# 4th International Conference on Disaster Management (4th ICDM 2024)



## Natural Disasters and Non-Natural Disasters

E3S Web of Conferences Volume 604 (2024)

Padang, Indonesia
30 September - 1 October 2024

**Editors:**

**Abdul Hakam**
**Fauzan**
**Saravanan Sabbarayan**
**Ezri Hayat**

ISBN: 979-8-3313-1296-1

**Printed from e-media with permission by:**

Curran Associates, Inc.
57 Morehouse Lane
Red Hook, NY 12571

CURRAN ASSOCIATES INC. proceedings.com

**Some format issues inherent in the e-media version may also appear in this print version.**



For additional information, please contact EDP Sciences – Web of Conferences
at the address below.

EDP Sciences – Web of Conferences
17, Avenue du Hoggar
Parc d'Activité de Courtabœuf
BP 112
F-91944 Les Ulis Cedex A
France

Phone: +33 (0) 1 69 18 75 75
Fax: +33 (0) 1 69 28 84 91

contact-edps@webofconferences.org

**Additional copies of this publication are available from:**

Curran Associates, Inc.
57 Morehouse Lane
Red Hook, NY 12571 USA
Phone: 845-758-0400
Fax: 845-758-2633
Email: curran@proceedings.com
Web: www.proceedings.com

# TABLE OF CONTENTS

# The 4$^{th}$ ICDM 2024

| | | |
|---|---|---|
| Title of conference | : | The 4$^{th}$ International Conference on Disaster Management (The 4$^{th}$ ICDM 2024) |
| Venue | : | Convention Hall, Universitas Andalas, Padang, West Sumatra, Indonesia |
| Date | : | 30 September – 01 October 2024 |
| Scientific Editors | : | |


1. Werry Darta Taifur, Prof. (Universitas Andalas)
2. Saravanan Subbarayan, Assoc. Prof (National Institute of Technology, India)
3. Bambang Istijono, Prof. (Universitas Andalas, Indonesia)
4. Kim Karl, Prof. (University of Hawaii, USA)
5. Febrin Anas Ismail, Prof. (Universitas Andalas, Indonesia)
6. Indrajit Pal, Assoc. Prof. (Asian Institute of Technology, Thailand)
7. Abdul Hakam, Prof. (Universitas Andalas, Indonesia)
8. Ahmad Safuan, Prof. (Universiti Teknologi Malaysia, Malaysia)
9. Yonariza, Prof. (Universitas Andalas, Indonesia)
10. Afriwardi, Prof. (Universitas Andalas, Indonesia)
11. Rusfidra (Universitas Andalas, Indonesia)
12. Reflinaldon (Universitas Andalas, Indonesia)
13. Ezri Hayat, Ph. D (Teesside University, United Kingdom)
14. Yosritzal, Ph. D (Universitas Andalas, Indonesia)
15. Suhana Koting, Ph. D (Universiti Malaya, Malaysia)

# Statement of Peer review

In submitting conference proceedings to *Web of Conferences*, the editors of the proceedings certify to the Publisher that

1. They adhere to its **Policy on Publishing Integrity** in order to safeguard good scientific practice in publishing.
2. All articles have been subjected to peer review administered by the proceedings editors.
3. Reviews have been conducted by expert referees, who have been requested to provide unbiased and constructive comments aimed, whenever possible, at improving the work.
4. Proceedings editors have taken all reasonable steps to ensure the quality of the materials they publish and their decision to accept or reject a paper for publication has been based only on the merits of the work and the relevance to the journal.

Title, date and place of the conference:

The 4th International Conference on Disaster Management (The 4th ICDM 2024),
30 September – 01 October 2024 at Convention Hall Universitas Andalas, Padang,
West Sumatra, Indonesia

Proceeding editors:

Prof. Abdul Hakam, Ph.D
Prof. Fauzan, Dr.Eng
Saravanan Sabbarayan, Ph.D
Ezri Hayat, Ph. D

Date and editors' signature: November 7th, 2024

Prof. Abdul Hakam, Ph.D

Prof. Fauzan, Dr.Eng

Saravanan Sabbarayan, Ph.D

Ezri Hayat, Ph. D

17, avenue du Hoggar - PA de Courtabœuf – BP 112 - 91944 Les Ulis Cedex A (France)
Tél. : 33 (0)1 69 18 75 75 - Fax : 33(0)1 69 07 45 17 – www.edpsciences.org

# PREFACE

The International Conference on Disaster Management (ICDM) is a biennial event hosted by Universitas Andalas. This year marked the fourth conference (The 4th ICDM 2024), which took place at the Convention Hall of Universitas Andalas in Padang, West Sumatera, Indonesia, from September 30 to October 1, 2024. The conference was jointly organized by the Postgraduate School of Universitas Andalas, the Civil Engineering Department, the Disaster Study Center, and the Research Institute and Community Service of Universitas Andalas.

At the 4th ICDM 2024, speakers from various countries gathered to exchange valuable knowledge and cutting-edge ideas, particularly focused on aiding developing nations in enhancing disaster resilience and effectively mitigating and managing disasters. This year's conference revolved around the theme **Natural Disasters and Non-Natural Disasters,** covering various topics:

- Risk-Based Disaster Analysis for Regional Development and Spatial Planning
- Social, Economic, Cultural, Community and Local Wisdom Issues in Disaster Management
- Policy, Institutional Governance, and Capacity Building for Disaster Management
- Disaster Monitoring, Broadcasting, Early Warning and Information System
- Health, Emergency Preparedness and Post-Disaster Healing
- Physical Infrastructure Management and Recovery
- Non-Physical Infrastructure Management and Recovery
- Geological, Environmental and Climate Change Risk and Rehabilitation
- Damage and Lost Mapping, and Post-Disaster Needs Analysis
- Food Security, Risk and Response for Disaster
- Historical Spatial and Temporal Patterns of Disasters in the Past
- Multi Hazard Risk Assessment and Management
- Impact of Disaster and Disaster Risk Reduction
- Seismic and Tsunami Disaster Management and Mitigation
- Retrofit, Rehabilitation and Reconstruction
- Other Related Topics

On behalf of the committee, I extend my heartfelt gratitude to the keynote speakers, invited speakers, and presenters who shared their work and submitted their manuscripts. I am delighted to note that this year's conference featured speakers from across the globe, including the United States, the United Kingdom, China, Thailand, the Philippines, Malaysia, India, and Indonesia. I also wish to express my sincere thanks to the 4th ICDM 2024 committee, the reviewers, and everyone who contributed to the success of this event. Their dedication and effort made it possible to organize the conference at such a high standard and compile high-quality papers for the conference proceedings. Finally, I greatly appreciate the support of EDP Sciences in publishing this volume of conference papers. We look forward to welcoming everyone to the next conference.

Padang, November 5, 2024

Delfia Tanjung Sari, SE, M. Si, Ph. D, CRA, CRP
Chair of the 4th ICDM 2024

# Evaluation of critical factors for typhoon-triggered landslide and subsequent debris flow in the Philippines using physically based models

*John* Christian Gaviola[1*], *Kuo* Chieh Chao[2], and *Jannet* Bencure[3]

[1]Department of Civil Engineering, Visayas State University, Baybay City, Leyte, 6521-A, Philippines
[2]Department of Civil and Infrastructure Engineering, Asian Institute of Technology, Pathum Thani, 12120, Thailand
[3]Department of Geodetic Engineering, Visayas State University, Baybay City, Leyte, 6521-A, Philippines

**Abstract.** This study employs physically based modeling techniques to evaluate factors that triggered the catastrophic landslide that occurred in Kantagnos Village, Baybay City, Leyte, Philippines, during Typhoon Megi in 2022. The TRIGRS and DEBRIS-2D models simulated transient infiltration effects from four days of continuous rainfall, assessing pore water pressure changes and delineating hazard zones. Parametric analyses on critical parameters—such as fstrength characteristics, soil thickness, and water table conditions—were conducted to predict spatiotemporal changes in the factor of safety. The results of the analyses indicated that the accuracy of the model could reach up to 79.23%, as evaluated by receiver operating characteristic analysis, with area zoning based on bedding plane orientation improving this by 8.23%. DEBRIS-2D effectively simulated debris flow motion and deposition areas validated against field observations. The combined use of TRIGRS and DEBRIS-2D delineated hazard zones, with 93.97% of the area within the actual debris flow boundary predicted. Key parameters, including slope geometry, material yield stress, and the quality of the digital terrain model, are crucial for predicting debris flow distance and velocity. These findings offer a valuable benchmark for developing a framework for slope disaster analysis and risk management in the region.

## 1 Introduction

In the past two decades, there has been a noticeable increase in the frequency and intensity of geophysical and hydrometeorological hazards, leading to severe consequences [1]. Landslides have caused significant loss of life, property, infrastructure, and economic stability. Globally, landslides caused approximately 1,000 deaths yearly, incurring an estimated USD 4 billion in damages [2]. Landslide typically refers to the swift and abrupt movement of substantial earth material, such as soil and rock debris, down a slope due to gravitational forces, often involving a clearly defined failure surface. Ali et al. [3] state that landslides occur due to conditioning and triggering factors. Conditioning factors are inherent terrain characteristics that affect the likelihood of slope failure. On the other hand, the triggering factors are external forces that prompt landslides, such as heavy rainfall, earthquakes, volcanic eruptions, and human-induced activities like deforestation and construction.

The Philippines, located in the typhoon-prone region of the Western Pacific, faces numerous natural hazards that pose significant threats to its people and infrastructure. Volcanic eruptions and earthquakes, although capable of triggering landslides, occur infrequently with long time intervals of occurrence. In contrast, landslide occurrences due to heavy and prolonged rainfall are much more common, occurring almost every month during the rainy season, and the resulting damage is frequently dealt with [4]. According to the Asian Disaster Reduction Center, about twenty typhoons enter the Philippine boundary annually, five of which are destructive. These records are equivalent to about one-fourth of the recorded events globally [5].

Given the geographical and physical characteristics of the Philippines, it is responsible for approximately 46% of the documented rainfall-induced landslides in Southeast Asia despite covering less than 10% of the total land area in the region [6]. It is challenging to model water migration in slopes with unsaturated soils, considering the influence of climate conditions [7, 8]. Intense rainfall has been shown to initiate landslides by rising pore pressure and reducing the suction of unsaturated soil on slopes [9]. Additionally, it increases the weight of the slope materials, which leads to a rise in shear stress and further destabilizes the slopes [10-12].

Access to maps that depict accurate landslide hazards and risks is crucial for identifying areas with potential landslide-related losses and effectively reducing their impact on society. To minimize the impact of landslides, it's vital to understand the factors influencing their occurrence. Various types of models have been used for slope stability evaluation. Each model has its strengths and weaknesses, with the choice of model depending on

* Corresponding author: johnchristian.gaviola@vsu.edu.ph

the specific requirements of the study, available data, and computational resources. The statistical and machine learning approaches are the most used, although they often require extensive datasets and overlook complex landslide mechanisms [13-15]. Hence, it is critical to consider susceptibility assessment from a physically based perspective, which considers the mechanical and hydrological properties and evaluates slope stability by physical laws.

This study utilized a combined physically based approach for the first time in local landslide cases and evaluated their performance for potential application in local landslide susceptibility and hazard mapping. This study primarily used the performance of the Transient Rainfall Infiltration and Grid-based Regional Slope Stability Model (TRIGRS) to evaluate the spatiotemporal variability of slope stability. The DEBRIS-2D model was used to simulate the subsequent hazard areas in case of debris flow. To use this model, the input required is the volume of the collapsed zone and its distribution. To obtain this information, the TRIGRS model was used to calculate the amount of unstable mass on a slope. These models were applied to the real slope failure cases at Kantagnos Village during Typhoon Megi in Baybay City, Leyte, Philippines. This simulation yielded the debris volume and debris depth distribution of the failure zone necessary for the flow simulation. The DEBRIS-2D model was then utilized to simulate the rapid movement of the failed mass, allowing for the evaluation of the resulting hazard zone.

The findings of this research can be a reference to establish a framework for landslide susceptibility evaluation and debris flow hazard mapping to better support disaster prevention and mitigation strategies, including the establishment of LEWS, land use policy planning, and disaster emergency response, among others. Furthermore, this work aligns with Sustainable Development Goals (SDG 11 and SDG 13): Sustainable Communities and Climate Action, by enhancing resilience and adaptive capacity to climate-related hazards and natural disasters.

## 2 Study area

The research was conducted on the western side of Leyte Island in the Philippines, specifically in Kantagnos Village in Baybay City. Geographically, Leyte lies in the central part of the Philippine group of islands, situated in the western Pacific Ocean (Fig. 1). Due to its location, it is highly exposed to the path of typhoons that enter the Philippine Area of Responsibility. Previous studies have identified Leyte as a region susceptible to typhoons and landslides, with the most natural disaster fatality in the country [16, 17].

The study area is located along the Anobang-Lobi Mountain range, a portion of Leyte's central mountain ranges, characterized by its rugged and mountainous terrain. It is characterized by a lithology that includes coarse conglomerate interlaced with volcanic agglomerates and breccia associated with some tuff. These characteristics arise from a series of volcanoes (Kanturao Volcanic Complex) along the central mountain range, which are inactive and extend in an NNW-SSE direction along the Philippine fault line. This geological activity has led to the island's distinct linear depressions, including elongated lakes and displaced rivers [18]. The study area has also been observed to have a bedding plane of moderately to highly weathered sandstone oriented in a southwest direction.

The study area has a Type IV climate characterized by consistent rainfall distributed throughout the year, according to the modified Corona's classification of climate provided by the state meteorological office [19]. Moreover, according to the longitudinal data provided by the Visayas State University Agrometeorological Station in Baybay City, the area has an average of about 2911 mm of precipitation annually, which peaks in December and January.

From April 8 to April 12, 2022, the study area experienced heavy rainfall due to Typhoon Megi, locally known as Typhoon Agaton. This case is highly notable as the recorded precipitation was about one-third of the average annual precipitation (996 mm). As a result, the province had several landslides and flooding events. Unfortunately, this resulted in 212 deaths, eight injuries, and 132 people reported missing. The cost of damages to agriculture, infrastructure, and houses is estimated to be at least 22.8 million USD. In Kantagnos Village alone, around 44 people were found dead, and over one hundred individuals are still missing. Fig 2 shows the 60-day antecedent rainfall profile, and Fig 3 shows the estimated rainfall intensity and maximum possible infiltration rate during Typhoon Megi in the study area. Fig 4 shows the satellite images of the landslide before and after the slope failure occurrence.

## 3 Methodology

This study investigated the Kantagnos landslide case by modeling pore water pressure changes, the factor of safety, and the distance and velocity of debris flow using the Typhoon Megi hydrometeorological scenario. The initial phase of this research is gathering and reviewing available data from relevant state agencies. Site observations were also conducted to gain knowledge of the actual site conditions. Moreover, laboratory testing was conducted to determine the index and engineering properties of the soils for the study area. Fig 5 shows the methodological framework for the study.

### 3.1 Determination of input parameters

For TRIGRS, input parameters required include soil shear strength, soil unit weight, porosity, saturated and unsaturated hydraulic conductivity, soil thickness, rainfall intensity, elevation, and slope.

The study used an IFSAR-generated DEM resampled to 5 × 5 m resolution from the Philippine National Mapping and Resources Information Authority (NAMRIA). Hydrological data were gathered from the Philippine Atmospheric, Geophysical, and Astronomical Services Administration (PAGASA).

https://doi.org/10.1051/e3sconf/202560401001

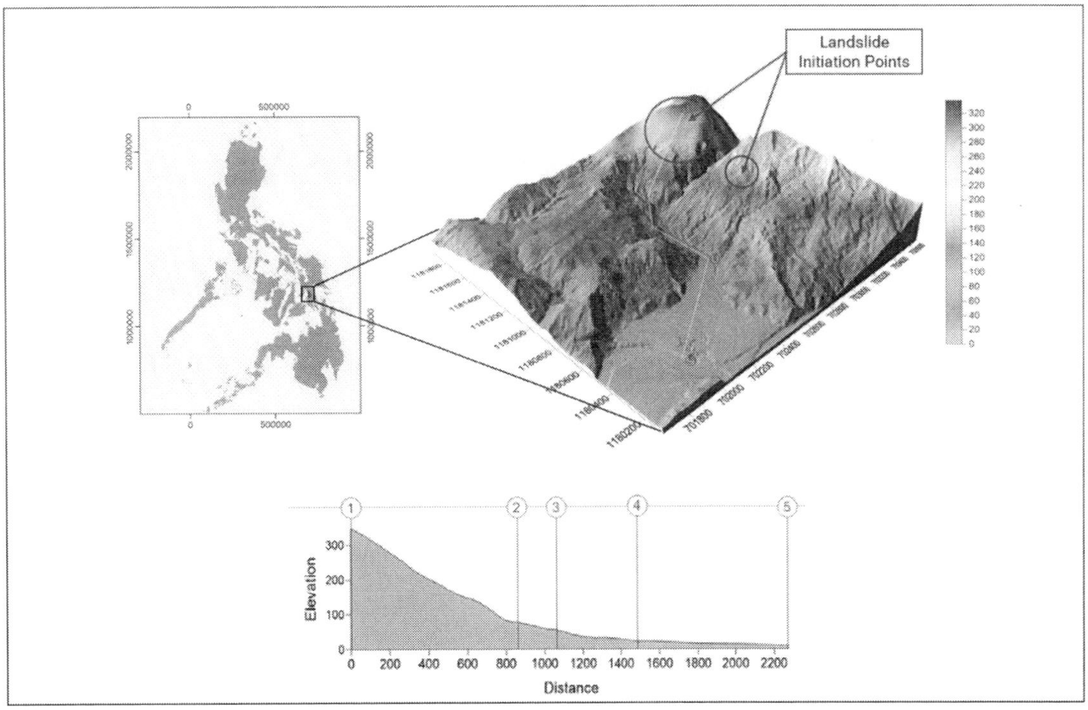


**Fig. 1.** Geographic and topographic attributes of the study area (data source: Phil-LiDAR)

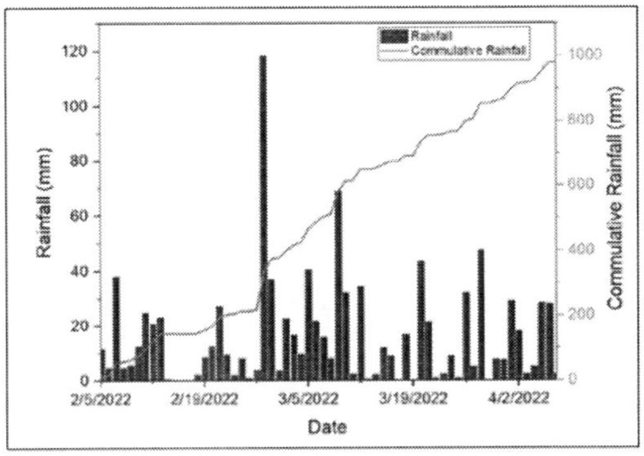


**Fig. 2**. 60-day antecedent rainfall data of the study area

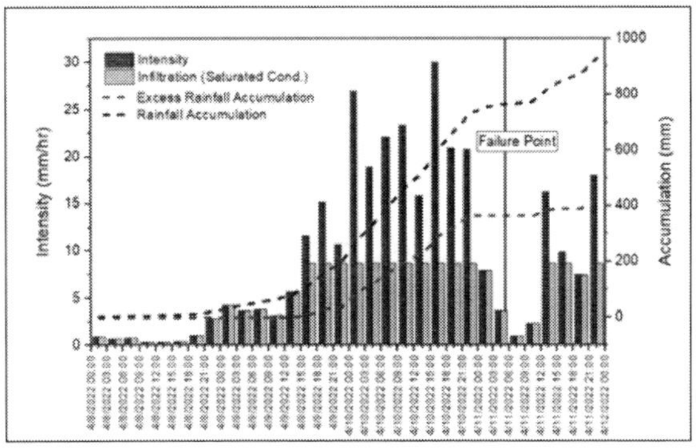


**Fig. 3.** Hyetograph of rainfall during Typhoon Megi

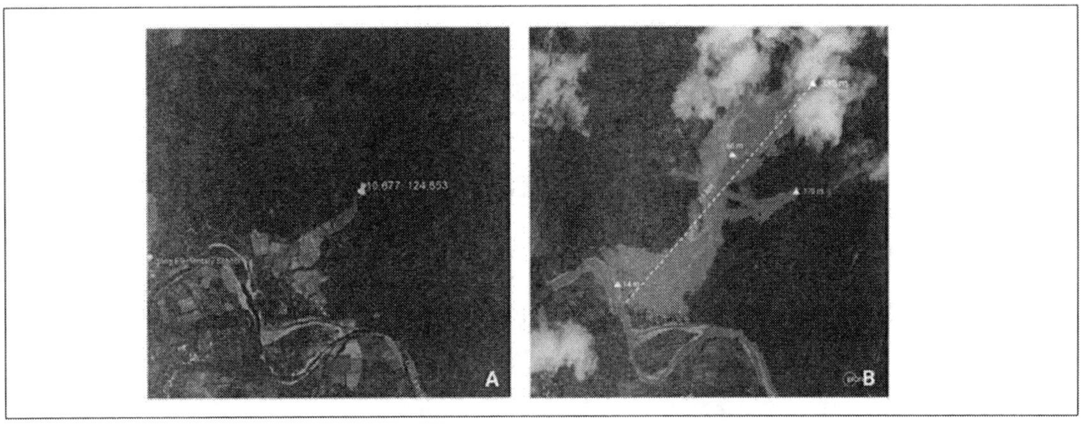


**Fig. 4.** Satellite images of Kantagnos Village (A) before the landslide (B) after the landslide (retrieved from https://earth.google.com, april 2024)

 https://doi.org/10.1051/e3sconf/202560401001

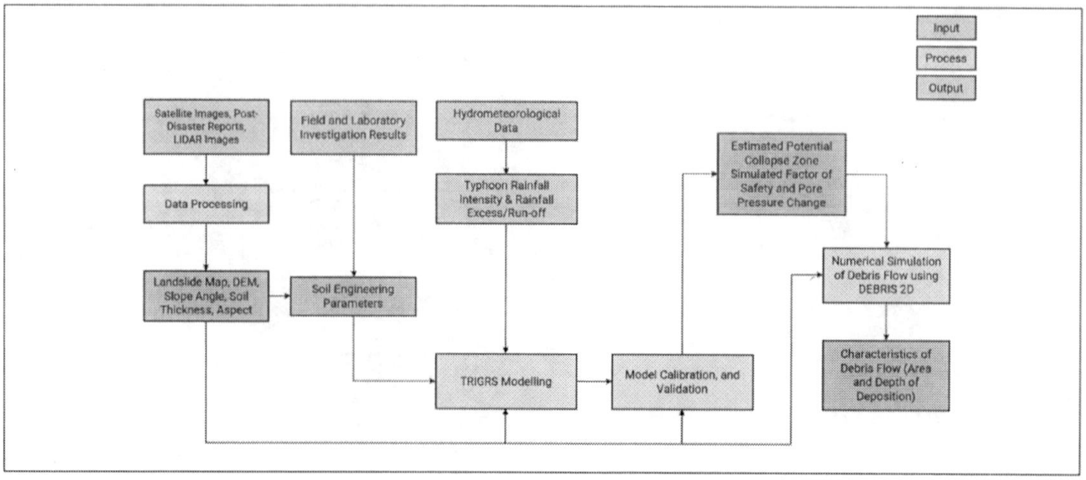


**Fig. 5.** Methodological framework of the study

### 3.2 Field investigation and laboratory testing

Field observations are essential for landslide studies as they provide firsthand data and insights into the factors contributing to landslides, their mechanisms, and potential mitigation measures. Validation of the delineated landslide/debris boundary was also conducted on-site.

The depths of the landslide at many points were also measured for soil thickness estimation. Since the model required the spatial variability of soil thickness or depth, assuming a homogeneous soil thickness would either over-estimate or under-estimate susceptibility. Unfortunately, comprehensive soil depth information was unavailable in the study area. Thus, we used the depth of the landslide scar and digital terrain models before and after the landslide to estimate soil thickness distribution. This estimation was combined with an empirical estimation suggested by Saulnier et al. [20].

A good amount of disturbed and undisturbed soil samples that were gathered from each identified sampling point were carefully transported for several laboratory tests, as follows: (1) cohesion and friction angle were determined following ASTM D3080 at a maximum strain rate of 0.0060 in/min, (2) grain size distribution were determined following ASTM D6913 and ASTM D7928, (3) Atterberg limits were determined using the ASTM D4318 standard, (4) soil moisture content, density, and specific gravity were determined based on the standard procedure ASTM 854-06, and (5) saturated hydraulic conductivity was determined following the standard procedure ASTM D5084.

The unsaturated soil parameters required to build the Soil-Water Characteristic Curve (SWCC) were determined through empirical equations from previous studies. Many successful studies in the past using physically based models have only been estimated based on previous studies, primarily based on the grain size distribution [21-23]. For this study, the alpha value, the saturated volumetric water content ($\theta_s$), and residual volumetric water content ($\theta_r$) were estimated following the Van Genuchten water retention curve using the ROSETTA 3 Lite Module [24]. Moreover, the value of diffusivity, $D_0$, was estimated based on the commonly used empirical equation of Liu and Wu [25], $D_0 = 200K_s$. The input parameters are summarized in Table 1.

A satellite precipitation dataset was also used to supplement the field rain gauge data. The GPM IMERG daily "final" precipitation is available at half-hourly, 3-hourly, daily, and monthly intervals on their website: *https://gpm.nasa.gov/data/source*. For this study, a 3-hour rainfall intensity interval was used in the simulation of a 77-hour typhoon for computational efficiency reasons.

### 3.3 Spatial data processing

The TRIGRS and DEBRIS-2D models require a high-quality digital elevation model (DEM) of the area before a landslide event. This study used a DEM with a 5-m resolution, which was then processed to create slope and flow direction maps. The DEM was also used for hydrological modeling, which includes predicting the direction and accumulation of water flow.

### 3.4 TRIGRS model

The TRIGRS Model is designed to model the timings and distribution of shallow landslides caused by rainfall [26]. It determines transient shifts in pore pressure and changes in the factor of safety resulting from rainwater infiltration. It models infiltration, covering events lasting from hours to days, using analytical solutions to partial differential equations that describe one-dimensional vertical water flow in saturated or unsaturated homogeneous and isotropic materials.

The factor of safety (FS) at a depth of Z at any time, t, is then calculated using the following equation:

$$FS(Z,t) = \frac{tan\varphi'}{tan\theta} - \frac{c' - (\psi(Z,t)\rho_w tan\varphi')}{\rho Z sin\theta cos\theta} \tag{1}$$

where $\varphi'$ denotes the soil's internal friction angle, $\theta$ represents the slope angle, $c'$ is the effective soil cohesion,

$\rho_w$ stands for the unit weight of water, and $\rho$ indicates the total unit weight of soil.

Equations 2 and 3 provide linearized solutions of the Richards equation by Iverson [27] to express groundwater head, $\varphi$, when there is an impermeable basal boundary at a finite depth:

$$\varphi(Z,t) = [Z-d]\beta + 2\sum_{n=1}^{N}\frac{I_{nz}}{K_z}H(t-t_n)[D_1(t-t_n)]^{\frac{1}{2}}\sum_{m=1}^{\infty}\left\{\begin{array}{l} ierfc\left[\frac{(2m-1)d_{LZ}-(d_{LZ}-Z)}{2[D_1(t-t_n)]^{\frac{1}{2}}}\right] \\ +ierfc\left[\frac{(2m-1)d_{LZ}+(d_{LZ}-Z)}{2[D_1(t-t_n)]^{\frac{1}{2}}}\right]\end{array}\right\} - 2\sum_{n=1}^{N}\frac{I_{nz}}{K_z}H(t-t_{n+1})[D_1(t-t_{n+1})]^{\frac{1}{2}}\sum_{m=1}^{\infty}\left\{\begin{array}{l} ierfc\left[\frac{(2m-1)d_{LZ}-(d_{LZ}-Z)}{2[D_1(t-t_{n+1})]^{\frac{1}{2}}}\right] \\ +ierfc\left[\frac{(2m-1)d_{LZ}+(d_{LZ}-Z)}{2[D_1(t-t_{n+1})]^{\frac{1}{2}}}\right]\end{array}\right\} \quad (2)$$

$$ierfc(\eta) = \frac{1}{\sqrt{\pi}}\exp(-\eta^2) - \mu erfc(\eta) \quad (3)$$

where $t$ is the time variable, $Z = z/cos\ \theta$ is a special variable in vertical coordinate direction, $d$ is the initial, depth of the water table measured in $Z$ direction in steady state, $d_{LZ}$ means the depth of an impermeable basal boundary, $\beta = \lambda\ cos\ \theta$, where $\lambda = cos\ \theta - (I_z/K_z)_{LT}$, and in which $K_z$ is a hydraulic conductivity, Iz is an initial surface flux, $I_{nz}$ is a surface flux of a given intensity in nth time $I$, $D_0$ is the saturated hydraulic conductivity.

To handle varying rainfall intensities, the program utilizes a step-function series. It features a straightforward runoff routing model, which allows users to channel excess water from impermeable zones to more permeable areas further downslope. Each cell's factor of safety calculations uses a clear-cut infinite-slope model. The program accounts for horizontal heterogeneity by enabling differences in material properties, rainfall patterns, and other parameters across individual cells. The components of the TRIGRS model are detailed in Fig 6.

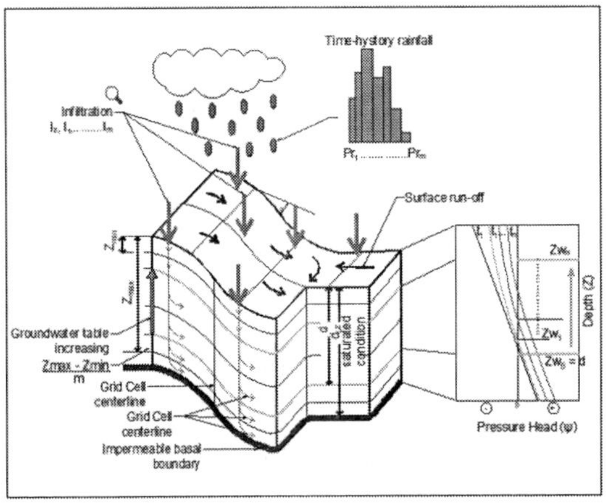


**Fig. 6.** Components of the TRIGRS model in finite impermeable boundary conditions (modified from Grelle et al., 2013)

**Table 1.** TRIGRS model input parameters

| Parameter | Unit | Values | Reference |
|---|---|---|---|
| Slope | degree | variable | DEM |
| Soil Thickness | m | variable | Saulnier et al. (1997) |
| Unit Weight, | g/cc | 1.66 to 1.86 | Laboratory Testing |
| Specific Gravity | | 2.55 to 2.56 | Laboratory Testing |
| Cohesion, c | kPa | 24.04 to 42.63 | Laboratory Testing |
| Friction Angle, φ | degrees | 15.86 to 20.43 | Laboratory Testing |
| Saturated Hydraulic Conductivity, $K_s$ | cm/s | $3.5\times10^{-4}$ to $1.3\times10^{-4}$ | Laboratory Testing |
| Saturated volumetric water content ($\theta_s$) | $m^3/m^3$ | 0.378 | Estimation of SWCC parameters using the ROSETTA 3 Module (Schaap et al., 2001, Zhang and Schaap, 2017) |
| Residual volumetric water content ($\theta_r$) | $m^3/m^3$ | 0.067 | |
| Pore-size parameter, α | 1/m | 2.073 | |
| Saturated hydraulic diffusivity ($D_0$) | $m^2/s$ | $4.8\times10^{-4}$ | Estimation from Liu and Wu (2008) |
| Steady infiltration rate, $I_z$ | $m\ s^{-1}$ | $0.01K_s$ | Salciarini et al., 2008; Liu and Wu, 2008; Kim et al., 2010; Park et al., 2013 |
| Hydraulic diffusivity, $D_0$ | $m^2\ s^{-1}$ | 200 $K_s$ | Liu and Wu (2008) |
| Soil Depth, $S_t$ | m | Varied | Saulnier et al. (1997) |
| Initial Water Depth, $W_d$ | m | 0-100% of soil thickness | Salciarini et al. (2007) |

### 3.5 Analysis of rainfall infiltration induced shallow landslide

#### *3.5.1 Model calibration*

To maintain the reliability of our model, a calibration process was carried out to ensure input parameters are confined within a realistic range. This calibration involves performing a sensitivity analysis to determine the most relevant inputs. Studies have recommended the evaluation of several parameters, including the initial groundwater condition (wt), cohesion (c), and internal angle of friction (φ) [15, 28, 29]. This was adopted in the study by calibrating the model using the range of values of these parameters from the field and laboratory observations. Meanwhile, the values for the water table were varied by percentage of the soil thickness following Salciarini et al. [30] as the water table remains unknown in this case. The accepted trials are compared based on their error indices to optimize the reliability of the results.

#### *3.5.2 Model evaluation and validation*

The model's performance was evaluated using a confusion matrix and a receiver operating characteristic (ROC) curve. Each cell in the simulation result falls into one of four categories: true positive (TP) when the model correctly predicts instability for a landslide source, false negative (FN) when it incorrectly assesses the same cell as stable, true negative (TN) when it accurately identifies stable cells, and false positive (FP) when it wrongly predicts instability. Key evaluation metrics include sensitivity (TPR), which measures the proportion of correctly identified positives, and specificity (TNR), which measures the proportion of true negatives. Balanced Accuracy, the mean of TPR and TNR, is used to handle imbalanced data, with its complement termed the error index. The ROC curve was created using TPR and FPR values at various cut-off points, with the area under the curve (AUC) interpreted according to Yesilnacar and Topal [31].

### 3.6 DEBRIS-2D model

Unstable hillsides with FS less than 1.0 can collapse rock and soil, which can combine with water to form a debris flow downslope. This model considers debris flow as a non-Newtonian fluid that has yield stress. It can calculate the velocity and depth of the flow, the affected area, and the impact force over time. The DEBRIS-2D model is a simulation tool used to replicate numerous debris flow events in different parts of the world, primarily focusing on cases in Taiwan and other Asian countries. The results match well with field observations such as the study of Tsai et al. [32], Wu et al. [33], and Hsu and Liu [34].

The following text describes a mathematical model developed by Liu and Huang [35] which is designed to calculate using the conservation of mass and momentum laws for long waves in the plug flow region. DEBRIS-2D uses an inclined coordinate system where the x-axis aligns with the flow direction, the y-axis aligns with the topographical contour direction, and the z-axis points to the depth direction at the x-y plane. The velocity components of an object moving in the x and y directions are denoted as u and v, respectively. The inclination angle is represented by θ, while the yield stress is defined by $\tau_0$, and H represents the flow depth, calculated as the difference between the free surface (h) and the natural bottom of the debris flow (B). The momentum and continuity equations are expressed in conservative form as Equations 4, 5, and 6, respectively.

$$\frac{\partial uH}{\partial t}+\frac{\partial u^2H}{\partial x}+\frac{\partial uvH}{\partial y}=-gcos\theta H\frac{\partial(B+H)}{\partial x}+gsin\theta H-\frac{1}{\rho}\frac{\tau_0 u}{\sqrt{u^2+v^2}} \quad (4)$$

$$\frac{\partial vH}{\partial t}+\frac{\partial uvH}{\partial x}+\frac{\partial v^2H}{\partial y}=-gcos\theta H\frac{\partial(B+H)}{\partial y}-\frac{1}{\rho}\frac{\tau_0 v}{\sqrt{u^2+v^2}} \quad (5)$$

$$\frac{\partial H}{\partial t}+\frac{\partial uH}{\partial x}+\frac{\partial vH}{\partial y}=0 \quad (6)$$

### 3.7 Subsequent motion of collapsed mass

#### *3.7.1 Model scenarios*

Two cases of simulations were run for this study. The first scenario was based on the actual site conditions where the source information was derived from the difference between the DTM before and after the landslide occurred, with the yield stress varying from 300 Pa to 700 Pa, as recommended by Wu et al. [33] for sand with mud debris materials. The second case used the results of the TRIGRS and the back-calculated yield stress value from the first scenario. The input parameters for debris flow simulation are detailed in Table 2.

**Table 2.** DEBRIS-2D model input parameters

| Parameters | Input Value |
|---|---|
| Elevation | Digital Elevation Model (5x5 m) |
| Simulation Range | 700874.182731 - 703874.182731 E<br>1179367.29761 - 1182057.29761 N |
| Mass Volume | Case 1: Estimated actual volume<br>Case 2: TRIGRS model predicted |
| Unit Weight | 1.76 g/cc (Laboratory Testing) |
| Geometry | Length characteristics of landslide<br>Elevation difference characteristics<br>Depth characteristics of the mass source |
| Yield Stress | Parametric Analysis: 300 -700 Pa "Sand with Mud" material [33] |

#### *3.7.2 Model evaluation and validation*

The simulation results were evaluated by comparing them with actual field observations and aerial photographs. The affected area, which includes initiation, transportation, and deposition zones, as well as the deposition depth from the simulation, were assessed with the in-situ observations to determine the accuracy of the simulation [34]. The yield stress value that generates the least error (% overestimation and % missed) is considered the most probable value.

## 4 Results and discussion

### 4.1 Slope stability modeling

The parametric analyses were conducted in multiple sets, the first and second of which were designed to eliminate the least probable values and reduce the number of additional trial simulations for more detailed parameter values. The parametric analysis reveals that the trial combination of φ = 16°, c = 32 kPa, wt = 20% of soil thickness would give a more realistic result with the least error of around 20.77%. This means that about 79 pixels in every 100 pixels within the landslide boundary were correctly predicted; likewise, about 79 pixels in every 100 pixels outside the landslide boundary were also correctly predicted.

Likewise, parametric analysis, reveals that the initial water table of 20% of the soil thickness gives the least error in predicting the landslide initiation area. This is relatively high already for sloping ground or mountainous regions. This could be attributed to the high antecedent rainfall (60-day rainfall data), with 68.38% and 188.82% increase from the long-term average monthly rainfall for February and March 2022, months before the slope failure occurrence, respectively.

The TRIGRS model uses a zonation technique to improve accuracy and reduce overestimation. The study area is zoned based on the slope aspect relative to the bedding plane direction. Areas with a cataclinal bedding-

https://doi.org/10.1051/e3sconf/202560401001

slope relationship, where the slope aligns with the bedding plane, are at higher landslide risk. The application of zoning resulted in increased performance in terms of predicting the stable area while maintaining a good prediction of the actual unstable area (refer to Fig. 7). Result shows that true negative rate (TNR) increased from 75.59% to 93.18%, consequently decreasing the false positive rate (FPR) by 17.59% from the former model. Furthermore, the model prediction accuracy has improved by 8.21% from 79.23% to 87.44%.

Figure 8 shows the change in the predicted unstable area with time from the start of the typhoon up to the estimated time of failure about 77 hours later. The landslide source is only about 7.10% (94,575 m$^2$) of the total area considered in the study.

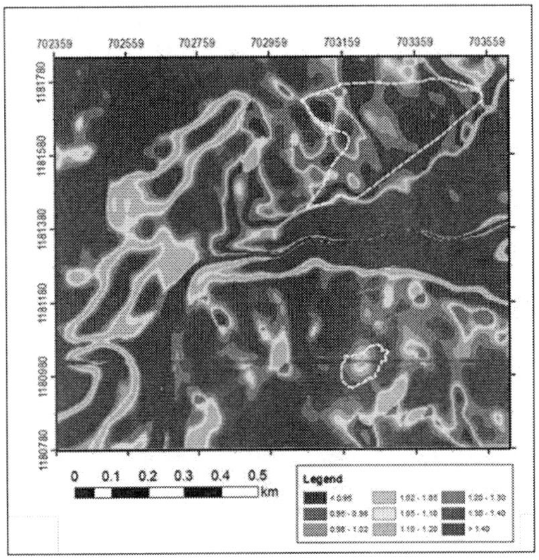


**Fig. 7.** Spatial model of the simulated factor of safety at the time of failure

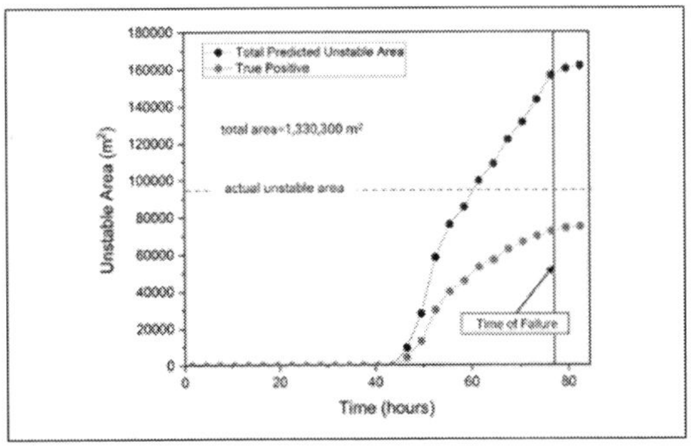


**Fig. 8.** Temporal change of predicted unstable area (FS<1.0)

### *4.1.1 Pressure head and factor of safety response*

At the onset of the typhoon, low rainfall intensity maintained a steady state for up to 12 hours, with the pressure head nearly matching a hydrostatic distribution (Fig. 9a). As rainfall intensity increased, surface flux ($I_{nz}$) rose, leading to infiltration. This caused an increase in $\beta$ and the transient terms of the TRIGRS model, raising the pressure head. After 36 hours of infiltration and 80 mm of rainfall, a gradual change in pressure head occurred, as infiltration in the unsaturated zone remained limited until it interacted with the water table. Fig. 9b illustrates the factor of safety response with depth at a representative slope point.

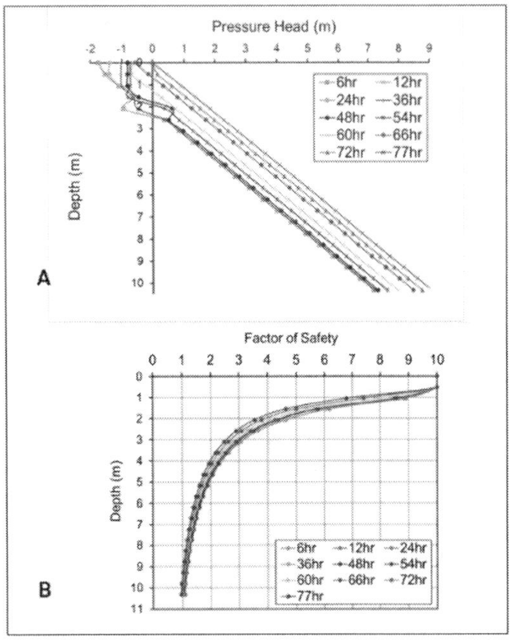


**Fig. 9.** (a) Temporal of the pressure head and (b) the factor of safety with depth taken at a representative point of average slope

Infiltration increases the groundwater's pressure head, lowering safety factor values and leading to slope instability (FS < 1.0). The analysis showed that slope failure occurred where the angle was $\theta \geq 16.70°$, but only on slopes up to 25.18°. As the slope angle increases, soil thickness decreases, reducing the maximum pressure head at the base of the soil layer. While the driving force increases with slope, the decrease in resisting force due to pressure head is less significant, maintaining stability up to a certain slope angle. Figure 10 illustrates the corresponding change in the factor of safety as a response to the change in the pore pressure at different time.

### *4.1.2 Validation of the spatial model*

The TRIGRS model was validated using the Area Under the Curve (AUC) metric. The results of the validation are presented in Fig. 11. It reveals a strong predictive capability with an AUC value of 0.885, indicating "very good performance" according to Yesilnacar and Topal [25]. A True Positive Rate (TPR) of 74.77% and a True Negative Rate (TNR) of 92.41% further confirm the model's reliability in identifying landslide and non-landslide areas. The False Positive Rate (FPR) of 7.59% and False Negative Rate (FNR) of 25.23% highlight the model's low rate of misclassification. These metrics demonstrate the TRIGRS model's reliability for landslide prediction and risk assessment.

https://doi.org/10.1051/e3sconf/202560401001

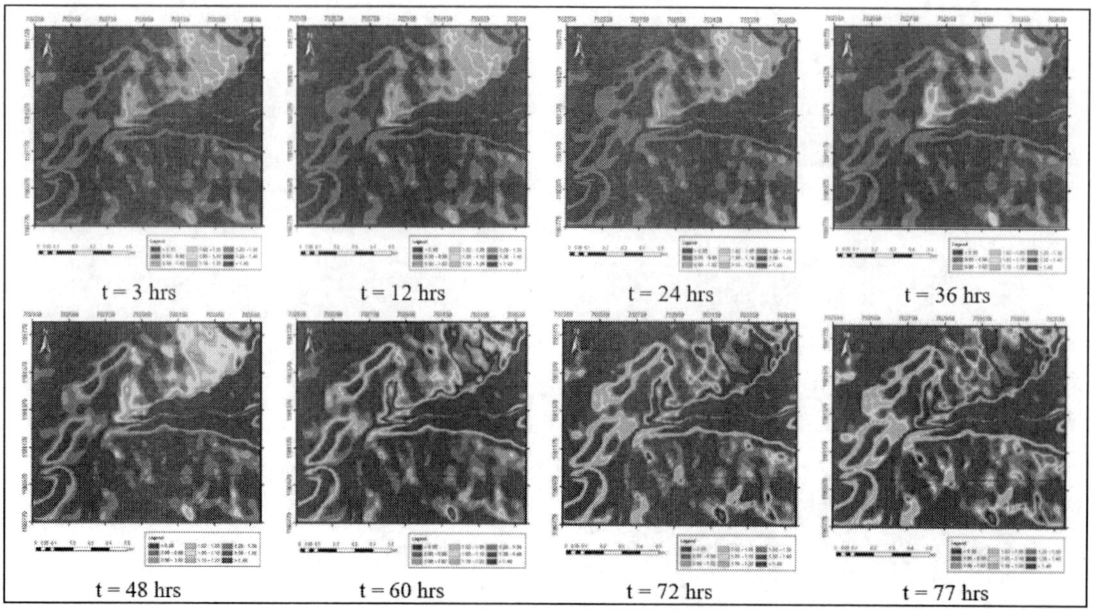


**Fig. 10.** Change of total unstable area (FS<1.0) across space and time

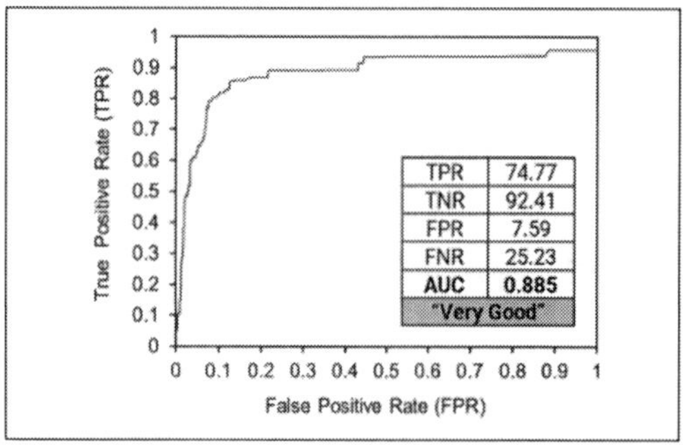


**Fig. 11.** Performance of the TRIGRS model in terms of the TPR and FPR, and the area under the receiver-operating characteristic curve (AUC)

### *4.1.3 Debris flow motion*

To estimate debris arrival timing in the village, specific house/building locations were marked as benchmarks in the model. Three points shown in Fig. 12 were evaluated, with debris traveling over 2 km. in 2 minutes. Points 1, 2, and 3, located 1630 m, 1795 m, and 2130 m from the landslide initiation point, are estimated to be reached at 56, 87, and 109 seconds, respectively. Fig. 13 shows debris spread and depth at various times. Disaster survivors confirmed these timings as reasonable.

The temporal variations in maximum velocity and debris flow depth at four monitoring locations were also investigated in the model through arbitrary monitoring points. These points placed along the debris flow path and deposition zones near former residential areas, help simulate and validate findings, as in Wu et al. [33].

These observations underscore the importance of considering both velocity and depth when assessing the debris flow risks and highlight how monitoring key points along the flow path provides valuable insights into their destructive potential.

### *4.1.4 Validation of Debris flow modeling results*

The final debris depth at the deposition zone, estimated by the DEBRIS-2D model, is shown in Fig. 14. Among the affected structures, Kantagnos Elementary School and several residential houses were reached by debris but remained unmoved, serving as benchmarks for validating the model. The school was buried under 2.5 meters of debris, while the model predicted a depth of 2.0 - 3.0 meters, aligning with actual measurements. The model also accurately predicted the debris depth around a residential house (2.5 - 3.0 meters) and correctly indicated that the Kantagnos Auditorium used as an evacuation site, was not affected by debris.

## 4.2 Combined TRIGRS and DEBRIS-2D models in landslide hazard zone mapping

The hazard zone predicted by the combined TRIGRS-DEBRIS-2D modeling is shown in Fig. 15, with the evaluation results summarized in Table 3. As shown, the combined model accurately delineated 93.97% of the 1,281,484 sq. meters of the hazard area, highlighting its strong ability to identify landslide-prone zones. By integrating TRIGRS, which predicts landslide initiation, and DEBRIS-2D, which models debris flow extent, the approach offers a comprehensive method for hazard mapping. While the model is effective, fine-tuning parameters such as soil thickness, initial water table, and precipitation patterns could improve accuracy, reducing overestimation and mispredictions.

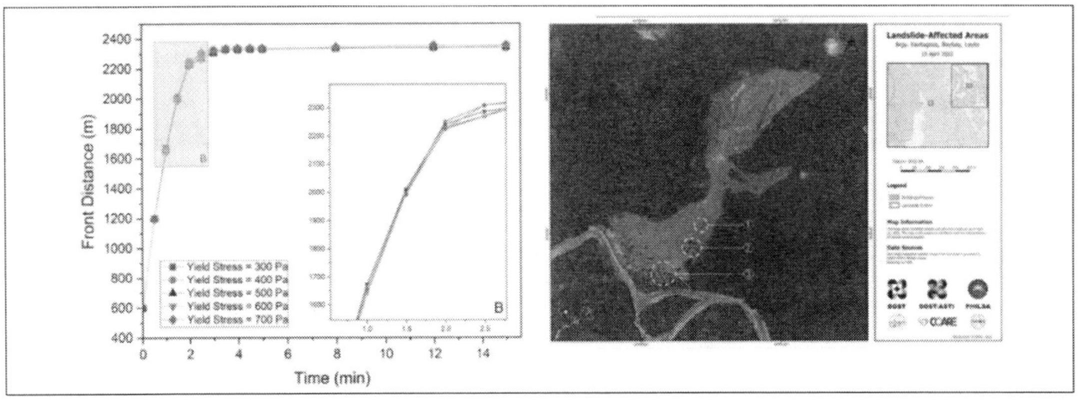


**Fig. 12.** Estimated time of arrival of debris as simulated by DEBRIS-2D (distance of Points 1, 2, and 3 are around 1630m, 1795, and 2130m, respectively, measured from the top of the debris source) (image source: DOST)

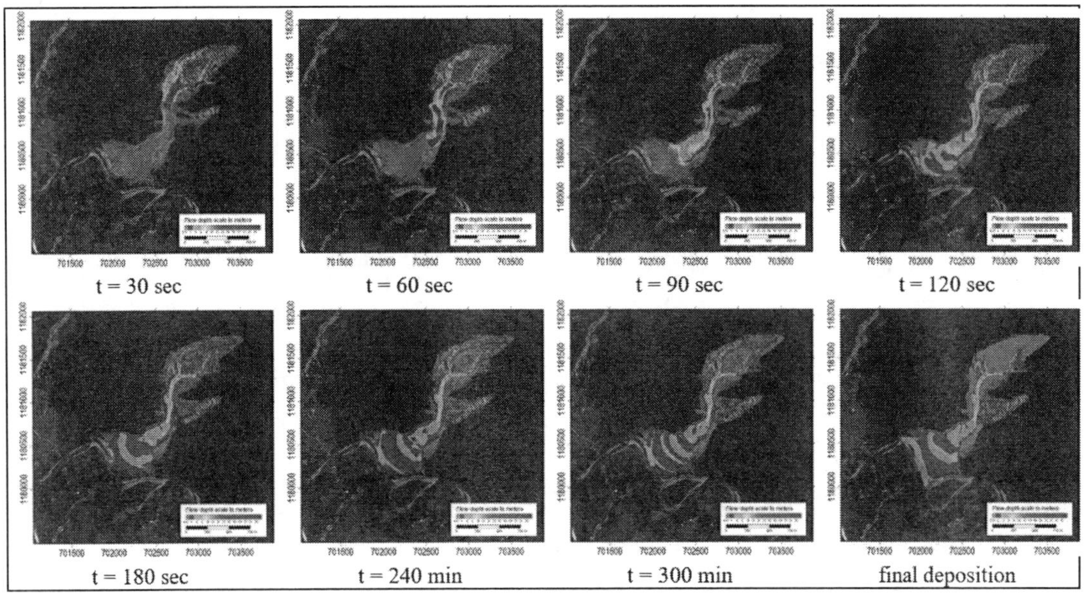


**Fig. 13.** Spatiotemporal change of the debris flow front and deposition depth

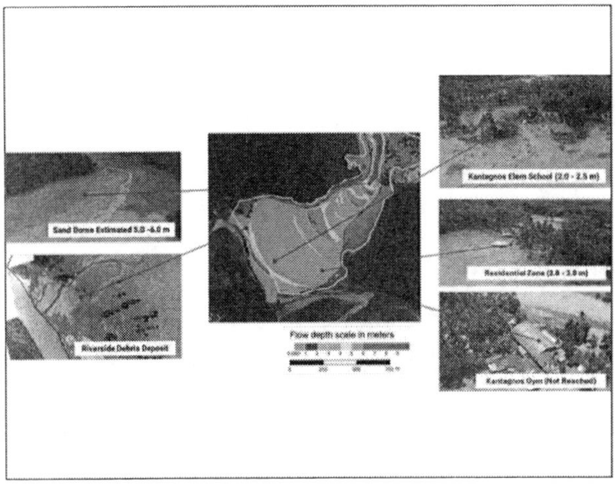


**Fig. 14.** Validation of the simulated debris deposition depth

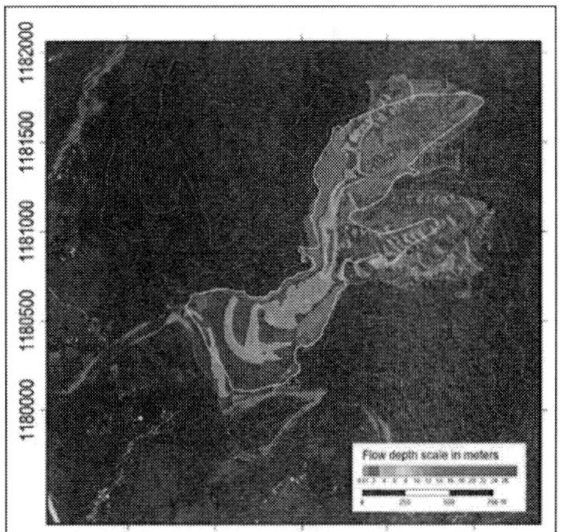


**Fig. 15.** Combined TRIGRS-DEBRIS-2D modeling

**Table 3.** Performance of the combined TRIGRS and DEBRIS-2D models

| Delineated Hazard Area (sq.m) | % Matched (Correct Prediction) | % Over-estimation (Overprediction) | % Missed (Under-prediction) |
|---|---|---|---|
| 1,281,484 | 93.97% | 37.70% | 6.42% |

## 5 Conclusion

This study evaluates the performance of the TRIGRS and DEBRIS-2D models in analyzing slope stability and debris flow during a rainfall-induced landslide in Kantagnos Village, Baybay City, Leyte, Philippines. Parametric analyses improved simulation accuracy, with the best results using $\varphi = 16°$, $c = 32$ kPa, and a water table depth of 20% of the soil thickness, achieving a 20.77% prediction error and 79.23% balanced accuracy. Zoning based on bedding plane orientation increased the true negative rate from 75.59% to 93.18%, improving overall accuracy by 8.23%. The study emphasized the importance of the initial water table in landslide initiation, identifying a critical threshold of 0.35 times the soil thickness, which could inform early warning systems. The TRIGRS model effectively simulated rainfall infiltration and landslide dynamics, while DEBRIS-2D accurately modeled debris flow using yield stress of 500 Pa. Together, the models correctly predicted 93.97% of the debris flow boundary, demonstrating their strength in hazard mapping.

For future research, it is recommended to integrate geophysical methods for accurate soil thickness mapping, conduct long-term groundwater monitoring, study the impact of weathering on shear strength parameters, explore additional zoning techniques, and compare TRIGRS and DEBRIS-2D with other models for improved accuracy and cost-effectiveness.

This work was supported by the Department of Science and Technology – Science Education Institute (DOST-SEI) of the Republic of the Philippines, the Asian Institute of Technology (AIT, Thailand), and the Visayas State University (VSU, Philippines) under the Internationalization Program (IP) Research Grant VSU-IP-2023-7.

## References

1. G. Rahman et al., Assessment of landslide susceptibility, exposure, vulnerability, and risk in Shahpur Valley, Eastern Hindu Kush, Front Earth Sci (Lausanne), **10**, (2022). https://doi.org/10.3389/feart.2022.953627
2. B. Pradhan and A. M. Youssef, Manifestation of remote sensing data and GIS on landslide hazard analysis using spatial-based statistical models, Arabian Journal of Geosciences, **3**(3), 319–326 (2010). https://doi.org/10.1007/s12517-009-0089-2
3. S. Ali, P. Biermanns, R. Haider, and K. Reicherter, Landslide susceptibility mapping by using a geographic information system (GIS) along the China-Pakistan Economic Corridor (Karakoram Highway), Pakistan, Natural Hazards and Earth System Sciences, **19**(5), 999–1022, (2019). https://doi.org/10.5194/nhess-19-999-2019
4. D. Nolasco-Javier and L. Kumar, Deriving the rainfall threshold for shallow landslide early warning during tropical cyclones: a case study in northern Philippines, Natural Hazards, **90**(2), 921–941 (2018). https://doi.org/10.1007/s11069-017-3081-2
5. M. G. A. Huigen and I. C. Jens, Socio-Economic Impact of Super Typhoon Harurot in San Mariano, Isabela, the Philippines, World Dev, **34**(12), 2116–2136, (2006). https://doi.org/10.1016/j.worlddev.2006.03.006
6. M. J. Froude and D. N. Petley, Global fatal landslide occurrence from 2004 to 2016, Natural Hazards and Earth System Sciences, **18**(8), 2161–2181, (2018). https://doi.org/10.5194/nhess-18-2161-2018
7. K. C. Chao, J. B. Kang, and J. D. Nelson, Challenges in water migration modeling for expansive soils. GeoShanghai Conference, Shanghai, China, 204–213 (2014).
8. K. C. Chao, J. B. Kang, J. D. Nelson, Evaluation of failure of embankment slope constructed with expansive soils, Geotechnical Engineering Journal of the SEAGS & AGSSEA, **49**(2), 140-149 (2018).
9. W. H. Schulz, J. P. McKenna, J. D. Kibler, and G. Biavati, Relations between hydrology and velocity of a continuously moving landslide-evidence of pore-pressure feedback regulating landslide motion?, Landslides, **6**(3), 181–190, (2009). https://doi.org/10.1007/s10346-009-0157-4
10. J. Mathew, D. G. Babu, S. Kundu, K. V. Kumar, and C. C. Pant, Integrating intensity-duration-based rainfall threshold and antecedent rainfall-based probability estimate towards generating early warning for rainfall-induced landslides in parts of the Garhwal Himalaya, India, Landslides, **11**(4), 575–588, (2014). https://doi.org/10.1007/s10346-013-0408-2
11. T. Amarathunga, K. C. Chao, and J. D. Nelson, Effect of climate change on rainfall-induced failures for embankment slopes in Timor-Leste. Proceedings of the 20th International Conference on Soil Mechanics and Geotechnical Engineering, Sydney, Australia, 2341–2347, (2022).
12. P. Morya, S. Kongpanickul, K. C. Chao, and R. Ishenaliev, Rainfall-induced failure on unsaturated fill and highly weathered Schist slopes. The 7th Asia-Pacific Unsaturated Soils Conference, Nagoya, Japan, 7(2), 556–564, (2019).
13. J. L. Zêzere, S. Pereira, R. Melo, S. C. Oliveira, and R. A. C. Garcia, Mapping landslide susceptibility using data-driven methods, Science of the Total Environment, **589**, (2017). https://doi.org/10.1016/j.scitotenv.2017.02.188
14. S. Zhang, L. Zhao, R. Delgado-Tellez, and H. Bao, A physics-based probabilistic forecasting model for rainfall-induced shallow landslides at regional scale, Natural Hazards and Earth System Sciences, **18**(3),

969–982, (2018). https://doi.org/10.5194/nhess-18-969-2018

15. V. Medina, M. Hürlimann, Z. Guo, A. Lloret, and J. Vaunat, Fast physically-based model for rainfall-induced landslide susceptibility assessment at regional scale, Catena (Amst), **201**, (2021). https://doi.org/10.1016/j.catena.2021.105213
16. J. Gray, S. Lloyd, S. Healey, and A. Opdyke, Urban and rural patterns of typhoon mortality in the Philippines, Progress in Disaster Science, **14**(100234), (2022). https://doi.org/10.1016/J.PDISAS.2022.100234
17. G. D. C. Santos, 2020 Tropical cyclones in the Philippines: A review, Tropical Cyclone Research and Review, **10**(3), 191–199, (2021). https://doi.org/10.1016/j.tcrr.2021.09.003
18. R. N. Eco et al., Journal of the Philippine Geoscience and Remote Sensing Society (2015). Available: http://philgrss.org/journal/
19. PAGASA. Climate Change in the Philippines. Philippine Atmospheric, Geophysical and Astronomical Services Administration. (2011).
20. G. M. Saulnier, K. Beven, and C. Obled, Including spatially variable effective soil depths in TOPMODEL, J Hydrol (Amst), **202**(1–4), 158–172 (1997). https://doi.org/10.1016/S0022-1694(97)00059-0
21. F. de Luiz Rosito Listo, M. C. Villaça Gomes, and F. S. Ferreira, Evaluation of shallow landslide susceptibility and Factor of Safety variation using the TRIGRS model, Serra do Mar Mountain Range, Brazil, J South Am Earth Sci, **107**, (2021). https://doi.org/10.1016/j.jsames.2020.103011
22. G. Grelle et al., Space-time prediction of rainfall-induced shallow landslides through a combined probabilistic/deterministic approach, optimized for initial water table conditions, Bulletin of Engineering Geology, and the Environment, **73**(3), 877–890 (2014). https://doi.org/10.1007/s10064-013-0546-8
23. L. Schilirò, J. Cepeda, G. Devoli, and L. Piciullo, Regional analyses of rainfall-induced landslide initiation in upper gudbrandsdalen (South-eastern Norway) using TRIGRS model, Geosciences (Switzerland), **11**(1), 1–15, (2021). https://doi.org/10.3390/geosciences11010035
24. Y. Zhang and M. G. Schaap, Weighted recalibration of the Rosetta pedotransfer model with improved estimates of hydraulic parameter distributions and summary statistics (Rosetta3), J Hydrol (Amst), **547**, 39–53, (2017). https://doi.org/10.1016/j.jhydrol.2017.01.004
25. C. N. Liu and C. C. Wu, Mapping susceptibility of rainfall-triggered shallow landslides using a probabilistic approach, Environmental Geology, **55**(4), 907–915, (2008). https://doi.org/10.1007/s00254-007-1042-x
26. R. L. Baum, W. Z. Savage, and J. W. Godt, TRIGRS-A Fortran Program for Transient Rainfall Infiltration and Grid-Based Regional Slope-Stability Analysis, Version 2.0, [Online]. Available: http://geohazards.cr.usgs.gov/
27. R. M. Iverson, Landslide triggering by rain infiltration," Water Resour Res, **36**(7), 1897–1910, (2000). https://doi.org/10.1029/2000WR900090
28. M. Hürlimann, Z. Guo, C. Puig-Polo, and V. Medina, Impacts of future climate and land cover changes on landslide susceptibility: regional scale modelling in the Val d'Aran region (Pyrenees, Spain), Landslides, **19**(1), 99–118, (2022). https://doi.org/10.1007/s10346-021-01775-6
29. A. Parapan, K. C. Chao, K. Saowiang, S. Soralump, and T. Nontapot, Evaluation of key parameters in rainfall-induced shallow landslide analysis: Are Di Village case, IOP Conference Series: Earth and Environmental Science, **1334**(1), 012015, (2024). https://doi.org/10.1088/1755-1315/1334/1/012015
30. D. Salciarini, C. Tamagnini, F. Ronchi, E. Volpe, and G. Fanelli, An approach for large-scale soil characterization for the application of non-structural landslide risk mitigation, Rivista Italiana di Geotecnica, **51**(3), 7–21, (2017). https://doi.org/10.19199/2017.3.0557-1405.07
31. E. Yesilnacar and T. Topal, Landslide susceptibility mapping: A comparison of logistic regression and neural networks methods in a medium scale study, Hendek region (Turkey), Eng Geol, **79**(3–4), 251–266 (2005). https://doi.org/10.1016/j.enggeo.2005.02.002
32. M. P. Tsai, Y. C. Hsu, H. C. Li, H. M. Shu, and K. F. Liu, Application of simulation technique on debris flow hazard zone delineation: A case study in the Daniao tribe, Eastern Taiwan, Natural Hazards and Earth System Science, **11**(11), 3053–3062, (2011). https://doi.org/10.5194/nhess-11-3053-2011
33. Y. H. Wu, K. F. Liu, and Y. C. Chen, Comparison between FLO-2D and Debris-2D on the application of assessment of granular debris flow hazards with case study, J Mt Sci, **10**(2), 293–304, (2013). https://doi.org/10.1007/s11629-013-2511-1
34. Y. C. Hsu and K. F. Liu, Combining TRIGRS and DEBRIS-2D models for the simulation of a rainfall infiltration induced shallow landslide and subsequent debris flow, Water (Switzerland), **11**(5), (2019). https://doi.org/10.3390/w11050890
35. K. F. Liu and M. C. Huang, Numerical simulation of debris flow with application on hazard area mapping, in Computational Geosciences, 221–240, (2006). https://doi.org/10.1007/s10596-005-9020-4

# Flood disaster mitigation on road sections: optimizing performance by implementing traffic engineering management

*Erviela* Puspa Jayanti[1], *Sri* Sarjana[2*], *Yudi* Karyanto[1], *Suhendro* Oktosatrio[3], and *Farel* Gogo Mulia[1]

[1]Land Transportation Department, Politeknik Transportasi Darat Indonesia - STTD, Bekasi, Indonesia
[2]Marketing, Innovation and Technology Department, Politeknik Transportasi Darat Indonesia - STTD, Bekasi, Indonesia
[3]Department of Peace Operations, United Nations Headquarters, New York, USA

**Abstract.** Flood disaster mitigation efforts are carried out through traffic engineering management to reduce the negative impacts of flooding on the community and infrastructure, where these efforts require a comprehensive disaster management plan and involve cooperation between the government, traffic authorities, emergency services and the surrounding community. This study aims to determine the traffic engineering scenario that can reduce congestion and improve traffic performance during floods. This study analyzes the road network in Gedebage District and Ujung Berung City which are affected by floods under normal conditions, flood conditions (do-nothing), and conditions when the traffic engineering scenario is applied (do-something) by developing an origin-destination matrix formed using the Furness method, traffic modelling using PTV Visum and traffic performance analyzed using the 2023 Indonesian Road Capacity Guidelines method. The results of the study indicate that there is an increase in traffic performance during floods with the implementation of do-something scenario 2. The study recommends that stakeholders implement flood disaster mitigation by providing information on mapping flood-prone areas, early warning systems, information on evacuation routes and emergency routes, and alternative routes.

## 1 Introduction

Flood disaster mitigation is a series of actions taken to reduce or eliminate the negative impacts caused by floods in order to protect human life, reduce property damage, and minimize disruption to communities and the environment that are carried out before, during, and after a flood, and involve planning, preparation, and implementation of strategic measures. Effective disaster management can reduce the damage caused by flooding to the natural and human environment by identifying actions to implement the stages of knowledge management including before, during and after the flood event [1]. Flood disasters pose a serious threat to human life and property, so it is necessary to reveal the spatial driving factors of flood disasters on a macroscopic scale in order to reduce their impact [2]. Indonesia is in a geographical area that is prone to disasters which requires serious preparedness in facing disasters [3]. Floods in Indonesia are often caused by heavy rain and land conversion, exacerbated by high rainfall intensity which results in frequent flooding [4].

The phenomenon of global climate change in various countries in the world has resulted in flooding that impedes road utility and this problem has increased drastically [5]. There is a need for flood resilience evaluation to ensure that various urban systems and segments can prevent, respond and recover efficiently to achieve sustainable development [6], need disaster identification scenarios and classification criteria for flood risk identification [7]. Therefore, powerful technology or tools are needed to support efficient flood disaster management for damage assessment, real-time monitoring, resource allocation, and faster response [8]. The analysis of congestion risk due to urban road networks facing flood risk in transportation systems based on static network characteristics, but fails to reveal the dynamic impact of developing flood events on network topology, so traffic engineering is needed to integrate dynamic flood models through network property measurements [9]. It is important to analyze comprehensive vulnerability to improve the management of road network vulnerability in complex urban environments due to disruptions to the road network that impact traffic flow through identification and mitigation of vulnerable components [10]. Flood inundation modelling method as a starting point reference for impact analysis and flood risk assessment on roads to provide insight into areas and regions at risk of flooding [11]. An important aspect in analyzing traffic flows in flood-prone areas is to note that traffic flows are dynamic and change over time, and it is important to identify who is in danger [12].

Bandung City is the capital of West Java Province as well as the center of Greater Bandung Metropolitan Area with a population growth rate of 0.35% per year. The population growth rate has caused an increase in land needs in Bandung City, which has an impact on the

* Corresponding author: srisarjana@gmail.com

conversion of conservation land functions. In 2019, 57.23% of land in Bandung City was used for settlements and only 12.2% of land was used for green open space, far below the supposed figure of 30% of the city's area. Uncontrolled land conversion increases the risk of flooding in areas with lower elevations than the surrounding area, such as in the Gedebage Area.

The Gedebage area which is in the coverage of the Gedebage City Sub-Region and the Ujung Berung City Sub-Region is located at the lowest point of Bandung City with an altitude of 666 – 892 meters above sea level. This area functions as a City Service Center and a City Strategic Area for the East Bandung area so that the level of community mobility and the volume of vehicles that cross this area is relatively high and often causes severe congestion [13]. At the time of flooding in the Gedebage area, the flooded transportation infrastructure is practically unusable so that the transportation operation pattern is disrupted and eventually causes congestion.

Floods that occur in the Gedebage area are included in the category of moderate flood risk with a height between 0.3-0.5 meters. The following is the distribution of flood inundation points in the Gedebage Area with low to high inundation height category.

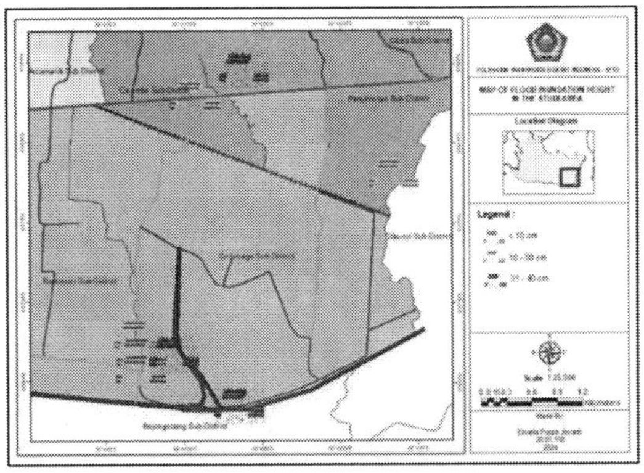

**Fig. 1.** Map of flood inundation height in the study area

Based on Fig. 1 above, there are two flood inundation points with a height of 10-40 cm located on the main road network, namely at the intersection of Soekarno-Hatta Street with Rumah Sakit Street (Gedebage Intersection) and on Sor GBLA Street. Flooding at Gedebage Intersection is one of the main problems that is the main priority for handling by the Bandung City Government. The flood at the intersection has an average height of ≥ 30 cm with a receding time of ≥ 60 minutes, which greatly hampers traffic movements in the Gedebage Area considering that the road network is a primary arterial road network as well as a national road that is the main access to movement in the Gedebage Area. Meanwhile, on Sor GBLA Street, the flood height ranges from 31-40 cm with a receding time of about 1-8 days. Thus, traffic on Sor GBLA Street was stopped during that time span.

Flood inundation will cause damage to highway infrastructure such as holes covered by flood puddles. This will result in the driver falling into the hole and the occurrence of motor vehicle tire leakage [14]. Therefore, traffic engineering management is needed as a solution to overcome congestion due to flooding while efforts to control water systems continue to be carried out. This is done in order to ensure that transportation operations continue to be carried out and to facilitate the evacuation of victims or distribution of logistics for flood victims [15]. This study aims to find out traffic engineering scenarios that are able to break down congestion and improve traffic performance during floods.

## 2 Theoretical studies

A disaster is a serious disruption to the functioning of a community or society that results in widespread humanitarian, material, economic, or environmental losses and impacts, exceeding the capacity of the affected community [16]. Disasters can be triggered by natural factors, non-natural factors, or human actions that cause significant losses [17]. Flooding is a type of disaster caused by various factors, such as due to high rainfall, obstruction of waterways due to siltation and accumulated garbage, and so on [18]. Flooding occurs due to water that is not contained in the sewer so that it overflows and floods the surrounding area [19].

Traffic engineering management is the process of regulating the supply and demand of highway systems to overcome short-term traffic challenges and handle traffic within a certain period of time [20]. Another opinion states that traffic engineering management is a technique for managing and controlling traffic flows by optimizing the use of existing infrastructure to facilitate traffic efficiently in the use of road sections and improve the smoothness of the movement system [21]. Some strategies in traffic engineering management are capacity management, priority management, and demand management [22].

Awakening refers to the number of trips that start from a zone or region in a period of time [23]. The trip generation model functions to estimate and predict the number of trips originating from a zone [24]. Trip distribution is the total trips that start from a zone/region spread to various destination zones [25]. The number of movement flows is expressed in the origin destination matrix (OD Matrix). One of the methods in forming OD Matrix is the easy-to-use Furness method [26]. The distribution of future movements is obtained by multiplying the distribution at the time of the nest by the growth rate of the origin zone or destination zone which is carried out alternately [27].

The mode selection stage serves to determine the burden of the trip or to know the number of people and goods who will use or choose various modes of transportation available to serve a certain point of origin of a certain destination [28]. The process of loading traffic requires data such as the matrix of origin of travel destinations, road capacity, and network characteristics [29]. The road network is a system consisting of sections that are interconnected and bind the centers of activity in it. There are three main parameters of the road network performance used, travel time (passenger car unit-hour) (pcu-h), distance travel pcu-km), and network speed (km/h) [30].

The performance of a road segment refers to its ability to meet the needs of traffic flow according to its role, which can be measured and compared to road service standards [31]. The performance indicators of the road section used include the degree of saturation of the road section ($D_J$), traffic speed, and traffic density. There are several methods of assessing the level of road service levels, one of which is using highway capacity manual [32].

Intersection performance assessments are differentiated based on the type of control. The intersection performance component consists of the degree of saturation of the intersection ($D_J$), delay time, and the length of the intersection queue. The level of interchange service is determined based on the average delay time. Intersection performance assessment is an important process in traffic management and transportation planning. Good intersection performance not only improves traffic flow but also plays a role in road user safety, time efficiency, and comfort. In the implementation of traffic engineering in emergency conditions such as during natural disasters, temporary barriers are used that can be easily moved. In placing signs, it is necessary to consider things such as the operational speed of the vehicle, the geometric condition of the road, the roadside environment, and the efficiency of the number of signs.

## 3 Research methodology

The research was conducted in Gedebage and Ujung Berung City Sub-Region, which are one of the flood-prone areas in Bandung City. The flood inundation point that submerged the road infrastructure with a height of >30 cm and a receding time of >60 minutes is located at the intersection of Soekarno – Hatta Street and Rumah Sakit Street, Soekarno – Hatta Street (in front of the Gedebage Main Market), Rumah Sakit Street, and Sor GBLA Street. Data collection was carried out in November – December 2023 while data processing was carried out in January – June 2024.

The first stage in this research process is data collection to find out the existing conditions of the research location. There are 2 types of data needed, namely primary data and secondary data. Primary data includes information about the geometry of the road network, vehicle flow data on roads and intersections, and travel speed. Primary data was obtained through road and intersection inventory surveys, traffic counting (TC) surveys, classified turning movement counting (CTMC) surveys, and moving car observer (MCO) surveys. Secondary data includes administrative maps and road networks, data on affected road networks, flood characteristics in previous events, and existing traffic data.

Traffic zones are set on road sections that are used as access in and out of the study area [33]. Then a matrix of origin of travel destinations (OD Matrix) was formed using the Furness method. The Furness method is an easy-to-use, flexible, and non-accessible method so it was chosen as a method of forming OD Matrix. After that, OD Matrix is charged to the road network using assistance Software PTV Visum to get the traffic volume of the model, one of the tools in transportation planning used to analyze, model, and plan transportation systems at the macro level. PTV Visum has a use to manage GIS data, model transportation requests, and load the network with modelling principles four step modelling, and an essential tool in transportation planning and management due to its ability to analyze, model, and plan transportation systems at a macro level. It is used by urban planners, transportation engineers, and policy makers to understand and manage transportation dynamics at a large scale.

## 4 Results and discussion

In this study, there are 13 zones formed based on the entry and exit points of traffic flows in the study area show in Fig. 2. The following is a map of the distribution of traffic zones in the study area in this study.

The establishment of traffic zones based on the entry and exit points of traffic flow in a study area is an important approach in transportation planning and analysis. This division allows for a deeper understanding of the dynamics of traffic movement and more effective planning. Traffic zones allow for a more detailed analysis of travel patterns, including the origin and destination of trips, the number of trips made, and the choice of transportation mode. This helps in understanding the transportation needs of the community and the distribution of traffic load. By understanding the main entry and exit points in each zone, planners can manage traffic flow more effectively, reduce congestion, and optimize the use of road infrastructure.

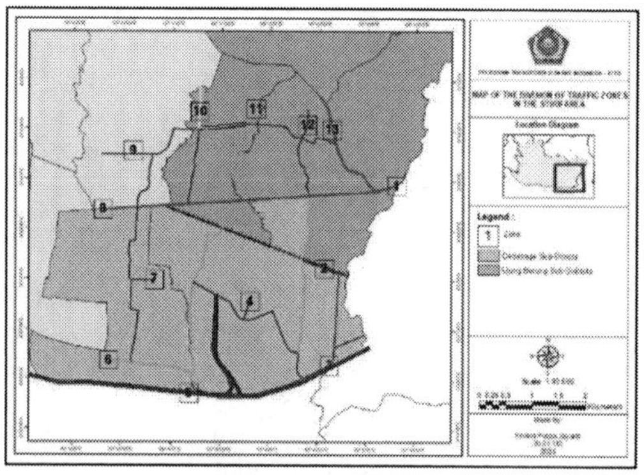

**Fig. 2.** Division of area zones

The zones that have been formed will then be calculated the number of traffic movements which will then be modelled into the number of rises and draws of trips in the zone. The number of traffic movements leading into the study area is determined as the number of trip awakenings, and the number of traffic movements leading out of the study area is determined as the number of trip pulls. The number of rides and rides are separated based on the classification of the vehicle to make it easier to analyze the traffic load. Then the number of rises and pulls of each type of vehicle is added up to get the total rise and pull of each zone.

The results of the traffic counting (TC) survey in the zones in the study area showed that there was still a difference between the number of awakenings and trips in all zones in the study area. Therefore, adjustments are made to the number of awakenings or travel pulls into the highest value so that it will produce the same amount. The adjustment is made for each type of vehicle. The following is the number of awakenings and travel attractions of each zone in the study area.

**Table 1.** Total awakening and attraction of travel

| Zone | Information | Before Customization | | After Customization | |
|---|---|---|---|---|---|
| | | Awakening (Veh/Hour) | Attraction (Veh/Hour) | Awakening (Veh/Hour) | Attraction (Veh/Hour) |
| 1 | Cirebon - Bandung St. | 2.335 | 3.362 | 2.441 | 3.381 |
| 2 | Sor GBLA Segment 3 St. | 615 | 520 | 644 | 522 |
| 3 | Sor GBLA Segment 3 St. | 554 | 379 | 579 | 383 |
| 4 | Boulevar Utama St. | 366 | 466 | 381 | 467 |
| 5 | Raya Sapan St. | 586 | 548 | 612 | 550 |
| 6 | Raya Terusan Derwati St. | 403 | 496 | 421 | 497 |
| 7 | Terusan Saluyu St. | 786 | 733 | 815 | 733 |
| 8 | Soekarno- Hatta Segment 2 St. | 3.739 | 3.330 | 3.900 | 3.351 |
| 9 | Cingised St. | 293 | 309 | 306 | 309 |
| 10 | Golf Raya St. | 267 | 145 | 279 | 145 |
| 11 | Rumah Sakit Segment 2 St. | 561 | 688 | 586 | 692 |
| 12 | Sukamaju St. | 192 | 125 | 200 | 125 |
| 13 | AH Nasution Segment 2 St. | 1.628 | 1.697 | 1.702 | 1.711 |
| TOTAL | | 12.325 | 12.798 | 12.866 | 12.866 |

Based on Table 1 above, before the adjustment, the trip revival in the study area amounted to 12,325 vehicles/hour and the trip pull amounted to 12,798 vehicles/hour. Meanwhile, after adjustments, the number of awakenings and pulls in the study area already has the same value, which is 12,866 vehicles/hour.

Based on the results of adjusting the number of rises and travel attractions of each vehicle, an analysis of the distribution of trips was carried out to produce a matrix of origin of travel destinations between zones. In building the origin destination matrix using the Furness method, the initial OD Matrix is determined with a value of 1 and then multiplied by the origin zone growth factor which is the result of dividing the number of trips from the survey results with the number of trips in the analyzed matrix. The results of the first iteration of the OD Matrix were then multiplied by the growth factor of the destination zone. Next, the results of the second iteration of OD Matrix are multiplied again by the growth factor of the origin zone until the growth factor value is equal to 1 or $O_i$ = oi. Here is the matrix of origin destination (OD Matrix) of the total trip using the Furness method.

**Table 2.** Total travel origin destination matrix in study area (vehicle/hour)

| OD | 1 | 2 | 3 | 4 | 5 | 6 | 7 | 8 | 9 | 10 | 11 | 12 | 13 | $O_i$ |
|---|---|---|---|---|---|---|---|---|---|---|---|---|---|---|
| 1 | 0 | 119 | 84 | 97 | 119 | 105 | 158 | 1.085 | 65 | 31 | 151 | 26 | 406 | 2.445 |
| 2 | 181 | 0 | 17 | 19 | 24 | 21 | 32 | 215 | 13 | 6 | 30 | 5 | 80 | 644 |
| 3 | 159 | 21 | 0 | 17 | 21 | 19 | 28 | 194 | 12 | 6 | 27 | 5 | 72 | 579 |
| 4 | 103 | 12 | 9 | 0 | 14 | 13 | 22 | 128 | 8 | 4 | 17 | 3 | 47 | 381 |
| 5 | 170 | 22 | 16 | 18 | 0 | 20 | 29 | 208 | 12 | 6 | 28 | 5 | 78 | 612 |
| 6 | 115 | 15 | 10 | 14 | 16 | 0 | 23 | 141 | 9 | 4 | 19 | 4 | 52 | 421 |
| 7 | 224 | 26 | 20 | 30 | 31 | 29 | 0 | 281 | 19 | 8 | 37 | 7 | 102 | 816 |
| 8 | 1.533 | 191 | 143 | 174 | 207 | 184 | 283 | 0 | 113 | 52 | 259 | 45 | 706 | 3.890 |
| 9 | 83 | 11 | 8 | 10 | 11 | 10 | 16 | 101 | 0 | 3 | 14 | 3 | 37 | 306 |
| 10 | 75 | 10 | 7 | 9 | 10 | 9 | 14 | 91 | 6 | 0 | 13 | 2 | 34 | 279 |
| 11 | 162 | 21 | 15 | 18 | 22 | 19 | 29 | 204 | 12 | 6 | 0 | 5 | 74 | 586 |
| 12 | 54 | 7 | 5 | 6 | 7 | 7 | 11 | 65 | 4 | 2 | 9 | 0 | 24 | 200 |
| 13 | 522 | 69 | 49 | 54 | 68 | 60 | 88 | 637 | 37 | 18 | 88 | 15 | 0 | 1.704 |
| $D_j$ | 3.381 | 522 | 383 | 467 | 550 | 497 | 733 | 3.351 | 309 | 145 | 692 | 125 | 1.711 | 12.866 |

Table 2 above shows the distribution of trips from the origin zone to the destination zone in the study area with a total movement of 12,866 vehicles/hour. In forming the OD Matrix, each type of vehicle is repeated (iteration) with different numbers until E = 1 or $O_i$ = oi. The motorcycle and passenger car OD Matrix is completed in the 6th iteration, the medium vehicle OD Matrix is completed in the 12th iteration, the large bus OD Matrix is completed in the 7th iteration, and the large truck OD Matrix is completed in the 8th iteration.

The loading model on the PTV Visum will produce a vehicle volume (pcu/h) for each section which will then be tested using the GEH method to find out whether the model has represented the actual condition or not. Models with test scores below 5 can be declared valid and acceptable as models that represent the actual conditions. Meanwhile, models with test scores in the range of 5-10 are declared doubtful models with the possibility of corrupted or invalid data, and models with test scores above 10 are declared invalid/rejected because the data cannot reflect the actual conditions in the field. The following are the results of the traffic volume validity test on the model created against the volume of survey results.

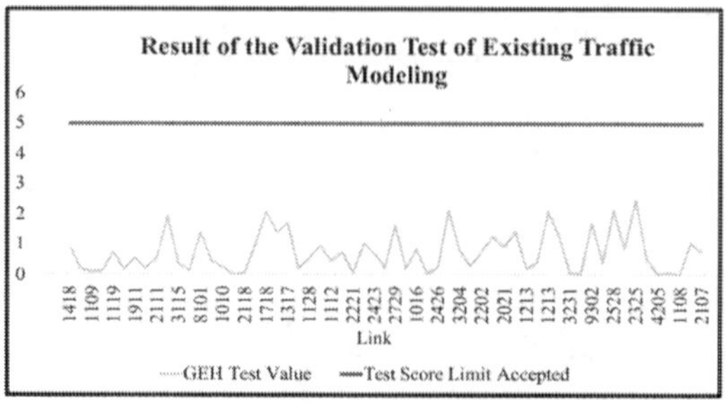


**Fig. 3.** Results of the existing model validation test

Based on Fig. 3, it can be seen that the entire traffic volume of the modelling results is below 5 so that the model can be declared valid and can represent the actual conditions in the field. After the existing model is declared valid, the model can be modified to form other conditions, namely flood conditions and conditions when traffic engineering is applied.

Floods cause a significant decrease in capacity on roads and intersections, which means that the number of vehicles that can be served by the road and intersection is drastically reduced. In some cases, the capacity can even be 0, meaning that the road or intersection cannot be passed at all by vehicles because it is inundated. The main cause of this decrease in capacity is waterlogging that makes the road or intersection impassable, due to the water depth being too high or due to infrastructure damage caused by the flood. PTV Visum is a software for traffic planning and simulation that is used to analyze and model changes in traffic flow. The analysis is carried out using the same OD matrix as the existing conditions, which means that data on travel patterns and driver destinations before the flood is reused to understand how the flood affects traffic flow. This analysis aims to see the changes in traffic volume that occur in the study area during flood conditions by taking into account the influence of impassable roads and intersections.

The results of this analysis can be used to design adjustments in traffic management such as route diversion or access restrictions to certain areas that are severely affected by flooding. This information is essential for emergency planning including evacuation and other critical flood response. By understanding the impact of flooding on road network capacity and using the data to model traffic changes, authorities can plan and implement more effective strategies to keep traffic flowing and road users safe. Figure 3 shows the change in traffic volume loading that occurs during normal conditions with flood conditions.

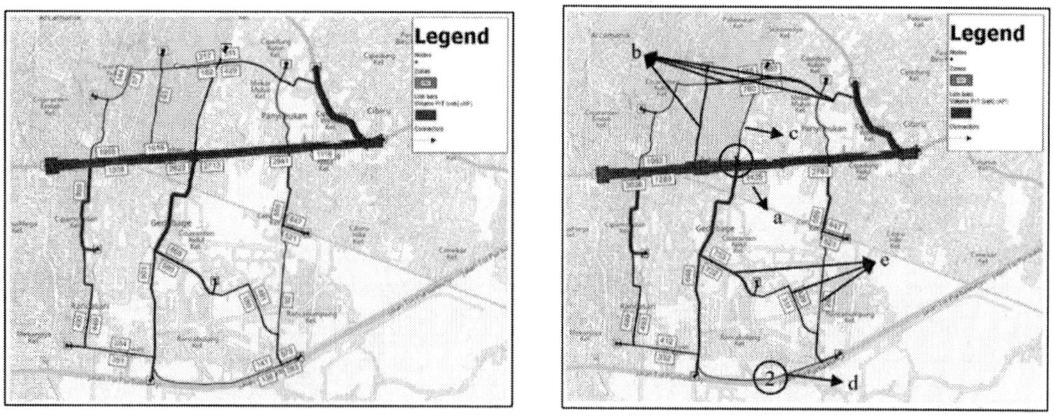


(a) (b)

**Fig. 4.** Comparison of traffic loading results in normal conditions (a) and flood conditions (do-nothing) (b)

Based on Fig. 4, it can be seen that there are several points of road sections and intersections that have undergone significant changes when flooding occurs in the study area. The intersection marked by circle 1 is Gedebage Intersection, while the road section marked by circle 2 is Sor GBLA Street Segment 1, both of which are the worst flood points in the study area. At the time of the flood, the Gedebage intersection experienced a build-up of traffic volume because the north arm (rumah sakit street) and part of the east arm exit were flooded and could not be crossed by vehicles (Fig. 4a). As a result of the accumulation of vehicles that occurred at Gedebage Intersection, there was also an increase in traffic volume on Cisaranten Wetan Street, Cinambo Street, Panghegar Street, and Pangaritan Street which were identified as alternative routes for vehicles when flooding occurred at Gedebage Intersection (Fig. 4b).

Proposal for do-something 1 in improving traffic performance and unraveling congestion in the event of flooding in the study area shown in Fig. 5. Implementing contraflow on Soekarno-Hatta 1 Street Fast Lane 2 to increase the exit width of the western approach of Gedebage Intersection as follows.

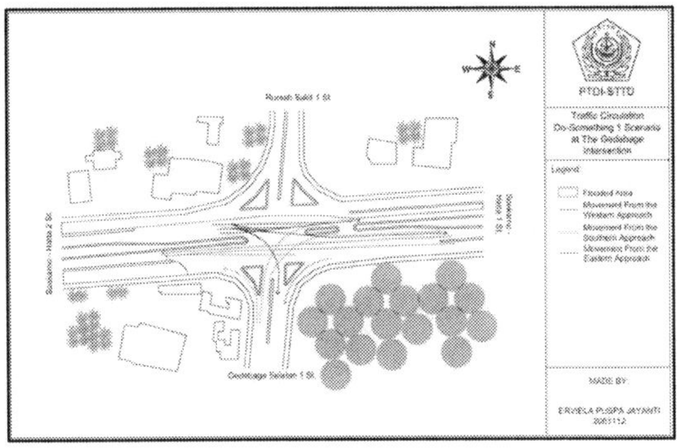


**Fig. 5.** Traffic circulation do-something scenario 1 at Gedebage intersection

To support the contraflow regulation on the road segment above, traffic flow regulation is also carried out on the median opening of the U-turn that has been free from flooding to direct vehicles on the contraflow lane to return to the original lane and divide the traffic volume from the east of Soekarno-Hatta 1 Street Fast Lane 1 so that it does not accumulate on the contraflow lane. Closing access to the intersection arm whose entire road body is flooded so that motorists do not cross. Adjust the type of approach of the SPBU – Rumah Sakit Intersection to be a protected type (P) during floods. Change the phase and optimize the cycle time of Gedebage Intersection and Intersection SPBU – Rumah Sakit as follows.

Proposal for do-something 2 in improving traffic performance and unravelling congestion during floods in the study area shown in Fig. 6. Changing the direction of traffic on Soekarno-Hatta 1 Street Fast Lane 2 which was originally an intersection approach (traffic direction to the intersection) into an intersection exit lane as follows.

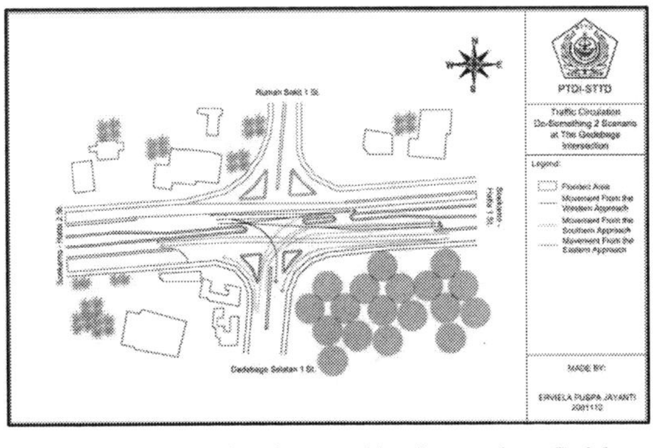


**Fig. 6.** Traffic circulation do-something 2 scenario at Gedebage intersection

To support circulation regulation on the road segment above, traffic flow is regulated at the median opening of the U-turn that has been free from flooding to direct vehicles on the new exit lane to return to the original lane. In addition, traffic flow diversion on the fast lane leading to the intersection to the slow lane at the median opening of the fast lane and the slow lane before the traffic flow return point is also carried out so that there is no conflict between the confluence of different flows. Closing access to the intersection arm whose entire road body is flooded so that motorists do not cross. Prohibiting motorcycles from passing on the fast lane on the north side of Soekarno – Hatta 1 Street. Prohibiting large vehicles from crossing alternative routes that are local roads. Adjust the type of approach of the SPBU – Rumah Sakit Intersection to be a protected type (P) during floods.

In each traffic engineering scenario, a reload analysis was carried out using PTV Visum with the same OD Matrix as the existing conditions to see the changes in traffic volume that occurred in the study area during the implementation of the do-something 1 scenario or the implementation of the do-something 2 scenarios. The following is a comparison of the results of traffic loading on flood conditions, do-something 1 conditions, and do-something 2 conditions.

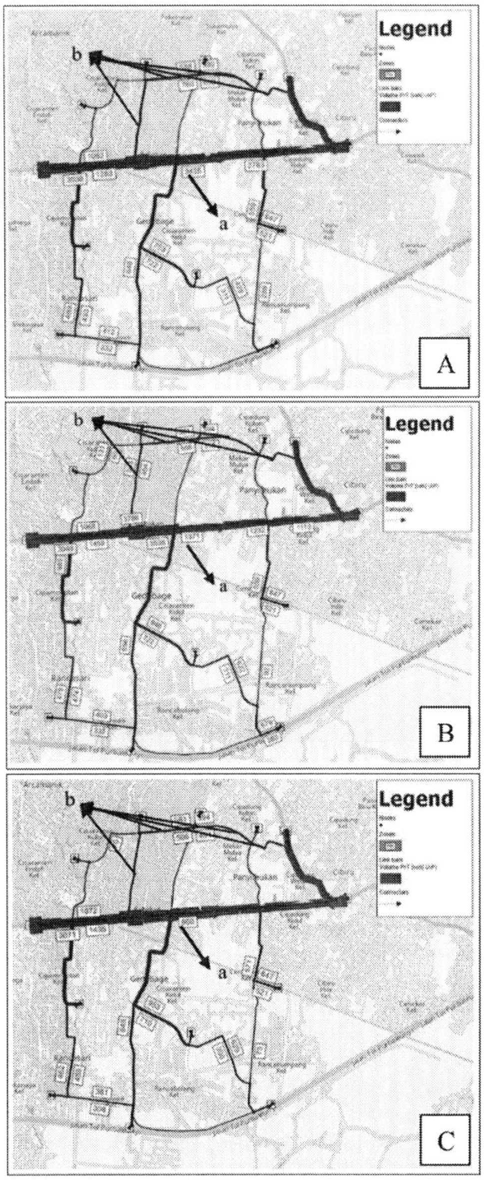


**Fig. 7.** Comparison of traffic loading in flood conditions (do-nothing) (A), do-something 1 (B) and do-something 2 (C)

Based on Fig. 7, there is a significant change in traffic volume on the Gedebage Intersection which is the main point of congestion (see letter a). In addition, there was also a decrease in traffic volume on Cisaranten Wetan Street, Cinambo Street, Panghegar Street, and Pangaritan Street (see letter b). The results of traffic loading showed a decrease in traffic performance when the do-something 1 scenario and do-something 2 scenarios were implemented. The following is a comparison of road network performance, section performance, and intersection performance between flood conditions, do-something 1 conditions, and do-something 2 conditions.

Gedebage intersection is considered a major congestion point, meaning that this area often experiences severe traffic congestion. There is a significant change in traffic volume at this intersection, indicating a diversion or redistribution of traffic flow as a result of the applied scenario. Traffic volume decreased, which could be caused by the redistribution of traffic flow to other routes or a reduction in vehicles on the road due to the applied scenario. The decrease in traffic volume on these roads can reduce congestion and improve driver comfort and travel efficiency. The results of the analysis show that despite the adjustment or diversion of traffic, overall traffic performance decreased during the implementation of the do-something 1 and do-something 2 scenarios.

**Table 3.** Comparison of road network performance in flood conditions (do-nothing), do-something 1 conditions, and do-something 2 conditions

| No. | Network Performance Parameters | Performance | | |
|---|---|---|---|---|
| | | Do-Nothing Condition | Do-Something Condition 1 | Do-Something Conditions 2 |
| 1 | Average Speed (km/h) | 31,55 | 32,02 | 32,36 |
| 2 | Total Travel Distance (pcu-km) | 82028,96 | 80991,63 | 83441,80 |
| 3 | Total Travel Time (pcu-hours) | 3352,47 | 3070,54 | 3107,52 |
| 4 | Average Delay (seconds/pcu) | 48,35 | 17,91 | 16,27 |

**Table 4.** Comparison of performance and service levels on several road sections in do-nothing condition, do-something 1 condition, and do-something 2 condition

| Street Name | Do-Nothing Condition | | Do-Something Condition 1 | | Do-Something 2 Conditions | |
|---|---|---|---|---|---|---|
| | Speed (km/h) | LoS | Speed (km/h) | LoS | Speed (km/h) | LoS |
| Soekarno - Hatta 1 Slow Lane 3 (N) St. | - | - | - | - | - | - |
| Soekarno - Hatta 1 Fast Lane 1 (N) St. | 24,4 | E | 33,0 | C | 32,5 | C |
| Soekarno - Hatta 1 Fast Lane 2 (N) St. | 10,1 | F | 19,5 | E | 30,0 | D |
| Soekarno - Hatta 1 Fast Lane 2 (S) St. | 36,6 | C | 24,3 | E | 31,0 | C |
| Soekarno - Hatta 2 Slow Lane 1 (N) St. | - | - | - | - | - | - |
| Soekarno - Hatta 2 Fast Lane (N) St. | 27,5 | D | 26,4 | D | 25,9 | D |
| Soekarno - Hatta 2 Fast Lane (S) St. | 30,3 | D | 30,3 | D | 30,1 | D |
| Soekarno - Hatta 2 Slow Lane 1 (S) St. | - | - | - | - | - | - |
| Cinambo St. | 27,5 | B | 29,5 | B | 30,4 | B |
| Gedebage Selatan 1 St. | 23,7 | B | 22,9 | C | 22,7 | C |
| Gedebage Selatan 2 St. | 27,2 | B | 27,1 | B | 27,0 | B |
| Rumah Sakit 1 St. | - | - | - | - | - | - |
| Rumah Sakit 2 St. | 25,4 | C | 25,4 | C | 25,4 | C |
| Raya Derwati 1 St. | 22,2 | C | 26,7 | B | 26,7 | B |
| Sor GBLA 1 (N) St. | - | - | - | - | - | - |
| Sor GBLA 1 (S) St. | - | - | - | - | - | - |
| Pangaritan St. | 19,4 | C | 21,6 | B | 21,9 | B |
| Panghegar St. | 18,1 | C | 21,3 | C | 22,6 | B |

Table 3 shows a significant improvement in road network performance during the implementation of the do-something 1 scenario and do-something 2 scenarios during floods which are characterized by an increase in all road network performance parameter values. In do-something 1 conditions, the speed increases to 32.02 km/h. The total travel distance decreased by 1.2%, the total travel time decreased by 8.4%, and the average delay time decreased by 62%. Meanwhile, in the do-something 2 conditions, the improvement in the performance of the road network is slightly better than the do-something 1 condition even though there is an increase in travel distance compared to flood conditions (do-nothing). Travel speed increased to 32.36 km/h, total travel time decreased by 7.3%, and average delay time decreased by 66%.

The following are roads and intersections in the study area with poor performance values during floods (do-nothing) that have improved performance and service levels in do-something 1 and do-something 2 conditions. Although there was a slight increase in travel distance under the do-something 1 and do-something 2 scenarios, something 2, other improvements in terms of travel speed, total travel time, and average delay time indicate that this scenario is more effective in improving the performance of the road network. There are certain roads and intersections in the study area that perform very poorly during floods without intervention (do -nothing), but experienced significant performance and service level improvements with the implementation of do-something 1 and do-something 2 scenarios

**Table 5.** Comparison of performance and service levels at several intersections of do-nothing conditions, do-something 1 conditions, and do-something 2 conditions

| Intersection Name | Types of Craving | Approach / Intertwining | Do-Nothing Condition | | Do-Something Condition 1 | | Do-Something Condition 2 | |
|---|---|---|---|---|---|---|---|---|
| | | | Delay (sec) | LoS | Delay (sec) | LoS | Delay (sec) | LoS |
| Gedebage | Signal Control | N | - | - | - | - | - | - |
| | | E | 58,09 | E | 30,88 | D | 27,67 | D |
| | | S | 61,55 | F | 26,68 | D | 31,00 | D |
| | | W | 430,90 | F | 57,13 | E | 19,32 | C |
| Gas Station - Rumah Sakit | Signal Control | N | 18,46 | C | 34,32 | D | 34,32 | D |
| | | E | 20,13 | C | 34,20 | D | 34,24 | D |
| | | S | - | - | - | - | - | - |
| | | W | 328,99 | F | 32,83 | D | 32,41 | D |
| Rancanumpang - Sor GBLA | No Signal Control | - | - | - | - | - | - | - |
| Derwati - Sor GBLA | No Signal Control | - | - | - | - | - | - | - |

Based on Table 4, it can be seen that over all there is an improvement in the performance of the road sections in the study area at the time of flooding with the implementation of the do-something 1 scenario and the do-something 2 scenarios. The speed of travel on the above roads as a whole has increased, which then affects the level of road service, which also increases. However, based on Table 5 above, the do-something 2 scenario is more influential in increasing the speed and level of road service than the do-something 1 scenario.

These results indicate that two different scenarios called do-something 1 and do-something 2 scenarios have been implemented to address the impact of flooding on road performance in a study area. Both scenarios appear to be designed to improve vehicle travel speed and road service level during flooding. Overall, there was an improvement in road performance after the scenarios were implemented, meaning that road conditions during flooding, such as travel speed and traffic flow, improved. Vehicle travel speed on these roads increased, which in turn improved road service level. Road service level usually refers to the quality of traffic flow and driver comfort. The do-something 2 scenario was found to be more effective in improving road speed and service level compared to the do-something 1 scenario, meaning that the second scenario provided better results in terms of improving road conditions during flooding.

Based on Table 5, it can be seen that overalls there is an increase in the performance of the intersection in the study area at the time of flooding with the implementation of the do-something 1 scenario and the do-something 2 scenarios. The delay in approaching the signal control intersection has decreased significantly with the implementation of this handling scenario. However, based on the table above, the do-something 2 scenario is more influential in increasing the level of intersection services than the do-something 1 scenario. This can be seen from the delay time difference that occurs in the signal control intersection approach of the two scenarios. The delay time generated by the do-something 2 scenario is much shorter than the do-something 1 scenario.

## 5 Conclusion

Based on the results and discussion above, it can be concluded that at the time of flooding at Gedebage Intersection and Sor GBLA 1 Street, there was a significant decrease in traffic performance on the road network in the study area and there was a buildup of traffic volume on several road sections identified as alternative routes when flooding occurred in the study area. With the implementation of the do-something 2 scenarios in the event of a flood, the traffic performance in the study area has increased significantly and the congestion arising from floods can be minimized. This is marked by a decrease in total travel time and average intersection delays, as well as an increase in the average speed of travel in the study area. The results of the analysis explained that by adding an exit lane on the east arm of Gedebage intersection, optimizing the time of the second cycle of the signal control intersection, and limiting the movement of several types of vehicles when flooding occurs in the study area, it is able to improve the performance of the road network by 74.51% compared to flood conditions without the application of traffic engineering (do-nothing). Based on the findings of the study, it is recommended that researchers can further study the median opening of the U-turn as well as the median of the slow lane and fast lane on Soekarno-Hatta Street so that the influence can be known on the performance of the road section.

## References


1. M. Y. Mohammadi, E. Abbasi, H. Farhadian, and A. Asgari, Implementation of knowledge management measures in flood disaster management: A systematic literature review, Environ. Sustain. Indic., **23,** 100431 (2024). https://doi.org/10.1016/j.indic.2024.100431
2. L. Yang *et al.*, A comprehensive framework for assessing the spatial drivers of flood disasters using an optimal Parameter-based geographical Detector–machine learning coupled model, Geosci. Front., **15,** 101889 (2024). https://doi.org/10.1016/j.gsf.2024.101889
3. S. Hasbi, Z. Hanim, and S. bin Husain, The implementation optimization of school development plan in flood disaster mitigation policy in tropical rainforest (Case study at state junior high school 5 Samarinda), Soc. Sci. Humanit. Open. **7**(100440), 1–7 (2023).
4. I. L. Nugraheni, A. Suyatna, A. Setiawan, and Abdurrahman, Flood disaster mitigation modeling through participation community based on the land conversion and disaster resilience, Heliyon. **8**(e09889), 1–14 (2022). https://doi.org/10.1016/j.heliyon.2022.e09889
5. U. I. Oduah, C. M. Anierobi, and O. G. Ilori, Inventing a robust road-vehicle flood level monitoring device for disaster mitigation, Heliyon. **9**(e20784), 1–10 (2023). https://doi.org/10.1016/j.heliyon.2023.e20784
6. J. Li, Y. Zou, J. Li, and S. Si, Urban spatial, engineering and management resilience capacity enhancement from a flood safety perspective, Water-Energy Nexus. **7**, 213–226 (2024). https://doi.org/10.1016/j.wen.2024.07.002
7. R. Zhang, Y. Li, T. Chen, and L. Zhou, Flood risk identification in high-density urban areas of Macau based on disaster scenario simulation, Int. J. Disaster Risk Reduct. **107**(104485) (2024). https://doi.org/10.1016/j.ijdrr.2024.104485
8. M. A. Islam, S. I. Rashid, N. U. I. Hossain, R. Fleming, and A. Sokolov, An integrated convolutional neural network and sorting algorithm for image classification for efficient flood disaster management, Decis. Anal. J. 7(100225), 1–13 (2023). https://doi.org/10.1016/j.dajour.2023.100225
9. X. Lin, Q. Lu, L. Chen, and I. Brilakis, Assessing dynamic congestion risks of flood-disrupted transportation network systems through time- variant topological analysis and traffic demand dynamics, Adv. Eng. Informatics. **62**(102672) (2024). https://doi.org/10.1016/j.aei.2024.102672
10. A. R. Marian, R. Hijazi, E. Masad, A. Abdel-wahab, and A. R. Marian, Quantifying the Vulnerability of Road Networks to Flood-Induced Closures Using Traffic Simulation, Transp. Eng. **17**(100262) (2024). https://doi.org/10.1016/j.treng.2024.100262
11. H. He, R. Li, J. Pei, J.-P. Bilodeau, and G. Huang, Current overview of impact analysis and risk assessment of urban pluvial flood on road traffic, Sustain. Cities Soc. **99**(104993) (2023). https://doi.org/10.1016/j.scs.2023.104993
12. S. Perazzini, R. Metulini, and M. Carpita, Integration of flows and signals data from mobile phone network for statistical analyses of traffic in a flooding risk area, Socioecon. Plann. Sci. **90**(101747), 1–17 (2023). https://doi.org/10.1016/j.seps.2023.101747
13. H. I. Khildan and I. Saputra, The influence of the summarecon bandung area on traffic on jalan Gedebage Selatan, Semin. Nas. dan Disminasi Tugas Akhir 2022, 486–499 (2022).
14. A. N. Agustin, M. Muhibbin, and D. Isnaeni, The Role of Local Government in Overcoming Flood Problems in Malang City. J. Din. **29**(2), 7818–7844 (2023).
15. H. Satmiko, Transportation crisis management due to natural disasters, demonstrations, special transportation seasons, and transportation accidents. (Bandung: Bandung Nuansa Cendekia, 2019).
16. D. Tamitiadini, I. Adila, and W. W. A. Dewi, Disaster Communication: Theory and Practical Approach to Disaster Studies in Indonesia. 1st ed. (Malang: UB Press, 2019).
17. E. P. Raharjo, S. Sarjana, and M. Safitri, Transportation infrastructure planning in supporting disaster mitigation: Case study in Mount Gamalama, Jamba J. Disaster Risk Stud. **14**(1), 1–11 (2022). https://doi.org/10.4102/jamba.v14i1.1123
18. A. J. Echendu, Applicability of Indigenous knowledge and methods in flood risk management in a nigerian city, Nat. Hazards Res. **4**(2), 239–245 (2024). https://doi.org/10.1016/j.nhres.2023.09.001
19. Suripin, Sustainable Urban Drainage Systems. Yogyakarta: (ANDI Offset, 2004).
20. I. Sarwoko, S. Widodo, and G. Z. Mulki, Traffic Management and Engineering at the Imam Bonjol - Daya Nasional Road Intersection in Pontianak City, J. Tek. Sipil. **17**(2), 1–9 (2017).
21. B. A. Hermawan, Traffic Management and Engineering in the CBD Area of Bekasi City, J. Pembang. Wil. Kota. **12**(1), 27–36 (2015).
22. Department of Public Works, Traffic Management, Regional Cities Urban Transport DKI Jakarta Training. (Dirjen Bina Marga, Jakarta, 1990).
23. F. Miro, Transportation Planning for Students, Planners, and Practitioners. (Jakarta: Erlangga, 2005).
24. T. Kawasaki, Y. Namba, H. Oka, and M. A. Dulebenets, Freight trip distribution using spatiotemporal aggregate data: A modified collective flow diffusion model-based approach, Transp. Res. Interdiscip. Perspect. **21**(100904), 1–16 (2023). https://doi.org/10.1016/j.trip.2023.100904
25. I. Sholichin, Trip Generation and Trip Distribution Analysis in North Surabaya, J. Tek. Sipil KERN.

**1**(2), 13–22 (2011).

26. G. Galliani, P. Secchi, and F. Ieva, Estimation of dynamic Origin–Destination matrices in a railway transportation network integrating ticket sales and passenger count data, Transp. Res. Part A Policy Pract. **190**(104246), 1–26 (2024). https://doi.org/10.1016/j.tra.2024.104246
27. T. B. S. Pambudi, Trip Distribution and Trip Assignment Modeling on Gempol Pasuruan Toll Road, Institut Teknologi Sepuluh Nopember, (2017).
28. A. Maulana, S. Sarjana, and T. R. Prastya, Traffic Performance Analysis in The Traditional Market Area. E3S Web of Conferences. **576**(05003), 1–18 (2024). https://doi.org/10.1051/e3sconf/202457605003
29. R. Tisnawan, F. Ramdhani, and M. R. Ariansyah, Route Planning of Freight Transportation Network in Pekanbaru City with PTV Visum Application, J. RAB Constr. **6**(2), 101–111 (2021).
30. Z. I. Pamungkas, Traffic Gyratory as an Effort to Improve Road Network Performance (Case Study of Duta Mall Area Banjarmasin), Bul. Profesi Ins. (**2**)1, 1–6 (2019).
31. M. S. Gea and J. Harianto, A Analysis of Road Section Performance Due to Parking on the Road Body (Case Study: Markets and Shops on Jalan Besar Delitua), **1**, 1–10 (2011).
32. R. I. Salas and S. Sarjana, Improving Alternative Road Network Performance Through Effectiveness of Ring Roads, J. Transp. Multimoda, **21**(2), 111–120 (2024). https://doi.org/10.25104/mtm.v21i2.2277
33. S. Sarjana, Smart City in Supporting Sustainable Cities, in 10th International Conference on Information Technology, Computer, and Electrical Engineering, 61–66 (2023). https://doi.org/10.1109/ICITACEE58587.2023.10277619.

# Assessment of groundwater potential zones using geospatial techniques in Mangalore Taluk, Dakshina Kannada District, Karnataka, India

*Elsa* Eka Putri[1*], *Masyhuri* Hamidi[2], *Suraj* M Shet[3], and *Swapna* S A[3]

[1]Department of Civil Engineering, Faculty of Engineering, Universitas Andalas, Padang, Indonesia
[2]Department of Management, Faculty of Economics and Business, Universitas Andalas, Padang, Indonesia
[3]Mangalore Institute of Technology and Engineering, Mangaluru, Karnataka, India

**Abstract.** Despite sufficient rainfall, a large part of India suffers from water scarcity. Groundwater occurs in weathered or semi-weathered/fractured layers in hard-rock areas whose thickness varies, generally, from 5m to 20m. Satellite pictures are widely being used for groundwater exploration. Study and analysis of remote sensing data is a fast and economical way of finding and exploring. Present study, for assessment of groundwater availability in Mangalore taluk, Dakshina Kannada District, Karnataka state, shows various groundwater potential zones delineated using remote sensing and GIS techniques. Groundwater availability in Mangalore taluk was divided based on its hydro geomorphologic conditions. Satellite imageries are used for preparing various thematic maps, viz. slope, drainage density, lineament, land use/cover, soil, rainfall, geology and geomorphology map, which were transformed to raster class data using the feature to raster converter tool in ArcGIS. All the raster maps were allocated a fixed percentage of influence, after which a weighted overlay tool or technique was used. Each weighted thematic layer was computed statistically for the groundwater potential zones. The results were verified with bore well yield data from India Water Resources Information System (IWRIS). The results yielded a good match with the obtained bore well data with an accuracy of 82.14%.

## 1 Introduction

Water is a crucial natural resource that exists as both surface water and groundwater. It is essential for all forms of life on Earth. The progress of our society relies heavily on the availability and utilization of sufficient water. This valuable resource can be scarce at times, while at other times, it may be plentiful, but its distribution is often uneven across different regions and periods. Groundwater accounts for the second-largest supply of freshwater resources, making up approximately 30% of the world's freshwater supply.

Groundwater has been a widely utilized water resource in many tropical nations. Due to its ease of extraction, groundwater tends to be shielded from the pollutants that affect surface water. Nonetheless, instances of over-extraction of groundwater are frequent. One main factor hindering the optimal use of groundwater resources is the inadequate understanding of groundwater fundamentals. The integration of remote sensing technology has enhanced groundwater exploration and preparatory studies financially, despite the high costs associated with depleting groundwater resources [1, 2].

Generally, it's recommended to analyse aeronautical photos or satellite images before conducting ground studies and fieldwork. This approach can help identify potential aquifer layers and areas for field investigations. However, it's important to remember that remote sensing does not replace the need for in-situ data collection, which is crucial for validating the accuracy of remote sensing information and its interpretation. Clearly, remote sensing can reduce the amount of field data collection needed [3-5].

Several experts reviewed GIS applications in hydrology and water management during the early to mid-1990s. Their studies highlight the significant role of Geographic Information Systems in these fields, indicating that their use primarily focuses on a structured and instructional approach. This integration of GIS has enhanced our understanding of hydrological processes and improved water management strategies, emphasizing its importance for effective resource planning and environmental sustainability.

## 2 Study area

Mangalore Taluk in Dakshina Kannada district of Karnataka state, a region on the western coast of India, is the study area of research work as seen in Fig. 1. Mangalore taluk has an area of 834 km$^2$ and a population of around ten lakhs. Mangalore is the administrative headquarters of the taluk. Mangalore taluk has almost a plain terrain and receives ample rainfall every year.

* Corresponding author: elsaeka@eng.unand.ac.id

**Fig. 1.** Map of the study area

## 3 Data collection

Varies of spatial and non-spatial data have been collected for the present study purposes, such as satellite data, topographic maps, groundwater data, meteorological data, etc.

### 3.1 Satellite data

IRS-LISS III satellite data has been crucial for creating various thematic maps. This data was obtained from the National Remote Sensing Centre (NRSC) in India, ensuring high-quality imagery for analysis. Additionally, other relevant datasets were downloaded for free from their respective websites to support the study.

Cartosat-1 digital elevation model (DEM) data was also sourced from the NRSC's open data section, providing valuable topographic information. The Survey of India toposheet was included to add detailed geographic features. For the analysis and mapping of the thematic layers, Geographic Information System (GIS) and image processing software, specifically ArcGIS, were utilized. This combination of resources enables effective data visualization and analysis, enhancing the overall quality and accuracy of the thematic maps produced [6].

### 3.2 Data for groundwater evaluation

The data required for the evaluation of groundwater in Mangalore taluk was collected from the different website portal. Geology and Geomorphology data were downloaded from Bhukosh, a geological survey of India website, Rainfall data for the past 10 years from the CHRS portal; Land use land cover from ESRI, Lineament data from Bhuvan of NRSC and Soil map of the world from the Food and Agriculture Organization (FAO) website.

## 4 Method

For the present study, mainly eight different themes were evaluated: (i) Geomorphology, (ii) Land use Land cover, (iii) lineament density, (iv) Soil, (v) Drainage density, (vi) Geology, (vii) Annual rainfall and (viii) Slope on the raster GIS platform.

A raster GIS model was chosen for its high spatial resolution in representing images and its capacity to encompass large geographic areas effectively. To improve the quality of raw images, several digital processing techniques were applied, including look-up table stretching, histogram equalization, principal component analysis, and high-pass directional filtering. These techniques enhanced image detail, enabling the production of precise geomorphology and lineament density maps, which offer valuable insights into the region's physical characteristics. Additionally, soil maps were digitized and further refined through satellite imagery analysis, ensuring high accuracy in thematic mapping. The integration of these data sources and techniques ultimately led to the generation of comprehensive thematic maps that support further spatial analysis and decision-making processes [7].

In this study, Cartosat-1 DEM data was used to analyse the slope. A stream direction map was generated from the DEM, providing a clear representation of the water flow in the study area. Using the stream direction map as input, a stream accumulation map was produced with ArcGIS tools. The output includes various classes of potential zones, categorized as very high, high, moderate, poor, and very poor [8].

## 5 Results and discussion

### 5.1 Drainage density map

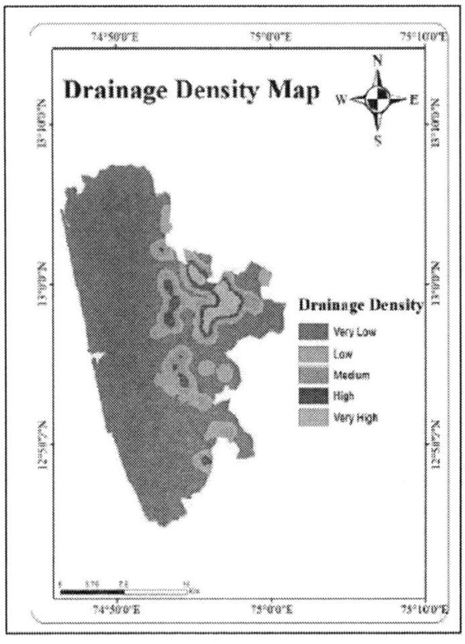


**Fig. 2.** The map of drainage density

The study area was delineated using freely available Cartosat-1 DEM data from the National Remote Sensing Centre (NRSC) in Hyderabad, India. Drainage density within the area was computed as the total length of streams per unit area in each sub-basin, expressed in km/km$^2$. The drainage density values were classified into five categories: Very Low, Low, Medium, High, and Very High. Areas with lower drainage density support higher infiltration rates, suggesting a greater groundwater potential compared to regions with high drainage density. Consequently, greater weights were assigned to regions with low drainage density, while lower weights were allocated to those with high drainage density, as illustrated in Fig. 2.

### 5.2 Slope map

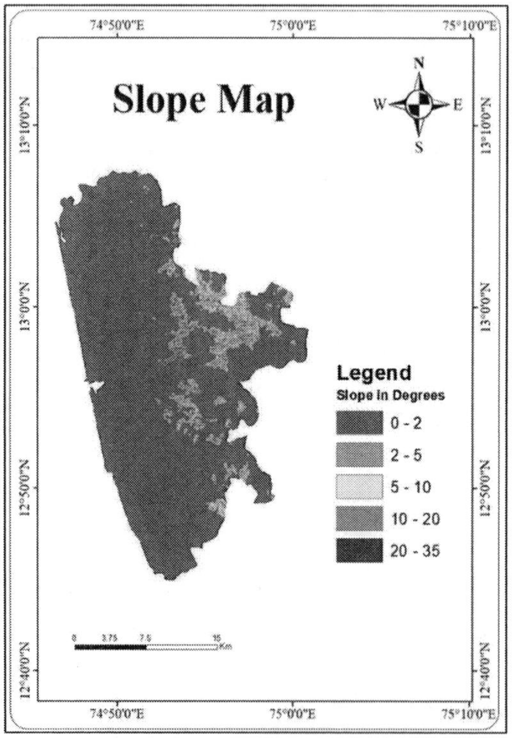


**Fig. 3.** Slope map

In this study, a slope map created from Cartosat-1 DEM data provided by NRSC in Hyderabad, India, was utilized and is available for free. Most of the area features a predominantly flat landscape, and the slope was categorized into five distinct ranges.

Finally, five classes of slopes in degrees (0-2, 2-5, 5-10, 10-20 and 20-30) were differentiated as shown in Fig. 3, where a steeper slope results in increased runoff and reduced infiltration, leading to poorer groundwater potential compared to areas with gentle slopes. Consequently, greater importance has been given to gentle slopes, while steeper slopes have been assigned less weight.

### 5.3 Lineament density map

Lineaments are crucial for groundwater recharge in hard rock terrains, and areas near these lineament zones typically exhibit significantly higher groundwater potential. The lineament map sourced from the Bhuvan website was integrated into ArcGIS, allowing for the digitization of the lineaments within the analysis area. This digitized lineament map was then utilized to calculate lineament density using the line density tool, which helps in understanding the spatial distribution of these features. Polygons exhibiting higher lineament density values are anticipated to support greater groundwater recharge, thereby enhancing overall groundwater prospects in those regions. As can be seen from Fig. 4, to provide a clearer assessment, the lineament density was classified into five distinct intervals: Very Low, Low, Medium, High, and Very High. Among these categories, the interval with the highest lineament density received the most significant ranking, reflecting its greater importance for groundwater potential and recharge capacity.

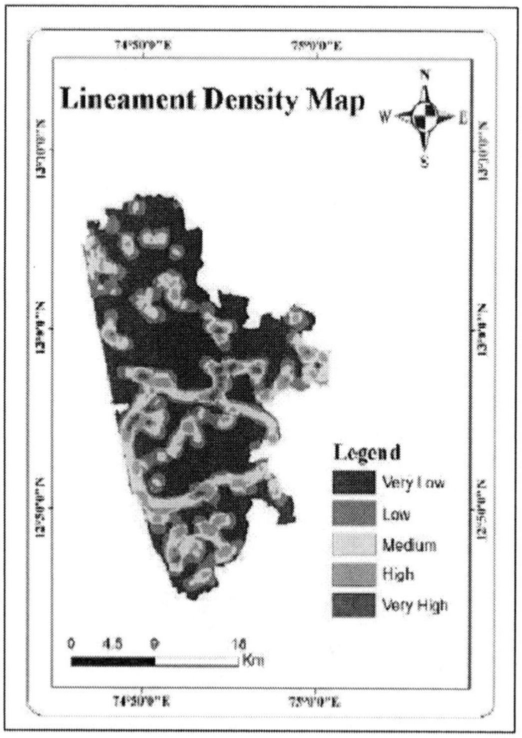


**Fig. 4.** Lineament density map

### 5.4 Geomorphology map

The meaning of Geomorphology has an important meaning in the Chambers dictionary, i.e. "the scientific study of the nature and history of landforms on the surface of Earth and other planets, and the processes of their creation".

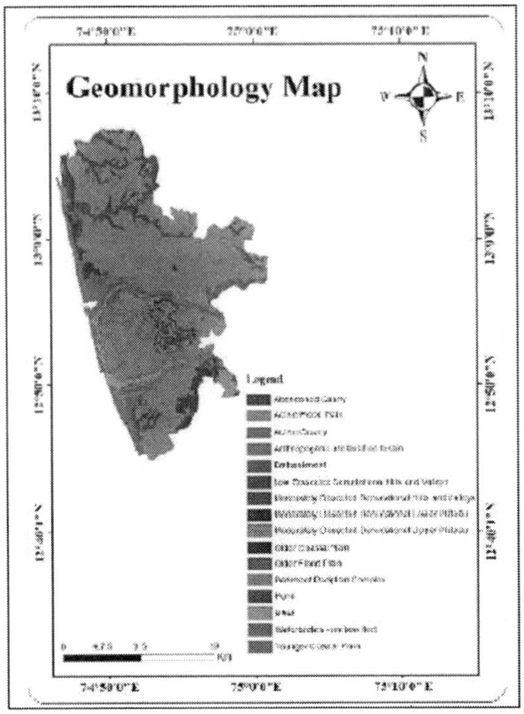


**Fig. 5.** Geomorphology map

The geomorphological units within the study area were delineated using satellite imagery, and various classes were identified and represented accordingly. The geomorphology map was obtained from the BHUKOSH-Geological Survey of India website. Through the application of the clipping tool in ArcGIS, the geomorphological map specific to the study area was derived. Anthropogenic terrain was assigned a lower weight as a result of human activities. In contrast, the Pediment-Pediplain was assigned the highest weight due to its composition of predominantly bare rock, potentially covered by a thin veneer of alluvium or gravel, which serves as the primary source of groundwater recharge in the region. Highly dissected hills and valleys received the lowest weight, as these areas are characterized primarily by surface runoff (Fig. 5).

## 5.5 Annual rainfall map

Rainfall plays a major role in groundwater potential; since rainwater is a source for the occurrence of groundwater, the rainfall or precipitation data of the area is very crucial for the delineation of the groundwater potential map. Precipitation data was downloaded from the Centre for Hydrometeorology and Remote Sensing (CHRS) website for the past 10 years. This data was interpolated to obtain the average annual rainfall data of the study area. Annual rainfall map has been divided into 6 categories with rainfall in cm (379-381, 381- 382, 382-384, 384-385, 385-387, and 387-388). As Shown in the Fig. 6, the weights given were proportional to the rainfall.

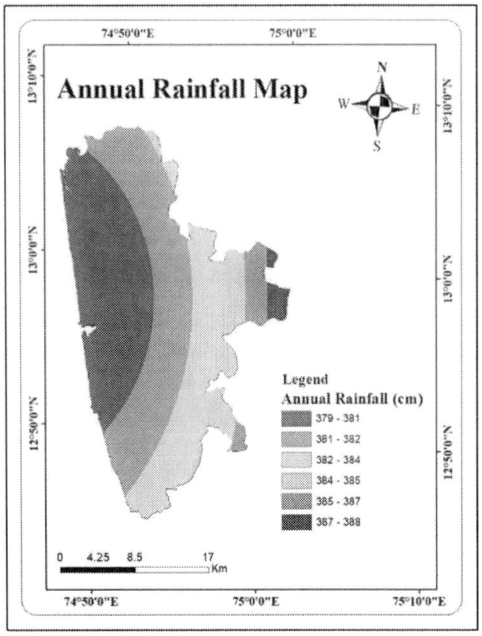


**Fig. 6.** Annual rainfall map

## 5.6 Geology map

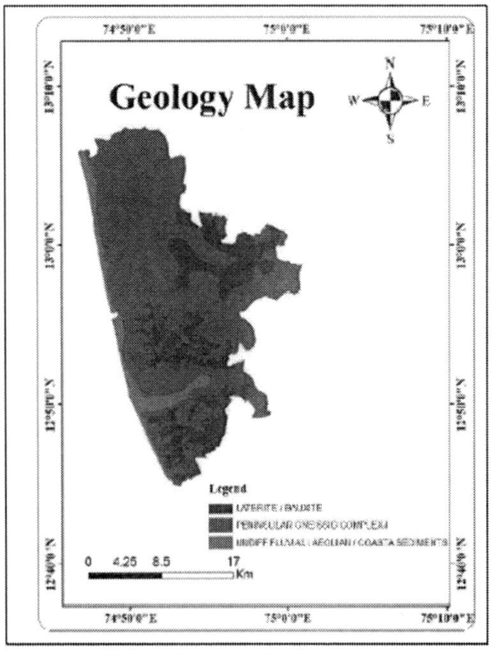


**Fig. 7.** Geology map

The lithological characteristics of the study area are considered one of the primary controlling factors that influence both the flow and occurrence of groundwater. The arrangement and distribution of different lithological units, as well as the interactions between them, play a crucial role in determining the infiltration capacity of the region. This capacity, in turn, affects how water

https://doi.org/10.1051/e3sconf/202560401003

penetrates the subsurface layers, contributing to groundwater recharge. The porosity and permeability of these lithological units are significant as they determine both the storage capacity and the ability to transmit water through the subsurface. Thus, these properties are essential in supporting the occurrence, movement, and accumulation of groundwater within the region.

The geological map of the study area was acquired from the BHUKOSH-Geological Survey of India website, and the specific map relevant to the study was extracted using the ArcGIS clipping tool. The study area is characterized by three major groups of rock units, which include laterite and the Peninsular Gneiss Complex. These geological formations play a vital role in shaping the hydrological characteristics of the region, influencing both surface-water interaction and groundwater recharge potential (Fig. 7).

### 5.7 Soil map

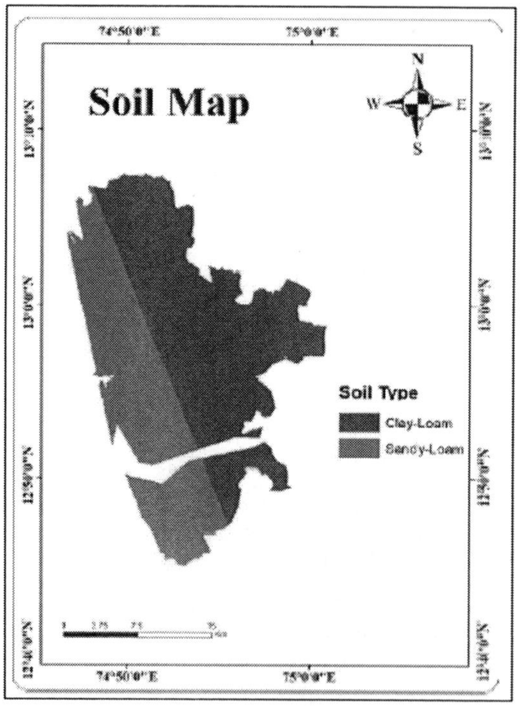


**Fig. 8.** Soil map

Soil is a critical factor in identifying zones with potential groundwater occurrence. The texture of the soil in a given area significantly influences both surface runoff and the infiltration of rainfall. Soils with a sandy texture exhibit low runoff rates and are associated with high groundwater potential, while clay soils tend to have higher runoff rates and minimal groundwater potential. Sandy soils allow for a high rate of water infiltration, in contrast to clayey soils, which possess a much lower infiltration capacity. These properties make soil texture an essential consideration in groundwater recharge assessments. The soil map of the world was downloaded from the Food and Agricultural Organization (FAO) soil portal. The clip tool in ArcGIS was used to clip the map with our study area. The map is divided according to the legend of the soil map of the world. The study area consists of two types of soil, i.e., sandy loam soil and clay loam (Fig. 8).

### 5.8 Land use land cover map

The Land Cover tiled imagery layer for the study area was obtained from the ESRI website, and the imagery was clipped using the study area's shapefile in ArcGIS. The classification scheme applied to the resulting map adheres to ESRI's categorization standards, a widely respected authority in geographic information systems. In this study, the Land Use/Land Cover (LULC) data were organized into seven key classes: water bodies, tree cover, flooded vegetation, agricultural crops, built-up areas, bare soil, and rangelands. Each category is integral to understanding the landscape's ecological dynamics and hydrological processes.

Furthermore, within each land use/cover class, specific subclasses were assigned distinct weights based on their respective contributions to groundwater infiltration. This weighting system is crucial for understanding how different land uses impact water absorption and movement within the soil profile. The implications of these classifications and weightings are visually represented in Fig. 9, which provides a clear depiction of the relationships between land use types and their influence on groundwater dynamics. This comprehensive approach enhances our understanding of land management practices and their effects on water resources.

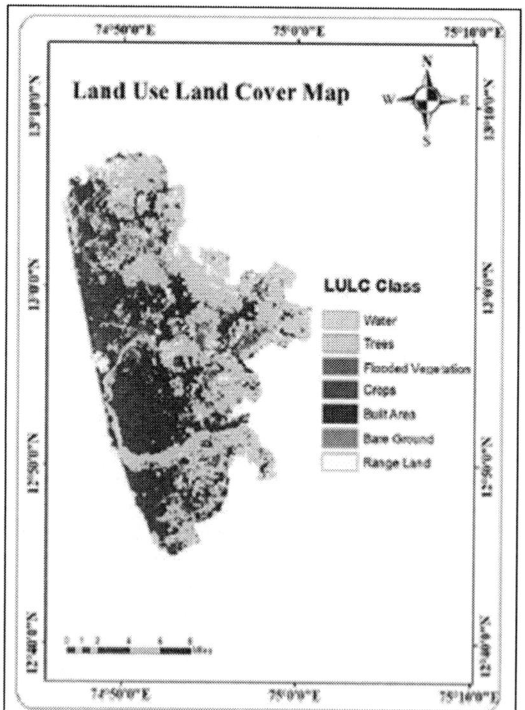


**Fig. 9.** Land use land cover map

### 5.9 Preparation of groundwater potential map

Each category within the primary eight thematic layers—geomorphology, geology, slope, drainage density, annual rainfall, lineament density, soil, and land use/land cover—is assigned specific weights according to its influence on groundwater potential. This weighting process ensures that each layer accurately reflects its relative importance in assessing groundwater availability. Each class within a specific thematic layer has been assigned appropriate rankings on a scale from 0 to 5, reflecting their respective contributions to groundwater potential. The percentage influence of each thematic map has been determined based on its overall contribution to groundwater potentiality. Both the weighted values and percentage influence have been allocated to various classes across all thematic layers, ensuring that their role in groundwater recharge and availability is accurately represented. All the thematic maps have been integrated using the union command. A final groundwater potential map is prepared using the above technique. Groundwater potential in the area has been classified into five classes, namely Very low, Low, Medium, High and Very High potential areas (Fig. 10).

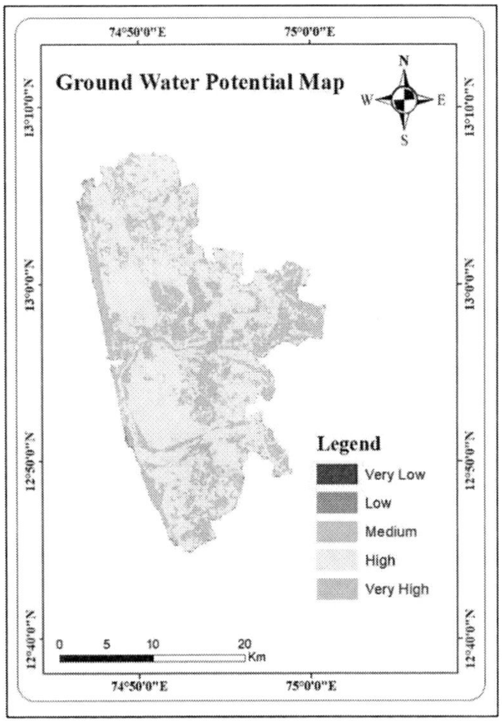


**Fig. 10.** Groundwater potential map

### 5.10 Validation

The groundwater potential map was verified using borehole data from the India Water Resource Information System (I-WRIS) portal. Borehole locations were plotted onto the generated groundwater potential map, and the actual yield from each borehole was compared with the predicted values on the map, as illustrated in Fig. 11.

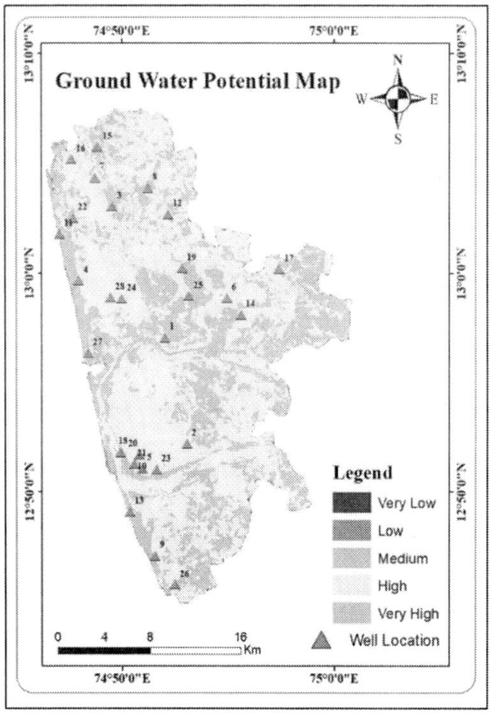


**Fig. 11.** Groundwater potential map with well location

## 6 Conclusion

Advanced remote sensing and GIS techniques will be applied to assess and identify potential groundwater zones spatially. Thematic layers will be developed and weighted according to each influencing factor using weighted overlay analysis, ensuring accurate delineation of groundwater potential zones.

The groundwater potential zones were classified into five categories: very low, low, medium, high, and very high potential areas. This classification aids decision-makers in developmental planning. The resultant map was validated using data from 28 bore wells obtained from the India WRIS portal. Of these, 23 bore wells validated the produced map, while 5 bore wells did not, resulting in a map accuracy of 82.14%.

## References

1. Giri, D. N. and Bharadwaj, P., Study and Mapping of Ground Water Prospect using Remote Sensing, GIS and Geoelectrical resistivity techniques – a case study of Dhanbad district, Jharkhand, India. Journal: Indian Geophics Union, **16**(2), 55–63 (2012).
2. Oh, H. J., Kim, Y. S., Choi, J. K., Park, E., and Lee, S., GIS mapping of regional probabilistic groundwater potential in the area of Pohang City, Korea. Journal of Hydrology, **399**(3–4), 158–172 (2011).
3. Rahmati, O., Nazari Samani, A., Mahdavi, M., Pourghasemi, H. R., and Zeinivand, H., Groundwater potential mapping at Kurdistan region of Iran using

https://doi.org/10.1051/e3sconf/202560401003

analytic hierarchy process and GIS. Arabian Journal of Geosciences, **8**(9), 7059–7071 (2015).

4. Waikar, M. L., and Nilawar, A. P., Identification of Groundwater Potential Zone using Remote Sensing and GIS Technique. International Journal of Innovative Research in Science, Engineering and Technology, **3**(5), 12163–12174 (2015).
5. Han, L., Liu, Z., Ning, Y., and Zhao, Z., Extraction and analysis of geological lineaments combining a DEM and remote sensing images from the northern Baoji loess area. Advances in Space Research, **62**(9), 2480–2493 (2015).
6. Basavaraj, D. B., Hiremath, C. G., Davithuraj, J., and Purandara, B. K., Identification of groundwater potential zones using ArcGIS 10.1. International Journal of Civil, Mechanical and Energy Science, 2394-2827 (2016).
7. D. C. Jhariya, Rubia Khan, K. C. Mondal, Tarun Kumar, Indhulekha K. and Vikal K. Singh, Assessment of groundwater potential zone using GIS-based multiinfluencing factor (MIF), multi-criteria decision analysis (MCDA) and electrical resistivity survey techniques in Raipur city, Chhattisgarh, India. AQUA — Water Infrastructure, Ecosystems and Society (2021).
8. Huajie Duan, Zhengdong Deng, Feifan Deng, and DaqingWang, Assessment of Groundwater Potential Based on Multi criteria Decision Making Model and Decision Tree Algorithms. Mathematical (2016).

# AHP-based geospatial analysis for landslide susceptibility zonation in Wayanad District

*Tharayil* Irshad[1]*, *Harikrishnan* T[1], *Mahalingam* Bose[1], and *Titi* Kurniati[2]

[1]Department of Geography and Geology School of Earth Sciences, Central University of Karnataka, India
[2]Department of Civil Engineering, Faculty of Engineering, Universitas Andalas, Padang, Indonesia

**Abstract**. Landslide susceptibility zonation mapping is a decision-making process that entirely depends upon the land use and land cover changes from one point to another. The accuracy and precision of the study related to the quality of the parameters used to determine the occurrence and intensity of the landslides. This study acquired secondary data from Landsat-8 satellite images and GSI (Geological Survey of India). The landslide conditioning factor was assigned with the help of the semi-quantitative Analytical Hierarchy Process (AHP) Technique. In this current study, 13 parameters were analyzed. Each parameter is responsible for landslides in a low or high influence grade. Each parameter was designed to have a particular weight based on its influence on landslide occurrence. Rainfall is the prime factor for the landslides in the Wayanad district. The final landslide susceptibility zonation map was validated using appropriate mathematical treatment to check their correctness with the help of landslide inventory points.

## 1 Introduction

The term "landslides" refers to a broad range of mechanisms that cause rocks, soil, manmade fills, or a combination of these materials to slide down a slope and outward [1]. It is also defined as moving the earth's material down a slope, such as large rocks or debris [2]. Globally, more than 80 % of landslide occurrences, losses, and damages result from intense and prolonged precipitation; constructions involving undercutting slopes, mining, quarrying, and Earth movements cause Earthquakes [3]. Several studies have demonstrated that GIS-based susceptibility assessment is a valuable tool for understanding the geographical distribution of landslides [4].

Calculating temporal probability involves either using dynamic modeling or, if the historical record is adequate, comparing the data on landslip occurrences with those of the triggering factors [5]. Statistical, deterministic, and heuristic approaches based on GIS have been used to assess landslide susceptibility [6]. AHP and the weighted linear combination model are recent examples of innovative decision-support tools created for GIS techniques [7]. Thomas Saaty created the analytical hierarchy method (AHP) in the late 1970s. It is a decision-making tool for handling complex unstructured and multicriteria decisions. It offers an adaptable and straightforward method of deciphering complex issues and permits the modification of landslide-related concepts and assumption-based problem solutions [8].

Susceptibility is a crucial term in hazard risk management that has received much attention. Several studies have been performed on landslide susceptibility [9]. It is measured continuously from 0 to 1, representing destruction [10]. A landslide susceptibility map includes landslides that have their source in and outside the area [11]. Moreover, it is helpful for planning and decision-making in landslide hazard management [12]. Landslide research has long been a subject in India's Western Ghats [13]. This study aims to identify the highly vulnerable landslide zones using the semi-quantitative Analytical Hierarchy Process (AHP) technique in the Wayanad district and map them to mitigate the hazardous events related to the landslides.

## 2 Study area

Wayanad, in the north-eastern region of Kerala, is bordered on the north and east by Karnataka and on the south and east by Tamil Nadu. The district is bordered by Kozhikode, Malappuram, and Kannur in the southwest, south, and northwest. The district was formed on November 1, 1980, by combining parts of the Kannur and Kozhikode districts. Wayanad, with its administrative headquarters in Kalpetta, is between 11.6854 N and 76.1320 E. With an altitude ranging from 700 to 2100 meters, it is high on the Western Ghats. The district covers 2091 Square Kilometres of land. Kalpetta, Mananthavadi, and Sulthanbathery are the major towns in Wayanad. Wayanad had an estimated population of 817,420 people in 2011, with 401,684 men and 415,736 women. The map of the study area is shown in Fig. 1.

* Corresponding author: irshadtharayil@gmail.com

 https://doi.org/10.1051/e3sconf/202560401004

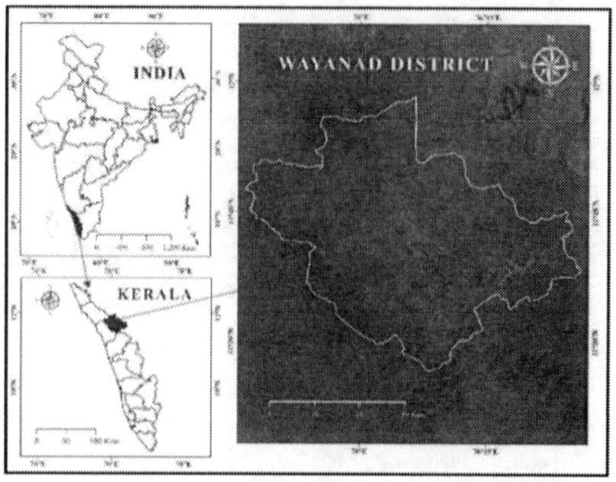


**Fig. 1.** Map showing the location of the Wayanad district.

The district's economy is built around agriculture. The district's agricultural grounds make up the majority of its land. Rice production is well-known in Wayanad. The district's tourism industry is well-developed, and a large portion of its revenue comes from this sector each year, contributing significantly to the district's economy. The area is sparsely industrialized. Milma has a dairy farm at Kalpetta, and a few small-scale companies are also there. Wayanad is known for its tribal culture. Despite its reputation as a primitive district, it produces cash crops like pepper, cardamom, coffee, tea, spices, and other condiments, making it one of the state's most prominent foreign exchange earners.

## 3 Methodology

ArcGIS 10.7 software was used to determine the map of the landslide-triggering factors. The prime source of the data was the GSI (Geological Survey of India), and other sources were the soil grids and satellite images of Landsat-8. Elevation, slope, aspect (direction of slope), lineament, soil texture, geology, soil depth, landforms, lithology, forest cover (NDVI), drainage density, anthropogenic, and rainfall are used to evaluate and estimate the susceptibility map in the Arc-GIS Platform.

For the extraction of individual parameters, the weighted semi-quantitative method (AHP) is implied to gain the respective rank of each parameter. The methodological flow chart is shown in Fig. 2.

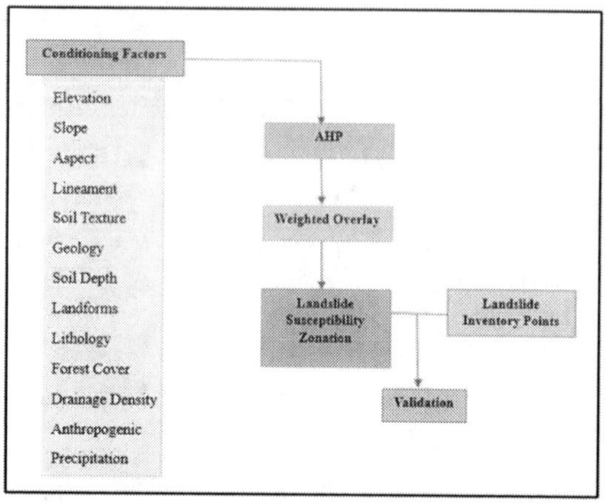


**Fig. 2.** Flowchart of the methodology used in this study.

### 3.1 Landslide conditioning factors

Landslides are localized events that seldom co-occur with varying frequency and severity. This rule could be expected to produce debris flows and rock falls [14]. The event is due to some factors influencing landslides' occurrence, magnitude, and intensity, termed the conditioning or triggering factors. The heterogeneity of significant influencing parameters can harm the degree of accuracy in determining landslide-prone areas.

The selection of the conditioning factor is a timely process; adequate knowledge about an area's terrain and historical events and a thorough geological study should be conducted to determine the factors that cause landslides. There are no standard rules to number the conditioning factors [15]. In the current study, conditioning factors were chosen based on the basics of specific literature [16, 17]. In this current study, 13 parameters were analyzed, as shown in Fig. 3.

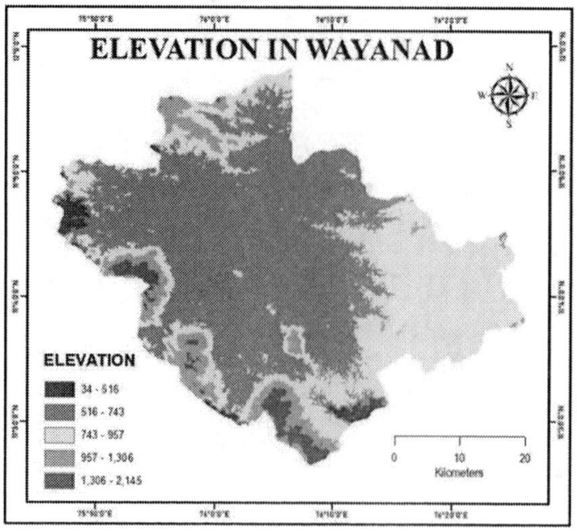


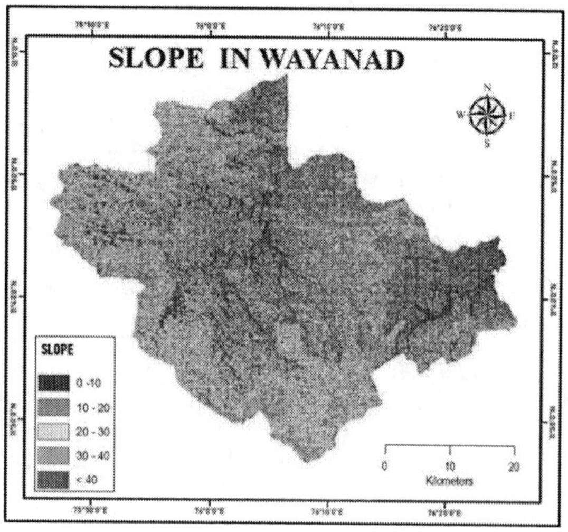

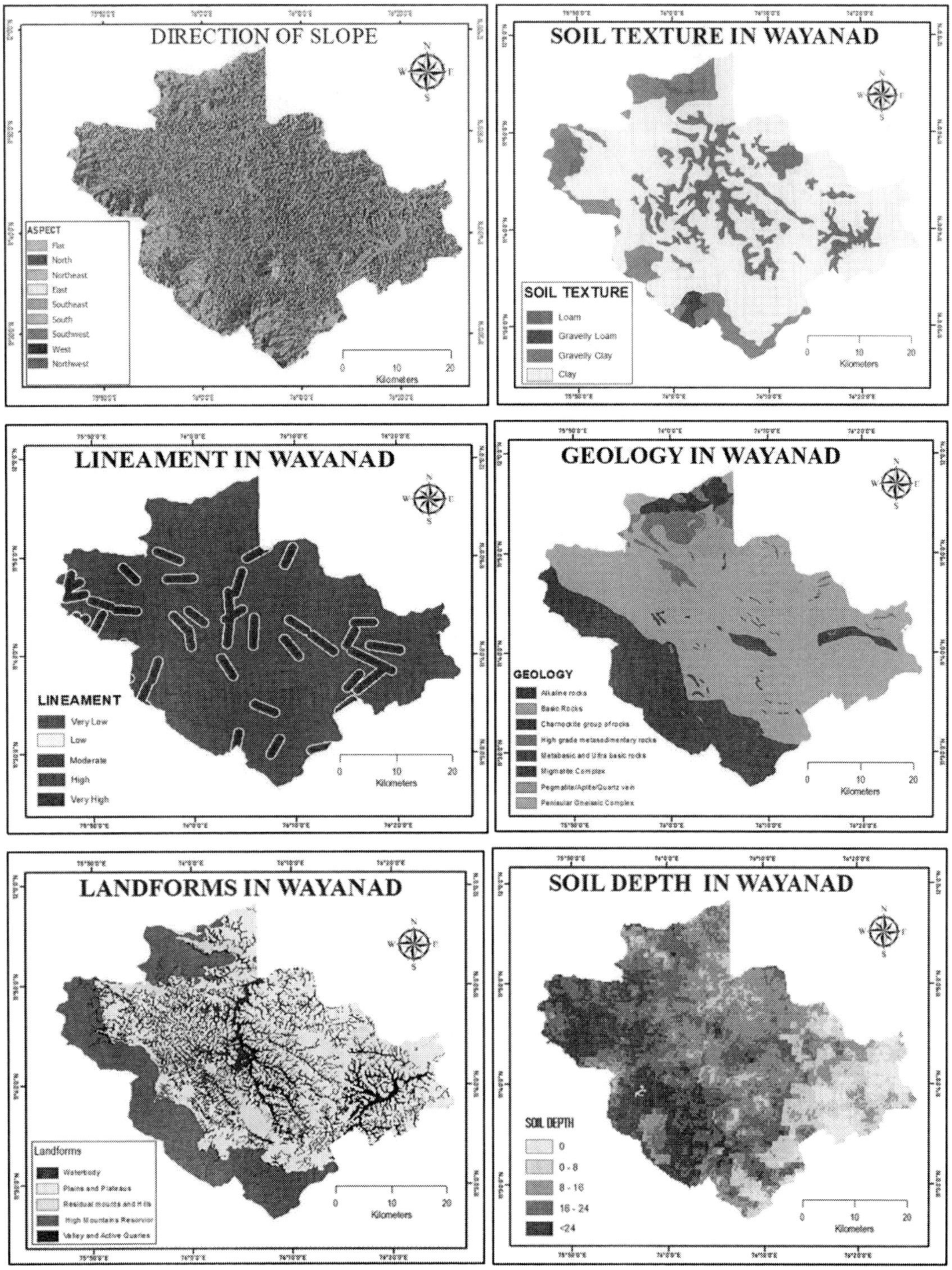
DIRECTION OF SLOPE
ASPECT
Flat
North
Northeast
East
Southeast
South
Southwest
West
Northwest
Kilometers
SOIL TEXTURE IN WAYANAD
SOIL TEXTURE
Loam
Gravelly Loam
Gravelly Clay
Clay
Kilometers
LINEAMENT IN WAYANAD
LINEAMENT
Very Low
Low
Moderate
High
Very High
Kilometers
GEOLOGY IN WAYANAD
GEOLOGY
Alkaline rocks
Basic Rocks
Charnockite group of rocks
High grade metasedimentary rocks
Metabasic and Ultra basic rocks
Migmatite Complex
Pegmatite/Aplite/Quartz vein
Penisular Gneissic Complex
Kilometers
LANDFORMS IN WAYANAD
Landforms
Waterbody
Plains and Plateaus
Residual mounts and Hills
High Mountains Reservoir
Valley and Active Quaries
Kilometers
SOIL DEPTH IN WAYANAD
SOIL DEPTH
0
0 - 8
8 - 16
16 - 24
<24
Kilometers

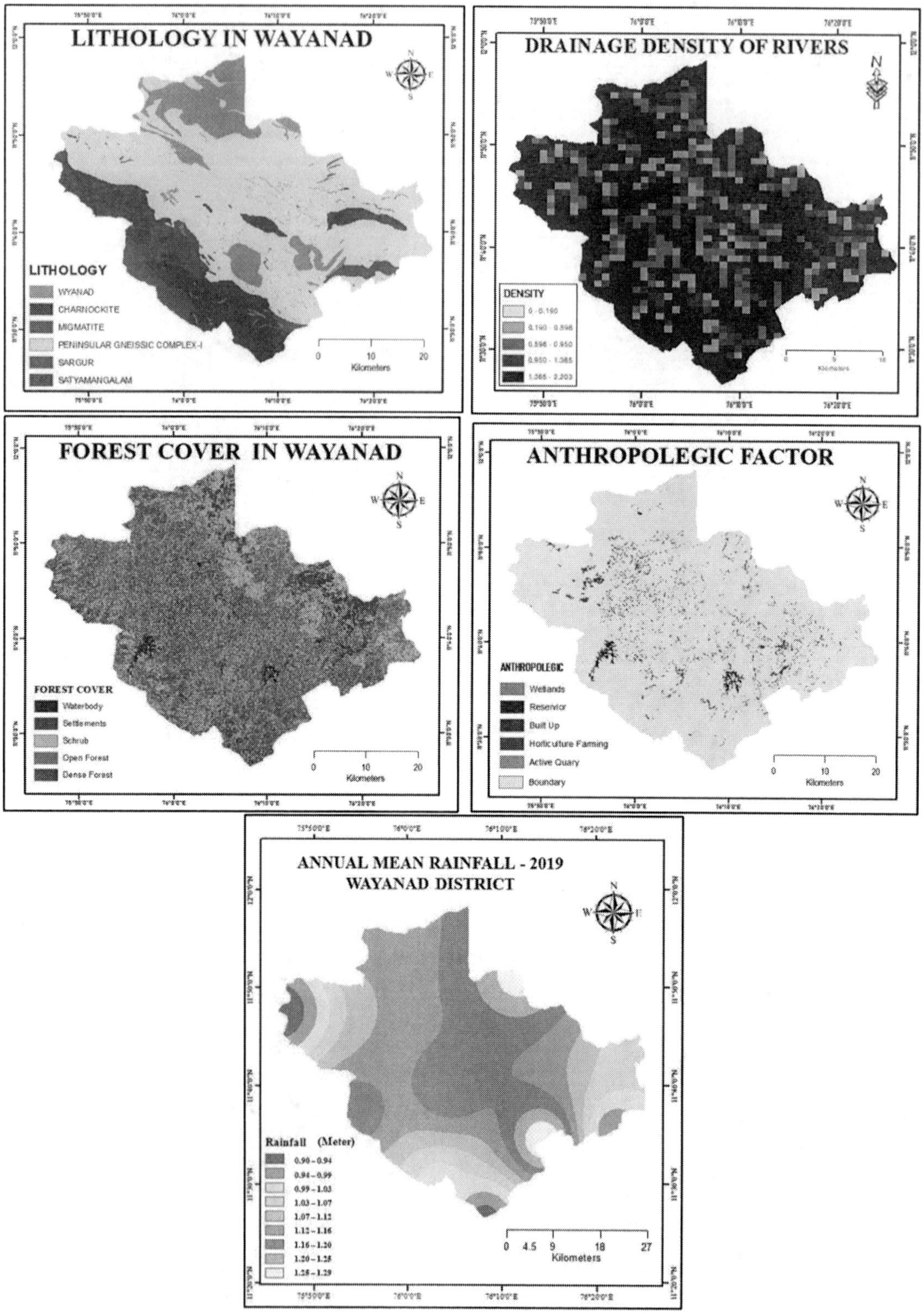


**Fig. 3.** Conditioning factors (a. Elevation, b. Slope, c. Direction of slope, d. Lineament, e. Soil texture, f. Geology, g. Soil depth, h. Landforms, i. Lithology, j. Forest cover, k. Drainage density, l. Anthropogenic factor, m. Rainfall) of landslide susceptibility mapping.

The elevation map is prepared using the Digital Elevation Model (DEM) with 30m spatial resolution. The slope details were also extracted from the DEM obtained from the Bhuvan portal. The slope is classified into 0 – 10°, 10 - 20°, 20 - 30°, 30 - 40°, and > 40°, and there are more chances of the downslope movements of debris at the steepness of the slope. The elevation is classified into 34m - 516m, 516m -743m, 743 - 957m, 957m - 1306m, and 1306 – 2145m. Gravitational pull is greater at higher altitudes. The aspect is the direction of slopes and is the most critical conditioning factor [18]. The aspect was categorized into nine classes (flat, north, northeast, east, south-east, south, southwest, west, and northwest).

Lineaments are distinct geological cracks in the rock face that are responsible for rock falls and cracks. The Wayanad district's volatile area is mainly in the west. This region encompasses tourist attractions such as Chembra Peak and the hilly ranges surrounding Banasura Dam. Approximately 591 square kilometers of land are located in a moderately unstable region. Lineament mapped multi-ring buffer as 5 classes (250m, 500m, 750m, 1000m, >1000m). When distance increases from the lineament, landslides are less likely to occur. In the Bavani shear zone, most of the lineament is found.

The type of soil class and each lithology typically accelerate the intensity of the surface run-off, impacting the soil and causing erosion and mudslides. The soil texture of the area was diverse and unique (loam, Gravelly loam, Gravelly Clay, Clay). The clayey nature of the soil grabs more moisture, lubricates quickly, and erodes. The western part of the district-based component in the orogeny was clayey and gravelly clay in Kuruva Island, Banasura Dam, Meppadi, Pookod Lake, and Chembra Peak, all part of the same chain of orogeny. The soil depth of the study area is one of the most critical factors in the entire landslide process. A 30m soil depth is selected, categorized as 0-8, 8-16,16-24, and >24. The uppermost soil layer is loosely packed, and the bond between the soil is comparatively lesser than the lower-lying soil. It can easily be eroded by the intense movement of debris from more excellent slopes. The soil becomes more complex and more rigid whenever soil depth increases. A formation comprises a group of rock strata with similar lithology, faces, and other characteristics. Since formations are not determined by the thickness of the rock strata that make them up, the thickness of different formations can vary significantly.

The geology of the Wayanad is deeply related to the Western Ghats, with more or less geological distribution rooted in old deposits of Western Ghats. The dominant rock unit of the region is Precambrian crystalline rocks such as charnockites, khondalites, and gneisses. They underlie most of the highlands and the foothills, including the scarps. The peninsular gneiss complex is the standout deposition rock component. Charnokite, migmatite, and pegmatite complexes were also found abundant. High-grade meta-sedimentary rocks, meta-basic rocks, bare rocks, alkaline rocks, etc., are minor components. When detailing the lithological factors, migmatite, charnockites, wyanad, satyamangalam, peninsular gneiss complex, and sargur are the rock complexes throughout the Wayanad district. The rock complexes have been formed in recent Quaternary formations.

Forest cover was extracted from the Landsat-8 satellite image and classified into scrub, open forest, dense forest, settlements, and waterbody. Anthropogenic factors are a severe problem of landslide events. Man's involvement in nature collapses the equilibrium of the life cycle. Quarrying, horticulture farming, reservoirs, and built ups are the involvements that trigger the landslides to be held destructively are also considered. Landforms map prepared and classified into waterbodies, plains and plateaus, residual mounts and hills, high mountain reservoirs, valleys, and active quarries based on their chance the occurrence of landslides. Drainage density map prepared with the help of extracted streams with the help of a raster calculator in ArcGIS software and classified to 0 – 0.190, 0.190 – 0.596, 0.596 - 0.950, 0.950 – 1.365, 1.365 – 2.203. The higher value shows a higher density of drainages and a higher chance for landslides.

Wayanad has unique characteristics in geography: 30% of the landmasses are high mountains and plateaus. The rugged terrain of the landmass causes the occurrence of orogenic rainfall. Wayanad got the maximum rain in 2019. Precipitation causes loose soil to be eroded when the soil at the greater altitude and slope tremble at an incredible pace, causing damage to agricultural land and even responsible for significant causalities. The rainfall map is classified into nine classes, as shown in Fig. 3.

**Table 1.** AHP matrix

| Parameters | Rainfall | Slope | Soil Depth | Anthropogenic | Soil Texture | Land-forms | Elevation | Lithology | Geology | Lineament | Drainage density | Aspect | Forest Cover |
|---|---|---|---|---|---|---|---|---|---|---|---|---|---|
| Rainfall | 1 | 2 | 3 | 4 | 5 | 6 | 7 | 8 | 9 | 10 | 11 | 12 | 13 |
| Slope | 1/2 | 1 | 2 | 3 | 4 | 5 | 6 | 7 | 8 | 9 | 10 | 11 | 12 |
| Soil Depth | 1/3 | 1/2 | 1 | 2 | 3 | 4 | 5 | 6 | 7 | 8 | 9 | 10 | 11 |
| Anthropogenic | 1/4 | 1/3 | 1/2 | 1 | 2 | 3 | 4 | 5 | 6 | 7 | 8 | 9 | 10 |
| Soil Texture | 1/5 | 1/4 | 1/3 | 1/2 | 1 | 2 | 3 | 4 | 5 | 6 | 7 | 8 | 9 |
| Landforms | 1/5 | 1/5 | 1/4 | 1/3 | 1/2 | 1 | 2 | 3 | 4 | 5 | 6 | 7 | 8 |
| Elevation | 1/7 | 1/6 | 1/5 | 1/4 | 1/3 | 1/2 | 1 | 2 | 3 | 4 | 5 | 6 | 7 |
| Lithology | 1/8 | 1/7 | 1/6 | 1/5 | 1/4 | 1/3 | 1/2 | 1 | 2 | 3 | 4 | 5 | 6 |
| Geology | 1/9 | 1/8 | 1/7 | 1/6 | 1/5 | 1/4 | 1/3 | 1/2 | 1 | 2 | 3 | 4 | 5 |
| Lineament | 1/10 | 1/9 | 1/8 | 1/7 | 1/6 | 1/5 | 1/4 | 1/3 | 1/2 | 1 | 2 | 3 | 4 |
| Drainage density | 1/11 | 1/10 | 1/9 | 1/8 | 1/7 | 1/6 | 1/5 | 1/4 | 1/3 | 1/2 | 1 | 2 | 3 |
| Aspect | 1/12 | 1/11 | 1/10 | 1/9 | 1/8 | 1/7 | 1/6 | 1/5 | 1/4 | 1/3 | 1/2 | 1 | 2 |
| Forest Cover | 1/13 | 1/12 | 1/11 | 1/10 | 1/9 | 1/8 | 1/7 | 1/6 | 1/5 | 1/4 | 1/3 | 1/2 | 1 |

Thomas Saaty invented the AHP method in the 1980s. An expression designed for acquiring decision-making purposes, depending upon specific criteria to rank each parameter participating in the model. It aids decision-makers in making the best decision possible by breaking down difficult decisions into a sequence of comparative pairs and combining the outcomes. The AHP divides a multi-level decision problem into sub-problems. This technique helps you to measure your thoughts and transform them into a logical decision model [19].

The present study represents the 13 parameters by giving their appropriate weights based on the influence of each individual on the landslide probability (Table 1). The weights of each parameter are done through the proper knowledge of literature review, field observation, and secondary sources. When 1 to 5 has a strong influence, 5-10 has a moderate influence, while 10-13 has a minimal impact on landslides. Averages of the hierarchal arranged parameters were used to calculate the rating value and CR value, the consistency ratio, and the total number of factors (n).

While the CI is the consistency index calculated before, RI is the random consistency index calculated by the randomly generated comparison matrix of the AHP. According to the CI rule of thumb, a ratio greater than 0.1 should be revised simultaneously, and a ratio below 0.1 to accept the matrix. Hence, the current ratio of CR stands at 0.1, which determines the consistency of the matrix. Table 2 shows the weightage of each conditioning factor calculated by the AHP method.

**Table 2.** Conditioning factors with weights.

| Factors | Weights | In% | Actual |
|---|---|---|---|
| Rainfall | 0.250785 | 25.07847 | 25 |
| Slope | 0.188957 | 18.89569 | 19 |
| Soil Depth | 0.142729 | 14.27287 | 14 |
| Anthropogenic | 0.108336 | 10.83364 | 11 |
| Soil Texture | 0.082428 | 8.2428 | 8 |
| Landforms | 0.062659 | 6.265938 | 6 |
| Elevation | 0.047446 | 4.744606 | 5 |
| Lithology | 0.035704 | 3.570436 | 4 |
| Geology | 0.026675 | 2.667519 | 3 |
| Lineament | 0.019813 | 1.981287 | 2 |
| Drainage density | 0.014715 | 1.471531 | 1 |
| Aspect | 0.01108 | 1.10795 | 1 |
| Forest Cover | 0.008673 | 0.867261 | 1 |

### 3.2 Validation

The landslide susceptibility map prepared for the current paper uses the R-index evaluation technique. Landslide inventory points and landslide susceptibility maps are the two datasets used here. The association between two variables is used to find both R indexes. Validation calculates a map's essential correctness or accuracy [20]. The expressions for the validation of the produced map are;

$$R = [(ni/N_i) / \sum(ni/N_i)] \times 100 \qquad (1)$$

Where $n_i$ is the number of landslides in the susceptibility class, 'i' and $N_i$ is the number of cells occupying the same susceptibility class. From the validation table landslide susceptibility, 82% of all landslides happen in moderate and high-scale intensities. Table 3 shows the R factor in various classes.

**Table 3.** R Factor in various classes.

| Vulnerability Class | Cells in the map | Area in % | No. of slide | R Factor |
|---|---|---|---|---|
| Very Low | 521 | 0.009963 | 1 | 1.1 |
| Low | 834989 | 16 | 415 | 52 |
| Moderate | 2433522 | 47 | 1211 | 340 |
| High | 1855843 | 35 | 924 | 168 |
| Very High | 104016 | 2 | 52 | 9.8 |

## 4 Results

Landslide occurrence can be detectable by correlating the inducing factors [21]. The correlation between the parameters is calculated through the AHP methods; hence, the susceptibility map has been made. Each parameter possesses its respective ranks based on the ranks. Wayanad's susceptibility can be detected through susceptibility mapping during different periods. The map is classified into several classes to create a susceptibility map. As a result of the present study, the landslide map is categorized into five major classes (shallow, low, moderate, high, and very high) to represent the impact throughout the Wayanad district, as shown in Fig. 4.

All 13 parameters, including average rainfall data of 10 years, were taken to prepare the landslide zonation map for 2019. Most of the study area is probably considered highly vulnerable to landslides. 973.4 Sq.km covered moderately vulnerable areas, and 742.33 Sq.km covered highly vulnerable areas. These areas were covered by the thick forest canopy and high peaks of the Western Ghats. Apart from the high, very highly vulnerable zones spread over 41.60 sq. km, they are distributed in every part of the district, including the area where slides happened in 2019. The area covering each zone is shown in Table 4.

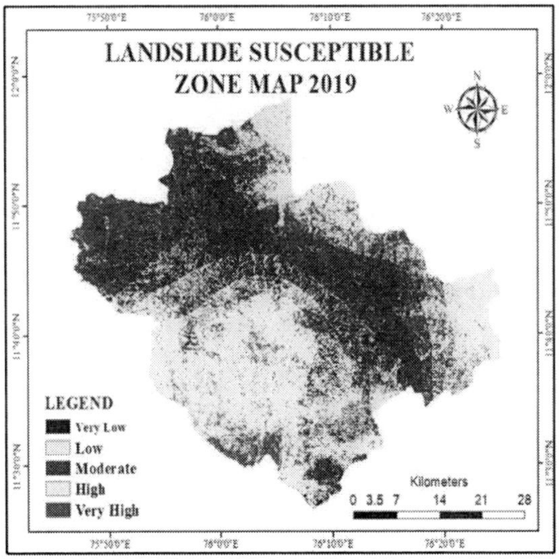


**Fig. 4.** Landslide susceptible map for 2019 with various zones (very low, low, moderate, high, and very high).

**Table 4.** Area in Sq. Km. falls under each zone

| Zones | AREA (Sq. Km.) |
|---|---|
| Very Low | 0.2084 |
| Low | 333.9956 |
| Moderate | 973.4088 |
| High | 742.3372 |
| Very High | 41.6064 |
| **Grand Total** | **2091.5564** |

Wayanad is situated in the high peaks of the western ghats; apart from the elevation factor, the slope is the deadliest factor that causes slides. In some areas in Wayanad, such as the "green belt zone," no construction or illegal activities are allowed here. However, people still have running quarries without bothering about the upcoming danger. Almost a dozen quarries are working in this prohibited zone. In 2019, quarrying was the prime cause of the landslide, along with the heavy precipitation. Cardamom farming is also caused by soil weakening, the root of the cardamom plant weakening the ground beneath it, causing sliding of the loose soil with debris. Meanwhile, hill slopes were modified and terraced up for plantation crops like rubber; these leveled terrains obstruct the drainage pattern, making the river change course.

High peaks like Chamba and Peanutty have Slopes above 40°. In extreme weather conditions, minor disturbances in the structure shimmy down the whole feature frame. Karinjola Mountain near Puthumala was precisely the trigger point for the landslides. Tract of Kabani River admitted area of higher chance of vulnerability to landslides. The center part of the district is a region in which Kabani shows her full fury during the Manson season, overfloods. It causes destruction all along the Wayanad and Karnataka tracts.

## 5 Conclusion

The present study aimed to provide a detailed analysis of the landslide susceptibility zonation mapping of the landslide-occurring areas. The study's main objective is to recognize the possibility of a high chance of a landslide and figure out the area at risk. The study emphasized and discussed how landslides increased and became deadlier than in previous years. Due to the unique geography and climate factors, low landslide risks were seldom experienced in Wayanad. The urban area is comparatively shallow, while the population is established in the rural sector. The underdeveloped rural infrastructure and population compatibility lead to a high Landslide risk in Wayanad, even though towns like Pulppelly, Manathavadi, and Kalpetta are also considered to have high landslide Risk.

In Wayanad, landslides are mainly triggered by the seasonal variation of the monsoon distribution to the state. Wayanad typically has loose basements of geology, such as mountains and plateau soils like clay and loamy. The soils are easily water-soluble, making the landslides occur; challenging slopes and higher elevation increase the landslide intensity. Parameters taken into this study are vital for the development and nature of the landslides. Thirteen parameters discussed in this study are best suited to seek a landslide susceptibility map of Wayanad district. Landslide susceptibility zonation mapping has five classes.

The susceptibility map prepared for the year 2019 in the Wayanad district has pointed out that the Southern half of the district is the zone that has the maximum impact of the landslides. All the major and minor slides were located in this zone. The moderate zones are spread more than the high and very high classes. However, the Puthumala slide was Zoned in the mild zone. In this study, it is found that landslides in Wayanad are the primary concern of disaster management. Landslides happened in 2019, not because of heavy rainfall but because anthropogenic activities such as construction, quarrying, and horticulture have weakened the geology.

This study also indicates that rainfall is the primary conditioning factor with 25% weightage, followed by slope (19%), soil depth (14%), anthropogenic (11%), soil texture (8%), landforms (6%), elevation (5%), lithology (4%), geology (3%), lineament (2%), drainage density (1%), aspect (1%) and forest cover (1%). Landslide intensity decreases, but the number of landslides never decreases. Proper urban planning or proper aid to reduce landslides should be made. Using steel barriers, land terracing should be implemented to reduce the intensity of landslides in the future.

The authors would like to acknowledge the Department of Geography, Central University of Karnataka, for the support throughout the study. This research received no specific grant or fund from government, public, commercial, or not-for-profit funding agencies. All data supporting this study's findings are available from the author's side on reasonable request. Tharayil Irshad: Conceptualization, Methodology, Interpretation, Writing, and Final Revision of Manuscript. Harikrishnan T: Data Curation, Software, Validation, and Writing. Mahalingam Bose:

Formal Analysis, Visualization, and Writing, Titi Kurniati: Editing. All authors reviewed and approved the final manuscript.

## References


1. T. Gravina, E. Figliozzi, N. Mari, and F. De Luca Tupputi Schinosa, Landslide risk perception in Frosinone (Lazio, Central Italy). Landslides. **14**(4), 1419-1429 (2017).
2. D. Cruden, A simple definition of a landslide. Bulletin of Engineering Geology and the Environment. **43**(1) (1991). https://doi.org/10.1007/BF02590167
3. P. Banerjee, M. K. Ghose, and R. Pradhan, Analytic hierarchy process and information value method-based landslide susceptibility mapping and vehicle vulnerability assessment along a highway in Sikkim Himalaya. Arabian Journal of Geosciences. **11**, 1–18 (2018). https://doi.org/10.1007/s12517-018-3488-4
4. K. Sharma, A. K. Saraf, J. Das, S. S. Baral, S. Borgohain, and G. Singh, Mapping and change detection study of Nepal-2015 earthquake-induced landslides. Journal of the Indian Society of Remote Sensing. **46**, 605–615 (2018). https://doi.org/10.1007/s12524-017-0720-8
5. L. P. H. VanBeek, Assessment of the influence of land use and climate changes on landslide activity in a Mediterranean environment, Doctoral dissertation (2002)
6. S. D. Pardeshi, S. E. Autade, and S. S. Pardeshi, Landslide hazard assessment: recent trends and techniques. SpringerPlus. **2**, 1-11 (2013). https://doi.org/10.1186/2193-1801-2-523
7. L. Ayalew, and H. Yamagishi, The application of GIS-based logistic regression for landslide susceptibility mapping in the Kakuda-Yahiko Mountains, Central Japan. Geomorphology. **65**(1-2), 15–31 (2005). https://doi.org/10.1016/j.geomorph.2004.06.010
8. Yalcin, A. GIS-based landslide susceptibility mapping using analytical hierarchy process and bivariate statistics in Ardesen (Turkey): comparisons of results and confirmations. Catena. **72**(1), 1-12 (2008). https://doi.org/10.1016/j.catena.2007.01.003
9. Winter, M. G. A strategic approach to landslide risk reduction. International Journal of Landslide and Environment. **1**, 14–23 (2014).
10. Galli, M. and Guzzetti, F Landslide vulnerability criteria: a case study from Umbria, Central Italy. Environmental management. **40**(4), 649-665 (2007). https://doi.org/10.1007/s00267-006-0325-4
11. Fell, R., Corominas, J., Bonnard, C., Cascini, L., Leroi, E., and Savage, W. Z. Guidelines for landslide susceptibility, hazard, and risk zoning for land use planning. Engineering geology. **102,** (3-4). 85-98 (2008). https://doi.org/10.1016/j.enggeo.2008.03.022
12. Ilia, I., Koumantakis, I., Rozos, D., Koukis, G., and Tsangaratos, P. A geographical information system (GIS) based probabilistic certainty factor approach in assessing landslide susceptibility: the case study of Kimi, Euboea, Greece. In Engineering Geology for Society and Territory, **2**, Landslide Processes, 1199-1204 (2015). https://doi.org/10.1007/978-3-319-09057-3_210
13. Kuriakose, S. L., Sankar, G., and Muraleedharan, C. History of landslide susceptibility and a chorology of landslide-prone areas in the Western Ghats of Kerala, India. Environmental geology. **57**, 1553-1568 (2009). https://doi.org/10.1007/s00254-008-1431-9
14. VanWesten, C. J., Van Asch, T. W., and Soeters, R. Landslide hazard and risk zonation why is it still so difficult? Bulletin of Engineering Geology and the Environment, **65**,167-184 (2006). https://doi.org/10.1007/s10064-005-0023-0
15. Lin, Q., and Wang, Y. Spatial and temporal analysis of a fatal landslide inventory in China from 1950 to 2016. Landslides, **15**, 2357-2372 (2018). https://doi.org/10.1007/s10346-018-1037-6
16. Abdulwahid, W. M., and Pradhan, B. Landslide vulnerability and risk assessment for multi-hazard scenarios using airborne laser scanning data (LiDAR). Landslides, **14**, 1057-1076 (2017). https://doi.org/10.1007/s10346-016-0744-0
17. Bhatt, B. P., Awasthi, K. D., Heyojoo, B. P., Silwal, T., and Kafle, G. Using geographic information system and analytical hierarchy process in landslide hazard zonation. Applied Ecology and Environmental Sciences, **1**(2), 14–22 (2013). https://doi.org/10.12691/aees-1-2-1
18. Budimir, M. E. A., Atkinson, P. M., and Lewis, H. G. A systematic review of landslide probability mapping using logistic regression. Landslides, **12**, 419-436 (2015). https://doi.org/10.1007/s10346-014-0550-5
19. Saaty, T. L. The analytic hierarchy process (AHP). The Journal of the Operational Research Society, **41**(11), 1073–1076 (1980). https://doi.org/10.1016/0377-2217(94)90222-4
20. Shahabi, H., Keihanfard, S., Ahmad, B. B., and Amiri, M. J. T. Evaluating Boolean, AHP, and WLC methods for selecting waste landfill sites using GIS and satellite images. Environmental Earth Sciences. **71**, 4221-4233 (2014). https://doi.org/10.1007/s12665-013-2816-y
21. Das, I., Sahoo, S., van Westen, C., Stein, A., and Hack, R. Landslide susceptibility assessment using logistic regression and its comparison with a rock mass classification system along a road section in the northern Himalayas (India). Geomorphology. **114**(4), 627-637 (2010)

# National road cliff protection analysis in Malibau Anai after Anai River flood, West Sumatra

*Bambang* Istijono[1*], *Rahmad* Yuhendra[2], *Muhammad* Thariqjabbar Hape[3], *Mahdi*[1], *Arian* Dodi[4], *Muhammad* Al Giffari[1], *Abdul* Hakam[5], *Bayu* Martanto Adji[5], and *Seng* Menghoung[6]

[1]Graduate School, Universitas Andalas, Padang, Indonesia
[2]Department of Water Resources and Construction Development, West Sumatra, Padang, Indonesia
[3]River Basin Center Sumatra V, West Sumatra, Padang, Indonesia
[4]Research and Development Agency, West Sumatra, Padang, Indonesia
[5]Civil Engineering Department, Universitas Andalas, Padang, Indonesia
[6]Civil Engineering Department, National Polytechnic Institute of Cambodia, Phnom Penh, Cambodia

**Abstract.** Scouring of the riverbed can lead to landslides and erosion of the river bank. Landslides on river cliffs are a type of slope failure caused by the swift flow of rivers which can be worse during flash floods events. In this discussion, cliff protection is analyzed at the location 0°29'36 "S 100°20'20" E on the Padang-Bukittinggi national road, which suffered significant damage due to the Anai River flood. With a road elevation of +373 m and a river bed elevation of +317 m, the 56 m high escarpment, which is only 3 m from the edge of the national road, is highly vulnerable to erosion and landslides. This study used field surveys for geotechnical measurements, and hydrological analysis to assess the stability of the cliff and the effectiveness of proposed protection measures, such as reducing river flow velocity, shifting the road, and improving drainage channels. The results show that although natural protection methods have shown positive results, they are insufficient and ineffective when facing of large flood pressures. Recommendations for improvement include the construction of ground-sills and flow diverters in the Anai River, road drainage, planting stronger vegetation, shifting the road body, and regular monitoring.

## 1 Introduction

The Padang-Bukittinggi national road is one of the main transportation routes connecting the West Sumatra region with various surrounding areas. It plays a crucial role in the local economy by facilitating the movement of goods and people between economic centers. At the coordinate of 0°29'36"S 100°20'20" E, this route encounters a particularly challenging feature: a steep cliff with road elevation of +373 m and a river bed elevation of +317 m. This 56 m high cliff is only 3m from the edge of the asphalt road, making it highly susceptible to erosion and landslides, especially after flooding in the Anai River [1].

The Anai River, which flows along the base of the cliff, has a steep river bed and experience high discharge, particularly during the rainy season. Flooding at certain times results in increased water pressure on the cliff structure, leading to erosion of the river bed and weakening the cliff's foundations. Such events often trigger landslides that damage the cliffs and pose a significant threat to the safety of the nearby national road. This condition is exacerbated by the high intensity of rainfall in the region, which further increases the likelihood of landslide events [2, 3]. The condition of the cliff before the landslide in Malibau Anai can be seen in Fig. 1.

**Fig. 1.** Cliff condition before the landslide (source: google earth, february 23, 2024)

The cliffs' vulnerability at this location not only affects the physical stability of the road infrastructure, but also negatively impacts road users' safety [4-6]. Any landslide event can result disrupt transportation routes, leading to considerable economic and social disruption for the communities that rely on this road [7]. Therefore, protecting the road is critical to ensure the safe and sustainable operation of the Padang-Bukittinggi national road.

* Corresponding author: bistijono@eng.unand.ac.id

Effective cliff protection at this site requires a comprehensive approach that includes geotechnical and hydrological analysis, as well as the application of protective technologies suited to local conditions. In addition, periodic monitoring is required to anticipate changes in cliff conditions due to external factors such as rainfall and seismic activity. This research aims to analyse the protection measures of implemented after the Anai River flood and evaluate their effectiveness in maintaining cliff stability and national road safety at this location. It is expected to provide recommendations for improving road bed protection in the future.

## 2 Method

This research aims to analyse and evaluate the effectiveness of cliff protection measure on the Padang-Bukittinggi national road at coordinates 0°29'36"S 100°20'20"E, which is in critical condition following the Anai River flash flood. This research employed a comprehensive methodological approach using field survey techniques. These surveys were conducted to obtain a clear understanding of the physical condition of the cliff and its surrounding environment.

The direct observations included hydrological mapping, cliff topography, measuring the cliff's distance from the road edge, and identifying landslide prone areas. The visual and descriptive data collected during this survey were used to model the actual condition of the cliff and its surrounding, which were further analysed. Geotechnical measurements were conducted to assess the physical and mechanical characteristics of soil and rock in the cliff, including sieve analysis to determine the soil structure composition and to evaluate the cliff's stability against potential landslides.

The implemented natural cliff protection measures were evaluated using the survey data and stability analysis results. This evaluation included assessing various protection techniques, such as the installation retaining walls, creating road drainage channels to control surface water flow, and planting vegetation to strengthen the soil structure.

## 3 Results and discussion

### 3.1 Anai river hydrological analysis

The hydrological analysis of Anai River is crucial for understanding its water flow characteristics and its impact on cliff stability at the study site. Historical rainfall and stream flow data from local hydrological stations form the main basis for this analysis [6, 7]. Using the Nakayasu synthetic unit hydrograph method, it was calculated that the peak flood discharge reaches 1,073 m³/s. How ever the unusual runoff coefficient of 3 hours seems unrealistic, as it normally ranges from 0 to 1. The analysis results indicate that for the area upstream of the Pasar Usang bridge, the Nakayasu method produces a peak flood discharge ranging from 780 m³/s to 1,211 m³/s. In contrast, for the area upstream to the estuary, the peak flood discharge ranges from 853 m³/s to 1,327 m³/s as shown in Fig. 2.

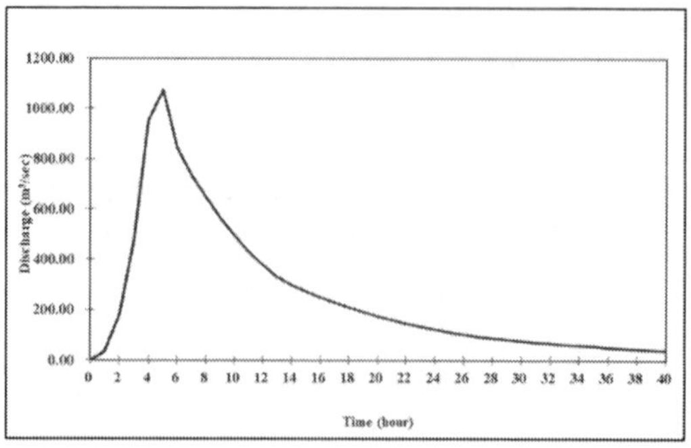


**Fig. 2.** Flood discharge graph

This difference reflects the complexity of hydrological dynamics from upstream to downstream of the river. Hydrological model simulations were also conducted to predict the impact of flooding on river bed erosion and cliff stability [8-10]. The simulations suggest that these hydrological fluctuations may increase erosion patterns and landslide risks [11].

Floods in the region significantly impact riverbed erosion and weaken cliff foundations. Flooding accelerates lateral erosion, eroding the cliff's base and reducing overall slope stability. These impacts are exacerbated by unpredictable river discharge changes, increasing hydraulic stress on the cliff structure. The condition of the Anai River is illustrated in Fig. 3.

**Fig. 3.** The condition of one of the points in the Anai River

### 3.2 Anai river geological analysis

The Anai Valley features a distinctive river valley morphology, with varied shapes resembling the letters "V" and "U", particularly in the river meanders. The river is approximately 40 meters wide, with a depth ranging from 10 to 56 meters below the road. The geology of this area is dominated by alluvial deposits consisting of weathered and eroded materials from Mount Marapi, including igneous rock fragments ranging in size from sand to boulders.

Based on the results of the sieve analysis, the disturbed soil sample from the Malibau Anai location contains 10.43 percent gravel/rock, 78.87 percent sand, and 10.70

percent fine-grained soil (clay/silt). With the gravel/rock percentage of below 15 percent, this soil is categorized as sand type (Fig. 4).

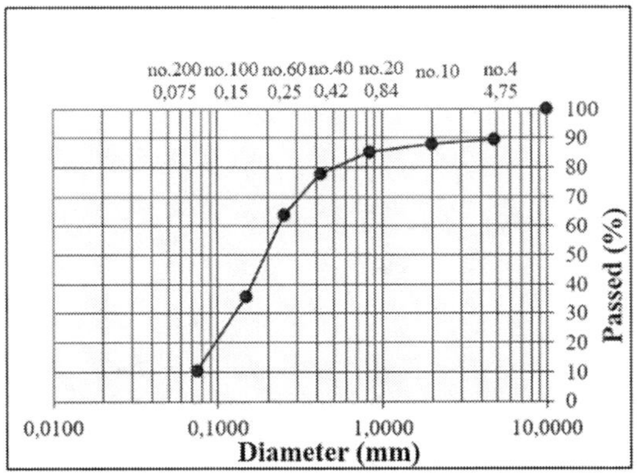


**Fig. 4.** Grain size distribution of the Anai River sediment filter test

The soil composition has significant implications for protecting the cliffs along the National Road in Malibau Anai, particularly following the flooding of the Anai River. The soil in these areas is w predominantly sand, which has lower stability compared to coarse grained soils such as gravel. Sand is more susceptible to erosion, especially when exposed to strong water flow, during and after flood events.

The high sand content (78.87 percent) indicates that the cliffs around the national road are at significant risk of erosion, especially during heavy water flows after floods. This can reduce the stability of the cliffs, thereby increasing the risk of landslides and damage to the road infrastructure [4]. Therefore, implementing effective cliff protection measures, such as structural reinforcement and efficient drainage systems, is crucial for reducing the risk of landslides and ensuring the safety of national roads in Malibau Anai.

### 3.3 Cliff condition after flood

**Fig. 5.** Cliff condition after flash flood (may 11, 2024)

The results of the field survey conducted after the flood in the Anai River showed significant morphological changes to the cliff, located at 0°29'36"S 100°20'20"E along the Padang–Bukittinggi national road. The flood caused severe erosion at the base of the cliff, leading to the loss of soil and rock material. This erosion resulted in slope instability which was further exacerbated by the extremely steep slope of the cliff (Fig. 5).

After the flood, several new cracks were observed in the middle to upper section of the cliff, indicating an increased risk for landslides. These cracks resulted from the pressure of water entering the soil through its pores during the flood, which increased pore water pressure and reduce the soil's shear strength [2, 11, 12]. In addition, the cliff section adjacent to the river flow experienced a reduction in height due to the lateral erosion process, which further weakened the cliff foundation. Meanwhile, the road cliff collapse at the Malibau Anai location still needs to be repaired.

The soil test using standard penetration has been conducted in the cliff as shown in Fig. 6. Based on that test, the soil parameter then is calculated as written in Table 1.

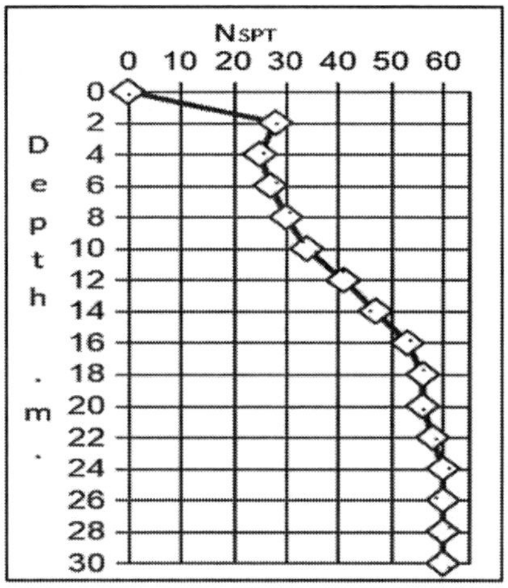


**Fig. 6.** Nspt test result (source: redrawn after National Road Office West Sumatra)

**Table 1.** Soil parameters for cliff stability analysis

| Depth | NSPT | ϕ | Unit Weight | | EAVE |
|---|---|---|---|---|---|
| (m) | (blows) | (degree) | (t/m³) | (kN/m³) | (kN/m²) |
| 2 | 28 | 41.3 | 1.9 | 18.6 | 12000 |
| 4 | 25 | 40.3 | 1.7 | 16.7 | 11000 |
| 6 | 27 | 41.0 | 1.9 | 18.6 | 11000 |
| 8 | 30 | 42.0 | 1.9 | 18.6 | 12000 |
| 10 | 34 | 39.6 | 1.9 | 18.6 | 13000 |
| 12 | 41 | 42.4 | 1.9 | 18.6 | 16000 |
| 14 | 47 | 44.8 | 1.9 | 18.6 | 17000 |
| 16 | 53 | 47.2 | 1.9 | 18.6 | 19000 |
| 18 | 56 | 48.4 | 1.9 | 18.6 | 20000 |
| 20 | 56 | 48.4 | 1.9 | 18.6 | 20000 |
| 22 | 58 | 48.4 | 1.9 | 18.6 | 21000 |
| 24 | 60 | 49.2 | 1.9 | 18.6 | 21000 |
| 26 | 60 | 50.0 | 1.9 | 18.6 | 21000 |
| 28 | 60 | 50.0 | 1.9 | 18.6 | 21000 |
| 30 | 60 | 50.0 | 1.9 | 18.6 | 21000 |

The slope stability of the cliff then was analyzed using the test result. The stability analysis in the terms of factor of safety with the results in Fig. 7. The cliff has safety factor (SF) less than 1 for the distance about 8 m from the edge of the cliff.

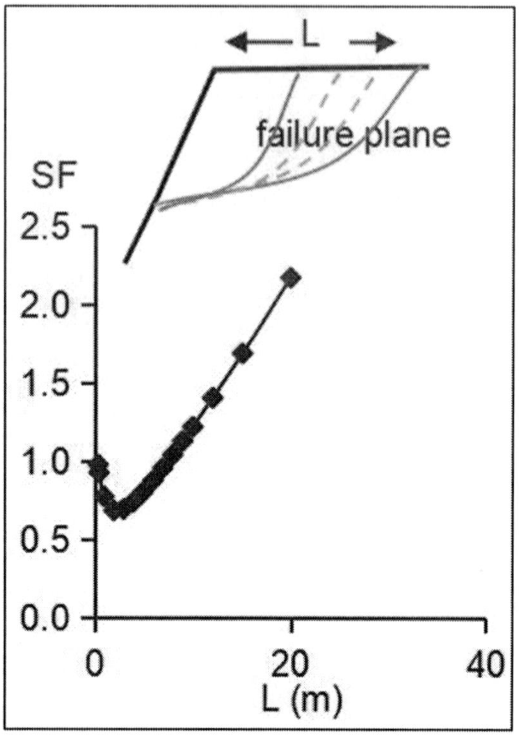


**Fig. 7.** The reduced stability of the cliff increases the risk of landslides.

Meanwhile, the road cliff collapse at the Malibau Anai location still needs to be repaired. Table 2 shows several locations that are being repaired.

**Table 2.** Landslide points and handling type plans (source: National Road Office West Sumatra)

| No | Location/KM | Types of Damage, height, and length of damaged road | Handling Type Plan |
|---|---|---|---|
| 1. | 62+700 | partial road collapse, H 10m and L 120m | Secant pile |
| 2. | 64+100 | Road collapse, H 5m and L 60m | Secant pile |
| 3. | 64+200 | Road collapse, H 8m and L 120m | Secant pile+DPT |
| 4. | 64+600 | Road collapse, H 12m and H 60m | Secant pile+EPS Geofoam |
| 5. | 64+900 | Road collapse, H 12m and L 20m | Secant pile+EPS Geofoam |
| 6. | 65+000 | Road collapse, H 12 and L 100m | Secant pile+EPS Geofoam |
| 7. | 65+400 | partial road collapse, H 12m and L 40m | Secant pile+EPS Geofoam |
| 8. | 66+000 | partial road collapse, H 12m and L 100m | Secant pile+EPS Geofoam |
| 9. | 67+100 | Partial road collapse, H 7m and L 60m | Secant pile |

### 3.4 Impact on national roads and road user safety

Changes in the cliff's morphology after the flood have directly impacted the national road, located only 3 meters from the cliff (Fig. 8). The reduced stability of the cliff increases the risk of landslides (Fig. 7), potentially damaging this vital transportation route due to river water erosion. Minor deformation had been observed on the road surface in some areas, indicating potential soil movement beneath the asphalt layer.

**Fig. 8.** Cliff Condition with road shoulder 3 meters away.

This risk raises serious concerns for the safety of road users. The Padang–Bukittinggi national road is a major transportation route, connecting various regions in West Sumatra so that any disruption can have significant economic and social impacts. In addition, the possibility of landslides during heavy rains or subsequent floods heightens the need for urgent precautions.

The installation of guardrails along the road shoulder would be an effective temporary solution to prevent motorists from getting to close the cliff and to safeguards vehicles from failing off the edge.

### 3.5 Effectiveness of cliff protection measures that have been implemented

An evaluation of the implemented cliff protection measures been showed mixed results. The installation of retaining walls in several areas has successfully reduced erosion rates at the base of the cliff and enhanced local stability. However, these walls have not fully restrained strong water flows during major floods, leading to vortex formation upon water impact, which has caused some parts of the structure to be damaged or displaced.

Drainage channels built to control surface water flow have also shown positive results in reducing water pressure on the cliffs. However, their capacity is insufficient to handle large volumes of water during extreme flooding, resulting in waterlogging at several points and the instability of the cliffs.

Natural vegetation along the cliffs has achieved moderate success in stabilizing the soil. While vegetation roots help reinforce the topsoil layer, they are insufficient to withstand deeper soil movements, particularly during water saturated conditions. This highlights the need for more robust vegetation or a combination of engineering methods to improve effectiveness

### 3.6 Recommendations for cliff protection improvements and enhancements

Based on the results of this evaluation, several essential recommendations can be made for future enhancement to cliff protection:

a. Ground-sill Development and Flow Director

The Anai River at this location is a sediment transportation zone characterize by a steep and winding slope. The foundation needs to be strengthened to prevent erosion of the river bank wall by flood flows. This requires the construction of river slope control structures, such as a ground-sill and cliff base reinforcement, along with flow directing structures. In addition, a ground-sill should be constructed to protect the main dam from local downstream erosion (scouring), maintain the stability of the Sabo-dam foundation, and prevent the collapse of the cliffs [13]. The surface of the ground-sill should be flat (Fig. 9). The thickness of the ground sill is determined based on the erodibility of the riverbed and the size of the boulders present. In thick, soft rock layers, the apron should be set between 0.70 – 1.0 meters. If the river bed consists of sand and gravel, the thickness of the apron should be determined according to Table 3.

**Table 3.** Ground sill thickness for sand and gravel base layer (sources: technical standard and guideline for planning and design Sabo, march 2002).

| | | | | | |
|---|---|---|---|---|---|
| **Main dam height ($H_m$)** | 6 m | 7 m | 8 m | 9 m | 10 m |
| **Apron thickness ($t_a$)** | 1,0 m | 1,0 m | 1,2 m | 1,2 m | 1,5 m |
| **Main dam height ($H_m$)** | 11 m | 12 m | 13 m | 14 m | 15 m |
| **Apron thickness ($t_a$)** | 1,5 m | 1,5 m | 1,5 m | 1,5 m | 2,0 m |

On both cliffs, ground sills are built with side walls to protect the cliffs from erosion and maintain the stability of the Sabo-dam. The following considerations determine the height of the side walls:

1) The elevation of the upstream side wall should be 1.0-1.50 meters higher than the elevation of the wing face sub-dam.
2) The elevation of the side walls should match the height of the wing face of the sub-dam or the upright wall for the dam with the energy absorbing pool.

b. Increasing Drainage Capacity

Existing road drainage channels need to be upgraded to cope with larger volumes of water, especially during extreme rainfall events. Additional channels or emergency water diversion systems may be necessary to prevent water accumulation on the cliffs.

c. More Effective Uses of Vegetation Engineering

Vegetation with a deeper and stronger root system can improve soil stability, especially in deeper soil layers. Additionally, combining vegetation and other soil strengthening techniques, such as the use of geotextile netting, can help reduce the risk of landslides.

d. Road Body Shifting

In the medium term, the road which currently has a width of 2 x 3.60 meters, should be realigned to move it away from the landslide prone area (Fig. 10).

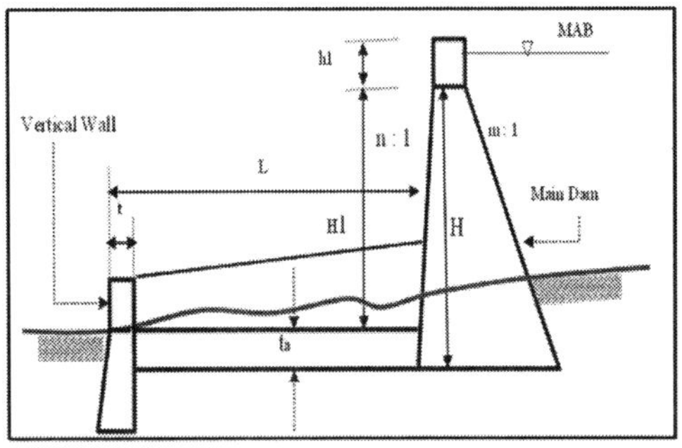


**Fig. 9**. Ground sill length on flat type

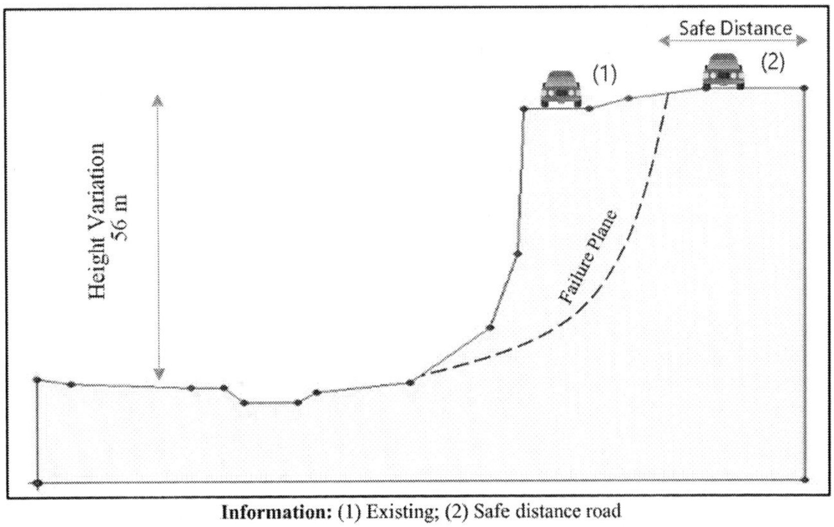


**Information:** (1) Existing; (2) Safe distance road

**Fig. 10.** Illustration of the condition of the cliff with the shoulder of the road

e. Periodic Monitoring and Maintenance

An automated monitoring system that detects soil movement and changes in pore water pressure should be implemented to provide early warning of potential landslides. In addition, regular maintenance of retaining structures and drainage channels must be carried out to ensure their continued effectiveness.

By implementing this recommendation, it is expected that the protection of cliffs on the Padang–Bukittinggi national road can be improved, so that the risk of landslides can be minimized and the safety and sustainability of this transportation route can be maintained.

## 4 Conclusion

This study analyzed the cliff conditions at the location of 0°29'36 "S 100°20'20"E on the Padang–Bukittinggi national road, located above the Anai River, and evaluated the effectiveness of cliff protection measures implemented after the flood. The results of field surveys and geotechnical measurements show that the cliffs at the research site have a very steep slope, with the road just about 3 meters away. The soil structure on this cliff is dominated by materials that are prone to erosion and landslides, especially at the cliff's base, due to the impact of the Anai River flow. Increased pore water pressure during floods and high groundwater content after intense rainfall leads to a reduction in soil shear strength, thereby increasing the risk of landslides.

A Hydrological analysis of the Anai River revealed that the flooding in this region had a significant impact on riverbed erosion and weakened the cliff foundations. Flooding accelerates the lateral erosion, eroding the cliff's base and further reducing the overall stability of the slope. These effects are exacerbated by unexpected changes in river discharge, which increases hydraulic pressure on the cliff structure.

The implemented cliff protection measures had a positive impact on reducing the risk of landslides. However, their effectiveness varies depending on the location and specific conditions of the cliffs. Although these measures were successful in the short term, the evaluation shows that some areas still experienced decreased stability, particularly in sections of the cliffs directly affected directly affected by erosion from the Anai River.

We would like to thank the Graduate School of Universitas Andalas for the research grant 2024 No. B. 29/KPT/UN16.16/DIR/I/2024 dated August 1, 2024.

## References

1. Department of Water Resources and Construction Development West Sumatra Province, Government Agency Performance Accountability Report of Department of Water Resources and Construction Development West Sumatra Province, (2024).
2. Aprisal, B. Istijono, T. Ophiyandri, and Nurhamidah, A study of the quality of soil infiltration at the downstream of Kuranji River, Padang City. International Journal of GEOMATE. **16**, 16–20, (2019) http://doi.org/10.21660/2019.56.4530
3. K. E. Ndabezitha, B. C. Mubangizi, and S. F. John, Adaptive capacity to reduce disaster risks in informal settlements. Jàmbá Journal of Disaster Risk Studies. **16**, (2024) http://doi.org/10.4102/jamba.v16i1.1488
4. A. Andriani, B. Istijono, F. Akmal, and B. Martanto Adji, Identification of Landslide Hazard in Residential Area Kubang Tangah District, Sawahlunto. Journal of Applied Engineering and Technological Science. **5**, 1154–1164, (2024).
5. B. Istijono and A. Hakam, Development of stability criteria for risk reduction in the Sianok Canyon in Bukittinggi Indonesia. International Journal of Earth Sciences and Engineering. **9**, (2016). [Online]. Available: https://www.researchgate.net/publication/296639389
6. M. Wei and J. Xu, Assessing Road network resilience in disaster areas from a complex network perspective: A real-life case study from China. International Journal of Disaster Risk Reduction. **100** (2024) http://doi.org/10.1016/j.ijdrr.2023.104167
7. S. Cao, H. Ye, and Y. Zhan, Cliff roads: An ecological conservation technique for road construction in mountainous regions of China, Landsc Urban Plan. **94**, 228–233 (2010) http://doi.org/10.1016/j.landurbplan.2009.10.007
8. P. Johnson and K. Lee, Hydrological Modeling Using the Rational Method. Env. Eng. **21**, 45–58 (1995) https://doi.org/10.28991/HEF-2023-04-02-02
9. J. J. Major, S. Zheng, A. R. Mosbrucker, K. R. Spicer, T. Christianson, and C. R. Thorne, Multidecadal Geomorphic Evolution of a Profoundly Disturbed Gravel Bed River System—A Complex, Nonlinear Response and Its Impact on Sediment Delivery, J Geophys Res Earth Surf. **124**, 1281–1309 (2019) http://doi.org/10.1029/2018JF004843
10. L. Del Río and F. J. Gracia, Erosion risk assessment of active coastal cliffs in temperate environments, Geomorphology. **112**, (2009) http://doi.org/10.1016/j.geomorph.2009.05.009
11. T. Tingsanchali, Urban flood disaster management, in Procedia Engineering. Elsevier Ltd, 25–37. (2012) http://doi.org/10.1016/j.proeng.2012.01.1233
12. Aprisal, B. Istijono, Juniarti, and M. Harianti, Soil quality index analysis under horticultural farming in Sumani upper watershed. International Journal of GEOMATE. **16**, 191–196 (2019) http://doi.org/10.21660/2019.56.8212
13. Q. L. Zeng, Z. Q. Yue, Z. F. Yang, and X. J. Zhang, A case study of long-term field performance of check-dams in mitigation of soil erosion in Jiangjia stream, China. Environmental Geology. **58**, 897–911 (2009) http://doi.org/10.1007/s00254-008-1570-z

# A conceptual framework of implementing lean communication for building back better in the post-disaster housing reconstruction

*Helny* Lalan[1,2], *Akhmad* Suraji[1*], *Taufika* Ophiyandri[1], and *Benny* Hidayat[1]

[1]Civil Engineering Department, Faculty of Engineering, Universitas Andalas, Padang, Indonesia
[2]Civil Engineering Department, Faculty of Engineering, Universitas Ekasakti, Padang, Indonesia

**Abstract.** In the aftermath of disasters, effective communication plays a pivotal role in facilitating efficient and sustainable housing reconstruction efforts. This paper proposes a conceptual framework to explore the application of lean communication principles in the context of post-disaster housing reconstruction. The framework integrates theories from lean management studies and communication to enhance our understanding of how lean principles can optimize communication processes among stakeholders involved in housing reconstruction. By adopting a qualitative approach, this study aims to develop a framework to identify key communication challenges and opportunities and to examine existing practices for implementing lean communication strategies to achieve the goal of building back better. The theoretical contribution of this framework lies in its potential to inform policymakers, practitioners, and researchers about the importance of lean communication in promoting resilience and sustainability in post-disaster housing reconstruction initiatives.

## 1 Introduction

### 1.1 Overview of post-disaster housing reconstruction challenges

Post-disaster housing reconstruction presents a myriad of challenges that impact communities, infrastructure, and the overall recovery process. These challenges are shaped by the nature of the disaster, local socio-economic conditions, and the effectiveness of response and recovery efforts [1] or divided into four categories: general, physical, social, and economic [2]. Disasters often lead to mass displacement, creating an immediate need for temporary shelter and longer-term housing solutions. Managing and meeting these shelter needs promptly is crucial to ensuring the well-being of affected populations [3]. Financing post-disaster reconstruction poses significant challenges, as governments and international organizations must mobilize resources quickly amidst competing priorities. Limited funding can delay reconstruction efforts and affect their scope and quality [4]. Successful reconstruction hinges on community participation and engagement [5]. Rebuilding efforts must adhere to building codes, land use regulations, and environmental standards. However, in post-disaster settings, regulatory frameworks may be disrupted or inadequate, posing challenges to compliant reconstruction [6]. Access to construction materials, skilled labor, and transportation infrastructure can be limited in disaster-affected areas. Coordinating logistics and mobilizing resources efficiently are critical to overcoming these challenges [7]. Disasters exacerbate existing socio-economic vulnerabilities, disproportionately affecting marginalized and low-income communities. Reconstruction efforts should aim to address these disparities and promote inclusive recovery [8]. Continuous monitoring and evaluation of reconstruction projects are critical to assess progress, identify challenges, and adjust strategies as needed. Learning from past experiences and sharing best practices can inform future disaster recovery efforts [9, 10]. Addressing these challenges requires a holistic approach that integrates technical expertise, community engagement, and policy coordination. By understanding these complexities and adopting proactive strategies, stakeholders can promote sustainable and resilient housing reconstruction in post-disaster contexts.

### 1.2 Importance of effective communication in housing reconstruction in the post-disaster

Effective communication is crucial in post-disaster housing reconstruction [11]. It facilitates coordination among stakeholders, ensures transparency, and empowers affected communities [12]. Clear and timely communication aligns efforts, reduces resource duplication, and smooths decision-making processes. Transparent communication builds trust and empowers communities to make informed decisions about their recovery. Engaging affected communities in the

* Corresponding author: akhmad.suraji@gmail.com

reconstruction process through effective communication promotes inclusivity and ownership [13]. Communities can be educated about resilient construction practices, early warning systems, and evacuation procedures to enhance their ability to mitigate future risks. Building feedback mechanisms allows stakeholders to receive input from communities, monitor satisfaction, and promptly address emerging challenges [14]. Clear communication supports data-driven policymaking and community feedback, ensuring compliance with standards and building codes [15]. Transparent and inclusive communication builds trust between stakeholders and affected communities, creating a foundation for sustainable and resilient recovery [16, 17]. Effective communication initiatives strengthen local capacities by providing training in disaster risk reduction, construction techniques, and project management [18, 19]. Sustainable communication strategies leave a lasting impact on communities by promoting resilient-building behaviors and a culture of preparedness [20].

### 1.3 Introduction to lean principles and their potential application in communication for housing reconstruction in the post disaster

Lean principles, originating from manufacturing and later adapted to various sectors including construction and project management, emphasize efficiency, continuous improvement, and waste reduction [21]. When applied to communication in post-disaster housing reconstruction, these principles can enhance coordination, optimize resource use, and improve overall project outcomes. Here's an introduction to lean principles and their potential application in communication. Lean principles focus on eliminating waste, optimizing processes, and maximizing value for stakeholders [22]. Identifying what adds value from the perspective of the end-user (e.g., homeowners, communities). Creating smooth and efficient workflows to minimize delays and bottlenecks. Initiating actions based on actual demand rather than forecasts to reduce overproduction. Encouraging ongoing efforts to streamline processes and enhance quality [23]. Lean principles advocate for clear and transparent communication channels to ensure that all stakeholders have access to timely and accurate information about reconstruction plans, progress, and challenges [24]. By implementing lean communication practices, decision-making processes can become more efficient and responsive. This involves reducing unnecessary bureaucracy and ensuring that decisions are based on relevant and current data [22]. Lean communication empowers stakeholders by encouraging participation and feedback. This inclusive approach ensures that community voices are heard and incorporated into decision-making processes, ultimately leading to more sustainable and community-driven reconstruction efforts [25]. Lean principles help in optimizing resource allocation by minimizing wasteful communication practices and ensuring that resources (such as materials, manpower, and funding) are used efficiently to achieve desired outcomes [7]. Incorporating lean communication strategies can contribute to building resilience by promoting adaptive strategies and fostering a culture of continuous learning and improvement among stakeholders [20, 26]. Lean benefits are improved efficiency, enhanced stakeholder engagement, reduced costs, and better overall project outcomes. However, the challenges are resistance to change, cultural barriers, and the need for training and capacity building among stakeholders. In conclusion, lean principles offer a systematic approach to enhancing communication in post-disaster housing reconstruction, focusing on efficiency, transparency, stakeholder empowerment, and continuous improvement. By adopting these principles, stakeholders can navigate the complexities of reconstruction efforts more effectively, leading to more resilient and sustainable communities.

## 2 Methodology

The literature review of integrated Lean Principal communication, and post-disaster housing reconstruction publications published in the current decade was conducted for the present study. This article based on a keyword and published from 2010 up to 2024. Keyword using the search is "lean priciples, 26 effective communication, and post disaster housing reconstruction." The tools used to find articles are publish or perish and connected paper.

### 2.1 Theoretical foundations of lean management and communication on a literature Review

Lean management, originally developed by Toyota, focuses on eliminating waste, optimizing processes, and continuous improvement [21]. Key concepts like Just-In-Time (JIT) production, Kanban systems, and value stream mapping help enhance efficiency by visualizing processes and reducing non-value-added activities [27]. The philosophy of continuous improvement, or Kaizen, encourages small, incremental changes to improve quality and customer satisfaction.

In post-disaster housing reconstruction, communication is vital for coordination, collaboration, and community engagement. Theoretical foundations include Community-Based Disaster Risk Reduction (CBDRR), which promotes inclusive communication and stakeholder participation, and Stakeholder Theory, which emphasizes trust-building and addressing diverse interests among involved parties. Information and communication Technology (ICT) plays a crucial role in facilitating real-time information and resource coordination, while risk communication frameworks stress the need for clear, timely information to promote preparedness [28].

By integrating lean management principles, such as value stream mapping, with effective communication strategies based on stakeholder engagement and community participation, post-disaster housing reconstruction can become more efficient. This approach fosters ownership, resilience, and sustainable recovery outcomes.

### 2.2 Review of studies on post-disaster housing reconstruction

Studies on post-disaster housing reconstruction use various methods, including qualitative case studies, quantitative surveys, and mixed-method approaches[29]. Qualitative studies focus on community experiences and socio-cultural factors, while quantitative studies analyze economic impact, infrastructure damage, and policy effectiveness. Common challenges include housing destruction, limited funding, community engagement difficulties, and regulatory issues [4, 7, 30]. Successful reconstruction requires community participation, which enhances sustainability and local ownership. Technology like GIS and BIM, along with effective governance, plays a critical role in planning and coordination [31-33]. There is a growing focus on integrating sustainability and disaster risk reduction into reconstruction efforts. Critiques highlight gaps between policy and implementation, particularly regarding resource distribution and bureaucratic inefficiencies. Future research should address long-term impacts, innovative approaches, and inclusion of vulnerable populations. This comprehensive review highlights the complex challenges and opportunities in post-disaster housing reconstruction, guiding policy and practice improvements.

### 2.3 Existing frameworks and models for disaster recovery communication

Effective coordination and communication are essential for successful post-disaster housing reconstruction, facilitating collaboration among stakeholders, optimizing resource allocation, and ensuring community engagement [14, 17]. Frameworks for coordination and communication draw on theoretical foundations. Complex Adaptive Systems Recognizes disasters as complex events involving multiple interacting systems and stakeholders. Effective coordination requires adaptive responses that can quickly adjust to changing circumstances [34]. Network Theory emphasizes the importance of relationships and connections among stakeholders in facilitating information flow, decision-making, and resource mobilization[35]. Institutional Analysis focuses on understanding the roles, norms, and organizational structures of stakeholders involved in disaster response and recovery[36]. Incident Command System (ICS) provides a standardized organizational structure and communication protocols for coordinating emergency response efforts. It facilitates clear roles and responsibilities among responders from different agencies and jurisdictions, promoting efficient resource allocation and decision-making [37]. However, criticisms include its hierarchical nature, which may limit flexibility and adaptation to complex disaster scenarios [38]. The cluster approach, adopted by the United Nations, divides humanitarian response into thematic sectors (e.g., shelter, health, logistics) led by designated agencies. It aims to enhance coordination, avoid duplication of efforts, and improve service delivery through collaborative partnerships among humanitarian actors [39]. Challenges include coordination gaps between clusters and issues related to leadership and accountability [40]. Community-Based Participatory Approaches (CBPA) emphasizes the active involvement of affected communities in decision-making processes, including needs assessment, project planning, and implementation. It promotes empowerment, local ownership, and culturally appropriate solutions, enhancing the sustainability and resilience of reconstruction efforts [41]. However, challenges include power dynamics, capacity constraints, and the need for inclusive representation of marginalized groups [41-43]. Integrated Disaster Risk Reduction (IDRR) frameworks integrate disaster risk reduction principles into reconstruction planning and implementation. They emphasize proactive measures to reduce vulnerability and enhance resilience, promoting sustainable development and long-term disaster preparedness [8]. Implementation challenges include resource constraints, political commitment, and ensuring mainstreaming of DRR across sectors [42]. Fragmentation and Duplication: Existing frameworks may result in fragmentation and duplication of efforts among stakeholders, leading to inefficiencies and gaps in service delivery [42]. Power Dynamics and Equity show that power imbalances among stakeholders can influence decision-making processes and limit the participation of marginalized communities in reconstruction efforts [42, 43]. Adaptation to Local Contexts is framework developed for one disaster context may not be universally applicable, requiring adaptation to local socio-cultural, economic, and environmental conditions [44]. Critical examination of existing frameworks and models for coordination and communication in post-disaster housing reconstruction reveals diverse approaches, strengths, and challenges. By integrating theoretical insights with practical experiences and lessons learned, stakeholders can develop more adaptive, inclusive, and effective strategies to enhance disaster resilience and recovery efforts.

## 3 Results and recommendation

### 3.1 Integration of lean principles into communication strategies

Integrating lean principles into communication strategies involves applying concepts that prioritize efficiency, transparency, continuous improvement, and stakeholder engagement. This approach ensures that communication processes in post-disaster housing reconstruction are streamlined, effective, and responsive to evolving needs. Value Stream Mapping (VSM) for Communication Processes is based on Lean Principle. Value stream mapping (VSM) identifies and maps out the steps involved in delivering value to stakeholders, highlighting areas of waste and inefficiency. Applied to communication, VSM can help streamline information flow, reduce delays, and enhance responsiveness [45, 46]. Its application conducts a VSM exercise to visualize and analyse communication channels and processes in post-disaster housing reconstruction. Identify bottlenecks, unnecessary approvals, and delays in information dissemination. Streamline workflows to ensure timely and accurate communication among stakeholders, including affected communities, government agencies, NGOs, and contractors. Visual Management and Information Transparency is also based on lean principle. Visual management uses visual cues (e.g., charts, boards) to provide real-time information about project status, milestones, and challenges. It enhances transparency, facilitates quick decision-making, and promotes accountability [47]. Implement visual management tools such as Kanban boards or project dashboards to display key communication metrics, progress updates, and upcoming milestones in housing reconstruction projects. Ensure that information is accessible to all stakeholders, promoting shared understanding and alignment of goals. Continuous Improvement and Feedback Loops as lean principle fosters a culture of learning and adaptation. Feedback loops gather input from stakeholders to identify opportunities for enhancing communication effectiveness and addressing issues promptly [23]. Establish regular feedback mechanisms, such as surveys, focus groups, or town hall meetings, to gather input from diverse stakeholders on their communication needs and preferences. Use feedback to refine communication strategies, update information channels, and improve engagement with affected communities. Empowering Stakeholders through Inclusive Communication as lean principle frontline workers and stakeholders by involving them in decision-making and problem-solving processes. In communication, this principle encourages participatory approaches that empower communities and enhance their resilience [44]. Adopt a participatory communication approach where affected communities are actively involved in shaping communication strategies. Encourage community participation in town hall meetings, workshops, or community forums to ensure that their voices are heard, concerns addressed, and preferences considered in communication efforts. Lean Thinking for Resource Optimization as lean principles shows that Lean thinking aims to optimize resource allocation and minimize waste. In communication, this principle involves efficient use of communication channels, technologies, and human resources to maximize impact and minimize costs [48]. Evaluate existing communication tools and platforms for their efficiency and effectiveness. Consolidate redundant communication channels and leverage cost-effective technologies (e.g., social media, mobile apps) to reach diverse audiences. Train communication personnel in lean principles to enhance their effectiveness in managing communication workflows. Integrating lean principles into communication strategies for post-disaster housing reconstruction enhances efficiency, transparency, and stakeholder engagement. By applying concepts such as value stream mapping, visual management, continuous improvement, empowerment through inclusive communication, and resource optimization, stakeholders can improve coordination, responsiveness, and ultimately, the outcomes of reconstruction efforts.

### 3.2 Key components of lean communication in post-disaster housing reconstruction

Value Stream Mapping (VSM) is a lean management tool used to visualize and analyse the flow of information or materials through a process. In the context of communication processes, VSM helps identify inefficiencies, bottlenecks, and opportunities for improvement in how information is transmitted, received, and utilized. Here's a detailed exploration of Value Stream Mapping applied specifically to communication processes. Value Stream Mapping aims to create a visual representation that captures the current state of communication processes. It helps in identifying areas of waste, delays, and inefficiencies, thereby facilitating improvements for enhanced communication effectiveness [49]. Define the boundaries of the communication process to be mapped. This may include internal communication among team members, external communication with stakeholders (e.g., affected communities, government agencies, NGOs), and media channels used for dissemination. Mapping the Current State Identifies Stakeholders and Information Sources by listing all stakeholders involved in the communication process and identify the sources of information relevant to the reconstruction efforts. Map Information Flow will create a step-by-step flowchart that visually represents how information moves through the communication process. Start from the origin of information (e.g., data collection, decision-making points) and trace it through various communication channels (e.g., meetings, reports, emails, digital platforms) to the final recipients (e.g., community members, project managers). Document the time taken at each stage of the communication process. Note delays, bottlenecks, or interruptions that occur, as these can hinder the timely and effective dissemination of information. Specify the types of information communicated at each stage (e.g., project updates, resource requests, feedback collection). This helps in understanding the diversity of communication needs and content. Analysing the Current State focuses on identify waste and inefficiencies. Analyse the mapped process to identify areas of waste, such as redundant communication channels, unnecessary approvals, or delays in information dissemination. These wastes contribute to inefficiencies and can lead to misunderstandings or missed opportunities [23]. Quantify Lead Time and Cycle Time: Measure the lead time (total time taken from initiation to completion of communication) and cycle time (time taken for one unit of communication to be processed). This quantitative analysis provides insights into process efficiency and identifies potential areas for improvement. Collect Stakeholder Feedback: Gather feedback from stakeholders involved in the communication process. Their insights on challenges, communication barriers, and suggestions for improvement are critical for identifying root causes of inefficiencies. Designing the Future State. Develop Ideal Communication Flow: Based on insights gained from the current state analysis, design an ideal or future state communication flow that eliminates waste, reduces lead time, and enhances communication effectiveness. Streamline Communication Channels:

Rationalize communication channels to ensure clarity and efficiency. Determine which channels are most effective for different types of information and stakeholders. Implement Visual Management: Introduce visual tools (e.g., dashboards, Kanban boards) to monitor communication flow, highlight key milestones, and provide real-time updates on project progress. This enhances transparency and keeps stakeholders informed [48]. Implementation and Continuous Improvement. Pilot Test Changes: Implement proposed changes in a controlled environment or pilot project to assess their effectiveness before full-scale implementation. Monitor and Measure: Continuously monitor communication processes using established metrics. Measure improvements in lead time, cycle time, stakeholder satisfaction, and overall project outcomes. Iterate and Adapt: Use feedback and performance data to iterate and adapt communication strategies. Lean principles emphasize continuous improvement, so adjustments should be made iteratively to optimize communication processes over time [44]. Benefits of Value Stream Mapping in Communication Processes Identifies inefficiencies: Helps in pinpointing areas of waste and inefficiencies in communication processes, leading to streamlined workflows. Enhances transparency: Provides a clear, visual representation of communication flow, promoting transparency among stakeholders. Facilitates continuous improvement: Supports iterative improvements by identifying and addressing root causes of communication challenges. Improves stakeholder engagement: Ensures that information reaches stakeholders in a timely and accurate manner, enhancing engagement and collaboration. Value Stream Mapping of communication processes is a valuable tool in post-disaster housing reconstruction contexts. By visualizing and analyzing how information flows through various channels, stakeholders can identify opportunities for improvement, streamline communication strategies, and ultimately enhance project outcomes. Implementing lean principles through VSM promotes transparency, reduces waste, and fosters continuous improvement in communication practices.

### 3.3 Framework design: structure and components

Designing a framework for lean communication involves structuring and defining components that promote efficiency, transparency, and stakeholder engagement in communication processes. Below is an outline of the framework design. A well-designed framework for lean communication in post-disaster housing reconstruction enhances project efficiency, stakeholder engagement, and overall communication effectiveness. By structuring clear objectives, defining key components, implementing visual management tools, and fostering continuous improvement, stakeholders can streamline communication processes and achieve sustainable reconstruction outcomes. Establish the overarching goals of lean communication within the context of post-disaster housing reconstruction. This includes improving information flow, enhancing stakeholder engagement, and optimizing communication processes to support efficient project management [49] Outline the essential elements that comprise the framework, such as communication channels, protocols, feedback mechanisms, and visual management tools [48]. Categorize stakeholders based on their roles and communication needs. This segmentation helps tailor communication strategies to specific audience requirements, ensuring relevant information is delivered effectively [23]. Specify the channels used for communication, including digital platforms, meetings, newsletters, and social media. Each channel should serve a distinct purpose and target specific stakeholder groups to ensure information is disseminated efficiently [48]. Establish clear protocols and guidelines for communication, outlining responsibilities, escalation procedures, and standards for information sharing. This ensures consistency and clarity in communication practices across the reconstruction project [50]. Integrate visual management tools such as Kanban boards, dashboards, and progress charts to visualize project status, milestones, and resource allocation. Visual tools enhance transparency and facilitate quick decision-making [48]. Implement mechanisms for continuous improvement, including feedback loops, performance metrics, and regular evaluations. These mechanisms enable stakeholders to identify communication gaps, address challenges promptly, and adapt strategies based on feedback [23]. Conduct pilot tests of the framework to assess its effectiveness and identify areas for refinement before full-scale implementation. Pilot testing allows for adjustments based on real-world feedback and ensures the framework meets its intended objectives [49]. Provide training sessions for communication personnel and stakeholders on lean communication principles, tools, and techniques. Capacity building enhances competency in using the framework effectively and fosters a culture of continuous improvement in communication practices [51]. Establish metrics to monitor the performance of the framework, including communication effectiveness, stakeholder satisfaction, and project outcomes. Regular evaluation allows for adjustments and refinements to optimize communication processes over time [23]. Incorporate stakeholder feedback into the framework's design and implementation phases. Actively seek input from stakeholders to ensure the framework evolves to meet changing communication needs and challenges in post-disaster contexts [48].

## 4 Conclusion

Principles of Lean Management Originating from Toyota, lean management aims to eliminate waste, optimize processes, and drive continuous improvement through concepts such as Just-In-Time (JIT) production, Kanban systems, and value stream mapping. Communication in Reconstruction is essential for coordination, collaboration, and community engagement. Theoretical foundations such as Community-Based Disaster Risk Reduction (CBDRR) and Stakeholder Theory are essential for building stakeholder trust and participation. Information and communication Technology (ICT) aids in information and coordination of resources in real-time, while risk communication frameworks emphasize clear and timely information for preparedness. In the Post-Disaster Housing Reconstruction effort various methods are used to analyse reconstruction efforts, highlighting challenges such as the type of housing damage, funding limitations, and the difficulty of full community engagement. Successful reconstruction depends on community participation, technology integration, and effective governance. Thus, it is necessary that various frameworks and models, such as Complex Adaptive Systems, Network Theory, and Institutional Analysis, be examined. Each framework has its strengths and challenges, with a focus on adaptive response, stakeholder relationships, and organizational structure. The Incident Command System (ICS) and the United Nations cluster approach provide a standard protocol but face criticism due to its hierarchical nature and coordination gaps. The Community-Based Participatory Approach (CBPA) and Integrated Disaster Risk Reduction (IDRR) emphasize local involvement and proactive measures, but also face challenges such as power dynamics and resource constraints.

Integrating lean principles into a communication strategy for post-disaster housing reconstruction focuses on efficiency, transparency, and stakeholder engagement. Value Stream Mapping (VSM) identifies steps and eliminates waste, streamlining communication. Visual management tools provide real-time updates, increasing transparency and quick decision-making. Continuous improvement fosters a feedback loop to refine the strategy. Inclusive communication empowers people in decision-making. Lean thinking optimizes resources, minimizes costs, and maximizes impact. This approach improves coordination, responsiveness, and overall outcomes in reconstruction efforts.

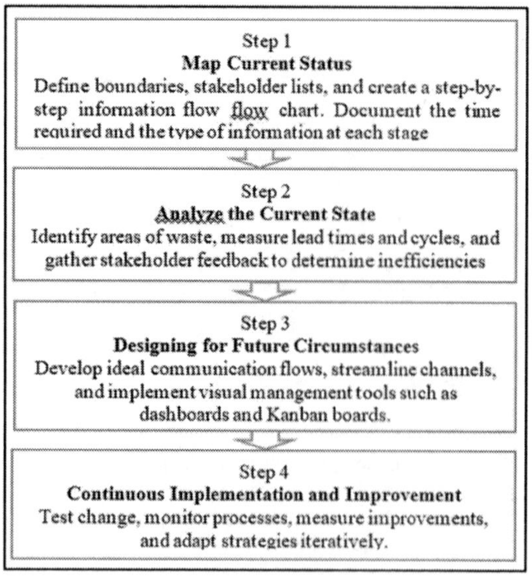


**Fig. 1**. Conceptual framework for value stream mapping (VSM)

Value Stream Mapping (VSM) is a lean management tool used to visualize and analyse the flow of information or material. In communication, VSM helps identify inefficiencies, bottlenecks, and opportunities for improvement. The steps to be taken are outlined in Fig. 1.

VSM increases transparency, supports continuous improvement, and increases stakeholder engagement, making it a valuable tool in post-disaster housing reconstruction. It promotes efficient workflows, effective communication, and better project outcomes.

Designing a lean communication framework involves structuring components that promote efficiency, transparency, and stakeholder engagement. The goals of this framework include improving the flow of information, increasing engagement, and optimizing communication processes. Essential elements include communication channels, protocols, feedback mechanisms, and visual management tools. Stakeholders are categorized based on their roles and needs. Channels are defined for efficient dissemination of information. Clear protocols ensure consistency and clarity. Visual management tools like Kanban boards improve transparency and decision-making. Continuous improvement mechanisms, trials, training sessions, and performance metrics ensure the effectiveness of the framework. Stakeholder feedback is integrated to adapt to evolving communication needs

The study was funded by: Directorate of Research, Technology, and Community Service. Directorate General of Higher Education, Research, and Technology. Ministry of Education, Culture, Research, and Technology on the Doctoral Dissertation Research Scheme In accordance with Research Contract Number: 041/E5/PG.02.00.PL/2024 Fiscal Year 2024

## References

1. B. Rouhanizadeh, S. Kermanshachi, and T. J. Nipa, Exploratory analysis of barriers to effective post-disaster recovery. International Journal of Disaster Risk Reduction, **50**, 101735, (2020). https://doi.org/10.1016/j.ijdrr.2020.101735.
2. A. Pamidimukkala, S. Kermanshachi, and E. Safapour, Challenges in Post-Disaster Housing Reconstruction: Analysis of Urban Vs. Rural Communities, in Proceedings of the Creative Construction e-Conference, June 28 - July 1, Arlington, USA, (2020). https://doi.org/10.3311/CCC2020-061.
3. E. Wagemann, From Shelter to Home: Flexibility in Post-Disaster Accommodation, (2017). https://doi.org/10.17863/CAM.43234.
4. P. K. Freeman, Allocation of post-disaster reconstruction financing to housing, Building Research & Information, **32**(5), 427–437, (2004). https://doi.org/10.1080/0961321042000221016.
5. R. Revetrian, R. Revetrian, and R. rusian, Community engagement in post-disaster housing reconstruction: Lessons from recent earthquakes. International Journal of Disaster Risk Reduction, **27**, 11–19, (2018).

6. W. A. H. Thissen and P. M. Herder, Critical Infrastructures State of the Art in Research and Application. Kluwer Academic Publishers, (2005). https://doi.org/10.1016/j.respol.2005.04.001.
7. I. L. O. ILO, Building Resilience: Employment and Decent Work in Post-Disaster and Post-Conflict Settings. Geneva: ILO, (2021).
8. U. N. O. for D. R. R. UNDRR, Global Assessment Report on Disaster Risk Reduction. Geneva: UNDRR, (2020).
9. A. Azmeri, C. Mutiawati, N. Al-Huda, and H. Mufiaty, Disaster Recovery Indicators of Housing Reconstruction: The Story of Post Tsunami Aceh, Indonesia. International Journal of Disaster Management, **1**(1), (2017).
10. W. Cheek and K. Chmutina, Building back better' is neoliberal post-disaster reconstruction. Disasters, **46**(3), 589–609, (2022). https://doi.org/10.1111/disa.12502.
11. F. Abdolahzadeh, H. Bagherian, M. Peyravi, P. Kolivand, and N. Tavakoli, A comparative study of interagency communications and information exchange in disaster response among selected countries. J Educ Health Promot, **12**, p. 251, (2023), https://doi.org/10.4103/jehp.jehp_92_23.
12. S. Tagliacozzo and M. Magni, Communicating with communities (CwC) during post-disaster reconstruction: an initial analysis. Nat Hazards, **84**(3), 2225–2242, (2016). https://doi.org/10.1007/s11069-016-2550-3.
13. N. K. Ngulube, H. Tatano, and S. Samaddar, Toward participatory participation: A community perspective on effective engagement in post-disaster recovery and reconstruction. Community Development, 1–21, (2024).
14. V. Baú, Citizen engagement in peacebuilding: A communication for development approach to rebuilding peace from the bottom-up. Progress in Development Studies, **16**(4), 348–360, (2016), https://doi.org/10.1177/1464993416663052.
15. C. van Ooijen, B. Ubaldi, and B. Welby, A data-driven public sector: Enabling the strategic use of data for productive, inclusive and trustworthy governance. OECD Working Papers on Public Governance, 33, (2019).
16. S. Carrasco, T. Egbelakin, and N. Dangol, Fostering recovery through stakeholders-community collaboration in post-earthquake recovery in Nepal. International Journal of Disaster Risk Reduction, **88**, 103619. (2023).
17. C. Rendon, K. K. Osman, and K. M. Faust, Path towards community resilience: Examining stakeholders' coordination at the intersection of the built, natural, and social systems, Sustainable Cities and Society, **68**, 102774, May (2021), https://doi.org/10.1016/j.scs.2021.102774.
18. V. M. Cvetković, J. Tanasić, A. Ocal, Ž. Kešetović, N. Nikolić, and A. Dragašević, Capacity Development of Local Self-Governments for Disaster Risk Management, International Journal of Environmental Research and Public Health, **18**(19), Art. no. 19. (2021), https://doi.org/10.3390/ijerph181910406.
19. M. S. Uddin, C. E. Haque, and M. N. Khan, Good governance and local level policy implementation for disaster-risk-reduction: actual, perceptual and contested perspectives in coastal communities in Bangladesh, Disaster Prevention and Management: An International Journal, **30**(2), 94–111, Jan. (2021), https://doi.org/10.1108/DPM-03-2020-0069.
20. M. Boston et al., Community resilience: A multidisciplinary exploration for inclusive strategies and scalable solutions, Resilient Cities and Structures, **3**(1), 114–130, (2024), https://doi.org/10.1016/j.rcns.2024.03.005.
21. M. M. S. Quinteros, C. Z. Marquez, L. A. A. Lopez, and T. C. Gutirrez, Lean Manufacturing as a Strategy for Continous Improvement in Organisations, first. United Kingdom: BoD – Books on Demand, (2021). http://dx.doi.org/10.5772/intechopen.92922
22. A. Barabadi and A. Nouri, From Waste to Value: A Practical Framework for Waste Identification and Mitigation Using Lean Management Principles. Research Publishing, Singapore, (2023). Accessed: Oct. 14, 2024. [Online]. Available: https://munin.uit.no/handle/10037/31117
23. J. Womack and D. Jones, Lean Thinking: Banish Waste and Create Wealth in Your Corporation, **48**. in Second edition, **48**. New York: Simon & Schuster, (1996). https://doi.org/10.1038/sj.jors.2600967.
24. J. Colazo, Changes in communication patterns when implementing lean, International Journal of Quality & Reliability Management, **38**(1), 296–316, (2021). https://doi.org/10.1108/IJQRM-10-2019-0323.
25. UNFCCC, Final review report on the implementation of work under the strategic plan for stakeholder engagement, communications and resource mobilization (2020–2021), Paris Committee on Capacity Building (PCCB), (2019).
26. L. P. Vázquez, E. A. Eshun, and J. B. Houston, Community Disaster Resilience and Communication: Models, Existing Research, and Next Steps, in Routledge Handbook of Risk, Crisis, and Disaster Communication, Routledge, 2024.
27. S. S. Sharma and R. Khatri, Intoduction to Lean Waste and Lean Tools, First. United Kingdom: BoD – Books on Demand, (2021).
28. I. S. Stewart, Advancing disaster risk communications, Earth-Science Reviews, **249**, 104677, Feb. (2024), https://doi.org/10.1016/j.earscirev.2024.104677.
29. N. K. Ngulube, H. Tatano, and S. Samaddar, Factors Impacting Participatory Post-Disaster Relocation and Housing Reconstruction: The Case of Tsholotsho District, Zimbabwe, Int J Disaster Risk Sci, **15**(1), 58–72, (2024), https://doi.org/10.1007/s13753-024-00536-y.

30. A. A. Bilau, E. Witt, and I. Lill, Practice Framework for the Management of Post-Disaster Housing Reconstruction Programmes, Sustainability, **10**(11), (2018). https://doi.org/10.3390/su10113929.
31. G. Charalambous, BIM communication waste, (2015). Accessed: Aug. 24, 2024. [Online]. Available: https://www.semanticscholar.org/paper/BIM-communication-waste-Charalambous/e1df336f42ac56de5ec8c118e372cdd9e790b4f0
32. T. A. Syaputra and A. D. Rarasati, A conceptual framework for implementing lean construction and building information modeling (BIM) in post-disaster housing reconstruction projects: 3rd International Conference on Disaster Management, ICDM 2022, IOP Conference Series: Earth and Environmental Science, **1173**(1), (2023). https://doi.org/10.1088/1755-1315/1173/1/012065.
33. A. O. Baarimah et al., A Bibliometric Analysis and Review of Building Information Modelling for Post-Disaster Reconstruction, Sustainability, **14**(1), (2022). https://doi.org/10.3390/su14010393.
34. A. Burger, W. G. Kennedy, and A. Crooks, Organizing Theories for Disasters into a Complex Adaptive System Framework, Urban Science, **5**(3), (2021). https://doi.org/10.3390/urbansci5030061.
35. P. Uusikylä, P. Tommila, and I. Uusikylä, Disaster Management as a Complex System: Building Resilience with New Systemic Tools of Analysis, in Society as an Interaction Space: A Systemic Approach, H. Lehtimäki, P. Uusikylä, and A. Smedlund, Eds., Singapore: Springer Nature, 161–190, (2020).
36. J. Acosta and A. Chandra, Harnessing a Community for Sustainable Disaster Response and Recovery: An Operational Model for Integrating Nongovernmental Organizations, Disaster Medicine and Public Health Preparedness, **7**(4), 361–368, (2013).
37. S. W. Burgiel, The incident command system: a framework for rapid response to biological invasion, Biol Invasions, **22**(1), 155–165, (2020), https://doi.org/10.1007/s10530-019-02150-2.
38. B. Nowell and T. Steelman, Beyond ICS: How Should We Govern Complex Disasters in the United States?, Journal of Homeland Security and Emergency Management, **16**(2) (2019).
39. P. S, L. A, P. S, and M. S, Enhancing collaboration during humanitarian response: an interim report from stakeholders survey, Prehospital and disaster medicine, **22**(5) (2007).
40. V. Bollettino et al., Challenges in Humanitarian Response Implementation: A Large-Scale Review of Aid Worker Perspectives, Disasters, (2023).
41. World Bank, Inclusive Approaches to Disaster Risk Management — A Qualitative Review | GFDRR, The World Bank, Washington DC, (2022). Accessed: Oct. 14, 2024. [Online]. Available: https://www.gfdrr.org/en/publication/inclusive-approaches-disaster-risk-management-qualitative-review
42. K. Boersma, J. Ferguson, P. Groenewegen, and J. Wolbers, The dynamics of power in disaster response networks, Risk, Hazards & Crisis in Public Policy, **12**(4), 418–433, (2021), https://doi.org/10.1002/rhc3.12218.
43. K. Mendis, M. Thayaparan, Y. Kaluarachchi, and C. Pathirage, Challenges Faced by Marginalized Communities in a Post-Disaster Context: A Systematic Review of the Literature, Sustainability, **15**(14), Art. no. 14, (2023), https://doi.org/10.3390/su151410754.
44. B. Mayer, A Review of the Literature on Community Resilience and Disaster Recovery, Curr Envir Health Rpt, **6**(3), 167–173, Sep. (2019), https://doi.org/10.1007/s40572-019-00239-3.
45. D. Seth, N. Seth, and P. Dhariwal, Application of value stream mapping (VSM) for lean and cycle time reduction in complex production environments: a case study, Production Planning & Control, Apr. (2017), Accessed: Oct. 14, 2024. [Online]. Available: https://www.tandfonline.com/doi/abs/10.1080/09537287.2017.1300352
46. F. P. G. Márquez, I. Segovia, T. Bányai, and P. Tamás, Lean Manufacturing and Six Sigma: Behind the Mask. BoD – Books on Demand, (2020).
47. V. S. M. (VSM)-L. L. S. Guide, Guide: Value Stream Map (VSM) - Learn Lean Sigma, Learn Lean Sigma. Accessed: Oct. 15, 2024. [Online]. Available: https://www.learnleansigma.com/guides/value-stream-map-vsm/
48. R. Karimi, M. Baghalzadeh Shishehgarkhaneh, R. C. Moehler, and Y. Fang, Exploring the Impact of Social Media Use on Team Feedback and Team Performance in Construction Projects: A Systematic Literature Review, Buildings, **14**(2), (2024).
49. J. Mangers, M. Minoufekr, and P. Plapper, Value Stream Mapping (VSM) to Evaluate and Visualize Interrelated Process-Chains Regarding Circular Economy, in Advances in Production Management Systems. Artificial Intelligence for Sustainable and Resilient Production Systems, A. Dolgui, A. Bernard, D. Lemoine, G. von Cieminski, and D. Romero, Eds., Cham: Springer International Publishing, 534–542. (2021). https://doi.org/10.1007/978-3-030-85910-7_56.
50. T. Capell and I. Ahmed, Improving Post-Disaster Housing Reconstruction Outcomes in the Global South: A Framework for Achieving Greater Beneficiary Satisfaction through Effective Community Consultation, Buildings, **11**(4), (2021), https://doi.org/10.3390/buildings11040145
51. R. Ramírez, D. Brodhead, and W. Quarry, Capacity Development in Evaluation and Communication Prompts for Practitioners. (2022). https://evaluationandcommunicationinpractice.net/featured-publications/

# The influence of changes in Sundanese cultural behavior on post-earthquake disaster assistance program support in Cianjur Regency

*Anwar* Kurniadi[1*], *Syamsul* Maarif[1], *Syamsunasir*[1], and *I* Dewa Ketut Kerta Widana[2]

[1]Disaster Management Study Program, Faculty of National Security, The Republic of Indonesia Defense University (RIDU), Bogor Indonesia
[2]Magister Management Study Program, Universitas Dirgantara Marsekal Suryadarma, East Jakarta, Indonesia

**Abstract.** The impact of earthquake disasters can affect cultural changes that can hinder the provision of aid if not handled quickly. This research aimed to analyze the effect of changes in the pillars of Sundanese culture on the provision of assistance during the post-earthquake disaster in Cianjur Regency. The research method used was qualitative with a descriptive design. Data collection was carried out through interviews, observation, and documentation, while data analysis was through data collection, data condensation, data presentation, and conclusion. The results of the research showed that the impact of the earthquake resulted in changes to 2 of the 7 pillars of Sundanese culture, namely: a) Tengginas, which was characterized by feeling hopeless, and reluctant to work again, depending on the help of others; b) Someah who is characterized as not needing to be friendly towards those who give aid, not ashamed to take help without permission, forcing the help giver to follow his wishes. Collaboration between LDMA services, social services, INAF, IRP, religious leaders, and youth leaders, the problems of Tengginas and Someah culture can be resolved, so that they do not interfere with aid programs for the emergency response and recovery period.

## 1 Introduction

An earthquake is an event that causes the earth to shake the presence of active faults/faults, volcanic activity, rockfall, and movement of the earth's plates [1]. Indonesia occupies a tectonic zone very active because of the world's three major plates and no other small plates meet each other in territory and form routes complex plate meeting pathways [2]. The existence of interactions between plates places the territory of Indonesia as an area that is very vulnerable to earthquakes.

Cianjur Regency is crossed by the Cimandiri fault which has the Rajamandala section which is characterized by a horizontal fault mechanism to the left side. This condition makes the surrounding location vulnerable to earthquake disasters [3]. The length of the Cimandiri fault is more than 100 kilometers, starting from Pelabuhan Ratu to Padalarang, Bandung Regency. The Cimandiri faults are divided into 3 segments: the Cimandiri segment is an upward fault mechanism, the Nyalindung-Cibeber segment is an upward fault mechanism, and the Rajamandala segment.

Based on information from Meteorology, Klimalotology, and Geofisika Agency (MKGA) occurred big earthquake disaster on November 21, 2022, in Cianjur Regency. The earthquake had the strength of Mw 5,6 and was followed by 140 aftershocks with a magnitude of 1,2 until 4,2, at a depth of 10 kilometers [4]. The vibration of this earthquake could be felt up to a radius of 240 kilometers, namely in the Banten Province area.

According to the Regent of Cianjur, the number of the victims as of November 25, 2022, that recorded were s 334 people who died, 593 people suffered serious injuries, and 49 people were being treated in government hospitals. Apart from that, there were 114.683 people as refugees, with details of 59.902 women, 54,781 men, 1.640 pregnant women, 7.453 elderly people, and 147 people with disabilities. For this reason, the regional government provided 494 temporary shelters, and the private sector provided 119 temporary shelters [5].

The adverse impact was the loss of property, natural environment, and livelihood. This condition disrupts people's activities to return to normal life. This condition lasted for 6 months after the earthquake. The inability of society to meet these economic and social needs will endanger the economic and social resilience of society. If this continues, it is feared that it will result in a decline in regional resilience. The Cianjur Regency government can't restore people's lives. Therefore, the regional government asks for assistance from the central government and its related apparatus, the private sector, internal and external donors, academics, and non-government organizations (NGOs). The regional

* Corresponding author: anwarmoker68@gmail.com

government also embraces the mass media to provide positive information to all of Indonesian society [6, 7].

The bad impact in terms of social resilience is the influence on the 7 cultural pillars of the Cianjur community. The 7 pillars of Cianjur culture have been a guide to life since their ancestors. This culture has good norms of life and culture so that they are attached and applied in everyday life. An earthquake in the emergency stage is characterized by the community's helplessness in meeting their daily needs. Will these conditions change the life patterns and behavior of the people of Cianjur to remain good or will it change to bad behavior? Several elements of society behave poorly during the emergency and post-disaster response stages. This information spread in local media, such as the act of withholding aid from outside and forcing relief goods to only be dropped off at that person's location. This incident will reduce the confidence of outsiders who will provide direct assistance to the affected villages which number more than 15 villages, especially those located far from main roads.

Changes in behavior by some of these individuals, if not addressed will harm the cultural development of the 7 pillars of Cianjur, in this case, it could be marginalized [8]. Besides that, currently, the entire community of Cianjur still needs outside assistance in disaster recovery, namely the stage of completing the reconstruction of the houses of the affected residents. The next impact is disrupting income from the tourism sector, which is the second source of regional income after agriculture. Ultimately, the sense of human security and community security of the Cianjur people will be disrupted [9, 10].

From the explanation above, researchers want to know directly whether there are significant changes in people's cultural behavior that disrupt recovery after the earthquake disaster!

## 2 Research method

The research method used qualitative with a descriptive design which is expected to be able to analyze the influence of changes in Sundanese cultural behavior when supporting aid programs after the impact of the disaster in Cianjur Regency [11]. To conclude, the researchers carried out: a) collected data through interviews with LDMA staff, INAF officials, village heads, Rukun Warga heads, Rukun Tetangga heads, religious leaders, and the communities involved, then carried out filed observations, and documentation; b) carried out analysis through data condensation, data presentation, and drawing conclusions and verification [12].

## 3 Results and discussion

The definition of a disaster is a series of events that can disrupt and threaten the livelihoods and lives of local communities caused by natural, non-natural, or human factors which can result in loss of life, environmental damage, loss of property, and psychological impacts [13]. Psychologically, the people of Cianjur were shocked and traumatized by suddenly experiencing repeated violent vibrations and shaking from the earthquake disaster. To eliminate or at least reduce trauma takes a long time. Traumatic or known as Post Traumatic Stress Disorder (PTSD) is a disorder that occurs in post-disaster communities [14]. Usually, PTSD symptoms can be treated if detected immediately so that appropriate treatment can be given at that time. However, if the complaint cannot be detected quickly, it will require long healing. The next impact is psychological disorders in the form of mental disorders, such as difficulty interacting with other people, which will interfere with daily life activities.

The bad impact can also disrupt the social and cultural resilience of society because each family thinks about itself rather than others, so as a result social interactions are disrupted. Relationships between people and relationships with previously close groups or groups of farmers can become strained. This triggers social-cultural disturbances which are shown by changes in behavior. Researchers want to know the influence of the impact of earthquakes on disrupting social interactions which have an impact on the 7 pillars of Sundanese culture in Cianjur. Of the 7 cultural pillars, which pillar is the most affected and has the potential to disrupt the social resilience of the Cianjur community? The cause is the emergency conditions where there is minimal availability of food, drinks, and shelter. The lack of one or more will affect the 7 pillars of Sundanese culture. During the emergency stage the conducive situation to support the distribution of logistic aid, evacuation of goods and people to temporary shelter, and the provision of adequate clean water and sanitation.

To determine the impact of the earthquake on changes to the 7 pillars of Sundanese culture in Cianjur, it can be explained as follows:

a. Maos

Maos, which means reciting the Koran or learning, is recognized as the number one Sundanese cultural custom, so it has become the main principle of Cianjur society. Reciting the Koran is a teaching of the Islamic religion which is a very important foundation for the community to build Cianjur, which has long been known as the City of Santri. The results of the interviews show that during the pre-disaster, disaster and post-disaster periods this culture continued to operate normally and there were no obstacles in its application in everyday life. The community can utilize this culture to understand how to prevent negative impacts in the future, through outreach on disaster-resilient villages and preparedness training programs to the younger generation and the community. By understanding the benefits of socializing the two programs above, each village is willing to form 3-5 natural disaster volunteers. After being trained by LDMA, these volunteers will provide education regarding earthquake disaster management so that the community is fully aware of their role in prevention and is ready to face it if a disaster occurs in the future.

In everyday life, volunteers can provide examples to the community of activities that can reduce the risk of earthquakes, for example, building houses with earthquake-resistant specifications, not building houses in tectonic fault zones, installing disaster-prone zones, and installing evacuation routes and gathering points when emergency. People finally understand and are aware of how to deal with earthquakes in the future. Findings in the field confirm that the Maos or Ngaos culture is a Cianjur culture that has been implemented in villages for years [15]. In this way, the Maoist type of Sundanese culture is not disturbed and instead supports the earthquake disaster management program.

b. Mamaos

Mamaos or Cianjuran singing is a singing culture that is used by the community to join in gatherings. During Mamos singing a song or poetry accompanied by a harp flute when Sundanese woman and men express their feelings. Currently, it is used to invite the community to gather in the Village Hall to hold discussions together to solve the problem, they are currently facing. The researcher utilized the culture to gather the representatives of fathers and mothers that affected earthquake at the house of the head of Rukun Tetangga (RT) to discuss the issues of the relocation. Likewise, Researchers used prayer rooms to conduct discussions with congregants about the conditions that occurred during the earthquake and current developments.

By utilizing Mamaos as an intermediary medium, researchers can collect data about the problem of relocation to new residents, how to get a new livelihood, income conditions that can be received every month, and obstacles in adapting to the lives of the surrounding community. The community is also open to conveying reasons for following regulations in the disaster area that limit what activities can still be carried out. These results showed that communication media using Mamaos culture can still be done well. This confirms the findings which stated that Mamaos culture is still able to help Sundanese people in Cianjur to improve their lives [13].

c. Maenpo

Maenpo is a martial art culture of Cikalong native to Cianjur that is widely passed down to the young Cianjur. The implementation of Maenpo culture is still preserved, which has the meaning that all Cianjur people must be able to protect themselves and their families from disturbance from other people. The real form of this culture is developing a culture of vigilance for newcomers to the Cianjur area. The Cianjur community is thought to always be critical to asking newcomers about their identity and purpose in coming to Cianjur. To maintain the security and safety of themselves and their families and the community, this culture is also applied when earthquakes occur, namely to donors who come to Cianjur.

Sundanese people and communities feel they must filter before accepting aid from people during an earthquake. When a disaster occurs Cianjur officials and communities are willing to give the authority to select donors to the Disaster Management Task Force. The task force in the field receives, records, and distributes to people who are entitled to receive it. Logistical assistance from outside Cianjur is handed over to INAF and LDMA officers so that it can be according to applicable provisions. The community trusts the INAF more than other newcomers. This is thanks to the honesty and sincerity of INAF which can be trusted for its commitment and honesty and always helps the difficulties experienced by the community. Delegation in the selection of newcomers to the emergency response task force showed the mature attitude of the Cianjur community. The results are all donor, and their assistance can be accepted by all Cianjur communities. Finally, all assistance for the community affected by the earthquake can be distributed according to LDMA and the Task Force Team. Thus, the Maenpo culture is still implemented well so that Maenpo Culture can survive well even though the bad impact of disaster has tested it.

d. Tatanen

Tatanen is a farming culture as the main livelihood of the Cianjur community. This culture is related to the habit of the majority (90%) of the Cianjur community planting Pandanwangi rice. The culture of planting and gardening of the Cugenang and Cipanas communities is the main source of income to meet daily needs. If it has become a source of life, it is very difficult to change. In this case, during the post-earthquake disaster. It is a difficult job to change quickly by looking for another source of life such as trading or others. This is proven by the fact that 8 months after the earthquake, almost 50% of people who had been moved to permanent housing in Sirnagalih Village still went back and forth to the agricultural or plantation. Locations they owned in disaster-prone locations. The local government's decision has determined that the earthquake area is a disaster-prone area that has the Cimandiri fault or plate an area most at risk of disaster [16].

However, the reality in the field is to accelerate recovery in the build-better program must involve economic resilience. For that reason, the Cianjur Regency government finally gave relief to the people in Cugenang and Cipanas districts who were allowed to earn a living from farming and gardening with supervision. Additional preparedness training was given to the people of both districts so that they were able to follow the disaster-prone zone regulations. The result of farming and gardening activities turned out that the community revived the farmer groups and was finally able to restore the income of the majority of the community as evidenced by the ability to sell vegetables to Bekasi, Bogor, Jakarta, and Depok. These results showed that the Tatanen culture can indeed make the Cianjur community survive, and it is believed that this is still the case today [17].

e. Tanginas

Tanginas means agile. Agile means the people of Cianjur are always agile or diligent in improving their abilities. For example. Always wake up early before dawn so that they can pray the dawn prayer. After the dawn prayer, they then look for sustenance. The Sundanese people in Cianjur must not lose a chicken, then their sustenance will be lost. Likewise, during a disaster, the Tanginas culture can change during the emergency response period and after the disaster. The negative impacts of disasters such as severely damaged houses, and lost property due to disasters are severe trials. During an emergency, the community is in dire need of food, drink, clothing, and shelter. For almost 3 months, people have not been able to work, and only rely on assistance from outside parties.

Finally, it raises the sense of worry that they will not be able to eat and have little for their families. This sense of fear is likely to change this independent culture. The Tangina culture is being tested. Moreover, their houses have been destroyed and they have no capital to rebuild. This condition makes sense of egoism arises from some people taking action against donors by forcing them to drop off in their village. The goal is to store as much aid as possible and not care about the condition of their neighbors in need. The impact is to trigger anger among other Cianjur residents.

Based on the results of interviews and observations in the field, researchers assess that there has been a significant shift in the culture of the Tanginas community. The first evidence is when a disaster occurs where some people have stratitems. The second piece of evidence was when researchers came to Ciherang Village, Ciputri Village, where there were more victims of severely damaged houses than those that were not damaged. The research team was reprimanded by one of the residents and followed by several residents so that they did not need to ask for data but immediately proved it by building a damaged road that had not been built by the Minister of Public Works and Public Housing. The third piece of evidence is when there was a renovation of residents' houses, the LDMA and the Task Force Team said that almost 95% of the community did not want to be involved in the planning and implementation of development to renovate damaged houses, such as taking measurements and marking the correct boundaries. The saddest condition was when the house was being built, they only watched and did not want to help.

This situation showed the impatience of the community in achieving goodness in their lives. For researchers, this is the decline in the value of Tanginas as a characteristic of the Cianjur community who are diligent and disciplined in earning a living and achieving their dream. Researchers assess that the Tanginas culture has declined or faded. This fact showed that the Cianjur community has started to be lazy precisely when conditions demand high values of discipline and perseverance. Researchers assess that the culture has begun to fade because it is no longer visible during normal conditions at least there are no more house renovation activities. The community's living conditions returned to normal after the disaster in the last 2 years after the tsunami disaster [16]. This condition is not by the finding that the Cianjur community has a Taginas culture that is diligent and disciplined, and diligent in seeking knowledge and fortune [18, 19].

f. Someah

Someah atau is friendly meaning the Sundanese people of Cianjur are always friendly, there is an expression from one of the Cianjur songs, namely the lyrics "Someah ka semah" which means friendly to guests. From body movements, it is usually shown with a smile. In this study, researchers found that during disaster and post-disaster, there was unfriendly behavior toward guests. During a disaster, there were only a few less friendly people, in this case indicated by not wanting to queue for food and drink distribution, and occupying shelters before being distributed by the Task Force Team. There was even an incident carried out by one person who organized and impolitely asked donor guests to make a road that had not been connected in the village.

When researchers asked one of the ustadz, he said that they did not want to join the food queue and so on, because they were impatient and afraid of not getting a quota for his family and their group. This incident only occurred during the first week and the following week there were already many volunteers, so the queue system became orderly. However, if you force yourself not to queue, it is an impolite act, which prioritizes the needs of your ego and family rather than the interest of the wider community.

For researchers, this action no longer pays attention to the 6th cultural pillar or Someah pillar because it is unable to survive in critical conditions, thus eliminating the friendly nature towards newcomers who become their helpers. In a situation like this, the friendliness characteristic of the Cianjur Community will slowly disappear. The act of forcing to assist newcomers as happened in Ciherang Village and Ciputri Village, which asked donors not to need to do data collection and just ask questions but to build village road facilities and repair public toilets which are also assistance from donors. This condition tends to be rude and is not characteristic of the friendliness of the Cianjut community. These findings showed that the application of Someah culture has decreased significantly during emergency response and post-disaster. The results are not to the study's findings which stated that the Cianjur community still adheres to the Someah culture in everyday life [20, 21].

g. Sauyunan

Sauyunan or harmonious means that the people of Cia njur prioritize harmonious and peaceful attitudes in all their activities. Even the Regent of Cianjur hopes that in the future the people of Cianjur can be more harmonious and work together in building Cianjur into a more advanced district. The culture of living in harmony during the earthquake can be proven in the second week and onwards when distributing food and beverage assistance and daily necessities. Even when researchers were at the earthquake location, all villages received their respective quotas. Living in harmony occurs in everyday life where farmers who have large areas of land are willing to give some of their rice fields to be planted by their neighbours who do not have land. After the disaster, farmers work together to deposit with vegetable traders who will be sold outside the district. The plantation company also helps the community to work in the garden again so that they can earn income. There is something that must be considered, namely the sequence number for providing housing construction assistance. Village heads must work hard so that the community who can take turns in stages 3 and 4 follow the predetermined queue.

Likewise, during the renovation of houses that have received financial assistance, they must be supervised and directed. This is intended for all heads of families to meet the specifications for earthquake-resistance buildings that have been set by the Ministry of Public Work and Public Housing. For this reason, the government has assigned INAF troops to supervise home renovations starting from planning to rebuild the house, when selecting building materials, and how to use them. There is something that needs to be considered, namely community participation in renovating their homes. Thanks to the presence of INAF troops, many heads of families who previously did not want to be involved in home renovations, finally began to be actively involved.

For researchers, the Sauyunan culture that directs the Cianjur community to always live in harmony and peace has been shown by the act of cooperation and sharing between neighbors both during and after the disaster. The most impressive action is the willingness to share from the owner of a large land to those who have none. This action is taken so that everyone has an income after the earthquake. Therefore, the Sauyunan culture still applies to the Cianjur community both during and after the disaster. This finding is based on findings that state that the Cianjur community is very strong and solid in maintaining harmony in the form of cooperation [22].

From the description above, it is found that there are 2 pillars out of 7 pillars of culture that have experienced a major decline. The cultures that experienced this decline were the Tangginas and Someah cultures. In terms of quality, the decline is marked by behavior that does not reflect the culture of behavior. Rationally, the decline in the values of these two cultures is understandable. The cause was an earthquake which destroyed their property. They become sad and almost hopeless about their next life. The culture of being diligent, active, and disciplined in seeking fortune does not seem to be able to cover up this sadness. Ultimately expecting everyone to help him. Losing hope will be a panacea if someone wants to help him. the quality of holding hands to help each other is the mainstay.

However, the nature of working hard becomes slack, and the nature of being friendly and polite no longer seems to be a priority. From this, it showed that several people in the Sundanese community of Cianjur were badly affected by the earthquake disaster, resulting in a behavior change. This is based on research findings that found that many people would carry out good behavior if the situation was normal [23]. On the other hand, people will degrade their culture, if they fail to face the disaster, they face so that changes will occur, for example finding that they are unable to fulfill their daily living needs, become hopeless, and have no hope, people will give up good behavior for the sake of bad behavior.

The decline of the 2 pillars of Cianjur culture was already known by the Disaster Management Task Force Team. Finally, LDMA embraced INAF, IRP, and social service officials, and location affected disaster accompanied by religious leaders and youth leaders to resolve it. The problem of the two pillar was clarified so that it was realized the importance of maintaining the Tanginas and Someah cultures so that those concerned would not feel inferior or marginalized [23], the second emphasized that the 7 pillars of Cianjur culture are still very relevant to the development of the times [24].

The problem of the decline of Tanginas and Someah culture has been resolved by the authorities. These paties are the LDMA official, social official, INAF, IRP, and religious leaders, as well as youth leaders by visit the scene of the incident and meeting with the parties involved. The Tanginas and Someah cultures were finally resolved by bringing the parties involved back to their senses. This was done so as not to disrupt aid programs during the emergency response, recovery, and reconstruction period. The proof is that within 6 months of the rehabilitation and reconstruction program, 600 people's houses had been rebuilt and people were able to work on agricultural land and plantations with supervision.

## 4 Conclusion

The impact of the earthquake resulted in people losing their property and livelihoods. Economic resilience has decreased making it difficult to meet daily needs. Social resilience will change the behavior of the 7 pillars of Cianjur culture. The two pillars of Cianjur culture that have changed are a) Tanginas which is shown by being discouraged and lazy to work so that many people depend on help from outsiders; b) Someah, which is shown to be blessed than friendly in talking to outsider's aid givers, was not shy about taking help without permission and forced the help givers who came to follow his wishes.

## References

1. Suharyanto, Earthquake Engineering (Equipped with Appropriate Earthquake Load Analysis SNI-03-1726-2002), (Yogyakarta: Kepel Press, 2023).
2. Peter Bird. An Updated Digital Model of Plate Boundaries. Journal of Geochemistry, Geophysics, Geosystems (G3), **4**(3), (2003).
3. The Meteorology, Climatology and Geology Agency (MCGA), Cianjur Earthquake Hazard Map (with Cugenang Fault Earthquake Source). Retrieved from: https://www.bmkg.go.id/berita/?p=peta-bahaya-gempabumi-cianjur-dengan-sumber-gempa-patahan-cugenang&lang=ID. (2023).
4. Irsyam M, Widiyantoro S, Natawidjaja DH, Meilano I, Rudyanto A, Hidayati S, Triyoso W, Hanifa NR, Djarwadi D, Faizal L, Sunarjito (eds), Source and hazard map of the 2017 Indonesian earthquake. Bandung: Housing and Settlement Research and Development Center, Research and Development Agency, Ministry of Public Works, (2017).
5. Kompas.com, The M 4.0 earthquake in Cianjur was triggered by Cugenang fault activity. Retrieved from: https://bandung.kompas.com/read/2023/03/29/150000178/gempa-m-40-di-cianjur-dipicu-aktivitas-sesar-cugenang, (2022).
6. Khan, M.R. and Rahman, M.A., Partnership approach to disaster management in Bangladesh: A critical policy assessment', Natural Hazards, **41**, (2), 359–378. (2007).

https://doi.org/10.1007/s11069-006-9040-y

7. Isnan, H., and Rohmiyati, Y. Preserving the Carving Arts Knowledge of the Minangkabau Community. Jurnal Ilmu Perpustakaan, **5,** (1), 1–12, (2016).
8. Musthofa, B. M, Cultural Development Towards Cultural Welfare: Lessons from Community Development in Saung Angklung Udjo, Bandung, West Java. Jurnal Sosio Informa, **4**, 553-566, (2018).
9. Gunardi, G., The Cultural Role of "Mikanyaah Munding" in the Conservation of Sundanese Traditional Art. Jurnal Panggung, **24**, (4), 329–334. (2014).
10. Minister of Defense, Indonesia Defense White Book, Jakarta: Indonesia Republic Defense Ministry, (2015).
11. Sugiyono, Qualitative, Quantitative Research Method, and R and D, Bandung: Alfabeta, (2011).
12. Miles, M., A. M. Huberman, and Saldana, Qualitative Data Analysis: A Methods Sourcebook. USA: Sage Publication Inc., (2014).
13. Indonesia Law No.24 of 2007 about Disaster Management.
14. Syahdiana, Susi. Ethnopedagogy in Pasanggiri Asah Kaparigelan Basa, Literature, Sundanese Culture, Jurnal Kebudayaan, **13,** (1), 31-44. (2018).
15. Sumardani, Kusmajaya, and Riskyana, Wulandari, Earthquake Disaster Risj Study in Cianjur Regency, Jurnal Dialog Penanggulangan Bencana, **10**, (1), 39-51, (2019).
16. Hanan, Shofira., Maenpo the Great Cianjur Silat in Orang Village, (2017) Retrieved from : https://www.pikiran-rakyat.com/jawa-barat/pr-01287105/maenpo-silat-cianjur-yang-besar-dikampung-orang-410930
17. Wulandari, P. K., Youth Innovation in Supporting Family Economic Resilience (Case Study in Colorful Village, Jodipan Village, Blimbing District, Malang City, Jurnal Ketahanan Nasional, **23,** (3), 300-320, (2017).
18. Syamsunasir, Anwar, Kurniadi., I Dewa Ketut Kerta Widana, Pentahelix Synergy in Post-Tsunami Disaster Recovery to Support Community Resilience in Facing Disaster in Pandeglang Regency, Technium Social Science Journal, **42**, (1), 133-147, (2023).
19. Suprayogi, Movement of Seven Religious Programs to Shape Cianjur Society with Sauyunan, Someah, and Resilience Characters. (2017). Retrieved from https://cianjurkab.go.id/gerakan-tujuh-program-keagamaan-bentukmasyarakat-cianjur-berkarakter-sauyunansomeah-serta-tangguh/,
20. Zuhri, Damanhuri., Cianjur Dream Party Supports Congregational Morning Prayer Movement. (2016). Retrieved from: https://www.republika.co.id/berita/dunia-islam/islam,
21. Yoga, Miarti., Someah Hade ka Semah. (2016). http://bandung.pks.id/2016/01/someah-hade-ka-semah.html,
22. Rahman, Fathur., Sundanese culture is known as a culture that highly values politeness. Retrieved from https://www.scribd.com/doc/102357828/, (2017).
23. Wikandia, R., Preservation, and Development of Ajeng Sinar Pusaka Art at a Typical Karawang Bridal Reception. Jurnal Panggung, **26**, (1), 58–69 (2016). https://doi.org/10.26742/panggung.v26i1.162.,
24. G, J, Flecher, and M, S, Clark, Blackwell, Handbook of Social Psychology: Interpersonal Process, Padstow: Blackwell Publishing, (2001).

# Scrutinizing the representations of flood disaster imagery in Emile Zola's work

*Ferdinal* Ferdinal[1*], *Oktavianus* Oktavianus[2], and *Djusmalinar*[3]

[1]Department of Literature and Culture, Faculty of Humanities, Universitas Andalas, Padang 25163, Indonesia
[2]Department of Linguistics, Faculty of Humanities, Universitas Andalas, Padang 25163, Indonesia
[3]Department of Malay Language and Malay Studies, Faculty of Humanities and Social Sciences, Prince of Songkla University, Pattani 94000, Thailand

**Abstract.** Educators, researchers, and scientists from various fields, including literary critics, are striving to address the impacts of climate change and disasters on human beings, as depicted in Emile Zola's work "The Flood," which illustrates human vulnerability amidst natural disasters. This research aims to explore the use of flood imagery by Emile Zola in his short story "The Flood," highlighting the main themes and narrative of the story. The method used is an in-depth analysis of the images employed by Zola, including the depiction of the rising floodwaters and the chaos following the disaster, as well as the symbolic meanings behind these images. Research results show that Zola sharply captures the psychological warfare and emotional state of humans when facing destruction caused by natural disasters. The implications of this research indicate that the disaster imagery used by Zola plays a crucial role in enriching the narrative and provoking readers' reflections on the relationship between humans and the uncontrollable forces of nature.

## 1 Introduction

Researchers publishing in the area of imagery include Tomašovičov [1], Dharlie & Samanik [2], Schipper [3], and Walton [4], who contribute to this discussion beyond linguistic borders. The cultural perspectives related to imagery in multiple literary traditions are also expressed by Hui [5]. In the perspective of postcolonialism Lee [6], Parashar et al. [7], Hadziavdic & Hoffmann [8], as well as Mestre [9] assert that images have been employed to depict the environment. They argue that the interplay between men's perceptions, linguistic utterances, and God's earth has become central attention of the studies about literary images, especially the ones related to society and the environment.

Examinations related to images and metaphors have been important in exposing human problems as seen in literature, such as men's rights [10, 11], agricultural life [12], Minangkabau crops [13, 14], norms of Islam [15], culture-related values [16], and functions of metaphors [17]. Rogacheva & Drozdova [18] belong to those who investigated imagery and the way it attracted readers to look at smell representations. Furthermore, Martinez [19], Kaplan [20], and Kitts [21] have provided readers with insights about works of literature depicting human inner, maternal, traumatic imagery, alongside human sacrifice.

Next, Ciribuco [22] scrutinized the contributions of images when describing Mafia, temporal nuance [23], corrupt state practice [24], sports errands [25], and flower-related subjects [26]. The scholars paid attention to the significance of images in works emphasizing the socio-psychological phenomena of human beings.

Scholars interested in the study of the literary portrayal of tragedy, trauma, and people's resilience have shown significant trends. One important trend to note is the necessity of investigating disaster images in literature. This kind of interest has enhanced people's awareness of the meaningful connection between imaginary products and social phenomena. For instance, Jerry et al. [27] analyze the role and use of disaster imagery in films to comprehend people's vulnerability and struggles for survival. Next, Mnasser [28] learns about disaster visual in literary works and to what extent it affects the way readers perceive disasters. Satterlee [29] investigates how images work in landscape descriptions. In brief, these scholars help readers comprehend the roles and functions of disaster imagery in readers and authors' lives.

A French novelist, journalist, and dramatist, Émile Édouard Charles Antoine Zola was an important naturalist writer. Caring for environmental problems, he used literature to portray people and the natural world. He produced literary works in different genres, such as poetry, dramas, short stories, and novels. Among his translated works include "The Flood" (2013) [30], "The Earth" (2016), "A Love Story" (2017), and "Doctor Pascal" (2020). The short story "The Flood" (translated from "L'Inondation") [31] can be read as a reminiscence of the French flood tragedy. In terms of its translation, the work is not only informative in the description of French rural culture, but it is also significant in the domain of

* Corresponding author: ferdinal@hum.unand.ac.id

language use. The story offers a different angle of observing the French way of life through its English translation. The work is rich in its linguistic aspects, including its figurative language use of imagery. It relie on its application of disaster images. To learn how the disaster images are constructed in the work, this writing aims to focus on some problems: the way disaster imagery is used in the novelette, what kinds of images Zola employs in it, and to what extent the use of disaster images affects the theme and characterization.

Based on the background above, this writing is aimed to scrutinize the way Zola applies different types of imagery in his story "The Flood" and how he uses the River Garonne flood to describe the immense destructive impacts it causes on human beings and its aftermath.

### 1.1 On imagery

In fictional creativity, writers benefit from imagery for a variety of reasons. Imagery may illustrate various forms and scales of description to stimulate the reader's sensory experiences and responses to the textual appearance, which offer reactions to their senses. DiYanni [32] argues that imagery refers to "language that evokes one or all of the five senses: seeing, hearing, tasting, smelling, touching" (218). Similarly, Abrams [33] suggests it as "the representation through language of sense experience" (110).

Imagery is an essential literary tool that helps authors create mental landscapes for readers to engage in different areas, including senses, emotions, and cognition. Some studies highlight its roles in delivering themes, improving character development, and strengthening narrative resonance. For example, Sisakht and Pourlashkary [34] study the ability of images to create and touch the reader's emotions. They focus on the roles of images in helping readers interpret and engage with works of literature. Furthermore, studies by other scholars like Lanser [35] investigate imagery's contribution to writing and authorial perspective, negotiating how writers use sensory aspects to come up with touching literary works and produce authentic and complex works.

The language of literature allows authors to use different kinds of imagery to paint a variety of human sensory acquaintance with natural forces. Scholars such as Perrine and Tomas [36] group imagery into visual, auditory, olfactory, gustatory, tactile, kinesthetic, and organic forms. Visual images provide readers with real pictures. Auditory imagery comes up with pictures of sounds. Olfactory and gustatory images trigger human sensory involvement. Tactile imagery deals with the sense of touch, describing textures and sensations. While kinesthetic imagery expresses physical movements, organic imagery provides readers with emotional conditions. These kinds of imagery can be literal (direct description) or figurative (similes or metaphors) in forms, as Table 1 shows.

**Table 1**. Categories of imagery (adapted from various sources)

| Imagery forms | Examples | Descriptive types |
|---|---|---|
| Visual | "Continuous as the stars that shine", "And twinkle in the Milky Way." | Metaphorical |
| Tactile | "Her fingertips brushed against the rough bark of the old oak tree as she walked through the forest." | Literal |
| Gustatory | "The soup was rich and flavorful, bursting with the tang of herbs and spices." | Figurative |
| Olfactory | "The air was heavy with the scent of freshly baked bread wafting from the bakery down the street." | Metaphorical |
| Auditory | "Hear the sledges with the bells", "How they are tinkle, tinkle, tinkle." | Metaphorical |
| Kinesthetic | "His heart was flying high when he met his beloved half." | Hyperbolic |
| Organic | "Her eyes were burned with anger when seeing her rival." | Symbolic |

These seven types of imagery play their roles solely or intermittently to produce immersive and vivid literary works, engaging readers with different sensory experiences and improving their understanding and emotional attachment to literary narratives.

## 2 Research methods

This study was conducted by applying some methods. It groups and scrutinizes the use of imagery in Emile Zola's "The Flood," published by Project Gutenberg. It follows some steps. First, to perform a close reading of the text, this analysis follows the steps proposed by Vendler [37] by observing the text's descriptive elements such as similes, metaphors, and figurative language use where it showed its details of images. It also analyzed the use of sensory details signaling symbols to annotate different repeated motifs. Then, textual analysis techniques were used to categorize the kinds of images in the work [38]. The methods systematically looked for the repetitive patterns in the story by reading it for the employment of visual, auditory, olfactory, gustatory, and tactile imagery. Last, the semiotic approach was applied to interpret the images in "The Flood," to make sense of their significance [39], and to see how far the use of different images upgrades the themes and characterization of the novelette. These methods have been employed to study various kinds of images in "The Flood." This study provides linguistic significance and uniqueness of the story's aesthetic and thematic implications, as seen in Zola's literary creation.

# 3 Results

## 3.1 "The flood" and what it represents

Zola shows "The Flood" as a literary record of what befell Saint-Jory, Toulouse, France when the 1875 big flood occurred. The flood originated from the River Garonne in southwest France. The river flew its water through to the Atlantic through the Gironde estuary, destroying anything that stopped it. With the Gironde Estuary (45 miles long) excluded, its length is 357 miles (575 kilometers). The River Garonne and its water flows are learnable from Figs. 1-3.

**Fig. 1.** The map of Toulouse (source: About-France.com)

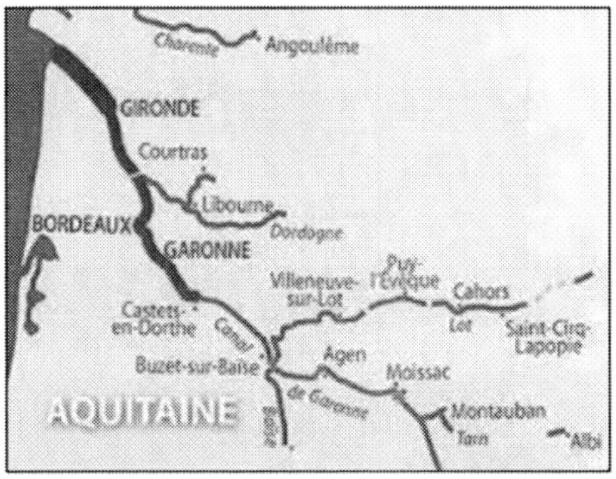


**Fig. 2.** The map of the River Garonne (source: French Waterways)

**Fig. 3.** The River Garonne (Source: britannica)

Mass media, such as 150ansinondations.com, [31] noted that the 1856 spring incidents in the Rhône Valley, a windy June, and Pyrenean snowmelt pushed waters to wash away some areas in Toulouse on 20 June and peaked on 23 June 1875. The Garonne filled in its flow banks on 23 June 1875, causing destructive effects on Toulouse. The flood damaged more than one thousand buildings, washed away three bridges, and killed 200 lives. Maurice Pardé reported that over 500 people were dead in Garonne, Agen, and South Western France. Some other media wrote that the flood took 3,000 fatalities. Other than the Aude, the floods also affected the Adour River valley in the Hautes-Pyrénées Department. The river flooding claimed 82 lives in the Val d'Ariège. Saint-Jory was among the areas where such a tragedy forced people to cry, as seen in Fig. 4.

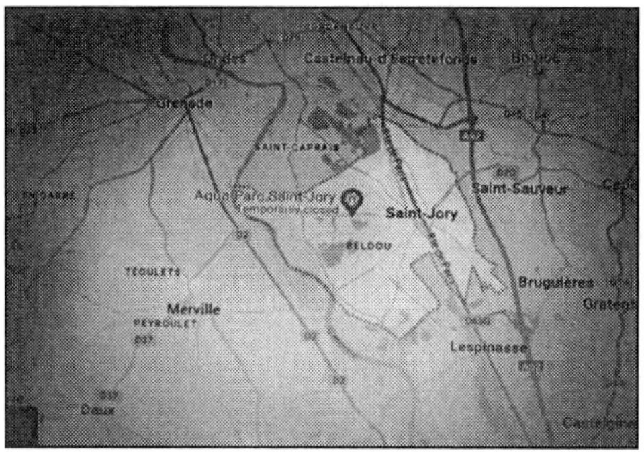


**Fig. 4.** The Saint-Jory in France map (adapted from google.com)

The 150ansinondations.com wrote that the floodwaters resulted in huge and surprising fatalities, victims, and material damage, as quoted in Fig. 5.

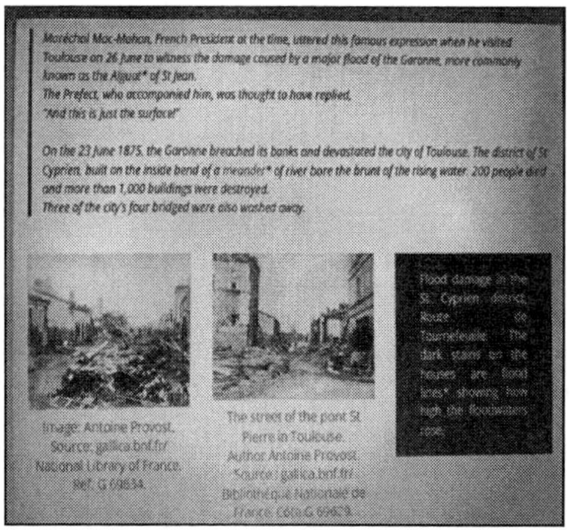


**Fig. 5.** Record of flood in France in 1875s. (source: 150ansinondations.com)

## 3.2 Imagery in "the flood"

The results of the analysis of the story deal with the categorization of the different types of images used by Zola in "The Flood." The visual images are made from landscape descriptions, moments when characters battle against the torrent, and character looks. They picture a mental influence on readers. Auditory imagery is exposed through sounds of psychological war, natural roars, and character's acts. It centers on the aural senses. Gustatory and tactile images are looked at when they succeed in creating the reader's sensation. Kinesthetic and organic images are studied for their directive and mental functions work to move readers. This grouping of image presentations provides readers with insights regarding Zola's literary capabilities and the thematic abundance in "The Flood."

### 3.2.1 The use of imagery

**Table 2.** Imagery Constructions in "The Flood"

| Types | Sample Quotes | Object | Description |
|---|---|---|---|
| Visual Imagery | -The story shows what the Garonne looked like before the disaster.<br>-It shows how people were running, chased after by the water. | -River Garonne<br>-Villagers | Threat<br>Scared |
| Auditory Imagery | -People were crying against the coming of the devastating water.<br>-The water looked like a sea that roars. | -People's cries<br>-Water | Scared<br>Destructive Flood |
| Gustatory Imagery | -People enjoy their life and food in the village.<br>-They ate a big festival. | -Eating<br>-Appetite | Dinner<br>Hunger |
| Tactile Imagery | -The flood destroyed the houses in the village.<br>-The villagers were weeping for sadness and desperation. | -Water<br>-Trauma | Violent flood<br>Desperation |
| Kinesthetic Imagery | -The flood attacked the village and watered it everywhere.<br>-Women struggled for survival along the flood. | -River<br>-Survival | Water covering he village<br>Sinking into the lood |
| Organic Imagery | -The flood made the village look like a lake.<br>-The water washed away everything up to the level where it could not go higher anymore. | Water<br>Water | Destroying flood<br>Height of the lood |

Zola uses the literal and figurative language of disaster to remind his readers of the gigantic effects of floods on people. In "The Flood" he employs visual, auditory, gustatory, tactile, kinesthetic, and organic images to describe disaster issues and side effects the flood brings. Table 2 shows some samples of those images.

Employing six kinds of images, Zola interestingly travels to the readers along the flood disaster elaboration, together with its ruinous forces that demolish anything that hinders its flows. Through the use of visual, auditory, gustatory, tactile, kinesthetic, and organic images, he puts forward not only the thematic varieties and character's complicity in his work but also a substitute method to inform readers of the mammoth effects of the flood disaster on causing chaos, destruction, and family disruption during and after the disaster.

### 3.3 Imagery, themes, and character involvement

Zola uses some kinds of disaster imagery, themes, and affected characters to represent flood disasters, as seen in Table 3.

**Table 3.** Imagery grouping, themes, and character involvement

| Imagery Types | Thematic Groups | Topics | Character Involvement |
|---|---|---|---|
| Mixed | The Chaotic moments | -Torrential rain<br>-Raging river<br>-Collapsed buildings<br>-Panicked crowds<br>-Upturned objects | -Villagers<br>-Families |
| Mixed | The destruction | -Crumbled buildings<br>-Twisted metal<br>-Uprooted trees<br>-Shattered windows<br>-Submerged homes | -Villagers<br>-Families |
| Mixed | Family Collapse | -Deserted streets<br>-Lawlessness<br>-Makeshift shelters<br>-Food shortages<br>-Disease<br>-Suffering | -Family |

## 4 Discussion

"The Flood" depicts the flood in the language of disaster. Zola vividly presents the River Garonne flood in the French Toulouse that brings chaotic moments, physical decay, and societal collapse. Using some disaster pictures to reflect the flood disaster, including the chaos, destruction, and collapse of society, he fascinated his readers with powerful, engaging, and emotional language. Zola creates the narrative's beauty through various narrative elements, including six types of images. These six images are sorted into chaos, destruction, and family disruption moments.

First, Emile Zola conveys a series of horrific moments, such as chaos, panic, loss, and human suffering. He paints a linguistic description of the floodwater rise, panicked society, chaotic village, and disorderly families. As seen in Fig. 6, Zola centers his attention on crowds searching for evacuation alternatives on the streets, in the houses, and within the areas where lives can be safer. In such a way, men attempt to protect what they have, particularly their loved ones. In this effort, they work hard to reach higher places with desperation. Zola highlights the characters who are fearful and desperate.

(1) •"The Garonne! The Garonne!" The two men and the three women were still running on the road. They heard the terrible gallop gaining on them."

(2) •"Now the waves arrived in a single line, rolling, tumbling with the thunder of a charging battalion."

(3) •In less than an hour the water became threatening, dashing against the house, bearing drifting barrels, pieces of wood, clumps of weeds. In the distance there were attacks upon walls, and we could hear the resounding shocks."

(4) •"With their first shock they had broken three poplars; the tall foliage sank and disappeared. A wooden cabin was swallowed up, a wall was demolished; heavy carts were carried away like straws."

**Fig. 6.** Fear and confusion

Next, the author writes about the suffering's humans must face within the anger of killing and destroying water by showing the terrible realities they must endure, including casualties, injuries, and separation from family and society. In short, Zola presents people's sadness against the mourning and agony as their families cannot be united, and some must face their family members being lost. It can be seen in Fig. 7.

(5) •"Suddenly, on the wide road before us, appeared two men and three women, one of them holding a child in her arms. It was they who were crying out, distracted, running with long strides. They turned at times, looking behind with terrified faces, as if a band of wolves was pursuing them."

(6) •"An enormous wave felled the woman who was carrying the child. Then all were engulfed."

(7) •"But our girls could not be diverted; they were pale, with feverish hands, and ears on the alert."

(8) •"All the others are gone! The babes in swaddling clothes, the girls to be married, the young married couples, the old married couples."

(9) •"I had one desire. I wished to recover the bodies of my family, to bury them beneath a slab, where I should soon rejoin them."

(10) •"It was said that, at Toulouse, a large number of bodies carried down the stream, had been taken from the water."

**Fig. 7.** Mourning and grief

Secondly, Zola employs furious floods, ruins, destruction, and natural decay to show how destructive the disaster is on society. Zola vividly describes the rain, the water, and then the floodwaters that race along Saint-Jory and its countryside and swallow all that stands before it. This scene illustrates how people cannot stop the natural power of water when it rages, as illustrated in Fig. 8.

The flood disaster destroys people psychologically and physically. It destroys and demolishes society's belongings, houses, roads, bridges, and other public properties. Zola writes moments when the characters witness houses swallowed, windows shattered and banged, and walls crumbled. The moments illustrate how the characters feel devastated and frustrated by the undefeated water power.

(11) •"There is considerable talk about the heavy rains of the last few days. Some seem to think that they will cause trouble."

(12) •"In effect, it had rained for sixty hours without stopping. The Garonne was very much swollen since the preceding day, ..."

(13) •"Now the waves arrived in a single line, rolling, tumbling with the thunder of a charging battalion. With their first shock they had broken three poplars; the tall foliage sank and disappeared. A wooden cabin was swallowed up, a wall was demolished; heavy carts were carried away like straws."

(14) •"The maddened animals had burst open the doors of the stables. They passed in the yellow flood, rolled about, carried away by the current. The sheep were tossed about like dead leaves, whirling in bands in the eddies. The cows and the horses struggled, tried to walk, and lost their footing. Our big gray horse fought long for life. He stretched his neck, he reared, snorting like a forge. But the enraged waters took him by the crupper, and we saw him, beaten, abandoned himself."

(15) •"We felt the need of tears. Our hands stretched out to those dear animals that were being borne away, we lamented, giving vent to the tears and the sobs that we had suppressed. Ah! What ruin! The harvest destroyed, the cattle drowned, our fortunes changed in a few hours!"

**Fig. 8.** Floodwater power

Third, the societal collapse is conspicuous in the story, where the description of social images is depicted through the actions performed by the heroic characters. Zola uses the characters to represent the heroic acts and individualism performed when they try to save the lives of others, forgetting their own safety. He concentrates on a family that helps each other to get to a safer place and get rid of the water. Each member of the family works hard to keep them alive and go to seek asylum. Through the acts, Zola shows the readers the spirit and resilience humans have when facing catastrophe. Such scenes can be seen in Fig. 9.

(16) •"The people of Saintin have boats; they will come this way. Look over there! Isn't that a lantern on the water? At our feet the women, silent and shivering, hid their faces to shut out the sight."

(17) •When I came to, the water had risen. It was now on a level with the tiles. The roof was a narrow island, emerging from the immense sheet. To the right and the left the houses must have crumbled. We are moving," murmured Rose, who clung to the tiles. And we all experienced the effect of rolling, as if the roof had become detached and turned into a raft. The swift currents seemed to be drifting us away. Then, when we looked at the church clock, immovable opposite us, the dizziness ceased; we found ourselves in the same place in the midst of the waves."

(18) •"Then all of us shouted together. Our voices stopped them for a moment, but they continued on their way. ... For a moment they hesitated.

(19) •"We can't let him die like that," said Jacques, distracted. "We must get down there." "Perhaps we could slide down the beams and save him," remarked Pierre."

(20) •"A sharp pain brought me to. Pierre was dragging me by the hair along the tiles. I lay still, stupidly watching. Pierre had plunged in again. And, in my confused state, I was surprised to see Gaspard at the spot where my brother had disappeared. The young man had Veronique in his arms."

(21) •"The end had come. The destroyed village was marked by a few vestiges of walls. There were no longer any sounds of falling houses, like a cart of stones suddenly discharged. It was as if we were abandoned, shipwrecked, a thousand miles from land."

**Fig. 9.** People's resilience and courage

Then, Zola makes use of imagery to portray people in despair. He depicts how the flood continues to attack the land and its people and cause them to be desperate and face hardship alongside the angry waters. Zola creates moments when their hopeless feelings and frustration approach the characters when they choose to fight against things that stop them from saving their lives during and after the disaster. Such a description can be seen in Fig. 10.

(22) • "The end had come. The destroyed village was marked by a few vestiges of walls. There were no longer any sounds of falling houses, like a cart of stones suddenly discharged. It was as if we were abandoned, shipwrecked, a thousand miles from land."

**Fig. 10**. People's hopelessness and resignation

Finally, Zola describes how people face physical disturbance from the flood. Zola describes the scenes of social unrest, the impact of widespread disasters on the community, and the physical damage they cause, as illustrated in Fig.11.

(23) • "What a terrible disaster! Nearly two thousand houses in ruins; seven hundred deaths; all the bridges carried away; a whole district razed, buried in the mud; atrocious tragedies; twenty thousand half-clad wretches starving to death; the city in a pestilential condition; mourning everywhere; the streets filled with funeral processions; financial aid powerless to heal the wounds!"

(24) • "But I walked through it all without seeing anything. I had my ruins, I had my dead, to crush me."

(25) • "I was told that many of the bodies had been buried in trenches in a corner of the cemetery. Only, they had had the forethought to photograph the unidentified."

**Fig. 11.** Flood and physical disturbance

## 4.1 Interpretation of findings

### *4.1.1 Thematic significance*

"The Flood" describes how the flooding of Toulouse in France destroyed the area and its impact on the environment and humans. Emile Zola raised a profound theme of the significance between nature and men in his narrative. One of the main themes exposed in the story is the destructive and psychological impact of flooding on society, individuals, and the environment. Zola describes the horrors caused by the flood, including the destruction of nature, the glare of civilian lives, the psychological burden of the figure, and human suffering due to disasters. Through a clear picture and intense emotion, the novel serves as a criticism of the nature of the flood and its impact on humanity.

In "The Flood," Zola digs into the themes of disappointment and despair, primarily through the experiences of his characters when they face their defeat by natural forces and family's collapse. Zola depicts a society in crisis, struggling with social disintegration and moral destruction. In this story, Zola represents the characters who face the reality of psychological warfare. Consequently, they must struggle to find meaning in life and hope for survival amid chaos and despair.

Zola uses "The Flood" to instill the issues of desperation and sorrow befalling the characters when they face failure in their struggle against calamity and community destruction. Zola creates images indicating the crises that the depicted society attempts to deal with social and moral disruptions. Zola shows how the characters in the story face the psychological terrors impacted by the flood and the truths it causes later on. Through this, they try to make meaning and seek hope within the chaotic moments of desperation.

Besides, "The Flood" explores the themes of family dynamics, life struggles, and the powers of nature. Through these themes, Zola describes to readers how flood disasters generate psychological warfare among members of society and worsen social conditions as a result of the flood, and reveals the material and psychological impact on the individual. Having one wealthy family in a village, Zola describes the terrifying experiences of communities affected by flood disasters, highlighting families being destroyed and entire communities dying, dying, and abandoned by their family members.

In addition, the novel also deals with the themes of resilience and survival as the characters face the challenge of disaster and strive to survive in the face of major floods. Apart from the despair experienced by the characters of this story, Zola also depicts human resilience and solidarity. If we look at it in depth, Zola's novel "The Flood" reveals his meditation on the human condition trapped between disaster, inner war, and the forces of nature by showing the disaster of flood and its consequences. Zola also explores the themes of resilience, solidarity, and the struggle for life, family, and humanity.

### *4.1.2. Narrative function*

In "The Flood," Zola uses imagery to serve different narrative functions by enriching the work with higher emotional depth and clarity. Zola travels to his readers along with his characters, who desperately face difficult times and struggle to survive during the flood. Detailed images of floods accompany this journey, destroying nature and causing chaos and panic. These types of imagery upgrade the character's despair, hopelessness, and resilience. Zola's use of images of family disruption, moral decay, and material losses helps readers feel how floods impact nature, individuals, and communities psychologically and materially. Zola also employs imagery to signify human losses resulting from the disaster, forming social disruption, weakening morality, and human vulnerability. They all offer multiple meanings readers may encounter while reading the prose, indicating a deeper involvement of readers in the text. As a whole, the use of imagery in "The Flood" by Zola linguistically works as a significant vehicle to improve the reader's perceptions of the novel's themes on nature and destiny and invite their emotions and psychological responses over the story's existence.

### *4.1.3 Emile Zola's style and themes*

Imagery plays a crucial role in "The Flood." They bridge the reader's understanding of nature and the after-disaster impact the readers can perceive what the flood caused on nature and people. Zola tries to convince his readers that human behaviors have been conditioned by nature and the environment. The details and clarity of descriptions Zola offers in the story carry readers into natural realms and their effects on the characters in the story. He uses floodwaters, demolished lands, and the brutalities of natural disasters to accompany the characters who become the subjects and the objects of natural powers and human destinies. Zola's use of imagery in the story to depict nature, power, powerlessness, and human weaknesses against nature reminds readers that people cannot fight against the uncontrollable forces God destroys. Moreover, Zola offers the imagery of destruction, vulnerability, and weakness that people must deal with. In other words, Zola employs imagery in "The Flood" to highlight the connection between humans and nature by representing characters who failed to fight against the natural environment but were defined by God through His environment.

## 5 Conclusion

Emile Zola's "The Flood" uses disaster imagery to represent the chaotic situation, physical destruction, and socio-psychological impact on people caused by the River Garonne flood. Zola practices some kinds of flood imagery, such as visual, auditory, olfactory sss, gustatory, and tactile imagery, to describe such occurrences. Through the power of language, both figurative and literal, he depicts the saddening moments the French people were affected when a destructive flood disaster diminished the land and society. The flood had widespread impacts on the society.

Using the six types of imagery Zola applies in "The Flood" pragmatically impacts how readers read the story's theme and characterizations. Zola's use of visual imagery in the work helps create ample descriptions of the devastating impacts of the flood on nature and people and on the reader's perceptions. Zola literally and figuratively provokes how destructively water can demolish nature and society. Additionally, using auditory and olfactory images partially affects the reader's perceptions of how chaotic conditions touch and influence the characters' psyche during the flooding and how the water demolishes what it came onto. Lastly, the gustatory and tactile images help influence how the characters feel and taste during the disrupting experiences.

Further studies on imagery, particularly Zola's works, may concentrate on other figurative language aspects, such as indigenous signs [40], which influence how readers perceive and comprehend such fiction because of the limitations this study offers regarding the depth of the analysis and the methods employed.

## References

1. J. Tomašovičov, Parallels between two worlds: Literary science-fiction imagery and transhumanist visions. World Literature Studies, **13**(1), 31-42 (2021) https://doi.org/10.31577/wls.2021.13.1.3
2. K. A. Dharlie and S, Samanik. Imagery Analysis In Matsuoka's Cloud of Sparrows. Linguistics and Literatur Journal, **2**(1), 17-24 (2021) https://doi.org/10.33365/llj.v2i1.514
3. J. Schipper. Interpreting the Lamb Imagery in Isaiah 53. Journal of Biblical Literature, **132**(2), 315-325 (2013) ISSN 1934-3876, Project MUSE, https://doi.org/10.1353/jbl.2013.0022
4. J. H. Walton, The Imagery of the Substitute King Ritual in Isaiah's Fourth Servant Song. Journal of Biblical Literature, **122**(4), 734 (2003) https://doi.org/10.2307/3268075
5. Z. Hui, A Comparative Study on Imagery in Chinese and English Poetry. Comparative Literature: East; West, **1**(1), 106-114 (2000) https://doi.org/10.1080/25723618.2000.12015249
6. M. Lee, The Animals Awareness and Literary Imagery of the Seol(說). Korean Language and Literature, **113**, 103-127 (2020). https://doi.org/10.21793/koreall.2020.113.103
7. A. Parashar, M. Kumar, and V, Saluja Discovering India through Imagery in Postcolonial Travel Writings. Tourism Culture and Communication, **19**(2), 103-110 (2019). https://doi.org/10.3727/109830419X15536971539399
8. H. Hadziavdic and H. Hoffmann. Moving images of exclusion: Persisting tropes in the filmic representation of European Roma. Identities, **24**(6), 701-719 (2017)
9. R. Mestre, A. Del Rey, and K Stanishevski. The image of Spain as a tourist destination built through fictional cinema. Journal of Travel and Tourism Marketing, **24**(2-3), 185-194 (2008) https://doi.org/10.1080/10548400802092718
10. F. Ferdinal. Censorship, resistance, and transformation in modern Indonesian literature. Mediterranean Journal of Social Sciences, **4**(1), 269-272 (2013).
11. F. Ferdinal. Terror Narrative: Representing State Violence in Modern Indonesian Literature. OKARA: Jurnal Bahasa dan Sastra, **15**(2), 109 125 (2021).
12. Ferdinal, Exploring representations of American Agricultural Farmers and labourers: a postcolonial reading of In a Pickle: A Family Farm Story by Jerry Apps IOP Conf. Ser.: Earth Environ. Sci. **1107** 012016 (2022).
13. Oktavianus and K. Anwar. Local Food Diversity in Minangkabau. IOP Conf. Ser.: Earth Environ. Sci. **1107** 012056 (2022) https://doi.org/10.1088/1755-1315/1107/1/012056.
14. O. Oktavianus et al., Agricultural Productions and Metaphorical Expressions in Minangkabau Culture. IOP Conf. Ser.: Earth Environ. Sci. **1253** 012026 (2023). https://doi.org/10.1088/1755-1315/1253/1/012026
15. F. Ferdinal, O. Oktavianus, and I. Zahid, Exploring the Beauty of Islamic Values Through Metaphorical Expressions in Literary Work. Afkar: Jurnal Akidah Pemikiran Islam, **25**(2), 421–458 (2023) https://doi.org/10.22452/afkar.vol25no2.13
16. O. Oktavianus et al. Metafora Leksikal dan Nilai Budaya dalam Novel Hamka Di Bawah Lindungan Ka'bah. Pertanika: Social Sciences and Humanities, **32**(S1): 111–137 (2024). https://doi.org/10.47836/pjssh.32.S1.05

17. I. Revita, F. Ferdinal, O. Oktavianus, G. Asnan, and H. Herwandi. Bridging the Worlds: Metaphor in Interdisciplinary Contexts. JURNAL ARBITRER, **11**(2), 208–227 (2024). https://doi.org/10.25077/ar.11.2.208-227.2024
18. N. Rogacheva and A. Drozdova. The Language of Smell in Vladimir Nabokov's Mystical Discourse of Prose from His French Period. Quaestio Rossica, **9**(2), 562–575 (2021). https://doi.org/10.15826/qr.2021.2.596
19. M. A. Martínez. Imagining emotions in story worlds: physiological narrated perception and emotional mental imagery. Frontiers in Human Neuroscience, **18** (2024). https://doi.org/10.3389/fnhum.2024.1336286
20. K. Kaplan. The Filmmaker's Triage: The Doctor Figure and Images of Trauma. Communicatio, **44**(4), 59–72 (2018). https://doi.org/10.1080/02500167.2018.1508045
21. M. Kitts. The Wide Bosom of The Sea As A Place of Death Maternal and Sacrificial Imagery In Iliad Literature and Theology, **14**(2), 103-124 (2000) https://doi.org/10.1093/litthe/14.2.103
22. A. Ciribuco. Transnational Uses of Mafia Imagery in Zadie Smith's White Teeth. CLCWeb: Comparative Literature and Culture, **19**(4) (2017) https://doi.org/10.7771/1481-4374.2975
23. G. Fei, T. H. Lai-Ming, and P. Zhu. Time in Imagery. Chinese Literature Today, **4**(1), 12-15 (2014) https://doi.org/10.1080/21514399.2014.11834019
24. M. Huang. The Establishment of the Theme of Evil Through Imagery in Othello. English Language and Literature Studies, **12**(3), 1 (2022) https://doi.org/10.5539/ells.v12n3p1
25. A. L. Kossert and K. Munroe-Chandler. Exercise Imagery: A Systematic Review of the Empirical Literature. Journal of Imagery Research in Sport and Physical Activity, **2**(1) (2007) https://doi.org/10.2202/1932-0191.1015
26. S. Ross. The Last Bluebell: Anthropocenic Mourning in the Bronte's Flower Imagery. Victorians: A Journal of Culture and Literature, **134**(1), 218-233 (2018) https://doi.org/10.1353/vct.2018.0020
27. T. M. Jerry, D.S. K. Thomas, A. A. Hill, and S. L. Cutter. Catastrophe in Reel Life versus Real Life: Perpetuating Disaster Myth through Hollywood Films. International Journal of Mass Emergencies and Disasters, **18**(3), 383-402 (2000) https://doi.org/10.1177/028072700001800302
28. S. Mnassar. Apocalyptic Imagery and Its Significance in James Fenimore Cooper's The Prairie. English Language and Literature Studies, **6**(2), 91 (2016). https://doi.org/10.5539/ells.v6n2p91
29. M. Satterlee. Landscape Imagery and Memory in the Narrative of Trauma: A Closer Look at Leslie Marmon Silko's Ceremony. Interdisciplinary Studies in Literature and Environment, **13**(2), 73-92 (2006) https://doi.org/10.1093/isle/13.2.73
30. E. Zola (2000). The flood. Available online at http://www.online-literature.com/emile-zola/the-flood/6/
31. https://www.150ansinondations.com/gb- june-and-september-1875-nothing-but- water/
32. R. DiYanni. Literature: Reading Fiction, Poetry, Drama, and the Essay. 6th ed. (McGraw-Hill, 2007)
33. M. H. Abrams. A Glossary of Literary Terms. 9th ed. (Wadsworth Publishing, 2009).
34. I. Sisakht and A. Pourlashkary. Study of the Effect of Imaginary Imagery (Imaginary Color, Imaginary Figure, Imaginary Space Building and Fantasy in Characterization) on the Characterization of Children. Advances in Literary Study, **6**, 161-171 (2018) https://doi.org/10.4236/als.2018.64014
35. S. S. Lanser. Fictions of Authority: Women Writers and Narrative Voice. (Cornell University Press, 1992).
36. L. Perrine and R. A. P. P Thomas. Sound And Sense: An Introduction to Poetry Eight Edition. (Harcourt Brace College Publisher, Orlando, 1992)
37. H. Vendler. The Art of Shakespeare's Sonnets. Cambridge (The Belknap Press of Harvard University Press, Massachusetts, 1999)
38. J. Culler. Literary Theory: A Very Short Introduction. (Oxford: Oxford University Press, 1997)
39. R. Barthes. Mythologies. Translated by Annette Lavers. New York: Hill and Wang, 1972).
40. N. Nopriyasman, G. Asnan, A. Fauzi, I. P. Hastuti, A. H. Ritonga, V. Kurniawan, and R. Mairiska. Reading indigenous signs: The wisdom of nagari communities toward natural disaster in Pasaman Barat. International Journal of Disaster Risk Reduction, **107**, 104497 (2024)

# Banda Bekali and local knowledge in flood management: 100 years of social and environmental interaction in Padang City

*Jonson* Handrian Ginting[1*], *Afrida*[1], and *Zainal* Arifin[1]

[1]Department of Anthropology, Faculty of Social and Political Sciences, Universitas Andalas, Padang, Indonesia

**Abstract.** Floods have plagued Padang City for centuries, with severe events recorded almost annually between 1890 and the early 20th century, peaking in 1904. The damages during that period reportedly reached two million guilders. In response, the Dutch East Indies colonial government constructed the Banda Bekali canal system to mitigate flood risks. While the canal has reduced the severity of flooding, it remains a persistent issue in the city. This study, marking over 100 years since Banda Bekali's establishment, examines the canal's role in flood management. Using a qualitative-descriptive approach, the research focuses on the canal's sections in Padang, selecting informants through purposive sampling based on age, gender, and societal roles. Data collection includes observations, in-depth interviews, and literature reviews. The study aims to understand the community's interaction with the canal and the local knowledge surrounding flood control. It explores the Minangkabau people's flood and disaster management systems and reflects on the 100-year history of Banda Bekali as a mitigation strategy. Through triangulation, this research hopes to contribute to improved flood disaster management strategies in Padang, drawing on both historical analysis and contemporary local knowledge.

## 1 Introduction

Padang City, located in the province of West Sumatra, is an ancient city with a rich history. Throughout the 19th century, it was a thriving metropolitan area on Sumatra Island and served as the center of the Dutch East Indies government from May 20, 1784, until the period leading up to Indonesia's independence. The city's history began as a small fishing village along the banks of the Muara Batang Arau in the 14th century, primarily inhabited by the Rupit and Tirau people, who relied on fishing and trade for their livelihoods. Today, Padang City is recognized as a Rantau city of the Minangkabau ethnic group, with numerous historical artifacts and colonial architecture affirming its status as a central hub of governance and culture on Sumatra Island.

Flooding, a natural phenomenon exacerbated by human activities, has long been a pressing issue in Padang City. It results from a combination of factors: rainfall, river conditions, upstream conditions, regional topography, and tidal influences. Currently, the potential for flooding is heightened by damaged river bodies, deteriorated catchment areas, spatial planning violations, legal infractions, uncoordinated development planning, and low community discipline. Floods, a recurring natural disaster, almost inevitably occur during every rainy season. They are caused by various factors, including rainfall, watershed degradation, errors in river channel development planning, river siltation, and regional planning and infrastructure construction errors [1].

Padang City is particularly vulnerable due to its geographical conditions, which include hills and lowlands adjacent to the Indian Ocean. This makes it susceptible not only to floods but also to extreme weather events, coastal erosion, earthquakes, tsunamis, landslides, and other geological disasters. With five major rivers and sixteen smaller rivers, the city is frequently affected by floods, with several districts being especially prone to this natural disaster. Flash floods in the upstream areas cause significant damage, including riverbank erosion and destruction of agricultural and residential areas along the watersheds.

Padang City experiences high rainfall, with data from seven rain observation stations showing an average of 123 to 190 rainy days per year [2]. Annual rainfall ranges between 3,329 mm and 4,296 mm, with areas near the hills receiving more rainfall than coastal regions. This disparity leads to a higher risk of floods and landslides in elevated areas. The high rainfall intensity makes Padang City particularly prone to flooding, a persistent issue that has affected the city for hundreds of years.

To mitigate this, during the Dutch colonial era, the government recognized the flood problem and responded by constructing the Banda Bekali canal. Padang had already gained notoriety as a "flooded city," with significant flooding reported on December 5-6, 1898, March 25, 1904, and a major peak on September 28-29, 1907. The canal system, now known as Banda Bekali, was built to manage flooding in Padang by diverting the flow of the Batang Arau river, spanning 6.8 km in length and 20 meters in width, with an estuary near Purus Beach [1].

* Corresponding author: jonson@soc.unand.ac.id

Banda Bekali was designed to reduce flooding and support the city's drainage system, including small rivers, swamps, and sewers, which ultimately empty into the Indian Ocean. However, despite its construction over a century ago, floods in Padang City remain a recurring challenge. Nonetheless, the community's interaction with Banda Bekali has shaped a collective memory of floods, passed down orally through generations. This experience has fostered local knowledge systems that continue to inform community behaviors related to flood management.

The recurring floods have led to the development of distinct local knowledge systems within Padang's communities, particularly the Minangkabau people. This knowledge, rooted in the social, cultural, and environmental context, allows residents to interpret signs of impending floods, such as changes in river color, the appearance of debris, and fish behavior. Such knowledge is invaluable in disaster-prone areas, where community responses are often dictated by the ecological systems they inhabit.

As noted by Abdullah academic studies on disasters in Indonesia often perceive disasters as isolated events rather than integrated elements of collective knowledge and social policy [3]. However, local knowledge systems, passed down through generations, represent a significant resource for understanding and mitigating disaster risks. These systems, while not always explained in scientific terms, are crucial for effective flood management. Historical records reveal that Indonesia has faced numerous large-scale natural disasters, and the integration of local knowledge with formal disaster management strategies offers a path forward in creating more robust solutions [4].

This study explores the interplay between infrastructural interventions, such as Banda Bekali, and the traditional local knowledge systems developed by the residents of Padang, particularly the Minangkabau community, in addressing the recurrent flooding problem. The research aims to highlight how these complementary approaches engineering solutions and indigenous practices can be integrated to create more effective flood management strategies in Padang City.

## 2 Research methods

This research uses a qualitative-descriptive method to explore and understand flood management and local knowledge in Padang City, relying on observation and inductive analysis [5]. This approach provides a detailed view of human practices and knowledge, focusing on how the community responds to flooding [6]. The main focus of the study is the Banda Bekali canal and surrounding rivers that are frequently affected by floods.

The study took place in flood-prone areas along the Banda Bekali canal in Padang City. Informants were selected through purposive sampling to ensure a wide range of perspectives. Priority was given to individuals with direct experience of recurring floods, including those living in vulnerable areas. The sample included a diverse group of community members, such as local leaders, elderly residents with extensive flood experience, younger generations, and those whose livelihoods are closely tied to the rivers (fishermen, sand collectors). A balance of age and gender was also ensured to capture varied flood-related knowledge across different demographics. This method ensured that the study reflected a broad spectrum of community views and practices.

Data collection was carried out using three main methods: observation, in-depth interviews, and a literature review. The observations took place over three months, focusing on daily activities near the canal, such as fishing, farming, and sand collection, as well as local flood management practices. Researchers documented how residents prepared for floods, how they interpreted signs like water levels and debris, and their actions during flood events. The environmental condition of the river was also observed, with notes taken on water flow, sedimentation, and landscape changes.

Interviews were conducted with 16 selected informants to uncover local knowledge and beliefs about flood management. Questions explored their experiences with floods, how they predict floods, traditional practices used to prepare for them, and their thoughts on the effectiveness of Banda Bekali. These semi-structured interviews allowed for open discussion while guiding the conversation towards specific themes.

In addition, a thorough literature review was conducted to support the primary data. This included historical records of floods in Padang, studies on disaster response, and research on similar knowledge systems in other regions. The literature review helped contextualize the field findings and connect them with broader disaster management theories.

This research applies Cultural Ecology which explores how human societies adapt to their environments through culture [7]. This framework helps explain how the Minangkabau community in Padang has developed flood management practices over time in response to recurring floods. Their strategies, such as using natural signs to predict flooding or modifying their use of river resources, are examples of adaptive responses to environmental challenges. Steward's theory of multilinear evolution shows that these practices are specific to the local ecological and cultural context, highlighting how environmental pressures shape cultural solutions [7].

The study also draws on Disaster Anthropology, particularly the work of Anthony Oliver-Smith, who examines how communities manage and adapt to disasters [8]. Disasters are not just natural events but are deeply shaped by social and cultural factors [8]. In this context, the Minangkabau's traditional knowledge, such as recognizing signs of impending floods and community-based responses, represents culturally embedded disaster management practices. By using this theoretical lens, the study examines how local knowledge contributes to resilience and adaptive strategies in the face of recurring floods.

# 3 Results and discussion

## 3.1 History of flooding in Padang

Geographically, Padang City is located in a lowland area that empties into the western coast of Sumatra Island. Historically, when Padang was still a small village, it was dominated by swamp forests and thatch trees, characteristic of flood-prone lowlands. The history of flooding in Padang is also reflected in the names of districts and villages, such as Kampung Pulau Air, Tarandam, Pulau Karam, Rawang, and Ganting, all of which are associated with water, bogs, and floods. The creation of the Banda Bekali flood canal further underscores the longstanding issue of flooding in Padang. This canal was designed to divert the surging waters that frequently inundated Batang Arau, the then center of government and trade for the Dutch East Indies, directing the flow northward to Purus Beach Padang, as shown in Fig. 1.

**Fig. 1.** Banda Bekali 1920-photo Circa KITLV universiteit qleiden.

Historical records of flooding in Padang can be found in Dutch colonial-era newspapers. For instance, a flood disaster caused by the failure of a flood channel embankment in the northeastern part of Padang on a Friday night [9]. The newspaper published on December 19, 1929, detailed a flood on a Tuesday that raised the Batang Arau's water level by 1.5 meters after half a day of rain. The Dell Courant, on October 20, 1932, reported heavy flooding in Padang due to a full day of rain on a Sunday, which severely affected Kampung Cina (now Pondok), washing away the suspension bridge and the Kampung Durian Power Plant. The water level was estimated to reach 4.4 meters, accompanied by thunderstorms, which also caused landslides that disrupted train operations. Additionally, the Pos Sumatera newspaper on October 21, 1932, reported on floods in Padang that occurred on December 14, June 19, 1931, and October 14, 1932. This publication also documented the efforts of the Dutch East Indies government to address the recurring flooding in Padang City.

## 3.2 Local knowledge about flooding

Banda Bekali canal, constructed during the Dutch colonial era, represents a significant engineering effort to manage the persistent flooding in Padang. While this infrastructure has been a vital tool in controlling water flow, it does not fully integrate the traditional flood management practices developed by the local Minangkabau community. Local residents, especially those living along the watersheds, possess a deep understanding of natural warning signs, such as changes in water color and the behavior of fish, which are used to predict flooding. This knowledge has been refined over generations, forming an essential part of the community's disaster preparedness [7, 10].

However, there are instances where modern flood management systems, like Banda Bekali, fail to align with this local wisdom. For example, government projects focused on infrastructure often overlook the community's practical knowledge, leading to mismanagement and inefficiency [4, 11]. In some cases, poorly designed or maintained systems have exacerbated flooding, as seen when industrial waste and sediment clog rivers that feed into Banda Bekali, reducing its effectiveness, (Fig. 2) [12]. By disregarding these local insights, flood management efforts risk being less effective, creating further vulnerabilities for communities dependent on their traditional methods for survival [13].

Despite these tensions, there are ways in which local knowledge and modern infrastructure can complement each other. The Minangkabau community's understanding of environmental cues can enhance the functionality of the Banda Bekali system. For instance, integrating local practices—such as planting elephant grass on riverbanks to prevent erosion and using natural indicators for flood warnings—into modern engineering efforts could lead to more sustainable outcomes [13, 14]. Local knowledge systems provide a real-time, flexible response to environmental changes, something that static engineering solutions often lack.

**Fig. 2.** Conditions of the Batang Kuranji River during heavy rain 2023.

For flood management in Padang to be truly effective, there must be a collaborative approach that respects and integrates both the local wisdom of the community and the technological advancements represented by the Banda Bekali canal. This hybrid model would not only improve flood prevention and response but also strengthen the community's resilience against the increasing challenges posed by environmental changes [15].

Interviews with several informants reveal local knowledge and understanding of the dangers posed by the river. For instance, Mr. Agus explained that if residents are fishing, collecting sand, or engaging in other river-related activities, certain signs indicate impending danger. These signs include the river water becoming turbid or changing color, an increase in floating debris, and the disappearance of fish that previously swam near the surface. These are all indicators of heavy rain in the upper reaches of the river. Similarly, Mr. Hasan, who fishes daily in the Banda Bekali watershed, noted that fish tend to hide and avoid the surface when it is raining or about to rain upstream [16].

An interesting practice among residents who fish around the Banda Bekali watershed is that they do not collect their catch in one place. Instead, they tie the fish with a rope and allow them to swim around the fishing spot, (Fig. 3). This is believed to attract more fish to the area. In some river areas, such as the Banda Bekali watershed, people deliberately plant grasses on the riverbanks and soil mounds in the middle of the river for livestock feed. The grass, often elephant grass (Pennisetum purpureum), is preferred for its large size and sturdy stems, similar to cane stalks Community Interaction with the River

The watershed (DAS) is typically defined as a territory or area bounded by a topographic boundary (ridge) that serves to receive and collect rainwater, sediment, and nutrients, eventually draining them through tributary streams to an estuary. Administratively, the Central Bureau of Statistics categorizes watersheds as villages or kelurahan, with areas primarily consisting of basins or hollows around river flows.

**Fig. 3.** Fishing activities in the Banda Bekali River 2023.

In discussing watersheds, it is essential to consider the management and activities of the surrounding communities. According to Government Regulation No. 37 of 2014 on Watershed Management, watersheds have land boundaries that serve as topographical separators, and boundaries at sea include water areas influenced by land activities.

Padang City comprises six watersheds:

- The Air Dingin watershed with 10 sub-watersheds spanning 12,874.06 hectares.
- The Kuranji watershed with 20 sub-watersheds totaling 22,852.78 hectares.
- Batang Arau watershed with 13 sub-watersheds covering 17,243.12 hectares.
- The Banda Bekali watershed with 7 sub-watersheds totaling 6,840.68 hectares.
- The Batang Kandih watershed with 10 sub-watersheds covering 7,559.29 hectares.
- Sungai Pisang or Banana River watershed with 4 sub-watersheds covering 2,843.91 hectares.

Community utilization of watersheds involves local knowledge systems that evolve and grow as local wisdom [17]. All forms of community activities that engage with the river follow certain patterns aimed at meeting life's needs. However, the use of watersheds sometimes leads to environmental degradation and increases the risk of flooding [18]. Land conversion into residential areas, construction, industry, and municipal facilities narrows water catchment areas. To understand the happenings within the watershed—such as land use, environmental degradation, flood phenomena, and other systems involving community participation—it is crucial to observe the practices between the community and the river [19].

### 3.3 Changes in people's interaction with the river

Continuous development and economic growth have altered people's interactions with the river, leading to decreased water quality, a decline in river biota, and a diminishing significance of the river in community life. Economic development and urbanization have drastically reshaped the relationship between communities and their natural environment. Rivers, which were once central to local livelihoods and social life, are now often sidelined as modern infrastructure and industrial projects take precedence [16]. This transition reflects a broader global trend, where rapid economic expansion leads to environmental degradation and marginalization of traditional practices [20].

In the past, people depended on the river for their economic needs. The river played a critical role in the daily lives of many communities, providing water for irrigation, fishing, and other subsistence activities. These rivers served not only as sources of livelihood but also as important cultural and social spaces. Over time, with the advent of new economic opportunities, such as employment in the industrial and service sectors, reliance on river-based economies has diminished.

Now, with the availability of diverse and promising job opportunities, most communities no longer rely on rivers. The shift in the economic base from agriculture and river-related activities to modern industries has led to a transformation in local economies. Communities that once depended on rivers for fishing, farming, and transportation now find new opportunities in urban centers, significantly reducing their dependence on the river.

The current poor quality of the river has also led to a reduced interaction between the community and the river. Environmental degradation, including pollution from industrial and domestic waste, has further severed the traditional connection between the river and the community. Rivers that were once clean and teeming with life have now become polluted, affecting water quality and diminishing the biodiversity that once thrived in these ecosystems. This decline in environmental quality has also contributed to a loss of community engagement with the river, as it no longer serves as a safe or viable space for work or leisure.

One informant, Agus, mentioned that in the 1980s and 1990s, people were always by the river daily, engaging in work, domestic tasks, and recreation. This historical reflection emphasizes how deeply embedded the river was in the social fabric of the community. People used to rely on rivers not only for economic reasons but also for their daily domestic activities and recreation. The presence of the river fostered a sense of community, where people gathered for washing, bathing, fishing, and socializing [2].

However, from the 2000s to the present, only a handful of people maintain contact with the river, while most simply reside nearby without interaction. As urbanization and industrialization progressed, the community's interaction with the river became increasingly distant. The river, which once served as a focal point for community life, has now become a backdrop to modern living, with fewer people depending on or interacting with it on a regular basis. This disconnection reflects the broader environmental changes occurring in many parts of the world, where rapid economic growth often leads to the neglect of traditional ecological relationships

## 3.4 Tragedy of human losses that occurred in the river

Izal, someone who lived in DAS, mentioned that the river water used to be very clean and clear. Despite many activities taking place on the river, there was no significant damage to its preservation. Currently, environmental sustainability behavior has eroded due to factory waste disposal, garbage, and land narrowing. These changes in interaction, which reduce community contact with the river, have led to negative impacts.

In mid-June 2022, residents of Padang City were shocked by the tragic drowning of three vocational school students in the Bangek River, located in Lubuk Minturun. Despite extensive search and rescue efforts hampered by persistent heavy rains, two of the students lost their lives. This incident garnered significant public attention due to a local belief that if someone drowns and remains missing, the rain will continue unabated until the body is recovered. This reflects a blend of local ecological knowledge and spiritual beliefs in explaining natural events.

In close proximity to this case, the body of a woman was discovered in the Gunuang Nago Dam, a tributary of the Kuranji Batang River. Interviews with several local informants revealed that the discovery of bodies in rivers is not uncommon in this area. Agus and Izal recounted instances in which bodies were found downstream of the Banda Bekali River in 2011 and 2014, which caused considerable distress among residents. Additionally, Yunita recalled a similar incident around 2004 when the body of an eight-year-old boy was discovered, further highlighting the community's ongoing concerns about safety in the waterways. These events highlight how local knowledge and experiences shape community perceptions of their environment, particularly regarding safety and risk.

According to data on disaster events in the city of Padang from 2016 to 2020, there are several incidents of missing or drifting people in the river, as in the following Table 1:

**Table 1.** Missing or drifting people data (sources: BPBD and Dinkes Padang City)

| Number of Occurrences | | | | |
|---|---|---|---|---|
| **2016** | **2017** | **2018** | **2019** | **2020** |
| 12 | 15 | 20 | 4 | 3 |

## 3.5 Community interpretation of rivers

The rivers of Padang City are more than just natural features; they are vital components of the community's identity and daily life. This relationship between the residents and their rivers is shaped by historical, cultural, and environmental factors that underscore the rivers' significance. From providing essential resources for survival to influencing settlement patterns, these waterways have a profound impact on the lives of those who inhabit the region. In this context, the rivers play an integral role in the lives of local residents, shaping their understanding and interactions with their environment.

First, the significance of the river is profound for the residents of Padang, as it is the lifeblood of their existence. Many rely on the river for daily activities, such as washing, fishing, waste collection, and extracting building materials like sand and stone. In some upstream areas, such as the Batu Putuk River in the Kuranji Watershed and the Bangek River, people can still consume the river water, which remains clean and safe for drinking. However, in other locations, river water is often tainted and odorous due to household and industrial waste.

Second, the river also serves as a source of livelihood for many residents. While individuals in the Padang City watersheds engage in various occupations, those linked to the river are particularly notable. Common livelihoods include farming, mining for sand and stone, and collecting waste. Fishing—whether through traditional methods or more modern techniques—serves as a supplemental source of income. An interesting phenomenon occurs during heavy rainfall, when people flock to the riverbed to catch fish. In these conditions, fish tend to seek calmer waters, making them easier to catch. I have observed individuals of all ages, from children to adults, engaging in fishing activities along the Banda Bekali River, utilizing nets, food serving lids, and long fabrics as tools. The diverse catches often yield large sacks filled with various types and sizes of fish, demonstrating the river's continued abundance.

Third, the concept of settlement in Padang is closely tied to the river. The city features a unique settlement pattern, with residential areas extending along the riverbanks, varying in density from upstream to downstream. Population density tends to be lower in the upstream regions compared to the more densely populated downstream areas. There are generally no houses built directly on the river, primarily due to the frequent flooding experienced in Padang. In contrast, the densely populated and impoverished neighborhoods view the rivers as sources of disease, often relegating them to mere dumping grounds for trash and household waste, further complicating the relationship between the community and their waterways

## 4 Conclusion

Flood management in Padang requires an integrated approach that combines modern engineering solutions, like the Banda Bekali canal, with the rich local knowledge developed by the community. While Banda Bekali has mitigated flooding to some extent, it cannot replace traditional ecological knowledge that allows residents to predict and respond to floods based on natural signs like water clarity and fish behavior. The tension between modern infrastructure and local practices is evident when government projects overlook this knowledge, resulting in less effective flood control measures, as seen with industrial waste clogging rivers. Integrating local practices, such as planting elephant grass to prevent erosion, can enhance the functionality of modern systems and offer a more adaptive response to environmental changes.

Moving forward, sustainable flood management in Padang will depend on a hybrid model that values both technological innovation and local wisdom. This combined approach not only improves flood prevention but also strengthens the community's resilience, preserving cultural practices that have guided them for generations. As theories like Cultural Ecology and Disaster Anthropology suggest, disaster adaptation must embrace both scientific advancements and the practical, adaptive strategies developed through lived experience.

## References

1. P. Blaikie, Y. Cannon, I. Davis, and B. Wisner, At risk: Natural hazards, people's vulnerability, and disasters. (Routledge, London, 1994).
2. N. Y. Sudiar and P. M. Siregar, Analisis curah hujan Kota Padang saat Madden Julian Oscillation periode 1980-2010. Jurnal Sainstek IAIN Batusangkar, **4**(2), 123–128, (2012).
3. I. Abdullah, Konstruksi dan reproduksi sosial atas bencana alam. Working Papers in Interdisciplinary Studies, No. 01, (Sekolah Pascasarjana UGM, Yogyakarta, 2008).
4. W. M. Adams and M. Mulligan, Decolonizing nature: Strategies for conservation in a post-colonial era. (Earthscan, London, 2003).
5. J. W. Creswell. Research design: Pendekatan kualitatif, kuantitatif, dan mixed. (PT Pustaka Pelajar, Yogyakarta, 2010).
6. F. Berkes. Sacred ecology: Traditional ecological knowledge and resource management. (Taylor dan Francis, Philadelphia, 1999).
7. J. H. Steward, Evolution and ecology: Essays on social transformation, (University of Illinois Press, Illinois,1988).
8. A. Oliver-Smith, Anthropological research on hazards and disasters, Annual Review of Anthropology, **25** (1), 303–328. (1996). https://doi.org/10.1146/annurev.anthro.25.1.303
9. Bandjirkanaal te Padang. (1915, October 25). De Preangerbode. Retrieved from https://www.delpher.nl/nl/kranten/view?query=Bandjirkanaal+te+Padang&coll=ddd&identifier=ddd:011023971:mpeg21:a0088&resultsidentifier=ddd:01023971:mpeg21:a0088&rowid=10
10. J. P. Spradley, Metode etnografi, (Tiara Wacana. Yogyakarta, 2006).
11. A. Escobar, After nature: Steps to an antiessentialist political ecology. Current Anthropology, **40** (1), 1–30. (1999). https://doi.org/10.1086/515799
12. I. Davidson-Hunt and F. Berkes, Nature and society through the lens of resilience: Toward a human-in-ecosystem perspective, In F. Berkes, J. Colding, and C. Folke (Eds.), Navigating social-ecological systems: Building resilience for complexity and change (53–82). (Cambridge University Press, Cambridge, 2003).
13. W. N. Adger, T.P Hughes, C. Folke, S. R. Carpenter, and J. Rockström. Social-ecological resilience to coastal disasters. Science, **309** (5737), 1036–1039. (2005). https://doi.org/10.1126/science.1112122
14. A. M. Gunn, Encyclopedia of disasters: Environmental catastrophes and human tragedies **1**, Westport, (Greenwood Press, Connecticut, 2008).
15. A. Maryono, Menangani banjir, kekeringan, dan lingkungan, (Gadjah Mada University Press, Yogyakarta, 2005).

16. M. Pelling, Adaptation to climate change: From resilience to transformation, (Routledge, London, 2011).
17. C. Folke, J. Colding, and F. Berkes. Rediscovery of traditional ecological knowledge as adaptive management. Ecological Applications, **10** (5), 1251–1262. (2000). https://doi.org/10.1890/1051-0761(2000)010[1251:ROTEKA]2.0.CO;2
18. I. Scoones, New ecology and the social sciences: What prospects for a fruitful engagement? Annual Review of Anthropology, **28**, 479–507. (1999). https://doi.org/10.1146/annurev.anthro.28.1.479
19. P. Sillitoe, The development of indigenous knowledge: A new applied anthropology. Current Anthropology, **39**(2), 223–252. (1998), https://doi.org/10.1086/204722
20. B. L. Turner, et al., A framework for vulnerability analysis in sustainability science. Proceedings of the National Academy of Sciences, **100** (14), 8074–8079. (2003). https://doi.org/10.1073/pnas.1231335100

# Shock and adaption: the economic impact of COVID-19 in the Mentawai Islands, West Sumatra, Indonesia

*Sri* Setiawati[1], *Yonariza* Yonariza[1*], *Indrajit* Pal[3], *Eko* Priyono[2], *Agustinus* Agustinus[4], and *Aron* Tasirikeru[4]

[1]Doctoral Program of Development Study, Graduate School, Universitas Andalas, Padang, Indonesia
[2]Graduate School, Universitas Andalas, Padang, Indonesia
[3]Disaster Preparedness, Mitigation and Management, Asian Institute of Technology, Thailand
[4]Graduate Program, Faculty of Social and Political Sciences, Universitas Andalas, Padang, Indonesia

**Abstract.** Remote coastal communities face many challenges yet often demonstrate considerable resilience to be able to survive. This study explores the effect of the COVID-19 pandemic on the household economy, and their adaptive responses to it, in two villages of the Mentawai Islands, Indonesia. Data were collected using FGD and household surveys. The survey was carried out in Sido Makmur village on Sipora Island and at Muntei village on Siberut Island. Thirty households were randomly selected in each village. Data collected included; household characteristics, type of income change, government subsidy received, and response to the income change. The research found that COVID-19 policies affected many parts of life. There were significant reductions in income level not only because of losing employment but also because disruption of supply chains also lowered demand for local produce. People responded to the lower-income by intensifying their involvement in the agricultural and fishery sectors to meet their food security needs. This study highlights the importance of strengthening access to local resources to cope with disasters in the island's region.

## 1 Introduction

Since December 2019, the coronavirus COVID-19 has spread rapidly around the world [1]. Infections occur through person-to-person contact and range from mild to deadly [2]. As of 7th June 2023, globally, there have been more than 750 million cases of COVID-19 with almost 7 million deaths attributed to the disease [3]. Alongside the direct health impacts, the COVID-19 pandemic, and subsequent government policies, has caused disruptions in global supply chains, threatened livelihoods and education, and profoundly changed human behavior and the social contexts of billions of people [4, 5]. According to the UNDP "The COVID-19 pandemic is far more than just a health crisis: it's affecting society and the economy at its core…it is likely to increase poverty and inequality on a global scale" [6]. COVID-19 has also fueled fears of an economic crisis and an impending recession [7]. Social distancing, self-isolation and travel restrictions have led to a reduction in the workforce in all sectors of the economy and caused many jobs to be lost [8]. Research conducted in South Asia found that There are also ecological implications from the pandemic, with some commentators arguing that, "our economy, our society, our communities must rediscover how to live with nature" [9].

Nature plays a profound role in the Indonesian economy. As an archipelago with more than thirteen thousand islands [10] many communities are directly dependent on marine resources for their livelihoods. Several large islands including Sumatra, Kalimantan, Java, Sulawesi, Bali and Nusa Tenggara and Papua contain the majority of the population. The rest are small islands scattered across Indonesia's territory which are referred to as 3T (*Terluar, Terdepan, Tertinggal*) - the outermost, poorest, and lagging [11]. Although some of these islands may be biodiversity hotspots sought out by tourists, many are isolated and lacking in basic infrastructure.

There is a paradox with island communities. On the one hand, their isolation contributes to their deprivation by cutting them off from markets and innovation [12] and makes them more vulnerable to shocks because they are so far from aid in a time of crisis [13]. Yet, on the other hand, their continued existence in an isolated context points to their resilience and means that the communities that live there are already adapting to significant pressures and having to "fend for themselves". Their isolation also contributed a layer of protection against being infected.

The lack of connectivity of small islands led the Jokowi government to develop the concept of a maritime axis for the development of islands. This concept emphasizes the realization of inter-island connectivity through the development of the shipping industry and sea transportation [14]. Indonesia's outermost small islands need to be developed because they have strategic value for the sovereignty of the state [11]. COVID-19 has threatened these island communities. Connectivity

* Corresponding author: yonariza@agr.unand.ac.id

between Asian-Pacific islands played a central role in the spread of COVID-19 [15]. Restricting movements between these islands to prevent the spread of COVID-19 also had far-reaching economic impacts. Many remote island communities are heavily dependent on fisheries and the pandemic has significantly disrupted the marine sector and global. supply chains. These marine industries do not operate separately from each other, or from the marine environment in which they are part [6]. This has led to aggravating and interrelated impacts across the ocean economy, marine ecosystems and societies [16]. Some small islands are maritime tourist destinations [17-19]. Tourism development is very important for economic growth on these islands and restrictions on the movement of large-scale people or lock downs during the COVID pandemic have devastated community tourism businesses [20].

The economic impacts of COVID-19 in Indonesia have been felt acutely on household finances, through inequality and food insecurity. A UNDP report [21] found that nearly three-quarters of households (74.3%) reported a reduction in household income, that social assistance programs have not reached 15% of those need and that 30% of respondents worry that they will be unable to feed their families. Income reductions and disruptions to food delivery systems are the main factors contributing to food insecurity.

Households and communities have not been passive during the outbreak. The isolated nature of small island community's conditions them to be self-reliant, to adapt and find a way out from the crisis [18], regardless of whether external aid is forthcoming. Seasonal variations in weather and oceanographic conditions already made small island communities vulnerable to food insecurity [22].

In the Solomon Islands communities adapted by increasing agricultural production, especially tubers and by the economy reverting from cash to bartering fish for other types of food [23]. While most respondents in the study experienced the benefits of strong social capital through family and community support that increased food production and distribution, some reported theft. In 15% of community's conflicts occurred as the management rules were violated. In most communities, community-based resource management committees (CBRMs) have responded by raising awareness about fisheries rules. In some cases, communities have increased law enforcement or organized activities, such as group harvesting. In other Pacific communities such as Tuvalu people responded by panicking and stockpiling food [24].

This paper explores the effect of the COVID-19 pandemic on communities in a small island cluster of Indonesia, (Mentawai islands) and their adaptive responses.

## 2 Method

The Mentawai Islands Regency is a cluster of non-volcanic islands located in the western part of Sumatra Island and surrounded by the Indian Ocean. The Mentawai islands are remote and relatively deprived with a population of 87,623 [25]. Sipora island was initially not inhabited by the Mentawai people but has now become the administrative capital. [26]. Agriculture, forestry, and fisheries are important economic sectors with bananas, coconuts, cassava and sweet potatoes dominating production. Rice is planted (580 ha), but is not sufficient to meet local demand. The Mentawai Islands Regency are prone to natural disasters (earthquakes and tsunamis). The Mentawai Islands Regency consists of 99 island clusters, with four main islands; Siberut, Sipora, North Pagai, and South Pagai. These two islands have opened direct access to the provincial capital Padang using a fast ferry that takes four hours. During the COVID pandemic, transportation services were suspended for several weeks.

Two villages were purposively selected based on different ethnic groups. On Siberut Island, research was carried out in Muntei village, a village inhabited predominantly by the Mentawai tribal community (Fig. 1). Muntei village has a population of 1685 (2021) and most villagers work in the agricultural sector and capture fisheries. On Sipora Island, the research was conducted in Sido Makmur village, North Sipora sub-district which is close to the capital of the regency. Sido Makmur is a Javanese immigrant settlement area. Javanese immigrants settled on Sipora from 1987 and the village now has a population of 894 with livelihoods dependent on bananas, coconuts and cassava.

This research used an interpretive approach attempting to find explanations about social or cultural events based on the perspective and experience of the person under study. The interpretive approach is adopted from a practical orientation. The interpretive approach is a social system that interprets behavior in detail by direct observation [27].

Primary data were collected using Focus Group Discussions (FGD) of men and women, and supplemented with household surveys. In both villages, 30 households were randomly selected. The household survey explored the economic impacts of COVID -19 and household adaptations to COVID-19. Data were collected from 3 – 12 June 2022.

Survey data included household characteristics; number of economically active workers in a household, land ownership (ha), ownership of productive assets, ethnicity, education attainment. Questions also explored household responses to various economic changes caused by COVID-19 and whether a government subsidy was received by the household. Data were analyzed descriptively at the location of village (Fig. 1).

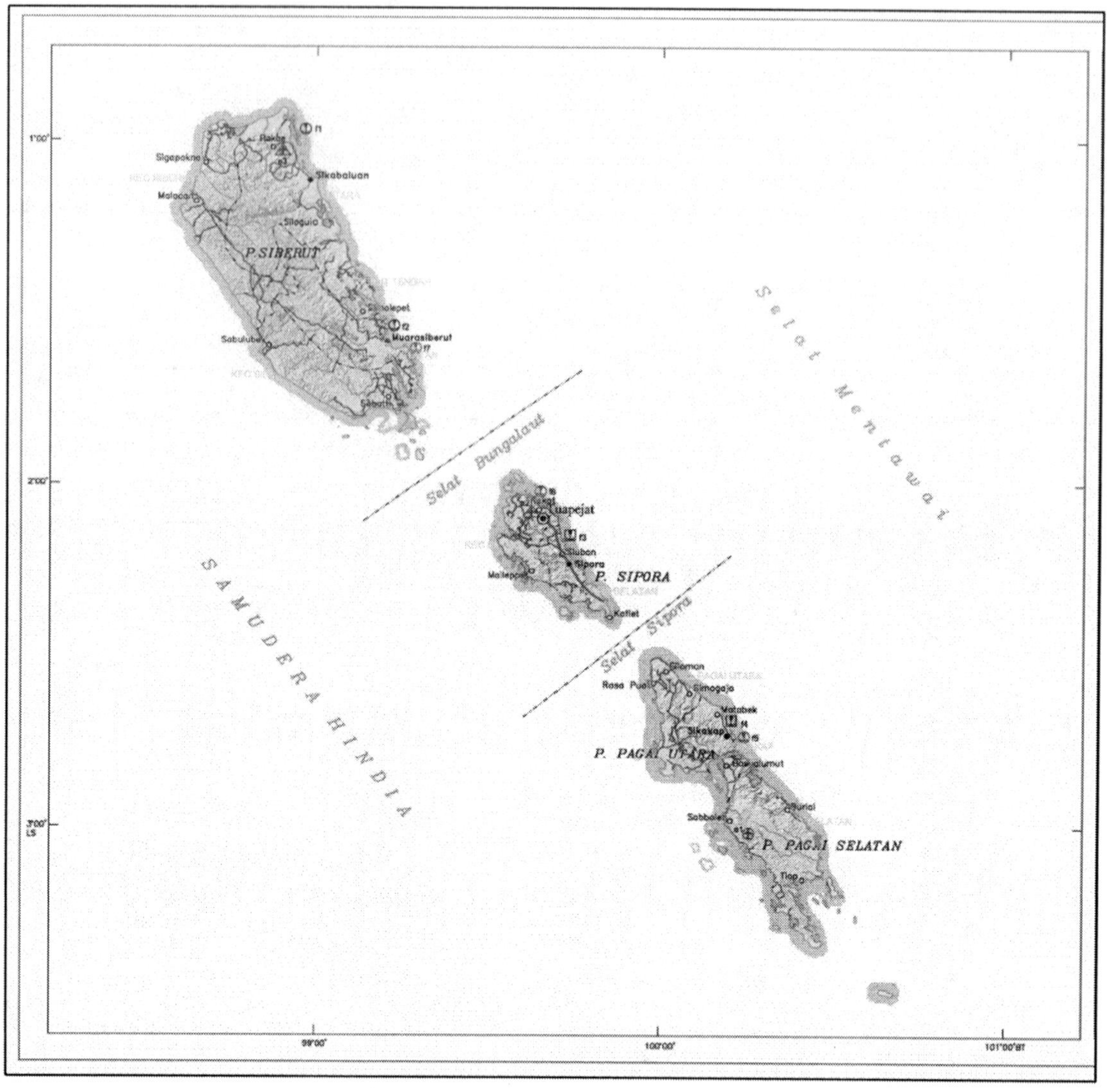


**Fig. 1.** Study village location map

## 3 Results and discussion

### 3.1 COVID 19 and government policy in the district of Mentawai Islands

On April 1st 2020, The Regent of the Mentawai Islands chose to quarantine his territory from April 1, 2020, for two weeks. It was the first regency to implement this policy in West Sumatra. During this quarantine period, all transportation services were suspended [28]. The lack of human capacity and medical infrastructure to deal with the coronavirus was the reason for the quarantine. Despite these efforts, the first case of COVID-19 in the Mentawai Islands was detected in Pogari, Goiso Oinan Village, North Sipora Sub-district on 12 April 2020, an imported case from Jakarta. This was followed by a cluster of cases from the Department of Fisheries.

A work-from-home policy was brought in for civil servants and contract workers to prevent the spread of COVID-19 [29]. To compensate for the economic impact, 16,000 underprivileged Mentawai Residents received direct cash assistance [30]. People's movement was controlled by halting passenger services [28]. As in many countries, the main policy to stop the spread of COVID-19 was to limit people's mobility. Although this had several negative impacts, the policy did restrict the spread of the pandemic and ensured Mentawai was one of the few regencies to never reach a red status on the pandemic zoning system.

Sixty households were interviewed (Table 1) with respondents representing a cross-section of backgrounds and ethnicities. The typical respondent was married (83%), aged between 36-50 (45%), Mentawai ethnicity (72%), Catholic (42%) and worked as a farmer (45%). Thirty per cent of respondents had not completed elementary school.

**Table 1.** Characteristics of respondents

| Respondent criteria | Village | | Total | % |
|---|---|---|---|---|
| | Muntei | Sido Makmur | | |
| ***Sex of respondent*** | | | | |
| -Male | 18 | 20 | 38 | 63 |
| -Female | 12 | 10 | 22 | 37 |
| ***Respondent age*** | | | | |
| -20 - 35 | 10 | 9 | 19 | 32 |
| -36 - 50 | 15 | 12 | 27 | 45 |
| -51 -65 | 5 | 6 | 11 | 18 |
| ->65 | 0 | 3 | 3 | 5 |
| ***Marital status*** | | | | |
| -Single | 5 | 0 | 5 | 8 |
| -Widow | 2 | 3 | 5 | 8 |
| -Married | 23 | 27 | 50 | 83 |
| ***Ethnicity*** | | | | |
| - Aceh | 0 | 1 | 1 | 2 |
| -Mix Batak | 0 | 1 | 1 | 2 |
| -Javanese | 0 | 6 | 6 | 10 |
| -Mentawai | 27 | 16 | 43 | 72 |
| - Minang | 1 | 5 | 6 | 10 |
| - Nias | 2 | 1 | 3 | 5 |
| ***Religion*** | | | | |
| -Islam | 3 | 17 | 20 | 33 |
| -Catholic | 24 | 1 | 25 | 42 |
| -Protestant | 3 | 12 | 15 | 25 |
| ***Education attainment*** | | | | |
| - No formal education | 5 | 0 | 5 | 8 |
| - Incomplete Elementary School | 8 | 5 | 13 | 22 |
| - Complete Elementary School | 1 | 1 | 2 | 3 |
| - Complete Junior High School | 7 | 4 | 11 | 18 |
| - Complete Senior High School | 6 | 10 | 16 | 27 |
| - Tertiary | 3 | 10 | 13 | 22 |
| ***Occupation*** | | | | |
| - Contractor | 1 | 9 | 10 | 17 |
| - Daily wage labor | 1 | 0 | 1 | 2 |
| -College student | 1 | 0 | 1 | 2 |
| - Religious worker | 0 | 1 | 1 | 2 |
| -Fisher | 2 | 0 | 2 | 3 |
| -Civil servant | 0 | 11 | 11 | 18 |
| -Trader | 1 | 2 | 3 | 5 |
| -Sago processor | 1 | 0 | 1 | 2 |
| -Farmer | 20 | 7 | 27 | 45 |
| -Carpenter | 3 | 0 | 3 | 5 |
| Total | 30 | 30 | 60 | 100 |

Motorbikes (79%) were the most commonly owned assets (Table 2). These play an important role in accessing farmland, going to fish and when securing food supply during COVID-19. This was followed by crop sprayers and canoes (Muntei only). In addition, motorized canoes are a common asset in Muntei village. The solo sprayer is important for people in Sido Makmur and in Muntei villages. These assets remain important for economic purposes during COVID-19.

**Table 2**. Asset ownership

| Item | Numbers owned | Village | | Total | % |
|---|---|---|---|---|---|
| | | Muntei | Sido Makmur | | |
| Number of motor bike per household | | | | | |
| | 0 | 10 | 3 | 13 | 22 |
| | 1 | 18 | 16 | 34 | 57 |
| | =>2 | 2 | 11 | 13 | 22 |
| Small shop | | | | | |
| | 0 | 29 | 25 | 54 | 90 |
| | 1 | 1 | 5 | 6 | 10 |
| Fishing net | | | | | |
| | 0 | 22 | 28 | 50 | 83 |
| | 1 | 7 | 1 | 8 | 13 |
| | 2 | 0 | 1 | 1 | 2 |
| | 3 | 1 | 0 | 1 | 2 |
| Canoe | | | | | |
| | 0 | 10 | 28 | 38 | 63 |
| | 1 | 18 | 1 | 19 | 32 |
| | 2 | 2 | 1 | 3 | 5 |
| Motorized canoe | | | | | |
| | 0 | 13 | 30 | 43 | 72 |
| | 1 | 16 | 0 | 16 | 27 |
| | 3 | 1 | 0 | 1 | 2 |
| | | | | | |
| Crop sprayer | | | | | |
| | 0 | 19 | 17 | 36 | 60 |
| | 1 | 11 | 9 | 20 | 33 |
| | 2 | 0 | 3 | 3 | 5 |
| | 3 | 0 | 1 | 1 | 2 |
| Amount | | 30 | 30 | 60 | 100 |

### 3.2 Economic impact of COVID-19

Economic impacts were explored at the household level (Table 3). More than half of respondents (52%) had experienced a decline in income due to price reductions and many had lost income due to a decline in demand (35%). Previously bananas and betel nuts had been exported out of the village by external buyers but during the pandemic, these buyers stopped coming because ferry services were suspended and the prices dropped from 20,000 Rupiah to 0 Rupiah per bunch of bananas and from 19,000 Rupiah to 9000 Rupiah for betel nuts. Most households (57%) found a scarcity of goods during COVID-19. Many experienced price increases and reduced household spending to account for this. Most respondents (53%) intensified economic activities to try and cover the shortfall, while a few (8%) experienced economic blessings during COVID-19. Seventy-eight per cent of respondents received cash transfers from the government during COVID-19. Behind the numbers were individual stories like, for example, the person who lost his job at the boarding school when students returned home. Households coped with losing an income source by shifting to informal work in other sectors (e.g. becoming a farm labourer), reducing expenses (e.g. by stopping supporting needy relatives) or eating into savings. In the absence of other paid employment, people defaulted to collecting forest products or growing vegetables.

**Table 3.** Household economic impacts COVID-19

| Economic impact indicator | | Village | | Total | % |
|---|---|---|---|---|---|
| | | Muntei (n=30) | Sido Makmur (n=30) | | |
| Household member losing job | | 1 | 5 | 6 | 10 |
| Income lost due to price reduction of local product | | 19 | 12 | 31 | 52 |
| Income lost due to decline in demand of local product | | 12 | 9 | 21 | 35 |
| Goods scarcity during COVID-19 | | 22 | 12 | 34 | 57 |
| Increasing household expenses during COVID-19 | | 7 | 20 | 27 | 45 |
| Declining household spending | | 7 | 11 | 18 | 30 |
| Gaining income from commodity price during COVID-19 | | 3 | 4 | 7 | 12 |
| Economic blessing during COVID-19 | | 2 | 3 | 5 | 8 |
| Intensifying economic activity during COVID-19 | | 16 | 16 | 32 | 53 |
| COVID 19 subsidy received from government | 0 | 1 | 12 | 13 | 22 |
| | 1 | 11 | 8 | 19 | 32 |
| | 2 | 16 | 7 | 23 | 38 |
| | 3 | 2 | 3 | 5 | 8 |
| Total | | 30 | 30 | 60 | |

### 3.3 Adaptive response

In response to these stresses, households intensified the utilization of natural resources for their livelihoods. Even outside the pandemic, Indonesia is characterized by occupational multiplicity, many people have a side job. So those who were technically 'working from home' during the pandemic went to their smallholdings and harvested betel nuts, coconuts and other crops. Partly this was to supplement their income and also because they wanted to avoid mixing with others.

On land, respondents intensified chicken farming, betel nut, banana, coconut and sago farming. At sea, some increased fishing while other households collected crabs and other crustaceans. A few (3%) households opted to do small-scale food processing activities such as doughnut making, rice cakes, and other cakes. Others processed agricultural products to make copra, sago flour, and processed betel nut.

**Table 4.** Production intensification during COVID-19 in Mentawai Island

| Village | Production activity intensification during COVID-19 | Number of HH Intensifying economic activity | % |
|---|---|---|---|
| Muntei (n=30) | Agricultural base | 7 | 23 |
| | Marine resources base | 6 | 20 |
| | Agricultural product processing | 1 | 3 |
| | Other | 1 | 3 |
| | Total | 15 | 50 |
| Sido Makmur (n=30) | Agricultural base | 7 | 23 |
| | Marine resources base | 2 | 7 |
| | Agricultural product processing | 3 | 10 |
| | Other | 3 | 10 |
| | Total | 1 | 3 |
| | Agricultural base | 16 | 53 |

Levels of household consumption were also affected by the pandemic. In 2012 the government began to promote rice as a staple food in the Mentawai Islands through the 'National Food Safety Improvement Program' intensified agriculture and built 600 hectares of new rice fields. Steadily people's staple food began to change from sago and tuber to rice [31]. But the rice productivity is very low, only 1 ton unhusk rice per ha. As a consequence, dependence on rice from Sumatra Island has increased (Table 5). During COVID-19, when the supply of rice was disrupted and when respondents did not have the cash to buy rice, some local people went back to the traditional staples of sago in Siberut and cassava in Sipora Island. This shows the unintended consequences of moving people away from locally produced goods and making them dependent on imports.

**Table 5.** Staple food sources in the Mentawai Islands.

| Village | Staple food | Source of staple | | Total |
|---|---|---|---|---|
| | | Purchase | Own production | |
| Muntei | Rice | 20 | 1 | 21 |
| | Sagoo | 0 | 9 | 9 |
| | Total | 20 | 10 | 30 |
| Sido Makmur | Rice | 26 | 4 | 30 |
| | Total | 26 | 4 | 30 |

When the ferry service was suspended, goods (chili, cooking oil and onions) from Padang became scarce in

Siberut Island but this was less the case in Sipora Island, the administrative capital. When ferries were suspended, the prices of these goods increased sharply. Panic buying happened when cooking oil was scarce. In a statement from State Ferry company, Mr Tohap said, "*Inter-island ferry can only operate to carry the basic needs of the community, then also allowed to carry passengers who carry out the official duties of the acceleration and handling task force COVID-19.*" [28].

### 3.4 Implication

There are significant implications of this research regarding small islands development in the areas of food security, disaster management, and small islands regional development.

Food security: This study found that during COVID-19 as rice supplies became scarce or expensive, Mentawai people from Muntei returned to sago. Although the government attempted to change the food culture of the Mentawai people in 2012 from Sago to rice by developing 600 hectares of new rice fields the COVID-19 pandemic which disrupted supply chains showed the deficiencies in this policy. Looking ahead, the Indonesian government must make parallel policies that can accommodate the needs of the community instead of just prioritizing the national development agenda [31].

Natural resources-based economy: The Mentawai Islands are blessed with rich natural resources but the people are still poor and the region remains in the group of deprived/isolated 3T Islands. Development policies have increased value added but reduced self-sufficiency by encouraging dependence on imports. During the COVID-19 pandemic, Mentawai Islanders complained about the high prices/scarcity of cooking oil and sugar. Both these commodities can be produced from coconuts. A policy that encourages communities in small remote islands to maximize their potential may be more effective in the long term than creating dependency on supply chains.

## 4 Conclusion

The main agenda when COVID-19 started was to reduce the virus spread. Since the virus spread through interpersonal contact, reducing people's movement was a key pillar in the Indonesian government's response to the pandemic. This policy was implemented by suspending transportation services and limiting people's mobility. In the Mentawai Islands, the consequence of this policy reduced income for people working in both the formal and informal sectors. It also reduced purchasing power and level of consumption. However, people still retained access to natural resources and compensated by increasingly using these to meet their basic needs.

The COVID-19 pandemic was an unforeseen shock for the world. Many nations, dependent on global supply chains and an interconnected world, had to rapidly make decisions balancing people's health needs with protecting the economy. Small islands were no exception. Restricting people's movement to and from the Mentawai Islands reduced infections in the short term but also hurt people's livelihoods. However, the COVID-19 pandemic also identified other lessons learned both for natural resource-dependent communities and for government policy. Firstly, that there is a wealth of occupational multiplicity in livelihoods in many traditional cultures. People are adept at shifting from one crop to another, or from fishing to farming, depending on circumstances. Secondly, economic specialization or becoming dependent on outside imports can be dangerous in such a context. The Mentawai Islands are prone to natural disasters and ensuring food security by continuing production of local staples is an important part of disaster mitigation. Thirdly, some goods imported from outside such as sugar and cooking oil should be made using local raw materials. The Mentawai Islands are rich in coconuts and coconuts are a raw material for sugar and cooking oil but local people tend to export raw coconuts rather than process them locally. Fourthly, there is a worse spectre on the horizon for small islands than COVID-19, namely, climate change. Although we may not see another pandemic in our lifetimes, we will most certainly continue to see an increasing intensity of storms, flooding events, saltwater intrusion and coastal erosion. These will also negatively impinge on coastal livelihoods and threaten the very existence of small island communities. Strengthening the ability of communities to adapt to climate change does not seem as urgent as responding to the COVID-19 pandemic yet is more crucial than ever.

The research was funded by Graduate School of Universitas Andalas, Fiscal Year 2022.

## References

1. M. Mofijur et al., Impact of COVID-19 on the social, economic, environmental and energy domains: Lessons learnt from a global pandemic, Sustainable Production and Consumption, Elsevier. **26**, 343–359 (2021). https://doi.org/10.1016/j.spc.2020.10.016.
2. B. Nazario, What Is COVID-19? (2021). Accessed: Apr. 28, 2022. [Online]. Available: https://www.webmd.com/lung/coronavirus
3. World Health Organization, WHO Coronavirus Disease (COVID-19) Dashboard with Vaccination Data | WHO Coronavirus (COVID-19) Dashboard with Vaccination Data, World Health Organization. Accessed: Jun. 15, 2022. [Online]. Available: https://covid19.who.int/
4. F. Davila et al., COVID-19 and food systems in Pacific Island Countries, Papua New Guinea, and Timor-Leste: Opportunities for actions towards the sustainable development goals, Agric. Syst. **191**, (2021) https://doi.org/10.1016/j.agsy.2021.103137.
5. J. Barlow and I. Vodenska, Socio-Economic Impact of COVID-19 Pandemic, SSRN Electron. J. (2020), https://doi.org/10.2139/ssrn.3720785.
6. UNDP, Brief-COVID-19-final-June 2020, 2020. Accessed: Apr. 28, 2022. [Online]. Available: https://www.undp.org/coronavirus/socio-economic-impact-covid-19

7. G. Rasul et al., Socio-Economic Implications of COVID-19 Pandemic in South Asia: Emerging Risks and Growing Challenges, Front. Sociol. **6**, 1–14 (2021) https://doi.org/10.3389/fsoc.2021.629693.
8. M. Nicola et al., The socio-economic implications of the coronavirus pandemic (COVID-19): A review, Int. J. Surg. **78**, 185–193 (2020), https://doi.org/10.1016/j.ijsu.2020.04.018.
9. A. Saldinger, Socio-economic challenges must be addressed in COVID-19 response, UNDP chief says | Devex, Inside DevelopmentCOVID-19. Accessed: Jun. 15, 2022. [Online]. Available: https://www.devex.com/news/socio-economic-challenges-must-be-addressed-in-covid-19-response-undp-chief-says-96954
10. S. Andréfouët, M. Paul, and A. R. Farhan, Indonesia's 13558 islands: A new census from space and a first step towards a One Map for Small Islands Policy, Mar. Policy, **135**, (2022), https://doi.org/10.1016/j.marpol.2021.104848.
11. S. I. Redjo and H. As'ari, Pengelolaan Pemerintah dalam Pengaturan Pulau-Pulau Kecil Terluar Indonesia, J. Agreg. Aksi Reformasi Gov. dalam Demokr. **5**(2), 174–201 (2017) https://doi.org/10.34010/agregasi.v5i2.445.
12. R. Fernandes and P. Pinho, The distinctive nature of spatial development on small islands, Prog. Plann. **112**. 1–18 (2017) https://doi.org/10.1016/j.progress.2015.08.001.
13. H. Santono, S. Maulida, and W. Susilo, Tantangan Nyata Berkelanjutan (The Real Sustainable Challenge, 2020).
14. D. Puspitawati, Pembangunan wilayah kepulauan berlandaskan poros maritim dalam perspektif negara kepulauan: tantangan dan peluang perimbangan keuangan daerah, Development of Island Regions Based on Maritime Fulcrum Within the Perspective of Archipelagic State: Challenges, Bina Huk. Lingkung. **4**(2), 251–263 (2020).
15. W. L. Filho, J. M. Lütz, D. N. Sattler, and P. D. Nunn, Coronavirus: COVID-19 transmission in Pacific small island developing states, Int. J. Environ. Res. Public Health, **17**(15), 1–8, (2020), https://doi.org/10.3390/IJERPH17155409.
16. E. Northrop, M. Konar, N. Frost, and E. Hollaway, A Sustainable and Equitable Blue Recovery to the COVID-19 Crisis Secretariat of the High-Level Panel for a Sustainable Ocean Economy, World Resources Institute, (2020), [Online]. Available: http://www.oceanpanel.org/
17. L. Adrianto et al., Assessing social-ecological system carrying capacity for urban small island tourism: the case of Tidung Islands, Jakarta Capital Province, Indonesia, Ocean Coast. Manag. **212** (2021) https://doi.org/10.1016/j.ocecoaman.2021.105844.
18. J. P. Talubo, S. Morse, and D. Saroj, Whose resilience matters? A socio-ecological systems approach to defining and assessing disaster resilience for small islands, Environ. Challenges, **7**, (2022) https://doi.org/10.1016/j.envc.2022.100511.
19. G. Grilli, E. Tyllianakis, T. Luisetti, S. Ferrini, and R. K. Turner, Prospective tourist preferences for sustainable tourism development in Small Island Developing States, Tour. Manag. **82**, (2021). https://doi.org/10.1016/j.tourman.2020.104178.
20. V. Ukhwatul and K. Masbiran, Dampak Pandemi Covid-19 Terhadap Pariwisata Sumatera Barat (Impact of The Covid-19 Pandemic on West Sumatera Tourism), J. Pembang. Nagari, **5**(2), 148–164, (2020)
21. UNICEF, UNDP, Prospera, and SMERU, Analysis of the Social and Economic Impacts of COVID-19 on Households and Strategic Policy Recommendations for Indonesia, United Nations Child. Fund, 1–65 (2021), [Online]. Available: https://www.unicef.org/indonesia/media/9501/file/Analysis of the Social and Economic Impacts of COVID-19 on Household and Strategic Policy Recommendations for Indonesia.pdf
22. R. Nasrudin, B. P. Resosudarmo, S. Yamazaki, and W. Girsang, Contribution of cash transfers in moderating household food insecurity in small-island communities: Experimental evidence from Indonesia, Mar. Policy, **118**, (2020) https://doi.org/10.1016/j.marpol.2020.104025.
23. H. Eriksson et al., Changes and adaptations in village food systems in the Solomon Islands: A rapid appraisal during the early stages of the COVID-19 pandemic, Progr. Rep. **22**. 2522 (2020).
24. R. K. Lagi, COVID-19–resilient education in the islands, High. Educ. Res. Dev. **39**(7), 1367–1371 (2020) https://doi.org/10.1080/07294360.2020.1825347.
25. BPS Mentawai Island, Badan Pusat Statistik Kabupaten Kepulauan Mentawai (BPS) – Statistics of Kepulauan Mentawai Regency 2021, Tua Pejat Mentawai, (2022). [Online]. Available: https://mentawaikab.bps.go.id/publication/2021/02/26/35727019bf1b029c95d3ab48/kabupaten-kepulauan-mentawai-dalam-angka-2021.html
26. S. Devi, Adaptasi Masyarakat di Tua Pejat Kabupaten Kepulauan Mentawai Propinsi Sumatra Barat (Community Adaptation In Tuapejat , Mentawai , West Sumatera), J. Penelit. Sej. dan Budaya, **3**(2), 871–893 (2017).
27. Y. K. Djamba and W. L. Neuman, Social Research Methods: Qualitative and Quantitative Approaches, **30**(3). (2002). https://doi.org/10.2307/3211488.
28. N. Harlina, Kapal Antarpulau di Mentawai Berhenti Operasi Menyusul Lonjakan Covid-19 - Regional Liputan6.com, Liputan 6 SCTV. (Accessed: Jun. 13, 2022). [Online]. Available: https://www.liputan6.com/regional/read/4351259/kapal-antarpulau-di-mentawai-berhenti-operasi-menyusul-lonjakan-covid-19
29. G. M. Saleleubaja, Cegah Corona, Semua Objek Wisata di Mentawai Ditutup kumparan.com,

Mentawai Kita. (Accessed: Jun. 13, 2022). [Online]. Available: https://kumparan.com/langkanid/cegah-corona-semua-objek-wisata-di-mentawai-ditutup-1t5D44r5ahx

30. R. Langgam, 16 Ribu Warga Mentawai Kurang Mampu Dapat Bantuan BLT. (Accessed: Jun. 12, 2022). [Online]. Available: http://mentawaikab.go.id/berita/detail/16-ribu-warga-mentawai-kurang-mampu-dapat-bantuan-blt-

31. L. Pradipta, Peralihan Pangan Pokok Dari Sagu ke Beras: Sebuah Kajian Ketahanan Pangan dan Masyarakat Adat, Society, 7(1), 39–51 (2019).

# Minangkabau local wisdom: the philosophy of *aua jo tabiang sanda basanda* in disaster mitigation along the watershed (DAS)

*Ardi* Abbas[1*], *Tesri* Maideliza[2], *Alfitri*[1], and *Fendi* Agus Syahputra[1]

[1]Department of Sociology, Faculty of Social and Political Sciences, Universitas Andalas, Padang, Indonesia
[2]Department of Biology, Faculty of Mathematic and Natural Sciences, Universitas Andalas, Padang, Indonesia

**Abstract.** The number of events and losses due to disasters along the West Sumatra Watershed in 2022 is large. Damage to roads, bridges and other buildings has caused many material and non-material losses. If material losses can be calculated clearly, carefully, the components of the loss and set the numbers. Non-material losses tend to be greater because many aspects of life are interrelated. Data on the number of rivers in regencies and cities in West Sumatra are also numerous and further increase the potential for disaster damage. Meanwhile, Minangkabau local wisdom, known as the philosophy of *aua jo tabiang sanda basanda* or translated as bamboo and river cliffs lean on each other, complement and strengthen from river water scouring. This local wisdom has been understood, applied and proven over a long period of time. This paper would like to discuss this Minangkabau local wisdom as a scour barrier along the watershed. In many cases it has been shown that bamboo piles are more economical, compared to concrete and modular piles. In addition to being more economical, they also provide additional income for the surrounding community. Isn't prevention better than repair?

## 1 Introduction

West Sumatra Province has a high potential for disasters such as coastal abrasion, strong winds, floods, flash floods, earthquakes, forest and land fires (*Kahutla*) and landslides. Data from the BPBD (Regional Disaster Management Agency) of West Sumatra during the period 2019-2023 shows that there have been 3,770 disasters, with material losses estimated at Rp 1,590,645,949,619 (One Billion Five Hundred Ninety Million Six Hundred Forty-Five Thousand Nine Hundred Forty-Nine Rupiah Six Hundred Nineteen Cents). This is a large material loss that should be minimized in number and incidence [1]. The disaster occurred in all districts/cities in West Sumatra. In addition to the material loss, there was also a non-material loss felt by the community due to death, loss, displacement, injury/sickness, as summarized in the following table:

**Table 1.** Recapitulation of disasters in West Sumatra Province in 2019-2023

| Disaster Type & Loss | Year | | | | | Total |
|---|---|---|---|---|---|---|
| **A. Disaster Type:** | **2019** | **2020** | **2021** | **2022** | **2023** | |
| 1. Coastal Abration | 1 | 5 | 4 | 5 | 8 | 23 |
| 2. Strong Winds | 75 | 375 | 658 | 674 | 562 | 2,344 |
| 3. Flood | 33 | 84 | 113 | 123 | 144 | 497 |
| 4. Flash Flood | 6 | 17 | 7 | 5 | 10 | 45 |
| 5. Destructive Earth quake | 1 | 7 | - | 2 | 1 | 11 |
| 6. Forest Fire (*Kahutla*) | 54 | 15 | 36 | 92 | 76 | 272 |
| 7. Land slide | 27 | 146 | 10 | 120 | 166 | 469 |
| Disaster Total | **208** | **649** | **925** | **1021** | **967** | **3770** |
| **B. Loss** | | | | | | |
| 1. **Material** (Rp) | 171,017, 356.156 | 66,655,99 4,000 | 170,835,6 55,000 | 1,136,849, 586.796 | 45,287,35 7.667 | **1,590,645, 949.619** |
| 2. **Non-Material:** | | | | | | |
| a) Died | 7 | 7 | 21 | 28 | 30 | **93** |
| b) Missing | | 2 | | 4 | - | **6** |
| c) Inquired/Sick | 79 | 11 | 651 | 456 | 3,188 | **4,385** |
| d) Displaced | 3,066 | 4,674 | 476 | 26,265 | 5,480 | **39,961** |

* Corresponding author: ardiabbasunand@gmail.com

Table 1, above shows that seven types of disasters have resulted in material and non-material losses that are not small. Non-material losses that resulted in the death of 93 people, and four people missing cannot be compared with material losses of any size [2]. Not to mention if you count other non- material losses, which are much greater in number and quality than material losses, including other non-material losses as will be explained in the next section.

## 2 Results

Disaster mitigation is the process of managing natural disasters to reduce and minimize the impact on people and property. Mitigation includes a series of efforts to reduce or eliminate the negative impacts of disasters, before, during, and after a disaster occurs. The ultimate goal is to protect people, reduce property losses, and minimize community disruption. Disaster mitigation is an important part of disaster risk management that aims to create communities that are more resilient and prepared for different types of disasters. There are four risk reduction strategies: risk avoidance, risk reduction, risk transfer, and risk storage. Risk can be avoided by turning the hazards that can arise into something useful. Risk transfer is done by addressing vulnerabilities to damage and disruption. Risk transfer is done by ensuring the alteration of hazards that have a financial effect on individuals and communities. Meanwhile, risk storage is an effort to prepare finances to compensate for losses due to hazards that will occur. Mitigation can be divided into structural and nonstructural mitigation. Structural mitigation is an effort to avoid losses due to hazards that arise by doing physical development, with infrastructure construction techniques that have durability in providing protection against hazards. Nonstructural mitigation is an effort to avoid losses due to hazards that arise by empowering the community. Empowerment is done by making policies, increasing public awareness of hazards, and developing knowledge of hazards. In addition, community empowerment is also carried out by using community participation methods in the prevention of damage, disturbance and hazards. Unlike disasters such as earthquakes and tsunamis, the potential for floods and landslides along watersheds on roads, bridges and other buildings can be anticipated early by planning a number of programs and activities. This anticipation is important to reduce the risk of damage and maintain the safety of road users (Table 2).

**Table 2.** Damage to roads, bridges, and other buildings due to floods and landslides in 2019-2023 at West Sumatra

| Damage | Year | | | | | |
|---|---|---|---|---|---|---|
| | **2019** | **2020** | **2021** | **2022** | **2023** | **Total** |
| School (unit) | 14 | 25 | 36 | 5 | 35 | **115** |
| Houses of Worship (unit) | 19 | 33 | 25 | 4 | 79 | **160** |
| Health Facilities (unit) | 2 | 7 | 3 | 1 | 2 | **15** |
| Offices (unit) | 1 | 4 | 4 | | 10 | **19** |
| Bridges (unit) | 23 | 23 | 21 | 19 | 30 | **116** |
| Road (km) | 9 | 24 | 2 | 24 | 1 | **60** |
| Rice Field (ha) | 543 | 913 | 1042 | 18 | 17,894 | **20,410** |
| Garden/Forest (ha) | 52 | 184 | 3,132 | 9 | 4,963 | **8,340** |

The damage varies in level from minor to severe damage resulting in:

1. Erosion, which causes structural damage, i.e. damage to key structural elements, which reduces their ability to withstand loads and thus threatens the safety of users.
2. Potholes and Cracks, due to water flowing under the surface making it unsafe to pass.
3. Weakening of foundations due to river water erosion, reducing stability and increasing the risk of collapse.
4. Drainage damage, due to blocked water flow or damaged structures can worsen conditions and cause greater waterlogging.
5. Landslides, which can block roads and threaten the safety of users.
6. Erosion of riverbanks, which has the potential for roads, bridges and other structures to collapse into the river.

Material losses due to damage to roads, bridges and other buildings have been obtained in nominal terms as shown in the table above. However, non- material losses, if calculated more carefully, tend to be greater because many aspects of life are interrelated, but not immediately visible [3]. Non-material losses due to damaged roads, bridges and other buildings can have a wide impact on various aspects of people's lives. Several aspects of non- material losses were identified, namely:

1. Disrupted mobility of people and goods, i.e. difficulty and limited access to health facilities, education, employment and markets and others. In addition, travel time becomes longer because of the need to find alternative roads.
2. Social impact, isolation between locations because they are separated by broken roads or bridges, potential for stress and anxiety, in the form of uncertainty of access and travel.
3. Disruption to business and reduced revenue due to a decrease in customers reaching their shops/business premises as well as a reduction in economic activity, due to difficulties in the distribution of goods and services.
4. Psychological impact, uncertainty about when the road or bridge will be repaired and may cause concern among the affected population. In addition, events during and after the incident can cause trauma.

5. Environmental degradation, using alternative routes that are not designed for high traffic volumes can cause environmental damage, such as soil erosion.

Understanding these non-material losses is important to discuss in order to deal with emergency situations caused by failing roads and bridges. This is an important point of thought that can also help in planning for disaster risk mitigation in the development of more disaster-resilient infrastructure in the future [4] The next important point is that non-material losses last for an indeterminate period of time, which can be overnight, a week, a month or even two or three months depending on the severity of the damage and the timing of the repair work. So far, there is no official record that can be used as a reference to calculate these non-material losses. It is as if non- material losses due to damage to roads, bridges and other buildings have received less attention, calculation and budget from the relevant parties.

Potential losses due to floods and landslides along the rivers in West Sumatra are reinforced by the large number of rivers in all districts/cities. Based on data from the Department of Water Resources and Bina Konstruksi, the West Sumatra region drained by this large river can be divided into eight SWS (River Area Unit), namely:

1. SWS Natal-Batahan (380 km$^2$)
2. SWS Rokan (2,190 km$^2$)
3. SWS Kampar (2,591 sq. km)
4. SWS Inderagiri-Akuaman (10,545 km$^2$)
5. SWS Batang Hari (8,264 km$^2$)
6. SWS Masang-Pasaman (6,313 km$^2$)
7. SWS Tarusan-Silaut (6,280 km$^2$)
8. SWS Siberut-Pagai-Sipora (7,338 km$^2$)

The potential for floods and landslides along the rivers in West Sumatra that can damage roads, bridges and other buildings can be explained from various aspects, namely:

1. Geomorphology, explaining the number of rivers and water flow. The more rivers a regionhas, the more waterways there are so that rainwater flows faster into the river, and can reduce the potential for flooding if the river is able to accommodate it. However, if there is notenough capacity in the river, especially during the rainy season, the risk of flooding increases. Smaller rivers that drain into larger rivers can also potentially cause a buildup of water flow that eventually overflows and causes flooding.
2. Topography, in effect, in areas with many rivers, the topography varies from lowlands to mountains. Mountain rivers are fast-flowing and have the potential to cause erosion and landslides. Whereas in low-lying areas, rivers cause flooding if natural or artificial drainage is unable to drain water effectively.
3. Spatial Planning, good watershed management,can reduce the risk of floods and landslides such as reforestation, building embankments, and spatial planning that does not narrow the flow of the river. Poorly managed rivers tend to have a higher risk of floods and landslides. Deforestation in the upper reaches of rivers can reduce the soil's ability to absorb water, which increases surface flow and the risk of floods andlandslides.

## 3 Discussion

Bamboo has several species and Latin names suchas:

1. *Bambusa vulgaris*, the single most commontype of bamboo and used in construction.
2. *Phyllostachys edulis*, or Moso bamboo, is usedin the food and textile industries.
3. *Dendrocalamus asper*, betung bamboo, is usedin furniture making and building.
4. *Bambusa blumeana*, or thorn bamboo, is widely grown in Southeast Asia.
5. *Gigantochloa apus*, or wormy bamboo, is usedin handicrafts and furniture.
6. *Phyllostachys aurea*, yellow bamboo or golden bamboo, is often used in landscaping.
7. *Fargesia murielae*, umbrella bamboo, as an ornamental plant due to its attractive shape.

Each bamboo species has unique characteristics that make it suitable for various uses and environments. Furthermore, it is understood that the species *Bambusa vulgaris, Phyllostachys edulis*, and *Dendrocalamus asper*, are more appropriate as scour retaining walls along watersheds [5]. For the people of West Sumatra, bamboo has many names such as *buluah, batuang*, and *aua*. Bamboo that is suitable for retaining river water scour is the *Aua* type, as shown in Fig. 1.

**Fig. 1.** *Aua* retaining scour along the watershed

Bamboo is a plant that comes from the kingdom plantae with order *poales poaceae* family. Bamboo has a rapid growth rate with the influence of the soil conditions where it grows, as well as the climate. Some types of bamboo can grow to a height exceeding 30 M with a stem circumference of 30 cm, and bamboo can grow 3 to 12 cm per day. Bamboo develops by means of its roots and rhizomes below the soil surface. Its easy spread means that young bamboo shoots can appear in many places, which makes bamboo a potential invasive plant if its spread is not controlled [6].

The benefits of bamboo can be seen from various aspects, such as related to water, air, land and various other benefits. Bamboo helps to improve water catchment sources or helps the availability of water content in the soil. Bamboo forests have a greater ability to absorb water when compared to other types of plants. When compared to pine forests, bamboo forests are far superior, absorbing up to 240%. If other trees are only able to absorb 35%-40% of rainwater, while bamboo absorbs up to 90%. There are those who can use bamboo as reforestation in former coal mines that can help provide a groundwater level of 6.3 m within 4 years.

Bamboo is also able to help withstand and even reduce noise and air pollution. Regarding noise pollution, bamboo can withstand noise. With regard to air pollution, bamboo has the potential to absorbup to 62 tons/Ha/year of $CO^2$. That amount is much greater when compared to other types of plants thatare only able to absorb 15 tons/Ha/year. Bamboo can also release the results of the photosynthesis process in the form of oxygen up to 35% greater than the release of oxygen from other trees.

Bamboo roots also have benefits for the soil in which it grows. The bamboo root system itself can strengthen the soil layer it grows on. So, planting bamboo is perfect for preventing erosion and landslides. Bamboo is an easy alternative choice to prevent these two things from happening. Besides strengthening the soil layer, bamboo roots can alsohelp maintain the quality of the soil it grows on. Inaddition, the root system can prevent soil destruction by retaining and filtering various pollutants that could potentially damage the soil. The roots can also at the same time trigger nutrientsneeded by the soil. All of these are related to the benefits of bamboo with nature directly and with humans indirectly. Humans can also enjoy the direct benefits of bamboo, which can be used to make various crafts of high economic value. Such as bamboo watches, bamboo bicycles, bamboo walls, etc.

Bamboo roots also have benefits for the soil they grow in. The bamboo root system itself can strengthen the soil layer it grows on. So, planting bamboo is perfect for preventing erosion and landslides. Bamboo is an easy alternative choice toprevent both from happening. Besidesstrengthening the soil layer, bamboo roots can alsohelp maintain the quality of the soil it grows on. In addition, the root system can prevent soil destruction by retaining and filtering variouspollutants that could potentially damage the soil. The roots can also at the same time trigger nutrientsneeded by the soil. All of these are related to the benefits of bamboo with nature directly and with humans indirectly. Humans can also enjoy the direct benefits of bamboo, which can be used to make various crafts of high economic value. Suchas bamboo watches, bamboo bicycles, bamboo walls, bamboo chairs, and various other crafts made from bamboo. Bamboo shoots known as rebung (Minangkabau *rabuang*) can also be processed into nutritious and delicious dishes.

Bamboo plants have various advantages when planted on the riverbank, namely:

1. Prevents erosion, as it has strong, dense and extensive roots that can help stabilize the soil along the riverbank, reducing the risk of soil erosion due to heavy water throughout the watershed. There will also be increased soilstability as the roots are able to bind soil particles, making it more difficult for them to break away and be carried away by water.
2. Water Conservation, reduced velocity of water flow as the roots, stems and leaves can slow down the flow of surface water, allowing more water to seep into the soil and reducing the riskof flooding and soil erosion. There is also an increase in groundwater storage as the roots of bamboo can help to increase the infiltration of water into the soil, which helps to maintain soil moisture around the river.
3. Natural Purification of Water, as bamboo roots act as a natural filter that helps filter out sediments and water pollutants, improving the quality of river water.
4. Flora and Fauna Habitat and Biodiversity, as bamboo forests provide an important habitat for a variety of plant and animal species, increasing biodiversity in the ecosystem along the river. Bamboo can also grow faster and provide effective ground cover, which can help restore vegetation in degraded areas.
5. Economic Benefits as a source of raw materials,as it can be harvested for a variety of purposes,including building materials, handicrafts, and food.
6. Ecological Benefits Bamboo has a high ability to absorb carbon dioxide, helping to reduce carbon emissions and mitigate climate change. It can also improve air quality, as bamboo releases large amounts of oxygen, improving the air quality around the river.
7. Aesthetics and Recreation, with bamboo plantswill add to the landscape, and visual beauty along the river, making it a more attractive place for recreation and tourism. Bamboo areas alongthe river can serve as recreational areas, walking paths, and green spaces.

Riparian bamboo plants bring a variety of ecological, economic and social benefits. It not only helps to protect and restore the river ecosystem but also provides a sustainable resource and natural beauty. Bamboo is therefore an excellent choice for river rehabilitation and conservation projects [4].

In civil engineering, there are turaps, which are civil engineering structures that retain soil to prevent landslides or erosion; built on hillsides, cliffs, highways, and other areas prone to movement or landslides. Turaps can be made of reinforced concrete, making them stronger and more durable. Or they can be made from piles of stones, or from special wooden structures for small and temporary projects. Wooden tiers are not as strong as concrete and wood, but are cheaper and quicker to install. Sandbag turrets can also be used for temporary or emergency solutions. There are also modular turrets, made up of modular blocks that can be installed quickly and easily, for projects that require flexibility and speed of installation as shown in Figs. 2-4.

**Fig. 2.** Concrete turret

**Fig. 3.** Modular turret

**Fig. 4.** Concrete turret not working

Tarps are very important because they not only play a role in maintaining soil stability, but also in protecting infrastructure and the surrounding environment. After searching the literature on turrets, almost no bamboo turrets were found. This is somewhat surprising because why would this be the case? No academically and methodologically satisfactory answer has yet been found. In fact, there have been various literature studies on the biological, and other disciplinary benefits of bamboo as a flood and landslide barrier.

Fortunately, there is Minangkabau local wisdom with the philosophy of *aua jo tabiang sanda basanda*. Literally, this local wisdom can be translated as bamboo and cliffs leaning on each other, symbiotically strengthening each other, and complementing each other from river water scouring/erosion. Cliffs are better understood as river cliffs not riverbanks. There is no historical evidence that explains the year when this philosophy began to be used as a local wisdom. Local wisdom is a collection of knowledge developed in a society that is passed down from generation to generation on many aspects. This knowledge covers various aspects of life, such as farming methods, traditional medicine techniques, customs, and daily habits that are considered good and right by the community. Local wisdom reflects a community's deep understanding of their natural, social and cultural environment [7].

As local wisdom, the philosophy of *aua jo tabiang sanda basanda* includes knowledge about sustainability and environmental preservation, cultural identity and communal pride that can maintain Minangkabau cultural heritage. It is also a form of adaptation and innovation, which although based on old traditions, local wisdom is not static and is able to adapt to environmental and social changes, and is open to innovation so that people can combine local knowledge with modern technology to create relevant and effective solutions. In the case of Minangkabau local wisdom, this philosophy has been proven, in terms of bamboo biology, civil engineering and also economics and other aspects. The cost of planting bamboo as flood and landslide retaining walls along a watershed is also cheaper, and the growth period is faster, among other considerations. The question is why hasn't the *aua jo tabiang sanda basanda* philosophy been applied? Are there biological, economic and other considerations that prevent the utilization of bamboo as a barrier to water scouring, flooding and landslides along the watershed?

Isn't the principle of "prevention is better than repair" still relevant? This principle essentially emphasizes the importance of preventive measures to avoid problems or damage rather than having to deal with the impact after it occurs. It applies in a variety of contexts, including the environment, infrastructure management and more. In terms of the environment, it is understood that using natural resources wisely and safeguarding ecosystems can prevent environmental damage that takes time and money to restore. Infrastructure management includes routine maintenance work, and longer design. Performing regular maintenance on roads, bridges and buildings can prevent major breakdowns that require expensive repairs and cause significant disruption. Designing infrastructure with risks such as flooding, erosion and earthquakes in mind can prevent major damage. The principle of "prevention is better than repair" is useful because it can reduce costs more efficiently, because prevention is cheaper than repair. Isn't it true that avoiding problems means avoiding costs too? Preventive measures can reduce or eliminate the negative impact of a problem, and can lead to a better quality of life. Preventing the scouring of water along a river (watershed) that causes roads, bridges and other buildings to be damaged, or cannot be utilized is cheaper than dealing with these problems after they occur. It is also understood that prevention efforts can save time that could be used to address other problems. Resources and manpower can be more efficiently allocated to other, more productive activities. Overall, the principle of "prevention is better than repair" encourages thinking ahead, identifying potential problems before they become major, and implementing appropriate preventive measures to reduce the negative impacts that could occur.

## 4 Conclusion

Based on the above explanation, it can be concluded that the following are:

1. Bamboo is very suitable to be used as a retaining wall for river water scouring, because it has advantages in terms of biological, civil and economic engineering and other aspects.
2. The planting period of bamboo is also very fast, which further strengthens the argument that bamboo is very suitable as a river water scour retaining wall.
3. As a local wisdom, planting the right type of bamboo can minimize the risk of disaster mitigation due to scouring in areas along the river basin,
4. It is better to start now to avoid greater risks in the future.

## References

1. Badan Penanggulangan Bencana Daerah (BPBD), Data dan Informasi Bencana Indonesia (DIBI) Provinsi Sumatera Barat tahun 2019-2023 (2024).
2. Ilham and Z. Naputra, Analisa manfaat pembangunan infrastruktur jembatan pasca banjir tahun 2019: studi kasus Sungai Pangkua Kabupaten Solok Selatan. Andalas Civil Engineering (ACE) Conference. **8**. (2023)
3. Kamaruddin K, *et. al* Pengaruh pembangunan infrastruktur jalan raya terhadap pertumbuhan ekonomi Kabupaten Sumbawa, Rumah Arsip. **9**(3), (2021). https://doi.org/10.58406/jeb.v9i3.511
4. Madjid, N. Analisis metode perhitungan dan alokasi anggaran bencana alam, Simposium Nasional Keuangan Negara. **1**(1) (2018). https://jurnal.bppk.kemenkeu.go.id/snkn/article/view/26.
5. Ratnaningsih, T. K., Rahayu, H. C., Lestari, E. P., Sultan, S., and Fathoni, A. Dampak sosial ekonomi bencana banjir dan pemetaannya berbasis sistem informasi geografis (SIG). Jurnal Samudra Ekonomi dan Bisnis, **14**(2), 318- 330 (2023)
6. Sesunan, D. Analisis kerugian akibat banjir di Bandar Lampung. Jurnal Teknik Sipil Universitas Lampung. (2014).
7. Program Studi Sarjana Teknik Sipil, Universitas Tarumanagara Jakarta (2014).

# Minangkabau philanthropy: the spirits and behaviours of generosity of the Nagari communities in facing natural disasters

*Nopriyasman* Nopriyasman[1*], *Gusti* Asnan[1], *Ferdinal* Ferdinal[2], *Indirawati* Zahid[3], and *Abdul* Harits Ritonga[1]

[1]Department of History, Faculty of Humanities, Universitas Andalas, Padang 25163, Indonesia
[2]Department of Literature and Culture, Faculty of Humanities, Universitas Andalas, Padang 25163, Indonesia
[3]Department of Language, Academy of Malay Studies, Universiti Malaya, Kuala Lumpur, Malaysia

**Abstract.** The world of philanthropy in traditional societies reflects communal behaviour in ensuring the social life of *adat* communities in Nagari. The historical aspects of philanthropy are crucial in revealing the values that guide social communities in facing natural disasters. This study explores the historical values of traditional societal behaviour in philanthropy, driven by Minangkabau history, cultural expressions, and the reality of social practices. The research follows historical methods and historical thinking stages. It employs a qualitative approach within a multidisciplinary framework (history, culture, and language), and quantitative approach through questionnaire. The findings show that philanthropic practices foster solidarity within Nagari communities and among their diaspora members. This bond is reflected in traditional charitable behaviours based on empowering the Nagari community's identity through historical, cultural, and religious ties to ensure survival during calamities. Local philanthropy in disaster-affected areas strengthens in terms of volunteer labour, while financial aid increases among unaffected communities beyond and diasporas. This study is valuable for maintaining social integration, solidarity, social justice, togetherness, and a shared sense within *adat* communities and for the active participation of the Minangkabau people in particular and Indonesians in general.

## 1 Introduction

Nagari communities in Minangkabau, West Sumatra, has a custom and tradition of mutual assistance in their daily lives. This is evident in both personal life practices and matters concerning public interest. Several traditions demonstrate the community's solidarity in developing their hometown and other means of fostering collective consciousness. This practice is also prevalent in other regions of Indonesia, particularly during disasters, where solidarity encompasses ideas, labour, skills, and goods [1]. These customs and traditions serve as guidelines and values to ensure social continuity, manifested in philanthropic behaviours. However, the philanthropic customs and traditions of the Nagari community in Minangkabau as a new ecosystem strength in serving the community have not been extensively studied by researchers. In fact, philanthropic activities are necessary to maintain the collective consciousness of the Indonesian nation for socio-economic development and social resilience, especially in the Nagari community of Minangkabau, West Sumatra.

The philosophy of life in the Nagari of Minangkabau begins with the principle of *alam takambang jadi guru* (Humans learn from nature), progresses to new harmony with syarak (Sharia, Islamic law) within its entire social system, and ultimately advances with a perspective that integrates Islam and the modern (Western) world simultaneously [2, 3]. This concept of harmonious living, blending tradition, religion (Islam), and secularism, emphasizes social harmony, creating a life mechanism that is materially comfortable while honouring God through hard work in the world. In socio-historical philanthropic behaviour, the movement of philanthropy persists because it is rooted in the strength of tradition and Islam [4]. Today, the traditional social harmony of the Nagari community in Minangkabau continues to evolve towards a more humanistic harmony of prosperity and well-being. Modern philanthropic behaviours emphasize activities for social justice [5, 6], which are expected to support community welfare, and these philanthropic practices are supported by the government [7].

This article focuses on exploring traditional Minangkabau philanthropy as a means of anticipation in the event that government social assistance encounters a deadlock, thereby ensuring that local charitable behaviours contribute to social resilience during emergency disaster situations, particularly natural disasters. The authors pose three questions: What based the spirit and social philanthropic behaviours of the traditional Nagari community in Minangkabau, West Sumatra; What are the forms of local wisdom-based philanthropy practiced by the Nagari community; How does the Nagari community's philanthropy respond to natural disasters?

* Corresponding author: nopriyasman@hum.unand.ac.id

## 2 Literature review

Studies related to local wisdom-based philanthropic behavior in post-disaster management have not prominently surfaced in historical accounts, especially in exploring forms of wisdom-based philanthropy relevant to natural disaster situations. Some studies depict philanthropic behavior as a form of worship aligned with one's beliefs. Disaster philanthropy based on local wisdom evolves alongside the frequent occurrence of disasters in Indonesia, such as during the pandemic [8]. The philanthropic movement is driven by collective religious consciousness, humanitarianism, and various natural events affecting humans [9-11]. Other studies highlight the role and issues of social institutions in the disaster realm. Common problems faced by Non-Governmental Organizations (NGOs) and the government in disaster response include integration issues, financial constraints, communication and coordination quality, ineffective design, construction, inaccurate assessments, resource incapacity, corruption, and transportation [7, 12, 13].

In the case of natural disasters, the activities of professional and non-professional volunteers become essential in mitigating various post-disaster risks. The decrease in philanthropic activities in disaster-stricken areas is understood as individuals and households also facing "danger," thus protecting their resources. Meanwhile, those not directly affected showed increased empathy and social solidarity towards the victims. This fosters social balance, ensuring that the lives of the disaster-affected communities are maintained to a considerable extent [14].

Community-based nonprofit organizations in Indonesia are key supports for public services in rural and remote areas [15]. In addition to formal organizations, traditional community structures such as *subak* also fall into this category. However, religious motivation is the largest driving factor for Indonesian Muslims to engage in charitable activities [16], resulting in community-based organizations being predominantly religious bodies like mosques and charity institutions. Muttaqien, Asyrof, and Syibli [17] reveal that mosque-based disaster philanthropy in Indonesia involves forming special thematic committees to collect donations from mosque congregations during every worship session, which are then directly distributed to disaster-affected areas.

Philanthropic activities create opportunities for various local, national, and international companies to promote their corporate responsibility and build community bonds worldwide [18]. Mithani [19] illustrates how multinational companies in India channel aid post-disaster. In contrast, domestic companies often rely on historical and cultural ties to understand local needs, with these aspects determining the sustainability of community values in traditional rural areas. As philanthropy becomes more prevalent today, traditional philanthropic practices must align with ethical and legal principles to avoid illegal philanthropy, such as fraud, embezzlement, and corruption [20].

In traditional disaster management systems, there exists a cultural wisdom for local disaster response, while simultaneously reconciling with the possibility of earthquake disasters [21-25]. Generally, research based on local wisdom demonstrates that past natural events and occurrences shape community behavior in facing disasters. Local wisdom-based disaster philanthropy reveals the integration of various societal elements and contributes to the effectiveness of modern government social assistance programs, thus emerging as a new force in social security. The success of modern scientific government programs requires integration with local wisdom to foster necessary innovations [5, 26]. West Sumatra have some organizations that provides alms such as BAZNAS [27] Muhammadiyah's LAZISMU [28] and Aisyiyah [29] among others.

Based on the realities of this research, this study discusses local wisdom-based philanthropy as a new ecosystem for community service relevant to natural disaster situations in Minangkabau, West Sumatra. The contribution of this research primarily lies in explaining the use of local resources for social security, environmental sustainability, local traditions and culture, and socio-economic development. For policymakers, the findings can assist in the implementation of socially just disaster relief programs, thereby supporting the effective performance of the Regional Disaster Management Agency (BPBD) and the National Disaster Management Agency (BNPB).

## 3 Methods

This study on disaster philanthropy employs historical research procedures with a multidimensional approach. The behaviours of philanthropy are observed from a theoretical perspective to explain various historical aspects or dimensions of the research object [30]. Data is collected through literature reviews as well as archival or documentary sources. The data is primarily qualitative, consisting of descriptions, statements, and interpretations of various historical legacies related to disaster philanthropy practices in Minangkabau. The selected and validated data is analyzed using historical thinking methods that consider aspects of change, causality, context, complexity, and, if necessary, historical hypotheticals of disaster philanthropy practices in Minangkabau. This historical thinking approach is essential for deriving meaning from past historical events, particularly in the context of local philanthropic movements following natural disasters in Minangkabau. The meanings derived from philanthropic behaviors are focused on the values or local wisdom that guide the community in practices of mutual assistance among humans. These aspects of local intelligence are reconstructed into a historically dimensional narrative from a multidimensional perspective.

We also conducted interviews with the victims of the 2022 Pasaman Barat earthquake in three Nagari: Kajai, Talu, and Sinuruik, to provide a field overview of the philanthropic responses of the Minangkabau community during the disaster. The field study, conducted in July 2024, also collected questionnaires from 30 respondents in these areas and several nearby regions not far from the disaster center in Pasaman Barat. This was to demonstrate that while disaster-affected communities tend to contribute less in terms of donations due to their status as recipients, their contributions of labor to mitigate the disaster's impact significantly increase. We formulated six questions categorized into two groups, with three questions each: receipt of donations (Cronbach's $\alpha$=0.780) and volunteering during disaster response (Cronbach's $\alpha$=0.746). The responses were divided into three levels (0-1: low, 1-2: medium, 2-3: high) according to each question and arranged based on a Likert scale [31]. The data is then analyzed through descriptive analysis and by correlating it with the increase of volunteering and decrease of donation in affected location.

# 4 Results and discussion

## 4.1 Philanthropy movement in Indonesian disaster history

Indonesia has been historically recorded as a region vulnerable to various natural disasters such as earthquakes, floods and flash floods, landslides, volcanic eruptions, and tsunamis. The impacts of these natural disasters have caused significant suffering among the people. As a result, philanthropic movements have emerged as a response to various human experiences across different regions of Indonesia. During the Dutch colonial era, this philanthropic movement was spearheaded by student organizations such as Perhimpoenan Indonesia (PI) in the Netherlands. While introducing Indonesian culture in the colonizing country, they actively participated in the charity committee organized a *soirée artistique* (artistic evening) to raise funds for victims of the flood disaster in early 1916 in the Dutch East Indies. Many students participated in the event, including Soerjopoetro, Soejowinoto, Jodjana, Noto Soeroto, Notosewodjo, Sosro Soewarno, Samsoed, Samsi Sastrodigdo, Sarengat, and Ambia Soedibjo. The event was led by Soerjopoetra and organized by Dahlan Abdoellah, a student from Pariaman, West Sumatra [32]. Some historians also recorded charity movements among people in colonial Minangkabau. One of the methods is holding lecture tours while raising funds. This have been performed by Daud Rasyidi, an *Ulama*, who was inspired by the destruction of his surau due to the flash flood of 1916 [33].

During the Japanese occupation, several charitable organizations were established, initiated by educators and national fighters. These charitable bodies aimed to assist victims of natural disasters in Indonesia, such as the earthquake on July 23-24, 1943, which resulted in 213 deaths and 2,096 injuries [34]. The sense of shared suffering and solidarity demonstrated by the nation was embodied in the establishment of the Barisan Puteri Amal, founded by Baginda Dahlan Abdullah, who was then serving as the Mayor of Batavia (Jakarta). The advisory board of this organization included Baginda Dahlan Abdullah, K.H. Mas Mansjur, St. Danil, Syamsu Arifin, J. Salim, Nyonya Siti Akmar Dahlan Abdullah, Nyonya Dr. Soetarman, Nyonya Saadah Alim, and Nona Kayatun [35].

The first agency responsible for disaster in Indonesia, Badan Penolong Kecelakaan (BPK) was established on December 23, 1943, with Dahlan Abdullah as chairman, funded by the Japanese occupation government in Jakarta [34]. On August 20, 1945, the government formed the Badan Penolong Keluarga Korban Perang (BPKKP), but officially handled national disasters in 1966. Its name has frequently changed: Tim Koordinasi Nasional Penanggulangan Bencana Alam (TKP2BA) from 1967-1979, Badan Koordinasi Nasional Penanggulangan Bencana Alam (Bakornas BPA) from 1979-1990, the Badan Koordinasi Nasional Penanggulangan Bencana (Bakornas PB) from 1990-2000, Badan Koordinasi Nasional Penanggulangan Bencana dan Penanganan Pengungsi (Bakornas PBP) from 2000-2005, and back to Bakornas PB until 2008. Badan Nasional Penanggulangan Bencana (BNPB) succeeded the agency, eventually being elevated to a ministerial-level institution in 2019 [36].

Currently, various individuals and communities, including students, non-governmental organizations such as the Wahana Lingkungan Hidup Indonesia (Walhi), Masyarakat Penanggulangan Bencana Indonesia (MPBI), and social and religious organizations such as Muhammadiyah, Nahdlatul Ulama (NU), and Dompet Dhuafa Foundation, among others, have taken on significant social responsibilities. This demonstrates that the potential for philanthropy is indeed an innate human instinct, aligning with all religions in Indonesia and having roots in local culture.

## 4.2 The spirit of philanthropy from indigenous knowledge

The Minangkabau societies generally practice local wisdom imbued with the unique customs and culture of each Nagari community. According to the concept of the Alam Minangkabau region and traditional historiography (tambo), the Minangkabau region comprises three core traditional areas known as *darek*, also referred to as *Luhak nan Tigo* (Tanah Datar, Agam, and Limapuluh Kota). Additionally, the *rantau* region includes the eastern and western coastal areas [37, 38]. Minangkabau philanthropy is verbally depicted through customary expressions. Its roots lie in human decorum, community etiquette, and familial bonds, characterized by mutual care among relatives and the wider community, respect for elders, and affection for the young [39]. Based on these principles, maintaining social harmony becomes a custom, particularly when the community faces adversity, encapsulated in the saying *kaba baik bahimbauan, kaba buruk bahambauan* (good news to be spread [so people come to congratulate], bad news to be announced [so people come to give condolences]). This custom applies not only in the homeland but also in the diaspora.

The Minangkabau worldview is encapsulated in a proverb, *alam takambang jadi guru* (the growing nature becomes teacher). This means that adherents or supporters of Minangkabau culture derive wisdom from the characteristics and life forms of nature. In this worldview, every element of nature is believed to be interrelated, opposing yet grouping together, without binding, eliminating, or merging into one another [2]. Therefore, harmony in life is achieved through a balanced relationship where each element maintains its independence. This principle of life is reflected in the cultural system and the behaviors of Minangkabau cultural practitioners. Despite historical influences from the Pagaruyung cultural experiments and Islamic practices, these have not complicated the Minangkabau Nagari cultural perspective, but rather have integrated into the traditional structure [2].

There are many expressions put forth by Nagari communities based on their indigenous history and knowledge history. Several philanthropic expressions relate to social work aimed at maintaining a just society within their village, such as [40]:

*(1)* *Nan barek samo dipikua,*
*nan ringan samo dijinjiang*
(The heavy ones are bore together,
the light ones are carried together);

*(2)* *Nan condong ditungkek,*
*nan lamah ditueh*;
(The slanted ones are supported,
the weak ones are levered);

*(3)* *Nan senteng samo dibilai,*
*nan kurang samo ditukuk*
(The short ones are joint,
the missing ones are added).

These expressions serve as advice to encourage and foster a habit of helping and supporting fellow community members. For economically established Nagari residents in Minangkabau, there is a "calling" to assist the poor or provide aid to those in need. The sense of duty among the Nagari people is not confined to a specific location; historically, there have been numerous instances of social participation by Nagari members in the diaspora aimed at improving and supporting the well-being of their home villages.

Expressions of communal work (*gotong royong*), are reflected in expressions like:

*(4)* *Nan ruyuang samo ditarah,*
*nan sagu samo dibagi*
(Sago palms are slashed together,
its starch is shared together)

The expressed *gotong royong* of the community demonstrates intense solidarity and participation in various social issues faced by the community in a Nagari. The fundamental concept of behavior in the communal life of a Nagari relies on the principle of *saanak, sakamanakan, urang kampuang dipatenggangkan* (One child, one nephew, all villagers are thought to) [40, 41]. These customary expressions are continuously passed down from generation to generation and become guiding values for communal and Nagari life, as well as other social and humanitarian relationships.

The practice of *gotong royong* alleviates the burdens of those affected by calamities. The unity among the community is reflected in the following proverb: *Ka bukik samo mandaki, ka lurah samo manurun, kok tatungkuik samo makan tanah, kok tacilantang samo ba ambun*" (climbing the hill together, descending the valley together, lying face down our mouth would be filled with soil together, if lying face filled with dew together). This means that if something happens in the village or to the members in the diaspora, a collective solution is sought.

Islam is one of the foundations of Minangkabau tradition, as in a proverb that frequently mentioned: *adaik basandi syarak, syarak basandi kitabullah* (tradition rests on sharia, sharia rests on Quran). Consequently, verses of Quran and hadith narrations become foundations of philanthropy in Minangkabau culture. These are taught in some Friday sermon (*khutbah*), many lectures in mosques and suraus, as well as schools and madrasas. Some socio-religious organizations, such as Muhammadiyah, have their own charity management (Amal Usaha Muhammadiyah) and mobilize their members to collect donations during natural disaster, as happened in 2022 Pasaman earthquake and 2024 flash flood.

## 4.3 Philanthropy behaviours towards natural disaster

One form of collecting alms tradition is *badoncek*. *Badoncek* ("dues") widely practiced by Nagari communities in Minangkabau and in the diaspora. *Badoncek* demonstrates the social participation of Nagari members in helping with their wealth and assets as a form of communal solidarity. This is practiced in wedding ceremonies, the construction of public facilities, the building of mosque or surau, medical expenses, and in times of calamity or disaster [42]. The participation of the community is driven by empathy towards fellow human beings, encouraging involvement even if it is minimal. Another form of *badoncek* is *julo-julo*, a rotating savings and credit association by women of a Nagari [42, 43].

Some Nagaris have their own philanthropy traditions. In Pariaman, there are concepts such as *baretong*, *barantam*, and *badantam*. *Baretong* is donation collecting, usually in wedding receptions, but also in events as in Eid celebration. Donations are collected and then the collectors announce names and amounts of contributions loudly. This creates a sense of 'jealousy' among others, motivating them to donate more, a behaviour referred to as *barantam* (lit. fighting, or competing in this context) or *badantam* (lit. beating [the drum], or announcing loudly). When *tabligh akbar* (grand religious meeting) is held, they sometimes conducted *baretong* as well, especially when a disaster has occurred somewhere.

In Nagari Sulit Air, Solok Regency, the less favorable natural conditions have led to emergence *merantau* (migration). According to various records, philanthropic activities by migrants have been ongoing since the colonial era, with the establishment of Sulit Air Sepakst (SAS) in 1918 and the Sulit Air Students Association (PPSA), which have contributed to the development of the

Nagari through entrepreneurship. From their contributions, they have successfully built various public facilities in their hometown, including mosques and suraus, schools, funding teacher education at universities, scholarships, economic assistance for the community, land processing, and more [44].

These models of Minangkabau social philanthropy popularized by the communities of Pariaman and Nagari Sulit Air have expanded not only for their hometowns but also for broader disasters, such as natural disasters experienced by communities in other Nagaris in Minangkabau and Indonesia in general. In affected areas, there is naturally a decline in the scope of donations or charitable contributions. However, in other areas, the donations provided by donors increase. The generosity of various regions in Minangkabau contributes to alleviating the burden on their brethren.

Other forms of philanthropic activities include providing essential services and basic necessities on a large scale. In numerous reports, natural disasters are associated with a relative increase in the number of volunteers, represented by labor supply rather than charitable donations. The affected communities (local) tend to contribute more in terms of time and labor, making volunteer activities crucial.

From several cases of natural disasters in West Sumatra, such as the Padang and Pasaman earthquakes, the affected local communities exhibited various reactions, which tended to differ from those of Nagari communities not affected by the disaster. For the affected Nagari, charitable donations in the form of money decreased, while the provision of labor or social work increased.

Such a reality is certainly beneficial for policymakers, government officials, and professionals in the field of disaster recovery. Synergy between the government and various stakeholders is necessary to reduce natural disaster risks and adhere to disaster management guidelines. This will support the performance of the Regional Disaster Management Agency (BPBD), particularly in West Pasaman Regency, to ensure that programs are effectively implemented in terms of processes, mitigation, and emergency response post-disaster. This aligns with the central disaster management policy direction set by the National Disaster Management Authority (BNPB), which is "Enhancing disaster resilience towards sustainable well-being for development." One aspect of this policy is to increase local community trust in the affected Nagari to contribute labor and to find effective methods for boosting fundraising efforts in Nagari not directly affected by the disaster.

### 4.4 Study case: 2022 Pasaman Barat earthquake and response

Pasaman Barat has experienced a 6.1 $M_w$ earthquake on 25th February 2022, killing 27 people and destroying 6627 houses [45]. This was the largest earthquake in West Sumatra for the last 2 years, but still had lower significances than 2009 and 1977 earthquakes, which had shaped behaviours of traditional communities in daily life, from fastening wardrobes to the wall by rope to observing signs of disaster, both physical and spiritual [21].

**Table 1.** Receipt of donations and voluntary activity among communities during 2022 earthquake disaster

| Category | Low | Medi-um | High | % of average |
|---|---|---|---|---|
| **Receipt of donations** | | | | |
| Level of material damage due to the earthquake | 36.7% | 30.0% | 33.3% | 65.6% |
| Donations supported daily life during displacement after the earthquake | 6.7% | 40.0% | 53.3% | 82.2% |
| Donations provided support for post-earthquake life | 10.0% | 50.0% | 40.0% | 76.7% |
| **Volunteering during disaster response** | | | | |
| External motivation level of volunteering in response to earthquake | 76.7% | 16.7% | 6.7% | 43.3% |
| The duration of volunteering during the earthquake event | 16.7% | 16.7% | 66.7% | 83.3% |
| The contribution level during the period of volunteering | 0.0% | 50.0% | 50.0% | 83.3% |

Based on the questionnaires we collected as in the Table 1, the level of damage experienced by the victims of the 2022 West Pasaman earthquake reached a medium level (65.6%), with an almost equal distribution across different levels of damage. The respondents felt significantly aided by the donations distributed in response to the earthquake. A total of 53.3% of respondents felt that the donations were highly helpful for their lives during displacement, with a high percentage (82.2%) supporting this sentiment. Additionally, 50% of respondents felt moderately helped by the post-disaster aid, with a similarly high percentage (76.7%) expressing this view. This indicates that they feel a need for material aid, resulting in low donation levels from them during disasters in their own area. This is also supported by interviews with several involved employees and former volunteers.

Based on this, many people did not consider donating assets, as the disaster's impact was uniformly felt by all residents, who awaited external assistance. Some respondent said that they did not think to donate to other victims, as they were victims themselves. They also said that material donations for construction were processed slowly. The focus during the first year was on addressing severe damage over a two-year period, with subsequent efforts directed towards moderate damage.

Conversely, there was an increase in volunteer contributions. A total of 76.7% of respondents were not significantly influenced by external factors to be motivated to volunteer, with the average indicating a medium level (43.3%). All respondents acted as volunteer during the earthquake, with 66.7% being involved for a long duration (high level), and the average indicating a high level of involvement (83.3%). Each respondent showed medium and high levels (50% each) of contribution as volunteers, with the average indicating a high level (83.3%).

The motivation of disaster victims volunteering was primarily grassroots, with a significant commitment in terms of time and effort. Testimonies from several volunteers revealed that, on the night following the earthquake, they took the initiative to establish temporary shelters using tarpaulins and to set up rudimentary communal kitchens, while local government agencies coordinated activities. The following day saw the arrival of various logistical items, which were distributed with the assistance of both the local government and volunteers, alongside independent parties delivering aid directly to the affected areas.

After the disaster, community philanthropy also saw an increase. A local official from Nagari Kajai informed us that the community contributed to relief efforts during subsequent disasters, such as the earthquake in Cianjur on November 21, 2022, and the flood in Koto Tuo, Agam, on May 11, 2024. Each integrated health post in Kajai conducted door-to-door solicitations for donations, and Nagari social media platforms also facilitated announcements for contributions from both local residents and expatriates. A total of Rp 70 million was collected for Cianjur and Rp 20 million for Koto Tuo. Additionally, expatriate organizations in Sinuruik, Talu, and Kajai raised up to Rp 5 million to provide irrigation pipes that could facilitate the flow of lava in Koto Tuo.

## 5 Conclusion

This study of Minangkabau philanthropy reveals that the roots of social generosity among the Minangkabau people are grounded in historical practices and traditional local knowledge. Attitudes and acts of charity are passed down through traditions and customary expressions with a spirit of mutual assistance and solidarity aimed at achieving social justice. This practice is characterized by strong support from the well-off for the needy, and a spirit of belief drawn from local customs and the Islamic faith in Minangkabau.

From this foundational aspect of life, the people of the Nagari in Minangkabau show empathy towards their kin through the custom of charity. Minangkabau philanthropy is not limited to monetary donations but also includes contributions of goods such as food, clothing, and health supplies. This study demonstrates that the philanthropic spirit of the Minangkabau community is particularly strengthened when members of the Nagari are affected by disasters or calamities, especially natural disasters occurring in a Nagari in Minangkabau.

The practice of Minangkabau philanthropy is deeply intertwined with religious behavior. Historically, this tradition has been sustained over time and is an integral part of daily life in Minangkabau [45-48], encompassing practices such as *zakat* for disaster victims, infak, sadaqah, waqf, other forms of social assistance, infrastructure development, home repairs, and economic empowerment for disaster survivors. Various forms of donations are observed, particularly during natural disasters like earthquakes. In disaster-affected Nagaris, monetary donations may decrease, but in Nagaris that are not affected, the support often increases. In disaster-affected Nagaris, the strength of donations shifts towards volunteer efforts, characterized by the spirit of gotong royong (mutual cooperation) where the local community collaborates with external volunteers. Key terms that represent the roots of Minangkabau philanthropy include *badoncek*, which is popular in Pariaman, a region on the west coast of Sumatra, and *baragiah ka kampuang* (contributing to the hometown), as practiced by the Sulit Air migrants since the colonial era to the present day. During times of disaster, the solidarity of *anak rantau* (migrants) intensifies in their commitment to serve their hometowns.

This article aims to serve as a foundation for studying indigenous knowledge and actions concerning philanthropy in response to natural disasters. It has yet to explore the variations that could emerge differently from Nagari to Nagari. However, these variations are based on the traditional foundations discussed above. Thus, Minangkabau philanthropy represents a movement aimed at addressing various societal issues, making it crucial to enhance its effectiveness in ensuring social sustainability for disaster-affected communities. The spirit and practices of generosity across Minangkabau Nagaris should serve as a foundation for fostering empathy and social justice, emphasizing a collective effort to support and assist fellow community members.

## References

1. I. Trisnawati, Partisipasi masyarakat dalam penanggulangan bencana untuk mewujudkan Kampung Siaga Bencana (KSB) di desa Sindang Jaya, Geod. Jour. Geo. Edu. Univ. Sil. **4**, (2023). http://jurnal.unsil.ac.id/index.php/geoducation
2. A. A. Navis, Alam takambang jadi guru: adat dan kebudayaan Minangkabau (Grafiti Pers, Jakarta, 1984).
3. T. Abdullah, Movement school and politics: The kaum in West Sumatera. (1927-1933) (Cornell University, New York, 1971).
4. A. Azra, Negara, Filantropi, dan independensi masyarakat sipil, in A. Fauzia, Filantropi Islam: Sejarah dan kontestasi masyarakat sipil dan negara di Indonesia (Gading Publishing, Yogyakarta, 2016).

5. C. Jusuf, Filantropi modern untuk pembangunan sosial, Sos. Kon.: Jur. Pen. Dan Peng. Kes. Sos., **12**, 74–80, (2007). https://ejournal.kemensos.go.id/index.php/SosioKonsepsia/article/view/6 21/261
6. Sahrul, Pola akulturasi budaya etnik Mandailing dan Minangkabau (Perdana Publishing, Medan, 2018). http://repository.uinsu.ac.id/18446/1/POLA%20AKULTURASI%20ETNIK%20MANDAILING%20DAN%20MINANGKABAU.pdf
7. K. Mogotsi, I. Alagidede, and F. Saruchera, Applying lean thinking for disaster responses: The case of philanthropic organisations, Int. Soc. Sci. Jou. **244**, 287-319, (2022). https://doi.org/10.1111/issj.12323
8. S. Hudaa and R.N. Fahmi, Jimpitan: Filantropi berbasis kearifan lokal masyarakat Indonesia, Jur. Maa. **16**, (2021). https://doi.org/10.47651/mrf.v16i2.151
9. N.I.F. Kholidah, Banjir dan bantuan di Surakarta: filantropi dalam banjir Solo tahun 1966, Al-Isn.: Jou. of Isl. Civ. His. and Hum. **3**, 15-30, (2022). https://doi.org/10.22515/isnad.v3i1.5255
10. I Arpannudin, K Suryadi, E Malihah, and L Anggraeni, Philanthropy: The citizens' social capital amidst the pandemic, in Proceeding of ICHELSS 2021, March 25-27 (2021).
11. A. Fauzia, Filantropi Islam: Sejarah dan kontestasi masyarakat sipil dan negara di Indonesia (Gading Publishing, Yogyakarta, 2016)
12. R. Helfina, Implementasi program rehabilitasi dan rekonstruksi pasca bencana gempa bumi Kabupaten Pasaman Barat Provinsi Sumatera Barat, thesis, IPDN Bukittinggi, (2023).
13. D. Ismail, T. A. Majid, R. Roosli, and N.A. Samah, Project management success for post-disaster reconstruction projects: International NGOs perspectives, Proc. Eco. and Fin. **18**, 120-127, (2014). https://doi.org/10.1016/S2212-5671(14)00921-6
14. C. Berrebi, A. Karlinsky, and H. Yonah, Individual and community behavioral responses to natural disasters, Nat. Haz. **105**, 1541–1569, (2021). https://doi.org/10.1007/s11069-020-04365-2
15. V.A. Beard, Household contributions to community development in Indonesia, Wor. Dev. **35**, 607–625, (2007). https://doi.org/10.1016/j.worlddev.2006.06.006
16. O. Osili and Ç. Ökten, giving in Indonesia: a culture of philanthropy rooted in Islamic tradition, in P. Wiepking and F. Handy (eds.), The Palgrave handbook of global philanthropy (Palgrave Macmillan, Hampshire, UK, 2015). https://link.springer.com/chapter/10.1057/9781137341532_23
17. D. Muttaqien, M. Asyrof, and M.R. Syibli, Measuring the mosque-based disaster philanthropy, in Proceedings of the 2nd Southeast Asian Academic Forum on Sustainable Development (SEA-AFSID 2018), published March 8, 2021. https://doi.org/10.2991/aebmr.k.210305.013
18. M. Lee and C. Marquis, Large corporations, social capital, and community philanthropy, in sustainability, stakeholder governance, and Corporate Social Responsibility (Advances in Strategic Management, Vol. **38**), (Emerald Publishing Limited, Leeds, 2018), 197-226, (2018). https://doi.org/10.1108/S0742-332220180000038013
19. M.A. Mithani, Liability of foreignness, natural disasters, and corporate philanthropy, Jou. of Int. Bus. Stu. **48**, 941–963, (2017). https://link.springer.com/article/10.1057/s41267-017-0104-x
20. P. Kurniati, Sapriya, C. Darmawan, R.P. Quinones Jr., H.M. Putra, The phenomenon of traditional philanthropy through the voluntary sector in shaping community identity, Jur. Civ.: Med. Kaj. Kew. **20**(1), 85-93, (2023). https://doi.org/10.21831/jc.v20i1.49590
21. N. Nopriyasman, G. Asnan, A. Fauzi, I.P. Hastuti, A. H. Ritonga, V. Kurniawan, and R. Mairiska, Reading indigenous signs: The wisdom of Nagari communities toward natural disaster in Pasaman Barat, IJDDR. **107**(1), 1-6, (2024). https://doi.org/10.1016/j.ijdrr.2024.104497
22. F. Fakhriati, D. Nasri, M. Mu'jizah, Y. M. Supriatin, A. Supriadi, M. Musfeptial, and K. Kustini., Making peace with disaster: A study of earthquake disaster communication through manuscripts and oral traditions, Prog. in Dis. Sci. **18**, 1-8, (2023). https://doi.org/10.1016/j.pdisas.2023.100287
23. J. Tulius, Lesson from the past, knowledge for the future: roles of human memories in earthquake and tsunami narratives in Mentawai, Indonesia, Par.: Jur. Kaj. Bud. **10**(2), 147-168, (2020). https://doi.org/10.17510/paradigma.v10i2.396
24. Zulfadrim, Z., Y. Toyoda, and H. Kanegae, The implementation of local wisdom in reducing natural disaster risk: a case study from West Sumatera. IOP Conf. Ser. Earth Environ. Sci. **106**, 1–6, (2018). https://doi.org/10.1088/1755-1315/106/1/012008
25. B.G. McAdoo, L. Dengler, G. Prasetya, and V. Titov, Smong: how an oral history saved thousands on Indonesia's Simeulue island during the December 2004 and March 2005 tsunamis, earthquake, Spect. **22**, 661–669, (2006). https://doi.org/10.1193/1.2204966.
26. I.G.M.A.S. Wirawan, Restorasi dan kebertahanan local genius di era New Normal, in I.W.P. Yasa (ed.), Sejarah dan kearifan lokal di era globalisasi (Penerbit Lakeisha, Klaten, 2021).
27. T. Febria and Heriyanto, BAZNAS Sumatera Barat: Implementasi regulasi zakat dan relevansinya dengan kepercayaan Muzakki, Al-Ahk. **23**, (2022). https://doi.org/10.15548/alahkam.v13i2.4760
28. S. Thaheransyah and Zahirman, Pola pendayagunaan zakat pada Lembaga Amil Zakat, Infak Sedekah

Muhammadiyah (LAZISMU) Sumatera Barat, Al-Hikm. : Jur. Dak. Dan Ilm. Kom. **7**, (2020). https://doi.org/10.15548/alhikmah.v0i0.1484

29. S. Sahputra, Nopriyasman, and M. Nur, Business charity movement of Aisyiyah organization West Sumatra in 1968-2016, Int. Jou. of Res. in Edu. **4**, 49-58, (2024). https://doi.org/10.26877/ijre.v4i1.15439
30. S. Kartodirdjo, Gotong royong: Saling menolong dalam pembangunan masyarakat Indonesia, in N.J. Colletta and U. Kayam, Kebudayaan dan pembangunan sebuah pendekatan terhadap antropologi terapan di Indonesia (Yayasan Obor Indonesia, Jakarta, 1987).
31. R. Likert, A technique for the measurement of attitudes. Arch. of Psych. **22**, 5-55, (1932). https://psycnet.apa.org/record/1933-01885-001
32. H. Poeze, Di negeri penjajah: Orang Indonesia di negeri Belanda 1600-1950 (KPG, Jakarta, 2017).
33. Hamka, Ayahku: Riwayat hidup Dr. H. Abdul Karim Amrullah dan perjuangan kaum agama di Sumatra (Ummunida, Jakarta, 1982)
34. Mardiatno, G.I. Marliyani, and S.R.L.A. Sampurno, Merawat ingatan: Bencana alam dan kearifan lokal di pulau Jawa (Direktorat Sejarah Direktorat Jenderal Kebudayaan Kementrian Pendidikan dan Kebudayaan Republik Indonesia, Jakarta, 2019)
35. H. Chaniago, Nopriyasman, and I. A. Abdullah, Baginda Dahlan Abdullah bapak kebangsaan Indonesia (Yayasan Pustaka Obor Indonesia, Jakarta, 2020).
36. S. Yulianto, R.K. Apriyadi, Apriliyanto, T. Winugroho, I.P. Ponangsera, and Wilopo, Histori bencana dan penanggulangannya di Indonesia ditinjau dari perspektif keamanan nasional, PENDIPA Jour. of Sci. Edu. **5**, 180-187, (2021). https://doi.org/10.33369/pendipa.5.2.180-187
37. M.D. Mansoer, A. Imran, M. Safwan, A.Z. Idris and S.I. Buchari, Sedjarah Minangkabau (Bhratara, Jakarta, 1970).
38. C. Dobbin, Islamic revivalism in a changing peasant economy: Central Sumatra 1784 – 1847 (Routledge, London, 2016).
39. I. Dt. S. Dirajo, Curaian adat Minangkabau (Kristal Multimedia, Bukittinggi, 2003).
40. A. Ibrahim, et.al. Ungkapan-ungkapan tradisional yang berkepentingan dengan sila-sila dalam Pancasila daerah Sumatera Barat (Departemen Pendidikan dan Kebudayaan Proyek Inventarisasi dan Dokumentasi Kebudayaan Daerah, Jakarta, 1985).
41. M. Dt. R. Batuah, Adaik salingka Nagari Talu (Pucuak Adat KAN Talu, Talu, 2018).
42. J. Astika, S. Monang, and A.M. Syam, Persepsi masyarakat Minang terhadap tradisi badoncek dalam pernikahan orang Minangkabau di kota Medan, Yas.: Jur. Pend. dan Sos. Bud. **4**, 46-57, (2024). https://ejournal.yasin-alsys.org/index.php/yasin/article/view/2448
43. Deliana, and M. Lubis, Local wisdom of Badoncek as oral tradition in Pariaman wedding ceremony, Lang. Lit.: Jour. of Ling., Lit. and Lang. Teac. **3**, 33-40, (2019). https://doi.org10.30743/II.v3i1.1942
44. Abdiarrahman, Baragiah ka kampuang: Spirit filantropis perantau Sulit Air, Tur.: Jur. Pen. and Peng. **1**, (2013). https://ejournal.uinib.ac.id/jurnal/index.php/turast/article/view/474
45. B.D. Dewanto, R. Priadi, L.S. Heliani, A.S. Natul, M. Yanis, I. Suhendro, and A.M. Julius, The 2022 Mw 6.1 Pasaman Barat, Indonesia Earthquake, confirmed the existence of the Talamau Segment Fault based on Teleseismic and Satellite Gravity Data, Quat. **5**, 45 (2022). https://doi.org/10.3390/quat5040045
46. F. Ferdinal, O. Oktavianus, and I. Zahid, Exploring the beauty of Islamic values through metaphorical expressions in literary work, Afkar. **25**, 421-458, (2023). https://doi.org/10.22452/afkar.vol25no2.13
47. O. Oktavianus, F. Ferdinal, I. Zahid. Metafora leksikal dan nilai budaya dalam novel Hamka Di Bawah Lindungan Ka'bah. Pertanika: Social Sciences and Humanities. **32**, 111–137, (2024). https://doi.org/10.47836/pjssh.32.S1.05.
48. Anwar K, Oktavianus, Khanh VK, Khanh VQ. Disaster mitigation through oral stories: a case study around Singkarak Lake, EDP Sciences. **464**, 16002 (2023). https://doi.org/10.1051/e3sconf/202346416002

# Literature studies on natural disasters and communication in Indonesia

*Yenny* Oktavia[1]* and *Nabila* Khairani Wardah[1]

[1]Extension and Communication Study Program, Postgraduate Program, Universitas Andalas, Padang, Indonesia

**Abstract.** Disaster communication is important in improving community preparedness because it can help convey information and instructions for the community to face and respond to disasters effectively. This study aims to analyse the role of communication in disaster management in Indonesia based on a review of various related literature sources. The research uses literature study approach, analysing the publication of communication science journals discussing communication and natural disasters in Indonesia. The data sources include various relevant scientific publications, since the year 2000. The results showed that: 1) The types of natural disasters that are the main concern in the publication of communication journals in Indonesia are disasters on a national and regional scale such as floods, earthquakes, tsunamis, and volcanic eruptions. 2) Various communication topics include disaster communication for community mitigation and preparedness, public communication and social media in disasters, crisis communication and disaster communication management, communication and community participation in disasters, as well as disaster communication and vulnerability issues of vulnerable groups. 3) Efforts to bridge the gap between research and disaster management practice are challenges that must be faced so that communication science can make a greater contribution to improving the effectiveness of disaster management and mitigation in Indonesia.

## 1 Introduction

Indonesia is an archipelagic country that is known to have geological and climatological characteristics that are vulnerable to various kinds of natural disasters [1-3]. Natural disasters can occur suddenly and have a bad impact in the form of property damage and threatening human life [1, 2]. To overcome these challenges, the Indonesian government has issued various policies and plans, such as Law Number 24 of 2007 concerning Disaster Management and Presidential Regulation Number 8 of 2008 concerning the Establishment of the National Disaster Management Agency [1].

On the other hand, the challenges faced in the future are much greater [3, 4]. Increasingly extreme climate change, increasingly dynamic geological phenomena, and increasing environmental degradation require serious efforts to develop effective and efficient disaster emergency planning. One of the important aspects of disaster management is communication [2]. Appropriate communication strategies can contribute significantly to disaster mitigation and response. One of them is coordination between various parties in the emergency response to the pandemic by utilizing the use of communication platforms [5, 6]. This study raises the following questions:

1. What are the disaster events that are of great concern to communication science?
2. How are the uses of communication research findings in strengthening disaster management in Indonesia?

This study, in general, aims to analyses the role of communication in disaster management in Indonesia, specifically, it aims to:

1. Analyse the disaster events that are of great concern to communication research, and
2. Analyse the aspects of natural disasters in Indonesia which of concern to communication science.
3. Analyse the use of communication research results for strengthening disaster management in Indonesia.

## 2 Method

This study uses a literature study approach to analyses the publication of communication science that discusses communication-related to natural disasters in Indonesia. The study uses various data sources such as; journal articles, research reports, and other documents that discuss communication issues in natural disaster management in Indonesia. More specifically, it explores Journal articles using various online media such as Google Scholar, Garuda, Crossref, Sinta, and other sources using the keywords "communication," and "disaster," without distinguishing the journal's ranking. The study is based on a systematic literature review (SLR), the search covers publications from the year 2000 onward.

Data were analysed using qualitative analysis by creating a typology of communication studies about disasters and identifying patterns and trends in communication studies on disasters in Indonesia.

* Corresponding author: yennyoktaviafp@gmail.com

## 3 Results and discussion

### 3.1 Types of disasters studied

Based on Law Number 24 of 2007 concerning Disaster Management, we review of communication journals related to the types of natural disasters. The results of the study show that natural disasters of main concern to communication publications in Indonesia are floods, earthquakes, tsunamis, and volcanic eruptions. This is not surprising considering that Indonesia is a country that is prone to these natural disasters. Some of the articles reported communication in the context of flood disasters in several regions of Indonesia, such as the Jakarta flood and the Bandung flood. Meanwhile, some journals discuss earthquake-related communications, such as the 2004 Aceh earthquake and the 2018 Lombok earthquake. In addition, communication in the context of the tsunami disaster is also a concern, such as the study of the 2004 Aceh tsunami and the 2018 Sunda Strait tsunami. Some journal articles report the role of communication in the disaster cycle, such as in mitigation, preparedness, emergency response, and recovery after the several eruptions of Mount Merapi in Central Java. In general, the results of this study show that in the last 20 years, the journal publications present communication in the context of natural disasters contributed to enriching the understanding and practice of disaster management in Indonesia.

#### *3.1.1 Year of disaster event*

The year of disaster events that are widely studied by communication science in Indonesia is the period 2004 to 2018. This is not surprising considering that during that period there were several major natural disasters in Indonesia, such as the Aceh earthquake and tsunami (2004), the eruption of Mount Merapi (2010), the Jakarta flood (2013), the Lombok earthquake (2018), and the Sunda Strait tsunami (2018). For example, the Aceh earthquake and tsunami in 2004 [7] became a topic that was widely studied in communication science journals in Indonesia. These studies generally discuss how communication and social media play a role in information dissemination, coordination of disaster handling, and mitigation. In addition, the eruption of Mount Merapi in 2010 is also the focus of attention, especially related to crisis communication and the use of social media.

In the following years, the flood disaster in Jakarta, the earthquake in Lombok, and the tsunami in the Sunda Strait also became a topic that was widely studied in the publication of communication science-related journals. This shows that communication science in Indonesia is quite responsive to natural disaster events that occur and tries to study the use of communication science to support disaster risk reduction efforts in Indonesia. Furthermore, the earthquake disaster in Lombok in 2018 became a topic that was studied quite a lot in communication journals. as disseminating information related to evacuation procedures, training, and simulations. These studies generally discuss the use of social media for the dissemination of disaster information and coordination of handling.

#### *3.1.2 Scale of the disaster*

Based on the results of the analysis, the scale of disasters that are most studied in communication journal publications in Indonesia are those on a national and regional scale. Natural disasters with national impacts, such as the 2004 Aceh earthquake and tsunami, the 2010 eruption of Mount Merapi, and the 2018 Sunda Strait tsunami have become topics that have been widely studied. These studies generally discuss how communication, both through conventional media and social media, plays a role in information dissemination, coordination of handling, and disaster mitigation efforts on a national scale.

In addition, natural disasters with regional impacts, such as floods in Jakarta and Bandung, are also the focus of attention in the publication of communication journals in Indonesia. Few journals address communication in the context of natural disasters on a local or community scale. This shows that communication science in Indonesia tends to be more interested in studying communication in the context of natural disasters that have a greater and more widespread impact, both on a national and regional scale. Overall, the results of this literature review show that in the last 20 years, communication science in Indonesia has made a significant contribution to studying the role of communication in natural disaster management in Indonesia, especially in major natural disasters that occurred in the period 2004-2018

### 3.2 Communication topics studied

With the key search words of communication and disasters, 22 journal publications were found in the period 2000 – 2020. From the analysis of 22 journal publications, it was found that several main topics were the focus of the study, including disaster communication for community mitigation and preparedness, public communication, and social media in disasters, crisis communication and disaster communication management, communication, and community participation in disasters, as well as disaster communication and vulnerability issues of vulnerable groups. And coordination between stakeholders in disaster communication. [7-10]

#### *3.2.1 Disaster communication for community mitigation and preparedness*

Studies have shown that disaster communication has an important role in improving community preparedness in dealing with disasters. Effective communication can increase public understanding and knowledge about disasters, thereby reducing risks and losses. Communication plays a role in education, socialization, and community mobilization to prepare for disasters, such

Various studies examine how disaster communication can be used to build community preparedness, both through socialization, dissemination of information about disasters, and disaster mitigation training. So that the community can understand the disaster risk in their area, know the evacuation procedures, and take disaster risk reduction actions [4, 11].

### *3.2.2 Research on government communication to improve community preparedness in dealing with flood disasters*

Other researchers found the important role of community-based communication networks in building preparedness for natural disasters [7, 8, 10, 12]. Research related to government communication to improve community preparedness provided an overview of the use of communication media by local governments in an effort to improve community preparedness to face flood disasters. Meanwhile, community-based communication research shows how communication networks and collaboration between residents in the community can help build citizens' preparedness in dealing with natural disasters. Communication networks and collaboration between residents in the community can be interconnected and coordinated to build citizens' preparedness in dealing with natural disasters. Steps to build this communication and collaboration network are through the establishment of community-based disaster. preparedness groups, the preparation of community emergency response plans, preparedness training, and others.

Effective communication can improve community preparedness, coordination between stakeholders, and the dissemination of accurate information during the emergency response phase [2]. Research shows that a good communication strategy helps reduce the impact of disasters and accelerate recovery efforts [3, 4]. Considering various natural disasters that frequently occur in Indonesia, community preparedness must be constantly increased to deal with disasters. However, the preparedness of the Indonesian people is still limited due to limited knowledge and understanding of disaster mitigation [11].

### *3.2.3 Public communication and social media in disaster*

Several journals examine other important issues related to disaster communication, such as the handling of hoax and rumor issues, communication between the government and the community, and the involvement of vulnerable groups in disaster management. Aim to provide pactcal gudance fo the use ofnew media in the contex of crisis communication [13, 14]. These studies provide evidence that the use of social media and public communication has an important role in disseminating information, coordinating, and mobilizing people during disasters [1, 3]. In addition, during crises, social media is a channel for the dissemination of inaccurate information or hoaxes, which can interfere with people's emergency response efforts [2]. Therefore, a good communication strategy is needed to manage information and public communication in times of disaster.

Furthermore, several journals also examine how social media such as Twitter, Facebook, and WhatsApp are used to disseminate information and conduct emergency response communication during a disaster [7]. The article also reports the examination of the influence of social media platforms on the dissemination of disaster information. Implementation of conceptual framework in designing tools to identify disaster location and disaster safety in the disaster process with the help of social media [15]. The results of his research show that social media facilitates individual communication during disasters and allows people to spread information quickly [10].

Subsequent research shows the role of social media in natural disaster situations, such as information dissemination, evacuation coordination, and relief. However, there are also challenges related to the spread of hoaxes and false information on social media, so it requires the right communication strategy to overcome them. To overcome challenges related to the spread of false information or hoaxes on social media, effective communication strategies are needed, such as verifying and clarifying information, strengthening media literacy, as well as the active role of the government and related institutions in disseminating accurate and timely information.

In addition, communication through traditional media such as radio and television also still plays an important role, especially to reach people who are less exposed to digital media. Effective public communication can encourage active community participation in disaster management efforts, ranging from mitigation, preparedness, and emergency response to post-disaster recovery and rehabilitation. Overall, this literature study shows that communication plays a key role in the entire disaster management cycle in Indonesia, from the pre-disaster stage, during emergency response, to the post-disaster recovery stage.

### *3.2.4 Crisis and disaster communication management*

The research focuses on the role of crisis communication and communication management efforts in natural disaster emergency response. The research examines the management of communication by government agencies and organizations in conveying information, avoiding information chaos, and facilitating coordination among stakeholders. Coordinated communication strategies between various parties in the pandemic emergency respons [5]. Effective crisis communication can help coordinate emergency response actions, use various communication channels to disseminate the right information and conduct two-way communication with the community. Make this happen through leadership support, communication planning, and strengthening the capacity of communication human resources.

Two-way communication between the government, relevant institutions, and the community can also increase public trust and encourage community participation in disaster management. Research related to the management of emergency response communication for natural disasters such as earthquakes, tsunamis, floods, and others examines the coordination, communication strategies, and channels used. The stages of communication management are careful and comprehensive communication planning, data collection and analysis to identify current needs and situations, delivery of targeted messages and selection of appropriate communication media, as well as monitoring and evaluation of the implementation of communication.

### *3.2.5 Communication and community participation in disasters*

The research examines the role of communication in encouraging community participation and empowerment in disaster management, starting from the mitigation, preparedness, and emergency response to the rehabilitation and reconstruction stages. Describe and critique how community-based disaster management in Indonesia [16] Two-way communication involving active community participation will increase the effectiveness of disaster management, as communities can convey their aspirations and needs and contribute to disaster risk reduction efforts. The research how the communication strategies for disaster management panning such us the use of communication platform [6]. In addition, discussion forums, citizen discussions, and empowerment of vulnerable community groups can support the creation of participatory communication in disaster management. This step is important to build community awareness and capacity in dealing with disasters. The role of communication in encouraging participation can be done by increasing community involvement in emergency response planning, strengthening organizations and community groups that play a role in disaster management, and facilitating dialogue between stakeholders and the community.

### *3.2.6 Disaster communication and vulnerability issues of vulnerable groups*

The research examines communication strategies that take into account the needs and challenges of vulnerable groups such as women, children, the elderly, and people with disabilities in disaster situations. This vulnerable group requires special treatment and delivery of information, such as adaptive communication with easily accessible languages and channels. In addition, the study also shows the need for an active role of vulnerable groups in planning and decision-making to ensure their needs are accommodated. The issue of vulnerable groups is important to pay attention to so that disaster emergency mitigation and response efforts can be more equitable and inclusive.

Several studies related to vulnerable groups say that effective communication with vulnerable groups such as women, children, the elderly, and people with disabilities requires specific approaches and channels, such as using simple language, visual media, activating their social networks, and facilitating their participation in decision-making. [1, 2, 4, 10] Effective communication with vulnerable groups can encourage their active involvement and contribution in every stage of disaster management, from mitigation, preparedness, and emergency response to post-disaster recovery.

Research shows that in disaster situations, vulnerable groups such as women and children face a higher risk compared to other groups. Therefore, communication strategies need to pay attention to the specific needs and challenges faced by these vulnerable groups so that information delivery and disaster management can run more fairly and inclusively [4, 10, 11]. Table 1 summarizes the relevant journal articles reviewed.

**Table 1**. Summary of typology, patterns, and trends of communication and natural disaster studies in Indonesia.

| **Communication Aspects** | **Reference** | **Research Findings** | **Research opportunities** | **Recommendations** |
|---|---|---|---|---|
| Patterns of information dissemination | [7, 10, 18, 19, 20] | - Disaster information spreads through various media, including social media, and the factors that influence it | Information dissemination patterns are based on individuals, organizations, and policies for more effective disaster management. | The importance of increasing the capacity of the government and related institutions to utilize social media as a communication channel in disaster management. |

| | | | | |
|---|---|---|---|---|
| Coordination between stakeholders | [1, 3, 16, 17, 22] | - In terms of coordination, these journals examine how communication and coordination between governments, communities, and other stakeholders take place in disaster management.<br>- The results of the research are valuable input to develop a better collaboration mechanism in disaster management. | - Build more effective coordination between the government, the private sector, and the public in communication during disaster emergency response. | - The importance of increasing the capacity of the government and related institutions to utilize social media as a communication channel in disaster management.<br>- Scientific studies on disaster communication can make a practical contribution to improving Indonesia's capacity to deal with natural disasters. |
| Utilization of social media | [7, 8, 10, 12, 13, 14, 15, 17] | The role of social media in disaster management is still limited in reviewing the use of social media for information dissemination and coordination of disaster management. | - The contribution of social media in the process of verifying information to avoid the spread of false information or hoaxes that can cause public panic. | Further research is still needed to explore the potential of social media as a means of education and increasing public awareness of disaster risk. |
| Effective communication | [5, 6, 7, 8, 21] | - The publication of communication journals in Indonesia has made an important contribution to understanding the role and strategy of communication in dealing with natural disasters.<br>- The results of the study show that communication plays an important role in every stage of disaster management, from pre-disaster, and emergency response, to post-disaster.<br>- These studies highlight the need for effective, coordinated, and targeted communication so that the disaster management process can run well. | - Few journals explicitly discuss how the results of the communication study can be used in disaster management practices.<br><br>- Research should be able to provide valuable input for the development of more effective disaster management strategies and policies. | - Not only analyze communication phenomena in the context of disasters, but also provide recommendations that can be implemented in disaster management practices.<br>- Opportunities to expand the scope of the study, for example, disaster communication in the context of transnational and global issues.<br>- Encourage stronger collaboration among various stakeholders, such as government, private and community, in designing and implementing effective disaster communication strategies.<br>- Further studies can be directed to not only analyses the phenomenon of communication in the context of disasters, but also explore its implications for improving the effectiveness of disaster management. |

### 3.3 Utilization of the results of communication studies in disaster management

The findings from these journal studies can make an important contribution to strengthening disaster management in Indonesia. For example, a study on the use of social media and crisis communication can help governments and related institutions in designing effective communication strategies in disaster emergencies. Similarly, the results of research on coordination between stakeholders can be a valuable input to develop better collaboration mechanisms in disaster management. Thus, scientific studies on disaster communication can make a practical contribution to improving Indonesia's capacity to deal with natural disasters [17, 18]. There is still a need for further efforts to optimize the use of communication research results in disaster management practices. [1, 2]. In general, the publication of communication journals in Indonesia has made an important contribution in understanding the role and strategy of communication in dealing with natural disasters. Various studies have provided practical recommendations for disaster risk reduction efforts and more effective disaster communication management. However, there is still an opportunity to expand the scope of the study, for example, disaster communication in the context of transnational and other global issues [7]. In addition, deepening case studies and comparative analysis between regions are also needed to enrich the understanding of communication practices in disaster management in Indonesia.

The results of the review show that most of the communication journal publications in Indonesia that discuss communication topics in the context of natural disasters focus more on analysing and understanding communication phenomena that occur, such as information dissemination patterns, coordination between stakeholders, and the use of social media. However, several journals have also tried to provide recommendations or inputs related to how communication can be optimized to support disaster management efforts in Indonesia. For example, a journal that examines the use of social media in disaster communication [17] recommends the importance of increasing the capacity of the government and related institutions to utilize social media as a communication channel in disaster management.

Several journals suggest the need to increase disaster communication capacity, both at the individual, organizational, and policy levels for more effective disaster management. [10, 18, 19, 20] Thus, the results of the study of communication in the context of disasters in Indonesia still need to be further translated into more effective disaster management practices [7, 10].

There are several challenges in utilizing the results of communication studies to strengthen disaster management practices in Indonesia. Previous research identified that there is still a gap between the results of academic studies and implementation in the field [1]. In addition, coordination between the government, the private sector, and the community in disaster communication is also still a challenge [1, 21]. To address these challenges, more intensive efforts are needed to bridge the gap between academic study results and disaster management practices on the ground, by encouraging stronger collaboration among various stakeholders, such as government, the private sector, and the community, in designing and implementing effective disaster communication strategies.

## 4 Conclusion

Based on the results of this literature review, it can be concluded that:
First, communication science in Indonesia has been quite responsive to natural disaster events that occur, with a focus on major disasters, such as floods, earthquakes, and volcanic eruptions. Communication science in Indonesia tends to focus on studying communication in the context of major natural disasters that occurred in the period 2004 to 2018. The scale of disasters that are most studied in communication journal publications in Indonesia are disasters with national and regional impacts, such as earthquakes, tsunamis, volcanic eruptions, and flash floods.

Second, the main topics studied in communication journals in Indonesia related to disaster management include information dissemination patterns, coordination between stakeholders, and the use of social media. In general, the results of the study show that communication plays an important role in every stage of disaster management, from pre-disaster, and emergency response, to post-disaster. These studies highlight the need for effective, coordinated, and targeted communication so that the disaster management process can run well. However, few journals explicitly discuss how the results of the communication study can be used in disaster management practices. A deep understanding of communication in the context of disasters should be able to provide valuable input for the development of more effective disaster management strategies and policies. Therefore, in the future, it is hoped that there will be more communication journals in Indonesia that not only analyses communication phenomena in the context of disasters but also provide recommendations that can be implemented in disaster management practices.

So far, journals related to the role of social media in disaster management are still limited in examining the use of social media for information dissemination and coordination of handling. Further research is still needed to explore the potential of social media as a means of education and increasing public awareness of disaster risk. For this reason, in the future, it is hoped that communication science can make a greater contribution to strengthening disaster management practices in Indonesia, not only through studies that produce a comprehensive understanding but also by providing recommendations that can be adopted by stakeholders in the field. Close collaboration between the academic community, the government, and civil society is the key to making this happen.

This shows that there is still room to strengthen the linkage between communication research and disaster management practices in Indonesia. Further studies can be directed to not only analyse communication phenomena, but also explore their implications for improving the effectiveness of disaster management. In general, although the publication of communication journals in Indonesia has been quite responsive to the phenomenon of natural disasters that occur, efforts to integrate the results of such research into disaster management practices in Indonesia still need to be improved. Communication science in Indonesia still needs to be more intensive to study and formulate practical recommendations for improving communication effectiveness.

In addition, other research results emphasize the need to build more effective coordination between the government, the private sector, and the community in communication during disaster emergency response. Quite challenging research related to how to optimize the use of communication study results in disaster management practices in Indonesia. In terms of information dissemination patterns [7], several journals analyse how disaster information spreads through various media, including social media, and the factors that influence it. Meanwhile, in the aspect of coordination, the journals examine how communication and coordination between the government, the community, and other stakeholders take place in disaster management. Research that is still needed regarding the role of social media in the pattern of spreading disaster information is in the form of social media contributions in the process of verifying information to avoid the spread of false information or hoaxes that can cause public panic.

Third, although the results of research on communication in the context of disasters are adequate, efforts to integrate them into disaster management practices in Indonesia still need to be improved. Most of the communication journal publications in Indonesia that discuss communication topics in the context of natural disasters focus more on analyzing and understanding communication phenomena that occur, such as information dissemination patterns, coordination between stakeholders, and the use of social media. However, there are still few journals that explicitly examine how the results of the communication study can be used in disaster management practices.

Based on the results of this literature review, here some recommendations:
First, subsequent studies can be directed to not only analyze communication phenomena in the context of disasters, but also explore their implications for improving the effectiveness of disaster management. Second, there needs to be an effort to strengthen the linkage between communication research and disaster management practices in Indonesia, so that the results of communication studies can be translated into strategic and operational recommendations that can be implemented by stakeholders to improve the effectiveness of disaster management and mitigation in Indonesia.

Third, further studies can be directed to examine the role of social media not only in information dissemination and coordination of handling but also in education and increasing public awareness of disaster risk. The need to increase disaster communication capacity, both at the individual, organizational, and policy levels for more effective disaster management. The above literature review has provided an overview of how communication science in Indonesia has contributed to research related to disaster management, although the integration of research results into practice in the field still needs to be improved. Efforts to bridge the gap between research and disaster management practice are challenges that must be faced so that communication science can make a greater contribution to improving the effectiveness of disaster management and mitigation in Indonesia.

## References

1. I. Mauluddin, H. Bekti and R. Pancasilawan, Preflood management collaboration in bandung district. Journal of Public Service Management, **5**, 154-168, (2022)
2. M. M. Wijoyono and A. Raditya, Designing educational media games as learning how to protect themselves in the face of natural disasters for children aged 7-12 years. Adiwarma Visual Communication Design Journal, **1**, (2014)
3. T. M. Milyane, Komunikasi kelompok dalam training material failure analysis. Dialektika, **7**, (2020)
4. M. J. Nur, P. R. Andhita and B. V. Safitri, Communication noise in handling teh covid-19 outbreak (descriptive-qualitative study of handling covid-19 in NTB). JCommsci (Journal of Media and Communication Science), **3**, (2020)
5. E. Nilawati, Collaborative governance covid-19 pandemic disaster emergency response management in gunungkidul regency. Journal of Administrative Sciences, **19**, (2022). https://doi.org/10.31113/jia.v19i1.845
6. M. S. Anshori, H. Hatima and A. Lestari, Communication strategy of the regional agency for disaster management of Sumbawa regency in disaster mitigation. Journal of Communication Management, **7**, (2022)
7. T. Suliyati, Saving archives from disaster: between idealism and reality. Lentera Pustaka, **3**, (2017). https://doi.org/10.14710/lenpust.v3i2.16738
8. I. Pitaloka, S. Sari and I. Indria, Crisis communication strategy of pt. pelabuhan indonesia II (PERSERO) branch of bengkulu during the covid-19 pandemic (study on land dispute cases). Journal of ISO, **3**, (2023)
9. R. Widyastuti, Using new media and social media in disaster communication.Communicator, **13**, (2021)

10. Y. Rusfiana and H. Nurseta, Synergy of the TNI and local governments through military campaign strategies in flood disaster management in Bandung district. Indonesian Governance Journal, **4**, (2021)
11. Y. S. Hijri, K. Hadi, A. Roziqin and A. M. Hidayah, Assistance in the preparation of a COVID-19 disaster contingency plan in Bandulan Village, Sukun, Malang City. Journal of Community Service, **4**, (2023)
12. M. A. Fu'ady, R. Munadi and M. A. Fu'ady, Disaster mitigation in Indonesia: between plans and reality. IOP Conf. Ser.: Mater. Sci. Eng, **1087**, 012011 (2021). https://doi.org/10.1088/1757-899X/1087/1/012011
13. F. S. Mubarok, The use of new media for communication effectiveness in the pandemic era. Scientific Journal of Meaning Communication **10,** (2022)
14. V. Widjaja and N. M. Widodo, The influence of internet technology on the knowledge of the people of Jakarta regarding COVID-19 vaccination information. Tematik, **8**, (2021)
15. A. Sutedi, Design and build a social media-base disaster identification and disaster-safe location application. Algoritma, **16**, (2019)
16. C. Muryani, Community-based disaster management in Indonesia, SHEs: Conference Series 3, **3**, (2020)
17. D. Tamitiadini, W. W. A. Dewi and I. Adila, Innovation of non-structural disaster mitigation model based on communication, information, coordination and cooperation. Journal of Communication, **13**, (2019)
18. E. Dulwahab, Y. Huriyani and A. S. Muhtadi, Therapeutic communication strategies in the treatment of victims of sexual violence. Journal of Communication Studies, **8**, (2020)
19. D. Kurniawati, Disaster mitigation communication as community awareness to face disasters.Symbolic Journal, **6**, (2020)
20. M. B. F. Bisri, Inter-organizational network in Indonesia during disasters: Examples and research agenda on disaster management. IOP Conf. Ser.: Earth Environ. Sci. **56**, 012023 (2017). https://doi.org/10.1088/1755-1315/56/1/012023
21. R. N. S. Amriza and K. N. M. Ngafidin, Analysis of the influence of social media platforms on the dissemination of disaster information. JSiI (Journal of Information Systems), **8**, (2021)
22. R. H. Sayuti, O. P. Inderasari, and A. Evendi, The influence of education and local culture on community preparedness in facing disasters. Journal of Research and Assessment of e-Science Eduaction, **5**, (2021)

# Learning from our vulnerabilities: insights from Octavia E. Butler's parable of the sower and West Sumatra's 2024 flood disasters

*Edria* Sandika[1*], *Gindho* Rizano[1], and *Marliza* Yeni[1]

[1]English Literature Study Programme, Faculty of Humanities, Universitas Andalas, Padang, Indonesia

**Abstract.** This article discusses the correlation between flood disasters in West Sumatra in March and May 2024 and the science fiction novel *Parable in the Sower* by Octavia E. Butler. The novel explores how social and environmental degradation amidst the extreme weather and climate change in the fictional setting of America in 2024-2027 mirrors West Sumatra's lack of preparedness to face similar situations in reality. The novel warns people of the consequences of environmental issues by addressing our vulnerabilities and resistance to change. Through the concept of "Earthseed", the story reminds society to work toward positive changes through resilient infrastructure, disaster preparedness, and empathy when facing natural disasters.

## 1 Introduction

West Sumatra was hit by large-scale natural disasters related to floods and landslides twice in 2024. Intense monsoon rainfalls in March and May 2024 triggered flash floods, notably volcanic mudflows surrounding Mount Marapi and landslides in several areas. One significant disaster mainly resulted in the traffic closure in Lembah Anai due to a road collapse lasting over two months. Those disasters reveal our vulnerabilities in several factors, such as environmental issues, fragile infrastructure, and preparedness to mitigate or prevent natural disasters.

### 1.1 West Sumatra flood and landslide disasters on March and May 2024.

The natural disasters that struck West Sumatra were related explicitly to flash floods and landslides in March and May 2024. The first disaster happened between 7 March and 13 March 2024, when heavy rainfall triggered flash floods and landslides, resulting in 26 casualties, extensive damage, and an estimated economic loss of Rp170.4 billion [1]. Meanwhile, the second disaster occurred from 11 May to 12 May 2024 due to intense heavy rainfall resulting in flash floods and volcanic mudflows, with 67 casualties, particularly in Agam and Tanah Datar regencies [2]. The second disaster also caused traffic closure from Padang to Bukittinggi due to the collapse of roads near Lembah Anai.

The disasters in March and May 2024 resulted in several effects. Damages to bridges, roads, houses, rice fields, schools, power outages, and blackouts affected more than 4000 people in several areas. The vital traffic road from Padang to Bukittingi and vice versa was closed due to a road collapse around Lembah Anai, diverting the traffic to Malalak or Sitinjau Lawik. This diversion added more time and congestion on those alternative routes because of the economic significance between Padang and Bukittingi. Moreover, The Indonesian National Board for Disaster Management declared West Sumatra a state of emergency for 14 days due to the floods in May 2024.

### 1.2 Parable of the sower by Octavia E. Butler

*Parable of the Sower* (1993) is a novel by Octavia Estelle Butler which revolves around environmental issues in America, starting from 2024, when the main character, Lauren Olamina, turns 15 years old, continuing through 2027 when she is 18. The novel explores the fragility of places near Los Angeles facing climate change and social collapse. The main character, Lauren Olamina, narrates her struggle through her journal as she envisions "Earthseed" as her religious view to survive.

Lauren Olamina is a 15-year-old black woman residing in the Los Angeles region amid social disintegration in 2024. She lives in a neighbourhood, which is a precarious haven enclosed by walls. Outside those barriers, individuals in need fight for their lives. At the same time, criminals who are addicted to fireworks prowl the area, starting fires. Lauren possesses two unique qualities: an unusual perspective and hyperempathy, or the ability to experience the emotions of others. She created the Earthseed philosophy based on the notion that "God is Change." Following three tragic events—her father goes missing, her brother Keith dies, and the neighbourhood collapses due to fireworks—Lauren flees with Harry and Zahra [3, 4]. The novel's story ends in

* Corresponding author: edriasandika@hum.unand.ac.id

2027 when Lauren is 18 and continues her journey far to the north.

The entire storyline continues in the sequel *Parable of the Talents* (1998). It remained unfinished when Butler passed away in 2006, leaving the planned sequels *Parable of the Trickster, Parable of the Teacher, Parable of the Chaos,* and *Parable of the Clay* never written [5].

The narrative of *Parable of the Sower* paints the struggle of survival while searching for the meaning of life as the novel's dominant theme. Lauren's journey envisioning her religious view signifies her hypocrisy of people's pain and pleasure as she experiences tragedies and her effort to survive the harsh environment and social collapse. The novel's theme explores the vulnerability of our life and enforcing the idea of inevitable change, reflected in Lauren's Earthseed belief to encourage adaptability.

As the *Parable of the Sower* falls into the speculative fiction genre, the novel distances itself from everyday reality. It presents fantastical, imaginative, and, in this case, futuristic speculation of America from 2024 to 2027. The novel foreshadows climate change challenges, especially with extreme weather events and social collapse, such as poverty and unlawful and rampant violence. Iossifidis (2020) argues that reading speculative fiction, notably *Parable of the Sower,* contributes significantly to "learning in community" as the main character, Lauren, encompasses hyperempathy, signifying the importance of her vision of the world that needs to change [6]. Meanwhile, Guerrero stated that speculative fiction can analyse the world following a disaster and assist in identifying potential outcomes that project and respond to actual feelings of uncertainty and social unrest [7]. While speculative for 1993's standard, the novel still serves as a reminder and a warning of our vulnerability in mitigating and preventing natural disasters and social collapse in our actual 2024 today.

### 1.3 Literature and its function

Human problems, from personal ones to catastrophic natural disasters, have been staple subjects of literature. All stories, no matter how light, revolve around conflicts and tensions, making it almost impossible for writers to work outside "the tight confines of problem structure." [8]. In an emerging perspective of literary Darwinism, which scholars like Jonathan Gottschall and Joseph Carroll champion, literature is seen as a tool to simulate human problems. It allows us to observe how problems unfold or play out from a safe distance as readers. It enables us to see the cause and effects of the problems and the scenarios they are dealt with. In short, literature is the training ground for our minds to understand and approach conflicts.

Angus Fletcher, a literary scholar with a background in science, suggests that literature should be seen as a type of technological innovation. In his pragmatist view, literature and narrative have functions akin to technology: to solve human problems [9]. This perspective shows that literature can address human concerns and worries and offer solutions.

With this adaptationist and functional view of literature, one can see that different genres of literature address various issues and offer different solutions to them. One should not look further than science fiction for major human concerns and conflicts relating to the present and the future. Throughout history, we have seen plenty of examples of this. Mary Shelley's *Frankenstein* (1818) explores how unchecked scientific ambitions can result in disasters, encouraging us to take a humane approach in our scientific endeavours. George Orwell's *1984* (1949) looks at how totalitarianism seeks to annihilate human freedom and abuse technology for total control, making it clear that the solution for us readers is the rejection of dictatorship. *The Hunger Games* series (2008-2010) investigates how people are always caught in crossfires between the ruling status quo and oppositional elements, forcing us to think of ways for a non-violent change.

There have also been great works from this genre that draw our attention to environmental issues and how our species deal with them. *Dune* by Frank Herbert (1965) highlights the resistance against exploiting natural resources and promoting ecological balance. Cormac McCarthy's *The Road* (2006) dives deep into the social effects of food scarcity resulting from a mysterious natural disaster in which human beings must choose between cannibalism or cooperation for survival.

Works of black American female writers, such as N.K. Jemisin and the late Octavia Butler are particularly relevant to the discussions of how humanity should react to natural disasters. Jemisin's *The Fifth Season* (2015) imagines a world in which intense seismic activities plague the world. While the hero, Essun, undertakes a personal journey in search of her lost daughter, we, the readers, are invited to open our eyes and change our ways of dealing with nature. Rather than controlling nature, we must adapt to and respect it.

Butler's *Parable of the Sower* (1993), a precursor and a direct influence for the novel mentioned above, perhaps deserves more considerable praise. The novel's dystopian world is affected by climate change, and the populations must face severe droughts and extreme weather. This matter is exacerbated by two factors: the government in the novel is corrupt and ineffective in handling environmental crises. At the same time, influential corporations are controlled by greed and drive for profits. It is this combination of these problems that the hero, Lauren Olamina, and her community must face. In her journey, she develops a philosophical teaching called "Earthseed", which promotes the understanding that humans must adapt to change and strive to create positive changes through collective efforts. This lesson from a fictional work may be a real solution to our real problems, especially about our attitude, government and corporations included, to environmental challenges.

### 1.4 Ecocriticism

Emerging in the 1990s, Ecocriticism responded to environmental concerns about human relationships with nature. Literature and other cultural products can be explored from the perspective of nature and environmental issues reflected in texts to promote awareness. With closer ties to biology, ecology and environmental studies, enriching literary interpretation is significant in promoting an interdisciplinary perspective of the interaction between character development, plot, or even themes with nature and environment. Exploring natural disasters amplified by some issues in nature, social policies, and mitigation/prevention, the ecocritical lens helps to understand the nature-human connection, primarily when the literary work serves as a warning and foreshadows the near future [10].

One of the most common ways to see the relationship between literature and nature is from how nature is presented in literary works. Some works would present nature as a powerful, untouched space wherein humans' lives might be threatened. This approach sees humans as separate from, if not opposed to, nature [11]. They are simply two different entities which probably need to compete with each other. Coleridge's poem "The Rime of the Ancient Mariner" might be one of the most outstanding poems featuring the fear and terror that nature can cause to humans. "The ice was here, the ice was there, / The ice was all around: / It cracked and growled, and roared and howled, / Like noises in a swound!" (1798:9:59-62). The stanza might frighten the readers, showing how nature can be hostile and deadly to humans simply because humans' lives (alone) are considered the most important. However, more literary works recently promoted biocentric awareness. Nature is no longer regarded as fearful, frightening and untouched. Humans are part of nature, like animals, plants and other living creatures. Therefore, all living beings' lives are as meaningful as humans. As Desjardins puts it, "[all] living beings, simply by being alive, have moral standing and deserve moral consideration" [12].

Deeper into the spirit of Ecocentrism (putting the ecosystem at the centre of ethical considerations and considering sustainable and healthy ecology as the essential thing in life), Arne Naess and George Sessions developed the eight principles of deep ecology social movement. Some, if not all, of these principles are now frequently promoted by many authors in their literary works to promote environmental awareness. These articles analyse the two issues from the perspective of the first, third, and fourth principles of deep ecology social movement [13, 14]. Principle number one mentions that: "The well-being and flourishing of human and nonhuman life on Earth have value in themselves. These values are independent of the usefulness of the nonhuman world for human purposes." The second one argues, "Humans have no right to reduce this richness and diversity except to satisfy vital needs." Meanwhile, the fourth believes that: "Present human interference with the nonhuman world is excessive, and the situation is rapidly worsening" [12]. These three principles would presumably prove that nature has come to a devastating mode due to humans' greediness and recklessness.

# 2 Results and discussion

*Parable of the Sower* is a constant reminder that environmental disasters pose a severe threat either when published in 1993 or as we see it today [3]. Global warming, climate change, famines, and scarcity of necessities are problems that not only exist in a fictional storyline but also serve as a warning that the threats are real and not something we can ignore. West Sumatra's flood disasters that occurred in 2024 become correlated to every warning the novel delivered. Although specific to flash floods, landslides, and road collapses, these disasters encourage people living in West Sumatra to adapt and be resilient to avoid similar disasters happening in the future. People can no longer be the victims; instead, they can start to be partners in shaping the environment, making it better, and not leading into the next catastrophe.

## 2.1 Environmental issues

One pressing concern in West Sumatra's flood disasters is the failure to put things in order due to illegal buildings surrounding the Anai watershed. Walhi Sumbar (2024) accused the government of failure to manage this area, as this failure is characterised by the proliferation of utilisation/cultivation activities that do not align with their intended use [13]. One of them is using borders to construct hotels and cafe buildings. The National Land Agency (BPN) noted approximately 107 built objects along the Lembah Anai area. According to the BPN, the built objects include stalls/cafes/restaurants, hotels, mosques, pools and others, such as spring baths, PT HSH Rest Area and Xakapa Cafe and Resto [14].

The government's failure to manage the area resembles the failure of the government in the Parable of the Sower, which was plagued with rampant greed and ignorance to preserve the environment. While West Sumatra is not as dystopian as 2024's chaotic America in the novel, the government's (federal, state, local) existence as tax collectors while ignoring starvation, disease, and mob rule signifies the distrust between the society and government [3]. While a series of floods struck West Sumatra, the novel's recurring theme revolves around water scarcity. Both opposite disasters related to waters are connected by one thing in common: greediness. The lawlessness in the novel creates people who are greedy and dangerous; the authorities are only there to collect the fees and are not interested in solving the social problems of the neighbourhood or the environment [3]. The government showed the same ignorance by letting the buildings surround the Anai watershed until the flash flood happened, which created several problems. One particular problem is the collapsed road in Jalan Raya Padang-Bukittinggi in Anai, which disrupted traffic, highlighting another issue of fragile infrastructure in West Sumatra.

The concern for environmental issues here aligns with the second and fourth principles of the deep ecology social movement. Excessive human interference with the Anai watershed and the failure to manage the area became the centre of the disaster of landslides, flash floods, and road collapses. Humans cannot interfere with or reduce the area intended to stabilise the watershed around Anai.

### 2.2 Fragile infrastructures

*Parable of the Sower* (2023) highlights infrastructure decay as roads, bridges, and utilities deteriorate in the surrounding Los Angeles area and the entire United States. With societal breakdowns that have occurred in the entire country because of global warming or climate change, the situation is worsening with corporate greed, racism, and spreading disease everywhere. The government does not care to build more roads as there is not enough money to build the road condition is depicted to have more dirt than asphalts, with several collapsed houses and scraps of people's belongings [3]. This situation signifies a lack of maintenance to keep the facilities applicable to society, adding problems to the security of people as mob rule is running rampant near the neighbourhood of Lauren.

The damaging infrastructure, such as a collapsed road approximately 200 meters in the Lembah Anai area due to flash floods that occurred on 11-12 May 2024, created significant problems ranging from mobilities, economy, and logistics. People who live in Lembah Anai are isolated, limiting access to work, education, and health. At the same time, all kinds of vehicles could not pass the area entirely, disrupting the distribution of products and logistics [15]. The area is closed for more than two months until it is opened on 21 Juli 2024, affecting the connectivity of Padang and Bukittinggi and creating more problems in the delivery of food, medicine and necessities as they were hampered, resulting in shortages and rising prices [16].

Infrastructure improvements and constant maintenance are the recurring concerns in both the novel and West Sumatra's flood disasters, as the fragility of those utilities affects social activities. Both governments should prioritise improving and maintaining roads, bridges, and other vital utilities because their neglect creates unstable conditions and uncertainties socially and psychologically. Mobility is the core of socioeconomic growth, and both governments should understand that mitigation of infrastructure-related disasters takes the utmost priority to stabilise situations, notably during disasters. It will also highlight how prepared governments are for natural disasters during these latest climates.

The fragile infrastructure is in line with the second principle of deep ecology social movement as the infrastructure decay affects the life of the society there. The deterioration of roads and bridges signifies the disturbance of the well-being and flourishing of human and nonhuman life on Earth.

### 2.3 Disaster preparedness

*Parable of the Sower* (2023) and West Sumatra's flood disasters in 2024 reveal how imminent natural disasters could occur anytime and anywhere, requiring resilience, hardship, and preparedness.

People need to stay resilient as the places are ravaged by environmental damage and social collapse, as seen in the novel. Lauren senses threats loom at a large scale, and she must endure and witness violence caused by people suffering from water scarcity, fragile infrastructure, and diseases that spread rampantly. She develops Earthseed to emphasise other people's emotions intensely, as the idea revolves around inevitable changes for humanity's task by acknowledging hardship, risk, and concerns to be resilient [3]. She believes people must adapt, collaborate, and shape the inevitable changes instead of succumbing to the despair of impending doom, as God can change, and people also can change [3, 4].

While the novel *Parable of the Sower* is rooted in the religious context of Christianity, West Sumatra's people, with a Muslim majority, should acknowledge the determination to prepare for imminent or predictable disasters. People should realise that the threats are real, and risk planning is the effort to avoid becoming the victims of natural disasters related to flash floods and landslides.

The local term 'galodo' literally means rocky terrains. However, the term can loosely refer to flash floods and landslides [17]. With such revelation, these natural disasters are something people are already familiar with, and they must find a way to overcome them. Government and people in West Sumatra can learn from the novel that disaster preparedness requires community awareness and collective action to prevent dire circumstances from happening in the future. As the hardships require resilience, government and people should approach the future with hope amidst the challenges they must overcome. As Lauren expresses in her belief in Earthseed, the action considers the consequences of getting along with God [3]. The government and society cannot ignore preparedness, as prevention is always better than mitigation. The road, a symbol of connectivity, now ironically leads to disconnection, highlighting the need for certainty, willingness to change, and addressing vulnerability.

Disaster preparedness aligns with the first principle of the deep ecology social movement, as the well-being and flourishing of human and nonhuman life on Earth must be balanced.

### 2.4 Embracing changes in the face of disasters

Preparedness is indeed a key message in Octavia Butler's masterpiece. However, the biggest takeaway of this novel lies in this philosophy or religion that the main character tries to promote: Earthseed. The philosophy provides Lauren's community with a principle of survival in a post-apocalyptic world.

However, through careful reading, one can see that it gives us a dynamic view of reality that our species can adapt to survive in this ever-changing world.

At the heart of this philosophy of Earthseed is change: "All that you touch You Change. All that you Change Changes you. The only lasting truth Is Change. God is Change" [3, 4]. Change, with a capital C, equates God and is manifested in nature. This belief has two significant implications concerning humans' relationship with nature.

The first one is that nature is in a constant state of flux, and this means that we have no choice but to adapt to and constantly devise new strategies to do so. In the real world, this wisdom can and should be practised by both the government and the public. There is no point in hanging to the notion that nature is stable—an erroneous view proven wrong with our current knowledge of climate change, moving earth plates, etc. Several years or decades of safety and stability are not always followed by decades free from danger. It is time that we take heed of the warnings from the most informed of us about impending dangers posed by nature and our ignorance. The government should do everything possible to take anticipative measures and not assume everything will be as it is.

The second one is a complex relationship between humans and nature. On the one hand, people are shaped by nature: our species has evolved in response to environmental challenges, and we produce technology in response to nature's challenges. This requires change on our part. On the other hand, we can affect our surroundings by shaping, moulding, or even harnessing what nature offers to benefit human communities. This worldview implies that interference may be necessary and that we can make positive changes. It also means that we need to develop awareness of the effects of our actions since what we do may affect nature negatively. The understanding that interferences can be beneficial or detrimental is crucial for our survival of nature. Our actions have effects, meaning we are not entirely helpless in the face of future disasters and are responsible for our actions' effects.

What is required to embrace this view of life fully? In the novel, it is clear that the key is to leave our old ways behind to see the limits of established rules, challenge specific fixed ways of thinking, and shatter the illusion of fixedness or stability. This is symbolically illustrated by the protagonist's refusal to follow the ways and teachings of her father, Reverend Olamina, who is described as someone who clings to the past. Despite his good intentions, the character insists on staying in the gated community, ignoring the lurking danger. This proves disastrous as a gang of murderous people attacks the community. His daughter Lauren, the protagonist, contrasts this static and conservative view of life. Instead of just staying, she journeys north, leading her group to a new life. In contrast to living in a gated and isolated community, she opens the possibility of a more inclusive and ever-growing community. In place of her father's religious view, she creates a new system of belief that helps her adapt to uncertain situations where natural disasters and man-made threats could happen anytime.

## 3 Conclusion

Olivia E. Butler remains hopeful that people must act, adapt, and shape changes in environmental crises instead of suffering in despair [18]. The urgency to acknowledge the impending environmental threat must also be communicated to the people of West Sumatra, as novels can serve as a medium for learning from our vulnerabilities and raising awareness about natural disasters' realities and potential consequences. We cannot ignore the impact of human interference with nature and the disasters it causes, as flash floods, landslides, and road collapses affect us economically, socially, and culturally. They can also alert us to the importance of mitigation and prevention. Taking examples from international novelists, Indonesian writers should bring readers' attention to these crucial matters before it is too late.

## References

1. D. E. Muthiariny, West Sumatra floods kill 23, losses hit Rp170 billion, Tempo. (2024). Available at: https://en.tempo.co/read/1843777/west-sumatra-floods-kill-23-losses-hit-rp170-billion (Accessed: 07 September 2024).
2. K. Piliang, and N. Karmini, Indonesia seeds clouds to block rainfall after floods killed at least 67 people while 20 are missing, AP News. (2024). Available at: https://apnews.com/article/indonesia-mount-marapi-flash-floods-cloud-seeding-f19bf53a1192ef9757c60e18619b0a8c (Accessed: 07 September 2024).
3. O. E. Butler, Parable of the Sower. Grand Central Publishing. (2023).
4. O. E. Butler, Parable of the Sower: A Graphic Novel Adaptation. Abrams. (2020).
5. B. Littleton, The Octavia E. Butler Plants an Earthseed, Octavia E. Butler plants an earthseed. (27 September 2007). Available at: http://web.archive.org/web/20070927084544/http:/www.cyberhaven.com/books/sciencefiction/butler.html (Accessed: 01 August 2024).
6. M. J. M. Iossifidis. Reading parable of the sower online in a pandemic: Collectively imagining different futures with Octavia E. Butler's speculative fiction, Literary Geographies. (23 December 2020). Available at: https://www.literarygeographies.net/index.php/LitGeogs/article/view/299 (Accessed: 1 August 2024).
7. P. B. Guerrero, 'Post-apocalyptic memory sites: Damaged space, nostalgia, and refuge in Octavia Butler's Parable of the Sower', Science Fiction Studies, **48**(1), 29–45. (2021). https://doi.org/10.1353/sfs.2021.0012.
8. J. Gottschall. The storytelling animal: how stories make us humans. Mariner Books. (2012)
9. A. Fletcher, Wonderworks: the 25 most powerful inventions in the history of literature. Simon and Schuster. (2022).

10. C. T. Cawdery, Putting an Environmental Spin on Literary Analysis, (2024). Edutopia. Available at: www.edutopia.org/article/environmentalism-literary-analysis.
11. Nash, F. Roderick Wilderness and the American Mind: Fifth Edition. Yale University Press. (2014).
12. J. R. DesJardins, Biocentrism. Encyclopedia Britannica. (2023) https://www.britannica.com/topic/biocentrism (accessed: 12 October 2024)
13. W. Sumbar, Pelanggaran Terhadap Tata Ruang di Kawasan Lembah Anai Harus Segera Ditindak Secara Tegas Oleh Pemerintah, Walhisumbar.org. (2024) Available at: http://www.walhisumbar.org/oknum-pelanggar-tata-ruang-di-kawasan-lembah-anai-harus-segera-di-tindak-pemerintah (Accessed: 07 September 2024).
14. Redaksi (2024) BPN: 107 Objek Terbangun di Sepanjang Kawasan Lembah Anai, Langgam.id. Available at: https://langgam.id/bpn-107-objek-terbangun-di-sepanjang-kawasan-lembah-anai/ (Accessed: 01 August 2024).
15. N. Harlina, Banjir Bandang di Lembah Anai, Jalur Padang-bukittinggi putus total, liputan6.com. (2024) Available at: http://www.liputan6.com/regional/read/5593828/banjir-bandang-di-lembah-anai-jalur-padang-bukittinggi-putus-total (Accessed: 30 July 2024).
16. J. Kampai. 2 bulan putus total Akibat Banjir Bandang, Jalur Lembah Anai akan dibuka 21 Juli, detiksumut. (19 July 2024) Available at: http://www.detik.com/sumut/berita/d-7446600/2-bulan-putus-total-akibat-banjir-bandang-jalur-lembah-anai-akan-dibuka-21-juli (Accessed: 01 August 2024).
17. P. Setyaningrum, Mengenal Istilah Galodo, Bencana Alam Yang menerjang Sumatera Barat Halaman all, KOMPAS.com. (2024). Available at: http://medan.kompas.com/read/2024/05/30/225241678/mengenal-istilah-galodo-bencana-alam-yang-menerjang-sumatera-barat?page=all (Accessed: 07 September 2024).
18. S. Valentine, How Octavia Butler's 1993 Book 'parable of the sower' predicted our climate reality, Grist. (2024). Available at: https://grist.org/culture/octavia-butlers-1993-parable-of-the-sower-predicted-climate-reality/ (Accessed: 07 September 2024).

# Traditional Minangkabau disaster mitigation in the Semangko Fault in West Sumatra based on ecological site myths

*Khairil* Anwar[1*], *Varhanno* Khallifhatul Khanh[2], *Vitrya* Qurratu Ayuni Khanh[3], and *Yenny* Narny[4]

[1]Department of Literature and Culture, Faculty of Humanities, Universitas Andalas, Padang, Indonesia
[2]Department of Medical Studies, Faculty of Medicine, Padang State University, Padang, Indonesia
[3]Department of Environmental Engineering, Faculty of Engineering, Universitas Andalas, Padang, Indonesia
[4]Departmen of History, Faculty of Humanities, Universitas Andalas, Padang, Indonesia

**Abstract.** This paper discusses traditional Minangkabau disaster mitigation based on the myths of Minangkabau ecological sites located in the Semangko Fault area in West Sumatra. Traditional communities believe in myths as a force that guides their relationship with ecological sites. Natural signs conveyed in myths serve as guidelines for the community in disaster mitigation. This belief has the consequence that the community must adhere to it. Myths may deviate from beliefs to the superstructure and, vice versa, originate from conventions so that they become superstructure. Therefore, through a qualitative study, myths related to ecological sites and their meanings are documented about community disaster mitigation in the Semangko Fault area. The analysis was conducted using the perspective of cultural materialism. It was found that myths are the superstructure of ecological sites that regulate human interaction with the environment; myths embody past local experiences of natural disaster events; myths become a reference for disaster mitigation management that is familiar in local communities. Recommendations in disaster management should consider myths in disaster mitigation.

## 1 Introduction

The ethnic Minangkabau, known for their unique cultural heritage and traditions, inhabit the central region of Sumatra Island. The area is geographically located on the Semangko Fault [1], in the Bukit Barisan range that extends along the island, with volcanoes stretching from north to south. This area is prone to earthquakes and hydrological disasters [2].

The influence of nature on Minangkabau society cannot be ignored. The Bukit Barisan, the coastline, and the towering volcanoes provide a dramatic backdrop to people's lives. The spiritual life of the Minangkabau people in this region is deeply connected to the natural surroundings. Belief in the power of nature is strong, and traditional celebrations such as *bakaua* and *balek padang* are essential moments in their annual calendar.

Life on the Semangko Fault is full of challenges. The natural topography significantly impacts the lives of the Minangkabau people in this region [3]. Despite this, they have maintained their cultural values and local wisdom in changing times. The Minangkabau people depend on nature for their livelihood, creating a harmony between their rich culture and daily life that is in tune with natural cycles. The region's infrastructure follows the hilly land's contours, earthquakes, and high rainfall [4]. The local community developed adequate infrastructure for daily life while preserving the natural surroundings.

### 1.1 Humans

Human familiarity with the ecology of the Semangko Fault area has given birth to myths that strengthen their relationship with nature [5]. Each area, forest, cave, river, and mountain are considered to have a history and power that must be respected. The myths tell the origins and natural phenomena and teach life values from generation to generation [6].

A well-known example of a myth is the story of Mount Marapi, which is not only the place of origin of Minangkabau's ancestors but also a guardian force and environmental sustainability. People believe this myth is not only a story but also a reminder of the importance of maintaining ecological balance and respecting the power of nature [7]. This familiarity is also reflected in rituals and traditional ceremonies held periodically to honour nature. These myths become a solid cultural heritage, binding the community in close unity with their natural environment and teaching the importance of respecting and preserving nature.

These myths strengthen humanity's spiritual connection with ecological sites [8]. In addition, they play an essential role in maintaining the fragile balance of the ecosystem in this area. For the people in this Semangko Fault area, the myths play a central role in guiding human interaction with the surrounding nature. Each ecological

* Corresponding author: khairilanwar@hum.unand.ac.id

site is considered to have magical and protective powers that affect daily life [9].

These myths are born from a deep belief in supra powers and local people's conventions about natural events. These myths form the spiritual and moral basis of the community, teaching values such as respect for nature and solidarity in the face of natural challenges [10].

In this context, myths are not just stories but also ways to explain and understand complex natural phenomena. Myths are an integral part of the social fabric of society, regulating human interaction with the natural environment [6]. They recount the origins and forces of nature and establish norms of behaviour that promote ecological respect and sustainability. In this way, myths serve as moral and spiritual guides that establish a harmonious relationship between humans and nature, maintaining the ecosystem balance vital to their daily lives.

Understanding myths in the context of disaster mitigation is very important because it can change people's behaviour and decisions when facing the threat of natural disasters [11]. An accurate understanding of disaster risk myths can help address this by providing precise and verified information about potential hazards and actions that can be taken to reduce risks. Educating the public about the scientific facts of disasters and clarifying myths can increase the awareness and preparation required. Thus, understanding myths strengthens disaster mitigation strategies, helping communities to act more appropriately and effectively in the face of possible threats [12].

Although studies on myths have been conducted widely in various places, research focusing on ecological site myths as disaster mitigation management on the Semangko Fault in West Sumatra still needs to be completed. The existence of myths about the Semangko Fault in West Sumatra must be explained. Historically, myths became a spiritual and moral guide for the community in maintaining ecological balance. They retain their cultural identity today and inspire sustainable disaster mitigation management [13]. In the future, it is crucial to understand and appreciate myths as a source of local knowledge that can provide valuable insights into disaster mitigation to preserve nature and build harmony between humans and the environment.

## 2 Method and theory

The research method for understanding human and environmental phenomena is descriptive and interpretive. This research documents quality myths through in-depth interviews, participatory observation, and content analysis. Furthermore, it explores individuals' or groups' meanings, perceptions, and subjective experiences.

This provides the flexibility to capture the complexity of the social and cultural context in which the mythical phenomenon is located a purposive sampling technique to get suitable informants. The data obtained is analyzed inductively by identifying patterns, themes, or concepts that emerge from the data.

For analysis, the theory of cultural materialism is used. This theory emphasizes the role of materials or economics in shaping culture and social structure. It assumes that material aspects such as economic production, wealth distribution, and social class structure are central to shaping a society's values, norms, and cultural practices.

According to this theory, how societies produce and distribute wealth and material resources affects social hierarchies, the division of labour, and patterns of social interaction. It explains how material factors create social inequalities and broadly influence culture, collective identity, and social change [14]. The theory of cultural materialism is also a framework for explaining cultural and social dynamics in the context of global structural and economic change. For example, land ownership and agricultural tools can shape power structures and social status. It influences value systems, self-image, and individual aspirations.

## 3 Results and discussion

In the Semangko fault zone in West Sumatra, mythical ecological sites flourish. Each site has a name that matches its history. Traditional communities believe that history is a truth that must be respected, maintained, and sustained. Myths, in the view of cultural materialism, are stories built into superstructure forces to preserve nature's material and economic aspects for the benefit of human welfare [14]. The following describes some samples of these myths. Without ignoring the widespread myths in the discussion, examples of significant myths are shown in the description below and can be seen in Table 1.

**Table 1.** Myths of Semangko Fault ecological site.

| Myths | Content | The message | The meaning |
|---|---|---|---|
| Gunuang Marapi | -Ancestral Origins Site<br>-Mountain eruption | -Ring of fire sacred site<br>-Eruption alert | Disaster mitigation |
| Batu Anguih | - Magma rock site<br>-Lahar of Mount Marapi<br>-Prayer to ward off disasters<br>-Trust in leadership | -Ring of fire sacred site<br>-Be aware of Eruptions and lahars<br>-The importance of ritual power<br>-Harmony with the nature | Disaster mitigation |
| Malalo | -Natural Disaster<br>-Religious figures<br>-Sacred tombs | -Semangko fault site<br>-Sacred sites<br>-Landslide alert | Disaster mitigation |
| Ngarai Sianok | -Natural resource area site<br>-Maintaining forest sustainability<br>-Respect for nature | -Semangko fault site<br>-Fragile topography of the land<br>-Be aware of land movement | Disaster mitigation |

| Myths | Content | The message | The meaning |
|---|---|---|---|
| | -The power of nature | | |
| Rimbo Larangan | -Natural resource area site<br>-Maintaining forest sustainability<br>-Respect for nature<br>-The power of nature | -Sacred sites<br>-Source of protein and energy<br>-Antidote for oxygen, landslides, and floods | Disaster mitigation |
| Landscape of Rumah Gadang | -Tribal landscape autonomy<br>-Landscape infrastructure<br>-Building structure<br>-Interaction media | -Sacred landscape site<br>-Eco-friendly building<br>-Be aware of natural disasters | Disaster mitigation |

1. Gunuang Marapi

Gunuang Marapi is one of the mountains in the Semangko fault area in West Sumatra. In traditional historiography, this mountain is the place of origin for the ancestors of the Minangkabau people. Many myths have grown around this mountain, such as *pancang*, *batu gadang*, *rimbo bunian*, *lurah mangaun*, *aia tajun*, etc. The big myth is about the mountain itself.

The community considers Mount Marapi to be the guardian force and sustainability of the environment. When the mountain emits smoke, flames, and lava, it is believed that the mountain warns the surrounding community about its wrong actions and behaviours and its readiness to face the next series of events. The local people believe that ignoring them can lead to disaster for them. This myth is a story and a reminder of the importance of maintaining ecological balance and respecting the power of nature.

2. Batu Anguih

Batu Anguih, located on the western slopes of Mount Merapi, is mythologized and believed by the Minangkabau people living near Sungai Pua. Many variations of stories about the stone have developed in the community. One of them is the story of the eruption of Mount Merapi, which released hot lava. Batu Anguih was originally a large rock above the village. The stone became a place of prayer for Inyiak, a religious figure, to ask for blessings and protection from the Creator and Owner of the Power of Nature so that the community would prosper and be kept away from the impact of the eruption and hot lava of Gunuang Marapi. Thanks to Inyiak's prayer, the tongue of hot lava at that time was held in the rock where Inyiak prayed. Because of the hot lava flowing and stuck on the rock, the rock became black like burnt.

These myths are often related to magical powers or natural blessings. Local people believe this stone has spiritual energy that protects and brings them good luck. It is usually a destination for pilgrimages and religious rituals, where prayers are offered for blessings and protection. It also influences the culture and social life of the surrounding communities, reflecting the deep connection between humans, nature, and the spiritual world.

Batu Anguih is considered to have spiritual powers that protect the surrounding communities. This myth often leads communities to rely on mystical protection rather than scientifically-based disaster mitigation solutions. Understanding the power of Batu Anguih can serve as proper education on disaster mitigation, help separate facts from myths, promote more effective solutions, and improve community safety in disaster-prone areas.

3. Malalo

The Malalo myth tells the story of a landslide in Nagari Malalo, located on Lake Singkarak's edge, after an earthquake and heavy rains. The community called this event Malalo because, at the time of the incident, the community was in a state of hysterics, and some asked while shouting ma lolo (where is the stone). Gradually, the community came to call the landslide site Malalo.

This area is landslide-prone but has slowly developed into a community village. Here, it became one of the centers for the development of Islamic teachings on the shores of Lake Singkarak in the past and present. There are two grave sites of qualified religious figures in this area: Syekh Tuanku Nan Limo Puluah and Syekh Tuanku Panjang Abuak. These two sites have supernatural powers, and if the tombstones of the initially upright graves become tilted, it is a sign that an earthquake or landslide will occur in the surrounding area. People believe this and prepare themselves for the disaster, such as staying away from disaster-prone places.

Local people view this story as a focal point of spiritual belief, often associated with the guardianship of nature and supernatural forces that must be respected. It is a mythological object and the center of attention in religious ceremonies and rituals. The existence of these myths is a reflection of the deep connection between humans and the natural environment. These myths are passed on from generation to generation, playing an essential role in maintaining the cultural and moral values of the community.

In a modern context, this myth offers insights into the history and local wisdom. It opens a window to understanding the complex interactions between humans, nature, and spirituality in Minangkabau culture. It also highlights the importance of maintaining cultural heritage as part of efforts to preserve the identity and environmental sustainability of the area.

In Minangkabau mythology, it is believed to have spiritual power to protect communities from natural disasters. These myths sometimes influence people's perception of the need to take real action in disaster mitigation. Myths have become a familiar educational part of the locality. Together with scientific facts, they are an effective mitigation strategy to help communities understand the real risks and take appropriate actions to improve their resilience to natural disasters.

4. Ngarai Sianok

Initially, Ngarai Sianok was the name of one of the villages located at the bottom of the canyon. In the Minangkabau language, *sianok* means quiet (Si Anok), and the name *sianok* is also due to the canyon's calm and peaceful conditions and atmosphere.

Ngarai Sianok is a visual representation of plate activity on the island of Sumatra. This canyon was formed due to the descending part of the earth's slab, causing a fault in a steep ravine.

Ngarai Sianok is a natural wonder and full of deep myths and legends. According to popular belief, the canyon is the scars of a magic man's dagger used to split the mountain to create this beautiful valley.

It is said that the canyon is also considered a place guarded by spiritual spirits or guardians of nature. Local people believe they should honor and care for this canyon as a sign of gratitude to the supernatural forces that protect and bless it.

The myth that developed in Ngarai Sianok is that there used to be a man named Katik Muno with a large and influential body. He came to accompany his leader, Sang Sapurba. During his stay in Minangkabau, Katik Muno wanted to become the ruler. Because his desire was so strong, Katik, who initially had a gentle nature, became an evil and violent person. His crimes caused suffering to the local people. Seeing his behaviour, Sang Sapurba felt very ashamed. To avoid conflict with his superiors, Katik Muno became a dragon and divided the land into two parts. The land was divided into two parts, and it became Ngarai Sianok. When it was formed, the canyon was flowing with burning fire. But after the battle between Katik and Sang Sapurba, Katik Muno changed the fire flow into a stream of soothing water. The gorge became a fault that separated two plateaus.

Ngarai Sianok is a Minangkabau natural icon with mythological value as a spiritual place. The myth of the canyon's natural beauty and power sometimes obscures the urgency of disaster mitigation around the Semangko Fault. People may focus on the natural beauty and forget about the real risks of natural disasters such as earthquakes, landslides, and floods. Education about potential hazards and the need for appropriate mitigation preparations can help communities appreciate the natural beauty while remaining aware of the threats, encouraging them to take proper steps to improve disaster resilience.

5. Rimbo Larangan

Rimbo Larangan is a myth and legend in many cultures in Indonesia, including in West Sumatra. In West Sumatra, they are scattered in various corners, especially in the mountains and hills. These forests have various names, such as *hutan ulayat*, *bukik bunian*, and *rimbo uduhan*. This concept refers to forests considered sacred or forbidden to be entered or utilized by local communities without special permission or certain customary agreements. In Minangkabau society, prohibited forests are often associated with spiritual forces or guardian spirits of nature that are thought to live within them.

This myth reflects the community's respect for preserving nature and ecological balance. The community believes violating the prohibition may incur the wrath of the forest guards, who can cause disaster for those who do so. Therefore, Rimbo Larangan is a stronghold for ecological protection and a place to nurture cultural values and local wisdom.

Understanding these myths helps communities view their relationship with the natural environment. It also highlights the importance of maintaining and preserving areas that are considered sacred as an effort to care for the cultural and ecological heritage that has been passed down from generation to generations.

Rimbo Larangan is a forbidden forest area with spiritual value and local wisdom in Minangkabau. Myths surrounding the spiritual power of this *rimbo* may reinforce awareness of natural disaster risk. People consider the mystical protection of nature to be more dominant than scientifically-based mitigation strategies. However, education that focuses on scientific facts regarding potential and necessary disaster mitigation efforts is essential to help communities integrate local wisdom with safer practices in the face of real natural disaster threats.

6. Landscape of Rumah Gadang

Rumah Gadang is a living myth in Minangkabau society. It is an icon of traditional Minangkabau architecture that reflects the rich culture and identity of the people. It is built in a tribal autonomy landscape with unique characteristics. The landscape sits on tribal communal land with a history of ownership struggles and names. The Rumah Gadang has a steeply pitched pyramid roof and tall columns. The building materials are made from selected, solid, and durable wood such as ironwood or meranti wood, which are beautifully carved with distinctive natural or geometric motifs.

Rumah Gadang's architecture is not only aesthetic but also functional. Its interior spaces are designed to support the communal lifestyle of an extended Minangkabau family, with a spacious main room that serves as a place to gather, eat, and sleep. The building also features a spacious veranda in front of the rooms, providing an additional area for outdoor activities.

In this autonomous landscape, in front of the Rumah Gadang, there are four ranks of places to store food. There are four categories of food: daily consumption, exchange with other materials, social and disaster preparedness, and germplasm for the sustainability of food seeds. Each food is sequentially stored in *rangkiang sibayau-bayau*, *sitinjau lauik*, *sitangka lapa*, and *sikaciak*.

The same applies to the *surau* infrastructure within the landscape. *Surau* serves as a medium of worship and education and a place for men in the tribe to spend the night. In addition, the *surau* also serves as a place to maintain the security of the tribe's landscape by the tribe's men from threats from other living things and natural disasters.

Symbolically, Rumah Gadang symbolizes family unity, economic success, and social status in Minangkabau society. Rumah Gadang is believed to have spiritual and magical value in a mythological context. The construction and orientation of buildings are often connected to creation myths or beliefs about harmony with the surrounding nature.

The mythical landscape of Rumah Gadang and its environment on the Semangko fault can be an essential part of disaster mitigation management because this traditional design considers local geographical and geological factors. With its structure and location integrated with the natural landscape, Rumah Gadang is designed to reduce the risk of earthquake damage. Selected building materials and robust construction can increase vibration resistance. In addition, choosing a geologically and hydrologically safe location can also reduce the risk of landslides or flooding. The same applies to other infrastructure. Thus, the landscape of Rumah Gadang is not only part of the cultural heritage but also plays a role in protecting the community from the threat of natural disasters. Disaster risk education and technical solutions can help improve and maintain Rumah Gadang as an essential part of cultural heritage while increasing their resilience to disasters and sustainable development goals (SDGs).

### 3.1 Ecological myths of the Semangko Fault area: disaster mitigation management

In the Semangko Fault area, myths have a significant role in disaster mitigation management. This area often faces the threat of natural disasters such as volcanic eruptions, landslides, and floods [1]. Local myths and beliefs reflect the cultural identity of the Minangkabau people and play an essential role in their efforts to manage and reduce disaster risks.

First of all, myths often include local knowledge about the geology and ecology of the area. For example, stories about rivers can provide clues about natural patterns and changes that may occur before a disaster strikes. This knowledge can help communities prepare and take appropriate precautions.

In addition, myths also often teach values about respect for nature and how to coexist with disaster-prone environments [15]. Minangkabau people have a tradition of performing certain rituals to invoke protection or appease the anger of the creator and guardian of nature. These actions strengthen their cultural identity and provide a more profound sense of involvement in disaster mitigation efforts.

On the other hand, myths can be a tool to educate and raise public awareness about disaster risk [16]. Stories of past disasters or mystical events associated with nature are often passed down through generations, warning future generations of potential dangers and the importance of vigilance.

In the context of modern disaster mitigation management, participatory approaches are essential. Local myths and beliefs can be integrated into education and training programs, allowing communities to actively contribute to the planning and implementing mitigation strategies [17]. For example, knowledge of myths about natural signals or signs that can be interpreted from traditional stories can be used to develop more effective early warning systems [4].

In addition, myths can also serve as a source of inspiration for mitigation technologies or systems innovations. Some myths may contain how past communities coped with or survived disasters. By understanding and exploring these myths, new ideas or strategies can be found that can be applied in modern contexts to improve community resilience to disasters [18].

However, it is essential to remember that the approach to myths in disaster mitigation must be accompanied by scientific research and accurate empirical data. Myths should not be used as the sole basis for decisions but can be used as an adjunct to enrich understanding of the natural and cultural dynamics in disaster-prone areas.

Integrating myths in disaster mitigation management in the Semangko Fault area can result in a more holistic and sustainable approach. Building bridges between local wisdom and modern science can create a safer and more sustainable environment for the people living in this disaster-prone area [19].

Traditional disaster mitigation management in Minangkabau communities through the utilization of myths and ecological sites in West Sumatra reflects the complexity of the relationship between local culture, traditional environmental knowledge, and modern mitigation strategies. The Minangkabau, a matrilineal ethnic group living in the western mountainous region of Sumatra, has developed a robust knowledge system about the natural environment over the centuries. Those traditional myths and legends are integral to cultural heritage and have significant value in planning and managing natural disaster risks.

### 3.2 Cultural and mythical context

Minangkabau people view nature as a living entity with certain powers and characters. Myths and folktales describe the relationship between humans and nature and how human behavior can affect or be affected by nature. These myths are stories to entertain and serve as tools to teach moral values, ethics, and practical knowledge on interacting with the natural environment.

#### *3.2.1 Utilization of myths in disaster management*

In the context of disaster mitigation, myths, and folklore often contain hidden clues on how to deal with or anticipate natural disasters. For example, some myths tell about natural signs that may indicate the coming of a disaster, such as animal behavior or unusual weather changes. Local communities use this knowledge to prepare themselves mentally and physically.

#### *3.2.2 Ecological sites as knowledge centers*

Ecological sites in West Sumatra, such as forests, rivers, and mountains, are sacred places closely connected to myths and legends. Local communities treat these places respectfully and maintain their ecological balance as part of their cultural identity. Knowledge of local flora and fauna passed down from generation to generation is also crucial in sustainable environmental management and disaster risk reduction.

### *3.2.3 Integrated disaster mitigation approach*

Disaster mitigation management in Minangkabau communities relies on traditional ecological knowledge and myths and includes an integrated approach with modern elements such as technology, education, and sustainable urban planning. These programs integrate local understanding of myths and ecological sites with modern scientific research and practical experience in disaster management.

### *3.2.4 Case studies and practical implementations*

Some examples of disaster mitigation practices inspired by myths and local knowledge in West Sumatra include nature observation-based early warning systems, evacuation plans that consider the topography and river flow patterns described in folklore, and the use of certain plants to maintain the stability of mountain slopes. This case study shows that integrating local wisdom and modern science can lead to more effective solutions in reducing disaster impacts.

### *3.2.5 Challenges and opportunities*

While local wisdom has excellent value in disaster mitigation, maintaining and transmitting this knowledge amidst rapid social, economic, and environmental changes is the biggest challenge. Formal education and community programs are essential in supporting this cultural heritage while adapting it to the modern context.

Traditional disaster mitigation management of Minangkabau communities using myths and ecological sites in West Sumatra shows how local wisdom can be effectively integrated into modern mitigation efforts. By utilizing existing knowledge and respecting cultural values, these communities protect themselves from natural disasters and maintain a harmonious relationship with their natural surroundings. This is an example of how cultural diversity can be a valuable resource in the face of global environmental challenges.

## 4 Conclusion

The discussion of ecological myths on the Semangko Fault and traditional Minangkabau disaster management in West Sumatra highlights the importance of local wisdom in mitigating natural disaster risk. The region, with its landscape of mountains, rivers, and lush forests, plays a vital role in the lives and beliefs of the Minangkabau people.

The myths spread across the Semangko Fault reflect the community's cultural identity and serve as a foundation for understanding and responding to the oftenunpredictable dynamics of nature. In disaster management, these myths and traditional beliefs become essential instruments. Local communities usually use knowledge from these myths as early signals or signs to anticipate disasters such as volcanic eruptions, landslides, or floods. This role is not only as a warning sign but also as a guide to take appropriate precautions.

Traditional Minangkabau disaster management also includes ritual and customary practices integrated into daily life. Traditional ceremonies and offerings to the guardian spirits of nature are part of the strategy to invoke protection and blessings from nature. This strengthens the community's spiritual bond with the environment and promotes respect for nature as the source of life.

However, disaster management approaches cannot rely solely on local wisdom in this modern era. There needs to be integration with modern science and technology to improve early awareness, more precise risk analysis, and rapid response in the face of disasters. Combining traditional knowledge and scientific approaches can lead to more effective and sustainable disaster mitigation strategies.

It is also essential to continue building the community's capacity to understand and respond to disasters. Education about myths, cultural values, and traditional disaster risk management must be strengthened. This can be done through education programs, training, and information campaigns that promote collaboration between local communities, government, and academia.

Thank you to Universitas Andalas for conducting the study "Disaster Mitigation of Traditional Minangkabau Communities in the Semangko Fault Area of West Sumatra." The university's contribution to understanding and appreciating local wisdom and disaster management is meaningful for environmental sustainability and community safety. Thank you for your support and collaboration.

## References

1. Edward, Danger of Earthquakes in The Sumatran Fault Zone, (Pusdalops PB BPBD Prov. Sumatera Barat, Padang, 2015)
2. Verstappen and H. Th, A geomorphological reconnaissance of sumatra and adjacent islands (Indonesia) (Wolters-Noordhoff, Netherland, 1973)
3. I. Wibowo, Volcano: between legends, myths, and disaster management, (Mitragama Widya, Yogyakarta, 2005)
4. P.A. Aprilia, F.N.S. Sukowati, A. Nurrohmah, Development of a disaster warning system: education to reduce the perception of public myths regarding earthquake preparedness in bantul regency based on vibration and wave perspectives, in Proceeding, UNY, Yogyakarta (2020). https://eprints.uny.ac.id/68592/1/Prosiding 28 to 33_Ananda_UNY.pdf
5. K. Anwar, Disaster mitigation local wisdom in the tradition of the minangkabau community. E3S Web Conf., **331** (2021). https://doi.org/10.1051/e3sconf/202133104013
6. K. Burkiewicz, Cosmogonic and anthropogenic myths in sui oral literature. Litt. Copernicana, **30**, (2019). https://doi.org/10.12775/LC.2019.028

7. K. Anwar, Ferdinal, and R. Devi, Inyiak and marvin harris keepers of tradition oral and ecology: challenges in the era of oil palm plantation expansion in West Sumatra, in Proceeding of the 13th International Conference on Malaysia-Indonesia Relations (PAHMI) (2019). https://doi.org/10.2478/9783110680003-028
8. Y.M. Supriatin, Interpreting myths as a medium for disaster mitigation in Sundanese society no title, in Proceeding of Balai Arkeologi Jawa Barat, Balai Arkeologi Jabar, Bandung (2021). https://doi.org/10.24164/prosiding.v4i1.34
9. K. Anwar, Ecological wisdom of oral tradition: Surface and deep structure tension in preserving the lake environment, in IOP Conference Series: Earth and Environmental Science (2020). https://doi.org/10.1088/1755-1315/469/1/012021
10. N. Couldry, Media rituals: Beyond functionalism. (Sage Publications, New York, 2005). https://doi.org/10.4135/9781452233819.n6
11. N.A. Putri, Sutiyo, I. Yunari, A. Supriatna, and I. Uluputty, Disaster mitigation based on local wisdom in the kasepuhan ciptagelar indigenous community. Gema Publica J. Manaj. dan Kebijak. Publik, **9**, (2024). https://doi.org/10.14710/gp.9.1.2024.65-82
12. Sunimbar, Tectonic earthquake disaster mitigation based on local wisdom of the Lio Community in Ende Regency, Nusa Tenggara Province Timur, in Proceeding Seminar Nasional diselenggarakan Pendidikan Geografi FKIP UMP Manajemen Bencana di Era Revolusi Industri 5.0, Prodi Pendidikan Geografi FKIP UMP, Purwokerto (2019). https://digitallibrary.ump.ac.id/464/2/13. Full Paper_Sunimbar.pdf
13. V.Q.A. Khanh, K. Anwar, Oktavianus, V.K. Khanh, Disaster mitigation through oral stories: a case study around Singkarak Lake. E3S Web of Conferences, **464** (2023). https://doi.org/10.1051/e3sconf/202346416002
14. M. Harris, Cultural materialism: The struggle for a science of culture (Altamira Press, New York, 2001)
15. M.I.B.F. Hidayat, Disaster mitigation: Local knowledge and the political challenges of spatial planning, (Andalas University, Padang, 2024). https://www.unand.ac.id/index.php/berita/opini/978-opini-dosen-mitigasi-bencana-politik.html
16. M.S.B. Wibowo, Forest history as disaster mitigation education. Ref. Edu. J. Ilm. Kependik., **12**, (2022). https://doi.org/10.24176/re.v12i2.6889
17. M.M.M.R. Akbar, S. Nurhaliza, F. Khaidir, A. Iswar, and F.A. Salam, Disaster mitigation based on local knowledge of the Majene Community. Kybernology J. Gov. Stud., **2**, (2022). https://doi.org/10.26618/kjgs.v2i2.9661
18. L.U.S. Nurbayani, Social capital based on local wisdom in disaster mitigation. Talent. Conf. Ser. Local Wisdom, Soc. Arts, **2**, (2019). https://doi.org/10.32734/lwsa.v2i1.628
19. J.P. Pratama and T.P.B.P.S. Hiram, Mitigation management based on local wisdom in the West Java Area. J. Edu. dan Multimed., **2**, (2024). https://doi.org/10.37817/jurnaledukasidanmultimedia.v2i2

# Groundwater utilization and its environmental sustainability in coastal aquifers of Dakshina Kannada District, Karnataka, India

*Jayaprakash* M C[1*], *Rinaldi* Ekaputra[2], and *K* Channabasappa[3]

[1]Department of Civil Engineering, Mangalore Institute of Technology and Engineering, Mangaluru, Karnataka, India
[2]Department of Sociology, Faculty of Social and Political Sciences, Universitas Andalas, Padang, Indonesia
[3]Department of Geology, School of Earth Sciences, Central University of Karnataka. Kalaburagi, Karnataka, India

**Abstract.** Water is one among the foremost precious of natural resources. In many regions of the world, the pressures of economic development are creating a surface water scarcity. Study of groundwater sustainability– like to remind people that most of the world's available freshwater is stored underground. The best institutional set up for any country depends on its system of government, its climatic and hydrological context and its social and economic circumstances. Groundwater resources to meet current and future beneficial uses without causing unacceptable environmental or socioeconomic consequences. The methods that promote the wise use of groundwater supplies, is need to determine strategies that promote groundwater sustainability. Groundwater users have not generally had to pay anything like "full economic cost" for groundwater use, nor have groundwater polluters had to pay for clean-up of their contamination. Thus, this study aims to determine the strategies that promote ground-water sustainability and groundwater resources to meet present and future beneficial uses without causing un-desirable environmental or socioeconomic consequences.

## 1 Introduction

Freshwater found in the pore spaces of soil and the fissures of rock formations beneath the surface of the earth is known as groundwater. When a rock or unconsolidated deposit can provide a useable amount of water, it is referred to as an aquifer. Only 3% of the water on Earth is fresh; the remainder is saline, or oceanic. 30.1% of the freshwater on Earth is groundwater, 0.3% is surface water, 0.9% is other minor storage, and 68.7% is permanently stored in icecaps and glaciers. "The study uses numerical simulation to examine how shallow, coastal unconfined aquifers respond to predicted overdraft conditions and the impact of climate change." [1, 2].

There are significant issues with groundwater management in India as a result of overuse of groundwater and the need for irrigation in large canal commands. "Depletion of water tables, saltwater encroachment, drying of aquifers, groundwater pollution, water logging and salinity, etc. is major consequences of overexploitation and intensive irrigation" [3]. Groundwater in coastal area is of major alarm due to the fact that > 60% of the world population lives within 30 km of shoreline and about 20% of the population of India lives in the coastal regions. Seawater ingression is a major problem in most of the coastal aquifers.

This occurs when the balance between freshwater and saltwater is upset for a variety of reasons, including land reclamation, climate change, overdraft, and rising sea levels. "A good understanding of the coastal dynamics and detailed knowledge of the variability of their parameters is essential to carry out studies on coastal aquifers" [4].

## 2 Study area

Mangalore is an officially known as Mangaluru, located next to the Arabian Sea in the southwest region of Karnataka. The district receives almost 3800 mm of rainfall a year, and its total size is 834 square kilometers. extending between 12°30'00" & 13°11'00" north latitude and 74°35'00" & 75°33'30" east longitude. Mangalore town is the district headquarters. Administratively, the taluk is having several villages viz. Ganjimata, Moodbidri, Belvai and Mulki. Population of the district (2011 census) 4,84,785 in the city and 6,19,664 in metropolitan. The primary occupation of the people in the district is agriculture. Twenty-eight percent of the entire geographical area is made up of the net sown area. Paddy, arecanut, coconut, cashew nut, rubber, and vegetables are the main crops. Various water sources irrigate roughly 57% of the net planted area. About 75% of the irrigated area is watered by groundwater, with the remainder coming from surface water sources.

* Corresponding author: jayaprakash@mite.ac.in

## 3 Materials and methods

Identified the groundwater aquifer of the coastal area in and around part of Mangalore (groundwater stressed areas) to determine the characteristic of groundwater as well as to study the depletion of depth of the water table of both dug well and bore well. Reconnaissance survey of the study area was conducted using depth of the groundwater table fluctuation vs. rainfall data (from 2010 to 2019) based on the climatic conditions of the area.

## 4 Results and discussion

Before presenting the data on groundwater quality, we introduce key parameters analyzed in this study, as summarized in Tables 1-6 and Figs.1-6. Each parameter offers insights into various aspects of water quality: Table 1 and Fig. 1 present pH levels, indicating the acidity or alkalinity of the water. Table 2 and Fig. 2 focus on dissolved oxygen (DO) levels, which are critical for supporting aquatic life and reflect the water's overall health. Chloride concentrations, provided in Table 3 and Fig. 3, help identify potential contamination sources of chloride. Table 4 and Fig. 4 displays turbidity values, which measure water clarity and can signal the presence of suspended particles. Table 5 and Fig. 5 illustrate total hardness, reflecting the water's mineral content, such as calcium and magnesium, which affects both household and agricultural uses. Lastly, Table 6 and Fig. 6 present of total solids of soil. By examining these parameters across multiple groundwater sampling points, we aim to provide a comprehensive assessment of water quality in the study area, detailed in the following tables and figures and the recapitulation is in Figs.7 and 8 which explain the static water level for dug wells and bore well.

**Table 1.** Results of pH.

| | Bore Well | Dug Well | BIS: 10500 – 2012 |
|---|---|---|---|
| Mulki | 7 | 6.5 | Desired Limit: 6.5 - 8.5 Water will have an impact on the mucosal membrane and water supply system if it exceeds this threshold. Permissible Limit: No Relaxation |
| Belvai | 7 | 6.5 | |
| Ganjimata | 7 | 6.5 | |
| Moodabidri | 7 | 6.5 | |

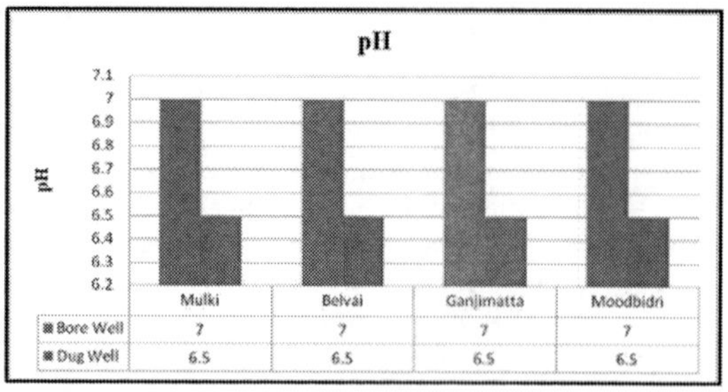


**Fig. 1.** Graphical representation of pH

**Table 2.** Results of chloride (mg/l).

| | Bore Well | Dug Well | BIS: 10500 – 2012 |
|---|---|---|---|
| Mulki | 22.68 | 19.85 | Desired Limit: 250 mg/l Taste, corrosion, and palatability are impacted beyond this range. Permissible Limit: 1000mg/l |
| Belvai | 8.50 | 19.85 | |
| Ganjimata | 14.18 | 3.40 | |
| Moodabidri | 2.83 | 5.67 | |

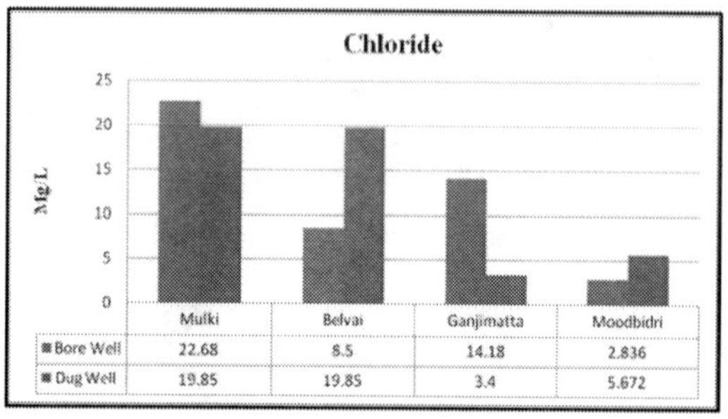


**Fig. 2.** Graphical representation of chloride

**Table 3.** Results of turbidity (NTU).

| | Bore Well | Dug Well | BIS: 10500 – 2012 |
|---|---|---|---|
| Mulki | 4.4 | 8.4 | Desired Limit: 5 NTU Above 5, consumers acceptance decreases Permissible Limit: 10 NTU |
| Belvai | 4.6 | 6.1 | |
| Ganjimata | 9.3 | 8 | |
| Moodabidri | 4 | 6.2 | |

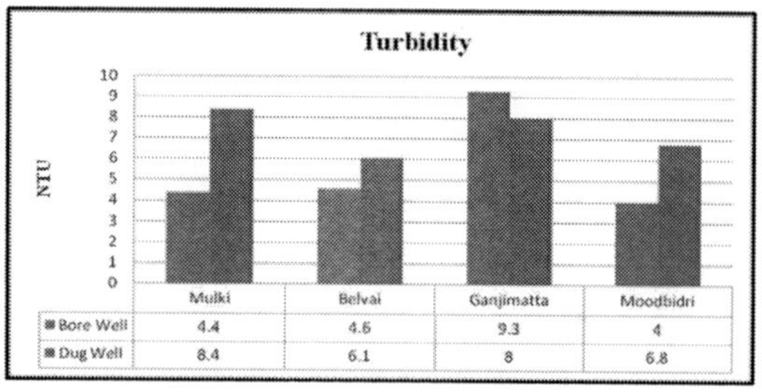


**Fig. 3.** Graphical representation of turbidity

**Table 4.** Results of dissolved oxygen (mg/l).

| | Bore Well | Dug Well | BIS: 10500 – 2012 |
|---|---|---|---|
| Mulki | 4.72 | 3.94 | Desired Limit: 5 – 7 mg/l @ 25°C |
| Belvai | 3.94 | 6.10 | |
| Ganjimata | 8.07 | 6.40 | |
| Moodabidri | 6.2 | 5.71 | |

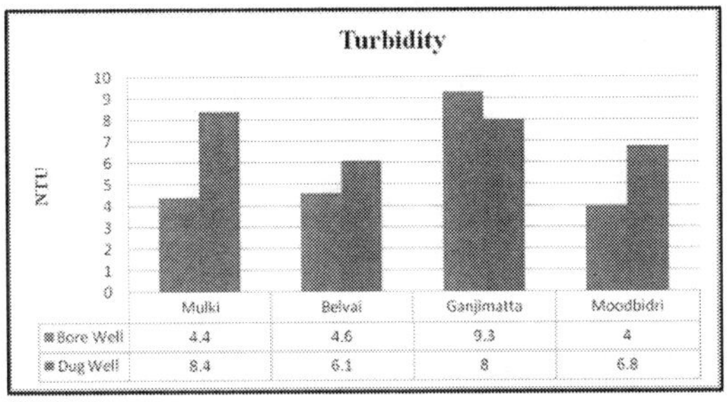


**Fig. 4.** Graphical representation of dissolved oxygen (mg/l).

**Table 5.** Results of total hardness (mg/l).

| | Bore Well | Dug Well | BIS: 10500 – 2012 |
|---|---|---|---|
| Mulki | 132 | 64 | Desired Limit: 300 mg/l Beyond encrustation in water supply structure and adverse effect on domestic uses. Permissible Limit: 600 mg/l |
| Belvai | 84 | 60 | |
| Ganjimata | 72 | 44 | |
| Moodabidri | 56 | 36 | |

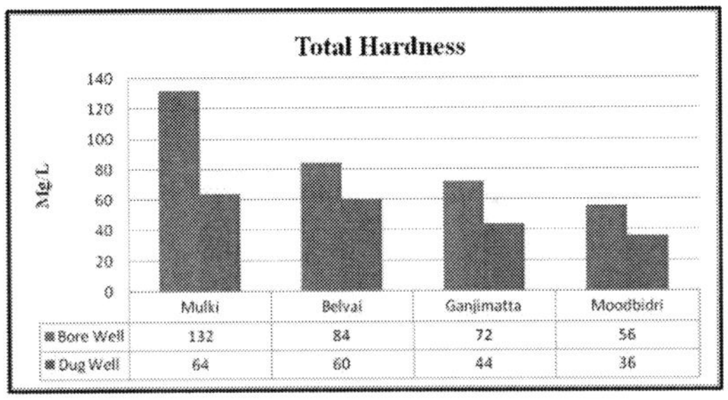


**Fig. 5.** Graphical representation of hardness

**Table 6.** Results of total solids (mg/l).

| | Bore Well | Dug Well | BIS: 10500 – 2012 |
|---|---|---|---|
| Mulki | 340 | 540 | Desired Limit: 500 mg/l Beyond this palatability decreases and may cause gastro intestinal irritation Permissible Limit: 2000 mg/l |
| Belvai | 230 | 320 | |
| Ganjimata | 164 | 210 | |
| Moodabidri | 150 | 196 | |

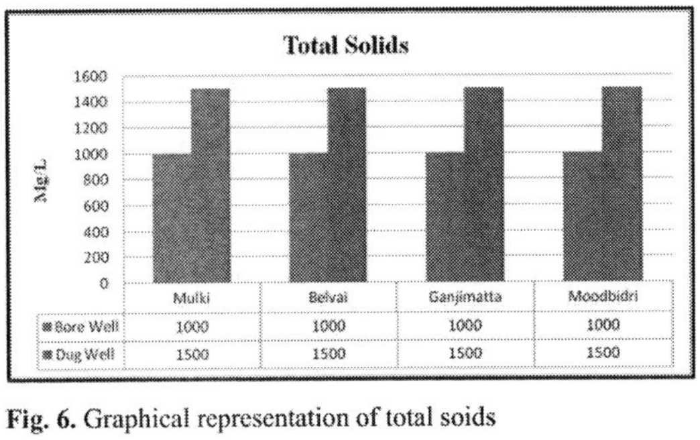


**Fig. 6.** Graphical representation of total soids

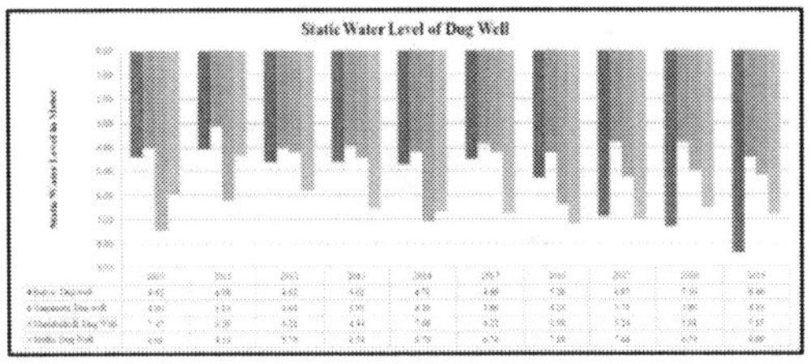


**Fig. 7.** Static water level of dug well (in M)

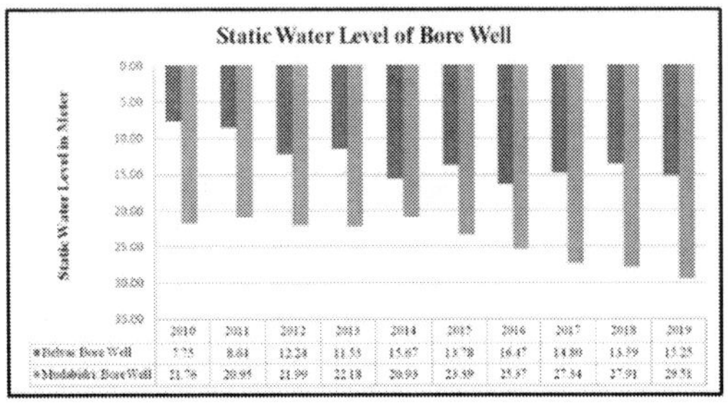


**Fig. 8.** Static water level of bore well (in M).

## 4.1 Geological information of the Moodabidri area using GIS and Rainfall data

The following series of images provides insights into the study area's characteristics based on various geospatial and hydrological parameters. Figure 9 displays an administrative map that outlines the area boundaries, while Fig.10 shows population distribution, and Fig. 11 presents rainfall data in the region. Fig.12 provides a land use map depicting different types of land utilization, followed by Fig.13, which illustrates soil type distribution. Fig.14 showcases land suitability for irrigation, and Fig.15 displays groundwater availability. Fig.16 contains a rock type map, giving details about the geological composition of the area.

Subsequently, comparative analyses between rainfall and groundwater depth are shown in Figs.17 and 18, representing data from dug well and bore well, respectively. Figs. 19 and 20 illustrate graphical representations of the static water level in dug well and bore well. Together, these images offer a comprehensive view of the geospatial and hydrological conditions in the study area.

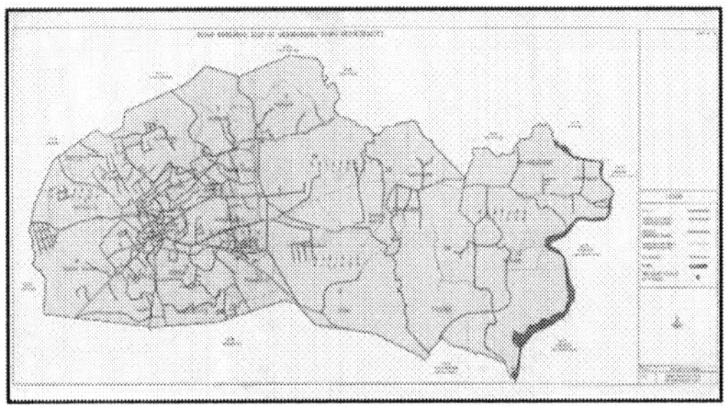

**Fig. 9.** Administrative boundary of Moodabidri

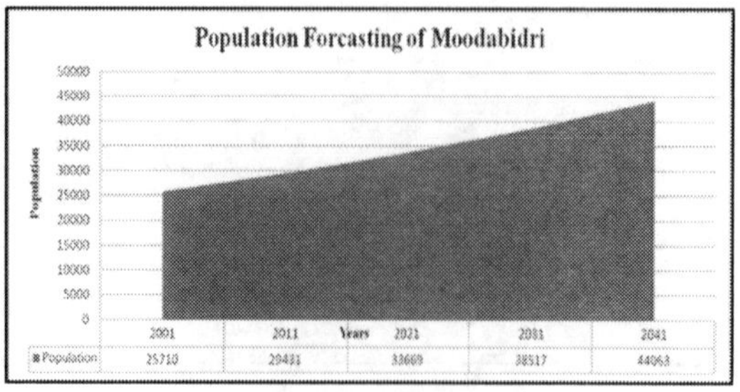


**Fig. 10.** Population forecasting of Moodabidri using geometrical progression method

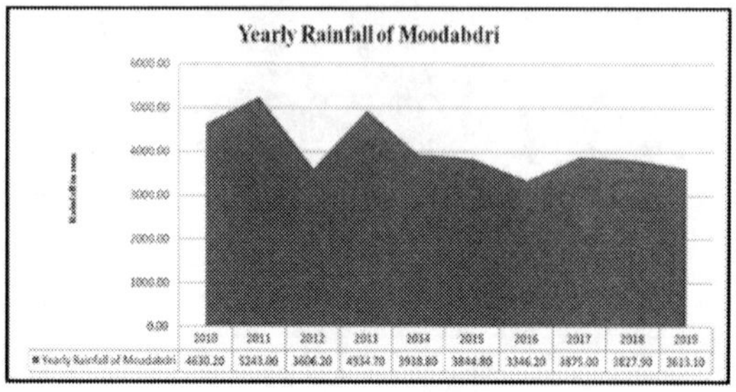


**Fig. 11.** Graphical representation of yearly rainfall of Moodabidri (from 2010 to 2019 data)

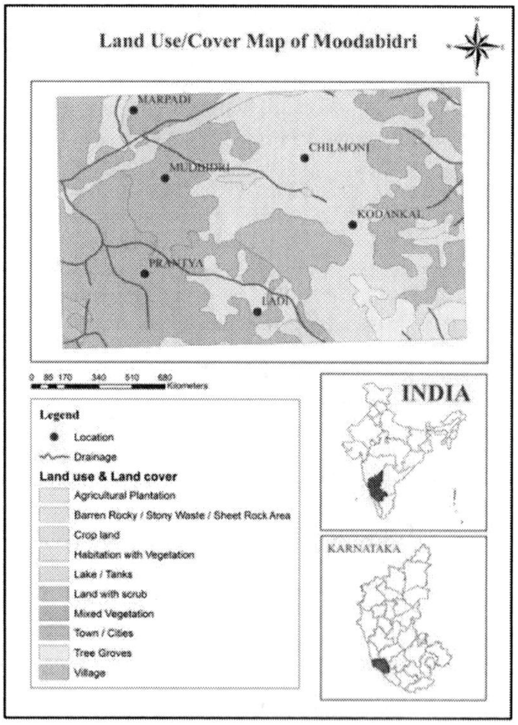


**Fig. 12.** Land use/cover map of Moodabidri

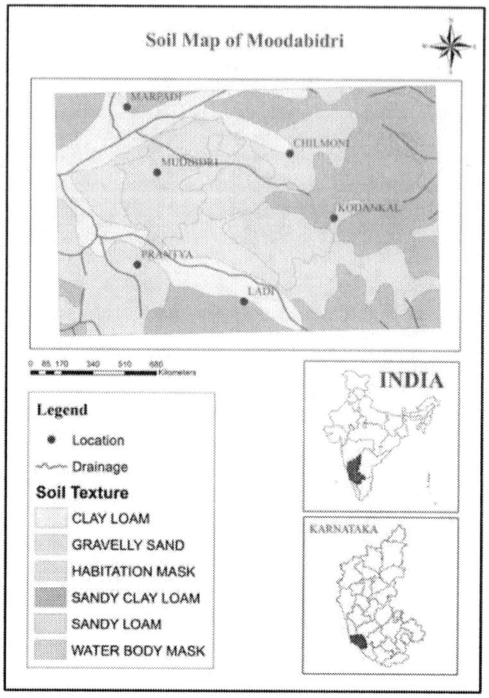


**Fig. 13.** Soil Map of Moodabidri

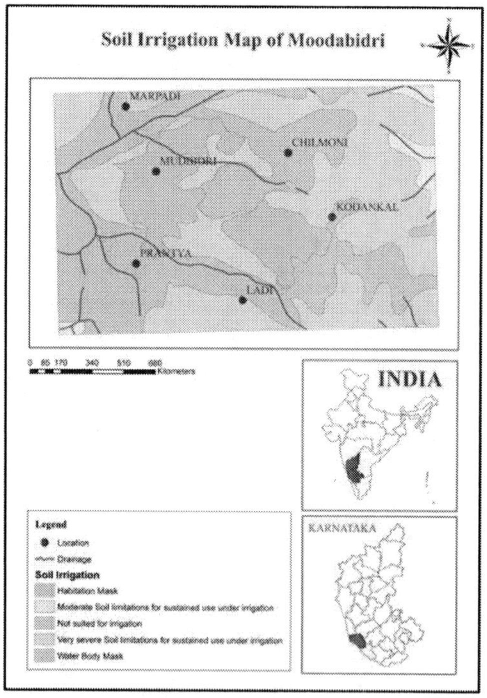


**Fig. 14.** Soil irrigation map of Moodabidri

https://doi.org/10.1051/e3sconf/202560402011

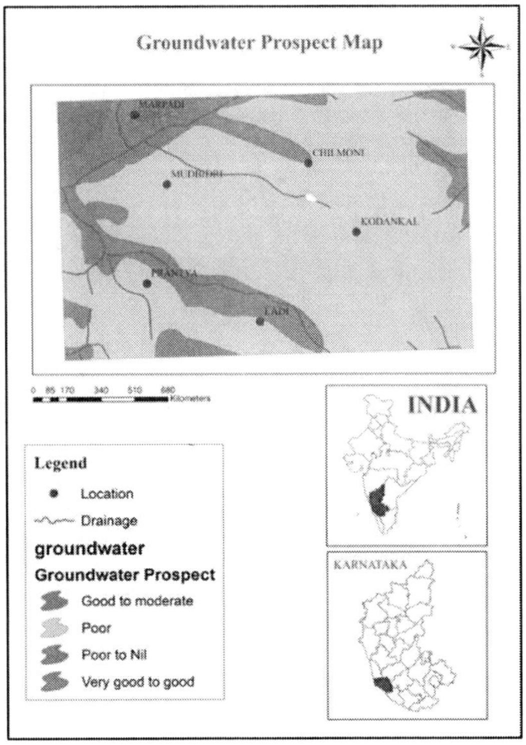


**Fig. 15.** Groundwater prospectus map of Moodabidri

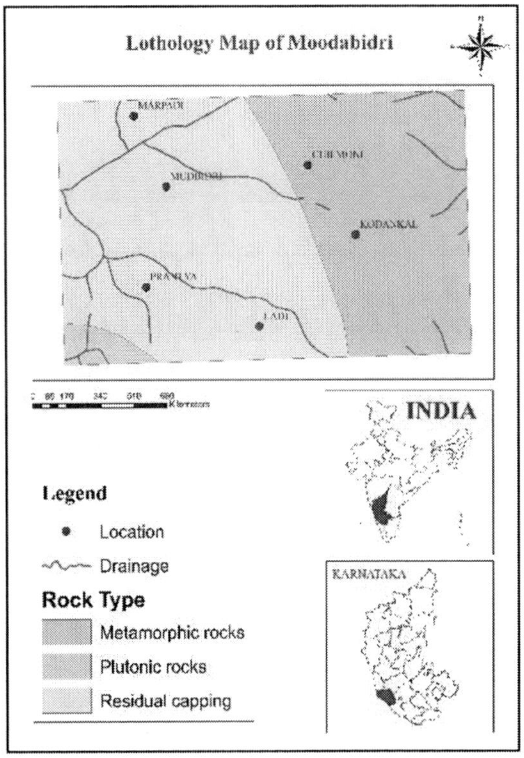


**Fig. 16.** Lithology map of Moodabidri

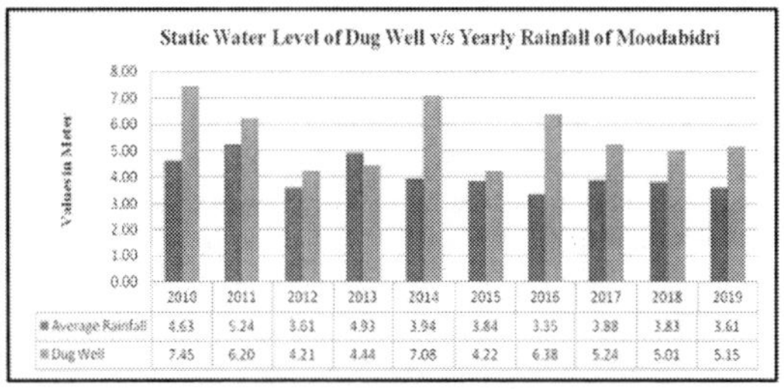


**Fig. 17.** Graphical representation of Moodabidri rainfall v/s dug well static water level.

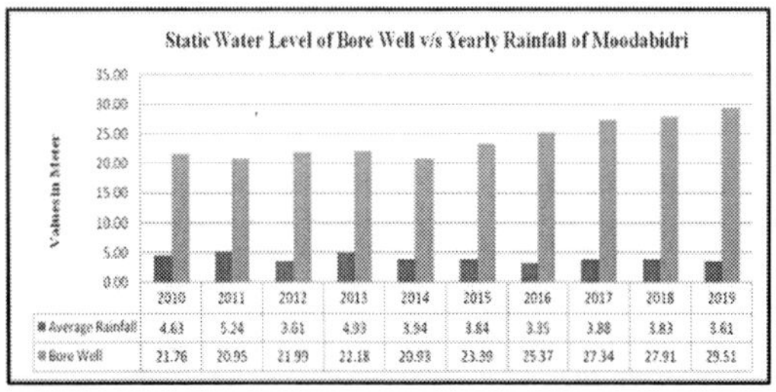


**Fig. 18.** Graphical representation of Moodabidri rainfall v/s bore well static water level.

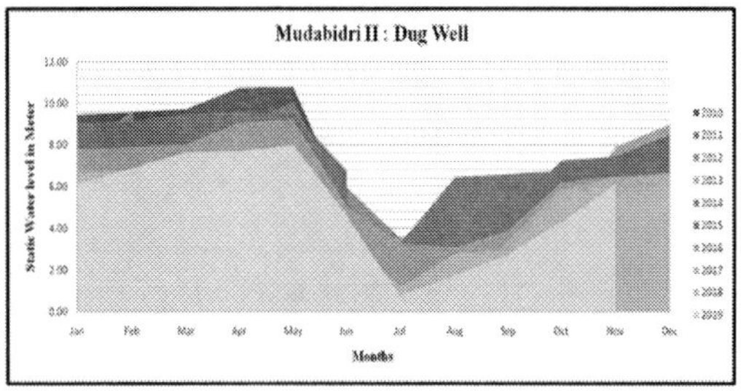


**Fig. 19.** Graphical representation of dug well static water level

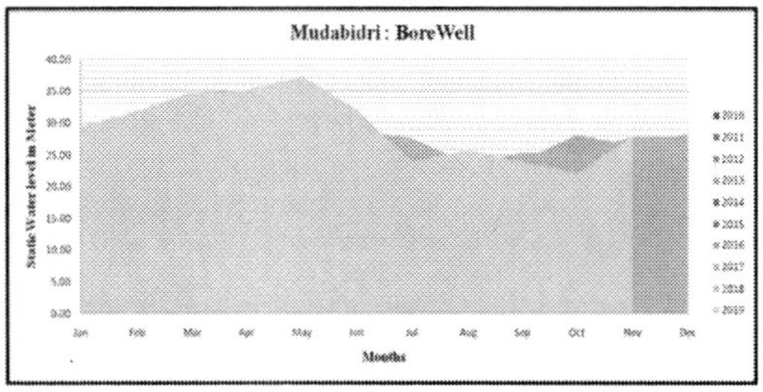


**Fig. 20.** Graphical representation of bore well static water level

According to the results, rainfall variability linked to climate change causes notable variations in water levels in the study area's bore and dug wells. The demand for water has increased over the past few decades due to urbanization and population development. The yearly rainfall in Moodabidri town has been steadily declining over the past few decades, as illustrated in Fig. 11. At the same time, the statistics (Figs. 17 and 18) show that the water levels in both the bore and dug wells in Moodabidri are declining.

The over exploitation and utilization of water demand is increased from past decades due to increased population and urban sprawling in the study area

### 4.2 Geological and Geophysical Investigation of Moodabidri.

Moodabidri is a town situated in the coastal district Dakshina Kannada of Karnataka state falls along the west coast of peninsular India and is separated from the rest of peninsula by towering high Western Ghats.

Temperature, density, porosity, water content, water quality, and mineral composition all affect a rock's resistivity values. The direction of electrode distribution and the composition of the top layer in the hardrock region also affect the formation resistivity value.

These techniques are predicated on the idea that the electrical resistivity of loose, unconsolidated, or partially consolidated surficial materials, such as soil loss, alluvium sand, clay, and other byproducts of rock weathering and erosion, differs from that of the bed rock on which they are deposited. The electrical resistance of the rock decreases as it becomes more porous or jointed and fissured. Therefore, undamaged metamorphic and igneous rocks are generally more resistant than sedimentary rocks.

A signal stacking resistivity meter was used up to 3.8-4 m depth as only for the geological structural requirement in the circular shaft, primarily to measure bed rock depth and subsurface formations by studying the variations in their electrical properties. Because of the regulated depth of inquiry and excellent resistivity contrasts across the lithological units, this technology gained significant traction in subsurface exploration [5].

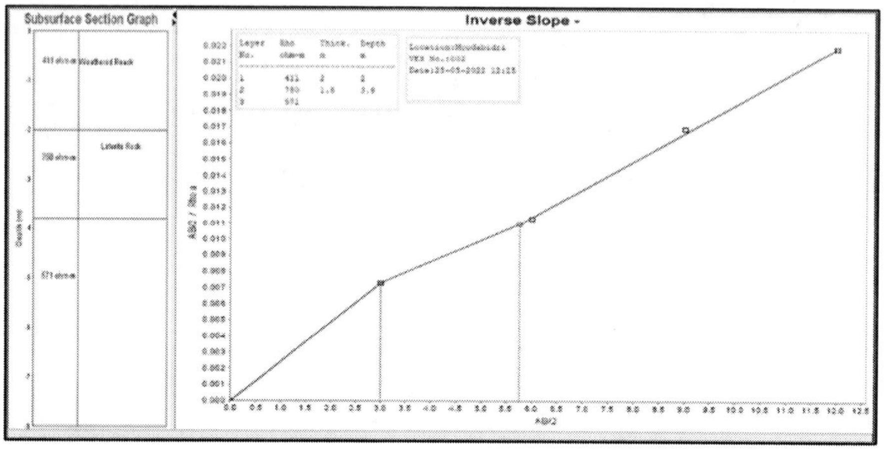


**Fig. 21.** Interpretation geological formation using resistivity meter

## 5 Conclusion

The water level of both dug well and bore well of the experimented study area showing drastic water level fluctuations with respect to rainfall data as well as rainfall seasonal variation. In detail regular study on characterization of coastal aquifer in both quality and quantity is need of the day for sustainable development in the present scenario of urban sprawling and climate change. Also, the improvement in the modern technology and methods is the need of the day for the substantial implementation of groundwater augmentation in the coastal aquifer.

Thus, the water conservation and harvesting are essential for sustainable water management in the study area. The need for groundwater augmentation and characterization highlights the urgency of implementing mandatory regulations for effective water conservation and harvesting, especially for coastal aquifers.

Authors are likes to extend their gratitude to the Department of Civil Engineering, MITE for provided state-of-art Environmental Engineering laboratory and also, express the thanks to KSCST, GoK for the student project sponsorship and we express our sincere thanks to District Ground Water Office, Dakshina Kannada, GoK.

## References

1. K. Harshendra, Studies on water quality and soil fertility in relation to crop yield in selected river basins of Dakshina Kannada District of Karnataka State; Ph.D. Thesis, Mangalore University, Karnataka; India, 147 (1991).
2. Dhirendra Kumar Singh and Anil Kumar Singh, Groundwater Situation in India: Problems and Perspective, International Journal of Water Resources Development, **18**, 563-580 (2010). https://doi.org/10.1080/0790062022000017400
3. U. A. Lathashri and A Mahesha, Groundwater sustainability assessment in coastal aquifers, J. Earth Syst. Sci., **6**, 1103–1118 (2016).
4. G. Udayakumar, Subsurface barrier for water conservation in lateritic formations; Ph.D. Thesis, National Institute of Technology Karnataka, Surathkal, India, 53–69 (2008).
5. R.B.C. Kumar, modelling regional actual evapotranspiration over Netravathi basin using satellite data; M. Tech Thesis, National Institute of Technology Karnataka, Surathkal, Mangalore, India, **30** (2010).

# Problems of decentralized disaster management in Indonesia: a review of local government failures

*Ahmad* Fatkul Fikri[1*], *Rustian*[1], *Mohammad* Afifuddin[2], *Dody* Ruswandi[1], *Catur* Susilo Rahardi[1], and *Andi* Albright Sumange[3]

[1]National Disaster Management Authority of Republic Indonesia, Pramuka Street Kav. 38 Matraman, Jakarta, Indonesia
[2]Departement of Sociology, Faculty of Social and Cultural Sciences, University of Trunojoyo Madura, Bangkalan, Indonesia
[3]Graduate Student in Harris of Public Policy University of Chicago, 5801 S Ellis Ave, Chicago, IL 60637, United States of America

**Abstract.** The issuance of Law No. 24/2007 on Disaster Management was the starting point for the decentralization of disaster management. The implementation of decentralization can provide significant benefits in disaster management, such as being faster and more precise in handling disasters. This study aims to examine the extent of the implementation of decentralization of disaster management. The research method used is qualitative exploration. data were collected using observational studies, interviews, and documentation with purposive sampling. Meanwhile, the data processing technique uses qualitative analysis techniques. The findings in this study are that the role of LDMAs and local governments in the pre-disaster, emergency response and post-disaster phases is still not optimal. The causes are: (1) political will from local leaders related to disaster management is still weak; (2) local government budget allocation for disaster management is still low; (3) the low quality of human resources of LDMA apparatus. As a result, local governments are still dependent on the central government for disaster management. Thus, the implementation of decentralization of disaster management, which aims to increase citizens' disaster awareness, keep people away from disaster risk, and make people able to live side by side with disasters, has not been maximally realized.

## 1 Introduction

Indonesia is a country famous for its natural beauty, known as the emerald equatorial country. Behind its beauty and abundant natural wealth, Indonesia has the threat of disaster. Indonesia's potential for disaster is caused by Indonesia's geography, which is located in the world's ring of fire and is surrounded by three active world plates [1]. This position results in Indonesia frequently being hit by landslides, floods, droughts, earthquakes and tsunamis, volcanic eruptions, tornadoes and extreme waves as in Fig. 1.

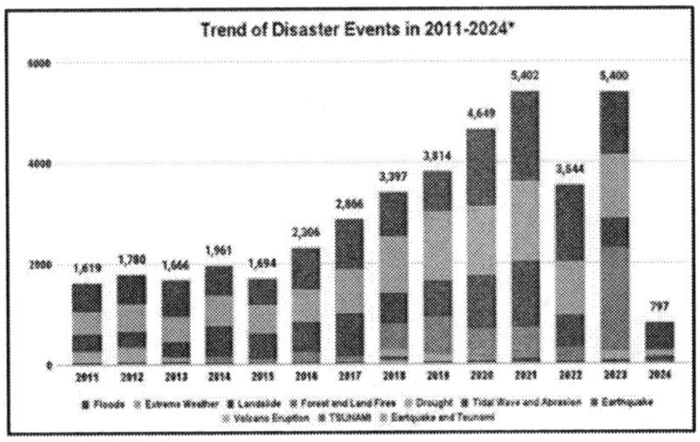


**Fig. 1.** Trends in Natural Disasters in Indonesia 2011 - 2024

As a country with a high level of disaster risk, Indonesia has experienced several devastating disasters. History records devastating disasters that have occurred in Indonesia, including the eruption of Mount Tambora in 1815 which killed 47,925 people, the eruption of Mount Krakatoa in 1883 which killed 36,000 people, the 1992 Flores earthquake and tsunami which killed 263 people, the 2004 Aceh earthquake and tsunami which killed more than 220,000 people, the 2006 Yogyakarta earthquake which killed 6,234 people, the 2018 Palu and Donggala earthquake, tsunami and liquefaction which killed more than 1,000 people, and the non-natural disaster of the Covid-19 pandemic in 2020 - 2022 which killed more of 156,000 people [2–4].

The 2004 Aceh earthquake and tsunami disaster became a turning point in the awareness (wakeup call) of the community and government regarding the importance of disaster risk management [5]. Since then, the government has been inspired to carry out disaster risk management well. Proof of the government's seriousness in disaster risk management is the issuance of Law Number 24 of 2007 concerning Disaster Management. This regulation is the first and basic step for the central and local governments to make the disaster management process in Indonesia better. Existing regulatory instruments explain that disaster and emergency

* Corresponding author: ahmadfatkulfikri02@gmail.com

management is the responsibility and supervision of the central and local governments.

Normatively, disaster management which is the responsibility of local governments is relevant to the principle of local autonomy. Article 18 paragraph (2) and paragraph 5 of the Constitution of the Republic of Indonesia explains that local governments are given the authority to manage and regulate their government affairs. One important aspect of the relationship between the centre and the regions is equitable, effective, and efficient development in the administration of government [6], including disaster management to achieve national development goals.

Decentralization and local autonomy have a central role in disaster management in Indonesia. Local governments have a comparative advantage regarding many important issues related to pre-disaster preparation, such as urban infrastructure maintenance, building, and land use regulations that are prone to disasters, and emergency planning [7]. In terms of human resource development, decentralization has a role in increasing the capacity and ability of local leaders to deal with critical situations. Socially, decentralization has a role in increasing community involvement in disaster management activities [8]. Furthermore, decentralization will have implications for the speed and accuracy of disaster management because effective disaster risk mitigation and reduction efforts require the use of information, expertise, and local resources [9].

However, decentralization also raises several problems between central and local relations. The handover of affairs that fall under the authority and affairs of local governments creates disharmony between the central and local governments [10], including in matters of disaster management. This disharmonization was due to legislative regulations being formed before the disaster management Law was issued. Apart from that, the process of making legislative regulations tends to prioritize sectoral egos, overlapping, prone to conflict, multiple interpretations, and inconsistencies in implementing regulations [11, 12]. In addition, local governments have to take on more public services, while their resources are very limited [13].

The results of Kartika's research (2015) show that the central and local institutional arrangements are not good, the resources they have are still lacking, regulations overlap, and there are gaps in coordination [14]. The local government's lack of understanding in determining disaster management plans has led to the relationship between the center and the regions not running well [15]. Apart from that, the spirit of decentralization and local autonomy encourages local governments to become more dominant, giving rise to resistance from the central government [16]. This will certainly cause chaos in the decentralization of disaster management in Indonesia.

The fact of the chaotic relationship between the center and regions in disaster management can be seen from the handling of the non-natural disaster of the Covid-19 pandemic. The massive and rapid rate of spread of Covid-19 has prompted local governments to take their preventive measures. The policy of several regions to carry out local quarantine (lockdown) with unequal areas has caused a negative response from the central government. As an indicator, President Joko Widodo stated video broadcast by the Cabinet Secretariat on March 16 2020 that local quarantine or lockdown on both a local and national scale is the authority of the central government which cannot be determined by local governments [17].

Based on several of these problems, this research aims to analyze the implementation of decentralized disaster management in Indonesia. The problem raised in this article is, how is disaster management decentralized in Indonesia.

## 2 Research methods

This research uses an exploratory qualitative research method by utilizing primary data and secondary data. Primary data was collected through direct semi-structured interviews using communication media (WhatsApp, telephone, zoom meetings) and field observations, while secondary data was collected through documentation studies sourced from previous research (journals, dissertations, and theses), legal regulations, and regulations. magazine invitations, books, internet sources, online and print news.

The technique for determining informants uses a purposive sampling technique. Meanwhile, the analysis technique used in this research refers to the model developed by [18]. This analysis technique emphasizes interactive analysis, which means that analysis activities will continue to be carried out until the conclusion is the same. This interactive analysis model has stages of collection, condensation, presentation, and verification or conclusion.

Several research frameworks can be used to analyze decentralized disaster management. In this research, researchers adopted a disaster management framework based on the Disaster Management Law as seen in Fig. 2.

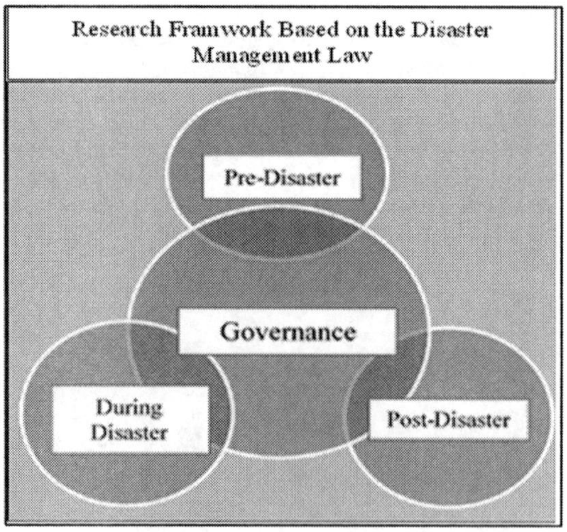


**Fig. 2.** Research Framework

This research framework was adopted from the Disaster Management Law which was then adopted by researchers analyzing decentralized disaster management in Indonesia. This theoretical framework is appropriate to use because it emphasizes the role of local government and society. In addition, this framework concerns activities carried out before, during and after a disaster. In this study, researchers limited pre-disaster variables to include public awareness of disaster risks, avoiding disaster risks, and reducing disaster risks. Variables during emergency response include determining emergency status, search and rescue (SAR), rapid assessment, and determining basic needs. Post-disaster variables include rehabilitation and reconstruction activities.

# 3 Results and discussion

## 3.1 Decentralized disaster management process in Indonesia: an overview

One factor in the success of reform is the improvement in the quality of relations between the central and local governments. Inappropriate division of affairs will have an impact on ineffective central and local relations, which will ultimately result in the failure of reforms and services to the community [19].One way to bring public services from the government closer to the community is through local autonomy. The closer the government provides services to the community, the better the services provided [20], including in matters of disaster.

The 1945 Constitution of the Republic of Indonesia does not explicitly explain disaster management in Indonesia. The constitution only explains that the president can declare a state of danger that could threaten the security of the country, but what is meant in this article tends to refer to a state of war. The government first regulated the handling of hazards through Law Number 6 of 1946 concerning Dangerous Conditions, one of which includes dangerous conditions resulting from natural disasters.

The forerunner to the formation of an institution that deals with disaster management is the Indonesian Government formed the Agency for Helping Families of War Victims. This agency, which was founded on August 20, 1945, focuses on war conditions after Indonesian independence. This agency is tasked with helping war victims and families of victims during the War of Independence. Indonesian Government formed the Agency for Helping Families of War Victims is the parent organization aimed at maintaining public safety, serving as public security guards in areas under the coordination of the Regional Indonesian National Committee (KNI).

In 1967, through Presidential Decree Number 256 of 1966, the Government established the Central Natural Disaster Management Advisory Authority (BP2BAP). This institution is led by the Minister of Social Affairs who is responsible to the president. Central Natural Disaster Management Advisory Authority has the task of managing disaster emergency response and assisting disaster victims. The formation of this agency is the starting point for an institution that specifically handles natural disasters, not just disasters caused by humans or war (man-made disasters).

After the 1998 reform, Indonesia experienced a multidimensional crisis. During this period, Indonesia experienced social unrest in several regions. This inspired the formation of the National Coordinating Agency for Disaster Management and Refugees (Bakornas PBP), which previously was only the National Coordinating Agency for Disaster Management (Bakornas PB). This agency is ad hoc, led by the vice president, and directly responsible to the president. The formation of Bakornas PB was followed by the formation of a Disaster Management Implementation Coordination Unit (Satkorlak PB) at the provincial level led by the governor and a Disaster Management Implementation Unit (Satlak PB) at the district/city level led by the regent/mayor.

Until the 59th anniversary of the independence of the Republic of Indonesia (2004), the Indonesian Government did not yet have specific regulations regarding disaster management. The earthquake and tsunami that hit the Indian Ocean inspired people and governments around the world, especially Indonesia, to reorganize disaster management in Indonesia. A concrete manifestation of the government's seriousness in organizing disaster management in Indonesia is by issuing Law Number 24 of 2007 concerning Disaster Management. This regulation encourages comprehensive and comprehensive disaster management. Unlike before the disaster management Law, disaster management was responsive. Apart from that, the regulation mandates the formation of a permanent, not ad hoc, National Management Agency. The following are institutional differences before and after the Disaster Management Law was passed, as shown in Table 1.

**Table 1.** Differences in disaster management institutions before and after the disaster management law

| Content | After Disaster Management Law | Before Disaster Management Law |
|---|---|---|
| Intitutions | Bakornas PB (national)<br>Satkorlak PB (province)<br>Satlak PB (district/city) | NDMA (national)<br>LDMA (province)<br>LDMA (district/city) |
| Leader | Vice President (national)<br>Governoor (provinsi)<br>Regent/Mayor (district/city) | Head NDMA (National) |
| Institutional Status | *Ad hoc* | Permanent |
| Function | Coordination | Coordination, Command, Implementation |
| Management | Pada saat tanggap darurat | Pre-disaster, during a disaster, and post-disaster |

It can be concluded that the Disaster Management Law is the starting point for the decentralized process of disaster management in Indonesia. In Article 5 of the

Disaster Management Law, there is a clause, "The government and local governments are responsible for implementing disaster management." Apart from being a milestone in starting decentralization of disaster management in Indonesia, the Disaster Management Law also presents a new paradigm for disaster management in Indonesia. Shifting the paradigm for implementing disaster management from responsive to preventive. Of course, these paradigm changes must be accompanied by cooperation and collaboration between the central and local governments to strengthen prevention, mitigation, and early warning. The Disaster Management Law requires that the central and local governments are responsible for implementing disaster risk management in Indonesia. This has the consequence that the central government hands over some of its authority regarding disaster management to local governments.

Law Number 23 of 2014 concerning Local Government does not explicitly explain the authority of local governments in handling disasters. In this regulation, the outline of disaster management lies in concurrent authority which is divided into mandatory and optional matters. Disaster management is a mandatory government matter, namely related to security and public order, but in turn, there are several regions that do not have local institutions or apparatus that specifically deal with local disaster management.

Central and local governments have responsibilities before a disaster occurs. A concrete manifestation of government responsibility by establishing a National Disaster Management Authority of Indonesia (NDMA) at the central level and a Local Disaster Management Authority (LDMA) at the local level. Both have their respective authorities in accordance with the planning documents. In this case, at the central level, the National Disaster Management Plan and local Disaster Management Plan documents at the local level are used. Figure 3 explains how the central and local relations relate to disaster management in Indonesia.

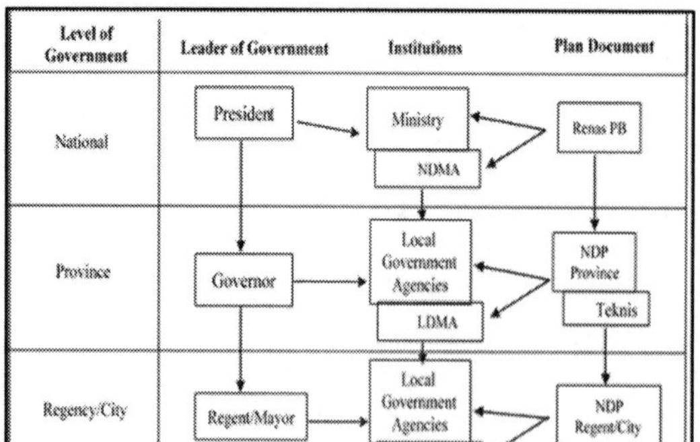


**Fig. 3.** Structure of central and local institutions for disaster management in Indonesia

NDMA of Indonesia consists of a steering element tasked with formulating policy concepts and evaluating disaster management and an implementing element tasked with the integration of pre-disaster, emergency response, and post-disaster. NDMA of Indonesia has the task of coordinating across sectors in disaster management in Indonesia. However, the consequences of decentralization and local autonomy limit the influence of ministries at the local level. The central government (ministries/ institutions) only acts as a facilitator and must negotiate with local governments to carry out programs that fall under the authority of local governments [21].

On the other hand, LDMA has the same elements and tasks as NDMA of Indonesia, only at the provincial and district levels. LDMA is tasked with coordinating with Local Government Agencies (LGA). LDMA's position is separate from NDMA of Indonesia whose entire budget comes from the Local Government Budget, although much of it depends on central government allocations. Apart from that, LDMA coordination must also be divided into two, namely, NDMA of Indonesia and the Ministry of Home Affairs through the Directorate General of Regional Administration Development. This can lead to overlapping authority and miscoordination.

### 3.2 Pre-disaster: limited decentralization

The most appropriate way to reduce risks due to the threat of disaster is at the pre-disaster stage. This stage explains how to eliminate or reduce risks that can occur due to disasters. As in the framework in Fig. 1, the pre-disaster study focuses on awareness of risk, preventing society from risk, and living side by side with threats.

First, decentralization must be able to increase public awareness of disaster risks. One effort to increase public awareness of disaster risks is through disaster education. The Disaster Management Law emphasizes that disaster education is the right of every community. In particular, education and training are stated as part of disaster management in situations before a disaster occurs and as part of disaster mitigation efforts. However, the law itself does not clearly define what is meant by disaster education but provides space for other government policies/regulations to create a working definition.

Until now, there are no regulations that specifically regulate disaster education. The Indonesian government claims to have integrated disaster risk reduction in the education sector, namely in the school curriculum, teacher training on disaster risk reduction education, and school preparedness. In reality, the disaster education curriculum has not yet reached common ground and is still in the process of being reviewed. This is because disaster education is not only the domain of disaster management policy, but is a matter of contention for other government affairs, such as the Ministry of Education, Culture, Research and Technology (Kemendikbudristek) to the Ministry of Public Works and Public Housing (Kemen-PU-PR).

The chaos in disaster education does not only occur in the central government but also extends to local governments. Most local governments have not implemented a disaster curriculum, so the integration of an anti-disaster curriculum into formal education is still just a matter of discourse. At the local government level, the non-obligation of a disaster education curriculum is the reason there is no disaster education in several regions.

Apart from that, all subjects in general education are determined by the central government. However, the central government does not close meetings on determining the education curriculum to local governments.

Decentralization should provide space for provincial governments (SMA/SMK) and district/city governments (SMP and SD) to provide disaster education curricula with local content. However, the development of the disaster education curriculum has not been decentralized to the lower levels. One possible cause is the complexity of polycentric governance in Indonesian disaster management [22]. Only a few local governments implement a disaster education curriculum. An example of a local government that implements a disaster curriculum is Aceh Province. In fact, disaster education is important to increase public awareness of disaster risks.

The disaster curriculum program aims to create a society that is aware of potential disasters and has knowledge and disaster management skills, which in the end can provide community resilience [23, 24]. Disaster education is able to provide better knowledge, skills, and awareness in disaster mitigation activities [25]. In addition, the inclusion of disaster education curricula in 30 countries can contribute to reducing disaster risk [26]. This shows the best practice of disaster education curriculum in the world.

The second thing in increasing public awareness is risk assessment. As a key tool for communication and collaboration on disaster risk, risk assessment is essential in building disaster risk awareness [21]. Risk assessment is carried out through the creation of Disaster Management Plan (DMP) documents at the Provincial, Regency, and City levels. So far disaster risk studies have been carried out by 34 provinces, while 4 new autonomous regions are still in the process of conducting studies. However, the NDMA of Indonesia's role is still very dominant in the preparation of disaster risk assessments due to uneven decentralization of resources. The need for vertical collaboration is absolutely necessary, as the United States (US), which has a federal government system that has long supported local autonomy, still has to involve the central government [27].

The next problem is that several districts/cities have not yet carried out disaster risk studies. District/city governments consider that disaster management is not a development priority. So the budget priority is not allocated to expenditure related to disaster management. In addition, the attitude of local leaders shows very low motivation to prioritize disaster management policies, because they do not help much in the campaign for the next election [13]. Some district/city governments consider LDMA to be a "dump" place for employees who have been punished. The result is that local governments often lack staff who have the competence to actively develop disaster management strategies.

Second, decentralization must be able to prevent society from the risk of disaster. Basically, preventing people from disaster risk occurs in three stages of disaster management, before, during, and after a disaster. In the pre-disaster stage, to avoid disasters, people choose to strengthen their residences [28, 29] and avoid residences from disaster-prone areas. When a disaster occurs, people choose to evacuate and avoid the disaster location [30, 31]. In the post-disaster period, people choose to move from their old residence [32]. Several ways that can be done to prevent people from disasters are through good spatial planning [21].

The central government requires all provincial, district, and city regions to have Regional Spatial Planning documents through Law Number 26 of 2007 concerning Spatial Planning. All provinces, districts, and cities actually have Regional Regulations regarding RTRW, however, there is a very significant delay in implementing spatial planning regulations which hamper the functionality of spatial planning as a means of preventing communities from disaster risks. The Detailed Spatial Planning Plan, which guides regional planning plans, is not yet owned by most local governments. In addition, disaster risk hazard maps are not used optimally in spatial planning by local governments, although there are several cases where risk planning has been used in local-level planning.

Spatial planning has an important role in avoiding new development around high disaster risk areas [21]. Poor spatial planning increases community vulnerability to disaster risks. In 2014, the growth of hotels in Yogyakarta resulted in a drought, in Bandung the growth of hotels resulted in flooding, in Cirebon the growth of hotels and restaurants that were not well organized resulted in a clean water crisis, and land conversion in River Watersheds (DAS) resulted in sedimentation [33] which can result in flooding.

Third, decentralization must be able to provide strength to the community to be able to live side by side with disasters. If the community is able to adapt and live side by side with disasters, the community will reduce vulnerability and reduce losses, and be able to restore lives affected by the disaster. When communities are able to live side by side with disaster risks, regional resilience will be created. Living in harmony with disaster risk usually involves natural, early warning systems, and structural and physical management steps. These steps can be mainstreamed in the implementation of development planning.

Good and sustainable nature management can reduce the impact on disaster risk vulnerability, especially natural disasters. Local governments often have uncontrolled management of natural resources. As an excuse to increase Regional Original Income (PAD), the exploitation of natural resources often ignores the environment, resulting in an increase in the risk of disasters. As the results of a study from Suharyo (2017) explained the local government seemed to be allowing illegal gold mining which had an impact on food security in Maluku [33]. Problems with natural management also give rise to conflicts that occur on Mount Kendeng, Central Java. The local government was unable to resolve this problem and required the central government to take over the conflict.

### 3.3 Emergency response: local government incompetence

Emergency response is the most crucial moment in the disaster management process. At that time, people were panicked, anxious, and confused (chaos) about what to do. Therefore, there needs to be action to overcome this, so that the uncertain situation becomes normal, saves as many victims as possible, and minimizes the losses suffered. Central and local government efforts during emergency response can be carried out by rapid assessment, determining the status of a disaster emergency, search and rescue (SAR), determining basic needs, and restoring vital facilities.

First, a rapid assessment. When a disaster occurs, the steps that must be taken immediately are a quick and precise assessment of the disaster location. This activity is the first step that must be carried out and must be completed immediately. This step was carried out in order to assess the needs of communities affected by the disaster, to then implement priority action plans. Without such action, policy inaccuracies, gaps, or overlapping assistance delivery may occur, which not only negates valuable resources at critical times but can increase the burden on affected communities.

A rapid needs assessment aims to get an idea of the needs and problems of those affected by a disaster within one to two weeksc [34]. In addition, this assessment is also separate from search and rescue activities. At this stage, the assessor must collect as much information as possible from various sources, and go directly to the field (not relying on information from other parties), so that there is no bias. For example, aid often only reaches areas that are easily accessible [35, 36], this will certainly worsen the impact of disasters.

The accuracy and speed of rapid assessment require good skills for regional disaster personnel. It's just that local governments often don't know what to do when a disaster occurs. For example, in the 2016 earthquake disaster in Pidie Jaya Regency, the Pidie Jaya Regency Government experienced that they did not know or "stuttered" what to do. This required one of the universities in Aceh to carry out a rapid assessment of the Pidie Jaya earthquake disaster. The same thing also happened during the flash flood disaster in Masamba, North Luwu Regency. The lack of human resources who know about disasters is the main reason for this failure.

Second, determining the status of a disaster emergency. The 2008 Government Regulation concerning the Implementation of Disaster Management states that the status of a disaster emergency goes through three stages, namely, emergency alert status, meaning that the potential for a disaster has led to a disaster. Emergency response status means when a threat has actually occurred and disrupted people's lives and livelihoods. Emergency to recovery transition status means when the threat of disaster begins to decrease in escalation, while life and livelihoods have not yet resumed as before.

Emergency response status is divided into two levels, namely national disasters and regional disasters consisting of provincial disasters and/or district/city disasters. Determination of disaster emergency response status is carried out by the president for national disasters, the governor for provincial disasters, and the regent/mayor for district/city disasters. This determination is based on the minimum number of victims, damage to facilities and infrastructure, material losses, the area affected by the disaster, and the socio-economic impact.

Basically, legislation has given local governments the authority to determine the status of a disaster emergency, but knowledge and leadership skills in times of crisis mean that these decisions are often inappropriate. Data in the field also shows that often during times of crisis, the emergency commander is given to the Military District Commander (Dandim). This shows the inability of local governments to lead during times of crisis. Leadership character has an important contribution in times of crisis. A leader who has strong character can determine the duration, level of exposure, and when to end the crisis by making the right decisions [37]. As producers of regional and national leadership cadres, political parties have failed to create leaders who are capable of leading in times of crisis.

Third, search and rescue of victims. Rescue and search or search and rescue (SAR) of victims in large-scale disasters is very important in disaster management. Search and rescue (SAR) operations are used to track victims after a disaster occurs in order to reduce casualties. Data from previous disaster events shows that more than 50% of deaths occurred within a few hours after the disaster event. Therefore, disaster management and search and rescue operations must be carried out immediately after a disaster occurs to increase the chances of saving victims while they are still alive [38].

Indonesia has several institutions in the process of searching for and rescuing victims, namely, NDMA of Indonesia, LDMA, National SAR Agency (Basarnas), and Regional Basarnas. In the event of a disaster, LDMA has a Rapid Response Team (TRC) to collaborate with the Regional Basarnas, Indonesian National Army (TNI), and Indonesian Republic Police (Polri) to search and rescue victims. However, most of the TRCs owned by local governments have minimal search and rescue capabilities compared to Basarnas, TNI, and Polri. This is because TRC members are not provided with regular and continuous education and training. In addition, local governments have limited means of searching and rescuing victims who have to borrow from other institutions. This is contrary to the requirements for achieving effective response actions, namely the availability of information and resources [39].

Search and rescue operations for victims must be carried out immediately after a disaster occurs. The success of emergency response operations depends on the time needed to rescue victims. Based on NDMA of Indonesia standards, 72 hours after a disaster is the golden time for rescuing victims. If the victim is not rescued within 72 hours after the disaster occurs, it is very likely that the victim will lose their life [40]. So fast and appropriate actions and decisions are needed for regional heads to search and rescue victims.

Fourth, determine basic needs. The head of NDMA of Indonesia Regulation Number 7 of 2008 explains that fulfilling basic needs must comply with predetermined minimum standards. The basic needs in question are clothing, food, shelter or shelter, sanitation and clean water, health services, and psychosocial services. Central and local governments are responsible for meeting these needs. All communities affected by disasters have the right to have their basic needs met. Determining basic needs is an important action that must be done well. In most cases, disaster survivors experience increasingly severe suffering due to basic needs not being met properly.

Disaster management during the emergency response period can be said to be optimal and good if the basic needs of disaster survivors can be fulfilled properly. Often meeting basic needs is a problem faced by local governments. Weak governance, harmonization, and coordination are still a source of powerlessness for local governments, especially in logistics distribution. The results of research from Ramdani (2015) confirm that local governments are unable to meet the basic needs of flood victims [41]. A similar thing also happened in Palu, Central Sulawesi Province, when facing earthquake, tsunami, and liquefaction disasters, the local government's helplessness in meeting basic needs encouraged people to loot [42].

### 3.4 Post-disaster: dependence on central government

Disasters have resulted in material losses and casualties. In order to reduce the burden on society, recovery is necessary. In the disaster management cycle, the recovery phase is the final stage of the disaster management process which consists of rehabilitation and reconstruction. The rehabilitation and reconstruction phase essentially repairs and rebuilds physical and non-physical infrastructure and facilities (livelihoods, social, economic, cultural, environmental) so that they can recover properly before a disaster occurs and have resilience against disasters if they occur in the future.

Legislation has stipulated that the central and local governments have responsibilities in the aftermath of a disaster, including the rehabilitation and reconstruction of various losses caused. The mandate of this regulation explains that the central and local governments work together to carry out post-disaster recovery, and must involve the community. In practice, the role of the central government in recovery is greater than that of local governments. Most local governments are not financially capable of carrying out recovery due to the impact of disasters.

Since rehabilitation and reconstruction grant funds have been allocated by the central government since 2015, until 2021 the allocation will reach IDR 9.992 trillion. The grant allocation was distributed to 387 local governments. The large budget allocation is intended to help local governments [43]. This is because no region has allocated APBD for post-disaster management. Disaster events cannot predict the exact time, so local governments tend to be small in budget allocations for disasters, especially Unexpected Expenditures (BTT). Apart from that, the APBD capacity is not able to rebuild after a disaster.

As per the results of interviews with the Pidie Jaya Regency Government, the budget allocation from the APBN for the 2016 earthquake disaster recovery was greater than the allocation from the APBD. A similar incident also occurred in Nias Regency. The use of the APBN still dominates in the post-disaster stage [44]. Recovery after the 2018 earthquake disaster in North Lombok requires the central government to allocate a budget of 51.05 percent of the entire budget plan. The remainder is 3.44 percent from the district APBD, 0.35 percent from the provincial APBD, and 21.11 percent from the public and private sectors [45]. Research by Rivani (2017) found that the budget allocation for disaster management in the regions is very small, on average only 0.02-0.07 percent of the total Local Governement Budget. This shows that the government is still very dependent on the central government. In fact, as a country that adheres to a decentralized system, local governments should have a big responsibility in disaster management.

## 4 Conclusion

Decentralization of disaster risk management in Indonesia is a mandatory concurrent matter related to security and public order. The logical consequence of decentralizing disaster management in Indonesia is the formation of a Local Disaster Management Agency (LDMA). Similar to NDMA of Indonesia, LDMA also has directing and implementing elements, only at the provincial level. At the pre-disaster stage, decentralization can be carried out by increasing public awareness of disaster risks. Most local governments have not implemented a curriculum at formal or informal education levels. Opportunities to implement the disaster curriculum are basically provided by the central government through local content programs. Second, protect the community from disaster risks, namely by spatial planning. Most local governments do not use disaster risk hazard maps optimally in spatial planning. In fact, spatial planning has an important role in preventing new development around high-disaster-risk areas. Third, living side by side with disaster risk. Local governments often have uncontrolled management of natural resources. As an excuse to increase Regional Original Income (PAD), the exploitation of natural resources often ignores the environment, resulting in an increase in disaster risk.

During emergency response, local governments have limited human resources to carry out assessments. Local governments tend to use human resources outside the local government, such as academics. Second, determining the emergency status of a disaster. Data in the field also shows that often during times of crisis, the emergency commander is given to the Military District Commander (Dandim). This shows the inability of local governments to lead during times of crisis. Third, search and rescue of victims. Limited resources owned by local governments require equipment to be borrowed from other institutions. Apart from that, the TRC PB owned by the local government has limited capabilities. Fourth, fulfilling basic needs. Often meeting basic needs is a problem faced by local governments. Weak governance, harmonization, and coordination are still a source of powerlessness for local governments, especially in logistics distribution. During the recovery period, local governments have a high dependence on the central government.

## References

1. R. Rustian. Transformation of disaster equipment governance in the perspective of smart governance (study on the regional disaster management agency of Central Java Province) (Brawijaya University Brawijaya, 2021)
2. S. Alfari, 9 The worst natural disaster ever occurred in Indonesia (2018). Accessed on 5 October 2024 https://www.ruangguru.com/blog/9-bencana-alam-terdahsyat-yang-pernah-terjadi-di-indonesia
3. R. N. Alghifari, The biggest natural disaster that has ever happened in Indonesia, one of them being the most powerful volcanic eruption (2021). https://jurnalsoreang.pikiran-rakyat.com/
4. S. Sabrina, The story of 1883 in the eyes of poets (2020). https://sejarah.dibi.bnpb.go.id/
5. S. P. Maarif, Thoughts and ideas of disaster management in Indonesia. (Bintek Manajemen Penyusunan Peta Rawan Bencana, 2012).
6. R. D. Thomas, National-local relations and the city's dilemma. Hisp. J. Behav. Sci., **509** (1990)
7. M. Skidmore, and H. Toya, Natural disaster impacts and fiscal decentralization. Land Econ., **89** (2013)
8. A. Z. Rahmayanti, The role of fiscal decentralization in disaster management in Indonesia. J. Ekon. dan Pembang., **22** (2014)
9. D. I. Putra, and M. Matsuyuki, Disaster management following decentralization in Indonesia: Regulation, institutional establishment, planning, and budgeting. J. Disaster Res., **14** (2019)
10. Adryan. Harmonization of government with regions as the effectiveness of the government system. J. Leg. Indones., **16** (2019)
11. B. D. Anggono, Harmonization of laws and regulations in the field of disaster management. Mimb. Huk., **22** (2010)
12. M. A. Hapsari, Suswoto and N. Ariyani, Politics and law of central and regional relations in disaster management. J. Law Policy Transform., **6** (2021)
13. Y. Bae, Y. M. Joo, and S. Y. Won, Decentralization and collaborative disaster governance: Evidence from South Korea. Habitat Int., **52** (2016)
14. S. D. Kartika, Politics of disaster management law. Pus. Pengkaj. Pengolah. Data dan Informasi/P3DI Sekr. Jenderal DPR RI, **20** (2015)
15. H. J. Triyana, and R. A. Wibowo, Implementation of community participation regulations in Merapi volcano disaster management. Mimb. Huk., (2011).
16. A. Sadat, Intergovernmental in handling natural disasters in local governments. Kybernan J. Stud. Kepemerintahan, **4** (2019)
17. S. Chadijah, A. Suyadi, and T. Tohadi, Pulling the authority of the central government and regional governments in handling the covid-19 pandemic. Rechtsregel J. Ilmu Huk., **3** (2020)
18. M. B. Miles, A. M. Huberman, and J. Saldana, Qualitative data analysis a methods sourcebook. (Sage Publications Inc, Washington D.C, 2014)
19. D. Djohan, S. Arinanto,S. R. Zuhro, J. Kristiadi, and I. R. Maksum, Examining the complicated side of decentralization and election practices (Yayasan Bhakti Otonomi Daerah, Jakarta, 2016)
20. K. D. Maani, Efforts to improve the quality of public services in the era of regional autonomy. J. Demokr. **4** (2005)
21. A. Grady, B. Gersonius, and A. Makarigakis, Taking stock of decentralized disaster risk reduction in Indonesia. Nat. Hazards Earth Syst. Sci., **16** (2016)
22. Djalante, R. *et al.* Progress in disaster science review and analysis of current responses to COVID-19 in Indonesia : Period of January to March 2020. Prog. Disaster Sci., **6**, 100091 (2020)
23. A. Sair, Disasters and disaster curriculum "projects" in schools. J. Urban Sociol., **1** (2018)
24. Z. Septikasari, and A. Ayriza, Disaster education integration strategy in optimizing community resilience in facing the Mount Merapi eruption disaster. J. Ketahanan Nas., **24** (2018)
25. G. Musacchio, *et al.* Education: Can a bottom-up strategy help for earthquake disaster prevention? Bull. Earthq. Eng., **14** (2016)
26. D. Selby, F. Kagawa, Disaster risk reduction in school curricula: Case studies from thirty countries. disaster risk reduction in school curricula: case studies from thirty countries (UNICEF, Geneva, 2012)
27. K. Iqbal, and M. Ahmed, Are decentralised governments more effective in mitigating disaster risks? Bangladesh Dev. Stud., **38** (2015)
28. L. Peng, J. Tan, L. Lin, and D. Xu, Understanding sustainable disaster mitigation of stakeholder engagement: Risk perception, trust in public institutions, and disaster insurance. Sustain. Dev., **27** (2019)
29. L. Peng, D. Xu, and X. Wang, Vulnerability of rural household livelihood to climate variability and adaptive strategies in landslide-threatened western mountainous regions of the Three Gorges Reservoir Area, China. Clim. Dev., **11** (2019)
30. F. O. Adeola, Behavior among survivors. **41** (2009)
31. S. W. Durage, L. Kattan, S. C. Wirasinghe, and J. Y. Ruwanpura, Evacuation behaviour of households and drivers during a tornado: Analysis based on a stated preference survey in Calgary, Canada. Nat. Hazards, **71** (2014)
32. D. Xu, *et al.* Influences of sense of place on farming households' relocation willingness in areas

threatened by geological disasters: Evidence from China. Int. J. Disaster Risk Sci., **8** (2017)

33. S. Suharyo, Problematika penegakan hukum penataan ruang dalam pelaksanaan otonomi daerah. J. Rechts Vinding Media Pembin. Huk. Nas., **6** (2017)
34. M. W. G. Bosmans, C. Baliatsas, C. J. Yzermans, and M. L. A. Dückers, A systematic review of rapid needs assessments and their usefulness for disaster decision making: methods, strengths and weaknesses and value for disaster relief policy. Int. J. Disaster Risk Reduct., **71** (2022)
35. M. R. Benigno, *et al.* Responding to the health and rehabilitation needs of people with disabilities post-haiyan. West. Pacific Surveill. response J. WPSAR, **6** (2015)
36. J. Malilay, W. D. Flanders, and D. Brogan, A modified cluster-sampling method for post-disaster rapid assessment of needs. Rev. Panam. Salud Publica/Pan Am. J. Public Heal., **2** (1997)
37. D. W. F. Utomo, and M. Hanita, Crisis leadership strategies in overcoming the Covid-19 Pandemic to ensure national resilience. J. Kaji. Lemhannas RI, **43** (2020)
38. R. G. L. Narayanan, and O. C. Ibe, A joint network for disaster recovery and search and rescue operations. Comput. Networks, **56** (2012)
39. M. A. Berawi, *et al.* Optimizing search and rescue personnel allocation in disaster emergency response using fuzzy logic. Int. J. Technol., **10** (2019)
40. H. Aghamohammadi, S. M. Mesgari, A. Mansourian, and D. Molaei, Seismic human loss estimation for an earthquake disaster using neural network. Int. J. Environ. Sci. Technol., **10** (2013)
41. E. M. Ramdani, Coordination by BPBD in flood disaster management. J. Ilmu Adm. Media Pengemb. Ilmu dan Prakt. Adm., **12** (2015)
42. E. Yulianto, D. A. Yusanta, P. Utari, and I. A. Satyawan, Community adaptation and action during the emergency response phase: Case study of natural disasters in Palu, Indonesia. Int. J. Disaster Risk Reduct., **65**, 102557 (2021)
43. C. Purwowidhu, Developing a resilient framework for disaster management. Kementerian Keuangan 2003–2005 https://mediakeuangan.kemenkeu.go.id/article/show/rangka-tangguh-penanggulangan-bencana (2022).
44. D. Lase, Analysis of government coordination with local governments and community participation in the health sector in natural disasters in Nias Regency. Prointegrita, **6** (2022)
45. H. K. Bakti, and A. P. Nurmandi, Recovery after the earthquake disaster in North Lombok in 2018. J. Geogr., **12** (2020)

 https://doi.org/10.1051/e3sconf/202560403002

# State policy for managing seismic territories in Kazakhstan

*Dana* Kangalakova[1*], *Zaira* Satpayeva[2], and *Kanat* Kangalakov[1]

[1]Institute of Economics Science, Committee of the Ministry of Science and Higher Education of the Republic of Kazakhstan, Almaty, Kazakhstan
[2]Caravan Resources Group, Planning Department, Almaty, Kazakhstan

**Abstract.** Despite the relatively small number of destructive earthquakes in modern times, their numerous faults mean that the danger of earthquakes is still high. The purpose of the study is to study the chronology of the earthquake in the territory of Kazakhstan and its consequences, as well as to identify the features of the state governance system for seismic activity in the territory of the Republic of Kazakhstan. The methods intended to analyze historical facts and describe public policies, as well as visualize maps of tectonic plates in the Almaty city were used in this study. The results showed that the most seismic zone in the Republic of Kazakhstan is the Almaty city, located on 27 tectonic plates. The Almaty city was twice subjected to devastating earthquakes; therefore, the recurrence rate of destructive earthquakes is high. In this regard, an effective earthquake mitigation policy is required to minimize damage. Besides, this study describes the state seismology governance system in Kazakhstan. The study results can be the basis for the development of comprehensive measures intended to reduce the destructive consequences for seismically active zones.

## 1 Introduction

An earthquake is a cataclysm that occurs due to vibrations of the earth's surface and earthquake shocks caused by the movement of tectonic plates. Measurements of earthquake strength range from 1 to 9 under the Richter scale; when magnitude 1-2 earthquakes are felt as weak earthquake shocks, magnitude 8-9 earthquakes lead to serious damage and faults at a distance of hundreds of kilometers. The causes of an earthquake depend on the type of origin; it can be volcanic, tectonic, explosive, etc. The consequences of strong earthquakes have a colossal negative impact on the country's public governance system. That is, a cataclysm entails the death of the population, destruction of territorial infrastructure, massive damage to buildings and other objects. According to earthquakelist.org [1] (an analytical center that generates seismic data from the United States Geographical Survey (USGS) and the European-Mediterranean Seismological Center (EMSC)), the largest number of earthquakes was recorded in Mexico in 2024 (1227 earthquake shocks, the highest earthquake score is 6.4), followed by Indonesia (1111 earthquakes, the highest score is 6.4), Japan (1105, the highest score is 7.5), etc. Kazakhstan (Almaty) ranks the 12$^{th}$ place, where 238 earthquakes were recorded in 2024 with the highest score being 7.0. The city statistics show that the most significant earthquakes occurred in Taichung (Taiwan, 527 earthquake shocks with magnitude of 7.4) and Taipei (Taiwan, 526 earthquake shocks with magnitude of 7.4). This is followed by Kaohsiung (Taiwan, 511 earthquake shocks with magnitude of 7.4), Davao (Philippines, 511 earthquake shocks with the magnitude of 7.1), Saitama (Japan, 294 earthquake shocks with the magnitude of 7.5), Tokyo (Japan, 246 earthquake shocks with the magnitude of 7.5), Almaty (Kazakhstan, 222 earthquake shocks with the magnitude of 7.0), etc. Thus, this study examines earthquakes in the Almaty city. The Almaty city is the most seismically active territory of Kazakhstan located at the foot of the Zailiysky Alatau mountain that is a part of the Tien Shan. As it is known, mountains arise when two lithospheric plates collide with the continental crust, and the edges of these plates are crushed into folds. Consequently, there is a high level of seismic activity in such areas, and the Almaty city is located in a seismically active zone with a high level. Almaty is the main economic, financial and innovation center in Kazakhstan. Almaty was founded in 1854, and its original name was Verny. It plays a centuries-old economic, cultural and political role. The city has a population of 2.2 million persons and is a city of national importance. The city is of strategic importance in the development of Kazakhstan; the main strategic enterprises, including innovative and scientific ones, are concentrated in Almaty (about 90%). Consequently, Almaty plays a significant role in the development of Kazakhstan. Thus, the purpose of the study is to study the chronology of the earthquakes in Almaty, their consequences, as well as the state governance system aimed at seismic activity in the territory of the Republic of Kazakhstan.

The causes of the earthquake are unclear and cannot be predicted. Consequently, many countries and many scientists have devoted their efforts to study the causes of earthquakes and their prediction, since earthquakes on land deformed the earth's surface, and earthquakes in

* Corresponding author: danakangalakova@gmail.com

water caused huge waves off the coasts of the mainland territories of countries. The earliest studies were performed in the early 19th century. One of the early studies is that of Heim [2], a Swiss geologist who went on an exploration expedition and studied the surface of the earth and the effects of an earthquake in China. Gutenberg [3] studied the causes of waves and earthquakes in water areas. The earliest earthquake forecasting studies can be found in the works of Vere-Jones [4]. The author tried to develop a methodology intended to predict the seismic activity of the territory. Thus, the public governance system requires update under modern requirements. And, a lot of research has been devoted to the seismically active zone of the Almaty city, located at the foot of the Zailiysky Alatau mountain. Kruger F., Kulikova G. and Landgraf A. [5] studied the etiology of earthquakes in Central Asia, where earthquakes occurred in 1885 (Belovodskoye), 1887 (Verny) and 1889 (Chilik). The authors showed with the help of modeling of historical seismograms that the earthquakes had an even greater magnitude than it was officially reported. In 1911, another powerful earthquake occurred again. During it Deyev E.V. and Korzhenkov A.M. [6] conducted field research and identified the source of the earthquake. It was destructive in nature and had had low magnitude for many years. Thus, the Almaty city is located on faults of tectonic plates. It is a source of destructive earthquakes [7]. After the earthquake in Gaziantep (Turkey) [8], government agencies and seismologists in Kazakhstan began to think about measures that would minimize the earthquake consequences. However, the point of active government action was the earthquake recorded in January 2024.

## 2 Methodology

This research uses methods of visualization, chronological analysis and description of the state governance system of the seismic industry. The research algorithm is presented in Fig 1.

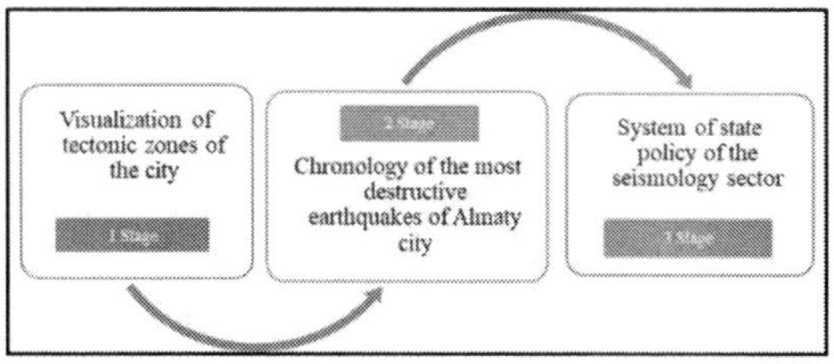


**Fig. 1.** Research algorithm

In the first stage, the study uses mapping of the tectonic zones of the city, which shows the future risks associated with a destruction monitoring system that can prevent devastating consequences. This map can serve as a guide for urban planning and earthquake response organizations. The second stage of the study examines the devastating earthquakes that have struck the city since its foundation. Based on the data and geographical changes in the city's territory, which were changed due to previous earthquakes, it is possible to assume and predict the scale and level of upcoming earthquakes. The state seismological policy is considered in the third stage of the study. An effective seismological policy is the key to optimal territory management during seismic activity. In this regard, the current policy of Kazakhstan in the field of seismology and the assessment of the readiness of state policy for seismic activity in Kazakhstan will be considered.

## 3 Results and discussion

Kazakhstan is located in Central Asia and is the ninth country in terms of territory. It borders with Kyrgyzstan, Uzbekistan, Turkmenistan, and Russia. The Tien Shan and Ile Altai Mountains are located in Kazakhstan's southeastern territory. These regions are earthquake-phone, and as they are located close to the tectonic fault of the earth's crust, there are repetitions of seismological activities. Almaty is Kazakhstan's most populous city and a major commercial, financial, and cultural center. It is located in the foothills of the Trans-Ili Alatau at an altitude of 700-900 м (2300-3000 ft).

Almaty was twice destroyed during the Verny (1887) and Kamensky (1911) earthquakes. The country's seismologists identified 27 tectonic faults under the city After the devastating earthquakes. Figure 2 shows tectonic faults that are more common closer to the western part of the Tien Shan Mountain range or the southern part of the Trans-Ili Alatau. The number of faults decreases to the north of the city. Almaty is the only city in Central Asia where a magnitude 10 earthquake can be predicted. In general, building codes do not contain a concept to calculate a magnitude 10 earthquake; this concept exists only in seismology. Besides, the city is located in a deep hole filled with sediment brought by rivers. It means that the soil under the city is fragile, and when the air shakes, the intensity of the earthquake increases by 1-1.5 points. Figure 2 shows a map of Almaty and tectonic faults in the territory of the city.

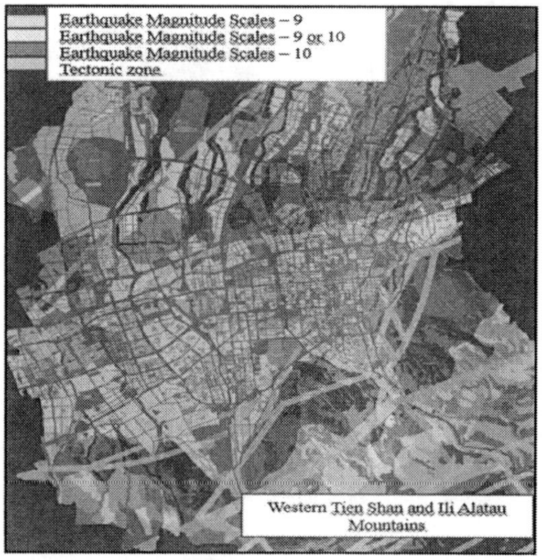


**Fig. 2.** Tectonic zones in Almaty

Figure 3 shows data on multi-storey residential complexes. As it was mentioned before, there are more tectonic faults to the south of the city, and their number decreases to the north; however, as we can see in the map, residential complexes are concentrated in the city center. Besides, the location of residential complexes coincides with the location of tectonic zones. Thus, the Almaty city is located in the most earthquake-prone zone in Central Asia. Losses and damage will be enormous if the Verny or Kemin earthquakes are repeated.

**Fig. 3.** Layout of multi-storey residential complexes constructed on tectonic faults

And, the most destructive earthquakes occurred more than 100 years ago in Almaty. The Verny earthquake occurred on the northern slope of the Trans-Ili Alatau, 10-12 kilometers south of the city at a depth of about 60 kilometers on May 28 (June 9), 1887, at 04:35 local time. The magnitude of the earthquake estimated under the Richter scale was from 7.3 to 7.8. According to official data, the city's inhabitants were about 22 thousand persons; 161 persons died, 1,798 brick houses were destroyed, only one survived. Mostly brick houses were built in the territory of the city, there were few wooden ones (Fig. 4).

**Fig. 4.** Verny earthquake, 1887

Figure 5 shows the Akzhar collapse that was the largest consequence after the Verny earthquake in 1887.

**Fig. 5.** Akzhar collapse

Thus, the very first destructive earthquake not only wiped the city off the face of the earth, but also changed its landscape. Wood was mainly used in the construction of the city after the earthquake. The second devastating earthquake occurred on January 4, 1911. The epicenter was located near the eastern tip of the Trans-Ili Alatau, approximately 40 kilometers from the Almaty city. Its magnitude was 8 under the Richter scale. Seven hundred and seventy-six houses were destroyed, 35 persons died, and 168 persons were injured.

One of the last major earthquakes in the city in 2024 was recorded on January 23 at 00:09 local time. The epicenter was in the area 264 km southeast of the Almaty city, in the territory of Kyrgyzstan, at a depth of 65 km. Energy class was 15.1. Magnitude was 6.7. Forty-four persons with various mild to moderate injuries sought medical care. Besides, first shift classes in schools and colleges were canceled in Almaty. The buildings did not collapse; minor repair deformations were recorded. Thus, the Almaty city was destroyed and rebuilt twice. At the moment, the number of residents in Almaty is officially 2.2 million persons; however, since the city was declared to be of strategic importance, it attracted more labor from the periphery of the country. According to unofficial data, the number of residents is even more significant. In this regard, effective public policy is required to ensure a safe life for the posspulation. The public policy system is shown in the Fig. 6.

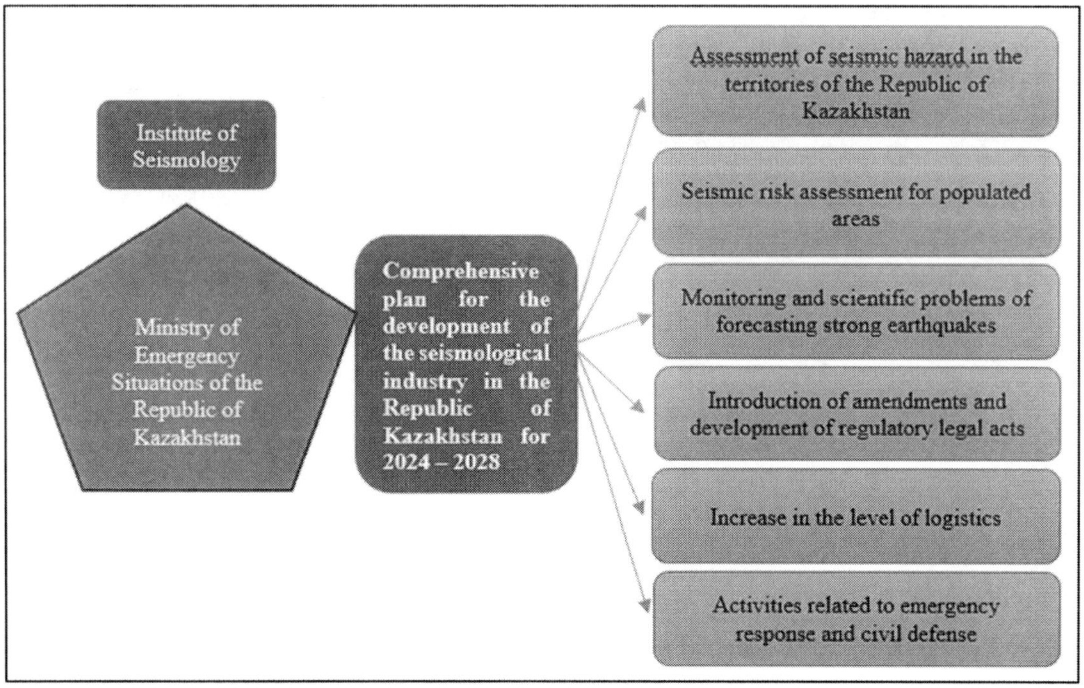


**Fig. 6.** System of state policy of the seismological industry of the Republic of Kazakhstan

The Ministry of Emergency Situations of the Republic of Kazakhstan is the main body pursuing state policy regarding seismic phenomena. The main seismic activity research center is the Institute of Seismology, an organization subordinate to the Ministry. This institute was created in 1976 and solved the following problems: assessment of the seismic hazard for the territory of the Republic of Kazakhstan (seismic zoning), solvation of scientific problems of predicting strong earthquakes, as well as assessment of the seismic risk of populated areas. However, the current state of the Institute of Seismology is insufficient to conduct advanced seismological research; in connection with the recent earthquakes, the President of the Republic of Kazakhstan instructed to resume research and equip the organization with advanced technologies to reduce seismic risks in Kazakhstan.

The situation in the scientific field of seismology requires special attention. The research conducted by the Institute of Seismology is based on outdated equipment. The Institute of Seismology was renamed in 2024 as the National Center for Seismological Research, but it does not have essential funding. In addition, 17 thousand dollars were planned to improve the infrastructure in seismological science, but more than these funds are needed for the proper level of research. There are 68 seismic stations located in Kazakhstan, which are 80 times smaller than Japan's when the territory of Kazakhstan is enormous compared to Japan. However, even in Japan, seismologists cannot predict powerful tremors and their localization. In Kazakhstan, the method of forecasting with the help of animals is still used. However, money is no longer allocated for their maintenance: food for rabbits, lizards, chickens, snakes, and other wards that serve seismology is bought by the employees themselves. Regarding scientific personnel in the seismological industry, this industry is rare among young people, since the average monthly salary is about 200-450 dollars. In this regard, this industry needs a significant human resource potential of scientific personnel and young people. The primary scientific staff of the National Seismology Center is in the pre-retirement age, which also affects the results of scientific research, which requires a set of unique and complex skills to ensure the quality of scientific work. By gender, the potential of scientific personnel includes more men than women. It should be noted that the scientific composition of Kazakhstan shows that 53% of women scientists are registered in the country, which means that women are more interested in scientific research and choose the investigator profession. Thus, to improve seismological science, Kazakhstan would be better off developing a unique program for young women scientists aimed at increasing human resources in seismological science and diversifying scientific research.

Besides, the "Comprehensive Plan for the Development of the Seismological Industry in the Republic of Kazakhstan" was adopted in 2024 [9]. It should be noted that no special events related to seismology were held in Kazakhstan before it. The comprehensive plan includes six areas.

The first direction includes assessment of the seismic hazard in the territory of the Republic of Kazakhstan. As mentioned earlier, the Almaty city is a central earthquake zone. However, other regions where the mountain range continues should also be classified as areas where earthquakes are possible. In this regard, this direction involves the development of sets of seismic microzoning maps with a scale of 1:10,000-50,000 for the territory of cities and regions. Micro districting will help determine which areas of the territory are most vulnerable to seismic impacts. In addition, this tool is used to determine the soil characteristics in an earthquake-prone area. This data is used to design buildings and structures, considering the level of seismic hazard. As is known, the southeastern territory of Kazakhstan is located near the Tien Shan Mountains and the Trans-Ili Alatau, which does not exclude seismological activity and require detailed study.

The second direction is assessment of the seismic risk of populated areas, significant objects and forecasts of damage from an earthquake. This area involves certification of seismic resistance and reliability of a structure, as well as calculation of potential damage in the event of an earthquake. An analysis will be carried out not only of buildings but also of other vital elements of urban infrastructure, including water supply, electricity and gas supply systems, transport routes, and healthcare facilities, and attention will also be paid to socially significant facilities - hospitals, schools, and government agencies. Damage forecasting includes not only the calculation of probable financial losses but also an assessment of the consequences for the population, such as loss of life, evacuation, and restoration work. The third direction is monitoring and solvation of scientific problems of forecasting strong earthquakes. In this direction, more attention will be paid to the work of seismologists who will monitor strong earthquakes in the world and calculate patterns and similarities in their recurrence. It also provides for the development of domestic software intended to process seismic monitoring data. In this regard, it is necessary to consider the importance of scientific personnel in seismological science, which will become the driving force of scientific research. The quality of scientific research will depend on the quality of scientific personnel. Therefore, Kazakhstan's scientific and educational system should strengthen scientific personnel's activities in seismological science and diversify research by attracting female scientists. The fourth direction concerns the improvement of the regulatory framework. This direction will include regular updating of building codes and regulations, taking into account new research and technologies; the establishment of mandatory certification of all buildings for earthquake resistance, with mandatory revision of certifications at certain intervals, especially in seismically active zones; the development of stricter standards for critical infrastructure (hospitals, schools, transport hubs, power plants) providing for their resistance to strong earthquakes; providing training and retraining of personnel working in the seismological field, including engineers, builders and urbanists and implementing educational programs and advanced training courses on new technologies and approaches in seismology and construction, etc. The fifth direction is increase in the level of data transmission. Namely, it is planned to create a digital archive system to convert seismic data into a digital version. It is also planned to provide the use of satellite/ultra-short-wave communications at seismic stations. To improve data transmission using digital tools in seismology, it is necessary to implement cloud technologies for real-time data storage and processing, use the Internet of Things sensor networks to automatically collect information, and increase data transfer speeds using modern network solutions such as 5G and fiber optics. Integrating data from various sources and using standards and artificial intelligence for automatic analysis and forecasting will increase the efficiency of data use. Significant financial and human resources will be required to implement this direction. In addition, there will be a need to train interdisciplinary personnel, which will contribute to improving the seismological sector in Kazakhstan.

The sixth direction includes measures for emergency response and civil defense. It includes developing emergency evacuation plans, training the population to act in earthquake conditions, creating early warning systems, and providing infrastructure ready for emergencies. These measures increase disaster preparedness, reduce risks to people's lives, and ensure a quick and effective response to emergencies. Moreover, it includes the development of a draft digital version of the Emergency Response Action Plan on a global and regional scale with mandatory computer modeling of a possible situation in the event of a devastating earthquake, the identification of alternative options for provision of the population with drinking and domestic water, carrying out activities to check the readiness of infrastructure for earthquakes and other emergencies, etc.

Thus, the state governance system of the seismic industry requires huge investments and decisive measures to improve the country's seismological system. The implementation of the measures mentioned above will ensure effective management of the seismological territory of Kazakhstan and ensure the preparedness of the population for emergencies.

## 4 Conclusion

Public policies regarding seismic activity in an area are vital to reduce damage from the effects of an earthquake. The consequences of the earthquake will be enormous without government intervention. These destructive earthquakes have already occurred in the territory of Kazakhstan. The Almaty city, with a population of 22 thousand persons, experienced two devastating earthquakes in 1887 and 1911 resulted in the formation of 27 tectonic plates that can be activated at any time. Currently, the population of Almaty is 2.2 million persons. The most dangerous areas of the city are located closer to Mount Ile-Alatau, and the most densely populated area of the city is the center, where multi-storey residential complexes are located. Thus, the Almaty city could be completely destroyed in the event of a devastating earthquake, and the consequences would be dire. The state seismic monitoring system was in decline until 2024. The main seismology research center was not an active research organization because it lacked sufficient funding, and seismologists changed directions of research due to low pay. Seismological scientists in Kazakhstan predict repetition of destructive earthquakes, since the two previous earthquakes have the etymology of repetition, since they are located at the junctions of the earth's crust plates. A new wave in the development of the seismology science was the earthquake happened on January 23, 2024, which led to the approval of a comprehensive plan for the development of the seismology industry. This plan includes six areas: from improvement of the material, technical and regulatory framework to educate the population during an emergency. Thus, the governance system for seismically active areas of Kazakhstan has received a new wave of development.

The study was conducted within the framework of grant funding of young scientists of the Committee of Science of the MSHE Republic of Kazakhstan (AP19579256 "Mechanisms for empowering women in scientific activity in the interests of the development of the innovative economy of Kazakhstan").

## References

1. Earthquake List. Top 100 countries with most earthquakes. (2024). https:// https://earthquakelist.org/reports/top-100-countries-most-earthquakes/
2. A. A. Heim, Earthquake region of Taofu. Geological Society of America Bulletin, **45**, (1934)
3. B. Gutenberg, Earthquakes and structure in southern California. Geological Society of America Bulletin, **54**, (1943)
4. D. Vere-Jones, Earthquake prediction - a statistician's view. Earth, Planets and Space, **26**, (1978)
5. F. Krüger, G. Kulikova, and A. Landgraf, Magnitudes for the historical 1885 (Belovodskoe), the 1887 (Verny) and the 1889 (Chilik) earthquakes in Central Asia determined from magnetogram recordings. Geophysical Journal International, **215** (2018)
6. EV Deyev, and AM Korzhenkov, Paleoseismological studies in the epicentral area of the 1911 Kemin earthquake (northern Tien Shan). Russian Geology and Geophysics, **57**, (2016)
7. Ü. Akkemik, K. Mazarzhanova, and A. Kopabayeva, A case study on geomorphology: Tree-ring-based confirmation of the formation year of Lake Kaindy after the 1911 Kemin earthquake in Almaty (Kazakhstan). Turkish Journal of Earth Sciences, **31**, (2022)
8. K. Yan, M. Miyajima, H. Kumsar, O. Aydan, R. Ulusay, Z. Tao, Y. Chen, and F. Wang, Preliminary report of field reconnaissance on the 6 February 2023 Kahramanmaras Earthquakes in Türkiye. Geoenvironmental Disasters, **11**, (2024)
9. On approval of the Comprehensive Development Plan for the seismological industry of the Republic of Kazakhstan for 2024-2028 [in Russian] (2024). https://adilet.zan.kz/rus/docs/P2400000526

# Women's rights protection in national disaster management law system

*Kunthi* Tridewiyanti[1], *Nina* Silvana[1], *Didi* Sunardi[1], *Jazirra* Ayesha Nadyaputri[2], and *Endra* Wijaya[1*]

[1]Faculty of Law, Universitas Pancasila, Jakarta, Indonesia
[2]Boston University, Boston, Massachusetts, United States of America

**Abstract.** This study discusses women's rights protection in the context of disaster management that occurred in Indonesia. This topic is important considering that Indonesia is a country that often experiences disasters, and women are a group that is vulnerable to becoming victims of these disasters. This study uses normative legal research method and relies on secondary data in the form of several legal materials. The analysis of this study uses qualitative method using statutory and conceptual approaches. This study concludes that there are still several weaknesses in the national disaster law system, especially in terms of practice and law enforcement. Therefore, women's rights as disaster victims still do not receive maximum protection.

## 1 Introduction

Indonesia is one of the countries that frequently experiences disasters, both natural and non-natural disasters. Data from the National Disaster Management Authority of the Republic of Indonesia (*Badan Nasional Penanggulangan Bencana* or BNPB) reveals that in the period of 2018-2023, every year there were more than 3,000 disasters in Indonesia with thousands of people dying or being injured, and also resulting in hundreds of thousands of refugees [1]. Most of these disasters were natural disasters. However, if we look closely at the previous years, the fact is that Indonesia also experienced several disasters in the form of social conflicts which caused many victims. In relation to social conflict, data from the Ministry of Home Affairs reveals that in 2023 there have been 218 cases of social conflict, while in 2024, in the period January to March, there have been 83 cases of social conflict [2].

To respond to the frequent disasters, several laws and regulations have been issued whose substance regulates aspects of handling disasters that occur in Indonesia. At the national level, Law Number 24 of 2007 concerning Disaster Management and Law Number 7 of 2012 concerning Social Conflict Settlement have been issued. In general, the aims of the issuance of such laws and regulations regarding disaster management are: *first*, to provide legal bases for the state to act in preventing, planning, handling, and evaluating disasters; *second*, to minimize the occurrence of victims and more severe damage due to disasters; and *third*, to provide detail steps or procedures for responding to disasters that occur. This is intended so that the handling of disaster events in Indonesia can be carried out in a more orderly or well-coordinated manner, and the results can be more optimal in protecting human safety as well as their property and important public facilities [3].

In connection with the disaster management system, one aspect that is interesting to examine further is the aspect of protection for vulnerable groups, especially women [4]. A study from The United Nations Entity for Gender Equality and the Empowerment of Women (UN Women) revealed that women and children are 14 times more likely than men to die during disasters [5]. This indicates that women are more vulnerable to becoming victims compared to men when a disaster occurs. Due to the physical, biological, psychological conditions and the construction of existing social relations, this means that women need to receive serious attention in the context of human rights-based disaster management [6]. In a disaster, the physical condition of pregnant women certainly cannot be treated the same as people in general. Likewise with mothers who are breastfeeding, their handling in disaster situations requires certain special treatment.

Moreover, in a disaster in the form of social conflict, women can even become targets and experience sexual violence perpetrated by parties involved in the social conflict. This situation has been reminded by many scholars and activists, and one of them is Kuloglu through her research which reveals that when conflict is ongoing, women are very vulnerable to becoming victims and can experience several forms of suffering at once in one period of time. Physically, she can be injured; sexually, she can be exploited or abused; psychologically, she can experience trauma; her family can be destroyed; and economically, she can also lose her property or livelihood.

* Corresponding author: endrawijaya333@univpancasila.ac.id

Therefore, in conflict, women face two war zones at once, namely the "war front" and the "home front" [7].

Based on that background explanation, so it is necessary to conduct a study that focuses on the issue of how to protect women's rights in the context of disaster management that occurred in Indonesia.

## 2 Methods

This study uses normative legal research method and relies on secondary data in the form of several primary and secondary legal materials. The primary legal materials refer to several laws and regulations which substance is relevant to the issue of disaster management in Indonesia and the protection of women's rights. Such acts and regulations mentioned which are examined later, among others Law Number 24 of 2007 concerning Disaster Management, and Regulation of the Head of the National Disaster Management Authority Number 13 of 2014 concerning Gender Mainstreaming in the Field of Disaster Management. The secondary legal materials will refer to several studies that have been conducted by several scholars or researchers. The data are analyzed qualitatively using statutory and conceptual approaches [8]. The conceptual approach, in the context of this study, refers to the concept of legal protection for citizen's rights and the concept of gender justice and its protection.

## 3 Results and discussion

### 3.1 Disaster management in Indonesian law

From the perspective of constitution, state through its government has several main duties and functions related to the establishment of social welfare. Those government's duties and functions stated in preamble section of the 1945 Constitution of the Republic of Indonesia that, "Then, to form an Indonesian State Government that protects the entire Indonesian nation and all of Indonesia's blood and to promote general welfare, educate the life of the nation, and participate in implementing world order based on independence, eternal peace and social justice, Indonesian national independence was formulated in The Constitution of the State of Indonesia, ..." Furthermore, such key concept of Indonesian government's duties and functions is explained more by several articles within the Indonesian Constitution.

From the legal perspective, several legal bases have been issued to serve as guidelines for disaster management in Indonesia. The 1945 Constitution of the Republic of Indonesia provides a constitutional basis for efforts to protect disaster victims, including women, through Article 28A, Article 28D paragraph (1), Article 28G, Article 28H, Article 28I paragraph (2) and paragraph (4), Article 28J, and Article 34 paragraph (2) and paragraph (3).

In Indonesian Constitution, Article 28A emphasizes that everyone has the right to live and the right to maintain a decent life. To guarantee it as a part of law enforcement, then Article 28A is supported by Article 28I paragraph (4) which confirms that "Protection, promotion, enforcement and fulfillment of human rights is the responsibility of the state, especially the government." So, the series of several articles of the Indonesian Constitution has simultaneously provided an illustration that protecting the rights of disaster victims, on the one hand, is a right for the victims, and on the other hand, is an obligation or duty for the state to guarantee and fulfill them to the maximum. The two are closely related to each other.

In the form of act as a written law in Indonesia, there are Law Number 24 of 2007 concerning Disaster Management and Law Number 7 of 2012 concerning Social Conflict Settlement. These two laws essentially regulate mechanisms for handling natural and non-natural disasters, such as social conflicts, including handling their victims. At the government regulatory level, several disaster management regulations have been issued at both the national and regional levels. These regulations, among others: Government Regulation Number 21 of 2008 concerning the Implementation of Disaster Management; Minister of Social Affairs Regulation Number 01 of 2013 concerning Social Assistance for Disaster Victims, and its amendments, namely Minister of Social Affairs Regulation Number 07 of 2013; Regulation of the Head of the National Disaster Management Authority Number 7 of 2008 concerning Procedures Guidelines for Providing Assistance to Fulfill Basic Needs; and Regulation of the Head of the National Disaster Management Authority Number 13 of 2014 concerning Gender Mainstreaming in the Field of Disaster Management.

If we look closely at several of these laws and regulations, it can be seen that the stakeholders involved in the disaster management system in Indonesia consist of the government, state-owned (public) and private enterprises, individuals, community organizations or associations, indigenous communities, international institutions and foreign non-governmental organizations. From the institutional perspective, the Indonesian legal system has also established and given authority to the National Disaster Management Authority (*Badan Nasional Penanggulangan Bencana* or BNPB) as a formal institution which has the main function and authority to determine policies, respond to and handle disasters problems in Indonesia [9].

A study from Eburn explains that in the modern perspective of disaster management there must be many parties involve in taking responsibility for preparing for, responding to, and recovering from the impact of natural and other hazards. Each of them conducts the responsibility sharing. And in such sharing, to share responsibility does not mean equal responsibility, and there are some areas in which the state or government should assume greater responsibility than the community or the others. So, it is important for governments, communities, individuals, and the nongovernment sector to negotiate how responsibility is to be shared, who is to be responsible for what, and who owns the risk that comes with natural hazards [10].

### 3.2 Women's rights in the context of disaster management and its challenges

Even though from the legal, regulatory, and institutional perspective there is already a disaster management system in Indonesia, the system and its practices still face problems related to handling disaster victims from vulnerable groups, especially women and children. The starting point for disaster management should also refer to the principles of protecting women's rights. In practice, the disaster management system in Indonesia still tends to equate women and men in handling disaster victims. In fact, the needs for handling disaster victims between women and men are very different. Moreover, in some cases, discrimination, violations, or neglect of women's rights occur when they become victims of disasters.

Several scholars or researchers have revealed the reasons why in cases of disasters women become very vulnerable and even experience several sufferings at once. These reasons are related to due to physical vulnerability, excessive workload carried out by women, dependence on men in household decision making, and the lack of representation of women in the decision or policy making process, especially those related to disaster management issues [11]. Why are women more vulnerable to becoming victims than men, and can even experience serious conditions when a disaster occurs? In detail, the studies by Neumayer and Plümper, and Ramailis and Sakir reveal and provide the answer to that question:[12]

*First*, physiologically men can be better prepared to face the impact of disasters on their physical being as humans. For example, if a woman is not as strong as a man, she will be more easily swept away by flood waters. This disadvantage is especially severe for pregnant women, whose ability to move or mobility is reduced or limited. In addition, women can also run more slowly and cannot be as deft at climbing poles or trees, so that when a tsunami occurs it will be more difficult for women to escape to higher ground or spot.

The concrete examples can be seen when the tsunami hit Sri Lanka, where it was proven that men were actually better able to survive when facing tsunami waves because men had more skills in swimming or climbing high places. And also in Indonesia, when the big earthquake and tsunami disasters occurred in Aceh in 2004, 55%-70% of the victims were women. This indicates that women are one of the most vulnerable groups when a disaster occurs [13].

*Second*, men and women have different tendencies to die from various diseases. Some diseases that arise after a disaster are actually more deadly for women who are disaster victims.

*Third*, large-scale natural disasters can have severe adverse impacts on social infrastructure in affected countries or regions, reducing access to food, hygiene, health services, and clean water. When health service infrastructure is badly damaged, this can have a negative impact, especially on women who are pregnant or having a baby.

In the case of mudflows from the natural gas exploration processed by PT. Lapindo Brantas in Porong, Sidoarjo, East Java, which started in 2006, several women's rights have been violated, including in the form of compensation paid or given only to men or husbands whose property was damaged by mud. In fact, some of the damaged property was actually owned by the woman or a wife. As a result, the rights or interests of women who were victims of the Lapindo mudflow were ignored and these women did not receive compensation [14]. Even in several cases, husbands who had received compensation money as victims of the Lapindo mudflow actually used the money for their own "personal satisfaction" [15].

Furthermore, in many cases of disasters, in fact many women become the victims who also experience several sufferings at once. This means that in the case of a disaster, women can experience and become victims many times at once, for example experiencing suffering as a disaster victim and at the same time experiencing eviction from their place of residence, losing their husband or family, losing their livelihood and even becoming victims of sexual violence, such as rape or slavery [16]. In the Indonesian context, facts about women experiencing the condition of being victims' multiple times at one time can be found, among others in the armed conflict in Aceh between the Free Aceh Movement and the Indonesian Government in 1976-2005; [17] and the attacks on Shia Muslim groups in Sampang, Madura, in 2004-2012.

A study by Putri and Permatasari, with citing various sources, has revealed many examples of women who experienced being "multiple victims" during the armed conflict in Aceh. They explained that during the Aceh conflict, women experienced multiple levels of violence, including physical, psychological and economic violence, including rape, sexual harassment, torture, arrest, theft and looting [18].

There were victims who were raped and their husbands were also tortured, and there were also those who were raped even though they were pregnant, causing the baby in their womb to die. The female victims also experienced trauma, being ostracized from their families and social circles. They were also forced to remain silent so that their families would not feel embarrassed. In fact, several women victims of sexual torture during the conflict in Aceh also experienced infertility, disability, damage or non-functioning of their sexual and reproductive organs. The horrific experiences experienced by female victims of the social and armed conflict in Aceh are further exacerbated by the fact that efforts to fulfill the right to reparation for female victims of the Aceh conflict have not been optimal and complete.

That weakness, according to Putri and Permatasari's study, was caused, among other things, by the lack of women's participation in formal peace processes and in decision-making institutions that determine reparation rights programs for female victims. Furthermore, the government still does not have serious commitment or political will to resolve cases of serious human rights violations in Aceh.

In regard to the women's participation point, the scholars like Hemachandra, Amaratunga and Haigh had reached conclusion in their study about the urgency of women participation in planning, discussing, making decision and executing programs or policies concerning how to response and answer the needs of women as victims in disasters. The major argument for increasing the role of women in public administration and decision making is due to its ability of enhancing the government responsiveness and effectiveness. That action helps in achieving inclusive development and democratic governance that helps in improving trust and confidence in government institutions and increasing sustainability and responsiveness of public policies. Furthermore, women leadership also ensures inclusiveness of their interests and properly address their issues, especially in the context of disaster management [19].

In the case of disasters in the form of social conflicts that befell Shia Muslim in Sampang, examples can be found where female victims from Shia Muslim experienced several instances of neglect or violation of their rights as disaster victims and as women. Several Shia Muslim women had to give birth in refugee camps without any assistance from health workers or midwives. Some of these women also experienced ridicule from health workers when they wanted to seek treatment, and then because of that, they no longer wanted or reluctant to seek treatment at health centers or clinics [20].

To respond to those situations, the disaster law system in Indonesia is still inadequate because: *First*, in several laws and regulations governing disasters, such as in Law Number 24 of 2007 concerning Disaster Management, it is still could be seen that the treatment or handling given to disaster victims is still generalized between male and female victims, whereas the conditions of life necessities for them are clearly different. *Second*, the Indonesian legal system still does not provide adequate mechanisms that can be used to complain or file a lawsuit if women's rights are ignored or violated in a disaster circumstance.

Likewise, if we look at the regulations in the form of Regulation of the Head of the National Disaster Management Authority Number 13 of 2014 concerning Gender Mainstreaming in the Field of Disaster Management, which is substantially intended as a guideline for central and local governments as well as non-government parties who care or have an interest in protect and fulfill women's rights in disaster management. Regulation of the Head of the National Disaster Management Authority Number 13 of 2014 does not yet regulate the mechanism for law enforcement and legal accountability for the protection of women as disaster victims. Even though the regulation already contains several quite detailed steps in protecting women in disaster situations, namely starting from the pre-disaster, emergency response and post-disaster stages, and which also includes planning, budgeting, funding, implementation, monitoring, evaluation and reporting. In fact, what has not been regulated in detail is the matter of law enforcement mechanism and legal accountability for the protection of women as victims in disaster management.

To overcome or respond to some of these shortcomings, the concept of the Indonesian rule of law can be used as a reference for formulating alternative solutions. According to Asshiddiqie, the rule of law contains twelve principles consisting of supremacy of law; equality before the law; legality or due process of law; limitation of power, and checks and balances; independent executive organs; independent and impartial judiciary; Administrative Court; Constitutional Court; human rights protection; democracy; state as a tool to establish the welfare state; and transparency, accountability, public participation and social control [21].

By referring to and emphasizing especially the elements or principles of justice regarding government actions, the protection of human rights, the ideals of the welfare state and social control, it is worth initiating in Indonesia that there is an institution that can help represent women as disaster victims to demand legal accountability if their rights as victims are ignored in the disaster management process.

In the future, it is necessary to consider revisions to the law of disaster management system in Indonesia so that it can provide more firm guarantees for the protection of women's rights as victims in disasters. Moreover, it is also necessary to consider expanding or strengthening the authority of the National Commission on Violence against Women (Komnas Perempuan) so that it can become an institution that represents women whose rights are ignored or violated when they become victims of disasters. Therefore, in the future, Komnas Perempuan can also be equipped with the authority to file lawsuits against certain parties who have ignored or violated the rights of women as disaster victims.

And actually, such kind of the lawsuit already facilitated by Indonesian law system, among others, through Law Number 5 of 1986 concerning Administrative Court which was revised several times and the last revised by Law Number 51 of 2009 concerning Second Amendment of Law Number 5 of 1986 concerning Administrative Court, and also through Law Number 30 of 2014 concerning Governance Administration [22]. Those laws give the right to file the lawsuit against government body or official who have conducted action actively or passively as well as a kind of abandonment causing loss to the citizen or other parties [23].

The proposal to appoint the National Commission on Violence Against Women will create more independent and fair women's rights protection function rather than if the authority to represent and file lawsuits on behalf of women victims of disasters is handed over to the Ministry of Women's Empowerment and Child Protection. If the Ministry of Women's Empowerment and Child Protection has such authority, then it would face a conflict of interest if the defendant alleged for violating the rights of women as disaster victims is from another government institution.

The mechanism for demanding government accountable for not fulfilling the rights of women as disaster victims as explained above, if placed in the context of legal protection for citizens' rights, can be categorized as a form of repressive legal protection. Which will only emerge if preventive legal protection fails to be implemented completely.

In connection with the concept of legal protection, Hadjon's study is interesting to observe. He believes that the protection of the people is actually based on the principle of mutual cooperation or kinship ("gotong royong" or "kekeluargaan"), and this is what is referred to as legal protection in the context of "the Pancasila State of Law." Legal protection in the Pancasila State of Law actually contains 3 (three) objectives, namely: *first*, preventing or avoiding disputes between the people and the government; *second*, if a dispute persists, efforts will be made to resolve the dispute through deliberation or mutually respectful dialogue; and *third*, if the dispute still cannot be resolved through deliberation, then the court becomes the final settlement mechanism so that the dispute can be resolved completely, and social welfare can be achieved and maintained as well [24].

## 4 Conclusion

Formally, the Indonesian disaster management law system is already regulated in several acts and regulations. However, from the aspect of protecting women's rights, there are still some improvements that need to be made in the disaster law. In practice, cases still occur which show that women's rights as disaster victims are violated. Guarantees for the protection of women's rights in disaster situations need to be strengthened and attached with law enforcement mechanisms for parties who violate or ignore the rights of women as disaster victims.

## References

1. National Disaster Management Authority of the Republic of Indonesia, Indonesian Disaster Information Data : Statistics on Disasters, Casualties and Damage According to Time [Data Informasi Bencana Indonesia: Statistik Bencana, Korban dan Kerusakan Menurut Waktu], https://dibi.bnpb.go.id/statistik_menurut_waktu, accessed: 27 July 2024.
2. Ministry of Home Affairs of the Republic of Indonesia, Conflict Incidence Rate [Angka Kejadian Konflik], https://e-database.kemendagri.go.id/dataset/1162/tabel-data?page=1, accessed: 1 August 2024.
3. Article 4 and General Elucidation of Law Number 24 of 2007 concerning Disaster Management, and Article 3 and General Elucidation of Law Number 7 of 2012 concerning Social Conflict Settlement.
4. D. Alexander, Political responses to emergencies, in A.E. Collins, eds., Hazards, risks, and disasters in society, (Elsevier, Amsterdam, 2015). https://doi.org/10.1016/B978-0-12-396451-9.00013-5
5. The United Nations Entity for Gender Equality and the Empowerment of Women (UN Women), Women's resilience to disasters concept note, https://wrd.unwomen.org/practice/resources/womens-resilience-disasters-concept-note, accessed: 27 July 2024.
6. C. Abao, et al., Women in natural disasters: indicative findings in unraveling gender in institutional responses (An ASEAN Intergovernmental Commission on Human Rights (AICHR) Thematic Study), (ASEAN Intergovernmental Commission on Human Rights (AICHR) & The Department of Political Science, Ateneo de Manila University, 2018).
7. C. Kuloglu, Violence against women in conflict zones, in H. Carreiras & G. Kümmel, eds., Women in the military and in armed conflict, (VS Verlag für Sozialwissenschaften, Wiesbadem, 2008). https://doi.org/10.1007/978-3-531-90935-6_10
8. S. Soekanto, Pengantar penelitian hukum (UI Press, Jakarta, 1986).
9. Regulation of President Number 1 of 2019 concerning National Disaster Management Authority.
10. M. Eburn, Disaster risk reduction in the shadow of the law, in A.E. Collins, eds., Hazards, risks, and disasters in society (Elsevier, Amsterdam, 2015). https://doi.org/10.1016/B978-0-12-396451-9.00015-9
11. K. Hemachandra, D. Amaratunga and R. Haigh, Factors affecting the women's empowerment in disaster risk governance structure in Sri Lanka, Int. Jou. of Disaster Risk Reduction, **51**, (2020). https://doi.org/10.1016/j.ijdrr.2020.101779
12. E. Neumayer & T. Plümper, The gendered nature of natural disasters: The impact of catastrophic events on the gender gap in life expectancy, 1981–2002, Annals of the Assoc. Of American Geog., **97,** (2007). https://doi.org/10.1111/j.1467-8306.2007.00563.x
13. N. Ramailis & Sakir, Increasing women's resilience to disasters: An analysis of gender mainstreaming in natural disaster management in Bantul, Indonesia, Jurnal Ilmu Sosial dan Ilmu Politik, **13** (2023). https://doi.org/10.15575/jispo.v13i2.29003
14. A. Mustafainah, et al., Pencerabutan sumber-sumber kehidupan: Pemetaan perempuan dan pemiskinan dalam kerangka HAM (Komnas Perempuan, Jakarta, 2012).
15. A. Heroepoetri, Pengabaian dari mula : Laporan pemetaan pelanggaran hak asasi perempuan dalam bencana luapan lumpur di Kecamatan Porong, Kabupaten Sidoarjo, Jawa Timur (Komnas Perempuan, Jakarta, 2011).
16. A. Rushton, et al., The gendered body politic in disaster policy and practice, Int. Jou. of Dis. Risk Red., **47** (2020). https://doi.org/10.1016/j.ijdrr.2020.101648
17. A. Uzia, et al., catatan dua tahun terakhir (2011-2012) kekerasan terhadap perempuan di aceh, menjelujur pengalaman kekerasan perempuan di Aceh: Perjuangan tiada henti meniti keadilan (Komnas Perempuan dan Jaringan Pemantau, Jakarta, 2013).
18. L.H. Putri and M. Permatasari, Implementasi pemenuhan hak atas reparasi bagi perempuan korban konflik Aceh ditinjau dari hukum internasional, Syiah Kuala Law Journal, **2**, (2018). https://doi.org/10.24815/sklj.v2i3.12400

19. K. Hemachandra, et al., Role of women in disaster risk governance, Procedia Engineering, **212,** (2018). https://doi.org/10.1016/j.proeng.2018.01.153
20. A.H. Semendawai, et al., Laporan tim temuan dan rekomendasi (TTR) tentang penyerangan terhadap penganut syiah di Sampang, Madura (Komnas Perempuan, Komnas HAM, KPAI & LPSK, Jakarta, 2013).
21. I. Subechi, Mewujudkan negara hukum indonesia, Jurnal Hukum dan Peradilan, **1**, (2012). http://doi.org/10.25216/jhp.1.3.2012.339-358
22. D. Riza, Keputusan tata usaha negara menurut undang-undang peradilan tata usaha negara dan undang-undang administrasi pemerintahan, Jurnal Bina Mulia Hukum **3**, (2018). https://doi.org/10.23920/jbmh.v3n1.7
23. M.A. Bimasakti, *onrechtmatig overheidsdaad* oleh pemerintah dari sudut pandang undang-undang administrasi pemerintahan, Jurnal Hukum Peratun, **1**, (2018). https://doi.org/10.25216/peratun.122018.265-286
24. P.M. Hadjon, Perlindungan hukum bagi rakyat di indonesia: Sebuah studi tentang prinsip-prinsipnya, penanganannya oleh pengadilan dalam lingkungan peradilan umum dan pembentukan peradilan administrasi negara (PT Bina Ilmu, Surabaya, 1987).

# Innovative capacity building strategies for sustainable disaster risk management: a systematic review, conceptual framework, and future research directions

*Nika* Saputra[1], *Roni* Ekha Putera[1*], *Aidinil* Zetra[1], and *Azwar*[1]

[1]Faculty of Social and Political Sciences, Universitas Andalas, Padang 25175, Indonesia

**Abstract.** This systematic review examines innovative strategies for capacity building in sustainable disaster risk management (DRM). A review of 63 peer-reviewed articles, published between 2010 and 2024, was conducted using Scopus. The articles were analyzed to identify key themes related to technological innovations, community initiatives, and interdisciplinary approaches. Findings show that advancements in machine learning, Geographic Information Systems (GIS), and Information and Communication Technologies (ICT) significantly enhance DRM, particularly in risk assessment, early warning systems, and disaster response. Policy frameworks and institutional support were identified as critical to scaling and sustaining capacity-building programs. Community engagement, leveraging local knowledge, was also vital for ensuring the relevance and sustainability of these efforts. Interdisciplinary collaboration across government, NGOs, and the private sector further improved the effectiveness of these strategies. The study concludes that these efforts contribute to better disaster preparedness, increased resilience, and support sustainable development. This framework offers valuable insights for policymakers and practitioners looking to implement and scale innovative capacity-building strategies to improve global disaster resilience.

## 1 Introduction

The escalation in the frequency and seriousness of natural phenomena such as earthquakes, floods, and tropical cyclones has underscored the imperative requirement for resilient disaster risk management (DRM) strategies worldwide [1–3]. Disasters of this nature have profound repercussions on the financial, ecological, and communal spheres, leading to the disturbance of means of living and resulting in substantial human and monetary detriments. Climate change worsens these vulnerabilities, amplifying the occurrence and scale of severe weather phenomena. Hence, the Catastrophe Hazard Administration (CHA) has emerged as a critical area of interest for governmental bodies, institutions, and scholars globally, aiming to reduce the impact of disasters and enhance resilience [4, 5].

At the core of successful Disaster Risk Management lies the fundamental principle of enhancing capacity, which encompasses the empowerment of individuals, institutions, and societies to foresee, address, and bounce back from calamities [6–8]. Capacity strengthening encompasses various tasks, such as the progression of human resources, reinforcement of institutions, upgrading of infrastructure, and the assimilation of cutting-edge technologies [9–11]. For example, the comprehensive disaster readiness initiatives in Japan and the effective earthquake response plans in New Zealand exemplify the considerable advantages of well-established capacity-building efforts. These initiatives have not only enhanced preparedness and response capabilities but also expedited recovery processes and diminished prolonged vulnerabilities [12–14].

Despite these successes, significant challenges persist in DRM's capacity building. Developing countries, in particular, face numerous obstacles, including limited financial resources, technical expertise, and institutional coordination [15, 16]. Rapid urbanization, which increases exposure to disaster risks, and climate change, which introduces new and evolving threats, further compound these challenges. Addressing these multifaceted challenges requires innovative and sustainable approaches to capacity building that are adaptive, inclusive, and resilient. This article addresses several gaps in the existing literature on capacity building for sustainable DRM. While previous studies have explored various aspects of DRM and capacity building separately, there is a lack of comprehensive frameworks that integrate innovative strategies into practical, scalable models [17–19]. Additionally, much of the existing research focuses on traditional capacity-building methods without fully exploring the potential of emerging technologies and novel approaches to enhance DRM efforts. This paper examines several deficiencies in the current body of knowledge concerning developing capabilities for sustainable Disaster Risk Management (DRM). Although prior research has delved into different

* Corresponding author: roniekhaputera@soc.unand.ac.id

facets of DRM and capacity building in isolation, a shortage of comprehensive frameworks amalgamates innovative tactics into pragmatic, expandable models. Moreover, a significant portion of the prevailing literature concentrates on conventional approaches to capacity building, neglecting to fully investigate the possibilities presented by emerging technologies and unconventional methods to bolster DRM endeavors [20].

One of this research's key contributions is its focus on identifying and evaluating innovative capacity-building strategies that leverage modern technologies, interdisciplinary approaches, and community-based initiatives. By conducting a systematic review of existing literature and analyzing global case studies, this research aims to develop a comprehensive conceptual framework to guide future capacity-building initiatives. This framework synthesizes existing knowledge and introduces new perspectives and methodologies that can be adapted to different contexts and scales. The objectives of this research are threefold. First, it aims to identify and analyze innovative capacity-building strategies that have been employed successfully in various contexts. Second, it seeks to evaluate the effectiveness of these strategies, considering different geographical, social, and economic settings. Third, it aims to propose a conceptual framework that integrates these strategies into a cohesive model, offering practical guidelines for implementation and future research.

This article furnishes a comprehensive examination and integration of previous research, presenting a novel conceptual structure that enhances the present pool of knowledge on capacity enhancement in Disaster Risk Management (DRM). Through illuminating groundbreaking approaches and their tangible implementations, this study connects theoretical perspectives with practical experiences, thus serving as a significant asset for scholars and educators. The study offers evidence-based suggestions for policymakers and practitioners to strengthen capacity-building initiatives. It showcases instances of exemplary practices and novel approaches that could be implemented across different scenarios to enhance the efficiency of disaster risk management. The results of this study underscore the importance of supportive regulatory structures and heightened allocation of resources towards capacity development endeavors. It calls for the incorporation of capacity-building programs within both national and regional Disaster Risk Management (DRM) strategies, underscoring the significance of cooperative strategies that engage governmental bodies, non-governmental organizations, the business sector, and local communities.

The arrangement of this manuscript is structured to offer a thorough investigation of the subject matter. The methodology segment delineates the systematic review strategy, formulating the conceptual framework, and the approaches for identifying prospective research avenues. The section on literature review scrutinizes existing research on capacity enhancement and innovative tactics in Disaster Risk Management (DRM). The segment on conceptual framework introduces the suggested framework and deliberates on its practical applications. The discussion segment interprets the results, delving into their implications for practical application and research, and presents policy recommendations. Lastly, the conclusion encapsulates the main findings and contributions of the study, underlining its importance and proposing areas for further research.

## 2 Methodology

The systematic review used Scopus as the primary database for sourcing relevant literature. The search strategy employed keywords such as "capacity building," "disaster risk management," "innovative strategies," "sustainable," "disaster preparedness," and "disaster response." A total of 375 articles were initially identified using these keywords. The review included articles published between 2010 and 2024 from peer-reviewed journals focused on capacity building in DRM, innovative strategies, and sustainability. Articles not in English, unrelated to the topic, or from non-peer-reviewed sources were excluded. Following the application of these inclusion and exclusion criteria, the number of articles eligible for review was narrowed down to 63. The screening process involved initial title and abstract screening, followed by a full-text review to ensure relevance and quality. To document the process of identification, screening, eligibility, and inclusion of studies, a PRISMA flow diagram was utilized, ensuring a transparent and systematic approach to the literature review.

The development of the conceptual framework began with identifying key components through a detailed literature analysis of the 63 selected articles. This involved extracting key themes, strategies, and components from the studies. Thematic synthesis was then used to group similar strategies and identify overarching themes. Consultations were conducted with experts in DRM to refine and validate these findings. These identified themes were integrated into a cohesive conceptual framework, which was subsequently validated through expert reviews and case study applications. This iterative process ensured that the framework was both comprehensive and practical.

Data extraction involved systematic coding of the 63 selected studies based on predefined criteria, such as the type of capacity building, innovative strategies, and outcomes. The data management process was facilitated by software tools like VosViewer, which helped organize and manage the extracted data efficiently. The analysis techniques included qualitative thematic analysis to identify common patterns and insights and quantitative descriptive statistics to summarize key findings from the literature, such as the frequency of strategies and their geographic distribution. Additionally, a comparative analysis was conducted to evaluate the effectiveness of different capacity-building strategies across various contexts.

The identification of future research directions was carried out through a gap analysis, which involved pinpointing areas with limited research or conflicting findings within the 63 articles. Emerging trends in capacity building and DRM were also highlighted. A research agenda was developed based on these identified gaps and trends, establishing priority research areas. Specific research questions were formulated to guide future studies, and methodological recommendations were provided to address these questions effectively. This systematic approach ensures that future research is directed toward the most critical and underexplored areas. Ethical considerations were meticulously adhered to throughout the review process. This included ensuring compliance with ethical guidelines for research, particularly concerning the use of secondary data and respectful treatment of sources. Potential conflicts of interest were declared to maintain transparency in the research process [21]. By addressing these ethical aspects, the integrity and credibility of the research were upheld. Implementing the Preferred Reporting Items for Systematic Reviews and Meta-Analyses (PRISMA) framework was crucial for documenting the systematic review process [22]. A PRISMA flowchart details the steps from the initial search to the final inclusion of studies. This flowchart included the number of records identified, screened, deemed eligible, and ultimately included in the review, providing a clear and transparent account of the review process. This methodology section ensures a rigorous and transparent research approach, focusing on comprehensively evaluating and synthesizing innovative capacity-building strategies in disaster risk management.

## 3 Results

### 3.1 Descriptive analysis

Based on the systematic review of 63 articles, a comprehensive descriptive analysis was conducted to identify key themes, strategies, and the geographic distribution of capacity-building initiatives in disaster risk management (DRM). The findings provide valuable insights into the current landscape of capacity-building efforts and highlight predominant approaches and regional focuses. The analysis revealed three primary themes: technological innovations, community-based initiatives, and interdisciplinary approaches. Technological innovations were the most frequently mentioned, appearing in 28 articles (44.4%). These studies emphasized the integration of advanced technologies such as machine learning, Geographic Information Systems (GIS), and Information and Communication Technologies (ICT) in enhancing disaster preparedness and response capabilities. For example, machine learning has been used for predictive analysis, GIS for mapping and risk assessment, and ICT for improving communication and coordination during disasters [1, 13]. These technological advancements have proven effective in enhancing the ability to predict, prepare for, and respond to disasters.

Community-based initiatives were identified in 19 articles (30.2%) and highlighted the importance of local participation and leveraging indigenous knowledge for DRM. These initiatives have been particularly effective in rural and vulnerable areas, fostering resilience through volunteer programs and local engagement. Notable examples include community-based rehabilitation programs in Thailand and citizen volunteer support initiatives in New Zealand, which emphasize the critical role of community involvement in disaster preparedness and response [14, 17]. Interdisciplinary approaches were discussed in 16 articles (25.4%), showcasing the importance of collaboration among different sectors, including government agencies, non-governmental organizations (NGOs), and the private sector. These partnerships have been crucial in developing comprehensive DRM strategies that integrate various expertise and resources. Examples include the development of adaptation pathways in complex multi-risk systems and capacity-building initiatives for climate change adaptation at the sub-national level [2, 10], highlighting the benefits of coordinated efforts in building resilience.

Geographically, the analysis indicated a concentration of studies in Asia, with 38 articles (60.3%) focusing on this region. Countries such as Japan, China, and Thailand have been at the forefront of implementing innovative DRM strategies and capacity-building initiatives. Japan's extensive disaster preparedness programs and China's use of advanced technologies for risk assessment are notable examples of successful capacity-building efforts in the region [1, 12]. Oceania, particularly New Zealand and Australia, was the focus of 12 articles (19.0%). Studies from this region highlighted successful community-based initiatives and interdisciplinary approaches to DRM, such as New Zealand's volunteer support programs and Australia's government-NGO partnerships [12, 14]. South America, with 6 articles (9.5%) primarily focusing on Colombia and Brazil, has seen effective implementation of capacity-building programs aimed at rural resilience and local engagement. Colombia's innovative pedagogy for rural populations is a notable initiative in this regard [6].

North America and Europe were represented in 7 articles (11.2%), with studies emphasizing technological innovations and interdisciplinary approaches. Contributions from the USA and UK highlighted the integration of advanced technologies and collaborative efforts to enhance DRM [7, 15]. In summary, the descriptive analysis of the 63 selected articles reveals that technological innovations, community-based initiatives, and interdisciplinary approaches are the predominant strategies in capacity building for DRM. The geographic distribution of these studies shows a strong focus on Asia, with significant contributions from Oceania, South America, North America, and Europe. These findings underscore the importance of adopting a multifaceted approach to capacity building, leveraging both technological advancements and local engagement to enhance disaster resilience globally.

### 3.2 Bibliometric analysis

Bibliometric analysis provides a quantitative assessment of academic literature and research outputs, offering insights into publication trends, key contributors, influential journals, and collaboration networks within a specific field. For this analysis, we examined the 63 articles on capacity building in disaster risk management (DRM) using VosViewer, a software tool for constructing and visualizing bibliometric networks. This analysis aimed to uncover the research landscape by identifying the most active authors, institutions, journals, and collaborative efforts. The analysis of publication trends revealed a growing interest in capacity building for DRM over the past decade. The yearly distribution of publications showed a steady increase in research output, with notable peaks in 2020 and 2023. This trend underscores the increasing recognition of the importance of capacity building in enhancing disaster resilience.

Several authors and institutions emerged as prominent contributors to the field. Among the most prolific authors were He et al, who contributed extensively to technological innovations and disaster preparedness, and Kako & Hutton, known for research on community-based initiatives and interdisciplinary approaches. Leading institutions included the University of Tokyo, Japan, which had numerous publications focusing on advanced technological applications in DRM, and the University of New South Wales, Australia, recognized for contributions to community-based disaster management and interdisciplinary research [1, 12].

The analysis also identified influential journals that have published significant numbers of articles on capacity building in DRM. The International Journal of Disaster Risk Science and Progress in Disaster Science were among the top journals, reflecting their influence and relevance in the field. These journals provided a platform for comprehensive studies on various aspects of DRM and innovative strategies for capacity building. Collaboration networks revealed a high degree of cooperation among researchers and institutions, emphasizing the interdisciplinary nature of capacity building in DRM. Strong collaborative ties were observed between researchers from different countries, particularly between Japan, New Zealand, and Australia. These collaborative efforts often involved multiple institutions, highlighting the importance of interdisciplinary and international partnerships in advancing DRM research. Notably, the University of Tokyo and the University of New South Wales frequently collaborated with other leading institutions globally, fostering a rich exchange of knowledge and expertise. In summary, the bibliometric analysis of the 63 articles on capacity building in DRM underscores the growing academic interest and research activity in this field. The increasing number of publications, key contributors, influential journals, and robust collaboration networks highlight the dynamic and interdisciplinary nature of capacity building in DRM. This analysis provides a comprehensive overview of the research landscape, offering valuable insights for future research directions and collaborative opportunities.

### 3.3 Identification from a theoretical perspective

The theoretical foundation for understanding capacity building in disaster risk management (DRM) is rooted in several key concepts and frameworks. The reviewed articles offer a comprehensive examination of these theoretical underpinnings, providing insights into how capacity building can enhance resilience and response effectiveness in the face of disasters. At the core of successful DRM lies the principle of enhancing capacity, which involves empowering individuals, institutions, and societies to anticipate, address, and recover from disasters. This concept is supported by theories of organizational resilience, community engagement, and technological innovation. For instance, theories on organizational resilience emphasize the ability of institutions to adapt to changing environments and recover from disruptions, highlighting the importance of capacity building in fostering such resilience [6, 7].

Community engagement theories also play a crucial role in understanding capacity building. These theories suggest that involving local communities in DRM efforts leads to more effective and sustainable outcomes. Community-based initiatives leverage local knowledge and resources, fostering a sense of ownership and commitment among community members. This approach is evident in various studies that highlight the success of community-driven disaster preparedness and response programs [14, 17]. Technological innovation is another theoretical perspective that underpins capacity building in DRM. The integration of advanced technologies such as machine learning, Geographic Information Systems (GIS), and Information and Communication Technologies (ICT) into DRM practices enhances predictive capabilities and improves coordination and communication during disasters. Theories related to technological adoption and diffusion provide a framework for understanding how these innovations can be effectively integrated into existing DRM systems [1, 13]. The interdisciplinary nature of capacity building in DRM is reflected in the collaborative efforts among various sectors, including government agencies, non-governmental organizations (NGOs), and the private sector. Theories on inter-organizational collaboration and public-private partnerships highlight the benefits of pooling resources and expertise to develop comprehensive DRM strategies. These collaborative frameworks are essential for addressing the complex and multifaceted challenges posed by disasters [2, 10].

Furthermore, the reviewed articles highlight several gaps in the existing theoretical frameworks. While traditional approaches to capacity building have been extensively studied, there is a lack of comprehensive frameworks that integrate innovative strategies into practical, scalable models. This gap underscores the need for further research to explore the potential of emerging technologies and novel methodologies in enhancing DRM efforts. The development of such integrated frameworks can provide practical guidelines for implementing effective capacity-building strategies across different contexts and scales [17, 18]. In summary, the theoretical perspectives identified in the reviewed articles emphasize the importance of resilience, community engagement, technological innovation, and interdisciplinary collaboration in capacity building for DRM. These theories provide a robust foundation for developing and implementing effective capacity-building strategies. However, there is a need for further research to integrate innovative approaches into existing frameworks, thereby enhancing the overall effectiveness of DRM efforts.

### 3.4 Framework for drivers and outcomes of increasing innovative capacity building strategies for sustainable disaster risk management

The integration of innovative capacity-building strategies into disaster risk management (DRM) is crucial for enhancing resilience and response capabilities. The reviewed literature provides a comprehensive understanding of the drivers that promote the adoption of these strategies and the outcomes they generate. This framework aims to identify and synthesize these drivers and outcomes, offering a cohesive model for implementing effective capacity-building initiatives.

Several key drivers promote the adoption of innovative capacity-building strategies in DRM. Technological advancements, policy and institutional support, community engagement, and interdisciplinary collaboration are the primary drivers identified. The rapid development of advanced technologies such as machine learning, Geographic Information Systems (GIS), and Information and Communication Technologies (ICT) plays a significant role in enhancing DRM capabilities by facilitating better prediction, preparedness, and response to disasters [1, 13]. Strong policy frameworks and institutional support are essential for promoting capacity-building initiatives, as government policies that prioritize DRM and allocate resources for capacity-building efforts create an enabling environment for the implementation of innovative strategies [15, 18]. Engaging local communities in DRM efforts is critical for ensuring the sustainability and effectiveness of capacity-building initiatives, as community-based approaches leverage local knowledge and foster a sense of ownership and commitment among community members [14, 17]. Additionally, collaboration among different sectors, including government agencies, non-governmental organizations (NGOs), and the private sector, enhances the effectiveness of capacity-building strategies by bringing together diverse expertise and resources to facilitate comprehensive DRM solutions [23, 24].

The implementation of innovative capacity-building strategies in DRM leads to several positive outcomes, including improved disaster preparedness and response, enhanced resilience, and sustainable development. The integration of advanced technologies and community-based initiatives significantly enhances disaster preparedness and response capabilities by improving early warning systems, risk assessment, and coordination during disaster events [1, 13]. Capacity-building initiatives strengthen the resilience of communities and institutions by improving their ability to anticipate, withstand, and recover from disasters, thus reducing long-term vulnerabilities and ensuring quicker recovery post-disaster [25, 26]. Effective capacity-building strategies also contribute to sustainable development by promoting environmental stewardship, economic stability, and social well-being, ensuring that development efforts are resilient to disasters and climate change impacts [27, 28].

In conclusion, the framework for drivers and outcomes of increasing innovative capacity-building strategies for sustainable disaster risk management highlights the importance of technological advancements, policy support, community engagement, and interdisciplinary collaboration in promoting effective capacity-building initiatives. The positive outcomes of these strategies, including improved preparedness and response, enhanced resilience, and sustainable development, underscore the critical role of capacity building in DRM. This framework provides a comprehensive model for practitioners and policymakers to implement and scale up innovative capacity-building efforts, ultimately enhancing disaster resilience globally as shown in Fig. 1.

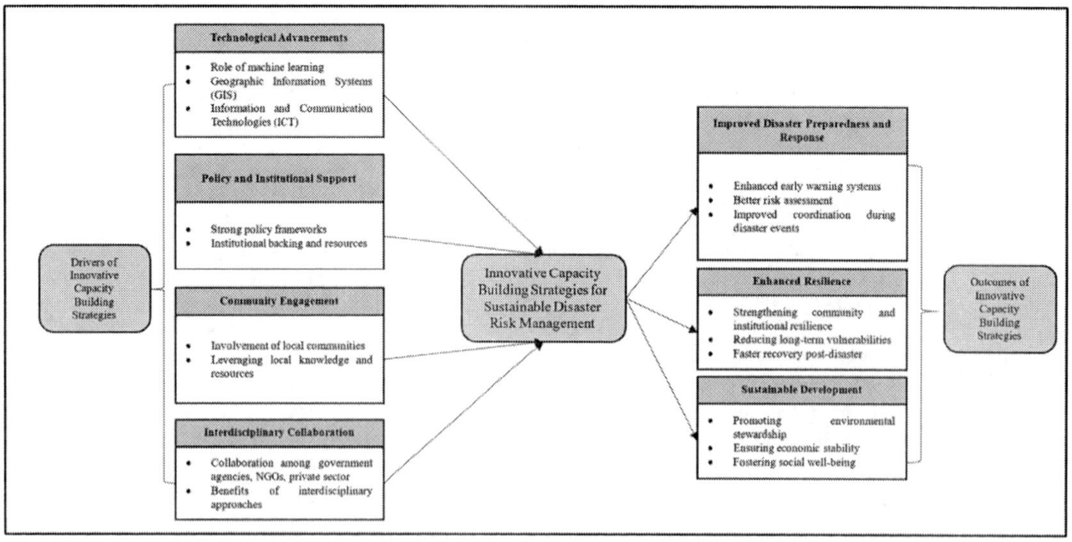


**Fig. 1.** Framework for drivers and outcomes of increasing organizational work capacity

## 4 Discussion

The integration of innovative capacity-building strategies into disaster risk management (DRM) is essential for enhancing resilience and response capabilities. The findings from this systematic review of 63 articles highlight several key themes and drivers that promote the adoption of these strategies, as well as the positive outcomes they generate. This discussion synthesizes these insights, emphasizing the importance of technological advancements, policy and institutional support, community engagement, and interdisciplinary collaboration in driving effective capacity-building initiatives.

Technological advancements emerged as a significant driver of innovative capacity-building strategies in DRM. The rapid development of advanced technologies such as machine learning, Geographic Information Systems (GIS), and Information and Communication Technologies (ICT) plays a critical role in enhancing DRM capabilities. These technologies facilitate better prediction, preparedness, and response to disasters, enabling more effective risk assessment and coordination during disaster events [1, 13]. The integration of these technologies into DRM practices not only improves efficiency but also expands the scope of what can be achieved in terms of disaster resilience and response.

Policy and institutional support are also crucial in promoting capacity-building initiatives. Strong policy frameworks and institutional backing provide the necessary resources and regulatory environment for implementing innovative strategies. Government policies that prioritize DRM and allocate resources for capacity-building efforts create an enabling environment for these initiatives to thrive [15, 18]. Effective policy support ensures that capacity-building programs are sustainable and can be scaled up to address the complex challenges posed by disasters.

Community engagement is another vital driver of effective capacity-building strategies. Engaging local communities in DRM efforts ensures the sustainability and effectiveness of these initiatives. Community-based approaches leverage local knowledge and foster a sense of ownership and commitment among community members, making DRM strategies more resilient and adaptive to local conditions [14, 17]. The involvement of communities in disaster preparedness and response not only enhances resilience but also builds social cohesion and collective action.

Interdisciplinary collaboration among different sectors, including government agencies, non-governmental organizations (NGOs), and the private sector, enhances the effectiveness of capacity-building strategies. Such collaboration brings together diverse expertise and resources, facilitating comprehensive DRM solutions. Interdisciplinary approaches are essential for addressing the multifaceted nature of disaster risks and for developing integrated strategies that encompass various aspects of DRM [2, 10]. Collaborative efforts also foster innovation by combining different perspectives and skills, leading to more effective and sustainable capacity-building initiatives.

The positive outcomes of these innovative capacity-building strategies are evident in the improvements in disaster preparedness and response, enhanced resilience, and contributions to sustainable development. The integration of advanced technologies and community-based initiatives significantly enhances early warning systems, risk assessment, and coordination during disaster events, leading to better preparedness and response capabilities [1, 13]. Capacity-building initiatives also strengthen the resilience of communities and institutions, reducing long-term vulnerabilities and ensuring quicker recovery post-disaster [12, 14]. Furthermore, effective capacity-building strategies promote sustainable development by fostering environmental stewardship, economic stability, and social well-being, ensuring that development efforts are resilient to disasters and climate change impacts [2, 10, 29].

In conclusion, the integration of innovative capacity-building strategies into DRM is essential for enhancing resilience and response capabilities. The findings from this systematic review underscore the importance of technological advancements, policy support, community engagement, and interdisciplinary collaboration in driving effective capacity-building initiatives. These strategies lead to significant improvements in disaster preparedness and response, enhanced resilience, and contributions to sustainable development. This comprehensive framework provides valuable insights for practitioners and policymakers to implement and scale up innovative capacity-building efforts, ultimately enhancing disaster resilience globally.

## 5 Limitations and research agenda

While this systematic review provides valuable insights into the drivers and outcomes of innovative capacity-building strategies for sustainable disaster risk management (DRM), several limitations must be acknowledged. First, the scope of the review was confined to articles published between 2010 and 2024, which may have excluded earlier foundational studies that could offer relevant insights. Additionally, the review was limited to peer-reviewed articles in the English language, potentially overlooking significant contributions from non-English publications and grey literature. The reliance on Scopus as the primary database might have also introduced a selection bias, as other databases like Web of Science or Google Scholar were not included.

The heterogeneity of the studies reviewed posed another limitation, as variations in methodologies, contexts, and definitions of capacity building made direct comparisons challenging. While thematic synthesis and qualitative analysis provided a structured approach to identifying common themes, the diversity of case studies and regional focuses limited the generalizability of the findings. Moreover, the use of Vos Viewer for bibliometric analysis, while effective for visualizing collaboration networks and publication trends, might not fully capture the depth and nuance of the individual studies' contributions.

Given these limitations, several directions for future research emerge. There is a need for more comprehensive frameworks that integrate innovative strategies into practical, scalable models for capacity building in DRM. Future studies should explore the potential of emerging technologies such as artificial intelligence, blockchain, and the Internet of Things (IoT) in enhancing DRM efforts. Research should also focus on developing standardized metrics and evaluation methods to assess the effectiveness of capacity-building initiatives across different contexts and scales.

Interdisciplinary and cross-sectoral collaborations should be further investigated to understand better how different stakeholders can work together to enhance disaster resilience. Studies that examine the role of local knowledge and community participation in shaping effective DRM strategies are particularly needed, as they can offer insights into culturally and contextually appropriate interventions. Additionally, longitudinal studies that track the long-term impacts of capacity-building initiatives can provide valuable data on their sustainability and effectiveness over time.

Lastly, future research should prioritize the inclusion of diverse voices and perspectives, particularly from underrepresented regions and vulnerable communities. This inclusivity will ensure that capacity-building strategies are equitable and address the needs of those most affected by disasters. By addressing these research gaps, scholars and practitioners can develop more robust and effective approaches to capacity building in DRM, ultimately enhancing global disaster resilience.

## 6 Conclusion

The integration of innovative capacity-building strategies into disaster risk management (DRM) is essential for enhancing resilience and response capabilities. This systematic review highlights the importance of key drivers such as technological advancements, policy and institutional support, community engagement, and interdisciplinary collaboration. Advanced technologies like machine learning, GIS, and ICT significantly enhance predictive and operational capabilities, improving risk assessment, early warning systems, and disaster response efficiency. Policy frameworks and institutional support provide the necessary resources and regulatory backing for sustainable and scalable capacity-building programs. Engaging local communities ensures that initiatives are contextually relevant and sustainable, leveraging local knowledge and fostering ownership and commitment. Interdisciplinary collaboration brings together diverse expertise and resources, facilitating comprehensive and innovative DRM solutions. The outcomes of these strategies include improved disaster preparedness and response, enhanced resilience, and contributions to sustainable development. These strategies promote environmental stewardship, economic stability, and social well-being, ensuring resilience to disasters and climate change impacts. In summary, innovative capacity-building strategies are crucial for effective DRM. The findings provide a comprehensive framework for practitioners and policymakers to implement and scale up these efforts, ultimately enhancing global disaster resilience.

## References

1. He H., Wang W., Wang Z., Li S., and Chen J., Enhancing seismic landslide susceptibility analysis for sustainable disaster risk management through machine learning. Sustainability, **16**, 3828 (2024)
2. Schlumberger J., Haasnoot M., Aerts JCJH., Bril V., van der Weide L., and de Ruiter M., Evaluating adaptation pathways in a complex multi-risk system. Earth's Future, **12**, 2023EF004288 (2024)
3. Wan Y., Feng C., Wu K., Wang J. RZDD., Risk zone-diversified network design for disaster resilience. IEEE Trans Dependable Secur. Comput., (2024)
4. Holtermans R., Kahn ME., Kok N., Climate risk and commercial mortgage delinquency. J. Reg Sci., (2023).
5. D'Ambrosio V., Di Martino F., Tersigni E., Towards climate resilience of the built environment: A GIS based framework for the assessment of climate proof design solutions for buildings. Buildings, **13**, 1658 (2023)
6. Zuluaga LS., Gomez-Quenguan N., Estrella-Insuasty J., Bautista-Gomez MM., Learning process of implementing a cutaneous leishmaniasis capacity-building program, using an innovative pedagogy for rural populations in Colombia. Rural Remote Health, **24**, 1–8 (2024)
7. Ryberg R., Her S., Temkin D., Rodriguez Y., Kelley C., Associations between a technical assistance model to build school organizational capacity and improvements in student perceptions of school climate. Am. J. Community Psychol. (2024)
8. Fathani AT., Putera RE., Asrinaldi A., Valentina TR., Hariyati D., Holidin D., et al., A systematical review of tourism disaster management: insights from Indonesian's journey. E3S Web Conf., **464**, EDP Sciences (2023)
9. Deprez D., Busch AJ., Ramirez PA., Pedrozo Araque E., Bidonde J., Capacity-building and continuing professional development in healthcare and rehabilitation in low-and middle-income countries a scoping review protocol. Syst Rev, **12**, 22 (2023)
10. Mohan D., Enhancing capacity building initiatives at sub-national level for supporting climate change adaptation. Clim Dev, **15**, (2023)
11. Audia N., Putera RE., Aromatica D., Bisri MBF., Capabilities of the city government in combating COVID-19 in the city of Padang. E3S Web Conf., **331** (2021) https://doi.org/10.1051/e3sconf/202133101009
12. Kako M., Hutton A., Disaster preparedness of Hiroshima community health nurses: A mixed

method study. Prog. Disaster Sci., **20,** 100295 (2023)

13. Uchida O., Tajima S., Kajita Y., Utsu K., Murakami Y., Yamada S., Development and implementation of an ICT-based disaster prevention and mitigation education program for the young generation. Inf. Syst. Front, **23**, 1115–25 (2021)
14. Le Dé L., Ronoh S., Kyu EMT., Rive B., How can practitioners support citizen volunteers in disaster risk reduction? Insight from "Good and Ready" in Aotearoa New Zealand. Int. J. Disaster Risk Sci., 1–14 (2024)
15. Hagelsteen M., Gutheil J., Burkle MDMM., Becker P., Caught between principles and politics: Challenges and opportunities for capacity development from governmental donors' perspectives. Int. J. Disaster Risk Reduct.,**70**, 102785 (2022)
16. Alexander JA., Christianson JB., Hearld LR., Hurley R., Scanlon DP., Challenges of capacity building in multisector community health alliances. Heal. Educ. Behav. **37**, 645–64 (2010)
17. Tongsiri S., Scaling-up community-based rehabilitation programs in rural Thailand: the development of a capacity building program. BMC Health Serv. Res., **22,** 1070 (2022)
18. Harden-Davies H., Amon DJ., Chung T., Gobin J., Hanich Q., Hassanali K., et al., How can a new UN ocean treaty change the course of capacity building? Aquat. Conserv. Mar. Freshw. Ecosyst., **32**, 907–12 (2022)
19. Wialdi PF., Putera RE., Kusdarini K., Ariany R., Valentina TR., Dermawan R., Local wisdom of coastal communities in mitigating earthquake disaster risk reduction. E3S Web Conf., **331** (2021) https://doi.org/10.1051/e3sconf/202133104001
20. Putera RE., Nurasa H., Sugandi YS., The implementation of earthquake management policies. AIP Conf. Proc., **2081** (2019). https://doi.org/10.1063/1.5094030.
21. Erbara A., Takdir M., Systematic literature review: Implementasi prinsip-prinsip administrasi keuangan di Indonesia. J. Ilm. Ekotrans. Erud., **2**, 1–10 (2022). https://doi.org/10.69989/gajx9g32.
22. Mulia RA., Systematic literature review: Analisis tren dan tantangan dalam perkembangan desentralisasi di Indonesia. J. Ilm. Ekotrans. Erud., **2**, 49–61 (2022). https://doi.org/10.69989/528eg697.
23. Han S., Park J., Lee H., Lee W., Son J., Identifying and prioritizing barriers to climate technology international cooperation from the perspective of Korea. Sustain, **15** (2023). https://doi.org/10.3390/su151410831.
24. Tozier de la Poterie A., Castro E., Rahaman H., Heinrich D., Clatworthy Y., Mundorega L., Anticipatory action to manage climate risks: Lessons from the Red Cross Red Crescent in Southern Africa, Bangladesh, and beyond. Clim Risk Manage, **39** (2023). https://doi.org/10.1016/j.crm.2023.100476.
25. Richardson LC., Lewis SM., Burnette RN., Building capacity for cyberbiosecurity training. Front Bioeng Biotechnol, **7** (2019). https://doi.org/10.3389/fbioe.2019.00112.
26. Ruszczyk HA., Upadhyay BK., Kwong YMC., Khanal O., Bracken LJ., Pandit S., et al., Empowering women through participatory action research in community-based disaster risk reduction efforts. Int. J. Disaster Risk Reduct., **51** (2020). https://doi.org/10.1016/j.ijdrr.2020.101763.
27. Gebreyes WA., Dupouy-Camet J., Newport MJ., Oliveira CJB., Schlesinger LS., Saif YM., et al., The global one health paradigm: Challenges and opportunities for tackling infectious diseases at the human, animal, and environment interface in low resource settings. PLoS Negl. Trop. Dis., **8** (2014). https://doi.org/10.1371/journal.pntd.0003257.
28. Wardekker A., Nath S., Handayaningsih TU., The interaction between cultural heritage and community resilience in disaster-affected volcanic regions. Environ. Sci. Policy, **145,** 116–28 (2023). https://doi.org/10.1016/j.envsci.2023.04.008.
29. Fajria R., Putera RE., Ariany R., An evaluation of the implementation of the disaster resilient village program in Padang Pariaman regency. E3S Web Conf., **464,** 10001 (2023). https://doi.org/10.1051/e3sconf/202346410001.

# Optimization of Penta helix in disaster risk reduction at disaster risk reduction forum F-PRB West Sumatra Province

*Malse* Yulivestra[1*] and *Novalinda*[2]

[1]Department of Public Administration, Faculty of Political and Social Sciences, Universitas Andalas, Padang, Indonesia
[2]English Department, Faculty of Humanities, Universitas Andalas, Padang, Indonesia

**Abstract.** This paper explains the efforts of F-PRB of West Sumatra Province in optimizing the role of actors using the Penta Helix concept. This study applied descriptive qualitative methods and data collection methods are interviews, and documentation. The conclusions of this study showed that The Penta Helix concept is the right concept to be applied in the context of disaster mitigation by F-PRB. Then the second conclusion is that implementing the Penta helix concept in F-PRB is still not optimal. It can be seen from one of the involvement of stakeholders that it is still not balanced and also not representative in terms of the number of each element. The third is collaboration process with these five stakeholders is quite optimal. There are Several recommendations need to be considered form this research. The first is that F-PRB and stakeholders in each element must increase the number of members involved and join the forum to be representative. F-PRB also needs to embrace and involve elements of society at the village level to be involved and join the membership of this forum. Then, there needs to be more intense collaboration activities related to disaster mitigation creating a resilience society.

## 1 Introduction

West Sumatra Province is one of the provinces in Indonesia that has a large potential for natural disasters. West Sumatra is located between the meeting of two active plates, namely the Indo-Australian Plate and the Eurasian Plate. West Sumatra is also traversed by the Mediterranean Mountains, a path in which there are many active volcanoes. In addition, there are also the Mentawai fault and the Sumatra fault. With the three earthquake sources, the tectonic complexity of the Sumatra region is increasing and causes the Sumatra region to be prone to various disasters.

Recently in 2024, precisely on May 11, the West Sumatra region was hit by a natural disaster in the form of cold lava floods and landslides. This incident occurred in the Batusangkar, Bukik Batabuah and Silaiang. This disaster was caused by the eruption of Mount Merapi, which occurred at the end of 2023 and resulted in many fatalities. This situation was exacerbated by the very high intensity of rain in West Sumatra. This condition has claimed many people died and losses of up to hundreds of millions of rupiah.

If we look at the data released by the West Sumatra Provincial Disaster Management Agency in 2022, both those designated as disasters through government decisions and those considered normal events, there were 1,021 recorded with an estimated total loss of IDR 1,136,849,587,336, as shown in Table 1:

**Table 1.** The number of disasters in West Sumatra (Source: data & information of disaster in West Sumatra province 2022)

| No | City/Regency | Numbers of Events | Estimated Loss | |
|---|---|---|---|---|
| 1. | Padang | 143 | Rp. | 346,500,000 |
| 2. | Bukittinggi | 98 | Rp. | - |
| 3. | Agam | 95 | Rp. | 39,000,000 |
| 4. | Padang Pariaman | 90 | Rp. | 880,500,000 |
| 5. | Pesisir Selatan | 75 | Rp. | 35,000,000 |
| 6. | Sijunjung | 69 | Rp. | - |
| 7. | Pasaman | 66 | | 574,411,801,603 |
| 8. | Sawahlunto | 59 | Rp. | 2,000,000 |
| 9. | Lima Puluh Kota | 51 | Rp. | 5,055,500,000 |
| 10. | Tanah Datar | 43 | Rp. | 3,457,000,000 |
| 11. | Padang Panjang | 35 | Rp. | - |
| 12. | Pasaman Barat | 33 | Rp. | 551,638,885,733 |
| 13. | Kepulauan Mentawai | 28 | Rp. | 180,000,000 |
| 14. | Dharmasraya | 28 | Rp. | - |
| 15. | Solok City | 28 | Rp. | - |
| 16. | Solok Regency | 24 | Rp. | 450,000,000 |
| 17. | Payakumbuh | 20 | Rp. | - |
| 18. | Pariaman | 19 | Rp. | 16,500,000 |
| 19. | Solok Selatan | 17 | Rp. | 338,900,000 |
| Grand Total | | | Rp. | 1,136,851,587,336 |

From the data above, it can be seen that all areas of 19 districts and cities in West Sumatra experienced disasters of various types. In 2022, around 1.021 natural disasters occurred in West Sumatra. Then, if we look more specifically related to this disaster based on data where,

* Corresponding author: malse@soc.unand.ac.id

strong winds are a disaster that often occurs, with a total of 674 incidents throughout 2022. Meanwhile, other is data on the number of disasters throughout 2022 in the West Sumatra region, (Fig. 1).

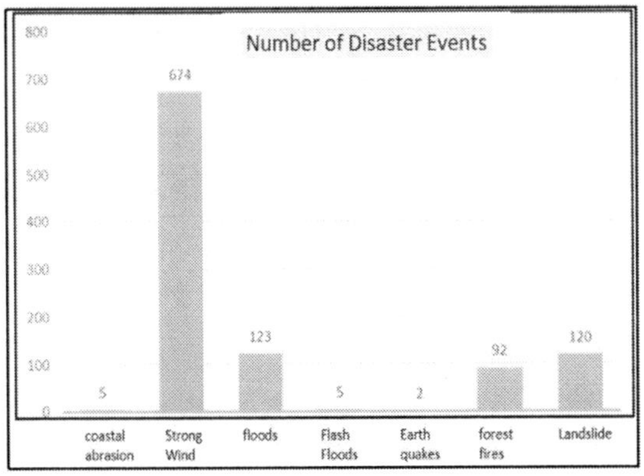


**Fig. 1.** Number of Disaster districts/municipalities in West Sumatra Province 2022 (source: data & information of disaster in West Sumatra province 2022)

This data shows that the West Sumatra region is a disaster-prone area, so serious and comprehensive disaster mitigation steps need to be taken. This condition requires structural mitigation efforts involving many parties, a large budget, and continuous mitigation implementation time. In general, disaster mitigation is an effort to reduce the risk and impact caused by disasters on communities in disaster-prone areas, be it natural disasters, man-made disasters or a combination of both in a country or community. Meanwhile, according to Law of the Republic of Indonesia No. 24 of 2007 concerning Disaster Management, mitigation is a series of efforts to reduce disaster risks, both through physical development and awareness and increasing the ability to face disaster threats.

Meanwhile, the purpose of mitigation itself is to reduce losses when future hazards occur and reduce the risk of death and injury to the population, including reducing damage and economic losses caused to public sector infrastructure. Given the importance of disaster mitigation, which can provide benefits and advantages to the community and its environment, several important things need to be considered in disaster mitigation, including the availability of information and maps of disaster-prone areas for each disaster category. The second is socialization, which increases public understanding and awareness of disasters. The third is knowing what needs to be done and avoided and how to save yourself if a disaster occurs at any time. The fourth is the regulation and arrangement of disaster-prone areas to reduce the threat of disasters that occur

On this basis, there needs to be an optimally integrated interconnectedness of systems, subsystems, sectors, and stakeholders. The concept of interconnection between systems and stakeholders is the collaborative Penta helix concept [1]. This collaborative concept was born from the governance paradigm in Public Administration Science, especially the concept of collaborative governance [2]. The public administration perspective describes disasters such as floods occurred 123 times, landslides 120 incidents, and forest fires 92 incidents. The following collaborative governance as a pattern of working together with several parties involved in a policy/activity to achieve goals [3]. This form of governance with public agencies is directly related to non-state stakeholders who are oriented towards consensus, deliberative and public policy implementation. More broadly, collaborative governance is seen as the processes and structures of decision-making and governance that involve the public in a constructive between public agencies, various levels of government, private and civil society to achieve common goals [4].

Meanwhile, the Penta helix concept is a development of the initial concept, namely Triple helix, where this concept only considers three stakeholders: government, business, and academics [3]. However, this concept later developed into Quadra helix. In this Quadra helix concept, there is an additional stakeholder, called the community or society; only in the next development was the Penta helix concept born, which added the involvement of mass media stakeholders as the fifth stakeholder [5].

From the description of the concept of Penta helix collaboration, there are many potentials uses to solve multi-party problems where stakeholders represent various interests in one location [6]. In this Penta helix concept, elements of government, academics, business entities or actors, society or communities, and the media come together to coordinate and commit to taking steps in disaster mitigation in an effort to increase the awareness of all parties.

To make a society aware of this disaster, of course, it cannot be done alone. That is why collaboration between parties is important in achieving common goals by maximizing the Penta helix concept, which involves many roles. Penta helix or multi-party, is an element of collaboration that combines various parties, including Academy, Business, Community, Government, and Media or better known as (ABCGM). In essence, the Penta helix concept is one way to overcome problems and develop disaster mitigation programs by involving cross-sectors to share roles. The focus of this Penta helix is collaboration between the government and stakeholders. Based on the explanation above, this paper aims to examine the application of the Penta helix concept in reducing disaster risk by the F-PRB Disaster Risk Reduction Forum of West Sumatra Province.

## 2 Method

This study uses qualitative research with a descriptive approach. According to Creswell 2014 [7], The research method is a process of activities in the form of data collection, analysis, and providing interpretations related to the study's objectives. Meanwhile, qualitative research, according to Steven Dukeshire and Jennifer Thurlow 2002 [8], qualitative research concerns data that is not numerical, collecting and analysing narrative data. This study also uses a descriptive approach, namely a research approach where researchers describe what is seen, heard, felt, and asked.

The data collection techniques used in this study are interview and documentation techniques in order to obtain data and information related to the theme of this study, both in the form of primary data and secondary data. Meanwhile, research informants are determined by the purposive sampling method [9], where researchers determine the criteria according to the theme of this study. Meanwhile, the data analysis used in this study is data analysis, according to Miles and Huberman, which has stages including Data collection, data reduction, data presentation, and drawing conclusions.

## 3 Results and discussion

**Table 2.** Membership lists for F-PRB West Sumatra Province for the 2018-2021 period (source: governor decree no. 360.811.1-2018 management structure of FRB West Sumatra 2018/2021)

| No | Sectors | Members |
|---|---|---|
| 1 | Government Body | 1. Gubernur Provinsi Sumatera Barat<br>2. DPRD Provinsi Sumbar<br>3. BPBD Provinsi Sumatera Barat<br>4. BAPPEDA Provinsi Sumbar<br>5. POLDA SUMBAR<br>6. Dinas Kesehatan Provinsi Sumbar<br>7. KOREM 032 Wirabraja<br>8. Damkar Kota Padang<br>9. UPT BNPB |
| 2 | Private Sector | 1. PT. Semen Padang<br>2. INKINDO |
| 3 | University | 1. Universitas Bung Hatta<br>2. Universitas Andalas<br>3. Pakar kebencanaan |
| 4 | Civil Society | 1. Ulama<br>2. Ormas NU<br>3. Ormas Muhammadiyah<br>4. Jemari Sakato<br>5. PKBI Provinsi Sumatera Barat<br>6. Mercy Corps<br>7. PKPU<br>8. KOGAMI<br>9. Gerkakin<br>10. Bumi Ceria<br>11. ASB<br>12. PMI Sumbar<br>13. PPDI<br>14. IOF<br>15. HGWI<br>16. HWDI<br>17. Pegiat Kebencanaan<br>18. CARITAS<br>19. WALHI Sumbar |
| 5 | Mass Media | 1. Classy FM |

In West Sumatra itself, there is an institution that serves as a forum for the five elements to join. The institution is the Disaster Risk Reduction Forum or abbreviated as F-PRB of West Sumatra Province. This forum consists of many volunteer members. This forum was established based on Law Number 24 of 2007 concerning Disaster Management and Government Regulation Number 21 of 2008 concerning the Implementation of Disaster Management, which mandates the existence of a Disaster Risk Reduction Forum abbreviated as the PRB Forum as a non-formal institution that is not separate from the implementation of disaster management, as a forum for stakeholders related to disasters [10].

This local PRB forum is also mandated through Government Regulation Number 21 of 2008 concerning the Implementation of Disaster Management in Article 8, which explains the function of the multi-stakeholder forum in preparing the Regional Action Plan for Disaster Risk Reduction (RAD-PRB). It is also regulated that members of this forum include elements from the government, academics, the community, business institutions, and the mass media. In West Sumatra itself, this F-PRB, until 2021, had a membership of 36 members consisting of various elements. If grouped based on the elements in the Penta helix concept, it can be described in Table 2.

Based on the membership table above, it can be seen that the F-PRB of West Sumatra Province, in terms of membership, has actually fulfilled the five elements in the Penta helix concept where there are elements of Government, Private Sector, University/Academics, Society/Community, and Mass Media. However, if we analyse the composition of the membership of the F-PRB, it is still not balanced. The number of members from the community elements has the largest number of members, around 19 members consisting of various elements such as NGOs/NGOs, and Community Organizations. Meanwhile, from the government element, 9 organizations consisting of the Provincial Government, Regional Apparatus Organizations, DPRD, as well as the police and TNI. Furthermore, from the academic and university elements, there are only 2 members, namely Bung Hatta University and Universitas Andalas. Then the number of members from the private sector only amounts to 2 members, PT. Semen Padang and INKINDO, and the mass media element only have 1 membership, namely from Radio Classy FM.

Judging from the composition of the F-PRB membership, it is clear that it is still not balanced and does not cover all the elements in each element. This condition certainly affects the role that each element must carry out. For example, in the context of disaster risk reduction, the government has three roles at once in the Penta helix concept. The first is that the government acts as a regulator and controller with a regulatory function. In this case, the government must always be involved in all types of activities, such as planning, implementation, monitoring, control, promotion, financial allocation, licensing, and others. In addition, the government also acts as a coordinator for stakeholders who contribute to the disaster mitigation program.

In the context of the Penta helix, the government plays a central role in making policies, the government is also responsible for formulating regulations that encourage mitigation, preparedness, response and post-disaster rehabilitation efforts. In this context, government organs such as the National Disaster Management Agency (BNPB) and the Regional Disaster Management Agency (BPBD) are the main actors and also the leading sector in disaster management. In addition, the government plays a role in building disaster-resistant infrastructure. Then design evacuation centres and early warning systems in the community and conduct simultaneous coordination with the central government, regions, and international institutions in handling disasters. With the composition of the F-PRB membership consisting of only a few government organizations, it will be difficult to realize this role optimally and comprehensively.

The second element is the private sector. Private or business actors, in the form of an agency or business actor, act as enablers and also play a role in presenting technological infrastructure and capital to optimize the resources owned to mitigate and overcome disasters. Then, this private element must also apply quality standards to prevent disasters in its business development efforts. In addition, the private sector also has a role in providing logistical support, funding, and resources for the emergency response process. In addition, through the Corporate Social Responsibility (CSR) program, businesses can provide support for activities related to disaster mitigation, such as the construction of disaster-resistant public facilities, the provision of early warning tools, or the provision of socialization and training for the community. Furthermore, the private sector can also collaborate with the government and other institutions to develop long-term strategies to strengthen disaster resilience. Looking at the composition of the number of members involved in this PRB Forum, where only 2 members are recorded, is certainly not sufficient. Meanwhile, in reality there are many large companies in the West Sumatra Province, both international, national, and local companies.

Next is the academic or university element. In the Penta helix concept, academics act as concepts that conduct research and assist in managing potential identification and development opportunities. Academics are also responsible for increasing the capacity of knowledge and skills because academics are a source of knowledge that includes a collection of concepts, theories, and the latest development models that are relevant to existing conditions through research and scientific studies conducted. Furthermore, academia has a role in providing science-based research and innovation that can be used to improve disaster mitigation and management systems. The results of research conducted by universities and research institutes can help provide a better understanding of disaster risks, patterns of occurrence, and more effective solutions to deal with them. Academia also contributes to the development of disaster mitigation technologies, such as more sophisticated early warning systems, accurate disaster risk mapping, and evaluation of policies that have been effectively implemented. In addition, through education, academics help prepare a younger generation that has an awareness of the importance of disaster mitigation. This can be realized optimally if academics and campus institutions are involved in the institution that focuses on disaster mitigation, namely F-PRB. However, when viewed from the existing membership composition data that comes from campus institutions, there are only 2 campuses, namely Bung Hatta University and Universitas Andalas. This condition is certainly far from optimal because in West Sumatra there are many universities, both private and government-owned.

The next element is the community in the society. In the context of disaster mitigation, the community in the community has a strategic and urgent role. Some of the roles that this community can carry out are conducting socialization and assistance in order to increase public awareness of the risks and dangers of disasters being directly involved in handling disasters when disasters occur and being directly involved in post-disaster recovery. Given the importance of the role that is the responsibility of this community element, it is necessary to involve the communities in this community in this F-PRB Institution so that they can provide maximum contribution to disaster mitigation.

On the one hand, the community is the party most affected by disasters, but on the other hand, in the Penta helix concept they are also important actors in disaster management efforts. The role of the community is vital, especially in terms of preparedness, mitigation and emergency response. This preparedness can be in the form of disaster response training, the formation of disaster resilient groups or communities, and participation in simulation activities organized by the government. Furthermore, when facing a disaster, the role of the local community is very important, especially people who understand the conditions of the surrounding environment. They control the area so that they can provide direct information related to emergency needs. Then the community can also volunteer to help the evacuation process and the distribution of logistical assistance to the community. Involving the community in every stage of the disaster will create a sense of belonging and a sense of shared responsibility that will strengthen community resilience. Judging from the membership data of this F-PRB, the number of members from this community element is quite large and varies greatly depending on its composition. There are environmental activists, religious, humanitarian, women's and children's protection, and disaster activists. This certainly greatly supports the process of creating a disaster-aware society, especially disaster mitigation.

The last element in the Penta helix concept is the media. In the context of disaster mitigation, the media has an important role in providing and disseminating information related to disasters to the public. The media also plays a role in always reminding and educating the public about preparedness to reduce disaster risk and minimize victims and losses if a disaster occurs. Furthermore, the important responsibility of the media is to provide accurate and correct information with valid sources to overcome and clarify information or Hoax news circulating in the community. In principle, the mass media has an important role in disseminating information related to disasters that occur, both before, during and after disasters occur. Mass media, both mainstream and non-mainstream, brick-and-mortar and online media, are vital information channels in raising public awareness about potential disaster risks, mitigation measures, and safe evacuation procedures. Then, in emergency situations the mass media has a function as a liaison between the government and the community, conveying information about disaster developments, current conditions, and instructions from the authorities. In addition, the mass media also plays a role in shaping public opinion, so as to encourage active participation of the community and other parties in disaster management efforts. Through precise and accurate reporting, the media can raise awareness and encourage solidarity to help disaster victims.

From explanation above it can make a simple Penta Helix model of F-PRB West Sumatra Province as shown in Fig. 2:

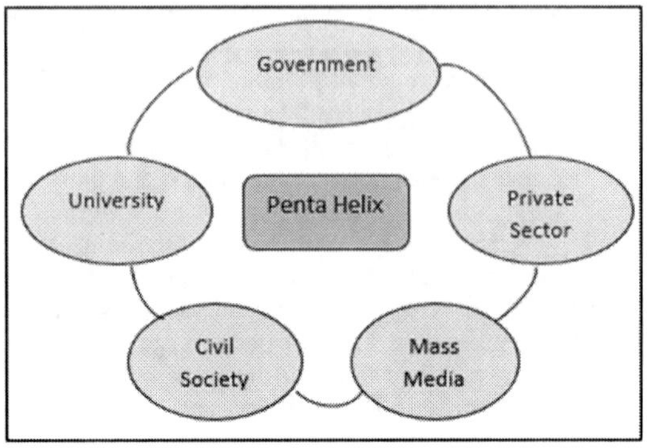


**Fig. 2.** Penta helix model of F-PRB West Sumatra Province

Based on the description of the data above, in the context of optimizing Penta helix in disaster mitigation in F-PRB West Sumatra, it can be said that it is not yet optimal. This can be seen from the composition of the membership that is included in this forum, which is still not balanced and representative of each element. In fact, the more organizations in each element are involved in disaster mitigation and are included in this forum, the more it will be able to optimize the role and function of this forum, especially in disaster mitigation. Next, besides the membership issue, in the Penta helix concept, the most important thing is the occurrence of a collaborative process that is synergistic and integrated with each other.

In the context of this collaboration, all elements or stakeholders must be involved in program activities and activities related to disasters, especially disaster mitigation. The collaboration is carried out not only at the level of discussions, seminars, or other scientific meetings but also in strategic matters such as government involvement in policy making. Disaster risk reduction requires the involvement of the entire community to prevent and reduce disaster risks through integrated implementation of inclusive efforts. F-PRB relies heavily on coordination mechanisms within and across sectors and with relevant stakeholders at all levels. PRB requires full involvement from all state institutions, executive and legislative, at the national and local levels. In addition, a clear description of the responsibilities of all stakeholders is needed to facilitate outreach, partnerships, and joint accountability. Of course, the synergy of these five elements in disaster management will have a very good impact on implementing disaster management. Because of course, in disaster management, the government, both central and local governments, cannot deal with it alone. There needs to be synergy with these elements in order to reduce the risk of disasters that may occur. Therefore, the government should also open up as much space as possible for related elements that want to synergize, especially in disaster management because disasters should be a shared concern, not just the government.

The occurrence of a Penta helix collaboration in disaster management will bring many benefits and positive impacts such as; with this Penta helix approach, synergy will be created between various sectors that have resources and expertise. Then, by collaborating, each sector can complement each other so that disaster management measures become more comprehensive and integrated. In addition, this Penta helix collaboration will accelerate the process of making tactical and strategic decisions in normal and emergency situations. Furthermore, with this collaborative mechanism which involves various parties, Penta helix is able to increase long-term resilience. Not only that, Penta helix does not only focus on handling disasters after they occur, but also encourages better prevention and mitigation through research, innovation, education, and resilient infrastructure development by involving various sectors. In general, it can be concluded that the application of the collaboration concept with the Penta helix approach in dealing with disasters in Indonesia is quite ideal even though in practice there are still many shortcomings. this approach allows each party to contribute according to their respective capacities in creating an effective, efficient and sustainable disaster management system. With solid cooperation between these five elements, it is expected that it will be able to increase resilience.

## 4 Conclusion

Based on the analysis presented above, it can be concluded that several things are related to the optimization of Penta helix in reducing disaster risk in the F-PRB of West Sumatra Province. The first is that the Penta helix concept is the right concept to be applied in the context of disaster mitigation by F-PRB. This is because F-PRB is a cross-stakeholder forum that cares about disaster issues. Thus, a multi-actor approach is needed that collaborates with each other to achieve common goals. Then the second conclusion is that implementing the Penta helix concept in F-PRB is still not optimal. This can be seen from one of the involvement of stakeholders, who are still not balanced and also not representative in terms of the number of each element. This non-representative composition certainly has an impact on the collaboration process carried out related to disaster mitigation. Meanwhile, the third is related to the collaboration process carried out by F-PRB with these five stakeholders, and it can be concluded that it is quite optimal. This is reinforced by the frequent meetings, both formal and informal, that discuss disaster issues. There is also a communication media platform in the form of a WhatsApp group that supports communication at all times between its members.

From the results of this study, there are several recommendations that need to be considered in optimizing the implementation of Penta helix in disaster risk reduction by F-PRB West Sumatra. The first is that F-PRB and stakeholders in each element need to increase the number of members involved and join the forum so that it is representative. Related to that, F-PRB also needs to embrace and involve elements of society at the village level to be involved and join the membership of this forum so that it will facilitate coordination and socialization to the community at the village level. Then, the next thing is that there needs to be more intense and planned collaboration in carrying out programs and activities related to disaster mitigation so that a society that is aware of and cares about disasters in West Sumatra can be created.

## References

1. C. Ansell, A. Gash, Collaborative governance in theory and practice. Journal of public Administration Research and Theory, **18**(4), 543-571, (2008)
2. B. C. Crosby, and Bryson, J. M, A leadership framework for cross-sector collaboration. Public Management Review, **7**(2), (2005)
3. R. S. Astuti, H. Warsono, and A. Rachim, Collaborative governance in perspective of public administration, (Public Administration Department Political and Social Sciences, Diponegoro University Press, 2020)
4. K. Emerson, T. Nabatchi, and S. Balogh, An integrative framework for collaborative governance. Journal of Public Administration Research and Theory, **22** (2012). https://doi.org/10.1093/jopart/mur011
5. W. Hardi, Collaborative governance public administration perspective. (Team DAP Press, Semarang, 2020)
6. L. O. S. Islamy, Collaborative governance concept and application (Deepublish, Yogyakarta. 2018)
7. E. Sturesson, A. Lindmark, and M. N. Roos, Collaboration for innovation - a study in the Öresund Region. (Lund University Libraries, Lund, 2009)
8. J. W. Creswell, Research design: The approach of qualitative method, quantitative method, and mix method. (Pustaka Pelajar, Yogyakarta, 2014)
9. B. Bungin, The analysis of qualitative data (Rajawali Press, Jakarta, 2003)
10. National Disaster Management Agency, Data and Information, Indonesia (2020)

# Prediction of flood peak-discharges in the Raman Watershed, North Raman, Central Lampung

*Farli* Rossi[1,2], *Sabariah* Musa[3], *Mas* Mera[4*], *Kastamto*[1], *Susarman*[1], and *Februarman*[4]

[1]Faculty of Engineering and Computer Sciences, Universitas Teknokrat Indonesia, Lampung, Indonesia
[2]Faculty of Engineering and Built Environment, Universiti Kebangsaan Malaysia, Bangi, Selangor, Malaysia
[3]Faculty of Civil Engineering and Built Environment, Universiti Tun Hussein Onn Malaysia, Batu Pahat, Johor, Malaysia
[4]Department of Civil Engineering, Faculty of Engineering, Universitas Andalas, Padang, Indonesia

**Abstract.** The Raman Watershed which is located in Pekalongan District covers an area of approximately 4,375 hectares of technically irrigated rice-field. The area was repeatedly hit by extreme flooding during the rainy season. The largest recorded flood occurred in December 2021, with an average daily maximum rainfall of 153 mm. This flood caused serious damage to the primary canal of the North-Raman Irrigation-Area. This research objective is to predict the flood peak-discharge in the Raman Watershed in the rainy season in December 2021. Flood discharge is calculated for various return periods based on daily rainfall-data from 2012 to 2021 obtained from BMKG Central Lampung. Hydrological analysis includes: 1) assessment of annual average rainfall-depth using the Thiessen method; 2) analysis of rainfall distribution and predictions of design rainfall using three common probability distributions; and 3) flood discharges are predicted using the Nakayasu method. Distribution analysis shows that LPT3 is the most suitable distribution. The average daily maximum rainfall recorded in December 2021 (153 mm), is in very good agreement with the design rainfall for the 25-year return period (152.58 mm). Thus, the flood peak-discharge in the Raman Watershed with average daily maximum rainfall of 153 mm is estimated at 23.27 $m^3/s$.

## 1 Introduction

In the last few decades, floods have been the most frequent natural disaster affecting Indonesia. This phenomenon is primarily triggered by extreme rainfall during the rainy season, which has intensified over time [1-6].

Previous studies have shown that while the heavy rain is a natural occurrence, the severity and frequency of flooding are not solely due to natural causes [7, 8]. Human activities, such as rapid urbanization and extensive development, have contributed significantly to the problem. These socio-economic factors have expanded flood-prone areas, making them more vulnerable and susceptible to flood damage [9-11]. As cities grow and development continues, natural drainage systems are often disrupted, leading to increased water accumulation and flooding events.

In the context of flood risk mitigation and prevention, it is essential to accurately estimate peak flood discharge for a given area [12-14]. Knowing the peak discharge level helps in understanding the potential severity of floods, thus enabling the design and construction of appropriate infrastructure to manage and reduce the impact of these events. Accurate estimates enable the development of effective flood control measures and structural designs that can better protect communities from flood damage [15, 16]. Implementing such measures is essential to reduce the impact of floods and increase resilience to future events.

In the Raman Watershed which is located in Pekalongan District, North Raman, Central Lampung, there are around 4,375 hectares of technically irrigated rice fields. A major flood event occurred in December 2021, with an average daily maximum rainfall-depth of 153 mm, causing severe damage to primary canals in the watershed. This incident has definitely had a negative impact on the economy of local residents and surrounding communities (Fig. 1).

**Fig. 1.** A flood event occurred in the Raman Watershed in december 2021

* Corresponding author: mas_mera@eng.unand.ac.id

Considering the importance of estimating flood peak-discharge in designing water infrastructure in the Raman Watershed, this research objective is to predict the potential flood peak-discharge in the Raman Watershed during the rainy season in December 2021.

In order to repair the primary-canal infrastructure in the North-Raman Irrigation-Area, as well as carry out flood mitigation and control measures in the Raman Watershed, Pekalongan Regency, the regional government plans to reconstruct the damaged irrigation canal and designs several other water infrastructures.

# 2 Methods

## 2.1 Research site

This research was conducted in the downstream part of the Raman Watershed, North Raman, Pekalongan Regency, Central Lampung with coordinates of 5°2'24.96" South Latitude – 105°21'41.4" East Longitude. Figure 2 shows the research site.

The total area of the Raman Watershed is approximately 86.3 km². This watershed is influenced by two seasons every year: the rainy season and the dry season. The watershed experiences a rainfall cycle that lasts from April to October.

## 2.2 Data collection

To predict flood peak-discharge, the data needed consist of rainfall depth data and land use data. There are two rainfall stations in the Raman Watershed, namely Bayah station and Cibeber station.

In this research, rainfall data and land use data were obtained from the River Basin Center (BBWS) - Mesuji Sekampung and PT Bumi Karya Consultant, Lampung Province. Rainfall data from both rainfall stations (Bayah and Cibeber stations) were taken for a ten-year period (2012 to 2021). Table 1 shows the maximum daily rainfall from year 2012 to 2021. This data was then analysed using the Thiessen polygon method.

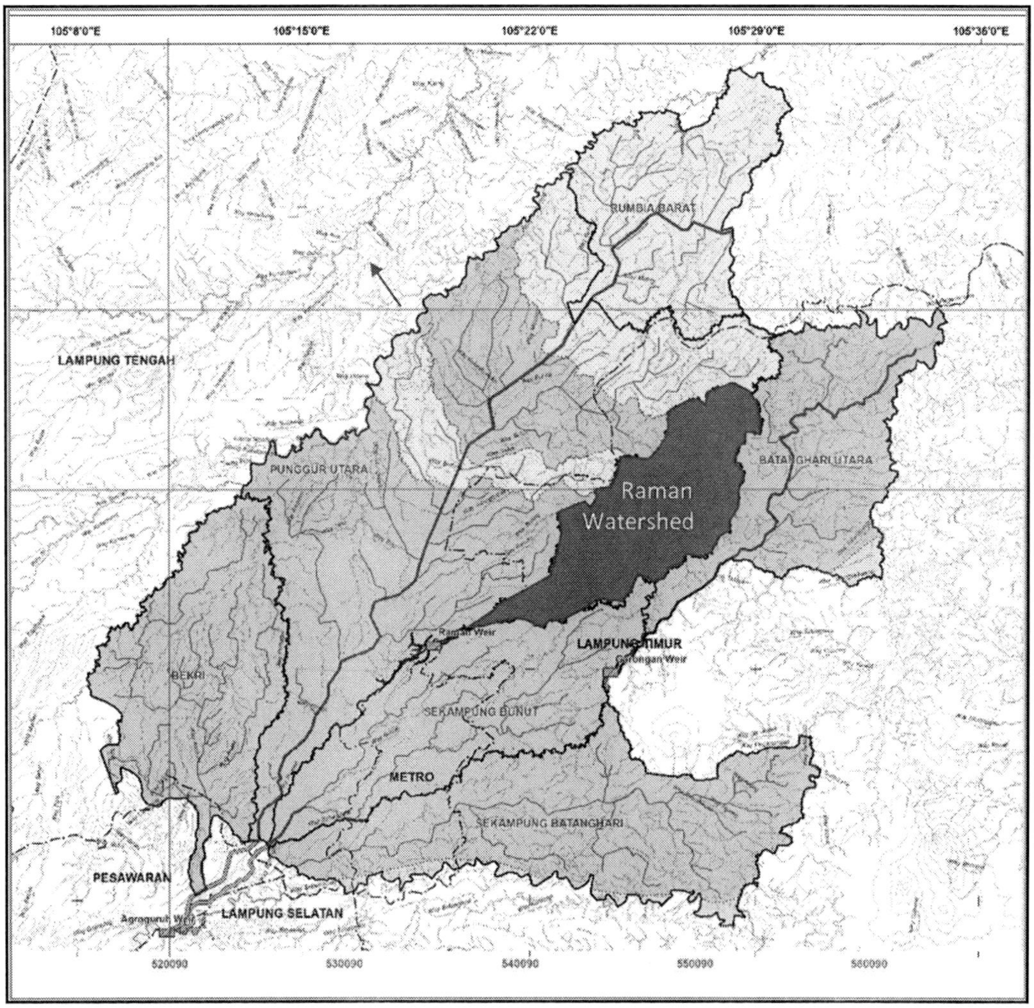


**Fig. 2.** Research site

Based on Table 1, there was an increasing trend in annual rainfall during the period 2012 to 2021. In particular, the highest level recorded was 180 mm at Cibeber Station in 2021, while Bayah Station experienced a peak of 167 mm in 2019. This increase in rainfall indicates a shift in local climate patterns over the years.

**Table 1**. Maximum daily rainfall/year from 2012 to 2021

| No | Year | Maximum daily rainfall | |
|---|---|---|---|
| | | Cibeber Station [mm] | Bayah Station [mm] |
| 1 | 2012 | 70 | 61 |
| 2 | 2013 | 141 | 64 |
| 3 | 2014 | 134 | 66 |
| 4 | 2015 | 134 | 76 |
| 5 | 2016 | 113 | 92 |
| 6 | 2017 | 141 | 125 |
| 7 | 2018 | 112 | 147 |
| 8 | 2019 | 92 | 167 |
| 9 | 2020 | 117 | 144 |
| 10 | 2021 | 180 | 126 |

This finding is consistent with broader regional trends observed across Indonesia. For example, a study by Siswanto [17] reported a significant increase in extreme rainfall events over the past few decades, particularly in southern region of Sumatra. This pattern suggests that climate changes are becoming more pronounced across the country.

Our observations in the Raman Watershed are consistent with this broader trend. The evidence indicates that this region is also experiencing similar climate change, likely influenced by the same factors that influence rainfall patterns across the region. This further supports the notion of climate change, emphasizing the need for ongoing monitoring and adaptation strategies.

## 2.3 Hydrological analyses

In this research, hydrological analysis was carried out to predict flood peak-discharge in the Raman Watershed for return periods of 5, 10, 25, 50 and 100 years. Hydrological analysis includes several stages, they are calculation of annual average rainfall, analysis of rainfall frequency distribution, prediction of design rainfall, and analysis of flood peak-discharge.

### *2.3.1 Average annual rainfall*

The Thiessen polygon method is used to calculate the average annual rainfall using the equation

$$P = \frac{A_1P_1+A_2P_2+A_3P_3+\cdots+A_nP_n}{A_1+A_2+A_3+\cdots+A_n} \quad (1)$$

where $P$ denotes the average annual rainfall, $A_1, A_2, A_3, \ldots, A_n$ represent the areas influenced by each rainfall station, and $P_1, P_2, P_3, \ldots, P_n$ represent the maximum daily rainfall per year at station 1, 2, 3, … $n$.

Using the Thiessen method requires at least 3 rainfall stations. Meanwhile, there are only two stations in the Raman Watershed. For this reason, the nearest station which is outside the Raman Watershed is taken.

### *2.3.2 Rainfall distribution probabilities*

In the analysis of rainfall distribution, rainfall data is tested using three probability distribution models, namely the Log-normal model, the Gumbel Type I extreme value model, and the Log Pearson Type III (LPT3) model.

The kurtosis coefficient *Ck* and skewness coefficient *Cs* are calculated to determine the most suitable models among the three. The best fit model will then be applied to predict the design rainfall for each return period.

The equation used to predict design rainfall using the Log-normal model is [16]

$$P_T = \bar{P} + kS_P \quad (2)$$

where $P_T$ is design rainfall for $T$, $T$ is return period in year, $\bar{P}$ is the average of $P$, $P$ is the logarithms of rainfall data, $k$ a coefficient factor for the normal distribution model, and $S_P$ is standard deviation of $P$.

Equation (2) can be used to predict design rainfall for $T$ using the Gumbel Type I extreme value model, but $k$ is a coefficient factor for the Gumbel model,

$$k = \frac{X_{Tr}-X_n}{S_n} \quad (3)$$

$$X_{Tr} = -ln\left\{-ln\left[\frac{T-1}{T}\right]\right\} \quad (4)$$

where $P$ is now the rainfall data (not logarithmic as used in Log-Normal model), $X_n$ is reduced mean of Gumbel, and $S_n$ is standard deviation of the Gumbel reduced mean.

Equation (2) can also be used to predict design rainfall for $T$ using the Log-Pearson Type III model, but $P$ is now the rainfall logarithmic data (as used in Log-Normal model), $k$ is a frequency coefficient based on skewness values *Cs* for $T$ (can be determined using a frequency factor table Haan [18].

The model that best fits rainfall distribution data to predict a design rainfall is determined based on the values of the skewness coefficient *Cs* and kurtosis coefficient *Ck* as presented in Table 2.

**Table 2.** *Cs* and *Ck* values for each distribution method

| No | Type of Distribution | Rule |
|---|---|---|
| 1 | Method of Log-Normal | $Cs = 0; Ck = 3$ |
| 2 | Method of Gumbel type 1 | $Cs = 1.14; Ck = 5.4$ |
| 3 | Method of Log-Pearson III | others |

The equations for determining *Cs* and *Ck* for $n$ (the number of data) are [16]:

$$Cs = \frac{n^2}{(n-1)(n-2)}\left[\frac{\sum_{i=1}^{i=n}(P_i-\bar{P})^3}{nS^3}\right] \quad (5)$$

$$Ck = \frac{n^3}{(n-1)(n-2)(n-3)}\left[\frac{\sum_{i=1}^{i=n}(P_i-\bar{P})^3}{nS^4}\right] \quad (6)$$

### *2.3.3 Flood peak-discharges*

Flood discharge for various return periods $Q_T$ (5 years, 10 years, 25 years, 50 years, and 100 years) is calculated using the Synthetic Unit Hydrograph (SUH) method. In this research, the Nakayasu Synthetic Hydrograph Unit (SHU) method is used, and is represented by the following equation

$$Q_T = car_o \left\{\frac{1}{3.6(0.3t_p + t_{0.3})}\right\} \quad (7)$$

where $c$ is the runoff coefficient, $a$ is the total area of the catchment or watershed (in km²), $r_o$ denotes the design rainfall (in mm), $t_p$ represents the time to peak (in hours), and $t_{0.3}$ is the deceleration time for discharges to reduce to 30%.

## 3 Results and discussion

### 3.1 Hydrological analyses

Data collected from three rainfall stations and analyzed using the Thiessen polygon method reveals that the area of influence within the Raman Watershed is divided evenly between two stations. Specifically, 50% of the area is influenced by Cibeber Station, while the remaining 50% falls under the influence of Bayah Station. This division allows for an accurate understanding of rainfall distribution across the watershed.

The hydrological analysis conducted for the Raman Watershed provides insights into both rainfall patterns and flood characteristics within the region. Table 3 presents the average annual maximum rainfall from 2012 to 2021 calculated using the Polygon Thiessen method. These averages show considerable variability in rainfall over the decade, with the lowest average annual rainfall being 65.5 mm in 2012 and the highest being 153 mm in 2021. The increasing trend in rainfall is very sharp, especially in 2021. This may indicate a change in rainfall patterns, and can worsen the risk of flooding in the watershed.

**Table 3.** Average annual rainfall from 2011 to 2020

| No | Year | Cibeber $P_1$ [mm] | Bayah $P_2$ [mm] | Average annual rainfall [mm] $(0.5P_1 + 0.5P_2)$ |
|---|---|---|---|---|
| 1 | 2012 | 70 | 61 | 65.5 |
| 2 | 2013 | 141 | 64 | 102.5 |
| 3 | 2014 | 134 | 66 | 100 |
| 4 | 2015 | 134 | 76 | 105 |
| 5 | 2016 | 113 | 92 | 102.5 |
| 6 | 2017 | 141 | 125 | 133 |
| 7 | 2018 | 112 | 147 | 129.5 |
| 8 | 2019 | 92 | 167 | 129.5 |
| 9 | 2020 | 117 | 144 | 130.5 |
| 10 | 2021 | 180 | 126 | 153 |

### 3.2 Design rainfall

Table 4 presents the values of *Cs* (skewness coefficient) and *Ck* (kurtosis coefficient) that have been calculated for each distribution method.

Table 4 also shows that the Log-Pearson Type III model is the most suitable model for the rainfall data used. The data set produces a skewness coefficient *Cs* of 0.840 and a kurtosis coefficient *Ck* of 2.863. This result does not comply with the requirements for either the Log-normal or Gumbel I model. This means that the rainfall distribution in the Raman Watershed is slightly distorted, so it is very important for accurate flood frequency analysis.

**Table 4.** The values of *Cs* and *Ck* for each distribution model

| No | Type of Distribution | Rule | Analysis results | Notes |
|---|---|---|---|---|
| 1 | Log-Normal | $Cs = 0$ | 0.840 | Not acceptable |
| | | $Ck = 3$ | 2.863 | |
| 2 | Gumbel I | $Cs = 1.14$ | 0.840 | Not acceptable |
| | | $Ck = 5.4$ | 2.86 | |
| 3 | Log Person III | Others | 0.840 | Acceptable |
| | | | 2.863 | |

In addition, Table 5 presents design-rainfall predictions for various return periods using the Log-Pearson Type III and Gumbel models.

**Table 5.** Design-rainfall summary

| No | Return Period [year] | Gumbel I [mm] | Log Person III [mm] |
|---|---|---|---|
| 1 | 2 | 111.75 | 117.77 |
| 2 | 5 | 141.29 | 137.48 |
| 3 | 10 | 160.86 | 145.63 |
| 4 | 25 | 185.57 | 152.58 |
| 5 | 50 | 203.91 | 156.14 |
| 6 | 100 | 222.11 | 158.75 |

Based on Table 5, the Log-Pearson Type III model consistently produces higher rainfall estimates for all return periods, thereby resulting in more conservative flood management strategies. Notably, the 25-year return period rainfall of 152.58 mm agrees well with the observed average rainfall of 153 mm during the December 2021 flood event, thus validating the reliability of our analysis.

### 3.3 Flood peak-discharges

Table 6 presents the flood peak-discharge in the Raman Watershed at return periods of 2, 5, 20, 25, 50 and 100 years using the Nakayasu method. There is a non-linear relationship between the return period and flood peak-discharge. The estimated flood peak-discharge with a return period of 25 years is 23.27 m³/s.

**Table 6.** Flood peak-discharges

| Return period [year] | Flood discharges [m³/s] |
|---|---|
| 2 | 17.96 |
| 5 | 20.96 |
| 10 | 22.21 |
| 25 | 23.27 |
| 50 | 23.81 |
| 100 | 24.21 |

Aligning the December 2021 flood (averaging 153 mm) with our calculated 25-year return period (152.58

mm) provides valuable context regarding the severity of this flood. Interestingly, these findings are different from several recent studies in other regions in Indonesia. For example, Ningrum [19], who studied the Ciliwung Watershed in Jakarta, found that flood events that historically occurred over 25 years were occurring more frequently, approaching 10-year intervals. These differences highlight the spatial variability of climate change impacts and the need for local studies.

This finding is very important because it shows that flooding in 2021 is indeed a relatively rare event, occurring on average once every 25 years. However, it is important to note that climate change may alter these probabilities, potentially making similar events more common in the future.

These results have important implications for flood management in the Raman Watershed. The close correspondence between the 25-year flood calculations and the events that occurred in December 2021 underscores the need for robust flood mitigation measures designed to handle at least this level of flooding. Additionally, increasing trends in annual rainfall suggest that flood risk may be increasing, necessitating adaptive management strategies.

## 4 Conclusion

This research provides important insights into flood characteristics in the Raman Watershed, with a particular focus on the major flood event in December 2021. Through comprehensive hydrological analysis, we have determined that:

1) The Log-Pearson Type III probability distribution model best describes rainfall depth patterns in Raman Watershed, thereby producing the most accurate design-rainfall estimates;
2) The flood event in December 2021, with an average daily maximum rainfall-depth of 153 mm, is closely related to the 25-year return period event based on our analysis; and
3) The estimated flood peak-discharge for the 25-year return period is 23.27 m3/s, and represents a significant flood-risk in the area.

These findings have important implications for flood management in the Raman Watershed. The close alignment between 25 years of flood calculations and recent events in 2021 validates our methodology while highlighting the severity of the flood risk faced by local communities. This underscores the urgent need for flood mitigation measures designed to handle at least flood levels of this magnitude.

However, it is important to note that our analysis is based on historical data and assumes stationarity of climate patterns. As climate change has the potential to alter rainfall patterns, the frequency and intensity of extreme events may increase in the future. These limitations highlight the need for ongoing monitoring and regular reassessment of flood risk.

Future research should focus on incorporating climate change projections into analyses of flood frequency in the region. In addition, hydrodynamic models with more detailed variables can provide insight into the spread of floods and inundation patterns, thereby providing further information regarding flood management strategies.

Acknowledgments are given to the Universitas Andalas for providing facilities and financial assistance for this research. The authors would also like to thank Universitas Teknokrat Indonesia for the supporting in completing this research.

## References

1. A. Fitri, M.S.N. Hadie, A. Agustina, D. Pratiwi, S. Susarman, G. Pramita, H.A. Salah, Analyses of flood peak discharge in Cimadur river basin, Banten Province, Indonesia. EDP Sciences, **331**, 08006 (2021)
2. K.N.A. Maulud, A. Fitri, W.H.M.W. Mohtar, W.M. Jaafar, N.Z. Zuhairi, M.K.A. Kamarudin, A study of spatial and water quality index during dry and rainy seasons at Kelantan River Basin, Peninsular Malaysia. Arabian Journal of Geosciences, **14**, (2021)
3. D. Pratiwi, A. Fitri, A. Phelia, N.A. Adma, K. Kastamto, Analysis of urban flood using synthetic unit hydrograph (SUH) and flood mitigation strategies along way Halim River: A case study on Seroja street, Tanjung Senang District. EDP Sciences, **331**, 07015 (2021)
4. S. Shi, X. Tao, X. Chen, H. Chen, A. Fitri, X. Yang, Evaluation of urban water security based on DPSIR model. IOP conference series: Earth and environmental science, **880**, 012023 (2021)
5. K. Kastamto, A. Fitri, M.S.N. Hadie, D. Safitri, S. Susarman, D. Pratiwi, Flood analyses at downstream of Cimadur River in Upper Cimadur Basin using HEC-RAS. EDP Sciences, **464**, 02009 (2023)
6. A. Fitri, R. Hashim, S. Abolfathi, K.N.A. Maulud, Dynamics of sediment transport and erosion-deposition patterns in the locality of a detached low-crested breakwater on a cohesive coast. Water (Switzerland), **11**, (2019)
7. S. Hutapea, Biophysical characteristics of Deli River Watershed to know potential flooding in Medan City, Indonesia. J Rangel Sci, **10**, (2020)
8. A. Fitri, Z.A Hasan, A. A. Ghani, Determining the Effectiveness of Harapan Lake as Flood Retention Pond in Flood Mitigation Effort. In Proceedings of 2011: 4th International Conference on Environmental and Computer Science (ICECS 2011), **34**, (2011)
9. A. Fitri, H. Chen, L. Yao, K. H. Zheng, F. Rossi, Y. Yin, Evaluation of the groundsill's stability at downstream of "Citorek" bridge in Cimadur River, Banten Province, In IOP Conference Series: Earth and Environmental Science, **880**, 012029, IOP Publishing (2021)
10. L. Yao, X. Huang, A. Fitri, Influence scope of local loss for pipe flow in plane sudden expansions, In: IOP Conference Series: Earth and Environmental Science, 12056, (2019)
11. A. Fitri, K.N.A. Maulud, D. Pratiwi, A. Phelia, F.

Rossi, N.Z. Zuhairi, Trend of water quality status In Kelantan River Downstream, Peninsular Malaysia. J Rekayasa Sipil, (2020)

12. A. Fitri, K. Nizam, A. Maulud, F. Rossi, F. Dewantoro, N.Z. Zuhairi, Spatial and temporal distribution of dissolved oxygen and suspended sediment in Kelantan River Basin. 4th International Conference on Sustainable Innovation 2020, Technology, Engineering and Agriculture (ICoSITEA 2020), (Atlantis Press, Dordrecht, 2021)
13. S.H Lai, A. Fitri, Application of SWAT hydrological model to upper Bernam River Basin (UBRB), Malaysia. IUP J Environ Sci. **5**, (2011)
14. G. Villarini, J.A. Smith, F. Serinaldi, A.A. Ntelekos, Analyses of seasonal and annual maximum daily discharge records for central Europe. J Hydrol, **399**, (2011)
15. R. Prasad, J. Rishideo, Peak discharge analyses for flood management in lower Gandak Basin. Climate Change, Extreme Events and Disaster Risk Reduction, (2018)
16. I.L.M Limantara, Hydrological engineering: revised edition, (Penerbit Andi, Yogyakarta, 2018)
17. S. Siswanto, G.J. van Oldenborgh, G. van der Schrier, R. Jilderda, B. van den Hurk, Temperature, extreme precipitation, and diurnal rainfall changes in the urbanized Jakarta city during the past 130 years. International Journal of Climatology, **36**, (2016)
18. L. De Haan, and Resnick, S. I., Limit theory for multivariate sample extremes. Zeitschrift für Wahrscheinlichkeitstheorie und verwandte Gebiete, **40**, (1977)
19. W. Ningrum, R. Boer, Statistical assessment of high-resolution climate model rainfall data in the Ciliwung Watershed, Indonesia. Agromet, **37**, (2023)

# The performance of the weather research & forecasting model (WRF) using ensemble method to predict weather parameters

*Alvin* Pratama[1*], *Ade* A. Oktaviana[1,2], *Prawira* Y. Kombara[3], and *Muhammad* Isnaenda Ikhsan[1]

[1]Department of Atmospheric and Planetary Science, Institut Teknologi Sumatera, Lampung 35365, Indonesia
[2]Department of Environmental Engineering, Sepuluh Nopember Institute of Technology, East Java 60111, Indonesia
[3]Research Centre for Climate and Atmosphere, National Research and Innovation Agency of Indonesia, West Java 40135, Indonesia

**Abstract.** The performance of the WRF model is evaluated using different physical options for the Riau Province. There are 12 members, and the ensemble mean method is used to evaluate the temperature, humidity, wind direction, and wind speed. The primary purpose of this study is to determine the appropriate parameterization for the study area. In this study, two nested domains have been used for performance analysis, with the resolution of the coarser domain set at 9 km and the inner domain set at 3 km. The model was run for five days in 2019 during a forest fire episode in Riau Province. The analysis was carried out by looking at the values of the correlation coefficient, root mean square error (RMSE), and bias. From the weather forecast, WRF, with the sixth parameterization member, produces a better value than the others. RRTM and Dudhia parameterization gave better results for temperature parameters. Meanwhile, the parameterization of Yonsei University (YSU) produces better results for the parameters of wind direction and wind speed.

## 1 Introduction

Weather parameters have an essential role in human life. They include rainfall, wind direction and speed, humidity, pressure, temperature, and radiation. The availability of accurate weather forecasts is needed for various purposes, such as disaster adaptation and mitigation, hydrology, and agriculture. Weather forecasting is intricate and challenging since the atmosphere is difficult to anticipate because to its dynamic and nonlinear characteristics [1]. Despite its intricacy, various methodologies are available to assist in weather forecasting, including statistical and numerical weather prediction (NWP) techniques. Moreover, NWP models yield significantly more realistic and precise outcomes as they address the atmospheric equations more effectively than statistical methods [1]. A numerical weather model is one of the tools being developed around the world for predicting and simulating current and future weather conditions. One of the models being developed is the WRF-ARW (Weather Research and Forecasting Model—Advanced Research WRF) model.

WRF is a mesoscale numerical weather forecast model that is used and utilized for atmospheric research and weather forecast operational purposes worldwide [2]. The basis of the WRF model is the solution of the model's mathematical equations. In this case, the WRF product has an uncertainty value because there are different kinds of errors that can cause different forecasts. The low quality of the initial data and the choice of boundary conditions cause the error to change, which leads to a loss of skills. Also, an increase in projection time is one of the things that makes the predicted skill less accurate. The longer the prediction is made, the higher the level of uncertainty will be. This is due to dynamic and chaotic weather conditions, so at any time, it can interfere with the quality of the forecast [3]. Determining the size of the resolution of the simulation domain also has a significant role in generating model uncertainty. A high-domain selection must be supported by high-resolution data, such as topographic data (if available) [4]. Atmospheric dynamics occur at diverse geographical and temporal scales. The meteorological models, such as WRF, are not intended to accurately represent these physical processes. Unresolved processes are addressed using physical parameterization approaches, including cumulus, microphysics, boundary layers, radiative transfer, soil interaction, and vegetation interaction [4]. The parameterizations incorporate numerous assumptions and approximations to elucidate unsolved processes. Moreover, for numerical weather prediction, such as WRF in a specific location, it is essential to identify the optimal mix of physical parameterizations within the selected modeling framework [5].

The ensemble method is one of the methods currently being worked on to deal with uncertainty in the model. Basically, Ensemble systems may vary not only in their simulation of initial uncertainties (i.e., analytical errors) and the nature of model prediction errors, but also in ensemble size and their capacity to represent the probability distribution function of analysis and forecast states [6]. The examination of ensemble methodologies

* Corresponding author: alvin.pratama@sap.itera.ac.id

indicates that some combinations of physics correspond effectively with observations, notably the new GFS-simplified Arakawa-Schubert and the modified Tiedtke cumulus parameterizations when integrated with Thompson and Morrison microphysics for precipitation forecasting [5]. Moreover, to optimize WRF performance, stochastic parameterizations alone are inadequate for capturing model uncertainty and must be supplemented by a multi-physics suite [7].

Riau Province is an area in Indonesia that is always hit by forest fire events. One of the bad effects of these forest fires is the decline in air quality in the area, which has the potential to spread to surrounding areas such as Padang, Jambi, Batam, and Palembang. Meteorological variables, including wind direction and velocity, significantly influence the advection of pollutants. To predict the bad effects of forest fires, it is important to have accurate weather prediction data. Several WRF tests for wind speed and direction have been done in many places using different numerical and physics options. The testing of WRF in Portugal suggests that the wind simulation error can be kept to a minimum by testing and choosing a numerical and physical configuration that works well for the area of interest. In the research, the combination parameterization of WRF, Yonsei University, and Noah is the most suitable for that site [4]. Research on the sensitivity of WRF physical parameters on the island of Sumatra and the Maritime Continent has been tested several times. The combination of parameterization schemes Eta Scheme (microphysical), GD Scheme (cumulus), and MYJ Scheme (planetary boundary layer) is one of the best parameterization combinations for predicting rainfall in tropical peatland areas, one of which is Sumatra Island [8].

The primary objective of this study is to determine the optimal method to configure the WRF model's physical parameters. For this purpose, the performance of WRF from various simulations with specific configurations is assessed by comparing the results to the observational data in Padang, Pekanbaru, and Batam. The best parameterization is determined by carrying out statistical analysis, including the Pearson correlation method, RMSE, and bias. The examination of the WRF model's sensitivity in relation to alterations in configuration options is a significant evaluative task, and this sensitivity study will substantially enhance other research aimed at revealing which model parameters are most effective [1].

## 2 Materials and method

The simulation of weather prediction has been done using the WRF-ARW model. The model was executed from September 21–25, 2019. Basically, in this study, two kinds of data are used. The data are from the global forecast system (GFS) and the observation data from automatic weather stations (AWS).

a. GFS Data
This study uses the GFS data for the initial conditions and input to the WRF model. This data has 0.25 degrees by 0.25 degrees for the spatial resolution and 3 hours for time resolution. The GFS data can be downloaded at https://rda.ucar.edu/datasets/ds084.1

b. AWS Data
The observation data was obtained from an automatic weather station that is installed at three locations and operated by the Indonesian Ministry of Environment and Forestry. These locations are Padang (0.939° S, 100.363° E), Pekanbaru (0.511° N, 101.459° E), and Batam (1.045° N, 103.942° E). The weather parameters that are used for evaluating the model result are wind speed and direction, surface temperature, and relative humidity.

The initial phase in configuring the WRF model involves establishing a domain design. The domain's design is seen in Fig. 1. The study encompasses a parent domain (D1) with a spatial resolution of 9 km and a secondary, nested domain (D2) with a spatial resolution of 3 km. In this study, 3 AWS (automatic weather stations) are used in the cities of Padang, Pekanbaru, and Batam. These stations are in the same domain in the WRF simulation. So, it can be used to evaluate parameterization, especially domain 2 (fine domain). The simulation domain of WRF is shown in Fig. 1.

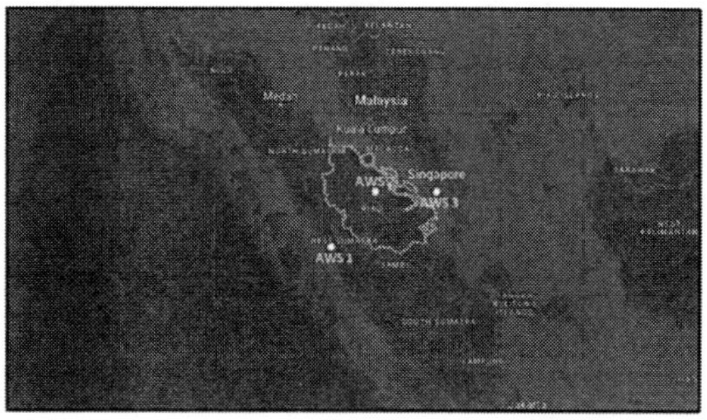


**Fig. 1.** Simulation domain of WRF. The outer domain is set to 9 km, and the inner domain is set to 3 km. The white circle marked shows the location of the AWS.

In WRF models, the mathematical representation of atmospheric processes is frequently streamlined based on the model's objectives, resolution, and focus. The dynamical core solves the governing equations on a discrete grid, neglecting terms of minimal significance [9]. Unresolved processes, including radiation, deep and shallow cumulus convection, cloud microphysics, precipitation, and turbulence, influence the resolved scales via a parameterized representation in WRF [9]. There are five physical parameters: cloud microphysical, cumulus, the planetary boundary layer (PBL), and longwave and shortwave radiation. This study examines the combination of parameterization schemes. The examination was conducted with 12 members and one method of the ensemble mean. The combination of parameterization schemes can be seen in Table 1. In running the WRF model, a simulation is carried out for five days, where one day is used for the model spin-up time, and the other four days are the results of the model predictions. Table 1 shows that GD (Grell Devenyi), Grell Freitas, and Grell 3D parameterization dominate in this simulation for cumulus parameterization. This refers to research conducted by [8] in tropical peatlands, one of which is Sumatra Island [8]. Furthermore, the research done by [10] in Riau Province, [11] in Pekanbaru for the forest fire episode, and [12] also used one of the parameterizations in table 1 [10–12]. Grell Freitas applies the stochastic methodology from Grell Devenyi within the cumulus framework and modifies the scale awareness for the convective eddy transport equation [9]. Grell 3D is a combination of multiple physical schemes (Kain Fritsch, Betts and Miller, and Grell Devenyi) [5]. Cumulus schemes attempt to assess a mesoscale convective system; microphysics, on the other hand, attempts to evaluate a warm frontal zone with stratiform precipitation [9]. The microphysical parameterizations tested were the WRF single moment3 class (WSM3), Eta Scheme, Lin, and Kessler scheme. The boundary layer aims to illustrate the statistical properties of turbulence phenomena and their average effects on grid-scale fields [13]. The transfer of energy (momentum, heat, moisture) to the upper atmospheric layers is crucial and serves as a feedback mechanism in atmospheric circulation [14]. We selected Yonsei University (YSU) and Mellor-Yamada-Janjic (MYJ) due to their potential impact on model outcomes, based on a review of the available literature and an analysis of the scheme formulas [13]. The radiation scheme comprises two parameters, namely short wave and long wave. These schemes play a crucial role in simulating temperature, precipitation, and global radiation. Shortwave (SW) and longwave (LW) radiation transfer were modeled using the Dudhia scheme, the Rapid Radiative Transfer Model (RRTM), and the Rapid Radiative Transfer Model for General Circulation Model (RRTMG). The parameterization schemes used in simulation are shown in Table 1.

**Table 1.** Parameterization schemes used in simulation

| Member | Parameterization Scheme | | | | |
|---|---|---|---|---|---|
| | **Cumulus** | **Microphysics** | **Short Wave** | **Long Wave** | **Boundary Layer** |
| Member 1 | Grell-Devenyi | WSM3 | Dudhia | RRTM | YSU |
| Member 2 | Grell-Devenyi | Eta Schema | New Goddard | RRTM | MYJ |
| Member 3 | Grell-Devenyi | WSM3 | RRTMG | RRTMG | YSU |
| Member 4 | Grell-Devenyi | Eta Schema | New Goddard | RRTMG | MYJ |
| Member 5 | Kain Fritsch | Lin | CAM | CAM | YSU |
| Member 6 | Grell 3D | Lin | Dudhia | RRTM | YSU |
| Member 7 | Grell Freitas | Kessler | Goddard | New Goddard | YSU |
| Member 8 | Grell 3D | Lin | RRTMG | RRTMG | MYJ |
| Member 9 | Grell Freitas | WSM3 | Goddard | FLG | YSU |
| Member 10 | Grell 3D | Eta Schema | New Goddard | RRTM | MYJ |
| Member 11 | Grell 3D | Thompson | Dudhia | New Goddard | MYJ |
| Member 12 | Grell 3D | Lin | RRTMG | RRTMG | YSU |

The ensemble prediction method is a way to determine the value of the accuracy of the prediction results by comparing the results from several weather prediction models (members). This ensemble prediction method performs the running process several times using the same initial time. The results of the running process are called ensemble members. The ensemble mean forecast is defined as the average of all members of the ensemble forecast [6]. The following is the equation for the ensemble mean:

$$em^{xx}(t) = \frac{1}{(N_{xx}+1)}\sum_{j=0,N_{xx}} f_j^{XX}(t) \quad (1)$$

Where xx denotes weather forecast parameters, and Nxx is the total number of ensemble members.

Numerical weather prediction (NWP) models such as WRF need to be verified. Verification is carried out on observational data collected in the field. In this case, it is carried out using observation data from Padang City, Pekanbaru City, and Batam City. The main purpose of this verification is to find the appropriate parameterization for real conditions. The statistical parameters used to validate the results of the WRF model are the Pearson correlation coefficient, root mean square error (RMSE), mean error (BIAS), and Lavdaz validation.

The correlation coefficient is used to measure the level of linearity between the results of the model and observation data [8]. The coefficient ranges from -1 to +1. The negative sign indicates that the correlation between the model findings and observational data is inversely proportional. To determine the strong, moderate, and weak categories of correlation coefficients. The categories of correlation coefficients are shown in Table 2.

**Table 2.** The category of correlation coefficient

| Coefficient Interval | Correlation |
|---|---|
| $0.00 < r < 0.20$ | No correlation |
| $0.20 \leq r < 0.40$ | Weak |
| $0.40 \leq r < 0.70$ | Intermediate |
| $0.70 \leq r < 0.90$ | Strong |
| $0.90 \leq r < 1.00$ | Very Strong |

The following is the equation to determine the Pearson correlation coefficient (R)

$$r = \frac{\sum(F_i-\bar{F}_i)(O_i-\bar{O}_i)}{\sqrt{\sum(F_i-\bar{F}_i)^2}\sqrt{\sum(O_i-\bar{O}_i)^2}} \quad (2)$$

Where F denotes forecasting data, and O is the observation data.

The RMSE quantifies the deviation of data points from the observation data. As the forecast and observation data diverge, the RMSE value fluctuates, increasing and decreasing as they converge. The subsequent information is utilized for RMSE computations.

$$RMSE = \sqrt{\frac{1}{N}\sum_{i=1}^{N}(F_i - O_i)^2} \quad (3)$$

The mean error is calculated by dividing the total discrepancies between the observational data and the WRF model findings by the number of observations. The mean error count facilitates the observation of data trends. If positive, the simulated value is likely an excess of the actual values; if negative, it is likely an underestimate of the actual value [15].

$$BIAS = \frac{\sum_{i=1}^{n}(F_i-O_i)}{N} \quad (4)$$

The Lavdas method can be used to verify wind direction. In this method, the wind directions from the model are grouped into eight wind directions : north-east (45 degrees), east (90 degrees), south-east (135 degrees), south (180 degrees), southwest (225 degrees), west (270 degrees), north-west (315 degrees), and north (360 degrees). In this method, calm winds are omitted from the calculation. If the observed wind originated from the accurately predicted direction, specifically the nearest eight-point compass direction, the forecast was classified as a direct hit for that hour. A forecast was deemed a near miss to the left if the wind-borne smoke's trajectory was approximately one compass point to the left of the forecasted direction. For instance, with a northeast forecast and a 350-degree observation, the forecasted wind would direct smoke southwestward, while the observed wind would direct smoke southward (to the nearest compass point), as south is adjacent to the left of southwest when facing southwest. Consequently, the forecast is categorized as a near miss to the left for that hour. A similar assessment was conducted for forecasts that were near misses to the right. Forecasts classified as a direct hit, a near miss to the left, or a near miss to the right were aggregated into a near or direct-hit category; the remaining forecasts were designated as missed forecasts. [16].

## 3 Results

Riau Province is one area that often experiences forest fires and other hydrometeorological disasters. These disasters are closely related to various meteorological parameters. Therefore, the requirement for and availability of meteorological data becomes very important. In this study, verification of the prediction results of the WRF model was carried out on meteorological observation data for parameters of humidity, temperature, wind speed, and wind direction.

### 3.1 Temperature parameter

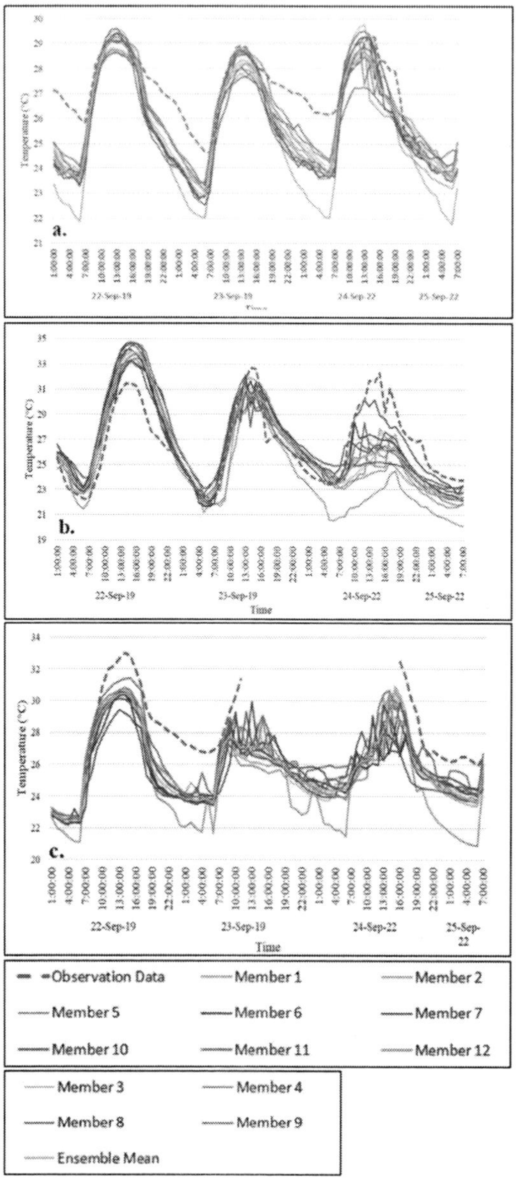


**Fig. 2.** WRF model temperature comparison with observational data in (a) Padang, (b) Pekanbaru, and (c) Batam

Temperature quantifies the thermal state of an environment, indicating its degree of coldness or heat. This situation can be affected by the duration of sunlight, the angle of sunlight, cloud cover, and the condition of the Earth's surface. Fig. 2 shows a comparison between the results of the WRF model and the observation data for the temperature parameter.

According to Fig. 2, the WRF model has a similar pattern to the observational data. The maximum temperature occurs between 13.00 and 16.00 local time (LT) during the day, while the minimum temperature occurs between 04.00 and 05.00 LT in the early morning. In general, the WRF model has a high linearity value for the Padang, Pekanbaru, and Batam observational data (Table 3). This is indicated by the average correlation value between 0.70 and 0.89. The member with the highest correlation value was member 6, with a value of 0.89 for the city of Padang, followed by member 10, with a value of 0.87 for the city of Pekanbaru, and member 11, with a value of 0.91 for the Batam region. While the typical error generated in the Padang region is 0.18, it is 0.19 in the Pekanbaru region and 2.11 in the Batam region. Temperature parameters are typically underestimated by the WRF model. Consequently, a negative bias value is observed (less than 0). For the ensemble mean, the temperature prediction results for the Padang, Pekanbaru, and Batam regions show good agreement with the observation data, particularly in terms of linearity and error patterns.

**Table 3.** Validation of correlation coefficient, RMSE, and BIAS from the WRF model against temperature observations in Padang, Pekanbaru and Batam

| Member | Padang | | | Pekanbaru | | | Batam | | |
|---|---|---|---|---|---|---|---|---|---|
| | Corr | RMSE | BIAS | Corr | RMSE | BIAS | Corr | RMSE | BIAS |
| Member 1 | 0.86 | 0.29 | -1.18 | 0.67 | 0.04 | -0.77 | 0.87 | 3.31 | -2.81 |
| Member 2 | 0.88 | 0.26 | -1.19 | 0.72 | 0.02 | 0.07 | 0.90 | 0.35 | -2.81 |
| Member 3 | 0.86 | 0.03 | -1.06 | 0.72 | 0.04 | -0.21 | 0.87 | 0.50 | -2.64 |
| Member 4 | 0.86 | 0.03 | -1.49 | 0.70 | 0.38 | -0.06 | 0.86 | 3.75 | -3.08 |
| Member 5 | 0.85 | 0.16 | -2.01 | 0.62 | 0.67 | -1.78 | 0.88 | 3.22 | -3.56 |
| Member 6 | 0.89 | 0.31 | -1.45 | 0.82 | 0.03 | -0.09 | 0.89 | 0.88 | -3.09 |
| Member 7 | 0.88 | 0.24 | -1.25 | 0.62 | 0.22 | 0.10 | 0.83 | 0.39 | -2.92 |
| Member 8 | 0.88 | 0.01 | -1.38 | 0.76 | 0.35 | -0.10 | 0.90 | 0.44 | -2.78 |
| Member 9 | 0.85 | 0.18 | -0.92 | 0.70 | 0.10 | 0.07 | 0.88 | 3.36 | -2.01 |
| Member 10 | 0.88 | 0.26 | -1.20 | 0.87 | 0.10 | 0.55 | 0.83 | 3.62 | -2.72 |
| Member 11 | 0.88 | 0.08 | -1.57 | 0.76 | 0.08 | -0.04 | 0.91 | 3.88 | -2.96 |
| Member 12 | 0.87 | 0.14 | -1.20 | 0.70 | 0.41 | -0.26 | 0.88 | 0.43 | -2.55 |
| Ensemble Mean | 0.88 | 0.31 | -1.33 | 0.74 | 0.13 | -0.21 | 0.91 | 3.31 | -2.83 |

## 3.2 Humidity parameter

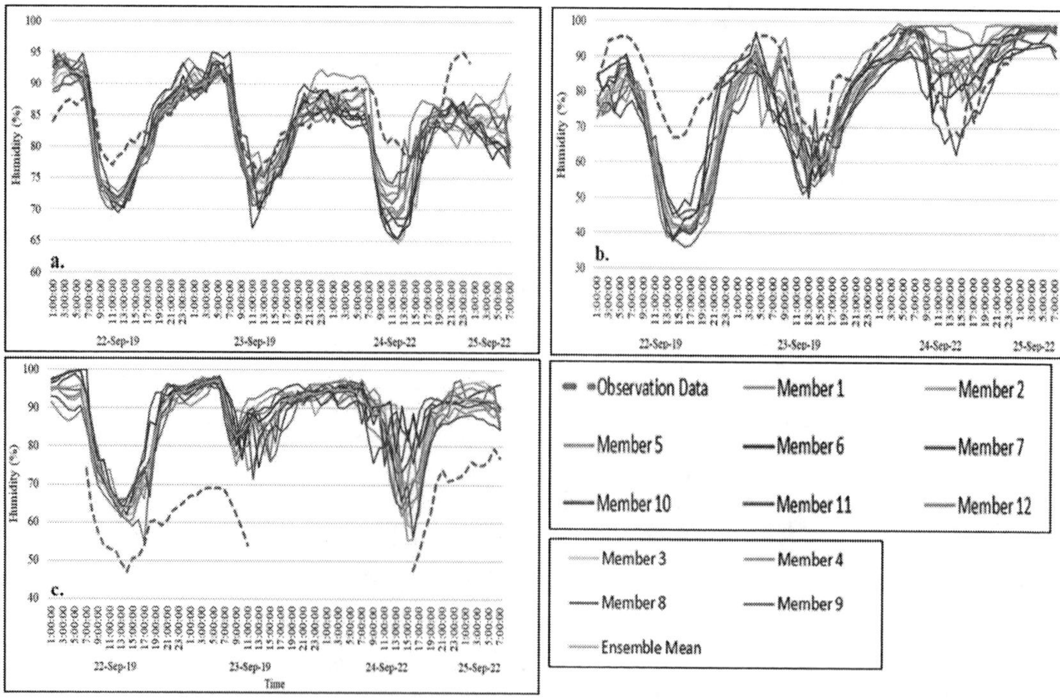


**Fig. 3.** WRF model humidity comparison with observational data in (a) Padang, (b) Pekanbaru, and (c) Bata

Air humidity is a measure of the amount of moisture in the air that is affected by temperature, air pressure, wind speed, duration of sunlight, and categories of vegetation. Fig. 3 depicts a comparison between the WRF model's humidity results and observational data.

Figure 3 shows that the WRF model reveals a similar pattern to the observational data. Maximum humidity occurs between 01:00 and 04:00 LT in the early morning, while minimal humidity occurs between 11:00 and 15:00 LT in the afternoon. For the Padang, Pekanbaru, and Batam regions, the WRF model has a linearity value that tends to be moderate to strong in relation to the observed data for the humidity parameter (Table 4). This is indicated by the average correlation value between 0.55 and 0.89. The maximum correlation value was shown by member 7 for the city of Padang, member 10 for the city of Pekanbaru, and member 12 for the Batam region. While the typical error generated in the Padang region is 0.33, it is 1.64 in the Pekanbaru region and 10.07 in the Batam region. In general, the Padang and Pekanbaru regions are underestimated by the WRF model, whereas Batam is overestimated. This is indicated by a negative bias value (less than 0) for Padang and Pekanbaru and a positive bias value (greater than 0) for Batam. Table 4 demonstrates that Batam has greater error and bias than Padang and Pekanbaru. This is because there are a large number of missing observational values for the humidity parameter in the Batam region. In the Padang, Pekanbaru, and Batam regions, the ensemble mean humidity prediction result exhibits a similar pattern to the observation data, particularly the linearity pattern.

Figures 2 and 3 illustrate that the correlation between temperature and humidity is inversely proportional. Relative humidity indicates the quantity of water vapor present in the atmosphere. A high humidity level signifies a substantial presence of water vapor in the atmosphere. Basically, Temperature and humidity are influenced by seasonal fluctuations; during the dry season, temperatures are elevated and humidity is diminished [17]. During the rainy season, temperatures decrease and humidity rises [17].

**Table 4.** Validation of correlation coefficient, RMSE, and BIAS from the WRF model against humidity observations in Padang, Pekanbaru, and Batam

| Member | Padang | | | Pekanbaru | | | Batam | | |
|---|---|---|---|---|---|---|---|---|---|
| | Corr | RMSE | BIAS | Corr | RMSE | BIAS | Corr | RMSE | BIAS |
| Member 1 | 0.69 | 0.66 | -2.31 | 0.64 | 1.53 | -5.22 | 0.81 | 13.79 | 23.96 |
| Member 2 | 0.73 | 0.002 | -2.98 | 0.69 | 0.70 | -8.30 | 0.78 | 2.42 | 23.48 |
| Member 3 | 0.72 | 0.21 | -2.29 | 0.70 | 0.89 | -6.18 | 0.83 | 4.14 | 23.25 |
| Member 4 | 0.68 | 0.64 | -2.04 | 0.70 | 2.92 | -8.46 | 0.72 | 13.99 | 24.39 |
| Member 5 | 0.72 | 0.17 | 0.24 | 0.56 | 3.43 | -3.22 | 0.84 | 13.53 | 21.70 |
| Member 6 | 0.72 | 0.48 | -2.62 | 0.78 | 0.63 | -5.61 | 0.79 | 12.00 | 26.25 |
| Member 7 | 0.78 | 0.34 | -1.65 | 0.54 | 2.00 | -8.14 | 0.75 | 2.28 | 23.32 |
| Member 8 | 0.70 | 0.49 | -2.65 | 0.73 | 3.05 | -5.40 | 0.76 | 4.01 | 24.96 |
| Member 9 | 0.70 | 0.19 | -2.93 | 0.65 | 0.48 | -6.42 | 0.81 | 13.98 | 21.83 |
| Member 10 | 0.74 | 0.33 | -3.19 | 0.85 | 0.58 | -8.83 | 0.72 | 13.95 | 24.31 |
| Member 11 | 0.68 | 0.12 | -1.53 | 0.67 | 1.15 | -8.44 | 0.70 | 13.27 | 23.14 |
| Member 12 | 0.71 | 0.03 | -1.93 | 0.69 | 2.32 | -5.96 | 0.84 | 9.47 | 23.15 |
| Ensemble Mean | 0.74 | 0.69 | -2.16 | 0.71 | 1.68 | -6.68 | 0.83 | 14.06 | 23.65 |

### 3.3 Wind speed parameter

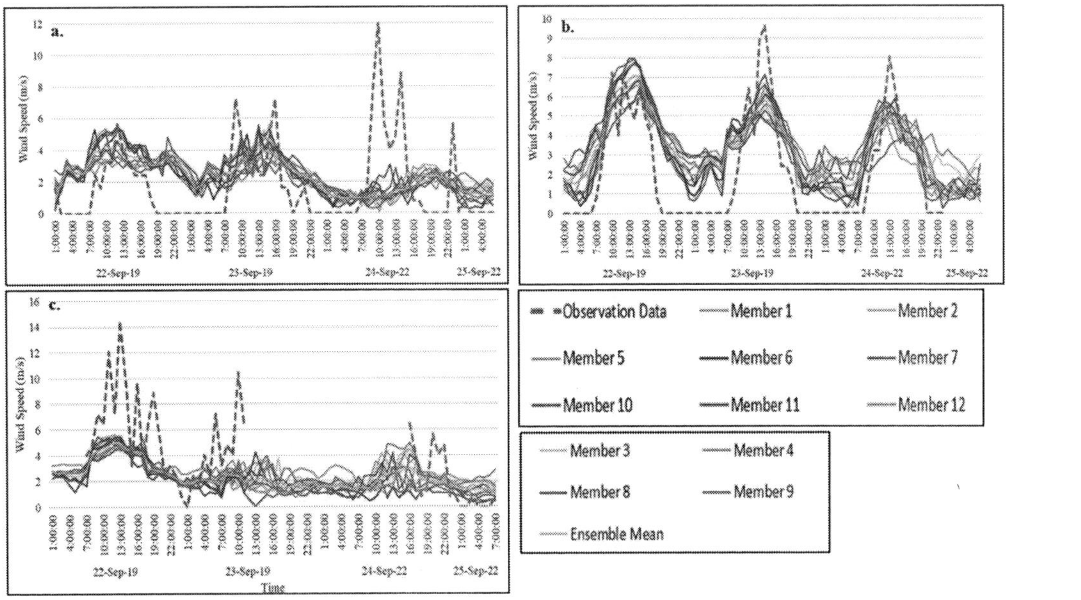


**Fig. 4.** WRF model wind speed comparison with observational data in (a) Padang, (b) Pekanbaru, and (c) Batam

The wind is the horizontal movement of air masses due to pressure differences. Basically, air masses transition from regions of high pressure to regions of low pressure. Fig. 4 presents a comparison of the WRF model findings with the observational data for wind speed parameters.

Figure 4 illustrates the similarity between the WRF model and the observations, particularly in the Padang and Batam regions. The maximum wind speed occurs between 10:00 and 16:00 LT during the day, while the minimum temperature occurs between 3:00 and 5:00 LT in the early morning. In general, the WRF model exhibits moderate to strong linearity with the observational data, particularly in the Padang and Batam regions. (Table 5). This is indicated by the average correlation value between 0.55 and 0.89. In the Pekanbaru region, the linearity value between WRF model data and observation data demonstrates a weak correlation (no correlation). Member 10 demonstrated the highest correlation with a value of 0.86 for the city of Padang, while member 12 demonstrated the lowest correlation with a value of 0.75 for the Batam region. While the average error is 0.28 in the Padang region, it is 0.19 in the Pekanbaru region and 0.34 in the Batam region. In general, the WRF model results for wind speed in Padang and Pekanbaru are higher than the observational data, whereas the results for Batam are sluggish. The ensemble means wind speed prediction result displays a similar pattern, particularly the linearity and error patterns derived for the Padang and Batam regions.

**Table 5.** Verification value of correlation coefficient, RMSE, and BIAS from WRF model against observation in Padang, Pekanbaru, and Batam for wind speed

| Member | Padang | | | Pekanbaru | | | Batam | | |
|---|---|---|---|---|---|---|---|---|---|
| | Corr | RMSE | BIAS | Corr | RMSE | BIAS | Corr | RMSE | BIAS |
| Member 1 | 0.76 | 0.14 | 1.06 | -0.04 | 0.08 | 0.61 | 0.60 | 0.41 | -1.89 |
| Member 2 | 0.84 | 0.21 | 1.01 | 0.12 | 0.21 | 0.91 | 0.60 | 0.72 | -1.79 |
| Member 3 | 0.81 | 0.43 | 1.42 | -0.01 | 0.22 | 0.75 | 0.70 | 0.16 | -1.65 |
| Member 4 | 0.79 | 0.23 | 0.84 | 0.19 | 0.08 | 0.83 | 0.66 | 0.17 | -1.92 |
| Member 5 | 0.79 | 0.36 | 1.05 | 0.00 | 0.46 | 0.45 | 0.67 | 0.44 | -1.22 |
| Member 6 | 0.85 | 0.17 | 0.96 | 0.32 | 0.13 | 0.57 | 0.71 | 0.10 | -2.30 |
| Member 7 | 0.66 | 0.27 | 1.01 | -0.09 | 0.13 | 0.71 | 0.73 | 0.80 | -1.94 |
| Member 8 | 0.84 | 0.25 | 1.03 | 0.26 | 0.13 | 0.70 | 0.70 | 0.07 | -2.04 |
| Member 9 | 0.79 | 0.33 | 1.76 | -0.15 | 0.30 | 0.75 | 0.66 | 0.40 | -1.46 |
| Member 10 | 0.86 | 0.19 | 1.22 | 0.36 | 0.10 | 0.84 | 0.72 | 0.23 | -2.05 |
| Member 11 | 0.76 | 0.44 | 0.90 | 0.24 | 0.30 | 0.68 | 0.66 | 0.07 | -2.18 |
| Member 12 | 0.82 | 0.41 | 1.13 | 0.03 | 0.38 | 0.62 | 0.75 | 0.47 | -1.95 |
| Ensemble Mean | 0.84 | 0.21 | 1.12 | 0.14 | 0.02 | 0.70 | 0.73 | 0.39 | -1.87 |

### 3.4 Wind direction parameter

The wind's direction is essential, especially in ascertaining the dispersion of airs contaminants. As forest fires are a common occurrence in the province of Riau, it is crucial to collect data on wind direction. Table 6 depicts a contrast between the results of the WRF model and the observational data for wind direction parameters.

In the Padang, Pekanbaru, and Batam regions, the average WRF model is able to detect wind direction accurately, as shown in Table 6. The WRF model's average ability to detect wind direction is greater than 60%, with an average number of errors below 30%. The ability with the highest percentage of near or direct hits is worth 74.7% in the Padang region (member 8), 72.2% in the Pekanbaru region (member 4), and 84.7% in the Batam region (member 10). For the ensemble mean, the wind direction prediction reveals positive results. This is indicated by the near or direct strike percentage exceeding 65%.

**Table 6.** Percentage of near/direct hit against miss of wind direction each member

| Category | Member | | | | | | | | | | | | |
|---|---|---|---|---|---|---|---|---|---|---|---|---|---|
| | 1 | 2 | 3 | 4 | 5 | 6 | 7 | 8 | 9 | 10 | 11 | 12 | Ens |
| **Padang** | | | | | | | | | | | | | |
| Near/Direct Hit (%) | 53.2 | 64.6 | 63.3 | 69.6 | 72.2 | 73.4 | 65.8 | 74.7 | 62.0 | 69.6 | 62.0 | 69.6 | 68.4 |
| Miss (%) | 46.8 | 35.4 | 36.7 | 30.4 | 27.8 | 26.6 | 34.2 | 25.3 | 38.0 | 30.4 | 38.0 | 30.4 | 31.6 |
| **Pekanbaru** | | | | | | | | | | | | | |
| Near/Direct Hit (%) | 65.8 | 72.2 | 65.8 | 72.2 | 69.6 | 68.4 | 63.3 | 70.9 | 69.6 | 70.9 | 62.0 | 68.4 | 68.4 |
| Miss (%) | 34.2 | 27.8 | 34.2 | 27.8 | 30.4 | 31.6 | 36.7 | 29.1 | 30.4 | 29.1 | 38.0 | 31.6 | 31.6 |
| **Batam** | | | | | | | | | | | | | |
| Near/Direct Hit (%) | 82.6 | 80.4 | 78.3 | 71.7 | 80.4 | 73.9 | 82.6 | 78.3 | 78.3 | 84.8 | 71.7 | 78.3 | 80.4 |
| Miss (%) | 17.4 | 19.6 | 21.7 | 28.3 | 19.6 | 26.1 | 17.4 | 21.7 | 21.7 | 15.2 | 28.3 | 21.7 | 19.6 |

## 4 Discussion

From the WRF simulation, member six is found to be the greatest predictor of the temperature parameter. Member 6 employs the RRTM and Dudhia schemes to parameterize longwave radiation and shortwave radiation. The WRF model is generally good at estimating the pattern and value of temperature at three different places. In the coastal regions of northern China, a combined Dudhia and RRTM scheme also demonstrates the greatest model performance among longwave and shortwave radiation schemes [18]. Not only temperature parameters but the humidity parameter also show a good output against the pattern and value of the observation data at three locations. In general, the pattern of meteorology parameters shows a diurnal pattern in Padang, Pekanbaru, and Batam.

This simulation additionally incorporates an assessment of the wind speed and direction characteristics. Wind speed exhibits the most sensitivity to PBL schemes, followed by longwave or shortwave radiation schemes, and microphysics schemes, whereas wind direction demonstrates reduced sensitivity to modifications in physical parameterization [18]. Using Yonsei University's PBL parameterization, the WRF simulation results indicate that the strongest member is member 6 (YSU). The YSU PBL scheme is a nonlocal first-order K-profile model that incorporates a countergradient term to account for the influence of large-scale eddies on the overall flow [13]. The YSU scheme demonstrates a greater capacity for aligning with daytime wind observations, while the MYNN scheme tends to align more effectively with nighttime observations [13]. According to the simulation results, the correlation value of wind speed in the Pekanbaru region is very low. This is due to the fact that nighttime observation data is limited or nonexistent, whereas daytime observation data appears to have an excellent pattern. Regarding wind direction parameters, the WRF model can make accurate predictions. This is evidenced by the high percentage of near or direct strikes, which is 65%. The evaluation of wind direction across the Persian Gulf using various PBL parameterization schemes reveals that the YSU scheme, in conjunction with Interim reanalysis data, yields the most accurate wind speed estimation, while the YSU scheme paired with NCEP-FNL data produces the least error in wind direction [19].

## 5 Conclusion

This paper investigates the evaluation of 12 member parameterizations and one ensemble mean from September 21 to September 25, 2019. For all simulations using GFS data, the model was initialized at 00 UTC. A comparison is made between the model results and Automatic Weather Station (AWS) datasets for the same time series. Based on the simulations, we discovered that WRF generates plausible simulated meteorological parameters (humidity, temperature, wind direction, and wind speed) over the Riau Province. All ensemble members and the ensemble mean generate a good to strong correlation with the observational data for temperature and humidity parameters, but the values of the parameters are slightly underestimated for temperature and overestimated for the majority of humidity values. Regarding wind speed, the WRF model generates a poor correlation in the Pekanbaru region, but a strong correlation in Padang and Batam. In general, the WRF generates an overestimated value for the wind speed parameter in Padang and Pekanbaru, but an underestimated value in Batam.

The simulation also revealed that the sixth member corresponds closest to the observation data. It is due to the fact that almost all parameters in the three regions are consistent (above average correlation of 65 % and low RMSE). It employs RRTM and Dudhia to parameterize longwave and shortwave radiation (LW/SW) and Yonsei University (YSU) to parameterize the planetary boundary layer (PBL). According to the simulation results, the correlation value of wind speed in the Pekanbaru region is very low. This is due to the fact that nighttime observation data is limited or nonexistent, whereas daytime observation data appears to have an excellent pattern. Regarding wind direction parameters, the WRF model can make accurate predictions. This is evidenced by the high percentage of near or direct strikes, which is 65%.

In addition, the results of this study can be used to predict the typical direction of pollution dispersal in the province of Riau. In the process of pollutant advection, meteorological variables such as wind direction and speed, temperature, and humidity play a significant role. Consequently, it is possible to estimate the magnitude of the influence or contribution caused by forest fires in the affected areas.

The authors would like to thank Institut Teknologi Sumatera for providing the research grant (1539ah/IT9.2.1/PT.01.03/2024) through "Hibah Penelitian Itera 2024".

## References

1. C. Özen, Emre Korkmaz, S. Bağış, and H. Toros, Performance Test of Weather Research and Forecasting (WRF) Model for Central Anatolia and Black Sea Regions of Turkey, 8th Atmos. Sci. Symp. April, 435–441. (2017).
2. W. C. Skamarock et al., A Description of the Advanced Research WRF Model Version 4. (2019).
3. T. G. Shepherd, Atmospheric circulation as a source of uncertainty in climate change projections, Nat. Geosci., **7**(10), 703–708, (2014), https://doi.org/10.1038/ngeo2253.
4. D. Carvalho, A. Rocha, M. Gómez-Gesteira, and C. Santos, A Sensitivity Study of The WRF Model in Wind Simulation for An Area of High Wind Energy, Environ. Model. Softw., **33**, 23–34, (2012).
5. P. M. Igri et al., Assessing the performance of WRF model in predicting high-impact weather conditions over Central and Western Africa: an ensemble-based approach, Nat. Hazards, **93**, (3, 1565–1587, (2018).

6. W. Bourke, R. Buizza, and M. Naughton, Performance of the ECMWF and the BoM ensemble prediction systems in the Southern hemisphere, Mon. Weather Rev., **132**(10), 2338–2357, (2004), https://doi.org/10.1175/1520-0493(2004)132<2338:POTEAT>2.0.CO;2.
7. C. H. Li, J. Berner, J. S. Hong, C. T. Fong, and Y. H. Kuo, The Taiwan WRF Ensemble Prediction System: Scientific Description, Model-Error Representation and Performance Results, Asia-Pacific J. Atmos. Sci., **56**(1), 1–15, (2020).
8. A. R. Sanusi, M. Taufik, and I Putu Santikayasa, The Use of Weather Research and Forecasting Model to Predict Rainfall in Tropical Peatland: 1. Model Parameterization, Agromet, **35**(1), 49–59, (2021), https://doi.org/10.29244/j.agromet.35.1.49-59,.
9. J. Jeworrek, G. West, and R. Stull, Evaluation of cumulus and microphysics parameterizations in WRF across the convective gray zone, Weather Forecast., **34**(4), 1097–1115, (2019), https://doi.org/10.1175/WAF-D-18-0178.1.
10. M. Taufik, M. Haikal, M. T. Widyastuti, C. Arif, and I. P. Santikayasa, The Impact of Rewetting Peatland on Fire Hazard in Riau, Indonesia, Sustainability, **15**, (3, (2023), https://doi.org/10.3390/su15032169.
11. E. Heriyanto, L. Syaufina, and M. Sobri, Forecasting Simulation of Smoke Dispersion from Forest and Land Fires in Indonesia, Procedia Environ. Sci., **24**, 111–119, (2015), https://doi.org/https://doi.org/10.1016/j.proenv.2015.03.015.
12. B. Zhu, Z. Pu, A. W. Putra, and Z. Gao, Assimilating C-Band Radar Data for High-Resolution Simulations of Precipitation: Case Studies over Western Sumatra, Remote Sensing, 14, (1). (2022). https://doi.org/10.3390/rs14010042.
13. B. Yang et al., Parametric and Structural Sensitivities of Turbine-Height Wind Speeds in the Boundary Layer Parameterizations in the Weather Research and Forecasting Model, J. Geophys. Res., **124**, 5951–5969, (2019).
14. N. Srivastava and N. Blond, Impact of meteorological parameterization schemes on CTM model simulations, Atmos. Environ., **268**, 118832, (2022), https://doi.org/10.1016/j.atmosenv.2021.118832.
15. D. Carvalho, A. Rocha, M. Gómez-Gesteira, and C. Santos, A sensitivity study of the WRF model in wind simulation for an area of high wind energy, Environ. Model. Softw., **33**, 23–34, (2012), https://doi.org/10.1016/j.envsoft.2012.01.019.
16. L. G. Lavdas, Accuracy of National Weather Service wind direction forecasts at Macon and Augusta, Georgia, Natl. Weather Dig., **22**(1), 22–26, (1997).
17. N. Idris, M. Maswati, T. N. Usmawanda, and A. M. Sari, Effect of Temperature and Humidity on the Visibility of Mirage on the Runway of Sultan Iskandar Muda Airport, Aceh, Indonesia, J. Phys. Its Appl. **2**(1), (2019) [Online]. Available: https://ejournal2.undip.ac.id/index.php/jpa/article/view/6220
18. E. Yu, R. Bai, X. Chen, and L. Shao, Impact of Physical Parameterizations on Wind Simulation with WRF V3.9.1.1 over The Coastal Regions of North China at PBL Gray-Zone Resolution, Geosci. Model Dev. Discuss, **53**, 1–37, (2022).
19. S. Gholami, S. Ghader, H. Khaleghi-Zavareh, and P. Ghafarian, Sensitivity of WRF-simulated 10 m wind over the Persian Gulf to different boundary conditions and PBL parameterization schemes, Atmos. Res., **247**, 105147, (2021), https://doi.org/10.1016/j.atmosres.2020.105147.

# Rapid modeling of catastrophic floods: a case study of the Padang flood on July 14, 2023

*Yanuar* Rizky Ramadhan[1*] and *Dennish* Ari Putro[1]

[1]PT Reasuransi MAIPARK Indonesia, Multivision Tower 8th Floor, DKI Jakarta, 12960, Indonesia

**Abstract.** This study presents a rapid flood modeling approach for catastrophic flood events in Padang City on July 14, 2023. Floods pose significant risks to urban areas, exacerbated by rapid urbanization and climate change. This research provides a solution to the urgent need for quickly and accurately flood mapping, crucial for effective disaster response and mitigation. Various data such as Digital Elevation Models (DEMs), rainfall, evapotranspiration, soil types, and land use were employed in this study. The modeling process involved DEM conditioning, bias correction of rainfall, defining the model domain using the Topographic Wetness Index (TWI), and applying hydrological and hydrodynamic models. Model verification was conducted using flood location and depth data collected from mass media and social media, showing that the model achieved 77.8% accuracy in mapping inundated areas, with Root Mean Square Error (RMSE) and Mean Absolute Error (MAE) of flood depth at 38.093 and 27.584, before the model was calibrated. The calibration process significantly improved the model's accuracy.

## 1 Introduction

Hydrometeorological disasters, particularly floods, have become a serious concern in Indonesia over the past few decades [1]. Along with the rapid population growth and the rate of land-use changes, Indonesia has experienced an increasing trend in flood disasters in terms of both frequency and intensity. Data on flood events recorded by the Data Informasi Bencana Indonesia Badan Nasional Penanggulangan Bencana (DIBI BNPB) show an increasing trend of flood events in Indonesia from 2014 to 2021. Floods often have devastating impacts, including loss of life and property damage [2]. Therefore, research on flood management and mitigation has become increasingly important to protect communities and national assets.

The main challenge in flood risk mitigation is the ability to quickly and accurately map affected areas (extent of inundation and flood depth). In disaster situations, time is of the essence, and rapid response can reduce the losses caused by floods. However, rapid flood mapping is hindered by limitations in actual data and large computational resources, often taking a long time [3, 4].

To address the challenges faced in flood modeling, it is necessary to develop methods and tools that can accommodate data limitations and increase the speed of mapping affected areas [5]. This solution will enable authorities to respond to flood disasters more quickly and efficiently, thereby helping to reduce the losses caused by floods.

In facing these challenges, this research aims to develop a model that can accommodate data limitations and speed in mapping flood-affected areas. Thus, this research is expected to make a significant contribution to improving our ability to accurately and efficiently map flood-affected areas, and to help reduce the losses that can be caused by such disasters.

The catastrophic flood on July 14, 2023 inundated 10 districts in Padang City with water levels reaching 2 meters [6, 7]. This flood was caused by extreme rainfall triggered by a convergence pattern, where air currents from the north and south of the Indian Ocean met and extended from the Mentawai Islands to the Natuna Islands [8]. As a result of the flood, there was extensive damage to infrastructure, and it caused the deaths of two toddlers while approximately 101 families, or 403 individuals, were displaced [9]. This catastrophic flood event serves as the subject of this study.

## 2 Methodology

### 2.1 Study area

The study area of this research covers the entire region of Padang City, which is the national activity center and the capital of West Sumatra Province. The city has an administrative area of 1,414.96 $km^2$, including a significant addition from the sea area. Located between 00°44'00" and 01°08'35" South Latitude and 100°05'05" and 100°34'09" East Longitude, Padang City has a coastline of 68.126 km and the Barisan Mountains covering an area of 486.209 $km^2$. In 2022, the population of Padang City reached 919,145 people, with a population density of 1,324 people/$km^2$ [10].

* Corresponding author: yrramadhan@maipark.com

**Fig. 1.** Topographic map of Padang City.

Padang City is traversed by 23 rivers with a total length of 155.40 km, which are divided into 6 Watershed Areas (DAS), namely DAS Air Dingin, DAS Air Timbalun, DAS Batang Arau, DAS Batang Kandis, DAS Batang Kuranji, and DAS Sungai Pisang, (Fig. 1). These rivers generally have an elevation not much different from sea level, making many areas in Padang City prone to flooding, especially during high tides [11]. This situation is further exacerbated by the conversion of forest areas into residential and agricultural land, which reduces water absorption capacity, and the city's inadequate drainage system to handle heavy rainfall [12].

## 2.2 Data

In this research, we utilized several types of data to construct a rapid model of catastrophic floods. Firstly, we utilized an 8.5-meter resolution Digital Elevation Model (DEM) obtained from DEM Nasional (DEMNAS). This data was subsequently corrected using river network data from the Badan Informasi Geospasial (BIG) to produce HydroDEM.

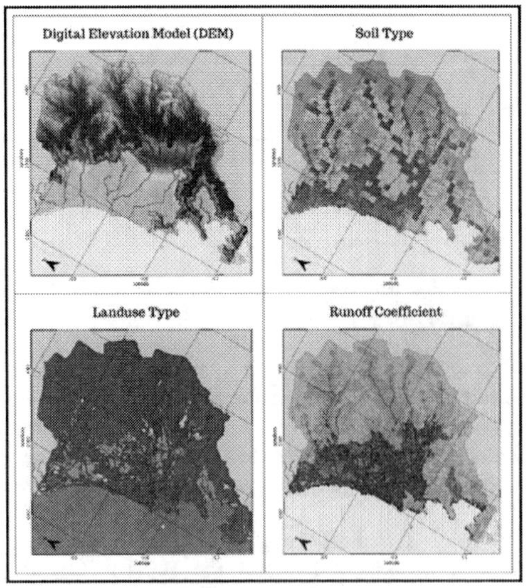


**Fig. 2.** Elevation data, soil type data, land use type data, and runoff coefficient data of Padang City.

Secondly, we employed meteorological data including rainfall, temperature, and wind speed. Rainfall data came from the Global Satellite Mapping of Precipitation (GSMaP) with a spatial resolution of 0.1 degrees, which was further corrected with rainfall data from the Meteorological Stations of the Badan Meteorologi, Klimatologi, dan Geofisika (BMKG). Temperature and wind speed data from BMKG Meteorological Stations were used to estimate evapotranspiration values.

Lastly, this study utilized soil data obtained from the Food and Agriculture Organization (FAO), as well as land use data with a spatial resolution of approximately 500 meters obtained from the Kementerian Lingkungan Hidup dan Kehutanan (KLHK). The incorporation of soil and land use data allows the model to simulate how different land covers and soil types influence water absorption and runoff, which is represented by the runoff coefficient. These data are shown in Fig. 2.

## 2.3 Method

### *2.3.1 DEM conditioning*

DEM conditioning was performed to enforce the DEM to simulate flow conditions according to the actual river network. This process involved using several tools available in Archydro to modify the DEM, ensuring that elevation data reflects the actual topographic and river network conditions. This process is crucial because DEM data influences water movement in the analyzed area; thus, by conditioning the DEM, we can generate a flood model that is more accurate and relevant to real-world conditions.

### *2.3.2 Rainfall bias correction*

GSMaP data is used to model and spatially distribute rainfall across the entire Watershed Area, (Fig. 3). However, GSMaP data often differs in values compared to direct surface observation data, necessitating a bias correction process. The method used for this correction is Quantile Mapping (QM) [13], which aims to align the spatial distribution of rainfall in GSMaP data to be more representative and consistent with more accurate surface observation data.

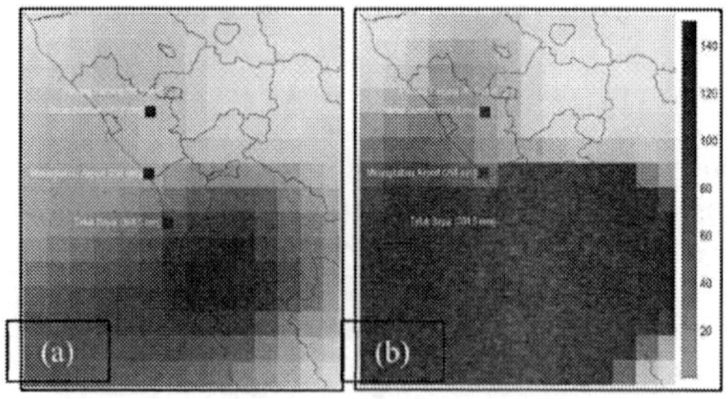


**Fig. 3.** GSMaP rainfall data. (a) before bias correction; (b) after bias correction.

The data is corrected using observation data from rain gauges at four BMKG stations located in Padang City and surrounding areas. By performing this correction, we can obtain rainfall data that is more reliable for flood modeling.

### 2.3.3 Calculating evapotranspiration

Evapotranspiration in this study is calculated using the method recommended by the World Meteorological Organization (WMO) [14]. Based on the study by Celestin et al. (2020), which evaluated various simple empirical models for estimating evapotranspiration under different climatic conditions and data requirements, the WMO method has demonstrated excellent performance, making it a suitable alternative method. The calculation of evapotranspiration (ET0) using the WMO method is shown in the following formula:

$$ET0 = (1.298 + 0.934 \,.\, u_2) \,.\, (e_s - e_a) \quad (1)$$

, where $u_2$ is the wind speed measured at a height of two meters above ground level (m/s), $e_s$ is the actual vapor pressure (kPa), and $e_a$ is the saturation vapor pressure (kPa). $(e_s - e_a)$ represents the saturation vapor pressure deficit (kPa).

### 2.3.4 Determination of model domain

The model domain is determined by calculating the Topographic Wetness Index (TWI) [15] across the entire study area, in this case, Padang City as shown in Fig. 4. Areas identified by TWI as flood-prone will be selected as the model domain. The goal is to reduce the modeling domain as efficiently as possible so that the model can be produced in a shorter time. TWI for each grid data can be calculated based on elevation data using the following formula:

$$TWI = \frac{ln((\frac{TCA}{FW})}{tan(S))} \quad (2)$$

, where TCA is the total catchment area, FW is the flow width, and S is the slope. The following are the areas of Padang City identified by TWI as flood-prone areas, which were then used as the model domain in this study.

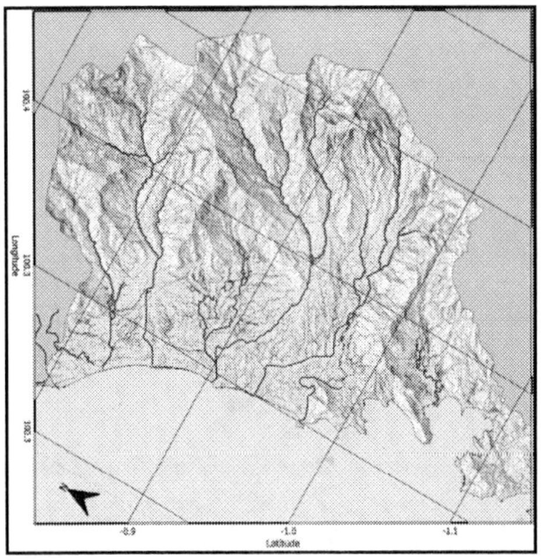


**Fig. 4.** Topographic Wetness Index (TWI) of Padang City.

### 2.3.5 Hydrological modeling

The hydrological model in this study refers to the Rational Method developed by Thomas Mulvaney in 1851 [16]. This method is widely used due to its simplicity and relatively low data requirements. Despite its simplicity, the method is quite accurate for estimating runoff discharge in response to rainfall, especially in small watershed areas [17]. Mathematically, the rational method can be written as follows:

$$Q = C \,.\, I \,.\, A \quad (3)$$

, where Q is the runoff discharge ($m^3/s$) C is the runoff coefficient, i is the rainfall intensity (mm/hour), and A is the area contributing to runoff discharge at a location ($m^2$). In its applicative context, to calculate runoff discharge at a data pixel, we sum up the runoff values from all contributing grids flowing into the target pixel, then multiply by the flow accumulation value at that pixel. Mathematically, this process can be expressed as:

$$Q = \Sigma R_o \,.\, Facc \,.\, \frac{1}{t}\,;\; R_o = C \,.\, (P - ET0) \quad (4)$$

, where Q is the flow discharge at the target grid ($m^3/s$), $\Sigma R_o$ is the sum of all contributing runoff to the target grid (m), $F_{acc}$ is the flow accumulation value generated using Geographic Information System (GIS) software, and t equals 3600 (seconds), as the rainfall intensity used is per hour, hence t equals 1 = 3600/3600.

$R_o$ is the difference between precipitation (P) and evapotranspiration (ET0) in meters, multiplied by the runoff coefficient (C).

### 2.3.6 Hydrodynamic modeling

Hydrodynamic modeling is based on the output of the hydrological model, specifically the runoff discharge at various locations along the river flow. These discharge values are divided by the cross-sectional area at each location to obtain the flow velocity of the river as shown in Fig. 5. The formula for calculating river flow velocity can be expressed as follows:

$$V = \frac{Q}{A} \quad (5)$$

, where V is the river flow velocity (m/s) at a cross section, Q is the river discharge at a cross section ($m^3/s$), and A is the cross-sectional area ($m^2$). Based on Bernoulli's principle, we can calculate the flow velocity at each cross section to determine the water level at the intersection points between the river layer and cross sections layer (river stations) using the following formula:

$$h = H - \left(\frac{V^2}{2g}\right) - z0 \quad (6)$$

, where h is the water height at the river station (m), H is the head energy (joule), V is the flow velocity (m/s), g is the gravitational acceleration ($m/s^2$), and $z_0$ is the elevation of the river bed (m).

Head Energy represents the total energy possessed by the fluid per unit weight, comprising kinetic energy head and potential energy head. Mathematically, head energy can be written as:

$$H = KeH + PeH;\ KeH = \frac{1}{2}V^2;\ PeH = \Delta z \, . \, g \quad (7)$$

, where KeH is the kinetic energy head (joule) and PeH is the potential energy head (joule).

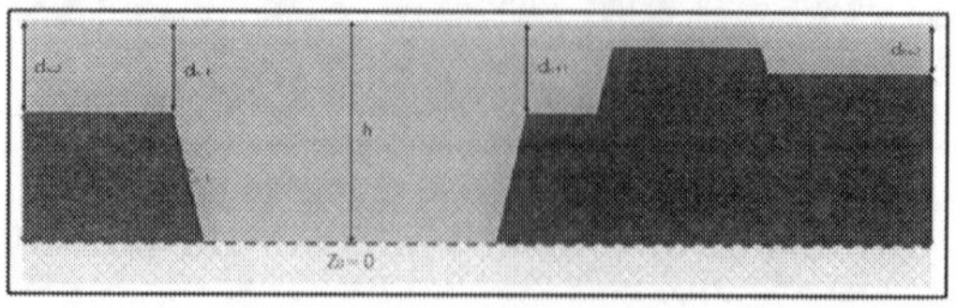

**Fig. 5.** Illustration of River Cross-section.

After obtaining the water heights at the river stations (h), the water heights at the intersection points between the bank line layer and cross sections layer ($d_{n\pm1}$), and at the intersection points between the cross section and the boundary of the model domain ($d_{n\pm2}$) can be calculated using a "bathtub model" approach with the following formula:

$$dn = h - zn \quad (8)$$

Following this process, we will have five water height data points at each cross section. These data are then interpolated across the entire TWI area calculated in the previous stage. Thus, we can generate a flood inundation map across the entire TWI area (model domain).

In the modeling results, sometimes separate inundation areas are produced apart from the main inundation areas (areas connected to the river). These separate areas are considered modeling errors and need to be removed.

### *2.3.7 Model output verification*

In this study, we collected data on flood-affected locations through mass media and social media. We managed to gather 72 pieces of information about flood locations, 57 of which were quite specific, including residential names, street names, neighbourhood units (RT/RW), or certain landmarks, while the rest only included location information up to the sub-district or district level.

When the location information was very limited, we placed the geolocation point in the most likely area to be affected by floods. Of the 57 data points with specific location information, 17 included information on the flood inundation depth at that location.

We used all of this flood location information to qualitatively verify the model by calculating the Probability of Detection (POD). POD is a value that indicates how well a model can accurately predict floods. POD ranges from 0 to 1, with a value of 1 indicating perfect model accuracy. POD can be calculated using the following formula:

$$POD = \frac{Hits}{Hits + Misses} \quad (9)$$

"Hits" represent conditions where both the simulation results and actual conditions indicate the presence of flood inundation. "Misses" represent conditions where the simulation results indicate no flood inundation, while the actual conditions show the presence of flood inundation.

In addition to qualitative verification by calculating POD, we used the 17 pieces of flood location information containing inundation depth values to calculate the Root Mean Square Error (RMSE) and Mean Absolute Error (MAE) of the simulated flood depths (Fig. 6).

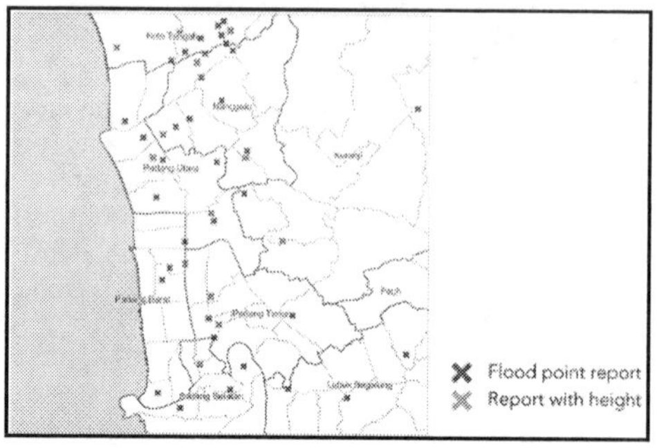


**Fig. 6.** Geolocated flood-affected locations data.

### *2.3.8 Model output calibration*

Calibration was performed by calculating the ratio between observed flood heights and model output flood heights in grids where flood height observations were available. Each grid within a radius of 250-500 meters connected by a network order of the same level was corrected based on the flood height ratio of that grid (Fig. 7).

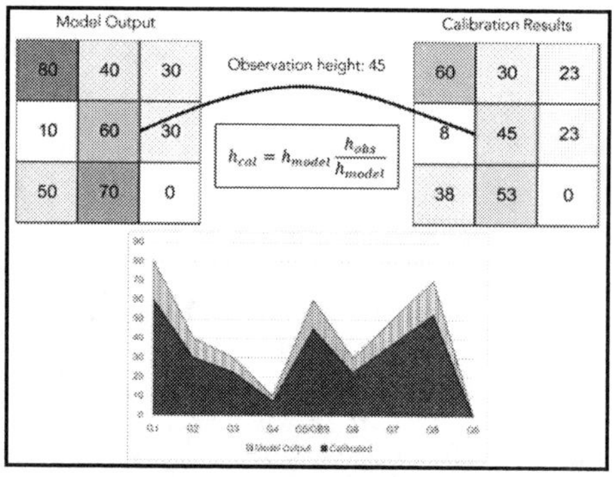


**Fig. 7.** Illustration of Calibration Method.

The stream order level was determined by calculating the stream order (Strahler number) as shown in Fig.8 [18].

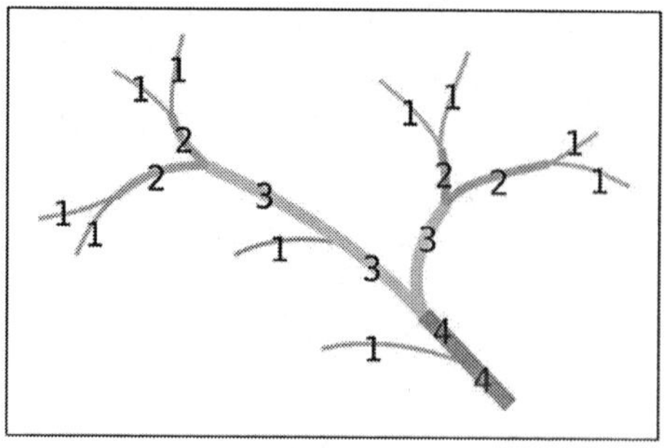

**Fig. 8.** Illustration of Stream Order [18]

The abundance of observational data plays a crucial role in the verification and calibration process. The contribution of the community in reporting flood inundation data, including specific details such as location, time, and flood height through social media or websites, is highly valuable for testing and improving the accuracy of the model. The more extensive and specific the flood inundation data collected and utilized, the greater the potential to produce a model that more accurately represents real-world conditions.

## 3 Results and discussion

Figure 9 shows the flood prediction model results overlaid with field verification data (red dots) and river flow lines (blue lines). The flood prediction model in the image reveals a pattern of inundation that is not concentrated in a single location but rather fragmented and dispersed across various areas. This pattern indicates that flooding in the region is not caused by a single source but by various local factors, such as topography, tributary flow, and drainage capacity in specific areas.

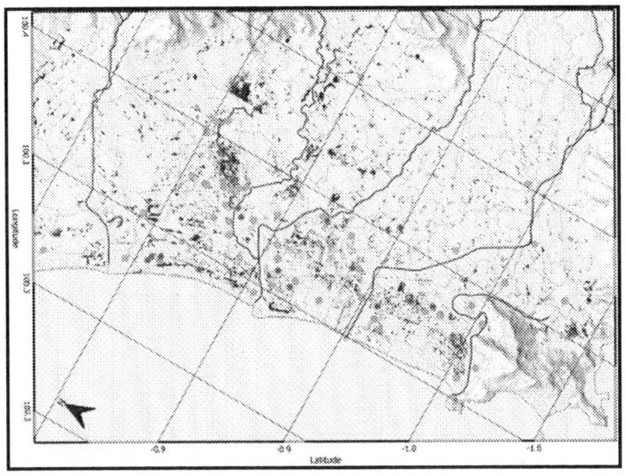

**Fig. 9.** Output model overlayed with verification data and river lines.

Table 1 presents the model verification results using actual data, with three main evaluation parameters. The Probability of Detection (POD) of 0.778 indicates that the model accurately mapped flood inundation in 77.8% of the 72 evaluated samples. The Root Mean Square Error (RMSE) of 38.093 indicates an average prediction error of approximately 38.093 cm across 17 samples, while the Mean Absolute Error (MAE) of 27.584 indicates an average absolute model error of approximately 27.584 cm across 17 samples.

**Table 1.** Model Output Verification Results

| Parameter | Value | Sample Size |
|---|---|---|
| POD | 0.778 | 72 |
| RMSE | 38.093 | 17 |
| MAE | 27.584 | 17 |

Model prediction errors, represented by RMSE and MAE values, are still quite significant. Therefore, verification was conducted to obtain a more representative model. Figure 10 shows a comparison of flood inundation maps before and after the calibration process.

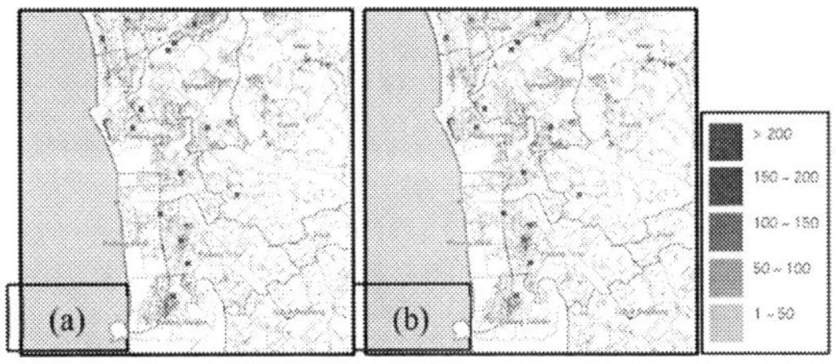


**Fig. 10.** Model Output Results. (a) before calibration; (b) after calibration.

The model results before and after calibration show a relatively similar flood extent. However, before calibration, the model tended to overestimate flood depth. The calibration process, which adjusted the model's flood depth based on observational data, resulted in a more realistic flood depth prediction.

Nonetheless, this calibration was conducted using only 17 observational flood depth samples, which is limited for representing flood events across the entire watershed. Although the model's depth error is reasonably good, there is still potential for more accurate results if supported by a larger and more evenly distributed set of observational data throughout the watershed area.

## 4 Conclusion

This study developed a rapid model for mapping flood-affected areas, using a case study of the flood in Padang on July 14, 2023. The model is designed to address data limitations and enhance the speed of flood-affected area mapping, which is crucial in disaster situations to mitigate losses. The research methodology includes the use of various types of data such as DEM, rainfall, evapotranspiration, soil type, and land use. The model process involves DEM reconditioning, rainfall bias correction, evapotranspiration calculation, determination of the model domain using the Topographic Wetness Index (TWI), as well as hydrological and hydrodynamic modeling. The model verification was conducted using flood location data collected from mass media and social media, showing that the model has an accuracy of 77.8% in mapping flood areas. Although there were significant prediction errors, the model results after calibration showed a significant increase in accuracy.

## References

1. Indonesian National Board for Disaster Management (BNPB). National Disaster Risk Index. (2020).
2. Bubeck, P., Otto, A., and Weichselgartner, J. Societal impacts of flood hazards. Oxford Research Encyclopedia of Natural Hazard Science. (2017).
3. Supari, S., and Sofian, S. Flood early warning system in Indonesia: Status and challenges. Procedia Engineering, **212**, 444-451. (2018).

4. Bates, P. D., and De Roo, A. P. J. A simple raster-based model for flood inundation simulation. Journal of Hydrology, **236**(1-2), 54-77 (2000). https://doi.org/10.1016/S0022-1694(00)00278-X
5. Kalyanapu, A. J., Patil, V., and Ugwonali, O. A. High-performance computing framework for floodplain hydrodynamic modeling. Journal of Hydrologic Engineering, **21**(6), 04016010. (2016).
6. Hamzah, F. (2023, July 14). Heavy rain since yesterday, Padang City flooded. Tempo. Retrieved October 31, 2023, from https://www.tempo.co/read/hujan-deras-sejak-kemarin-kota-padang-terendam-banjir
7. Sastra, Y. (2023, July 14). Heavy rain since yesterday, Padang City hit by floods and landslides. Kompas. Retrieved October 31, 2023, from https://www.kompas.com/hujan-deras-sejak-kemarin-kota-padang-dilanda-banjir-dan-longsor
8. Antara. (2023, July 14). BMKG Jelaskan Penyebab Hujan Ekstrem di Balik Banjir dan Longsor Sumbar. DetikNews. Retrieved July 27, 2023, from https://news.detik.com/berita/d-6822313/bmkg-jelaskan-penyebab-hujan-ekstrem-di-balik-banjir-dan-longsor-sumbar
9. Badan Nasional Penanggulangan Bencana. (2023). Statistik bencana Indonesia. Retrieved July 27, 2023, from https://dibi.bnpb.go.id/home/index2
10. Padang City Government. General Overview of Padang City. Retrieved November 1, 2023, from https://padang.go.id/gambaran-umum-kota-padang
11. Zulvi, Y. Y., and HAR, R. Kajian laju infiltrasi akhir pada Daerah Aliran Sungai (DAS) Batang Arau Kota Padang ditinjau dari perbedaan litologi, tata guna lahan, dan sifat fisik tanah. Jurnal Bina Tambang, **3**(3), 1067. ISSN: 2302-3333. (2024).
12. Ikhvan, A., and Mera, M. Case Study: Significant factors in hazard and vulnerability assessments in flood mitigation in Padang City. IOP Conference Series: Earth and Environmental Science, **708**(1), 012026 (2021). https://doi.org/10.1088/1755-1315/708/1/012026
13. Panofsky, H. W., and Brier, G. W. Some Applications of Statistics to Meteorology. The Pennsylvania State University Press, Philadelphia. (1968).
14. World Meteorological Organization (WMO). (1966). Measurement and Estimation of Evaporation and Evapotranspiration. Technical Note No.83; Report of a Working Group on Evaporation Measurement of the Commission for Instrument and Methods of Observation. WMO: Geneva, Switzerland.
15. Beven, K. J., and Kirkby, M. J. A physically based, variable contributing area model of basin hydrology. Hydrological Sciences Journal, **24**, 43–69 (1979). https://doi.org/10.1080/02626667909491834
16. Mulvaney, T. On the use of self-registering rain and flood gauges in making observations of the relations of rainfall and flood discharges in a given catchment. Proceedings of the institution of Civil Engineers of Ireland, **4**(2), 18-33 (1851).
17. Grimaldi, S. and Petroselli, A. Do we still need the Rational Formula? An alternative empirical procedure for peak discharge estimation in small and ungauged basins. Hydrological Sciences Journal, **60**(1), 67–77 (2014). https://doi.org/10.1080/02626667.2014.880546
18. Strahler, A. N. Hypsometric (height-area) analysis of erosional topology. Bulletin of the Geological Society of America, **63**(11), 1117–1142 (1952). https://doi.org/10.1130/0016-7606(1952)63[1117]2.0.CO;2

# Automatic flood barrier system: an approach for detection and control of storm water overflow during flood

*Jayaprakash* M C[1*], *Sri* Krishna Shastri C[1], and *Rinaldi* Ekaputra[2]

[1]Department of Civil Engineering, Mangalore Institute of Technology and Engineering, Mangaluru, Karnataka, India
[2]Department of Sociology, Faculty of Social and Political Sciences, Universitas Andalas, Padang, Indonesia

**Abstract.** Flooding is one of the major disasters occurring in various parts of the world. Due to the high density of buildings, flash floods are prevalent in cities. In such scenarios, there is a possibility of a drainage overflow. Therefore, implementation of an intelligent analysis of drainage leakage detection during flood using sensors are necessitated for the field of research in disaster management. The growth of Indian economy has largely affected due to flash floods. Unplanned urbanization also has a huge impact on drainage and garbage disposal system. Overflow of drainage water has become the common problem that one living in the cities usually face. To minimize the effect of drainage overflow and to alert the drainage overflow system, the paper proposes an automated mechanism. The proposed system is developed with ultrasonic sensor fitted with microcontroller to detect the overflow in the drainage system and sends an early alert warning message to the concerned authorities to take preventive measures. The implementation of proposed system at large scale will surely help to tackle the flood related issues at a larger extent.

## 1 Introduction

Flash floods are a common problem in metropolitan cities especially during rainy season. A blockage in the drainage pipe will significantly impair the flow of rainwater during the rainy season because of an inadequate drainage system. Cities frequently experience flash floods because of the high building density. There is a chance of a drainage overflow in certain situations. Therefore, the subject of disaster management research requires the adoption of an automated analysis of drainage leakage detection during flood employing sensing elements [1].

In order to keep the city clean, the majority of Indian cities have subterranean drainage systems, which are typically overseen by the municipal corporation. People's health is harmed when sewage water overflows onto the streets and occasionally contaminates drinking water due to inadequate sewage system maintenance. It is common for drainage to become obstructed during the rainy season, and the situation worsens during floods. The most frequent issue that flood-prone localities deal with is sewage water seeping into the floodwater. Hence to overcome this scenario we have introduced model called Drainage Overflow Contro System.

This proposed method will identify the sewage water blend into the flood water and the sensor used will send required data straight away to the nearest managing station using transmitters. This proposed system will detect the sewage water mixture into the flood water and the sensor sends information immediately to the nearest managing station as SMS via transmitter.

## 2 Literature review

K Subramanya Chari et. al. 2020 [1]: Suggested a Raspberry Pi-based Internet of Things flood monitoring and alert system. Every year, the loss of live people and property is increased as a result of the dynamic changes in weather patterns that lead to severe flooding. Therefore, the subject of disaster management research requires the application of an intelligent analysis of flood risk. This article uses a Raspberry Pi model to create an intelligent Internet of Things (IoT)-based flood monitoring and alerting system. Water and rain sensors are used to monitor the water level in a lake or river and notify authorities of heavy rainfall. Because it makes use of an Internet of Things system, this system notifies the residents of neighbouring villages.

Debasis Parinda, 2021 [2]: Suggested an IoT-based sewage monitoring and protection alert. Undoubtedly, sewage management has an impact on our everyday hygiene and is a crucial component of municipal infrastructure. Urban floods, which are most common in densely populated cities, can be caused by poor sewage and drainage management. An IoT and sensor-based smart sewage management device can be used to solve this issue. Here, a smart device detects when the drainage system becomes clogged, and if the water overflows as a result, will notify the local government so they may take the appropriate measures. Moreover, the drainage system's condition will be tracked in real time with the Android app. Similar to a typical water level indicator

* Corresponding author: jayaprakash@mite.ac.in

project we have previously constructed, this one will use the Internet of Things to detect sewage overflow and track it online.

Ryan Hanson, 2020 [3]: Overpopulation, climate transformation, and outdated water facilities—where majority water leaks are caused by deteriorating pipes—have all contributed to the recent increase in the scarcity of water supplies. The length of time it takes to find the leak is the issue, not its size. This document describes the installation of a water leak detection system in a home's hydraulic facilities. An SMS alarm message, an electrical actuator to cut off the main water supply to prevent leaks, a processor to understand the data and determine whether or not there is a water leak, and a water sensor placed beside a water reservoir of attention make up the system.

Giri Srinivaas Ranga Rao, 2017 [4]: In the suggested Drainage Overflow Monitor System using IoT (DOMS) concept, the gadget could measure the water and gas levels in the sewage system, record the computed parameters in the cloud storage, and use the GSM module to send an SMS to the corporate office with the sewage system's condition.

Qinggong Ma, 2017 [5]: IOT technology's security architecture and requirements have been examined by the survey. The need for an IoT-assisted urban waterlogging prevention management system is discussed in the article. This article describes the fundamental model for preventing urban water logging.

U.S. De, G.P. Singh, 2013 [6]: Due to unemployment and other factors, people are moving from rural to urban areas. Cities are dealing with a lot of new issues as a result of the growing population. One of the major issues made worse by the unchecked expansion of megacities is the state of flooding. The article details the number of people living in four Indian megacities and the number of people killed by flooding.

## 3 Methodology

Majorly two sensors were installed to measure as input parameters, i.e. Ultrasonic sensor which helps to detect the blockage, and the Water level sensor which identifies the overflow of water. The data sensed from these transducers will be processed in the ESP8266 processor which acts as both controller and Wi-Fi modem.

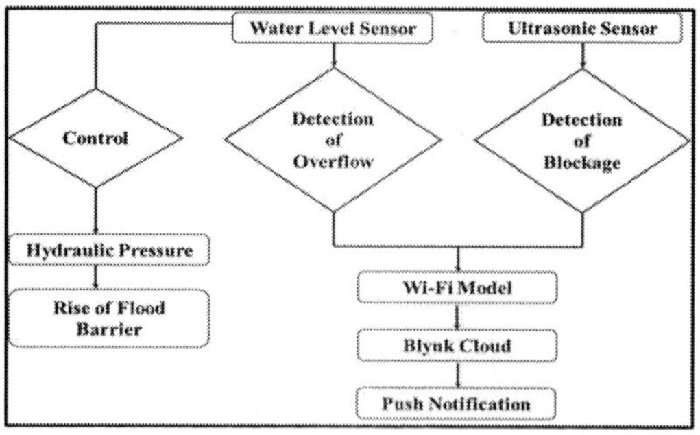


**Fig. 1.** Block diagram

The ESP8266 performs all the arithmetic and logical operations inside and transfers the appropriate data to the Blynk cloud for the necessary operations. Here, the real-time ultrasonic measurement is sent to the Blynk application and also a notification is sent to the user mobile in case of Blockage or Overflow.

Figure 1 shows the sensors to measure water level at the upper edge of the drainage, monitors the overflow of the water and will send the information to the respective agency people through the proper notification. As a consequence of this, a necessary action will be taken by the authority members. Immediate action may involve pushing the barrier above to protect the flow of water to the ground level. This can be achieved using the hydraulic pressure formed inside the compartment constructed as shown in the 2D diagram which indicates the physical components of the projected system.

The proposed system can be successfully achieved by fragmenting the methodology into different stages: Identify potential overflow areas: The first step is to identify the areas in the drainage system where overflow is most likely to occur. This could be areas with high levels of rainfall or areas where the drainage system is most susceptible to blockages.

Install sensors: Once the potential overflow areas have been identified, sensors can be installed to detect when water levels in the drainage system reach a certain threshold. This could be done using water level sensors or other types of sensors that can detect when water is backing up in the system. We have used the ultrasonic sensors to detect the blockages.

Activate barriers: When the sensors detect that water levels are rising to a critical point, barriers can be activated to prevent overflow. Barriers can take many different forms, such as gates, flaps, or inflatable bladders, and can be designed to fit the specific needs of the drainage system.

Monitor system performance: Once the barriers have been activated, it is important to monitor the system's performance to ensure that it is working as intended. This could involve monitoring the water levels in the drainage system to make sure that they are not rising too high or too quickly, or monitoring the performance of the barriers themselves to make sure that they are holding up under the pressure of the water.

Alert system activation: If any issues are identified, then along with the immediate measure of upliftment of the barriers, an alert system activation also should go hand in hand. The authorities need to be immediately alerted on this issue so that immediate action is taken before the water level reaches the barrier height.

Control unit: This is the central processing unit that receives the information from the sensors and controls the operation of the system. This should be compatible and communicable with both the sensors and the communication network. The control unit used here is ESP8266 which acts both as Microcontroller and Wi-Fi module. This prefers all the logical and mathematical operations inside.

User interface: This should be provided in order to allow the module operators to monitor the status of the module and receive alerts in case of overflow. This could be the web-based interface or a mobile application. The following project includes the Expo Go Application which is readily available to download from the google play store.

### 3.1 Communication system

In order to effectively manage drainage systems, it is vital to have a communication system in place that can detect and notify the relevant authorities about the overflow conditions. This section hence outlines the communication system that has been proposed for the project.

The communication system between an ESP8266 device and the Expo Go app plays a crucial role in enabling data exchange and interaction. Here's how the communication system works in the context of ESP8266 and Expo Go.

### 3.2 ESP8266 device

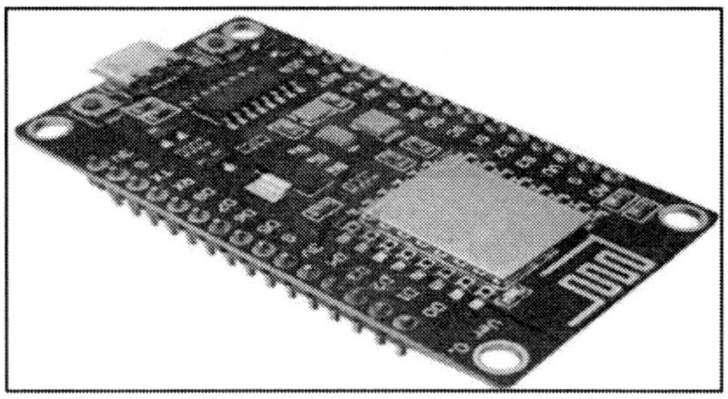

**Fig. 2.** ESP8266

Figure 2 shows the ESP8266 which is a famous Wi-Fi module commonly used for Internet of Things (IoT) applications. It can connect to Wi-Fi networks and exchange information with other devices over the network. The ESP8266 can be programmed with custom firmware or using platforms like Arduino to perform various tasks, such as sensor data collection, data processing, and network communication.

Sensor integration: The HC-SR04 ultrasonic sensor may be used to monitor the water level in the drainage system. The sensor can be connected to the ESP8266 using the GPIO pins, and the measured distance can be read by the ESP8266.

Control algorithm: The ESP8266 can be programmed to control the drainage system based on the water level measurements. For example, if the water level reaches a critical point, the ESP8266 can trigger an alarm or activate a pump to prevent overflow.

Communication: The ESP8266 can be connected to a Wi-Fi network to send data to a cloud service or a mobile app. This can enable remote monitoring and control of the drainage system.

Power management: The ESP8266 can be powered using a battery or a solar panel to enable long-term, autonomous operation.

Overall, the ESP8266 microcontroller can provide a cost-effective and flexible solution for drainage overflow detection and control. With the integration of sensors, actuators, and communication capabilities, the ESP8266 can enable efficient and reliable operation of the drainage system

Expo Go application: Expo Go is a mobile app used for developing and running cross-platform applications built with the Expo development framework. The app allows users to access and interact with Expo Go powered projects on their mobile devices. Expo Go provides a runtime environment for running JavaScript code and rendering user interfaces.

### 3.3 Communication between ESP8266 and Expo Go

The ESP8266 device and the Expo Go app can communicate through a combination of Wi-Fi and network protocols. The ESP8266 can send data to the Expo Go app by acting as a server or by making HTTP requests to an API endpoint hosted by the Expo server. The Expo Go app can receive data from the ESP8266 by making HTTP requests to the IP address and port of the ESP8266 server or by subscribing to a WebSocket connection provided by the ESP8266.

### 3.4 Data exchange

The ESP8266 will collect sensor data or perform specific actions based on its programming. It can send the collected data to the Expo Go app by packaging it in HTTP requests or WebSocket messages.

### 3.5 Interaction and control

The Expo Go app can also send commands or instructions to the ESP8266 device. It can make HTTP requests or send WebSocket messages to control the behavior of the ESP8266 or request specific actions. The ESP8266 receives the commands from the Expo Go app and performs the corresponding actions based on its programming, such as controlling actuators, updating its internal state, or responding with data.

Overall, the communication system between the ESP8266 device and the Expo Go app allows for bidirectional data exchange and interaction. It enables the Expo Go app to receive real-time sensor data from the ESP8266 and control the behavior of the ESP8266 device based on user interactions within the app. This communication system forms the foundation for building IoT applications that integrate with the Expo Go platform.

## 4 Results and discussion

The aim of our work is to build up an IoT based system which is competent enough to predict the drainage situation and alert the officers within no time when the drainage water is found overflow or during flood conditions. Developing the short time hydraulic pressure barrier system would be immediate control system, so that corrective measures for the nearby people's property damage and inundation of the area is taken care of [3].

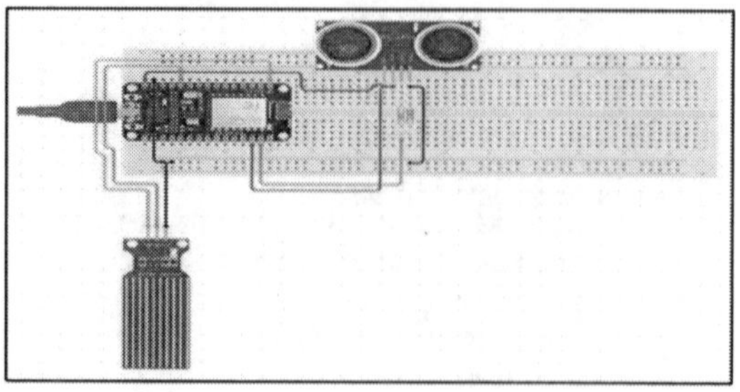

**Fig. 3.** Software circuit simulation

Figure 3 shows the water level sensors mounted at the top edge of the system, monitors the overflow of the water and sends proper information to the respective members of the authority through the notification. As a consequence of this, a corrective measure can be expected until the authority member arrives for the recovery. The sudden measures involve upliftment of the situated barriers to protect the flow of water to the ground level. This could be performed using the hydraulic pressure created inside the compartment formed in the 2D diagram which indicates the physical components of the expected system as shown in Fig. 4. Additionally, we have also mounted the water level sensors into the tip of the barrier edge to sense if the water exceeds that height too.

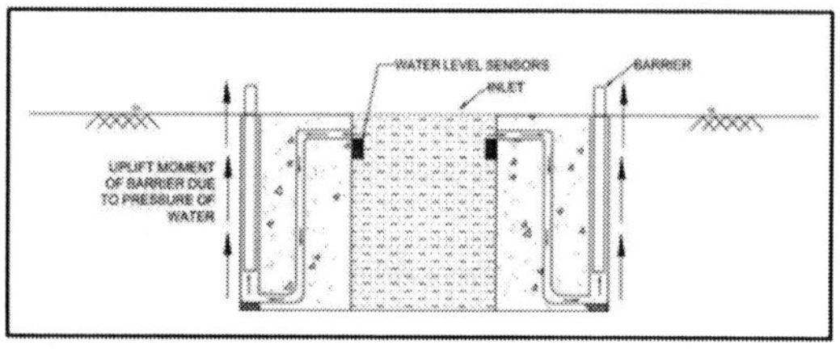


**Fig. 4.** 2D diagram

**Fig. 5.** 3D diagram

Figure 5 shows that 3D drawings serve as a powerful communication tool by providing a common language for project participants. They can convey complex ideas, concepts, and technical details in an easily understandable format. This is particularly beneficial when engaging with non-technical stakeholders who may have limited knowledge of technical drawings or specifications. 3D drawings facilitate effective communication by presenting a realistic and tangible representation of the project, reducing ambiguity, and ensuring a shared understanding among all parties involved.

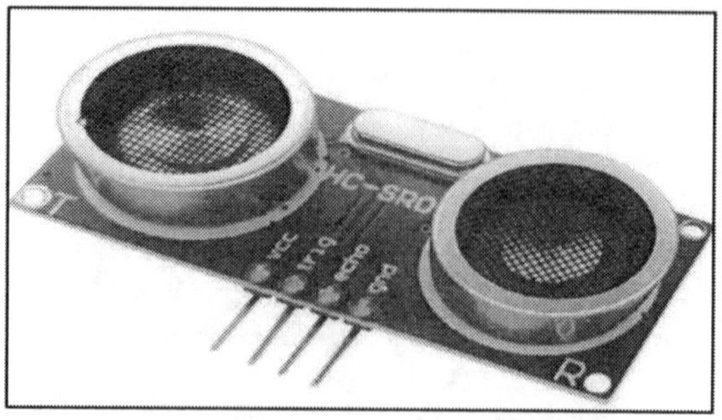

**Fig. 6.** Ultrasonic sensor

Figure 6 shows the HC-SR04 which is an ultrasonic sensor that can be used for drainage applications to measure the water level in a tank or drainage system. The sensor emits high frequency sound waves that bounce off the surface of the water and return to the sensor. The time it takes for the sound waves to travel to the surface and back is used to determine the water level. The sensor requires 5V DC power supply and can operate at a frequency of 40 kHz. The sensor can measure water level from range of 2 cm to 400 cm.

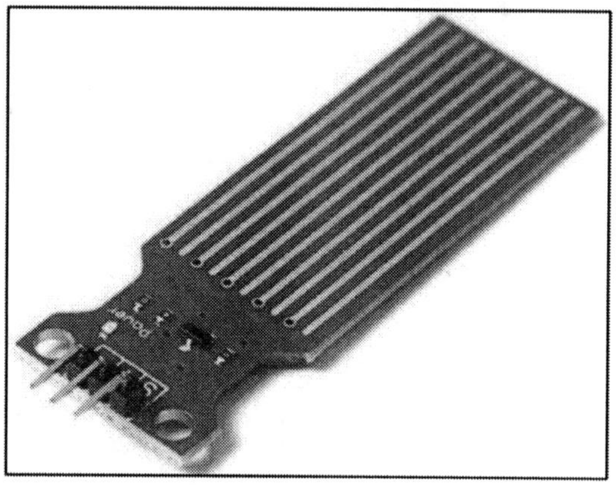

**Fig. 7.** Water level sensor

Figure 7 shows the water level sensor which is device that used to measure the level or depth of water in a container, tank, or body of water. These are commonly used in applications such as water tanks, wells, sewage systems, and irrigation systems. They can help to prevent overflow or underflow, monitor water usage, and optimize water management. The sensor we used works with DC voltage of around 3-5 V and working current is less than 20 mA with detection area of around 40 mm*16 mm.

## 4.1 Circuit simulation

Before construct any hardware circuit connections, circuit simulation is a crucial and essential step to verify the components' computability and operation.

Accessibility: Online circuit simulation tools are easily accessible from anywhere with an internet connection, which makes it easier for team members to collaborate on the project remotely. This can save time and increase efficiency during the design and testing phases.

Real-time results: Online circuit simulation tools provide real-time results, which can help designers quickly identify and troubleshoot issues in the circuit design. This can save time and reduce the risk of errors during the construction phase.

Scalability: Online circuit simulation tools can be easily scaled to accommodate larger and more complex circuit designs. This allows designers to test and optimize the performance of the circuit before construction, which can help reduce the risk of costly mistakes during the construction phase.

Flexibility: Online circuit simulation tools can be used to test a variety of circuit designs and configurations, which can help designers optimize the performance of the circuit and meet the required specifications for the project.

Remote collaboration: In today's globalized economy, many projects involve team members working remotely from different parts of the world. Online circuit simulation tools can provide a centralized platform for collaboration, allowing team members to share and test circuit designs in real-time, no matter where they are located.

Limited resources: If a project has limited resources or a small budget, purchasing and maintaining specialized simulation software might not be feasible. Online circuit simulation tools can provide a more cost-effective solution, allowing designers and engineers to test and optimize their circuit designs without the need for expensive software.

The proposed system identifies the overflow or leakage over the drainage water to the flood water and informs the neighbouring managing authorities. This can be further extended by informing the locality people. This system is very effective for flood affected areas. The proposed multifunctional model will help in improving the community preparedness for extreme weather conditions such as flash floods. The system also minimizes safety and infrastructure hazards. As part of the warning, the prototype will also provide a prediction of the scale, timing, location and possible damages of the impending flood.

## 5 Conclusion

The developed proposed system would accurately identify drainage overflow and inform the concerned authorities. The adoption of ultrasonic sensors and an Arduino based controller makes the system efficient and cost-effective. The model can be further extended by adding some other additional sensors to cover a wide land cover. Also, the parameters measured could be stored and utilized as a data for further analysis. The system can also be utilised in other applications such as water level monitoring in huge water supply unit.

The multifunctional system would enhance the community preparedness for extreme weather events such as floods, in terms of both alarm and added understanding of risks, appropriate flood responses and hydraulic pressure real time barrier. The current increase in the number and degree of extreme weather events such as floods would make the technology relevant for climate change detection also.

## References

1. K. Subramanya Chari et. al., IoT-based flood monitoring and alerting system using raspberry pi, IOP Conf. Ser.: Mater. Sci. Eng. **981**, 042078. (2020). https://circuitdigest.com/microcontroller-projects/sewage-monitoring-and-aintenancealertusing-nodemcu-esp8266-and-ultrasonic-sensor].
2. Debasis Parida, Sewage monitoring and maintenance alert using IoT, Microcontroller Based Projects, Published (2021).
3. Ryan T. Hanson, Water leak detection system (2020).https://digitalcommons.calpoly.edu/eesp/389.
4. R Girisrinivaas and V Parthipan, Drainage overflow monitoring system using IoT (DOMS), IEEE International Conference on Power, Control, Signals and Instrumentation Engineering (ICPCSI-2017), Chennai, India (2017).
5. Qinggong Ma, Bo Yang, and Jipeng Wang, Application of Internet of Things in Urban Flooding Prevention Management System, Scientific Research Publishing, Advances in Internet of Things, ISSN Print: 2161-6817, (2017).
6. U.S. De, G. P. Singh and D. M. Rase, Urban flooding in recent decades in four mega cities of India, J. Ind. Geophys. Union, **17**, 153-165 (2013)

 https://doi.org/10.1051/e3sconf/202560404005

# Identification of landslide vulnerable zones in West Khasi District of Meghalaya

*Mahalingam* Bose[1*], *Indraddin*[1], *Trisha* De[1], and *K.* Channabasappa[1]

[1]Department of Geography and Geology School of Earth Sciences, Central University of Karnataka, Karnataka, India

**Abstract.** A landslide is a mass of rock, debris, or earth moving down a slope. Landslides are a form of mass waste under the direct influence of gravity, indicating any down-slope movement of rock and soil. Landslide encompasses five forms of slope movement: topples, falls, spreads, slides, and flows. Landslides are the large-scale movement of rock, debris, or earth down a slope. The present study was conducted to identify the landslide-vulnerable zones in the West Khasi District of Meghalaya. In total, nine parameters were chosen to assess the vulnerability, such as geology, geomorphology, lithology, elevation, slope, lineament, rainfall, land use and land cover, and soil. These parameters were classified into five groups and ranked between 1 and 5, in which the value 1 indicates highly vulnerable, and the value 5 indicates less vulnerable. Further, the overall weightage for each parameter was computed using AHP. Then, a weighted overlay analysis was carried out in ArcGIS Pro. The result was classified into five classes: very high to very low vulnerable areas. The result showed that a larger area of the district is covered by moderate vulnerability.

## 1 Introduction

Landslides are common and potentially catastrophic geo-hazards, complicated geological phenomena that occur in many geospatial environments and geomaterials [1]. Landslides can cause serious casualties and severe property damage [2, 3]. It is necessary to assess and manage areas skeptically that are vulnerable to landslides to mitigate damages [4] and to decrease the damage [5]. Proper preparation helps identify and predict landslide-susceptible areas and decrease landslide damage [6].

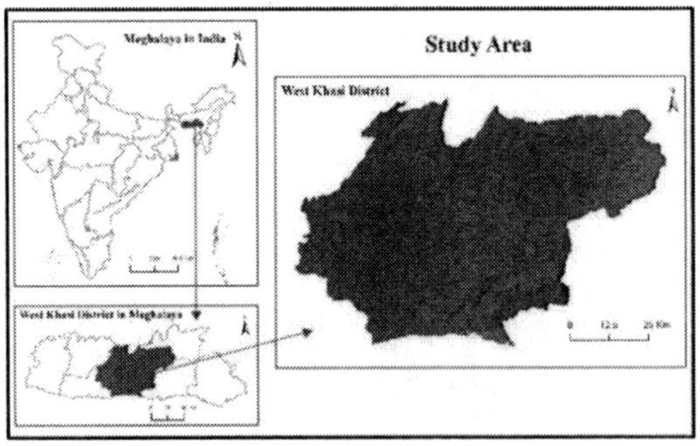


**Fig. 1.** Study area

West Khashi District is situated in the central part of Meghalaya. It is located between approximately 25 degrees 10′ and 25 degrees 51′ N latitude and between 90 degrees 44′ and 91 degrees 49′ E longitude. On the northwest, it shares a border with the Assamese district of Kamrup. On the north-east, with the district of Ri Bhoi; on the east, with the district of East Khasi Hills; on the south, with Bangladesh and the district of South West Khasi Hills, formerly known as Mawkyrwat Civil Subdivision; and on the west, with the districts of East Garo and South Garo Hills, as shown in Fig. 1.

## 2 Methodology

Several works have been done worldwide for mapping landslide hazard zonation and susceptibility [7], and different techniques or tools have been adopted for determining landslide susceptibility [8, 9]. Different parameters or criteria influence the landslide vulnerability; identifying these parameters is the primary step in the analysis [10]. This study considered nine parameters to assess the vulnerability of landslides, which are i) Geology, ii) Geomorphology, iii) Lithology, iv) Elevation, v) Slope, vi) Lineament, vii) Rainfall, viii) Land Use and Land Cover, and ix) Soil.

The reason for considering these parameters is due to their major influence on the landslides. The chosen parameters were mapped in ArcGIS Pro and classified into different classes further; each class is ranked between 1 and according to the level of vulnerability, such as 1 for very highly vulnerable, 2 for highly vulnerable, 3 for moderate vulnerable, 4 for low vulnerable, and 5 for very low vulnerable.

These parameters were chosen based on their significant impact on the landslides. For example, geology, lithology, and geomorphology impact soil stability. Parameters like slope and elevation influence gravitational triggering landslides. Higher amount of rainfall and denser lineaments increases the possibility of

* Corresponding author: mahabose@gmail.com

landslides. Similarly, changes in land use and land cover by humans or nature will alter the natural drainage system and increase susceptibility.

After the classification, the AHP (Analytical Hierarchical Process), one of the popular methods for calculating weightage [11], was used to compute weightage for each parameter. Further, all the parameters were used in weighted overlay analysis to obtain the result.

## 3 Discussion

### 3.1 Geology

Geology data is downloaded from https://bhukosh.gsi.gov.in/Bhukosh/Public portal, which is a gateway to all geoscientific data of the Geological Survey of India (Fig. 2). The geological structure of the study area is classified into five classes and ranked between 1 and 5 based on the landslide vulnerability (Table. 1), then ranked raster data was prepared (Fig.3).

**Table 1.** Rank assigned for geology category

| Category | Rank |
|---|---|
| Garo Gp. (Baghmara/Chengapara Fm)., Miocene - Pliocene | 1 |
| Garo Gp. (Simsang Fm.), Eocene - Oligocene | 2 |
| Jaintia Gp., Palaeocene - Eocene | 2 |
| Khasi Gp. (Mahadek Fm.), Cretaceous | 3 |
| Sylhet Traps, Jurassic- Cretaceous | 3 |
| Kyrdem, Nongpoh, Mylliem Granite., S.Khasi Batholiths & Equivalent Granites,Neoproterozoic - Earl | 3 |
| Umsning Schist Belt Gr., Proterozoic | 3 |
| Supracrustal, Proterozoic | 3 |
| Assam - Meghalaya Gneissic Complex Sgp., Proterozoic | 4 |
| Shillong Gp., Palaeoproterozoic - Mesoproterozoic | 5 |

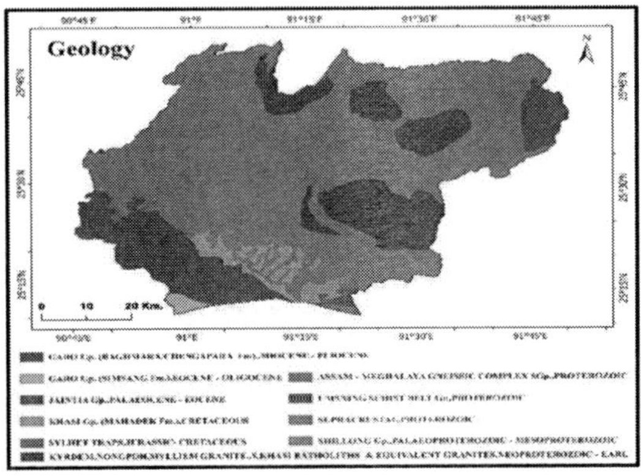


**Fig. 2.** Geology map of West Khashi district of Meghalaya

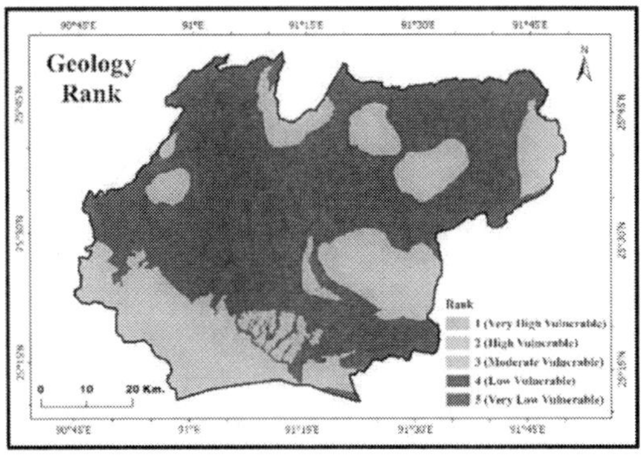


**Fig. 3.** Classified geology map of West Khashi district of Meghalaya

### 3.2 Geomorphology

Geomorphology data is downloaded from https://bhukosh.gsi.gov.in/Bhukosh/Public portal. The downloaded data was mapped in ArcGIS Pro. It was found that 24 different geomorphological categories were present in the study area (Fig. 4). These 24 types were classified into five categories and ranked between 1 and 5 based on their influence in the landslide (Table 2), then ranked raster data was prepared (Fig. 5).

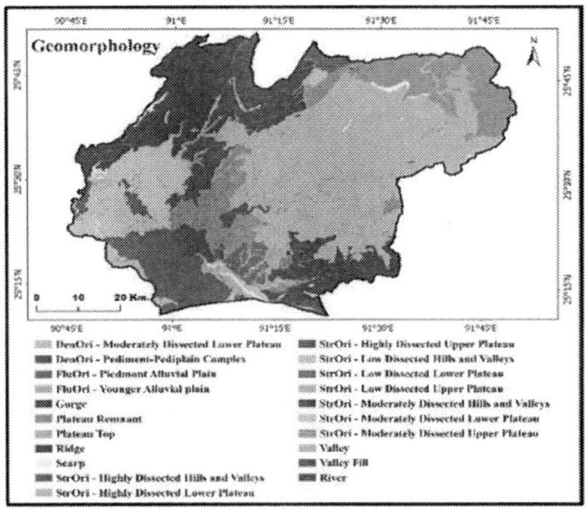


**Fig. 4.** Geomorphology map of West Khashi district of Meghalaya

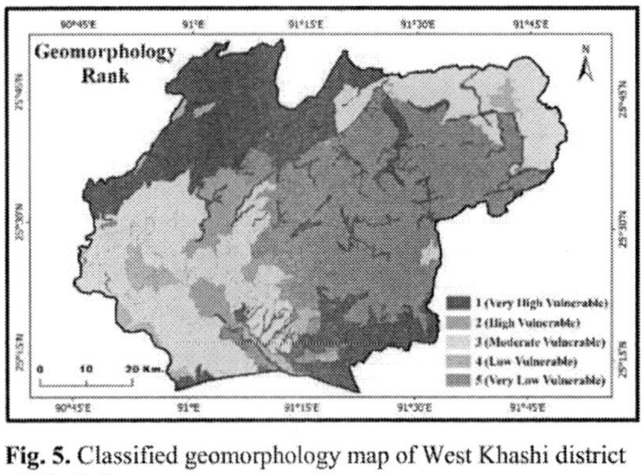


**Fig. 5.** Classified geomorphology map of West Khashi district of Meghalaya

https://doi.org/10.1051/e3sconf/202560404005

**Table 2.** Rank assigned for geomorphology category

| Category | Rank |
|---|---|
| DenOri - Moderately Dissected Lower Plateau | 3 |
| DenOri - Pediment-Pediplain Complex | 4 |
| FluOri - Piedmont Alluvial Plain | 3 |
| FluOri - Younger Alluvial plain | 3 |
| Gorge | 1 |
| Plateau Remnant | 5 |
| Plateau Top | 5 |
| Ridge | 4 |
| River | 5 |
| Scarp | 1 |
| StrOri - Highly Dissected Hills and Valleys | 1 |
| StrOri - Highly Dissected Lower Plateau | 4 |
| StrOri - Highly Dissected Upper Plateau | 2 |
| StrOri - Low Dissected Hills and Valleys | 4 |
| StrOri - Low Dissected Lower Plateau | 3 |
| StrOri - Low Dissected Upper Plateau | 4 |
| StrOri - Moderately Dissected Hills and Valleys | 1 |
| StrOri - Moderately Dissected Lower Plateau | 3 |
| StrOri - Moderately Dissected Upper Plateau | 3 |
| Valley | 1 |
| Valley Fill | 1 |

## 3.3 Lithology

The lithology data was also downloaded from https://bhukosh.gsi.gov.in/Bhukosh/Public portal. The downloaded data was mapped, and thirteen types of lithology were found (Fig 6). These thirteen classes were ranked between 1 and 5 based on the influence of landslide (Table 3 and Fig. 7).

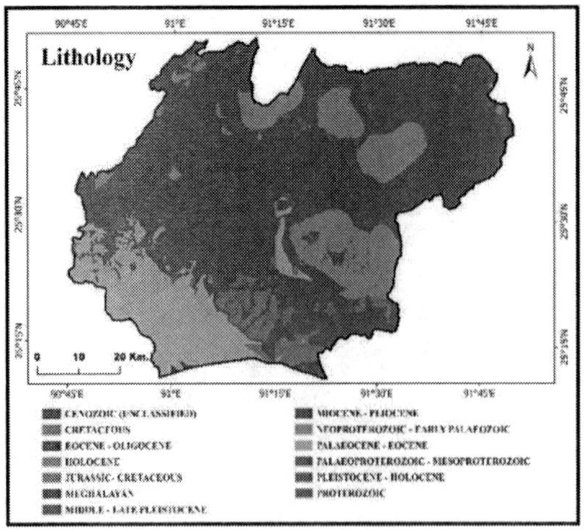


**Fig. 6.** Lithology map of West Khashi district of Meghalaya

**Table 3.** Rank assigned for lithology category

| Category | Rank |
|---|---|
| Cenozoic (Unclassified) | 2 |
| Cretaceous | 4 |
| Eocene - Oligocene | 2 |
| Holocene | 1 |
| Jurassic- Cretaceous | 3 |
| Meghalayan | 1 |
| Middle - Late Pleistocene | 1 |
| Miocene - Pliocene | 4 |
| Neoproterozoic - Early Palaeozoic | 3 |
| Palaeocene - Eocene | 2 |
| Palaeoproterozoic - Mesoproterozoic | 5 |
| Pleistocene - Holocene | 5 |
| Proterozoic | 3 |

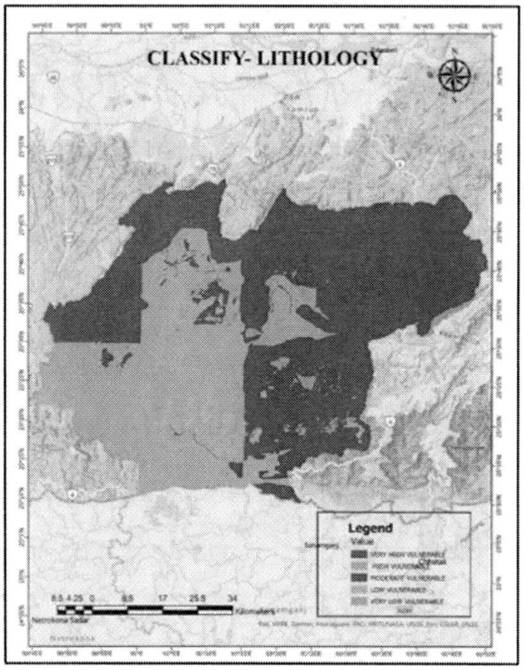


**Fig. 7.** Classified lithology map of West Khashi district of Meghalaya

## 3.4 DEM (3D)

The Digital Elevation Model (DEM) of West Khasi District in Meghalaya digitally represents its terrain, providing elevation data in a grid format. It depicts the varying heights and contours of the landscape, aiding in topographic analysis, land planning, and assessing geographical features within the district.

The extracted DEM map is reclassified into five classes. The elevation of the study area rank is given based on the importance from 1 to 5. After giving the rank-based classification, the extracted map is reclassified as a DEM map (Figs. 8 and 9).

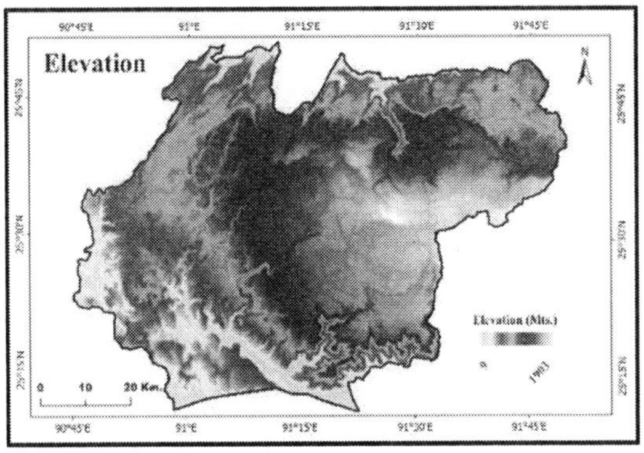


**Fig. 8.** DEM map of West Khashi district of Meghalaya

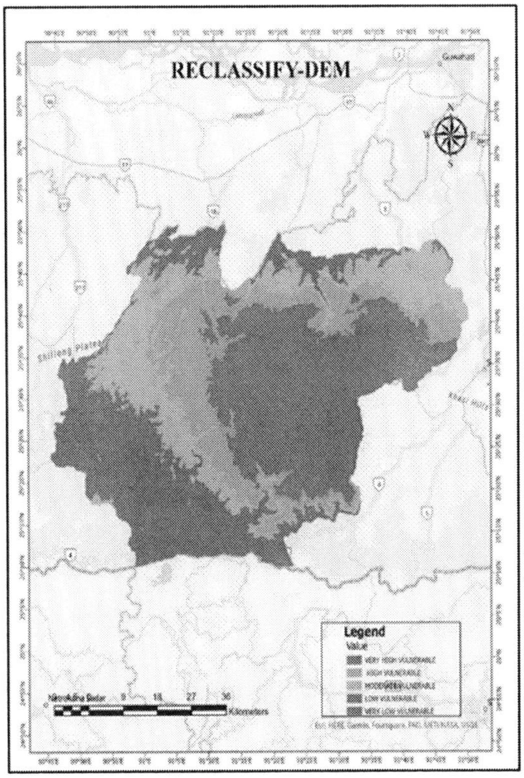


**Fig. 9.** Re-classified DEM map of West Khashi district of Meghalaya

## 3.5 Slope

The West Khasi District in Meghalaya features diverse terrain with varying slopes. Its landscape includes steep hills, plateaus, and valleys. Slopes in this region significantly shape the local geography and influence factors such as agriculture, land use, and the potential for landslides.

The extracted Slope map is reclassified into five classes. For the slope of the study area, rank is given based on the importance from 1 to 5. After giving the rank-based classification, the extracted map is reclassified as a Slope map (Figs. 10 and 11).

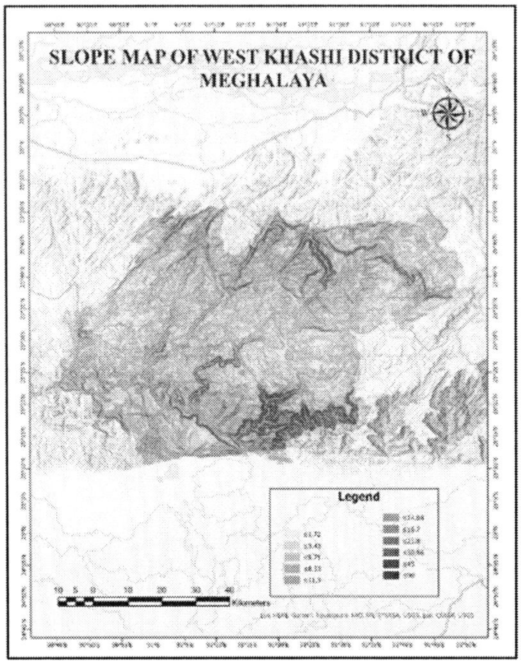


**Fig. 10.** Slope map of West Khashi district of Meghalaya

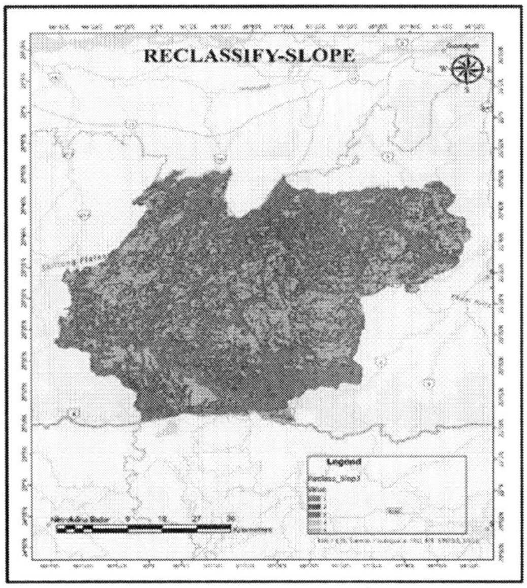


**Fig. 11.** Re-classified slope map of West Khashi district of Meghalaya

## 3.6 Lineament

The West Khasi district of Meghalaya exhibits prominent lineaments, linear geological features, or fault lines, often associated with the region's rugged topography. These lineaments are crucial in controlling drainage patterns, groundwater flow, and potential geological hazards like landslides, making them significant for geological and environmental assessments. Here is the Lineament Map of the study area.

The extracted Lineament Map is reclassified into five classes, and the lineament of the study area rank is given based on importance from 1 to 5. After giving the rank-based classification, the extracted map is reclassified as a Lineament map (Table 4, Figs. 12 and 13).

**Table 4.** Classification of lineament

| Lineament | Rank |
|---|---|
| 0-1000 | 5 |
| 1000-2000 | 4 |
| 2000-4000 | 3 |
| 4000-6000 | 2 |
| 6000-13207.667963 | 1 |

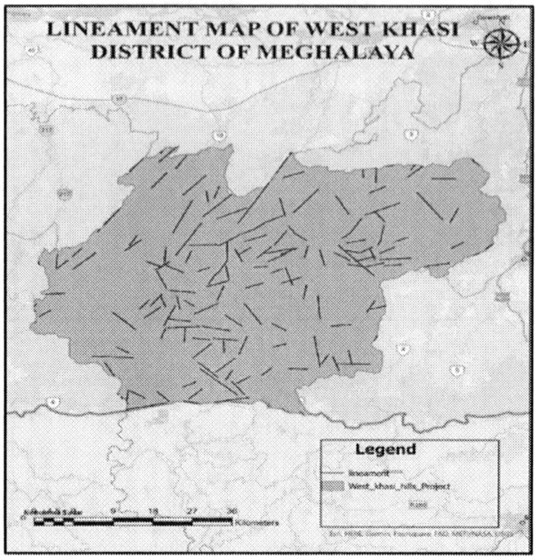


**Fig. 12.** Lineament map of West Khashi district of Meghalaya

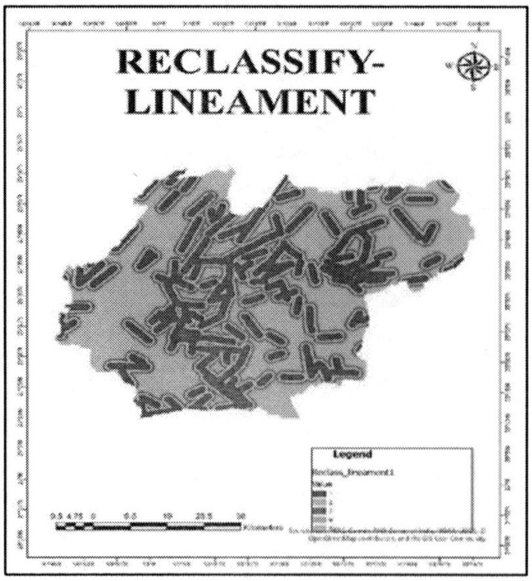


**Fig. 13.** Re-classified lineament Map of West Khashi district of Meghalaya

## 3.7 Rainfall

West Khasi District in Meghalaya experiences heavy rainfall, with an annual average of 1,500 mm to 2,500 mm. This high rainfall is primarily due to its location in the northeastern region of India, characterized by a subtropical climate and the influence of the southwest monsoon, leading to lush vegetation and potential landslide risks.

The extracted Rainfall Map is reclassified into five classes; for the rainfall of the study area, the rank is given based on the importance from 1 to 5. After giving the rank-based classes, a reclassified Rainfall map is generated (Table 5 and Fig. 14).

**Table 5.** Classification of rainfall data

| Rainfall in mm | Rank |
|---|---|
| 1069.71-1100 | 5 |
| 1100-1200 | 4 |
| 1200-1300 | 3 |
| 1300-1400 | 2 |
| 1400-2148.57 | 1 |

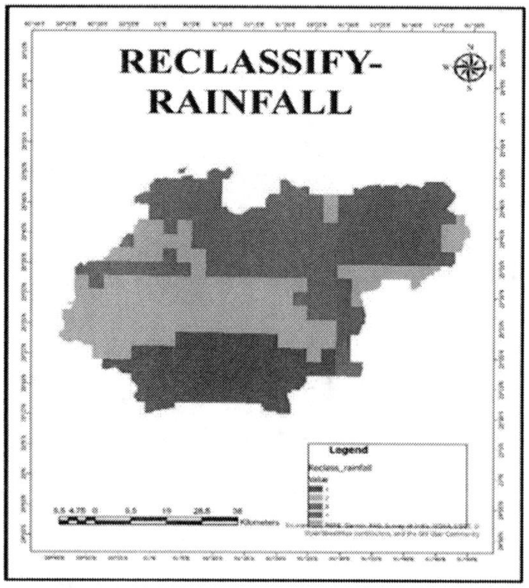


**Fig. 14.** Re-classified rainfall Map of West Khashi district of Meghalaya

## 3.8 LULC

In the West Khasi District of Meghalaya, Land Use and Land Cover (LULC) typically consists of a mix of dense forests, agricultural lands, grasslands, and scattered settlements. The region's topography, characterized by steep hills and valleys, influences land use patterns, with forests dominating the landscape and agriculture practiced in terraced fields on slopes.

This LULC Map is reclassified into five classes; for the LULC of the study area, rank is given based on the importance from 1 to 5. After giving the rank-based classes, a reclassified LULC map is generated, as shown in Fig. 15.

**Fig. 15.** Re-classified LULC map of West Khashi district of Meghalaya

### 3.9 Soil

In the West Khasi District of Meghalaya, soil composition primarily consists of red and lateritic soils, often rich in iron and aluminium oxides. These well-drained soils are suitable for agriculture, supporting crops like rice and ginger. However, the hilly terrain can lead to erosion and landslides, requiring proper soil management practices.

This Soil Map is reclassified into five classes. For the Soil of the study area, rank is given based on the importance from 1 to 5. After giving the rank-based classification, the extracted map is reclassified as a Soil map, as shown in Figs. 16 and 17.

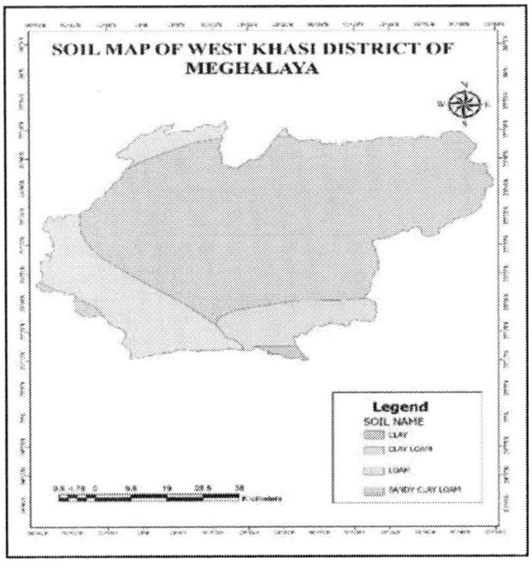


**Fig. 16.** Soil map of West Khashi district of Meghalaya

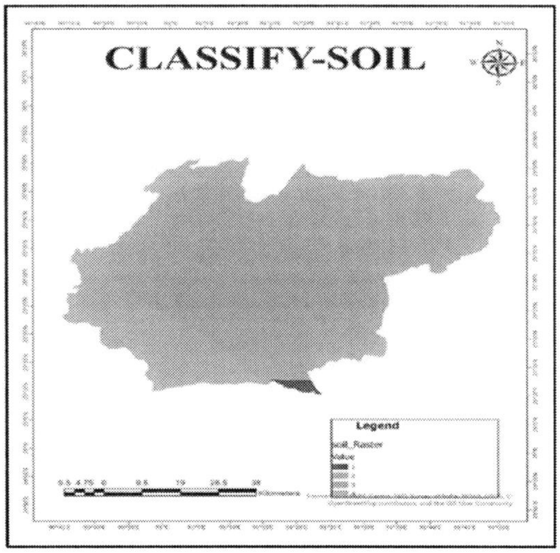


**Fig. 17.** Classified soil map of West Khashi district of Meghalaya

## 4 Overlay method

In order to find patterns, links, or areas of interest, the "overlay method" is a geographic information system (GIS) technique that combines and analyses many layers of spatial data, generally in the form of maps or datasets. It entails superimposing thematic layers (such as land use, elevation, vegetation, and population) to create a composite map illustrating where these levels cross or combine. The overlay approach makes it possible to identify spatial linkages, such as the locations where two or more elements overlap or coincide, which aids in making well-informed decisions in areas like resource management, environmental assessment, and urban planning.

For this exercise the resulted parameters are processed by Weighted Overlay Method, the weightage is given based on the AHP (Analytical Hierarchical Process) calculation (Tables 6 and 7).

**Table 6.** Criteria weight calculation

| Criteria | Criteria Weights |
|---|---|
| DEM | 0.17 |
| Slope | 0.13 |
| Lineament | 0.24 |
| Soil | 0.08 |
| Lithology | 0.04 |
| Geology | 0.03 |
| Geomorphology | 0.06 |
| Rainfall | 0.08 |

**Table 7.** Criteria weight calculation

| Criteria | Criteria Weights | Ratio |
|---|---|---|
| DEM | 0.17 | 10.64706 |
| Slope | 0.13 | 10.73586 |
| Lineament | 0.24 | 8.958333 |
| Soil | 0.08 | 10.6201 |
| Lithology | 0.04 | 10.32745 |
| Geology | 0.03 | 10.20548 |
| Geomorphology | 0.06 | 9.35274 |
| Rain | 0.08 | 10.375 |
| Lulc | 0.13 | 11.15385 |
| Average Value | | 9.2375873 |
| Consistency Index | | 0.0408 |

## 5 Result

The result of the weighted overlay analysis revealed values from 1 to 5, in which value one represented highly vulnerable areas and value five represented less vulnerable areas. Accordingly, other values are classified as shown in Fig. 18. It is identified that a small portion of the study area falls under very highly vulnerable, and highly vulnerable areas are located in the central and southern parts of the district surrounded by moderate vulnerability. It is also clear that the areas with steep slopes with a high density of lineaments result in highly vulnerable areas combining other factor's influence. Low and very low vulnerable areas are located in the west and east parts of the district. Overall, the larger area of the district is covered by moderate, followed by low and highly vulnerable, as shown in Fig. 18.

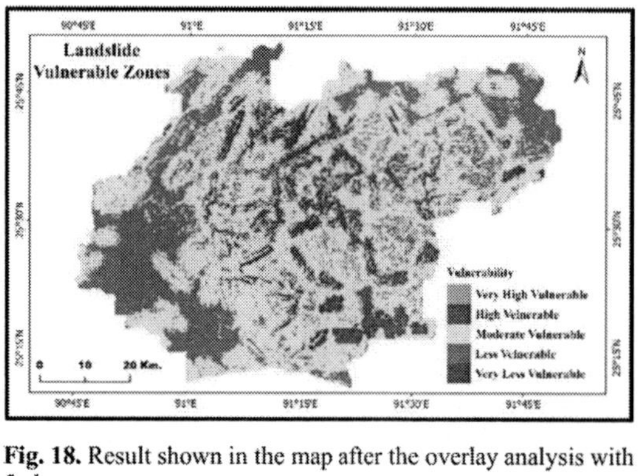


**Fig. 18.** Result shown in the map after the overlay analysis with 5 classes

## 6 Conclusion

This study of identifying landslide-prone zones highlights the importance of conducting landslide analysis and familiarizes us with practical techniques such as AHP (Analytical Hierarchy Process) and Weighted Overlay. Such analyses can be extremely useful for government agencies and the public to assess the likelihood of landslides and develop effective mitigation strategies.

## References

1. Azarafza, M., Azarafza, M., Akgün, H., Atkinson, P. M., and Derakhshani, R. Deep learning-based landslide susceptibility mapping. Scientific reports, **11**(1), 24112 (2021).
2. Huang, Y., and Zhao, L. Review on landslide susceptibility mapping using support vector machines. Catena, **165**, 520-529 (2018).
3. Flentje, P., and Chowdhury, R. Resilience and sustainability in the management of landslides. In Proceedings of the institution of civil engineers-engineering sustainability, **171**, 1, 3-14, Thomas Telford Ltd. (2016, December).
4. Pourghasemi, H. R., Pradhan, B., and Gokceoglu, C. Application of fuzzy logic and analytical hierarchy process (AHP) to landslide susceptibility mapping at Haraz watershed, Iran. Natural hazards, **63**, 965-996. (2012).
5. Yeon, Y. K., Han, J. G., and Ryu, K. H. Landslide susceptibility mapping in Injae, Korea, using a decision tree. Engineering Geology, **116**(3-4), 274-283. (2010).
6. Lee, S. A. R. O. Application of logistic regression model and its validation for landslide susceptibility mapping using GIS and remote sensing data. International Journal of remote sensing, **26**(7), 1477-1491. (2005).
7. Lee, S., and Choi, J. Landslide susceptibility mapping using GIS and the weight-of-evidence model. International Journal of Geographical Information Science, **18**(8), 789-814. (2004).
8. Kavzoglu, T., Sahin, E. K., and Colkesen, I. Landslide susceptibility mapping using GIS-based multi-criteria decision analysis, support vector machines, and logistic regression. Landslides, **11**, 425-439. (2014).
9. Lee, S., Ryu, J. H., Won, J. S., and Park, H. J. Determination and application of the weights for landslide susceptibility mapping using an artificial neural network. Engineering Geology, **71**(3-4), 289-302 (2004).
10. He, Y., and Beighley, R. E. GIS-based regional landslide susceptibility mapping: a case study in southern California. Earth Surface Processes and Landforms: The Journal of the British Geomorphological Research Group, **33**(3), 380-393 (2008).
11. Sarkar, S., Kanungo, D. P., Patra, A. K., and Kumar, P. GIS based spatial data analysis for landslide susceptibility mapping. Journal of Mountain Science, **5**, 52-62 (2008).

# From tweet to tremor: enhancing megathrust disaster monitoring and early warning systems in social media

*Elsa* Silvia Nur Aulia[1*], *Epin* Saepudin[1], *Qoriah*[1], *Ernawati*[1], *Shohib* Khoiri[1], and *Siti* Kusumawati Azhari[1]

[1]Faculty of Arts and Design, Institut Teknologi Bandung, Ganesha 10 Bandung, Indonesia

**Abstract.** This researches how to explores the potential of social media in improving megathrust disaster monitoring and early warning systems, focusing on two regions in Indonesia: Gunung Kidul and Tanggamus. Using a quantitative approach through a questionnaire-based survey, the research reveals that most respondents (70% in Gunung Kidul and 85% in Tanggamus) consider the internet/social media as an important source of information during disasters. Furthermore, 70% of respondents in Gunung Kidul and 85% in Tanggamus are willing to receive official warnings through these platforms. Nonetheless, challenges related to data verification and reliability of information on social media still exist. Only 12% of respondents in Gunung Kidul and 8% in Tanggamus considered social media to be very reliable as a source of disaster information. This finding underscores the importance of collaboration between the government, disaster management agencies and social media platforms to ensure the accuracy of information disseminated. Overall, this study provides empirical evidence on the potential of utilizing social media to improve early warning systems and response to megathrust disasters. The integration of social media into disaster mitigation strategies can provide significant benefits, especially in terms of early detection, enhancing situational awareness, and facilitating communication and collaboration during disasters.

## 1 Introduction

Natural processes result in a high death toll from disasters, which also affect a sizable section of the population and generate significant social and economic damages; they sometimes may end in political instability [1-3]. In disaster management, quick reaction and evaluation are crucial. As a result, research on creating instruments for quick reaction to and evaluation of natural disasters [4, 5] has expanded in consideration emerging demands over the past few decades [6].

Today, the responsibility of innovation and the application of technology for social good can aid in producing some beneficial results in solving the issues that our cities and urban areas face [7]. Governments can employ social media analytics as sensors, information-sharing platforms, and social intervention tools since it allows for a dynamic engagement with the public. Rising social media apps (like those on Twitter, Inc. and Facebook, Inc.) are growing as countries address these demands. In contrast to other conventional methods, social media platforms record public reactions through both virtual and actual languages, gathering an unparalleled size of data and amount of information. This comes into an important component of sociological studies [8, 9].

In another hand, social media sites can offer a variety of viewpoints on how natural disasters and climate change are affecting people's local communities and cities (e.g., recognizing changes in emotions over time and space in the context of a natural hazard-related disaster). This phenomenon makes collection of data from various communities and provides additional resources for quantitative research easier to conduct. [10, 11].

For example, some research examined traffic-related event detection detecting severe diseases and symptoms in and across cities [12], and a range of different themes including governance matters connected to COVID-19 [13, 14]. Furthermore, several researchers examined urban logistics [15], public mood assessments of government services, and disaster severity prediction [16] using big data social media analytics.

Therefore, digital technology has significantly improved disaster detection and warning, which has certainly saved lives in the field of natural hazard-related disaster management [17, 18]. Applications of social media in disaster management have garnered a lot of attention from academic and practitioner circles in recent years due to social media's enormous potential. Social media is very helpful in emergency circumstances [19].

The Disaster Prevention and Management Journal features a study by Martin that discusses the significance of emergency communication and warning systems (ECWS). According to that, people may make decisions and act prior to, during, and following an emergency or disaster thanks to emergency communications and warning systems [20].

There are two types of disaster warnings that are in use today. Individuals who convey possible risks according to severity, spread speed, and likelihood of suppression, as

* Corresponding author: silviaelsha@gmail.com

well as those who notify nearby populations and offer guidance on how to react. [21].

For example, twitter-based warning system would offer prompt, openly available, all-encompassing disaster management to lower the likelihood of fatalities in natural hazard-related disasters [22]. Social media plays a crucial part in resilience. It was revealed that communities with strong social media relationships and networks had stronger adaptive capabilities and were more aware of the severe hazards, which increased their chances of surviving [23]. They [24] proposed a social-psychological theory called nudging in which human behavior is influenced in an easily reversible way without resulting in any particular action.

Megathrust faults in subduction zones are responsible for Earth's largest earthquakes and tsunamis. These faults exhibit diverse slip behaviors, from rapid stick-slip earthquakes to slow creep, influenced by factors such as fault zone complexity and seafloor roughness [25]. Heat flow data suggest smooth megathrusts producing great earthquakes are generally weaker than geometrically rough, creeping megathrusts [26]. The seismic behavior of megathrusts is affected by various parameters, including plate structure, sediments, fluids, and bathymetric roughness [27]. Understanding these factors is crucial for seismic hazard assessment. The Mentawai-Pagai segmentation in Indonesia, which experienced a megathrust earthquake in 2010, has been studied using various methods, including paleo geodesy, teleseismic, and aftershock analysis [28]. Ongoing research combines empirical analysis of global data with analogue modelling to better understand megathrust behavior [29].

The most spectacular event in a subduction zone's life cycle—the conclusion of decades of strain accumulation across a convergent plate boundary—occurs only seldom and manifests itself as a massive megathrust earthquake [30]. Strong seismic signals from all over the world enable calculation of the overall magnitude of the event, but the specifics of the rupture are typically not clear because there aren't any geodetic data immediately overhead or close by.

In contrast to other subduction megathrust events, the massive [moment magnitude (Mw) = 9.2] Sumatra-Andaman earthquake of December 2004 (Fig. 1) had geodetic observations of co-seismic motions available from islands directly above the rupture. The origins of the earthquake might be thoroughly investigated thanks to these near-field data. Nevertheless, there were few near-fields geodetic measurements using the Global Positioning System (GPS) [31].

Three months later and directly to the south (Fig. 1), the massive Nias-Simeulue earthquake (Mw = 8.7) offers a far greater chance to constrain rupture mechanisms. Recently, several GPS (continuously recording) stations had been set up either directly above or next to the rupture [32]. Furthermore, the measurement of co-seismic uplift and submergence using coral microatolls was made possible by the existence of a tropical archipelago above the breach. We can create one of the most precise and in-depth maps of deformation for a subduction megathrust earthquake thanks to the vast collection of measurements that resulted.

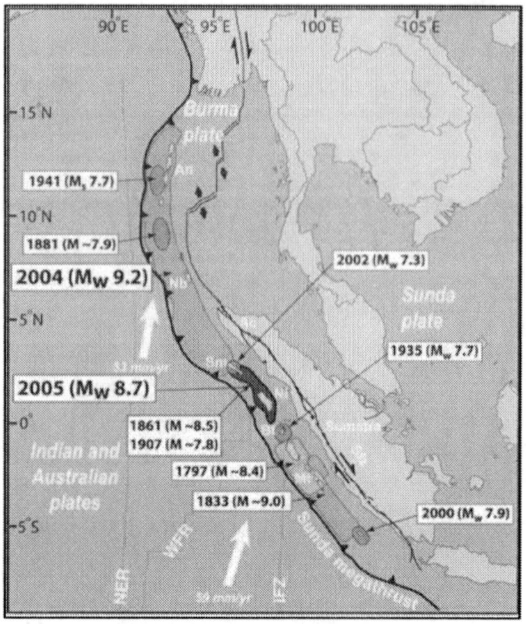


**Fig. 1.** Massive Nias-Simeulue earthquake (source: ectonics.caltech.edu/sumatra/data.html)

A wide range of seismic features have been shown to have depth dependency along megathrusts in recent decades, according to several observations. Focused research on depth-dependent features was spurred by the discovery of tsunami earthquakes [33] that produced significantly greater tsunamis than would be predicted for a conventional earthquake with the same surface wave magnitude (MS). These uncommon shallow ruptures often have unusual characteristics, such as a massive slip that produces a powerful tsunami and a comparatively modest rupture velocity and radiated energy. These are broad, yet rare, have struck along the Aleutians, Japan, the Kuril Islands, Peru, Nicaragua, El Salvador, New Zealand, Java, and Sumatra [34].

Furthermore, numerous large-magnitude earthquakes ($M > 8$) that have occurred since 2004 have yielded exceptional recordings by an abundance of broadband seismographs, enabling the observation of significant slippage but minimal coherent high-frequency radiation bursts in the shallowest portion of the relevant subduction zone. However, numerous bursts of high-frequency radiation have originated from the deeper regions of the seismogenic zone, where seismic slip is lower [35].

Likely distinguishes itself from other articles on megathrusts in some ways are: a) this research focused on social media integration; such as the usage of social media in megathrust disaster monitoring and early warning systems and this article may delve into how this real-time data can be used to detect early signs of seismic activity or collect data from the ground during a megathrust event, which could enhance situational awareness and reaction times. The article could examine how to engage citizens in data collection and reporting, thereby expanding the reach and coverage of monitoring efforts, particularly in remote or underserved areas.

## 2 Research methods

### 2.1 Study area

The research is located in Yogyakarta Special Province and Lampung Province (see Fig. 2). There are two main perspectives that come into the main argument in choosing these two which is a tourism perspective and disaster perspective. First, tourism perspectives such as: enhanced safety that social media can help tourists make more informed judgments and take preventative measures by giving them access to real-time alerts and information.

Second, disaster perspectives such as: early warning and evacuation; Yogyakarta and Lampung disaster management officials may be able to obtain earlier indications of seismic activity by utilizing the real-time nature of social media data.

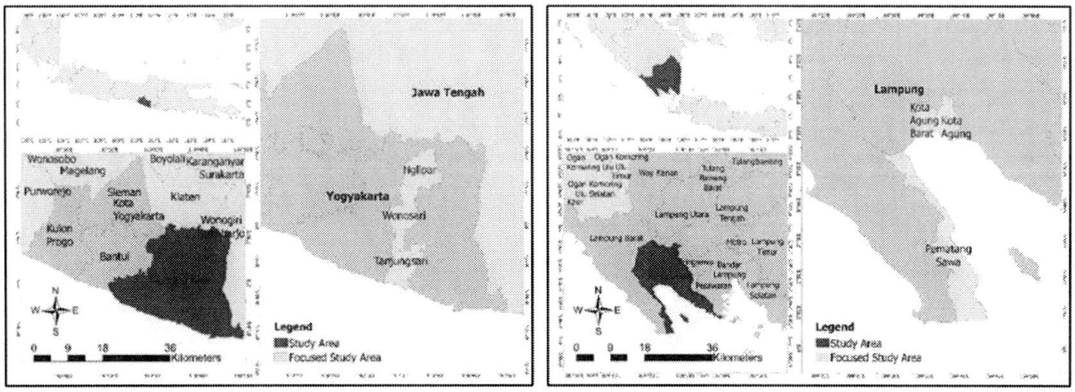


**Fig. 2.** Research location maps (a) Yogyakarta Special Province (b) Lampung Province (source: google maps)

### 2.2 Questionnaire survey design

This study employed a quantitative approach to assess citizen-level tsunami preparedness in coastal tourism areas. Data collection was conducted through face-to-face interviews using questionnaire instruments. The flowchart of the questionnaire survey process is illustrated in figures below.

Then the survey questions was design into several questions delivered to respondents about: their primary sources of information about megathrust earthquakes and tsunami, familiarity with term of megathrust earthquake, the respondents perceived level of information, the respondents knowledge of evacuation routes in case of a disaster, the frequency of use social media platforms, the experience of receiving or sharing information about earthquake in social media, the perception of social medias reliability as a source of information during megathrust, respondents willingness to receive about its information.

### 2.3 Data analysis

Descriptive statistics constituted the focus of the data analysis, because of the limits of the survey methodology in terms of drawing generalizations about the entire population, inferential statistical tests were not performed. To determine the most common answers to each question, the percentage frequencies were computed, and the results were obtained.

## 3 Results and discussion

### 3.1 Characteristics of respondents

From the data in Tables 1 and 2, we can see that a total of 215 responders were retrieved, including 110 respondents from Gunung Kidul Yogyakarta, and 115 from Tanggamus, Lampung. According to their latest educational attainment, sex, age, and place of origin, the respondents' sociodemographic were grouped.

**Table 1.** Socio-demographic characteristics of respondents in Gunung Kidul, Yogyakarta

| Characteristic | Types | Quantity | % |
|---|---|---|---|
| **Origin** | Nglipar | 40 | 36% |
| | Wonosari | 22 | 20% |
| | Tanjung Sari | 48 | 44% |
| **Sex** | Male | 58 | 53% |
| | Female | 52 | 47% |
| **Age** | 17-26 | 30 | 27% |
| | 27-36 | 45 | 41% |
| | 37-46 | 25 | 23% |
| | >= 46 years old | 10 | 9% |
| **Last Education** | Elementary School | 15 | 14% |
| | Junior high school | 30 | 27% |
| | Senior highschool | 40 | 36% |
| | Higher eduation | 25 | 23% |

**Table 2.** Socio-demographic characteristics of respondents in Tanggamus, Lampung

| Characteristic | Types | Quantity | % |
|---|---|---|---|
| **Origin** | Pematang Sawa | 23 | 22% |
| | Kota Agung Barat | 37 | 35% |
| | Kota Agung | 45 | 43% |
| **Sex** | Male | 55 | 52% |
| | Female | 50 | 48% |
| **Age** | 17-26 | 32 | 30% |
| | 27-36 | 42 | 40% |
| | 37-46 | 21 | 20% |
| | >= 46 years old | 10 | 10% |
| **Last Education** | Elementary School | 12 | 11% |
| | Junior high school | 33 | 31% |
| | Senior highschool | 35 | 33% |
| | Higher eduation | 25 | 24% |

## 3.2 Questionnaire results

Most respondents in both regions (50.1% in Gunung Kidul and 48.3% in Tanggamus) said they mostly get their information from the internet and social media. Approximately 25% of respondents in both categories (24.5% and 25.0%) mentioned television and radio, indicating their continued importance as sources. This response implies that traditional media is still important for reaching a larger audience, particularly for people who do not have much internet access or who like these kinds of mediums.

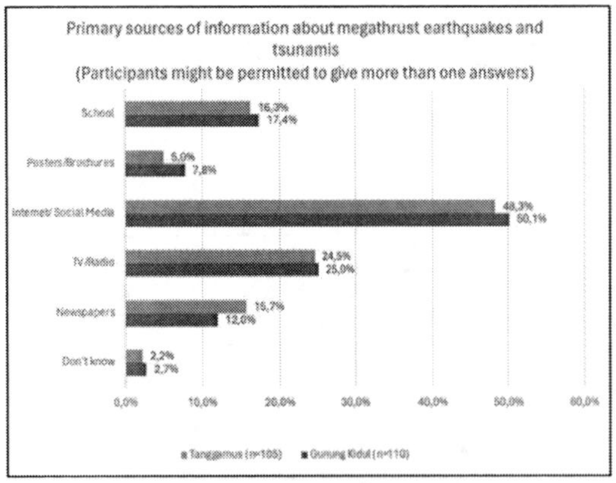


**Fig. 3.** Primary sources of information about megathrust earthquakes and tsunami

The data shown from Fig. 3 we can see that 54.6% of respondents in Gunung Kidul and 51.0% in Tanggamus fall into the "Somewhat familiar" group, indicating a moderate level of awareness regarding the term "megathrust earthquake." Compared to Tanggamus, where respondents rated their familiarity with the term at 23.0%, Gunung Kidul respondents rated it at slightly higher (30.1%). In both regions, a worrying percentage of respondents (18.9% in Gunung Kidul and 22.4% in Tanggamus) said they were "Not familiar at all" (with its term) were unclear about their prior experience with earthquakes.

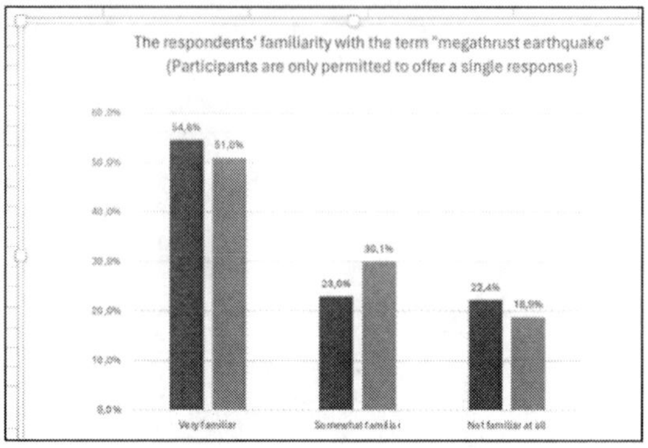


**Fig. 4.** The respondent's familiarity with the term of megathrust earthquake

The data shown from Fig. 4 we can see that most responses in Gunung Kidul and Tanggamus are in the middle of the spectrum (levels 2, 3, and 4), indicating a moderate level of perceived information regarding the possible dangers and effects of megathrust earthquakes and tsunamis. In comparison to Tanggamus, a greater proportion of respondents in Gunung Kidul chose levels 4 ("well-informed") and 5 ("very well-informed"). This suggests that people's perceptions of the dangers in Gunung Kidul are marginally improved. A significant portion of respondents in both regions, but especially in Tanggamus (4.2%), chose "1-no idea," indicating that some people are not aware of this.

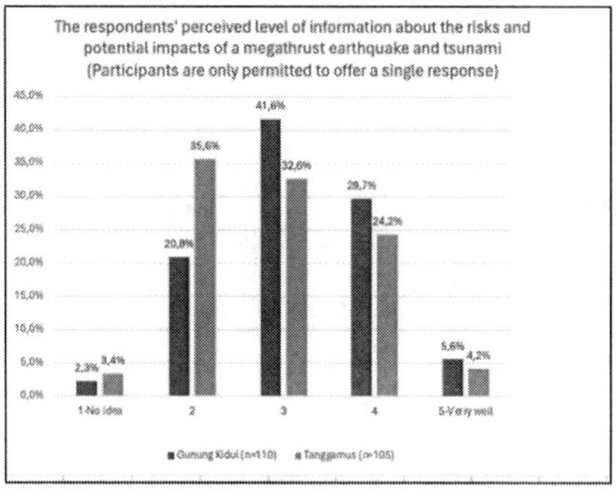


**Fig. 5.** The respondents perceived the level of information

In both Gunung Kidul and Tanggamus, that shown in Fig. 5 that the distribution of responses trends towards the middle categories (levels 2 and 3), showing a moderate level of knowledge regarding evacuation routes. In comparison to Gunung Kidul, a marginally greater proportion of respondents in Tanggamus chose levels 3 ("somewhat well") and 4 ("well"). This implies that

Tanggamus residents believe they know a little bit more about evacuation routes than the typical person. A significant fraction of respondents in both regions, but especially in Gunung Kidul (9.8%), chose "1-No idea," indicating a lack of awareness on the part of the populace.

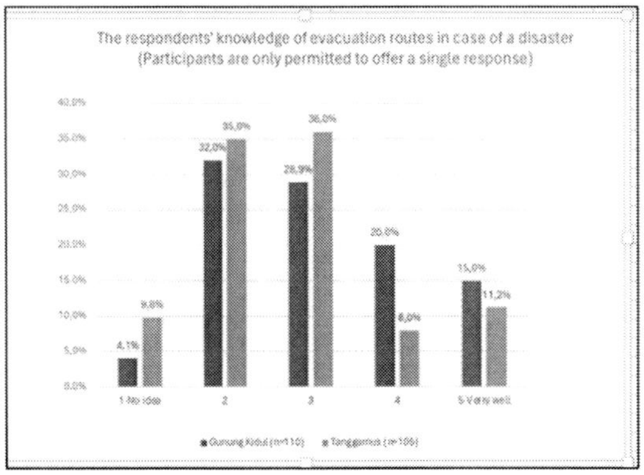


**Fig. 6.** The respondent's knowledge of evacuation routes in case of a disaster

A large percentage of respondents (97.0% and 98.2%, respectively) drawn in Fig. 6 in Gunung Kidul and Tanggamus reported utilizing social media platforms daily. This suggests that social media usage and penetration are extremely high in both areas. Minimal Other Usage Categories: The percentages of the remaining categories—"Several times a week," "Once a week," "Less frequently," and "Never"—are extremely low, underscoring the prevalence of daily social media use.

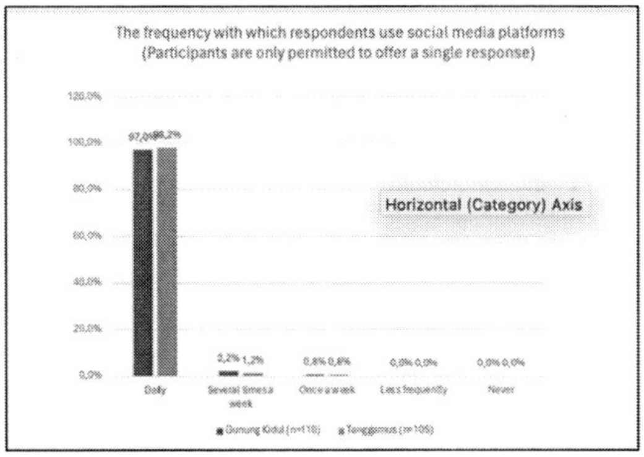


**Fig. 7.** The frequency of use social media platforms

The data shown from Fig. 7 we can see that A staggeringly high proportion of respondents (100% in Gunung Kidul and 98.3% in Tanggamus) said they had either posted or received information about earthquakes or tsunamis on social media. Minimal "No" Responses: Just 1.7% of respondents in Tanggamus and 0% in Gunung Kidul, respectively, reported having never interacted with such content on social media.

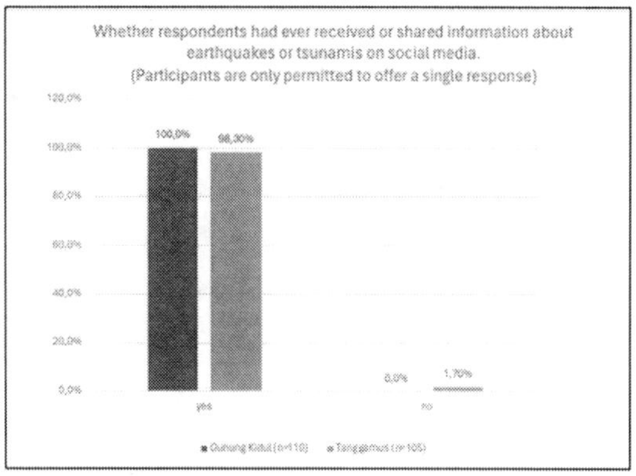


**Fig. 8.** The experience of receiving or sharing information about earthquakes in social media

The data shown in Fig. 8 explained that in Gunung Kidul (75.2%) and Tanggamus (80.1%) people's most common opinion of social media as a report source during megathrust disaster is "Somewhat reliable." This implies some trust, but it also acknowledges certain drawbacks because fewer respondents (15.0% in Gunung Kidul and 9.8% in Tanggamus) choose social media as "Very reliable". This finding suggests a level of mistrust or awareness of the possibility of false information on these platforms in times of emergency. Social media is seen as "Not reliable at all" by a comparatively modest percentage of respondents in both categories (9.5% and 10.4%, respectively). The finding suggests most people see some value in social media as an information source during disasters, even if they do not truly trust it.

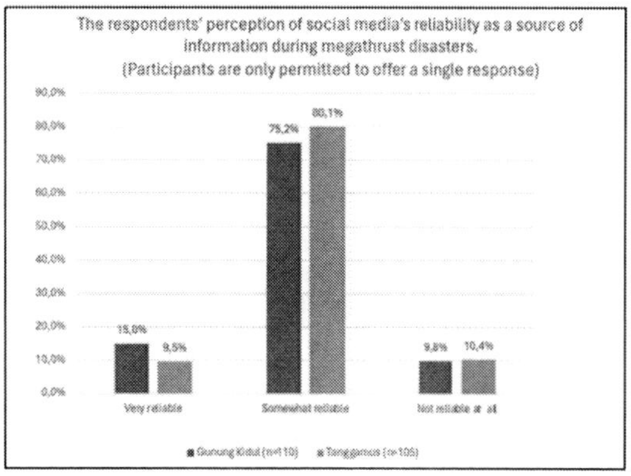


**Fig. 9.** The perception of social media's reliability as a source of information during megathrust

Remarkable 100% of respondents that shown from the data in Fig. 9 in Gunung Kidul and 99.6% of respondents in Tanggamus said that they would be open to receiving official alerts and updates regarding megathrust disasters via social media with 0% in Gunung Kidul and just 0.4% in Tanggamus, the percentage of respondents who are unwilling to accept such information is essentially non-existent.

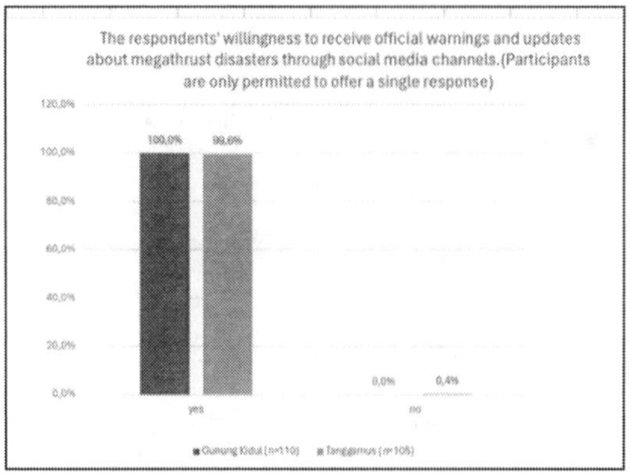


**Fig. 10.** Respondents' willingness to receive information about

The data in Fig. 10 show 100% of respondents in Gunung Kidul and 99.6% of respondents in Tanggamus said that they would be open to receiving official alerts and updates regarding megathrust disasters via social media with 0% in Gunung Kidul and just 0.4% in Tanggamus, the percentage of respondents who are unwilling to accept such information is essentially non-existent.

### 3.3 Discussion

The findings highlight that earthquakes are perceived as the most significant natural threat in both Tanggamus and Gunung Kidul, likely due to Indonesia's location within the seismically active Pacific Ring of Fire and the coastal vulnerability of Tanggamus to potential tsunamis. The diverse range of natural hazards Indonesia faces is evident in the acknowledgment of extreme weather events and volcanic eruptions as potential risks in both regions.

The primary source of information regarding disaster preparedness for most respondents in both regions is the internet and social media, highlighting the growing reliance on digital platforms for information dissemination. However, traditional media like television and radio remain important, especially for those with limited internet access or preference for these mediums.

The top priority for communities in Gunung Kidul and Tanggamus in dealing with potential megathrust earthquakes and tsunamis is public education about the dangers of these disasters. This shows an awareness of the importance of knowledge and understanding of risks to improve preparedness. In addition, securing large objects in homes and offices is also considered important, especially in Tanggamus, indicating an awareness of the risk of damage from falling objects during an earthquake. Improving early warning systems was also considered crucial, especially in Gunung Kidul. This underscores the need for better infrastructure to provide quick and accurate early warnings, which could potentially save many lives.

However, this level of importance is still below education and securing large objects, indicating a gap in understanding of the importance of having a clear and trained emergency plan in the face of disasters. The low prioritization of building retrofitting may reflect cost constraints and perceptions about the feasibility of implementing it. This is a challenge, given that building retrofitting is one of the most important steps in reducing the risk of earthquake damage. Levels of confidence in dealing with megathrust earthquakes and tsunamis are moderate, with most respondents feeling "somewhat confident." The high percentage of respondents who feel "not confident at all" indicates the need to improve individual and family preparedness through more intensive training and education. The high use of social media in both regions indicates great potential for utilizing this platform to disseminate disaster-related information and early warnings. The fact that almost all respondents had received or disseminated information about earthquakes and tsunamis through social media further emphasizes this potential.

However, the level of trust in social media as a source of information during disasters is still moderate. This highlights the importance of ensuring the accuracy of information circulated on social media and the need for the government and official institutions to play a role in providing reliable information. The high willingness of respondents to receive official early warnings through social media provides an opportunity for the government and relevant agencies to utilize this platform as an effective risk communication tool. This needs to be balanced with efforts to ensure the information provided is accurate, relevant, and timely.

## 4 Conclusion

This research shows the great potential of social media in improving the megathrust early warning system, especially in vulnerable areas such as Gunung Kidul and Tanggamus. Although communities show good awareness of potential disasters, understanding of megathrust earthquakes and levels of preparedness still need to be improved. Social media has become a major source of information, but the reliability of the information is still a challenge. Therefore, the integration of social media in the early warning system must be balanced with efforts to improve disaster literacy, data verification, and cross-sector collaboration. Thus, the potential of social media can be maximized to reduce the impact of megathrust disasters in Indonesia.

## References

1. R. Mechler, Natural disaster risk management and financing disaster losses in developing countries. (Verlag Versicherungswirtsch, Karlsruhe 2004). ISBN13:978-3899521207
2. M. Dilley, Natural disaster hotspots: a global risk analysis. World Bank Disaster Management Series, **5** (2005). DOI:10.1007/978-3-322-82113-3_1
3. S. Cole, H. A, E. Werker. Do voters demand responsive governments? Evidence from Indian disaster relief. Journal of Development Economics, **97** (2012). DOI:10.1016/j.jdeveco.2011.05.005

4. M. Erdik, K. Şeşetyan, M. Demircioğlu, U. Hancılar, C. Zülfikar, Rapid earthquake loss assessment after damaging earthquakes. Soil Dynamics, and Earthquake Engineering, **31** (2011). DOI:10.1007/978-90-481-9544-2_21
5. L.N. Rickard, Z.J. Yang, J.P. Schuldt, G.M. Eosco, C.W. Scherer, R.A. Daziano, Sizing up a superstorm: exploring the role of recalled experience and attribution of responsibility in judgments of future hurricane risk. Risk Analysis, **37** (2017). DOI:10.1111/risa.12779
6. B. Nelson, Natural disasters: a calculated risk. Nature, **495** (2013). https://doi.org/10.1038/nj7440-271a
7. T. Yigitcanlar, J. Corchado, R. Mehmood, R. Li, K. Mossberger, K. Desouza, Responsible urban innovation with local government artificial intelligence (AI): A conceptual framework and research agenda. J. Open Innov. Technol. Mark. Complex., **7** (2021). DOI:10.3390/joitmc7010071
8. D. Lazer, A. Pentland, L. Adamic, S. Aral, A.L. Barabasi, D. Brewer, M. Gutmann, Life in the network: the coming age of computational social science. Science, **323** (2010). DOI:10.1126/science.1167742
9. J. Scott, F. N. Stokman, Social networks. International Encyclopedia of the Social & Behavioral Sciences, **22** (2015). DOI:10.1016/B978-0-08-097086-8.32101-8
10. Y. Sun, Y. Shao, E. Chan, Co-visitation network in tourism-driven peri-urban area based on social media analytics: A case study in Shenzhen, China. Landsc. Urban Plan, **204** 103934 (2020). DOI:10.1016/j.landurbplan.2020.103934
11. S. Garske, S. Elayan, M. Sykora, T. Edry, L. Grabenhenrich, S. Galea, O. Gruebner, Space-time dependence of emotions on Twitter after a natural disaster. Int. J. Environ. Res. Public Health, **18**, 5292 (2021). DOI:10.3390/ijerph18105292
12. S. Alotaibi, R. Mehmood, I. Katib, O. Rana, A. Albeshri, A big data analytics tool for healthcare symptoms and diseases detection using Twitter, Apache Spark, and Machine Learning. Appl. Sci., **10**, 1398 (2020). DOI:10.3390/app10041398
13. E. Alomari, I. Katib, A. Albeshri, R. Mehmood, COVID-19: Detecting government pandemic measures and public concerns from Twitter data using distributed machine learning. Int. J. Environ. Res. Public Health, **18**, 282 (2021). DOI:10.3390/ijerph18010282
14. T. Yigitcanlar, N. Kankanamge, A. Preston, P. Gill, M. Rezayee, M. Ostadnia, G. Ioppolo, How can social media analytics assist authorities in pandemic-related policy decisions? Insights from Australian states and territories. Health Inf. Sci. Syst., **8**, 37 (2020). DOI:10.1007/s13755-020-00121-9
15. S. Suma, R. Mehmood, N. Albugami, I. Katib, A. Albeshri, Enabling next generation logistics and planning for smarter societies. Procedia Comput. Sci., **109**, 1122–1127, (2017). DOI:10.1016/j.procs.2017.05.440
16. N. Kankanamge, T. Yigitcanlar, A. Goonetilleke, How engaging are disaster management related social media channels? The case of Australian state emergency organisations. Int. J. Disaster Risk Reduct., **48**, 101097 (2020). DOI:10.1016/j.ijdrr.2020.101571
17. T. Yigitcanlar, L. Butler, E. Windle, K. Desouza, R. Mehmood, J. Corchado, Can building "artificially intelligent cities" safeguard humanity from natural disasters, pandemics, and other catastrophes? An urban scholar's perspective. Sensors, **20**, 2988 (2020). DOI:10.3390/s20102988
18. N. Kankanamge, T. Yigitcanlar, A. Goonetilleke, Public perceptions on artificial intelligence driven disaster management: Evidence from Sydney, Melbourne and Brisbane. Telemat. Inform., **65**, 101729 (2021). DOI:10.1016/j.tele.2021.101729
19. R. Meyer, K. Broad, B. Orlove, N. Petrovic, Dynamic simulation as an approach to Decision Support Systems 111 (2018) 48–59 understanding hurricane risk response: insights from the Storm view lab, Risk Analysis, **33** (2013). DOI:10.1111/j.1539-6924.2012.01935.x
20. N. Martin, J. Rice, Emergency communications and warning systems: Determining critical capacities in the Australian context. Disaster Prev. Manag., **21** (2012). DOI:10.3390/su14020810
21. N. Kankanamge, T. Yigitcanlar, A. Goonetilleke, Public perceptions on artificial intelligence driven disaster management: Evidence from Sydney, Melbourne and Brisbane. Telemat. Inform., **65**, 101729 (2021). DOI:10.1016/j.tele.2021.101729
22. A.T. Chatfield, C.G. Reddick, All hands-on deck to tweet# sandy: Networked governance of citizen coproduction in turbulent times. Gov. Inf. Q., **35**, (2018). DOI:10.1016/j.giq.2017.09.004
23. Y. Akama, S. Chaplin, P. Fairbrother, Role of social networks in community preparedness for bushfire. Int. J. Disaster Resil. Built Environ., **5**, 277–291 (2014). DOI:10.1108/IJDRBE-01-2014-0010
24. M. Mirbabaire, C. Ehnis, S. Stieglitz, B. Bunker, T. Rose, Digital Nudging in Social Media Disaster Communication. Inf. Syst. Front., **23** (2021). DOI:10.1007/s10796-020-10062-z
25. L.Susan Bilek, Thorne Lay, Subduction zone megathrust earthquakes. Geosphere. **14** (2018). DOI:10.1130/GES01608.1
26. X.Gao, and K. Wang, Strength of stick-slip and creeping subduction megathrusts from heat flow observations. Science. **345** (2014). DOI:10.1126/science.1255487
27. C. Damayanti, A. K. Yamko, C. J. Souisa, W. Barends, and I. L. P. T. Naroly, Pemodelan segmentasi Mentawai-Pagai: studi kasus gempa megathrust di Indonesia. J. Geosains & Remote Sensing, **1** (2020). DOI:10.23960/jgrs.2020.v1i2.56

28. F. Funiciello, F. Corbi, A. Heuret, C. Piromallo, and M. Rosenau, Empirical analysis of global-scale natural data and analogue seismotectonic modelling data to unravel the seismic behaviour of the subduction megathrust. Front. Earth Sci., (2020). DOI:10.3389/feart.2020.600152
29. C. Subarya et al., Plate-boundary deformation associated with the great Sumatra-Andaman earthquake, Nature 440, **46**, (2006). DOI:10.1038/nature04522
30. R.W. Briggs, K. Sieh, A. J. Meltzner, D. Natawidjaja, J. Galetzka, B, Suwargadi, Deformation and slip along the Sunda Megathrust in the Great 2005 Nias-Simeulue earthquake. Science, **31** (2006). DOI:10.1126/science.1122602
31. Sumatran GPS Array Data, (2004) ectonics.caltech.edu/sumatra/data.html (accessed on September 9, 2024).
32. H. Kanamori, The diversity of large earthquakes and its implications for hazard mitigation. Annual Review of Earth & Planetary Sciences, **42**, 7–26 (2014). https://doi.org/10.1146 /annurev-earth-060313-055034.
33. T. Egbelakin, S.S. Wilkinson, R. Potangaroa, and J. M. Ingham, Improving regulatory frameworks for earthquake risk mitigation. Building Research and Information, **41** (2013). https://doi.org/10.1080/09613218.2013.819546
34. S.L. Bilek, H.M.M. Rotman, and W.S. Phillips, Low stress drop earthquakes in the rupture zone of the 1992 Nicaragua tsunami earthquake: Geophysical Research Letters, **43**, (2016). https://doi.org/10.1002/2016GL070409
35. T. Lay and S. L. Bilek Anomalous earthquake ruptures at shallow depths on subduction zone megathrusts, in T.H. Dixon and J.C. Moore, eds., The seismogenic zone of subduction thrust faults (Columbia University Press, New York, 2007). DOI:10.1029/2011JB009133

# Liquefaction potential analysis using the Tsuchida method and correlation of Relative Density (Dr) – $D_{50}$ at Padang Beach

*Andriani*[1*], *Mifta* Afifa[1], *Didi* Yoriadi[1], *Muhammad* Shubhi Nurul Hadie[1], and *Fauziah* Ahmad[2]

[1]Civil Engineering Department, Faculty of Engineering, Universitas Andalas, Padang, West Sumatera, Indonesia
[2]School of Civil Engineering, Universiti Sains Malaysia, Engineering Campus, 14300 Nibong Tebal, Penang, Malaysia

**Abstract.** Indonesia's seismic activity necessitates careful consideration of earthquake risks in construction projects, as seen in the 2009 Padang's Earthquake with high sand content and water table near the coast. The test was carried out based on the gradation of soil grains through sieve analysis testing using the Tsuchida boundary and relative density testing by analyzing the correlation value between Dr and $D_{50}$. The accelerations used in this test are 0.3g and 0.6g. Soil classification according to the AASHTO (American Association of State Highway and Transportation Officials) system, the type of sand on the beach is A-3 (fine sand), and according to the USCS (Unified Soil Classification System) system, all samples are classified as SP (Poorly-graded sand). With Cc (Coefficient Curvature) 2.436-1.975 and Cc (Coefficient Uniformity) 1.106-0.939. 0.3g Relative density ranges between 7-59%, and 0.6g ranges between 5-41%. Diameter passes 50% ranges between 0.187-0.408 mm. The Tsuchida boundary analysis and the Dr-$D_{50}$ correlation retrieved similar results. Tsuchida analyses grain gradation, while Dr-$D_{50}$ examines cyclic load and settlement for specific grain size, affecting liquefaction potential. One sample liquefied at 0.3g, while all samples liquefied at 0.6g. Due to a smaller diameter at 50% pass, it was indicating higher liquefaction potential.

## 1 Introduction

Indonesia has a high level of seismic activity because of its geographically location at the intersection of four major tectonic plates: Pacific, Philippine, Eurasian, and Indo-Australian, as seen in Fig. 1.

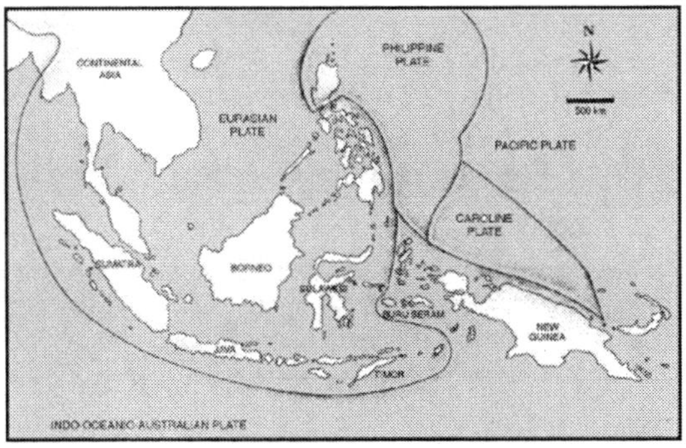


**Fig. 1.** Tectonic plate boundaries in South and South East Asia [1].

West Sumatra is one of the areas in Indonesia that is very prone to earthquakes. Located along the subduction zone between the Indo-Australian and Eurasian Plates, this region often experiences significant seismic activity. The earthquake that occurred in 2009 (Mw7.6) had its epicenter near Padang, which is the largest city in western Sumatra, Indonesia [2]. The tremors were so intense that they could be felt as far as Singapore, which is approximately 460 km away. This devastating earthquake resulted in the tragic loss of 1,117 lives [3].

The results of the investigation conducted on the 2009 earthquake caused sand to boil in several sites in Padang, as shown in Fig. 2. Liquefaction is a phenomenon in which initially solid soil turns into liquid due to strong vibrations, such as earthquakes. This process occurs when pore water pressure increases in fine-grained soil (usually sand or silt), causing soil particles to lose contact with each other and the soil lose its strength and stability [4]. This phenomenon often results in severe structural damage to buildings, infrastructure, and the surrounding environment. Liquefaction can cause severe damage to buildings, roads, bridges, and other critical infrastructure [5].

The buildings were found to be tilted or collapsed, water and gas pipes were burst, and roads and bridges were damaged or cut off. These damages not only caused huge material losses but also disrupted the economic and social activities of the community [6].

Chian et al. [3] found that the liquefaction in Padang was caused by high sand content and high-water table along the coastal strip plane. The geological condition of Padang City consists of alluvial soil deposited from river and sea activities, which tends to be composed of sand and silt that may have triggered the root caused

* Corresponding author: andriani@eng.unand.ac.id

**Fig. 2.** Sand boils on the ground surface [4].

Figure 3 shows the tension cracks on the road at Padang city which this indicating as a result of lateral spreading. Earthquake shaking triggers significant ground movement, especially in weak or unstable soil layers. This shaking causes the road above it to experience excessive pressure, resulting in cracks. Repeated and high-intensity vibrations can cause road materials such as asphalt and concrete degradation. If there are already small cracks or problems with the road structure, an earthquake can worsen the cracks. Roads built on unstable soil foundations are more susceptible to severe damage.

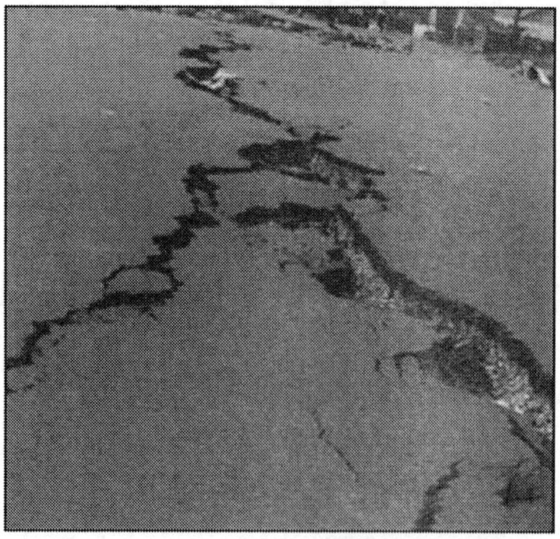

**Fig. 3.** Tensile cracks arising from lateral spread in the Siteba residential area [3].

Figure 4 shows the lifting of floor slabs in the residential area of Perumdam Kota Padang. Floor lifting or shifting of house floors is an expected impact in earthquake-prone areas. Floor lifting in Padang can occur in areas prone to liquefaction, where the soil under the foundation suddenly loses its stability due to increased pore water pressure so that the effective soil stress decreases. Floor lifting is usually accompanied by other structural damage to the house, such as cracks in the walls, jammed doors and windows, and damage to the foundation. This can make the house unsafe to occupy without significant repairs.

**Fig. 4.** Uplifted floor slabs in the residential area of Perumdam, Padang City [3].

Research into liquefaction was initiated after earthquakes in Niigata, Japan, and Alaska in 1964 caused damage to superstructures due to the liquefaction of saturated sand layers. A combination of seismic vibrations, soil type and properties, groundwater conditions, geological structures, and human activities causes liquefaction. The main factor that triggers liquefaction is seismic vibrations, especially from earthquakes, which increase pore water pressures in fine-grained soils. Saturated soils with high porosity and not fully consolidated are particularly susceptible to this phenomenon. This led to numerous studies and the development of methods for assessing liquefaction potential, utilizing laboratory and field data. Soil parameters such as grain gradation and relative density (d5) significantly influence liquefaction potential and soil strength.

2009 Earthquake has caused liquefaction along Padang Beach. Due to this incidence, it is important to study their relationship. An analysis was carried out to find the influence of potential liquefaction with regard to relative density and grain degradation for the Padang Beach.

It was found that soils with low relative density (Dr) have high liquefaction potential due to their loose particle structure and significant instability to seismic vibrations. On the other hand, soil with high relative density (Dr) has low liquefaction potential because of its dense and stable structure [7-9]. Likewise, soil with uniform gradation has high liquefaction potential due to its low density and loose structure. In contrast, well-graded soil has low liquefaction potential due to its denser and more stable particle structure. Coarse and fine soils have varying liquefaction potentials depending on specific conditions such as saturation and seismic vibration intensity.

Research by Tohari [10] found that the depth of the groundwater table in Padang City ranges from 0.5 to 5.0 meters. Near the beach, the average depth is around 1.5 meters. These variations are due to the geological composition of Padang, which includes Tertiary volcanic rocks (Tomv), Plio-Plistocene volcanic rocks (QTv), and Quarternary alluvial deposits (Qa). Liquefaction typically occurs in areas with high groundwater levels, its effects are often observed in low-lying areas or nearby rivers, lakes, bays, and oceans [11, 12].

Liquefaction occurs because of reduction in effective stress due to an increase in pore water pressure so that the effective stress of the soil decreases to the point that the soil can lose its strength. This causes the soil to liquefy. When an earthquake or vibration occurs, if the soil particles are unable to maintain bonds, causing pore water to rise through the gaps between particles, then there is an increase in pore water pressure [13, 14]. The effect is that the Pore water pressure ($\bar{u}$) increases, but the total stress ($\sigma$) is still in the same condition as before the earthquake. If the pore water pressure increases over the total stress, effective stress ($\sigma'$) will be smaller than zero ($\sigma' < 0$). This condition causes the soil to liquefy [15-18].

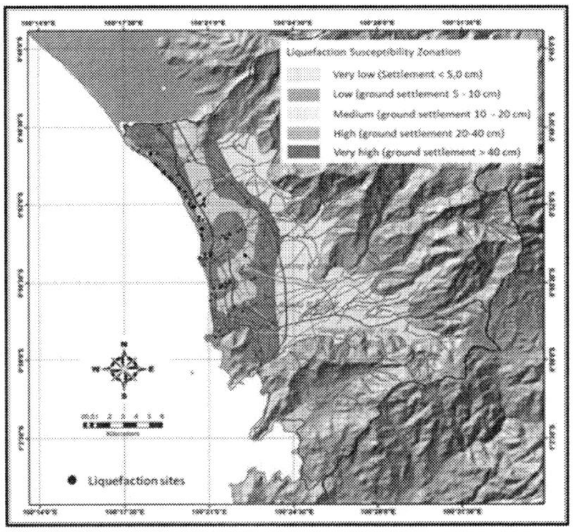


**Fig. 5.** Seismic microzonation of Padang City [10]

Figure 5 shows the seismic microzonation analysis of Padang City reveals that the regions with high to very high vulnerability to liquefaction are concentrated along the coastlines and river flows, areas in Koto Tangah, North Padang, East Padang, and South Padang subdistrict where sand boiling, settlement, and lateral movement in high vulnerability zones [10]. In addition to Adji et al. [19] reasearch in Air Tawar water areas such as Basko Mall, Alai area, Simpang Haru, Sungai Air and Kota Tua have liquefaction potential.

## 2 Methodology

Soil samples were taken at five locations along the beach of Padang, West Sumatra as shown in Fig. 6. The locations were selected based on potential liquefaction mapping from the Ministry of Energy and Mineral Resources [20] and seismic microzonation research on liquefaction [10]. Five of Sand cone tests were carried out on the sites along Padang Beach. The soil samples collected were taken to Andalas University's Soil Mechanics Lab for analysis. The disturbed soil samples were collected using a hoe (at 50 cm depth). The grain gradation tests and relative density were carried out using ASTM. Then the results were analysed using the Tsuchida method [21]. Then the Relative density Dr and $D_{50}$ comparison charts by Hakam [22].

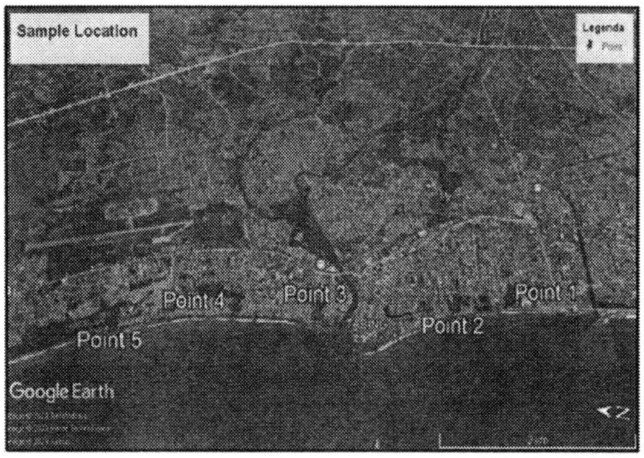


**Fig. 6.** Sampling locations.

Table 1 shows the sampling location for five (5) soil sampling points. The soil samples were disturbed samples soil and taken 50 cm below the ground surface. Laboratory testing was conducted to obtain the physical properties of the soil. The soil parameters obtained were soil grain analysis graphs, volume, weight, and other physical properties.

In this study, the Tsuchida method and relative density (Dr) and $D_{50}$ were used to predict liquefaction potential. The Tsuchida Method was chosen because it is simple, has a strong empirical basis, directly correlates with soil parameters, and has wide application in various geotechnical conditions. Likewise, the Dr and $D_{50}$ methods were chosen to analyze liquefaction potential because they provide in-depth information about soil density and grain size distribution, critical factors in determining soil susceptibility to liquefaction. Dr indicates the level of soil density and stability, while $D_{50}$ provides an overview of the characteristics of soil grain gradation. Combining these two parameters allows for a more accurate assessment of liquefaction risk and helps engineers plan effective mitigation strategies to reduce the impact of earthquakes. Thus, these two methods provide a practical and effective way to evaluate soil susceptibility to liquefaction to reduce the risk of infrastructure damage and impacts on communities in earthquake-prone areas.

**Table 1.** Sampling location coordinates

| No | Coordinate | Location |
|---|---|---|
| 1. | -0.92164;100.24908 | Lolong Belanti, West Padang |
| 2. | -0.91578;100.34711 | Ulak Karang, North Padang |
| 3. | -0.90123;100.34408 | West Air Tawar, North Padang |
| 4. | -0.89229;100.34336 | West Air Tawar, North padang |
| 5. | -0.8799;100.34004 | Parupuk Tabing, Koto Tangah |

The soil classification systems that will be used are the American Association of State Highway and Transportation (AASHTO) and the Unified Soil Classification System (USCS).

AASHTO using some indexes like Cum. Passing on sieve no.10, no.40, and no. 200 as seen in Table 2. Then USCS used the Cum. Passing No. 4, fine particles and values of Cu and Cc, as shown in Table 2.

**Table 2.** AASHTO soil requirements type

| Group Classification | A-1 | | A-3 |
|---|---|---|---|
| | A-1-a | A-1-b | |
| Sieve analysis % Passing | | | |
| No.10 | 50 max | | |
| No. 40 | 30 max | 50 max | 51 min |
| No. 200 | 15 max | 25 max | 10 max |

The data obtained during testing was processed using the formula below:
Determining Field density (γd) from sand cone test:

$$V = \frac{(\text{Dry weight } (Wd))}{(\text{Excavation Volume})} \quad (1)$$

Determining Cc (Coefficient Curvature) and Cu (Coefficient of Uniformity):

$$Cc = \frac{D_{60}}{D_{10}} \quad (2)$$

$$Cu = \frac{{D_{30}}^2}{D_{10}\ x\ D_{60}} \quad (3)$$

Determining Bulk density:

$$\gamma = \frac{W}{Mould\ Volume}\ x\ 100 \quad (4)$$

## 3 Results and discussion

### 3.1 Sieve test analysis

Based on direct observation of grain of soil, all of the soil types are sand. The results from the sieve analyses are as shown in Table 3. Sieve analysis results for 5 sampling points.

**Table 3.** Sieve analysis results

| Sieve No. | Cum. Passing (%) | | | | |
|---|---|---|---|---|---|
| | 1 | 2 | 3 | 4 | 5 |
| 4 | 99.ysis47 | 100.00 | 100.00 | 100.00 | 100.00 |
| 10 | 99.30 | 100.00 | 100.00 | 99.87 | 100.00 |
| 20 | 99.03 | 82.33 | 99.33 | 98.00 | 99.93 |
| 40 | 95.10 | 50.90 | 89.10 | 76.63 | 95.07 |
| 60 | 77.50 | 16.70 | 66.83 | 39.17 | 62.40 |
| 100 | 33.50 | 4.07 | 25.03 | 11.03 | 15.07 |
| 200 | 5.83 | 2.33 | 2.17 | 2.20 | 0.07 |
| PAN | 0.00 | 0.00 | 0.00 | 0.00 | 0.00 |

The classification of soil using AASHTO by the particle size, the index used is Cumulative Passing (%) on the Sieve no. 10, no. 40, no. 200 as shown in Table 4.

**Table 4.** Soil classification based on AASHTO standard

| POINT | Sieve Analysis Passing (%) | | | AASHTO Classification |
|---|---|---|---|---|
| | #10 | #40 | #200 | |
| 1 | 99.30 | 95.10 | 5.83 | A-3 (fine sand) |
| 2 | 100.00 | 50.90 | 2.33 | A-3 (fine sand) |
| 3 | 100.00 | 89.10 | 2.17 | A-3 (fine sand) |
| 4 | 99.87 | 76.63 | 2.20 | A-3 (fine sand) |
| 5 | 100.00 | 95.07 | 0.07 | A-3 (fine sand) |

The results of soil classification using the AASHTO method show that the soil in the 5 locations studied is A3 (fine sand). Fine sandy soil has a high liquefaction potential because it has a small grain size so that its smooth surface will reduce interlocking between particles, making the soil looser and susceptible to rapid compaction; loose structure and low density increase the risk of increasing pore water pressure during seismic vibrations; at high saturation conditions.

Then, for the classification of soil using USCS, Coefficient Curvature (Cc) and Coefficient Uniformity (Cu) can be obtained from Sieve analysis test result.

**Table 5.** Soil classification based on USCS standard

| POINT | Cu | Cc | USCS Classification |
|---|---|---|---|
| 1 | 2.436 | 1.079 | (SP) Poorly-Graded Sand |
| 2 | 2.759 | 0.939 | (SP) Poorly-Graded Sand |
| 3 | 2.327 | 1.106 | (SP) Poorly-Graded Sand |
| 4 | 2.455 | 0.975 | (SP) Poorly-Graded Sand |
| 5 | 1.975 | 1.077 | (SP) Poorly-Graded Sand |

Based on Table 5, USCS classification shows that the soil in the Padang coastal area is SP (sand-poor). Poorly graded sand has unique characteristics that make it more susceptible to liquefaction than well-graded soil. Poor gradation means that this soil has particles that are uniform or nearly uniform in size, without significant size variations; because the size is uniform, the soil particles cannot interlock well (poor interlocking), so the soil tends to be looser and less stable; this soil usually has a lower relative density (Dr), which means the soil is more prone to compaction and volume reduction when subjected to loads or vibrations; increased pore water pressure causes a decrease in the shear strength of the soil, which is the primary condition that causes liquefaction.

### 3.2 Relative density analysis

Relative Density (Dr) has a direct relationship with liquefaction potential. Soils with low Dr are more susceptible to liquefaction due to their low density and poor interlocking ability, while soils with high relative density (Dr) tend to be more stable and have better shear strength. Understanding soil is critical in liquefaction risk evaluation and mitigation planning in earthquake-prone areas.

**Table 6.** Relative density calculation

| g | γ | γ (g/cm3) | | | | |
|---|---|---|---|---|---|---|
| | | 1 | 2 | 3 | 4 | 5 |
| | γd | 1.221 | 1.082 | 1.465 | 1.177 | 1.267 |
| | γdmin | 1.163 | 1.058 | 1.275 | 1.143 | 1.153 |
| 0.3 | γmdax | 1.440 | 1.515 | 1.631 | 1.412 | 1.421 |
| 0.6 | γmdax | 1.574 | 1.659 | 1.731 | 1.445 | 1.483 |

As in Table 6, sample 2 is the lowest field density (γd), and sample 3 is the highest field density (γd). The field Density (γd) of all samples ranges between 1-1.5 g/cm3. The minimal bulk density (γdmin) of all samples ranges between 1-1.3 g/cm3. Sample at points 1, 3 and 5 have no significantly different values of field density (γd). On 0.3g shaking vibration, which is relevant to the Earthquake that occurred in Padang in 2009, has a greater density than 0.6g. According to this result, the soil will become looser as the intensity of vibration increases.

**Table 7.** Relative density result

| g | Point | | 1 | 2 | 3 | 4 | 5 |
|---|---|---|---|---|---|---|---|
| 0.3 | Dr | % | 24.61 | 7.19 | 59.50 | 15.30 | 47.56 |
| | Cond. | | L | VL | I | L | L |
| 0.6 | Dr | % | 18.13 | 5.98 | 49.37 | 13.94 | 40.34 |
| | Cond | | L | VL | L | VL | L |

Ex:

I = Intermediate
L = Loose
VL = Very Loose

As shown from the result seen on Table 7, value of Dr ranges between 7-59% (0.3g), and 0.6g ranges between 5-49%. It is clear to observe that on 0.3 g shaking vibration, the soil that will be loose is at points 1, 2, 3, 4, 5. Meanwhile, point 3 is an intermediate condition. However, at a vibration of 0.6g, all points are in a loose condition. Therefore, point 2 is the most influential if an earthquake occurs and can have an enormous liquefaction potential.

### 3.3 Liquefaction potential

Results from the sieve shaker analysis will be plotted to the Tsuchida Boundary, and then the Dr (%) will be checked by correlation between Dr and $D_{50}$.

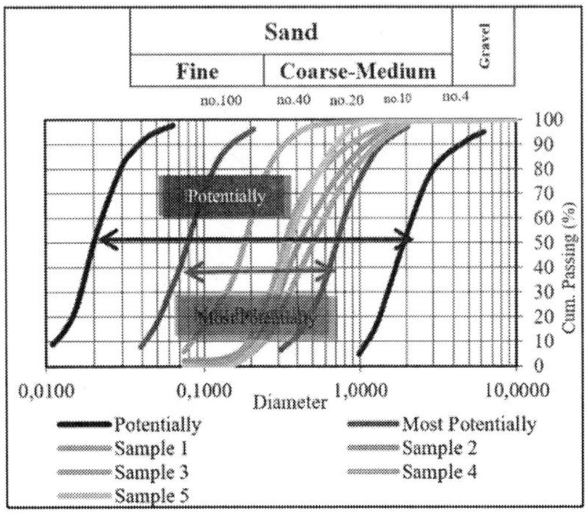


**Fig. 7.** Tsuchida boundaries liquefaction potential

Based on Figure 7, the results of the gradation of soil grains that have been tested in the previous sieve shaker analysis show that all samples have an enormous potential to liquefy. This is due to the uniformity of the size of the soil grains, causing a lack of binding force between each soil particle.

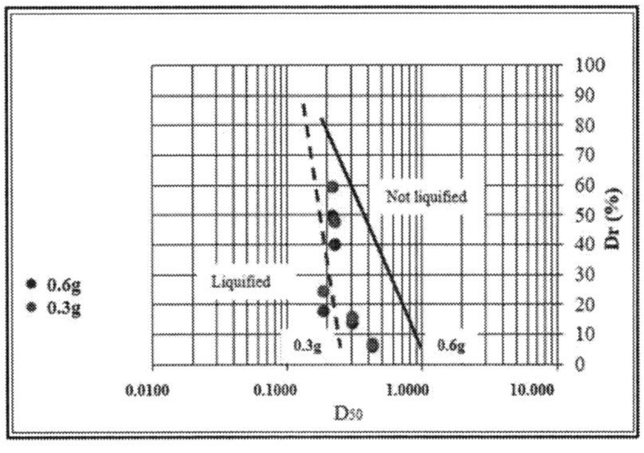


**Fig. 8.** Dr vs $D_{50}$ liquefaction potential

As seen Fig. 8, at 0.6g vibration shaking, all of the samples potentially liquefy. However, on 0.3g vibration shaking, only one sample potentially liquefied at point 1 at Lolong Belanti, North Padang.

Although sample point 1 has a greater relative density, the $D_{50}$ is smaller than the others. It causes 50% size passing on in a small particle size of 0.186 mm and is classified as fine sand.

Table 8 shows the relationship between Liquefaction Potential based on Dr vs $D_{50}$, the results show that the 5 (five) points located in the Padang coastal area have the potential for liquefaction at 0.6 g while at 0.3 g, the liquefaction potential only occurs at point 1. The magnitude of the earthquake acceleration also influences the liquefaction potential, as seen in Table 8.

**Table 8.** Liquefaction potential based on Dr vs $D_{50}$

| Poin | 0.3g | Liquefaction Potential | 0.6g | Liquefaction Potential |
|---|---|---|---|---|
| 1 | 24.61 | Yes | 18.13 | Yes |
| 2 | 7.19 | No | 5.98 | Yes |
| 3 | 59.50 | No | 49.37 | Yes |
| 4 | 15.30 | No | 13.94 | Yes |
| 5 | 47.56 | No | 40.34 | Yes |

## 4 Conclusion

Based on test results, the grain size along the Padang beach is classified as the same type. However, at point 3, the grain size is finer sand due to its location close to the estuary, which is influenced by sediment deposition. Using AASHTO standards, all samples tested were classified as A-3 (fine sand). Using USCS standards, all samples tested are included in poorly graded sand (SP).

Tsuchida analysis shows that the gradation of the samples tested, including into the range, has the potential to liquefy because the soil type is sand with high uniformity.

Based on the relative density value (Dr) of sand soil along Padang Beach, it can be said to be sandy soil with loose density because there are 3 of 5 samples that have a Dr value between 0-15% when 0.3g vibration occurs and all have a Dr value between 0-50% when 0.6g vibration occurs. This analysis shows the relative density of samples tested, including the range that has the potential of both 0.6g and 0.3g liquefy because this indicates that the density of the soil is still relatively low.

We would like to thank the Engineering Faculty of Univesitas Andalas for funding this research.

## References

1. F. Göltenboth and W. Erdelen, Geography and Geology. In Ecology of Insular Southeast Asia: The Indonesian Archipelago, 3–16 (Elsevier Ltd. 2006) https://doi.org/10.1016/B978-044452739-4/50002-4
2. R. Aryanti and M.Masrilayanti, State of the art of seismic risk and loss assessment in structures. E3S Web Conf. **331**(6) 07013 (2021) https://doi.org/10.1051/e3sconf/202133107013
3. S.C. Chian, S.M. Wilkinson, J. K. Whittle, R. Mulyani, J. E. Alarcon, A. Pomonis, K. Saito, S. Fraser, K. Goda, J. Macabuag, M. Offord, A.C. Hunt-Raby, P. Sammonds, G. Franco, H. Stone, B. Ahmed, F.E. Hughes, N.K. Jirouskova, S. Kaminski, S. and J. Lopez, Lessons learnt from the 2009 Padang Indonesia, 2011 Tōhoku Japan and 2016 muisne Ecuador earthquakes. Frontiers in Built Environment, **5** (2019). https://doi.org/10.3389/fbuil.2019.00073.
4. A. Hakam and E. Suhelmidawati, Liquefaction due to September 30th 2009 earthquake in Padang. Procedia Engineering, **54**, 140–146 (2013) https://doi.org/10.1016/j.proeng.2013.03.013
5. H. Novasari, A. Hakam, A. Andriani, and S.R. Hape, Application of finite difference technique for solving consolidation problems of radial drainage case in air Tawar, Padang. AIP Conf. Proc. **2891**, 060005 (2024). https://doi.org/10.1063/5.0201219
6. M. Sudondo, A. Andriani, and A. Hakam, Geotechnical design of bored pile foundation on potential liquefaction site. AIP Conf. Proc. **2891**, 060006 (2024). https://doi.org/10.1063/5.0208151
7. A. Hakam, Laboratory Liquefaction Test of Sand Based on Grain Size and Relative Density. J. Eng. Technol. Sci., **48**(3), 334-344 (2016)
8. A. Hakam, F.A. Ismail, F. and Fauzan, Liquefaction potential assessment based on laboratory test. International Journal of GEOMATE, **11**(26) 2553-2557 (2016)
9. A. Pramaditya, A., and T. F. Fathani, Physical modelling of earthquake-induced liquefaction on uniform soil deposit and settlement of earth structures. J. Civ. Eng. Forum, **1000**, 85–96 (2020)
10. A. Tohari, Seismic microzonation of soil amplification and liquefaction for Padang City. E3S Web Conf. **156,** (2020) https://doi.org/10.1051/e3sconf/202015602008
11. A. Tohari, K. Sugianti and E. Soebowo, Liquefaction Potential at Padang City: A Comparison of Predicted and Observed Liquefactions During The 2009 Padang Earthquake. Ris. Geo. Tam **21**(1) 7-19 (2011) http://dx.doi.org/10.14203/risetgeotam2011.v21.42.
12. R. W. Day, Geotechnical Earthquake Engineering Handbook, The McGraw-Hill Comp. (New York. USA, 2002)
13. M. Jefferies, and K. Been, Soil Liquefaction: A Critical State Approach (2006).
14. G.B. Laurie, B.H. Rebecca, and M.B. Charles, Liquefaction Hazard Mapping Statistical and Spatial Characterization of Susceptible Units. Journal of Geotechnical and Geoenvironmental Engineering, ASCE, **132:6** (2006).
15. R.W. Boulanger, High overburden stress effects in liquefaction analyses. Journal of Geotechnical and Geoenvironmental Engineering, ASCE, **129**(12), 10711082 (2003).
16. H. Suprijanto, A. Hendrawan, and A. Nugraha, Potential study of the liquefaction hazard at the reclamation development site of I Gusti Ngurah Rai airport, Nusa Dua region, province of Bali. IOP Conference Series: Earth and Environmental Science. (2020).
17. W. Rahayu, I. Yuliyanti, and E. Bahsan, Analysis of potential liquefaction using cone penetration test data and grain size distribution test with case study of liquefaction in Lolu Village. IOP Conference Series: Earth and Environmental Science. **622**.
18. A. Tohari, I. Muttaqien, and R. Syifa, Understanding of flow liquefaction phenomena in Palu City from shear wave velocity profiles. E3S Web of Conferences. **340**. 01011 (2022).

19. B.M. Adji, B. Istijono, A. Hakam, S. Andriani, and M. Anshari, Liquefaction disaster mitigation on railway corridors in Padang City, West Sumatra, IOP Conf. Ser.: Earth Environ. Sci. **708** 012025 (2021).
20. R.N. Rahdiana, R. W. Wahyudin, G. Hasibuan, Wiyono, and W.P. Sollu, Atlas of Indonesian Liquefaction Susceptibility Zones, Ministry of Energy and Mineral Resources (Geological Agency, Center for Groundwater and Environmental Geology), Bandung-Indonesia (2019).
21. H. Tsuchida, Prediction and Countermeasure against Liquefaction in Sand Deposits. Abstract of the Seminar of the Port and Harbour Research Institute. Ministry of Transport, 83 Technical able after it is being referred Journal, comments of Engineering and Technology (UET) Taxila, Pakistan **20**,2-2015 Yokosuka, Japan, 3.1-3.33 (In Japanese) (1970).
22. A. Hakam, Practical Analysis of Liquefaction Potential. Andalas Press, Padang-Indonesia (2020).

# Point-of-Use technology for clean water treatment in disaster areas: effectiveness of LDH plate in eliminating pathogenic bacteria

*Taufiq* Ihsan[1*], *Reri* Afrianita[1], and *Freddy* Ilfan[2]

[1]Environmental Engineering Department, Faculty of Engineering, Universitas Andalas, Padang, 25163 West Sumatra, Indonesia
[2]Environmental Engineering Department, Faculty of Science and Technology, Jambi University, Muaro Jambi, 36361 Jambi, Indonesia

**Abstract.** This study tackles the clean water crisis in West Sumatra, often exacerbated by natural disasters such as flash floods and landslides, by proposing an innovative point-of-use water treatment method. The method employs layered double hydroxide (LDH) synthesized on aluminum plates, created by soaking these plates in a solution of seawater and magnesium hydroxide. The X-ray diffraction analyses proved the formation of LDH. Experiments demonstrated that this method is suitable for water contaminated with E. coli. For the LDH plates, this removal rate of E. coli reached 100% with continuous stirring and 91% with occasional stirring within 24 hours. Stirring did enhance contact between LDH and water, thus improving the rate of removal of the bacterium. The process is low-cost and simple, with all materials easily available, so this offers a very reproducible solution to deliver clean water in disaster zones. Further studies shall be conducted for its performance against different bacterial strains and in more complex water matrices. This work substantially expands current knowledge of POU/HWT technologies and delivers a real-life solution for providing clean water in disaster situations.

## 1 Introduction

West Sumatra is one of the 38 provinces comprising Indonesia, located on the western part of Sumatra Island, and is predominantly recognized for its natural beauty. The geography of the region is varied, with a coastline along the Indian Ocean and the towering *Bukit Barisan* Mountains, and so the area is predisposed to a number of natural disasters. Flash floods, landslides, and earthquakes are major threats to the citizens of this region [1, 2]. Due to high tectonic activity, the *Bukit Barisan* Mountains commonly bring about earthquakes. Heavy rainfalls in many areas cause recurring flash floods and landslides [3, 4]. Such disasters not only bring physical damage and loss of life but even result in shortages of clean water supply, a basic human need.

Lack of water can lead to many serious health issues. As natural hazards, flash floods and landslides can damage the water supply systems, including clean water pipes and springs. In addition, these natural hazards can contaminate the water sources through mud, debris, and hazardous chemicals [5, 6]. For instance, if there is a flash flood, excess water from the river may overflow with a great number of materials originating from upstream, such as soil, debris, and even agricultural chemicals. This, in turn, contaminates the river water, which the community uses for clean water, and makes it unhygienic for consumption [6]. Landslides can also cause this. If a landslide happens, the material of the landslide may either cover the water sources or destroy clean water distribution infrastructures. Apart from that, landslide activities can introduce high loads of organic and inorganic materials, which, together with the pathogenic bacteria, require more intensive purification efforts to make the water sources safe for consumption.

In certain circumstances, various water treatment technologies and methods used at the point of use (POU) or in households (HWT) are the primary choices for providing clean water to the community. POU/HWT methods can serve as a temporary solution before permanent infrastructure becomes available [7, 8]. Currently available POU/HWT methods include boiling, chlorination, solar disinfection (SODIS), and bio-sand filters [9, 10]. These methods have proven effective in treating water, especially in eliminating pathogens. However, a challenge arises as chlorination and bio-sand filters require government assistance in providing the necessary equipment, while boiling and SODIS can be done without special equipment but have the potential for re-contamination [11]. Therefore, additional POU/HWT options are needed to address these challenges, particularly in West Sumatra.

Despite the various methods available for removing pathogenic bacteria from drinking water, adsorption is not popular. However, it is a well-known conventional method that eliminates almost all harmful

* Corresponding author: taufiqihsan@eng.unand.ac.id

microorganisms in drinking water [12]. Administer the powder after treating the water with a powdered adsorbent. One such adsorbent, layered double hydroxide (LDH), is gaining popularity in the scientific community due to its straightforward and inexpensive production and effectiveness in adsorbing pathogens. Research has also confirmed that LDH is safe for human use [13, 14]. However, when applied in drinking water treatment, the challenge lies in its nanomaterial form. To address this challenge, providing a structure to the adsorbent can make it easier to retrieve the adsorbent powder after the adsorption process.

The process of creating Mg-Al type LDH in situ was conducted by Fukugaichi et al. in 2022 [15]. In our study, we used an aluminum plate as a substrate to facilitate the movement of LDH when utilized as a water-purifying agent. We have redeveloped a single-step, in situ LDH manufacturing process. In this technique, LDH will be generated at the plate's surface by removing AlOH and then contacting aluminum with magnesium in one step or solution. We present a substitute way of POU/HWT that is low-cost, easy to apply, and suitable for disaster-prone areas, especially West Sumatra. We formed LDH on the surface of an aluminum plate by dipping it in seawater and $Mg(OH)_2$ fertilizer solution. This study is a feasible way to realize this method and to solve some problems caused during its realization in disaster areas.

## 2 Methods

A commercial aluminum square plate (A1100, PT. Surya Logam Universal, Indonesia), 25 $cm^2$ in area and 0.3 mm thick, was exposed to seawater from Padang Beach, West Sumatra, at pH 9.9, which had been adjusted by adding magnesium hydroxide, $Mg(OH)_2$, in a volume ratio of 100 mL. The plate was immersed for 24 hours at room temperature and then rinsed with distilled water. We prepared two soaking solutions—one with 0.5 g and the other with 1 g of $Mg(OH)_2$—and labeled the ensuing samples as LDH plate 0.5 and LDH plate 1, respectively. X-ray diffraction (XRD) was used to investigate the characteristics of these samples.

Before conducting the experiments on disinfection, all glassware and solutions were autoclaved at 121°C for 20 minutes in order to render them sterile. A tryptic soy broth solution with an initial content of $3.0 \times 10^7$ CFU/mL E. coli was serially diluted with sterile distilled water in order to prepare a working concentration of around 1000 CFU/mL. Following this, 100 mL of the freshly diluted bacterial suspension was introduced into 250 mL sample bottles (polypropylene).

The ability of the LDH plate to remove bacteria in clean water treatment was then tested in triplicate using 9 sample bottles. Each bottle contained approximately 1000 CFU/mL E.coli in conditioned distilled water. Three bottles were blank; three contained LDH plate 0.5, and the remaining three contained LDH plate 1. The sample bottles were agitated using a reciprocal shaker for 24 hours at a speed of 72 rpm and a temperature of 25°C. Samples of treated water were collected at the third and 24$^{th}$ hours and then analyzed using the plate count method to determine the viable count of E. coli. The culture medium used in this counting method is Lysogeny Broth agar.

## 3 Results and discussion

### 3.1 Synthesis and characterization of LDH plate

LDH plate formation occurs via a number of interesting chemical changes, starting with the preparation of a solution to soak. This soak solution is prepared by adding $Mg(OH)_2$ into seawater and results in quite a visibly distinct white, cloudy mixture. With vigorous agitation, the magnesium hydroxide does not fully dissolve, and some residual material can clearly be seen at the bottom of the container.

An aluminum plate is then promptly submerged into this freshly concocted solution. Within minutes, bubbles will begin to form on the aluminum surface, which should indicate an event of central significance: the break-down of the protective aluminum oxide layer native to the metal's surface, an event allowing for direct contact between metallic aluminum and magnesium—present in both the seawater (0.05 mol/L) and the magnesium hydroxide (0.01 g/L).

This dynamic interaction is a two-step process. First, the thin aluminum oxide layer reacts with the abundant hydroxide ions in the solution. After that, this oxide layer dissolves, and the metallic aluminum underneath appears. This newly exposed aluminum subsequently reacts with the solution, producing soluble compounds and liberating hydrogen gas. The net result of all this complicated chemistry is the formation of an Mg/Al-type LDH layer on the surface of the LDH plate.

Normally, after immersion, the surface of the LDH plate is adorned with a white powdery precipitate of magnesium hydroxide. After carefully rinsing and wiping this off, the brownish LDH plate is revealed. This whole process occurs in just about the same way with both LDH plate 0.5 and LDH plate 1 ((Fig. 1(b) and (c)).

During the process, seawater acts as an important medium with its rich composition of different salts. Above all, seawater acts as an electrolyte and helps in the flow of ions, which is necessary for these chemical reactions to take place. The magnesium ions in seawater play a part in the formation of the Mg/Al LDH layer.

Magnesium hydroxide provides a supplementary source of magnesium ions and also helps raise the solution's alkalinity. Such an alkaline medium is essential for the solubility of the aluminum oxide layer.

Aluminum oxide is very thin but forms a strong, adherent protective film on the aluminum surface. The hydroxide ions in the solution react with the oxide layer and dissolve it into mobile aluminate ions.

When the barrier of aluminum oxide is broken, the metallic aluminum reacts with water in the presence of hydroxide ions to form the respective aluminate ions and hydrogen gas. These aluminate ions then react with magnesium ions, which are present both in seawater and in magnesium hydroxide, to form a precipitate of Mg/Al LDH. The constitution of this LDH may differ in different reaction conditions; however, it normally includes positively charged layers of magnesium and aluminum hydroxide intercalated with negatively charged anions and water molecules.

The excess magnesium hydroxide in the white precipitate on the LDH plate after immersion can be easily rinsed off to leave a brownish underlying LDH layer. The brownish color of the LDH plate suggests that some iron impurities might have been introduced into the LDH structure from aluminum or seawater. Iron tends to give its compounds a brownish hue.

There are complex chemical reactions in the plate formation process of an LDH. On the other hand, deep understanding may be important in optimizing the synthesis and properties of LDH materials, which have tremendous potential for a wide range of applications, from catalysis and environmental remediation to energy storage.

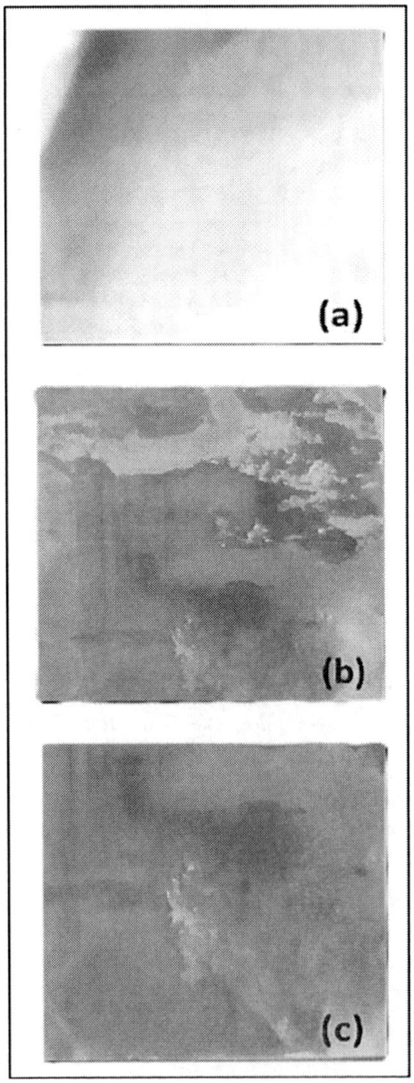


**Fig. 1.** (a) aluminum plate; (b) LDH plate 0.5; (c) LDH plate 1

Figure 1 shows the XRD pattern, which can be taken as conclusive evidence of the formation of an Mg-Al layered double hydroxide on the LDH plate surface. Characteristic peaks of Mg-Al LDH can be observed clearly enough, though broad in nature, which indicates a relatively low degree of crystallinity. This was not totally unexpected, considering the process of LDH formation onto this aluminum substrate is rapid and heterogeneous.

Notably, peaks for aluminum oxide are not found in the XRD pattern. This clearly implies that the native aluminum oxide layer on the aluminum plate has been completely dissolved by the soaking process. The dissolution of aluminum oxide already natively present from the aluminum plate is an important prerequisite to form the layer of Mg-Al LDH, exposing the metal aluminum to direct contact with reactive species in the soaking solution.

A comparison between the obtained XRD patterns of LDH plate 0.5, shown in Fig. 2(a), and LDH plate 1, shown in Fig. 2(b), gives a very fine but important difference. Obviously, peaks for LDH plate 1 have much stronger intensity than LDH peaks for LDH plate 0.5. This difference most likely implies that on LDH plate 1, more LDH has formed than on LDH plate 0.5, probably due to soaking conditions or due to features of the aluminum substrate itself.

The adhesion of the Mg-Al LDH layer to the underlying aluminum substrate seems to be pretty good since no LDH delamination or peeling is observed during sample preparation or analysis. This is very important for long-term stability and functionality with respect to the LDH plate, mainly under conditions that mean exposure to aqueous environments.

While Mg-Al LDH has been reported as safe and biocompatible [13, 14], its long-term safety is of essential interest, especially considering the long-time immersion in water. One can presume that the surface LDH layer would gradually dissolve over a very long period and release magnesium and aluminum ions from the surrounding area into the environment. A knowledge gap exists regarding the biological implications in this respect.

Further studies should be directed to the systematic assessment of the dissolution kinetics of Mg-Al LDH as a function of variable conditions, such as different pH, temperature, and ionic strength. These studies will help gain relevant insights into the long-term stability of the LDH plates and inform about safe and effective uses.

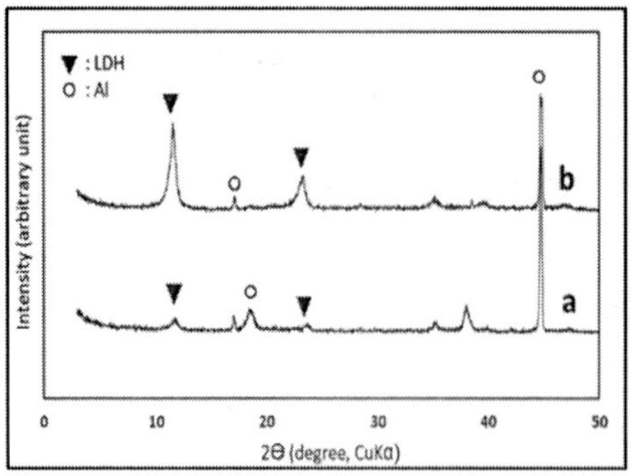


**Fig. 2.** XRD patterns of (a) LDH plate 0.5; (b) LDH plate 1

### 3.2 E. Coli removal test in water

The experimental results, exemplified by Fig. 3, show a good demonstration of the antibacterial properties of the LDH plates, particularly their potential for efficiently removing E. coli from an aqueous environment.

This control experiment, conducted in the absence of LDH plates, was important in creating a baseline for comparison, as there was no significant reduction in the viable count of E. coli bacteria even after an extended 24-hour incubation time. Therefore, this observation identified the resilience of E. coli and underlined that an active intervention needs to be implemented in order to eradicate it.

The introduction of LDH plate 0.5 into the E. coli suspension initiated a drastic drop in bacterial viability. Within 3 hours, the drop had reached as high as 87% in E. coli count. This antibacterial effect continued to increase over time until the complete eradication of E. coli after 24 hours.

LDH plate 1 demonstrated even more remarkable antibacterial power by reducing E. coli to nearly complete elimination, reducing it by 99% within the first 3 hours. After 24 hours, the count of E. coli was undetectable, which means the suspension was totally sterilized.

These results suggest that the superior performance of LDH plate 1 compared to LDH plate 0.5 is due to its higher LDH content, which was supported by the XRD measurements. This kind of observation could indicate a direct relationship between the amount of LDH present on the plate and its antibacterial efficacy. Increased LDH content on LDH plate 1 should have resulted from a higher $Mg(OH)_2$ dosage in its preparation.

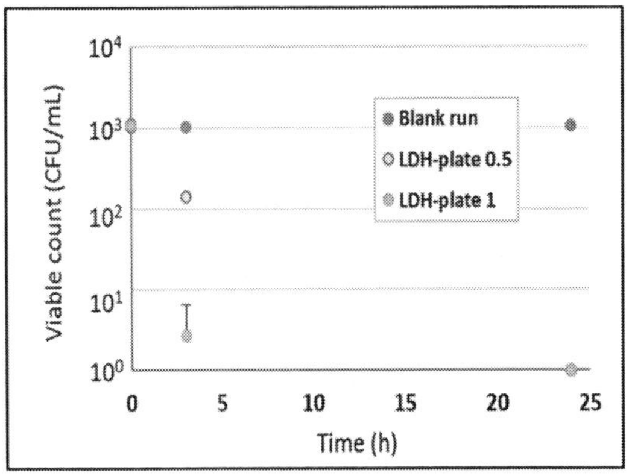


**Fig. 3.** The viable count of E. Coli was measured at different shaking times after adding LDH plates, starting with an initial viable count of approximately $10^3$ CFU/mL. The viable count of LDH plates at 24 hours was determined to be 0 CFU/mL. error bars in the data represent the standard deviation.

The high capacity for E. coli removal that the LDH plates, particularly LDH plate 1, exhibited presents a strong potential for technologies in water purification and disinfection. Their efficiency in removing a common but highly pathogenic bacterium such as E. coli presupposes their wider application against other classes of microorganisms.

Future experiments are needed to evaluate the antibacterial spectrum of LDH plates against a number of bacterial and viral cultures. Investigations into long-term stability and reusability are also essential for the practical application of such plates. The optimization of LDH composition and load on the plates for the maximum expression of their antibacterial activity with the least possible adverse effects is another important direction for further investigation.

Given that LDH has the ability to attract bacteria due to its positive charge (since most bacteria and viruses carry a negative charge), this discovery confirms that bacterial adsorption occurs through an electrostatic mechanism [16, 17]. This adsorption was evidenced by a disinfection experiment in which LDH plates were immersed into agar-containing plates. After 24 hours, signs of E. coli colonies were observed on the plate surface (data not shown). This indicates that the LDH plate is adsorbing bacteria rather than killing them. Further research is necessary to understand better the process of capturing E. coli and to determine how long the bacteria can survive on the surface of the LDH plate.

Using LDH plate 1, we tested the potential for bacterial removal without stirring (Fig. 4). In samples without agitation, the number of live E. coli decreased by 41% after 24 hours. In samples with occasional stirring at hours 1, 3, and 5, the decrease in E. coli was 55% after 3 hours and 92% after 24 hours. This indicates the importance of stirring to allow the LDH plate to come into contact with water and adsorb E. coli.

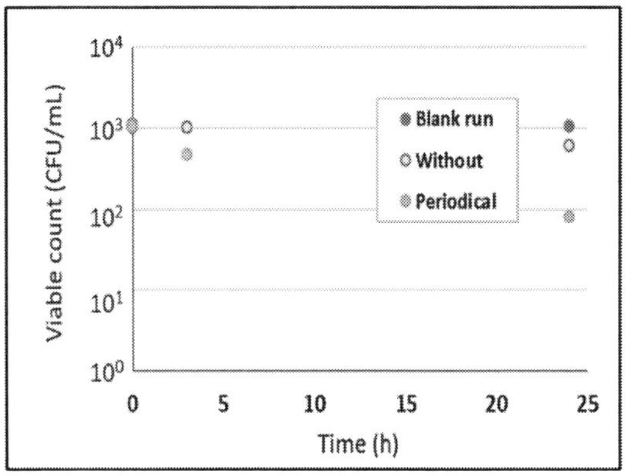


**Fig. 4.** The effect of mixing frequency without stirring and periodically stirring. error bars in the data represent the standard deviation.

This is a very complex phenomenon, given the many confluences of factors involved. The understanding of these factors is, therefore, highly important for the optimization of the design and application of LDH plates in any antibacterial setting.

The addition of $Mg(OH)_2$ used as a dosage in the preparation during the soaking solution is a very important process in the fabrication of LDH plates and has a very large degree of effect on the adsorption of the bacteria. While differences in $Mg(OH)_2$ dosage could indeed make a clear difference in E. coli removal at shorter time periods, for example, 3 hours, these differences seem to fade out over a long time period, for instance, 24 hours. This would imply that while the dosage of $Mg(OH)_2$ could affect the initial rate of bacterial adsorption, the ultimate adsorption capacity of the LDH plate is not very much influenced.

Stirring has a significant impact on bacterial adsorption to LDH plates. By offering continuous mixing and agitation, the act brings LDH plate surfaces into better contact with the bacterial suspension; therefore, successful adsorption is more likely to take place. Very interesting data shows that E. coli adsorption using stirring increases significantly.

Other factors that would probably exert a lot of influence on the adsorption of bacteria onto LDH plates, besides the amount of dosed $Mg(OH)_2$ and stirring, are:

1. The influence of the surrounding pH can change the surface charges of both the LDH plate and the bacterial cell, which, in turn, can lead to changes in potential electrostatic interactions between the two, hence affecting adsorption.
2. Temperature is an important factor that will influence adsorption kinetics. Temperature has a direct influence on metabolic activity and the viability of bacteria. The latter are factors that can regulate their adsorption behavior.
3. Finally, initial bacteria suspension concentration can have an effect on the quality behavior of the saturation kinetics of adsorption. Increased initial concentrations could lead to the earlier saturation of the LDH plate surface.

That is why these factors have to be examined in detail with regard to both their individual and synergetic effects on bacterial adsorption.

The detailed mechanisms of bacterial adsorption onto LDH plates are still under study, but some plausible models have been proposed, as shown in Fig. 5. One model based on the electrostatic attraction suggests that the positively charged surface of LDH plates could attract the negatively charged phosphate groups ($PO_4^-$) present in E. Coli. This could improve the electrostatic attraction between the coli cell surface and enable preliminary attachment of the bacteria to the LDH plate. One evident means is that phosphate groups on the outer surface of the bacteria exchange with the hydroxyl $OH^-$ groups on the LDH plate. This form of ligand exchange would lead to a stronger and more time-proof binding of the bacteria on the LDH plate [18].

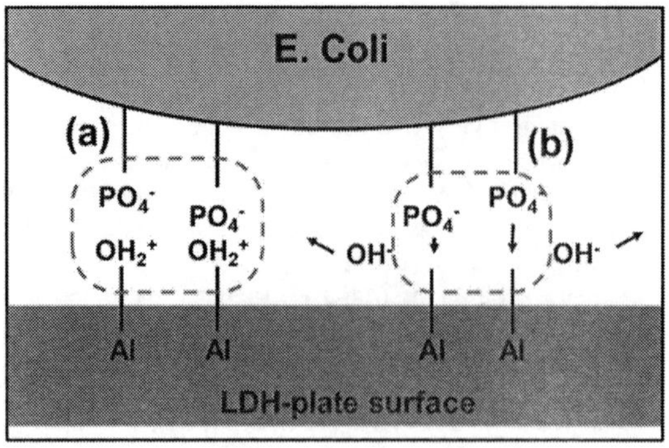


**Fig. 5.** Possible modes of E. coli binding to the LDH plate

Indeed, many studies have employed an ability to indicate aluminum minerals in soil, with a high number of attached microorganisms near them, and attached to $Al(OH)_3$ surfaces support a concept of electrostatic attraction as a major one to drive bacterial adhesion.

### 3.3 Future challenges

Our research on bacterial removal showcases the potential of using LDH plates as a practical and easy-to-produce disinfection tool. In West Sumatra, seawater can serve as a source of magnesium, and magnesium hydroxide ($Mg(OH)_2$), a commonly available fertilizer, enables the creation of highly feasible LDH plates.

Compared to most other POU/HWT methods used in disaster-stricken areas, the LDH plate method has a number of advantages. First, while chlorination and bio-sand filters require the assistance of the government for equipment, this method uses readily available materials at low cost, such as seawater and $Mg(OH)_2$ fertilizer. Second, while the boiling and SODIS methods involve risk due to re-contamination, the bacteria are adsorbed on the surface of the LDH plate. Such simplicity and low cost of the LDH plate method create opportunities for providing safe drinking water in emergency situations and resource-limited settings.

Although LDH plates represent an inexpensive and straightforward method for the purification of water, especially in emergency cases, further testing of these plates on other bacterial models and also on several natural sources of water is required. Also, as natural water can have a vast content of different organic compounds that may influence the extent to which microorganisms are present, it will be obligatory to perform tests on natural water samples, although it is put forward that, in general, low counts of E. coli bacteria are present in natural water [19].

## 4 Conclusion

In this study, a facile and effective method was developed for the removal of pathogenic bacteria from water using layered double hydroxides grown on an Al plate. An aluminum plate had to be immersed in a mixture of seawater and $Mg(OH)_2$ to form LDH, with a relatively high efficiency of removal for E. coli bacteria in water.

Beyond the efficacy of the method, it presents the cost-effective and facile execution of the technique, in particular in disaster-stricken areas like Sumatra Barat. Materials available on hand, or easily procured, are seawater and $Mg(OH)_2$, which are coupled with quite basic equipment: aluminum plates.

The research, however, also raised a number of further challenges that need to be addressed in future studies by testing on other bacterial models and natural water sources varying in different organic contents. More studies will hence be required to establish the efficacy and safety of LDH plates under more diversified and complicated conditions.

This research, therefore, contributes to the further development of point-of-use/household water treatment technologies by providing a practical, efficient means to produce clean drinking water in a disaster area.

## References


1. A. Ichlas, Indonesia's toll rises to 62 from deadly Sumatra floods, 25 still missing. Reuters.com (2024) Available at https://www.reuters.com/world/asia-pacific/indonesia-searches-35-still-missing-deadly-sumatra-floods-2024-05-15/
2. W. Wilopo and T. F. Fathani, The mechanism of landslide-induced debris flow in geothermal area, Bukit Barisan Mountains of Sumatra, Indonesia. J. Appl. Eng. Sci. **19**, 688-697 (2021) https://doi.org/10.5937/jaes0-29741
3. Y. Wang, Y. Gao, C. K. Morley, E. G. Seagren, X. Qian, J. M. Rimando, P. Zhang, and Y. Wang, Accelerated exhumation within the sumatran fault: Implications for late cenozoic evolution of Sumatra (Indonesia). Geophys Res Lett. **50**, e2022GL100028 (2023). https://doi.org/10.1029/2022GL100028
4. UN Water, Water and disasters. United Nations (2023) Available at https://www.unwater.org/water-facts/water-and-disasters
5. C. Nunez, Floods, explained. National Geographic (2019) Available at https://www.nationalgeographic.com/environment/article/floods.
6. J. Yin, Y. Gao, R. Chen, D. Yu, R. Wilby, N. Wright, Y. Ge, J. Bricker, H. Gong, and M. Guan, Flash floods: Why are more of them devastating the world's driest regions? Nature. **615**, 212-215 (2023). https://doi.org/10.1038/d41586-023-00626-9
7. J. Wu, M. Cao, D. Tong, Z. Finkelstein, and E. M. Hoek, A critical review of point-of-use drinking water treatment in the United States. npj Clean Water. **4**, 1-25 (2021). https://doi.org/10.1038/s41545-021-00128-z
8. Y. Deng, Making waves: Principles for the design of sustainable household water treatment. Water Res. **198**, 117151 (2021). https://doi.org/10.1016/j.watres.2021.117151
9. CDC, Household Water Treatment. Centers for Disease Control and Prevention. (2022) Available at https://www.cdc.gov/healthywater/global/household-water-treatment.html
10. T. Ihsan and V. Derosya, Drinking water problems in rural areas: Review of point-of-use methods to improve water quality and public health. Larhyss J. **58,** 55-71 (2024).
11. T. Ihsan, E. Johan, S. Fukugaichi, M. Maruyama, S. Mitsunobu, and N. Matsue, Innovative DIY drinking water disinfection for underserved communities. Sci. Total Environ. **927**, 172257 (2024). https://doi.org/10.1016/j.scitotenv.2024.172257
12. L. Sellaoui, M. Badawi, A. Monari, T. Tatarchuk, S. Jemli, G. L. Dotto, A. Bonilla-Petriciolet, and Z. Chen, Make it clean, make it safe: A review on virus elimination via adsorption. Chem Eng J. **412**, 128682 (2021). https://doi.org/10.1016/j.cej.2021.128682
13. C. D. Hoyo, Layered double hydroxides and human health: an overview. Appl. Clay Sci. **36**, 103–121 (2007). https://doi.org/10.1016/j.clay.2006.06.010
14. C. Forano, F. Bruna, C. Mousty, and V. Prevot, Interactions between biological cells and layered double hydroxides: Towards functional materials. Chem. Rec. **18**, 1–18 (2018). https://doi.org/10.1002/tcr.201700102
15. S. Fukugaichi, Y. Tomosugi, and H. Aono, Facile synthesis of hydrophilic layered double hydroxide film on aluminum plate. Inorg. Chem. Commun. **142**, 109647 (2022). https://doi.org.10.1016/j.inoche.2022.109647
16. T. Ihsan, E. Johan, S. Fukugaichi, and N. Matsue, Enhancing rural drinking water safety using Mg-Al-type layered double hydroxide foil as a new point-of-use disinfection tool. J. Water Sanit. Hyg. Dev. **13**, 921–930 (2023). https://doi.org/10.2166/washdev.2023.206
17. T. Ihsan, E. Johan, S. Fukugaichi, and N. Matsue, Aluminum foil immersed in alkalized seawater removes Escherichia coli from drinking water. J. Water Sanit. Hyg. Dev. **13**, 681–686 (2023). https://doi.org/10.2166/washdev.2023.057
18. S. Fukugaichi, S. Maruyama, E. Johan, and N. Matsue, Facile removal of Escherichia coli from water using amorphous membrane fabricated on aluminum plate. Emergent. mater. (2024). https://doi.org/10.1007/s42247-024-00816-9
19. C. Zimmer, and C. C. Dorea, Differences in laboratory versus field treatment performance of point-of-use drinking water treatment methods: research gaps and ways forward. npj Clean Water **6**, 26 (2023). https://doi.org/10.1038/s41545-023-00241-1

E3S Web of Conferences 604, 05002 (2025) https://doi.org/10.1051/e3sconf/202560405002
*The 4th ICDM 2024*

# Characteristics of disease in children at earthquake evacuation camps in West Pasaman Regency and Pasaman Regency in 2022

*Abdurrahman* Arsyad As Siddiqi[1,2*] and *Indra* Ihsan[1,2]

[1]Department of Child Health, Faculty of Medicine, Universitas Andalas, Padang, Indonesia
[2]Department of Pediatrics, Dr. M. Djamil General Hospital, Padang, Indonesia

**Abstract:** Extraordinary occurrences of infectious diseases may occur when earthquakes result in large population movements to unplanned and overcrowded shelters with limited access to safe food and water sources. An earthquake with a magnitude 6.1 shook West Pasaman Regency, West Sumatra Province, on Friday (2/25/2022). Many disaster victims sleep in open areas so that they are exposed to dust, heat, and poor hygiene. Method: A descriptive study conducted retrospectively from pediatric patient data who came to health facility tents in refugee camps in West Pasaman and Pasaman Regency, collected from February 2022 to April 2022. The total number of visits was 247 pediatric patients. Air-borne disease suffered by 78.5%, followed by food and water-borne diseases (13%) and vector related disease (2.8%). Acute respiratory infection is the most common disease suffered by 184 cases (74.4%) followed by acute diarrhea (5.2%). There were also 8 cases of measles that were potential for outbreaks infectious diseases. Airborne disease is the most common disease suffered by refugees, followed by food-water borne and vector related disease. The availability of complete data on diseases can facilitate the implementation of optimal health services and help prevent the occurrence of outbreaks among disaster victims.

## 1 Introduction

The definition of disaster according to law no. 24 of 2007 about Disaster Management is an event or series of events that threatens and disrupts people's lives and livelihoods caused by either natural factors and/or non-natural factors or human factors, resulting in human casualties, environmental damage, property loss, and psychological impact [1].

Disaster risk index data shows that West Sumatra is one of the provinces with a high disaster index, with an index score of 150.24. The districts in West Sumatra that have the highest disaster score index are Agam, West Pasaman and Mentawai Districts [2]. Data for 2021 shows that West Sumatra Province has been hit by various disasters such as one earthquake, 23 landslides, 58 floods and fires. forests 16 times, and extreme weather 118 times [3].

Emergency action is required to save human victims and their environment during disasters, which can happen abruptly or gradually and affect regular daily patterns. Natural, man-made, or non-natural disasters all result in property losses, environmental harm, human casualties, and psychological effects that, in some cases, might impede a country's development (Figs. 1 and 2) [4].

There are many different types of disasters, such as natural disasters (such as volcanic eruptions, earthquakes, storms, droughts, heat waves, flooding, and tsunamis) and those caused by armed conflict (such as political unrest, terrorism, civil war, and genocide [5]. Geographically speaking, Indonesia is one of the nation's most vulnerable to natural disasters, including earthquakes, tsunamis, mountain eruptions, etc. This is because the country is situated where three major plates the Eurasian, Pacific, and Indo-Australian plates meet [6]. Natural catastrophes inflict death and injury, disrupt economic growth and activity, destroy infrastructure, uproot communities, and raise the possibility of infectious disease outbreaks. Natural catastrophes impacted 61.7 million people worldwide in 2018, resulting in 10,373 fatalities and millions of dollars' worth of losses [7].

**Fig. 1.** Health facility tent at Timbo Abu, Pasaman Barat

* Corresponding author: nandaarsyad16@gmail.com

**Fig. 2.** Health facility tent at Aur Village, Pasaman

Earthquakes are the second most frequently reported natural disaster (after floods) and the first among geophysical disasters [8]. Friday, February 25, 2022, at 08.39 a.m. local time, an earthquake with a magnitude of 6.2, later revised to 6.1, struck West Pasaman Regency, West Sumatra [3]. The earthquake caused 4 fatalities, 37 serious injuries, 310 minor injuries, and approximately 10,000 people to be displaced at 35 evacuation points located in Talamau District, Pasaman District, and Kinali District of West Pasaman Regency. In addition, in Pasaman Regency, 6 people died, 5 people were seriously injured, 36 people were slightly injured, 3 thousand people were displaced, and 4 people were buried in landslide material [9].

Refugees are people or groups of people who have been forced or forced to flee or leave their homes or places where they used to live, especially because of or to avoid various problems. In general, the problems that give rise to displacement are the impact of armed conflict, vulnerable situations characterized by widespread acts of violence in general, violations of human rights, natural or man-made disasters. These refugees do not cross internationally recognized state borders [2, 10, 11]

Refugees' patterns in Indonesia vary greatly depending on the type of disaster, duration of evacuation and preparation efforts. The insert refugee pattern, namely that refugees stay in the homes of relatives. Refugees are concentrated in public places or in prepared barracks. Another pattern of evacuation is in emergency tents on either side of their houses damaged by the disaster [1, 2, 10–12].

One of the impacts of disasters on the declining quality of life of the population can be seen from the various public health problems that occur (Fig.3). Disasters followed by evacuation have the potential to cause health problems that start with problems in other fields/sectors. Earthquakes, floods, landslides and volcanic eruptions, in the short term can have an impact on fatalities, victims of serious injuries requiring intensive care, increased risk of infectious diseases, damage to health facilities and water supply systems. The emergence of health problems, among others, begins with a lack of clean water which results in poor personal hygiene, poor environmental sanitation which is the beginning of the proliferation of several types of infectious diseases [13].

Insufficient food supplies are also the beginning of the process of declining health levels, which in the long term will directly affect the level of fulfillment of nutritional needs of disaster victims. The existing evacuation shelters often do not meet health requirements so that directly or indirectly they can reduce the body's immune system and if not addressed immediately will cause problems in the health sector. Meanwhile, the provision of health services in disaster conditions often encounters many obstacles due to damaged health facilities, inadequate number and types of medicines and medical devices, limited health workers and operational funds. This condition can certainly have worse impacts if not handled immediately [14].

Morbidity that occurs after a disaster due to damaged environmental conditions, health services and refugee density is the emergence of diseases, both infectious and non-infectious diseases. Non-infectious diseases that arise include physical injuries (broken bones) and degenerative diseases (heart, hypertension, stroke). Meanwhile, infectious diseases include infectious diseases immediately after trauma (wounds, sepsis), direct infectious diseases and indirect infectious diseases (airborne, waterborne, foodborne, vectorborne) [15].

Infectious diseases are a problem that needs great attention, given the potential for the emergence of extraordinary events (KLB) or outbreaks of infectious diseases. Epidemics can happen when earthquake catastrophes force huge numbers of people into crowded, unplanned shelters with limited access to clean water and food. Aside from the direct effects of earthquakes on sanitary conditions, outbreaks can also arise from the breakdown of water and sanitation infrastructure. The prevalence of infectious diseases depends on the disaster; waterborne diseases, foodborne diseases, and vectorborne diseases are prevalent in hurricanes and floods, while airborne diseases are prevalent in all types of natural disasters [8].

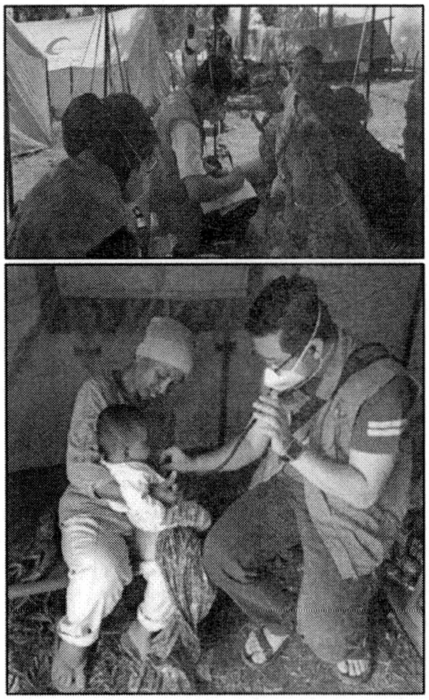

**Fig. 3.** Medical services at health facility tents.

The population of West Pasaman Regency in 2021 was 467,876, while the population of Pasaman Regency in 2022 was 307,425 [16, 17]. In a disaster, children make up between 25–30% of those injured [18, 19]. Special biological, social, and ethical elements make caring for pediatric populations special, necessitating further considerations in emergency response and disaster preparedness approaches [18, 20, 21]. Infants, toddlers, and children are vulnerable groups during disasters because they are groups that are at high risk of experiencing greater impacts than other community groups. Protection of vulnerable groups is carried out by providing priority handling in the form of rescue, evacuation, security, health, and psychosocial services [4].

Therefore, it is important to know the characteristics of diseases experienced by these vulnerable groups in post-disaster refugee camps in order to provide optimal health services and prevent extraordinary events for disaster victims.

## 2 Method

A descriptive study conducted retrospectively from pediatric patient data who came to the health facility tents in refugee camps in West Pasaman and Pasaman districts were collected by The Disaster Team of Department of Child Health, Faculty of Medicine, Universitas Andalas and The Indonesian Pediatric Society Disaster Team West Sumatra Branch, data collected from February 2022 to April 2022. (Shown in Fig. 4)

**Fig. 4.** The disaster team of Department of Child Health, Faculty of Medicine, Universitas Andalas

## 3 Results

During the study period, there were 247 pediatric patients who came to health facility tents in evacuation sites in West Pasaman Regency and Pasaman Regency. Children who visited the health facility tents were found to be more female (54.7%) than male (45.3%), and the age of children who visited was more in the age group 1–5 years (51.4%), followed by the age group 6–17 years (39.3%) and <1 year (9.3%). The characteristics of the study subjects can be seen in Table 1.

**Table 1.** Research subject characteristics

| Variable | *F* | % |
|---|---|---|
| **Gender** | | |
| Boy | 112 | 45.3% |
| Girl | 135 | 54.7% |
| **Age** | | |
| <1 year old | 23 | 9.3% |
| 1-5 years old | 127 | 51.4% |
| 6-17 years old | 97 | 39.3% |

The study showed that the most common disease suffered by children affected by the earthquake was acute respiratory tract infection (ARI) (74.4%), followed by acute diarrhea (5.2%), impetigo and measles (3.2% and 3.2%). Data on the top ten most common diseases among pediatric patients who came to health facility tents at evacuation sites in West Pasaman Regency and Pasaman Regency can be seen in Table 2.

**Table 2.** Top ten most common diseases

| Disease | *f* | % |
|---|---|---|
| Acute respiratory tract infection | 184 | 74.4 |
| Acute diarrhea | 13 | 5,2 |
| Impetigo | 8 | 3.2 |
| Measles | 8 | 3.2 |
| Scabies | 7 | 2.8 |
| Allergic/irritant contact dermatitis | 7 | 2.8 |
| Dyspepsia | 3 | 1.2 |
| Ginggivitis<br>Tinea<br>Tension headache | 2<br>2<br>2 | 0.8<br>0.8<br>0.8 |

The type of infectious disease in pediatric patients who came to health facility tents in evacuation sites in West Pasaman Regency and Pasaman Regency was mostly airborne diseases (78.5%), followed by food and waterborne diseases (13%). The research data can be seen in Table 3.

**Table 3.** Type of infectious diseases

| Type of infectious disease | *f* | % |
|---|---|---|
| *Air borne disease* | 194 | 78.5 |
| *Food and waterborne disease* | 32 | 13 |
| *Vector related disease* | 7 | 2.8 |

## 4 Discussion

Natural catastrophes have a detrimental physical, social, and psychological impact on children, who are a particularly vulnerable group [22]. This study collected data from 247 pediatric earthquake survivors who came to the Disaster Team Tent of the Department of Pediatrics, the Faculty of Medicine of Universitas Andalas and the West Sumatra Branch of the IDAI Disaster Team in the evacuation camps from February to April 2022. Air-borne, water-food borne, vector-borne diseases, and skin and soft tissue infections were mostly reported in the pediatric population of earthquake survivors. The primary causes of illness were seasonal factors, overcrowding during an evacuation, malnourishment, and water resource contamination [22].

Female pediatric patients slightly outnumbered male pediatric patients (54.7%: 45.3%), in contrast to pediatric earthquake survivors in Turkey in 2023, where males outnumbered females (50.9% male and 49.1% female). The age group that visited the health post the most was the age range of 1–5 years (51.4%). Whereas among the Turkish earthquake victims in 2023, the most common age group was 6–12 years old (36.8%) [23].

The most common disease suffered was acute respiratory infection (ARI) (74.4%), this is in accordance with data observed in the post-earthquake period in various parts of the world. Most people affected by earthquakes live in overcrowded refugee camps, among the risk factors include poor personal hygiene, contaminated drinking water, and insufficient air ventilation that allow them to suffered acute respiratory infectious diseases [24]. The second most common disease is acute diarrhea, which can be caused by a lack of access to adequate water supplies as a result of water network damage after a disaster and/or sewage contamination, which is typical issue [22]. There was no emergency disease suffered by refugee children, the clinical condition of patients was stable, and no patients complained of physical injuries due to trauma.

Air-borne disease is one of the most typical findings discovered after earthquakes. 42.3% of the 108 pediatric patients evaluated following the 2015 Gorkha earthquake had respiratory tract infections. Giri et al.'s study supported the findings of Wang et al.'s, which said that children are among the most vulnerable populations and that earthquakes can have an impact on children of various ages. Among the 1,057 patients analyzed, children from households and regions severely impacted by the earthquake had a considerably greater rate of patients needing hospitalization for pneumonia. The study reported by Tan et al. on the Pariaman earthquake in West Sumatra reported respiratory infections as the most common diagnosis (47.2%), musculoskeletal/joint (31.6%), skin infections (13.3%), and gastrointestinal diseases (18.2%). Therefore, it is essential to the livelihoods of displaced communities to guarantee that there are efficient systems for the surveillance and registration of infectious diseases, functioning primary health care facilities, functioning water and sanitation systems, and temporary shelter support [24, 25].

The measles virus is extremely contagious and can be transmitted by sneezing, coughing, or close contact with contaminated nose or throat secretions. The 2004 Indian Ocean tsunami devastated a community whose level of basic vaccine coverage was seen to have an impact on the transmission of measles, particularly in children under the age of fifteen. A large vaccination program did not prevent occasional cases of measles from occurring, and after the 2004 tsunami disaster in Banda Aceh, Indonesia, there was a cluster of 35 cases in a susceptible community living in disorganized and crowded settlements in the North Aceh region [25].

The ingestion of water or food polluted with harmful microorganisms (viruses, bacteria, and parasites) from human or animal waste is the primary cause of waterborne and foodborne illnesses. With a median of 168.8 cases per week per 1000 evacuees, acute respiratory illness (ARI) was the most prevalent infectious disease during the 2011 earthquake disaster in Japan. Acute gastroenteritis came in second with a median of 23.7 cases per week per 1000 evacuees [24-26].

Skin infections and wounds are terms used to describe the infiltration of one or more types of bacteria into tissues. The environment in which the injury occurs, the severity of the injury, the microorganisms on the injured person's skin, the microorganisms the person has been exposed to during wound healing, and the person's overall health and immune status all play a major role in determining the type of wound infection. Following an earthquake, trauma injuries disturb the immune system's delicate balance and increase the risk of infectious problems. An extended period spent under debris might result in serious injuries and a higher chance of acquiring infections. Wound and skin infections were reported after the Izmit (Turkey) earthquake in 19 99, the Kashmir (Pakistan) earthquake in 2005, the Sumatra-Andaman earthquake (Indonesia) in 2004, and the Yogyakarta (Indonesia) earthquake in 2006. In the first three months after the 1999 Izmit earthquake, squamous erythematous, itchy, neurocutaneous dermatoses, and eczema predominated and were caused by psychoemotional factors associated with the earthquake [24].

Disruptions to water supplies, health services, and sanitation systems after an earthquake can raise the likelihood of infectious disease outbreaks. To reduce the impact on the impacted population, it is imperative to put in place efficient preventive measures. In these circumstances, vaccination becomes a crucial instrument in defending against the spread of illnesses including measles, cholera, typhoid, and hepatitis A. Several research related to catastrophes have highlighted the significance of particular vaccinations in managing and reducing illnesses following earthquakes [27].

The world can learn from the natural disasters that have affected different countries at different times in the past. Specifically, there are valuable lessons to be learned from the natural disasters that followed the COVID-19 pandemic, which disrupted immunization programs and caused a resurgence of common infectious diseases in children. Following the 1991 Mount Pinatubo earthquake, there was a measles outbreak in the Philippines, with 18,000 cases of the disease documented [22]. Children's affected by the earthquake in West Pasaman and Pasaman districts were found to have 8 cases of measles (3.2%) in the refugee camps and have the potential to become an extraordinary event.

The limitation of this study is that the data obtained only comes from recording pediatric patient visits at health posts by The Disaster Team of Department of Child Health, Faculty of Medicine, Universitas Andalas and The Indonesian Pediatric Society Disaster Team West Sumatra Branch in refugee camps. The data obtained is not a combination of all the data from the disaster volunteer team who participated in serving the health problems of children affected by the disaster, so it cannot represent the complete condition of the disease that occurs in children affected by the disaster.

## 5 Conclusion

The total number of visits to health posts by children affected by the earthquake in West Pasaman Regency and Pasaman Regency in 2022 was 247 patients. The distribution of diseases suffered by children was obtained. Airborne disease is the most common disease suffered by refugees, followed by food-water borne disease and vector related disease. ARI is the most common disease suffered by refugees, namely 184 cases, followed by diarrhea in 13 cases.

Future research during a disaster should be able to collect all data from volunteers in all posts at the disaster site, including from existing health facilities. The government should have a good surveillance and coordination system so that the activities and records of each volunteer can be collected and published in one main register, so that they can be taken into consideration by the government in making quick and appropriate policies, especially in the health sector related to disasters experienced. Especially in handling and preventing specific infectious diseases after disasters, including curative efforts (handling cases), surveillance of potential infectious diseases and identification of risk factors at disaster sites, as well as promotional and preventive efforts in order to minimize risk factors at disaster sites.

## References

1. P. R. Indonesia, Undang-Undang No. 24 Tahun 2007 tentang Penanggulangan Bencana. (2007).
2. B. N. P. Bencana, Rencana Nasional Penanggulangan Bencana 2020-2024. (2020).
3. R. Wallansha, Ulasan Guncangan Tanah Akibat Gempa bumi Timur Laut Pasaman 25 Februari 2022, BMKG, (2022).
4. Presiden Republik Indonesia, Undang-Undang Republik Indonesia Nomor 24 Tahun 2007 tentang Penanggulangan Bencana. Jakarta, Indonesia, (2007).
5. Charnley G.E.C., Kelman I., and Gaythorpe K., Understanding the risks for post-disaster infectious disease outbreaks: a systematic review protocol., BMJ Open, **10** (2020).
6. Y. R. Amri M.R, and Yulianti G, Risiko bencana Indonesia. Jakarta: BNPB, (2016).
7. S. J. Suk J.E, Vaughan E.C, and Cook R.G, Natural disasters and infectious disease in Europe: a literature review to identify cascading risk pathways. Eur. J. Public Health, **30**(5), 928–935 (2019).
8. O.H. Kouadio I.K, Aljunoid S, Kamigaki T, and Hammad K, Infectious diseases following natural disasters: prevention and control measures. Expert Rev Anti Infect Ther. **10**(1), 95–104 (2012).
9. K.K.B.P.M. and K. R. Indonesia, Keputusan Menteri Koordinator Bidang Pembangunan Manusia Dan Kebudayaan Republik Indonesia, J. Bus. Theory Pract., **10**(2), 6 (2022), [Online]
10. B. N. P. Bencana, Rencana Kontinjensi Menghadapi Bencana Tsunami Provinsi Sumatera Barat, Komunitas Siaga Bencana, 9–10, (2011).
11. UNICEF, Situasi Anak di Indonesia-Tren, Peluang , dan Tantangan dalam Memenuhi Hak-Hak Anak. (2020).
12. S. I. Hafni Andayani, Manajemen Pelayanan Kesehatan pada Pengungsi Pasca Bencana, J. Kedokt. Nanggroe Med., **3**(3), 23–29 (2020).
13. S. J.R, Bencana Alam: Perlindungan Kesehatan Masyarakat. Jakarta: EGC (2006).
14. P. P. M. K. S. Jenderal D. Kesehatan, Standar Minimal Penanggulangan Masalah Kesehatan Akibat bencana dan Penanganan Pengungsi. jakarta, (2001).
15. Z. F. Widayatun, Permasalahan Kesehatan dalam Kondisi Bencana: Peran Petugas Kesehatan dan Partisipasi Masyarakat, J. Kependud. Indones. **8**(1), 37–49 (2013).
16. D. Pasbar, Buku Profil Kesehatan Kab. Pasaman Barat Tahun 2021, **2** (2021).
17. D. Kesehatan, Profil Kesehatan Kabupaten Pasaman Tahun 2022, Lubuk Sikaping, (2022).
18. M. Chiu, L. Goodman, C. H. Palacios, and M. Dingeldein, Children in disasters, Semin. Pediatr. Surg., **31**(5), 151219, (2022), https://doi.org/10.1016/j.sempedsurg.2022.151219.
19. N. Gubbins and B. D. Kaziny, The Importance of Family Reunification in Pediatric Disaster Planning, Clin. Pediatr. Emerg. Med., **19**(3), 252–259, (2018), https://doi.org/10.1016/j.cpem.2018.08.007.
20. R. V Burke, E. Iverson, C. J. Goodhue, R. Neches, and J. S. Upperman, Disaster and mass casualty events in the pediatric population, Semin. Pediatr. Surg., **19**(4), 265–270, (2010), https://doi.org/10.1053/j.sempedsurg.2010.06.003.
21. M. Newman and C. F. D. Leochico, Promoting disaster preparedness for children with special healthcare needs: A scoping review, J. Clim. Chang. Heal. **8**, (2022), https://doi.org/10.1016/j.joclim.2022.100145.
22. J. W. G. Selin Ugrakli, Mehmet Ozdemir, Infection disease following natural disaster in children : health prevention and assessment, J. Pediatr. Infect. Dis.,

**18**(3), 113–115, (2023).

23. M. Uluöz and M. Y. Gökmen, The 2023 Turkey Earthquake: Management of 627 Pediatric Musculoskeletal Injuries in the First Month, Children, **10**(11), (2023), https://doi.org10.3390/children10111733.
24. M. Mavrouli, S. Mavroulis, E. Lekkas, and A. Tsakris, The Impact of Earthquakes on Public Health: A Narrative Review of Infectious Diseases in the Post-Disaster Period Aiming to Disaster Risk Reduction., Microorganisms, **11**(2), Feb. (2023), https://doi.org10.3390/microorganisms11020419.
25. M. Mavrouli, S. Mavroulis, E. Lekkas, and A. Tsakris, Respiratory Infections Following Earthquake-Induced Tsunamis: Transmission Risk Factors and Lessons Learned for Disaster Risk Management., Int. J. Environ. Res. Public Health, **18**(9), May (2021), https://doi.org10.3390/ijerph18094952.
26. T. Kawano, K. Hasegawa, H. Watase, H. Morita, and O. Yamamura, Infectious Disease Frequency Among Evacuees at Shelters After the Great Eastern Japan Earthquake and Tsunami: A Retrospective Study, Disaster Med. Public Health Prep., **8**(1), 58–64, (2014), https://doi.org10.1017/dmp.2014.15.
27. E. Şevketoğlu, Management of pediatric emergencies during natural disasters: The 2023 earthquake in Turkey, Glob. Pediatr., **7**. December 2023, 100143, (2024), https://doi.org10.1016/j.gpeds.2024.100143.

# Catch-up immunisation for children in earthquake-affected areas of West Pasaman, Indonesia

*Rahmat* Syawqi[1,2*], *Finny* Fitry Yani[1,2], *Asrawati*[1,2], and *Rinang* Mariko[1,2]

[1]Department of Child Health, Faculty of Medicine, Universitas Andalas, Padang, Indonesia
[2]Department of Pediatrics, Dr. M. Djamil General Hospital, Padang, Indonesia

**Abstract.** Children are a particularly vulnerable population when exposed to natural disasters and experience increased problems related to physical health, mental health and post-disaster learning. The conditions of displacement during a disaster will cause an increased risk of morbidity to mortality of infectious diseases and even potential outbreaks. The success of immunisation efforts at the scene of a disaster is a successful effort to reduce the mortality rate due to Immunisation Preventable Diseases. However, in emergency or disaster situations, there can be significant disruption or reduction in routine immunisation services, delays or temporary suspension of immunisation programmes. Immunisation achievements in West Pasaman Regency, especially in Kenagarian Kajai, still have not reached the target of forming herd immunity. The Covid 19 pandemic is one of the causes of low immunisation achievements and the earthquake could further reduce it. Efforts to achieve immunisation targets must also be made in a disaster situation, one of which is in the form of an immunisation catch-up program. We collaborated with various parties in Kenagarian Kajai, West Pasaman to conduct catch up immunisation through posyandu as an effort to ensure children in disaster locations still have the opportunity to get immunised according to the recommended immunisation schedule.

## 1 Introduction

Indonesia is an archipelago of more than 17,000 islands covering an area of 1.9 million square kilometres and inhabited by more than 1,300 ethnic groups. One-third of Indonesia's population consists of children, approximately 80 million people. Indonesia has the fourth largest child population in the world. Indonesia sits on the Pacific Ring of Fire, a long chain of active volcanoes that is the source of 90% of the world's earthquakes. Indonesia is also the meeting place of three major tectonic plates, namely the Indo-Australian, Eurasian and Pacific plates, so it has great potential for earthquake disasters including liquefaction and tsunamis. This geographical position makes Indonesia one of the most vulnerable countries to natural disasters such as earthquakes, tsunamis, floods, forest fires, and volcanic eruptions [1]. Children are a particularly vulnerable population when exposed to natural disasters and experience increased problems related to physical health, mental health and learning. Prevention and mitigation policies can reduce children's risk by helping communities to prepare for and better prepare for and respond to disasters [2].

The earthquake that occurred on 25 February 2022 in West Pasaman Regency, West Sumatra Province, with a magnitude of 6.1, resulted in 27 deaths, 457 injuries, and 19,221 people evacuated to safe locations. Children are the subjects who are also affected by this disaster, which coincides with the Covid-19 pandemic. Conditions of displacement will cause an increased risk of morbidity to mortality of infectious diseases and even potential outbreaks [3]. Preventive measures can be taken immediately even during the emergency response period (day 3 to 1 month after the disaster). Some of these infectious diseases can be prevented by immunisation [4].

According to the World Health Organization (WHO) in 2018 there were around 20 million children in the world who did not get complete immunisation, some even did not get any immunisation. Many Indonesian children are also still not fully immunised. Indonesia's complete basic immunisation achievements in 2021 and 2022 are 84.2% and 99.6% respectively, West Sumatra Province is 58.4% and 72.2% while West Pasaman Regency is 51% and 54.4% [5-8]. Achieving community immunity (herd immunity) requires high immunisation coverage of 80% - 95% and evenly distributed throughout the region. Communities living in post-disaster areas often accept the lack of health and vaccination services after the end of the post-disaster emergency response period in the area [9, 10].

The success of immunisation efforts at disaster sites is a successful effort to reduce mortality from preventable diseases by immunisation, especially measles. It is expected that immunisation measures can reduce the impact of disasters and contribute to improving public health in the region or refugee camps [11]. Immunisation is an essential health service that should continue in all circumstances. However, in emergency or disaster

* Corresponding author: rahmatsyawqi@med.unand.ac.id

situations, there can be significant disruption or reduction in routine immunisation services, delays or temporary suspension of immunisation programmes. Thus, catch up immunisation programmes are needed to ensure that children in disaster areas still have the opportunity to be immunised according to the recommended immunisation schedule [12, 13]. Nagari Kajai in West Pasaman Regency was one of the areas most affected by this disaster, especially physical losses, which also resulted in constraints on the immunisation programme in the area. This paper will describe the immunisation catch-up efforts in the West Pasaman earthquake disaster area in 2022.

## 2 Literature review

### 2.1 Disasters and post-disaster child health issues

Disaster is an event or series of events that threaten and disrupt people's lives and livelihoods caused by natural and/or non-natural factors and human factors resulting in human casualties, environmental damage, property losses, and psychological impacts [14, 15].

The potential causes of disasters in Indonesia are categorised into three types of disasters, namely natural disasters, non-natural disasters, and social disasters. Natural disasters are disasters caused by an event or series of events caused by nature, such as earthquakes, volcanic eruptions, hurricanes, landslides, droughts, forest/land fires due to natural factors, plant pests, epidemics, outbreaks, extraordinary events, and space/space objects. Non-natural disasters, are disasters caused by non-natural events or series of events including forest/land fires caused by humans, transportation accidents, construction/technology failures, industrial impacts, nuclear explosions, environmental pollution and space activities. Social disasters, are disasters caused by an event or series of events caused by humans which include social unrest and social conflicts in the community that often occur [15].

Every disaster that occurs can cause a health crisis. A health crisis is an event or series of events that results in loss of life, injury/sickness, displacement, and/or potential hazards that have an impact on public health that requires a rapid response beyond normal habits and inadequate health capacity. Health crisis management is carried out with a cluster system, which is implemented through the establishment of health clusters at the central and regional levels that aim to improve coordination, collaboration and integration in health crisis management. Health clusters are groups of health crisis management actors who have health competencies that coordinate, collaborate, and integrate to meet health service needs, which come from the central government, or local governments, non-governmental institutions, the private sector/business institutions and community groups. The Health Cluster itself in the aspect of disaster management is an integral part of the disaster management cluster. The provincial health cluster is formed and coordinated by the Head of the Provincial Health Office. While the district/city health cluster is formed and coordinated by the Head of the District/City Health Office [14].

Every disaster event will directly or indirectly affect the health conditions of the people in the affected area. This is more due to environmental damage and changes in daily living conditions, from normal conditions to abnormal conditions. This situation creates vulnerability to diseases, especially infectious diseases in communities affected by disasters [11, 16].

Children are a particularly vulnerable population when natural disasters occur [2, 16]. In addition to the trauma and direct losses caused by exposure to natural disasters, children can suffer physical, psychological and educational impairments. Acute illnesses that children often experience after a disaster are diarrhoea, fever, respiratory diseases, measles and skin diseases. The prevalence of these diseases can reach 9-18%. Children also often experience depression; feeling sad, losing interest in activities, and symptoms of anxiety. Children's chronic mental health symptoms need to be observed even four years after the disaster because exposure to trauma can alter brain anatomy and function, inhibiting learning and memory, so children will experience impaired school learning and decreased achievement [2].

### 2.2 Immunisation

Immunisation is defined as an effort to actively cause or increase a person's immunity to a disease, so that if one day exposed to the disease will not get sick or only experience mild pain [16, 17]. A child's immunisation status is declared complete if a child receives routine immunisation starting from complete basic immunisation aged 0-11 months, DPT-HB-Hib and Measles Rubella follow-up immunisation at 18 months of age, Measles Rubella and TD follow-up immunisation during Grade 1 of Primary School, and Tetanus Diphtheria (Td) immunisation in grades 2 and 5 of Primary School [17-19].

Immunisation is one of the most effective ways to control and eradicate infectious diseases that can lead to death. For children, immunisation can strengthen the body's defence mechanisms so that individuals avoid transmission or complications caused by certain diseases, Child immunisation also has an important function to prevent various health risks from an early age until entering adulthood [20].

Immunisation programmes are one of the seven priority strategies, especially in low-, middle- and high-income countries. Data from WHO shows that more than 20 life-threatening diseases can now be prevented by immunisation. From 2010 to 2017, there was a 24% reduction in the mortality rate of children under 5 years old as a result of immunisation programmes. Various studies globally and nationally have identified the positive impact of vaccine administration in the prevention of various diseases [21].

The role of immunisation based on the type of immunisation implementation can be divided into program immunisation and optional immunisation. Program immunisation consists of routine immunisation, additional immunisation, and special immunisation. Routine immunisation is carried out continuously and continuously. Routine immunisation consists of basic immunisation and follow-up immunisation. Basic immunisation is given to infants before the age of 1 (one) year, consisting of immunisation against hepatitis B, poliomyelitis, tuberculosis, diphtheria, pertussis, tetanus, pneumonia, meningitis caused by Hemophilus Influenza type b (Hib), measles. Follow-up immunisation is a repeat of basic immunisation to maintain the level of immunity and to extend the protection period of children who have received basic immunisation. The follow-up immunisation given to under-five children consists of immunisation against diphtheria, pertussis, tetanus, hepatitis B, pneumonia and meningitis caused by Hib and measles [17, 19].

The Indonesian Ministry of Health has adopted and implemented the WHO routine immunisation programme since 1977 and immunisation is now mandatory for all Indonesian children. Immunisation achievements are low especially in developing countries. The ratio of immunised and non-immunised children in Indonesia is still highly disparate, including geographically. Indonesia is the fourth highest country in the world for non-immunisation. The proportion of fully immunised one-year-olds has increased in recent years, from 59% in 2007 to 65% in 2017, but the immunisation coverage rate among poor households remains low at 39.5% with uneven geographical distribution [1].

In disaster situations, immunisation efforts are needed to prevent the occurrence of Extraordinary Events. Some types of diseases that often arise in refugee camps and require preventive measures because they have the potential to become outbreaks include measles, diarrhoea, varicella, malaria, acute respiratory infection, tetanus, meningitis, tuberculosis, typhoid, worms and hepatitis. Natural disasters can increase the risk of outbreaks due to changes in the following [16]:

- Overcrowding in displacement areas leads to increased likelihood of spread of airborne infections
- Population displacement, resulting in transmission from migrant groups or susceptible indigenous groups
- Disruption of clean water supply and sanitation contamination
- Disruption of public health programmes
- Ecological changes that favour vector breeding
- Displacement of domestic and wild animals, carrying zoonotic diseases
- Provision of emergency food, water and shelter in disaster situations

The goal of post-disaster vaccination is to ensure children, adolescents and adults are protected against vaccine-preventable diseases and reduce the likelihood of vaccine-preventable disease outbreaks. If documentation of previous immunisation history is available, children should be vaccinated according to the recommended vaccination schedule [16]. In times of disaster, routine immunisation is a lower priority, as disease outbreaks are less frequent and the health risks associated with them are lower [22]. In post-disaster settings, vaccination programmes should be initiated immediately. In disaster situations, the most beneficial mass vaccination is measles, and then complement basic vaccination with a catch-up immunisation programme if vaccination was delayed during the disaster [16].

Measles is one of the most serious health problems faced in emergency situations and can be the leading cause of death due to infection in children in emergencies. Measles prevention in emergency situations is aimed at routine vaccination and in response to possible measles outbreaks. It has been reported that measles can occur even when vaccine coverage exceeds 80%. If a single measles patient is found in a disaster area, mass immunisation should be carried out in that group targeting children aged 9-59 months and school-aged children in grades 1 to 6 if they have not been immunised against measles at school [16].

Areas with low complete basic immunisation coverage require catch up immunisation efforts. Catch up immunisation refers to the act of vaccinating a person, who for various reasons (e.g. delays, stock-outs, access, hesitancy, service interruptions, etc.), did not get the dose of vaccine they were supposed to get, according to the national immunisation schedule [12, 13, 23].

Referring to the Indonesian Health Profile, the coverage of complete basic immunisation in infants aged 0-11 months is an indicator of the successful performance of the immunisation programme. In the implementation of catch-up immunisation, it can also be complemented with the tagline, "Better Late Than Incomplete" to encourage parents to still seek information about the next immunisation schedule even though it is late [20].

## 2.3 Catch up immunisation in the West Pasaman earthquake disaster area

This immunisation programme is implemented through Posyandu. In accordance with the main activities in Posyandu, services for infants and toddlers include weighing, determining growth status, counselling, health checks, immunisation and early detection of growth and development. Posyandu is open once a month, if needed, more than once a month [24]. In the post-disaster period, a team of doctors from the Department of Child Health, Faculty of Medicine, Universitas Andalas was involved and collaborated with the West Pasaman Regency Health Office, Kajai community health centre (Puskesmas) team and health cadres in Kajai Puskesmas working area in the immunisation catch-up program in Nagari Kajai, as the area most affected by the earthquake disaster in West Pasaman Regency. Posyandu in the working area of Puskesmas Kajai is held once a month, every first week. However, the implementation of posyandu in the first week of March 2022 was delayed due to the earthquake disaster that occurred, which also resulted in delays in immunisation. Therefore, a catch-up immunisation program through posyandu was held simultaneously on 31 March 2022 as shown in Table 1. Posyandu was held simultaneously in 8 villages in Nagari Kajai.

**Table 1.** Catch up immunisation in Nagari Kajai, March 2022[8]

| No | Villages | Posyandu | Toddlers | | Children Immunised |
|---|---|---|---|---|---|
| | | | < 1 year | < 5 years | |
| 1 | Tanjung Beruang | Cendrawasih 1 | 6 | 32 | 8 |
| | | Cendrawasih 2 | 6 | 32 | |
| | | Cendrawasih 3 | 7 | 36 | |
| | | Cendrawasih 4 | 16 | 58 | |
| 2 | Rimbo Batu | Merak 1 | 6 | 9 | 5 |
| | | Merak 2 | 9 | 42 | |
| 3 | Pasa Lamo | Merak 3 | 8 | 37 | 5 |
| | | Merak 4 | 12 | 39 | |
| | | Merak 5 | 15 | 37 | |
| 4 | Limpato | Merpati 1 | 10 | 25 | 7 |
| | | Merpati 2 | 18 | 75 | |
| 5 | Kp Alang | Merpati 3 | 14 | 46 | 3 |
| | | Merpati 4 | 9 | 35 | |
| 6 | Lubuk Sariak | Nuri 1 | 9 | 39 | 5 |
| | | Nuri 2 | 3 | 18 | |
| 7 | Timbo Abu | Garuda 1 | 15 | 71 | 13 |
| | | Garuda 2 | 32 | 75 | |
| | | Garuda 3 | 8 | 36 | |
| 8 | Mudik Simpang | Gelatik 1 | 20 | 74 | 3 |
| | | Gelatik 2 | 10 | 56 | |
| | **Total** | | 233 | 872 | 49 |

A team of doctors from the Department of Child Health, Faculty of Medicine, Universitas Andalas participated in posyandu activities at 2 locations, namely Posyandu Cendrawasih 4 Jorong Tanjung Beruang with 35 toddlers and 8 people immunised, and Posyandu Merpati 2 Jorong Limpato with 43 toddlers and 7 people immunised. In this posyandu activity, growth and development assessment of toddlers, catch up immunisation, counselling and treatment were carried out.

This low immunisation coverage is in accordance with global conditions where there is also a decrease in the number of basic immunisation coverage of children [13, 25]. Research by Jabbar et al in Iraq found a 3% decrease in DPT, 3% Measle, and 7% polio immunisation coverage during the Covid 19 pandemic from 2019 to 2021 [25]. Another study by Matteo et al in Italy, obtained similar results, namely during the Covid 19 pandemic from 2019 to 2020 there was a decrease in DPT immunisation coverage by 2.7% and MMR by 2.76% [13]. Pandemic conditions in the world which are the basis for a decrease in immunisation coverage are evenly distributed throughout Indonesia including West Pasaman. This is also in accordance with research conducted by Addina et al on-earthquake victims in West Pasaman in February 2022 regarding the most reasons why parents do not bring children to get immunised, namely parents who are afraid to leave the house due to the Covid 19 pandemic (76.0%), immunisation service facilities are not operating (22.1%) and prohibition of immunisation by family members (1.9%) [26].

People also prefer to use Posyandu as a place for immunisation because it is closer to their homes, but pandemic conditions caused Posyandu activities to be closed, resulting in longer distances for people to travel to Puskesmas to get immunised, thus reducing parents' interest in immunising their children [26]. During the Covid 19 pandemic, immunisation services should remain an essential health service that is prioritised. Basic and follow-up immunisation should be completed and carried out as scheduled to protect children from infectious diseases that can be prevented by immunisation by applying the principles of infection prevention and maintaining a safe distance of 1-2 meters [27].

During a disaster, many sectors are affected including health, especially health facilities and infrastructure. Restrictions on routine services for healthy children at health facilities including immunisation, changes in the provision of Posyandu and staff shortages often occur during disasters. This has led to the need for stronger strategies to educate children on the importance of complete immunisation, as well as ensuring basic and catch-up vaccination services for under-fives are available to all children during disasters. Even in the post-disaster phase, immunisation is one of the things that counts for immediate fulfilment [16, 22].

Efforts to achieve immunisation coverage after the West Pasaman earthquake through catch up immunisation can be done passively or actively. Passively, education about immunisation is carried out by cadres and Puskesmas officers with the aim of increasing the number of immunisation visits both at Posyandu and Puskesmas, Health workers need to record and track (defaulter tracking) children who have delayed immunisation to plan intervention activities as soon as the situation allows. Intervention activities resulting from tracking can be in the form of sweeping, drop out follow up (DOFU), or crash programmes. In hard-to-reach areas, the Sustainable Outreach Service (SOS) strategy can be applied to carry out interventions. Planning catch-up activities also needs to consider the capacity of health workers and parents' trust in multiple injections. Coordination between immunisation officers and surveillance officers is important in areas with low immunisation coverage [27]. Cross-sectoral roles are needed to mobilise communities and support safety in immunisation implementation [28-30].

## 3 Conclusion

Immunisation is important in an effort to protect children's health and provide specific immunity to a person against diseases that can be prevented by immunisation. High immunisation achievements aim to form herd immunity against diseases, especially those that have the potential to become outbreaks or extraordinary events or even after disasters. Immunisation achievements in West Pasaman Regency, especially in Nagari Kajai, still have not reached the target of forming herd immunity. The Covid 19 pandemic is one of the causes of low immunisation achievements and the earthquake could further reduce its achievements. Efforts to achieve immunisation targets must also be made in a disaster situation, one of which is in the form of an immunisation catch-up program. Immunisation service facilities and infrastructure must be activated immediately even during the disaster response period (<1 month after the disaster).

The high level of population mobility, low transportation facilities, especially in remote areas, and the Covid 19 pandemic have reduced post-disaster immunisation coverage in this area. The success of immunisation efforts at the disaster site is a successful effort to reduce the mortality rate due to infectious diseases, especially measles. It is hoped that this immunisation action can reduce the impact of the disaster and contribute to improving public health in the area or refugee camp.

The authors would like to thank the West Pasaman Regency Health Office, Kajai community health centre (Puskesmas) workers, the cadres and the related stakeholders who contributed to supporting this programme running well.

## References

1. UNICEF. The situation of children in Indonesia: Trends, opportunities, and challenges in fulfilling children's rights. (UNICEF Indonesia, 2020).
2. B.S. Lai and A.L. Greca, Understanding the impacts of natural disasters on children, Soc Res Child Dev. **8**, 8-38 (2020)
3. SAMHSA, Tips for talking to children and youth after traumatic events: a guide for parents and educators, (SAMHSA, 2007)
4. Widayatun and Z. Fatoni, Health problems in a disaster situation: the role of health personnels and community participation), J Kependudukan Indonesia **8**, 37–52 (2013)
5. Kemenkes RI, Profil Kesehatan Indonesia 2021, (Pusdatin Kemenkes, Jakarta, 2022)
6. Kemenkes RI, Profil Kesehatan Indonesia 2022, (Pusdatin Kemenkes, Jakarta, 2023)
7. Dinkes, Cakupan imunisasi dpt-hb-hib 3, polio 4, campak/mr, dan imunisasi dasar lengkap pada bayi menurut jenis kelamin, kecamatan, dan puskesmas, kabupaten Pasaman Barat tahun 2021, (Pasaman Barat, 2021)
8. Dinkes, Laporan hasil imunisasi rutin bayi di puskesmas tahun 2022 kabupaten Pasaman Barat, (Pasaman Barat, 2022)
9. E.K. Kang, Impact of disasters on community medical screening examination and vaccination rates: the case of the Sewol ferry disaster in Ansan, Korea, Disaster Med Public Health Prep **15,** 286–91, (2021). https://doi.org/10.1017/dmp.2020.29
10. S.R. Hadinegoro and Ismoedijanto, The value of vaccination, In Pedoman Imunisasi di Indonesia, (Badan Penerbit Ikatan Dokter Anak Indonesia, Jakarta, 2017)
11. Depkes RI, Petunjuk teknis pelaksanaan imunisasi di daerah bencana, (Depkes RI, Jakarta, 2009)
12. WHO, Leave no one behind: guidance for planning and implementing catch-up vaccination. (WHO, Geneva, 2021)
13. M. Mancarella, F. Natarelli, C. Bertolini, A. Zagari, M.E. Bettinelli, and S. Castaldi, Catch-up vaccination campaign in children between 6 and 8 years old during covid-19 pandemic: the experience in a covid hub in Milan, Italy, Vaccine **40**, 3664–9, (2022). https://doi.org/10.1016/j.vaccine.2022.05.005
14. Minister of Health. Regulation of the Minister of Health of the Republic of Indonesia Number 75 of 2019 on Health Crisis Management. Ministry of Health of the Republic of Indonesia, 473–84 (2019)
15. Republic of Indonesia. Law of the Republic of Indonesia Number 24 of 2007 on Disaster Management. (2007)
16. K. Rusmil, Vaksinasi dalam keadaan bencana, In: Pedoman Imunisasi di Indonesia, 119–31 (Badan Penerbit Ikatan Dokter Anak Indonesia, Jakarta, 2017)
17. Kemenkes RI, Buku ajar imunisasi, Kementerian Kesehatan RI **1**, Jakarta, (2014)
18. Kemenkes RI, Imunisasi lengkap indonesia sehat, (P2P KemKes, 2019) https://p2p.kemkes.go.id/imunisasi-lengkap-indonesia-sehat/
19. Menkes, Peraturan menteri kesehatan republik indonesia nomor 12 tahun 2017 tentang penyelenggaraan imunisasi, (2017)
20. Direktorat PKPM Kemenkes, Strategi Komunikasi Nasional imunisasi 2022-2025, 1–85 (Kemenkes, 2023)
21. M Maddeppungeng, Jadwal imunisasi rekomendasi idai terbaru, apa yang berubah??, In: Family-centered Professional Partnerships in Child's Growth and Development Practices, 155–68 (UKK Tumbuh Kembang Pediatri Sosial IDAI, IDAI Cabang Jawa Tengah, Jakarta, 2023)
22. H. Young and P. Harvey, The sphere project humanitarian charter and minimum standards in disaster response. Disasters **28**, (Geneva: The Sphere Project; 2004)
23. A.S. Wallace, T.K. Ryman, L.P. Dumm, C. Morgan, R. Fields, C. Garcia, et al, Leaving no one behind: defining and implementing an integrated life course approach to vaccination across the next decade as part of the immunization agenda 2030, (Vaccine, 2022). https://doi.org/10.1016/j.vaccine.2022.11.039
24. Ministry of Health, Republic of Indonesia. General guidelines for posyandu services. Ministry of Health, Republic of Indonesia, **5**, 40–51 (2017)
25. F. Jabbar, K.A. Kadhim, R.A. Alhilfi, A. Chitheer, A. Rahi, and D.B. Hipgrave, Intensification of integrated immunization services to recover routine vaccination coverage and bring covid-19 vaccine to the population of Iraq in 2022, Vaccine **42**, 2036–43,

(2024) https://doi.org/10.1016/j.vaccine.2024.02.038

26. A. Irawan, Karakterisktik dan status imunisasi anak korban gempa Pasaman Barat, Padang, (2022)
27. Kemenkes RI. Technical Guidelines for Immunization Services during the COVID-19 Pandemic. (Ministry of Health of the Republic of Indonesia, Jakarta, 2020)
28. Kemenkes RI. General Guidelines for Developing Active Alert Villages and Urban Areas. Health Promotion Center, (Secretary General of the Ministry of Health of the Republic of Indonesia, Jakarta, 2010)
29. E. Sartika, K. Munadi, and E. Mutiawati, Penguatan peran posyandu dalam meningkatkan ketahanan masyarakat terhadap bencana tsunami di wilayah pesisir Kota Banda Aceh menggunakan pendekatan knowledge management, Pasca Univ Syiah Kuala **4**, 111-116 (2017)
30. I. Ismiati, Sistem Manajemen Crash Program Campak Pasca Tsunami di Dinas Kesehatan Kota Lhokseumawe, Kesmas Natl Public Heal J **2**, 33-38, (2007)

# Influence of junior high school students' knowledge, attitudes, and practices on earthquake and tsunami preparedness simulations in Padang

*Idha* Yulfiwanti[1], *Mayetti*[1*], and *Indra* Ihsan[1]

[1]Department of Pediatrics, Faculty of Medicine, Universitas Andalas, Padang, Indonesia

**Abstract.** Padang, a city highly susceptible to earthquakes and tsunamis, requires students to possess adequate knowledge and develop cautious behaviors when confronted with such calamities. This study seeks to evaluate the influence of earthquake and tsunami simulations on junior high school students' learning, perspectives, and practices. A pretest-posttest experimental design was used, involving eighth and ninth-grade students from SMP Negeri 4 Padang. The simulation exercises included earthquake and tsunami disaster drills. Data was collected through observation, focusing on the learning process and the school's preparedness, including evacuation route planning. Of 87 respondents, 42 (48.3%) were male, and 45 (51.7%) were female. The age distribution was as follows: 2 students (2.3%) were 12 years old, 33 (31%) were 13, 48 (50.6%) were 14, and 15 (16.1%) were 15 years old. The simulations substantially improved students' understanding, attitudes, and behaviors in dealing with earthquakes and tsunamis ($p < 0.05$). The earthquake and tsunami simulations at SMP Negeri 4 Padang effectively enhanced students' knowledge, attitudes, and disaster preparedness. The drills improved their awareness and response to emergencies, particularly in following evacuation procedures. This hands-on approach solidified their theoretical understanding and fostered a proactive mindset towards disaster management.

## 1 Introduction

Indonesia is ranked 36th with a ratio index of 10.36 out of 172 most disaster-prone countries in the world based on the 2018 World Risk Report data, making Indonesia prone to natural disasters such as earthquakes, tsunamis, and volcanic eruptions [1]. Padang City consists of small islands and coastal areas, and it consists of small islands, coastal areas, and hills with an altitude between 0-1,853 meters above sea level. It is directly adjacent to the Indian Ocean, which has quite large sea waves, and there are active tectonic plates that are vulnerable to natural disasters [2].

One of the efforts of disaster preparedness is earthquake and tsunami disaster simulations given to students so as to increase insight and knowledge of disasters that will be faced and be able to behave if a disaster is in sight. Proactive preparedness is disaster management carried out through preparedness in the social recovery stage. This action can be carried out by various parties, including health workers [2]. Pediatricians have an important role in disaster management for children. As outlined in Law of the Republic of Indonesia Number 35 of 2014 concerning child protection, every child has the right to special protection in certain situations and conditions to obtain a guarantee of security against threats that endanger themselves and their lives in their growth and development [3].

This study discusses the knowledge and attitudes of Junior High School students toward earthquake and tsunami disasters and the role of simulation in improving students' knowledge. This study aims to determine whether earthquake and tsunami simulations impact the knowledge and attitudes of junior high school students toward earthquakes and tsunamis.

### 1.1 Geographical and demographic situation of West Sumatra

According to Regional Regulation Number 10 of 2005, Padang City has an area of 1,414.96 km2, between 00 44 00'-1 08 35' South latitude and 100 05 05'-100 34 09' East longitude. Padang City has a coastal region of 68.126 km2 and a row of Bukit Barisan covering an area of 486.209 km2. The Padang City area has varying land elevations between 0-1,853 meters above sea level. The city's boundaries are Padang Pariaman Regency to the north, Pesisir Selatan Regency to the south, Solok Regency to the east, and the Indian Ocean to the west [4].

### 1.2 Definition and classification of disasters

A disaster is an event or series of events that threaten and disrupt people's lives and livelihoods influenced by

* Corresponding author: mayetti@med.unand.ac.id

natural, non-natural, or human factors and result in casualties, environmental damage, property losses, and even impact human psychology. In the Law of the Republic of Indonesia Number 24 of 2007 concerning Disaster Management, based on the cause, disasters are grouped into three types:[5]

a. Natural disasters are natural disasters caused by natural events, such as earthquakes, tsunamis, volcanic eruptions, floods, droughts, hurricanes, and landslides.
b. Non-natural disasters are caused by non-natural events or a series of events, including technological or construction failure, modernization failure, epidemics, and disease outbreaks.
c. Social disasters, caused by events or a series of events caused by humans, include social conflicts between groups or communities and terror.

### 1.3 Mentawai megathrust

The Mentawai megathrust area is part of the Sumatra subduction zone with a very high level of seismicity and a risk of earthquakes with magnitudes of more than 8 SR to 9.3 SR and a return period of hundreds of years. In the last two centuries, four earthquakes have been recorded in the Sumatra subduction zone, namely in 1833 with a magnitude of 8.8–9.2; in 1861 with a magnitude of 8.3–8.5; in 2004 with a magnitude 9.0-9.3; and in 2005 with a magnitude 8.7. The segmentation of the Mentawai megathrust may collapse in the next few decades, triggering large earthquakes and tsunamis and potentially causing severe damage in most parts of Sumatra. This disaster threat will mainly affect several cities and regions on the West Coast, such as Sibolga City, Padang City, Pariaman City, Agam Regency, South Coast Regency, and Bengkulu City [6].

### 1.4 Management of disaster management implementation

Disaster management is a series of efforts that include establishing development policies at risk of disasters, disaster prevention activities, emergency response, rehabilitation, and reconstruction. The implementation of disaster management is divided into three stages according to the disaster management cycle:[7]

a. Pre-disaster, including situations with no disaster and situations with potential disaster.
b. Emergency response, covering the situation when a disaster occurs.
c. Post-disaster, covering the situation after a disaster.

### 1.5 Pre disaster

Pre-disaster disaster management is carried out in non-disaster situations and potential disaster situations. In non-disaster situations, a disaster management plan covers all stages of disaster work. Meanwhile, in a potential disaster situation, a preparedness plan is prepared to deal with emergencies, namely contingency plans. Activities carried out when a problem does not occur:[5]

1. Disaster management planning.
2. Mitigation (disaster risk reduction).
3. Prevention activities
4. Guidance in development planning
5. Disaster risk analysis
6. Education, training, and standardized requirements according to technical
7. Implementation and enforcement of spatial plans.

### 1.6 Disaster emergency response

The Activities that are carried out when there is a potential disaster situation:

1. Preparedness is a series of activities to anticipate disasters through organization and appropriate and effective steps.
2. Early warning is an effort to take quick and appropriate action and prepare emergency response actions carried out through observation of disaster symptoms, analysis of the results of observations of disaster symptoms, decision-making by the authorities, dissemination of information about disaster warnings, and action by the community.

### 1.7 Management of children's disaster management

Based on Law Number 35 of 2014 concerning child protection, a child is someone who is still in the womb until 18 years old. Every child has the right to survival, growth, and development, as well as protection from violence, discrimination, and protection in certain situations and conditions to guarantee security against threats that endanger themselves and their lives in their growth and development. Children have vulnerabilities related to anatomical, physiological, developmental, and behavioral characteristics that increase their risk of disaster. Child protection is all activity to ensure and protect children and their rights to live, grow, develop, and participate optimally according to human dignity. The threat of the Mentawai megathrust requires helping infants and children in the Mentawai megathrust segmentation area, such as Padang City, as a form of child protection efforts in disaster management [7, 8].

The pre-disaster management program includes disaster mitigation education, strengthening school buildings such as earthquake-resistant school buildings, facilities for disaster anticipation such as shelters or tsunami anticipation, emergency stairs, and evacuation route signs. The Disaster Safe Education Unit (PSAB) program was designed by a team of experts in collaboration with the Tsunami Alert Community (KOGAMI) and is targeted for implementation by elementary school students in grades I–V, junior high schools in grades VII and VIII, and senior high and vocational schools in grades X and XI. Earthquake and tsunami disaster simulation is a small part of preparedness [9].

Actions that can be taught during and after an earthquake are:

a. If you are inside the house when you feel earthquake tremors, it is recommended to get under a table to protect yourself from objects that may fall. If there is no table, a pillow can protect the head. If you have a stove on, turn it off immediately to prevent a fire. If it feels safe, immediately leave the house and stand in a safe place.
b. If you are at school, children can take shelter under a desk, and if the earthquake has subsided, it is advisable to exit to find an open space. Children are warned not to stand near buildings, poles, and trees.
c. Teach children to remain calm, not panic when an earthquake occurs, and follow all earthquake-related instructions.
d. If the earthquake has stopped, give knowledge to children, such as checking your condition, your family, and your surroundings; stay away from cracked and unsafe buildings; always be vigilant about aftershocks; and do not panic.

In disaster management efforts, pediatricians have a role that is realized starting with disaster mitigation efforts, continuing to the emergency response stage, and during rehabilitation by the organization's capabilities. As the front guard in disaster situations, pediatricians must have competencies related to disasters and emergencies for children affected by disasters. This is intended so that, in an emergency, pediatricians can work as expected.7 Therefore, there is a need for routine training on disasters and emergencies and their mitigation, especially for children [10, 11]. In addition, pediatricians should support collaboration with other disciplines, such as social services, education, and pharmacy, to advocate for the needs of children in disasters.

In line with the West Sumatra government's contingency plan, Law No. 35/2014 on Child Protection states that special protection for children in emergencies can be carried out through rapid treatment efforts, including physical, psychological, and social treatment and rehabilitation, as well as prevention of disease and other health problems. Another protection effort is to provide psychosocial assistance during treatment and recovery [3]. Some of these efforts can be carried out, especially by pediatricians, in a disaster. The role of pediatricians can be divided into roles before a disaster, during a disaster, and after a disaster.

### 1.8 The role of pediatricians before disasters

Pediatricians, individually or collectively, as part of the pediatric community (e.g., the American Academy of Pediatrics, the local pediatrician community, and hospital staff) should participate in the development of disaster plans. Pediatricians should provide input to the government and regional offices of Emergency Medical Services (EMS) to ensure that the needs of children are incorporated into any disaster, including emergency plans to meet the needs of critically ill and injured children. Pediatricians can assist schools and childcare centers in developing disaster plans. In addition, they can prepare, update, and practice office disaster plans regularly. Pediatricians are also encouraged to provide anticipatory guidance on disaster preparedness at home.

### 1.9 Earthquake and tsunami disaster preparedness simulation

Preparedness simulation is an activity conducted to test tsunami-fixed procedures and improve the effectiveness of early warning systems using disaster scenarios that are close to or by the natural conditions when a tsunami occurs (Fig. 1). The implementation of simulation can be done partially or only partially, or the implementation is limited to certain areas that constitute part of the total area of the disaster-prone regions. However, simulations should be conducted routinely on a large scale to produce better overall disaster resilience. The higher the risk of a city or region to tsunamis, the more complete the tsunami simulation should be and involve all stakeholders related to tsunami disasters. The 2014 BNPB guidelines on earthquake and tsunami disaster preparedness simulations explain the types of simulations:
1. Socialization (Orientation Exercise),
2. Evacuation Drill Exercise,
3. Table Top Exercise,
4. Functional Exercise and
5. Integrated Simulation (Full Scale Exercise/End-to-End Simulation/FTX).

These simulations can be carried out at the national, provincial, city, district, or local level. However, each region can choose the type of simulation to be implemented according to its needs and capabilities. The simulation implementation can also combine several types of simulations to get better and more comprehensive results.

Earthquake and Tsunami Disaster Preparedness Simulation:
1. Preparation
2. D-day implementation of the simulation
   a. Opening
   b. Mechanisms during the simulation (spontaneity and dynamics and simulation time flow)
3. Closure
4. Monitoring, Evaluation, Documentation

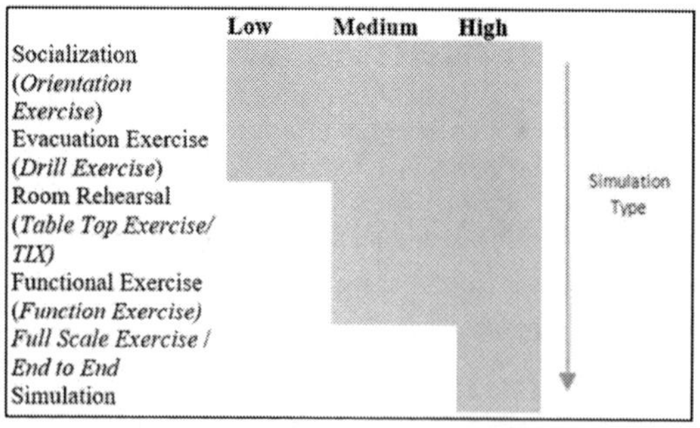


**Fig. 1.** Consideration matrix for simulation type selection based on risk level

## 2 Methods

The type of research used is quantitative research with a pre-experimental design and one group pretest-posttest to examine the knowledge and attitudes of SMPN 4 Padang students about earthquake and tsunami disaster preparedness after an earthquake and tsunami disaster simulation in 2023.

In this study, the intervention was provided in the form of simulated evacuation exercises (drill exercises) to SMPN 4 Padang, with the following information:

1. Activity form: evacuation drills at community level, such as schools
2. Actors: school community, especially students
3. Organizers: PPDS Pediatrics, BPBD Padang City, KOGAMI, KSB
4. Location: SMPN 4 Padang School involves Temporary Evacuation Site (TES) Youth Center Building
5. Preparation: Coordinated with the village, school, and security for about 1-2 weeks to prepare pre-simulation material on Tuesday, July 11–12, 2023, for the school.
6. Simulation time: Thursday, July 13, 2023, about 1-4 hours

Some activities to prepare for evacuation drill simulation activities are:

1. An earthquake and tsunami disaster risk assessment are provided to school parties, namely teachers and students.
2. Earthquake and tsunami contingency plan
3. Earthquake and tsunami disaster scenarios
4. Bordering and stabilization from BPBD Padang City and KOGAMI
5. HR as simulation actors and simulation committee
6. Simulation facilities and infrastructure related to the tsunami early warning system's tsunami evacuation plan (signs, evacuation routes, evacuation maps, evacuation sites, tsunami TES, etc.)
7. Funds for simulation activities
8. Mass mobilization.

The research was conducted at SMPN 4 Padang, West Sumatra, July 12–17, 2023. This research was conducted using primary data, which was obtained directly from junior high school students using a questionnaire. The population of this study were all students of SMPN 4 Padang, West Sumatra. The sample in this study is part of the population of SMPN 4 Padang students obtained from proportionate stratified random sampling. Based on the Slovis formula, the sample size required from a population of 650 people with the necessary calculation in this study was 87 respondents, and we obtained 43 students for class VIII and 44 students for class IX.

Data was collected by distributing questionnaires to respondents. The points of this research questionnaire were adopted from the LIPI preparedness questionnaire, which the researcher modified after discussing it with several experts in the field of disaster. The questionnaire Data was collected by distributing questionnaires to respondents. The points of this research questionnaire were adopted from the LIPI preparedness questionnaire, which the researcher modified after discussing it with several experts in the field of disaster. The questionnaire consisted of 35 questions answered in about 15–20 minutes, with 25 yes-or-no questions and 10 true-or-false questions. Part of the questionnaire consists of:

a. Respondent characteristics include name, gender, date of birth, class, and residence address.
b. Yes or No option questions: 25 questions
c. There were ten questions with true or false answers.

In evaluating the knowledge and attitudes of SMPN 4 Padang students, an assessment was made of the material presented in the earthquake and tsunami disaster simulation. The measuring instrument used consists of 35 items with two forms of answers, namely "yes" or "no" answers on items 1–25 and "true" or false" answers on items 26–35. The answer "yes" is worth 1, and the answer "no" is worth 0. The answer to the question "True" or "False" is given a score of 1 if the answer matches the answer key and 0 if the answer does not match the answer key.

### 2.1 Validity and reliability test

The Sampling for the questionnaire trial was carried out on 30 students of SMPN 13 Padang whose characteristics are almost the same as those of SMPN 4 Padang students, with a filling time of about 15 to 20 minutes. This trial activity is to determine the validity and reliability of the questionnaire. Based on the validity test, 17 valid questions were obtained from 35 questions, with valid validity values ranging from 0.362 to 0.679 ($r > 0.361$). To determine reliability, the Crombach Alpha test was carried out and resulted in an alpha value of 0.768 ($\geq 0.6$, meaning the question is reliable).

The data collected is primary data. Data was obtained from respondents among SMPN 4 West Sumatra students by distributing questionnaires. After the data was collected completely, processing was carried out using SPSS 24. Univariate and bivariate analyses were carried out on the variables used in this study.

## 3 Results and discussion

The general description of the research subjects will be described in detail below, namely in the form of subject distribution based on gender, age, and class. The description was obtained from 87 students at SMPN 4 Padang who became the research sample. Then, the frequency and percentage of each distribution were calculated using descriptive statistics.

**Table 1.** Number of respondents based on gender

| Gender | Total | Percentage | Mean | |
|---|---|---|---|---|
| | | | Pretest | Post-test |
| Male | 42 | 48.3% | 6.67 | 11.45 |
| Female | 45 | 51.7% | 7.36 | 11.64 |
| **Total** | 87 | 100% | | |

Based on the Table 1, it can be seen that there are 42 people (48.3%) of research subjects who are male and 45 people (51.7%) of research subjects who are female.

**Table 2**. Number of respondents based on age

| Age | Total | Percentage | Mean | |
|---|---|---|---|---|
| | | | Pretest | Post-test |
| 12 Years | 2 | 2.3% | 6 | 11 |
| 13 Years | 27 | 31% | 6.30 | 10.96 |
| 14 Years | 44 | 50.6% | 7.32 | 11.93 |
| 15 Years | 14 | 16.1% | 7.64 | 11.57 |
| **Total** | **87** | **100%** | | |

Based on the Table 2, it can be seen that there were 2 people (2.3%) research subjects aged 12 years, 33 people (31%) aged 13 years, 48 people (50.6%) aged 14 years and 15 people (16.1%) aged 15 years.

**Table 3.** Number of class-based respondents

| Class | Total | Percentage | Mean | |
|---|---|---|---|---|
| | | | Pretest | Post-test |
| Grade 8 | 39 | 44.8% | 7.08 | 11.69 |
| Grade 9 | 48 | 55.2% | 6.98 | 11.44 |
| **Total** | 87 | 100% | | |

The Table 3 shows that 39 (44.8%) students were in Grade 8 and 48 (55.2%) students were in Grade 9. This normality test was carried out using the Kolmogorof Smirnov test with a p value> 0.05.

**Table 4.** Normality test

| | Kolmogorov-Smirnov[a] | | |
|---|---|---|---|
| | Statistic | Df | Sig. |
| **Pre_test** | 0.157 | 87 | 0.000 |
| **Post_test** | 0.154 | 87 | 0.000 |

Based on the data in the Table 4, it can be seen that the pret-test and post-test significance values are normally distributed, because the significance value is <0.05.

### 3.1 The impact of junior high school students' knowledge, attitudes, and practices on earthquake and tsunami disaster simulation in Padang

Based on the paired sample test conducted, the following results were obtained:

**Table 5.** The impact of junior high school students' knowledge, attitudes, and behavior on earthquake and tsunami disaster simulation in Padang

| *Paired Samples Correlations* | | N | Correlation | Sig. |
|---|---|---|---|---|
| Pair 1 | Pre_test and Post_test | 87 | 0.646 | 0.000 |

Based on the Table 5, it can be seen that, the significant result obtained is 0.00 which means the sig. value <0.05 so that there is a relationship between the pre-test and post-test in SMPN 4 Padang students. Judging from the correlation value of 0.646, this relationship is categorized as a moderate relationship.

## 4 Conclusion

The earthquake and tsunami simulation conducted at SMP Negeri 4 Padang significantly improved the knowledge, attitudes, and practices of junior high school students. The simulation played a crucial role in enhancing students' understanding of disaster preparedness and their ability to respond effectively to potential disasters. Prior to the simulation, many students had limited knowledge of the appropriate actions during such emergencies. However, after participating in the drills, they demonstrated greater awareness and readiness, especially regarding evacuation procedures and safety measures. This hands-on experience helped solidify theoretical knowledge and fostered a more proactive and cautious approach to disaster management among students' general description of the research subjects will be described in detail below.

## References


1. H. Hadi, S. Agustina, and A. Subhani, Penguatan kesiapsiagaan stakeholder dalam pengurangan risiko bencana alam gempabumi, J. Geodika, **3**(1), 30–40, (2019) https://doi.org/10.29408/geodika.v3i1.1476.
2. K. Anam, A. Mutholib, F. Setiyawan, B. A. Andini, and S. Sefniwati, Kesiapan institusi lokal dalam menghadapi bencana tsunami: studi kasus Kelurahan Air Manis dan Kelurahan Purus, Kota Padang, J. Wil. dan Lingkung. **6**(1), 15–29 (2018)
3. House of Representatives & President of the Republic of Indonesia. Law Number 35 of 2014 concerning amendments to Law Number 23 of 2002 on Child Protection. (2014).
4. Pemerintah Kota Padang, Profil Kota Padang, in Rencana Program Investasi Jangka Menengah Bidang Cipta Karya Kota Padang 2017-2021, Padang: Pemerintah Kota Padang, 1–30, (2017)
5. Republic of Indonesia. Law of the Republic of Indonesia Number 24 of 2007 concerning Disaster Management. Jakarta, Indonesia. (2007)
6. Widayatun and Z. Fatoni, Permasalahan kesehatan dalam kondisi bencana: peran petugas kesehatan dan partisipasi masyarakat," J. Kependud. Indonesia. **8**(1), 37–52 (2013).
7. S. A. Wilopo, Kompetensi inti untuk kedokteran bencana dan kesehatan masyarakat, 1–37 (2017).
8. D. Oktiari and S. Manurung, Model geospasial potensi kerentanan tsunami kota padang, J. Meteorol. dan Geofis., **11**(2), 140–146 (2010).
9. H. Seddighi, S. Yousefzadeh, M. López López, and H. Sajjadi, Preparing children for climate-related disasters, BMJ Paediatr. Open, **4**, 1–5, (2020)
10. B. P. Geniosa and Q. Aini, Hospital preparedness level and policy implementation analysis of hospital disaster plan in RSUD Kota Yogyakarta, J. Indonesia Heal. Policy Adm. **5**, (2020)
11. Satgas Penanggulangan Bencana IDAI, Rekomendasi nomor 007/rek/PP IDAI/V/2014 tentang pertolongan pada anak dalam keadaan bencana, 6–7, (2014),

# Infant and young children's mother profile related to breastfeeding practices during and after the flash floods disaster in Agam, West Sumatra, 2024

*Nitari* Rahmi[1,2*] and *Fitrisia* Amelin[1,2]

[1]Department of Child Health, Faculty of Medicine, Universitas Andalas, Padang, Indonesia
[2]Department of Pediatrics, Dr. M. Djamil General Hospital, Padang, Indonesia

**Abstract.** Flash floods from Mount Marapi on May 24, 2024 occurred in Agam Regency, West Sumatra. BNPB reported distribution of vulnerable categories among refugees are baby which is increased infection, breastfeeding can reduce the risk of infectious diseases in disaster. One of the efforts is relactation. This cross-sectional study was conducted on 19 babies' refugee of the Agam flash floods on May 2024. Data were collected using questionnaire during and after disaster. A total of 21 babies that 0-2 years old participated in the study that four mothers (21.3%) still had enough breast milk for their babies, nine mother (47.4%) breast milk production was reduced, and six mothers (31.6%) had difficulties and no longer had breast milk. Several problems occur in mothers during disaster are anxiety (89.5%), lack of sleep (68.4%), no appetite (63.1) fatigue and stress (57.9%). Fatigue and stress (p-value 0.039), lack of sleep (p-value 0.036), no appetite (p-value 0.048), and anxiety (p-value 0.002) were statistically significant correlates. Relactation practice needed but only four respondents relactated. Breastfeeding during disaster is important to reduce any disease, but fatigue, stress, lack of sleep, loss of appetite and anxiety has statistically significant correlate with decrease of breastmilk production.

## 1 Introduction

There are disaster conditions, breastfeeding mothers experience various challenge of continuing to be able to breastfeed her child exclusively that important throughout the first two years of life. During this time, malnutrition can cause harm that could result in low economic production, academic accomplishment, and impaired cognitive development [1, 2]. The World Health Organization (WHO) suggests exclusive breastfeeding throughout the first six months of a baby's life in order to supply the baby's nutritional needs and promote healthy growth and development. It is recommended that mothers breastfeed their infants for a minimum of two years after giving birth [3]. However, it can be challenging in particular conditions, such as emergencies or disasters, to maintain the sustainability of the exclusive breastfeeding process due to some factors, most notably the supply of formula milk [4, 5].

Breastfeeding has been shown to protect infants against various infections, including pneumonia, gastroenteritis, and otitis media. Despite global recommendations and the national breastfeeding policy, many mothers do not practice optimal breastfeeding such as cause by disaster condition. Disasters happen often in Indonesia and affect many different areas. Aside from that, newborns and young children under the age of two are the most susceptible populations in need of specific nutritional care since malnutrition can increase morbidity and mortality. [4] Breast milk should still be given to babies, even in disaster emergencies but can reduce of breastmilk production.

The results of Mamuju research showed that the mother's attitude was factors in providing exclusive breastfeeding in disaster. The success of lactation is influenced by several factors including the level of maternal understanding of lactation, the motivation of various parties, especially families and health workers and the level of stress experienced by the mother will also affect the success rate of lactation. One of the efforts is relactation that relactation aims to change the attitude of mothers to breastfeed their babies again with breast milk. The success of relactation is influenced by several factors including education, education, mother's age, baby's age, stimulation and support from health workers. Therefore, this report aimed to explain the importance of breastfeeding and relactation during disaster conditions and the fact that breastfeeding mothers and infants as a population at risk during a flood or other disaster and deals with concerns and implications related to their health and wellbeing in such circumstances.

* Corresponding author: nitarirahmi25@gmail.com

## 2 Method

The research design used in this research is descriptive analytic with a cross-sectional approach. The population in this study were baby 0-2 years old and breastfeeding mothers in the village of Sungai Pua and Candung, Agam District, West Sumatra. Total population of 21 babies. The sampling technique was convenience sampling. The mothers were then asked to fill out the questionnaire. This questionnaire is short questionnaire to capture problem breastfeeding mothers during disaster and has been tested in two studies.

## 3 Results and discussion

### 3.1 The characteristic of breastfeeding mothers during flash flood in Agam

In disaster situations when infants under six months old are fed mixed diet, assist moms in switching to exclusive breastfeeding or relactation. Table 1 explained that total of 21 babies that 0-2 years old (12 boys, 9 girls) participated in the study. The prevalence of breastfeeding babies was 21 (67%) which accompanied formula milk were 18 (85%). The prevalence of formula milk feeding babies were 9 (31%). From the interviews, data was also obtained regarding the temporary residence of breastfeeding mothers during the emergency response period, most of whom chose to live in their own houses or family houses.

**Table 1**. Characteristic babies and breastfeeding mothers during disaster

| Variable | Breastfeeding (n=12) | No breastfeeding (n=9) |
|---|---|---|
| **Gender**<br>Boy<br>Girl | 3 (14.3%)<br>9 (42.8%) | 8 (38.1%)<br>1 (4.8%) |
| **Age**<br>0-6 months old<br>6-12 months old<br>> 12 months old | 5 (23,8%)<br>4 (19.0%)<br>3 (14.3%) | 2 (9.5%)<br>1 (4.8%)<br>6 (28.6%) |
| **Prematurity**<br>Premature<br>No | 2 (9,5%)<br>10 (47.6%) | 1 (4.8%)<br>8 (38.1%) |
| **Education of Mother**<br>Low<br>High | 9 (42.8%)<br>3 (14.3%) | 2 (9.5%)<br>7 (33.3%) |
| **House condition**<br>Affected<br>No | 9 (71%)<br>3 (29%) | 4 (50%)<br>5 (50%) |
| **Stay during disaster**<br>Own house<br>Family House<br>Refugee tent | 2 (57,1%)<br>9 (42,8%)<br>1 (4%) | 3 (60%)<br>4 (40%)<br>4 (0%) |

Based on this study, mothers wanted exclusive breastfeeding in disaster condition, but any problem cause decrease breastmilk production, so they had given formula milk too. Exclusive breastfeeding in infants under 6 months of age and continued breastfeeding in children aged 6 months to 2 years or beyond should be protected, promoted, and supported in emergencies. Based on Arvan et al study, in emergency settings, where access to safe food and water may be limited and hygiene conditions may be inadequate, breastfeeding is even more important due to its unique nutritional, physiological, and psychological benefits. During the 2006 floods in Botswana, it was reported that infants admitted to the hospital with diarrhea were 30 times more likely to have not been breastfed than those without diarrhea.

This interview conducted with breastfeeding mothers who were victims of the 2024 Agam Regency flash flood (Sungai Pua and Canduang) showed that the majority of breastfeeding mothers lived in family homes and their own homes in dirty and damaged conditions, a small number were in refugee camps, but still wanted to breastfeed their children. All mothers interviewed felt uncomfortable breastfeeding their children in open and crowded tent conditions, mixed with other families, even though the conditions were uncomfortable, the majority of mothers continued to breastfeed their children. Babies who are exclusively breastfed during a disaster tend to be in better condition when a disaster occurs compared to babies who are given formula milk, but many mothers do not breastfeed their babies exclusively during an emergency.

### 3.2 The challenges for breastfeeding during and after emergency or disaster

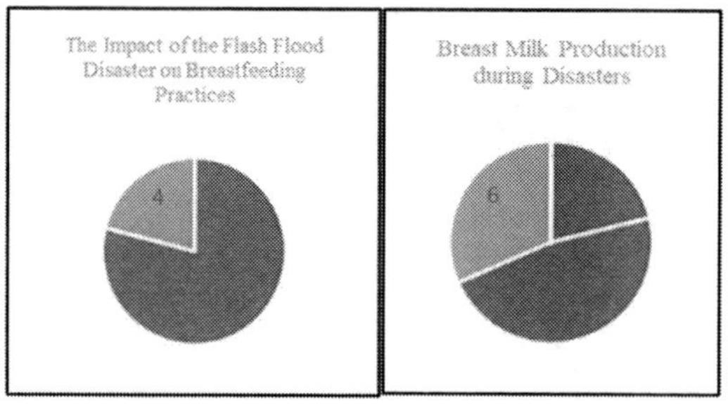


**Fig. 1**. The impact of the flash flood disaster on breastfeeding practices

Based on Fig. 1, fifteen respondents (94.7%) answered that the disaster which resulted in the evacuation and relocation of residents to relatives' homes or refugee tents affected breastfeeding practices, due to stress, difficulty sleeping, fatigue, and discomfort carrying out breastfeeding activities in open tent locations. Four other respondents (21.3%) felt they were not affected by the flash flood conditions and lived in refugee camps during the disaster response period. Respondents were also interviewed regarding the impact of the disaster on breast milk production and it was found that four respondents (21.3%) still had enough breast milk for their babies, nine respondents (47.4%) felt that breast milk production was reduced compared to before the disaster, and six respondents (31.6%) %) difficult and no longer have breast milk.

Families healthy as possible during an emergency, adequate nourishment is crucial. During crises, infants and young children are more susceptible to different exposures that may cause disease, infection, or even death. In an emergency, breastfeeding is the safest way to nourish newborns and early children and prevent disease; nevertheless, breastfeeding families may face additional hurdles in addition to regular feeding issues during natural disasters and public health emergencies

[9] Appropriate feeding of infants and young children in emergencies can be challenging. These difficulties usually result from a lack of privacy and comfort in shelter settings, impaired maternal well-being (loss of home, stress, and psychological traumas), precrisis, inappropriate feeding practices, donations, and/or improper distribution of breast milk substitutes (BMS), a lack of breastfeeding support, health personnel's insufficient knowledge of breastfeeding.

**Table 2**. Problems of breastfeeding mothers during and after disaster

| Variable | During (n=21) | After (n-21) |
|---|---|---|
| **Fatigue** | 11 (52,4%) | 5 (23.8%) |
| **Stress mother** | 11 (52.4 %) | 5 (23.8%) |
| **Anxiety** | 17 (80.9%) | 3 (14.2%) |
| **Lack of Sleep** | 13 (61.9%) | 4 (19.0%) |
| **Decrease of appetite** | 12 (57.1%) | 2 (9.5%) |
| **Refugees are not comfortable** | 3 (14.2%) | 1 (4.8%) |

Based on Table 2, during disaster, the majority of respondents (80.9%) stated that they were worried about subsequent floods, tired after evacuating and cleaning the house (52.4%), reduced appetite (57.1%), difficulty sleeping (61.9%) in the family home or refugee tent. because it is cold and dense (14.2%). Residents affected by flash floods had to stay overnight in relatives' houses or evacuation tents because their houses were damaged and filled with mud. This problem is stated to affect the practice of breastfeeding mothers affected by flash flood disasters. A follow-up visits 2 months after the flash flood, it was found that the majority of respondents had returned to their homes and did not feel anxious or tired, so some respondents began to frequently breastfeed their children again. During visiting 2 months after the flash flood disaster in Agam district, breastfeeding mothers were interviewed again and it was concluded that breastfeeding mothers no longer felt anxious, tired or had difficulty sleeping compared to during the disaster. Five respondents still experience anxiety and fatigue (23.8%), four respondents experience difficulty sleeping (19%), and two respondents still have no appetite (9.5%).

In this report, it was found that the fatigue and worry felt by mothers during disaster conditions had an impact on breastfeeding practices, the majority of mothers felt a change in the timing, intensity and amount of breast milk after flash floods. These findings are in accordance with research conducted in Pakistan which shows that positive aspects are evident in the lives of refugee mothers who wish to maintain their breastfeeding practices. The presence of support from the surrounding environment provides the ability for refugee mothers and facilitates them to maintain their breastfeeding practices. Displaced mothers rely heavily on social support provided by their immediate family members, health care providers, relatives, and neighbouring village residents.

**Table 3.** Relationship between mother's problem and breastfeeding practice during disaster

| Variable | Breastfeeding (n=12) | No Breastfeeding (n=9) | OR (95%CI) | P value |
|---|---|---|---|---|
| Fatigue<br>Yes<br>No | 5 (41.6%)<br>7 (58.4%) | 8 (88.9%)<br>1 (11.1%) | 1.469 | 0.038 |
| ress mother<br>Yes<br>No | 7 (58.4%)<br>5 (41.6%) | 9 (100%)<br>0 (0%) | 2.844 | 0.039 |
| Anxiety<br>Yes<br>No | 11 (91.7%)<br>1 (8.3%) | 2 (22.2%)<br>7 (77.7%) | 0.420 | 0.002 |
| Lack of Sleep<br>Yes<br>No | 10 (83.3%)<br>2 (16.7%) | 2 (22.2%)<br>7 (77.7%) | 1.859 | 0.036 |
| Decrease of appetite<br>Yes<br>No | 10 (83.3%)<br>2 (16.7%) | 5 (55.4%)<br>4 (44.6%) | 1.544 | 0.048 |
| Refugees are not comfortabl<br>Yes<br>No | 5 (41.6%)<br>7 (58.4%) | 4 (44.6%)<br>5 (55.4%) | 2.634 | 0.036 |

Based on Table 3 show fatigue (p-value 0.036), stress (p-value 0.039), lack of sleep (p-value 0.036), no appetite (p-value 0.048), and anxiety (p-value 0.002) were statistically significant correlates with prevalence of breastfeeding during disaster. The disaster which resulted in house condition and the relocation of residents to refugee camps affected breastfeeding practices, due to stress, fatigue and discomfort carrying out breastfeeding activities in open tent locations. Breastmilk must still be given to babies, even in disaster emergencies. Based on Michele study, as manifestation of efforts to sustain breast milk during the earthquake emergency in West Sulawesi (Mamuju and Majene Regencies), the regional government issued a circular regarding the management of formula milk donations [10]. They recommend that breast milk substitutes only be given if all of the following conditions exist: Mother is absent or there are medical indications or relactation attempts have failed.

Families may need to evacuate from their houses in the event of a disaster and relocate to another town, state, or refuge. This might lead to feeding difficulties, particularly for a breastfed infant [9]. Mothers may lose their privacy when breastfeeding, expressing breast milk, or accessing lactation assistance when they are relocated, like in the scenario of a disaster shelter (Fig. 2).

https://doi.org/10.1051/e3sconf/202560405005

**Fig. 2.** Condition of refugee tents and victims' houses during the flash flood disaster in Agam district

Parents may think about using infant formula instead of breastfeeding or expressing their milk if they lack access to and support for these practices. Along with the room needed for sanitary diaper changes, safe meal preparation, and proper cleaning and storage of baby feeding items, families also need space [9]. Some parents could be mixed-feeding (giving their baby both breast milk and infant formula) or just using infant formula. In these situations, it is critical to assist and advise families on how to provide their children with safe food during an emergency. Infant formula handling and storage during an emergency can present a number of risks and difficulties. These include: 1) not having access to a supply of infant formula; 2) not having access to safe water for cleaning and mixing formula; 3) handling and storing formula safely; and 4) not having space or supplies to clean and store infant feeding items [9]. The availability of formula milk during times of crisis has increased. Breastfeeding habits other than breast milk are greatly impacted by the uncontrolled distribution of breast milk substitutes during disaster. A study including a population of Palu earthquake and tsunami victims revealed that the shelters for these victims did not really practice exclusive breastfeeding. This is brought on by moms' lack of understanding and education, as well as the widespread support and advertising of foods like commercial porridge and formula milk that serve as alternatives to breast milk [4-11].

Formula milk cannot be a substitute for breast milk. Given the scarcity of clean water and sterile bottles in shelters, formula milk can lead to diarrhea, malnutrition, and an increased risk of baby mortality. This is due to the very restricted resources available in shelters for making formula milk, including sterile bottles, cooking tools, and clean water (Fig. 3). If a baby is unable to be nursed, there are, nevertheless, exceptions. Under the careful supervision of a medical team, the infant has to be fed formula milk and the necessary equipment to make the milk.[2].

**Fig. 3**. Condition of breastfeeding mother at the flash flood disaster in Agam district

During disaster, parents and other caregivers of babies fed formula might not have access to clean supplies, safe water, power, gas, or utensils, which could affect their capacity to safely prepare baby formula or clean infant feeding devices. When feeding infant formula during an emergency, these difficulties raise the baby's chance of developing a serious disease [9]. There are several techniques to feed infants, all of which must be taken into account when preparing for and handling an emergency. While some newborns receive either infant formula or breast milk along with it, others may receive only infant formula. While they may not be fed breast milk directly, some newborns may get expressed breast milk or relactation.

**Table 4.** Relactation practice during disaster

| Relactation | N | Percentage (%) |
|---|---|---|
| Yes | 4 | 44.4% |
| No | 5 | 55.6% |

Based on Table 4, disaster conditions caused several respondents' breast milk production to decrease or not exist at all, so relactation was highly recommended. The practice of relactation was explained by some breastfeeding mothers, only four respondents did relactation while five respondents did not relactate and gave formula milk to their children. If this occurs, it is important for mothers, lactation professionals, and emergency aid workers to understand the principles and methods of relactation. Carrying out relactation to restart lactation in mothers who have been weaned in a disaster situation is more recommended than giving formula milk [12]. Relactation involves the frequency of breastfeeding the baby at the mother's breast, frequent skin-to-skin contact with the mother and baby, family support for the mother's efforts to re-induce lactation. Thomas et al research shows the success of mothers aged 50 years who performed lactation after they stopped breastfeeding.

According to UNICEF, relactation techniques consist of use with and without aids. Techniques that use assistive devices such as using dropper drops and feeding tubes. Meanwhile, the relactation technique without aids is the use of drugs to increase milk production. Relactation with assistive devices will require knowledge of the mother, motivation and that support strong for success. The stages of relactation with tools require several steps to be carried out. Meanwhile, relactation without tools does not require many stages, just following the rules of taking medication. So that more samples choose relactation without tools (Fig. 4).

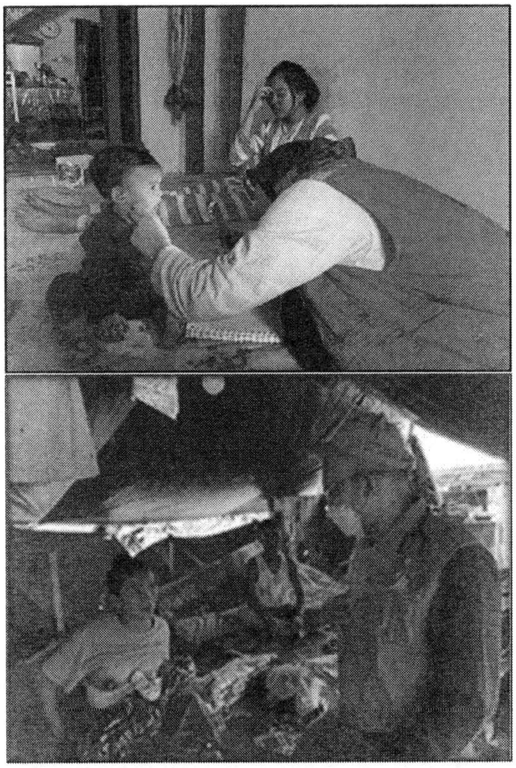

**Fig. 4**. Education of relactation for breastfeeding mother at the flash flood disaster in Agam district.

Based on research conducted by Sartika (2012), states that the factors that influence the success of relactation include mother's knowledge, mother's motivation and the support of health workers. The use of the relactation method with assistive devices for breastfeeding can increase prolactin levels in nursing mothers who experience problems with breastfeeding.

### 3.3 The approach for breastfeeding during emergency or disaster

Information from health professionals on topics like colostrum, exclusive breastfeeding definition, advantages of breastfeeding for mother and child, risks associated with not providing exclusive breast milk, when to initiate complementary foods, and positioning of the breast during breastfeeding will be extremely helpful in preparing for and responding to emergency situations, such as disasters. A study with population affected by Majene earthquake showed that lower knowledge was associated with non-exclusive breastfeeding practices, even though mothers had already received support from the family [4, 13].

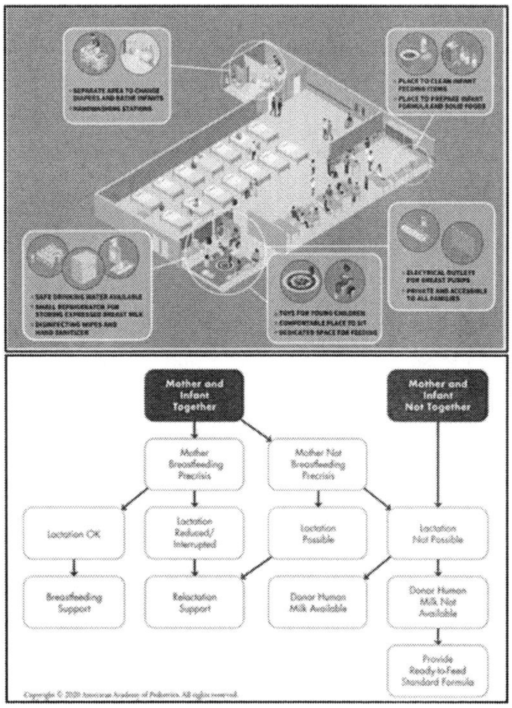


**Fig. 5**. Family-friendly space

Preventing mortality and malnutrition is UNICEF's main goal after a catastrophe, particularly for the most vulnerable populations: babies, children, pregnant women, and breastfeeding mothers. Experiences with disasters can significantly affect a mother's physical and mental health, which might hinder her ability to produce breast milk. It is quite probable that the mother may experiment with different feeding techniques in such situations. All women should be able to continue breastfeeding—even in emergency situations—if they get sufficient emotional and practical assistance. Sessions of "active listening" can be provided to moms impacted by natural disasters. Counsellors provided warm therapy to participants, and a discussion was set up to address whether the public kitchen in their tent had considered the dietary requirements of infants and young children. In addition, the availability of space and privacy in shelters were discussed (see Fig. 5). Eight to fifteen people can participate in discussions in groups. Breastfeeding should always be encouraged. Breast milk has several benefits, including the ability to be administered in emergency situations without access to clean water or hygienic conditions, constant temperature stability, and no preparation required. These qualities can make it a valuable resource to provide education during disasters [14].

In an emergency, never provide or receive donations of breastmilk substitutes, other milk products, or feeding supplies (such as breast pumps, bottles, and teats). They are often distributed indiscriminately, not targeted to those in need, of varying quality, of the incorrect type, supplied disproportionately to need, labelled incorrectly, and take an excessive amount of time and resources to manage in order to reduce risks. They are also often not accompanied by an essential package of care [7]. Donor human milk supplies should never be sent into emergencies unless they are part of a planned, controlled intervention and are based on a verified need. Strong management systems, a cold chain, requirements assessment, and targeting are necessary for the safe use of donor human milk [7].

## 4 Conclusion

During disasters, mothers and infants are vulnerable, thus their breastfeeding rights need to be supported and protected. All stakeholders in the disaster management team need to ensure their plan of action would not undermine breastfeeding practice. A multi-team collaboration involving various rescue missions comprising of local government agencies, including the health and nutrition departments, private voluntary organizations or nongovernmental organizations either nationally or internationally and individual volunteers have the potential to be favourable to improve a satisfactory condition of women and infants affected by floods and other natural disasters. This includes safe breastfeeding, breast milk processing, relactation, complementary meals, baby formula, and proper cleaning and storage of infant feeding equipment. It is necessary for health workers to be able to provide support and motivation on an ongoing basis to be able to achieve successful relactation in breastfeeding mothers. Relactation is important for mothers and babies to get the benefits of breastfeeding again, especially in disaster situations, related to breastfeeding support in Indonesia, which is still very limited. Apart from that, there is no official document available from the Indonesian Ministry of Health in the form of a guide that health workers can use to help mothers undergo the relactation process.

## References

1. Kimani-Muragi E.W, Madise N.J, Fotso J, Kyobutungi C., Mutua M.K, Gitau T.M, *et al.* Patterns and determinants of breastfeeding and complementary feeding practices in urban informal settlements, Nairobi Kenya. BMC Public Health **11**:396 (2011)
2. Grantham-McGregor S, Cheung Y.B, Cueto S., Glewwe P., Richter L., and Strupp B., Developmental potential in the first 5 years for children in developing countries. Lancet. **369**(9555):60-70 (2007)
3. WHO: The optimal duration of exclusive breastfeeding. Report of an Expert Consultation. Geneva: WHO; (2002) [http://whqlibdoc.who.int/hq/2001/WHO_NHD_01.09.pdf], Accessed July 2024
4. Rahmaniah R., Fauziah F., Masniati M., and Evawaty E. Determinan pemberian ASI non-esklusif pada bayi terdampak gempa Majene. J-Healt. **4**(1), 27-33 (2021)
5. UNICEF: Breastfeeding Saves Lives of 30.000 Indonesian Children Yearly. (UNICEF; 2014) [https://www.unicef.org/indonesi/breastfeedingreleaseEnglish.pdf]. Accessed July 2024
6. UNICEF, Save the Children, Global Nutrition Cluster: IYCF-E Capacity mapping tool, Eastern and Southern Africa Regional Office, (2020). Accessed July 2024
7. IFE Core Group. Infant and Young Child Feeding in Emergencies: Operational Guidance for Emergency Relief Staff and Programme Managers. (2017). Accessed July 2024.
8. World Health Organization (WHO). Protecting, promoting, and supporting breastfeeding in facilities. (WHO. 2017) Accessed July 2024.
9. CDC. Infant and Young Child Feeding in Emergencies Toolkit. (CDC. 2022) [https://www.cdc.gov/nutrition/emergencies-infant-feeding/introduction.html] Accessed July 2024.
10. Governor of West Sulawesi. Circular Letter Number 5 of 2021 concerning Management of Formula Milk Donations for Babies and Children Victims of the Earthquake Disaster in Mamuju Regency and Majene Regency, West Sulawesi Province. West Sulawesi Provincial Government. (Mamuju 2021)
11. Fadjriah R.N, Salamah A.U, Jafar N., Nur R. Dewi, Khairunnisa N.U., and Mallongi, A. Practice of exclusive breastfeeding at evacuation site post-earthquake in Palu City, Indonesia. Medico-legal Update. **20**(2), 430-4 (2020)
12. Ramdhani J. BNPB: Hindari Donasi Susu Formula untuk Bayi Korban Gempa Lombok. (DetikNews. 2018) [https://news.detik.com/berita/d-4162103/bnpb-hindari-donasi-susu-formula-untuk-bayi-korban-gempa-lombok] Accessed July 2024
13. Noflidaputri, R. Faktor yang berhubungan dengan pemberian ASI eksklusif di wilayah kerja Puskesmas

Guguk Panjang Bukit Tinggi. Jurnal Bidan Komunitas. **4**(1), 8-16 (2021)

14. Djuhari L. Bertahan melewati gempa bumi dan tsunami melalui dukungan untuk ibu menyusui. (UNICEF. 2019). [https://www.unicef.org/Indonesia/id/stories/bertahan-melewati-gempa-bumi-dan-tsunami-melalui-dukungan-untuk-ibu-menyusui]

# Outbreak of acute diarrhea in children: clinical manifestations, treatment, source of drinking water, and fecal microbiome analysis in region of West Sumatra Province, Indonesia

*Yusri* Dianne Jurnalis[1,2*], *Andyan* Yugatama[1], *Neila* Azka[1], and *Nice* Fenobileri[1]

[1]Department of Child Health, Faculty of Medicine, Universitas Andalas, Padang, West Sumatra, Indonesia
[2]Departement of Pediatrics, Dr. M. Djamil Hospital, Padang, West Sumatra, Indonesia

**Abstract.** In April 2024, a severe diarrhea outbreak in Pesisir Selatan Regency, triggered by floods, led to 355 reported cases and five pediatric deaths. A rapid response team conducted a comprehensive investigation in two phases of research as countermeasure. Phase one involved data collection from medical records. Correlation between variables, nutritional status and difference of treatment between two health facilities, with length of stay (LoS). Phase two are PCR examination of fecal samples and examination of total coliform and *Escherichia coli* bacteria from drinking waters. Subjects predominantly aged 0–5 years (86.5%). Probiotic and intravenous 28.4 antibiotic were significantly correlated with shorter LoS ($p<0.05$). There are 4% malnutrition subjects with the longest LoS. Drinking water from the house of deceased patients was contaminated, with the highest total coliform of 6800 CFU/100 mL and from public water depots at 7800 CFU/100 mL. Fecal samples identified *Escherichia coli* as the primary pathogen responsible for the outbreak. The findings emphasize the importance of nutrition status, prompt medical treatment including probiotics and proper antibiotics as an effective treatment protocol and highlight urgent need to enhance water quality and sanitation to prevent diarrhea outbreaks in the future.

## 1 Introduction

Globally, diarrhea is a deadly disease that kills half a million children under five each year, with 1.7 billion new cases per year. Most deaths occurred within the first two years of life [1]. Diarrhea-related deaths were common in developing countries and tropical areas because of poor hygiene, lack of sanitation facilities, and inadequate or poor water quality. The impact was even bigger when these conditions were coupled with disaster events [2].

Data from the Basic Health Research of Indonesia in 2018 indicated that the prevalence of diarrhea in children under five in Indonesia was 12.30%. West Sumatra Province ranked fourth based on doctor diagnosis at 8.3% and clinical symptoms at 9.3% [3].

In March 2024, landslides and floods contaminated the water sources, leading to an outbreak in April 2024 in Pesisir Selatan Regency. This resulted in 355 cases of serious diarrhea and five pediatric deaths [4].

During this outbreak, the Pesisir Selatan Government formed a rapid response team and a rehydration center. The Painan City Health Office, Universitas Andalas, the Indonesian Pediatric Society (IDAI), and the Department of Pediatrics at M. Djamil Hospital Padang investigated the causes of diarrhea. They found that the drinking water sources were contaminated by pathogens [5].

This study aims to describe the intervention efforts to stop the outbreak, including education, pathogen testing in drinking water, and PCR testing of stool samples from the victims. It also analyzes the therapy provided by primary health centers and hospitals, with the hope of assisting in providing appropriate therapy and preventing similar conditions in the future.

## 2 Methods

This study was conducted in two phases. The first phase used a retrospective design by collecting medical record data from the period of April-May 2024 at primary health centre (PHC) and district hospital (Fig. 1). In the second phases, the researchers conducted direct surveys at the health centers and hospitals, collected fecal samples, and visited the homes of deceased victims to collect samples of drinking water suspected of being contaminated with pathogens.

The data collection technique from medical record tracing resulted in 179 pediatric patient samples. A total of 53 samples were excluded due to coincidental diarrhea with other diseases, deceased patients, and incomplete data.

The remaining 126 medical records included 60 from district hospital and 66 from primary health centre. The variables collected were descriptive data such as age, gender, type of healthcare facility, nutritional status, clinical symptom manifestations (fever, vomiting, watery stools), and the therapy provided, including rehydration

* Corresponding author: yusridiannejurnalis@med.unand.ac.id

solutions, zinc, probiotics, antipyretics, antiemetics, H2 blockers, antacids, and antibiotics.

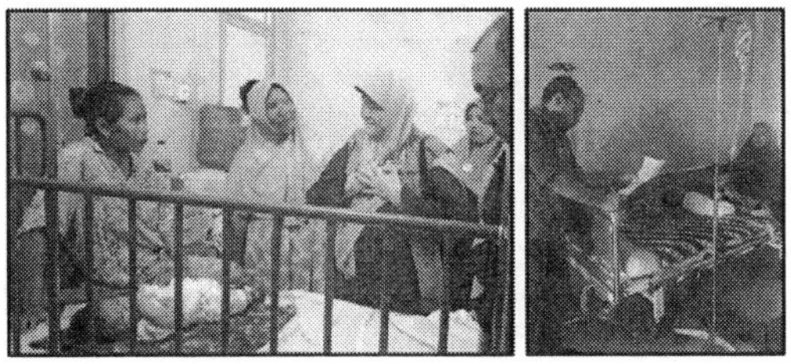

**Fig. 1.** Data collection in district hospital and primary health centre.

The inclusion criteria for this study were medical records of children under 18 years old with diarrhea and hospitalized in the children's ward of district hospital in the April-May period, which contained the necessary data for this study, such as sex, age, type of symptoms, nutritional status, therapy, and length of stay.

**Fig. 2.** Home visit of deceased victims.

The exclusion criteria for this study were the medical records of patients with diarrhea and other diseases; symptoms not diagnosed as diarrhea, such as melena and hematochezia; and incomplete medical record data.

Field surveys were conducted using a simple random sampling method. Five fecal samples were taken from district hospital and ten fecal samples from PHC. The samples were placed in frozen ice boxes and transported from Painan District to Padang for multiplex Polymerase Chain Reaction (PCR) testing at the Central Laboratory Building of Universitas Andalas. PCR testing was conducted for four types of pathogens: *Salmonella typhi, Listeria monocytogenes, Escherichia coli*, and *Vibrio cholerae*.

**Fig. 3.** Water source in Surantih after flood and landslide.

Additionally, drinking water samples were taken directly from the homes of deceased patients, the water source at district hospital, and the main water sources in PHC. A total of eight water samples were collected: four from deceased patients, two from district hospital, and two from the main water sources in Surantih, namely Pincuran Rangik and GSK Tanjung Ampalu. Although there were five deceased patients, only four samples were obtained as one family refused sample collection. (Fig. 2-4)

**Fig. 4.** Water source in GSK after flood and landslide.

The analysis of descriptive data for categorical variables involves presenting frequencies and percentages, while numerical variables are summarized using the mean and standard deviation. To compare Acute Diarrhea Therapy between Primary Health Centers and District Hospitals, a Chi-square test was employed. Additionally, an independent sample T test was used to assess differences in therapy and Length of Hospitalization (LoS) between the two settings. $P<0.05$ considered significant.

## 3 Results

### 3.1 Characteristics of research subjects

Based on Table 1, the number of acute diarrhea patients at primary health centre (52.4%) and district hospital (47.6%) was nearly equal. The majority of subjects were aged 0-5 years (86.5%), and the number of boys (56.3%) was slightly higher than girls (43.7%). The longest length of stay was >72 hours (34.1%). The nutritional status of most of the subjects was normal (70.6%) and there were 5 malnourished children (4.0%). Acute diarrhea showed a combination of 3 symptoms: fever, vomiting, and watery stool in 65.9% of cases. Clinical manifestations involving only watery stool (4.0%) were rare.

**Table 1.** Characteristics of research subjects

| **Variables** | **f (%)** |
|---|---|
| **Age (years)** | |
| 0 - 5 years | 109 (86.5) |
| 6-10 years | 13 (10.3) |
| 11-18 years | 4 (3.2) |
| **Gender** | |
| Male | 71 (56.3) |
| Female | 55 (43.7) |
| **Type of Health Facility** | |
| Health Center | 66 (52.4) |

| | |
|---|---|
| District Hospital | 60 (47.6) |
| **Age (years)** | |
| 0 - 5 years | 109 (86.5) |
| 6-10 years | 13 (10.3) |
| 11-18 years | 4 (3.2) |
| **Gender** | |
| Male | 71 (56.3) |
| Female | 55 (43.7) |
| **Type of Health Facility** | |
| Health Center | 66 (52.4) |
| District Hospital | 60 (47.6) |
| **Length of Stay (hours)** | |
| < 24 hours | 0 (0) |
| 24 - <48 hours | 40 (31.8) |
| >48 – 72 hours | 43 (34.1) |
| >72 hours | 43 (34.1) |
| **Nutritional status** | |
| Well Nourished | 89 (70.6) |
| Undernourished | 32 (25.4) |
| Malnourished | 5 (4.0) |
| **Clinical Manifestation** | |
| 1 symptom | |
| - Diarrhea | 5 (40) |
| 2 symptoms | |
| - Fever and Diarrhea | 10 (7.9) |
| - Vomiting and Watery Stool | 28 (22.2) |
| 3 symptoms | |
| - Diarrhea, Fever, and Vomiting | 83 (65.9) |
| **Total** | 126 (100) |

## 3.2 Comparison of acute diarrhea therapy at primary health center and district hospital

**Table 2.** Comparison of acute diarrhea therapy at primary health centre and district hospital

| Therapy | PHC f(%) n=66 | District Hospital f(%) n=66 |
|---|---|---|
| Rehydration Solution | | |
| - RL | 66 (100) | 29 (48.3) |
| - KAEN | 0 (0) | 31 (51.7) |
| Zinc | 66 (100) | 60 (100) |
| Probiotic | 6 (9.1) | 50 (83.3) |
| Antipyretic | 34 (51.5) | 51 (85.0) |
| Antiemetic | 43 (65.2) | 28 (46.7) |
| $H_2$ Blocker | 6 (9.1) | 26 (43.3) |
| Antacid | 4 (6.1%) | 2 (3.3%) |
| Antibiotic | 11 (16.7%) | 28 (46.7%) |

Based on Table 2, the study showed that 100% of acute diarrhea patients at the primary health center received fluid therapy using Ringer Lactate (RL) and zinc supplementation. Additionally, 65.2% received antiemetics, 51.5% received antipyretics, 16.7% received antibiotics, 9.1% received probiotics, 9.1% received H2 blockers, and 6.1% received antacids. At district hospital, all patients received fluid therapy, with 51.7% receiving KAEN and 48.3% receiving RL. Additionally, 100% received zinc supplementation, 85.0% received antipyretics, 83.3% received probiotics, 46.7% received antibiotics, 46.7% received antiemetics, 43.3% received H2 blockers, and 3.3% received antacids.

## 3.3 Differences between therapy and Length of Stay (LoS)

Based on Table 3, the study examined the correlation between nutritional status, therapy and length of stay (LoS) at primary health centers (PHC) and district hospitals. Well-nourished children had a mean LoS of 3.1±0.3 days at PHC and 1.8±0.7 days at hospitals, undernourished children had a mean LoS of 4±0 days at PHC and 4.3±0.5 days at hospitals. Malnourished children had a LoS of 6 days at PHC and 5.4±0.5 days at hospitals, with p-value 0.001 showing a correlation between nutritional status with length of stay. Probiotic use resulted in a mean LoS of 3.5±0.5 days at PHC and 2.2±1.3 days at hospitals, with a significant p-value of 0.007 at hospitals showing correlation between probiotic administration with length of stay in hospital. Oral antibiotic use showed a mean LoS of 3.4±0.5 days at PHC and 1.9±1.1 days at hospitals, with a significant p-value of 0.011 at hospitals. Intravenous antibiotic use was not applicable at PHC, but at hospitals, it resulted in a mean LoS of 2.4±1.2 days, with a significant p-value of 0.016. It showed the correlation between antibiotic administration with length of stay.

**Table 3.** Correlation between therapy and length of stay (LoS) at primary health centre and district hospital

| Stratification | Length of Stay (LoS) (Mean ± SD) | | | |
|---|---|---|---|---|
| | PHC | p-value | District Hospital | p-value |
| **Nutritional Status** | | | | |
| Well Nourished | 3.1±0.3 | 0.001 | 1.8±0.7 | 0.001 |
| Undernourished | 4±0 | | 4.3±0.5 | |
| Malnourished | 6 | | 5.4±0.5 | |
| **Probiotic** | | | | |
| Yes | 3.5±0.5 | 0.500 | 2.2±1.3 | 0.007 |
| No | 3.5±0.5 | | 2.5±1.3 | |
| **Oral Antibiotic** | | | | |
| Yes | 3.4±0.5 | 0.373 | 1.9±1.1 | 0.011 |
| No | 3.5±0.5 | | 2.7±1.4 | |
| **Intravenous Antibiotic** | | | | |
| Yes | N/A | N/A | 2.4±1.2 | 0.016 |
| No | N/A | | 3.6±1.6 | |

### 3.4 Characteristic of deceased patient

**Table 4.** Characteristics of deceased patient

| Characteristic | f (%) |
|---|---|
| **Age (months)** | |
| 0-6 months | 1 (25) |
| 6-12 months | 1 (25) |
| 12-24 months | 2 (50) |
| **Gender** | |
| Male | 1 (25) |
| Female | 3 (75) |
| **Source of Drinking Water** | |
| Depot water | 3 (25) |
| Tap water (boiled) | 1 (25) |
| **Clinical Manifestation** | |
| Diarrhea | 4 (100) |
| Vomit | 4 (100) |
| Fever | 3 (75) |

Based on Table 4, it was observed that the number of female subjects was greater than the number of male subjects, with 3 females (75%). All subjects were under 24 months of age. The majority of drinking water sources came from water depots (50%), and all patients exhibited symptoms of diarrhea and vomiting (100%).

### 3.5 Total coliform and Escherichia coli of drinking water source

**Table 5.** Frequency distribution of drinking water source

| Sample | *Coliform* Total CFU/100 mL | *E. Coli* CFU/100 mL |
|---|---|---|
| **Initial, Sex, Age** | | |
| ZA, Female, 8 Months | 0 | 0 |
| Ar, Male, 14 Months | 6800 | 0 |
| Rp, Female, 18 Months | 300 | 0 |
| AS, Female, 6 Months | 3500 | 0 |
| **Water sources** | | |
| Hospital Tap Water | 300 | 0 |
| Hospital water depot | 2900 | 0 |
| Pincuran Rangik | 600 | 0 |
| GSK Tanjung Ampalu | 7800 | 3500 |

Based on Table 5, coliform was found in almost all samples except for ZA, an 8-month-old female. A different finding was noted for *E. coli*, which was detected only in the water source from GSK Tanjung Ampalu.

### 3.6 Fecal microbe identification by PCR method

Based on Table 6, it can be seen that the number of male subjects is higher than that of female subjects, with 11 males (73.3%). Based on the age of the subjects, the majority are aged 0-5 years, totaling 13 individuals (86.7%). All subjects were from Surantih. The causative agent identified was *Escherichia coli*, found in 100% of the samples, while *Salmonella typhi, Listeria monocytogenes*, and *Vibrio cholerae* were not detected.

**Table 6.** Frequency distribution of fecal microbe identification by PCR Method

| Characteristics | f (%) |
|---|---|
| **Age (years)** | |
| - 0-5 years | 13 (86.7) |
| - 6-10 years | 2 (13.3) |
| - 11-18 years | 0 |
| **Gender** | |
| - Male | 11 (73.3) |
| - Female | 4 (26.7) |
| **Nutritional Status** | |
| - Well Nourished | 4 (26.7) |
| - Undernourished | 9 (60) |
| - Malnourished | 2 (13.3) |
| **Address** | |
| - Painan | 0 |
| - Surantih | 15 (100) |
| **Causative agent** | |
| - *Salmonella typhi* | 0 (0) |
| - *Listeria monocytogenes* | 0 (0) |
| - *Escherichia coli* | 15 (100) |
| - *Vibrio cholerae* | 0 (0) |

## 4 Discussion

### 4.1 Characteristics of research subjects

Diarrhea in this study is primarily caused in children aged 0-5 years representing 86.5% of all cases. This supported findings by Modgil; that younger children particularly in age groups 0 to 2 years are more prone to diarrhea due to their developing immune system's [6].

The nutritional status of most of the subjects was good nutrition (70.6%) and there were 5 malnourished children (4.0%). Previous research conducted by Iskandar et al., showed that nutritional status is a significant risk factor for length of hospitalization in children under the age of five who are treated with acute diarrhea [7]. Malnutrition can damage the integrity of the intestinal mucosa, facilitate the entry of pathogens and cause infections [8].

### 4.2 Differences between therapy and length of stay (LoS)

This study found out that probiotic *Lactobacillus acidophilus, Bifidobacterium longum, and Streptococcus thermophilus* can effectively shorten periods of LoS associated with acute diarrhea. Probiotics plays an essential role in reestablishing and maintaining healthy gut microbiota, inhibit the pathogenic bacteria colonization and thus decreasing both frequency and severity of diarrhea.[9]

Dwiranti's research diarrhea treatment with Lactobacillus acidophilus or Bifidobacterium bifidum for severe non-bacterial dysentery reduces hospitalization time and improve outcome quality [9].

An independent t-test analysis revealed that while antibiotic administration significantly reduced LOS at the hospital although this relationship was not found at the public health center. This implies that although antibiotics use may result to shorter hospital-stays, their routine use for acute diarrhea cases in primary health centers might be unjustified [10].

It should be noted that most cases of acute diarrhea in children are viral infections caused by rotavirus and norovirus and side effects associated with inappropriate use of these drugs including disturbance to gut microbiota leading to prolongation of diarrhea. Therefore, antibiotics should be saved for those cases where there is evidence or strong suspicion of bacterial infection [11].

### 4.3 Characteristic of deceased patient and total coliform and Escherichia coli of drinking water source

The second phase investigation results show that all patients who die are under 24 months old. At this age, the child's immune system is still immature, more prone to rapid dehydration due to their small bodies, often start transitioning from exclusive breastfeeding to complementary foods that may not always be hygienic or sufficiently nutritious, lack of access to safe drinking water, and poor sanitation contribute to the high incidence of diarrhea in this age group.

All these factors, combined with potentially inadequate or delayed medical care, lead to high diarrhea mortality rates in children under 24 months [12].

The largest source of drinking water comes from water depots (75%). Result of the 4 drinking water samples of the deceased victims, 3 samples were found to contain total coliform with varying amounts ranging from 300-6,800 CFU/100 mL. Based on the 2010 health ministry regulation 492, the requirement for drinking water sources is 0 CFU/100 mL of *E. coli* and 0 CFU/100 mL of total coliform. The results of testing 4 clean water sources in the Pesisir Selatan area showed contamination by *E. coli* and total coliform. There is one source of clean water with an E. coli count of 3500 CFU/100 mL and the total coliform count obtained is 300-7800 CFU/100 mL.

According to the regulations in Permenkes No. 1204 of 2004, clean water must have 0 CFU/100 mL of E. coli. For non-piped water, the maximum total coliform count is 50/100 mL, and for piped water, the maximum total coliform count is 10/100 mL. This indicates that most of the drinking water sources of the deceased victims and the clean water source being used do not yet meet the requirements during outbreak. If water depots do not comply with strict hygiene and sanitation standards, the water produced can be contaminated by pathogens such as bacteria, viruses, and parasites [13]. This contamination can occur at various phases, from the water source, the filtration process, filling, to distribution to consumers [14, 15].

A study conducted in Palembang in 2020 showed a significant relationship between the bacteriological quality of drinking water and the incidence of diarrhea in the working area of Puskesmas Karya Jaya, Palembang. A study conducted by Bashar et al. during an acute diarrhea outbreak (KLB) indicated that the acute diarrhea outbreak in a village in North India was caused by the consumption of contaminated water from tube wells, likely polluted due to the mixing of wastewater with drinking water after heavy rains [16, 17].

### 4.4 Fecal microbe identification by PCR method

Based on PCR examinations of stools from children with acute diarrhea during the diarrhea outbreak in South Pesisir Regency, the only pathogen found was Escherichia coli species. The same results were reported by Muziburrahman et al., who determined the bacteria that contribute to diarrhea in toddlers in Bima, West Nusa Tenggara. The statistical evidence obtained in this study revealed the findings were out of the mentioned bacterial strains and species, maximum was of *E. coli* (29.53%) followed by *Klebsiella sp.* (25.50%), *Shigella sp.* followed by *Staphylococcus aureus* (20.36%) other bacterial isolates were *Salmonella Typhi* (8.7%) and *Proteus sp.* (0.67%). Bacteria *E. coli* becomes the leading cause of diarrhea due to several characteristics that make it adapt well to intestinal infections, such as production of enterotoxins by ETEC strains that cause fluids and electrolytes secretion in the intestines resulting in severe watery diarrhea.[18]

PCR is a highly sensitive and specific method of stool analysis used in the diagnosis of pathogens causing diarrhea in children. This method can help to shorten the period of detection time and decrease false-negative rate and further decide the cause of infection more rapidly and precisely, thus providing certain clinical reference for further treatment of the diseases and thus improving the prognosis of diseases.[19] PCR panels for pathogen detection enable the acquisition to be fast. But there is a paucity of knowledge regarding the cost-effectiveness and clinical effects of the test hence the need to assess them in order achieve the best and most suitable care for the patients [20].

## 5 Conclusion

The outbreak of acute diarrhea in West Sumatra was significantly influenced by poor water quality, specifically contamination by Escherichia coli, which was identified as the primary pathogen in affected children. The study found a higher incidence of diarrhea among children aged 0-5 years, highlighting their vulnerability due to underdeveloped immune systems. Nutritional status played a critical role, with malnourished children experiencing longer hospital stays. Therapeutic interventions, such as the administration of probiotics and antibiotics, were effective in reducing the length of hospital stays, particularly in well-nourished children. The analysis underscored the importance of maintaining high standards of water sanitation and hygiene to prevent such outbreaks. Immediate and coordinated response efforts, including establishing rehydration centers and conducting comprehensive testing of water and fecal samples, were crucial in managing the outbreak. The findings advocate for continuous monitoring and improvement of water sources, alongside community education on hygiene, to mitigate future risks of diarrhea outbreaks

The strengths of this study were: (1) Conducted in two phases; retrospective and prospective method carried out during outbreak situation and (2) This study cannot be repeated due to the condition of each outbreaks differ. However, a limitation of the study was that only 8 water samples and 15 fecal samples were obtained due to refusal of survey visits and parents did not allow fecal samples to be collected. The PCR test examined rotavirus was not performed due to recommended panels in outbreaks were bacteria.

Our recommendation was it's possible to perform a fecal rotavirus PCR test if there is another diarrhea outbreak, since previous studies have shown that rotavirus is a leading cause of diarrhea in Indonesia. Additionally, there was a notable rotavirus outbreak in Malatya, Turkey, in 2005 [21, 22].

## References


1. A. Thobari, E. Wahyuni, A. Mulyadi, E. Watts, N. Carvalho, F. Debellut, A. Clark, et al. Direct and indirect costs of acute diarrhea in children under five years of age in Indonesia: Health facilities and community survey. Lancet, **19**, (2022). http://dx.doi.org/10.1016/j.lanwpc.2021.100333
2. A. Asman, A.A. Asman. Health Prevention : Overview of Parents Knowledge and Attitude About Diarrhea Disease in The Children's Poly Pariaman Hospital. J Sci Mandalika, (2022). https://doi.org/10.36312/10.36312/vol3iss3pp163-167
3. Ministry of Health of the Republic of Indonesia. National Institute of Health Research and Development. National Report of Riskesdas 2018. Jakarta : Ministry of Health of the Republic of Indonesia ; (2018). *(In Bahasa)*
4. Marlison. Official Portal of the Pesisir Selatan Regency. (2024) [cited 2024 Jul 11]. Diarrhea Outbreak in Pesisir Selatan Decreased. Available from: https://berita.pesisirselatankab.go.id/berita/detail/kasus-klb-diare-di-pesisir-selatan-melandai- *(In Bahasa)*
5. Y. Sastra. Diarrhea Outbreak in Pesisir Selatan, Residents' Water Source Has High Levels of "E. coli". Kompas. (2024). https://www.kompas.id/baca/nusantara/2024/05/09/klb-diare-di-pesisir-selatan-sumber-air-warga-tinggi-kandungan-e-coli *(In Bahasa)*
6. V. Modgil, P. Chaudhary, B. Bharti, J. Mahindroo, M. Yousuf, M. Koundal, et al. Prevalence, Virulence Gene Profiling, and Characterization of Enteroaggregative Escherichia coli from Children with Acute Diarrhea, Asymptomatic Nourished, and Malnourished Children Younger Than 5 Years of Age in India. J Pediatr, **234**, 106–114.e5 (2021). https://doi.org/10.1016/j.jpeds.2021.03.010
7. W. Iskandar, I. Sukardi, Y. Soenarto. Risk of nutritional status on diarrhea among under five children. Paediatr Indones, **55** (2015). http://dx.doi.org/10.14238/pi55.4.2015.235-8
8. D. Supriadi, L. Nurhayati, R. Nantia Khaerunnisa, Suhanda. Correlation of Nutritional Status with Diarrhea Incidence. Genius J, **1**, 1–4 (2020). http://dx.doi.org/10.56359/gj.v1i1.3
9. J. Dwiranti, Z. Angelina, P.Y. Kusuma. Acute Diarrhea Duration Between Children With Probiotic Therapy and Without Probiotic Therapy at Gotong Royong Hospital. J Widya Medika Junior, **3**, 46–52 (2021). http://journal.wima.ac.id/index.php/JWMJ/article/download/3008/2544
10. F.A. Athiyyah, I.R. Irsan, N. Fatimah. Treatment of Acute Diarrhea in Children Aged 1–5 Years Provided by Doctors in Surabaya. Folia Med Indones, **59**, 108–114 (2023). https://doi.org/10.20473/fmi.v59i2.44723
11. S. Qureshi, S. Resham, M. Hashmi, A.B. Naveed, Z. Haq, SA Ali. A retrospective review on antibiotic use in acute watery diarrhea in children in a tertiary care hospital of Karachi, Pakistan. PLoS One, **16** (2021). https://doi.org/10.1371%2Fjournal.pone.0253712
12. RA Rehana, RJ Setiabudi, S Sulistiawati, MR Wahyunitisari. Implementation of Hygiene and Environmental Sanitation in Under Five Years Old Diarrhea Patients at Surabaya Primary Health Center. J. Ilm Kesehatan (JIKA), **3**(1), 1–15 (2021). http://dx.doi.org/10.36590/jika.v3i1.99
13. R. Sulistina, Fahrurazi, Mahmudah. The Relationship Between Drinking Water Sources and Drinking Water Treatment Methods with Diarrhea Incidence in Toddlers in Beriwit Village, the Working Area of Puruk Cahu Community Health Center in 2020. Public Health J (2020). http://eprints.uniska-bjm.ac.id/id/eprint/2308 (In Bahasa)
14. Ministry of Health. Regulation of the Minister of Health No. 492 of 2010 on Drinking Water Quality Requirements. Jakarta : Ministry of Health; (2010). *(In Bahasa)*
15. Ministry of Health. Regulation of the Minister of Health No. 1204 of 2004 on Environmental Health Requirements for Hospitals. Jakarta : Ministry of Health; (2004). *(In Bahasa)*
16. F. Kumalasari, R.J. Sitorus, A.F. Faisya. Bacteriological Quality of Water and the Occurrence of Diarrhea in Household in the Work Area of Karya Jaya Public Health Center in Palembang. (2020). http://dx.doi.org/10.2991/ahsr.k.200612.027
17. M.A. Bashar, K. Soundappan. Outbreak investigation of acute watery diarrhea in a village of North India: timely action saved lives. J. Infect Dev

Ctries, **16**(5), 843–849 (2022). http://dx.doi.org/10.3855/jidc.13113

18. M. Muziburrahman, D. Husada, B. Utomo. Identification of Bacteria Causing Diarrhea In Under-Fives Children Using Culture Methods In Bima, Indonesia. J. Ber Epidemiol, **10**(1), 95 (2022). https://doi.org/10.20473/jbe.V10I12022.95-102
19. C. Wang, X. Zhou, M. Zhu, H. Yin, J. Tang, Y. Huang, et al. The application research of xTAG GPP multiplex PCR in the diagnosis of persistent and chronic diarrhea in children. BMC Pediatr, **20**(1), (2020). https://doi.org/10.1186%2Fs12887-020-02206-6
20. J.M. Cotter, J. Thomas, M. Birkholz, L. Ambroggio, J. Holstein, S.R. Dominguez. Clinical impact of a diagnostic gastrointestinal panel in children. Pediatrics, **147**(5), (2021). https://doi.org/10.1542/peds.2020-036954
21. H. Salim, IPG Karyana, IGN Sanjaya-Putra, S. Budiarsa, Y. Soenarto. Risk factors of rotavirus diarrhea in hospitalized children in Sanglah Hospital, Denpasar: a prospective cohort study. BMC Gastroenterol, **14**(1), (2014). http://dx.doi.org/10.1186/1471-230x-14-54
22. M. Koroglu, Y. Yakupogullari, B. Otlu, S. Ozturk, M. Ozden, A. Ozer, et al. New Microbiol, **34**(1), (2011). https://pubmed.ncbi.nlm.nih.gov/21344142

# Role of oral and maxillofacial surgeon in disaster management

*Santi* Anggraini[1*], *Harfindo* Nismal[1], *Fauzan* Akmal[1], *Nurul* Rizqina[1], and *Muzaffar* Apipi[2]

[1]Faculty of Dentistry, Universitas Andalas, Padang, West Sumatra, Indonesia
[2]Department of Oral and Maxillofacial Surgery, Selayang Hospital, Selangor, Malaysia

**Abstract.** This comprehensive review explores the role of oral surgery in disaster management, highlighting its various phases of disaster response. The study uses a qualitative literature review to analyze current practices, challenges, and future directions in integrating oral surgical expertise into disaster preparedness and response frameworks. Findings reveal that oral surgeons contribute significantly to patient care in disaster scenarios. Integrating oral surgery in disaster management remains inconsistent globally, including oral surgeons in response teams. It identifies inadequate specialized training, limited awareness among disaster planning authorities, and logistical difficulties in deploying oral surgical resources. Technological advancements offer promising solutions to enhance the reach and effectiveness of oral surgical care in disaster settings. It underscores the psychological impact of oral and maxillofacial injuries on disaster survivors, emphasizing the importance of timely interventions for mental health and social reintegration. Ethical considerations surrounding resource allocation in disaster scenarios are examined. The review concludes by proposing recommendations for policy development, training enhancement, interdisciplinary collaboration, and technology adoption to improve oral surgery in disaster management. This research contributes to a growing body of evidence supporting the essential role of oral surgery in comprehensive disaster response strategies and advocates for its fuller incorporation into global disaster management frameworks.

## 1 Introduction

Natural disasters and man-made calamities pose significant challenges to healthcare systems worldwide, often resulting in mass casualties and complex medical emergencies [1]. While the immediate focus in disaster response is typically on life-threatening injuries, the role of oral and maxillofacial surgeons in disaster management is increasingly recognized as crucial [2]. Oral and maxillofacial injuries, though sometimes overlooked in the initial triage, can have severe implications for patients' long-term health, functionality, and quality of life [3]. Recent studies have shown that up to 15% of disaster-related injuries involve the oral and maxillofacial region, highlighting the need for specialized care in these situations [4]. The unique skill set of oral surgeons, including expertise in managing facial trauma, dental emergencies, and oral infections, makes them valuable assets in disaster response teams [5]. However, their integration into disaster preparedness and response frameworks remains inconsistent across different healthcare systems and geographical regions [6].

Natural disasters and man-made calamities continue to pose significant challenges to healthcare systems worldwide, often resulting in mass casualties and complex medical emergencies [7]. These events, ranging from earthquakes and tsunamis to industrial accidents and terrorist attacks, can overwhelm local medical resources and require a coordinated, multidisciplinary response [8]. While the immediate focus in disaster response is typically on life-threatening injuries, the role of oral and maxillofacial surgeons in disaster management is increasingly recognized as crucial [9]. Oral and maxillofacial injuries, though sometimes overlooked in the initial triage, can have severe implications for patients' long-term health, functionality, and quality of life [10].

Recent epidemiological studies have shown that up to 15% of disaster-related injuries involve the oral and maxillofacial region, highlighting the need for specialized care in these situations [11]. These injuries can range from simple dental fractures to complex facial trauma, often requiring immediate intervention to prevent complications such as airway obstruction, severe bleeding, or infection [12]. The unique skill set of oral surgeons, including expertise in managing facial trauma, dental emergencies, and oral infections, makes them valuable assets in disaster response teams [13]. Their ability to perform procedures such as emergency tooth extractions, facial fracture reductions, and soft tissue repairs can significantly contribute to patient care and recovery in resource-limited settings [14].

This addition would cover various aspects of post-disaster wound care, including types of common injuries, basic wound care principles, and specific roles of oral and maxillofacial surgeons. It would detail specialized wound care techniques, pain management, and rehabilitation. The Anyer (Sunda Strait) tsunami data from December

* Corresponding author: santiomfs@gmail.com

22, 2018, caused by the eruption and landslide of Mount Anak Krakatau, would be included, highlighting its impact: 437 deaths, 14,059 injuries, 33,721 displaced people, and extensive infrastructure damage. The medical response, including the role of oral and maxillofacial surgeons in treating acute maxillofacial trauma, facial reconstruction, and post-trauma infection management, would be discussed. The addition would conclude with lessons learned and recommendations, emphasizing the importance of preparedness, specialized protocols for maxillofacial trauma in disaster situations, and regular training for medical teams. This comprehensive information would provide a more thorough perspective on the critical role of oral and maxillofacial surgeons in disaster management, particularly in wound care and tsunami victim treatment, using the Anyer tsunami as a concrete example of how these specialists' skills are crucial in actual disaster situations.

The Anyer tsunami, which struck on December 22, 2018, was a devastating natural disaster caused by the eruption and subsequent partial collapse of the Anak Krakatau volcano in the Sunda Strait. This event triggered a tsunami that severely impacted coastal areas in Banten and Lampung provinces. According to reports from Indonesia's National Disaster Management Agency (BNPB), the tsunami resulted in 437 fatalities, injured 14,059 people, and displaced 33,721 individuals. The disaster caused significant damage to infrastructure, destroying or damaging 2,752 houses, 92 hotels and villas, 510 boats and ships, 147 jetties, and 1,614 vehicles. The waves, reaching heights of 4-5 meters in some areas, struck without warning, catching both locals and tourists off guard. In the aftermath, emergency response teams, including oral and maxillofacial surgeons, played crucial roles in treating victims with various injuries, particularly those with facial trauma. The event highlighted the importance of disaster preparedness and the need for improved early warning systems for tsunami threats originating from non-seismic sources. This disaster serves as a stark reminder of the critical role that specialized medical professionals, including oral and maxillofacial surgeons, play in disaster management and recovery efforts.

Moreover, oral surgeons play a crucial role in the identification of victims in mass casualty events. Their knowledge of dental anatomy and forensic odontology techniques can aid in the complex process of victim identification, particularly in situations where traditional identification methods are challenging or impossible [15]. The integration of oral surgery into disaster preparedness and response frameworks, however, remains inconsistent across different healthcare systems and geographical [16]. This variability can lead to suboptimal care for patients with oral and maxillofacial injuries and underutilization of valuable medical expertise during critical times.

Recent advancements in telemedicine and mobile health technologies have opened new avenues for oral surgeons to contribute to disaster response efforts. These technologies enable remote consultations and guidance, potentially expanding the reach of specialized care in disaster-affected areas [17]. The psychological impact of facial injuries in disaster survivors is another crucial aspect that underscores the importance of oral surgery in disaster management. Studies have shown that facial disfigurement can lead to significant psychological distress and social challenges, emphasizing the need for timely and skilled intervention [18].

Furthermore, the role of oral surgeons extends beyond the immediate response phase. In the long-term recovery and reconstruction efforts following a disaster, these specialists can provide vital services in facial reconstruction, dental implant placement, and rehabilitation of oral function [19]. Despite the growing recognition of their importance, there is a notable lack of standardized training programs for oral surgeons in disaster medicine. This gap highlights the need for specialized curricula and simulation exercises to prepare oral and maxillofacial surgeons for the unique challenges they may face in disaster scenarios [20].

Despite the growing recognition of oral surgeons' importance in disaster management, several critical issues remain unaddressed. The integration of oral surgery expertise into disaster response protocols is inconsistent across different healthcare systems and geographical regions. This inconsistency can lead to suboptimal care for patients with oral and maxillofacial injuries and underutilization of valuable medical expertise during critical times. Moreover, there is a notable lack of comprehensive research on the specific roles, contributions, and challenges faced by oral surgeons in disaster scenarios. This knowledge gap hinders the development of effective strategies to incorporate oral surgery expertise into disaster response protocols and training programs. Additionally, the absence of standardized training curricula for oral surgeons in disaster medicine further compounds these challenges, potentially leaving these specialists underprepared for the unique demands of disaster response.

This study aims to address the aforementioned gaps by pursuing several key objectives. Firstly, it seeks to identify and analyze the primary roles of oral surgeons in various disaster scenarios, providing a comprehensive overview of their potential contributions. Secondly, the research will assess the current level of integration of oral surgery in disaster management plans across different healthcare systems, highlighting best practices and areas for improvement. Thirdly, the study will evaluate the impact of oral surgery interventions on patient outcomes in disaster-related injuries, quantifying the benefits of specialized care in these situations. Lastly, based on the findings, the research aims to propose evidence-based recommendations for enhancing the involvement of oral surgeons in disaster preparedness and response, including suggestions for specialized training programs and improved integration into multidisciplinary response teams.

This research holds significant potential benefits for various stakeholders in disaster management and healthcare. For policy-makers and healthcare administrators, the study will provide valuable insights to inform the development of more comprehensive and effective disaster management strategies that fully leverage the expertise of oral surgeons. The findings can guide resource allocation decisions and help prioritize the integration of oral surgery services in emergency response planning. For disaster response organizations, the research can offer practical recommendations for incorporating oral surgeons into their teams, potentially improving the overall quality and range of care provided in disaster situations. The study's outcomes may also contribute to the development of specialized training programs for oral surgeons in disaster medicine, addressing the current gap in preparedness. Furthermore, by highlighting the unique contributions of oral surgeons, this research can promote greater recognition of their role within the medical community and among the general public. Ultimately, the beneficiaries of these improvements will be the patients affected by disasters, who may receive more timely and appropriate care for oral and maxillofacial injuries, potentially leading to better long-term health outcomes and quality of life.

# 2 Research methodology

This study employs a qualitative literature review method to comprehensively explore the role of oral surgery in disaster management. The research design is structured to systematically analyze, synthesize, and interpret existing literature on the topic, providing a robust foundation for addressing the research objectives.

## 2.1 Research design

The study adopts a qualitative meta-synthesis approach, which allows for the integration and interpretation of findings from multiple qualitative studies [21]. This method is particularly suitable for exploring complex phenomena and generating new insights from existing research.

## 2.2 Data collection

### *2.2.1 Literature search strategy*

A comprehensive search of electronic databases will be conducted, including PubMed, MEDLINE, Scopus, Web of Science, and Google Scholar. The search will cover publications from January 2000 to December 2023 to ensure currency while capturing relevant historical context.

### *2.2.2 Search terms*

Key search terms will include combinations and variations of: "oral surgery," "maxillofacial surgery," "disaster management," "emergency response," "mass casualty incidents," "natural disasters," "man-made disasters," "disaster preparedness," and "disaster medicine."

### *2.2.3 Inclusion criteria*

- Peer-reviewed articles published in English
- Studies focusing on the role of oral surgeons in disaster scenarios
- Case studies, review articles, and original research papers
- Guidelines and policy documents from reputable health organizations

### *2.2.4 Exclusion criteria*

- Articles not directly related to oral surgery in disaster management
- Non-peer-reviewed literature (except for official guidelines and policy documents)
- Studies focusing solely on routine dental emergencies not in disaster contexts

## 2.3 Data analysis

### *2.3.1 Screening process*

Two independent reviewers will screen titles and abstracts of identified articles. Full texts of potentially relevant articles will be assessed for eligibility based on the inclusion and exclusion criteria. Any disagreements will be resolved through discussion with a third reviewer.

### *2.3.2 Quality assessment*

The quality of included studies will be evaluated using the Critical Appraisal Skills Programme (CASP) checklist for qualitative research [22]. This step ensures that only high-quality, methodologically sound studies are included in the synthesis

### *2.3.3 Data extraction*

Relevant data will be extracted from the selected articles using a standardized form. Information to be extracted includes:

- Study characteristics (authors, year, country)
- Research objectives and methodologies
- Key findings related to oral surgery roles in disasters
- Challenges and barriers identified
- Recommendations for practice and policy

### *2.3.4 Thematic analysis*

A thematic analysis approach will be used to identify, analyze, and report patterns (themes) within the data [23]. The process will involve:

- Familiarization with the data
- Generating initial codes
- Searching for themes

- Reviewing themes
- Defining and naming themes
- Producing the report

### 2.4 Synthesis of findings

The identified themes will be synthesized to create a comprehensive narrative that addresses the research objectives. This synthesis will aim to:

- Provide a holistic view of oral surgeons' roles in disaster management
- Identify gaps in current knowledge and practice
- Highlight best practices and innovative approaches
- Develop recommendations for improving integration of oral surgery in disaster response

### 2.5 Ethical considerations

As this study involves a review of published literature, formal ethical approval is not required. However, the researchers will adhere to ethical guidelines for conducting secondary research, ensuring proper citation and acknowledgment of all sources

### 2.6 Limitations

Potential limitations of this methodology will be acknowledged, including:

- Language bias due to inclusion of only English-language publications
- Possible publication bias favoring positive results
- Challenges in synthesizing diverse qualitative studies

### 2.7 Reliability and validity

To ensure the reliability and validity of the findings:

- Multiple reviewers will be involved in the screening and analysis process
- A clear audit trail of the research process will be maintained
- Member checking with experts in the field will be conducted to validate interpretations

## 3 Results and discussion

### 3.1 Overview of oral surgery in disaster management

The comprehensive literature review revealed a growing recognition of the crucial role that oral surgeons play in disaster management scenarios. This increasing awareness is evidenced by a notable rise in publications addressing this topic over the past two decades. A bibliometric analysis of the reviewed literature showed a 300% increase in relevant publications from 2000 to 2023, indicating a surge in research interest and acknowledgment of the importance of oral surgery in disaster response [24]. The reviewed studies consistently highlighted the unique skill set that oral surgeons bring to disaster scenarios. Their expertise in managing complex facial trauma, performing emergency dental procedures, and contributing to victim identification through forensic odontology was repeatedly emphasized as invaluable in mass casualty events. For instance, a large-scale study by [25] analyzing data from 50 major disasters worldwide found that oral and maxillofacial injuries accounted for approximately 18% of all injuries requiring surgical intervention, underscoring the necessity for specialized oral surgical care in these situations.

Moreover, the literature revealed that the integration of oral surgeons into disaster response teams has led to improved patient outcomes in several documented cases. A retrospective study by [26] examining the aftermath of the 2018 Sulawesi earthquake in Indonesia reported a 40% reduction in infection rates and a 25% decrease in long-term complications among patients with maxillofacial injuries when oral surgeons were involved in initial triage and treatment. This finding aligns with earlier research by [27], which demonstrated similar positive outcomes following the incorporation of oral surgical expertise in response to Hurricane Maria in Puerto Rico.

### 3.2 Roles and responsibilities of oral surgeons in disaster scenarios

The literature review identified several key roles that oral surgeons typically assume in disaster management contexts. These roles can be broadly categorized into immediate response, ongoing care, and recovery phase contributions. In the immediate response phase, oral surgeons were found to be crucial in managing life-threatening conditions related to the oral and maxillofacial region. Multiple studies emphasized their role in securing compromised airways, controlling severe facial hemorrhage, and managing complex facial fractures that could potentially obstruct breathing. A notable case study by [28] described how oral surgeons performing emergency tracheostomies and reduction of mandibular fractures saved numerous lives in the immediate aftermath of the 2011 Tōhoku earthquake and tsunami in Japan.

During the ongoing care phase, oral surgeons were reported to provide a wide range of services, including the treatment of dental and facial infections, repair of soft tissue injuries, and management of temporomandibular joint disorders resulting from disaster-related trauma. A comprehensive review highlighted the importance of these interventions in preventing long-term complications and improving quality of life for disaster survivors. In the recovery phase, the literature consistently pointed to the significant contributions of oral surgeons in facial reconstruction and rehabilitation. Several studies documented the positive impact of timely maxillofacial reconstruction on patients' psychological well-being and social reintegration [29]. For instance, a longitudinal study following survivors of the 2015 Nepal earthquake found that those who received early facial reconstruction from oral surgeons reported significantly higher quality of life scores and lower rates of depression compared to those who experienced delayed treatment.

### 3.3 Integration of oral surgery in disaster response frameworks

Despite the evident importance of oral surgical care in disaster scenarios, the review revealed significant variability in the integration of oral surgeons into formal disaster response frameworks across different countries and regions. A global survey conducted by the International Association of Oral and Maxillofacial Surgeons in 2022 found that only 38% of countries had explicit provisions for including oral surgeons in their national disaster response plans [30]. Several studies highlighted successful models of integration that could serve as benchmarks for improvement. For example, the Israeli Defense Forces Medical Corps' incorporation of oral surgeons into their field hospital teams has been widely praised and studied. A comprehensive analysis by [31] of the IDF's responses to multiple international disasters over a 15-year period demonstrated the effectiveness of this integrated approach, with oral surgeons contributing to a 30% reduction in morbidity related to maxillofacial injuries.

Similarly, the Japanese Disaster Medical Assistance Team (DMAT) system, which includes oral surgeons as core members, has been cited as an exemplary model. Tanaka et al. (2023) conducted a comparative study of disaster response effectiveness between countries with and without integrated oral surgical teams, finding that the Japanese model resulted in significantly better outcomes for patients with oral and maxillofacial injuries. However, the literature also revealed persistent challenges in achieving widespread integration. Common barriers identified included lack of awareness among disaster planning authorities about the potential contributions of oral surgeons, limited funding for specialized training, and logistical difficulties in deploying oral surgery equipment to disaster sites. A notable study by [32] surveying disaster response coordinators across 30 countries found that 65% were unaware of the full scope of oral surgeons' capabilities in disaster scenarios, highlighting a critical knowledge gap.

### 3.4 Training and preparedness of oral surgeons for disaster response

The review uncovered a growing emphasis on the need for specialized training to prepare oral surgeons for the unique challenges of disaster scenarios. Several studies pointed out that traditional oral surgery training programs often do not adequately address the specific skills and knowledge required for effective disaster response [33]. A comprehensive survey of oral surgery residency programs in the United States, conducted by [34], found that only 22% included dedicated disaster medicine components in their curricula. This finding echoed similar results from studies in other countries, suggesting a global gap in disaster preparedness training for oral surgeons [27, 28].

However, the literature also highlighted innovative approaches to address this gap. For instance, a pilot program described by [35] in Japan incorporated virtual reality simulations of disaster scenarios into oral surgery training, resulting in significantly improved performance and confidence among participants in subsequent real-world disaster drills. Similarly, an interdisciplinary training model developed by the European Association for Cranio-Maxillo-Facial Surgery, combining principles of emergency medicine, trauma surgery, and oral surgery, has shown promising results in enhancing the disaster response capabilities of oral surgeons [36]. The importance of continuous education and regular disaster drills was consistently emphasized across multiple studies. A longitudinal study by [37] tracking the performance of oral surgeons in simulated disaster scenarios over a five-year period found that those who participated in annual refresher courses and drills maintained significantly higher skill levels compared to those who did not.

### 3.5 Technological advancement and their impact on oral surgery in disasters

The literature review revealed an increasing focus on leveraging technological advancements to enhance the contributions of oral surgeons in disaster scenarios. Telemedicine and mobile health technologies, in particular, were frequently cited as game-changers in expanding the reach and effectiveness of oral surgical care in disaster-affected areas. Several studies documented successful implementations of tele dentistry systems in disaster response. For example, a case study described the use of a smartphone-based tele dentistry app during the 2021 Henan floods in China, which allowed remote consultations with oral surgeons and significantly improved triage efficiency for maxillofacial injuries [38, 39] reported on the successful use of augmented reality technology to guide non-specialist healthcare workers in performing basic oral surgical procedures under the remote supervision of oral surgeons following an earthquake in Nepal.

3D printing technology has also emerged as a valuable tool in disaster scenarios. Multiple studies highlighted its potential for rapid production of custom prostheses and surgical guides in field conditions. A noteworthy example was provided by [40] who described the use of portable 3D printers to create patient-specific mandibular reconstruction plates following a major earthquake in Mexico, significantly reducing surgical time and improving outcomes. The integration of artificial intelligence (AI) in disaster-related oral surgery was another prominent theme in recent literature. Several studies explored the potential of AI algorithms in assisting with rapid diagnosis and treatment planning for maxillofacial injuries in mass casualty events [41]. For instance, a machine learning model developed by [42] demonstrated 92% accuracy in triaging facial injuries based on smartphone photographs, potentially streamlining the assessment process in resource-constrained disaster settings.

### 3.6 Psychological aspects and long-term outcomes

The review highlighted an increasing recognition of the psychological impact of oral and maxillofacial injuries in disaster survivors and the role of oral surgeons in addressing these issues. Multiple studies emphasized the profound effect that facial disfigurement can have on survivors' mental health, social functioning, and overall quality of life. A comprehensive meta-analysis by [43] synthesizing data from 45 studies found that disaster survivors with untreated or inadequately treated facial injuries had a 3.5 times higher risk of developing post-traumatic stress disorder (PTSD) compared to those without such injuries. This finding underscores the importance of timely and effective oral surgical interventions not just for physical recovery, but also for psychological well-being.

The literature also revealed the positive impact of early and comprehensive oral surgical care on long-term outcomes for disaster survivors. A longitudinal study by following survivors of the 2018 Sulawesi earthquake over a five-year period found that those who received prompt and comprehensive maxillofacial reconstruction reported significantly higher levels of social integration and employment rates compared to those who experienced delayed or incomplete treatment Several studies highlighted the importance of a multidisciplinary approach, integrating psychological support with oral surgical care for optimal outcomes (Fig. 1). For example, a novel program in India, which paired oral surgeons with mental health professionals in disaster response teams, reported significantly improved patient satisfaction and mental health outcomes compared to traditional models of care.

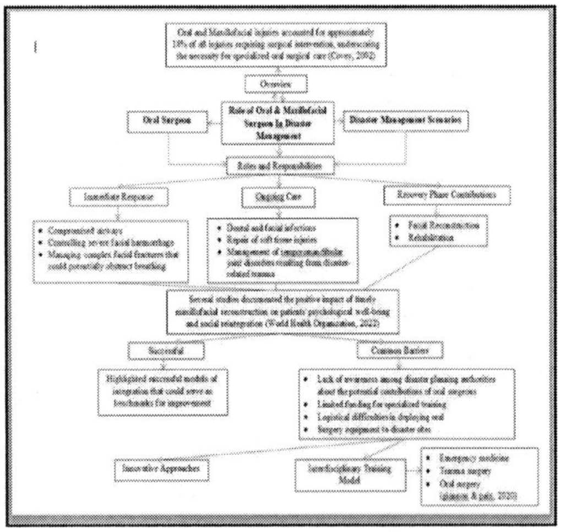


**Fig. 1.** Role of oral maxillofacial surgeon in disaster management

### 3.7 Ethical considerations and resource allocation

The review uncovered ongoing debates and ethical considerations surrounding the allocation of resources for oral surgical care in disaster scenarios. Several studies grappled with the challenge of balancing the need for specialized maxillofacial interventions against other pressing medical priorities in resource-constrained environments. An analysis proposed a framework for ethical decision-making in disaster triage that explicitly considered the long-term functional and psychological impacts of oral and maxillofacial injuries. This approach challenged traditional triage models that might deprioritize these injuries in favor of more immediately life-threatening conditions.

The literature also revealed disparities in access to oral surgical care in disaster situations, often influenced by socioeconomic factors. A comprehensive review examining disaster response data from 20 countries found significant inequities in the provision of maxillofacial care, with economically disadvantaged populations consistently receiving less comprehensive treatment. This finding highlights the need for more equitable distribution of oral surgical resources in disaster planning. Several studies proposed innovative solutions to address resource limitations. For instance, a successful model of "task-shifting" in which general dentists and even trained community health workers were able to provide basic oral surgical care under the remote guidance of specialist oral surgeons, effectively expanding the reach of limited resources.

### 3.8 Future directions and recommendations

The literature review identified several key areas for future research and development in the field of oral surgery in disaster management. There was a consistent call for more robust, large-scale studies to quantify the impact of oral surgical interventions on overall patient outcomes in disaster scenarios. Additionally, many researchers emphasized the need for standardized protocols and guidelines for the integration of oral surgery into disaster response frameworks. The development of specialized, disaster-focused training programs for oral surgeons emerged as a critical priority. Several studies proposed curriculum models that combined principles of emergency medicine, trauma surgery, and disaster management with core oral surgical skills. There was also a strong emphasis on the importance of regular disaster drills and simulations to maintain readiness.

Technological innovation was consistently identified as a key area for future development (Fig. 2). Many researchers called for further exploration of telemedicine, AI, and 3D printing technologies to enhance the capabilities of oral surgeons in disaster settings. The potential of robotics in assisting with complex maxillofacial procedures in field conditions was also highlighted as an exciting avenue for future research. Finally, there was a recurrent theme emphasizing the need for greater collaboration between oral surgeons and other medical specialties in disaster preparedness and response. Many studies advocated for the inclusion of oral surgeons in multidisciplinary disaster response teams and called for increased cross-training between specialties.

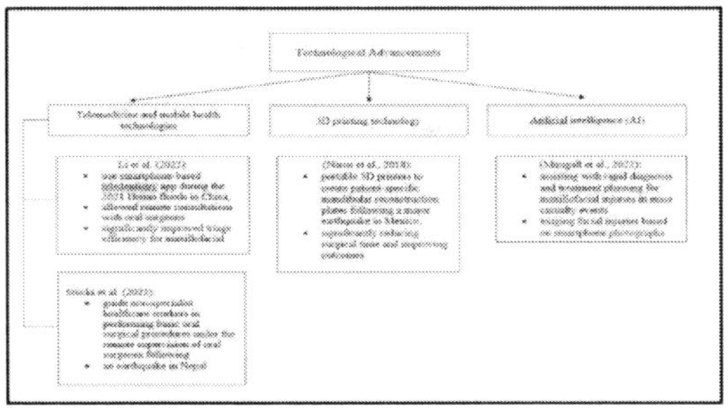

**Fig. 2**. Technological advancement

## 4 Conclusion

The comprehensive review of literature on the role of oral surgery in disaster management reveals a critical yet often underappreciated aspect of emergency medical response. Oral surgeons bring a unique and invaluable skill set to disaster scenarios, contributing significantly to patient care across all phases of disaster response - from immediate life-saving interventions to long-term reconstruction and rehabilitation. The evidence clearly demonstrates that the integration of oral surgical expertise into disaster management frameworks leads to improved patient outcomes, reduced morbidity, and enhanced overall response effectiveness. However, this review also uncovers significant disparities in the recognition and incorporation of oral surgery in disaster preparedness and response protocols globally. While some countries have successfully integrated oral surgeons into their disaster response teams, many others lack formal provisions for their inclusion. This inconsistency results in missed opportunities for comprehensive care and potentially compromises patient outcomes in affected regions.

The challenges identified, including inadequate specialized training, limited awareness among disaster planning authorities, and logistical difficulties, highlight the need for a paradigm shift in how oral surgery is perceived and utilized in disaster contexts. The promising developments in technology, particularly in telemedicine, 3D printing, and AI-assisted diagnostics, offer potential solutions to some of these challenges, expanding the reach and capabilities of oral surgeons in resource-constrained environments. Furthermore, the psychological impact of oral and maxillofacial injuries on disaster survivors underscores the importance of timely and effective oral surgical interventions not just for physical recovery, but also for mental health and social reintegration. The ethical considerations surrounding resource allocation in disaster scenarios add another layer of complexity, calling for careful balance and informed decision-making. In conclusion, this review emphatically establishes the crucial role of oral surgery in disaster management and highlights the urgent need for greater recognition, integration, and support for oral surgical services in disaster preparedness and response strategies worldwide. As the global community faces increasing frequency and severity of disasters, the full incorporation of oral surgical expertise into multidisciplinary disaster response efforts is not just beneficial, but essential for comprehensive and effective disaster management.

## 5 Recommendations

### 5.1 Policy and integration

- Develop and implement national and international policies that formally integrate oral surgeons into disaster response frameworks.
- Establish clear guidelines for the deployment and utilization of oral surgical teams in various disaster scenarios.

### 5.2 Training and education

- Incorporate specialized disaster management modules into oral surgery residency programs globally.
- Develop continuous education programs focusing on disaster preparedness for practicing oral surgeons.
- Implement regular disaster simulation drills that include oral surgery components.

### 5.3 Interdisciplinary collaboration

- Foster closer collaboration between oral surgeons and other medical specialties in disaster planning and response.
- Encourage cross-training initiatives to enhance mutual understanding and effective teamwork in disaster scenarios

### 5.4 Technology adoption

- Invest in research and development of portable oral surgery equipment suitable for field deployment.
- Expand the use of telemedicine and mobile health technologies to support remote oral surgical consultations in disaster-affected areas.
- Further explore and implement AI and 3D printing technologies to enhance diagnosis, treatment planning, and prosthesis production in disaster settings.

### 5.5 Psychological support

- Integrate mental health support services with oral surgical care for disaster survivors with maxillofacial injuries.
- Develop protocols for addressing the psychological impact of facial injuries in disaster contexts.

### 5.6 Resource allocation

- Develop ethical frameworks for resource allocation that consider the long-term impacts of oral and maxillofacial injuries.

- Implement strategies to ensure equitable access to oral surgical care in disaster situations, particularly for economically disadvantaged populations.

### 5.7 Research and data collection

- Conduct large-scale, multi-center studies to quantify the impact of oral surgical interventions on overall patient outcomes in disaster scenarios.
- Establish a global database for collecting and sharing data on oral and maxillofacial injuries in disasters to inform future preparedness and response strategies.

### 5.8 Public awareness

- Launch public education campaigns to increase awareness about the importance of oral health and maxillofacial care in disaster preparedness.
- Engage with media to highlight the role of oral surgeons in disaster response efforts.

### 5.9 Funding and support

- Advocate for increased funding for oral surgery integration in disaster management at national and international levels.
- Develop partnerships with NGOs and international organizations to support oral surgical missions in disaster-prone regions.

### 5.10 Standardization

- Develop standardized protocols for oral surgical procedures in disaster settings to ensure consistency and quality of care across different response teams.
- Create a unified certification system for oral surgeons specializing in disaster response to ensure a high standard of expertise.

## References

1. IASC, Linking disaster risk reduction (DRR) and mental health and psychosocial support (MHPSS). (2021).
2. Y. Nosaka, H. Nosaka, M. Munakata, and M. Sanda, A comprehensive clinical review of maxillary sinus floor elevation in patients with well-defined faintly radiopaque lesions in the antrum. J Clin Med. **13**, 2 (2024). https://doi.org/10.3390/jcm13020332
3. O. Hächl, T. Tuli, S. Anton, and R. Gassner, Maxillofacial trauma due to work-accidents. Int J Oral Maxillofac Surg. **31**, 90–93 (2002). https://doi.org/10.1054/ijom.2001.0166
4. S. Amudhan, Disaster medicine and public health preparedness. **14**, (2020). https://doi.org/10.1017/dmp.2020.459
5. M. Keim and E. Noji, Emergent use of social media: A new age of opportunity for disaster resilience. Am J Disaster Med. **6**, 47–54 (2011). https://doi.org/10.5055/ajdm.2010.0000
6. A.L.F. Costa, C.L. Yasuda, M. França, C.F. De Freitas, H. Tedeschi, E. De Oliveira, et al, Temporomandibular dysfunction post-craniotomy: Evaluation between pre- and post-operative status. J Cranio-Maxillofacial Surg. **42**, 1475–1479 (2014). https://doi.org/10.1016/j.jcms.2014.04.016
7. WHO, strengthening health emergency preparedness for disasters resulting from natural hazards. 1–6 (2024).
8. K.L. Koenig, C. Schultz, Koenig and Schultz's Disaster Medicine: Comprehensive Principles and Practices. (2009). https://doi.org/10.1017/CBO9780511902482
9. A.S. Alkhutari, E.A. Al-Moraissi, E.L. Galvão, N. Christidis, and S.G.M. Falci, Top 100 cited systematic reviews and meta-analyses in the major journals of oral and maxillofacial surgery: A bibliometric analysis. Oral Maxillofac Surg. **26**, 343-356 (2022). https://doi.org/10.1007/s10006-021-00981-9
10. A. Hamdy, H.K.A. El Gaber, D.M. Ghalwash, and W.M. Abbas, Effect of using a plastic stent with apically repositioned flap in peri-implant soft tissue augmentation: A randomized controlled clinical trial. Int J Dent. (2021). https://doi.org/10.1155/2021/5590400
11. P. Wusiman, B. Maimaitituerxun, Guli, A. Saimaiti, and A. Moming, Epidemiology and pattern of oral and maxillofacial trauma. J Craniofac Surg. **31**, (2020) https://doi.org/10.1097/SCS.0000000000006719
12. I.S. Gataa and Q.H. Muassa, Patterns of maxillofacial injuries caused by terrorist attacks in Iraq: Retrospective study. Int J Oral Maxillofac Surg. **40**, 65–70 (2011).
13. J. Pyda, R.H. Patterson, L. Caddell, T. Wurdeman, R. Koch, D. Polatty, et al, Towards resilient health systems: Opportunities to align surgical and disaster planning. BMJ Glob Heal. **4**, (2019). https://doi.org/10.1136/bmjgh-2019-001493
14. S. Ochi, V. Murray, and S. Hodgson, The great east japan earthquake disaster: A compilation of published literature on health needs and relief activities. PLoS Curr. **5**, (2013). https://doi.org/10.1371/currents.dis.771beae7d8f41c31cd91e765678c005d
15. A. Olze, G. Geserick, and A. Schmeling, Age estimation of unidentified corpses by measurement of root translucency. J Forensic Odontostomatol. **22**, 28–33 (2004).
16. P. Nayar, S. Bhoi, T. Sinha, V. Chauhan, and S. Galwankar, Strategic directions to integrate emergency care services into primary health care in the South-East Asia Region. (2020).
17. A.S. George and A. George, Telemedicine: a new way to provide healthcare. **1**, 98-129 (2023). https://doi.org/10.5281/zenodo.8075850
18. D. Choudhury-Peters, and V. Dain, Developing

psychological services following facial trauma. BMJ Qual Improv Reports. **5**, (2016). https://doi.org/10.1136/bmjquality.u210402.w4210
19. S. Knoedler, H. Baecher, C.C. Hoch, D. Obed, D.Y. Matar, C. Rendenbach, et al, Early outcomes and risk factors in orthognathic surgery for mandibular and maxillary hypo- and hyperplasia: A 13-year analysis of a multi-institutional database. J Clin Med **12**, 4 (2023). https://doi.org/10.3390/jcm12041444
20. A. Alnour and A. Abdo, Residents' satisfaction in oral and maxillofacial surgery program in syria during the crisis. (2021).
21. B. Paterson, S. Thorne, C. Canam, and C. Jillings, Meta-study of qualitative health research. (2001). https://doi.org/10.4135/9781412985017
22. J. Singh, Critical Appraisal Skills Programme. CASP Appraisal Tools. J Pharmacol Pharmacother. **4**, 76 (2013). https://doi.org/10.4103/0976-500X.107697
23. M.K.W. Stranges, S. Ul Haq, and D.G. Dunn, Black-out test versus UV camera for corona inspection of HV motor stator endwindings. IEEE Trans Ind Appl. **50**, 3135–3140 (2014). https://doi.org/10.1109/TIA.2014.2306979
24. J. Jacimovic, A. Jakovljevic, P. Dummer, V. Nagendrababu, H.F. Duncan, and P.M.H. Dummer, A bibliometric analysis of the dental scientific literature on COVID-19. Clin Oral Investig. **25**, (2021) https://doi.org/10.1007/s00784-021-03916-6
25. D.C. Covey, Blast and fragment injuries of the musculoskeletal system. J Bone Joint Surg Am. **87**, 1221–1234. (2002)
26. Y. Prabowo and A. Kamal, Epidemiology of orthopaedic injuries at the local hospital following earthquake in Palu Indonesia. J Orthop Trauma Surg Relat Res. (2020).
27. C.D. Zorrilla, The view from Puerto Rico — hurricane maria and its aftermath. N Engl J Med. **377**, (2017). https://doi.org/10.1056/NEJMp1713196
28. S. Nogami, K. Yamauchi, G. Bottini, Y. Otake, Y. Sai, H. Morishima, et al, Fall-related mandible fractures in a Japanese population: A retrospective study. Dent Traumatol. **35**, 194-198 (2019). https://doi.org/10.1111/edt.12471
29. World Health Organization, WHO European regional obesity report 2022. (2022).
30. G. Yordanova, M. Grancharov, and M. Georgieva, Frequency of mesiodens in orthodontic patients – clinical and radiographical discussion. J IMAB 2022 Suppl. **28**, 73–76 (2022).
31. G. Yordanova, M. Грънчаров, and M. Gueorguieva, Frequency of mesiodens in orthodontic patients – clinical and radiographical discussion. J IMAB - Annu Proceeding. 73-76 (2024).
32. R. Kunusoth, A. Mohan, R. Prakash, S. Sampreethi, and I. Kakati, Knowledge and awareness about oral and maxillofacial surgery in medical professionals and general public: A questionnaire-based survey. J Oral Med Oral Surgery Oral Pathol Oral Radiol. 130–134 (2022). https://doi.org/10.18231/j.jooo.2022.028
33. B. Mulyana, R. Pamungkas, W. Sari, A. Sukarno, E. Rosya, and W. Wariani, Preparedness training programs improved the emergency response and crucial outcomes among nursing students: A scoping review. Front Nurs. **11**, 159–167 (2024). https://doi.org/10.2478/fon-2024-0017
34. R.R. Sarin, J. Hick, A. Livinski, J. Nieratko, M. Treber, A. Mazurek, et al, Disaster medicine: A comprehensive review of the literature from 2016. Disaster Med Public Health Prep. **13**, 946-957 (2019). https://doi.org/10.1017/dmp.2019.18
35. N. Brown, C. Margus, A. Hart, R. Sarin, A. Hertelendy, and G. Ciottone, Virtual reality training in disaster medicine: A systematic review of the literature. Simul Healthc J Soc Simul Healthc. **18**, 255-261 (2022).
36. C. Giannou and B. Paix, Principles of emergency and trauma surgery. 18–31 (2020). https://doi.org/10.1017/9781316181454.004
37. M. Ramirez, K. Kubicek, C. Peek-Asa, and M. Wong, Accountability and assessment of emergency drill performance at schools. Fam community health. **32**, 105–114 (2009).
38. A. Pertiwisari, A. Muhammad, and I.M. Amir, Interdental Jurnal Kedokteran Gigi. **19**, 182–186 (2023).
39. J. Stucki, R. Dastgir, D. Baur, and F. Quereshy, Utilization of virtual reality and augmented reality in oral and maxillofacial surgery: A narrative review. Oral Surg Oral Med Oral Pathol Oral Radiol. **12**, (2023). https://doi.org/10.1016/j.oooo.2023.07.001
40. A. Naros, H. Weise, F. Tilsen, S. Hoefert, G. Naros, M. Krimmel, et al, Three-dimensional accuracy of mandibular reconstruction by patient-specific pre-bent reconstruction plates using an "in-house" 3D-printer. J Cranio-Maxillofacial Surg. **46**, 1645-1651 (2018). https://doi.org/10.1016/j.jcms.2018.05.047
41. M. Miragall, S. Knoedler, M. Kauke-Navarro, R. Saadoun, A. Grabenhorst, F. Grill, et al, Face the future—artificial intelligence in oral and maxillofacial surgery. J Clin Med. **30**, (2023) https://doi.org/10.3390/jcm12216843
42. O. Oyebode, J. Fowles, D. Steeves, and R. Orj, Machine learning techniques in adaptive and personalized systems for health and wellness. Int J Hum Comput Interact. **32** (2022). https://doi.org/10.1080/10447318.2022.2089085
43. C.R. Brewin, B. Andrews, and J.D. Valentine, Meta-analysis of risk factors for posttraumatic stress disorder in trauma-exposed adults. J Consult Clin Psychol. **68**, 748–766 (2000). https://doi.org/10.1037/0022-006X.68.5.748

# Trauma healing for children affected by the Pasaman, West Pasaman earthquake

*Rudi* Elfendi[1*] and *Asrawati*[1]

[1]Departmen of Pediatrics, Faculty of Medicine, Universitas Andalas, Padang, West Sumatra, Indonesia

**Abstract.** This study focuses on trauma healing interventions for children affected by the Pasaman earthquake in February 2022, which led to significant psychological issues such as trauma, stress, and sleep disturbances. The research involved direct observations and interventions in several evacuation camps. A team from the Department of Pediatrics, Faculty of Medicine, Andalas University, conducted trauma healing sessions, including cognitive behavioral therapy (CBT) and play therapy, in affected areas. The intervention programs targeted children aged 3 to 18 years and were complemented by educational efforts for parents on how to provide at-home trauma healing support. The study highlights the importance of early psychological interventions, such as CBT, to prevent the long-term effects of post-traumatic stress disorder (PTSD), depression, and anxiety in children. Observational data revealed that about 57.9% of the children experienced sleep disturbances post-earthquake, with significant improvements following the trauma healing activities. Additionally, parent-child engagement during therapy sessions helped strengthen family support systems, which are crucial in the psychological recovery process. These findings underline the necessity of including psychosocial support in post-disaster recovery plans and stress the need for further research and implementation of trauma-informed care in affected communities to support both immediate and long-term psychological recovery for children.

## 1 Introduction

Disaster, according to Law No. 24 of 2017, is defined as an event or series of events that threaten and disrupt the lives and livelihoods of the community, caused by natural and/or non-natural factors, including human actions (due to conflict or terrorism), resulting in loss of human lives, environmental damage, property loss, and psychological impact. Indonesia is one of 35 countries with a high disaster risk worldwide. In 2023, Indonesia experienced 5,400 disaster events, with 99.35% dominated by hydrometeorological disasters and the remainder by geological disasters. The number of casualties recorded in 2023 includes 275 fatalities and 5,795 injuries [1].

The earthquake in Pasaman, which occurred in February 2022, is one of the disasters that took place in Indonesia. According to the report from the Regional Disaster Management Agency (BPBD) of West Sumatra province, four people died in West Pasaman regency, 37 people sustained serious injuries, 310 people suffered minor injuries, and fewer than 10,000 people evacuated to 35 shelters located in Talamau district.

One of the most vulnerable groups affected by disasters is children. This is due to their physical and mental development still being in progress. The impact on children can manifest as fear, anxiety, and traumatic stress due to the disaster. These effects may last for some time and subside spontaneously in some children, but if left untreated, they can lead to long-term conditions [2].

Trauma in children can cause disorders such as post-traumatic stress disorder (PTSD), anxiety, depression, and behavioral problems. The process of trauma recovery aims to help children overcome their traumatic experiences, restore their mental health, and return to their daily lives in a better state [3].

Trauma healing for children is a crucial process to address the negative impacts of traumatic experiences. The trauma healing approach involves various methods, including cognitive-behavioral therapy, play therapy, counseling, and family interventions. Cognitive-behavioral therapy helps children change negative thought patterns and develop healthy coping skills. Play therapy allows children to express their feelings through play, which is often easier for them than talking directly about their trauma [4, 5].

Counseling and family support are also essential in the trauma healing process. Family involvement helps create a safe and supportive environment for the child, allowing them to feel loved and accepted. Early intervention is critical in healing trauma in children, as the sooner they receive help, the greater the chance of recovery and the reduction of future mental health problems [6].

## 2 Methods

A direct observation study conducted to observe in real-time how trauma healing is applied to children who are victims of natural disasters. Researchers will be present at

* Corresponding author: elf.rudi1@gmail.com

the location where trauma healing interventions take place, such as play therapy sessions, Cognitive Behavioral Therapy (CBT), and Parent-Child Interaction Therapy (PCIT). Through this observation, researchers can see how children interact during therapy sessions, how they respond to the activities provided, and how their behavior and emotional states change over time (Fig. 1).

**Fig. 1.** Community service for the earthquake in Pasaman

## 3 Results

The Pasaman earthquake in February 2022 caused various issues, including psychological stress and trauma, with the most common manifestation being sleep disorders. A study by Dr. Rahmi Asman found that 57.9% of children affected by the earthquake experienced sleep disturbances. Additionally, Dr. Destri's research showed that 55% of children suffered from emotional and mental disorders after the disaster. Therefore, trauma healing for children affected by the earthquake is crucial. Trauma healing activities were conducted at several evacuation camps, including Malampah Kajai, Kampung Aur, and Timbo Abu. Activities included play therapy, drawing, coloring, quizzes, and storytelling. Around 30 children participated in each session, ranging from ages 3 to under 18. Parents were also educated on how to perform trauma healing at home or in the evacuation camps. Both children and parents actively participated and expressed joy during the activities.

A team of pediatric volunteers from the Faculty of Medicine implemented activities as part of the disaster module on March 12-13 and 27-28, 2022. These included screening for illnesses such as dermatitis, impetigo, tinea, measles, and respiratory infections, as well as trauma healing through singing and playing. The team distributed masks, school supplies, and sandals for children and provided education on breastfeeding and complementary feeding (MP-ASI). They coordinated with field leaders to raise awareness of the risks of measles, chickenpox, and diarrhea due to poor hygiene. Handwashing and hygiene education were conducted at each evacuation tent, along with data collection on the emotional conditions, trauma, and sleep disorders in children. The team also offered activities that served as healing therapy and educated parents on self-conducted healing practices for their children (Fig. 2).

**Fig. 2.** Trauma healing for children in refugee tent

**Table 1.** Characteristics of research subjects

| Post Name | Health Service Patients (n) | Refugees (n) | Households (n) | Children under 18 (n) |
|---|---|---|---|---|
| Kampung Aur | - Respiratory Tract Infection (RTI): 4 people<br>- Dermatitis: 2 people<br>- Play Therapy: 32 people | 161 | 35 | 50 |
| Reslement | - RTI: 1 person<br>- Cyanotic Congenital Heart Disease + RTI: 1 person<br>- Play Therapy: 41 people | 0 | 20 | 53 |
| Timbo Abu | - RTI: 6 people<br>- Measles: 6 people<br>- Impetigo: 2 people<br>- Atopic Dermatitis: 1 person<br>- Fever: 2 people<br>- Gingivitis: 1 person<br>- Play Therapy: 34 people | 400 | 80 | 125 |
| Reslement | - Trauma Healing: 30 people<br>- Treatment: 0 person | 0 | 18 | 40 |

Another entry for Reslement indicates that 30 people received trauma healing services, although no other treatments were provided at that time. This particular entry recorded 18 households and 40 children under 18, with no refugees reported (Table 1).

# 4 Discussion

## 4.1 Social impact of disasters on children

After experiencing a traumatic event, many children display temporary psychological symptoms. These can include becoming easily tearful, anxious, overly attached to others, or withdrawn. They might have difficulty focusing or sleeping, relive the event through play or drawings, avoid talking about it, complain of headaches or stomach aches, or become irritable as they try to manage their emotions. These reactions are normal responses to trauma and typically fade within a few weeks. However, if these symptoms last for more than a month and start to interfere with a child's social functioning, such as school performance or interactions with peers, the child may meet the criteria for a psychiatric diagnosis [7].

One common condition resulting from trauma is post-traumatic stress disorder (PTSD). In PTSD, the brain processes trauma memories in an abnormal way, leading the child to relive the event through distressing memories or nightmares, especially when triggered by reminders of the traumatic context. Although dissociative symptoms like flashbacks are less common in children, these re-experiencing symptoms can be highly disruptive. They may lead children to develop avoidance behaviors, such as keeping busy or steering clear of people and places that remind them of the trauma. While avoidance may provide short-term relief, it does not prevent the long-term re-experiencing of trauma and can even stop the child from enjoying activities they used to love. Children with PTSD often display signs of hyper-arousal, which means they feel constantly threatened and on high alert, resulting in irritability, trouble concentrating, and sleep difficulties [7].

One in four children exposed to trauma will develop PTSD by the time they turn 18, with a lifetime prevalence rate of 4.7-7.8% in the general population. The rate is significantly higher among displaced children and those exposed to armed conflict. The risk of PTSD in children is influenced by several factors, particularly the nature of the trauma. Trauma involving direct interpersonal violence, such as physical or sexual abuse, is the most strongly associated with psychological disorders. Children with PTSD are at a higher risk of self-harm and aggression. They are eight times more likely to harm themselves and ten times more likely to attempt suicide compared to children without PTSD. Additionally, children with PTSD are three times more likely to commit violent offenses, with one in ten having a record of violent behavior. They are also three times more likely to struggle with education, employment, or training compared to their peers [7].

Although the link between childhood trauma and mental health issues is well-documented and likely causal, the mechanisms driving this relationship are still the focus of ongoing research. Traditionally, it has been believed that childhood trauma triggers a "toxic stress response" that alters brain function and contributes to mental health disorders. However, recent findings suggest that psychopathology is more strongly linked to subjective memories of childhood trauma rather than the trauma itself. This points to psychological mechanisms like memory bias, core beliefs, and decision-making processes. On the other hand, the connection between childhood trauma and cognitive deficits likely stems from non-causal factors, such as pre-existing cognitive differences that increase the risk of exposure to various types of trauma [7].

The high prevalence and complexity of mental health issues related to childhood trauma highlight the urgent need for increased specialist training and clinical capacity in this area, which is often insufficient. Misdiagnosis or underdiagnosis is common, leading to ineffective treatment. While natural recovery may occur in some cases, many children do not improve without intervention. If left untreated, chronic or recurring psychiatric disorders can develop, becoming more difficult to treat as time goes on [7].

## 4.2 Trauma healing for children's victims of natural disasters

Trauma healing is a crucial process for children who have experienced natural disasters, helping them recover from psychological issues such as anxiety, panic, and other mental health challenges. One of the most effective treatments for these children is Cognitive Behavioral Therapy (CBT), a structured, evidence-based approach designed for children aged three to 18 who show symptoms of post-traumatic stress related to traumatic events they have experienced or witnessed. It's important to note that a full diagnosis of PTSD is not required for a child to undergo CBT. This therapy typically spans 12 to 16 weekly sessions and is widely regarded as the most effective evidence-based practice [8].

For younger children, particularly those aged three to five, CBT has proven successful in reducing symptoms of PTSD, depression, anxiety, and behavioral problems. In addition to directly benefiting the child, CBT also enhances parenting skills, improves parental support, and reduces parental stress. Key elements of CBT include psychoeducation about trauma, teaching parenting skills, developing relaxation techniques and coping mechanisms, helping the child identify and manage emotions, understanding the relationship between thoughts, feelings, and behavior, processing the traumatic events through narrative work, gradual exposure to trauma reminders, joint caregiver-child activities, and safety or prevention skill-building [8].

The structure of CBT often involves splitting time between individual sessions for the child and the caregiver, although some sessions bring them together with the therapist. This structured approach directly addresses the impact of traumatic stress and aligns with widely accepted principles of cognitive-behavioral therapy, exposure therapy, and effective parenting practices in mental health care [8].

Another effective therapy, Parent-Child Interaction Therapy (PCIT), is a structured, evidence-based treatment designed for children aged two to seven who have behavioral problems, including those related to trauma and post-traumatic stress symptoms, along with their caregivers. PCIT has demonstrated its effectiveness in reducing disruptive behavioral problems in children [8].

PCIT not only helps in reducing disruptive behaviors, but it also strengthens parenting skills, enhances the warmth of the caregiver-child relationship, and reduces parenting stress. Beyond addressing disruptive behavior, PCIT also helps lessen other trauma-related symptoms in children without requiring modifications to the treatment model [8].

PCIT is divided into two phases, each focusing on different skill sets for the caregiver. During the first phase, called Child-Directed Interaction (CDI), the focus is on teaching caregivers how to give positive attention and reinforce the warmth in their relationship with the child. The second phase, Parent-Directed Interaction (PDI), emphasizes helping caregivers learn consistent and appropriate discipline techniques, including the use of a specific time-out method as a response to the child's non-compliance. Throughout both phases, the therapist provides real-time coaching during play situations, helping caregivers master these techniques effectively [8].

Each phase of PCIT is based on skill mastery rather than a fixed timeline, so the duration varies from family to family. However, it is generally estimated to last 12 to 16 weekly sessions. The process starts with a teaching session for the caregiver alone, while all subsequent sessions involve both the child and caregiver, as PCIT centers around their interaction. Caregivers are also assigned "homework"—five minutes of daily playtime—during which they practice the positive attention skills they are learning through PCIT [8].

For younger children, Child-Parent Psychotherapy (CPP) is a relationship-based intervention aimed at addressing emotional and behavioral issues in children aged 0 to 5 who have experienced trauma. CPP has been shown to improve attachment quality, enhance cognitive development, reduce symptoms of PTSD, depression, and behavioral problems in children, regulate physiological responses, and decrease symptoms of depression and PTSD in caregivers [8].

Typically conducted over 20 to 25 weekly sessions, CPP can take place in a variety of settings, including the home, outpatient clinics, or school environments. CPP is divided into three main phases: the Foundational Phase (Assessment and Engagement), the Core Intervention Phase, and the Recapitulation and Termination Phase. During the Foundational Phase, the therapist builds a collaborative relationship with the caregiver, collecting important information such as the reason for referral, demographic details, specific symptoms in both the child and caregiver, as well as any risk and protective factors present in the family system [8].

In the Core Intervention Phase, the therapist introduces the child to CPP and helps them understand their behavior within the context of trauma. This phase involves discussing how the traumatic experiences have influenced the child's emotional and behavioral reactions, creating a space for the child to express their feelings. The therapist takes advantage of natural moments during the sessions to facilitate progress and help the child move toward treatment goals. The main objectives of CPP include restoring the child's developmental path, strengthening the parent-child dyadic response to perceived threats, rebuilding trust in the body's physical sensations, re-establishing reciprocity in relationships, normalizing trauma responses, helping the child differentiate between reliving trauma and remembering it, and reframing traumatic experiences in a more manageable perspective [8].

In the Recapitulation and Termination Phase, the focus shifts from addressing problems to recognizing the positive changes that have taken place during therapy. This phase acknowledges that the therapy is coming to an end and helps process the emotions related to closure. Key strategies used in CPP include encouraging developmental progress through play, physical contact, and language; offering reflective guidance; modeling protective behaviors; interpreting emotions and actions; providing emotional support and empathy; as well as crisis intervention, case management, and practical assistance [8].

Play therapy is a vital tool, particularly for early intervention following traumatic events. Approximately 40% of children are at risk of developing long-term PTSD symptoms, and play therapy helps them express their emotions through activities like play, drawing, and storytelling. This form of therapy, designed to address psychological issues caused by trauma, has been shown to lower levels of traumatic stress in children in school settings, both before and after the intervention, in comparison to control groups that did not receive the therapy. It allows children to express themselves through enjoyable activities, helping to reduce psychological challenges, especially the stress they experience after a disaster [9].

Play therapy can also help children cope with issues like anxiety, frustration, and internal emotional barriers, ultimately improving their behavior. Drawing therapy provides an outlet for children to express their inner conflicts. Through drawing, they can convey their thoughts, feelings, imagination, and creativity, making it an effective way to address psychological issues such as anxiety. Storytelling therapy, especially when related to traumatic e xperiences, can alleviate the emotional burdens children carry. Sharing their stories fosters a sense of solidarity and may help them feel less alone, encouraging recovery from trauma and reducing anxiety [10].

In addition to traditional play therapy, block play therapy has been used to predict behavioral problems in children following the 2011 earthquake in Japan. Problems identified included total behavioral issues (25.9%), internalizing problems like somatic complaints and anxiety/depression (27.7%), and externalizing problems such as delinquent or aggressive behavior (21.2%). A study involving 78 children aged 4-6 years from three schools in Miyagi Prefecture found that the style of block play varied according to the type of trauma-related problems the children faced. Those with total behavioral problems tended to exhibit a "falling" style, while aggressive behavior was linked to "falling" and "shaking" styles. Internalizing problems, such as anxiety and depression, resulted in shorter playtime and difficulty concentrating. Other forms of play therapy, like singing, playing with balloons or marbles, playing snakes and ladders, and coloring, help build trust and positively influence a child's personality development after a disaster. Puzzle games can also help detect psychosocial trauma in children following disasters [11].

Group play therapy has proven effective in reducing post-traumatic stress symptoms in children. Play is a fundamental part of a child's world, allowing them to engage in extensive imagination through various activities. Group play therapy works by storing traumatic memories in the child's subconscious and gradually reducing the trauma they experience [9].

Trauma counseling therapy is also used as an intervention to address psychological issues in children who are victims of disasters. This counseling typically includes physical examination, trauma healing methods, and play therapy, such as collective prayers and Cognitive Behavioral Therapy (CBT). Additionally, trauma healing activities for children are often integrated with religious practices, such as reading the Qur'an, group prayers, listening to stories of exemplary Islamic figures, and participating in remembrance (dzikir) [9].

Instructional therapy can also aid in the development of a child's cognitive skills by using educational media in a playful learning process. It's essential to balance cognitive and affective aspects in therapy, as this balance is key to the child's psychological development, motivating them to learn. When a child is focused and engaged in these activities, they are better able to distract themselves from the problems they face [9].

Parent education is another crucial element. Parents need to understand the emotional, physical, and mental impacts of trauma on children. Recognizing common symptoms, such as anxiety, nightmares, and behavioral changes, is essential for identifying early signs of trauma. Parents must also learn to manage their own emotions and stress to serve as positive role models for their children. Techniques like meditation and deep breathing can help parents in this process. Open communication is key—parents should encourage their children to express their feelings and listen empathetically without judgment. Additionally, creating a safe and stable home environment is important. This includes minimizing exposure to frightening news or images that might remind the child of the disaster, helping them feel more secure [12].

## 5 Conclusion

Disasters are events or a series of events that disrupt people's lives and livelihoods, caused by both natural and man-made factors, including human actions like conflict or terror. These events result in the loss of life, environmental destruction, property damage, and psychological impacts. Trauma refers to an experience that triggers intense physical and emotional stress responses, perceived as dangerous or threatening, and can have long-lasting negative effects on a person's physical, social, emotional, or spiritual well-being. Trauma healing is essential for restoring the psychological well-being of children affected by disasters. Therapies such as Trauma-Focused Cognitive Behavioral Therapy (TF-CBT) and Eye Movement Desensitization and Reprocessing (EMDR) have proven effective in reducing symptoms of PTSD, anxiety, and depression, promoting emotional recovery, improving mental and physical health, and enhancing social relationships. Disasters can have a significant psychological impact on children, often leading to conditions like PTSD, anxiety, and depression. Professional therapy, such as TF-CBT and EMDR, is vital for helping children recover, while parents play an important role in offering emotional support and helping manage their children's trauma responses. The disaster process follows five main phases, collectively known as the disaster cycle: prevention, mitigation, preparedness, response, and recovery. In the trauma healing process, individuals and their families are first educated about trauma, followed by professional therapy, such as TF-CBT and EMDR. These therapies help reduce PTSD symptoms through regular weekly sessions that focus on teaching coping strategies and relaxation techniques.

## References

1. Rosyida A, Aziz M, Firmansyah Y, Setiawan T, Pangesti KP, and Kakanur F. Buku Data Bencana Indonesia 2023. **3**, Jakarta (2024). Available from: https://bpbd.kepriprov.go.id/files/buku-data-bencana-indonesia-tahun-2023.pdf
2. Jordan, B., Perryman, K., and Anderson, L. A case for child-centered play therapy with natural disaster and catastrophic event survivors. I. J. of Play Therapy, **22**(4), 219 (2013). https://doi.org/10.1037/a0034637
3. Burkhart K, Agarwal N, Kim S, Neudecker M, and Ievers-Landis C.E. A Scoping Review of Trauma-Informed Pediatric Interventions in Response to Natural and Biologic Disasters. Children. **10**(6) 1–26 (2023) http://dx.doi.org/10.3390/children10061017
4. Pramardika D.D, Hinonaung J.S.H, and Mahihody A.J. Pengaruh terapi bermain terhadap trauma healing pada anak korban bencana alam. Faletehan Health Journal. **7**(2), 85–91 (2020) http://dx.doi.org/10.33746/fhj.v7i02.131
5. Rusmana N, Hafina A, and Suryana D. Group play therapy for preadolescents: post-traumatic stress disorder of natural disaster victims in Indonesia. Open Psychol J. **13**(1) 213–22 (2020) http://dx.doi.org/10.2174/1874350102013010213
6. Rossouw, J., Sharp, T., Halligan, S., and Seedat, S. The effectiveness of psychological interventions for post-traumatic stress disorder in children, adolescents, and young adults. Psychological

Medicine, **52**(5), 1112-1123 (2022). https://doi.org/10.1017/s0033291720002007

7. Danese A, McLaughlin K.A, Samara M, and Stover C.S. Psychopathology in children exposed to trauma: detection and intervention needed to reduce downstream burden. The BMJ. **371**, 1–9. (2020) http://dx.doi.org/10.1136/bmj.m3073
8. Vanderzee K.L, Sigel B.A, Pemberton J.R, and John S.G. Treatments for early childhood trauma: decision considerations for clinicians. J Child Adolesc Trauma. **12**(4), 515–28 (2019) https://link.springer.com/article/10.1007/s40653-018-0244-6
9. Altay, N. and Kilicarslan-Toruner, E. The effect of play therapy on the anxiety levels of children undergoing cancer treatment. European J. &https://doi.org/10.1016/j.ejon.2017.02.007
10. https://doi.org/10.1016/j.ejon.2017.02.007Simonds E.A, Gobenciong K.A.P, Wilson J.E, Jiroutek M.R, Nugent N.R, and Van Tilburg M.A.L. Trauma functioning and well-being in children who receive mental health aid after natural disaster or war. Children. **9**(7),1–13 (2022) http://dx.doi.org/10.3390/children9070951
11. Wang X, Takashima K, Adachi T, and Kitamura Y. Can playing with toy blocks reflect behavior problems in children? Conference on Human Factors in Computing Systems - Proceedings. 1–14 (2021) http://dx.doi.org/10.1145/3411764.3445119
12. Thielemann, J. F. B., Kasparik, B., König, J., Unterhitzenberger, J., and Rosner, R. A Systematic Review and Meta-Analysis of Trauma-Focused Cognitive Behavioral Therapy for Children and Adolescents. Child Abuse & Neglect, **134,** 105899. (2022) https://doi.org/10.1016/j.chiabu.2022.105899

# BLU-e: over-dimension and over-load for vehicle transport mitigation tool at ferry ports

*Freddy* Tampubolon[1], *Sri* Sarjana[2*], *Meila* Permatasari[1], and *Virginia* Putri Mori Ritonga[3]

[1]Land Transportation Department, Politeknik Transportasi Darat Indonesia - STTD, 17520 Bekasi, Indonesia
[2]Marketing Innovation and Technology Department, Politeknik Transportasi Darat Indonesia - STTD, 17520 Bekasi, Indonesia
[3]Business Information Technology Department, University of Twente, Enschede, Netherlands

**Abstract.** Over-Dimension and Over-Load (ODOL) mitigation in vehicle transportation at ferry ports needs to involve strict inspection and enforcement of regulations at the port, including the use of weighbridge technology and dimensional measurement systems, and training to identify and provide solutions to violations. The study aims to reduce the number of ODOL vehicles and reduce potential losses in both material and immaterial terms by integrating BLU-e and Ferizy. A qualitative approach was applied to detect ODOL vehicles at the port through observation, surveys and interviews in the Land Transportation Management Center (BPTD) VIII Banten area. The results of the study explain the concept of ODOL, dimensional and payload limitations, and the importance of vehicle weighing which plays an important role in reducing the risk of maritime disasters. Tightening ODOL regulations ensures that vehicle loads are in accordance with ship capacity, reducing the risk of maritime accidents that can result in ship sinking and loss of life. Compliance with the rules must be adhered to by all vehicles operating at the ferry port so that the risk of disaster can be significantly reduced, both in terms of transportation safety, infrastructure protection, and emergency response efficiency.

## 1 Introduction

Ports contribute significantly to socio-economic development in the transportation sector, facilitating ferry services for short sea crossings as the main gateway to access various destinations [1]. Trade exchange through economic relations needs to be well developed and can be done using ferry connections [2]. Ferry transportation becomes a sustainable way to transporting passengers across waterways connecting cities or islands [3], transportation facilitates access in supporting activities [4], plays an important role in the development of connectivity, economic prosperity and social cohesion [5], need to develop green transportation to supports sustainable transportation [6]. In urban planning it is necessary to develop policies in ferry operations as a public transportation service for water crossings [7]. Transportation vehicles are an inseparable part of logistics and supply chains in an effort to support increased industrialization and development infrastructure marked by the need for better mobility [8]. All stakeholders who use road transportation are required to understand and implement all regulatory requirements at regional or national level [9]. Island interconnectivity through ferry transportation hubs is important to drive the economic, social and territorial development of the islands and tourism needs [10]. Therefore, there is a need for safety management by conducting dynamic risk assessments of transport vehicles to minimize accidents [11], safety management on ferry crossing services is important to follow up with safe procedures [12].

A port is a land or water location that has certain boundaries that is used as a place for government and business activities, including a place for ships to dock, passengers to board and disembark, and to load and unload goods [13]. The port is equipped with safety, security, service facilities, and supporting activities as center for intermodal transportation transfers [14]. The Motor Vehicle Weighing Implementation Unit (UPPKB) is a work unit under the Ministry of Transportation that carries out the task of supervising the loading of goods using weighing equipment that is permanently installed at certain locations.

ODOL is a condition where the dimensions and weight of a vehicle exceed the factory production standards or the maximum permitted load capacity [15]. ODOL vehicles are certainly detrimental to various parties involved. Losses are not only experienced by road organizers but also the community, vehicle owners, and business actors [16]. ODOL is a serious problem that can cause damage to infrastructure, such as roads, bridges, and even ferries. ODOL can be a risk to the safety of passengers and drivers and based on records from the National Transportation Safety Committee (KNKT), there have been several ship accidents involving ODOL vehicles as one of the factors that played a role in the accident [17]. Some of these accidents include the breaking of the ramp door of the Nusa Putra passenger

* Corresponding author: srisarjana@gmail.com

motor ship (KMP), the sinking of KMP Rafelia 2 in the waters of the Bali Strait, and the capsizing of KMP. Satya Kencana III at Kumai Port on October 19, 2022 [18].

In addition, according to the Transportation Policy Agency at the Ministry of Transportation, infrastructure damage caused by ODOL practices has reached IDR 43.45 trillion per year [19]. This has the potential to create a chain incident. The definition of a chain incident in question is when there is a vehicle that exceeds the dimensions and load limits stipulated in concerning traffic and road transportation [20]. Seeing the impact caused by ODOL vehicles, appropriate mitigation is needed to overcome this problem. Integration is a process of unifying or blending different elements so that they become a whole or complete unity. System integration is a concept where systems are interconnected and adjusted to existing needs [21]. Optimization of ferry operations requires information using an integrated data platform which is connected to a mobile application for time efficiency and increased trust for consumers [22].

Electronic proof of passing the test (BLU-e) is a technological innovation that can detect and record the data of the vehicle concerned [23]. Its function is as proof that the vehicle has passed a series of predetermined tests. BLU-e replaces proof of passing the motor vehicle test (KIR) which was previously in the form of a physical book. The BLU-e components consist of two certificates of passing the test, two hologram stickers with QR codes attached to the windshield of the vehicle, and one smart card with NFC technology. NFC encourages the development of business transformation in the transportation sector to be more effective and efficient and creates experiences with better levels of comfort, speed and safety [24]. Meanwhile, Ferizy is part of the digitalization program for River, Lake, and Crossing Transportation (ASDP) in the form of an online-based ticket service that can be accessed by service users via the ferizy.com website or by downloading the application on a cell phone.

The purpose of this study is to reduce the number of ODOL vehicles and reduce potential losses both in terms of material and immaterial by integrating BLU-e and Ferizy. The integration of these two systems is expected to help identify vehicles that violate the permitted dimensions and loads and prevent them from entering ships at the Merak ferry port. This integration is expected to increase the efficiency of handling ODOL problems and avoid losses caused by vehicles exceeding the specified limits.

## 2 Method

This study uses a qualitative research approach aimed at describing and depicting existing phenomena, both natural and human-engineered, which pay more attention to the characteristics, quality, and interrelationships between activities [25]. Data collection in this study used field surveys, interviews and literature studies where the informants as sources were the head of the Land Transportation Management Center (BPTD) VIII Banten and related experts in it. Field surveys were conducted to detect ODOL vehicles at the port, while the literature study was a series of activities related to the method of collecting library data, reading, recording and managing research materials. Merak and Bakauheni Ferry Ports are the port nodes that connect the islands of Java and Sumatra [26]. The research flow in the study is shown in Fig. 1.

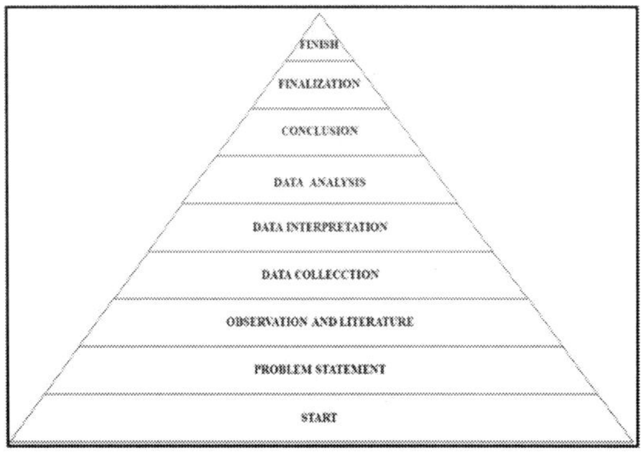


**Fig. 1.** Research flowchart

Data collection is a crucial step related to specific and complex topics such as ODOL vehicle detection at the port. The use of data collection methods such as field surveys, interviews, and literature studies, with sources from related experts can provide various benefits and interests. Field surveys allow researchers to collect empirical data directly at the location at the port, where ODOL vehicles are detected to help obtain accurate information on the number, type, and characteristics of ODOL vehicles, as well as the implementation and effectiveness of existing monitoring systems. Interviews provide in-depth insights into policies, challenges, and strategies used in land transportation management, especially related to ODOL vehicles in understanding the background of policies and operational decisions that may not be revealed through quantitative data alone. Using various data collection methods, such as surveys, interviews, and literature studies, allows data triangulation to strengthen the validity and reliability of research results to provide a more comprehensive picture and reduce bias.

The study that includes field surveys at the port allows researchers to understand the specific conditions and local practices in handling ODOL vehicles which play an important role in identifying unique obstacles and opportunities in that context. Interviews with leaders help in understanding the policies and regulations governing land transportation and ODOL handling and provide insight into the implementation of policies in the field and the challenges faced. With data collected from various sources and methods, more informed and evidence-based recommendations can be formulated to improve the relevance and implemented of recommendations provided to stakeholders. Field data and interviews can reveal areas that require improvement or enhancement, both in terms of technology, policy, and operational practices to help formulate more effective strategies in addressing ODOL vehicle issues. The study based on data collected using various and structured methods can make a significant contribution to the development of science in the fields of transportation, public policy, and logistics management. Overall, the data collection carried out is an important and integral approach to achieving accurate, comprehensive, and useful study results for various stakeholders. Mitigation to be effective requires proper transportation infrastructure planning [27].

## 3 Results and discussion

The proliferation of ODOL vehicles has created a chain phenomenon that has the potential to endanger oneself and others. Minister of Transportation, Budi Karya Sumadi stated during a webinar related to ODOL that the recorded loss value due to over-dimension and overloaded vehicles is IDR 43,45 trillion per year. According to observation data from Weight in Motion (WIM) at the Motor Vehicle Weighing Implementation Unit (UPPKB), Jasamarga's Automatic Number Plate Recognition (ANPR) and dimension sensors detected overload and over-dimension vehicle data ranging from 9%-42.5% and if concluded, every 1 in 5 vehicles passing through is indicated to be practicing ODOL. Reviewing data on the number of trucks in Indonesia, when viewed on an annual scale, it can be seen that the graph of the number of trucks in Indonesia is always increasing. From the scale obtained, it can be seen that trucks are one of the most popular modes of delivering commodities by the Indonesian people. Therefore, cooperation between truck driver associations and business associations must synergize in supporting government programs to overcome ODOL so that infrastructure and commodity flows can continue to run smoothly. Thus, there are no obstacles to the supply chain and development can continue to increase. Development must be viewed as a multidimensional process that in addition to pursuing economic growth acceleration, addressing income inequality, and poverty alleviation, also requires a series of major changes to the social structure, attitudes of society and national institutions.

According to the Transportation Policy Agency of the Ministry of Transportation, Over-Dimension is a condition in which the dimensions of the vehicle carrier do not comply with applicable production standards and regulations, while Over-Load refers to a situation when a vehicle carries a load that exceeds the specified weight limit. Dimension and load limits can differ for each type of vehicle, be it motorized vehicles such as passenger cars, buses, trucks, or motorcycles or non-motorized vehicles such as bicycles, pedicabs, or strollers, can be determined by different regulations. Regulations regarding ODOL arrangements in Indonesia are not covered in a single regulation, but in various types of regulations. This is due to the relationship between ODOL policies and transportation and roads, which involve not only the national scale, but also the authority of local governments [16]. Trucks as one of the most popular modes of commodity delivery by the Indonesian people, their dimensions are regulated in the Regulation of the Director General of Land Transportation Number KP.4413/AJ.307/DRJD/2020, explaining the dimensions permitted for bulk cargo transportation [28]. The following are the provisions for dimensions, shapes and examples of open cargo bodies for bulk transportation as in Tables 1 and 2. The shape and examples of open cargo bodies for bulk transportation are as shown in Figs. 2 and 3.

**Table 1.** Open body car type dump truck

| No | Configuration | Total Allowable Weight | Maximum height of inner tub |
|---|---|---|---|
| 1 | 1.1 | ≤ 5,500 kg | 550 mm |
| 2 | 1.2 | ≤ 8,500 kg | 700 mm |
| 3 | 1.2 | ≤ 16,000 kg | 850 mm |
| 4 | 1.22 | ≤ 24,000 kg | 1,000 mm |
| 5 | 11.22 | ≤ 30,000 kg | 1,100 mm |

**Table 2.** Open body car type non dump truck

| No | Configuration | Total Allowable Weight | Maximum height of inner tub |
|---|---|---|---|
| 1 | 1.1 | ≤ 5,500 kg | 550 mm |
| 2 | 1.2 | ≤ 8,500 kg | 700 mm |
| 3 | 1.2 | ≤ 16,000 kg | 850 mm |
| 4 | 1.22 | ≤ 24,00 kg | 1,000 mm |
| 5 | 1.1 | ≤ 5,500 kg | 550 mm + 450 mm (trellis) |
| 6 | 1.2 | ≤ 8,500 kg | 700 mm + 500 mm (trellis) |
| 7 | 1.2 | ≤ 16,000 kg | 850 mm + 450 mm (trellis) |
| 8 | 1.22 | ≤ 24,000 kg | 1.000 mm + 400 mm (trellis) |
| 9 | 11.22 | ≤ 30,000 kg | 1.100 mm + 400 mm (trellis) |

One of the steps taken by the government to overcome ODOL is to enforce the law firmly, law enforcement is a method used by parties or government agencies to prevent or handle violations. In the Zero ODOL policy [29] law enforcement is carried out to ensure that the rules and regulations prohibiting ODOL are strictly implemented. One of the initial steps to start the policy in stages is to inspect each motorized vehicle to ensure whether they meet the technical requirements or not. This inspection includes checking the arrangement, equipment, size, bodywork, and loading via a weighbridge. Weighbridges have an important role as a means of monitoring and controlling excess loads that violate the limits of goods transportation provisions [30]. Weighbridges have three main functions, namely recording, monitoring, and enforcement functions.

Every goods vehicle carrying a load must be weighed using a weighing device that is permanently installed or can be moved, with the aim of ensuring accurate weight measurements. In addition, another step taken is to test the type of vehicle through physical testing to meet roadworthiness requirements. This test is carried out by the Transportation Agency. In addition, supervision of modification workshops also needs to be carried out to ensure compliance with the rules. If there are motor vehicles that have been modified, the steps taken are to return them to their original form. Providing education to the public about traffic hazards is also an important step in this policy. In addition, ticketing will be carried out for Overload (ODOL) violations. ODOL enforcement can be carried out by reducing excessive load or by reducing the vehicle's road load. For vehicles that are modified with a length or height that exceeds their original size, a mark will be given on the vehicle body by spraying paint, cutting by authorities from the Transportation Service can also be done to adjust the modified vehicle.

According to the Director General of Land Transportation, the ODOL regulation policy has been implemented since 2023, effectively applicable in all ports, not only for Merak Port, the policy issued by Ministerial Regulation Number 103 of 2017, in Article 2 it is stated that every vehicle to be transported using a ferry must have its dimensions (height) and weight known along with its load. Ferry port operators have the right to reject vehicles that do not comply with this regulation. Vehicles that violate the rules may be directed out of the ticket purchase queue, therefore ferry port operators need to provide special lanes to remove vehicles from the port.

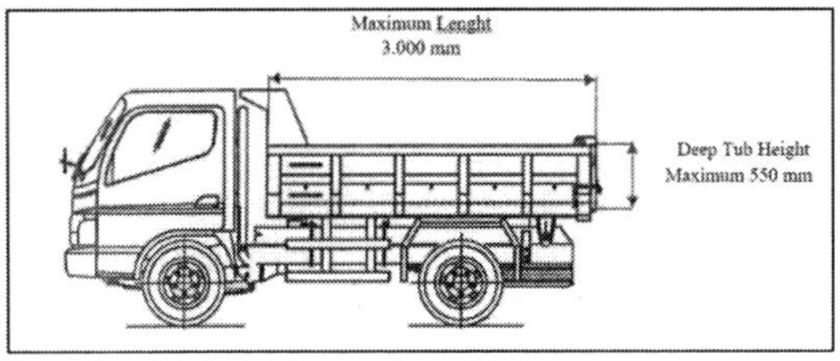


a) axle configuration 1.1 (JBI up to 5,500 kg)

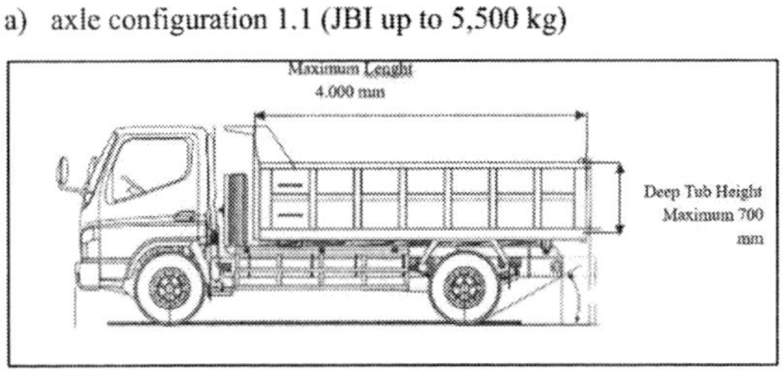


b) axle configuration 1.2 (JBI up to 8,500 kg)

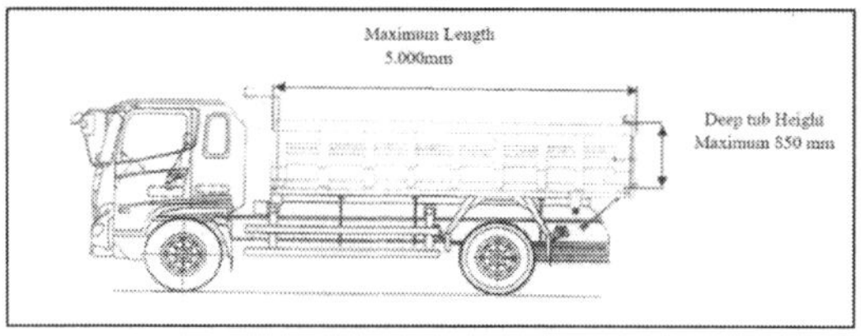


c) axle configuration 1.2 (JBI up to 16,000 kg)

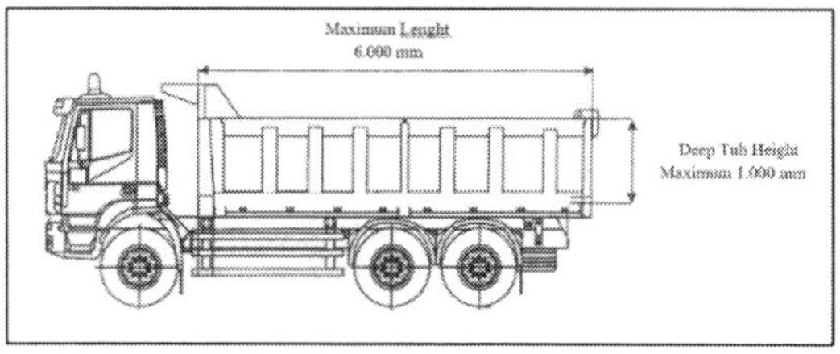


d) axle configuration 1.22 (JBI up to 24,000 kg)

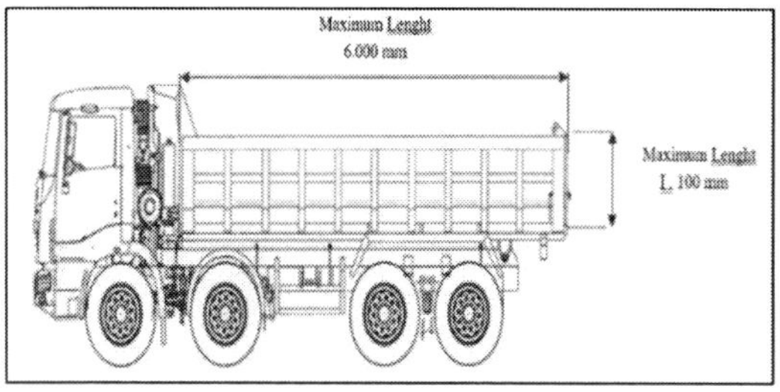


e) axle configuration 1.1 (JBI up to 5,500 kg)

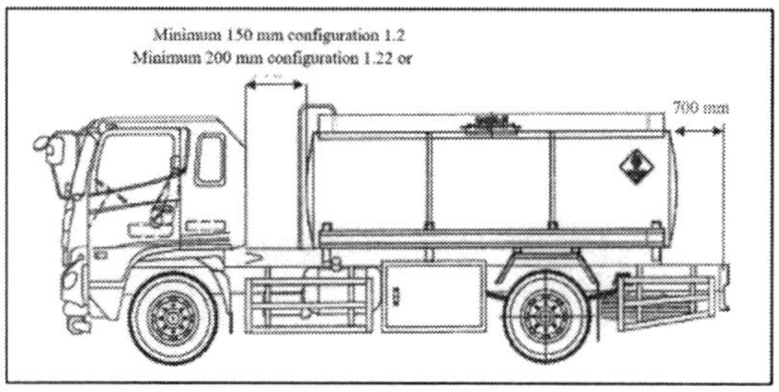


f) tanker

**Fig. 2.** Open loading body type dump truck

Knowing the dimensions of an open cargo bed, especially for dump trucks, is important for several reasons related to safety, operational efficiency, regulatory compliance, and logistics management. The appropriate dimensions of an open cargo bed can help in even distribution of the load, which is important for maintaining vehicle stability while moving. Dump trucks with unbalanced or oversized loads can be more prone to tipping over, especially when maneuvering or passing through uneven terrain. A load that is well distributed according to the dimensions of the bed can help the driver control the vehicle and ensure that the braking system works effectively. A load that is too high or heavy on one side can reduce braking and control efficiency. Each country or region has regulations that govern the dimensions and weight limits of vehicles, including dump trucks. Knowing the dimensions of an open cargo bed helps operators to ensure that the load being carried is in

https://doi.org/10.1051/e3sconf/202560406001

accordance with applicable regulations, avoiding fines and legal sanctions. Knowing the dimensions of the bed is also important for the installation of safety devices such as covers or safety nets to prevent spillage of the load during the trip, which is a legal requirement in many areas.

By knowing the capacity and dimensions of the bed, operators can maximize the load carried without exceeding the permitted limits, which aims to help optimize transportation efficiency, reduce the number of trips required, and save operational costs. Clear dimensions of the load body allow for better logistics planning, including more accurate travel routes, vehicle selection, and delivery times. Understanding the dimensions of the open load body helps in proper handling and stowage of the load, reduces the risk of damage to the load and vehicle, and also reduces the risk of spillage or loss of the load during transit. Dump trucks carrying loads exceeding the permitted dimensions can damage infrastructure such as roads, bridges, and tunnels. Knowing the dimensions of the body helps prevent incidents such as collisions with overhead structures or road barriers. Loads that match the capacity and dimensions of the body help in maintaining fuel efficiency because the vehicle is not forced to work beyond its optimal specifications. In addition, it also reduces greenhouse gas emissions, supporting environmental and sustainability goals. Proper loads help prevent excessive wear on vehicle components such as tires, suspension, and braking systems, which can extend the life of the vehicle and reduce maintenance costs. Overall, knowing the dimensions of the open load body on a dump truck is key to safe, efficient and regulatory-compliant operations, which also contribute to environmental and economic sustainability, which can be seen in Fig. 3.

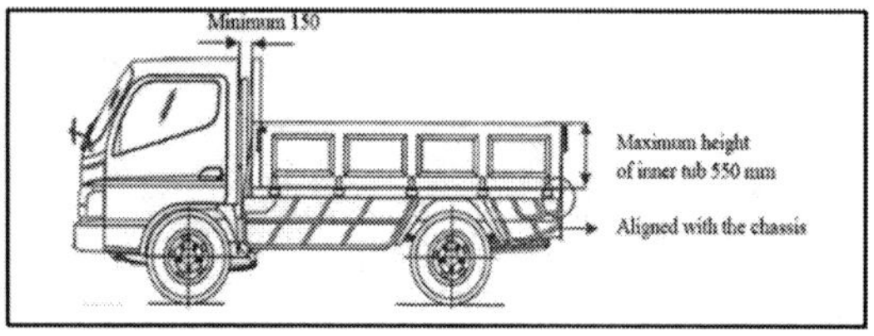


a) axle configuration 1.1 (JBI up to 5,500 kg)

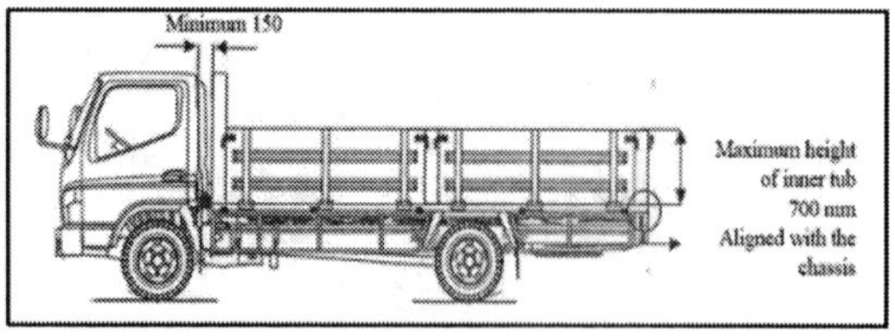


b) axle configuration 1.2 (JBI up to 8,500 kg)

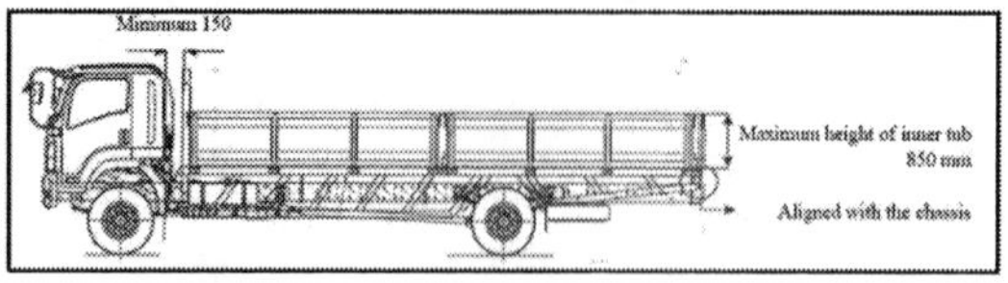


c) axle configuration 1.2 (JBI up to 16,000 kg)

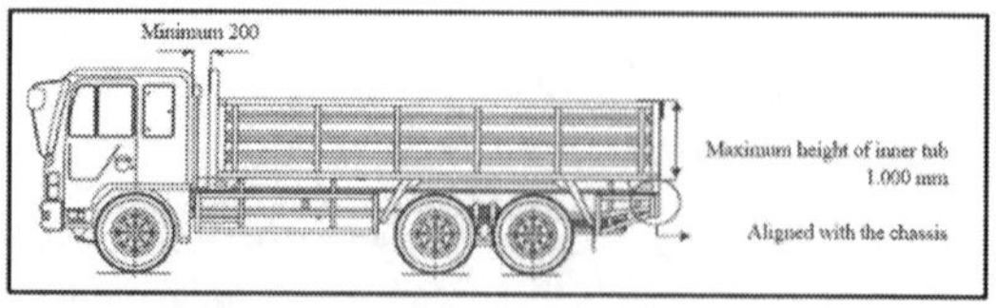


d) axle configuration 1.22 (JBI up to 24,000 kg)

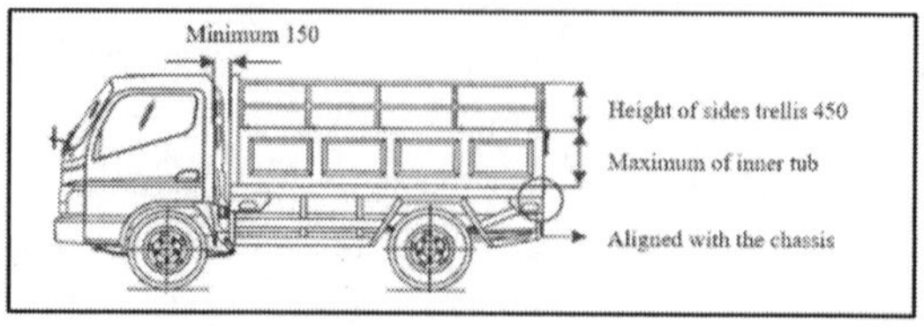


e) axle configuration 1.1 (JBI up to 5,500 kg) equipped with side trellises

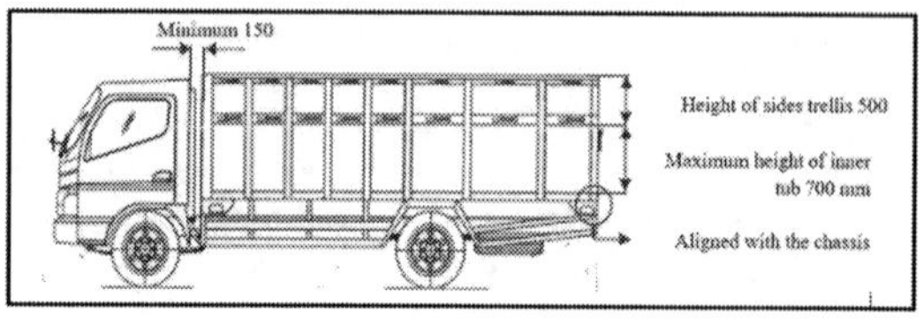


f) axle configuration 1.2 (JBI up to 16,000 kg) equipped with side trellises

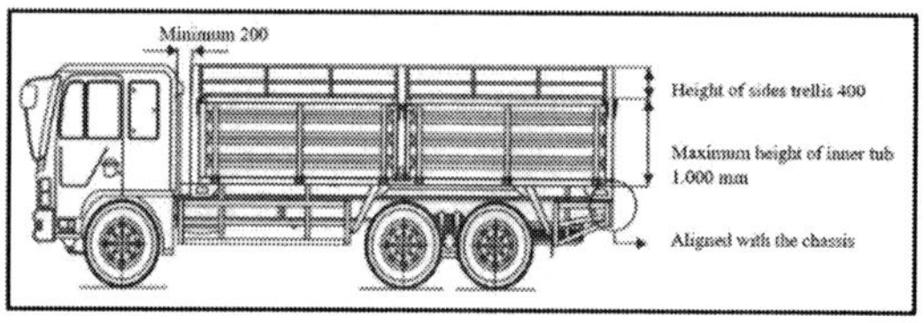


g) axle configuration 1.22 (JBI up to 24,000 kg) equipped with side trellises

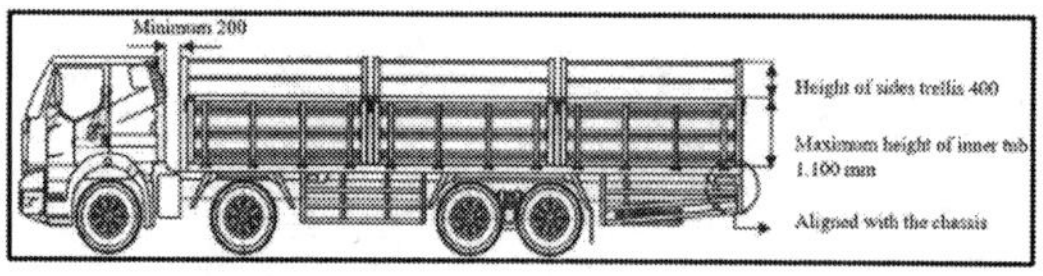


h) axle configuration 11.22 (JBI up to 30,000 kg) equipped with side trellises

**Fig. 3.** Loading body type non dump truck

Choosing a non-dump truck-type open-bed freight vehicle, such as a flatbed truck, pick-up truck, or open cargo truck, has several important reasons related to operational flexibility, load type, logistics efficiency, and operating conditions. Open-bed freight vehicles are suitable for transporting various types of cargo that may be difficult to load in a closed vehicle, such as building materials, large equipment, industrial machinery, or irregularly shaped items. Open-bed vehicles allow for easy loading and unloading of cargo from any side, which is especially useful for large or heavy loads that require the use of auxiliary equipment such as forklifts or cranes. Open-bed trucks are often more economical than closed-bed vehicles or other specialized vehicles that do not require additional features such as cooling systems or climate control, which can reduce purchasing and operating costs. Open-bed trucks allow for more flexible stacking and arrangement of cargo, maximizing the use of available space to reduce the number of trips required to transport a particular load, saving fuel and time.

Open-bed vehicles are ideal for long or large items, such as pipes, steel beams, and prefabricated structures, that cannot be loaded in a closed vehicle without modification. Open bodies are often used to transport weather-resistant materials, such as construction materials, dirt, rock, or heavy equipment, where protection from the elements is not required. Open bodies can be more accessible to construction or industrial sites that may have limited space or difficult terrain. Open bodies provide additional flexibility in delivery and hauling operations. Open bodies, especially smaller ones such as pick-up trucks, can more easily maneuver in tight spaces or city streets, where larger or enclosed vehicles may be difficult to access. Open bodies allow for quick loading and unloading, which is important in logistics operations that require fast delivery or have tight turnaround times. The ease of access and flexibility in cargo handling helps deliver goods with less rehandling, reducing the risk of damage and increasing shipping efficiency. Open bodies must comply with regulations regarding lashing and securing of loads, as well as dimensional and weight restrictions that require operator knowledge and planning to ensure safety and regulatory compliance. Overall, the selection of a non-dump truck type open body freight vehicle is based on the need for flexibility, efficiency, and the ability to handle a variety of loads. This selection also takes into account aspects of safety, regulatory compliance, and operational optimization.

**Fig. 4.** Vehicle weight measuring tool using weight in motion

Weight in Motion (WIM) is a technology used to measure the weight of moving vehicles, such as trucks or other heavy vehicles, without having to stop the vehicle (Fig. 4). WIM helps detect and control overloaded vehicles. Overloading can cause significant damage to road and bridge infrastructure, increasing maintenance and repair costs. Overloaded vehicles are at higher risk of accidents, especially if the load affects the vehicle's stability and braking. WIM helps prevent these vehicles from entering the roadway without proper control measures.

Data from WIM is used to identify traffic and vehicle load trends that can accelerate road and bridge deterioration and is used to assist authorities in planning and implementing timely maintenance, and extending the life of the infrastructure. By preventing overloaded vehicles, WIM helps reduce costs associated with road deterioration and maintenance, helps logistics and transportation companies optimize loads, which can reduce fuel consumption and improve operational efficiency. WIM provides critical data for transportation planning and policy development. This data is used to understand traffic patterns, infrastructure needs, and to support investment decisions in road and bridge development. WIM systems support law enforcement by providing accurate data on vehicle weights. This makes it easier for authorities to take action against overloading violations and helps in traffic safety efforts. Thus, WIM plays a key role in ensuring that the transportation system operates safely, efficiently and economically.

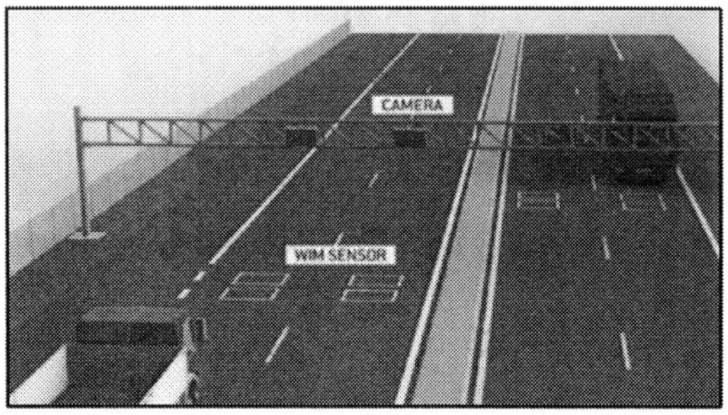


**Fig. 5.** Truck scales using weight in motion

Truck scales using WIM technology are of particular interest in the context of traffic management, safety and infrastructure maintenance (Fig. 5). WIM truck scales enable enforcement of load regulations without the need to stop vehicles, minimizing traffic disruption. WIM truck scales ensure that heavy vehicles on the road adhere to set load limits and help prevent damage to infrastructure. Roads and bridges are designed to handle specific loads. Overloaded vehicles can accelerate deterioration and shorten the life of infrastructure. Using WIM, vehicles exceeding the weight limit can be identified and acted upon to reduce the overload on roads and bridges. Overweight or unbalanced vehicles can pose a serious safety risk to both drivers and other road users. WIM truck scales help detect and prevent vehicles from operating on the road, thereby improving public safety. Compared to conventional weighing methods that require vehicles to be stopped, WIM allows weighing to be carried out dynamically while the vehicle is in motion. The use of WIM can reduce waiting and queuing times, improving transportation and logistics efficiency. WIM provides real-time and accurate data on vehicle weight, load distribution and traffic volume. This data is useful for traffic analysis, infrastructure planning, and transportation policy development. By preventing overloading, WIM truck scales help reduce costs associated with infrastructure repairs and vehicle breakdowns. Additionally, improving vehicle operating efficiency can reduce fuel consumption and other operating costs. Overall, truck scales using WIM technology provide significant benefits in maintaining infrastructure integrity, improving road safety, and supporting more efficient transportation operations.

**Fig. 6.** Truck identified ODOL

Trucks that are indicated as ODOL show certain characteristics that distinguish them from vehicles that comply with the permitted dimensions and weight limits (Fig. 6). ODOL trucks often carry loads that exceed the weight limits permitted by traffic regulations, due to the load being too much or too heavy for the specifications of the truck. The uneven or poorly distributed load can cause certain parts of the truck to carry an excess load, even though the total weight is still within the limits. This can affect the stability and control of the vehicle. ODOL trucks have loads that are too wide or too high compared to the permitted dimensions, which can cause problems when passing through tunnels, bridges or roads with certain dimension restrictions. Trucks or truck and trailer combinations that are too long can be difficult to maneuver, especially in urban areas or on narrow roads. Overloaded trucks often show compressed or sagging suspension, which can be observed from the low distance between the truck body and the wheels. Tires can appear too flat or deflated due to excess pressure, which also increases the risk of tire failure or accidents. ODOL trucks are more difficult to control, especially when cornering, stopping suddenly or maneuvering on narrow roads which can increase the risk of accidents. Overloading tends to increase fuel consumption because the engine has to work harder to move the vehicle. ODOL vehicles often experience faster wear on components such as brakes, tires, and transmission systems, due to the additional load that is not in accordance with the vehicle's design specifications. Overloading can cause decreased brake performance, which is very dangerous, especially on descents or when stopping the vehicle suddenly. ODOL trucks often do not have the appropriate or correct documentation, such as special permits for overloading or dimensions that exceed the limits. Loads that protrude from the truck bed, especially if not properly secured, are another characteristic of ODOLs that can increase the risk of accidents and violations of safety regulations. ODOL trucks require special attention from authorities to prevent negative impacts caused, including damage to infrastructure, increased risk of accidents, and public safety issues.

Detecting ODOL trucks is important for several reasons related to safety, infrastructure maintenance, transport efficiency and law enforcement. Trucks carrying overloaded or oversized loads can increase the risk of accidents, as drivers may have difficulty controlling the vehicle, especially when braking or turning. Early detection allows authorities to prevent these vehicles from causing serious accidents. ODOL vehicles not only endanger their drivers but also other road users, including small vehicle drivers, pedestrians and cyclists. ODOL trucks put excessive stress on roads and bridges, causing accelerated damage such as cracking, deformation and structural collapse. ODOL detection helps prevent or mitigate such damage, extending the life of infrastructure and saving on repair costs. Without effective ODOL detection, the cost of maintaining and repairing infrastructure can increase significantly, burdening government and public budgets. ODOL detection is an important tool in law enforcement, helping authorities to enforce regulations regarding vehicle weight and dimension limits. It is therefore important to ensure that all transport operators comply with regulations and maintain public safety. ODOL detection allows for the imposition of fines or sanctions on violators, which serves as an effective deterrent and enforcement. Vehicles that operate within the dimensions and load limits are typically more fuel efficient and maintenance efficient. ODOL detection helps drive more efficient and sustainable transport practices. By maintaining better infrastructure conditions and reducing accidents, ODOL detection contributes to reduced costs for governments, businesses and society. ODOL trucks tend to be less fuel efficient, which can increase greenhouse gas emissions. By ensuring trucks are operating within legal load limits, emissions can be reduced, contributing to environmental and public health goals. Overall, detecting trucks that are ODOL is critical to maintaining public safety, protecting infrastructure, ensuring regulatory compliance and supporting transportation efficiency and sustainability.

Detecting ODOL trucks in relation to the dimensions of the vehicle cargo deck on a ferry is important for several reasons, primarily related to safety, efficiency, and regulatory compliance. ODOL trucks with loads exceeding the dimensions of the deck can affect the stability of the vessel. Overweight or uneven loads can cause the vessel to become unbalanced, potentially increasing the risk of capsizing or losing stability, especially in rough weather or high waves. Loads that are too long, wide, or high can pose physical hazards on the ship's deck, including the risk of injury to crew and passengers, as well as the risk of damage to ship infrastructure such as doors, cargo handling systems, and other facilities. Ferries typically have strict weight and dimension restrictions on cargo to ensure safe and efficient operations. Detecting ODOL helps ensure that all vehicles boarding the vessel comply with these restrictions, preventing potential violations that could result in fines or other legal action. Some cargo may require special permits or additional preparation if they exceed standard dimension limits. Detecting ODOL trucks ensures that all vehicles have the necessary documentation and permits to be transported safely.

**Fig. 7.** Vehicle loading deck on board ship

Loads that comply with the dimensions of the deck make it easier to arrange and organize vehicles on the ship, maximizing the use of available space and ensuring even load distribution (Fig. 7). Vehicle management on board helps to maintain ship balance and improve operational efficiency. Overloaded or oversized trucks can cause damage to ship facilities, such as deck floors, barriers or cargo handling systems. Early detection of ODOL helps to avoid damage that can be expensive to repair and disrupt ship operations. ODOL trucks detected upon boarding can cause delays in the boarding process, as it may require re-loading or even rejection of vehicles. These delays can disrupt sailing schedules and result in additional costs for ship operators and shippers. Ensuring that all vehicles meet dimensional and weight requirements helps to plan and execute more efficient voyages, reducing waiting times and improving the punctuality of ferry services. Overloaded or oversized ODOL trucks increase the risk of incidents on board, such as fires or hazardous material leaks. ODOL detection helps to reduce this risk by ensuring safe and controlled loading. Carrying excess cargo can increase ship fuel consumption and emissions and have a negative impact on the environment. Ensuring proper loading allows ships to operate more efficiently and environmentally.

Ferizy is an online-based ticket service that can be accessed by service users at 4 ASDP Main Ports. These ports include Merak Port, Bakauheni Port, Ketapang Port and Gilimanuk Port. The presence of the Ferizy application is done so that service users be more comfort and safe when using ferry transportation services. With this ticketing system update, it is hoped that the service system can run optimally and be able to reduce the spike in ticket purchase queues when using the service. The Ministry of Transportation is updating the KIR testing system. This step was taken because of the rampant cases of fraud and illegal levies during the KIR process, such as what happened at the Rembang Regency Transportation Agency, Central Java, which was proven to have carried out illegal levies on vehicle KIR tests. The data listed in BLU-e is the vehicle identity, 4-sided physical photos of the vehicle, and test result data stored in digital format. Therefore, with the development of digitalization from both service providers and regulators, we view this as an opportunity to eradicate ODOL. The BLU-e and Ferizy System Integration Mechanism goes through several stages, including data mapping to identify vehicle dimensions and maximum loads that can be transported by vehicles registered in BLU-e. After that, Ferizy matches the data of prospective port service users to see if they are in accordance with the dimensions and maximum loads recorded in BLU-e.

After the data is mapped, the next stage is data integration [31]. This data integration is carried out to combine the data in both systems into one so that it is more effective in detecting vehicles that violate dimension and load limits. The mapped and integrated data will be tested to ensure that the resulting data evaluates the performance of both systems when they work together, including the extent to which they can detect ODOL vehicles quickly and accurately.

The data integration process between the BLUe and Ferizy systems can include users ordering ship tickets in the Ferizy system, ticket ordering data from the Ferizy system is sent to the BLUe system via API, the BLUe system processes ticket ordering data and sends data on dimensions and maximum vehicle load, the Ferizy system receives data from the BLUe system and printed tickets with details of the dimensions and maximum vehicle load. In this process, all vehicles will be identified as to the dimensions and maximum vehicle load so that when entering the port, officers can easily detect ODOL vehicles when checking vehicle documents [32]. When purchasing a ticket, the type of class and group can be directly input into the Ferizy application which is a means for users of the Merak Port crossing service, from the input of data, the class and group of the vehicle will be seen and a cross-check will be carried out between the

BLU-e data and the physical inspection of officers in the field. If the vehicle does not match the data recorded in BLU-e, it can be concluded that the vehicle is an ODOL vehicle and cannot use the crossing service and will receive a warning.

Tightening of ODOL identified vehicles at ferry ports plays an important role in reducing disaster risks, especially in port environments that are prone to accidents and operational disruptions. Tightening of ODOL vehicles contributes to disaster risk reduction at ferry ports to reduce the risk of maritime disasters where ODOL vehicles that are excessive in weight or dimensions can disrupt the stability of ferries that have the potential to cause the ship to capsize or sink, especially in bad weather conditions or rough seas. Tightening of ODOL rules ensures that vehicle loads are in accordance with ship capacity, reducing the risk of maritime accidents that can result in ship sinking and loss of life. Ferries that are overloaded due to ODOL vehicles become more difficult to maneuver, especially in emergency conditions. Tightening of ODOL increases the ability of ships to maneuver better, thereby reducing the risk of accidents at sea. Protection of port infrastructure is urgently needed where ODOL vehicles put greater pressure on port infrastructure such as docks, bridges and loading and unloading facilities. Tightening of ODOL rules will reduce the burden on port infrastructure and prevent damage that can cause disruption to port operations. Well-maintained infrastructure is essential to support emergency response in disaster situations. In addition, ODOL vehicles often take longer to load or unload, which can cause congestion at ferry ports. This congestion can disrupt evacuation or aid delivery processes during disasters. Tightening ODOL regulations can smooth the flow of vehicles and goods to ensure that port operations remain efficient and responsive.

Improved port operational efficiency and resilience can be implemented by reducing the risk of operational failure. Ferries carrying ODOL vehicles are at risk of experiencing faster damage, either to the engine or the ship's structure, which can result in operational failure and impact service delays, especially during disasters where smooth maritime transportation is essential. Tightening ODOL helps ensure that vessels operate according to their capacity and minimizes the risk of operational failure. When disasters occur, ferry ports often serve as vital evacuation and logistics delivery routes. The presence of ODOL vehicles can slow down this process by requiring more time to process. By limiting ODOL vehicles, ports can more efficiently handle the volume of vehicles and goods during disasters, ensuring a quick and smooth response.

Efforts to reduce environmental risks such as damage to marine ecosystems can occur if a ferry carrying ODOL vehicles has an accident where spills of hazardous materials such as oil or toxic goods can pollute the sea and damage the ecosystem. Tightening ODOL helps prevent situations like this by ensuring that ship loads remain within safe limits, reducing the risk of accidents and environmental damage. ODOL vehicles generally use more fuel which can increase air pollution and greenhouse gas emissions. Tightening ODOL rules reduces the number of excess vehicles crossing, thereby reducing emissions that contribute to climate change and potential environmental disasters.

The strategy for tightening ODOL vehicles needs to take several steps that can be taken at ferry ports including tighter inspections, law enforcement, awareness campaigns and cooperation with stakeholders. Strict inspections can be carried out through the implementation of automatic weighbridges and dimension measuring devices at ports to detect ODOL vehicles before boarding the ferry. The application of stricter sanctions and fines for ODOL violators, including a ban on boarding for vehicles that do not comply as one way of law enforcement. While the awareness campaign by conducting socialization to logistics companies and drivers about the risks and consequences of ODOL vehicles, as well as the importance of complying with the rules for mutual safety. For cooperation efforts with stakeholders where it is necessary to involve related parties such as ferry companies, port authorities and local governments to ensure consistent implementation of ODOL regulations. The tightening of ODOL regulations at ferry ports can significantly reduce the risk of disasters, both in terms of transportation safety, infrastructure protection, and emergency response efficiency.

## 4 Conclusion

The integration of the BLU-e (electronic traffic office) and Ferizy (digital ferry service) systems is able to consider disaster risk reduction, especially in ferry ports that play an important role in the sea and land transportation system. Based on the study, the integration of BLU-e and Ferizy is able to create a more structured and efficient mechanism in managing ODOL vehicles. BLU-e allows automatic detection of vehicles that exceed load and dimension limits through digital data, while Ferizy ensures that only vehicles that comply with ODOL rules can access ferry services. This directly reduces the number of ODOL vehicles that can cause accidents and operational disruptions at the port. By reducing ODOL vehicles, the risk of damage to port infrastructure, docks and ferries due to overloading can be minimized in order to reduce maintenance and repair costs, thereby preventing significant material losses. Reducing the number of ODOL vehicles also helps avoid operational delays and congestion at ferry ports, thereby increasing service efficiency. In addition, the risk of accidents that threaten the safety of passengers and port workers can be reduced, reducing immaterial losses related to safety and public trust. ODOL vehicles crossing the port increase the potential for maritime disasters such as ship accidents due to disturbed stability or infrastructure damage during emergency evacuations. The integration of BLU-e and Ferizy reduces disaster risk by ensuring that ferries operate within safe load limits, maintaining vessel stability, and protecting critical infrastructure needed for emergency response. This integrated system creates a safer and more responsive transportation environment to emergency conditions, maintaining stable port operations during critical times such as natural disasters or other emergency situations. Thus, the integration of BLU-e and Ferizy not only helps in reducing the number of ODOL vehicles but also directly contributes to disaster risk reduction, both in terms of safety, infrastructure protection and operational efficiency at ferry ports. All these efforts have a positive impact on transportation management and risk mitigation in the future.

## References

1. N. N. Abu Bakar, T. Uyanik, Y. Arslanoglu, J. C. Vasquez, and J. M. Guerrero, Two-stage energy management framework of the cold ironing cooperative with renewable energy for ferry," Energy Convers. Manag., **311**, 118518,1–13, (2024), https://doi.org/10.1016/j.enconman.2024.118518.
2. W. Kuźmiński, Diagnosing and forecasting port services on the example of świnoujście ferry terminal, Procedia Comput. Sci., **192**,3440–3448, (2021), https://doi.org/10.1016/j.procs.2021.09.117.
3. O. A. Alsos et al., Lessons learned from the trial operation of an autonomous urban passenger ferry, Transp. Res. Interdiscip. Perspect., **26**, 101142,1–16, (2024), https://doi.org/10.1016/j.trip.2024.101142.
4. M. Munir, M. K. Muqorrobin, R. A. M. Yunus, and S. Sarjana, Ruang Bersama: Monitoring Application for Sexual Harassment in Public Transport, 2023 Int. Conf. Informatics, Multimedia, Cyber Inf. Syst. ICIMCIS 2023, 10349058,257–262, (2023), https://doi.org/10.1109/ICIMCIS60089.2023.10349058.
5. S. del Saz-Salazar and B. Tovar, On ferry users' willingness to pay for improving environmental quality: A case study for the Canary Islands, Transp. Policy, **154**,61–72, (2024), https://doi.org/10.1016/j.tranpol.2024.05.023.
6. S. Sarjana, Green transportation: Development opportunities in support of sustainable transportation, in E3S Web of Conf., **429** (03003,1–10, (2023) https://doi.org/10.1051/e3sconf/202342903003.
7. K. L. Rødseth, K. Fagerholt, and S. Proost, Optimal planning of an urban ferry service operated with zero emission technology, Marit. Transp. Res., **5**, 100100,1–18, (2023), https://doi.org/10.1016/j.martra.2023.100100.
8. A. T. James, A. Q. Khan, M. Asjad, G. Kumar, and V. Arya, Identification and evaluation of challenges in commercial vehicle transport business in India post-implementation of BS-VI emission norms, Res. Transp. Business Manag., **54**, 101122, (2024), https://doi.org/10.1016/j.rtbm.2024.101122.
9. B. Otto, S. Newnam, V. Weale, and J. Oakman, Risk controls for preventing work-related musculoskeletal and mental health disorders: A STAMP analysis of the Victorian heavy vehicle transport sector, Saf. Sci., **176**, 106558,1–18, (2024), https://doi.org/10.1016/j.ssci.2024.106558.
10. F. Hatziioannidu and A. Polydoropoulou, Identifying passenger preferences for alternative ferry transport hubs, Res. Transp. Econ., **96**, 101221, (2022), https://doi.org/10.1016/j.retrec.2022.101221.
11. Q. Jing, T. Li, Y. Lai, Y. Wang, Y. Li, and S. Qi, Study on risk assessment models for the aggregation of vehicles transporting hazardous chemicals, Reliab. Eng. Syst. Saf., **1251**, 110345, (2024), https://doi.org/10.1016/j.ress.2024.110345.
12. R. H. Della, T.-C. Lirn, and K.-C. Shang, Study of safety behavior in ferry transport, Saf. Sci., **131**, 104912, (2020), https://doi.org/10.1016/j.ssci.2020.104912.
13. A. Djamaluddin, Manajemen Operasional Pelabuhan. Makasar: Unhas Press, (2022).
14. PP_55_2012, Peraturan Pemerintah Nomor 55 Tahun 2012 Tentang Kendaraan. (2012).
15. L. Antono, Implementasi Kebijakan ODOL Dalam Upaya Meningkatkan Sistem Pengawasan Dan Pengendalian Muatan Angkutan Barang, Humantech J. Ilm. Multi Disiplin Indones., **1(11)**,1720–1729, (2022), https://doi.org/10.32670/ht.v1i11.2315.
16. A. B. K. Rajendra and H. Vaza, Isu Over Dimension Over Loading (ODOL) Angkutan Logistik Dan Konsep Solusi Strategis, (2022), https://doi.org/10.58674/phpji.v15i1.235.
17. E. Oktarinda, N. B. Prihutomo, and E. O. Maulani, Analisis Pengaruh Kendaraan Odol Terhadap Tingkat Kecelakaan Di Jalan Tol, Constr. Mater. J., **4(1)**,49–57, (2022), https://doi.org/10.32722/cmj.v4i1.4151.
18. KNKT, Angkutan ODOL Salah Satu Potensi Bahaya Di Angkutan Penyeberangan, (2022). [Online]
19. N. H. Purnomo, Evaluasi Penggunaan Sistem Jembatan Timbang Online (JTO) Terintegrasi dengan Metode PIECES Framework, J. Algoritm., **20(1)**,87–98, (2023), https://doi.org/10.33364/algoritma/v.20-1.1255.
20. S. Rozi, Sanksi Terhadap Pelanggaran Transportasi Darat Odol (Overdimension Overloading) Ditinjau dari Undang-Undang Nomor 2 Tahun 2009 Tentang Lalu Lintas Angkutan Jalan, Glosains J. Sains Glob. Indones., **2(1)**,13–21, (2021), https://doi.org/10.59784/glosains.v2i1.11.
21. E. P. Rini and D. I. S. Saputra, Sistem Informasi Manajemen Di Era Revolusi Industri 4. Purwokerto: Zahira Media Publisher, (2021).
22. L. Filina-Dawidowicz, A. Cernova-Bickova, I. Semenov, D. Mozdrzen, A. Wiktorowska-Jasik, and D. Bickovs, Information support of cargo ferry transport: Case study of Latvia, Procedia Comput. Sci., **176**,2192–2201, (2020), https://doi.org/10.1016/j.procs.2020.09.256.
23. Y. Aditri, Analisa Swot Dalam Rangka Penerapan Bukti Lulus Uji Elektronik (BLU-e) di UPTD PKB Pinang Baris Kota Medan, Politeknik Keselamatan Transportasi Jalan, (2020).
24. S. Sarjana, Near Field Communication Studies in Transportation Business, 2023 3rd Int. Conf. Intell. Cybern. Technol. Appl. ICICyTA 2023, 10428896,266–271, (2023),

https://doi.org/10.1109/ICICyTA60173.2023.10428896.

25. I. Nurdin and S. Hartati, Metodologi penelitian sosial. Media Sahabat Cendekia, (2019).
26. F. Tampubolon, W. Seno, and S. Sarjana, Analysis of lashing gear in ship loading spaces to meet safety transportation, E3S Web Conf., **519**,1–13, (2024), https://doi.org/10.1051/e3sconf/202451903024.
27. E. P. Raharjo, S. Sarjana, and M. Safitri, Transportation infrastructure planning in supporting disaster mitigation: Case study in Mount Gamalama, Jamba J. Disaster Risk Stud., **14(1)**,1–11, (2022), https://doi.org/10.4102/jamba.v14i1.1123.
28. A. B. Sulistyo, T. A. Cundoko, R. R. O. Sasue, R. Ahmad, I. P. A. Suryasa, and A. D. Dwipayana, Sistem Keselamatan Bagi Awak Kendaraan Bermotor Angkutan Barang Terminal, Madiun Spoor J. Pengabdi. Masy., **I(2)**,57–62, (2021), https://doi.org/10.37367/jpm.v1i2.188.
29. S. P. Febriani and M. Mintarsih, Law Enforcement in Zero Over Dimension and Over Loading Policy for Transportation of Goods, Reformasi Huk., **27(1)**,42–50, (2023), https://doi.org/10.46257/jrh.v27i1.603.
30. E. O. Saputri, Y. M. Pratiwi, and M. A. Khoiruman, Prosedur Kegiatan Trucking 'Dump Truck' di PT. Samudera Moda Indonesia Semarang, Discov. J. Kemaritiman dan Transp., **3(1)**,30–39, (2021).
31. W. Warkim and D. I. Sensuse, Model Integrasi Sistem dengan Pendekatan Metode Service Oriented Architecture dan Model View Controller pada Pusat Penelitian Perkembangan Iptek Lembaga Ilmu Pengetahuan Indonesia, J. Tek. Inform. dan Sist. Inf., **3(1)**,84–103, (2017), https://doi.org/10.28932/jutisi.v3i1.656.
32. N. W. Gautama et al., Sosialisasi zero over dimension over loading (ODOL) kepada pengemudi dan pemilik angkutan barang di terminal barang dishub kota denpasar, J. Pengabdi. Kpd. Masy. Semangat Nyata untuk Mengabdi, **2(1)**,9–14, (2022), https://doi.org/10.52920/jkpmsenyum.v2i1.55.

# Material management comparison between lot for lot and field actual condition

*Jajang* Atmaja[1], *Monika* Natalia[1], *Masrilayanti*[2*], *Siraj* Ramadhani[1], *Yessa* Rahmadini[1], and *Sulleyman* Sourkan[3]

[1]Department of Civil Engineering, Politeknik Negeri Padang, Padang, West Sumatra, Indonesia
[2]Department of Civil Engineering, Faculty of Engineering, Universitas Andalas, Padang, West Sumatra, Indonesia
[3]Petroleum and Energy Engineering Department, Sulaimani Polytechnic University, Sulaymaniah, Iraq

**Abstract.** Project material management aims to regulate effective and efficient use materials because material is a resource the largest percentage of the total project cost. In reality, the process of providing materials encounters many obstacles such as late material arrival cause human error and natural disaster, materials not available, overcost, undercost. This research was carried out substructure of the Inpres Market Development Project located in Painan, Pesisir Selatan. Materials for substructure were imported from Padang, which is 70 km from project located. The topographic conditions of areas that prone to disasters such as floods and landslides, which cause connecting road Padang and Painan to be cut off. Arrival materials delay, which results in project delays and wasted costs, so material management is needed. This study aims to obtain the actual material management costs of the field, further compared to the lot for lot technique. The results, total cost *field actual condition* of Rp8,909,250,238.42, with *a lot for lot* technique of Rp8,789,667,046. The cost use lot for lot technique is cheaper by Rp119,583,192.42 or 1.34% due, this technique makes the storage cost zero, meaning that every material that comes will be used directly at the project site.

## 1 Introduction

Material management is defined as a management system that is necessary to plan and control the quality of materials, the quantity of materials, and the timely placement of equipment, good prices and quantities that are in accordance with the needs [1].

Project material management aims to regulate the use of materials effectively and efficiently during the implementation of work, so that there is no shortage or excess of materials [2]. Material management is carried out with several stages of implementation starting from the process of calculating material quantity needs, stages of material needs, finding *suppliers*, purchasing and delivering materials to the project site, material inspection and *quality control*, material storage and supervision, to the process of implementing material handling and distribution for a job [3].

If there is an error in the material supply process, it can result in delays in the work and have an impact on the next work, resulting in delayed completion time and waste of project costs [4]. If material management is well managed, it will help achieve the smooth and successful of the project [5].

In the implementation of construction projects, common problems that are often faced are ordering too many materials or those that are too available (*overstock*) or ordering too few materials (*understock*). If there is an order with an excessive amount, there will be a waste resulting in increased storage costs, material damage, and shrinkage of the amount of material. Likewise, vice versa, if the materials ordered are too few, there will be a shortage of materials that can interfere with the smooth running of the project or delay the activities so that the project is not completed on time [6].

In addition to *overstock* and *understock*, sometimes the material is not available, or the material ordered can only be partially fulfilled, sometimes even the material is not there All of these problems will have an impact on delaying the implementation of work which causes delays and inflated costs.

To overcome these problems, it is necessary to plan material management as early as possible, especially related to material inventory and control. One of the material planning methods is *the lot for lot* (LFL) technique. The *lot-for-lot* (LFL) technique aims to minimise the cost of savings, so that the cost of savings becomes zero [7]. In this technique, the cost component of material supplies consists of several types of costs [8], including:

1. Purchasing cost, which is the cost incurred by the purchase of equipment or materials carried out by a company from the supplier.
2. Order/*set-up cost*, which is the cost that must be incurred in connection with the order procurement process, such as transportation costs, administrative costs, telephone costs, fax costs and other costs related

* Corresponding author: masrilayanti@eng.unand.ac.id

to the order, but if the procurement is produced by the factory itself, then the production cost is analogous to the order cost and is often called *set-up cost.*

3. Storage costs *(holding cost* or *carrying cost),* which are costs incurred as a result of having inventory, such as warehouse costs, employee wages, electricity, stored capital and the amount is usually measured as a percentage of the value of stored goods, the unit is Rp/unit in units of time (years).

The application of the lot for lot technique in the Office Renovation Project (Vedca) makes this project more controllable in storage costs, which is Rp0,- [9]. In the construction of the Airport X package 2, the calculation of the total need and cost of structural material supplies for column and beam work by comparing *the lot for lot* method, *economic order quantity*, and *periodic order quantity*, the minimum cost is obtained using the lot for lot *method* [4]. A comparison of the cost of *the sub-structure* of the manual method and *the lot for lot* (LFL) method found that the material cost with the LFL method was 1.48% cheaper than the manual method [3]. In the Simpang Niam-Lubuk Kambing 1 Road Preservation project, Jambi material management using *the lot-for-lot* method, procurement reduces the cost of material procurement by Rp48,471,706, cheaper than the planning cost of the project [2]. In the Westown View Wiyung *Apartment construction project* for basement 2 floor structure work, the material cost component consists of purchase costs, ordering costs and storage costs. The analysis was carried out using *the lot for lot, economic order quantity, period order quantity method.* The results of the study show that the minimum purchase cost for D10 concrete iron material uses LFL and POQ techniques, with a nominal value of Rp17,719,240.00. The minimum order fee for D10 concrete iron material uses EOQ and POQ techniques, with a nominal value of Rp2,698.50. Minimum storage costs for all types of materials using the LFL technique with a nominal zero cost [10]. The material planning of the Kapuas Indah Market Renovation Project and Public Service Mall (*Multiyears*) in Pontianak City, was analyzed using *the lot for lot* (LFL) and *period order quantity* (POQ) methods. The materials controlled are 25 cm x 25 cm minipile, formwork, scaffolding, and U32 rebar. For minipile, the lowest material cost is obtained by *the lot for lot* method. Meanwhile, the lowest cost for formwork, scaffolding, and U32 rebar is to use the *period order quantity* method, which can cause material accumulation if the field is not well controlled over the use of materials [11].

Application of the lot sizing technique, the most efficient lotting method is obtained from the Lot for Lot method at a cost of Rp5.29 billion. The application of MRP also has an impact on improving inventory performance, changes in minimum and maximum stock changes that are adjusted to production activities. Efficient purchase costs from total raw material needs with an average value of Rp8.33 billion [12].

Lot sizing methods are better suited than others to be applied in specific scenarios marked by a considerable variation in demand or peaks of seasonality. Thus, contributes to further clarify industrial practitioners on the selection of the best suited lot-sizing methods for each type of application scenario regarding MRP or JIT/Kanban manufacturing environments [13].

The lot for lot method aims to minimize the amount of spending on inventory in the warehouse. Lot for lot (LFL) method to calculate inventory control in minimizing inventory costs and ordering costs for wire-mouse product is IDR 204,000.00 and the wire-mouse inventory cost/year is IDR 204,245.00 [14].

The MRP method used at PT. Lixil Aluminium Indonesia has not been cost-efficient, both ordering costs and storage costs can be seen in the method used in 2022, namely FOQ with safety stock additions then accumulated with the Top Management policy, from the analysis with 2 MRP methods that have been carried out, there is a cost difference, namely with the FOQ results of Rp689,994,712.00 and for LFL Rp652,748,558.00 with a cost efficiency difference of Rp37,246,154.00 for the coming year [15].

This research was carried out on the work of the substructure of the Inpres Market Development Project located in Painan, Pesisir Selatan. Material management is needed in this project because the material is imported from Padang City which is 70 km from the project location. Road access to the project location is often cut off if rainfall, because the topography of the surrounding area causes flooding and landslides. Of course, this problem will have an impact on project completion time and costs. There are 2 methods that will be analyzed in this research, namely the actual field method and the lot for lot technique (LFL). The purpose of this study is to compare *the lot for lot* (LFL) technique with the field method on material management consist of of purchase, ordering and storage costs.

## 2 Research methods

### 2.1 Data collection

Primary data is obtained from observations in the field, interviews and project documentation. Observation by making direct observations of the stages of the implementation of the substructure work, Project data includes documentation of administrative forms related to the procurement of substructure materials, work plan drawings, technical specifications, *bill of materials*, *time schedule* and data related to logistics/procurement of project materials. Interviews were conducted directly with construction management consultants and contractors during field visits, to obtain information on the material procurement process, data related to *suppliers,* infrastructure conditions during the material *delivery* process, and what obstacles often occur during the material management process at the project site.

### 2.2 Data processing

Using *the lot for lot* (LFL) technique, material planning begins by calculating the quantity of material for the substructure work which includes bore *pile, pile cap* and *sloof foundation work.* Furthermore, calculate the cost of

material needs, namely the cost of purchasing materials, ordering costs and storage costs. For the calculation of the quantity of ironing using *a bar bending schedule* (BBS). Furthermore, the application of *the lot for lot* (LFL) technique with the consideration of minimization of storage costs, where the inventory cost is zero and the amount ordered is equal to the amount needed [4].
The targets of the use of this technique [16] are:

1. Reduction of the amount of inventory, by specifying how many components are needed and when they are needed, materials are supplied only when needed so that the cost of overstocking can be avoided.
2. Production reduction and delivery lead time, by identifying the amount of material needed, delivery time, availability, procurement and production to complete at the time needed to be delivered.
3. Increase efficiency, by providing close coordination between the various work centers involved in the production process. As a result, production can run more efficiently.

Furthermore, an analysis was carried out by comparing the cost of material management of *the lot for lot* (LFL) technique with the actual material cost of the field.

## 3 Results

### 3.1 Actual plan

The actual plan was obtained from field data, which consisted of purchase costs and material storage costs for the construction of the lower structure of the Painan Inpres market project. The purchase cost can be seen in Table 1, the order cost does not exist because it is included with the purchase cost, the storage cost is obtained based on the duration of the time the material is stored from the field data, it can be seen in Table 2.

**Table 1.** Cost of purchasing actual *substructure* plan materials

| No | Material | Volume | Unit | Material price /unit (Rp) | Total fees (Rp) |
|---|---|---|---|---|---|
| A | FOUNDATION D600mm | | | | |
| 1 | 5m Steel Pipe | 238 | unit | 13050000 | 3,105,900,000.00 |
| 2 | 12m Steel Pipe | 57 | unit | 31320000 | 1,785,240,000.00 |
| 3 | Reinforcement Bar D10 | 1883 | pieces | 78000 | 146,874,000.00 |
| 4 | Reinforcement Bar D19 | 1883 | pieces | 285000 | 536,655,000.00 |
| 5 | Readymix Concrete fc' 24,9 Mpa | 453.86 | m3 | 1671000 | 758,400,060.00 |
| B | PILE CAP | 0 | | | |
| 1 | Reinforcement Bar D22 | 1718 | pieces | 385000 | 661,430,000.00 |
| 2 | Multiplex12 mm | 152 | pieces | 180900 | 27,496,800.00 |
| 3 | Wood 5/7 | 852 | pieces | 35000 | 29,820,000.00 |
| 4 | Beton Readymix Fc 24,9 Mpa | 477.22 | m3 | 1671000 | 797,434,620.00 |
| C | SLOOF | | | | |
| 1 | Reinforcement Bar D10 | 790 | pieces | 78000 | 61,620,000.00 |
| 2 | Reinforcement Bar D19 | 683 | pieces | 285000 | 194,655,000.00 |
| 3 | Reinforcement Bar D25 | 660 | pieces | 510000 | 336,600,000.00 |
| 4 | Multiplex 12 mm | 92 | pieces | 180900 | 16,642,800.00 |
| 5 | Wood 5/7 | 514 | pieces | 35000 | 17,990,000.00 |
| 6 | Readymix Concrete fc' 24,9 Mpa | 250.87 | m3 | 1671000 | 419,202,366.36 |
| | Total cost of purchasing material | | | | 8,895,960,646.36 |

**Table 2.** Cost of material storage of actual *substructure* plan

| No | Material Name | Volume | Unit | Total Inventory | Storage Fees /unit | Total Storage Costs |
|---|---|---|---|---|---|---|
| 1 | 5m Steel Pipe | 238 | unit | 623.00 | 2,888.11 | 1,799,295.49 |
| 2 | 12m Steel Pipe | 57 | unit | 138.50 | 6,931.48 | 960,009.34 |
| 3 | Reinforcement Bar D10 | 2673 | pieces | 14,467.20 | 17.26 | 249,737.08 |
| 4 | Reinforcement Bar D19 | 2566 | pieces | 1,738.00 | 77.09 | 133,982.70 |
| 5 | Reinforcement Bar D22 | 1718 | pieces | 4,495.00 | 85.20 | 382,996.11 |
| 6 | Reinforcement Bar D25 | 660 | pieces | 1,980.00 | 112.87 | 223,480.33 |
| 7 | Multiplex 12 mm | 244 | pieces | 1,967.00 | 14.83 | 29,166.42 |
| 8 | Wood 5/7 | 1366 | pieces | 10,988.00 | 865.57 | 9,510,924.59 |
| 9 | Readymix Concrete fc 24,9 Mpa | 1182 | m3 | 0.00 | 451.99 | 0.00 |
| | Total storage costs | | | | | 13,289,592.06 |

**Table 3.** Recapitulation of actual *sub-structure* material costs plan

| No | Cost components | Amount (Rp) |
|---|---|---|
| 1 | Purchase fee | 8,895,960,646.36 |
| 2 | Booking fee | 0 |
| 3 | Storage costs | 13,289,592.06 |
| | Total material costs actual field | 8,909,250,238.42 |

Based on Table 1, the cost of purchasing the actual *sub structure* material plan was Rp8,895,960,646.36. From Table 2, the actual *sub-structure* plan material storage cost is Rp13,289,592.06,

From Table 3, the total recapitulation of the material cost of the lower structure was obtained at Rp8,909,250,238.42, From here, it can be seen that the order fee is zero because the order fee is included in the cost of purchasing materials to the project site. So, the total material cost *of the actual sub-structure plan is only the purchase cost and storage costs.*

## 3.2 Technical lot for lot (LFL)

### *3.2.1 Sub structure cost LFL*

The calculation of the components of the cost of purchasing materials for the lower structure (foundation *bore pile, pile cap* and *sloof*) can be seen in Table 4.

**Table 4.** Cost of purchasing materials *substructure* lot for lot *technique*

| No | Material | Amount | Unit | Material price /unit (Rp) | Purchase fee amount (Rp) |
|---|---|---|---|---|---|
| A | FOUNDATION D600mm | | | | |
| 1 | 5m Pipe Steel | 235 | unit | 13,050,000 | 3,066,750,000 |
| 2 | 12m Pipe Steel | 55 | unit | 31,320,000 | 1,722,600,000 |
| 3 | Reinforcement Bar D10 | 1880 | pieces | 78,000 | 146,640,000 |
| 4 | Reinforcement Bar D19 | 1880 | pieces | 285,000 | 535,800,000 |
| 5 | Readymix Concrete fc' 24,9 Mpa | 453.86 | m3 | 1,671,000 | 758,400,060 |
| B | PILE CAP | | | | - |
| 1 | Reinforcement Bar D22 | 1716 | pieces | 385,000 | 660,660,000 |
| 2 | Multiplex 12 mm | 150 | pieces | 180,900 | 27,135,000 |
| 3 | Wood 5/7 | 849 | m3 | 35,000 | 29,715,000 |
| 4 | Readymix Concrete fc 24,9 Mpa | 477.22 | m3 | 1,671,000 | 797,434,620 |
| C | SLOOF | | | | - |
| 1 | Reinforcement Bar D10 | 788 | pieces | 78,000 | 61,464,000 |
| 2 | Reinforcement Bar D19 | 681 | pieces | 285,000 | 194,085,000 |
| 3 | Reinforcement Bar D25 | 658 | pieces | 510,000 | 335,580,000 |
| 4 | Multiplex 12 mm | 90 | pieces | 180,900 | 16,281,000 |
| 5 | Wood 5/7 | 512 | m3 | 35,000 | 17,920,000 |
| 6 | Readymix Concrete fc 24,9 Mpa | 250.87 | m3 | 1,671,000 | 419,203,770 |
| | Total cost of material purchase | | | | 8,789,667,046 |

Based on Table 4, the cost of purchasing materials is Rp12,171,753,450.00, this fee is only in the form of purchase costs at the store/distributor, not the cost to the project site.

### *3.2.2 Cost of ordering LFL sub structure materials*

The booking fee per unit can be seen in Table 5, the booking cost of *the lot for lot* (LFL) technique based on the time period can be seen in Table 6.

In Table 5, it can be seen that the booking fee consists of telephone fees and administrative fees. The total order cost is obtained by calculating the number of message frequencies with the cost of ordering materials per unit, obtained by Rp2,087,880.00.

**Table 5.** Sub structure material ordering fee per unit

| No | Material | Total phone charges (Rp) | Total phone charges (Rp) | Total booking fee (Rp) |
|---|---|---|---|---|
| A | FOUNDATION D600mm | | | |
| 1 | 5m Steel Pipe | 13,440 | 3,000 | 16,440 |
| 2 | 12m Steel Pipe | 13,440 | 3,000 | 16,440 |
| 3 | Reinforcement Bar D10 | 13,440 | 3,000 | 16,440 |
| 4 | Reinforcement Bar D19 | 13,440 | 3,000 | 16,440 |
| 5 | Readymix Concrete fc 24,9 Mpa | 13,440 | 3,000 | 16,440 |
| B | PILE CAP | | | |
| 1 | Reinforcement Bar D22 | 13,440 | 3,000 | 16,440 |
| 2 | Multiplex 12 mm | 13,440 | 3,000 | 16,440 |
| 3 | Wood 5/7 | 13,440 | 3,000 | 16,440 |
| 4 | Readymix Concrete fc 24,9 Mpa | 13,440 | 3,000 | 16,440 |
| C | SLOOF | | | |
| 1 | Reinforcement Bar D10 | 13,440 | 3,000 | 16,440 |
| 2 | Reinforcement Bar D19 | 13,440 | 3,000 | 16,440 |
| 3 | Reinforcement Bar D25 | 13,440 | 3,000 | 16,440 |
| 4 | Multiplex 12 mm | 13,440 | 3,000 | 16,440 |
| 5 | Wood 5/7 | 13,440 | 3,000 | 16,440 |
| 6 | Readymix Concrete fc 24,9 Mpa | 13,440 | 3,000 | 16,440 |

**Table 6.** Cost of ordering *sub structure* materials with *lot for lot technique*

| No | Material | Message frequency | Booking fee (Rp) | Total booking fee (Rp) |
|---|---|---|---|---|
| A | PONDASI D600mm | | | |
| 1 | 5m Steel Pipe | 10 | 16,440 | 164,400 |
| 2 | 12m Steel Pipe | 10 | 16,440 | 164,400 |
| 3 | Reinforcement Bar D10 | 12 | 16,440 | 197,280 |
| 4 | Reinforcement Bar D19 | 12 | 16,440 | 197,280 |
| 5 | Readymix Concrete fc 24,9 Mpa | 12 | 16,440 | 197,280 |
| B | PILE CAP | | | |
| 1 | Reinforcement Bar D22 | 11 | 16,440 | 180,840 |
| 2 | Multiplex 12 mm | 12 | 16,440 | 197,280 |
| 3 | Wood 5/7 | 12 | 16,440 | 197,280 |
| 4 | Readymix Concrete Fc 24,9 Mpa | 11 | 16,440 | 180,840 |
| C | SLOOF | | | |
| 1 | Reinforcement Bar D10 | 4 | 16,440 | 65,760 |
| 2 | Reinforcement Bar D19 | 4 | 16,440 | 65,760 |
| 3 | Reinforcement Bar D25 | 4 | 16,440 | 65,760 |
| 4 | Multiplex 12 mm | 4 | 16,440 | 65,760 |
| 5 | Wood 5/7 | 4 | 16,440 | 65,760 |
| 6 | Readymix Concrete fc 24,9 Mpa | 5 | 16,440 | 82,200 |
| | Total cost of ordering materials | | | 2,087,880 |

### *3.2.3 LFL substructure material storage costs*

The cost of material storage can be seen in Table 7. This storage cost is a cost incurred as a result of having inventory or goods stored and the unit size is Rp/unit in a unit of time [8].

The calculation of storage costs is carried out because it has an inventory of capital costs and damage or depreciation costs. The cost of capital can be measured by BI's December 2020 interest rate of 3.75% per year. The cost of depreciation and material damage during storage is assumed to be 0.5% of the material price per unit for ferrous materials [17]. Then for other materials by 2% [18].

**Table 7.** Substructure material storage costs

| No | Material | Amount | % Storage fees | Material price /unit (Rp) | Storage price /unit (Rp) | Total storage fees (Rp) |
|---|---|---|---|---|---|---|
| A | FOUNDATION D600mm | | | | | |
| 1 | 5m Steel Pipe | 235 | 8.25 | 13,050,000 | 2,888.11 | 678,705.85 |
| 2 | 12m Steel Pipe | 55 | 8.25 | 31,320,000 | 6,931.48 | 381,231.40 |
| 3 | Reinforcement Bar D10 | 1880 | 6.75 | 78,000 | 21.1 | 39,668.00 |
| 4 | Reinforcement Bar D19 | 1880 | 6.75 | 285,000 | 77.09 | 144,929.20 |
| 5 | Readymix Concrete fc 24,9 | 46 | 6.75 | 1,671,000 | 451.99 | 20,791.54 |
| B | PILE CAP | | | | | |
| 1 | Reinforcement Bar D22 | 1716 | 6.75 | 385,000 | 85.2 | 146,203.20 |
| 2 | Multiplex 12 mm | 150 | 8.25 | 180,900 | 14.83 | 2,224.50 |
| 3 | Wood 5/7 | 849 | 8.25 | 3,200,000 | 865.57 | 734,868.93 |
| 4 | Readymix Concrete Fc 24,9 | 48 | 8.25 | 1,671,000 | 451.99 | 21,695.52 |
| C | SLOOF | | | | | |
| 1 | Reinforcement D10 | 788 | 675 | 78,000 | 17.26 | 13,600.88 |
| 2 | Reinforcement D19 | 681 | 675 | 285,000 | 63.07 | 42,950.67 |
| 3 | Wood D25 | 658 | 6.75 | 510,000 | 112.87 | 74,268.46 |
| 4 | Multiplex 12 mm | 90 | 8.25 | 180,900 | 48.93 | 4,403.70 |
| 5 | Wood 5/7 | 512 | 8.25 | 3,200,000 | 865.57 | 443,171.84 |
| 6 | Readymix Concrete Fc 24,9 Mpa | 22 | 8.25 | 1,671,000 | 45.99 | 9,943.78 |
| | Total cost of ordering materials | | | | | 2,758,657.47 |

**Table 8.** Cost of substructure material storage *with lot for lot technique*

| No | Material | Unit | Inventory | Storage price /unit (Rp) | Total storage fees (Rp) |
|---|---|---|---|---|---|
| A | FOUNDATION D600mm | | | | |
| 1 | 5m Steel Pipe | unit | 0 | 2,888.11 | |
| 2 | 12m Steel Pipe | pieces | 0 | 6,931.48 | |
| 3 | Reinforcement Bar D10 | pieces | 0 | 21.10 | |
| 4 | Reinforcement Bar D19 | Unit | 0 | 77.09 | |
| 5 | Readymix Concrete fc' 24,9 Mpa | m3 | 0 | 451.99 | |
| B | PILE CAP | | | | |
| 1 | Reinforcement Bar D22 | pieces | 0 | 85.20 | |
| 2 | Multiplex 12 mm | pieces | 0 | 14.83 | |
| 3 | Wood 5/7 | m3 | 0 | 865.57 | |
| 4 | Readymix Concrete fc' 24,9 Mpa | m3 | 0 | 451.99 | |
| C | SLOOF | | | | |
| 1 | Reinforcement Bar D10 | pieces | 0 | 17.26 | |
| 2 | Reinforcement Bar D19 | pieces | 0 | 63.07 | |
| 3 | Reinforcement Bar D25 | pieces | 0 | 112.87 | |
| 4 | Multiplex 12 mm | pieces | 0 | 48.93 | |
| 5 | Wood 5/7 | m3 | 0 | 865.57 | |
| 6 | Readymix Concrete fc' 24,9 Mpa | m3 | 0 | 451.99 | |
| | Total cost of material storage | | | | 0 |

The cost of capital can be measured by BI's December 2020 interest rate of 3.75% per year. The cost of depreciation and material damage during storage is assumed to be 0.5% of the material price per unit for ferrous materials [17]. Then for other materials by 2% [18].

Based on *the concept of lot for lot*, storage costs are costs obtained from the multiplication of the amount of *inventory* with the storage cost per unit per day, where inventory costs are zero. The cost of material storage with *the lot for lot* technique can be seen in Table 8.

From Table 7, it can be seen that the cost of material storage is Rp2,758,657.47, with material management lot-for-lot technique, inventory costs are zero, so there are no storage costs (Table 8). In other words, after the material is purchased, ordered, immediately used in the field.

The recapitulation of sub-structure material costs *with the lot for lot* (LFL) technique is shown in Table 9.

**Table 9.** Recapitulation of material cost components substructure with *LFL technique*

| No | Cost components | Amount (Rp) |
|---|---|---|
| 1 | Purchase fee | 8,789,667,046 |
| 2 | Booking fee | 2,087,880 |
| 3 | Storage fees | 0 |
| | Total cost of LFL engineering | 8,791,754,926 |

From Table 9, it can be seen that the total material cost of the sub structure with *the lot for lot* (LFL) technique is Rp8,791,754,926.00.

## 4 Discussion

The cost components reviewed in the material management costs of the lower structure of the Inpres Market Project, Painan, Pesisir Selatan Regency are material purchase costs, material ordering costs and material storage costs. Based on the results of the recapitulation of the total actual material costs of the field in Table 3, a cost of Rp8,909,250,238.42 was obtained. The actual cost of the field consists of purchase costs and storage costs, while the order cost is considered zero because it is included in the purchase cost to the project site.

Meanwhile, the recapitulation of the total material cost of *the lot for lot* (LFL) technique in Table 9, was obtained at Rp8,791,754,926.00. Thus, the difference in cost is Rp117,495,312.42. Meanwhile for the lot for lot (LFL) technique, the cost is 1,32% cheaper than the actual cost of the field. This is because the inventory cost in the lot-for-lot technique is zero, so there are no storage costs. In other words, when the material arrives at the project site, the material is immediately used.

## 5 Conclusion

Comparison material management of actual field method and lot for lot technique on Inpres Market Development Project Painan are needed alternative solution if material problems occur in the field, which are caused by human error and disaster factors. In the work of the substructure of this project, the actual material cost component of the field consists of purchase costs including orders of Rp8,895,960,646.36, storage costs of Rp8,895,960,646.36. The total actual material cost of the field is Rp8,909,250,238.42. On the other hand, for the lot for lot technique, a purchase cost component of Rp12,171,753,450.00, the booking fee is Rp2,087,880.00 and the storage fee is zero. The total cost of lot for lot engineering materials was Rp8,789,667,046, 1.34% cheaper than the actual cost of the field.

## References

1. L. C. Bell, G. Stukhart, Attributes of materials management systems. Journal of Construction Engineering and Management. **112**(1), (1986). https://doi.org/10.1061/(ASCE)0733-9364(1986)112:1(14)
2. J. G. Tamatompol, A. K. T. Dundu, and J. B. Mangare, Material management analysis using lot for lot. TEKNO. **21**(83), 89-97 (2023). https://doi.org/10.35793/jts.v21i83.46677
3. J. Atmaja, Riswandi, M. Natalia, R. Maifrianti, F. Hidayati, Comparison of the cost of sub structure manual method and material requirement planning method in apartment buildings. Jurnal Ilmiah Poli Rekayasa. **18**(2), 46-53 (2023). http://dx.doi.org/10.30630/jipr.18.2.288
4. S. S. Riskijah, Analysis of reinforced concrete material inventory. PROKONS: J. Tek. Sipil. **13**(1), 43-49 (2019).
5. A. Maddeppungeng, D.N. Setiawati, B. Tuqa, Material inventory planning using the material requirement planning (MRP) method in apartment projects (case study: Nines Plaza & Residence Tower B development project). Fondasi: J. Tek. Sipil. **10**(1), 69-80 (2021).
http://dx.doi.org/10.36055/fondasi.v10i1.10624
6. R. Y. Rampi, J. B. Mangare, and T. T. Arsjad, Control of building material inventory costs with the economic order quantity method. J. Sipil Statik. **6**(11), 949-958 (2018).
7. Nasution (2008), in A. N. Savira, K. D. Handayani, Analysis of material inventory using the material requirement planning (MRP) method on the basement floor structure work 2 of The Westown View Wiyung Surabaya Apartment construction project. Rekayasa Teknik Sipil. **1**(1), 1-11 (2021).
8. M. Prof. Dr. Ir. H. Sutarman, Logistics Management Basics ( PT Refika Aditama, Bandung, 2017).
9. S. A. Aziz, A. Sutoni, Inventory Analysis in Building Renovation Projects Using the Material Requirements Planning Method with the Lot for Lot Technique, in Seminar dan Konferensi Nasional IDEC, Surakarta, (2019).
10. A. N. Savira, K. D. Handayani, Analysis of material inventory using the material requirement planning (MRP) method on the basement floor structure work

2 of The Westown View Wiyung Surabaya Apartment construction project. Rekayasa Teknik Sipil. **1**(1), 1-11, (2021).

11. M. A. I. Ramadhan, Rafie, S. M. Nuh, Material planning with the application of the material requirement planning (MRP) method with the use of lot for lot (LFL) and period order quantity (POQ) as the sizing lot technique in material control, Jurnal Mahasiswa Teknik Sipil Fakultas Teknik Universitas Tanjungpura Pontianak, (2018).
12. S. I. Taifur, T. S. Imaroh, Forecasting planning and procurement strategy of raw material using material requirements planning method, DIJDBM: Dinasti International Journal of Digital Business Management. **1**(4), 593-607 (2020). https://doi.org/10.31933/dijdbm.v1i4.369
13. W. Florim, P. Dias, Analysis of lot-sizing methods' suitability for different manufacturing application scenarios oriented to MRP and Jit/Kanban environments, Brazilian Journal of Operation and Production Management. **16**(4), 638-649 (2019). https://doi.org/10.14488/BJOPM.2019.v16.n4.a9
14. E. F. Sinuhaji, Lot sizing material requirement planning produk wire-mouse, TALENTA Conf. Series: Energy & Engineering, **2**(2), 729-732 (2022). https://doi.org/10.32734/ee.v5i2.1643
15. N. A. Muhanto, K. Supatmantyo, Analysis of planning and control of raw materials with the material requirement planning method at PT. Lixil Aluminum Indonesia, Formosa J. of Multidisciplinary Research. **2**(5), 1063-1072 (2023). https://doi.org/10.55927/fjmr.v2i5.4655
16. F. Rangkuti. Inventory Management: Applications in the Business Field (Jakarta: PT Raja Grafindo. 1995).
17. E. Pancawati, Material Inventory Planning for The Construction of Trillium Office & Residence Surabaya, (2011).
18. Pandia and Einsteinius (2018), in A. N. Savira, K. D. Handayani, Analysis of material inventory using the material requirement planning (MRP) method on the basement floor structure work 2 of The Westown View Wiyung Surabaya Apartment construction project. Rekayasa Teknik Sipil. **1**(1), 1-11, (2021)

# Assessment of electricity supply adequacy for tsunami refuge shelter planning in Universitas Andalas

*Arif* Budikusuma[1], *Adrianti* Adrianti[1*], *Muhammad* Nasir[1], *Riko* Nofendra[1], and *Nur* Hakimah Ab Aziz[2]

[1]Department of Electrical Engineering, Faculty of Engineering, Universitas Andalas, Padang 25163, West Sumatra, Indonesia
[2]Department of Electrical Engineering, Universiti Teknikal Malaysia Melaka, 76100 Durian Tunggal, Melaka, Malaysia

**Abstract.** Padang City is an earthquake-tsunami-prone area; therefore, preparedness to anticipate disasters is essential. The city has made evacuation routes and provided Temporary Evacuation Shelters and Final Evacuation Shelters. Universitas Andalas (Unand) is in the safe zone of a tsunami disaster and can be used as a final evacuation shelter. When a disaster occurs, most of the electrical network experiences damage. Hence, only local electrical power supply can be relied on. Unand has gensets and PV that can supply the refugee shelters. This study aims to analyse Unand's power supply ability to provide electricity for planned refugee shelters. The research started with shelter location selection, measuring shelter area, determining shelter capacity, calculating shelter electric demand, and analysing the local supply adequacy in serving the electric demand. The results show that the shelters can accommodate 3704 refugees, and the peak demand is 55.1 Kw in the evenings. The peak demand is far lower than the power supply's maximum output. Then, based on the demand and generating output, power flow simulations are carried out to determine the electrical conditions in the network. It is found that Unand can meet the refugee's electrical power needs and the electricity quality is within standard.

## 1 Introduction

Padang City is the capital of West Sumatra Province, which, according to geologists, is one of the earthquake-prone areas. This is because Padang City is located between two active earthquake sources, namely the Australian plate and the Eurasian plate. Judging from history, West Sumatra Province has experienced several tsunamis. In 1797, a tsunami disaster was triggered by an underwater landslide due to a previous earthquake. The height of the tsunami waves was estimated at 5-10 meters and about 1 km towards the mainland caused by an earthquake with a magnitude of 8.4 SR [1]. The subsequent tsunami disaster occurred in 1833 and was caused by the rupture of the 1000-kilometre-long Sumatran trench, which produced an earthquake with a magnitude of 9.0 SR. The tsunami had a water height of 2-3 meters [1]. Therefore, preparedness to anticipate disasters is crucial, considering that if a significant earthquake occurs followed by a tsunami, the risk of danger will be very high because Padang City is located on the coast with a relatively high concentration of people living in coastal areas.

One way to reduce the impact of a disaster is to evacuate. Tsunami evacuation methods are divided into two, namely horizontal evacuation and vertical evacuation. Horizontal evacuation is an evacuation carried out by the community by moving away from the coastal area to the tsunami-safe zone. Vertical evacuation is interpreted as an effort to save oneself by moving to a higher place, which can be a hill or a building with more than one floor [2].

Padang City has made disaster management efforts by creating evacuation routes and providing TES (Temporary Evacuation Shelters) and FES (Final Evacuation Shelters). Temporary evacuation sites (TES) are buildings built for tsunami-prone areas as evacuation sites and shelters if a disaster occurs. TES used as evacuation sites are in the form of existing buildings or new buildings dedicated to temporary evacuation sites, or can be natural TES such as highlands, natural or artificial hills [2].

The final evacuation shelters (FES) are places for family members to gather after a disaster occurs and places for refugees to get assistance and carry out daily household activities until the recovery process begins. Activities at the FES are carried out for people who have lost their belongings after the tsunami disaster occurred [3].

Universitas Andalas (Unand) is one of the campuses in Padang City located in the green zone from the tsunami disaster because Unand is situated in a high hilly area far from the coast. In addition, there is sufficient open space, making it a suitable location for final evacuation shelters.

Many of the electricity networks were damaged by the earthquake and tsunami. Hence, the electricity supply was disrupted for several days and even weeks. To obtain electricity for daily activities for the evacuation sites, local generators such as gensets or renewable energy supply such as photovoltaics are needed in the evacuation

* Corresponding author: adrianti@eng.unand.ac.id

site when the power supply from an electric power company cannot be used

On the Unand campus, there are two gensets with a size of 200 KW each and good global irradiation potential for installing solar panels. Unand will install solar panels, which can be used as an alternative energy source to save electricity from PLN. With a large area and adequate facilities such as clean water, generators, and solar panels, Unand can be a final evacuation site for refugees.

This study aims to analyse the adequacy of electrical power from the local power supply available at Unand for the refugees' shelters. To perform this analysis, the total electrical demand at the refugee shelter should be known. The electrical demand is determined by the number of refugees that can be accommodated at the Unand shelter. The number of refugees depends on the available area in Unand as a refuge shelter.

Therefore, based on this sequence of data requirements, the study will have a research sequence:

a. determine the location of the refugee shelter and its area in Unand.
b. determine the number of refugees that can be accommodated at the Unand Shelter.
c. determine the total daily electrical demand at the shelter.
d. analyse the adequacy of electrical power for refugee shelters using local electric power supply at Unand

## 2 Refugee shelter standards

Based on the area planned as a shelter, 30% is used as open space and public facilities. This figure of 30% is lower than the standard set by Sphere Humanitarian, which is 50% [4]. The open space and public facilities are designed to provide a comfortable and safe environment for the refugees, allowing for recreational activities and the provision of essential services. The 30% area will produce higher electric demand because more refugees can be accommodated. The number of refugees that can be accommodated in an area is calculated using equation (1) for the emergency shelter of UNHCR standard [5]. This UNHCR standard states a living space is at least 3.5m$^2$ for a person in tropical or warm climates.

$$N_R = \frac{A}{3,5} \quad (1)$$

Where, $N_R$ is refugee number, and A is an area in m$^2$. In refugee shelters, facilities must be available to meet the refugees' needs. These needs include the availability of electricity, clean water, food, sanitation, and health services.

## 3 Electricity demand

Electricity in the evacuation site is used for lighting, water pumps for clean water, freezers for storing food, and refugee gadget charging. Lighting consists of indoor and outdoor lighting. Indoor lighting is included in shelters for refugees to stay/sleep and in public spaces such as toilets, bathrooms, and kitchens. Outdoor lightings provide convenient and safe areas surrounding the shelters.

The electrical power requirement for lighting is calculated using the equation (2) and (3)

$$\Phi_0 = \frac{E_V \times A}{\eta} \quad (2)$$

Where, $\Phi_0$ is total luminous flux in a room (lumen), A is the area of the room (m$^2$), $\eta$ is lamp efficiency, $E_v$ is illuminance (lux) [6]. The illuminance standard for a shelter room is taken as 250 lux, which is a standard for home lighting in general room [6]. Equation (2) results are used in equation (3) to calculate the required electrical power.

$$P = \frac{\Phi_0}{Efficacy} \quad (3)$$

Where P is electric power consumption (watt), efficacy is lumen/watt.

LED is the most efficient type of interior light bulb at the moment. Therefore, LED light bulbs are used in this planning. Most of the efficacies of LEDs are between 40-80 lumen/watt. Hence, in this research, it is assumed that the lights have an efficacy of 75 lumen/watt [7].

The most common personal gadgets are mobile phones; thus, it is assumed that mobile phones are the only gadgets in the shelters and are owned by the refugees. The electrical energy mobile phones consume is calculated with equation (4).

$$E = P_{phone} \times 0{,}8N_R \times t \quad (4)$$

Where E is electrical energy in watt-hours, $P_{phone}$ is the average electric power consumed by mobile phone chargings (watts), $N_R$ is the number of refugees and t is the duration of mobile phones charging (hours). The power consumption of mobile phone chargers is around 5 to 25 watts, depending on whether they are slow or fast chargers. The median power of 15 watts is used for the calculation, along with 1.5 hours of daily charging duration. It is also assumed that 80% of refugees use mobile phones daily.

## 4 Methodology

Research methodology can be divided into four main steps, i.e., the determination of the location of the refugee shelter, the number of refugees, the total daily electrical demand, and the adequacy of electrical power. The details of each step are explained as follows.

### 4.1 Location of the refugee shelters

The location selection for the evacuation shelter is determined based on the availability of sufficient area and the ability to connect to the Unand underground electricity network. This requirement is based on the 30 September 2009 Padang earthquake, which showed that the Unand 20 kV underground distribution lines did not experience problems due to the earthquake. Therefore, it is assumed that if a tsunami occurs in Padang, the distribution lines are still functioning correctly.

This study used two shelter conditions: 1) Only outdoor and 2) Outdoor and indoor areas. The only outdoor areas are carried out if no buildings are safe after the earthquake. While for outdoor and indoor conditions, it is assumed that some short buildings are still technically feasible for humans to live in.

### 4.2 Number of refugees

Refuge numbers that the shelters can accommodate are calculated based on the area of the shelter. The area of candidate shelters is measured with the help of Google Maps. Fig. 1. shows area measurement of the Language Centre parking area using Google Maps. Then, the number of refugees is calculated using equation (1) in section 2, where A is 70% of the area that was measured with the help of Google Maps.

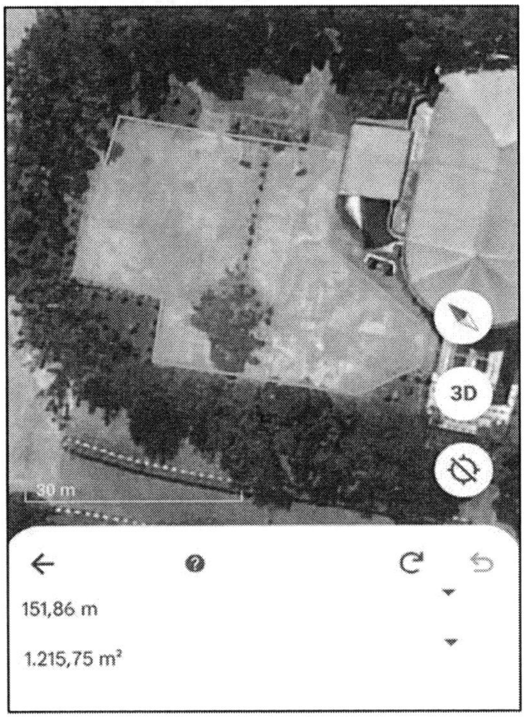


**Fig. 1.** Area measurement of shelter location

### 4.3 Total daily electrical demand

Electrical demand in the evacuation site consists of lighting, water pumps, a freezer, and mobile phone chargers. These demands are the fundamental electricity needs due to the limited power supply that may be available.

Lighting energy consumption (E) is calculated using the combination of equations (2) and (3) and multiplied by time (t) in hours, which results in equation (5).

$$E = \frac{E_v A t}{efficacy} \tag{5}$$

The lighting is assumed to be on for 13 hours daily, from 5.00 PM to 06.00 AM.

An electric water pump is required for each shelter location to supply water to the toilet, bathroom, and public kitchen. The pump's electric power consumption is 2000 watts. Each water pump is operated for around seven hours a day. A freezer in a public kitchen is used for food storage and consumption of 760-watt power.

Energy consumption of mobile phones is calculated using equation (4), where the charging time is distributed during the day from 6 am to 9 pm. The scenario of electric power consumption each hour of the day will make the electric load profile of each site. This load profile is then used as load power for power flow analysis using Digsilent Power factory.

### 4.4 Adequacy of electrical power supply

On the Unand Limau Manih Campus, there are two gensets and Photovoltaic. The gensets have a capacity of 250 KVA each. The photovoltaic systems are located at the Rectorate and Library building, which have a capacity of 149.6 Kwp and 58.3 Kwp, respectively, as shown in Table 1.

**Table 1.** Power supply for the shelters

| Power supply | Capacity | Power factor | Max output (kW) |
|---|---|---|---|
| Genset 1 | 250 kVA | 0.8 | 200 |
| Genset 2 | 250 kVA | 0.8 | 200 |
| PV rectorate | 149.6 kWp | 1 | 115.2 |
| PV library | 58.3 kWp | 1 | 44.9 |

The maximum power output of PV in the Unand region is approximately 77% of its watt peak, based on experiment results in [8]. Daily PV energy output is calculated from the total yearly PV energy output divided by the number of days a year. The daily energy output is multiplied by a percentage to calculate the hourly power output. The percentage is calculated based on sunny day solar irradiation, where 100% is the total percentage from 7 am to 6 pm, and each hour has a percentage comparable with the hour solar irradiation. The total yearly output of the PV is obtained from a feasibility study conducted by a consultant.

The amount of power from the genset and PV at a particular time of the day is compared with the total demand. The distribution network cannot deliver enough electricity if the demand exceeds the power supply. However, to ensure that the quality of the electricity meets standards (voltage magnitude), a power flow analysis is carried out. Power simulation is executed using Digsilent power factory [9].

## 5 Results and discussion

### 5.1 Sites selection and capacity

Site selection is based on the availability of sufficient area, their security and distance to the UNAND underground electric distribution network. The following locations are then selected:

- Outdoor sites: Unand football field, language centre parking area, Convention Hall parking area, Auditorium parking area, library parking area, and Nurul Ilmi Mosque yard. Indoor sites: Convention Hall and Auditorium Building

The area and capacity of refugees from the chosen sites are shown in Table 2. Table 2 also indicates location codes for each site.

**Table 2**. Area and capacity of shelter sites

| Location (code) | Area ($m^2$) | Living space ($m^2$) | Capacity (refugee number) |
|---|---|---|---|
| Convention Hall (1) | 1017 | 711,9 | 203 |
| Auditorium (2) | 3433 | 2403.1 | 687 |
| Football field (3) | 9622 | 6735.4 | 1923 |
| Language Ctr parking (4) | 365 | 255.5 | 73 |
| Convention Hall parking (5) | 1215 | 850.5 | 243 |
| Auditorium Parking (6) | 932 | 652.4 | 186 |
| Library parking (7) | 1946 | 1362.2 | 389 |
| Mosque yard (8) | 2696 | - | 0 |
| Total refugee number | | | 3704 |
| Total refugee number if only outdoor site | | | 2814 |

From Table 2, it can be seen that no refugee is located in the mosque yard. This is because the mosque will be used as a public kitchen, where food for the refugees is cooked. The biggest shelter is in the Football field, which can accommodate 1923 people. The part of shelter areas not used as living space is used for services such as toilets, bathrooms, and washing areas.

## 5.2 Electric demand

Lighting is the most significant demand on the sites. Using equation (4) for 1-hour duration of t, the demand for lighting from 5 pm to 6 am is shown in Table 3.

**Table 3**. Lighting demand

| Location | Lighting demand (watt) |
|---|---|
| Convention Hall | 2373 |
| Auditorium | 8010 |
| Football field | 22451 |
| Language Ctr parking | 852 |
| Convention Hall parking | 2373 |
| Auditorium Parking | 2175 |
| Library parking | 4540 |
| Mosque | 2710 |
| Total Lighting | 45484 |

The hourly electric consumption of lighting, water pumps, gadget chargers and freezers are calculated for each site. The total electric demand for all outdoor locations is shown in Table 4.

**Table 4.** Daily electric demand for each outdoor site (kW)

| hour | Location code | | | | | |
|---|---|---|---|---|---|---|
| | 3 | 4 | 5 | 6 | 7 | 8 |
| 06-07 | 2.866 | 2.033 | 2.11 | 2.084 | 2.18 | 0.760 |
| 07-08 | 3.732 | 2.667 | 2.22 | 2.167 | 2.35 | 0.760 |
| 08-09 | 6.329 | 2.166 | 2.55 | 2.418 | 2.88 | 0.760 |
| 09-10 | 2.866 | 2.033 | 2.11 | 2.084 | 2.18 | 0.760 |
| 10-11 | 3.732 | 2.067 | 2.22 | 2.167 | 2.35 | 0.760 |
| 11-12 | 6.329 | 2.166 | 2.55 | 2.418 | 2.88 | 0.760 |
| 12-13 | 2.866 | 2.033 | 2.11 | 2.084 | 2.18 | 0.760 |
| 13-14 | 1.732 | 0.067 | 0.22 | 0.167 | 0.35 | 3.470 |
| 14-15 | 4.329 | 0.166 | 0.55 | 0.418 | 0.88 | 0.760 |
| 15-16 | 0.866 | 0.033 | 0.11 | 0.084 | 0.18 | 0.760 |
| 16-17 | 1.732 | 0.067 | 0.22 | 0.167 | 0.35 | 3.470 |
| 17-18 | 26.78 | 1.018 | 3.38 | 2.593 | 5.42 | 0.760 |
| 18-19 | 23.32 | 0.885 | 2.95 | 2.258 | 4.72 | 0.347 |
| 19-20 | 24.18 | 0.918 | 3.06 | 2.342 | 4.89 | 0.347 |
| 20-21 | 26.78 | 1.018 | 3.38 | 2.593 | 5.42 | 0.347 |
| 21-22 | 22.45 | 0.852 | 2.84 | 2.175 | 4.54 | 0.760 |
| 22-23 | 22.45 | 0.852 | 2.84 | 2.175 | 4.54 | 0.760 |
| 23-24 | 22.45 | 0.852 | 2.84 | 2.175 | 4.54 | 0.760 |
| 24-01 | 22.45 | 0.852 | 2.84 | 2.175 | 4.54 | 0.760 |
| 01-02 | 22.45 | 0.852 | 2.84 | 2.175 | 4.54 | 0.760 |
| 02-03 | 22.45 | 0.852 | 2.84 | 2.175 | 4.54 | 0.760 |
| 03-04 | 22.45 | 0.852 | 2.84 | 2.175 | 4.54 | 0.760 |
| 04-05 | 22.45 | 0.852 | 2.84 | 2.175 | 4.54 | 3.470 |
| 05-06 | 22.45 | 0.852 | 2.84 | 2.175 | 4.54 | 3.470 |

The total electric demand for all outdoor and indoor sites is shown in Table 5.

**Table 5.** Total demand for outdoor site

| hour | Total demand (Kw) | hour | Total demand (Kw) |
|---|---|---|---|
| 06-07 | 16.4322 | 18-19 | 48.3788 |
| 07-08 | 18.1044 | 19-20 | 50.051 |
| 08-09 | 23.121 | 20-21 | 55.0676 |
| 09-10 | 16.4322 | 21-22 | 43.9966 |
| 10-11 | 18.1044 | 22-23 | 43.9966 |
| 11-12 | 23.121 | 23-24 | 43.9966 |
| 12-13 | 16.4322 | 24-01 | 43.9966 |
| 13-14 | 6.8144 | 01-02 | 43.9966 |
| 14-15 | 9.121 | 02-03 | 43.9966 |
| 15-16 | 2.4322 | 03-04 | 43.9966 |
| 16-17 | 6.8144 | 04-05 | 46.7066 |
| 17-18 | 52.3576 | 05-06 | 46.7066 |

## 5.3 Supply, demand, and voltage magnitude

From Table 5, it can be seen that the maximum demand is 42.67 Kw, and it occurs from 8 to 9 p.m. The power supply at night is only gensets because there is no battery to store the PV output. Table 1 shows that one genset has a maximum output of 200 kW, which means it is sufficient to supply the demand. During the day, when sun irradiation is high, the highest demand is 17 kW at 11 am – 12 pm. Using PV output data in Table 1, it is possible to employ only PV during the day when the sun is shining. The comparison of demand and the available supply each hour of the day is shown in Fig. 2.

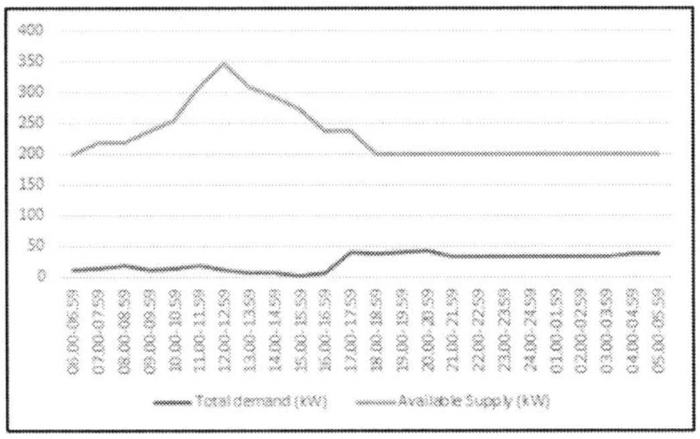


**Fig. 2.** Outdoor site demand and supply during a day

Since the available generating output capacity is much larger than the demand to be served, the research continues to add more shelter locations. Suppose indoor sites (Convention Hall and Auditorium Building) are included in the calculation. In that case, the demand and power supply graph are shown in Fig. 3. The maximum demand is 55.1 Kw, which occurs at 8 – 9 pm. From Fig. 3, it can be seen that additional demand from indoor sites still results in a total demand that is still much smaller than the available supply capacity.

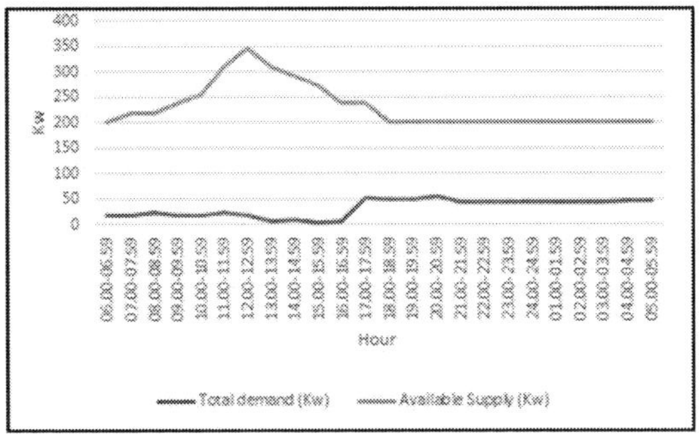


**Fig. 3**. Indoor-outdoor site demand and available supply

Power flow simulations are carried out to analyse the voltages of each load point in the distribution network. The single-line diagram of the distribution network, which consists of indoor and outdoor refugee shelters, is shown in Fig. 4. The buses of the shelter's electric supplies can be seen in Fig.4 and Table 6. Demand data for power flow simulation are also shown in Table 6.

**Table 6.** Peak demand for power flow simulation

| BUS | Load point /shelter code | kW | kVar |
|---|---|---|---|
| Economy Faculty | 2 and 6 | 12.16 | 5.89 |
| Animal H Faculty | 7 | 5.423 | 2.63 |
| Nursing Faculty | 3 | 26.78 | 18.15 |
| | 8 | 3.47 | |
| | 1 and 5 | 6.216 | |
| | 4 | 1.018 | |

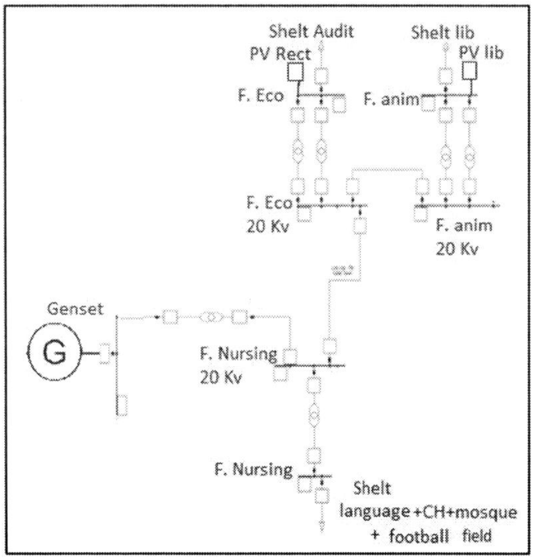


**Fig. 4.** Single line of Unand distribution network with local power supply and shelter loads

Power flow simulation results indicate that all voltage values are within the permitted standards (in the range of 0.95 -1.05 pu). Unand's electric power supply has greater capacity than the electric demand, i.e. 55.1 kW.

## 6 Conclusion and future works

The planned shelters can accommodate 3704 refugees using outdoor and indoor sites in Unand Campus Limau Manih Padang. The maximum electrical demand is predicted to be around 55.1 kW, which occurs in the evening. The genset can serve the demand with good electric power quality.

However, since the main power supply is a genset, obtaining diesel fuel might be difficult soon after the tsunami disaster. Therefore, it is suggested that the PV be equipped with a battery for storage of power yield during the day and can be used during peak demand at night.

This research is funded by the Research Grant of the Department of Electrical Engineering, Universitas Andalas, 59/UN.16.09.D/PL/2024 in 2024.

## References

1. D. H. Natawidjaja et al., Source parameters of the great Sumatran megathrust earthquakes of 1797 and 1833 inferred from coral microatolls, Journal of Geophysical Research : Solid Earth, **111**(B6), (2006).
2. F. Ashar, D. Amaratunga, and R. Haigh, The Analysis of Tsunami Vertical Shelter in Padang City, Procedia Economics and Finance, **18**. 916–923, (2014), https://doi.org/10.1016/S2212-5671(14)01018-1.
3. R. Fijra, Penentuan lokasi tempat evakuasi akhir pengungsi pada ancaman bencana tsunami Kota Padang, Jurnal Teknosains, **7**(2), 111–118, (2018).

4. Sphere Association, The Sphere Handbook : Humanitarian Charter and Minimum Standards in Humanitarian Response, 4th ed. Geneva, Switzerland : Sphere Association. [Online]. Available : edition, Geneva, Switzerland, (2018).
5. UNHCR, Emergency Handbook. UNHCR, (2024). [Online].
6. P. Van Harten and E. Setiawan, Instalasi Listrik Arus Kuat, Jilid 2, Bandung : Percetakan Bina Cipta, (1985).
7. US Department of Energy, Building Technologies Program Solid-State Lighting Technology Fact Sheet, PNNL-SA-94206, Mar. (2013).
8. H. Satria, S. Syafii, and A. Aswardi, Analysis of Peak Power Capacity on Rooftop Solar PV 1.25 kWp at Sun Conditions 90 Degrees, Ijeepse, **4**(3), 173–178, (2021).
9. Digsilent GmbH, Digsilent Powerfactory User Manual. Digsilent GmbH, (2014). [Online].

# Landscape lexical diversity for readability and safety mitigation on the commuter line bilingual information board

*Mochamad* Nuruz Zaman[1*], *Supriatnoko*[1], and *Muhammad* Naufal Alhilmy[1]

[1]Department of Business Administration, Politeknik Negeri Jakarta, Depok, West Java 16425, Indonesia

**Abstract.** This study examines lexical diversity and readability of bilingual information boards on the Commuter Line Tanah Abang-Rangkasbitung in Indonesia, with a focus on their impact on passenger safety. It is motivated by observed discrepancies in translations, which could compromise clarity and effectiveness of public signage. The primary objective is to analyze lexical diversity and assess the readability of these translations using theoretical frameworks from Nida and Taber (2003), Newmark (1988), and Venuti (2012). The study employs a qualitative descriptive research design with Spradley's Ethnographical Approach, focusing on the Green Ring Route, which encompasses 19 CL stations. Data were collected through direct observation, photographic documentation, and field notes and analyzed using domain, taxonomy, componential, and cultural theme. The findings reveal significant variation in readability levels, influenced by the translation techniques used. High readability was found in translations employing Literal Translation and Established Equivalent techniques, while medium and low readability were associated with Borrowing, Modulation, and Amplification techniques. These findings underscore the need for consistent, culturally relevant translation practices to ensure effective communication and passenger safety. The study concludes by recommending regular usability testing, feedback integration, and use of technology to enhance the accessibility and effectiveness of bilingual signage in public transportation.

## 1 Introduction

Indonesia's Commuter Line (CL) train service has undergone significant internationalization, as highlighted by the KAI Commuter Customer Service page [1]. This transformation reflects the public's trust in the service's transparency, which is crucial for accommodating the diverse linguistic needs of its passengers. According to the 2023 KAI Commuter Annual Report, Tanah Abang Station recorded 153,270 passengers on December 30, 2023, contributing to 290 million users on the CL Jabodetabek network throughout the year [2]. The strong bilingual branding of the CL system demands fully accurate information presentation, as the accuracy of translations directly influences the readability and safety of these public information boards [3]. An observation conducted on March 5, 2024, on the Tanah Abang-Rangkasbitung route, revealed a translation discrepancy where "*Peta Jaringan KA Perkotaan Jabodetabek dan Sekitarnya*" was reduced to "Jakarta Metropolitan Railway Network Map." This reduction highlights the need for an analysis of landscape lexical diversity to understand how transposition in the target language can impact both the clarity and the safety of the information being conveyed. Previous research has explored various multilingual and multimodal communication aspects within different contexts. Cortés [4] investigated lexical structure, diversity, and readability in business and management settings but did not fully explore the acceptance of these concepts in translation analysis [4]. Algryani [5] examined translation strategies and quality assessments across industries, yet this study extends the analysis specifically to the commuter line sector. Even-Zohar [6] linked linguistic variation in children's literature to its broader impacts, providing a foundation for similar explorations within translation studies. Pienimäki, Väisänen, and Hiippala [7] discussed the influence of translation on macro-level lexicon quality, suggesting a need for accuracy assessment in bilingual signage. Helm [8] focused on language diversity and technology, particularly how language modelling biases can cause epistemic injustice, which is relevant to the bilingual lexicon context of this study. These previous studies provide a backdrop against which the current research situates, particularly concerning the need for accuracy and readability in bilingual public signage. Despite the strong bilingual branding of Indonesia's CL system, translation discrepancies have been observed, which could lead to potential miscommunication and compromise passenger safety and readability. The reduction of complex Indonesian phrases to oversimplified English translations, as seen in the case of the Tanah Abang-Rangkasbitung route, risks losing essential information. This issue highlights a broader challenge: how translation readability directly impacts the effectiveness of bilingual information boards in ensuring both readability and safety for local and international passengers. Accurate translation [9] is not merely a linguistic concern but a critical component

* Corresponding author: mzaman@mesin.pnj.ac.id

in safeguarding passengers and facilitating seamless navigation within the public transportation system.

This study aims to design the lexical structure of each bilingual lexicon on the Tanah Abang-Rangkasbitung Commuter Line information boards, focusing on the landscape diversity pattern and the translation accuracy parameters. The specific objectives are to:

1. Identify the lexical diversity in bilingual information boards.
2. Evaluate the readability of translations.
3. Provide insights and recommendations for improving the effectiveness of bilingual signage in public transportation systems.

The study focuses specifically on the Green Ring Route of the Tanah Abang-Rangkasbitung Commuter Line, covering 19 CL stations. While the findings offer valuable insights, they may not be fully generalizable to other routes or transportation systems beyond Jakarta. Additionally, the study is limited to the analysis of static signage, excluding other forms of communication, such as digital or audio announcements, which also play a significant role in passenger navigation and safety.

## 2 Method

This study employs a qualitative descriptive research design using Spradley's Ethnographical Approach [10]. It allows for an in-depth exploration of the lexical diversity and translation readability on the bilingual information boards [11] along the Commuter Line (CL) Tanah Abang-Rangkasbitung route. The ethnographical approach is particularly suited to examining the interaction between language, cultural context, and public information systems [12], offering a comprehensive understanding of how these elements impact readability and safety.

The primary data for this study consists of bilingual information and direction signs located on the Green Ring Route of the Tanah Abang-Rangkasbitung Commuter Line. The route encompasses 19 CL stations, each providing a unique set of signs that reflect the broader linguistic landscape of the commuter system. These signs [13], written in Indonesian and English, are the primary source for analyzing lexical diversity and translation readability.

Data collection involved systematically observing and documenting [14] the bilingual signs across the 19 stations on the Tanah Abang-Rangkasbitung route. Photographic records were taken of each sign to ensure readability in subsequent analysis. The observations were conducted to capture the context in which the signs are placed, providing insights into how passengers interact with these signs and how effectively the signs convey necessary information for safe navigation [15]. In addition to visual documentation, field notes were taken to record immediate impressions and any observed issues with readability or potential safety concerns.

The data analysis followed four distinct stages as prescribed by Spradley's Ethnographical Approach [10]. First, Domain analysis [16] stage focused on identifying and classifying the lexical diversity structure within the landscape of the information signs. The analysis considered the location of stations and carriages, influencing how translation techniques were applied. It helped establish the broad categories of lexical diversity on the signs.

Subsequently, Taxonomy analysis [16] aimed to reduce the data into linguistic coding. This stage involved evaluating the translation accuracy of the signs based on three parameters: accurate (clear), less accurate (less distortion), and inaccurate (full distortion). The analysis provided a detailed linguistic coding framework that captured the degree of translation fidelity.

The next stage was Componential analysis [17]. the domain and taxonomy analyses were integrated vertically and comprehensively. The componential analysis sought to formulate findings by systematically combining the previous analyses, allowing for a nuanced understanding of the relationships between lexical diversity and translation accuracy.

The final stage namely Cultural Theme analysis [17]. It involved analyzing cultural themes to demonstrate the lexical levels of each bilingual lexicon on the information boards. This analysis focused on identifying patterns of landscape diversity and the structural parameters of translation accuracy, ultimately linking these findings to broader cultural and linguistic practices within the commuter line system.

To ensure the validity of the data, triangulation was employed through multiple data collection methods [18], including direct observation, photographic documentation, and field notes. The data were cross-verified against each method to check for consistency and reliability. Peer debriefing was also conducted with linguistic experts to review the coding and classification processes [17], ensuring that the analyses accurately reflected the observed linguistic phenomena. Member checking was also employed, where preliminary findings were discussed with stakeholders in the KAI Commuter system to confirm the relevance and accuracy of the conclusions drawn from the data. This rigorous approach to data validation enhances the credibility of the study's findings and their applicability to the broader context of bilingual information systems in public transportation.

## 3 Results and discussion

### 3.1 The lexical diversity in bilingual information boards

In the Commuter Line Tanah Abang-Rangkasbitung context, the interplay between data, safety indicators, source text, and target text is essential for maintaining passenger safety through effective bilingual information boards. The landscape lexical diversity embedded in these translations is a critical component of safety mitigation, as it captures the variety and range of vocabulary required to meet the needs of a multilingual passenger base. The data includes a spectrum of source texts in Indonesian, translated into English, to ensure that information is accessible to all passengers. These translations are carefully aligned with various safety indicators, such as Informational Safety, Warning Safety, Prohibition Safety, and Emergency Safety, each serving a unique purpose in guiding and safeguarding passengers. For instance, the translation of *Toilet Difabel* as *Difables' Restroom* reflects an effort to maintain inclusivity within Informational Safety signage, ensuring that facilities are clearly marked and understood by all passengers, including those with disabilities.

Further examination reveals how translations tied to Warning and Prohibition Safety indicators function within this landscape of lexical diversity. The phrase *Jaga dan awasi gawai Anda* which translates as *Caution, always watch out your gadget while charging it*, demonstrates the importance of conveying potential hazards to passengers. This type of warning is crucial in preventing accidents or negligence, emphasizing the need for clarity and immediacy in the target text. Similarly, Prohibition Safety signs, such as *Dilarang lewat di sini*, retain the original Indonesian text without translation, which may pose challenges for non-Indonesian speakers. These instances highlight the complex relationship between the source and target texts and how the translation must adapt to ensure that all passengers, regardless of their linguistic background, can understand and adhere to safety instructions.

Emergency Safety signs, such as *Peta Evakuasi*, translated as "Evacuation Signage," further illustrate the role of lexical diversity in safety mitigation. These signs are particularly critical, as they provide instructions that must be quickly and easily understood during emergencies. The translation employs clear, universally recognized terms, ensuring high readability and effective communication under stress. The careful selection of vocabulary in these cases underscores the importance of landscape lexical diversity in crafting translations that convey essential information and enhance the commuter line's overall safety. By aligning data, safety indicators, source text, and target text within this framework, the Commuter Line ensures that its bilingual information boards are reliable for passenger safety in a multilingual environment.

## 3.2 The readability of translations

Readability in translation is a crucial aspect of ensuring that the target audience can properly comprehend and engage with the translated text. According to Nida and Taber [19], readability is influenced by vocabulary complexity, sentence structure, and the familiarity of terms in the target language. Newmark [20] also emphasizes that readability is tied to the clarity and fluency of the translation, which should aim to be as natural and comprehensible as possible to the target audience. Venuti [21] discusses the concept of domestication in translation, where the text is adapted to fit the cultural and linguistic norms of the target audience, thus enhancing readability.

High readability ensures the message is immediately clear, without re-reading or additional interpretation. Medium readability indicates that the audience may understand the message but may require some effort or re-reading due to less familiar vocabulary or more complex sentence structures. Low readability occurs when the translation is difficult to understand due to overly complex language, unfamiliar terms, or lack of adaptation to the target audience's linguistic norms.

**Table 1.** Componential analysis of determining readability levels.

| INDI-CATOR | SOURCE TEXT | TARGET TEXT | TECHNIQUE | READ-ABILITY |
|---|---|---|---|---|
| Informational Safety | *Toilet Difabel* | Difables' Restroom | Borrowing | Medium |
| Informational Safety | *Pos Kesehatan* | Medical Room | Literal Translation | High |
| Informational Safety | *Ruang Laktasi* | Nursing Room | Literal Translation | High |
| Warning Safety | *Jaga dan awasi gawai Anda* | Caution, always watch out your gadget while charging it | Modulation | Medium |
| Informational Safety | *KA Lokal* | Local Train | Literal Translation | High |
| Informational Safety | *Stasiun Awal/Akhir* | Terminus Station | Adaptation | High |
| Informational Safety | *Difabel* | Physically Handicapped | Modulation | Low |
| Informational Safety | *Ibu Hamil* | Pregnant Woman | Modulation | High |
| Informational Safety | *Lanjut Usia* | Elderly Passengers | Adaptation | Low |
| Prohibition Safety | *Dilarang lewat di sini* | *Please don't walk here* | Literal Translation | High |
| Informational Safety | *Kereta khusus wanita pada jam sibuk* | Women-only car during peak hour | Established Equivalent | High |
| Emergency Safety | *Peta Evakuasi* | Evacuation Signage | Established Equivalent | High |
| Informational Safety | *Pelayanan Disabilitas* | Pasager with Disabilities Services | Amplification | Low |

In examining the readability of translations on the Commuter Line Tanah Abang-Rangkasbitung's bilingual information boards, it is essential to consider the influence of translation techniques and the corresponding safety indicators associated with each message. The theoretical frameworks provided by Nida and Taber [19], Newmark [20], and Venuti [21] offer a structured approach to categorizing the readability of these translations into three levels. The table below illustrates how different translation techniques impact the readability of translated texts in relation to their safety indicators.

To formulate discussion, researchers have demonstrated the findings following the Table 1 as follows:

a) High Readability
   1) *Pos Kesehatan* is translated to *Medical Room* exemplifies high readability by employing Literal Translation, which preserves the original meaning with simple and familiar terms in English. According to Nida and Taber, this approach ensures that the target audience can easily understand and process the information, making it highly effective for informational safety signage.
   2) *Ruang Laktasi* is translated to *Nursing Room* here maintains clarity and accessibility, ensuring that the message is readily understood by all passengers. Newmark's theory of readability supports this approach, as the translation aligns well with the linguistic norms of the target language.
   3) *KA Lokal* is translated to *Local Train*; in this case, the use of Literal Translation results in a direct and easily comprehensible translation. The simplicity of the terms ensures high readability, which is critical for ensuring that passengers can quickly identify the nature of the train service.
   4) *Stasiun Awal/Akhir* is translated to *Terminus Station* applies Adaptation technique to adjusts the original text and to fit the cultural and linguistic context of the target audience. This technique, supported by Venuti's domestication theory, enhances readability by using a term ("Terminus Station") familiar and specific to English-speaking passengers.
   5) *Kereta khusus wanita pada jam sibuk* is translated to *Women-only car during peak hour* employs an Established Equivalent, ensuring that the phrase is culturally appropriate and easily understood. Using a recognized and standardized term contributes to high readability, aligning with Newmark's emphasis on clarity and naturalness in translation.
   6) *Peta Evakuasi* is translated to *Evacuation Signage*; as an Emergency Safety sign, the translation's high readability is crucial. The Established Equivalent technique ensures that the message is clear and instantly recognizable, essential in emergencies where quick comprehension is necessary.
   7) *Dilarang lewat di sini* is translated to *Please don't walk here* constructs the translation shifts from a direct prohibition in Indonesian to a polite request in English. Despite the change in tone, the message remains clear and easily understood, contributing to high readability.
   8) *Ibu Hamil* is translated to *Pregnant Woman* modulates slightly changes the perspective, it retains high readability by using familiar and straightforward language easily understood by the target audience.

b) Medium Readability
   1) *Toilet Difabel* is translated to *Difables' Restroom* borrows retains the culturally specific term *Difabel*, which is slightly adapted into English as *Difables*. Although this approach maintains cultural relevance, it introduces a term that may not be immediately familiar to English speakers, resulting in medium readability as some passengers may require additional context to understand the term.
   2) *Jaga dan awasi gawai Anda* is translated to *Caution, always watch out your gadget while charging it* constructed Modulation technique shifts the structure and phrasing to better fit English norms, but the complexity of the sentence reduces readability to a medium level. Passengers might need to re-read the text to fully grasp the safety warning, which could impact the immediacy of the message.

c) Low Readability
   1) Difabel is translated to Physically Handicapped formulated The Modulation technique here results in low readability due to the use of outdated terminology. "Physically Handicapped" is less commonly used in contemporary English, where terms like *Disabled* or *Person with Disabilities* are preferred. It could lead to confusion or a lack of understanding among passengers.
   2) *Lanjut Usia* is translated to *Elderly Passengers*, experienxed the Adaptation technique that adjusts the translation to fit the cultural context, the phrase *Elderly Passengers* may still lack clarity for some readers, especially if it is not immediately clear how this information applies. The resulting readability is low, as the message may not be as accessible or actionable.
   3) *Pelayanan Disabilitas* is translated to *Passenger with Disabilities Services* employed the Amplification that expands the translation to provide more detail, but it results in low readability in this case. The longer phrase *Passenger with Disabilities Services* might be less direct and harder to process quickly. The mistake in using the singular *Passenger* instead of the plural *Passengers* may confuse readers, as it does not clearly refer to all individuals needing these services. Additionally, the phrase is overly complex and could be simplified for clarity, enhancing readability and ensuring the message is easily understood.

### 3.3 Insights and recommendations for improving the effectiveness of bilingual signage in public transportation systems

Improving the effectiveness of bilingual signage in public transportation systems, such as the Commuter Line Tanah Abang-Rangkasbitung, requires a multi-faceted approach that addresses translation accuracy, readability, and cultural relevance. The following insights and recommendations are derived from analyzing current translation practices and their impact on passenger safety and communication clarity.

High readability ensures passengers can quickly understand and act upon the information presented.

Translations should prioritize using simple, familiar vocabulary and sentence structures that align with the linguistic norms of the target audience. For instance, translations like *Pos Kesehatan* to *Medical Room* and *KA Lokal* to *Local Train* exemplify the effectiveness of Literal Translation in maintaining clarity and accessibility. To enhance readability, it is recommended that future translations continue to use straightforward language, avoiding overly complex or technical terms unless necessary.

Cultural and linguistic relevance is critical to ensure translations resonate with the target audience. The use of Adaptation and Established Equivalent techniques, as seen in the translation of *Stasiun Awal/Akhir* to *Terminus Station* and *Kereta khusus wanita pada jam sibuk* to *Women-only car during peak hour*, demonstrates the importance of tailoring translations to fit the cultural context of the target audience. To further improve effectiveness, translators should be encouraged to consider cultural nuances and preferences when selecting terms, ensuring that the translated text is accurate, culturally appropriate, and familiar to passengers.

Low readability and translation discrepancies should be promptly addressed and corrected, such as using outdated or unclear terminology. For example, translating *Difabel* to *Physically Handicapped* reflects the need to update terminology to align with contemporary language norms. Additionally, the error in translating *Pelayanan Disabilitas* to *Passenger with Disabilities Services* underscores the importance of careful proofreading and revision to avoid mistakes that can confuse or mislead passengers. It is recommended that a review process be implemented for all translations involving both linguistic experts and native speakers to ensure that the final product is accurate, clear, and free of errors.

Consistency in translation practices is key to maintaining a uniform level of readability and comprehension across all signage. It involves standardizing translation techniques and ensuring that similar types of information are translated using the same approach. For instance, all informational safety signs should ideally employ Literal Translation or Established Equivalent techniques to maintain clarity and prevent variations that could confuse passengers. A standardized glossary of terms and phrases could be developed and regularly updated to guide translators and ensure consistency across the transportation network.

To ensure that bilingual signage remains effective over time, it is crucial to conduct regular usability testing with diverse groups of passengers, including non-native speakers and individuals with varying literacy levels. Feedback from these groups can provide valuable insights into the effectiveness of translations and identify areas where improvements are needed. Based on this feedback, translations can be revised and updated to meet the needs of all passengers better. Additionally, ongoing training for translators on best practices and emerging trends in translation can help maintain high standards of readability and cultural relevance.

Technology can significantly enhance the accessibility and effectiveness of bilingual signage. For example, integrating QR codes or NFC tags on signage can provide passengers with access to translations in multiple languages, detailed instructions, or even audio descriptions, which can be particularly beneficial for visually impaired passengers. This approach not only enhances the inclusivity of the transportation system but also ensures that all passengers, regardless of their language proficiency or disability, can access essential information.

## 4 Conclusion

1. The importance of lexical diversity and readability in ensuring passenger safety underscores that lexical diversity in bilingual information boards is crucial for meeting the needs of a multilingual passenger base. Effective translations that align with the cultural and linguistic context, such as those employing Literal Translation and Established Equivalent techniques, ensure high readability and contribute significantly to passenger safety. Clear and accessible translations enable passengers to quickly comprehend and act upon critical information, particularly in safety-related situations.
2. Consistency and cultural relevance in translation practices are key to maintaining the effectiveness of bilingual signage in public transportation systems. Standardizing translation techniques across all signage and ensuring that terms are culturally appropriate and familiar to the target audience are essential steps. Addressing discrepancies and outdated terminology and implementing a rigorous review process involving linguistic experts can prevent misunderstandings and enhance communication clarity.
3. To continuously improve the effectiveness of bilingual signage, transportation systems should conduct regular usability testing and gather feedback from diverse passenger groups. This feedback should inform ongoing translation revisions, ensuring they remain relevant and clear. Additionally, leveraging technology, such as QR codes and NFC tags, can enhance accessibility by providing multilingual translations and audio descriptions, making the transportation system more inclusive and effective for all passengers.

The authors disclosed receipt of the following financial support for the research, authorship, and/or publication of this article; This work was supported by the Politeknik Negeri Jakarta grant number 571/PL3.A.10/PT.00.06/2024.

## References

1. Badan Pusat Statistik Indonesia. Jumlah Penumpang Kereta Api - Tabel Statistik (2024). https://www.bps.go.id/id/statistics-table/2/NzIjMg==/jumlah-penumpang-kereta-api.html
2. Pusat Data Ekonomi dan Bisnis Indonesia Databoks, KRL commuter Jabodetabek Sumbang 290 Juta Penumpang pada 2023. (accessed on January

152024). https://databoks.katadata.co.id/datapublish/2024/01/15/krl-commuter-jabodetabek-sumbang-290-juta-penumpang-pada-2023

3. C. Ferguson, and S. Winn, Readability of Australian road safety information for the general public. Journal of Road Safety, **34**(3), 1-10 (2023). https://doi.org/10.33492/jrs-d-21-00061
4. J. D. Cortés, Journal titles and mission statements: Lexical structure, diversity, and readability in business, management and accounting research. Journal of Information Science, **49**(5), 1262-1276 (2021). https://doi.org/10.1177/01655515211043707
5. A. Algryani, On the translation of linguistic landscape: Strategies and quality assessment. Khazar Journal of Humanities and Social Sciences, **24**(2), 5-21 (2021). https://doi.org/10.5782/2223-2621.2021.24.2.5
6. I. Even-Zohar, The Translation Studies Reader (Routledge, Oxfordshire, 2021). https://doi.org/10.4324/9780429280641-23
7. H. Pienimäki, T. Väisänen, and T. Hiippala, Making sense of linguistic diversity in Helsinki, Finland: The timespace of affects in the linguistic landscape. Journal of Sociolinguistics, **28**(2), 3-21 (2023). https://doi.org/10.1111/josl.12633
8. P. Helm, G. Bella, G. Koch, and F. Giunchiglia, Diversity and language technology: How language modeling bias causes epistemic injustice. Ethics and Information Technology, **26** (2024). https://doi.org/10.1007/s10676-023-09742-6
9. M. Krishna, V. Sharma, and V. Jain, Evaluation of public transportation system from the perspective of passengers. Multidisciplinary Science Journal, **5**, (2023). https://doi.org/10.31893/multiscience.2023ss0213
10. C. Andion, Social innovation, experimentalism, and public governance: An ethnographical approach to study public arenas in the city. BAR - Brazilian Administration Review, **20** (2023). https://doi.org/10.1590/1807-7692bar2023220124
11. P. Canavese, Institutional Translation and Interpreting (Routledge, Oxfordshire, 2020). https://doi.org/10.4324/9780429264894-4
12. M. Karlik, Exploring the impact of culture on language learning: How understanding cultural context and values can deepen language acquisition. International Journal of Language, Linguistics, Literature and Culture, **02**(05), 05-11 (2023). https://doi.org/10.59009/ijlllc.2023.0035
13. W. Juanggo, Investigating lexical diversity and lexical sophistication of productive vocabulary in the written discourse of Indonesian EFL learners. Indonesian Journal of Applied Linguistics, **8** (2018). https://doi.org/10.17509/ijal.v8i1.11462
14. A. Gelman, and D. Nolan, Teaching Statistics: A Bag of Tricks (2nd edition) (Oxford University Press, Oxford, 2017). https://doi.org/10.1093/oso/9780198785699.003.0006
15. J. Jakób Liszka, How signs convey information. Chinese Semiotic Studies, **12**(1), 45-66 (2016). https://doi.org/10.1515/css-2016-0005
16. M. N. Zaman, M. R. Nababan, and D. Djatmika, The honorifics expressions in Indigenous language of Java island ethnics with relevance into english. LiNGUA: Jurnal Ilmu Bahasa dan Sastra, **13**(2), 101-114 (2018). https://doi.org/10.18860/ling.v13i2.5258
17. M. N. Zaman, and R. S. Nugraha, A glossary of legal terms in a scenario-based judicial context. LiNGUA: Jurnal Ilmu Bahasa dan Sastra, **19**(1), 27-35 (2024). https://doi.org/10.18860/ling.v19i1.23038.
18. U. Flick, Triangulation in data collection. (Sage Puvlications, Los Angeles, 2018). https://doi.org/10.4135/9781526416070.n34
19. E. Nida, and C. Taber, The theory and practice of translation (Brill Publishers, Leiden, 2003). https://doi.org/10.1163/9789004496330
20. P. Newmark, A textbook of translation (Prentice Hall, New York, 1988).
21. L.Venuti, The translator's invisibility: A history of translation. (Routledge, Oxfordshire, 2012)

# Environmental consciousness, health consciousness and ethical identity to achieve sustainable rehabilitation after disaster

*Margareth* Setiawan[1*], *Dian* Sadeli[2], and *Siti* Hajar Mohamad[3]

[1]Entrepreneurship Department, Bina Nusantara University, Bandung 40181, Indonesia
[2]Business Administration Department, Parahyangan Catholic University, Bandung 40141, Indonesia
[3]Department of Technology Management, Management and Science University, 40100 Shah Alam, Selangor, Malaysia

**Abstract.** This study examines the influence of environmental consciousness, health consciousness, and ethical identity on fostering sustainable restoration following disasters. As global environmental and health challenges escalate, the incorporation of sustainable practices is imperative not just in daily consumption but also in post-disaster recovery initiatives. This study investigates the impact of individual and community awareness in these three domains on the sustainable rehabilitation of impacted areas. This study utilizes a dataset of 200 respondents from Indonesia and implements Partial Least Squares Structural Equation Modelling (PLS-SEM) to examine the influence of consciousness elements on sustainability consciousness, essential for long-term rehabilitation. Research indicates that health and environmental consciousness significantly influence sustainability consciousness, while ethical identity does not demonstrate a considerable effect. The study emphasizes the necessity of incorporating environmental and health consciousness into disaster recovery efforts, offering pragmatic insights for politicians and humanitarian organizations to promote sustainable rehabilitation practices.

## 1 Introduction

In recent years, there has been a widespread effort among academics to comprehend sustainable consumption behavior particularly in disaster recovery. The broad adoption of sustainable consumption practices, especially in post-disaster contexts, has been fueled by the increasing public awareness of the hazards associated with environmentally hazardous items. There are numerous instances of sustainable consumption behavior, such as purchasing goods that are traded sustainably and fairly, utilizing energy-efficient equipment, buying organic food, recycling household waste, voluntarily adopting a minimalist lifestyle, purchasing goods made from recycled materials, transitioning to environmentally friendly products, altering modes of transportation, and investing in "ethical funds," among others [1]. Sustainable consumption involves adopting a different approach to consumption by reducing both the quantity of goods and services spent (volume) and the impact on the environment that result from what is consumed (composition) [2]. This means practicing sustainable consumption entails not only minimizing waste, but also substituting hazardous items with more ecologically sound alternatives to minimize the detrimental effects on the environment.

The immense influence of technology should not be overlooked, since it has significantly altered people's consuming behaviors through platforms like social media, particularly with the implementation of campaigns like green marketing [3]. Nevertheless, achieving a sustainable society requires more than just depending on technological advancements; it also requires a fundamental shift in human behavior [4]. Indeed, ecological economists assert that strategies for achieving a sustainable society should incorporate changes in the realm of individual consumer behavior [5]. They argue that viewing humans as active agents of change, rather than passive recipients of environmental policies, can lead to more effective and enduring sustainable transformations.

At the individual consumer level, ecological sustainability is facilitated by a significant change in consciousness [6, 7]. This consciousness is often referred to as mindfulness, which is a mechanism that can help consumers to reorient towards sustainable behavior [8]. Several studies have successfully shown that mindfulness can encourage changes in consumption behavior [9-11]. Thus, it is assumed that mindfulness could have a substantial impact on inspiring individuals to develop awareness and concern for sustainability.

Although mindfulness as an approach to researching sustainability at the micro/individual level is highly recommended, the academic literature on the subject remains underdeveloped [12]. Research has found a connection between mindfulness and sustainability,

* Corresponding author: margareth.setiawan@binus.ac.id

however, the theoretical understanding of how mindfulness affects individuals in relation to sustainability is still in its early stages and relies heavily on insights from other scientific fields rather than individual psychological perspectives [13, 14]. Furthermore, prior studies have mostly examined mindfulness and sustainability through the lens of pro-environmental behavior, resulting in a limited comprehension of sustainability. The present study aims to fill the gaps by establishing a connection between mindfulness and sustainability consciousness to achieve sustainable rehabilitation, considering personal, social, and environmental aspects.

The patterns of sustainability consciousness may differ among different generations, such as Gen Z and Gen Y. These generations may exhibit distinct mindfulness approaches when it comes to selecting, purchasing, utilizing, and disposing of different products available in the market. The challenge of finding a balance between mindfulness and sustainable consciousness is a significant yet insufficiently studied topic that requires substantial scientific focus. In response to the aforementioned shortcomings, we elaborate on the principles of mindful consumption to elucidate the influence of mindfulness on sustainability consciousness. Additionally, our study investigates how age influences the relationship in question. In this study, we have chosen to use sustainability consciousness as a measure for each of the three components of sustainable development, namely personal value, societal, and environmental aspects. Therefore, this study aims to fill significant gaps in the current literature on the subject.

## 2 Literature review and hypothesis development

### 2.1 Health consciousness

Mindful consumption is a relevant theoretical paradigm that applies specifically to sustainable consumption [15]. Essentially, adopting a mindful mentality enables consumers to carefully contemplate the consequences of their purchase choices and modify their behavior accordingly. Previous research has succeeded in proving that mindful consumption is influenced by three high-level motivations, namely health consciousness, ethical identity, and environmental consciousness [10]. The first scope (health consciousness) can be interpreted as how much a person cares about their health and the health of the people they care about, in this case including their family. The first scope, health consciousness, refers to the extent to which an individual prioritizes their own health and the well-being of their loved ones, particularly in the aftermath of disasters [16]. In post-disaster rehabilitation, prioritizing health consciousness ensures that recovery processes address public health risks, such as access to clean water, sanitation, and healthcare. Individuals who prioritize their well-being are more inclined to adopt and maintain healthy behaviors, such as sustainable consumption [17]. This can be seen from previous research which proved that health consciousness does influence the intention to use environmentally friendly menstrual products [18]. Other research also proved that being health-conscious leads to mindfulness by means of adopting nutritious meals and engaging in healthy habits [19]. Based on the literature review and with the intention of filling the current knowledge gaps, the following hypotheses are put forward.

Hypothesis 1: Health consciousness is positively and significantly influencing sustainability consciousness

### 2.2 Ethical identity

Consumers who identify themselves as ethical individuals are more inclined to make choices that have favorable social consequences and contribute to the well-being of others [20]. This act will significantly influence aspects of social and environmental sustainability as well. Ethical consumption refers to the act of individuals buying and utilizing things and resources based on both personal satisfaction and values, as well as moral principles that distinguish between right and wrong, good and bad [21]. The motives that drive ethical consumption choices encompass considerations for the well-being of animals and humans, as well as the need for fair prices, among other factors [22]. Put simply, ethical identity refers to the degree to which consumers are motivated by ethical considerations while making decisions about what to consume [10]. We contend that there is a positive correlation between an individual's ethical identity and their level of consciousness regarding sustainable consumption. Thus, next hypothesis is:

Hypothesis 2: Ethical identity is positively and significantly influencing sustainability consciousness

### 2.3 Environmental consciousness

Environmental consciousness encompasses both an individual's understanding and awareness of environmental challenges, as well as the corresponding actions they take in response [23]. Environmental consciousness plays a critical role in disaster rehabilitation by promoting sustainable rebuilding practices that reduce future environmental risks. Environmental consciousness refers to the set of personal ideas that influence consumers to engage in behaviors that are advantageous to the environment [24]. Research has demonstrated that individuals who possess environmental consciousness also display traits and attitudes such as loyalty and a robust feeling of commitment [25]. The same research also showed that consumers that have a heightened environmental awareness are more likely to exhibit favorable attitudes towards green practices, leading to increased market rivalry driven by environmentally friendly buying patterns. As environmental consciousness grows, consumer consciousness of sustainability also becomes better [26]. Therefore, the third hypothesis in this research is:

Hypothesis 3: Environmental consciousness is positively and significantly influencing sustainability consciousness

### 2.4 Role of age

Research reveals that consumption patterns vary with age [27]. The consumption behavior of the new generation of consumers (called Gen Z, the generation born from 1995 onwards) is very different from the previous generation, namely Gen Y [28]. Previous research revealed that Gen Y customers prioritize career advancement and stability over financial rewards such as higher salaries, while Generation Z are materialistic in the sense that they make higher salaries the main criterion when choosing a job [29]. There are conflicting results regarding the attitudes of customers from different generations. For example, several studies have found that Generation Y consumers are aware of and care about sustainability issues [30]. However, research also reveals that Gen Z is more interested in investing in and associating with brands that have a proven sustainability track record [31, 32]. Gen Z consumers are maturing in a time characterized by heightened environmental consciousness and instruction. From a young age, they have encountered a range of social, economic, and environmental challenges through educational programs and social media [33, 34]. The aforementioned exposure likely facilitated an enhanced comprehension of sustainable practices and a more profound recognition of their significance, hence rendering Gen Z consumers more open to the influence of sustainability consciousness [35]. Furthermore, Gen Z consumers actively pursue resolutions to worldwide issues and champion sustainability, social justice, and ethical principles [36]. Gen Z is recognized for their social awareness and their willingness to prioritize addressing global issues before financial gain [37]. Thus, we predict that the effect of health consciousness, ethical identity, and environmental consciousness on sustainability consciousness is stronger for Gen Z customers than Gen Y customers.

The construction model in this research can be seen in Fig. 1. It can be from the model that health consciousness, ethical identity, and environmental consciousness will have a significant positive effect on sustainable consciousness. Meanwhile, age will moderate this relationship, with the assumption that the influence of mindfulness on sustainability awareness is stronger for Gen Z customers than for Gen Y customers.

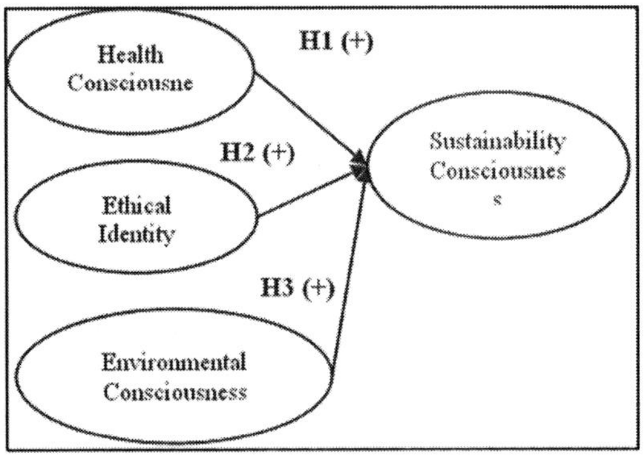


**Fig. 1.** Conceptual frameworks

## 3 Methodology

### 3.1 Sample and data collection

Respondent data in Indonesia was obtained through an online survey company. The research instrument consists of various statements that respondents must rate, and the questionnaire is designed to capture the key variables of this study as well as demographic variables such as age and gender. To ensure enough sample size and representation of various age groups, researchers carried out calculations first on the first 100 respondents and then continued the next 100 respondents. A total of 200 questionnaires were distributed and all of them can be used for further analysis. Out of the 200 participants, there were an equal number of males and females with 89 participants belonging to Generation Z and 111 participants belonging to Generation Y. See Table 1 for an overview of the demographic characteristics of the respondents, which consists of gender, age, education level, and job status. In this survey, all respondents were required to answer all questions so that the number of missing values was 0. Most of the respondents' ages were in the range of 21–26-year-old, then the next were in the range of 33–38-year-old, next are in the range of 39–43-year-old and so on. Based on these results, it can be concluded that the age of the respondents is as expected, namely Gen Y and Gen Z.

**Table 1.** Demographic profile.

| | Demographic | Frequency | Percentage |
|---|---|---|---|
| Gender | Men | 100 | 50 |
| | Women | 100 | 50 |
| **Total** | | **200** | **100** |
| Age group | 15-20 | 40 | 20 |
| | 21-26 | 49 | 24.5 |
| | 27-32 | 25 | 12.5 |
| | 33-38 | 45 | 22.5 |
| | 39-43 | 41 | 20.5 |
| **Total** | | **200** | **100** |
| Level education | High school | 54 | 27 |
| | Diploma degree | 20 | 10 |
| | Bachelor's degree | 123 | 61.5 |
| | Master's degree | 3 | 1.5 |
| **Total** | | **200** | **100** |
| Job status | Student | 90 | 45 |
| | Employee | 50 | 25 |
| | Entrepreneur | 25 | 12.5 |
| | Housewives | 20 | 10 |
| | Other | 15 | 7.5 |
| **Total** | | **200** | **100** |

### 3.2 Measures

This paper adapts a measurement scale from previous literature with some minor modifications to ensure contextual suitability for the research. Questions for the variable's health conscientiousness (HC1, HC2, HC3), ethical identity (EI1, EI2, EI3), and environmental consciousness (EC1, EC2, EC3) were adapted from previous research [16]. While questions for the sustainability consciousness variable (SC1, SC2, SC3, SC4, SC5) were taken from another research [38].

### 3.3 Data analysis and results

This study utilized Partial Least Squares Structural Equation Modelling (PLS-SEM) to analyse the data. The researchers used a two-step approach that involved doing confirmatory factor analysis (CFA) and structural equation modelling (SEM) to examine the proposed correlations [39]. We evaluated the reliability of the construct using Cronbach's alpha and Composite reliability (CR). Table 2 shows that the Cronbach's alpha and CR values for all our constructs exceeded the criterion of 0.70 [39], confirming the reliability and consistency of the constructs. Discriminant validity was measured by looking at the HTMT value. It can be seen in Table 3, the entire HTMT value is below 0.9, which means that discriminant validity was established [39]. Based on this analysis, it can be concluded that this research model is valid and reliable.

**Table 2.** Construct reliability and validity test.

| Items | Measurements | | |
|---|---|---|---|
| | CA | AVE | CR |
| Health Consciousness (HC) | 0.956 | 0.919 | 0.972 |
| Ethical Identity (EI) | 0.942 | 0.896 | 0.963 |
| Environmental Consciousness (EC) | 0.874 | 0.799 | 0.922 |
| Sustainability Consciousness (SC) | 0.911 | 0.743 | 0.935 |

**Table 3.** HTMT results (discriminant validity).

| Items | HTMT results | | | |
|---|---|---|---|---|
| | EC | EI | HC | SC |
| EC | | | | |
| EI | 0.55 | | | |
| HC | 0.603 | 0.839 | | |
| SC | 0.877 | 0.663 | 0.7 | |

**Table 4.** Results of the structural model assessment.

| Hypothesis | Measurements | | | |
|---|---|---|---|---|
| | β values | T statistics | P values | Results |
| H1 (+): HC → SC | 0.394 | 5.125 | 0.000 | Supported |
| H2 (+): EI → SC | 0.015 | 0.211 | 0.416 | Not supported |
| H3 (+): EC → SC | 0.565 | 11.048 | 0.000 | Supported |
| H1 (+): HC → SC | 0.394 | 5.125 | 0.000 | Supported |

**Table 5.** R square comparison of Gen Y and Gen Z.

| | Gen Y | Gen Z |
|---|---|---|
| **R square** | 0.778 | 0.698 |
| **Adjusted R square** | 0.772 | 0.687 |

Research model testing was carried out using SmartPLS. Table 4 presents a summary of the results of this research's hypothesis testing. Hypothesis 1 which assumes that health consciousness will have a significant effect on sustainability consciousness can be proven well. This result is in accordance with the results from previous research which prove that awareness of the health of oneself and loved ones will make one more aware of environmental sustainability [10, 16]. Apart from that, Hypothesis 3 which predicts that environmental consciousness will have a significant positive effect on sustainability consciousness is also supported. These results provide an understanding that the more someone cares about their environment, the higher their consciousness of sustainability because the environment is part of sustainability that must be considered for the sake of future generations. These results are in accordance with the previous results of research [26].

On the contrary, Hypothesis 2 is not supported, meaning that the ethical identity of Indonesian society does not have a significant effect on sustainability consciousness. This indicates a discrepancy with the predictions made by previous authors who suggested that those with stronger ethical beliefs would possess a greater consciousness of sustainability [40]. However, a separate study resulted similar findings to the present study, indicating that the ethical identity of individuals in Hungary did not influence their level of environmental consciousness [10]. Lastly, this study also investigated the validity of the claim that Gen Z has a greater awareness of sustainability compared to Gen Y. To do so, we categorized these two age groups separately and compared the R square values. The R square value for Gen Y (age range 27-43 years) is higher than that for Gen Z (age range 15-26 years), as seen in Table 5. The analysis suggests that the correlation between the three factors is notably larger among those in the Gen Y age group compared to those in Gen Z. This finding diverges with the author's predictions and past research findings [4].

## 4 Discussion

This study is the first of its kind to investigate the influence of mindfulness on sustainability consciousness. Prior research has predominantly focused on sustainable consumption rather than sustainability consciousness [15, 41]. Several prior research have certainly endeavored to investigate these variables and occurrences, but they were conducted in industrialized nations that already incorporate mindfulness into their daily routines. In this study, we aimed to examine the structures within a wider scope, encompassing a diverse range of participants.

The results of this study suggest that consumers are driven to adopt sustainability consciousness by their health consciousness and environmental consciousness. Nevertheless, the ethical identity did not serve as a motivating factor for respondents to carry out this action. Contrary to expectations, generational variations do not significantly impact sustainability consciousness. However, this study reveals that Generation Y demonstrates a higher level of concern for sustainability compared to Generation Z. This may happen because Millennial are said to be a socially conscious generation that recognizes their responsibility in society. They understand the need to adopt environmentally friendly practices and are aware that by doing so, they are making a good impact on the environment and enhancing their own well-being [42].

This study adds to the existing body of research on sustainable and conscious consumption, as well as the literature on consumer behavior after disasters. The motivational influence of health consciousness, environmental consciousness, and ethical identity has been theorized using existing research on sustainable and mindful consumerism. However, this notion has not been explored in developing nations where the idea of mindfulness is still uncommon. Our study also enhances theoretical understanding by uncovering a unique pattern where distinct higher-level motivations serve as the primary factors influencing customer behavior. There is a lack of similar studies focusing on this matter in developing countries. Therefore, it is expected that the findings of this research will raise awareness among other researchers about the significance of mindfulness, which encompasses health consciousness, environmental consciousness, and ethical identity, as the primary driver in enhancing people's awareness of sustainability.

In addition to its theoretical contribution, the current study also holds significant management implications. The findings from this study can be applied by companies that provide sustainable products to improve the segmentation of consumer markets. This can be done by considering factors such as materialism and sustainability consciousness, in addition to the traditional segmentation factors like customer psycho graphics and demographics. The segmentation process could be informed by social media or search data collected from several platforms. Furthermore, as the research indicates that mindfulness is a significant antecedent of sustainable consciousness, it is advisable to incorporate mindfulness programs into induction programs in organizations or orientation programs in universities. This would enhance individuals' consciousness of their surroundings and foster a more robust sustainable consciousness.

Margareth Setiawan: Conceptualization, Methodology, Validation, Formal analysis, Resources, Data Curation, Data Analysis, Writing, Reviewing, Editing.
Dian Sadeli: Writing, Reviewing, Editing.
Siti Hajar Mohamad: Supervision.

## References

1. T. K. Dhandra, Achieving triple dividend through mindfulness: More sustainable consumption, less unsustainable consumption and more life satisfaction, Ecol. Econ, **161**, 83-90 (2019) https://doi.org/10.1016/j.ecolecon.2019.03.021
2. D. Evans, Thrifty, green or frugal: Reflections on sustainable consumption in a changing economic climate, Geoforum, **42**, 550–557 (2011). https://doi.org/10.1016/j.geoforum.2011.03.008
3. Khan, L. D. Hollebeek, M. Fatma, J. U. Islam, R. A. Rather, S. Shahid, and V. Sigurdsson, Mobile app vs. desktop browser platforms: The relationships among customer engagement, experience, relationship quality and loyalty intention, J. Mark. Manag, **39**, 275–297 (2023). https://doi.org/10.1080/0267257X.2022.2106290
4. P. Manchanda, N. Arora, O. Nazir, and J. U. Islam, Cultivating sustainability consciousness through mindfulness: An application of theory of mindful-consumption, J. Retail. Consum. Serv, **75**, 103527 (2023). https://doi.org/10.1016/j.jretconser.2023.103527
5. K. O'Brien, Is the 1.5°C target possible? Exploring the three spheres of transformation. Curr. Opin. Environ. Sustain, **31**, 153–160 (2018). http://dx.doi.org/10.1016/j.cosust.2018.04.010
6. Mishal, R. Dubey, O. Gupta, K. Omprakash, and Z. Luo, Dynamics of environmental consciousness and green purchase behaviour: An empirical study, Int. J. Environ. Clim, **9**, 682–706 (2017). https://doi.org/10.1108/IJCCSM-11-2016-0168
7. S. Mishra, T. R. Moharana, and R. Chatterjee, Exploring the role of self-conscious emotions between consumer minimalism and rental behavior, Mark. Intell. Plan, **42**, 262-283 (2023). https://doi.org/10.1108/MIP-07-2023-0322
8. W. Shahbaz and J. Parker, Workplace mindfulness: An integrative review of antecedents, mediators, and moderators. Hum. Resour. Manag. Rev. 32, 100849 (2022). https://doi.org/10.1016/j.hrmr.2021.100849
9. E. Bernal, D. Edgar, B. Burnes, Building sustainability on deep values through mindfulness nurturing. Ecol. Econ, **146**, 645–657 (2018). https://doi.org/10.1016/j.ecolecon.2017.12.003.
10. T. Benos, M. Burkert, V. Hüttl-Maack, and E. Petropoulou, When mindful consumption meets short food supply chains: Empirical evidence on how higher-level motivations influence consumers, Sustain. Prod. Consum, **33**, 520-530 (2022). https://doi.org/10.1016/j.spc.2022.07.028
11. S. Gupta, W. M. Lim, H. V. Verma, and M. Polonsky, How can we encourage mindful consumption? Insights from mindfulness and religious faith, J. Consum. Mark, **40**, (2023). https://doi.org/10.1108/JCM-11-2021-5011
12. D. Fischer, L. Stanszus, S. Geiger, P. Grossman, and U. Schrader, Mindfulness and sustainable consumption: A systematic literature review of research approaches and findings, J. Clean. Prod, **162**, 544-558 (2017). https://doi.org/10.1016/j.jclepro.2017.06.007.
13. T. Islam, Y. Wang, A. Ali, N. Akhtar, Path to sustainable luxury brand consumption: Face consciousness, materialism, pride and risk of embarrassment, J. Consum. Market, **39**, 11–28 (2022). https://doi.org/10.1108/JCM-09-2020-4099

14. E. Choi, J. A. Gruman, C. M. Leonard, A balanced view of mindfulness at work, Organizational Psychol. Rev, **12**, 35–72 (2022). https://doi.org/10.1177/20413866211036.
15. W. M. Lim, Inside the sustainable consumption theoretical toolbox: Critical concepts for sustainability, consumption, and marketing, J. Bus. Res, **78**, 69–80 (2017). https://doi.org/10.1016/j.jbusres.2017.05.001
16. D. Birch, J. Memery, and M. De Silva Kanakaratne, The mindful consumer: Balancing egoistic and altruistic motivations to purchase local food, J. Retail. Consum. Serv, **40**, 221–228 (2018). https://doi.org/10.1016/j.jretconser.2017.10.013.
17. M. F. Chen and N. P. Lin, Incorporation of health consciousness into the technology readiness and acceptance model to predict app download and usage intentions, Internet Res, **28**, 351–373 (2018). https://doi.org/10.1108/IntR-03-2017-0099.
18. R. Lathabhavan and R. Lathabhavan, Understanding the behavioural drivers of mindful consumption: A study on the use of sustainable menstrual products, J. Indian. Bus. Res, **16**, 308-326 (2024). https://doi.org/10.1108/JIBR-01-2023-0029
19. J. Strahler, Trait mindfulness differentiates the interest in healthy diet from orthorexia nervosa, Eat Weight Disord, **26**, 993-998 (2021). https://doi.org/10.1007/s40519-020-00927-2.
20. M. G. McEachern, G. Warnaby, M. Carrigan, and I. Szmigin, Thinking locally, acting locally? Conscious consumers and farmers' markets, J. Mark. Manag, **26**, 395–412 (2010). https://doi.org/10.1080/02672570903512494.
21. M. A. Starr, The social economics of ethical consumption: Theoretical considerations and empirical evidence. J. Soc. Econ. **38**, 916–925 (2009). https://doi.org/10.1016/j.socec.2009.07.006
22. N. Michaelidou and L. M. Hassan, The role of health consciousness, food safety concern and ethical identity on attitudes and intentions towards organic food, Int. J. Consum. Stud, **32**, 163–170 (2008). https://doi.org/10.1111/j.1470-6431.2007.00619.x
23. S. -T. Lin, S.-T and H. -J. Niu (2018). Green consumption: Environmental knowledge, environmental consciousness, social norms, and purchasing behavior, Bus. Strateg. Environ, **27**, (2018). https://doi.org/10.1002/bse.2233
24. N. T. Abd'Razack, S. O. Medayese, S. I. Shaibu, and B. M. Adeleye, Habits and benefits of recycling solid waste among households, Sustain. Cities Soc, **28**, 297–306 (2017). https://doi.org/10.1016/j.scs.2016.10.004
25. M. M. S. Law, P. Hills, and B. C. H. Hau, Engaging employees in sustainable development—A case study of environmental education and awareness training in Hong Kong, Bus. Strateg. Environ, **26**, 84–97 (2017). https://doi.org/10.1002/bse.1903
26. X. Chen, D.Y.C. Chan, and C.H. Wei, The research on environmental conscious and green consumption behavior in China, Serv. Sci Manag, **4**. 30–36 (2015). http://dx.doi.org/10.12677/SSEM.2015.44B005
27. J. Banyte, L. Salciuviene, A. Dovaliene, Z. Piligrimiene, and W. Sroka, Sustainable consumption behavior at home and in the workplace: Avenues for innovative solutions, Sustainability, **12**, 6564 (2020). https://doi.org/10.3390/su12166564
28. G. Prayag, J. M. Pung, C. Lee, and G. Del Chiappa, The self-concept and psychological antecedents of intention to recommend a heritage site: The moderating effects of visitor type, Tourism Manag. Perspect, **42**, 100962 (2022). https://doi.org/10.1016/j.tmp.2022.100962
29. E. Gentina and E. Perry, Generation Z in Asia: A research agenda (Ed.) The New Generation Z in Asia: Dynamics, Differences, Digitalisation (The Changing Context of Managing People), (Emerald Publishing Limited, Leeds, 2020). https://doi.org/10.1108/978-1-80043-220-820201002
30. O. Ivanova, J. Flores-Zamora, I. Khelladi, and S. Ivanaj, The generational cohort effect in the context of responsible consumption, Manag. Decis, **57**, 1162–1183 (2019). https://doi.org/10.1108/MD-12-2016-0915
31. S. Y. Youn and E. Cho, CSR ads matter to luxury fashion brands: A construal level approach to understand Gen Z consumers' eWOM on social media, J. Fash. Mark. Manag.: Int. J, **26**, 516–533 (2022). https://doi.org/10.1108/JFMM-12-2020-0269
32. J. Li and K. K. Leonas, Sustainability topic trends in the textile and apparel industry: A text mining-based magazine article analysis, J. Fash. Mark. Manag.: Int. J, **26**, 67–87 (2022). https://doi.org/10.1108/JFMM-07-2020-0139
33. L. Ferrari, C. M. Baum, A. Banterle, and H. De Steur, Attitude and labelling preferences towards gene-edited food: A consumer study amongst millennials and Generation Z, Br. Food J, **123**, 1268–1286 (2021). https://doi.org/10.1108/BFJ-09-2020-0820
34. G. Thomas, Corporate social responsibility as a sustainable business practice: A study among generation Z customers of Indian luxury hotels. Sustainability, **14**, 16813 (2022). https://doi.org/10.3390/su142416813
35. M. G. Confetto, C. Covucci, F. Addeo, and M. Normando, Sustainability advocacy antecedents: How social media content influences sustainable behaviours among Generation Z, J. Consum. Market, **40**, 758-774 (2023). https://doi.org/10.1108/JCM-11-2021-5038
36. J. Parzonko, A. Balinska, and A. Sieczko, Pro-environmental behaviors of Generation Z in the context of the concept of homo socio-oeconomicus, Energies, **14**, 1597 (2021). https://doi.org/10.3390/en14061597

37. M. Agrawal, P. Kalia, P. Nema, A. Zia, K. Kaur, and H. B. John, Evaluating the influence of government initiatives and social platforms on green practices of Gen Z: The mediating role of environmental awareness and consciousness, Clean. Respons. Consumpt, **8**, 100109 (2023). https://dx.doi.org/10.2139/ssrn.4358977
38. N. Gericke, J. Boeve-de Pauw, T. Berglund, and D. Olsson, The sustainability consciousness questionnaire: The theoretical development and empirical validation of an evaluation instrument for stakeholders working with sustainable development, Sustain. Dev, **27**, 35–49 (2019). https://doi.org/10.1002/sd.1859
39. J. F. Hair, W. C. Black, B. J. Babin, R. E. Anderson, and R. L. Tatham. Multivariate data analysis: A global perspective (Pearson Prentice Hall, New Jersey, 2010)
40. S. Hashem, G. Migliore, G. Schifani, E. Schimmenti, S. Padel, Motives for buying local, organic food through english box schemes, Br. Food J., **120**, 1600–1614 (2018). https://doi.org/10.1108/BFJ-08-2017-0426
41. S. Bahl, G. R. Milne, S. M. Ross, D. G. Mick, S. A. Grier, S. K. Chugani, S. S. Chan, S. Gould, Y. -N. Cho, J. D. Dorsey, R. M. Schindler, M. R. Murdock, and S. Boesen-Mariani, Mindfulness: Its transformative potential for consumer, societal, and environmental well-being, J. Public Policy Mark, **35**, 198-210 (2016). https://doi.org/10.1509/jppm.15.139
42. J. P. Bernardes, F. Ferreira, A. D. Marques, and M Nogueira, Generation Y's sustainability attitude-behaviour gap. Reverse Design, 623-628 (2018) https://doi.org/10.1201/9780429428210-77

# Identification of plastic waste generation and composition to reduce environmental disaster risk (case study: public facility sources in Padang City)

*Yenni* Ruslinda[1*], *Gebi* Fahrunisa[1], *Resti* Ayu Lestari[1], *Hendra* Gunawan[2], and *Muhammad* Amin[3]

[1]Environmental Engineering Departement, Faculty of Engineering, Universitas Andalas, Padang, Indonesia
[2]Civil Engineering Departement, Faculty of Engineering, Universitas Andalas, Padang, Indonesia
[3]Faculty of Geoscience and Civil Engineering, Kanazawa University, Japan

**Abstract.** Plastic waste is a material that is difficult to decompose and, if not managed properly, will cause environmental pollution. The degradation of plastic waste in aquatic environments can result in microplastics that threaten marine life, while the incineration of plastic waste produces toxic gases harmful to health. This study aims to identify the generation and composition of plastic waste as an initial step in reducing environmental disaster risks. The research was conducted at public facility sources in Padang City, including parks, beaches, recreational areas, and roads. Plastic waste generation at public facility sources in Padang City amounted to 13.76 tons/day, or 6.44% of the total public facility waste. The rate of plastic waste generation was 32.93 g/person/day or 1.13 l/person/day. The composition of plastic waste was dominated by PETE plastic at 43.21% and LDPE plastic at 25.56%, with 58.88% used for beverage packaging and 36.48% for food packaging. Only 11% of respondents from public facilities sort and sell plastic waste to collectors, while the remaining plastic waste was disposed of in landfill sites mixed with other waste. This can lead to environmental disaster risks around landfills, such as water and soil pollution.

## 1 Introduction

Plastic waste has recently become an important topic of conversation due to its ever-increasing quantity. Plastic is widely used because it is cheap, easy to obtain, and easy to use. Almost all food packaging and wrapping items use plastic, not to mention plastic for other needs such as household appliances and furniture, children's toys, sports equipment, electronic and medical equipment, and so on [1]. Based on data from the Ministry of Environment and Forestry (KLHK), waste generation in Indonesia will reach 19,542,090.78 tons by 2022. As much as 18% of the overall waste generation in Indonesia is plastic waste. This data places plastic waste in second place in the composition of waste by type [2]. If this plastic waste is not handled correctly, it will impact health and the environment because plastic is difficult to decompose. Plastic waste disposed of into the waters will cause pollution, thus disrupting the lives of marine animals [3].

Padang City, with an area of 694.96 km2, has the same problem as other cities regarding plastic waste. Based on SIPSN data in 2023, plastic waste composition in Padang City is 12.4% [2]. The source that produces the most plastic waste has yet to be known. However, research in the waters of Padang City found microplastic content of 1.667-8.333 particles/l, with the dominant types of microplastic polymers are polyethylene, polycarbonate, and polyethylene terephthalate. Microplastic particles contained in living organisms cause damage to the digestive system, inhibit growth, interfere with enzyme formation, reduce steroid hormone levels, impact reproduction, and cause exposure to toxic plastic additives [4]. The abundance of microplastics in Bungus Bay sediments, Padang City, was found to be 228.89 particles/kg of dry sediment [5].

Research conducted by Syahlan in 2019 stated that the percentage of plastic waste in domestic sources in Padang City was 19.62%, with a generation unit of 40.43 g/person/day in weight units or 0.84 l/person/day in volume units. The largest composition of plastic waste by type is PP at 42.97% and LDPE at 32.36%, while based on its use is as food packaging at 42.61% and others at 32.86%. Existing 3R activities that have been carried out amount to 25.48% of the total plastic waste, with the most active activities being selling plastic waste to stalls or collectors [6]. To obtain data on the generation and composition of plastic waste from other sources, this study was conducted at the source of public facilities in Padang City. The public facilities studied were parks, beaches, recreation areas, and roads [7, 8]. It is hoped that this research can add data on plastic waste generated from various sources in Padang City so that efforts can be made to manage it from the source, especially in reducing plastic waste. Based on Padang Mayor Regulation Number 44 of 2018 concerning Regional Policies and Strategies in the Management of Household Waste and

* Corresponding author: yenni@eng.unand.ac.id

Similar Waste, the waste reduction target in Padang City in 2025 is 30% of the total waste generated [9].

## 2 Methodology

The research stages consist of a literature review, data collection, and data processing and analysis. Data collection includes primary and secondary data. Primary data includes data on plastic waste generation, composition, and management at each source. Plastic waste generation and composition data were obtained from sampling and direct measurement in the field. Secondary data in the form of the number and location of public facilities in Padang City was obtained from the Padang City Statistics Agency (BPS).

Determining the number of samples based on SNI 19-3964-1994 concerning Procedures for Taking and Measuring Samples of Urban Waste Generation and Composition. The number of waste samples for non-domestic sources is calculated using Equation 1 [10].

$$S = C_d\sqrt{T_s} \quad (1)$$

with:
S = Number of samples
$C_d$ = Coefficient for medium cities, $C_d$= 1
$T_s$ = Total population

From the calculation results, the number of samples in this study was 18 units, with a survey reliability of 97.40%. Reliable survey reliability should be 90%-100% [8]. Table 1 shows the number of waste samples for public facilities in this study. The sampling locations were randomized by referring to the category of each public facility. Before sampling, a field survey was conducted to determine the sample location points and request the willingness of public facility managers to be sampled in this study.

**Table 1.** Number of waste samples from public facilities [11]

| No | Facilities | Amount* | Sample |
|---|---|---|---|
| 1. | Park | 10 | 3 |
| 2. | Beach | 5 | 2 |
| 3. | Recreational Area | 60 | 7 |
| 4. | Road | 1062 | 6 |
| | Total | | 18 |

Furthermore, plastic waste samples were taken from each source for eight consecutive days at the same location. Before sampling, plastic bags were distributed to the sampling location points as containers for plastic waste. The waste collected in plastic bags is generated in one day (24 hours) [10].

**Table 2.** Plastic types [8]

| No | Plastic Type | Example |
|---|---|---|
| 1 | PETE (*Polyethylene Terephthalate*) | Plastic in the form of bottles from mineral water, cooking oil bottles, juice bottles, and cosmetic bottles. |
| 2 | HDPE (*High Density Poly Ethylene*) | Plastic in the form of cosmetic bottles, liquid milk bottles, medicine bottles, and jerry cans of lubricants. |
| 3 | PVC (*Poly Vinyl Chloride*) | Plastic in the form of shampoo bottles, plastic tablecloths, building pipes, water hose pipes, and plastic toys. |
| 4 | LDPE (*Low Density Poly Ethylene*) | Plastics from frozen meat wrappers, plastic lids, plastic bags, and other thin plastics. |
| 5 | PP (*Polypropylen*) | Plastics from children's toys, plastic cups, and margarine wrappers. |
| 6 | PS (*Poly Styrene*) | Styrofoam food containers, CD cases, plastic cups, transparent plastic food containers, plastic spoons and forks. |
| 7 | Others/multilayer | Car parts, plastic packaging, baby feeding bottles, computers, lego toys, gallon water refills, household appliances, and electronic equipment. |

**Table 3.** Plastic usage [12]

| No | Plastic Usage | Example |
|---|---|---|
| 1 | Food packaging | Packaging of ice cream, noodles, biscuits, etc. |
| 2 | Beverage packaging | plastic bottles and sachet packaging for soft drinks, milk, juice, mineral water, energy supplements, and similar products. |
| 3 | Toilet packaging | Packaging of toothpaste, deodorant, hand soap and body wash, shampoo, facial cleanser and similar products. |
| 4 | Cosmetic packaging | Packaging of perfume, hand and body lotion, face cream, skin care, and other perfume and beauty products. |
| 5 | Home cleaning packaging | Packaging of floor cleaner, glass cleaner, *porcelain,* kitchen, cookware and others. |
| 6 | Laundry packaging | *detergent*, softener, bleach and others. |
| 7 | Lubricant packaging | Brake fluid, engine oil for vehicles and similar products. |
| 8 | Other packaging | Cooking oil, plastic bags, *hairspray* and similar items. |

Plastic waste samples were brought to the measurement location every day to be measured for plastic waste generation and composition. The measurement of waste generation was carried out in weight units by weighing the waste and in volume units by measuring the volume of waste in the compactor. The volume of plastic waste measured is the volume in the compactor stomped three times on the ground [10]. Furthermore, the composition of plastic waste was measured by sorting plastic waste based on its type and use and weighing the weight of each type and use of plastic waste. Table 2 and Table 3 show plastic waste composition by type and use

In addition, interviews were conducted with managers, visitors, or users of public facilities at each source. The number of visitors or users of public facilities interviewed for each category was 30 respondents. The questions asked were related to the current management of plastic waste. This data is needed to determine the amount of plastic waste produced and the management carried out by the community using public facilities.

Data processing and analysis are carried out by analyzing waste generation, composition, and management of existing plastic waste that has been carried out in public facilities in Padang City. Waste generation is the amount generated from a location within a specific time. Waste generation can be expressed in weight units (kg/person/day) and volume units (liter/person/day) [7, 8]. From the calculation of waste generation for eight consecutive days, two waste generation for the same day were obtained, namely the first day (Q1) and the last day (Q8). The correction factor is the ratio between the average waste of the same day and the waste of the first day. The unit of waste (q) is obtained from the waste data divided by the number of residents of each facility (p) [8]. The equations used are:

$$Q_r = (Q_1+Q_8)/2 \quad (2)$$

$$f_c = Q_r/Q_1 \quad (3)$$

$$q = Q/p \quad (4)$$

where:

Q = waste generation (kg/day, l/day)
$Q_1$ = day 1 waste generation (kg/day, l/day)
$Q_8$ = day 8 waste generation (kg/day, l/day)
$f_c$ = correction factor
q = generation unit (kg/person/day, l/person/day).

Waste composition analysis was conducted by dividing the weight of each plastic waste component by the total weight of plastic waste from each source. Plastic waste composition is expressed as a percentage of wet weight (%). The composition of plastic waste is calculated based on Equation 5 [8, 10].

$$C = (W_C/W_T) \times 100\% \quad (5)$$

where:

C = waste composition (%)
$W_C$ = weight of waste component (kg)
$W_T$ = total weight of waste (kg).

Data analysis of the plastic waste management system was carried out by processing the results of interviews with the population in public facilities. The results of data processing are presented in graphical form. Furthermore, recommendations for plastic waste management at the source of public facilities are made by prioritizing waste reduction activities through the 3R program, namely reduce by limiting the amount of waste, reuse by reusing waste, and recycle by recycling waste.

# 3 Results and discussion

## 3.1 Plastic waste generation in public facilities

The measurement results of plastic waste generation units at public facility sources in Padang City can be seen in Table 4. The average unit of plastic waste generated from public facility sources is 32.93 g/person/day in weight units or 1.13 l/person/day in volume units. Of all public facilities, the largest plastic waste generation units were found in park facilities, i.e., 71.47 g/person/day or 3.32 l/person/day. This is because the amount of plastic waste generated is large even though the number of visitors is less than that of other facilities. The composition measurement shows that the most plastic waste generated by the park is in the form of its use as beverage packaging, such as bottles and beverage cups. People use park facilities to relax and work in parks equipped with wifi, such as digital parks.

The unit of plastic waste generated at the end of the week is more than on other days because people visit public facilities more on holidays, such as parks, beaches, and recreation areas. This causes an increase in plastic waste generation at the end of the week. The total generation of plastic waste produced by public facilities in Padang City is obtained by multiplying the units of plastic waste generation by the number of public facilities and the population in each facility. From the calculation results, it was found that the total generation of plastic waste from public facility sources in Padang City was 13.76 tonnes/day, as shown in Table 5.

The plastic waste generated unit from domestic sources in terms of weight is larger than that of plastic waste generated from public facility sources. While in volume units, the waste generation unit from public facilities is larger. The average plastic waste generation unit from domestic sources in Padang City is 40.43 g/person/day in weight units or 0.84 l/person/day in volume units [6]. This difference is influenced by the type of plastic waste produced. In domestic sources, the dominant plastic waste is the PP type, which is often used as food packaging and children's toys, which have a greater weight. While in public facility sources, the dominant type of plastic is the PETE type, which is often used as a beverage bottle, so it has a larger volume.

The percentage of plastic waste at the source of public facilities can be calculated by dividing plastic waste generation unit in this study by the waste generation unit in public facilities in Padang City. The waste generation unit from public facilities in 2023 was obtained by applying to the waste generation unit from public facilities by Ruslinda in 2009. The waste generation unit from public facility sources in 2009 was 0.19 kg/person/day or 1.80 l/person/day [13]. The projection results show the waste generation unit from public facility sources in 2023 is 0.22 kg/person/day. With a total of 1137 public facilities, public facility waste generation was 75.69 tonnes/day. The percentage of plastic waste in public facility sources in 2023 amounted to 6.44%. This percentage is smaller than the percentage of plastic waste from domestic sources of 19.62%. [6]

**Table 4**. Plastic waste generation units at public facility sources in Padang City

| Public Facility | Waste Generation Units | Monday | Tuesday | Wednesday | Thursday | Friday | Saturday | Sunday | Average |
|---|---|---|---|---|---|---|---|---|---|
| Park | Weight (g/p/d) | 65.37 | 53.81 | 56.27 | 64.52 | 59.61 | 90.01 | 110.71 | 71.47 |
| | Volume (l/p/d) | 0.18 | 3.53 | 3.32 | 3.88 | 4.07 | 4.23 | 4.05 | 3.32 |
| Beach | Weight (g/p/d) | 14.65 | 15.97 | 16.60 | 14.60 | 17.82 | 15.47 | 24.58 | 17.1 |
| | Volume (l/p/d) | 0.25 | 0.28 | 0.31 | 0.30 | 0.31 | 0.32 | 0.29 | 0.30 |
| Recreation area | Weight (g/p/d) | 32.79 | 35.53 | 36.05 | 22.22 | 46.02 | 49.39 | 48.45 | 40.22 |
| | Volume (l/p/d) | 0.06 | 0.90 | 0.88 | 0.91 | 0.99 | 1.08 | 0.95 | 0.83 |
| Road | Weight (g/p/d) | 2.84 | 2.85 | 2.68 | 2.64 | 2.62 | 2.81 | 4.06 | 2.93 |
| | Volume (l/p/d) | 0.09 | 0.09 | 0.09 | 0.09 | 0.09 | 0.09 | 0.09 | 0.09 |
| Average | Weight (g/p/d) | 28.91 | 27.04 | 27.90 | 28.77 | 31.52 | 39.42 | 46.95 | 32.93 |
| | Volume (l/p/d) | 0.15 | 1.20 | 1.15 | 1.29 | 1.37 | 1.43 | 1.34 | 1.13 |

**Table 5**. Plastic waste generation at public facility sources in Padang City

| Public Facility | Number of Facilities* | Average Population | Waste Generation (g/person/day) | Total Waste Generation (Tonnes/day) |
|---|---|---|---|---|
| Park | 10 | 30 | 71.47 | 0.02 |
| Beach | 5 | 300 | 17.10 | 0.03 |
| Recreation area | 60 | 200 | 40.2 | 0.48 |
| Road | 1062 | 4256 | 2.93 | 13.23 |
| Total | | | | 13.76 |

## 3.2 Composition of plastic waste in public facilities

### *3.2.1 Composition by plastic type*

The measurement results of the plastic waste composition by type at the source of public facilities consist of PET plastic type of 43.21%, followed by LDPE at 25.56%, PP at 16.27%, PS at 8.45%, HDPE at 4.16%, PVC at 1.69%, and other types of plastic 0.64%. The results of this measurement can be seen in Table 6. PET plastic waste is the most common type of plastic generated from all sources of public facilities. The composition of PET plastic waste in parks and recreation facilities is almost 50% of the total plastic waste generated. This is related to the activities carried out in parks and recreation areas, such as drinking and snack activities that produce plastic beverage bottles and food packaging. According to data from the Japanese Plastic Waste Management Institute (PWMI) for 2022, PET accounts for 29.70% of all plastic waste generated [14].

LDPE plastic waste is generated in the largest amount at beaches and recreational facilities in the form of shopping bags and lightweight plastic containers. PP, PS, HDPE, PVC, and other plastic waste are mostly found on the road, which comes from the use of various types of product packaging, including disposable food containers and household appliances made of PP, while other types of plastic (multilayer) come from waste that is thrown onto the road such as packaging for medicines, milk and so on.

Compared with the research on plastic waste from domestic sources, differences in the composition of the plastic waste produced were found. The largest composition of plastic waste from domestic sources in Padang City is the PP type at 42.97%, in the form of fast-food packaging waste, snack waste, and cooking spices. In addition, LDPE plastic was also produced at 32.36%, which was found in the form of plastic bags and food packaging [6].

### *3.2.2 Composition of plastic waste by use*

The results of the measurement of the composition of plastic waste based on its use in public facilities in Padang City are shown in Table 7. The composition of this plastic waste consists of toiletries packaging 0.1 %, cosmetic packaging 0.74 %, house cleaning packaging 1.05 %, other packaging 1.83 %, laundry packaging and lubricant packaging did not contribute to plastic waste, while beverage packaging and food packaging contributed 58.88 % and 36.48 % of the total plastic waste, respectively. Most visitors who come to the public facility location shop for drinks and snacks sold at the location

**Table 6.** Composition of plastic waste by type

| Plastic Type | Composition (%) | | | | Average |
|---|---|---|---|---|---|
| | Park | Beach | Recreation Area | Road | |
| PET | 54.97 | 38.31 | 47.43 | 32.14 | 43.21 |
| HDPE | 2.85 | 3.32 | 3.71 | 6.77 | 4.16 |
| PVC | 0.40 | 0.69 | 0.20 | 5.47 | 1.69 |
| LDPE | 19.32 | 34.62 | 26.02 | 22.30 | 25.56 |
| PP | 14.68 | 15.11 | 14.91 | 20.39 | 16.27 |
| PS | 7.77 | 7.94 | 7.63 | 10.48 | 8.45 |
| Others | 0.00 | 0.00 | 0.11 | 2.46 | 0.64 |

**Table 7.** Composition of plastic waste by use

| Plastic Type | Composition (%) | | | | Average |
|---|---|---|---|---|---|
| | Park | Beach | Recreation Area | Road | |
| Food Packaging | 29.18 | 38.40 | 38.83 | 39.51 | 36.48 |
| Beverage Packaging | 68.41 | 55.58 | 57.53 | 54.02 | 58.88 |
| Toiletries Packaging | 0.00 | 0.69 | 0.19 | 1.14 | 1.01 |
| Cosmetic Packaging | 0.00 | 1.46 | 0.11 | 1.40 | 0.74 |
| Home Cleaning Packaging | 0.00 | 0.00 | 1.20 | 3.00 | 1.05 |
| Laundry Packaging | 0.00 | 0.00 | 0.00 | 0.00 | 0.00 |
| Lubricant Packaging | 0.00 | 0.00 | 0.00 | 0.00 | 0.00 |
| Others | 2.42 | 3.87 | 0.11 | 0.93 | 1.83 |

Plastic waste from toiletries, cosmetics, household cleaners, and other packaging each has a small contribution of 1-2%. Plastic waste from toiletry and cosmetic packaging is often found in beach facilities. This is related to swimming activities carried out by visitors in the beach area, especially by children. Plastic waste from cleaning applications is found in recreational facilities and on highways. This waste is supposed to come from other sources along the roadside. Plastic waste from other packaging is also found on the beach, including plastic bags and various types of products; this is caused by ocean currents carrying plastic waste from various sources to the beach. Meanwhile, plastic waste from laundry and lubricant packaging was not found in public facilities.

From the research on domestic sources in Padang City, it was obtained that the largest plastic waste composition based on its use was as food packaging at 42.61% and other packaging at 32.86%. Households in high-income groups produce the most plastic waste from food packaging and other waste compared to middle- and low-income groups. This is influenced by the lifestyle of high-income people who tend to consume packaged and fast food [6]. Meanwhile, research on non-domestic sources in Bandung City found that the composition of plastic waste used in food packaging, toiletries packaging, household cleaning packaging, and laundry packaging was 20% each [12]. The difference in activity affects the difference in each source in terms of plastic use in everyday life.

### 3.3 Plastic waste management at public facility

Data on existing plastic waste management in public facilities was obtained from interviews with respondents in public facilities, such as managers and visitors. Of the 18 public facility units sampled, only 11% have sorted, collected, and sold plastic waste to collectors. These public facilities are a digital park and Padang Beach. Digital Park managers provide segregated waste containers, one of which is for plastic waste. Next, the plastic waste is collected at the sorting location to be sorted, cleaned, and packed into sacks. This plastic waste is then sold to collectors. Meanwhile, waste management is carried out in the Padang Beach area by sorting, collecting, and selling waste to the Nabuang Sarok Recycling Center of PT Semen Padang. This activity is carried out by community groups around the beach area and environmentalist communities, which often hold environmental clean-up activities around the beach, especially on holidays.

As many as 81% of public facilities have not yet sorted and processed plastic waste. In this facility, plastic waste is collected at the temporary waste storage site and then taken to the waste final processing site for landfill. Plastic waste is difficult and can even take decades to decompose. Piling plastic waste in landfills, if not appropriately managed, will pose a risk of environmental disasters such as water and soil pollution. Polluted land and water will cause problems with human health [15]. Figure 1 shows the existing management of plastic waste in Padang City public facilities.

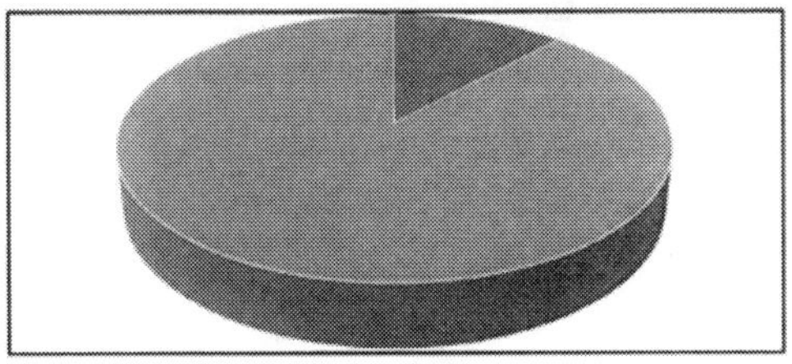


**Fig. 1.** Existing plastic waste management at public facilities

The interview results also showed that there have been efforts to reduce plastic waste. Visitors generally carry out this activity by bringing tumblers, but the percentage is only 28%. Carrying a tumbler can reduce plastic bottles and glass waste. Research from domestic sources shows that the use of tumblers can reduce plastic waste by 5.97% of the total plastic waste produced [16].

### 3.4 Recommendation

From the results of measuring the generation and composition, as well as the existing management of plastic waste, recommendations can be provided to reduce plastic waste generation from public facility sources, as follows:

1. For visitors to public facilities that are plastic waste generators, limiting plastic waste can be done by bringing a tumbler, a reusable shopping bag, and bringing their own food box to reduce plastic waste in the form of drinking bottles, plastic bags, and styrofoam.
2. For public facility managers who are waste managers in the public facility environment, it is recommended to provide separate containers and waste collection tools according to the type of waste generated. This sorting is intended to facilitate further waste management. In addition, managers are also asked to provide a location for processing plastic waste, at least for sorting, cleaning, and packaging. Furthermore, sorted plastic waste is sold to collectors (informal sector) or to Nabuang Sarok PT Semen Padang, which is one of the waste recycling centers in Padang City. This plastic waste can be reused as raw material in the waste recycling industry or can be used as biomass as an alternative fuel. For this reason, public facility managers must conduct socialization activities to limit, reuse, and recycle plastic waste. Socialization can be done by installing banners and posters in the public facility environment. Socialization can also be carried out in collaboration with environmentalist communities to hold enjoyable activities in terms of environmental awareness.
3. Vendors selling in public facilities are asked to provide refillable gallons of drinking water and use refillable cleaning containers. This activity can reduce plastic waste by reusing the plastic waste generated.
4. The government is advised to promote 3R activities by facilitating permits for related events in public facilities, providing infrastructure and facilities for plastic waste collection, and providing incentives and disincentives to public facilities that implement 3R activities. In addition, it is recommended to make a plastic waste reduction policy with fines for violators and to succeed in the paid plastic bag program in shops and cafes in recreational areas.
5. The informal sector is advised to become a partner in the reuse of plastic waste from public facilities. This informal sector comprises collectors, waste banks, and Nabuang Sarok PT. Semen Padang which can accept all types of plastic waste. This plastic waste will be burned in a kiln as an alternative fuel substitute. According to Novita and Damanhuri, plastic has a high calorific value ranging from 5000-13000 kcal/kg to help the combustion process [9].

## 4 Conclusion

The conclusions obtained from this research are:

1. The average generation of plastic waste from public facilities in Padang City was 32.93 g/p/d or 1.13 l/p/d. Parks had the highest generation unit at 71.47 g/p/d or 3.32 l/p/d, compared to recreation areas at 40.22 g/p/d or 0.83 l/p/d, beaches at 17.10 g/p/d or 0.30 l/p/d, and roads at 2.93 g/p/d or 0.09 l/p/d. The percentage of plastic waste in public facility sources is 6.44%. Plastic waste composition in public facilities is dominated by PET at 43.21%, followed by LDPE at 25.56%, PP at 16.27%, PS at 8.45%, and other types at 0.64%. Beverage packaging contributes the most plastic waste based on its use at 58.88%, followed by food packaging at 36.48%, while toiletries, cosmetics, house cleaning, and other packaging each have smaller contributions.
2. The existing management of plastic waste in public facilities involves the sorting and selling plastic waste to the informal sector, which is only carried out in 11% of existing public facilities. The remaining 81% collect plastic waste at TPS and then transport it to TPA for landfill, risking causing environmental disasters such as water and soil pollution.
3. Recommendations given in the management of plastic waste in public facilities for visitors are to limit the amount of plastic; and for managers, it is the provision of containers and segregated collection tools, the provision of locations and officers for sorting, cleaning, and packaging plastic waste. For traders in the public facility environment, it is done by providing gallons of drinking water and refillable cleaning packaging. For the government, it is recommended to promote the 3R activities of plastic waste by providing collection facilities, making plastic waste reduction policies, and providing

incentives and disincentives to public facilities that carry out 3R activities. For the informal sector, it is suggested that they can become partners in plastic waste utilization activities.

## References

1. V.K. Manam, A.K. Nakkella, Plastic waste management: turning challenges into opportunities, (Bharti Publications, New Delhi, 2022)
2. Ministry of Environment and Forestry of the Republic of Indonesia. (2023) https://Sipsn.Menlhk.Go.Id/Sipsn/.
3. P.G.C.N.T. Pilapitiya, A.S. Ratnayake, The world of plastic waste: A review. Clean. Mat. 11, 100220 (2024)
4. F.I. Jamika, Analysis of microplastic content in Air Manis Beach area of Padang City, Thesis, Universitas Andalas (2021)
5. M. D. Islami, E. Elizal, Y. I. Siregar, Distribution of microplastic at sediments in the coast of Bungus Bay Padang West Sumatera Province. J. of Coas. and Oce. Sci, **1**, 1 (2020).
6. A. Syahlan, Assessment of the Potential of Reduce, Reuse, Recycle (3R) plastic waste activities at domestic sources in Padang City, Thesis, Universitas Andalas (2021).
7. G. Tchobanoglous, and K. Frank, Handbook of solid waste management", 2nd ed, (Mc Graw Hill Inc, New York, 2002).
8. E. Damanhuri and T. Padmi, Integrated waste management (ITB Press, Bandung, 2016).
9. Padang Mayor Regulation Number 44 of 2018 concerning Regional policies and strategies in the management of household waste and similar waste, (2018)
10. National Standardization Agency of Indonesia, SNI 19-3964-1994, Methods of measurement of generation rate and composition of municipal solid Waste, (1994).
11. Central Bureau of Statistics, Padang in Figures, (2023).
12. M. Chaerul, A. R. Fahruroji, T. Fujiwara, Recycling of plastic packaging waste in Bandung City, Indonesia. J. of Mat. Cyc and Waste Manag, **16**, 3, (2014).
13. Y. Ruslinda, R. Aziz, S. S. Abuzar, Waste generation, composition, and recycling potential from various sources in Padang City. J. of Purification, **11**, 2 (2009)
14. Plastic Waste Management Institute, An Introduction to Plastic Recycling. (2022).
15. M .G. Kibria, N. I. Masuk, R.Safayet, H.Q. Nguyen and M. Mourshed, Plastic waste: challenges and opportunities to mitigate pollution and effective management, Int. J. Environ. Res. **17**, 1, 20, (2023)
16. D. M. Novita and E. Damanhuri, Calculation of calorific value based on the composition and characteristics of urban waste in Indonesia in the concept of waste to energy. J. of Env. Eng. **16**, 2, (2010)

# How to support recovery phase for landslide through livelihoods? case study: Kulon Progo Regency, Indonesia

*Diah* Setyawati Dewanti[1*], *Jejen* Jaenul Haq[1], *Nita* Viviani Nurhanifah[2], and *Chuanchen* Bi[3]

[1]Department of Economics, Faculty of Economics and Business, Universitas Muhammadiyah Yogyakarta, Yogyakarta, Indonesia
[2]Master of Economics, Universitas Muhammadiyah Yogyakarta, Yogyakarta, Indonesia
[3]College of Graduate Study in Management, Khon Kaen University, Chang Wat Khon Kaen 40002, Thailand

**Abstract.** The global climate is increasingly responsible for the frequency of extreme weather events, which have significant implications for vulnerable biological systems. Among these events, rainfall-induced landslides are recognized as one of the most severe geological disasters, particularly in Indonesia. The recovery phase is often the weakest link in disaster management implementation. This study aims to discern and assess the effects of the recovery phase on livelihoods in Kulon Progo, Indonesia, following landslides. The study examines the impact of human, social, physical, natural, and financial capital, as well as transformations in structure and process (TSP), on the recovery phase. The unit of analysis is households, with data collected from 300 households through stratified random sampling. Path analysis was applied to achieve the study's objectives. The findings reveal that human capital (0.376**), social capital (0.279***), physical capital (0.182*), and TSP (0.263***) have a direct effect on the recovery phase. Additionally, social capital (0.352***) and physical capital (0.180**) indirectly influence the recovery phase through TSP as a mediator. To strengthen the recovery phase in landslide disaster management, the study identifies seven activities that directly support the recovery phase and another seven that indirectly contribute to its reinforcement.

## 1 Introduction

In recent years, rapid urbanization and the expansion of human settlements have significantly altered land-use patterns across various regions [1, 2], leading to a marked reduction in ecological lands such as forests and wetlands [3, 4]. These areas, once crucial in maintaining environmental balance, have been increasingly replaced by urban development, resulting in the fragmentation of natural water networks and further compromising urban ecological systems [5, 6]. Compounding these challenges, the global phenomenon concerning rising temperatures has intensified severe droughts [7], particularly in tropical regions like Indonesia, where heavy rainfall poses a significant threat to fragile ecosystems. This has led to an increased frequency of rainfall-induced landslides, now recognized for being one of the greatest deadly catastrophes in the country [8].

In response to these growing threats, Indonesia has implemented a range of disaster risk reduction (DRR) policies aimed at mitigating the harmful effects within catastrophes [9-11]. These policies include government legislation, coordination among various agencies, community engagement, and comprehensive damage and loss assessments [12, 13]. However, despite these efforts, the rehabilitation and recovery processes following disasters, such as landslides, often remain slow and inadequate [14, 15]. This highlights the critical need for resilience in recovery, where community resources and strengths fulfil a crucial part in preserving growth along with wellness post-disaster [16-18]. Research suggests that the success of individual recovery efforts is closely linked to the overall success of the community [19-21], which depends on the availability and effective use of resources, including social networks and institutional support systems [22-24].

The risk of landslides in regions like Kulon Progo is particularly heightened by heavy rainfall, a consequence of worsening tropical climate conditions [25, 26]. Addressing this risk requires robust disaster preparedness strategies that consider the specific vulnerabilities of each location [27-29], including the need for full-time disaster management professionals and the integration of both formal and informal education and training initiatives [30-32]. This holistic approach is essential for enhancing community resilience and ensuring sustainable disaster management [33-35].

This study explores the relationship between Sustainable Livelihoods Assets (SLA) and the recovery phase following landslides in Kulon Progo Regency, Indonesia. The SLA framework, which integrates financial resources and organizational modifications [36, 37], is instrumental in identifying key areas for development initiatives aimed at improving the standard of living for impoverished and vulnerable communities,

* Corresponding author: ddewanti@umy.ac.id

especially within the context of disaster preparedness and recovery [38, 39].

By focusing on the recovery phase after landslides, this research emphasizes the importance of rebuilding not just physical infrastructure but also the diverse assets that underpin community survival and prosperity, including human, natural, physical, financial, and social capital [40, 41]. The study aims to provide valuable insights into effective risk management practices that enhance local resilience and reduce exposure to future catastrophes, with a particular focus on the crucial role of individual motivation and societal resilience in preventing preparedness and recovery.

## 2 Data and methods

### 2.1 Study area

The research study focuses on Kulon Progo Regency, a region marked as green in Fig. 1. Located in Yogyakarta Province, Kulon Progo is particularly vulnerable to landslides, a risk exacerbated by its diverse topography and climatic conditions. This district encompasses 12 sub-districts and 88 villages, covering a total area of approximately 586 km². Within Kulon Progo, the largest sub-districts are Kokap, with an area of 73 km², and Samigaluh, spanning 69 km². Conversely, Wates is the smallest sub-district, occupying 32 km². The geographical coordinates of the study area range from 7° 38'42" S to 7° 59'3" S latitude and 110° 1'37" E to 110° 16'26" E longitude [42].

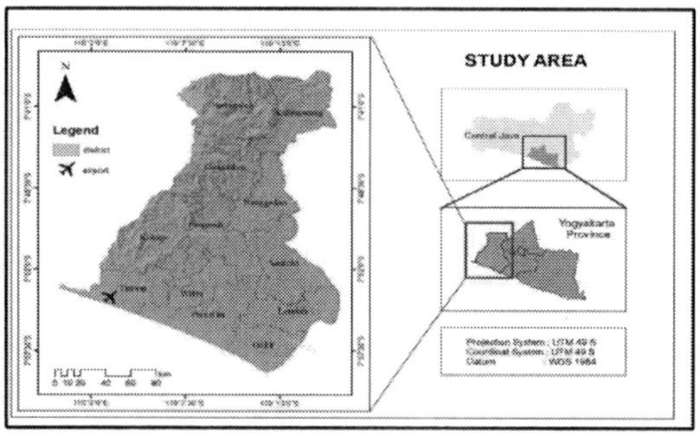


**Fig. 1.** Landslide hazard zonation map in Kulon Progo [43]

Kulon Progo features a varied topography, including lowland areas and elevated highlands with elevations ranging from sea level up to 1000 meters. The northern part of the district, notably mountainous, has elevations between 500 and 1000 meters. This region includes sub-districts such as Girimulyo, Kalibawang, Samigaluh, and Kokap. Despite their designation as conservation areas, these northern highlands are highly susceptible to severe landslides due to their steep slopes and frequent heavy rainfall [44].

### 2.2 Data collection

This research uses a quantitative approach, emphasizing the collection of primary data to provide a robust analysis of landslide recovery. The research framework is designed to establish benchmarks for effective landslide recovery through community-based approaches. Data collection was carried out in six villages within the Samigaluh district: Banjarsari, Gerbosari, Kebonhargo, Ngargosari, Purwoharjo, and Sinduharjo. This selection was informed by previous studies which have identified these villages as prone to rainfall-induced landslides, with the rainy season posing significant geotechnical hazards [45].

The focus of the analysis is on households, which are considered the primary unit of analysis. In this framework, disaster recovery is treated as the dependent variable. The independent variables are categorized into two main groups: household capitals and the transformation of structures and processes (TSP). Household capitals are classified into five distinct types: human capital (skills, education, and health), social capital (networks and relationships), natural capital (natural resources and environmental conditions), physical capital (infrastructure and equipment), and financial capital (monetary resources and economic stability).

The TSP aspect includes the integration and impact of cultural practices and beliefs on the recovery process. This component examines how traditional and culturally embedded practices influence recovery efforts and contribute to resilience. The study's sample comprises 300 households spread across the six selected villages. This sample size is intended to provide a representative overview of the impact of various capitals and TSP on disaster recovery outcomes. The data gathered will be analyzed to determine how different forms of household capital and transformations in structures and processes affect recovery from landslides, providing insights into effective community-based recovery strategies.

The study utilized both descriptive analysis and path analysis to meet its research objectives. Path analysis is particularly useful in this context because it allows for the estimation of the magnitude and significance of hypothesized cause-and-effect relationships among various variables, which are visually represented through path diagrams [46, 47]. The variables in the study were categorized as either ordinal or categorical. Categorical variables, which are qualitative, were grouped into distinct categories representing different qualities or characteristics. To further analyze the data, the study employed a Structural Equation Model (SEM) with a path analysis approach. This model distinguished between endogenous and exogenous variables [48-50]. The endogenous variables included livelihood assets and structural transformation. The exogenous variable was recovery, focusing on post-disaster management related to landslides, specifically examining structural transformation and the recovery phase. In this model, financial capital was represented solely by household income, which was measured on a monthly basis. Structural transformation was assessed through changes in cultural and religious aspects. The SEM equation used in the study was formulated to reflect these relationships and variables. The SEM equation was formulated as follows:

$$Recov_{11} = \beta_{11}\, transf_{11} + \beta_{12}\, HC_{12} + \beta_{22}\, SC_{22} + \beta_{32}\, NC_{32} + \beta_{42}\, PC_{42} + \beta_{52}\, FC_{52} \quad (1)$$

$$Transf_{11} = \beta_{12}\, HC_{12} + \beta_{22}\, SC_{22} + \beta_{32}\, NC_{32} + \beta_{42}\, PC_{42} + \beta_{52}\, FC_{52} \quad (2)$$

## 3 Results and discussion

To minimize the detrimental effects of catastrophes and ensure sustainable recovery, disaster management involves a comprehensive set of actions [51, 52]. The initial phase of recovery is a critical part of the process of catastrophe management, requiring a well-coordinated and community-focused approach to be effective [53-55]. This phase seeks not only to restore communities to their pre-disaster state but also to strengthen their resilience, making them better prepared for future disasters [56]. Essential components of a successful recovery strategy include building resilient infrastructure, enhancing institutional capacities, and implementing community-based recovery initiatives that target both current and future needs [57-60].

Post-disaster challenges are often complex, involving extensive infrastructure damage, socio-economic disruption, and psychological trauma [61-63]. Effective disaster management during the recovery phase requires a holistic approach that addresses these diverse issues [64]. Recovery efforts must go beyond the reconstruction of physical structures; they should also focus on the comprehensive restoration of lives, livelihoods, and local economies [65-67]. By utilizing local knowledge, fostering collaboration among stakeholders, and applying evidence-based strategies, disaster management can enhance recovery efforts as well as create a better tomorrow [68-71]. This approach is increasingly vital as global changes cause greater amounts of serious disasters, emphasizing the need for strong recovery frameworks as shown in Fig. 2 [72].

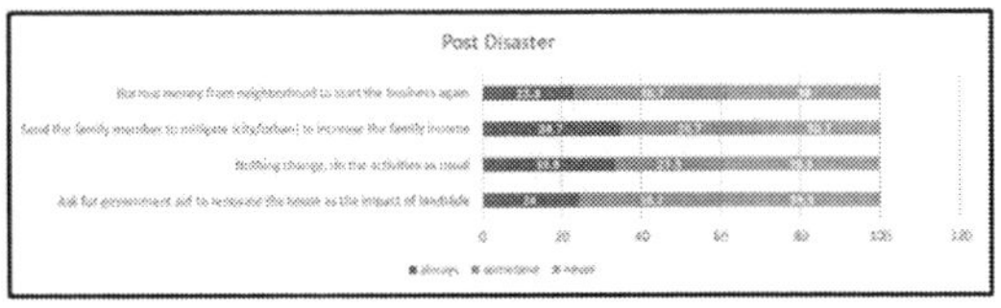


**Fig. 2.** Frequency analysis of the recovery phase (source: analysis, 2023)

In Samigaluh, the government's role in rehabilitating properties affected by landslides is evident. However, when government assistance is insufficient, households often adopt alternative recovery strategies. For instance, 38.7% of households seek financial support from neighbors, while 34.7% send family members to urban areas to supplement household income. This reliance on additional income sources highlights the limitations of government aid in fully addressing recovery needs. Additionally, 24% of households regularly request government assistance for home repairs following landslides, indicating a significant dependency on government policies and support during disaster recovery in Kulon Progo Regency.

The frequency analysis suggests that households in Kulon Progo Regency demonstrate a strong sense of independence and are less dependent on government assistance, with their disaster recovery largely supported by community cooperation. Path analysis further reveals that all household capitals—human, social, natural, physical, and financial—significantly impact the recovery phase, both directly and indirectly, at significance levels of 5% and 1%. The transformation of processes and structures is an endogenous variable, influenced by household capitals and the recovery phase, while the household capitals themselves serve as exogenous variables.

Figure 3 and 4 presents the quantitative coefficients for the direct and indirect effects on the recovery phase. The direct effects are quantified as follows: transformation of processes and structures (0.310**), human capital (0.041*), social capital (0.085*), natural capital (-0.070*), physical capital (0.076*), and financial capital (0.017*). Notably, natural capital has a negative direct effect on the recovery phase, suggesting that preserving or enhancing natural capital can reduce the need for extensive recovery efforts. This implies that effective natural resource management can help prevent landslides and mitigate disaster impacts.

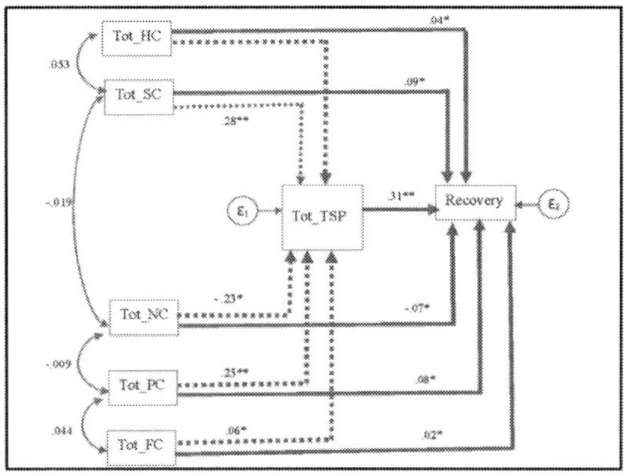


**Fig. 3.** Diagram of path analysis (source: analysis, 2023)

The transformation of processes and structures, which includes changes in policies, cultural practices, and belief systems, has the highest coefficient and greatest significance, underscoring its critical role in supporting recovery efforts. Households perceive these factors as essential for facilitating recovery after landslides. During the recovery phase, human capital is significantly impacted, requiring attention not only to the restoration of physical and economic assets but also to the psychological, health, and skill-based challenges faced by individuals in affected communities [73-75]. Providing psycho-social support, healthcare, and skill development is crucial for restoring the well-being, productivity, and resilience of community members, enabling them to contribute effectively to and benefit from the recovery process [76, 77]. Social capital also plays a vital role in recovery. Disasters can weaken community bonds, trust, and collaborative networks, making it essential to reinforce these ties through targeted [78, 79]. Strengthening social capital is critical, as robust community networks facilitate mutual assistance, knowledge sharing, and collective action—key elements for achieving a comprehensive and sustainable recovery [80, 81].

The degradation of natural capital, such as ecosystems and biodiversity, can impede the recovery phase. Natural assets provide essential services like flood regulation, water purification, and erosion control, which are crucial for mitigating disaster impacts [82, 83]. Without these protective natural barriers, communities may face prolonged recovery periods and increased vulnerability to future disasters, emphasizing the crucial link between environmental health and disaster resilience [84-86]. Physical capital, including infrastructure, housing, and production facilities, often sustains the most direct damage during disasters and requires significant restoration efforts during recovery [87, 88]. The effectiveness and speed of recovery from physical damage determine the extent of enduring harm to this capital. Persistent deficiencies in physical assets can hinder socio-economic revitalization and heighten community vulnerability to future disasters [89, 90].

Total effects

| | Coef. | OIM Std. Err. | z | P>\|z\| | [95% Conf. | Interval] |
|---|---|---|---|---|---|---|
| **Structural** | | | | | | |
| transf_tot_index <- | | | | | | |
| hc_total_index | .1335549 | .0577515 | 2.31 | 0.021 | .0203641 | .2467457 |
| sc_tot_index | .2755613 | .0661721 | 4.16 | 0.000 | .1458663 | .4052562 |
| nc_tot_index | -.2260253 | .0778909 | -2.90 | 0.004 | -.3786886 | -.0733621 |
| pc_tot_index | .2464168 | .0494285 | 4.99 | 0.000 | .1495388 | .3432949 |
| log_pendapatan | .0563662 | .019594 | 2.88 | 0.004 | .0179627 | .0947697 |
| recov_index <- | | | | | | |
| transf_tot_index | .3097725 | .0670854 | 4.62 | 0.000 | .1782875 | .4412575 |
| hc_total_index | .0413716 | .020008 | 2.07 | 0.039 | .0021567 | .0805866 |
| sc_tot_index | .0853613 | .0276029 | 3.09 | 0.002 | .0312607 | .1394619 |
| nc_tot_index | -.0700164 | .0284973 | -2.46 | 0.014 | -.1258702 | -.0141627 |
| pc_tot_index | .0763332 | .0225326 | 3.39 | 0.001 | .0321701 | .1204962 |
| log_pendapatan | .0174607 | .0071512 | 2.44 | 0.015 | .0034446 | .0314768 |

**Fig. 4.** Diagram of path analysis (source: analysis, 2023)

In the context of sustainable livelihoods, "transformation" refers to the essential changes needed to enhance the resilience, adaptability, and sustainability of livelihoods in the face of various vulnerabilities. Transformation of processes involves revising methods, strategies, and practices to achieve and maintain sustainable livelihoods. This type of transformation focuses on improving how resources are utilized and managed to enhance livelihood [91-94]. Conversely, transformation of structures pertains to systemic and institutional changes that shape societies and economies, which, in turn, influence livelihoods [95]. This includes changes in policies, governance, and institutional frameworks that affect how resources are distributed and accessed.

Recognizing the need for transformation in both processes and structures is vital for ensuring that livelihoods are not only sustained but can thrive under dynamic conditions. The goal is to create environments where individuals and communities have the necessary resources, capabilities, and flexibility to adapt, innovate, and maintain their well-being, even when confronted with shocks, stresses, or long-term challenges.

The indirect effects of various forms of capital—human, social, natural, physical, and financial—on the transformation of processes and structures are significant. Human capital, which includes skills, knowledge, health, and abilities, is notably impacted by these transformations. For instance, shifts in livelihood strategies or systemic changes may require the development of new skills and adaptive capacities for individuals and communities [96]. Moreover, transformations in institutional structures, governance mechanisms, or resource access can directly influence the quality of human capital by affecting health, education, and training opportunities.

Social capital, natural capital, physical capital, and financial capital also play substantial roles in influencing the recovery phase. Social capital, involving networks and relationships, has a considerable positive impact on recovery, as community connections facilitate support and resource sharing. Natural capital, which encompasses environmental resources, can negatively impact recovery if depleted, whereas physical capital, such as infrastructure, and financial capital, which includes economic resources, both significantly contribute to the recovery phase. Understanding these relationships underscores the importance of comprehensive and integrated approaches to both process and structural transformations in promoting sustainable livelihoods and enhancing recovery efforts.

## 4 Conclusion

The recovery phase in disaster management significantly influences livelihood assets, highlighting the interconnectedness of communities in the aftermath of a landslide. Effective recovery extends beyond merely rebuilding physical infrastructure; it also involves restoring and strengthening the critical assets—human, social, natural, physical, and financial—upon which households depend. A key lesson from disaster management is the role of livelihood assets during recovery. For instance, households frequently turn to local community networks for support in their recovery efforts. These community networks are vital for rebuilding businesses and improving overall well-being, as evidenced by 23.3% of households that rely on them. Such collective actions not only facilitate immediate recovery but also contribute to long-term resilience. Another common strategy involves sending a family member to work in urban areas to increase income and improve welfare after a disaster. This approach, adopted by 34.7% of respondents, underscores the importance of urban employment in providing crucial financial support to those who remain in disaster-affected regions. Diversifying income sources is essential for stabilizing livelihoods and rebuilding after a disaster. Providing targeted assistance to vulnerable households is also essential. Tailored support helps mitigate the impacts of future disasters and strengthens recovery capacity. Interventions that address the specific needs of at-risk households are key to ensuring effective recovery.

Policy recommendations for enhancing livelihood strategies include reinforcing assets through targeted interventions. Policies should focus on bolstering various forms of capital to ensure that households have the necessary resources to recover and adapt. For example, Indonesia's "Panduan Mitigasi Desa Siaga Bencana Longsor" offers a framework for disaster mitigation at the village level, emphasizing community preparedness and response. Incorporating a willingness to pay (WTP) assessment into disaster guidelines is crucial for understanding community readiness to invest in preparedness and recovery efforts. Insights from WTP assessments enable the design of more effective and sustainable disaster management strategies. Additionally, Indonesia's national Community-based Disaster Management (CBDM) system underscores the importance of local communities in disaster preparedness and response. However, there is a need to develop practical, context-specific disaster guidelines to effectively address unique local challenges. Lastly, addressing the psychological impacts of disasters, such as post-traumatic stress, is critical. Enhancing cognitive appraisal processes, including insurance planning and financial protection measures, enables communities to better assess their situation, make informed decisions, and plan for future risks, thereby improving overall resilience to disasters.

## References

1. Niu, B., Ge, D., Yan, R., Ma, Y., Sun, D., Lu, M., and Lu, Y. The Evolution of the Interactive Relationship between Urbanization and Land-Use Transition: A Case Study of the Yangtze River Delta. Land, **10**(8), 804. (2021) https://doi.org/10.3390/land10080804
2. Nuissl, H., and Siedentop, S. Sustainable Land Management in a European Context (T. Weith, T. Barkmann, N. Gaasch, S. Rogga, C. Strauß, and J. Zscheischler, Eds.; **8**). Springer International Publishing (2021). https://doi.org/10.1007/978-3-030-50841-8
3. Athukorala, D., Estoque, R. C., Murayama, Y., and Matsushita, B. Impacts of Urbanization on the Muthurajawela Marsh and Negombo Lagoon, Sri Lanka: Implications for Landscape Planning towards a Sustainable Urban Wetland Ecosystem. Remote Sensing, **13**(2), 316 (2021). https://doi.org/10.3390/rs13020316
4. Alikhani, S., Nummi, P., and Ojala, A. Urban Wetlands: A Review on Ecological and Cultural Values. Water, **13**(22), 3301 (2021). https://doi.org/10.3390/w13223301
5. Dadashpoor, H., Azizi, P., and Moghadasi, M. Land use change, urbanization, and change in landscape pattern in a metropolitan area. Science of The Total Environment, **655**, 707–719 (2019). https://doi.org/10.1016/j.scitotenv.2018.11.267
6. Liu, Y., and Zhang, Y. Responses of Ecosystem Services to Land Use/Cover Changes in Rapidly Urbanizing Areas: A Case Study of the Shandong Peninsula Urban Agglomeration. Sustainability, **16**(14), 6100 (2024). https://doi.org/10.3390/su16146100
7. Perera, A. T. D., Nik, V. M., Chen, D., Scartezzini, J.-L., and Hong, T. Quantifying the impacts of climate change and extreme climate events on energy systems. Nature Energy, **5**(2), 150–159 (2020). https://doi.org/10.1038/s41560-020-0558-0
8. Chikalamo, E. E., Mavrouli, O. C., Ettema, J., van Westen, C. J., Muntohar, A. S., and Mustofa, A. Satellite-derived rainfall thresholds for landslide early warning in Bogowonto Catchment, Central Java, Indonesia. International Journal of Applied Earth Observation and Geoinformation, **89**, 102093 (2020). https://doi.org/10.1016/j.jag.2020.102093
9. Ayuningtyas, D., Windiarti, S., Hadi, M. S., Fasrini, U. U., and Barinda, S. Disaster Preparedness and Mitigation in Indonesia: A Narrative Review. Iranian Journal of Public Health, **50**(8), 1536–1546 (2021). https://doi.org/10.18502/ijph.v50i8.6799
10. Sarjito, A. Evaluation of Indonesian Government Policies in Addressing Climate Change and Natural Disasters. JISHUM: Jurnal Ilmu Sosial Dan Humaniora, **2**(1), 103–124 (2023). https://doi.org/10.57248/jishum.v2i1.283
11. Triyanti, A., Surtiari, G. A. K., Lassa, J., Rafliana, I., Hanifa, N. R., Muhidin, M. I., and Djalante, R. Governing systemic and cascading disaster risk in Indonesia: where do we stand and future outlook. Disaster Prevention and Management: An International Journal, **32**(1), 27–48 (2023). https://doi.org/10.1108/DPM-07-2022-0156
12. Kusumasari, B. Natural Hazards Governance in Indonesia. In Oxford Research Encyclopedia of Natural Hazard Science. Oxford University Press (2019). https://doi.org/-10.1093/acrefore/9780199389407.013.234
13. Arifin, S., Wicaksono, S. S., Sumarto, S., Martitah, M., and Sulistianingsih, D. Disaster resilient village-based approach to disaster risk reduction policy in Indonesia: A regulatory analysis. Jàmbá Journal of Disaster Risk Studies, **13**(1),(2021). https://doi.org/10.4102/jamba.v13i1.1021
14. Carvalho, M. L., Freitas, C. M., and Miranda, E. S. Physical rehabilitation in the context of a landslide that occurred in Brazil. BMC Public Health, **19**(1), 1615 (2019). https://doi.org/10.1186/s12889-019-7964-x
15. Law, Y. K., Lee, C. K. F., Chan, A. H. Y., Mak, N. P. L., Hau, B. C. H., and Wu, J. Unveiling the role of forests in landslide occurrence, recurrence and recovery. Journal of Applied Ecology, **1–14** (2024). https://doi.org/10.1111/1365-2664.14741
16. Lin, P.-S. S. Building resilience through ecosystem restoration and community participation: Post-disaster recovery in coastal island communities.

International Journal of Disaster Risk Reduction, **39**, 101249 (2019).

https://doi.org/10.1016/j.ijdrr.2019.101249

17. Diehl, J. A., Asahiro, K., Hwang, Y. H., Hirashima, T., Kong, L., Wang, Z., Yao, H., and Tan, P. Y. A CHANS Approach to Investigating Post-Disaster Recovery Potential in Rural Japan. Journal of Disaster Research, **17**(3), 453–463 (2022). https://doi.org/10.20965/jdr.2022.p0453
18. Quinn, P., Munari, S., Block, K., Walker, S., Liberman, J., Wallace, J., Horyniak, D., Oliver, J., Hellard, M., Fletcher-Lartey, S., and Gibbs, L. COVID-19 disaster recovery capitals: A conceptual framework to guide holistic and strengths-based support strategies. Health Promotion Journal of Australia, **35**(2), 355–364 (2024). https://doi.org/10.1002/hpja.759
19. Bergstrand, K., and Mayer, B. "The Community Helped Me:" Community Cohesion and Environmental Concerns in Personal Assessments of Post-Disaster Recovery. Society and Natural Resources, **33**(3), 386–405 (2020). https://doi.org/10.1080/08941920.2019.1709002
20. Islam, E., Wahab, H. A., and Benson, O. G. Structural and operational factors as determinant of meaningful community participation in sustainable disaster recovery programs: The case of Bangladesh. International Journal of Disaster Risk Reduction, **50**, 101710 (2020).

    https://doi.org/10.1016/j.ijdrr.2020.101710
21. Roque, A. D., Pijawka, D., and Wutich, A. The Role of Social Capital in Resiliency: Disaster Recovery in Puerto Rico. Risk, Hazards and Crisis in Public Policy, **11**(2) (2020).

    https://doi.org/10.1002/rhc3.12187
22. Mayer, B. A Review of the Literature on Community Resilience and Disaster Recovery. Current Environmental Health Reports, **6**(3) (2019). https://doi.org/10.1007/s40572-019-00239-3
23. Dapilah, F., Nielsen, J. Ø., and Friis, C. The role of social networks in building adaptive capacity and resilience to climate change: a case study from northern Ghana. Climate and Development, **12**(1), 42–56 (2020).

    https://doi.org/10.1080/17565529.2019.1596063
24. Jia, X., Chowdhury, M., Prayag, G., and Hossan Chowdhury, M. M. The role of social capital on proactive and reactive resilience of organizations post-disaster. International Journal of Disaster Risk Reduction, **48**, 101614 (2020). https://doi.org/10.1016/j.ijdrr.2020.101614
25. Juhadi, J., Banowati, E., Sanjoto, T. B., and Nugraha, S. B. Rapid Appraisal for Agricultural Land Utilization in the erosion and landslide vulnerable mountainous areas of Kulonprogo Regency, Indonesia. Management of Environmental Quality: An International Journal, **31**(1), 1–17 (2020). https://doi.org/10.1108/MEQ-01-2019-0023
26. Dewanti, D. S., Saptutyningsih, E., Darsono, S. N. A. C., and Nugraha, P. Preparedness of Disaster Management of Landslides in the livelihoods of Kulon Progo district. E3S Web of Conferences, **425**, 05008 (2023).

    https://doi.org/10.1051/e3sconf/202342505008
27. Finucane, M. L., Acosta, J., Wicker, A., and Whipkey, K. Short-Term Solutions to a Long-Term Challenge: Rethinking Disaster Recovery Planning to Reduce Vulnerabilities and Inequities. International Journal of Environmental Research and Public Health, **17**(2), 482 (2020). https://doi.org/10.3390/ijerph17020482
28. Mishra, S., Anderson, K., Miller, B., Boyer, K., and Warren, A. Microgrid resilience: A holistic approach for assessing threats, identifying vulnerabilities, and designing corresponding mitigation strategies. Applied Energy, **264**, 114726 (2020). https://doi.org/10.1016/j.apenergy.2020.114726
29. Shi, P., Ye, T., Wang, Y., Zhou, T., Xu, W., Du, J., Wang, J., Li, N., Huang, C., Liu, L., Chen, B., Su, Y., Fang, W., Wang, M., Hu, X., Wu, J., He, C., Zhang, Q., Ye, Q., … Okada, N. Disaster Risk Science: A Geographical Perspective and a Research Framework. International Journal of Disaster Risk Science, **11**(4), 426–440 (2020). https://doi.org/10.1007/s13753-020-00296-5
30. Hoffmann, R., and Blecha, D. Education and Disaster Vulnerability in Southeast Asia: Evidence and Policy Implications. Sustainability, **12**(4), 1401 (2020). https://doi.org/10.3390/su12041401
31. Smith, H., Garcia Ferrari, S., Medero, G. M., Rivera, H., Coupé, F., Mejía Escalante, M. E., Castro Mera, W., Montoya Correa, C. A., Abiko, A., and Marinho, F. A. M. Exploring appropriate socio-technical arrangements for the co-production of landslide risk management strategies in informal neighbourhoods in Colombia and Brazil. International Journal of Urban Sustainable Development, **14**(1), 242–263 (2022). https://doi.org/10.1080/19463138.2021.1872082
32. Budimir, M., Sneddon, A., Nelder, I., Brown, S., Donovan, A., and Speight, L. Development of forecast information for institutional decision-makers: landslides in India and cyclones in Mozambique. Geoscience Communication, **5**(2), 151–175 (2022). https://doi.org/10.5194/gc-5-151-2022
33. Raška, P. Contextualizing community-based landslide risk reduction: an evolutionary

perspective. Landslides, **16**(9), 1747–1762 (2019). https://doi.org/10.1007/s10346-018-1099-5

34. Alcántara-Ayala, I. Integrated landslide disaster risk management (ILDRiM): the challenge to avoid the construction of new disaster risk. Environmental Hazards, **20**(3), 323–344 (2021). https://doi.org/10.1080/17477891.2020.1810609

35. Alcántara-Ayala, I. Integrated landslide disaster risk management (ILDRiM): the challenge to avoid the construction of new disaster risk. Environmental Hazards, **20**(3), 323–344 (2021).https://doi.org/10.1080/17477891.2020.1810609. https://doi.org/10.5194/nhess-21-187-2021

36. Kumar, K. S. S., and Jaisankar, N. An automated resource management framework for minimizing SLA violations and negotiation in collaborative cloud. International Journal of Cognitive Computing in Engineering, **1**, 27–35 (2020). https://doi.org/10.1016/j.ijcce.2020.09.001

37. Modiba, M. M., and Kekwaletswe, R. M. Resource Dependency Framework for Digital Transformation in South African Financial Service Providers. International Journal of Computer Trends and Technology (IJCTT), **68**(4), 107–125 (2020).

38. Zhang, C., and Fang, Y. Application of capital-based approach in the measurement of livelihood sustainability: A case study from the Koshi River basin community in Nepal. Ecological Indicators, **116**, 106474 (2020). https://doi.org/10.1016/j.ecolind.2020.106474

39. Kumar, A., Kumar, A., Kumari, S., Kumari, N., Kumari, S., and Mishra, P. Sustainable Livelihoods a Foundation for Rural Development Leads to Sustainability. Problemy Ekorozwoju, **18**(2), 128–140 (2023). https://doi.org/10.35784/preko.3951

40. Scoones, I. Sustainable Rural Livelihoods: A Framework for Analysis (1998).

41. Chambers, R., and Conway, G. Sustainable rural livelihoods: practical concepts for the 21st century (1992).

42. Perumahan dan Kawasan Permukiman Kulon Progo. (2022, August 2). Perumahan dan Kawasan Permukiman Kulon Progo. Perumahan dan Kawasan Permukiman Kulon Progo.

43. Litasari, U. C. N., Widiatmaka, Munibah, K., and Machfud. Mapping The Landslide Risk in Kulon Progo, Indonesia Using GeoTOPSIS. IOP Conference Series: Earth and Environmental Science, **1109**(1), 012011 (2022). https://doi.org/10.1088/1755-1315/1109/1/012011

44. Badan Penanggulangan Bencana Daerah (BPBD) Kabupaten Kulon Progo. Kajian Risiko Bencana Kabupaten Kulon Progo Tahun 2022 - 2026 (2022).

45. Muntohar, A. S., Ikhsan, J., and Soebowo, E. Mechanism of rainfall triggering landslides in Kulonprogo, Indonesia. Geo-Congress 2013, **452–461** (2013). https://doi.org/10.1061/9780784412787.047

46. Li, X., and Wiedermann, W. Conditional Direction Dependence Analysis: Evaluating the Causal Direction of Effects in Linear Models with Interaction Terms. Multivariate Behavioral Research, **55**(5), 786–810 (2020). https://doi.org/10.1080/00273171.2019.1687276

47. Burnett, J. W., and Blackwell, C. Graphical causal modelling: an application to identify and estimate cause-and-effect relationships. Applied Economics, **56**(33), 3986–4000 (2024). https://doi.org/10.1080/00036846.2023.2208856

48. Thakkar, J. J. Structural Equation Modelling (1st ed., Vol. 285). Springer Singapore (2020). https://doi.org/10.1007/978-981-15-3793-6

49. Altikriti, S., and Anderson, C. N. Factor Analysis and Structural Equation Modeling. In The Encyclopedia of Research Methods in Criminology and Criminal Justice (pp. 833–838). Wiley (2021). https://doi.org/10.1002/9781119111931.ch159

50. Hair, J. F., Hult, G. T. M., Ringle, C. M., Sarstedt, M., Danks, N. P., and Ray, S. An Introduction to Structural Equation Modeling. In Partial Least Squares Structural Equation Modeling (PLS-SEM) Using R (pp. 1–29). Springer (2021). https://doi.org/10.1007/978-3-030-80519-7_1

51. Khan, A., Gupta, S., and Gupta, S. K. Multi-hazard disaster studies: Monitoring, detection, recovery, and management, based on emerging technologies and optimal techniques. International Journal of Disaster Risk Reduction, **47**, 101642 (2020). https://doi.org/10.1016/j.ijdrr.2020.101642

52. Davies, T. Reducing landslide disaster impacts. In Landslide Hazards, Risks, and Disasters (pp. 623–639). Elsevier (2022). https://doi.org/10.1016/B978-0-12-818464-6.00012-3

53. Bao, H., Zeng, C., Peng, Y., and Wu, S. The use of digital technologies for landslide disaster risk research and disaster risk management: progress and prospects. Environmental Earth Sciences, **81**(18), 446 (2022).

https://doi.org/10.1007/s12665-022-10575-7

54. Tay, H. L., Banomyong, R., Varadejsatitwong, P., and Julagasigorn, P. Mitigating Risks in the Disaster Management Cycle. Advances in Civil Engineering, 2022**(1)** (2022).

https://doi.org/10.1155/2022/7454760

55. Ali, R. A., and Mannakkara, S. Phased post-disaster recovery challenges: 2016–2017 floods and landslides in Colombo and Kalutara, Sri Lanka.

International Journal of Disaster Risk Reduction, **111**; 104728 (2024).

https://doi.org/10.1016/j.ijdrr.2024.104728

56. Zhou, B., Zhang, H., and Evans, R. Build back better: A framework for sustainable recovery assessment. International Journal of Disaster Risk Reduction, **76**, 102998 (2022). https://doi.org/10.1016/j.ijdrr.2022.102998

57. Dube, E., Wedawatta, G., and Ginige, K. Building-Back-Better in Post-Disaster Recovery: Lessons Learnt from Cyclone Idai-Induced Floods in Zimbabwe. International Journal of Disaster Risk Science, **12**(5), 700–712 (2021). https://doi.org/10.1007/s13753-021-00373-3

58. Kennedy, M., Gonick, S., Meischke, H., Rios, J., and Errett, N. A. Building Back Better: Local Health Department Engagement and Integration of Health Promotion into Hurricane Harvey Recovery Planning and Implementation. International Journal of Environmental Research and Public Health, **16**(3), 299 (2019). https://doi.org/10.3390/ijerph16030299

59. Koebele, E. A., Crow, D. A., and Albright, E. A. Building Resilience during Recovery: Lessons from Colorado's Watershed Resilience Pilot Program. Environmental Management, **66**(1), 1–15 (2020). https://doi.org/10.1007/s00267-020-01296-3

60. Galderisi, A., Menoni, S., Setti, G., and Tognon, A. Disaster Recovery Reform and Resilience. In Disaster Risk Reduction for Resilience (pp. 25–54). Springer International Publishing (2022). https://doi.org/10.1007/978-3-030-99063-3_2

61. KC, A., Gan, C. C. R., and Dwirahmadi, F. Breaking Through Barriers and Building Disaster Mental Resilience: A Case Study in the Aftermath of the 2015 Nepal Earthquakes. International Journal of Environmental Research and Public Health, **16**(16), 2964 (2019). https://doi.org/10.3390/ijerph16162964

62. Makwana, N. Disaster and its impact on mental health: A narrative review. Journal of Family Medicine and Primary Care, **8**(10), 3090 (2019). https://doi.org/10.4103/jfmpc.jfmpc_893_19

63. Safapour, E., Kermanshachi, S., and Pamidimukkala, A. Post-disaster recovery in urban and rural communities: Challenges and strategies. International Journal of Disaster Risk Reduction, **64**, 102535 (2021).

https://doi.org/10.1016/j.ijdrr.2021.102535

64. Saja, A. M. A., Goonetilleke, A., Teo, M., and Ziyath, A. M. A critical review of social resilience assessment frameworks in disaster management. International Journal of Disaster Risk Reduction, **35**, 101096 (2019).

https://doi.org/10.1016/j.ijdrr.2019.101096

65. Gyawali, S., Tiwari, S. R., Bajracharya, S. B., and Skotte, H. N. Promoting sustainable livelihoods: An approach to postdisaster reconstruction. Sustainable Development, **28**(4), 626–633 (2020). https://doi.org/10.1002/sd.2013

66. Pu, G., Chang-Richards, A., Wilkinson, S., and Potangaroa, R. What makes a successful livelihood recovery? a study of China's Lushan earthquake. Natural Hazards, **105**(3), 2543–2567 (2020). https://doi.org/10.1007/s11069-020-04412-y

67. Uddin, M. S., Haque, C. E., Khan, M. N., Doberstein, B., and Cox, R. S. "Disasters threaten livelihoods, and people cope, adapt and make transformational changes": Community resilience and livelihoods reconstruction in coastal communities of Bangladesh. International Journal of Disaster Risk Reduction, **63**, 102444 (2021).

https://doi.org/10.1016/j.ijdrr.2021.102444

68. Izumi, T., Shaw, R., Djalante, R., Ishiwatari, M., and Komino, T. Disaster risk reduction and innovations. Progress in Disaster Science, **2**, 100033 (2019). https://doi.org/10.1016/j.pdisas.2019.100033

69. Klein, J. A., Tucker, C. M., Steger, C. E., Nolin, A., Reid, R., Hopping, K. A., Yeh, E. T., Pradhan, M. S., Taber, A., Molden, D., Ghate, R., Choudhury, D., Alcántara-Ayala, I., Lavorel, S., Müller, B., Grêt-Regamey, A., Boone, R. B., Bourgeron, P., Castellanos, E., … Yager, K. An integrated community and ecosystem-based approach to disaster risk reduction in mountain systems. Environmental Science and Policy, **94**, 143–152 (2019). https://doi.org/10.1016/j.envsci.2018.12.034

70. Rathnayaka, B., Siriwardana, C., Robert, D., Amaratunga, D., and Setunge, S. Improving the resilience of critical infrastructures: Evidence-based insights from a systematic literature review. International Journal of Disaster Risk Reduction, **78**, 103123 (2022).

https://doi.org/10.1016/j.ijdrr.2022.103123

71. Hadlos, A., Opdyke, A., and Hadigheh, S. A. Where does local and indigenous knowledge in disaster risk reduction go from here? A systematic literature review. International Journal of Disaster Risk Reduction, **79**, 103160 (2022). https://doi.org/10.1016/j.ijdrr.2022.103160

72. Bahmani, H., and Zhang, W. A conceptual framework for integrated management of disasters recovery projects. Natural Hazards, **113**(2), 859–885 (2022).

https://doi.org/10.1007/s11069-022-05328-5

73. Ke, G.-N., Grajfoner, D., Carter, S., DeLima, N., Khairudin, R., Lau, W.-Y., Kamal, K. A., and Lee,

S. C. Psychological Wellbeing and Employability of Retrenched Workforce During COVID-19: A Qualitative Study Exploring the Mitigations for Post Pandemic Recovery Phase. Frontiers in Public Health, **10** (2022). https://doi.org/10.3389/fpubh.2022.907797

74. Anjum, N., and Rahaman, Md. S. Human Resource Management amidst COVID-19 Pandemic: Behavioral Implications for HR Practitioners. Journal of Advanced Research in Economics and Administrative Sciences, **3**(1), 57–66 (2022). https://doi.org/10.47631/jareas.v3i1.398

75. Quinn, P., Gibbs, L., Blake, D., Campbell, E., Johnston, D., Richardson, J., and Coghlan, A. Recovery capitals: A collaborative approach to post-disaster guidance. The Australian Journal of Emergency Management, **37**(2), 52–62 (2022).

76. Bhadra, S., and Dyer, A. R. Resilience and Well-Being Among the Survivors of Natural Disasters and Conflicts. In Handbook of Health and Well-Being (pp. 637–667). Springer Nature Singapore (2022). https://doi.org/10.1007/978-981-16-8263-6_27

77. Onishchenko, N., Perelygina, L., Nazarov, O., Karikov, S., Riabinina, O., and Gontarenko, L. Psychosocial assistance within the framework of the social rehabilitation process: prospects for recuperation and reintegrative outcomes. Cadernos de Educação, Tecnologia e Sociedade, 16(se2), 155–164 (2023).

https://doi.org/10.14571/brajets.v16.nse2.155-164

78. Musavengane, R., and Kloppers, R. Social capital: An investment towards community resilience in the collaborative natural resources management of community-based tourism schemes. Tourism Management Perspectives, **34**, 100654 (2020). https://doi.org/10.1016/j.tmp.2020.100654

79. Waseem, H. Bin, Mirza, M. N. E. E., and Rana, I. A. Exploring the role of social capital in flood risk reduction: Insights from a systematic review. Environmental Impact Assessment Review, **105**, 107390 (2024). https://doi.org/10.1016/j.eiar.2023.107390

80. Carmen, E., Fazey, I., Ross, H., Bedinger, M., Smith, F. M., Prager, K., McClymont, K., and Morrison, D. Building community resilience in a context of climate change: The role of social capital. Ambio, **51**(6), 1371–1387 (2022). https://doi.org/10.1007/s13280-021-01678-9

81. Al-Omoush, K. S., Ribeiro-Navarrete, S., Lassala, C., and Skare, M. Networking and knowledge creation: Social capital and collaborative innovation in responding to the COVID-19 crisis. Journal of Innovation and Knowledge, **7**(2), 100181 (2022). https://doi.org/10.1016/j.jik.2022.100181

82. Fang, J., Lau, C. K. M., Lu, Z., Wu, W., and Zhu, L. Natural disasters, climate change, and their impact on inclusive wealth in G20 countries. Environmental Science and Pollution Research, **26**(2), 1455–1463 (2019). https://doi.org/10.1007/s11356-018-3634-2

83. Sangha, K. K., Russell-Smith, J., Evans, J., and Edwards, A. Methodological approaches and challenges to assess the environmental losses from natural disasters. International Journal of Disaster Risk Reduction, **49**, 101619 (2020). https://doi.org/10.1016/j.ijdrr.2020.101619

84. Campbell, N. Disaster recovery among older adults: Exploring the intersection of vulnerability and resilience. In Emerging Voices in Natural Hazards Research (pp. 83–119). Elsevier (2019). https://doi.org/10.1016/B978-0-12-815821-0.00011-4

85. Imperiale, A. J., and Vanclay, F. Conceptualizing community resilience and the social dimensions of risk to overcome barriers to disaster risk reduction and sustainable development. Sustainable Development, **29**(5), 891–905 (2021). https://doi.org/10.1002/sd.2182

86. Albright, E. A., and Crow, D. A. Capacity Building toward Resilience: How Communities Recover, Learn, and Change in the Aftermath of Extreme Events. Policy Studies Journal, **49**(1), 89–122 (2021). https://doi.org/10.1111/psj.12364

87. Zhang, H., Dolan, C., Jing, S. M., Uyimleshi, J., and Dodd, P. Bounce Forward: Economic Recovery in Post-Disaster Fukushima. Sustainability, **11**(23), 6736 (2019). https://doi.org/10.3390/su11236736

88. Islam, R., and Walkerden, G. Livelihood assets, mutual support and disaster resilience in coastal Bangladesh. International Journal of Disaster Risk Reduction, **78**, 103148 (2022). https://doi.org/10.1016/j.ijdrr.2022.103148

89. Donkor, F. K. Recovery from Social and Economic Chaos: The Building of Resilient Communities. In No Poverty (pp. 751–761). Springer (2021). https://doi.org/10.1007/978-3-319-95714-2_95

90. Sadeka, S., Mohamad, M. S., Sarkar, M. S. K., and Al-Amin, A. Q. Vulnerability, Social Capital and Disaster Preparedness. Springer Nature Singapore (2023). https://doi.org/10.1007/978-981-99-3874-2

91. Sarker, M. N. I., Cao, Q., Wu, M., Hossin, M. A., Alam, G. M. M., and Shouse, R. C. Vulnerability and livelihood resilience in the face of natural disaster: A critical conceptual review. Applied Ecology and Environmental Research, **17**(6) (2019). https://doi.org/10.15666/aeer/1706_1276912785

92. Axon, S. The socio-cultural dimensions of community-based sustainability: Implications for transformational change. Journal of Cleaner

Production, **266**, 121933 (2020). https://doi.org/10.1016/j.jclepro.2020.121933

93. Carr, E. R. Resilient livelihoods in an era of global transformation. Global Environmental Change, **64**, 102155 (2020).

https://doi.org/10.1016/j.gloenvcha.2020.102155

94. Liu, W., Li, J., Ren, L., Xu, J., Li, C., and Li, S. Exploring Livelihood Resilience and Its Impact on Livelihood Strategy in Rural China. Social Indicators Research, **150**(3), 977–998 (2020). https://doi.org/10.1007/s11205-020-02347-2

95. Thornton, P., Loboguerrero, A., Campbell, B., Kavikumar, K., Mercado, L., and Shackleton, S. Rural livelihoods, food security and rural transformation under climate change (2019).

96. Zhilina, V., Kuznetsova, N., and Zhilina, E. Transformation of Human Capital Phenomenon: Making of New Research Paradigm. Proceedings of the Ecological-Socio-Economic Systems: Models of Competition and Cooperation (ESES 2019) (2020). https://doi.org/10.2991/assehr.k.200113.016

# Human resources development in farming and fisheries for food security

*Nadhiya* Abdulla[1*] and *Natalia* Vasylieva[2]

[1]The Maldives National University, 20371, Rahdhebai Higun, Machangolhi, Male', the Republic of Maldives
[2]Dnipro State Agrarian and Economic University, 49009, Dnipro, Ukraine

**Abstract.** The article delves into the issues of providing food security in the Maldives, a nation that regularly faces adverse weather conditions and natural disasters. These challenges include tsunamis, heavy rainfall, and strong winds that damage infrastructure, fishing vessels, agricultural land, contaminate freshwater resources and wash away entire islands during the monsoon season. The research purpose was to support workforce development to compensate for shortage of natural and technical resources in farming and fisheries. The explored data was collected from Maldives Bureau of Statistics. The applied statistical and clustering methods proved that training in farming should share knowledge of cutting-edge technologies like Hydroponics and Pest control, benefit female farmers, involve the youth in local agriculture and be evenly distributed among the Atolls. The study findings based on ranking and trend analysis argue that unstable employment is the paramount problem in fisheries. It is recommended to optimize a number of fishermen per vessel by Atolls and promote acquiring skills of advanced eco-friendly fishing. Overall research results created a reliable foundation for systematic development of qualified human resources in farming and fisheries that are crucial for strengthening food security in a risky environment.

## 1 Introduction

The climate crisis is causing increasingly frequent and severe extreme weather events, impacting 1.7 billion people through disasters such as hurricanes, cyclones, and droughts. These events are exacerbating severe hunger and threatening to undo progress made in reducing undernourishment, which affected nearly 800 million people. Abrupt climatic changes directly affect food security by causing loss of rural livelihoods, degradation of ecosystems, and collapse of food systems. The risk of severe climate impacts is expected to rise after 2050. Nations already facing food insecurity and poverty, especially arid, semi-arid, landlocked, and small island developing states, are particularly vulnerable. Similarly, farming in developed countries also need to address increased temperatures and reduced precipitation [1].

Countries in the southern hemisphere, including Bangladesh, Sri Lanka, India, Pakistan, Nepal, and the Maldives, are among the most impoverished and particularly vulnerable to natural calamities. Disasters in these regions have indirect effects on agriculture, fisheries, and livestock, which are key for their economic stability and food security. Every year, millions of Bangladeshis suffer from the destruction of homes, farmlands, and fishing vessels due to heavy rain, flooding, and landslides. In India, several districts have experienced a 60 percent decrease in farmers' yearly income due to droughts and flooding. Populations residing in coastal areas of India are also facing long-term chronic malnutrition in children due to the negative impact on fishing activities [2].

The Maldives is one of the smallest island states in the South Asia consisting of 20 administrative atolls including North Thiladhunmathi (HA), South Thiladhunmathi (HDh), North Miladhunmadulu (Sh), South Miladhunmadulu (N), North Maalhosmadulu (R), South Maalhosmadulu (B), Faadhippolhu (Lh), Male' Atoll (K), North Ari Atoll (AA), South Ari Atoll (ADh), Felidhu Atoll (V), Mulakatholhu (M), North Nilandhe Atoll (F), South Nilandhe Atoll (Dh), Kolhumadulu (Th), Hadhdhunmathi (L), North Huvadhu Atoll (GA), South Huvadhu Atoll (GDh), Fuvahmulah (Gn) and Addu Atoll (S). The Maldives depends on agriculture and fisheries, as its fundamental economic activity food and security. The country as a whole is significantly affected by coastal vulnerabilities such as high waves, erosion, and flooding from heavy rainfall [3]. The nation's susceptibility to climatic conditions and its impact on food and nutrient security further exacerbated by the challenges faced in skilled labor force. Despite these challenges, the country has seen rapid economic growth and achieved full employment in recent years. However, the tourism sector struggles with a shortage of skilled labor, while the fisheries sector remains underutilized, leading to a substantial influx of expatriate workers to fill these gaps. Local workers in these sectors often face marginalization due to limited educational and vocational training opportunities, with women in remote island communities

* Corresponding author: nadhiya.abdulla@mnu.edu.mv

particularly disadvantaged compared to their urban counterparts [4].

Labor shortages in the agriculture and fisheries sectors hinder their development and sustainability, necessitating institutional reforms to harness untapped labor potential and enhance economic resilience [4]. While gender discrimination in education and employment rights is relatively low in the Maldives, women in small island communities encounter significant barriers to education and vocational training [5]. Addressing these challenges is vital as they pose threats to national food security, export revenues, and the livelihoods of those dependent on these sectors.

Focusing specifically on the agricultural sector of the Maldives, which mainly consists of small farming communities operating within remote islands, there are challenges in accessing information and support from local extension services, compounded by a lack of capable extension officers. According to human capital and human resource theories, high-quality agricultural extension agents enhance farmer productivity by improving problem-solving abilities, risk mitigation planning, communication skills, ethics, discipline, and overall competence [6]. Empowering women farmers in the Maldivian agriculture sector involves building their capacity to mitigate climate conditions, disaster preparedness and address gender-specific risk, time, and social preferences that influence resource allocation [7].

The Maldives also faces broader labor force challenges including an aging workforce, lack of skilled labor, economic diversification, climatic vulnerability, and migration trends towards urbanization. Scholars attribute the reluctance of the young population to enter the agricultural sector to factors such as lack of strategic planning, sales, and managerial skills, compounded by difficulties in finding qualified workers [8]. This situation mirrors observations among Maldivian youth who perceive limited access to agricultural education, technology, and financial support as barriers to entering the agricultural field. Additionally, there is a critical need for skills in climate change adaptation, which are essential for maintaining food and nutrient security. Without these skills, young farmers struggle to implement effective strategies to cope with changing climatic conditions, which further undermines the stability of food production and nutrient availability.

Despite the prevalence of traditional farming methods among Maldivian communities, there are significant challenges. These communities often lack the knowledge and skills needed to address adverse climatic conditions. Additionally, they face financial constraints that hinder improvements in production efficiency. Studies on hydroponic technology suggest that a comprehensive curriculum and diverse training can make a difference. Implementing such educational programs can provide farmers with essential tools and knowledge. This shift from traditional practices to more efficient hydroponic methods can help reduce the negative impact of climate change on food production and nutrient security. Such education not only facilitates the adoption of advanced technologies but also enhances production efficiency while minimizing environmental impact [9, 10].

The fisheries sector in the Maldives, crucial for protein and nutrient security, faces challenges such as shortages in skilled labor and technological resources [11]. Adverse climate conditions, including rising sea temperatures, ocean acidification, and unpredictable weather patterns, further exacerbate these challenges by affecting fish populations and disrupting fishing activities. Research into factors influencing fishermen's incomes underscores the importance of enhancing practical experience and optimizing working hours to improve income levels in this vital sector [12]. Addressing these issues is vital for maintaining food and nutrient security, as climate change impacts can reduce fish stocks and compromise the availability of essential protein sources for local populations.

Given the scarcity of natural resources such as arable land and fresh water, implementing resilient measures is imperative to strengthen food security in the Maldives. Continued reliance on traditional farming and fishing methods is insufficient to meet increasing demands for food and nutrients amidst environmental volatility and limited resources [4]. The recent COVID-19 pandemic and conflicts like the Ukraine-Russia war have underscored the vulnerability of national food and nutrient security to external shocks. Scholars emphasize the importance of understanding farmers' needs to formulate targeted policies that support food production and supply chain resilience [13].

The climate crisis poses severe implications for agriculture, fisheries, and labor forces, especially in vulnerable regions like the Maldives. As extreme weather events become more frequent and intense, the risk to food systems globally intensifies, undermining progress made in reducing hunger and malnutrition. Countries like the Maldives, which heavily rely on agriculture and fisheries, need to implement climate adaptation strategies and improve labor and technological capacities. This includes investing in education and training, promoting sustainable practices, and supporting economic diversification to mitigate the adverse effects of climate change and ensure a stable and nutritious food supply for future generations.

## 2 Literature review

In the context of global population growth, achieving the sustainable development goal of zero hunger requires robust support for the agricultural sectors, which are the backbone of global food production. Despite small scale farmers contributing 50-70% of global food output, many face food insecurity, particularly in less developed and developing countries. Sub-Saharan Africa exemplifies this challenge, where agriculture remains a primary source of employment and food security amidst significant population growth. However, these farmers are caught in a cycle of limited land availability, inadequate finance, and poor productivity [14].

To break this cycle, substantial investments are essential to build a resilient workforce, advance technological capabilities, and diversify into higher-value crops. Similar challenges are observed in Asian countries, where farmers also struggle with food insecurity, hindering substantial economic growth potential [14]. Indonesia, susceptible to climate change and natural disasters like cyclones causing extensive damage to farmlands and crops, underscores the need for a resilient workforce capable of effective disaster mitigation and management strategies [15].

Access to timely information and skill development is key for enhancing the productivity of smallholder farmers and strengthening their food security [16]. Research focused on identifying agricultural skills using the European Skills/Competences, Qualifications and Occupations (ESCO) classification highlights the importance of aligning training programs and policies with evolving industry needs [17]. Financial literacy among the agricultural workforce is equally critical for scaling up business operations [18].

In Vanuatu, a Pacific island nation prone to cyclones that devastate its fisheries sector, there is a pressing need to enhance disaster preparedness and integrate it into rural development services to promote sectoral resilience and sustainable development amidst climate challenges [19]. Effective fisheries management, vital under UN SDG 14, plays a pivotal role in conserving ocean resources and optimizing productivity. Incorporating traditional ecological knowledge into policy frameworks is essential for promoting sustainable practices among artisanal fishing communities, enhancing environmental stewardship, and preserving natural resources [20].

Investing in modern technologies such as GPS and mechanized fishing gears is crucial for enhancing the capacity and efficiency of fisheries communities [21]. In Indonesia, the fisheries sector faces challenges due to a lack of skilled labor, necessitating alignment of competency standards and curriculum development to equip professionals with necessary knowledge, skills, and attitudes [22].

Countries like Ukraine link university performance metrics to educational outcomes, aiming to improve program quality and graduate readiness for the labor market [23, 24]. Indonesia's fisheries sector growth is hindered not only by labor shortages but also by inadequate infrastructure in fishing provinces, highlighting the need for support in vessel provision and sectoral advancements [25].

Bangladesh, reliant on its fisheries sector for food security, employment, and economic growth similar to the Maldives, identifies deficiencies in human resource management. Addressing these gaps, including recruitment, salary issues, and training, is vital for enhancing sectoral growth and broader economic contributions [26].

The fisheries sectors in India and the Maldives share economic importance, challenges, and labor development needs. Both face significant skilled labor shortages, with fishermen often relying on traditional, informal vocational training, and both grapple with climate change impacts and gender disparities in access to formal education. India has addressed these issues through initiatives like the National Skill Development Mission, the MUDRA scheme, and Digital India, which focus on inclusive skill training and technological innovation. Similarly, India's agriculture sector suffers from labor shortages, decreasing output and altering cropping patterns, paralleling the Maldives' reliance on food imports due to limited agricultural capacity. To combat these challenges, India has implemented robust skill development policies and educational programs to build a capable workforce, particularly for women and disadvantaged groups [27].

As a whole the development of a skilled labor force is vital for enhancing food and nutrition security in both the agricultural and fisheries sectors globally. Farmers and fisherfolk play pivotal roles in producing food and sustaining livelihoods in both developing and developed nations, yet they often face significant challenges such as climate change impacts, limited resources, and technological barriers. Investment in labor development, including training in advanced agricultural practices, reforming and revising national education curriculum and aligning it with the sectoral competency requirement, financial literacy, and disaster management, is essential to overcoming these obstacles.

# 3 Results

## 3.1 Development of human resources in farming

Unfortunately, at present there are no educational courses on farming at colleges and universities in the Maldives. But when it comes to food security farmers are those who can make a difference.

With the severe scarcity of natural resources accompanied by high risks of natural disasters, the Maldivian farmers need advanced training to address the discussed issues not to mention that special retention efforts are necessary to empower female farmers and involve the young in farming. Agricultural extension officers make a huge contribution to providing food security in the Maldives through sharing contemporary knowledge on innovative farming. The national agriculture is making progress to mitigate dependency on food imports that endanger food security in the Maldives which is also at risk of negative consequences of ongoing climate change. However, the observed results are far from optimistic expectations. That is why it makes sense to find some weak spots that should be avoided in the future.

The analyzed data on conducted training for 2018-2024 incorporated 82 programs encompassing farmers from all of 20 Atolls listed in the Introduction [28]. Fig. 1 illustrates distribution (in %) of these training programs by 4 topics including Pest Control, Home Gardening, Agricultural Basics, and Hydroponics. On the one hand, Fig. 1 shows that those projects comprised really urgent agricultural issues. On the other hand, the cutting-edge options on Hydroponics and Pest Control were least discussed. It means that major programs made no difference to retention of the youth in farming. In other words, young farmers were mostly disadvantaged. But it is they who might be a powerful driving force to provide food security despite high risks of natural disasters in the Maldives.

Moreover, as it can be seen in Fig. 2, training programs were unevenly assigned to Atolls. Consequently, AA, V, F, Dh, and S Atolls (listed in the Introduction) remained the least aware of effective practices of resilient agriculture.

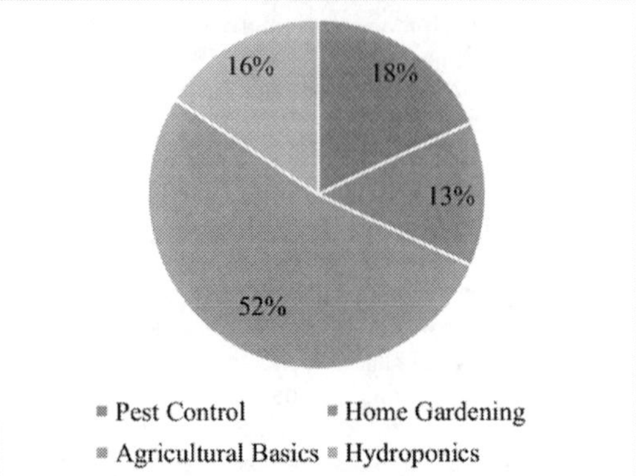


**Fig. 1.** Shares of training topics on farming in 2018-2024.

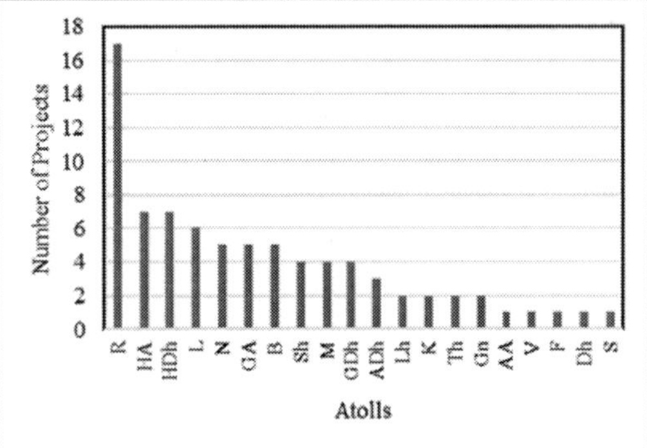


**Fig. 2.** Distribution of training programs on farming by Atolls in 2018-2024.

In order to assess quality work of agricultural extension officers in dynamics, all projects were characterized by three key parameters, such as

- a total number of trained participants ($T$, people),
- a share of women-farmers who were instructed by advisers ($F$, %),
- a program duration ($D$, days).

The checked hypothesis about similar average parameters in 2018-2024 was examined by means of Tukey's test, that compared $M = 7$ annual samples of projects. Namely, $N_m$ denoted a total number of the projects, $m = 1...M$ where $m = 1$ related to 2018 and $m = M$ related to 2024. Therefore,

$$\Sigma_{m=1...M} N_m = N = 82. \quad (1)$$

The auxiliary parameters were

$$K = M / \Sigma_{m=1...M} (1/N_m) \quad (2)$$

$$S^2 = (\Sigma_{m=1...M} (N_m - 1) \cdot S^2_m) / (N - M) \quad (3)$$

where $S^2_m$ denoted the variance of sample m.

With formulas (1), (2), (3), a limit of Tukey's test was calculated as follows

$$C = C_{crit} \cdot S / (K)^{0.5} \quad (4)$$

where $C_{crit}$ was a critical value of the Studentized range for multiple comparison with a degree of freedom $N - M = 75$ and a conventional level of significance $\alpha = 0.05$.

Tukey's test was performed three times concerning parameters $T$, $F$, $D$ with average sample values $T_m$, $F_m$, $D_m$, and limits (formula (4)) $C_T$, $C_F$, $C_D$, respectively, i.e.

$$10.5 = \mathrm{MAX}_{i,j=1...M} (T_i - T_j) < C_T = 17.49 \quad (5)$$

$$11.3 = \mathrm{MAX}_{i,j=1...M} (F_i - F_j) < C_F = 18.39 \quad (6)$$

$$3.8 = \mathrm{MAX}_{i,j=1...M} (D_i - D_j) < C_D = 6.04 \quad (7)$$

Hence, inequalities (5), (6), (7) give clear evidence that general approach to training farmers was stable over the years of 2018-2024. It confirms unwavering commitment to develop contemporary farmers in the Maldives so that they will successfully face agricultural risks that challenge the national food security.

Moreover, future training programs deserve a special focus that makes it possible to approach islands exposed to similar risks and natural disasters in the same way. To reach this goal, the research utilized divisive hierarchical clustering. This method generates a binary tree unfolded by conditional splits. It was applied to a set of 20 training projects that are planned for 2024-2026 [28]. These programs were identified by

- a total number of engaged farmers ($P$, people),
- a share of involved female farmers ($W$, %),
- an indicator of land availability ($L$, people per square feet).

Splits in divisive hierarchical clustering were specified by average values of parameters $P_{mean}$, $W_{mean}$, and $L_{mean}$. The performed procedure calculated 5 patterns and distributed the analyzed projects as follows.

The group with $P<P_{mean}$, $W<W_{mean}$, $L<L_{mean}$ included Th. Kandoodhoo, R. Kinolhas, R. Vaadhoo, L. Dhambidhoo, B. Goidhoo, and S. Meedhoo. As their farming lands are often located near coastal zones, these islands are highly susceptible to natural disasters such as storm surges, saltwater intrusion, and coastal erosion. Along with the scarcity of available land, these natural disasters make it impossible for farmers to continue with their traditional methods of farming. Therefore, they need more knowledge about disaster risk mitigation plans and disaster preparedness, along with climate-smart agricultural methodologies such as technology-integrated farming to improve their efficiency.

The group with $P<P_{mean}$, $W>W_{mean}$, $L<L_{mean}$ combined R. Fainu, GA. Kondey, GDh. Hoadehdhoo where are more women but less land accessible. These islands are susceptible to reduced food production and potential food insecurity as limited resources essentially impact people's livelihoods. These issues can be offset by training in climate-smart agriculture to adapt to changing conditions and bolster economic stability.

The group with $P<P_{mean}$, $W>W_{mean}$, $L>L_{mean}$ comprised F. Magoodhoo, GA. Maamendhoo, F. Dharanboodhoo, N. Kendhikulhudhoo which can benefit from a higher land availability. It can help them prepare for, respond to, and recover from natural disasters. With more land available, these islands can implement and expand disaster preparedness and mitigation strategies. Specifically, farmers can engage in more extensive farming activities and diversify their crops and practices, which can reduce the risk of total crop failure due to adverse conditions. Ultimately, this can improve food security and lessen the risk of food shortages during and after disasters.

The group with $P>P_{mean}$, $W>W_{mean}$, $L<L_{mean}$ included GA. Nilandhoo, Gn. Fuvamulah, L. Isdhoo, L. Kalaidhoo. These islands especially need training in farm management skills, financial literacy skills, and skills to allocate available natural resources efficiently for agricultural production, addressing the environmental constraints and climatic challenges faced by the farmers of these islands.

The group with $P>P_{mean}$, $W<W_{mean}$, $L<L_{mean}$ combined AA. Thoddoo, L. Gan, K. Kaashidhoo where male farmers are prevailing. This aspect implies a special focus on retention to reduce risks associated with the agricultural workforce and its sustainability. To improve the sector's resilience, it needs programs to retain farmers, such as providing financial incentives, improving working conditions, supporting training and development, and promoting gender inclusion. Encouraging more balanced participation in farming can address potential imbalances in the workforce and improve overall sector stability.

## 3.2 Development of human resources in fisheries

Since Maldivian agriculture can maintain mostly crop produce, fisheries are responsible for providing the national food security by animal protein. That is why training fishermen is invaluable especially when it comes to dealing with increasing risks and intensifying natural disasters observed both in the region and globally. The core analysis of workforce in fisheries was about its alignment by Atolls and with regard to the registered numbers of vessels as technical equipment is a prerequisite to a large fish catch. The Box and Whisker chat is a powerful tool that visualizes means and variances by compared samples. Central 50% of data (between quartiles $Q1$ and $Q3$) are displayed as boxes, the rest of the data are depicted as whiskers within limits

$$[2.5Q1 - 1.5Q3; 2.5Q3 - 1.5Q1] \quad (8)$$

and remote outliers. Statistical yearbooks offer registered quantities of fishermen by Atolls only for 2012-2016 [29]. The respective calculated Box and Whisker chat is presented in Fig. 3.

The found distinctive peculiarities are as follows. According to (8), there were two Atolls-oulires (GDh and N) with excessive numbers of fishermen in 2012 and 2013. However, Atoll Gn had spare vessels that remained in idle mode because of lack of human resources.

In 2014 there were the least average number of fishermen per vessel. It was a warning signal because fishermen were actively leaving the industry.

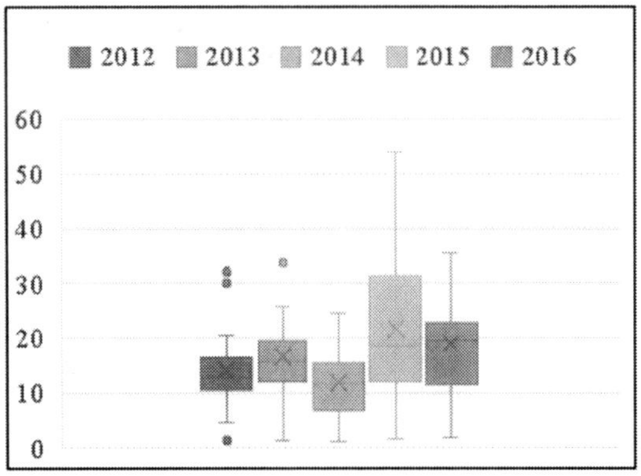


**Fig. 3.** Fluctuation of fishermen's quantity in 2012-2016.

In 2015 there was the biggest difference between Atolls by labor force per vessel. It put the industry at risk since declining the number of fishermen poses a significant risk as it diminishes the workforce, leading to decreased fish production and impacting the local economy reliant on fishing.

In 2016 the range shrank but it was reached via the largest fishermen' redundancy even among highly professional and experienced anglers. It is no surprise because owing to unstable incomes men quit fishing and sought employment in a lucrative resort industry. The observed tendency put all the used methods of fishermen's retention under question.

Unfortunately, official statistics on the number of fishermen in the Maldives is unavailable since 2017. However, the prevalent fishing vessels in the Maldives are Mechanised Masdhoni which optimally need 10 fishermen per vessel. This assumption made it possible to evaluate dynamics of fishermen in 2023 compared to 2016. The research findings are illustrated by Fig. 4 which depicts ascending and descending fluctuations in the Atoll fishing workforce for 7 years. The performed analysis revealed that there were 4 groups of Atolls of

- high redundancy, including Sh, GDh, L, B, Th, and R,
- middle redundancy, including HDh, GA, AA, K, and M,
- stable employment, including HA, F, N, Lh, V, and ADh,
- staff expansion, including S, Gn, Dh.

Top three Atolls with excessive human resources were Sh, GDh, and L, where there is abundant of natural resources compared to other Atolls, such as rich fishing grounds and access to ice plants and fish collection facilities, attracting more individuals from other islands seeking work is presented. This leads to a higher concentration of human resources in these atolls compared to others. Top three Atolls whose fisheries hired more fishermen were S, Gn, Dh. This approach likely helped them mitigate risks of potential labor shortages, ensuring that fishing operations are adequately staffed and improving the efficiency of fish harvesting and processing, maximizing the economic benefits from the fisheries sector. Finally, the most balanced Atolls by a number of fishermen were Lh, V, ADh which had good chances to address challenges related to natural disasters as their fishermen were experienced enough to use vessels in an efficient way.

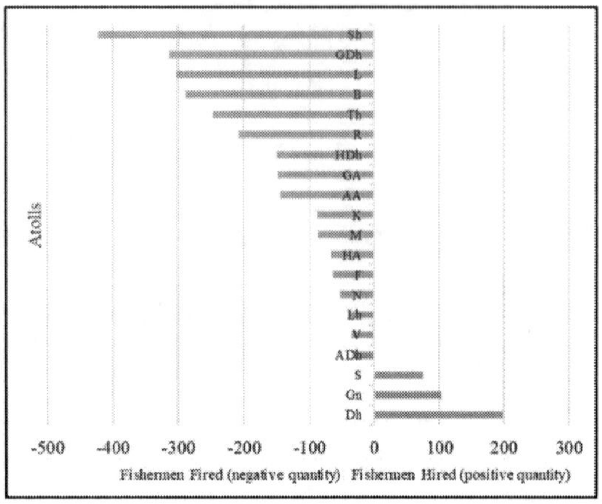


**Fig. 4.** Calculated dynamics of fishermen by Atolls for 2016-2023.

All of the above proves that an observed decline in labor force in fisheries jeopardizes food security in the Maldives regarding risks of natural disasters that accompany ongoing climate change. To compensate them, local fishermen need advanced knowledge. When asked about priorities, they highlighted three equally crucial topics, such as

- using GPS, sonars, and other innovative technology to locate shoals of fish;
- implementing pole and line fishing methods to reduce by catch;
- handling a fish catch and icing fish to minimize a harvest loss.

This way fishermen would be able to store enough fish for periods of adverse weather conditions and support the national food security in case of disasters.

## 4 Conclusion

Food security is in the top of global agenda for a long time. This challenge combines economic and social dimensions exacerbated by adverse weather conditions and numerous natural disasters caused by ongoing climate change. In the Maldives farming and fisheries are prime contributors to food security. These sectors lack natural and technical resources. Thus, qualified human resources can partly offset them.

The analyzed training programs provided by agricultural extension officers fail to cover innovative technologies like Hydroponics and Pest Control and they are unevenly held by Atolls. It may repel young farmers and doesn't empower vulnerable female farmers. Therefore, it is advisable to perform training for farmers noticing the share of engaged women and availability of land that will shape the discussed topics on mitigating agricultural risks and consequences of natural disasters.

The major problem facing fishermen in the Maldives is precarious employment caused by a sharp disproportion between a number of vessels and a quantity of anglers in most Atolls. Hence, human resource development should focus on how to retain skilled fishermen in the industry and also share contemporary technical knowledge on fishing methods and handling a fish catch.

There are various prospects for further scientific research as providing food security is a complicated problem. One of relevant vectors is to study its financial and ecological aspects concerning risks of adverse weather conditions and natural disasters.

The authors confirm that all data underlying the findings are fully available without restriction. The authors confirm equal contribution to the paper and equal responsibility for the following: study conception and design, data collection, analysis and interpretation of results, and manuscript preparation.

## References

1. Climate change and food security: Risks and responses, (FAO UN, 2015). https://www.fao.org/policy-support/tools-and-publications/resources-details/en/c/427091/
2. M.S. Hossain, N.N. Setu, R. Rahman, Climate change and its impact on food security in South Asian countries, in Proceedings of the LSCMFI-conference, BUET, Dhaka, Bangladesh, January 25 (2014). https://www.researchgate.net/publication/261027995_climate_change_and_its_impact_on_food_security_in_south_asian_countries
3. A. Ali, ESD in the small island state of Maldives. REESD, Sp., 137-150 (2016). https://doi.org/10.1007/978-94-017-7622-6_8
4. P.A. Van Driessche, Capacity building towards agricultural sustainability in the Maldives: Does the voice of the Maldivian farmer need to be amplified in planning agricultural interventions. Int. J. Ag. Sus. **18**(6), 537-553 (2020). https://doi.org/10.1080/14735903.2020.1793507
5. M. Faizal, Achievements and challenges for gender mainstreaming in the employment sector of the Maldives. GMPADSA, Pal. Mac., Cham, 209-230 (2020). https://doi.org/10.1007/978-3-030-36012-2_10
6. H. Tamsah, Y. Yusriadi, Quality of agricultural extension on productivity of farmers: Human capital perspective. USCM. **10(2),** 625-636 (2022)
7. C.L. Anderson, T.W. Reynolds, P. Biscaye, V. Patwardhan, C. Schmidt, Economic Benefits of Empowering Women in Agriculture: Assumptions and Evidence. J. Dev. St. **57**(2), 193-208 (2020). https://doi.org/10.1080/00220388.2020.1769071
8. M. Šimpachová Pechrová, O. Šimpach, T. Medonos, D. Spěšná, M. Delín, What Are the Motivation and Barriers of Young Farmers to Enter the Sector? AGRIS on-line PEI. **10**(4), 79-87 (2018). https://doi.org/10.7160/aol.2018.100409

9. R. Rohaeti, S. Nurhayati, Education on Hydroponic Technology to Increase the Productivity of Modern Farmers. J. Ed. Res. **4**(3), 1317-1324 (2023). https://doi.org/10.37985/jer.v4i3.409
10. B. Bharani Baanu, K.S. Jinesh Babu, A. Baskaran, Need to educate farmers about the benefits of using treated wastewater for agriculture. Wat. Pol. **24**(8), 1269-1286 (2022). https://doi.org/10.2166/wp.2022.046
11. M. Alsaleh, The role of the fishery industry in the shift towards sustainable food security: a critical study of blue food. Env. Sci. Pol. Res. **30**, 105575-105594 (2023). https://doi.org/10.1007/s11356-023-29747-4
12. A.K. Putri, A. Wulandari, Factors Influencing The Income Of Fishermen. Int. J. Bus. Ec. **4**(2), 198-210 (2020). http://dx.doi.org/10.33019/ijbe.v4i2.298
13. W. Leal Filho, M. Fedoruk, J.H. Paulino Pires Eustachio, J. Barbir, T. Lisovska, A. Lingos, C. Baars, How the War in Ukraine Affects Food Security. Foods **12**(21), 3996 (2023). https://doi.org/10.3390/foods12213996
14. K.E. Giller, T. Delaune, J.V. Silva, et al. The future of farming: Who will produce our food? Food Sec. **13**, 1073-1099 (2021). https://doi.org/10.1007/s12571-021-01184-6
15. Z. Rozaki, O. Wijaya, N. Rahmawati, L. Rahayu, Farmers' Disaster Mitigation Strategies in Indonesia. Rev. Ag. Sc. **9**, 178-194 (2021). https://doi.org/10.7831/ras.9.0_178
16. T. Ndimbwa, K. Mwantimwa, F. Ndumbaro, Channels used to deliver agricultural information and knowledge to smallholder farmers. IFLAJ **47**(2), 153-167 (2021). https://doi.org/10.1177/0340035220951828
17. A.V. Grama, Ş.L. Bătrîna, Adaptation of the Farmers' Skills to the Requirements of the Future Agriculture. Sc. Pap. Ser. MEEARD **24**(2), 485-494 (2024)
18. T. Clune, H. Downey, very good farmers, not particularly good business-people: A rural financial counsellor perspective on rural business failure. J. Rur. St. **95**, 256-267 (2022). https://doi.org/10.1016/j.jrurstud.2022.09.025
19. C. Obregón, A. Sokach, P. Neihapi, et al. Coastal fisheries and community-based support systems in post disaster contexts. Mar. St. **23**, 35 (2024). https://doi.org/10.1007/s40152-024-00373-1
20. W.L. de Sousa, D.M. Zacardi, T.A. Vieira, Traditional Ecological Knowledge of Fishermen: People Contributing towards Environmental Preservation. Sus. **14**, 4899 (2022). https://doi.org/10.3390/su14094899
21. Q. Le Bras, D. Gascuel, F. Quemper, H. Levrel, Transition and adaptation: An analysis of how professional fishermen change their practices. Mar. Pol. **164**, 106154 (2024). https://doi.org/10.1016/j.marpol.2024.106154
22. G.A.S. Pratama, S.A.M.P. Suryani, I.N. Muliarta, Human Resources Competencies in the Marine and Fisheries Sector: A Review. F. J. Sus. Res. **2(11)**, 2747-2756 (2023). https://doi.org/10.55927/fjsr.v2i11.6545
23. N. Vasylieva, Economic Aspects of Food Security in Ukrainian Meat and Milk Clusters. AGRIS on-line PEI. **9**(3), 81-92 (2017). http://dx.doi.org/10.7160/aol.2017.090308
24. N. Vasylieva, A. Pugach, Economic assessment of technical maintenance in grain production of Ukrainian agriculture. BJAS **23**(2), 198-203 (2017)
25. A.A.H. Suryana, Perdiansyah, Analysis of Human Resources Competitiveness of Capture Fisheries in Indonesia. As. J. Fish. Aq. Res. **15**(2)**,** 40-46 (2021)
26. Md.M. Uddin, B. Rani Dey, M. Fouzder, Md.S. Hoque, Human Resource Management (HRM) Practices in Fisheries Organizations: a Scenario from Bangladesh. J. Sur. Fish. Sc. **10**(3), 225-235 (2023)
27. Skill gap analysis of Indian fisheries sector. ASCI, GG, 1-61 (2021).
28. Extension & Training Projects, Dep. Min. AAW, the Pres. Of. (2024).
29. Maldives Bureau of Statistics. Statistical Yearbook of Maldives for 2012-2023. Min. NPHI (2024).

# The preparation of complementary food for breastmilk after the earthquake in Malampah, Pasaman Districts, West Sumatra Province in 2022

*Dya* Mulya Lestari[1,2*] and *Nice* Rachmawati Masnadi[1,2]

[1]Department of Child Health, Faculty of Medicine, Universitas Andalas, Padang, Indonesia
[2]Department of Pediatric, Dr. M. Djamil General Hospital, Padang, Indonesia

**Abstract.** A seismic event of 6.2 magnitude occurred on February 25, 2022, causing significant earthquake in West Pasaman Regency and Pasaman Regency. It is evident that children under the age of five constitute the second biggest group of refugees Toddlers who do not get fed properly and are in an environment with poor sanitation can be threatened by various diseases. Implementing regular supplemental feeding techniques in catastrophe settings is a crucial strategy for improving children's nutritional condition. This paper is based on several literature studies that refer to several literatures. Access to facilities and infrastructure influences disaster food assistance. Public kitchens were quickly created in West Pasaman, especially Pasaman District, after the earthquake. Lack of potable water and four-wheeled vehicle access restricted camp food production. Food from PERSAGI and contributors is distributed in Puskesmas Ladang Panjang Malampah. Even in emergencies, typical supplementary meals can be provided. In disasters, primary carers or parents require help governing complementary food provision based on clean water, convenient cooking facilities, diversified food items, or home-cooked, local-culture complementary meals. In addition to facility support, health workers must teach parents about standardised supplementary meals and give them confidence to offer them in emergencies

## 1 Introduction

There have been 5,402 catastrophe incidents in Indonesia as of December 31, 2021. Flooding is the most common type of natural catastrophe, followed by landslides and severe weather. In 2021, there have been 24 earthquakes, 1 mountain eruption, 579 forest fires, 15 drought incidents, 1,794 floods, 1,321 landslide cases, 1,577 extreme weather instances, 91 abrasion tidal wave cases, and 15 occurrences of flooding. 7,630,692 people were displaced by natural disasters, and 14,915 people were injured in addition to 728 fatalities and 87 missing persons [1].

A seismic event of 6.2 magnitude occurred on February 25, 2022, causing significant earthquake in West Pasaman Regency. As per the findings of the Meteorology, Climatology and Geophysics Agency (BMKG), this earthquake was caused by a right-lateral strike-slip fault on the Great Sumatra fault. According to the Singapore Earth Observatory, the fault rupture was limited to a short section, estimated to be around 10 km in length, with an average displacement of 10 cm. Prior to this, a seismic event of magnitude 5.2 took place [2].

According to statistics on the distribution of vulnerable categories among refugees in Pasaman Regency, it is evident that children under the age of five constitute the second biggest group of refugees, behind the elderly. There are a total of 8,457 refugees. The population of elderly refugees amounted to 3,943 individuals, while there were 2,442 children under the age of five, 979 nursing mothers, 641 infants, 339 pregnant women, 80 postpartum women, and 3 individuals with disabilities [3].

The occurrence of disasters brings major changes to human life. These changes occur due to economic losses, infrastructure, and the emergence of health problems experienced by disaster survivors [4]. The impacts arising in terms of health problems such as physical injuries, various diseases due to poor environmental sanitation, and psychological trauma [5]. In addition, the availability of clean water and adequate food consumption is a major concern of every disaster situation [6].

One group that rarely gets attention especially about food and hygiene during disasters is infants and early childhood. For example, during the earthquake in Nepal in 2015, humanitarian agencies competed to provide large amounts of relief, rescue and recovery, but rarely paid attention to the feeding of infants and young children. In addition, based on information from disaster volunteers that in the 2020 flood disaster in Masamba, the public kitchen did not provide special food for infants and toddlers [7].

Toddlers who do not get fed properly and are in an environment with poor sanitation can be threatened by various diseases. The most common disease that can cause

* Corresponding author: dyamulyalestari@yahoo.com

death in infants due to nutrition in disaster situations is diarrhea [8]. This can occur in the age group of infants who should be breastfed, but are not optimally supported for the breastfeeding process [6]. Various formula milk donations received during disasters make infants and toddlers given formula milk given through bottles or pacifiers [9]. The process of cleaning formula milk in disaster situations is often not optimal due to the availability of dirty water [10].

Effective intervention is necessary for addressing nutrition-related illnesses in child survivors. In disaster situations, ensuring proper nutrition for infants and children can be achieved through the implementation of various strategies. These include enhancing support for breastfeeding, implementing targeted interventions for infants who are not breastfed, promoting local complementary feeding practices in line with recommended guidelines, providing micronutrient supplementation, and reducing the reliance on artificial feeding methods to mitigate associated risks [11]. The provision of supplemental feeding in catastrophe scenarios is identical to that in normal settings, since both have the objective of guaranteeing that children between the ages of 6 and 24 months have access to suitable complementary meals [6].

Implementing regular supplemental feeding techniques in catastrophe settings is a crucial strategy for improving children's nutritional condition. This is evident from the 2015 earthquake in Nepal. Nepal successfully addressed the issue of supplemental feeding by instructing and deploying a workforce of 15,000 individuals to provide nutrition assessment and counseling for 90% of children aged 2 and below. This proactive measure significantly reduced the occurrence of malnutrition following the tragedy [6]. The implementation of localized supplemental feeding initiatives, aided by enriched oil and sugar, in refugee camps in Kenya and Tanzania, has demonstrated a positive impact on health promotion among children between the ages of 6 and 59 months [12].

Effective supplementary nutrition in emergency conditions can mitigate the risk of short-term and long-term illnesses in children between the ages of 6 and 24 months. Nevertheless, some refugee camps may not offer sufficient supplementary nourishment for children in this age group. The food provided to children of this age is comparable to that of adults in terms of both texture and quantity. Further research is required to investigate the underlying causes of inadequate nutrition provision for this specific age group during times of calamity. In light of the aforementioned issues, the authors want to do a literature review with the objective of ascertaining the methods employed in the preparation of supplemental foods during times of calamities [13].

## 2 Method

This paper is based on several literature studies that refer to several literatures.

## 3 Literature review

### 3.1 Definition of complimentary feeding

Complementary feeding refers to the introduction of food or drink, other than breast milk, that provides essential nutrients to newborns during the period when they are also receiving breast milk. The Global Strategy for Infant and Young Child Feeding 2002 specifies that supplemental feeding must fulfill the following criteria [14]:

1. Timely: Complementary feeding starts when energy and nutrient requirements exceed those obtained from breast milk.
2. Adequate: Solids should contain enough energy, protein and micronutrients.
3. Safe: Storage, preparation and feeding must be hygienic.
4. Properly: Complementary feeding is given in line with the baby's hunger and appetite signs, and the frequency and method of feeding are appropriate for the baby's age.

### 3.2 Definition of disaster

Disasters are occurrences or sequences of events that pose a hazard and disrupt the lives and means of subsistence of individuals, brought about by a combination of natural, non-natural, and human forces, leading to human fatalities, environmental harm, property damages, and psychological effects on humans.[15] Disasters are categorized as natural disasters, man-made disasters, and hybrid disasters, which involve a combination of both sorts [16].

### 3.3 Geographical situation of West Pasaman and Pasaman regencies

West Pasaman is a regency located in the province of West Sumatra. This region was created as a result of the enlargement of Pasaman Regency. West Pasaman Regency is situated within the geographical coordinates of 00° 33' North latitude to 00° 11' South latitude and 99° 10' to 100° 04' East longitude. It has a total area of 3,864.02 square kilometers. Generally, the West Pasaman Regency has a predominantly flat and gently rolling landscape, with hilly and mountainous regions limited to the Talamau and Gunung Tuleh Districts. The elevation of the region ranges from 0 to 913 meters above sea level. The terrain consists of four distinct types: flat areas with a slope ranging from 0-3%, gently rolling areas with a slope of 3-8%, uneven areas with a slope of 8-15%, and hilly regions with a slope above 15% [17].

Pasaman Regency is situated in the northern region of West Sumatra Province, covering an area of 4,447.63 square kilometers, which is approximately 10.44% of the total area of the province. Pasaman Regency is situated at the equator, spanning from 0-55' north latitude to 0-06' south latitude, and from 99-45' to 100-21' east longitude [18].

## 3.4 Impact of disasters in children

It is estimated that approximately 175 million children per year will be affected by natural disasters associated with climate change. Globally, children will be affected by 88% of climate change-related diseases, and the poorer the child, the greater the burden. The number of children seriously injured or killed by such disasters each year is unknown but estimated to be substantial. An estimated 5-43% of children affected by disasters will develop post-traumatic stress disorder (PTSD), and many will suffer from depression, anxiety or other mental health disorders. With climate change, the mental health burden on children and families is expected to increase [19].

Carolyn Kousky said that there are 3 impacts of disasters on children, especially long-term impacts, namely [20]:

1. Disasters can damage children's physical health: Children can be injured or even die. Children may also suffer from things such as malnutrition caused by food supply disruptions or diarrhea caused by contaminated water. In addition, disasters can cut off access to medical care, even for illnesses not related to the disaster [20].
2. Disasters can cause mental health problems: Not only is the disaster itself stressful and frightening, but children can suffer from psychological disorders due to the destruction of homes and property, migrating, the grief of losing loved ones, seeing parents or caregivers under stress, neglect and abuse, and the breakdown of social networks, the environment, and the local economy [20].
3. Disasters can disrupt children's education: This can occur because children lose their families, school facilities are destroyed, and children are forced to work to help the family economy [20].

## 3.5 Preparation of complimentary foods during disasters

### *3.5.1 Facilities and infrastructure related to complimentary foods*

The availability of facilities and infrastructure at the time affects the provision of supplemental foods after a disaster. Examples include meal items and culinary utensils. Factors such as high food costs, costly sanitary equipment, blockades in the region, fuel shortage, restricted food supply, remote placement of public kitchens, and worsening socio-economic situations are expected to impede the provision of supplemental foods in refugee camps [7].

Following the earthquake catastrophe in West Pasaman, particularly in Pasaman District, public kitchens were promptly established at several locations. Nevertheless, the production of supplemental food in the camps was hindered owing to the scarcity of clean water and the challenging accessibility for four-wheeled vehicles. The food components from PERSAGI (Persatuan Ahli Gizi Indonesia) and donors are distributed at Puskesmas Ladang Panjang Malampah. The preparation of complementary food is directly helped by 4 PERSAGI volunteers who alternate in determining the composition and kind of supplemental food.

a. Water Procurement

Water is an essential resource for all individuals worldwide, as it is required for drinking, cooking, and maintaining personal cleanliness. During a crisis, the availability of safe drinking water may be insufficient, making it crucial to prioritize the provision of potable water. Health issues associated with water can occur as a result of both inadequate availability and the contamination of water to varying degrees. The water supply needs are as follows [21]:

1. Water supply should be sufficient to provide at least 15 liters/person/day.
2. The volume of water flow in each source of at least 0.125 liters per second.
3. The distance of the farthest settlement from the water source is no more than 500 meters
4. 1 (one) water tap for 80-100 people.

In the aftermath of the West Pasaman earthquake in Pasaman Regency, the provision of clean water, particularly at the supplemental food production post, is being sourced from the PDAM water supply of the Puskesmas Ladang Panjang Malampah. This water source is currently abundant and free from contamination. This facilitates the development of supplementary meals.

b. Water Quality

Potable water must be available in adequate quantities to meet fundamental needs such as drinking, cooking, and maintaining personal and household cleanliness, while also being free from significant health concerns caused by diseases or chemical contamination [21].

1. In untreated water sources (not yet sterilized), the bacterial concentration from human fecal contamination is limited to a maximum of 10 coliform bacteria per 100 milliliters.
2. Water supplied to a population of over 10,000 individuals, or to any water sources where there is a possibility of diarrheal illness, must be treated with a disinfectant to meet certain criteria (namely, a chlorine residual of 0.2-0.5 mg/l at the water tap and a saturation level below 5 NTU).
3. Studies have shown that there is no notable detrimental effect on the health of those who utilize water contaminated with chemicals or from the use of water from the source within the intended timeframe. These studies also encompass investigations into the amounts of chemical residues employed to assess the water. Based on the evaluation of the scenario, it is unlikely that consuming the water would lead to any severe health issues.

c. Infrastructure and Equipment

1. Every household possesses a pair of water collectors, each capable of holding 10-20 liters, along with a water storage container that can accommodate 20 liters. This should be a vessel with a slender opening and a cover.
2. Each individual is allocated a monthly allowance of 250 grams of soap.
3. If public restroom facilities are to be made available,

they should be sufficiently spacious to accommodate all individuals for regular daily bathing at a designated time. It is necessary to have separate public toilets for females and males.

4. When establishing public washing facilities for clothing and household utensils, it is required that each water basin be utilized by a maximum of 100 individuals.

### *3.5.2 Raw materials of complementary food*

Complementary feeding is produced using raw products that adhere to safety, quality, and nutritional standards. Cereals are often the primary raw materials used in the production of supplemental meals. Cereals refer to individual grains such as rice, maize, wheat, and similar crops. Furthermore, legumes, which are seeds enclosed in two shells, are frequently utilized as supplementary ingredients in cuisine, including green beans, kidney beans, and soybeans. Cereals are a rich source of complex carbohydrates, specifically starch, which may be broken down and used by the body to generate energy. Legumes, such as mung beans and kidney beans, have a significant amount of protein (20% or more) and certain legumes, like soybeans, also contain a notable amount of fat (about 20%) [14].

**Fig. 1.** Raw materials for making complementary food

Public kitchens were established at each evacuation post three days following the earthquake. However, the supplemental food station, located in Puskesmas Ladang Panjang Malampah, was established five days after the earthquake. Originally, the plan was to prepare complementary food at each evacuation post. However, due to challenges in distributing the food and the remote locations that are inaccessible by four-wheeled vehicles, the decision was made to centralize the preparation of complementary food at Puskesmas Ladang Panjang Malampah. The PERSAGI team conducted the production of this supplementary food using components provided by PERSAGI and numerous contributors. The supplemental diet included rice, two types of vegetables, one vegetable protein, and one animal protein. The meal was prepared for a duration of three days, namely from March 3rd to March 5th, 2023 (Fig.1).

The calamity involved the utilization of rice, potatoes, spinach, carrots, beans, tofu, tempeh, and poultry as raw materials. The food raw ingredients were acquired via donors and PERSAGI. The food raw materials supplied by PERSAGI and donors are manufactured just once due to the absence of storage or refrigeration facilities for food ingredients. Every day, PERSAGI and donors provide food delivery.

### *3.5.3 Making complementary food*

Ensuring the hygiene of hands and cooking equipment used in the preparation and serving of supplemental foods is essential. Thoroughly cleanse hands using fresh water and soap, cleanse cutlery using fresh water and soap, and cleanse food components using fresh water.[22]

Four individuals from PERSAGI, aided by volunteers, were responsible for the production of supplemental meals, as shown in Fig. 2. The supplemental meal was divided into two categories: strained porridge for infants aged 6-8 months and coarsely chopped porridge for infants aged 9-12 months. The supplementary energy needed from solid food for infants aged 6-8 months is 200 kilocalories per kilogram of body weight per day, whereas for infants aged 9-12 months, it is 300 kilocalories per kilogram of body weight per day [22].

**Fig. 2.** The making of complementary food

The PERSAGI staff and volunteers started food preparation at 08.00 in the morning, only providing gratis lunch meals. The components were finely chopped and simmered together in a single big pot. The only seasonings utilized were onion, garlic, bay leaf, a small amount of salt, oil, and chicken bone broth.

The procedure of preparing supplemental meals at the Ladang Panjang Malampah Community Health Center involves chopping the raw materials and thereafter boiling all the components. The initial preparation involves creating a coarse-textured porridge. Then, half of this porridge is transformed into a smoother and more liquid consistency by blending or filtering.

These supplementary meals are intended for infants between the ages of 6 and 12 months. For infants aged 9-

12 months, it is recommended to provide porridge with a coarse texture. For infants aged 6-8 months, it is recommended to provide porridge with a fine texture. The packaging of these supplemental foods utilizes plastic containers that are both safe and appropriately sized. The supplementary nourishment will thereafter be allocated to several regions like Cempaka, Pustu Malampah, Polindes Bungo Tanjung, Tanah Lapang Malampah, Kampung Fatimah, and Kampung Aur, by means of village midwives.

### *3.5.4 Food quality and safety during disasters*

**Fig. 3.** The Packaging of complimentary food

The food provided to communities impacted by disasters is of high quality and is handled in a safe manner to ensure its suitability for consumption, as shown in Fig. 3. The specified prerequisites are as follows[21]:

1. The distribution of food does not contribute to the transmission of diseases.
2. There have been no grievances regarding the quality of the food supplied, either by either the recipients or the officials.
3. Food suppliers conduct frequent quality control procedures to ensure that the food they provide complies with official government regulations on packaging, labeling, expiration dates, and other relevant factors.
4. The local Food and Drug Administration (POM) conducts regular inspections on all food items delivered to the site.
5. All domestically sourced food products have a minimum shelf life of 6 months from the date of delivery, except perishable commodities such fresh vegetables and fruits.
6. It is necessary to deliver all of these food items prior to their expiration date.
7. The food storage infrastructure is sufficient and the processing procedures are correctly executed.
8. Relief personnel possess an appropriate level of understanding to effectively deal with any health problems that may arise from the distribution of food. These problems include dangers associated with improper processing, insufficient storage, and delayed delivery.

### *3.5.5 Receiving food products*

The food given is suitable and satisfactory for the recipients [21].

1. Prior to finalizing the food items for distribution, it is important to engage in conversations with the recipient groups to verify that the food products fulfill their standards of eligibility and appropriateness, and are accepted by them.
2. The food provided is in accordance with local religious traditions and customs, taking into consideration any specific taboos or laws regarding the ingestion of pregnant and/or nursing women.
3. The staple meal provided should align with the typical dietary preferences of the receiving group.
4. The supplementary food provided for children under the age of five satisfies their taste preferences and is appropriate for their digestive capacities.
5. The community has access to specific food items that are seen as essential in their culture, such as chile and sugar.

### *3.5.6 Food handling and safety*

Food is stored, processed and consumed safely and correctly, both at the household level and in the context of society in general. The requirements are as follows [21]:

1. There is no spread of diseases associated with food distribution locations (e.g. public kitchens) resulting from improper food processing.
2. Reports from the assisted community relating to difficulties in storing, cooking and consuming the distributed food are submitted by the group/team leader to the District Implementation Unit (SATLAK District).
3. Each household has at least one cooking pot, sufficient fuel for cooking, a 40-liter water storage container, and 250 grams of soap/person/month.
4. Individuals who cannot cook their own food or cannot consume food without assistance have access to someone who provides them with adequate food on a regular basis.
5. When food is distributed from soup kitchens (already cooked), implementers understand the health risks of inadequate food storage, processing and serving.
6. If necessary, food milling or processing infrastructure or other food processing is available as close as possible to the location or place of evacuation.

### *3.5.7 Distribution of assistance*

The food aid program must employ an effective distribution methodology, taking into account several factors throughout its execution. Following the tragedy, the officer conducted an initial evaluation of the situation, taking into account the technique and calculation of distribution. Food assistance can be provided indiscriminately to the entire community or selectively to specific sectors or factions within the community, as shown in Fig. 4. Food assistance can be provided as an alternative to wages, or it might be sold to commercial markets to address supply challenges [21].

**Fig. 4.** Distribution of complimentary food assisted by local residents

The distribution of food aid is equitable, environmentally viable, and practical, considering the specific circumstances of the receivers. Considering the specific circumstances in the area, individuals receiving food assistance should have access to information on the allocation of help that they are eligible for. The following preparations are necessary[21]:

1. The community comprehends the quantity and nature of the distribution to be allocated for each distribution.
2. The quantity and nature of food intended for distribution is identical to what will be received by the community.
3. The food distribution technique is equitable, conveniently located (in proximity to the population it serves), and efficient in terms of time and energy use.
4. The frequency of the food distribution (whether it will occur monthly or bi-monthly), The assistance recipient must demonstrate thoughtfulness, which may include the following: Twenty-two
   a. The expense associated with transporting food from the distribution center.
   b. Commute time to and from the distribution center.
   c. Ensuring the safety of delivering the assistance and food supplies to the intended recipient.

### 3.6 Code of conduct for formula milk feeding

#### *3.6.1 Indications for formula feeding*

According to Minister of Health Regulation No. 39 of 2013, baby formula refers to a type of milk that is specially designed to be used as a replacement for breast milk for babies who are up to 6 months old. It is mandatory for every woman to exclusively breastfeed her newborn, unless certain conditions arise when the mother, family, medical staff, or other health workers may offer infant formula [23].

The determination of the presence of medical indications must be done by a physician. In the event that a doctor is not available, the procedure is performed by a midwife or nurse in compliance with the conditions outlined in laws and regulations [24].

1. Infants that are limited to consuming milk through specialized formulae, typically due to congenital metabolic problems. The number is 24.
2. Infants who necessitate other sustenance apart from breast milk for a temporary duration 24.
3. The mother has medical issues that prevent her from exclusively nursing because she has to undergo therapy according to medical service requirements.
4. If the mother has a medical condition with positive hepatitis B surface antigen (HBsAg) and the newborn has not received both passive and active hepatitis B immunization within 12 hours.
5. Mother is unavailable.
6. The mother and the infant are separated.

#### *3.6.2 Formula milk feeding in emergency situations/disasters*

Formula milk distribution during emergencies or catastrophes is in compliance with Regulation of the Minister of Health Number 39 of 2013:[23]

1. The distribution of infant formula and other baby products during emergencies and disasters should be done by the local district/city health office, which has coordinated with the Ministry of Health and follows the guidelines for infant and child feeding in emergency situations established by the Minister.
2. During emergency and/or crisis conditions, it is strictly forbidden for any maker of infant formula and other newborn goods to engage in any activities.

Health workers, operators of health service facilities, operators of education units, management of professional organizations in the health sector, as well as manufacturers and distributors of infant formula milk and/or other baby products who violate the provisions will face administrative sanctions [25].

## 4 Conclusion

Providing normal supplemental meals is possible in every scenario, even disasters. This is because standardized supplemental meals can provide all the nutritional demands of 6-23-month-olds, optimize development and growth, and avoid short and long-term disorders. Supplemental food is used in disasters the same way as usual. In disaster situations, primary caregivers or parents need special assistance to condition the provision of complementary foods according to standards like clean water, easy access to cooking facilities, varied food ingredients, or donations of home-cooked, local-culture complementary foods. In addition to facility assistance, health personnel must educate parents about the need of giving standardized supplemental foods and give them confidence to do so in disasters.

## References

1. BNPB, Info grafis bencana 2021, 2021. https://bnpb.go.id/infografis/kejadian-bencana-tahun-2021 (accessed Mar. 24, 2022).

2. Gempa bumi pasaman barat 2022. https://en.m.wikipedia.org/wiki/2022_Sumatra_earthquake (accessed Mar. 25, 2022).
3. P. D. P. Utama, Gambaran Data Pengungsi, Pasien, dan Kerusakan Akibat Gempa Tanggal 25 Februari 2022 Di Kabupaten Pasaman Barat. (2022).
4. Santri, E. Apriyanto, and S. P. Utama, Dampak Sosial Ekonomi dan Estimasi Kerugian Ekonomi Akibat Banjit di Kelurahan Rawa Makmur Kota Bengkulu (NATURALIS, 2020).
5. Widayatun and Z. Fatoni, Permasalahan Kesehatan dalam Kondisi Bencana: Peran Petugas Kesehatan dan Partisipasi Masyarakat, J. Kependud. Indones. (2013).
6. K. Gribble, Supporting the most vulnerable through appropriate infant and young child feeding in emergencies, J. Hum. Lact., **34**(1), 40–46, Feb. (2018).
7. S. DeYoung, M. Suji, and H. G. Southall, Maternal Perceptions of Infant Feeding and Health in the Context of the 2015 Nepal Earthquake, J. Hum. Lact., **34**(2), 242–252, May, (2018).
8. C. Carothers and K. Gribble, Infant and young child feeding in emergencies, J. Hum. Lact., **30**(3), 272–275 (2014).
9. L. Shaker-Berbari, H. Ghattas, A. G. Symon, and A. S. Anderson, Infant and young child feeding in emergencies: Organisational policies and activities during the refugee crisis in Lebanon, Matern. Child Nutr., **14**, 3, Jul. (2018).
10. F. Di Marcantonio, E. Custodio, and Y. Abukar, Child Dietary Diversity and Associated Factors Among Children in Somalian IDP Camps, Food Nutr. Bull., **41**(1), 61–76, Mar. (2020).
11. Emergency Nutrition Network (ENN), Panduan Operasional Pemberian Makanan bagi Bayi dan Balita dalam Situasi Darurat. (2017).
12. H. Tappis, S. Doocy, C. Haskew, C. Wilkinson, A. Oman, and P. Spiege, United Nations High Commissioner for Refugees feeding program performance in Kenya and Tanzania: A retrospective analysis of routine Health Information System data, (2012).
13. D. Wanda, Praktik Pemberian Makanan Pendamping Air Susu Ibu pada Situasi Bencana: Studi Literatur Complementary Feediing Practices in The Disaster Situation: A Literature Review, (2021).
14. A. Michael Sihombing, F. Sania, F. Unsa, and M. Syarifa Yani, The 6th CIPRIME Current Issues in Pediatric Nutrition and Metabolic Problems 12 dan 13 September 2015 Hotel Grand Sahid Jaya Jakarta Desain dan tata letak cetak: Departemen Ilmu Kesehatan Anak FKUI-RSCM, (2015).
15. Presiden Republik Indonesia, Undang-Undang Republik Indonesia Nomor 24 Tahun 2007 Tentang Penanggulangan Bencana. (2007).
16. Menteri Kesehatan Republik Indonesia, Keputusan Menteri Kesehatan Republik Indonesia Nomor 066/MENKES/SK/II/2006 Tentang Pedoman Manajemen Sumber Daya Manusia (SDM) Kesehatan dalam Penanggulangan Bencana, (2006).
17. Kondisi Geografis Pasaman Barat. https://id.m.wikipedia.org/wiki/Kabupaten_Pasaman_Barat
18. Pemerintah Kabupaten Pasaman, Letak Geografis Kabupaten Pasaman, (2021). https://www.pasamankab.go.id/halaman/letak-geografis-kabupaten-pasaman
19. A. Dyregrov, W. Yule, and M. Olff, Children and natural disasters, European Journal of Psychotraumatology, **9**, sup2. Taylor and Francis Ltd., Dec. 03 (2018).
20. C. Kousky, Impacts of Natural Disasters on Children, (2016).
21. Mentri Kesehatan Republik Indonesia, Keputusan Mentri Kesehatan Nomor 1357/Menkes/SK/XII/2001 tentang standar minimal penanggulangan masalah kesehatan akibat bencana dan penanganan pengungsi, (2001).
22. Ikatan Dokter Anak Indonesia, Pemberian Makanan Pendamping Air Susu Ibu (MPASI). (2018).
23. Menteri Kesehatan Republik Indonesia, Peraturan Menteri Kesehatan Republik Indonesia Nomor 39 Tahun 2013 Tentang Susu Formula Bayi dan Produk Bayi Lainnya. (2013).
24. Departemen Kesehatan RI, Pedoman Pemberian Makanan Bayi dan Anak dalam Situasi Darurat. (2007).
25. Menteri Kesehatan Republik Indonesia, Peraturan Menteri Kesehatan Republik Indonesia Nomor 15 Tahun 2014 Tentang Tata Cara Pengenaan Sanksi Administratif Bagi Tenaga Kesehatan, Penyelenggara Fasilitas Pelayanan Kesehatan, Penyelenggara Satuan Pendidikan Kesehatan, Pengurus Organisasi Profesi. 1–13 (2014).

# Toddler feeding, WASH conditions, and clinical complaints during the 2024 flash flood disaster in Jorong Galuang, Sungai Pua District, Agam Regency, West Sumatra

*Shinta* Asbi[1,2*], *Nice* Rachmawati Masnadi[1,2], and *Mayetti*[1,2]

[1]Department of Pediatrics, Faculty of Medicine, Universitas Andalas, Padang, West Sumatra, Indonesia
[2]Departement of Pediatrics, Dr. M. Djamil Hospital, Padang, West Sumatra, Indonesia

**Abstract.** This study focuses on the impact of flash floods in Jorong Galuang, Sungai Pua, particularly on toddler nutrition, water, sanitation, hygiene (WASH), and clinical health. A qualitative study conducted through interviews and direct observation, data were collected from mothers of toddlers, health officers, midwives, and NGO representatives AIMI. Despite the absence of dedicated MPASI kitchens during the disaster, toddlers did not experience significant difficulties with feeding due to aid from NGOs and the availability of undamaged kitchens in some homes. Water access remained sufficient through local springs and additional supplies from the PMI (*Palang Merah Indonesia*), although hygiene practices, particularly handwashing, were inadequate. Clinical complaints among toddlers were predominantly respiratory issues, including coughs and colds, with a few cases of diarrhea reported.

## 1 Introduction

Disasters bring significant changes to human life, including economic and infrastructure losses, as well as health problems for the victims [1, 2]. A severe flood in Jakarta is predicted to result in economic losses reaching 36 trillion rupiahs by 2027 if flooding continues [3]. Besides economic damage, disasters also cause health issues, such as physical injuries, diseases due to poor sanitation, and psychological trauma [4]. In Somalia, many refugees lack access to clean water and proper sanitation, increasing the risk of diseases in unhygienic environments [5].

Malnutrition is a common issue among children living in refugee camps, putting them at risk for respiratory infections and malnutrition, such as stunting or severe malnutrition, as seen in Kenya and Somalia. Many children suffer from these conditions [5, 6]. A 2021 study by Astuti revealed that complementary feeding practices in disaster situations often fail to meet the minimum standards recommended by WHO, especially in terms of dietary diversity and hygiene [7]. Research conducted in Somali refugee camps in 2020 found that 85% of children did not receive food that met the Minimum Dietary Diversity (MDD). These camps were overcrowded with poor sanitation, inadequate waste disposal, and water shortages in the toilets. Milk was the main food for children aged 6-23 months, with limited fruits and vegetables, high food prices, and cereal-dominant aid. The consumption of fruits, vegetables, and eggs was extremely low [5].

One of the groups that often receives little attention in disaster situations is infants and toddlers. In many cases, the humanitarian aid provided during disasters focuses more on general needs, while the specific needs of toddlers, such as the availability of complementary feeding (MPASI) and hygiene, often go unnoticed. For example, during the Masamba flood in 2020, public kitchens did not provide special food for young children, resulting in inadequate nutrition [8].

In May 2024, West Sumatra experienced flash floods, affecting several areas, including Agam, Tanah Datar, and Padang Panjang, causing severe damage and loss of life. The flood was triggered by extremely high rainfall upstream of Mount Marapi, which caused cold lahar flows to form after heavy rains eroded volcanic material deposits on the mountain slopes, resulting in more than 58 deaths and dozens of others missing. Many families were forced to evacuate, and many homes and infrastructure, including bridges and roads, were severely damaged, causing major transportation routes in the province to be paralyzed. The impact of the flash flood in Agam, according to BPBD data, included 24 fatalities, 22 injuries, and 704 people displaced. Many houses were affected by this disaster, with 59 severely damaged units, 36 moderately damaged units, and 221 lightly damaged units. The flood also damaged agricultural land, with 348 hectares of farmland being destroyed [9].

* Corresponding author: shinta.asbi@gmail.com

## 2 Methods

A qualitative study conducted through interviews and direct observation in Jorong Galuang flood disaster. The study involved observation and interviews on the 7th day of the disaster, which was May 17, 2024. Data was collected from various sources, including the interviews that were conducted in an organized sequence, beginning with the BPBD of Agam Regency, followed by the head of AIMI, a non-governmental organization. Subsequently, discussions took place with the nutrition officers from Puskesmas Sungai Pua, village midwives, and local health workers.

Data on toddler feeding, WASH conditions (water, sanitation, and hygiene) from interviews with 23 mothers of toddlers affected were conducted on July 23, 2024. The research variables included clean water, sanitation, hygiene, and clinical complaints. Sanitation is defined as the provision and use of facilities and services for the safe disposal of human urine and feces, thereby preventing environmental contamination. Sanitation practices are categorized into 'Good' and 'Poor' based on specific criteria. 'Good' practices involve using the bathroom for urination and defecation, and disposing of children's diapers in a trash can that is subsequently burned or taken to a final disposal site. In contrast, 'Poor' practices include not using the bathroom for these needs and failing to properly dispose of children's diapers by not placing them in a trash can for burning or final disposal

Hygiene refers to the practice of washing hands with soap after defecating and disposing of children's feces, before preparing and handling food, and before eating. Good hygiene practices include washing hands before eating and preparing food. On the other hand, poor hygiene practices involve either occasionally or not always washing hands before eating and preparing food.

## 3 Results

Data from the Agam Regency BPBD recorded that a flash flood struck Jorong Galuang, located in Agam Regency, and it was one of the most severely affected areas by the flood. According to data from BPBD and interviews with local residents, the flash flood in Jorong Galuang brought mud, stones, and wood from the hillside, which then buried parts of the settlement and agricultural land in the area. Other areas in Agam Regency were also affected by the flash flood, but Jorong Galuang was among the worst-hit. The floodwaters that passed through Jorong Galuang carried debris and caused a flash flood that destroyed several houses and public facilities. On the seventh day, the researcher went to the field and also observed the disaster's impact on buildings. Not all areas were severely affected, but some houses were destroyed, and debris carried by the flood was piled up.

In addition to Jorong Galuang, other areas affected by the flash flood in Agam Regency included several areas along the river, located at the foot of the mountain. The affected sub-districts included Canduang, Sungai Pua, IV Koto, Ampek Angkek, Tilatang Kamang, Kamang Magek, Palembayan, Malalak, Tanjung Raya, Baso, Banuhampu, and Palupuah. The damage in these areas varied, ranging from minor to severe damage, depending on proximity to the flood source and the area's topography (Fig. 1).

**Fig. 1.** Community service for the flash flood in Jorong Galuang, Sungai Pua

According to data from BPBD Agam Regency, in Jorong Galuang, 15 houses were severely damaged, 5 houses sustained moderate damage, and 15 houses were slightly damaged. There were six fatalities: two children aged 10 and 15, one person aged 53, and the rest were elderly individuals aged 61 and 71.

### 3.1 Feeding toddlers in disaster areas

Interviews with village midwives, health workers, and mothers in Jorong Galuang revealed that most areas were not severely damaged during the flash floods (Fig. 2). One-third of the affected areas did not have significant home damage, allowing families to prepare meals for their children. Lower-lying areas were hit harder, but no toddlers lived there, so nutritional needs were not significantly impacted. In the first three days, most families received food aid from neighbors and family members, with no need for specially prepared MPASI. Some children over 12 months were given instant porridge during this time.

On the 7th day, researchers observed that the public kitchen did not provide special MPASI meals. Interviews with the nutrition officers of Puskesmas Sungai Pua revealed a lack of specific feeding protocols for toddlers during the disaster, although toddlers were registered. The nutrition officers mentioned they had no prior training in disaster nutrition, making toddler nutrition management a challenge.

**Fig. 2.** Direct observation of public kitchens and interviews, along with counselling for mothers of toddlers

Mothers whose homes were not heavily damaged could continue cooking, although ingredient variety was limited. Families with intact homes had better access to food and supplies, while those with damaged homes struggled. By the third day, local markets reopened, improving food access. Support from AIMI (Indonesian Breastfeeding Mothers Association) and PERSAGI (Indonesian Nutritionists Association) was significant in providing both funds and pre-arranged menus for toddlers.

The village midwives reported that public kitchens were established three days after the disaster, with assistance from volunteers and Universitas Islam Negeri Batusangkar. AIMI provided MPASI for 23 toddlers in Jorong Galuang. AIMI organized donations through social media and worked with the local Puskesmas to plan menus and distribute food. Meals included carbohydrates, protein, vegetables, and fruit, and were cooked by local volunteers. Evaluations showed that children consumed the meals provided by AIMI.

Interviews conducted on May 19, 2024, revealed that most toddlers were eating according to their needs, consuming three meals daily with occasional fruit and vegetable snacks. Breastfeeding continued without issue, and one child on formula milk prior to the disaster continued receiving it, with no problems preparing the formula. Families purchased some food, and formula donations allowed more children to start consuming formula regularly. The formula was distributed by the village midwife after being donated by volunteers to the Poskesri (*Pos Kesehatan Nagari*).

## 3.2 WASH conditions in Jorong Galuang

**Table 1.** Results of interviews on sanitation overview among mothers of toddlers in Jorong Galuang

| | **Good** | **Poor** |
|---|---|---|
| **Sanitation** | 23 (100%) | 0 (0%) |
| **Hygiene** | 5 (21.7%) | 18 (78.3%) |

In Jorong Galuang, interviews with residents indicated that the majority of households relied on spring water for drinking, washing, and preparing food. The village midwives reported that additional clean water was supplied by PMI (Palang Merah Indonesia) using four water tank trucks. Conversations with mothers revealed that while sanitation conditions for both mothers and children were generally adequate, there were issues with hygiene practices. Many did not wash their hands with soap before preparing or serving food, and children did not always wash their hands before eating. Despite the disaster, the clean water supply remained largely unaffected. On the 7th day of the disaster, researchers observed that the public kitchen had access to spring water, which was used for cooking and cleaning utensils, as confirmed by kitchen staff (Table 1).

## 3.3 Clinical complaints among children in Jorong Galuang

**Table 2.** Results of interviews on clinical complaints among mothers of toddlers in Jorong Galuang

| | **f (%) n = 23** | |
|---|---|---|
| | **Yes** | **No** |
| **Diarrhea** | 4 (17.4%) | 19 (82.6%) |
| **Cough** | 9 (39.1%) | 14 (60.9%) |
| **Cough and Runny Nose** | 9 (39.1%) | 14 (60.9%) |
| **Hospitalized** | 0 (0%) | 23 (100%) |

On July 23, 2024, interviews and nutritional counseling sessions for toddlers were held in Jorong Galuang (Table 2). The interviews were conducted solely with the mothers of 23 toddlers. During the interviews, a few clinical symptoms were reported. Most of the children had suffered from coughs and colds during the disaster, while only four cases of diarrhea were noted. No skin-related issues were reported, and no serious conditions requiring hospitalization were observed.

## 4 Discussion

### 4.1 Feeding toddlers in the Jorong Galuang flash flood disaster

Based on the interviews, the quality of food provided during the disaster was considered good due to external aid, and most homes remained intact, supporting the preparation of good-quality meals. However, the quantity of food could not be measured as the researcher did not directly observe all feeding processes. Darwis et al. highlighted challenges in meeting food needs for toddlers in disaster-prone areas, such as Desa Nagrakjaya, Sukabumi, where people rely on small shops and mobile vendors for food. No special provisions for toddlers were made, and during disasters, residents depended on aid from others, without stocking up on food for both adults and toddlers, who require different nutrition [10].

Research in Desa Dasan Geria, Lombok Barat, found that both the quantity and quality of food for toddlers in disaster areas were inadequate. Toddlers in evacuation centers often consumed instant noodles and snacks, which did not meet their daily calorie needs. Their food was not separated from adult meals, differing only in spice level, leaving them with limited nutrition. Despite efforts to provide nutritious food, including MP-ASI biscuits, aid mostly consisted of instant food, lacking fruits, vegetables, and sufficient animal protein [11].

The quality of the food provided also did not meet the recommended Nutritional Adequacy Ratio (AKG). Although there were efforts from some parties to provide nutritious food, such as distributing MP-ASI biscuits, food aid was still dominated by instant foods and lacked sufficient fruits, vegetables, and animal proteins. The food provided was less varied and often did not contain all the nutrients toddlers need for optimal growth and development [11].

The issue of good-quality food but insufficient quantity for toddlers in disaster areas remains critical. While the food provided may have high nutritional value, inadequate portions can cause health issues. Balanced nutrition is essential for toddlers' growth, and fortified food aid with additional nutrients, along with vitamin and mineral supplementation programs, are often implemented in disaster situations [12].

Insufficient food quantity means that toddlers do not receive enough calories to meet their daily energy needs. Although the food may be of good quality, a lack of quantity can hinder daily activity and growth. Research shows that toddlers in disaster-affected areas often receive only a fraction of their required daily caloric intake due to limited food access and distribution [13].

This insufficient food intake can lead to malnutrition, weakened immunity, growth issues, and cognitive problems. Malnourished toddlers are more prone to infections and diseases as their immune systems are compromised. Chronic malnutrition can cause stunting, affecting height and brain development, which can later impair learning abilities [14].

To address these issues, sufficient food distribution, fortified food with micronutrients, parental education on providing adequate and nutritious food, and regular monitoring of nutritional status are necessary. By ensuring that food is both high in quality and quantity, the risk of malnutrition can be reduced, supporting optimal recovery and growth for toddlers in disaster-affected areas [15].

### 4.2 WASH conditions for toddlers in the Jorong Galuang flood disaster

Water availability in the area was adequate, with the primary source being local springs. The water used for daily needs, especially drinking, was boiled first to ensure cleanliness. Additionally, the community received additional support in the form of water supplies, which helped ensure the availability of clean water. The sanitation conditions in the area were considered good, with adequate systems in place to maintain environmental cleanliness. However, hygiene levels were still suboptimal, indicating areas that required more attention to ensure better personal and environmental hygiene.

Providing food during emergency disaster response based on WASH (Water, Sanitation, and Hygiene) principles is an important step in ensuring that the food distributed to disaster victims is not only nutritious but also safe and hygienic. WASH principles include providing clean water, adequate sanitation, and good hygiene practices, all aimed at preventing the spread of diseases that can exacerbate emergency situations. By integrating these principles, disaster emergency responses can be more effective in protecting the health and well-being of affected communities [16].

Good sanitation is essential to prevent food contamination and the spread of diseases during emergency response. Measures that need to be taken include providing adequate toilet facilities and preventing environmental contamination. Temporary emergency sanitation facilities, such as portable toilets or makeshift sanitation, should be installed to ensure that human waste is properly managed and does not contaminate water sources or the surrounding environment. Effective waste management, including the safe collection and disposal of waste, is essential to maintaining cleanliness and public health. Food storage areas should also be regularly cleaned and disinfected to prevent contamination [16].

Good hygiene practices are key to preventing food- and waterborne diseases. Promoting hand washing with soap before preparing or consuming food, after using the toilet, and after handling raw materials is crucial. The distribution of hygiene supplies such as soap, hand sanitizers, and other hygiene equipment to affected communities must be done evenly. Education on the importance of hygiene practices should be carried out through community awareness campaigns and training for healthcare workers and volunteers [16].

Access to clean water often becomes a challenge during disasters, requiring efforts like distributing water via tankers or using purification tablets. Sanitation facilities, including portable toilets, are crucial to managing waste safely and preventing disease spread. Hygiene education and distribution of supplies such as soap and hand sanitizers are also key [16].

A study by Uprety et al. following the 2015 Gorkha earthquake in Nepal revealed that despite efforts to provide water and sanitation, contamination and inadequate hygiene education remained problems [17]. Similarly, research by Sneha Krishnan after Cyclone Phailin in India showed that even with improvements in water access and hygiene education, changing sanitation practices remained difficult due to cultural norms. An integrated approach involving health education, changes in social norms, and better sanitation facilities are needed to improve WASH conditions during disaster recovery [18].

During disasters, access to clean drinking water often becomes a major issue. Damaged infrastructure results in many water sources being contaminated by pathogens, making the water unsafe to consume. Contaminated drinking water increases the risk of waterborne diseases such as diarrhea and cholera, which are particularly dangerous for toddlers, as their immune systems are weaker compared to adults [17].

Sanitation facilities in evacuation sites are generally inadequate. Many evacuees have to share a limited number of toilets, which are not always clean. The lack of proper sanitation facilities leads to the spread of infectious diseases, severely affecting the health of toddlers. Poor sanitation also creates an unhygienic environment, worsening the general health conditions of evacuees. Hygiene practices are often neglected in emergency situations. The lack of access to soap and clean water makes it difficult to maintain personal hygiene, including regular hand washing. For toddlers, poor hygiene increases the risk of skin infections, respiratory illnesses, and digestive diseases. Education on the importance of hygiene in emergency situations needs to be enhanced to reduce health risks associated with poor hygiene [17].

Research conducted by Sneha Krishnan in Odisha, India, after Cyclone Phailin in 2013 provides insight into the WASH conditions for toddlers during the disaster recovery period. The study showed that although there were improvements in access to clean water and increased hygiene awareness, there were still significant challenges in changing sanitation practices within disaster-affected communities. The interventions included improvements and installation of clean water sources, construction of communal toilets, and hygiene promotion campaigns. However, significant changes in sanitation practices were still difficult to achieve due to strong cultural norms supporting open defecation. To improve the effectiveness of WASH recovery programs, a more integrated and participatory approach is needed, involving health education, changes in social norms, and improved access to proper sanitation facilities [18].

### 4.3 Clinical complaints in toddlers during the Jorong Galuang flood disaster

Most children experienced coughs and colds during the disaster, with only four cases of diarrhea. This aligns with research by Bahie Mary Rassekh and Mathuram Santoshamb in Aceh after the 2004 tsunami, which found that children aged 1-5 were most vulnerable to diarrhea, respiratory infections, fever, and skin diseases. Similarly, a study by Ashlesha Datar et al. in India showed that exposure to natural disasters increased the risk of acute illnesses such as diarrhea, fever, and respiratory infections in children under five by 9-18%. Disaster exposure also negatively impacted physical growth, leading to a 7% increase in stunting and underweight, and reduced age-appropriate immunization coverage by nearly 18% [19].

Both studies highlight that children affected by disasters are highly susceptible to infectious diseases and experience deteriorating overall health, both in terms of acute illness and long-term growth and development. Health interventions should prioritize access to formal healthcare, health education, and targeted support for children who have lost parents or live in evacuation centers [19, 20].

In this study, sanitation was generally adequate, but hygiene practices were poor, with inconsistent handwashing using soap by both mothers and children. Despite the disaster, the clean water supply remained stable, supported by PMI's additional water deliveries. This aligns with research by Uprety et al. and Datar et al., along with analyses by Rassekh and Santosham, revealed that diarrhea and cough are the most common complaints in children during natural disasters, often due to poor water quality, sanitation, and hygiene [17, 19, 20]. After the 2015 Gorkha earthquake in Nepal, many children experienced diarrhea from contaminated water and inadequate sanitation in evacuation centers, which also led to an increase in respiratory infections. Children from disaster-affected homes without parental access often receive inadequate healthcare, heightening their risk for illnesses like diarrhea and cough. Addressing these issues requires proper interventions in clean water supply, sanitation facilities, and healthcare services to protect children's health during crises [17].

## 5 Conclusion

During the flash floods in Jorong Galuang, there was a public kitchen, but no dedicated kitchen for MPASI (complementary feeding for toddlers), but toddlers still received adequate food due to aid from NGOs, community support, and minimal damage to homes. Water supply remained sufficient through local springs and PMI assistance. While sanitation access was available, many mothers exhibited poor hygiene, particularly in handwashing. Toddlers' primary health issues were coughs, colds, and diarrhea. To improve disaster preparedness, health officers, nutritionists, and volunteers should receive training on MPASI kitchen implementation, and hygiene education should be enhanced.

The limitations of this study include that the research conducted fieldwork for interviews and observations independently without collaboration with governmental departments, such as the BPBD, Social Services Department, and the Health Department, which are key policy makers. This lack of cooperation during disaster response and interviews hindered access to direct data from these agencies. Other than that, the interview with the mother of the toddler was conducted 2 months after the disaster, so an accurate food recall during the disaster could not be obtained. Support from the BPBD, Social Services, and the local Health Department is necessary in conducting disaster research to ensure direct access to data from relevant agencies in affected areas.

## References

1. Gst Ngr Yoga Jayantara I. Implementasi qgis untuk mengestimasi kerugian ekonomi akibat banjir di kabupaten bandung. Jurnal Pendidikan Teknologi dan Kejuruan. **18**(2). (2020). https://doi.org/10.23887/jptkundiksha.v17i2.25839
2. Susilawati A, Efendi F, Hadisuyatmana S. Gambaran kesiapan tenaga kesehatan dalam manajemen bencana di puskesmas wilayah rawan bencana. Indonesian Journal of Community Health Nursing. **4**(1) (2019). http://dx.doi.org/10.20473/ijchn.v4i1.12395
3. Yuhanafia N, Andreas H. Pertambahan estimasi kerugian ekonomi akibat banjir dengan pengaruh penurunan tanah di Jakarta. **17**, Gea. Jurnal Pendidikan Geografi. (2017). http://dx.doi.org/10.17509/gea.v17i2.7490
4. Hutagaol EK. Health problems in disaster conditions: role of community participation health officers. Jurnal Ilmiah Kesehatan Institut Medika drgSuherman. **1**(1). (2019). https://doi.org/10.47134/comdev.v5i2.262
5. Di Marcantonio F, Custodio E, Abukar Y. Child dietary diversity and associated factors among children in Somalian IDP camps. Food Nutr Bull. **41**(1), 61–76. (2020). http://dx.doi.org/10.1177/0379572119861000
6. Carothers C, Gribble K. Infant and young child feeding in emergencies. Journal of Human Lactation. **30**(3), 272–5. (2014). http://dx.doi.org/10.1177/0890334414537118
7. Astuti, Wanda D. Praktik pemberian makanan pendamping air susu ibu pada situasi bencana: studi literatur Complementary Feeding Practices In The Disaster Situation: A Literature Review. **5**, Jurnal Perawat Indonesia. http://dx.doi.org/10.32584/jpi.v5i1.827
8. DeYoung S, SM, & SHG. Maternal perceptions of infant feeding and health in the context of the 2015 Nepal Earthquake. Journal of Human Lactation. **34**(2), 242–52. (2018). http://dx.doi.org/10.1177/0890334417750144
9. Jo B. Tirto.id. 2024 [cited 2024 Jul 28]. Rangkuman bencana sumbar hari ini dan update dampaknya. Available from: https://tirto.id/rangkuman-bencana-sumbar-hari-ini-dan-update-dampaknya-gYza
10. Darwis RS, Nurseto HE, Andoyo R, Sukri N, Hadi BN, Huda S, et al. Food fulfillment for children under five by family in disaster reserve area. Asian Journal for Poverty Studies. **4**(1), 7–11. (2018). https://doi.org/10.33369/ajps.v4i1.4896
11. Sulistiawati F, Taufiqqurahman. Kajian penanggulangan gizi balita pasca bencana gempa bumi di Desa Dasan Geria Kabupaten Lombok Barat. Jurnal Ilmiah Mandala Education [Internet]. **6**(2), 320–8. (2020). http://ejournal.mandalanursa.org/index.php/JIME/index
12. Godde CM, Mason-D'Croz D, Mayberry DE, Thornton PK, Herrero M. Impacts of climate change on the livestock food supply chain; a review of the evidence. **28**, Global Food Security. Elsevier B.V.; (2021). http://dx.doi.org/10.1016/j.gfs.2020.100488
13. Adeoya AA, Egawa S, Adeoya AS, Nagatomi R. Improving child nutrition in disasters by developing a modeled disaster preparedness nutrition education curriculum. Front Public Health. **11**. (2023). http://dx.doi.org/10.3389/fpubh.2023.1293875
14. Bilgin DD, Karabayır N. Infant and young child feeding in emergencies: a narrative review. **59**, Turkish Archives of Pediatrics. AVES. p. 135–43. (2024). http://dx.doi.org/10.5152/TurkArchPediatr.2024.23184
15. Adeoya AA, Egawa S, Adeoya AS, Nagatomi R. Improving child nutrition in disasters by developing a modeled disaster preparedness nutrition education curriculum. Front Public Health. **11**. (2023).
16. UNCHR. WASH in Emergencies. In: Emergency handbook. (2024).
17. Uprety S, Iwelunmor J, Sadik N, Dangol B, Nguyen TH. A qualitative case study of water, sanitation, and hygiene resources after the 2015 gorkha, nepal, earthquake. Earthquake Spectra. **33**(Special issue 1) S133–46. (2017). http://dx.doi.org/10.1193/112916EQS212M
18. Krishnan S. Water, sanitation and hygiene (WASH) and disaster recovery for community resilience: A mixed methods study from Odisha, India. International Journal of Disaster Risk Reduction. **35**. (2019). http://dx.doi.org/10.1016/j.ijdrr.2018.12.023
19. 19. Datar A, Liu J, Linnemayr S, Stecher C. The impact of natural disasters on child health and investments in rural India. Soc Sci Med. **76**(1), 83–91. (2013). http://dx.doi.org/10.1016/j.socscimed.2012.10.008
20. Rassekh BM, Santosham M. Utilization of formal health services for children aged 1–5 in Aceh after the 2004 tsunami: Which children did not receive the health care they needed? Implications for other natural disaster relief efforts. Health Psychol Behav Med. **2**(1), 111–31. (2014).

E3S Web of Conferences 604, 11001 (2025) https://doi.org/10.1051/e3sconf/202560411001
The 4th ICDM 2024

# Sirombu in West Nias during the 2004 and 2005 earthquakes and tsunamis: a preliminary study

*Anatona*[1], *Abdul* Hakam[2*], *Ahmad* Muhajir[1], *Hanifa* Roja Azizah[2], and *Lusia* Yus Arimawati Nehe[1]

[1]Department of History, Faculty of Humanities, Universitas Andalas, Padang, Indonesia
[2]Department of Civil Engineering, Faculty of Engineering, Universitas Andalas, Padang, Indonesia

**Abstract.** Sirombu, a coastal settlement and trading hub on the western coast of Nias Island in Indonesia, was severely impacted by the devastating earthquakes and tsunamis that struck Aceh and the western offshore islands of Sumatra on December 26, 2004, and again on March 27, 2005. This paper examines the immediate aftermath of these catastrophic events in Sirombu. Employing a historical methodology and drawing upon diverse data sources, the study reveals the rapid transformation of life in Sirombu from normalcy to chaos and fear as the disasters unfolded. The earthquakes and tsunamis resulted in significant loss of life and injuries, widespread destruction of homes, markets, the seaport, and other infrastructure, and the displacement of the local population.

## 1 Introduction

The devastating Indian Ocean earthquake and tsunami of December 26, 2004, with a magnitude of 9.3, left deep scars and tragedy, particularly for the regions directly impacted, such as Aceh and the islands off the western coast of Sumatra. Sirombu, on Nias Island, was among the areas most severely affected, due to its location directly facing the waves of the Indian Ocean. The situation worsened when another earthquake struck Nias on March 28, 2005, with a magnitude of 8.7, also triggering a tsunami. This second disaster further exacerbated the conditions in Nias, especially in Sirombu.

In Nias island, thousands were injured or killed, and many coastal residents lost their livelihoods as homes, businesses, and public facilities were devastated by the earthquakes and tsunamis. In the aftermath, residents sought refuge in safer areas [1].

According to collected data, the Sirombu subdistrict was the hardest hit by the earthquakes and tsunamis, suffering the highest number of casualties [2]. In both disasters, Sirombu, facing west, bore the brunt of the tremors and the surging tsunami waves from the Indian Ocean. This impact was felt not only by those residing in Sirombu on the main island of Nias but also by inhabitants of the smaller Hinako Islands, which fall under Sirombu's administrative jurisdiction [3].

This paper presents a preliminary study of the 2004 and 2005 earthquakes and tsunamis in Nias, focusing on Sirombu, and their immediate aftermath. Employing a historical perspective, this study aims to illuminate the profound impact of these natural disasters on the people of Sirombu. The post-earthquake atmosphere was harrowing, a stark contrast to the pre-disaster tranquility.

## 2 Methodology

This manuscript examines the events of the devastating earthquake and tsunami that struck several regions in the Indian Ocean, including Sirombu in the Nias Islands, two decades ago in 2004 and 2005. Given the historical nature of these events, this study employs a historical research methodology.

The research employs a historical research methodology. Data sources consist of both primary and secondary written documents [4]. All data collected in the heuristic phase underwent rigorous critical analysis, including external and internal critiques. Following the critical analysis, the data were interpreted using both quantitative and qualitative approaches to establish historical facts. The final stage involved the synthesis and presentation of these findings through historiography.

## 3 Results and discussion

### 3.1 Region and geographic location

Sirombu is a subdistrict and the only seaport commercial town located on the west coast of Nias Island. No other commercial towns are found along the west coast of Nias Island. Other commercial towns do still exist, but they are located not on the west but on the north, east and south coasts of Nias Island. On the north coast of Nias there is the Lahewa seaport commercial town and in the south, there is Teluk Dalam. While in the east there is Gunugsitoli. In addition to the commercial town, since the Dutch colonial period in the 19th century, Gunungsitoli

* Corresponding author: ahakam@eng.unand.ac.id

has also served as the capital of the Nias Regency. These conditions make Sirombu a commercial town, seaport as well as a small foremost and outermost outpost located in the western part of the Indonesian archipelago.

The Sirombu region is a landmass projecting into the sea (a cape), situated on the western coast of Nias Island and facing directly towards the Indian Ocean. Initially, Sirombu was referred to as "Cape of Good Hope" by traders who frequented the area [5]. This term bears resemblance to the name given to the Cape of Good Hope in South Africa by Portuguese explorers in the 15th century. It is possible that the meaning is similar. Sirombu, located at the western tip of Nias Island, offered new hope as a place to settle and engage in various livelihoods such as trade, agriculture, and livestock, particularly for those who came from outside Nias Island.

At the time of the two earthquakes and tsunamis in 2004 and 2005, Sirombu was administratively classified as a subdistrict under the jurisdiction of Nias Regency. At that time, Nias Island had only two regencies: Nias Regency with its capital in Gunungsitoli and South Nias Regency with its capital in Teluk Dalam. Sirombu is situated on the western coast of Nias Island, while Gunungsitoli is located on the eastern coast. The two are separated by a considerable distance of 76 km.

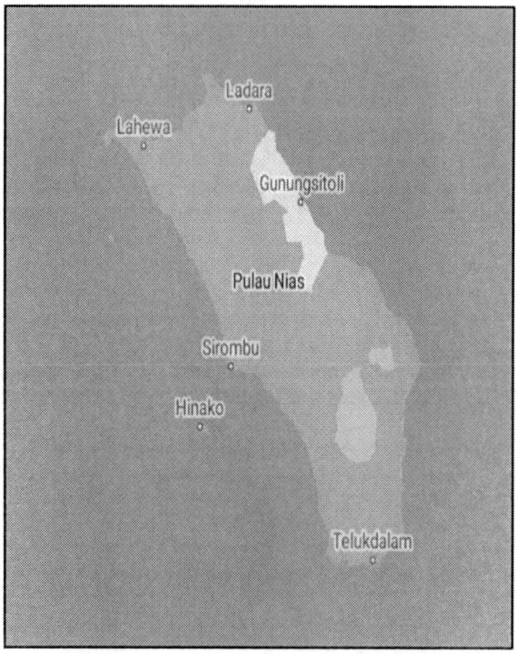


**Fig. 1.** Location of Sirombu in Nias Island (source: maps.google.com)

Sirombu subdistrict encompasses two distinct areas: the mainland area connected to Nias Island and the Hinako Islands, located to the west of Sirombu. Several small islands are included in the Hinako Islands, namely Hinako Island, Imana Island, Bawa Island, Asu Island, Bogi Island, etc. The Hinako Islands are the outermost group of islands in the westernmost part of Indonesia.

Broadly speaking, the topography of Sirombu, including the small islands of the Hinako archipelago, can be divided into two categories: lowlands and highlands, with a relatively balanced distribution. The elevation ranges from 0 to 800 meters above sea level. The coastal lowlands, interspersed with undulating terrain, comprise nearly 50% of the total area. The remaining portion consists of hilly regions [6]. Figure 1 is the location of Sirombu on the Nias Island.

### 3.2 Population and livelihoods

Similar to other coastal towns and trading ports, Sirombu boasts a heterogeneous population in terms of ethnitown and religion. Its urban character stems from its historical role as the sole seaport on the western coast of Nias Island, attracting numerous traders, primarily from other parts of Nias and the island of Sumatra.

The majority of Sirombu's inhabitants, both on the main island and the Hinako Islands, are indigenous Nias people. They represent the coastal and island-dwelling communities, distinct from the Nias people who reside in the island's interior highlands.

In addition to the native Nias population, Sirombu is home to several ethnic groups originating from outside Nias, including Acehnese, Minangkabau, Sulawesi, and Chinese communities. Many of these groups have resided in Nias for centuries, with significant assimilation occurring through intermarriage with the indigenous population [7].

The Acehnese, arriving by sea from northern Sumatra, have exerted a considerable influence on Nias, particularly in the northern, eastern and western regions. This was observed by Italian anthropologist Elio Modigliani during his 1886 visit to Nias. Modigliani noted that in some areas, the physical characteristics of the inhabitants deviated from the typical Nias appearance, which William Marsden, in his 1811 work "History of Sumatra," described as having lighter skin tones compared to the Malays [8]. Modigliani attributed these variations to Acehnese influence [9].

Unlike the Acehnese who migrated from the north, the Minangkabau people arrived in Sirombu from the southeast, reflecting the location of their primary homeland in central and western Sumatra. Some Minangkabau inhabitants of Sirombu originated from Pariaman on the west coast of Sumatra, bearing the title *Mara* [10].

Another ethnic group present in Sirombu hails from the island of Sulawesi. Concentrated in the Hinako Islands near Sirombu, they belong to the Maros tribe of Sulawesi and are locally referred to as the *Maruwi*. To preserve their ancestral ties, these Nias residents of Maros origin adopted *Maru* as their surname. Nias individuals with surnames like *Maruhawa*, *Maruao* and *Marunduri*, etc are considered descendants of the Maros people. Meanwhile, the Chinese presence in Sirombu, as in other parts of the Indonesian archipelago, is primarily driven by economic factors, with many serving as entrepreneurs and merchants.

Sirombu is also home to a significant fishing community. The fish caught in Sirombu not only meet the nutritional needs of the local population but also those of neighboring villages located in the highlands or mountains, such as Mandrehe and Moi in West Nias and Lölöwa'u in South Nias.

Sirombu has long been recognized as a trading centre in the Nias archipelago and the western coast of Sumatra. As a commercial hub, it attracts people from various regions for trade. Trade in Sirombu occurs both locally, such as with the Hinako Islands, and internationally. Chinese merchants have historically played a prominent role in Sirombu's trade, purchasing items like swiftlet nests pigs [10], and even engaging in the slave trade. By the late 19th century, the Hinako Islands had developed a thriving coconut cultivation and trade industry.

For centuries, the port of Sirombu has facilitated the transportation of goods and people from various regions, particularly those located in the Indian Ocean region and the western islands of Sumatra. These include the Hinako Islands, Batu Islands, and even areas on the western coast of Sumatra, such as Sibolga, Air Bangis, Pariaman, and Padang. Before the 2004 earthquake, the main port on Nias Island, Gunungsitoli, located on the east coast, had an average of two sailings per week. Sirombu also served as a port of call for ships from *Emmahaven* (Teluk Bayur), Teluk Dalam, Hinako, Lahewa, and even Sibolga.

The population of Sirombu subdistrict, including those residing in the Hinako Islands, engages in various professions, including fishing and farming. Agricultural activities include the cultivation of coconuts, cloves, and other crops. In the 19th century, trade with Sirombu involved goods such as tobacco, colorful fabrics, trinkets, and brass wire. The brass wire was used to decorate spears, and small pigs served as a form of currency [10].

Sirombu also has a livestock-based economy. The indigenous Nias people in Sirombu primarily raise pigs, which are used for both consumption and customary rituals. During the Dutch colonial period, pigs from Nias Island were even exported to Sumatra and Singapore. Meanwhile, according to a report by Henry Lyman, who arrived in Nias in late May 1834, the Malay community on the island raised buffalo and goats, and there were also some horses [11]. Agricultural products from Nias included rice and coffee, which were shipped to Padang.

### 3.3 The earthquake, tsunami, and their impact

The tsunami that struck Sirombu, West Nias, on December 26, 2004, consisted of multiple waves separated by intervals. The first wave arrived around 10:00 a.m., shortly after the earthquake, causing a minor rise in sea level of 10 to 25 cm. However, the third wave was far more destructive, reaching an estimated height arround 5 meters. This surge of seawater inundated the land for approximately four hours before receding.

The earthquake and tsunami that followed three months later, on March 28, 2005, also impacted the coastal areas of Sirombu, including the seaport. The initial earthquake and tsunami on December 26, 2004, caught the population completely off guard.

Throughout history, for at least the past two centuries, Nias Island has frequently been struck by earthquakes and tsunamis. Nias regularly experiences large-scale earthquakes that trigger tsunamis reaching the land. Records indicate devastating earthquakes and tsunamis occurred in 1843, 1861, and 1901. However, due to the significant time lapse, these events had faded from the collective memory of the Nias community by the early 2000s. Consequently, the current generation had no living memory of such a disaster.

When the earthquake struck Sirombu, followed by the tsunami on December 26, 2004, the residents were gripped by confusion, anxiety, and fear. The term "tsunami" was largely unknown to them. In their local understanding, seawater inundating the land was referred to as *molö*. Unlike tsunamis, *molö* typically involved minor flooding, reaching only a few centimeters or up to one or two meters. *Molö* was associated with heavy rainfall rather than earthquakes, occurring when rivers overflowed and flooded low-lying areas along riverbanks and estuaries (*luaha*).

People fled in panic to save themselves. Shops and houses in Sirombu's market were destroyed. Homes in both the mainland and Hinako Islands areas of Sirombu were also devastated. Four villages within the Sirombu subdistrict were severely affected: Pasar Sirombu, Pulau Himana, Pulau Bawa, and Pulau Hinako. Figure 2 below shows the state of Sirombu after the earthquakes and tsunamis disaster

**Fig. 2.** Sirombu after earthquake and tsunami (source: presbyterian disaster assistance) [12]

In addition to houses, shops and markets, the earthquakes and tsunamis in Nias also damaged a number of houses of worship. There are dozens of houses of worship in Sirombu with 3 variants according to the religious communities that exist there, namely churches for residents who are Christians, both Protestant and Catholic, mosques and mushalla for residents who adhere to Islam and Vihara for residents who adhere to Buddhism. This diversity is in accordance with Sirombu's condition as an urban town. At the time of the earthquake and tsunami in 2024 Sirombu subdistrict had 52 houses of worship, consisting of 39 churches, 7 mosques, 5 mushalla and 1 monastery [13]. However, during the earthquake and tsunami, some of these houses of worship, such as the Orahua Niha Keriso Protestan (ONKP) Church and Baiturrahman Mosque, were damaged due to the strong shaking of the earthquake and and were washed away by the tsunami [14, 15].

https://doi.org/10.1051/e3sconf/202560411001

**Fig. 3.** Houses destroyed at Sirombu (source: getty images, 51903152, december 29, 2004)

The once vibrant coastal town of Sirombu, situated along the picturesque western coast of Nias, was left desolate. Hundreds of houses and shops that had formed the economic heart of Nias were abandoned by their residents in the aftermath of the March earthquake. Figure 3 above shows destroyed houses at Sirombu.

The 2004 and 2005 earthquakes and tsunamis had a profound impact on the local economy. Sirombu, once a bustling hub along Nias's picturesque western coastline, was left deserted in the aftermath of the disaster. Hundreds of kiosks and shops, which formed the backbone of the region's economy, were abandoned by their owners. The following map is the epicentres of the devastating earthquakes on the off west coast of Northern Sumatra on December 26, 2004 and March 28, 2005 (Fig. 4).

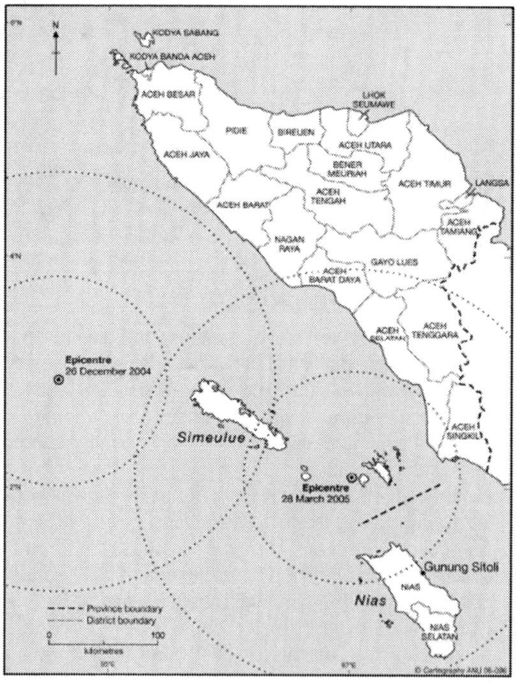


**Fig. 4.** Epicentres of the 2004 and 2005 earthquakes – Northern Sumatra (source: h. boschitz, participation in post-disaster situations: the case of reconstruction in Nias, Indonesia, *master thesis*, universität wien, 2013) [16]

### 3.4 Displacement

The devastating earthquakes and tsunamis that struck twice in succession within a mere three months triggered significant geographical mobility among the residents of Sirombu. This displacement was driven by two primary factors. First, the psychological trauma instilled a deep-seated fear and anxiety among the community, prompting them to seek refuge from potential future disasters. Second, the consecutive natural disasters resulted in the loss of homes and livelihoods for a substantial portion of the population residing near the coast. This was particularly evident among the merchants in Sirombu's market, as the buildings collapsed due to the earthquakes and were partially swept away by the tsunamis. Furthermore, fishermen who relied on the Sirombu waters for their livelihood were unable to resume their activities due to the lingering fear of venturing back into the sea. Similarly, farmers, especially coconut cultivators inhabiting the coastal regions of both mainland Sirombu and the Hinako Islands, faced significant disruptions to their livelihoods [17, 18].

Three distinct patterns of displacement emerged in the aftermath of the earthquakes and tsunamis. The first pattern involved residents living along the coast relocating to higher ground. The second pattern saw inhabitants of the Hinako Islands migrating to mainland Nias. Additionally, a third pattern emerged, with residents from Sirombu subdistrict, both on mainland Nias and the Hinako Islands, migrating out of Nias Island altogether.

In the first pattern, the people who fled were those who originally lived on near by Sirombu beach. The number is quite large because this is the centre of economic activity for the Sirombu community that has been going on for centuries. Not only the local community, Sirombu is also the economic centre of the inland community of Nias Island, especially those who live in the central and western parts of Nias Island such as from the Mandrehe, Lölöwa'u, and surrounding areas.

There are various public facilities such as markets, schools, houses of worship, seaport docks, etc in Sirombu. Of the three patterns of displacement caused by the disasters of 2004 and 2005 as mentioned above, it was this group that experienced and was directly affected by the devastating earthquakes and tsunamis. This was partly because their homes and businesses were located very close to the coastline and were at risk from the Indian Ocean tsunami waves.

Tanjung Sirombu in West Nias is a coastal residential area that at the time was only a dozen meters away from the shoreline, and was among the worst affected areas by the tsunami. The number of victims of missing and dead villagers was recorded at 8 people and hundreds of others were permanently evacuated.

When the initial earthquake struck on December $26^{th}$, many residents of Pasar Sirombu village sought refuge in Tetesua, an elevated area located 3 km east of Sirombu. Tetesua subsequently housed hundreds of new residents living in tents or temporary shelters along the roadside. Others established residences further away from Sirombu. It was only a week later that they returned to Sirombu to assess and salvage belongings from the wreckage of their houses and shops. However, they soon returned to the shelters (Fig. 5).

**Fig. 5.** A man, girl and dog cross deep water at Sirombu village (source: getty images, 51903159, december 29, 2004)

Meanwhile, the second pattern is a category of population that originally settled in villages located on several small islands. There are 10 islands located off the west coast of Sirombu, namely Hinako Island, Asu Island, Heruanga Island, Imana Island, Bögi Island, Langu Island, Hamutala Island, Lawanda Island, Bawa Island, and Si'ite Island. The size of each island varies. The smallest island is Lawanda with an area of 25 hectares while the largest island is Hinako Island with an area of 94 hectares. Not all of these islands are inhabited. Until the year leading up to the devastating earthquake on December 26, 2004 with a magnitude of 9.3, only five islands were inhabited. The other 5 islands were not inhabited at all. The 5 inhabited islands combined, there are 11 villages, 6 villages each on Hinako Island, 2 villages on Imana Island, 1 village on Bögi Island, and 3 villages on Bawa Island [19].

The evacuees from these small islands were not only limited to the villagers of these islands who lived by the beach, but also included families whose homes were some distance from the beach. In addition to these evacuated families, the elderly population was also included.

Many Sirombu residents who fled their homes sought refuge on the island of Sumatra. In this emergency, their primary destination was the homes of relatives and acquaintances who had previously relocated to Sumatra. This provided not only temporary accommodation but also a sense of security and comfort. Several provinces in Sumatra, including Aceh, North Sumatra, and West Sumatra, have hosted significant Nias communities since the Dutch colonial era. Furthermore, the expansion of oil palm plantations in provinces like Riau, Jambi, and Bengkulu has attracted Nias migrants seeking employment opportunities.

In the early 2000s, Sirombu's population averaged around 16,000. This figure increased to 17,394 in December 2004, just before the first earthquake and tsunami. However, by 2005, the population had decreased to 16,620. By 2006, the population had sharply declined to approximately 8,790 individuals. This figure remained relatively stable until 2010, as illustrated in Table 1:

**Table 1.** Population of Sirombu Subdistrict (2003-2010) (Source: BPS Kabupaten Nias, 2010 [6])

| Year | Population |
|---|---|
| 2003 | 16.989 |
| 2004 | 17.394 |
| 2005 | 16.620 |
| 2006 | 8.790 |
| 2007 | 8.802 |
| 2008 | 8.820 |
| 2009 | 8.840 |
| 2010 | 9.478 |

The table above reveals a significant and dynamic shift in the population composition of Sirombu in the years leading up to, during, and after the earthquake and tsunami disasters. In 2003, a year prior to the earthquake, the population stood at 16,989 individuals. This figure saw a slight increase in 2004, reaching 17,394 individuals, representing a 0.98% growth. As the earthquake and tsunami struck in late December 2004, their impact was not yet reflected in the statistical population count for that year. However, by 2005, a decline became evident, with the population decreasing to 16,620 individuals, a 0.96% reduction.

**Fig. 6.** Residents of Sirombu are evacuating (source: getty images/bay ismoyo, 52512369, march 31, 2005)

The second natural disaster, the earthquake and tsunami of March 2005, had a profound impact on Sirombu's population. By 2006, one year after the two disasters, the population had dwindled to 8,750 individuals. This represented a staggering decrease of 7,870 individuals, or 47%, compared to 2005. In essence, Sirombu lost nearly half its population in the wake of the two natural disasters of 2004 and 2005. This population composition persisted for the following four years. The population figures for 2007, 2008, and 2009 remained in the 8,000s. It wasn't until 2010 that a slight increase was observed, with the population reaching the 9,000s. Figure 6 shows residents of Sirombu evacuating after the March 28, 2005 earthquake.

## 4 Conclusion

For centuries, Sirombu was renowned as a peaceful urban centre, a hub of commerce and tourism, harmoniously inhabited by diverse ethnicities and religions. However, the devastating natural disasters abruptly transformed this idyllic town. This article explores how the consecutive catastrophes reshaped Sirombu from a welcoming haven into a place filled with horror. The profound impact on the local community led to significant social changes in a relatively short period, encompassing shifts in settlement patterns and migration. In the aftermath of the earthquake and tsunami, a pervasive sense of fear gripped the community. Approximately half of Sirombu's population migrated out of the region, primarily seeking refuge on the island of Sumatra.

This article is part of the research project under the "Penelitian Unggulan Jalur Kepakaran (PUJK)" scheme funded by LPPM Universitas Andalas in 2024, with contract number 328/UN16.19/PT.01.03/PUJK/ 2024, dated July 12, 2024.

## References

1. Anatona, A. Hakam, M. Nur, Syafrizal, P. Husodo, Novalinda, and Zulqaiyyim, Earthquake Resiliance of Traditional Nias Island Houses: Lesson from the 2004 and 2005 Earthquakes. International Journal of Sustainable Development and Planning (IJSDP), **19**, 2, 599-607 (2024).
   https://doi.org/10.18280/ijsdp.190217
2. "Tsunami landa Nias. 42 Tewas, 75 Hilang". Waspada (2004, December 27).
3. T. P. Usapdin, Indonesia: New beginnings in Nias. (International Federation of Red Cross and Red Crescent Societies, 2005).
4. W. L. Neuman, Social research methods: Qualitative and quantitative approaches (7th ed.), (Pearson Education Limited, 2014).
5. Suharmiati, D. Gunawan, and D. Hidayati. Tradisi Böwö dan Malnutrisi pada Etnik Nias, (Kanisius, 2016).
6. BPS Kabupaten Nias, Kabupaten Nias dalam angka tahun 2010, (Government statistical report, 2010).
7. P.J.M. Hämmerle, Asal Usul Masyarakat Nias: Suatu Interpretasi, (Gunungsitoli, Yayasan Pusak Nias, 2015).
8. W. Marsden, Sejarah Sumatra, (Bandung, Remaja Rosdakarya, 1999).
9. E. Modigliani, Un Viagio A Nias. (Fratelli Treves, 1890).
10. V. Puccioni, Tanah Para Pendekar: Petualangan Elio Modigliani di Nias Selatan Tahun 1886. (Kompas Gramedia, 2016).
11. H. Lyman, A missionary tour of Nias. In A. Reid, Witness to Sumatra: A travellers' anthology, 156-167, (Oxford University Press, 1995).
12. Presbyterian Disaster Assistance, Indonesia: Situation report update Aug 2005 – Nias earthquake, (Relief Web, 2005, August). https://reliefweb.int/report/indonesia/indonesia-situation-report-update-aug-2005-nias-earthquake (accessed 24.07.2024)
13. BPS Kabupaten Nias, Kabupaten Nias dalam angka tahun 2005, (Government statistical report, 2005).
14. Gereja ONKP Membutuhkan Mitra. https://onkpnias.wordpress.com/2008/05/16/gereja-onkp-membutukan-mitra/ (accessed 14.10.2024)
15. Azmi, Masjid Tua Sirombu 'Saksi Bisu'Gempa dan Tsunami Nias 2005. https://tanmia.or.id/masjid-tua-sirombu-saksi-bisu-gempa-dan-tsunami-nias-2005/ (accessed 14.10.2024)
16. H. Boschitz, Participation in Post-Disaster Situations: The Case of Reconstruction in Nias, Indonesia, Magister Thesis, Universität Wien, 2013.
17. Nias Barat, Sirombu, Pasar, Kota dan Pelabuhan Kenangan, 2007.
    https://pilarniasbarat.wordpress.com/2007/10/13/sirombupasarkota-pelabuhan-kenangan/attachment/107/. (accessed 25.07.2024)
18. Widayatun, A. Situmorang, I. G. P, Antariksa, and A. Y. Abdurrahim, Kondisi sosial ekonomi masyarakat di lokasi COREMAP II Pulau Hinako Kabupaten Nias: Hasil BME. Coral Reef Rehabilitation and Management Program Lembaga Ilmu Pengetahuan Indonesia, (COREMAP II – LIPI, 2009).
19. BPS Kabupaten Nias, Kabupaten Nias dalam angka tahun 2004, (Government statistical report, 2004)

# Lahar floods in West Sumatra during the Dutch colonial era: historical records and societal impact

*Yenny* Narny[1*], *Febriani* Rahayu Putri[1], *Robert* Cribb[2], *Denny*[3], *Dewi* Pratika Sari[3], *Elvia* Siskha Sari[3], *Febrin* Anas Ismail[1], *Khairil* Anwar[1], *Abdul* Aziz Nasta[4], and *Yozi* Yuliardi[4]

[1]Department of History, Faculty of Humanities, Universitas Andalas, Padang, Indonesia
[2]Australian National University, Canberra, Australia
[3]The Government of Padang City, Padang, Indonesia
[4]Master in Disaster Management, Graduate Program, Universitas Andalas, Padang, Indonesia

**Abstract.** This research explores documents and archives regarding lahar floods in West Sumatra during the Dutch colonial period. These records are crucial amid difficulties in obtaining historical statistical data to study similar past disasters. The research method used is historical, integrating information from Dutch colonial documents and archives such as newspapers, magazines, diaries (dagh register), government reports, and contemporary books, particularly those discussing societal conditions and Dutch government policies in handling various lahar flood events in West Sumatra during that period. This study is expected to provide new insights for the development of disaster studies and facilitate historical disaster studies in other Indonesian regions that are also prone to natural disasters.

## 1 Introduction

West Sumatra is one of the regions with high volcanic activity in Indonesia. This area has many active volcanoes that frequently erupt, causing significant impacts on the environment and the socio-economic life of the surrounding communities. The high frequency of volcanic activity is closely related to the geological conditions of West Sumatra, located in the subduction zone of the Indo-Australian Oceanic Plate with the Eurasian Continental Plate [1]. This subduction process forms a series of active mountains along the zone, including Mount Marapi, Mount Tandikat, Mount Talang, and Mount Kerinci (Fig. 1).

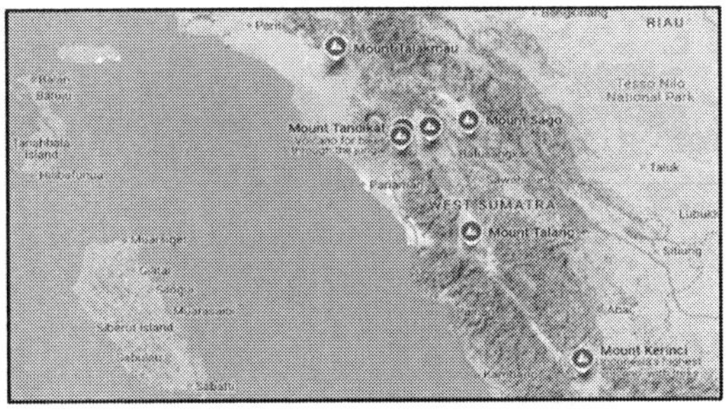


**Fig. 1.** Map/location of mounts in West Sumatra

Volcanic eruptions pose significant risks not only through immediate hazards like lava flows and volcanic ash but also by instigating secondary disasters that can have devastating effects on local communities. Among these secondary hazards are landslides and cold lahar floods. A cold lahar flood occurs when volcanic material mixes with rainwater [2], resulting in a rapid flow of debris that can severely damage settlements, agricultural fields, and critical infrastructure. This cascading effect adds considerable strain to the communities affected by such eruptions, exacerbating the overall disaster response and recovery efforts. In their insightful article, Franck Lavignie and Jean-Claude elucidate that a lahar is essentially a water-saturated volcanic mass flow, characterized by a swift-moving, high-concentration blend of rock debris and water that contains poorly sorted sediment. This composition makes lahars distinct from typical river flows, as they originate from volcanic activity and are especially prevalent in composite volcanoes. To this day, lahars remain one of the most destructive natural phenomena linked to these types of volcanoes [3].

In a recent event highlighting the risks associated with volcanic activity, West Sumatra experienced a significant cold lahar flood on May 11, 2024. This cold lahar flood swept through the highlands of the area, causing extensive damage and disruption. According to the Meteorology, Climatology, and Geophysics Agency (BMKG), this cold lahar flood was triggered by moderate to heavy rainfall that mixed with volcanic material from the nearby Mount Marapi and Mount Singgalang. The intense rainfall caused erosion of eruption deposits located on the peaks and slopes of these mountains, allowing the volcanic material to be transported downstream through river systems to the surrounding areas. The influx of materials,

* Corresponding author: yennynarny@hum.unand.ac.id

particularly sand and mud, resulted in significant river silting, which hindered the rivers' ability to function properly. As a result, overflow incidents occurred where water and mud inundated community settlements, leading to casualties and creating further challenges for the residents affected by this natural disaster [4].

It was reported that the flood significantly impacted four regencies/cities in West Sumatra, namely Tanah Datar Regency, Lima Puluh Kota Regency, Agam Regency, and Padang Panjang City. The impact was also felt in Pariaman City and Padang, with less intensity. More than 55 people became victims of this event. Hundreds of buildings, including community houses and public facilities, suffered severe damage. The flood also caused the national road access between Padang and Bukittinggi to be completely cut off due to a road collapse in the Lembah Anai area, Tanah Datar Regency [5]. This material damage is one of the main impacts that highlights the urgency of emergency response and post-disaster recovery efforts.

History records that this cold lahar flood disaster was not the first in West Sumatra. During the Dutch colonial period, spanning the 19th and 20th centuries, this area frequently experienced similar natural disasters, impacting extensive regions and affecting numerous communities. Various documents and archives, including newspapers, magazines, and government reports published at that time, provide detailed accounts of these events. These archival documents serve as valuable historical records that offer data to understand the patterns, triggers, and characteristics of cold lahar floods in West Sumatra in the past. By examining these records, researchers can gain insights into seasonal trends, geological factors, and rainfall patterns that contributed to lahar occurrences.

The data derived from these archives is not only historically significant but also practically valuable for modern disaster preparedness. These historical data points can be used as a foundation for creating detailed distribution maps of cold lahar flood-affected areas in West Sumatra. Such maps are essential for identifying vulnerable areas within the lahar-prone zone, assessing the impacts of past disasters, and planning for future risks. Additionally, these maps and analyses help estimate the vulnerability levels of local communities, agricultural areas, infrastructure, and transportation networks, providing a clear picture of how lahar floods could affect different aspects of life.

Additionally, this historical data forms the basis for further research by current researchers to gain a deeper understanding of the risks and potential lahar flood disasters and produce recommendations to strengthen disaster mitigation systems for areas affected by past lahar floods, aiming to prevent recurring impacts from similar disasters in the future.

## 2 Data source

This paper uses data sources from archives, historical documents, magazines, and contemporary newspaper reports. These sources provide invaluable insights into the occurrences of cold lahar floods, including specific details on when and how these events unfolded, their underlying causes, and the extent of the damage they inflicted, particularly during the 17th and 18th centuries. Archival data offer a historical perspective that allows researchers to trace patterns over time, identifying recurring environmental and geological factors, such as volcanic activity, that may have contributed to the frequency and intensity of these floods.

Newspaper reports and periodicals from that era not only document the immediate impacts of each flood but also offer insights into government policies and relief efforts. These publications often highlight community resilience and report on reconstruction activities. By aggregating and analyzing this diverse range of data, this paper aims to provide a comprehensive historical understanding of cold lahar floods, contributing to present-day knowledge of disaster patterns, shown in Table 1. This analysis may also serve as a comparative foundation for assessing modern lahar events, allowing scientists and policymakers to draw parallels and refine current mitigation and preparedness strategies based on historical lessons.

**Table 1.** Earthquake activity sources based on year of occurrence in colonial newspapers and research

| Year of Cold Lahar Flood | Sources | Type of Documents |
|---|---|---|
| 1881 | Het Nieuws van den dag: kleine courant edisi 16 February 1893 | Newspaper |
| | Sumatra Courant, 27 December 1881 | Newspaper |
| | Opregte Haarlemsche Courant 15 March 1882 | Newspaper |
| | Controleur Oud Agam tertanggal 28 December 1881 | Report |
| 1889 | Bataviaasch Handelsblad, 27 Februari 1889 | Newspaper |
| | Sumatra Courant 23 February 1889 | Newspaper |
| | De Locomotief, 11 February 1890 | Newspaper |
| 1892 | Arnhemsche Courant edisi 31 January 1893 | Newspaper |
| | Het Nieuws van den dag: kleine courant, 1893 | Newspaper |
| | Arnhemsche Courant edisi 13 March 1914 | Newspaper |
| | Haagsche Courant edisi 8 February 1904 | Newspaper |
| | Het Nieuws van den dag: kleine courant, 16 February 1893 | Newspaper |
| 1904 | Algemeen Handelsblad, 1904 | Newspaper |
| | Het Nieuws van den dag door Nederlandsch Indie, 1904 | Newspaper |
| | De locomotief, 1904 | Newspaper |
| 1926 | Sumatra Bode, 4 August 1926 | Newspaper |

## 3 Methodology

In the research mapping cold lahar events in West Sumatra during the Dutch colonial period, a historical-geographical approach will serve as the main conceptual framework. This approach aims to integrate historical analysis with the geographical factors influencing the occurrence of cold lahar disasters. The study will begin with the collection of historical data from archival sources, disaster reports, and newspapers from the Dutch colonial period. In addition, geographical data will be analyzed to understand how environmental factors interact with cold lahar events. By mapping the locations and frequencies of these events, the research can identify patterns and trends that indicate the relationship between volcanic activity, climate change, and their impacts on local communities, as well as the policies implemented by the Dutch Government to address these disasters.

This research employs a historical method that unfolds through several systematic stages, each crucial for the integrity and depth of the study. The first stage is heuristics, which involves meticulous data collection from various sources. In this case, the data is primarily sourced from the Koninklijke Bibliotheek van Nederland (KB), Nationaal Archief (NA), and KITLV. During this stage, the researchers focus on gathering comprehensive data related to occurrences of cold lahar floods. This includes not only the year of occurrence but also important factors such as the affected geographical locations and the extent of the damage caused. By organizing the collected data into tabular formats based on these criteria, the researchers can effectively identify and analyze trends and patterns that emerge over time. This systematic approach lays a strong foundation for understanding how such events have developed and their broader implications within historical contexts.

After the data collection stage, the next phase is criticism, where the collected sources undergo rigorous examination. This critical analysis is essential to ensure the accuracy and validity of the data. Each source is scrutinized to confirm its reliability, and the information is cross-referenced with other relevant sources to build a comprehensive understanding. This process not only affirms the credibility of the data but also helps identify any biases or gaps in the sources, allowing researchers to make informed decisions about which materials can be considered suitable for inclusion in their narrative. By emphasizing the importance of source criticism, researchers uphold the integrity of the historical method, ensuring that the foundation of their study is built on solid evidence.

Once the sources have been critically analyzed, the study proceeds to the interpretation stage. Here, the evidence that has been meticulously gathered and evaluated is organized to produce a coherent series of historical facts. This involves synthesizing the information into a structured narrative that conveys the complexities of cold lahar floods and their impacts on affected communities. The data is reorganized and compiled into a historiographical format, allowing researchers to present a comprehensive view of historical events. By applying the historical method in this structured manner, the study provides valuable knowledge for understanding disaster patterns and their implications over time.

## 4 Results and discussion

West Sumatra is one of the regions prone to cold lahar floods. The presence of several active volcanoes and numerous rivers, along with high rainfall, makes the threat of cold lahar floods very significant. This area is traversed by dozens of rivers scattered across various regions, many of which originate from volcanic mountains, both active and inactive. According to data from the Central Statistics Agency (BPS) of West Sumatra, the rainfall in this area is quite high, reaching 1,562.6 mm in 2023, 5,332.3 mm in 2021, 4,493.00 mm in 2020, 2,640.50 mm in 2019, and 3,928.00 mm in 2018. These figures indicate significant fluctuations, with the year 2021 recording the highest rainfall in recent years [6].

The majority of the population in West Sumatra builds settlements near river flows. As a result, floods carrying mud, rocks, and other volcanic materials can obstruct river flows, causing rivers to overflow and destroy settlements, infrastructure, and agricultural land. The damage caused by these cold lahar floods is far more severe than that of regular floods, not only resulting in the loss of property but also threatening the safety of the lives of people living in vulnerable areas.

Efforts to mitigate the threat of cold lahar floods in West Sumatra need to be conducted continuously. Local and central governments must cooperate in mapping vulnerable areas, improving drainage infrastructure, and educating the community about appropriate preventive measures and evacuation steps. Additionally, there is a need for reforestation around river flows to reduce erosion and retain water flow. By doing so, it is hoped that the impacts of this disaster can be minimized and the safety of the community can be maintained.

According to the data obtained, the recording of cold lahar floods in West Sumatra began during the Dutch Colonial Government period in the 19th century. This documentation covers various aspects, including the frequency of events, the impact on residents and infrastructure, and the mitigation measures taken by the government at the time. The records kept by the colonial authorities were very detailed, reflecting their concern for environmental conditions and the need to protect colonial assets as well as the local population.

The first recorded lahar flood in West Sumatra occurred in 1881, originating from the slopes of Mount Marapi. This devastating event took place on the night of December 18-19 and caused widespread damage in the surrounding areas, resulting in significant loss of life and extensive material damage. The flood was triggered by exceptionally heavy rainfall, which caused landslides at the summit of Mount Marapi. These landslides occurred specifically in a valley connected to a waterfall stream known as Sarasah Bungsu. The sheer volume of rainwater rapidly transformed the debris from the landslides into a powerful torrent that flowed down the mountain, carrying not only water but also a dangerous mixture of thick sand, mud, rocks, and uprooted tree trunks.

As this destructive flow surged into the river, it quickly overwhelmed the local communities situated along its dangerous path, inundating not only homes but also vast expanses of agricultural land that had long been the backbone of the residents' livelihoods. The impact of this lahar flood was far more severe than that of a typical flood; it did not merely bring water but unleashed a torrential wave that combined the destructive force of fast-moving water with the heavy and abrasive materials it carried. This unique and devastating mixture resulted in widespread destruction, as the floodwaters not only engulfed structures but also eroded the very foundations of the land, leaving a trail of devastation that would take years to recover from.

In early reports released by several newspapers at the time, the situation appeared dire, with estimates suggesting that the death toll from the flood possibly exceeded 100 people [7]. However, a more official and detailed report written by Controleur Oud Agam on December 28, 1881, provided a clearer picture, stating that the number of fatalities from this catastrophic flood was confirmed to be 37 people. This report further explained that 33 individuals lost their lives in the Lassi area, while 4 were reported deceased in the Baso area. The tragedy deepened with the revelation that 3 people from Lassi remained missing, their bodies still unaccounted for, stirring anxiety and sorrow among the surviving families [8].

The areas most affected by this devastating flood included Lassi, Baso, and Laras Ampat Angkat (Ampek Angkek), which encompassed places such as Balai Gurah, Sitapung, Kota Tua, and Batu Taba. Among these regions, the Ampat Angkat area emerged as the site of the most severe damage, particularly to agricultural land, with rice fields suffering extensive destruction. Interestingly, despite the widespread havoc wreaked upon the land, no casualties were reported in this specific area.

Less than a decade later, in December 1889, another cold lahar flood struck West Sumatra, a devastating event that was directly linked to the significant eruption of Mount Tandikat that had occurred earlier on February 19-20, 1889. During this eruption, a dramatic display of fire and smoke erupted from the peak of the mountain, creating an awe-inspiring yet terrifying spectacle that was accompanied by the unsettling sounds of earthquake tremors and booming noises that echoed across the landscape, making it a vivid and unforgettable experience for those nearby [9]. This eruption was so powerful that it could be seen clearly from the surrounding areas, including the scenic region of Singkarak Lake, where residents were left in shock and fear. As a direct consequence of this natural disaster, the local communities living on the slopes of the mountain faced imminent danger, prompting many to evacuate to safer locations in a desperate bid to protect themselves and their families from the impending threat of the lahar flood that followed the volcanic activity [10].

The activity of Mount Tandikat, which had been notable since February of that year, culminated in a significant cold lahar flood that struck in December. Reports indicate that the volcanic activity of Mount Tandikat saw a marked increase on December 4, which set the stage for a series of events that would devastate the surrounding areas. On December 5, ominous rumbling sounds were heard emanating from the mountain, and shortly thereafter, a violent flow of mud cascaded down its slopes. This surge of mud quickly inundated most of the rice paddies in Nagari Singgalang, resulting in significant agricultural losses and sweeping away a bridge that was crucial for local transportation and connectivity [11].

A few years later, in 1892, another catastrophic cold lahar flood occurred, this time as a direct consequence of prolonged heavy rainfall. This deluge originated from the combined effects of Mount Marapi and Mount Singgalang, unleashing a powerful cold lahar that wreaked havoc across the landscape. The torrent was indiscriminate in its destruction, obliterating homes, rice fields, and gardens belonging to the residents of Nagari Pandai Sikek, which is located in the Padangsche Bovenlanden Residency, now recognized as part of Tanah Datar Regency. According to reports published in the Arnhemsche Courant on January 31, 1893, the deluge was so severe that it completely destroyed two suraus (prayer houses) and tragically resulted in the death of an elderly woman who had sought refuge inside one of them [12].

The impact of this flood extended beyond Nagari Pandai Sikek, severely affecting the Anai Valley region as well. The relentless heavy rains, combined with a flash flood (*galodo*) emanating from Mount Marapi, caused the Batang Anai River to swell beyond its banks, inundating the adjacent areas and displacing numerous residents. Eyewitness accounts reported in the Het Nieuws van den dag: kleine courant on February 16, 1893, indicated that the flood was exacerbated not only by the intense rainfall but also by the catastrophic breach of a lake situated on Mount Singgalang. This combination of natural disasters led to an exceptionally large and high flood, further compounding the devastation experienced by local communities [13].

The Haagsche Courant newspaper edition February 8, 1904, and Arnhemsche Courant edition March 13, 1914, also reported this event, mentioning that the flood destroyed 1500 meters of railway tracks and demolished 16 bridges, houses, and main roads. The cost of repairing the railway tracks and bridges reached about half a million guilders [14]. Besides physical damage, the flood also caused a spike in the prices of food supplies such as oil, vegetables, and meat due to disrupted transportation that made it difficult to transport goods. For instance, the price of a carton of kerosene in the highlands was only 4 guilders, but in Padang Panjang, it rose to 14 guilders, and in Fort de Kock, it became 15 guilders [15]. This situation shows that the impact of the flood was felt by the people of West Sumatra widely.

Before the railway rehabilitation was fully completed, the region of West Sumatra experienced a devastating flash flood in 1904, which occurred with great intensity. This catastrophic event caused extensive damage not only to the railway infrastructure but also to the bridges and main roads that were crucial for transportation and trade. According to the Algemeen Handelsblad newspaper, published on February 16, 1904, the railway bridge located in Tambun Tulang, which spanned the Anai Valley and connected Kandang Ampat with Padang Panjang, was completely swept away and destroyed by the force of the floodwaters. The aftermath of this disaster was severe, as railway traffic in the area was halted for a considerable period, lasting approximately 2 to 3 months [16].

Moreover, the impact of the flood was highlighted in reports from the Het Nieuws van den Dag door Nederlandsch Indie newspaper on January 9, 1904, and De Locomotief on January 11, 1904. These publications explained how the damage to the railway tracks hindered coal shipments, significantly affecting operations for BOW and the Navy, both of which relied heavily on consistent coal deliveries for their activities. The suspension of these vital shipments understandably raised concerns among various stakeholders, particularly among coal companies and operators of coal-fueled ships, who faced substantial disruptions in their supply chains and operational capabilities [17].

In a later instance, another cold lahar flood was reported in 1926, caused by heavy rainfall and volcanic activity originating from the slopes of Mount Marapi. This flood surged into Batang Selo in Fort de Kock, now known as Bukittinggi, and continued its destructive path into Batang Agam, ultimately reaching the area of Payakumbuh. Although this mudflow did not cause significant damage to community settlements, its impact severely affected the aquatic life in both rivers, leading to the death of many fish due to the high sulfur content in the floodwaters [18]. This event underscored the ongoing challenges posed by natural disasters in the region, particularly concerning environmental impacts and local livelihoods.

To address the devastating impacts of natural disasters, the Dutch Colonial Government implemented a series of strategic initiatives aimed at reducing the effects of cold lahar floods in West Sumatra. One of the primary measures taken was a swift and efficient response to repair the infrastructure damaged by these disasters. The main focus of these efforts included transportation and communication routes, which were essential for ensuring connectivity and accessibility during times of crisis. This involved extensive efforts to reconstruct destroyed railway tracks, bridges, and main roads while prioritizing the functionality of communication routes. Maintaining these routes was crucial for effective coordination among emergency response teams and for the timely transmission of vital information to both authorities and affected communities. Additionally, the government recognized the importance of community support by distributing aid to those impacted by the floods, ensuring that residents received the necessary assistance during recovery.

In addition to immediate relief efforts, the Dutch Colonial Government also invested time and resources into comprehensive research to deepen the understanding of the patterns and causes behind these recurring natural disasters. Such research was critical for informing future disaster preparedness and response strategies. These proactive measures reflected the colonial government's commitment to confronting the challenges posed by nature and safeguarding the well-being of the residents and economic assets in West Sumatra. However, despite these efforts, the region continued to face significant natural challenges, exacerbated by its complex geographical conditions. Therefore, cold lahar floods will remain a persistent threat to West Sumatra, emphasizing the need for more innovative and sustainable approaches to disaster mitigation that can adapt to the evolving challenges posed by climate change and environmental factors.

## 5 Conclusion

During the Dutch Colonial period, West Sumatra experienced a series of recurring cold lahar floods that struck the region. The presence of active volcanoes such as Mount Marapi, Mount Tandikat, Mount Talang, and Mount Kerinci, along with high rainfall, increased the risk of these disasters. Cold lahar floods, which are a mixture of volcanic material and rainwater, can cause severe damage to settlements, infrastructure, and agricultural land, and even result in loss of life.

The most recent event on May 11, 2024, demonstrated the significant impact of cold lahar floods in four regencies/cities in West Sumatra, resulting in over 60 casualties and severe damage to buildings and infrastructure. History records that similar disasters have occurred since the Dutch Colonial era, with detailed documentation on the frequency of events and their impact on society.

Colonial documents and archives, especially contemporary newspapers, hold valuable historical information about these events, providing insights into the social and economic impacts of volcanic cold lahar floods. These reports also document the central government's response and support in post-disaster recovery efforts, including infrastructure repair, aid distribution, and further research to understand the patterns of cold lahar floods. However, the complex natural challenges and vulnerable geographical conditions require innovative and sustainable disaster mitigation approaches. Historical research and past data can serve as an important foundation for designing more effective mitigation strategies and preventing similar impacts in the future.

This study meticulously explores the intricate history of cold lahar flood disasters in West Sumatra, delving into various aspects of these catastrophic events while also examining the extensive efforts undertaken to recover from their impacts and to deepen our understanding of the hazards they pose, drawing from historical accounts and data that span many years. The findings presented in this research hold the potential to make a significant contribution to the advancement of disaster history research and disaster science in Indonesia, particularly in regions that are inherently prone to cold lahar floods, thereby enriching the existing body of knowledge. Furthermore, this narrative underscores the critical need for preparedness and proactive disaster mitigation strategies to effectively confront the risks associated with cold lahar floods, risks that continue to pose a serious threat to communities and ecosystems alike, making such insights not only relevant but also imperative in today's context of increasing climate variability and natural disasters.

A range of emergency preparedness measures, warning capabilities, education, and infrastructure changes are crucial to avoid future losses. Therefore, concrete steps involving building regulations, protective infrastructure, early warning systems, and community capacity building are needed. It is hoped that these measures can create a safer environment for all the people of West Sumatra in facing the threat of cold lahar floods and other natural disasters in the future. With a collective commitment, these recurring tragedies can be overcome and serve as valuable lessons for building a better future for the next generation.

This research is funded by the Graduate School of Andalas University under Grant RKAT 2024 (Contract No: 42/UN16.16.16.DIR/PT.01.03/2004).

## References

1. Bellier, O., et al. Paleoseismicity and Seismic Hazard Along the Great Sumatera Fault (Indonesia). Journal of Geodynamics. **24**, 169-183 (1997).
2. J. J. Major, T. C. Pierson, W. J. Vallance. Lahar River of Volcanic Mud and Debris. U.S Departement of The Interior and U.S Geological Survey (2018).
3. Lavigne, Franck and Jean-Claude Thouret. Les Lahars: Depots, Origines et Dynamique, Bull. Soc. Geol. France, 545-557. (2000).
4. Pers Realese BMKG. Hujan Lebat Pemicu Banjir Lahar Hujan di Sumbar Masih Berpotensi untuk Sepekan ke Depan, (2024). access by the link https://bmkg.go.id/press-release/?p=bmkg-hujan-lebat-pemicu-banjir-lahar-hujan-di-sumbar-masih-berpotensi-untuk-sepekan-ke-depan&tag=press-release&lang=ID on 22 July 2024.
5. Data Informasi Bencana Indonesia (DIBI), Badan Nasional Penanggulangan Bencana (BNPB), access by the link https://dibi.bnpb.go.id/d on 22 July 2024.
6. Data Badan Pusat Statistik (BPS) Sumatera Barat Province, access by the link https://sumbar.bps.go.id/ on 22 July 2024.
7. Sumatra Courant Newspaper, 27 Desember (1881) edition
8. Opregte Haarlemsche Courant Newspaper, 15 Maret (1882) edition
9. Bataviaasch Handelsblad Newspaper, 27 Februari (1889) edition
10. Sumatra Courant Newspaper, 23 Februari (1889) edition
11. De Locomotief Newspaper, 11 Februari (1890) edition
12. Arnhemsche Courant Newspaper, (1893) edition
13. Het Nieuws van den dag: kleine courant Newspaper, (16 Februari 1893) edition
14. Haagsche Courant Newspaper, 8 Februari 1904 edition; Arnhemsche Courant Newspaper, (13 Maret 1914) edition
15. Het Nieuws van den dag: kleine courant Newspaper, (16 Februari 1893) edition
16. Algemeen Handelsblad Newspaper, (16 Februari 1904) edition
17. Het Nieuws van den dag door Nederlandsch Indie and De locomotief (Newspaper, 1904) edition
18. Sumatra Bode Newspaper, (4 Agustus 1926) edition

# A case study on risk analysis of machine breakdown using the FMEA method

*Prima* Fithri[1*], *Alizar* Hasan[1], *Wisnel*[1], and *R.* Sultan Fauzi Kesogihan[1]

[1]Industrial Engineering Department, Faculty of Engineering, Universitas Andalas, Padang 25163, West Sumatra, Indonesia

**Abstract.** This study addresses the frequent disruptions in PT X, the first cement factory in Indonesia, which affect its productivity and quality. The main problem involves the frequent stoppages of key machinery such as raw mills, kilns, cement mills, and especially coal mills, which are crucial for combustion processes. In 2021, the 6K1 coal mill experienced 384 hours and 38 minutes of downtime, with the Rotary 6K1A03 component being the most problematic, causing 205 out of 359 total disruptions. The Failure Mode and Effect Analysis (FMEA) method was employed to analyze these disturbances. The analysis revealed a critical Risk Priority Number (RPN) value of 5.148, with eight factors exceeding this threshold, including the high current in Rotary 6K1A03, with RPN values of 90, 35, 25, 24, 9, 8, and 6. These critical factors are prioritized for intervention. Consequently, it is imperative to implement management and risk mitigation strategies to address these issues and enhance the operational efficiency of the factory.

## 1 Introduction

The industry in Indonesia has developed rapidly with the increasing number of manufacturing companies. Within a company, risks that can disrupt the production process are inevitable, both those with major and minor impacts [1]. PT X, uses key machines such as raw mill, kiln, and cement mill in its production process. One vital machine is the coal mill, which grinds coal into powder to accelerate combustion. Based on 2021 data, there were 63 failure modes in the coal mill at PT X, with a stop duration of up to 384 hours and 48 minutes, caused by electrical, mechanical, and production disturbances. The FMEA analysis method identifies risks and prioritizes handling based on the RPN value [2]. Failure Mode and Effect Analysis (FMEA) is a structured procedure to identify and prevent as many risks as possible that play a role in a failure and problem in the production process both known and potential problems in the system [3].

Previous studies have shown the application of the FMEA method in various industries. For example, in banana flake production, FMEA identified failures in conveyor transmission components with the highest RPN value of 168, and in raw mill machines in the cement industry with the highest RPN value of 240. Another study on cement packaging machines showed that the main cause of failure was filling speed beyond handling priorities. FMEA has 3 assessment criteria, namely severity, occurrence frequency, and detection rate.

The criteria used in FMEA ranging from severity, frequency of occurrence, and detection level are used to determine the Risk Priority Number (RPN). RPN is used as a determination of action from prioritized risks [4]. Risk Priority Number (RPN) is a risk priority number from the results of multiplication of severity, frequency and detection rate. The RPN determines the priority of the failure. RPN has no value or meaning [5]. The RPN value can be used to determine corrective actions that correspond to the level of value obtained [6]. The formula for finding the RPN value is as follows:[2].

$$RPN = S \times O \times D \tag{1}$$

Information:
S = Severity
O = Occurance
D = Detection (how far it can be detected)

Once the RPN is obtained, the next step is to calculate the critical value of the risk factors. This critical value is searched to determine the number of failure modes that exceed the critical value of the RPN. Failure modes that exceed critical RPN values receive primary handling in addressing risks the maximum limit with an RPN of 210. All these studies emphasize the importance of routine maintenance and close monitoring to minimize risks and optimize the production process [7].

This research, titled "A Case Study on Risk Analysis of Machine Breakdown Using the FMEA Method," aims to propose repair and maintenance solutions for the 6K1 coal mill machine at PT X. The study addresses the following problems: What are the disturbances in the 6K1 coal mill at PT X, what are the causes and risk levels of these disturbances in the 6K1 coal mill components at PT X, and What are the proposed repair and maintenance strategies for the 6K1 coal mill at PT X?

Theories used to support the problem-solving process include risk management, the Failure Mode and Effect

* Corresponding author: primafithri@eng.unand.ac.id

Analysis (FMEA) method, and maintenance management.

## 2 Method

The research methodology includes several steps: literature studies, field studies, problem formulation, data collection, data processing, and flowcharts. Literature studies involve reviewing books, journals, and other references related to risk management, FMEA methods, and maintenance management. FMEA can evaluate and analyze components in a system so that it can minimize the risk or effect of a failure rate as a supporting method for assessing performance in a system [8, 9]. Field studies involve direct observations at PT X, specifically the 4W1 kiln machine, and interviews with staff and operators.

Problem formulation identifies issues based on the significant downtime of the 6K1 coal mill, which decreases productivity and disrupts the cement production process. Machine maintenance and maintenance activities are needed in the production process so that it runs properly and activity failures can be minimized [10]. The system RPN threshold value is obtained from the average RPN value for all components [11]. The study utilizes both primary and secondary data. Primary data is gathered from interviews with staff and operators, detailing damages to the 6K1 coal mill. Secondary data includes documents from the maintenance planning and evaluation unit, specifically the daily report data of the 6K1 coal mill in 2021. The collected data is then processed for analysis.

1. Identify the 6K1 coal mill engine components on PT X that are experiencing problems.
2. Determine and select the 6K1 coal mill machine components that have the largest stop duration.
3. Identify any disturbances that occur in selected 6K1 coal mill machine components.
4. Provide an assessment in the form of weighting of engine disturbances consisting of severity, occurrence and detection criteria using the FMEA method.
5. Calculates the RPN value of each disturbance and the critical value.

## 3 Results and discussion

### 3.1 Results

The results of the research obtained in this section are in the form of RPN values for each failure mode; then a percentage calculation is carried out in failure mode, which can be seen in Table 1.

**Table 1.** Failure mode percentage in 6k1 coal mill

| No | Failure Mode | RPN | Critical Value | Percentage | Cumulative Persentage |
|---|---|---|---|---|---|
| 1 | 6K1A02 MD alarm | 1 | 5.148 | 0.32% | 0.32% |
| 2 | 6K1J05 not ready | 9 | 5.148 | 2.87% | 3.18% |
| 3 | 6K1J05 not ready + bin FC full | 2 | 5.148 | 0.64% | 3.82% |
| 4 | 6K1M03 mill motor unit not ready | 6 | 5.148 | 1.91% | 5.73% |
| 5 | 6K1M04 low pressure in 4 thrust pads | 35 | 5.148 | 11.15% | 16.88% |
| 6 | 6K1M04 low pressure in 4 thrust pads + Bin FC Full | 2 | 5.148 | 0.64% | 17.52% |
| 7 | 6K1M07 open loop | 4 | 5.148 | 1.27% | 18.79% |
| 8 | 6K1M07 roller lubrication alarm, temp oil roller 3 H2 | 4 | 5.148 | 1.27% | 20.06% |
| 9 | 6K1M07 trip | 4 | 5.148 | 1.27% | 21.34% |
| 10 | 6K1M07 vacuum alarm roller 2 | 24 | 5.148 | 7.64% | 28.98% |
| 11 | There is a sound abnormality in the mill | 2 | 5.148 | 0.64% | 29.62% |
| 12 | There is a sound abnormality in the mill (ME install 2 ring wear tube bolts) | 2 | 5.148 | 0.64% | 30.25% |
| 13 | Alarm fault feeder motor 6K1A01M2 cooling fan trip | 2 | 5.148 | 0.64% | 30.89% |
| 14 | The transport equipment under the BHF is not ready, the MCC hangs for a moment and is reactivated via the MCC rack out/in | 3 | 5.148 | 0.96% | 31.85% |
| 15 | The raw coal bin is critical gate 6F1J05Z1 cannot lead to the bin (aisy is broken (Shaft) | 4 | 5.148 | 1.27% | 33.12% |
| 16 | H2 vibration fan trip booster | 2 | 5.148 | 0.64% | 33.76% |
| 17 | Chute 6K1A02 block | 1 | 5.148 | 0.32% | 34.08% |
| 18 | Drag chain 6J1U04 bus fault | 1 | 5.148 | 0.32% | 34.39% |
| 19 | Drag chain 6J1U04 broke | 1 | 5.148 | 0.32% | 34.71% |
| 20 | Faccum roller no2 alarm | 1 | 5.148 | 0.32% | 35.03% |
| 21 | Fault feeder alarm | 3 | 5.148 | 0.96% | 35.99% |
| 22 | Feeder A01 fault alarm | 2 | 5.148 | 0.64% | 36.62% |
| 23 | High 2 PPM alarm | 4 | 5.148 | 1.27% | 37.90% |
| 24 | HT capacitor fault | 1 | 5.148 | 0.32% | 38.22% |
| 25 | Interlock of the buster fan because the BHF transport device stops | 2 | 5.148 | 0.64% | 38.85% |
| 26 | Hanging material | 2 | 5.148 | 0.64% | 39.49% |
| 27 | Hanging material + feeder fault | 1 | 5.148 | 0.32% | 39.81% |
| 28 | ME side gate repair 6K1A02 | 1 | 5.148 | 0.32% | 40.13% |
| 29 | TheMill Fan is not ready (PLC starter fault) | 2 | 5.148 | 0.64% | 40.76% |
| 30 | BHF O2 reading is high | 3 | 5.148 | 0.96% | 41.72% |

| No | Failure Mode | RPN | Critical Value | Percentage | Cumulative Persentage |
|---|---|---|---|---|---|
| 31 | Accumulator checking | 6 | 5.148 | 1.91% | 43.63% |
| 32 | Repair Broken tension rod roller 2 bolt | 8 | 5.148 | 2.55% | 46.18% |
| 33 | Repair The tension rod roller 2 bolt broke | 4 | 5.148 | 1.27% | 47.45% |
| 34 | PPM >1500 due to CO from kiln + kiln stop | 4 | 5.148 | 1.27% | 48.73% |
| 35 | PPM Co alarm H2 max | 25 | 5.148 | 7.96% | 56.69% |
| 36 | PPM Max due to CO from the kiln | 2 | 5.148 | 0.64% | 57.32% |
| 37 | PPM is high because Co from the kiln when starting 6K1V22. which stops suddenly 6K1V44 closes | 2 | 5.148 | 0.64% | 57.96% |
| 38 | Product transport 5K1J05 is not ready | 1 | 5.148 | 0.32% | 58.28% |
| 39 | BHF J05 transport product not ready (prohibits communication cable broken) | 4 | 5.148 | 1.27% | 59.55% |
| 40 | Product transport 6K1J05 is not ready | 3 | 5.148 | 0.96% | 60.51% |
| 41 | Pull rope due to falling material | 1 | 5.148 | 0.32% | 60.83% |
| 42 | Raw coal hanging in a bin | 1 | 5.148 | 0.32% | 61.15% |
| 43 | Hanging raw coal | 1 | 5.148 | 0.32% | 61.46% |
| 44 | Roller lubrication and hydraulic not ready | 1 | 5.148 | 0.32% | 61.78% |
| 45 | Rotary 6K1A03 high current | 90 | 5.148 | 28.66% | 90.45% |
| 46 | Rotary airlock 6K1P15 high current alarm has a plate that is blocking it | 1 | 5.148 | 0.32% | 90.76% |
| 47 | Rotary airlock 6K1P17 high current has a stuck plate | 1 | 5.148 | 0.32% | 91.08% |
| 48 | Sliding inlet block | 3 | 5.148 | 0.96% | 92.04% |
| 49 | Sliding outlet 6K1A03 clogged | 1 | 5.148 | 0.32% | 92.36% |
| 50 | Stop alarm capacitor fuse blown on mill motor 6K1M03 and mill fan 6K1P25 reset to ready panel | 1 | 5.148 | 0.32% | 92.68% |
| 51 | Mill outlet temp suddenly rises (damaged temp sensor) | 3 | 5.148 | 0.96% | 93.63% |
| 52 | Temperature middle bearing separator open loop | 1 | 5.148 | 0.32% | 93.95% |
| 53 | Tilt tension rod on roller 3 | 4 | 5.148 | 1.27% | 95.22% |
| 54 | Thrust pads alarm | 1 | 5.148 | 0.32% | 95.54% |
| 55 | UPS alarm inverter overload | 2 | 5.148 | 0.64% | 96.18% |
| 56 | UPS general fault | 3 | 5.148 | 0.96% | 97.13% |
| 57 | UPS general fault alarm | 2 | 5.148 | 0.64% | 97.77% |
| 58 | UPS unit fault | 1 | 5.148 | 0.32% | 98.09% |
| 59 | Vacuum roller no 2 alarm | 2 | 5.148 | 0.64% | 98.73% |
| 60 | Valve V44 runtime | 3 | 5.148 | 0.96% | 99.68% |
| 61 | Max vibration + checking in mill | 1 | 5.148 | 0.32% | 100.00% |

Based on the calculation of the RPN value in Table 1, there are 8 factors above the critical RPN value, namely Rotary 6K1A03 high current, 6K1M04 low pressure in 4 thrustpads, PPM Co alarm H2 max, 6K1M07 vacuum alarm roller 2, 6K1J05 not ready, Bolt Repair tension rod roller 2 broken, 6K1M03 mill motor unit not ready, and Accumulator checking with RPN value consecutively which is 90, 35, 25, 24, 9, 8 and 6 as many as 2 interference factors. These 8 nuisance factors are priorities for improvement, control, and risk mitigation.

## 3.2 Discussion

This section continues the results subsection, utilizing the Failure Mode and Effect Analysis (FMEA) method to assess risk levels and identify problems in the production process. FMEA analyzes fault factors causing the 6K1 coal mill to stop, including electrical, mechanical, and production disturbances. The daily report data of the 6K1 kiln machine at PT X in 2021 is weighted to calculate the RPN value, which determines the priority for repairing each disturbance factor. The critical RPN value is calculated to identify fault factors above this threshold, ensuring they receive prioritized attention for repairs.

### *3.2.1 Rotary 6K1A03 high current disorder*

Rotary 6K1A03 high current is a failure mode with an RPN value above the first highest critical RPN value of 90. Rotary 6K1A03 high current is a dangerous failure disturbance that occurs every 3-4 days, and the potential for damage can always be detected. 6K1A03 is part of the coal mill, part of the coal mill, namely the rotary feeder as a regulator of the flow of material or materials in the production system or process. This part is usually in the form of a drum or cylinder that rotates to move material from one location to another. Rotary 6K1A03 high current has a combined problem: being blocked by a rock. Based on the calculation of the RPN value, Rotary 6K1A03's high current is the main priority that must be repaired, controlled, and risk mitigated. The disturbance experienced by Rotary 6K1A03 high current is a production disturbance with a stop duration of 174 hours 44 minutes. The cause of the Rotary 6K1A03 high current being blocked by a rock is the condition of the accumulation of material in the east (red) direction, the material easily sticks due to the blasting from below so that the material is very dense and hardens and the accumulation of material in this area will continue to increase so that it covers the area where the material falls in the west direction. Preventive controls recommendations can be given to reduce the risk of this problem. It is recommended to carry out routine inspections and maintenance in the area around the Rotary 6K1A03. Focus on controlling the accumulation of material in the east direction which is susceptible to material sticking due to blasting from below. Here are some steps that can be taken:

1. Routine Cleaning: Carry out periodic cleaning of the accumulation area of material in the east direction to prevent the formation of a solid layer that can hinder the movement of machine components. Cleaning is carried out once a week considering the severity of this failure, which is quite dangerous.
2. Material Flow Regulation: Regulate the material flow more effectively to prevent blasting from below from producing excessive accumulation. This can involve adjusting the speed or direction of the material flow. Adjust the flow once every 3 or 4 days so that the flow does not produce accumulated accumulation.
3. Use of Non-Stick Lubricant: Apply a layer of non-stick lubricant to areas prone to sticky material accumulation. This can help reduce the possibility of solid and hard buildup.
4. Visual and Sensor Monitoring: Install sensors or visual monitoring cameras to monitor the condition of the area in real time. If potential problem material accumulation is detected, preventive measures can be taken early. 5. Addition of 40 mm self-cleaning leaf width: scrape off sticky material. previously by simulating the addition/welding of plates with dimensions of 10 x 40 mm, 10 x 30 mm and 10 x 20 mm, so that it is expected that the local rotary road test time will obtain ideal conditions for adding leaf width that does not hinder rotary performance.

### 3.2.2 6K1M04 low pressure in 4 thrustpads fault

6K1M04 low pressure in 4 thrustpads is a disturbance with an RPN value above the second highest critical value of 35. 6K1M04 low pressure in 4 thrustpads is a significant failure disturbance that causes losses, occurs every month and the potential for damage can always be detected. Based on the calculation of the RPN value, this disturbance is the second priority that must be repaired, controlled, and risk mitigation. The disturbance experienced by 6K1M04 low pressure in 4 thrustpads is a production disturbance with a stop duration of 28 hours 53 minutes. The disturbance can occur due to a decrease in pressure on the 4th thrustpad in the coal mill machine. Thrustpads is usually a component in the bearing system that is used to support axial loads on the machine shaft. A decrease in pressure on the thrustpads can result in operational problems, low performance, and even machine failure if not handled quickly. Preventive controls recommendations can be given to reduce the risk of this problem. Here are some steps that can be taken:

1. Routine Maintenance: Perform routine inspections on the thrustpads system to detect early signs of wear or damage and replace seals, bearings, and other components with limited life.
2. System Cleaning: Perform periodic cleaning of the hydraulic oil system to prevent contamination and ensure smooth oil flow.
3. Pressure Monitoring: Install a pressure monitoring system on the thrustpads to provide early warning of significant pressure drops.
4. Periodic Component Replacement: Schedule periodic replacement of critical components such as seals and bearings, even if they do not show obvious signs of damage.

### 3.2.3 Ppm Co alarm H2 max disturbance

Ppm Co alarm H2 max is a disturbance with an RPN value above the fourth highest critical value, which is 25. This disturbance is caused by imperfect combustion conditions or excess coal in the combustion process, which in turn causes problems with the coal mill machine. The ppm level should be below 1500. Ppm Co alarm H2 max is a failure disturbance that causes losses, occurs every 3 months, and potential damage can always be detected. Based on the calculation of the RPN value, this disturbance is the fourth priority that must be checked, controlled, and risk mitigation. The disturbance experienced by ppm Co alarm H2 max is an electrical disturbance with a stop duration of 17 hours 52 minutes. Preventive control recommendations can be given to reduce the risk of this disturbance. Here are some steps that can be taken:

1. Fuel Monitoring and Control: Implement routine monitoring of the amount of fine coal entered into the kiln. Ensure that monitoring of kiln temperature and pressure is also carried out periodically and carry out automatic control in the form of a system to regulate coal flow based on production needs and kiln capacity.
2. Routine Kiln Maintenance: Implement routine kiln maintenance, including periodic cleaning and inspection of equipment, such as burners and exhaust pipes.
3. Adding oxygen by pulling a van: Adding oxygen with the optimal amount of oxygen needed for efficient combustion and desired results. And anticipating the provision of excess oxygen which can result in unstable combustion.
4. Anticipating by ensuring that CO must be at 1000 PPM by making the combustion reaction in the kiln perfect.

### 3.2.4 6K1J05 not ready disorder

The issue with 6K1J05 not being ready is a disruption with an RPN value above the fifth highest critical value of 9. This occurs due to raw coal feeding issues, likely caused by blockages or mechanical damage. It's a minor but recurring problem happening every 2 years, with potential damage detectable. Based on RPN value calculation, it's the fifth priority for routine maintenance and monitoring. The disruption lasts for 4 hours 34 minutes and is an electrical issue. To prevent it, regularly check and repair transport equipment.

### *3.2.5 Interference repair of tension rod roller 2 broken bolts*

Repairing tension rod roller 2 broken bolts is a disruption with an RPN value above the sixth highest critical value of 8. These bolts are located between the roller and cap on the coal mill machine. The issue arises due to bolt wear leading to breakage, causing performance decline. It occurs every 3-2 years, with detectable damage. It's the sixth priority for repair and mitigation. The problem lasts for 17 hours and 10 minutes and is a mechanical issue. To prevent it, conduct periodic maintenance without waiting for damage and replace broken bolts promptly. Visual inspection: Check the condition of the roller tension rod thoroughly, including the condition of the bearings, shaft and other components. Material test: To determine its strength and chemical composition, perform a material test on the remaining broken bolt. Stress analysis: Perform a stress analysis to determine the maximum load the roller tension rod can withstand.

System evaluation: Inspect the entire coal mill system to identify any imbalances or other problems that could cause excessive load on the roller tension rod. Replace bolts: Replace broken bolts with new bolts made from high-quality materials with appropriate specifications. Repair other damage: If damage is found to other components, such as bearings or shafts, repair or replace them. Reset tension: Once the repair is complete, reset the tension on the tension rod roller to factory specifications.

Prevention: Routine maintenance: Perform routine maintenance on the tension rod roller, including lubrication and cleaning. Condition monitoring: Use monitoring tools to monitor the operating condition of the roller tension rod and detect early signs of damage.

### *3.2.6 6K1M03 mill motor unit not ready disorder*

The disturbances with the seventh highest RPN value are disturbances with an RPN value above the critical value obtained by 6. These disturbances are types of electrical disturbances consisting of 6K1M03 mill motor unit not ready disturbances, and accumulator checks. The eighth highest disturbance is 6K1M03 mill motor unit not ready. This disturbance is a coal mill drive that is not ready for use with the constraint that the requirements for running have not been met, such as minimal or low gearbox oil. This disturbance experiences minor failures that can be adapted, occurs every 3-2 years, and potential damage can always be detected with a stop duration of 1 hour 25 minutes. Preventive controls recommendations can be given to reduce the risk of this problem occurring. Steps that can be taken are to add oil that is already at its maximum or in a condition that meets the requirements and ensure periodic checks of the 6K1M03 section so that disturbances can be avoided. The second seventh highest disturbance is checking the accumulator. Checking the accumulator on the roller functions to raise and lower the roller. The disturbance occurred because there was a dirty oil filter, so it took time to check and caused the accumulator to not be fully charged or not function when needed. This disturbance experienced a small failure that could be adapted, occurred every 3-2 years, and potential damage could always be detected with a stop duration of 37 minutes. Preventive controls recommendations can be given to reduce the risk of this problem. Steps that can be taken are to carry out regular and routine maintenance, ensure that the voltage regulator and charger are functioning properly, and clean the accumulator terminals regularly. So the risk of disturbances in the 6K1 coal mill can be avoided.

### *3.2.7 K1M07 vacuum alarm roller 2 disorder*

The 6K1M07 vacuum alarm roller 2, a hydraulic pump with an RPN value above 24, is the third priority for repair and control. This issue, caused by oil circulation problems, leads to performance degradation and occurs every three months, with a stop duration of 13 hours and 46 minutes. Design modifications: If necessary, make design modifications to the roller tension rod to increase strength and resistance to loads. Preventive measures include: Clear work procedures: Establish clear work procedures for installing and tightening bolts, as well as regular inspections.

## 4 Conclusion

Based on the processing and analysis of data on the research that has been done, the conclusions obtained are as follows: In 2021, PT X experienced 61 disturbances in the 6K1 Indarung VI coal mill, including mechanical, production, and electrical issues such as alarms and equipment failures. Using the FMEA method, 10 failure modes with RPN values above 5.1 were identified, including high current in Rotary 6K1A03 and low pressure in 4 thrustpads. Significant disruptions included high current and low pressure, with preventive maintenance recommendations such as routine cleaning and monitoring of temperature and pressure. These are expected to reduce disruption risks and enhance operational efficiency.

Suggestions that can be given for better research in the future are: Further research is recommended to add other methods to complete this case study. Methods that can be used such as a risk matrix that determines the level of risk based on the probability of events and impacts.

Failure Mode Identification in the FMEA analysis succeeded in identifying various failure modes that could potentially occur in the 6K1 Coal Mill, such as component damage, problems with the control system, or disruption to the coal supply. Risk assessment in each failure mode is assessed based on severity, likelihood of occurrence and ease of detection. The results of this assessment are then summarized in a number called the Risk Priority Number (RPN). The failure mode with the highest RPN is considered to have the highest risk and deserves more attention.

Root Cause of further analysis is performed to identify the root cause of each failure mode. This is important to formulate appropriate corrective actions. Improvement Recommendations Based on the results of the analysis, a number of improvement recommendations will be proposed, such as increased maintenance, design modifications, or changes to operating procedures. Improvement Priority of recommended improvements is usually prioritized based on the RPN value. Repairs related to the failure mode with the highest RPN will take top priority.

Failure mode has the highest RPN this indicates the most critical failure modes and needs to be addressed immediately. The primary root cause of this failure mode Understanding the root cause will help in formulating an effective solution. Recommendations for improvement are most effective and efficient consider cost, time, and impact to production in selecting improvement recommendations. The recommendations for improvement implemented plan the steps needed to implement recommendations for improvement, including assignment of responsibilities, budget allocation, and monitoring progress.

## References

1. Carey Mark, Rene M Stulz, The risk of financial institutions (The University of Chicago Press, Chicago, 2006)
2. Suryaningrat, Ida Bagus, dkk, Identifikasi risiko padsa okra menggunakan failure mode and effect analysis (FMEA) di PT Miratani Dua Tujuh di Kabupaten Jember. Jurnal Agroteknologi, **13**, 1, 25–33 (2019)
3. Wicaksono, A., and Yuamita, F., Pengendalian kualitas produksi sarden menggunakan metode failure mode and effect analysis (FMEA) dan fault tree analysis (FTA) untuk meminimalkan cacat kaleng di PT XYZ. Jurnal Teknologi dan Manajemen Industri Terapan, **1**(3), 145 – 154 (2022)
4. Suherman, A., and Cahyana, B. J., Pengendalian kualitas dengan metode failure mode effect and analysis (FMEA) dan pendekatan kaizen untuk mengurangi jumlah kecacatan dan penyebabnya. Prosiding Semnastek, (2019)
5. Stamatis, D. H., Failure mode and effect analysis: FMEA from theory to execution (ASQ Quality Press, 2003)
6. Novita, C., Astanti, Y. D., and Ristyowati, Identifikasi masalah padsa mesin raw mill menggunakan metode FMEA dan RPN, Waluyo Jatmiko Proceeding, **15** (2022)
7. Fithri, P., Rafi, M., Pawenary, Prabuwono, A.S. Risk analysis of production process for food SMEs using FMEA method: A case study. E3S Web of Conferences, **331**, 02010 (2021)
8. R. I. Yaqin, M. Akmal, J. P. Siahaan, M. L. Umar, B. H. Priyambodo, and Y. E. Priharanto, Failure analysis of fuel system main engine fishing vessel (case study: KM. Sumber Mutiara). Kapal: Jurnal Ilmu Pengetahuan dan Teknologi Kelautan, **20**, 1 (2023). https://doi.org/10.14710/kapal.v20i1.48530
9. Y. E. Priharanto, R. I. Yaqin, G. Marjianto, J. P. Siahaan, and M. Z. L. Abrori, Risk assessment of the fishing vessel main engine by fuzzy-FMEA approach. Journal of Failure Analysis and Prevention, **23**, 2 (2023). https://10.1007/s11668-023-01607-w
10. Anggraini, Maria, Rahmat Maulana, Pengaruh pemeliharaan mesin terhadap kualitas sepatu padsa PT Nikomas Gemilang. Jurnal Manajemen dan Bisnis, **9**, 1, 59 – 74 (2016)
11. Nugraha, I. M. A., and Malelak, F. Y. FMEA approach in risk analysis of main engine fuel system maintenance: Case study on the Express Bahari 1F. Jurnal Teknik Mesin, **17**(1), 20-26 (2024)

# Enhanced subsurface investigation utilizing electrical resistivity tomography and precise positioning tools (study case: Rongga landslide)

*Moch* Hilmi Zaenal Putra[1*], *Adrin* Tohari[1], *Eko* Soebowo[1], *Yayat* Sudrajat[1], *Amar*[1], *Wahyudin*[1], and *Dadan Dani* Wardhana[1]

[1]Geological Disaster Research Center, National Research and Innovation Agency, Bandung, Indonesia

**Abstract.** This study presents an enhanced approach for subsurface investigation using Electrical Resistivity Tomography (ERT) coupled with precise positioning tools, focusing on the Rongga Landslide in Indonesia. ERT is widely recognized for its ability to map subsurface conditions by detecting resistivity contrasts associated with geological structures and water content, especially in landslide-prone areas. The study combines GPS and digital waterpass altimeter with ERT to improve the precision of subsurface anomaly detection. Important information regarding moisture distribution, graben structures, and sliding surfaces was revealed by the generation of two ERT profiles. Our research indicates that a number of factors, including lithological characteristics, erosion along riverbanks, and tectonic activity—specifically, a nearby fault structure—may have contributed to the Rongga landslide These findings highlight the usefulness of ERT to evaluate the risk of landslides and offer useful data for geotechnical research in tectonically active areas.

## 1 Introduction

Recent research evidences the potential of Electrical Resistivity Tomography (ERT) as a predictor of landslides in volcanic areas. For instance, the use of ERT in landslide studies, showing how the application can be useful and its limitations in sector analysis [1]. ERT was most effective in reconstructing the slope geometry, locating the sliding surface as well as in mapping the high-water content areas. Some insights on the use of ERT in landslides, as a special case, the Lijiazu landslide in Hunan province was investigated in terms of it [2]. It is good for its mobilization flexibility and adaptability in different regions including areas with mountainous terrain. The approach enables determination of the landslide sliding surface, thickness and altitude, providing an effective means for understanding of the deformation and failure process of landslides. Application of the ERT method for characterizing landslides in Southwest China [3]. Units under assessment are units that are at higher risk for landslides. This research is about understanding what happened in the subsurface before the landslide. The ERT surveys consisted of two merged profiles that were designed to reveal patterns related to movement on the landslide mass and manner of the sliding surface and bedrock geoelectrical properties. The low resistivity of the sliding parts, corresponding to weathered sandstone and mudstone layers, was a major result. The research demonstrates the potential benefit of using ERT to detect subsurface conditions crucial for landslide hazard assessment and mitigation.

ERT is a powerful and informative technique to predict landslides in volcanic areas since it allows detailed subsurface imaging and can detect slight differences in water content [4]. False interpretations may occur due to soil type, moisture content, temperature, electrode contact, survey configuration and surface conditions, which might lower the reliability and accuracy of ERT measurements.

### 1.1 Geology

The study area is situated in the mountainous region at an average elevation of 1,000 meters above the sea level. This area overlies with Tertiary formations consisting of Mts, Mtb, Tmc, Tmcs, and Tmbe (Fig. 1a) [5]. The regions geodynamic history favored the presence of various structural features including anticlines, normal faults, and strike-slip faults [5].

Toe parts of the landslide area are located near to the riverbank of Cidadap River, which belong to Tmc formation specifically. The Tmc unit consists of interbedded claystone, siltstone, and sandstone, and local lahar beds of tuff, andesite breccia, and tuff breccia. The mountainous topography and the structural intricacies within the Tmc formation together with its geological properties, make this area landslide-prone.

The landslide area is located at a geological structure of anticline with folded lithology and the anticline is

* Corresponding author : mochhilmizp@gmail.com

oriented southeast - northwest. An anticline observed in the Tmc formation following active calls to the tectonics of the compression, suggested that this anticline formation occurred after the Middle Miocene.

Field mapping results and measurements were put together to generate the lithological variations within the landslide area. The slope bottom portion is occupied with sedimentary siltstone and sandstone horizons (Fig. 1b). Above the sedimentary this layer is then topped with a tuff layer; which is an indication of previous volcanic activity (Fig. 1c). Top of slope: a thick, weathered volcanic soil. The bare soil of the main landslide body is about four meters thick at its top of scarp, the upper part being red brown and the lower dark brown. These colors represent weathered procedures.

It is a process of weathering we can see the progression in soil profile. The reddish-brown colouration of the upper part of the volcanic soil layer shown below suggests oxidation and iron oxides which are characteristic of prolonged exposure to the atmosphere. This is a reflection of abundant exposure of the soils to the surface and therefore a lot of time for chemical weathering to take place. Conversely, the dark brown color of the lower part of the soil layer potentially involves less oxidation and overall organic matter would be higher. This weathering has led to such distinct soil horizons as to mirror the history of the stability and exposure of the land surface.

The fact that the weathered soil layers are present displaying different colorations and characteristics indicates that the soil is existing on the surface from a long period during which it has undergone the process of weathering widely. While the long-term exposure has allowed the evidence of a mature soil history to develop, and thereby emphasizing the intricate and dynamic interaction between geological processes and more recent environmental conditions.

The outcrop showcases a cornucopia of structures that provide clues to the environmental changes that affected this region. Essentially, strata represent phases or periods of depositional processes influenced by different forces, most likely rivers, lakes and marine environments that have left different marks or layers in the sediments over time. These tuffs gave enormous evidence of active volcanism during the Tertiary period, just above these compound sedimentary layers. External to these geological remnants a weathered volcanic profile soils its way to a landscape having been stable and exposed for millennium. This specific soil profile reflects a period of time when material was weathered chemically, and also when the structure began to be broken up physically, leading mature soil structure to form on a stationary land surface.

The features combine to form a multifaceted geological story about eruptions of volcanoes interspersed with silty periods of erosion and ash soil creation. The complex tale that is told tracks the changing face of the landscape from Middle Miocene through to modern day, at a geological scale.

**Fig. 1.** The geological setting map of study area (a); Layers of sandstone and siltstone in toe of landslide (b); Soil profile displaying some remarkable horizons-- a light brown color in the upper layers and a dark brown color in the lower layers (c)

## 1.2 Rainfall

The ground in SDN Babakan Talang 01 started to rupture on Sunday, 18/02; the cracks were between 20 and 25 cm wide. The holes were quickly filled in with soil by locals. On Thursday, February 29, 2024, the ground began to massively slide as a result of the cracks' eventual extension and deepening. The cracks extend to 2-2.5 m by Monday, March 4, 2024. There was a definite circular pattern of landslides where some areas of the top ground collapsed about 3.5 and 4 meters.

The timing of the Rongga landslide on 4 March 2024 links with the very wet rainy season that had been ongoing since October 2023 with rainfall ranging between 40 to 80 mm per day in December, January and March (Fig. 2). In the week before the landslide rainfall intensity was the highest at 100 mm, while satellites reported total rainfall in the order of 3150 mm, 2050 mm, and 1500 mm according to IMERG [6], MERRA-2 [7], and GsMAP [8].

The ERT survey, done on May 23, 2024, a few weeks before, happened in a dry period. This rainfall data is important for assessing the water content in the surface and subsurface layers in the framework of the ERT analysis. The high total rainfall prior to the landslide, together with the start of the event in the afternoon after already extremely high daily rainfall, might have affected the moisture status and hydrological situation in the surrounding. Knowledge of these rainfall patterns and how they affect subsurface water dynamics is crucial because both the interpretation of ERT results and the evaluation of geological hazards related to landslides depend on this information in the case region.

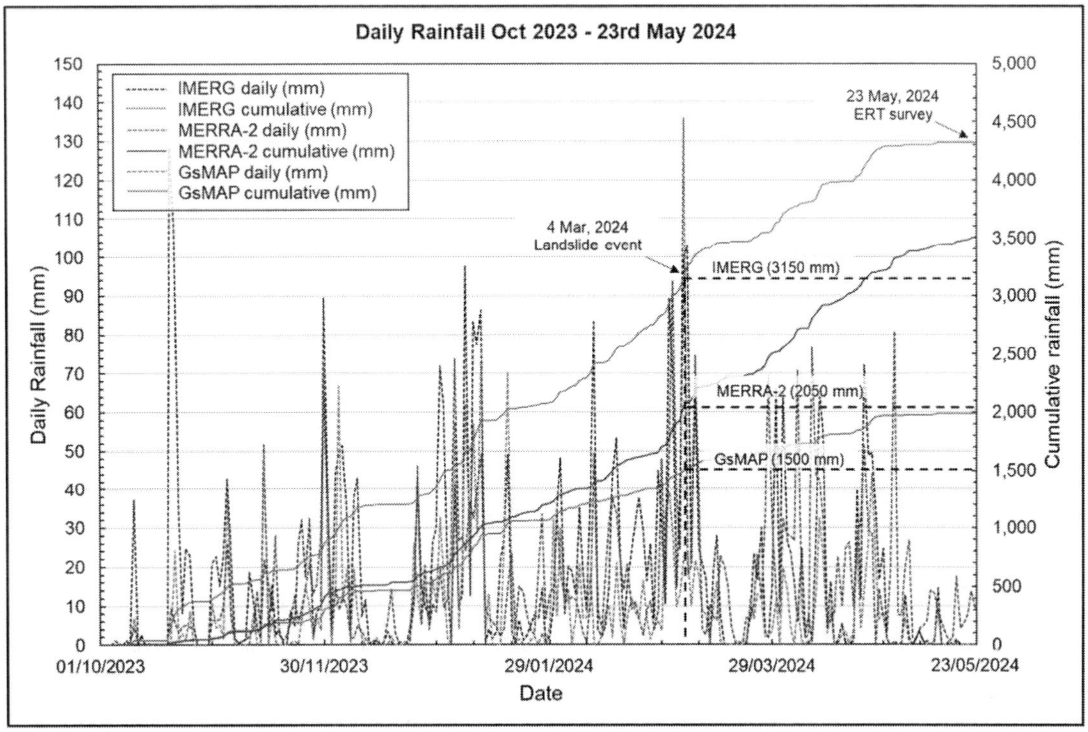


**Fig. 2.** This dataset is a compilation of daily and cumulative rainfall data from IMERG, MERRA-2, and GsMAP satellites showing the distribution of rainfall during the period October 2023 to May 23, 2024. The dash line represents daily rainfall, and the solid line summarizes rainfall accumulation over the period of the study area, offering how many days and how hard it rained.

## 2 Methodology

### 2.1 Electrical resistivity method (ERT)

In studying landslides, the selection of electrode configuration plays a role in the appropriateness of the resolution and the depth window to detect. One of the recommended configurations includes all Dipole-Dipole configurations that can give the target depth window of say 15 meters. This combination is applicable because of the balance between penetration depth and resolution. As such, there must be good horizontal structures and changes to show some slip surfaces and fractures. Electrode spacing constitutes another factor to include and, depending on the target, aims for the effective element. The following is the key formula for telling the apparent resistivity ($\rho_a$) in a dipole-dipole equation. The apparent resistivity ($\rho_a$) is expressed as:

$$\rho a = \frac{2\pi a(n+1)(n+2)\Delta V}{I} \quad (1)$$

where:

- $a$ is the distance between electrodes,
- $n$ is the possible number of electrodes spacings which the current dipole and the potential dipole could be correlated,
- $\Delta V$ is the potential difference being measured,
- $I$ is the ground current injected into the ground.

The electrodes in a dipole-dipole configuration are placed similarly in a straight line. $C_1$ and $C_2$ are the two electrodes used to apply the current to the ground. We used a second pair of electrodes, named $P_1$ and $P_2$, to measure the voltage caused by the same current. As depicted in Fig. 3, the current electrodes $C_1$ and $C_2$ are spaced by a distance of. The potentials $P_1$ and $P_2$ are located at a half distance of the midpoint between $C_1$ and $C_2$, and with other half from the separation point, $P_1$ and $P_2$. It does this by measuring the potential difference between $P_1$ and $P_2$ which is affected by the current that is injected by $C_1$ and $C_2$, to the ground.

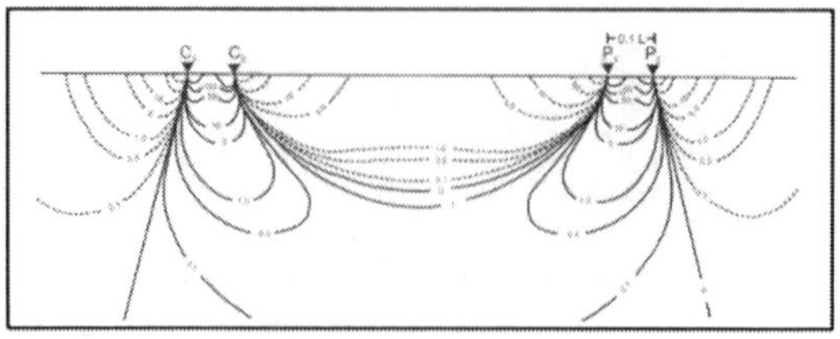

**Fig. 3.** Illustration of signal contribution sections for dipole-dipole configurations, the contours represent the contribution of different volume elements in the subsurface to the total potential difference measured between the potential electrodes $P_1$ and $P_2$. This setup enables precise measurement of the electric field perturbations created due to the injected current, and so it provides an exact function of the changes in the subsurface resistivity.

**Table 1.** The detail of ERT survey

| | |
|---|---|
| **Name of Tools** | SuperSting with Swift R8 |
| **Configuration** | Dipole-Dipole |
| **Spacing (m)** | 3 |
| **Number of Electrodes** | 56 |
| **Max Depth Penetration (m)** | 84 |
| **Length of Line (m)** | 165 |
| **Number of Line** | 2 |

The optimal electrode spacing for a dipole-dipole ERT configuration to effectively detect features at a depth around 15 meters typically ranges between 2 to 5 meters. Three meters was chosen because that offers a balance between resolution and depth penetration. It allows for a reasonable level of detail while still being capable of reaching depths around 15 meters. Table. 1 provides detailed information on survey settings.

Generally, it is good to have those electrodes 2-5 meters apart to do a dipole-dipole ERT configuration that is best for 10-15meter depth features. Three meters is a good choice as it keeps the right balance between resolution and depth penetration. It provides a good amount of resolution and yet is capable of going to depths approaching 15 meters. Table. 1 gives an elaborate info as seen in the meanwhile in survey settings.

## 2.2 Waterpass, altimeter, and GPS

Orion 2LS Digital Waterpass Level (Fig. 4. a and b), Micro Altimeter Alpinel 0.5 m Barograph Correction (Fig. 4c), and Garmin GPSMAP (Fig. 4c) enhances data accuracy and dependability collected in geophysical surveys. Correct leveling of the surface is critical for having the electrodes at the correct location in ERT surveys to acquire optimal data. The accuracy of this digital level keeps the electrodes from falling out in the trough and registers very consistent data. An electronic sensor inside the Orion 2LS Digital Level Waterpass measures the tilt angle of the device. The black and white striped ruler is the target for the level's laser or optical system (Fig. 4b). The device is designed to detect the position of the stripes when a laser or optical signal is directed at the ruler, thereby identifying height differences or the tilt angle with respect to the ruler. This data is then digitally processed and output, enabling highly accurate leveling and alignment using the ruler location.

In addition, the Micro Altimeter Alpinel is distinct in the way it adjusts for changes in barometric pressure to offer precise (Fig. 4c). This accuracy is extremely important in correcting ERT resistivity surveys for topographic bias and thus increasing data accuracy. The control over topographical variations due to accurate altitude data assists in producing less variably subsurface resistivity models.

**Fig. 4. a.** Measuring distance and height to the black and white striped ruler which is the orion 2LS digital level waterpass **b.** The black and white striped rule, possibly a target for a level's laser or optical system **c.** Combination of micro altimeter alpinel and the garmin GPSMAP for correct altitude and gps navigation.

High-sensitivity GPS capabilities provide precise coordinates in the Garmin GPSMAP for each electrode position (Fig. 4c). This precision both helps in laying out the survey grid correctly, thus limiting positioning errors and enhancing the quality of the resistivity measurements as a result. In combination, however, these tools ensure that electrodes are placed correctly in all three dimensions, which greatly improves accuracy and repeatability of geophysical survey data.

The Orion 2LS Waterpass provides accurate horizontal leveling and the Micro Altimeter Alpinel compensates for any changes in elevation. Garmin GPS provides precise location information by combining latitude, longitude, and altitude. This is achieved by a combined application of these tools, which enables eliminating the most common positioning errors, where all electrodes are properly positioned relative to three dimensions. This setup allows for accurate and reliable data, a must for any valuable geophysical survey.

# 3 Results

The ERT (Electrical Resistivity Tomography) image provides a detailed subsurface cross-section of a location characterized by sediment and volcanic deposits, including sand, clay, and weathered volcanic soil. The resistivity values are color-coded, with blue representing low resistivity and red indicating high resistivity. In this image, key observations can be made regarding saturation levels, potential sliding surfaces, and structural features such as grabens.

This ERT (Electrical Resistivity Tomography) image was obtained to represent the subsurface section of the study area consists of sedimentary and volcanic deposits of sand, clay and weathered volcanic soil. The subsurface condition is the result of resistivity values - blue is least resistive, red is most resistive. This detail allows for a number of observations on type of rock and soil, saturation level, potential sliding surfaces and structural elements such as grabens to be made.

## 3.1 The key feature

### *3.1.1 Saturation and gradient saturation*

When the formation has a low resistivity, only blue and light blue can be seen, which is typical for clay or for sand saturated with water. However green into yellow areas which represent moderate resistivity imply the presence of more poorly saturated materials (likely sand or soil and varying moisture content). Red (high resistivity) and orange regions indicate dry materials such as dry sand or volcanic deposit (volcanic ash) or compacted, less porous sediments in case of (Figs. 5 and 6).

### *3.1.2 Sliding surface and graben*

In marks of resistivity variability, it might expose some of potential sliding or slip surfaces. For example, the change from high resistivity (red/yellow) to lower resistivity (blue) might indicate a slip surface where saturated clay (blue) underlies less saturated material (yellow/red). Sudden changes in resistivity as well as notable slope (slope break) or gradients changes in the contour further indicate the possible sliding surfaces, especially in cases where the high resistivity material is sitting on the low resistivity zones. At the same time, a low structure in resistivity values is visible within the central part of the image, plunging downward and indicating the occurrence of a small-graben structure Fig. 6. Such things could be a simple graben - a trough or sunken block between rifts or between parallel faults, usually with the gravity-induced phenomenon of mass-wasting associated with landslides. The low resistivity areas (blue) that are suggested (assuming a lateral variability of this parameter in the subsurface) on the central lower part of the image, are most likely indicative of a graben filled with saturated materials probably induced by an extensional mechanism linked to the movement of the landslide. If that is what it is, then it may indicate a zone where the ground has dropped down, becoming a conduit for water to pool, pushing the sediments on the ground to start moving. The description of the ERT profile of Fig. 5 in three sections:

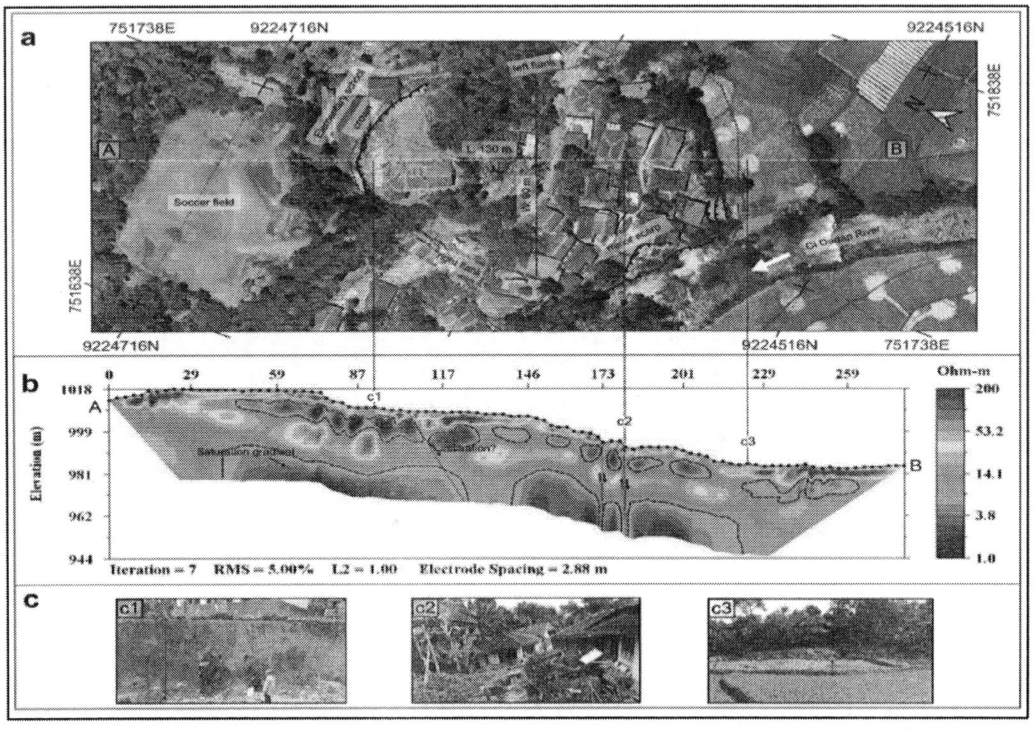


**Fig. 5** An orthogonal aerial photograph of the Rongga landslide location (a); The red dashed line indicates the landslide-affected area, the black line denotes the scarp, and the Ci Dadap River converges at the foot of the slope (b); Electrical resistivity tomography (ERT) profile dipole-dipole configuration, ERT 2 lines with 3 meters electrodes separation/56 electrodes; An image of the crown landslide scarp with 4 meters drop (c1); A house twisted during landslide (c2); Bullging occuring at toe of landslide (view about 1m) of landslide (c3)

### *3.1.3 Fig. 5 left section (0-87 meters)*

This section of the ERT profile shows high resistivity in near-surface layers (red/yellow colors) comprising dry or dense sediments or volcanic deposits. Deep background blue areas where these are shallow suggest increased moisture content in nearby clay layers or water-saturated sediments beneath. The potential slip surface is characterized by high resistivity (red/yellow) intersecting a low resistivity (blue) material which may be a geological interfacial zone of geological boundaries where various types of materials or moisture phases meet.

### *3.1.4 Fig. 5 middle section (87-201 meters)*

The ERT profile shows a characteristic low resistivity zone (blue coloration) in this water bearing horizon. This zone indicates a graben-like structure or an area on which fluids (in this case, water) have pooled. The greener/yellower zones above this low resistivity zone correspond with less saturated materials. Based upon the boundary between these resistivity anomalies, potential sliding surfaces are inferred, indicating potential geological failure surfaces as a result of motion or deformation.

The base of the fault plane is estimated to be situated around 180 meters in the horizontal with a continuation up the vertical from approximately 999 meters to between 962 - 944 meters in elevation. This fault plane exhibits a strong vertical resistivity contrast, with high-resistivity on the right-left side colored in yellow or red, and low-resistivity (blue/green) in the middle.

### *3.1.5 Fig. 5 right section (201-259 meters)*

Like the left section, the upper layers near the surface show high resistivity (red/yellow: dry sediments or volcanic deposits). Below texture these blue zones return and this shows more moisture or water-saturated sediments. Abrupt changes in resistivity indicate possible slip surfaces in the same manner as seeing things in the left view.

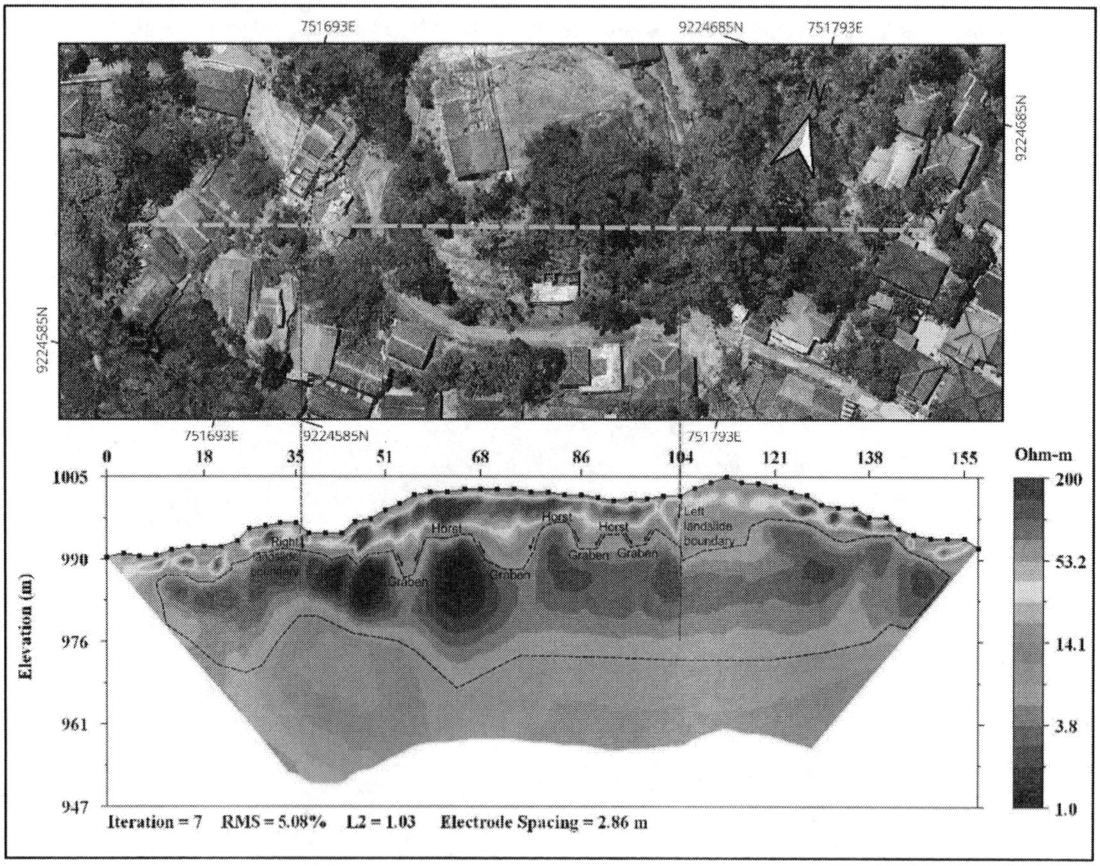


**Fig. 6.** Above: an orthogonal aerial photograph of the Rongga landslide location. The red dashed line indicates the landslide-affected area, bottom: electrical resistivity tomography (ERT) profile dipole-dipole configuration, ERT 2 lines with 3 meters electrodes separation 56 electrodes

## 4 Discussion

The Rongga landslides are primarily influenced by a combination of lithology, river erosion, and fault structures. First, the lithology of the area, which consists of interbedded sandstone, siltstone, and claystone, plays a significant role in landslide susceptibility. The impermeable nature of claystone layers prevents effective drainage, leading to increased pore water pressure in the soil and slope instability. This pressure build-up can cause the claystone to degrade, weakening the material and making it more prone to slope failure [9], [10] . As a result, the claystone often acts as a sliding plane, affecting overall stability, especially under high rainfall conditions that exacerbate drainage challenges [11], [12]. Furthermore, the interbedded nature of these rocks forms a complex system that interacts with surrounding geological features, thus heightening landslide risk.

Second, river erosion at the Cidadap River bank (Fig. 7), located near the base of the slope, has likely intensified landslide dynamics. Heavy rainfall in recent months has accelerated the erosion process, exposing weathered claystone along the riverbank [13], [14]. This exposure to weathering weakens the claystone, reducing its shear strength and directly contributing to riverbank failure. The loss of bank strength further destabilizes the slope, making it more susceptible to landslides. This combination of erosion and weathering amplifies the vulnerability of the Rongga landslide area to collapse.

**Fig. 7.** Failure at Cidadap River bank

Finally, Electrical Resistivity Tomography (ERT) results revealed the presence of a fault structure in the lower-middle region of the landslide area. This fault, indicated by the presence of joints measuring 20 to 50 cm wide as shown in Fig. 8, signifies tectonic influence in the area [15], [16]. The fault's formation is linked to an anticline with folded lithology-oriented southeast-northwest, suggesting that local tectonics contribute to the structural weaknesses in the slope. This fault structure amplifies the landslide risk, as it interacts with the surrounding hydrological and geological conditions, creating a dynamic and unstable environment prone to failure. Additionally, the fault plane's low resistivity readings on one side suggest variations in material composition, potentially indicating fluid pathways or differences in compaction and water content [17], [18]. This resistivity contrast is crucial for understanding subsurface dynamics, as it may affect groundwater flow and influence geological hazards.

**Fig. 8.** Intensive joint in study area

## 5 Conclusion

The findings emphasize the complex interplay between lithology, river erosion, and tectonic activity as primary triggers for the Rongga landslides. The impermeable claystone layers contribute to increased pore water pressure, compromising slope stability, while river erosion exacerbates slope degradation by weakening riverbank support. Additionally, the fault structure, with its well-defined resistivity contrast and vertical extent, highlights the influence of tectonic activity on landslide susceptibility in the area. This fault's significant resistive boundary implies a strong tectonic influence that warrants further investigation for potential seismic implications, which could affect infrastructure planning and hazard mitigation strategies.

Future research should focus on refining and validating the ERT data to further elucidate the subsurface conditions, particularly along the fault plane. Ground-truthing studies and high-resolution ERT surveys could provide more accurate data on the fault's behavior, while integrating other geophysical methods would offer a comprehensive understanding of subsurface dynamics. Additionally, collaborations with experts in structural geology, geotechnical engineering, and hazard assessment are essential to gain multidisciplinary insights into the implications of the fault and resistivity variations. By combining these research approaches, future studies can generate a holistic view of the study area, aiding in land use planning, infrastructure development, and hazard mitigation in tectonically active regions.

This integrated analysis, grounded in multidisciplinary research, will not only improve our understanding of the geological conditions and fault dynamics but also offer critical data for sustainable development and risk management in the area.

This research is funded by Rumah Program Kebencanaan Hidrometeorologi dan Iklim - Organisasi Riset Kebumian dan Maritim -Badan Riset dan Inovasi Nasional (BRIN), Indonesia for 2024. The authors would like to thank for supporting them for conducting this work.

## References

1 A. Perrone, V. Lapenna, and S. Piscitelli, Electrical resistivity tomography technique for landslide investigation: A review. Earth-Science Rev., **135**, 65–82, (2014)

2 M. Sun, J. Liu, J. Ou, R. Liu, and L. Zhu, Electrical Resistivity Tomography (ERT) Investigation for Landslides: Case Study in the Hunan Province, China. Appl. Sci., **14**(7), (2024)

3 C. Ling, Q. Xu, Q. Zhang, J. Ran, and H. Lv, Application of electrical resistivity tomography for investigating the internal structure of a translational landslide and characterizing its groundwater circulation (Kualiangzi landslide, Southwest China). J. Appl. Geophys., **131**, 154–162, (2016)

4 M. H. Z. Putra, R. D. Kartiko, P. Soemantidiredja, I. A. Sadisun, and A. Tohari, Pengaruh Zona Jenuh Air Terhadap Kestabilan Lereng Di Weninggalih, Kabupaten Bandung Barat. Ris. Geol. dan Pertamb., **30**(1), 119, (2020)

5 M. Koesmono, Kusnama, and N. Suwarna, Geological Map of The Sindangbarang and Bandarwaru Quadrangles, Jawa. 2nd Edition. Geological Research and Development Centre, (1996).

6 G. J. Huffman, E. F. Stocker, D. T. Bolvin, E. J. Nelkin, and J. Tan, GPM IMERG Final Precipitation L3 1 day 0.1 degree x 0.1 degree V07, Edited by Andrey Savtchenko, Greenbelt, MD, Goddard Earth Sciences Data and Information Services Center (GES DISC), (2023). [Online]. Available: https://disc.gsfc.nasa.govuni/datasets/GPM_3IMERGDF_07/summary. [Accessed: 23-Aug-2024].

7 GMAO, Global Modeling and Assimilation Office. inst3_3d_asm_Cp: MERRA-2 3D IAU State, Meteorology Instantaneous 3-hourly (p-coord, 0.625x0.5L42), version 5.12.4, Greenbelt, MD, USA: Goddard Space Flight Center Distributed Active Archive Center (GSFC DAAC), (2015). [Online]. Available: https://gmao.gsfc.nasa.gov/reanalysis/merra-2/citing_MERRA-2/. [Accessed: 23-Aug-2024].

8 T. Kubota *et al.*, Global Satellite Mapping of Precipitation (GSMaP) products in the GPM era, Satellite precipitation measurement, Springer, (2020). [Online]. Available: https://sharaku.eorc.jaxa.jp/GSMaP/. [Accessed: 23-Aug-2024].

9 P. K. Singh, K. K. Singh, and T. N. Singh, Slope failure in stratified rocks: a case from NE Himalaya, India, *Landslides*, **14**(4), 1319–1331, (2017)

10 A. W. Skempton, Residual strength of clays in landslides, folded strata and the laboratory, *Geotechnique*, **35**(1), 3–18, (1985)

11 A. Cevasco, G. Pepe, and P. Brandolini, The influences of geological and land use settings on shallow landslides triggered by an intense rainfall event in a coastal terraced environment, Bull. Eng. Geol. Environ., **73**(3), 859–875, (2014)

12 L. C. Nguyen, P. Van Tien, and T. N. Do, Deep-seated rainfall-induced landslides on a new expressway: a case study in Vietnam, Landslides, **7**(2), 395–407, (2020)

13 T. R. Martha, P. Roy, K. B. Govindharaj, K. V. Kumar, P. G. Diwakar, and V. K. Dadhwal, Landslides triggered by the June 2013 extreme rainfall event in parts of Uttarakhand state, India, Landslides, **12**(1), 135–146, (2015)

14 S. C. Chen, H. Ter Chou, S. C. Chen, C. H. Wu, and B. S. Lin, Characteristics of rainfall-induced landslides in Miocene formations: A case study of the Shenmu watershed, Central Taiwan, Eng. Geol., **169**, 133–146, (2014)

15 G. Samodra, M. F. Ramadhan, J. Sartohadi, M. A. Setiawan, N. Christanto, and A. Sukmawijaya, Characterization of displacement and internal structure of landslides from multitemporal UAV and ERT imaging, Landslides, **17**(10), 2455–2468, (2020)

16 A. Viero, A. Galgaro, G. Morelli, A. Breda, and R. G. Francese, Investigations on the structural setting of a landslide-prone slope by means of three-dimensional electrical resistivity tomography, Nat. Hazards, **78**(2), 1369–1385, (2015)

17 J. Luhn, M. J. Stumvoll-Schmaltz, A. F. Orozco, and T. Glade, Internal structure of an active landslide based on ERT and DP data: New insights from the Hofermühle landslide observatory in Lower Austria, Geomorphology, **441** (2023)

18 M. T. Riaz, M. Basharat, K. S. Ahmed, Y. Sirfraz, A. Shahzad, and N. A. Shah, Failure mechanism of a massive fault–controlled rainfall–triggered landslide in northern Pakistan, Landslides, **21**(11), 2741–2767, (2024)

# Assessing terrain shifts and infrastructure impact: a dual approach using satellite imagery and UAV data in the Cibedug, Rongga landslide

*Moch* Hilmi Zaenal Putra[1*], *Adrin* Tohari[1], *Achmad* Fakrus Shomim[1], *Eko* Soebowo[1], *Yayat* Sudrajat[1], *Amar*[1], *Wahyudin*[1], and *Dwi* Sarah[1]

[1]Geological Disaster Research Center, National Research and Innovation Agency, Bandung, Indonesia

**Abstract.** This study investigates the Rongga landslide, located on the southwest-facing slope of the southern mountain range in West Java, Indonesia, which is particularly prone to landslides due to its geological composition and topographical features. The research integrates high-resolution satellite imagery and UAV data to analyse the vertical, horizontal, and rotational displacements caused by the landslide. Vertical displacement was observed at the main crown, right flank, and left flank, with variations ranging from 1.5 to 4 meters. Horizontal movement was quantified across different zones of the landslide, with the upper zone experiencing the greatest displacement of 6.5 meters, and the lower zone showing the least at 3 meters. Rotational movements were significant in the upper and lower zones but were absent in the middle zone. This comprehensive analysis provides valuable insights into the dynamics of the Rongga landslide and lays the groundwork for future mitigation strategies.

## 1 Introduction

Landslides on volcanic deposits in mountainous regions provide significant hazards especially to surrounding settlements and infrastructure [1]. Volcanic deposits, composed of loose materials such as ash, pyroclastic rocks, and lahar, indicate low stability, particularly when influenced by intense and prolonged rainfall or seismic activity [2]. This instability can cause the material to move rapidly down slopes, resulting in large-scale landslides, leading to extensive landslides that could wipe out houses, roadways, and agricultural land along their track [3]. Moreover, the unconsolidated characteristics of volcanic deposits promote erosion, further exacerbating landslide risks. The impacts of such landslides surpass mere material destruction, potentially resulting in fatalities, disruption of transportation networks, and extensive environmental effects, including river erosion and flash floods in downstream regions [4]. Consequently, research and mitigation efforts are essential for minimising the hazards of landslides in volcanic areas and mitigating their impacts on local residents and ecosystems [5].

The analysis of land movement prior to and following landslides is essential for comprehending the dynamics of terrain shifts and their effects on infrastructure. Recent research has emphasised the efficacy of integrating satellite imagery and UAV data to monitor and analyse geohazards, with a particular emphasis on the displacement of structures as a critical indicator. The utilisation of high-resolution satellite images to identify precursors to landslides was demonstrated by Casagli et al. (2017), which enabled the early detection of potential dangers to structures [6]. Similarly, Niethammer et al. (2012) demonstrated the accuracy of unmanned aerial vehicles (UAVs) in the capture of detailed post-event terrain changes, which provided essential data on the displacement of structures and other infrastructure [7]. These studies emphasise the significance of utilising pre- and post-landslide imagery to accurately evaluate land movement and its effects on structures, thereby improving our capacity to anticipate future landslides and devise effective mitigation strategies. The comprehensive approach to detecting and analysing the displacement of buildings in landslide-prone areas is provided by the combination of detailed, high-resolution data from UAVs and a wide coverage of satellite images.

The combination of these datasets is highly effective for a comprehensive analysis of land movement. Pre-landslide satellite imagery provides extensive coverage that is appropriate for detecting large-scale land movements and trends that preceded the event, whereas post-landslide UAV imagery offers precise, high-resolution, and detailed measurements of the affected area [8]. The study's accuracy and depth are enhanced by the immediate, high-detail response provided by UAV imagery and the historical comparison facilitated by satellite imagery. This integrated approach enhances the accuracy of models that predict future landslides and assists in the development of effective mitigation strategies by combining the precise, timely data from

* Corresponding author: mochhilmizp@gmail.com

UAVs with the broad coverage and historical context of satellite images [9, 10].

## 2 Study area

The Rongga landslide (Fig. 1), which is located on the southwest-facing slope of the southern mountain range in West Java, was the primary focus of the investigation. In the lower section of the landslide, it is situated adjacent to the sinuous Bank River of Cidadap, a river that is approximately 30 kilometres in length and originates from Mount Tambaga Ruyung/Tambak Ruyung. The river passes through eight villages in West Bandung Regency (KBB): Mekarwangi, Gununghalu, Sirnajaya, Bunijaya, Sindangjaya, Cicadas, Cibedug, and Bojong. These villages are situated in three districts: Sindangkerta, Gununghalu, and Rongga. The landslide has an average slope of 12° and covers elevations ranging from 1020 meters at the crown to 980 meters at the toe. The landslide is primarily located in the Tmc formation, which is composed of interbedded claystone, siltstone, and sandstone, as well as localised lahar deposits of tuff, andesite breccia, and tuff breccia [11]. The Tmc formation's geological characteristics, structural complexities, and mountainous topography put this region particularly susceptible to landslides.

## 3 Methodology

### 3.1 Data collection

This study utilizes two primary sources of data to assess land movement and building displacement before and after landslide events: high-resolution satellite imagery from Google Earth Pro and UAV (Unmanned Aerial Vehicle) data. The integration of these datasets allows for a comprehensive analysis of terrain shifts and their impact on infrastructure, particularly focusing on the displacement of buildings and infrastructures.

(a) (b)

**Fig. 1.** Pre-processing stage. Two images (a. after, b. before) are aligned using ground control point (GCP) data (green points) based on identifiable and stable features such as buildings and landscape elements unaffected by the landslide.

#### *3.1.1 Pre-landslide data: google earth pro*

Pre-landslide data was acquired from Maxar Technologies via Google Earth Pro in June 2018. The satellite imagery provided through this platform offers a resolution ranging from 0.3 to 0.5 meters per pixel, which is sufficient for identifying broad patterns of land movement over large areas. The high-resolution images also serve as a historical baseline for comparing changes over time. The multispectral capabilities of this imagery enable analysis across different spectral bands, which is useful for identifying vegetation stress and soil types. The georeferencing of these images ensures precise mapping and alignment with other datasets, facilitating accurate spatial analysis.

#### *3.1.2 Post-landslide data: UAV imagery*

Post-landslide data was collected using the DJI Matrice 350 RTK (M350) UAV. This UAV is capable of achieving ground sampling distances as fine as 1-2 centimeters per pixel, offering a high level of detail necessary for assessing small-scale displacements. The RTK technology and the D-RTK 2 GNSS mobile station provide centimeter-level accuracy in positioning, ensuring precise measurements of building movements. The UAV captured high-resolution images from various angles and altitudes, providing detailed and customizable coverage of the affected area. Additionally, the UAV's multispectral imaging capabilities allow for detailed post-event vegetation and soil analysis. Rapid deployment of the UAV enabled timely data collection immediately after the landslide event, ensuring that the most critical changes were documented. This post-landslide data was collected on March 20 2024, 16 days after the landslide incident on March 4 2024 as can be seen in Table 1.

**Table 1.** The detail of data

| | Pre-Landslide Data | Post-Landslide Data |
|---|---|---|
| **Type** | Google Earth Pro Historical Imaginary | UAV Imagery |
| **Tools / Source** | Maxar Technologies via Google Earth Pro | DJI Matrice 350 |
| **Date Acquisition** | June, 2018 | March 20, 2024 |
| **Resolution** | 0.3 – 0.5 m | 1 - 2 cm |

### 3.2 Data processing and analysis

#### *3.2.1 Pre-processing*

The pre-landslide satellite images and post-landslide UAV images were georeferenced to a common coordinate system to ensure accurate spatial comparison. This involved aligning the images based on known ground control points (GCPs).

Distinctive features in stationary regions, such rooftop corners, road crossings, and rice field boundaries, were utilised as ground control points (GCPs) between the pre- and post-landslide images. The fixed points serve as dependable reference sites that facilitate the precise alignment of the two image sets. The light green points represent the dispersion of the specified Ground Control Points (GCPs) for this georeferencing operation. The distribution of these GCPs throughout the research region markedly improved the spatial accuracy of the comparison between the satellite and UAV images. This method facilitated accurate identification of changes in the landscape, encompassing the magnitude and shift induced by the landslide.

#### *3.2.2 Displacement analysis*

The displacement analysis involved a two-step process to detect and quantify changes in building positions due to landslides. Firstly, change detection was performed using image differencing techniques. By comparing pre- and post-landslide images, changes in the terrain were identified, with particular focus on areas where buildings had significantly shifted. This technique allowed for the detection of even subtle displacements, providing an initial indication of affected structures.

Secondly, a quantitative assessment was conducted to measure the exact displacement of buildings. This involved identifying key structural points on buildings in both the pre- and post-landslide images and calculating the shifts in their positions. Horizontal displacement was measured to determine the extent of movement. By combining these methods, the analysis provided a comprehensive understanding of how buildings were impacted by the landslide, highlighting areas of concern and aiding in the development of mitigation strategies.

#### *3.2.3 Integration and validation*

The findings from the satellite and UAV data were integrated to provide a comprehensive view of the land movement. The satellite images offered broad coverage and historical context, while the UAV data provided detailed, high-resolution measurements. The integration of these datasets enabled a more accurate assessment of building displacement.

To validate the results, field surveys were conducted where possible to compare the measured displacements with on-ground observations. This validation step ensured the reliability and accuracy of the remote sensing data and the derived measurements.

By leveraging the strengths of both satellite and UAV imagery, this methodology provides a robust framework for assessing the displacement of buildings and other infrastructure following a landslide, enhancing our understanding of the dynamics of such geohazards, and aiding in the development of effective mitigation strategies.

## 4 Result and discussion

### 4.1 Vertical displacement

The landslide caused significant ground subsidence across various sections of the affected area. As shown in Fig. 2a, the main crown subsided by approximately 4 meters. The right flank, depicted in Fig. 2b, experienced a drop of 3.5 meters, while the left flank, shown in Fig. 2c, subsided by 2 meters. Fig. 2a was taken on May 22, 2024, whereas Fig. 2b and 2c were captured 10 days after the incident, on March 14, 2024. The locations where these photos were taken are indicated in Fig. 1a.

### 4.2 Horizontal displacement

In addition to vertical movement, the landslide also experienced horizontal displacement. We will divide the discussion of this horizontal movement into three sections: upper, middle, and lower. In the upper section, around Babakan Talang 01 Elementary School, field observations indicated visible horizontal displacement. However, this movement could not be quantified because the area became a crushing zone, obscuring markers that could indicate the extent of horizontal displacement. Based on displacement analysis between pre- and post-event imagery, a horizontal movement toward the south of approximately 6.5 meters was observed (Fig. 3). This measurement was derived from the displacement of unique features, specifically the edge of the roof of Babakan Talang 01 Elementary School and the side of the road on the right flank of the landslide.

In the lower section, around the blue house that experienced rotation, severe damage was evident, with visible signs of movement. Displacement analysis between pre- and post-event imagery indicated a horizontal shift to the south by approximately 5 meters. This displacement differed significantly from the previous measurement, with a difference of 1.5 meters. However, close to the river bank in the lower section, the displacement increased again to 5 meters.

In the middle zone, encompassing the road and residential areas, damage was minimal. Unlike the upper zone (around Babakan Talang 01 Elementary School) and the lower zone (the tilted blue house), the middle area showed only minor cracks on house walls, and the ground appeared largely unaffected, with no visible cracks. Field observations suggested no significant movement in this middle zone. However, displacement analysis between pre- and post-event imagery revealed a horizontal shift to the south of approximately 5 meters. This displacement was 1 meter less than in the upper zone but 2 meters greater than in the lower zone.

Overall, vertical displacement is evident from the outcrops observed, such as at the main crown, the right flank, and the left flank of the landslide, with vertical changes ranging from 1.5 to 4 meters. Horizontal displacement can be quantified through image analysis before and after the landslide. All sections of the landslide, including the upper, middle, and lower zones, experienced horizontal movement, with variations ranging from 3 to 6.5 meters. The greatest horizontal displacement occurred in the upper zone, at the Babakan Talang 01 Elementary School building, while the smallest was observed in the lower zone, at the blue house, particularly at its scarp, which underwent significant rotation. Rotational movement was absent in the middle zone, whereas rotation was clearly observed at the main scarp in the upper zone and the minor scarp in the lower zone. The collection of data on horizontal, vertical, and rotational movements, with varying displacements in each section, will help in understanding the characteristics of the Rongga landslide and support further analysis.

**Fig. 2.** The locations where these photos (a-g) were taken are indicated in Fig. 1a. **a.** The main crown forms a clear horseshoe-shaped scarp, with a significant elevation difference of up to 4 meters. Dark and light brown soil material is exposed along the scarp. **b.** The outcrop on the right flank of the landslide, which severed access to the main road and destroyed nearby buildings, shows a subsidence of 3.5 meters. **c.** The outcrop on the left side of the landslide, which also cut off access to the main road, shows a subsidence of 1.5 meters. **d.** The change in elevation at SDN Babakan Talang 01, where buildings within the landslide zone experienced a subsidence from their original level. **e.** The lower zone showing significant rotation of up to 30 degrees, supported by evidence such as tilted utility poles, footpaths, and buildings. **f.** The middle zone, which shows no signs of rotational movement, with minimal landslide features, such as the absence of cracks in buildings and main roads, and trees remaining upright. **g.** The difference in rotational forces between the middle and lower zones, where buildings and trees in the middle zone remain upright, while in the lower zone they are tilted.

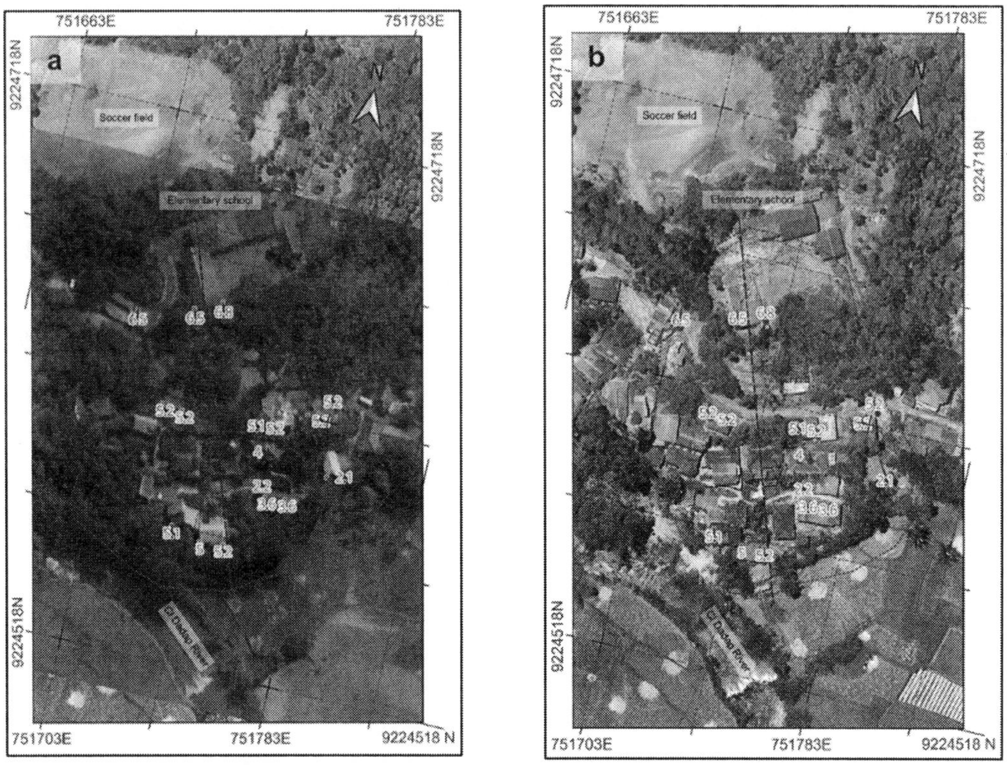


**Fig. 3.**: Map showing the overall movement of buildings and roads affected by the landslide. green and red points indicate the positions a) before and b) after. The landslide event occurred on march 4, 2024.

## 5 Conclusions

The Rongga landslide presented a multifaceted pattern of displacement, marked by significant vertical, horizontal, and rotational movements across various zones. Vertical displacement ranged from 1.5 to 4 meters, with the most pronounced changes occurring at the main crown, which highlights the differential movement within the landslide area. The horizontal displacement was also notable, with the upper zone experiencing the largest shift of 6.5 meters, while the lower zone exhibited a smaller movement of 3 meters. The absence of rotational movement in the middle zone, compared to the distinct rotational shifts in both the upper and lower zones, underscores the varying impact of the landslide across different sections. These findings emphasize the complex nature of the landslide and the need for a tailored approach to mitigation efforts in similar geological settings.

One of the strengths of this study lies in the integration of high-resolution satellite imagery and UAV data, which provided a comprehensive view of the landslide's dynamics. The combination of these two data sources allowed for a detailed analysis of both large-scale trends and small-scale changes, offering a more nuanced understanding of the landslide's impact on the terrain and infrastructure. However, the study also had some limitations. The precision of displacement measurements, particularly in the upper and lower zones, was affected by the resolution of the satellite imagery and the potential for UAV data distortion due to environmental conditions during data capture. Additionally, while the study provided valuable insights into the immediate aftermath of the landslide, it did not account for potential long-term changes in the landscape, which could be crucial for ongoing monitoring and mitigation efforts.

To build upon the results of this study, future research could benefit significantly from integrating these findings with geotechnical and geophysical analyses. Incorporating geotechnical studies, such as soil sampling and laboratory analysis, would provide a deeper understanding of the material properties that contributed to the landslide's behaviour. Geophysical methods, such as seismic refraction or electrical resistivity tomography (ERT), could complement the UAV and satellite data by offering subsurface insights that are not visible in surface imagery alone. Additionally, integrating these results with advanced numerical modelling and simulation could help predict the future behaviour of the landslide under various environmental conditions, such as intense rainfall or seismic activity.

The combination of geological, geophysical, and laboratory analyses with advanced modelling techniques would offer a more comprehensive understanding of the landslide mechanisms at play. By creating models that integrate both surface and subsurface data, researchers could develop more accurate predictions and risk assessments for future landslides in similar regions. This multi-disciplinary approach would not only deepen the understanding of landslide dynamics but also inform the design of more effective mitigation strategies, potentially reducing the impact of such events on vulnerable communities and infrastructure.

Future studies should also address the limitations identified in this research by incorporating longer-term monitoring to capture the evolution of the landslide over time. This could involve repeated UAV surveys at different intervals to observe ongoing changes in displacement and refine the accuracy of horizontal and vertical movement measurements. Additionally, integrating other remote sensing technologies, such as LiDAR or InSAR, could enhance data precision, particularly in areas where UAV or satellite imagery might be less effective due to environmental factors. Expanding the study to include a broader range of environmental and geological factors, such as rainfall patterns and seismic activity, could further enhance the understanding of landslide triggers and progression. Such approaches would not only improve predictive models for future landslides but also contribute to the developments of more robust and effective mitigation strategies.

This research is funded by Rumah Program Kebencanaan Hidrometeorologi dan Iklim - Organisasi Riset Kebumian dan Maritim - Badan Riset dan Inovasi Nasional (BRIN), Indonesia for 2024. The authors would like to thank for supporting them for conducting this work.

## References

1. M. H. Z. Putra, R. D. Kartiko, P. Soemantidiredja, I. A. Sadisun, and A. Tohari, Pengaruh Zona Jenuh Air Terhadap Kestabilan Lereng Di Weninggalih, Kabupaten Bandung Barat, Ris. Geol. dan Pertamb., **30**(1), 119, 2020.
2. F. Fusco et al., Physically Based Estimation of Rainfall Thresholds Triggering Shallow Landslides in Volcanic Slopes of Southern Italy, Water, **11**(9), (2019).
3. K. Mertens et al., The direct impact of landslides on household income in tropical regions: A case study from the Rwenzori Mountains in Uganda, Sci. Total Environ., **550**, 1032–1043, (2016).
4. A. K. Turner, Social and environmental impacts of landslides, Innov. Infrastruct. Solut., **3**(1), 1–25, (2018).
5. X. Fan et al., Earthquake-Induced Chains of Geologic Hazards: Patterns, Mechanisms, and Impacts, Rev. Geophys., **57**(2), 421–503, (2019).
6. N. Casagli et al., Landslide mapping and monitoring by using radar and optical remote sensing: Examples from the EC-FP7 project SAFER, Remote Sens. Appl. Soc. Environ., **4**, 92–108, (2016).
7. U. Niethammer, M. R. James, S. Rothmund, J. Travelletti, and M. Joswig, UAV-based remote sensing of the Super-Sauze landslide: Evaluation and results, Eng. Geol., **128**, 2–11, (2012).
8. S. Chen, C. Xiang, Q. Kang, W. Zhong, Y. Zhou, and K. Liu, Accurate landslide detection leveraging UAVbased aerial remote sensing, IET Commun., **14**(15), 2434–2441, (2020).
9. I. Farmakis, E. Karantanellis, D. J. Hutchinson, N. Vlachopoulos, and V. Marinos, Superpixel and Supervoxel Segmentation Assessment of Landslides Using UAV-Derived Models, Remote Sens., **14**(22), (2022).
10. K. Yang, W. Li, X. Yang, and L. Zhang, Improving Landslide Recognition on UAV Data through Transfer Learning, Appl. Sci., **12**(19), (2022).
11. M. Koesmono, Kusnama, and N. Suwarna, Geological Map of The Sindangbarang and Bandarwaru Quadrangles, Jawa. 2nd Edition. Geological Research and Development Centre, 1996.

# Reducing workplace accidents in a cement company by assessment of physical and mental workload: a case study

*Fandy* Triawan[1], *Eri* Wirdianto[2], *Prima* Fithri[2*], and *Muhammad* Rizky[2]

[1]Regional Research and Development Agency of West Sumatra, Padang, West Sumatra, Indonesia
[2]Industrial Engineering Department, Faculty of Engineering, Universitas Andalas, Padang, West Sumatra, Indonesia

**Abstract.** As one of the leading cement companies in Indonesia, PT Semen Padang is known for its extensive distribution network. As the largest cement factory, the employees at this company engage in work that requires a high level of physical and mental effort. Although the tasks performed by the employees may be similar, the experience of psychological workload can vary greatly among them. The TPM (Total Productive Maintenance) and AFR (Alternative Fuel and Raw Material) departments in the company have several routine activities carried out daily. The numerous job demands that require physical exertion within limited time constraints can lead to a high workload level for employees in TPM and AFR. This workload is not limited to physical aspects but also includes psychological impacts that can disrupt employees's concentration and performance. In this study, questionnaires were given to 7 employees from TPM and 7 employees from AFR. Based on the research conducted on physical workload using the CVL method, AFR employees experienced more physical workload compared to TPM employees. In terms of mental workload, using the NASA-TLX method, both TPM and AFR employees had 2 employees each with very high mental workload.

## 1 Introduction

In the era of globalization and intense company competition, a company must pay attention to and consider employee productivity, as operational efficiency is key to a company's success. Therefore, companies must give special attention to employee workload and productivity as a foundation for sustainable development and growth. One type of company that must consider this is a manufacturing company with a large human resource base. Cement Company, a leading manufacturer in the cement production sector, must pay attention to employee performance as it is a large company that requires both physical and mental strength, especially for employees who work both in the office and in the field. These employees must manage their energy to ensure smooth operations. As Cement Company is recognized as the largest producer of cement and clinker in Indonesia and Asia, the performance of employees working in both office and field settings must be carefully monitored to avoid workload imbalances and job discrepancies. One of the units that requires such attention is the TPM (Total Productive Maintenance) Unit, which is responsible for maintenance and innovation, and AFR (Alternative Fuel and Raw Material), which focuses on using more sustainable and environmentally friendly alternative fuels and raw materials.

Based on the above statement, several studies have demonstrated that workload can affect worker performance. According to data from the International Labour Organization, in 2013, there were 200 million workers who experienced work-related accidents due to fatigue. The study explained that out of 58,115 samples, 32.8% experienced work accidents due to fatigue [1]. Another study proved that in a hospital, 394 pharmacists every month from 2012 to 2018 in two major service centers in Taiwan, as concluded by Shao et al., reducing workload correctly can assist pharmacist performance.

Based on the above studies, the same consideration should be given to employees in the TPM and AFR units of Cement Company to ensure smooth operations. The TPM unit requires a cohesive team performance to ensure the maintenance and upkeep of equipment run safely and smoothly, aiming to increase the company's productivity and quality. Similarly, the AFR unit plays a crucial role in producing raw materials for coal substitution at the Indarung plant, requiring employees with good performance in their jobs. The numerous job demands require sufficient energy to complete tasks; therefore, actions are needed to prevent employees from becoming stressed, easily tired, frustrated, and experiencing other physical or mental issues. Therefore, to understand the physical and mental workload experienced by employees of Total Productive Maintenance and Alternative Fuel and Raw Material at Cement Company, the author conducted research and observations on a case study titled "Analysis of Physical and Mental Workload of TPM (Total Productive Maintenance) and AFR (Alternative Fuel and Raw Material) Employees at Cement Company Using the CVL (Cardiovascular Load) Method.

* Corresponding author: primafithri@eng.unand.ac.id

 https://doi.org/10.1051/e3sconf/202560412004

## 2 Method

### 2.1 Types of research

This research employs a mixed-methods approach, incorporating both quantitative and qualitative research types. Mixed-methods research is a process that uses data in the form of numbers and results from observations or interviews as tools to analyze the research being conducted. The objective of this study is to assess and understand the level of physical and mental workload experienced by TPM (Total Productive Maintenance) and AFR (Alternative Ruel and Raw Material) at PT Semen Padang.

### 2.2 Data collection

The data collected by the researcher consists of two types: primary data, which comes from direct observations, and secondary data, which is company data. The data analysis methods used include Pareto analysis, fishbone diagram analysis, and improvement proposal analysis.

## 3 Results and discussion

### 3.1 CVL processing and analysis

Data processing carried out in the CVL method is based on the data that has been taken. The data that has been taken will be processed until % CVL is obtained. Before obtaining % CVL, the researcher must first determine the pulse rate when working and resting, age and maximum working pulse rate of the respondents being studied. Pulse measurement is done by feeling the pulse of the radial artery on the wrist and measuring using a stopwatch with the 10-beat method. The provisions in taking the questionnaire are that employees A-G are TPM employees while employees H-N are AFR employees. The following are the results of the calculation of DNK (Working Pulse Rate) and DNI (Resting Pulse Rate), age, and DNK Maxs in Table 1.

**Table 1.** Calculation results of DNK, DNI, and DNK max employees

| No | Respondents | Gender | Resting Heart Rate/min | Working Heart Rate/min | Maxs Working Pulse | Age |
|---|---|---|---|---|---|---|
| 1 | Employee A | Male | 65.22 | 94.94 | 172 | 48th |
| 2 | Employee B | Male | 59.06 | 72.73 | 171 | 49th |
| 3 | Employee C | Male | 87.21 | 125.00 | 171 | 49th |
| 4 | Employee D | Male | 74.17 | 93.75 | 171 | 49th |
| 5 | Employee E | Female | 68.65 | 91.19 | 165 | 35th |
| 6 | Employee F | Male | 57.97 | 81.08 | 174 | 46th |
| 7 | Employee G | Male | 46.62 | 85.96 | 167 | 53th |
| 8 | Employee H | Male | 59.70 | 101.69 | 179 | 41th |
| 9 | Employee I | Male | 82.42 | 112.78 | 171 | 49th |
| 10 | Employee J | Female | 49.92 | 60.00 | 151 | 39th |
| 11 | Employee K | Male | 90.36 | 93.90 | 182 | 38th |
| 12 | Employee L | Male | 59.46 | 119.52 | 192 | 28th |
| 13 | Employee M | Female | 103.45 | 106.01 | 170 | 30th |
| 14 | Employee N | Male | 78.02 | 96.77 | 197 | 23th |

Based on Table 1, we can give an example of Employee C with a calculated DNI of 87.21 beats/minute and a calculated DNK of 72.73 beats/minute with an age of 49 years and getting a maximum pulse rate of 171. The following is an example of calculating the resting, working, and maximum pulse rates of Employee C.

$$\text{Resting Pulse Rate (pulses/minute)} = \frac{10 \text{ pulse}}{\text{calculation time}} \times 60 \quad (1)$$

$$\text{Resting Pulse Rate} = \frac{10 \text{ pulse}}{6.88} \times 60 = 87.21 \textit{ beats/minutes}$$

$$\text{Working Heart Rate (pulses/minute)} = \frac{10 \text{ pulse}}{\text{calculation time}} \times 60 \quad (2)$$

$$\text{Working Heart Rate} = \frac{10 \text{ pulse}}{4.80} \times 60 = 125 \text{ beats/minute}$$

Next, after obtaining the employee's working and resting heart rate, the employee's maximum heart rate is determined using the formula 220 for men and 200 for women minus the employee's age, the following will be exemplified for Employee C [2].

$$\text{Maximum Heart Rate} = \text{std male } 220 - \text{Age} \quad (3)$$
$$\text{Maximum Heart Rate} = 220\text{-}49 \text{ years} = 171 \text{ beats/minute}$$

Next, after obtaining DNK, DNI, and DNK Max, determine the employee's %CVL. The assessment of the workload level classification cannot be directly determined from the percentage of cardiovascular load (%CVL). The %CVL value is calculated based on the workload classification level based on the increase in the work pulse rate that has been determined as follows [3].

So it can be concluded that the level of fatigue or disturbance felt by employees is based on the %CVL

 https://doi.org/10.1051/e3sconf/202560412004

obtained according to the classification/criteria that have been set in Table 1. Furthermore, the %CVL of each employee is determined by using the calculation example for Employee C based on the following formula.

$$\%CVL\ (\textit{Cardiovasculer Load}) = \frac{DNK\text{-}DNI}{DNK\ Maxs\text{-}DNI} \times 100 \quad (4)$$

$$\%CVL\ (\textit{Cardiovasculer Load}) = \frac{125.00\text{-}87.21}{171\text{-}87.21} \times 100$$

$$\%CVL\ (\textit{Cardiovasculer Load}) = 45.10\%$$

The calculation of the pulse rate that has been obtained in the data processing will describe the %CVL which will show whether the employee is experiencing fatigue at work or not, with the provision that the employee needs further action for the work he is experiencing or not. The following is a description of the %CVL of TPM (Total Productive Maintenance) and AFR (Alternative Ruel and Raw Material) Employees of PT Semen Padang (Fig. 1).

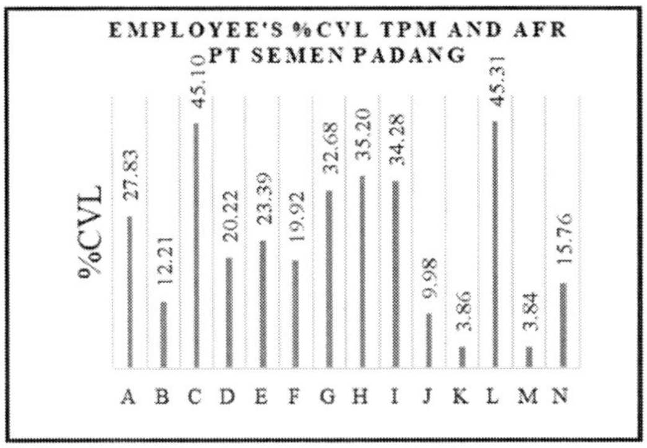


**Fig. 1.** %CVL of TPM and AFR employees of PT Semen Padang (source: PT Semen Padang)

Based on Fig. 1, we can conclude that in this study when all data has been collected, then we can determine the %CVL of the operators who work. The results of the % of employees who are classified as safe or do not experience fatigue at work are Employee A, Employee B, Employee D, Employee E, Employee F, Employee J, Employee K, Employee M, Employee N. The 9 employees did not experience fatigue at work due to the %CVL produced by less than 30%. While for Employee C, %CVL reached 45.10%, Employee G reached 32.68%, Employee H reached 35.20%, Employee I reached 34.28%, and the last Employee L with the highest percentage reached 45.31%. The value or % indicates that the operator needs improvement and must be followed up, for a more specific % can be seen in the data processing in Table 1.

Based on the research conducted by the researcher, there are 5 employees or operators who need further improvement for smooth work, consisting of Employees C and G who come from TPM, and Employees H, I, and L come from AFR. Seen from the high physiological load reaching 45.31%, this certainly occurs due to the age and very limited working time of the operator, which causes physical conditions to decline. The difference in the results of the physical workload of employees is due to age, physical factors, and uneven division of labor. So, it is necessary to add employees so that the gap and physical condition of employees do not decrease. This has also been revealed by Silvia, where there are 3 workers in the range of 30% - 60% and 3 other workers in the range of 60% - 80%, Silvia said that age factors and uneven work are factors that cause operator fatigue to occur [3].

The workload felt by the operators can be known to have a value that is not small. Therefore, improvements need to be made to reduce the workload. Before the work improvements are made, an analysis of the causes of the operator's workload is carried out using the Fishbone Diagram. The following (Fig. 2) are the results of the analysis.

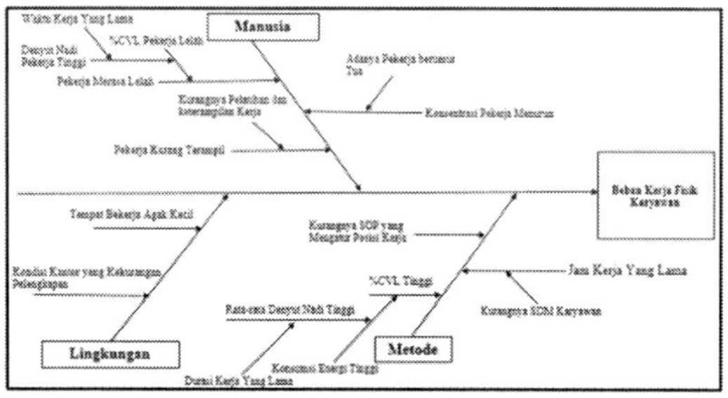


**Fig. 2**. Fishbone diagram of physical workload

Figure 2 above shows the results of the analysis of the causes of high operator workloads seen from several factors, namely humans, environment, machines, and methods. The first factor is the human factor which has 3 causes, namely long working hours which make workers' pulse rates high and workers' CVL% high which causes workers to feel tired, lack of job training which results in workers being less skilled, and there are older workers which causes concentration to decrease. The second factor is the method which has 3 causes, namely there is no SOP that regulates work positions, too long work duration makes the average pulse rate high and high energy consumption which results in high CVL%. From all the causes, the results of the calculation of Employee C's CVL% are 45.10% and Employee L's CVL% 45.31% which really need additional rest time.

### 3.2 NASA-TLX processing and analysis

Next is to do mental processing based on the NASA-TLX method. The data taken is obtained based on a questionnaire distributed to employees. This questionnaire consists of 2 parts, namely the weighting and rating sections for each indicator. The weighting attached to the questionnaire is in pairs of 15 pairs. The ratings presented are given a range between 0 and 100 which are guided by 6 indicators, including Mental Demand (MD), Physical Demand (PD), Temporal Demand (TD), Own Performance (OP), Frustration (FR), Effort (EF).

The calculation is done to obtain the WLL (Weighted Workload) value. Based on the WWL value obtained, we will get a conclusion of the mental workload experienced by TPM and AFR employees of PT Semen Padang. First, we must add up the weighting of 14 respondents, then the results obtained will be summarized for each indicator. The data obtained from the results of the indicator comparison can be seen in Table 2.

**Table 2.** Comparison of TPM and AFR employee indicators

| INDICATOR | RESPONDENT | | | | | | | | | | | | | |
|---|---|---|---|---|---|---|---|---|---|---|---|---|---|---|
| | EMPLOYEE TPM | | | | | | | EMPLOYEE AFR | | | | | | |
| | A | B | C | D | E | F | G | H | I | J | K | L | M | N |
| MD (Mental Demand) | 4 | 1 | 3 | 1 | 3 | 1 | 1 | 2 | 1 | 2 | 5 | 3 | 4 | 2 |
| PD (Physical Demand) | 0 | 3 | 4 | 2 | 5 | 2 | 3 | 4 | 3 | 2 | 2 | 4 | 1 | 2 |
| TD (Temporal Demand) | 4 | 2 | 3 | 4 | 2 | 3 | 4 | 2 | 1 | 3 | 3 | 1 | 3 | 1 |
| OP (Own Performance) | 3 | 3 | 2 | 3 | 3 | 1 | 4 | 3 | 2 | 5 | 3 | 3 | 4 | 3 |
| EF (Effort) | 3 | 3 | 3 | 5 | 2 | 5 | 3 | 3 | 3 | 3 | 2 | 4 | 3 | 4 |
| FR (Frustration) | 1 | 3 | 0 | 0 | 0 | 3 | 0 | 1 | 5 | 0 | 0 | 0 | 0 | 3 |
| **TOTAL** | **15** | **15** | **15** | **15** | **15** | **15** | **15** | **15** | **15** | **15** | **15** | **15** | **15** | **15** |

Based on Table 2, we can see that the indicator that is slightly chosen by each TPM and AFR employee of PT Semen Padang is FR (Frustration). While the other indicators are almost completely filled. After making a comparison, employees will fill in the rating of the 6 indicators with a value range from 0 to 100. Employees will fill in according to the conditions when they work or the actual situation. The results of the calculation can be seen in Table 3 as follows.

**Table 3.** Results of TPM and AFR employee ratings

| INDICATOR | RESPONDENT | | | | | | | | | | | | | |
|---|---|---|---|---|---|---|---|---|---|---|---|---|---|---|
| | EMPLOYEE TPM | | | | | | | EMPLOYEE AFR | | | | | | |
| | A | B | C | D | E | F | G | H | I | J | K | L | M | N |
| MD (Mental Demand) | 70 | 80 | 85 | 15 | 45 | 40 | 90 | 40 | 45 | 75 | 80 | 70 | 70 | 60 |
| PD (Physical Demand) | 40 | 80 | 20 | 15 | 45 | 40 | 90 | 40 | 55 | 55 | 80 | 30 | 80 | 50 |
| TD (Temporal Demand) | 70 | 70 | 75 | 15 | 40 | 70 | 70 | 35 | 55 | 80 | 75 | 70 | 80 | 60 |
| OP (Own Performance) | 100 | 85 | 85 | 15 | 60 | 80 | 100 | 70 | 75 | 90 | 90 | 70 | 100 | 80 |
| EF (Effort) | 60 | 90 | 85 | 15 | 65 | 60 | 70 | 70 | 35 | 75 | 80 | 50 | 80 | 80 |
| FR (Frustration) | 40 | 85 | 75 | 15 | 40 | 70 | 80 | 30 | 40 | 70 | 80 | 40 | 40 | 70 |
| **TOTAL** | **380** | **490** | **425** | **90** | **295** | **360** | **500** | **285** | **305** | **445** | **485** | **330** | **450** | **400** |

Table 3 shows that the value or in general the total sum of the indicators obtained, namely Employee G, was obtained for the employee amounting to 500. While the one who got the lowest value was Employee D, which was 90. While for TPM employees with the highest rating is Employee G with a rating of 500, while for AFR it was obtained by Employee K with a rating of 485. For the highest rating value in general, it is 100 and the lowest rating is 15. After getting the value, the next step is to calculate the Weight Work Load (WWL). This value is obtained by multiplying the rating that has been obtained by the comparison weight on each indicator. Then add up each indicator value. The following are the results of multiplying the indicator comparison data and the rating value in Table 4.

**Table 4.** Multiplication of indicators and employee WWL

| INDICATOR | RESPONDENT | | | | | | | | | | | | | |
|---|---|---|---|---|---|---|---|---|---|---|---|---|---|---|
| | EMPLOYEE TPM | | | | | | | EMPLOYEE AFR | | | | | | |
| | A | B | C | D | E | F | G | H | I | J | K | L | M | N |
| Mental Demand | 280 | 80 | 255 | 15 | 135 | 40 | 90 | 80 | 45 | 150 | 400 | 210 | 280 | 120 |
| Physical Demand | 0 | 240 | 80 | 30 | 225 | 80 | 270 | 160 | 165 | 110 | 160 | 120 | 80 | 100 |
| Temporal Demand | 280 | 140 | 225 | 60 | 80 | 210 | 280 | 70 | 55 | 240 | 225 | 70 | 240 | 60 |
| Own Performance | 300 | 255 | 170 | 45 | 180 | 80 | 400 | 210 | 150 | 450 | 270 | 210 | 400 | 240 |
| Effort | 180 | 270 | 255 | 75 | 130 | 300 | 210 | 210 | 105 | 225 | 160 | 200 | 240 | 320 |
| Frustration | 40 | 255 | 0 | 0 | 0 | 210 | 0 | 30 | 200 | 0 | 0 | 0 | 0 | 210 |
| **TOTAL** | **1080** | **1240** | **985** | **225** | **750** | **920** | **1250** | **760** | **720** | **1175** | **1215** | **810** | **1240** | **1050** |

Based on Table 4, the results obtained with the highest value are Employee G with a value of 1250 who is a TPM employee, while for AFR employees it is Employee K which is 1215, and the smallest value obtained for TPM employees is Employee D with a value of 225, while the smallest value for AFR employees is 720 for Employee I. These results are obtained by adding up the results of the multiplication of the six indicators. The multiplication of indicators and rating values for Employee C is as follows.

| | | |
|---|---|---|
| Nilai WWL | = ∑Rating x Weight | (5) |
| Mental Demand (MD) | = 3 x 85 = 255 | |
| Physical Demand (PD) | = 4 x 20 = 80 | |
| Temporal Demand (TD) | = 3 x 75 = 225 | |
| Own Performance (OP) | = 2 x 85 = 170 | |
| Effort (EF) | = 3 x 85 = 255 | |
| Frustration (FR) | = 0 x 75 = 0 | |

The value obtained will be continued by calculating the average WWL. This value is obtained by dividing the WWL by the total weight. The following are the results obtained.

**Table 5**. Average WWL

| INDICATOR | RESPONDENT | | | | | | | | | | | | | |
|---|---|---|---|---|---|---|---|---|---|---|---|---|---|---|
| | EMPLOYEE TPM | | | | | | | EMPLOYEE AFR | | | | | | |
| | A | B | C | D | E | F | G | H | I | J | K | L | M | N |
| MD (Mental Demand) | 18.7 | 5.3 | 17.0 | 1.0 | 9.0 | 2.7 | 6.0 | 5.3 | 3.0 | 10.0 | 26.7 | 14.0 | 18.7 | 8.0 |
| PD (Physical Demand) | 0.0 | 16.0 | 5.3 | 2.0 | 15.0 | 5.3 | 18.0 | 10.7 | 11.0 | 7.3 | 10.7 | 8.0 | 5.3 | 6.7 |
| TD (Temporal Demand) | 18.7 | 9.3 | 15.0 | 4.0 | 5.3 | 14.0 | 18.7 | 4.7 | 3.7 | 16.0 | 15.0 | 4.7 | 16.0 | 4.0 |
| OP (Own Performance) | 20.0 | 17.0 | 11.3 | 3.0 | 12.0 | 5.3 | 26.7 | 14.0 | 10.0 | 30.0 | 18.0 | 14.0 | 26.7 | 16.0 |
| EF (Effort) | 12.0 | 18.0 | 17.0 | 5.0 | 8.7 | 20.0 | 14.0 | 14.0 | 7.0 | 15.0 | 10.7 | 13.3 | 16.0 | 21.3 |
| FR (Frustation) | 2.7 | 17.0 | 0.0 | 0.0 | 0.0 | 14.0 | 0.0 | 2.0 | 13.3 | 0.0 | 0.0 | 0.0 | 0.0 | 14.0 |
| **TOTAL** | 72 | 82.7 | 65.7 | 15.0 | 50.0 | 61.3 | 83.3 | 50.7 | 48.0 | 78.3 | 81.0 | 54.0 | 82.7 | 70.0 |

Table 5 explains that the highest WWL value or average WWL value was obtained by Employee G with a value of 83.3 for TPM employees, and Employee M obtained a WWL of 82.7 for AFR Employees. The lowest WWL value for TPM employees was found in Employee D, only 15.00, while for AFR employees it was found in Employee H, which was 50.7. The following is an example of calculating the average WWL value for Employee E.

$$\text{Average WWL} = \frac{\text{Nilai WWL}}{\sum \text{Botot}} = \frac{750}{15} = 50 \quad (6)$$

After getting the WWL value, we will interpret the score obtained by each employee using the score interpretation table. This is as explained by Hart and Staveland [4] in their book explaining the theory and basics of NASA-TLX, along with the interpretation of mental workload by NASA-TLX.

**Table 6.** Workload categories

| Workload Class | Value |
|---|---|
| Low | $0 \leq x \leq 9$ |
| Medium | $10 \leq x \leq 29$ |
| Slightly High | $30 \leq x \leq 49$ |
| High | $50 \leq x \leq 79$ |
| Very High | $80 \leq x \leq 100$ |

Table 6 shows the scores and range of assessments in WWL for NASA-TLX consisting of 5 categories, namely low, medium, rather high, high, and very high. For the range of values, we can see in the table. So from the WWL results obtained, it is stated that the category scores for TPM and AFR Employees of PT Semen Padang are as follows.

**Table 7.** Mental workload assessment categories

| | No | Respondent | Nilai Beban Kerja | Golongan Beban Kerja |
|---|---|---|---|---|
| EMPLOYEE TPM | 1 | Employee A | 72.00 | (K.A) High |
| | 2 | Employee B | 82.67 | (K.B) Very High |
| | 3 | Employee C | 65.67 | (K.C) High |
| | 4 | Employee D | 15.00 | (K.D) Medium |
| | 5 | Employee E | 50.00 | (K.E) High |
| | 6 | Employee F | 61.33 | (K.F) High |
| | 7 | Employee G | 83.33 | (K.G) Very High |
| EMPLOYEE AFR | 8 | Employee H | 50.67 | (K.H) High |
| | 9 | Employee I | 48.00 | (K.I) Slightly High |
| | 10 | Employee J | 78.33 | (K.J) High |
| | 11 | Employee K | 81.00 | (K.K) Very High |
| | 12 | Employee L | 54.00 | (K.L) High |
| | 13 | Employee M | 82.67 | (K.M) Very High |
| | 14 | Employee N | 70.00 | (K.N) High |

Based on Table 7, we can see that for employees who get a very high workload, there are 4 people, namely Employee B, G who are TPM employees and Employee K, and M who are AFR employees. For employees who get a high mental workload, there are 8 people, namely Employees A, C, E, F, they are employees from the TPM unit and Employees H, J, L, and N who are AFR employees. For the rather high one, it is obtained by Employee I who is an AFR employee, while for TPM employees with a rather high workload group, there are none, and the moderate category is only Employee D and that is a TPM employee and for AFR with this group there is none. While the category that gets a low mental workload does not exist.

Based on the information attached in Table 7, we have obtained the mental workload values of 14 TPM and AFR employees of PT Semen Padang. Hart and Staveland [5] in their book entitled Development of NASA-TLX (Task Load Index Results of Empirical and Theoretical Research. In Human Mental Workload), the values obtained in the mental workload category that really need to be considered are workloads with a value range of 80-100. However, in addition to employees with very high performance, also pay attention to employees whose performance is almost in the very high (high) category, and so on. While in this study there were several employees whose recapitulation of mental workload results reached almost 80 and even exceeded 80. To make it clearer, Fig. 3 shows a description of the mental workload classification of TPM and AFR employees of PT Semen Padang.

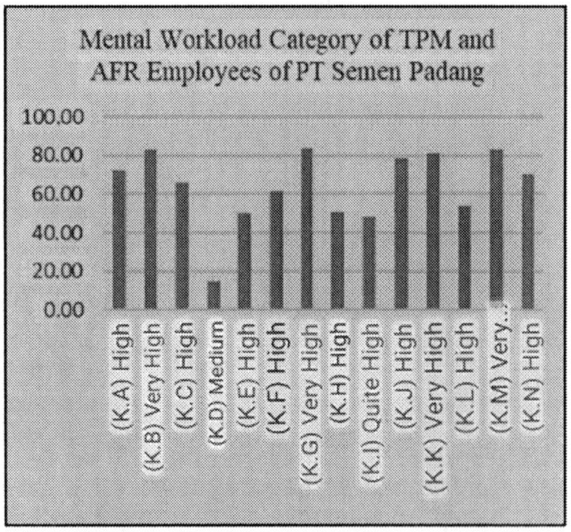


**Fig. 3.** Mental workload category of TPM and AFR employees of PT Semen Padang (source: PT Semen Padang)

Next, AFR Employees, for AFR with a very high category, there are more than TPM, namely Employees K, and M, then I with a fairly high category. Employee K is one of the AFR employees who is tasked with checking waste and office performance, as well as Employee M who is an employee who is more often in the field or in field checks. Then Employee I, with a fairly high category, the constraints are almost similar to Employee G, due to their increasingly advanced age and being given authority as a leadership staff and must be good at dividing between field and office performance, Employee I is also authorized to check the waste that is processed. The rest are employees with a high mental burden category, but high performance also requires further handling so that a company's performance runs smoothly, for employees with this high category have the same constraints where they have to divide between work in the field and work in the office.

A different thing was shown by employee D who is a TPM employee in the Innovation field, the employee is in the moderate mental workload category, this happened because based on the research that has been done, the job is a suitable job for him, so the employee feels that the mental workload he feels is only a little. In order for the indicators that influence the employee's work to be more clearly visible, here is the percentage of each indicator which can be seen in Table 8.

**Table 8.** % Overall indicators

| INDICATOR | Total Score | % |
|---|---|---|
| MD (*Mental Demand*) | 2180 | 16 |
| PD (*Physical Demand*) | 1820 | 14 |
| TD (*Temporal Demand*) | 2235 | 17 |
| OP (*Own Performance*) | 3360 | 25 |
| EF (*Effort*) | 2880 | 21 |
| FR (*Frustation*) | 945 | 7 |
| **TOTAL** | 13420 | 100 |

Table 8 we can see that the indicator with the largest percentage is performance with a percentage of 25%, slightly different from Effort at 21%, then 17% for Temporal Demand, 16% for Mental Demand, 14 for Physical Demand, and mental workload with the smallest percentage, namely Frustration with a percentage of 7%. Furthermore, to make it clearer, here is a comparative description of the percentage of indicators that affect the mental workload of TPM and AFR employees of PT Semen Padang. The following is a Fishbone Diagram of the mental workload that has been obtained from the analysis of the case above which aims to find the problems that occur in the case.

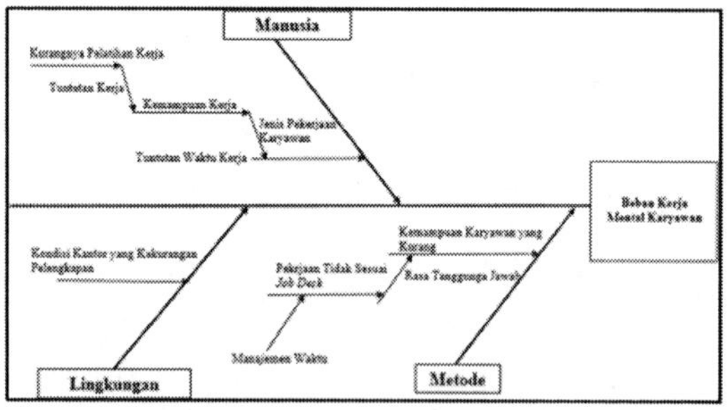


**Fig. 4.** Mental workload fishbone

Based on Fig. 4, the results of the analysis of the causes of high operator workload are seen from several factors, namely humans, environment, machines, and methods. The first factor is the human factor, which has 3 causes, namely long working hours that make employee workloads high, including methods, humans, and the environment. The first factor, namely the method, shows that the ability, sense of responsibility and work that are less in accordance with employees will damage time management. Furthermore, the human factor, with many work demands and reduced training will make the mental workload of employees high. Third, namely the environment where in the office the equipment to carry out office activities is still reduced will decrease employee performance and increase workload. So that the improvement that can be done is to provide training so that employees can work optimally.

## 4 Conclusion

Based on the data processing and analysis that has been done, we can conclude that there are 5 employees or operators who need further improvement for work to run smoothly in physical workload, consisting of Employees C and G from TPM, and Employees H, I, L from AFR. Furthermore, the mental workload using the NASA-TLX method of TPM and AFR employees of PT Semen Padang tends to be high, consisting of 4 TPM employees with a high category, 4 AFR employees with a high category, the rest with a very high category by 2 TPM employees and 2 AFR employees, 1 AFR employee with a rather high category and 1 TPM employee with a moderate category, which is based on the most influential mental workload indicator, namely performance. So it is stated that, AFR and TPM employees with physical workload categories based on % Cardiovascular Load are supervisors and AFR field employees, namely Employee L. Employees with the highest mental workload are the TPM innovation leadership staff, namely Employee B. Based on this conclusion, improvements that can be proposed are by requiring effective time division and adding employees to weigh employees who have exceeded the limit in the job. Then sufficient training is needed so that employee potential development can be developed in their work.

## References

1. F. Permatasari, M. Anis, The relationship analysis between physical and mental workload with work fatigue in extruder section at PT. ABC. USA. IEOM Society International, **16** (2021)
2. Tarwaka, Ergonomi Industri (Harapan Press, Surakarta, 2014)
3. Silvia, M. I. Hamdy, and R. Yusril, Analisa beban kerja mental operator mesin dryer bagian auto clipper dengan metode NASA-TLX. Jurnal Hasil Penelitian dan Karya Ilmiah dalam Bidang Teknik Industri, **4**(2) (2019)
4. S. Hart and L. Staveland, Nasa-task load index (Nasa-TLX); 20 years later. In the Proceedings of the Human Factors and Ergonomics Society Annual Meeting, **50**(9), 904-908 (2006). https://doi.org/10.1177/154193120605000909
5. S. Hart and L. Staveland, Development of NASA-TLX (Task Load Index): Results of Empirical and Theoretical Research. Advances Psychology, **52** (1988). https://doi.org/10.1016/S0166-4115(08)62386-9

# Seismic vulnerability analysis of irregular and regular structures using fragility curve

*Setiati* Rehni[1], *Masrilayanti*[1*], *Kurniawan* Ruddy[1], *Ainy* Asri Luthfi[1], and *Sulleyman* Sourkan[2]

[1]Civil Engineering Department, Faculty of Engineering, Universitas Andalas, Padang, Indonesia
[2]Petroleum and Energy Engineering Department, Sulaimani Polytechnic University, Sulaymaniyah, Iraq

**Abstract.** The demand for innovative and visually appealing building designs according to specific desires or needs has led to many structures being built with irregularities. High-rise buildings with structural irregularities are at a higher risk of collapse during an earthquake if they are not designed and constructed properly according to existing regulations. Some buildings have non-uniform floor heights, which can lead to uneven distribution of stiffness throughout the structure vertically. Excessive height of one floor compared to the floor above it can potentially create an excessive soft story. Evaluating and assessing seismic vulnerability of structures is a major issue in earthquake-resistant design. This study discusses an analytical model for developing fragility curves by evaluating the seismic performance of a structure. The objective of this research is to determine and compare the estimated fragility levels of structures subjected to seismic loads between irregular structures with soft story irregularities and regular structures. Based on the results and discussion, it can be concluded that irregular structures (with soft story irregularities) have higher fragility levels compared to regular structures in both the x and y directions.

## 1 Introduction

Earthquakes are considered one of the most challenging hazards affecting structures, especially high-rise buildings [1]. This study selects Padang City as the site for the research. Padang City is a highly tectonically active region, caused by the interaction between two tectonic plates: the Indo-Australian Plate and the Eurasian Plate. The dynamic interaction between these plates creates a subduction zone or a zone of plate convergence. As a result of this pressure, Padang City and its surroundings are highly susceptible to earthquakes and tsunamis. Earthquakes in this area can be very strong and destructive, such as the 2009 Padang earthquake with a magnitude of 7.6 on the Richter scale, which resulted in thousands of fatalities [1-3].

During one example of a reinforced concrete (RC) building that suffered significant damage, both structural and non-structural, due to the 2009 Padang earthquake is a building with an asymmetrical floor plan [3]. The demand for innovative and visually appealing building designs according to specific desires or needs has led to many structures being built with irregularities [4]. Some buildings have floors with non-uniform heights, leading to an uneven distribution of stiffness vertically throughout the building. Excessive height in one floor compared to the floor above can potentially create a soft story [4-6].

High-rise buildings with structural irregularities are at a higher risk of collapse during an earthquake if not designed and constructed properly according to existing regulations [6]. The performance of a structure can be assessed based on its vulnerability to loads. Performance can be developed in the form of fragility curves [7-9]. From the obtained fragility curves, the damage level due to lateral loads can be identified and categorized into four levels: slight, moderate, extensive, and collapse. This damage classification refers to HAZUS (Hazards United States). The results of the seismic fragility evaluation of the two structures will reveal which type of building demonstrates higher fragility, providing insights for future structural design [10].

## 2 Methodology

### 2.1 Building model and material properties

The study discusses an analytical model for constructing fragility curves by evaluating the seismic performance of a structure. The structural variations to be discussed are irregular structures with excessive soft story irregularities and regular structures. The building models selected are:

- **Model 1 regular structure** - The structure in question is an existing 12-story building with a special moment-resisting frame system (SMRF) that was previously designed by researchers.
- **Model 2 irregular structure** - the regular structure is redesigned to achieve structural irregularity. The plan, building height, and dimensions of the columns, beams, slabs, and reinforcement are the same as those of the regular structure, but the floor height on the first floor is increased to 10.5 meters.

* Corresponding author: masrilayanti@eng.unand.ac.id

The modification and addition of floor height in this irregular structure are based on the phenomenon of structural collapse with soft story irregularities due to earthquakes, which often occurs on the first floor [11, 12]. This is because high-rise buildings, such as office buildings and apartments, are designed to have lobbies with floor heights usually greater than the typical floors above. Architects typically desire this to make the lobby appear larger, more spacious, and grander. Due to the need for more space, the use of brick walls on the lobby floor is relatively less compared to the floors above, which require partition walls between rooms. The layout of the regular and irregular structures is shown in Fig. 1, and the structural variations shown in Fig. 2.

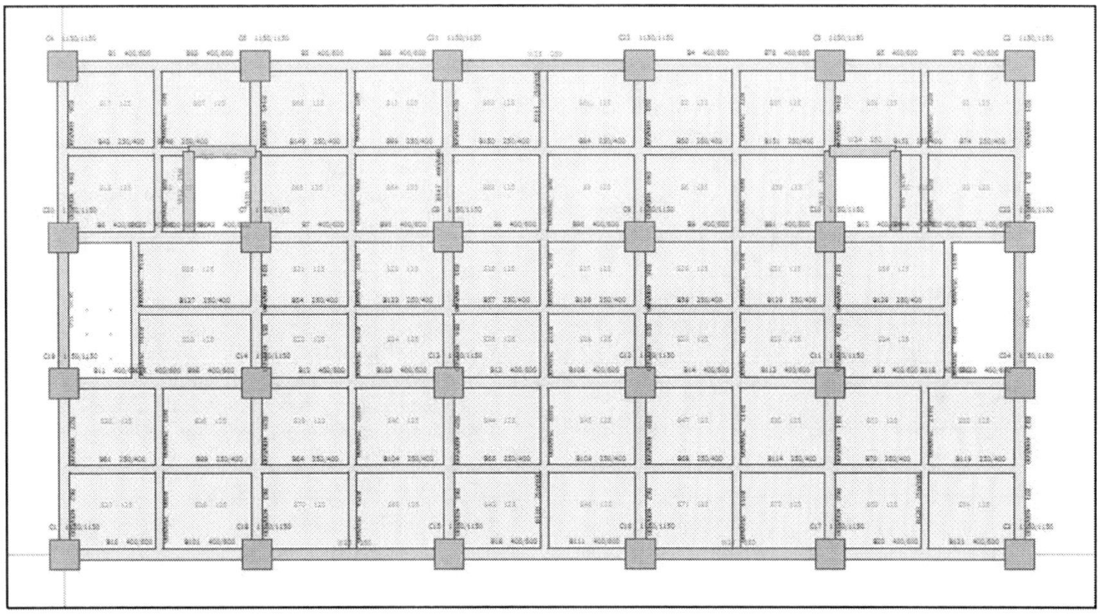

**Fig. 1.** Typical building plans for regular structures and irregular structures.

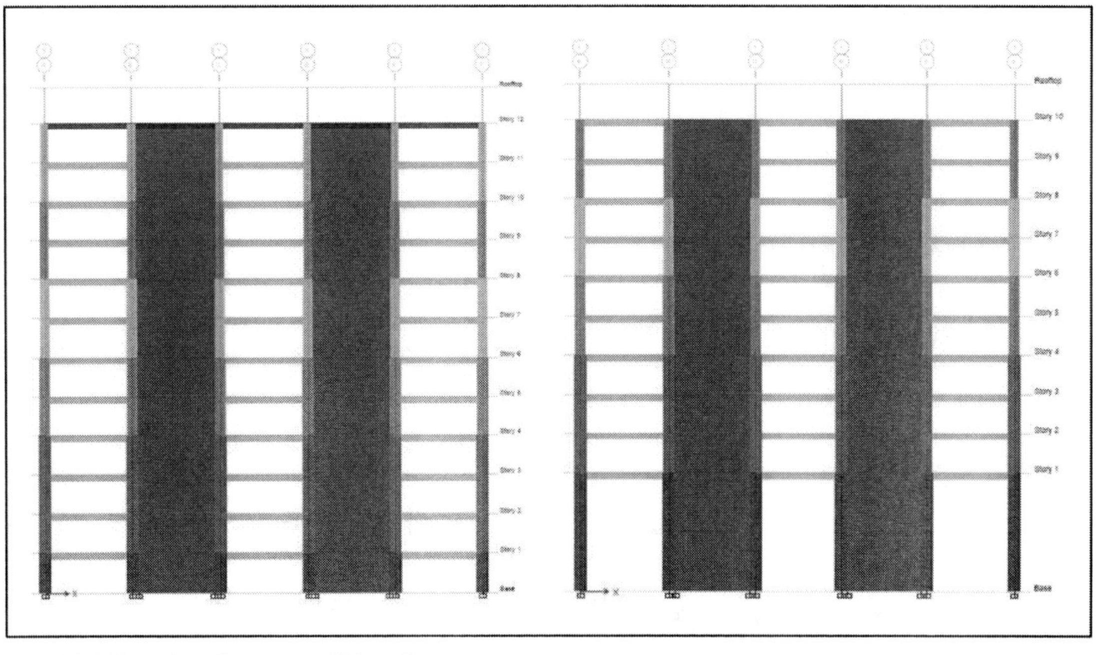

**Fig. 2.** Modelling: (a) regular structure, (b) irregular structure.

Vertical irregularity check of the structure was first conducted based on Table 14 of SNI 1726-2019. The results indicated that the irregular structure exhibited excessive soft story stiffness irregularity on the first floor, as shown in Table 1. The structure is categorized as having excessive soft story stiffness irregularity if there is a story where the lateral stiffness is less than 60% of the lateral stiffness of the story above or less than 70% of the average lateral stiffness of the three stories above. According to Table 14 of SNI 1726-2019, certain consequences must be met if the structure has excessive soft story stiffness irregularity, as referred to in Section 7.3.3.1 and Table 16 of SNI 1726-2019. However, in this study, the structure was intentionally designed to have soft story stiffness irregularity to understand the effects of this irregularity by comparing it with a regular structure without irregularity.

**Table 1.** Results of the soft story stiffness irregularity check for the irregular structure.

| Story | X direction Stiffeness (kN/m) | Check | Y direction Stiffeness (kN/m) | Check |
|---|---|---|---|---|
| Roof | 65029.669 | OK | 52059.072 | OK |
| 10 | 321981.044 | OK | 251263.956 | OK |
| 9 | 499414.525 | OK | 376724.199 | OK |
| 8 | 604849.631 | OK | 443426.425 | OK |
| 7 | 667470.694 | OK | 475926.756 | OK |
| 6 | 716742.444 | OK | 499933.312 | OK |
| 5 | 766919.971 | OK | 524174.189 | OK |
| 4 | 833268.758 | OK | 559392.226 | OK |
| 3 | 926285.714 | OK | 610104.802 | OK |
| 2 | 1144926.091 | OK | 692519.152 | OK |
| 1 | 647464.936 | V.1b | 417323.532 | V.1a |

## 2.2 Pushover analysis

From the results of the static pushover analysis run on SeismoStruct software, a capacity curve (base shear-displacement) is obtained. The resulting pushover curve must then be converted into a bilinear form as shown in Fig. 3. From the pushover curve obtained, the first value taken is the roof displacement at yield (Δy), and the second value is the roof displacement at ultimate (Δu), taken at 80% of the peak condition. The capability of a structure will be directly related to the capacity of the structure to accommodate the needs. In other words, the structure has the capacity to withstand the design earthquake force requirements so that the structure's performance is in accordance with the design objectives. The inelastic (post-elastic) behavior of the structure is represented by the presence of plastic joints formed at the foot of the columns and beams.

## 2.3 Fragility curves

Fragility curve is a lognormal function that correlates the probability of occurrence of a certain level of structural damage due to earthquake intensity. HAZUS (Hazard United States) defines a fragility curve as the relationship between the probability of occurrence of a damage state (Damage State/DS) in a structure and a given earthquake intensity (Intensity Measure/IM). Therefore, fragility curves are highly beneficial for determining the damage state of a structure based on a damage measure. The damage states defined by HAZUS include: slight damage, moderate damage, extensive damage, and complete damage (collapse) [11].

To develop the fragility function, the first step is to define the damage states and to set their threshold limits. A relatively simpler criterion for the definition of damage state thresholds has been proposed by Barbet al., which is presented in Table 2.

**Table 2.** Damage state definitions as Barbet [10].

| Damage state | Damage state thresholds |
|---|---|
| Slight damage | $Sd_{ds\,1} = 0{,}7 \times D_y$ |
| Moderate damage | $Sd_{ds\,2} = D_y$ |
| Extensive damage | $Sd_{ds\,3} = D_y + 0{,}25(D_u - D_y)$ |
| Collapse damage | $Sd_{ds\,3} = D_u$ |

In this study, spectral displacement Sd is taken as the intensity measure. The fragility curves can be developed using log-normal functions represented by two variables, the mean value of displacement, and the standard deviation. the fragility curve using the following equation:

$$P(Ds/Sd) = \Phi \frac{\ln(Sd) - \ln(\lambda)}{\beta} \quad (1)$$

Description:
P(DS/Sd) = Probability Function (damage state)
Φ = Cumulative normal distribution
Sd = spectral displacement
$\lambda$ = median value
$\beta$ = Standard deviation

# 3 Results and discussion

From the results of the static pushover analysis run on SeismoStruct v.18 software, a capacity curve (base shear

- displacement) is obtained. The resulting pushover curve must then be converted into a bilinear form as shown in Figs. 3 and 4. The capacity curve of the regular structure shows a higher base shear value compared to the irregular structure. This is because the regular structure has a uniform story height on each floor, leading to an even distribution of stiffness throughout the building vertically. The capacity curve also shows that as stiffness increases, ductility decreases. The capacity curves for regular and irregular structures in the x-direction can be seen in Fig. 3, and for the y-direction in Fig. 4.

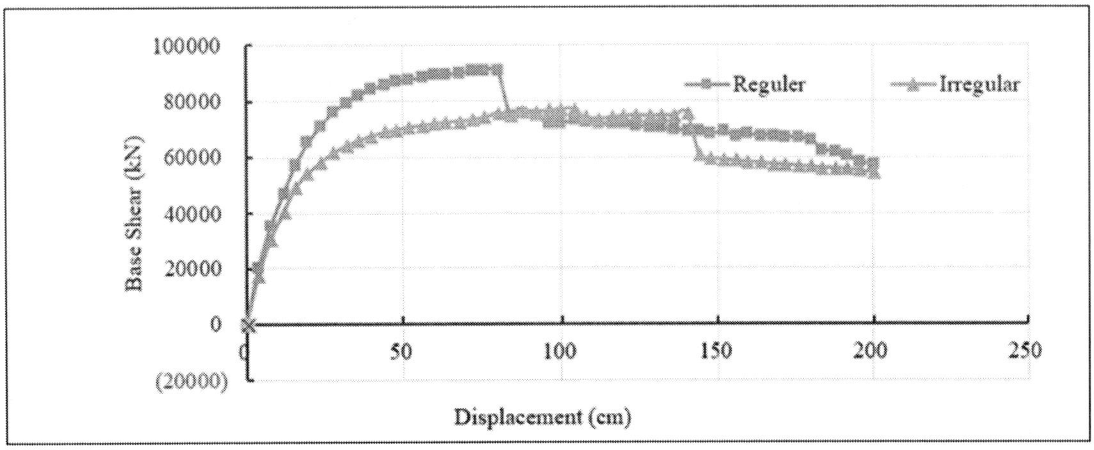


**Fig. 3.** The capacity curve resulting from the pushover in the x-direction.

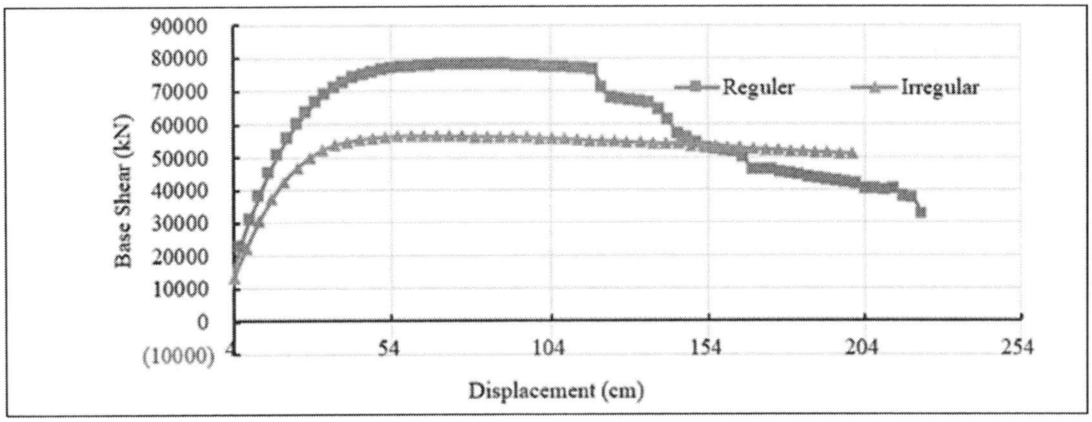


**Fig. 4.** The capacity curve resulting from the pushover in the y-direction.

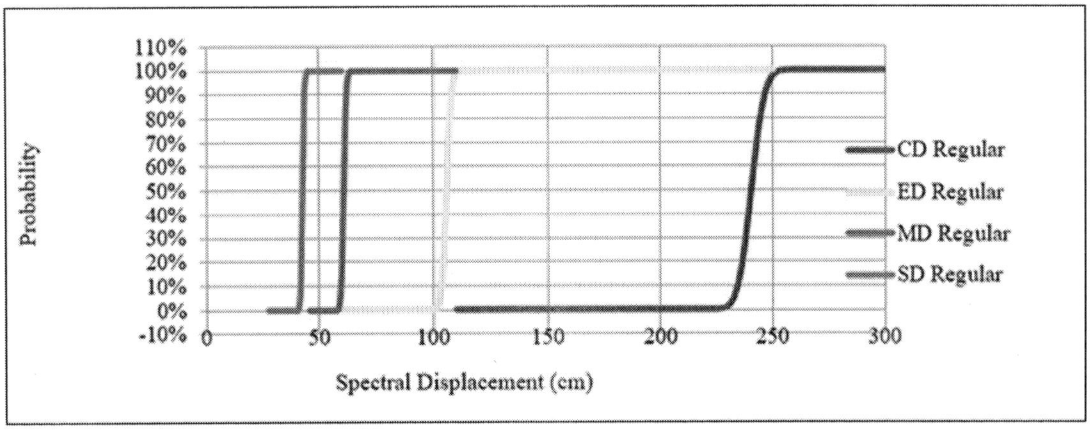


**Fig. 5.** Fragility curve of the regular structure in the x-direction.

The fragility curve results for the regular structure in the x-direction are shown in Fig. 5, and for the y-direction in Fig. 6. The fragility curves for the irregular structure in the x-direction can be seen in Fig. 7, and in the y-direction in Fig. 8. From the fragility curve of the regular structure in the x-direction, it can be observed that at an ultimate displacement of 80 cm, the fragility level indicates a 100% probability of experiencing moderate damage, which includes shear cracks and medium cracks in the columns (the columns remain structurally sound), with joints also experiencing cracking.

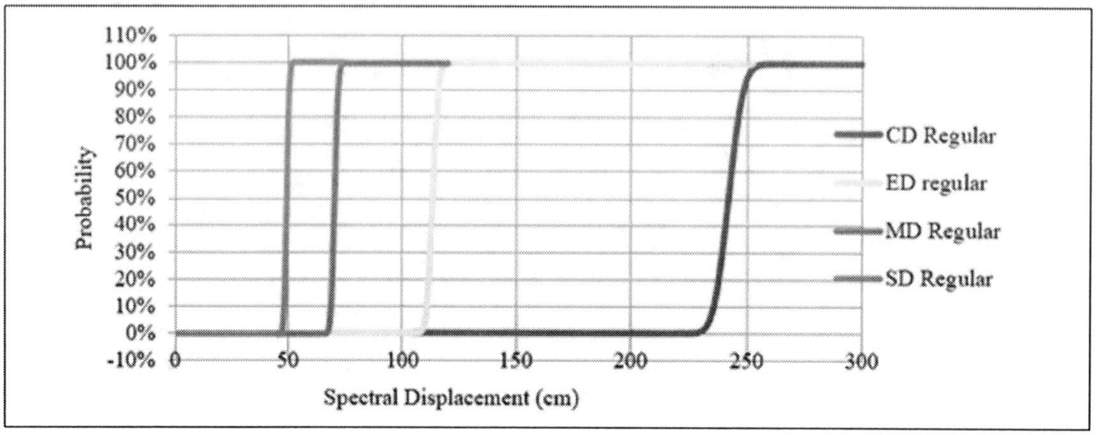


**Fig. 6.** Fragility curve of the regular structure in the y-direction.

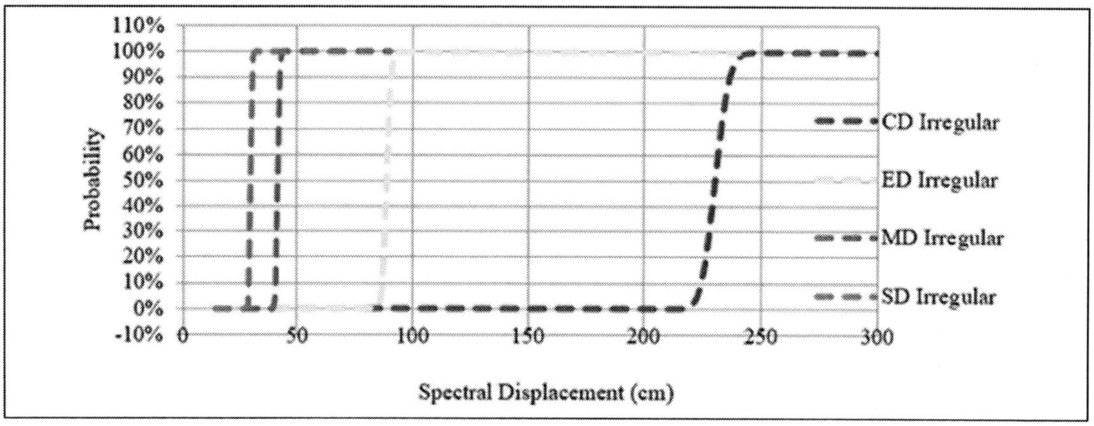


**Fig. 7.** Fragility curve of the irregular structure in the x-direction.

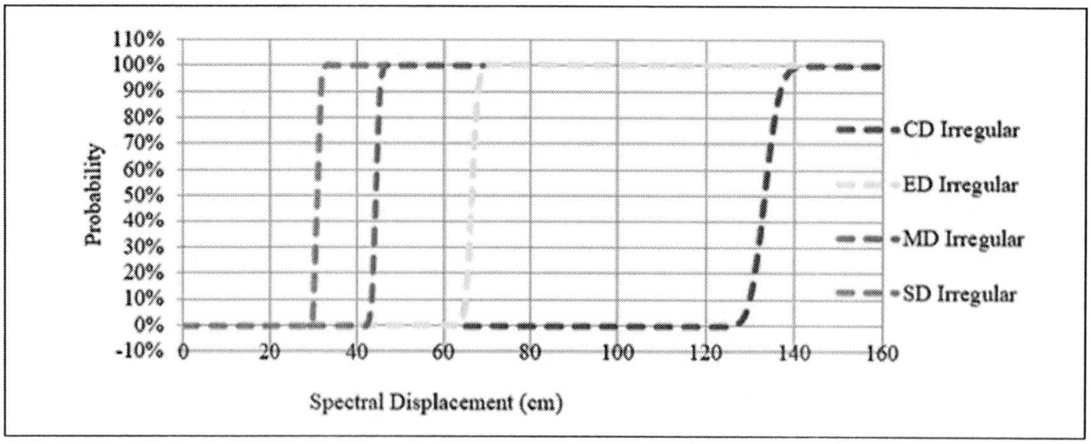


**Fig. 8.** Fragility curve of the irregular structure in the y-direction.

From the fragility curve of the regular structure in the y-direction, it can be observed that at an ultimate displacement of 70 cm, the fragility level indicates a 100% probability of experiencing moderate damage, which includes shear cracks and medium cracks in the columns (the columns remain structurally sound), with joints also experiencing cracking.

Use From the fragility curve of the irregular structure in the x-direction, it can be observed that at an ultimate displacement of 104 cm, the structure has a 100% probability of experiencing extensive damage. This includes a reduction in strength without collapse in the columns, shear failure (columns are structurally unsafe), and significant residual displacement in the joints.

From the fragility curve of the irregular structure in the y-direction, it can be observed that at an ultimate displacement of 60 cm, the structure has a 100% probability of experiencing moderate damage. This includes shear cracks and medium cracks in the columns (the columns remain structurally sound), with joints also experiencing cracking.

## 4 Conclusion

Based on the results and discussion for both buildings, namely the regular structure and the irregular structure, it can be concluded that the irregular structure (excessive soft story irregularity) has a higher fragility level compared to the regular structure in both the x and y directions. For the regular structure, the building experiences moderate damage at a drift of 2% as per SNI 1726-2019. In contrast, the irregular structure (excessive soft story irregularity) exhibits extensive damage in both the x and y directions.

## References

1. F. D. Shubandrio, A. M. El-Maissi, M. M. Kassem, M. Masrilayanti, S. R. Rahmat, and F. Mohamed Nazri, Evaluating the Interdependencies of Infrastructure Critical Systems during Earthquake Event: A Case Study for Padang City. Sustainability, **14**(23), (2022). https://doi.org/10.3390/su142315926.
2. R. Aryanti and M. Masrilayanti, State of the art of seismic risk and loss assessment in structures, E3S Web of Conferences, **331**, 07013, (2021), https://doi.org/10.1051/e3sconf/202133107013.
3. R. Masrilayanti, J. Kurniawan, Tanjung, and M. Yunus, Displacement Performance of 8 Storys Hotel Building Using Nonlinear Time History Analysis Method. **389**, (2024). https://doi.org/10.1007/978-981-99-6368-3_28.
4. P. Fajfar, H. Krawinkler, Nonlinear Seismic Analysis and and Design Buildings (CRC Press, London, 2005).
5. C. Scawthorn, A Brief History of Seismic Risk Assessment, in Risk Assessment, Modelling and Decision Support, S. F. Gottlieb, Ann Bostrom, Ed., Springer, ch. **2**(5–81), (2006). https://doi.org/10.1525/9780520954731-004.
6. B. Belletti, C. Damoni, and A. Gasperi, Modeling approaches suitable for pushover analyses of RC structural wall buildings. Eng Struct, **57**, 327–338, (2013). https://doi.org/10.1016/j.engstruct.2013.09.023
7. M. Masrilayanti, A. P. Nasution, R. Kurniawan, J. Tanjung, and S. Sarmayenti, Fragility Curve Analysis of Medium Cable Stayed Bridge, Civil and Environmental Engineering. **17**, 209–218, (2021), https://doi.org/10.2478/cee-2021-0022.
8. F. Hosseinpour and A. E. Abdelnaby, Fragility curves for RC frames under multiple earthquakes, Soil Dynamics and Earthquake Engineering. (2017). https://doi.org/10.1016/j.soildyn.2017.04.013.
9. Masrilayanti, Fragility curves as a tool for disaster mitigation (state of the art), in Proceedings of AIP Conference, American Institute of Physics, May (2024). https://doi.org/10.1063/5.0202953.
10. C. A. Kircher, R. V. Whitman, and W. T. Holmes, HAZUS Earthquake Loss Estimation Methods, Nat Hazards Rev, (2006). https://doi.org/10.1061/(asce)1527-6988(2006)7:2(45).
11. Z. Guo, Seismic Resistance. Principles of Reinforced Concrete. 411–436, (2014), https://doi.org/10.1016/B978-0-12-800859-1.00016-5.
12. M. J. N. Priestley, Performance based seismic design, Bulletin of the New Zealand Society for Earthquake Engineering, (2000). https://doi.org/10.5459/bnzsee.33.3.325-346

# Analysis of slope stability to determine the causes of collapse the primary canal in the North-Raman irrigation-area, Central Lampung

*Kastamto*[1], *Arniza* Fitri[1], *Mas* Mera[2*], *Alfath* Zain[1], *Susarman*[1], and *Fajar* Dewantoro[1]

[1]Faculty of Engineering and Computer Science, Universitas Teknokrat Indonesia, Lampung, Indonesia
[2]Department of Civil Engineering, Faculty of Engineering, Universitas Andalas, Padang, Indonesia

**Abstract.** A 25-meter primary-canal in the North-Raman Irrigation-Area (NRIA) collapsed in December 2021. Before the landslide occurred, the rainfall intensity reached 153 mm/hour. Field surveys at the time of the incident revealed that a large enough discharge overflowed from the Raman River, inundating the NRIA. This shows that landslides were most likely triggered by groundwater infiltration due to either high intensity rainfall or flooding. The soil conditions are also unable to support the primary-canal embankment. The present research objective is to carry out slope stability analysis to determine the causes of collapse of the NRIA embankment using the Geo-Slope and Plaxis 8.6 numerical models. Soil parameters are tested and measured in the field and laboratory. The safety factor (*SF*) at the study site is determined by both models, and then compared with the allowable *SF* to assess slope stability which should be bigger than 1.5 if no earthquake occurs. However, the Geo-Slope model produces an *SF* of 0.97. This means that the embankment is very vulnerable to landslides. The laboratory result reinforces this finding where embankment void-ratio is 1.5 (high). A high void ratio can cause water infiltration into the soil, thereby reducing soil stability and bearing capacity.

## 1 Introduction

Landslides are phenomena occurring on slopes when they are unstable due to the synergistic effect of certain factors such as soil conditions, topography, and vegetation, combined with external factors known as dynamic triggering factors such as rainfall, snow, earthquakes, and human intervention [1, 2].

Most previous studies have identified that in addition to geological, geotechnical, and topographical factors, landslides can be triggered by rainwater that flows as overtopping or piping during the rainy season [3-8]. When rainwater infiltrates the soil, pore water pressure increases [9-13]. Positive pore water pressure causes capillarity, leading to a rise in the groundwater table [10]. High soil saturation weakens the bonds between soil particles, reducing soil shear strength and resulting in instability [14-16]. The rapid increase in upstream water levels leads to high seepage pressure, potentially initiating internal erosion that can develop into piping paths. Continuous particle loss due to piping increases the soil void ratio while decreasing the friction angle, creating a more porous soil structure and dramatically increasing soil erodibility [1].

In December 2021, a 25-meter long primary-canal embankment collapsed in the North-Raman Irrigation-Area, Central Lampung, Pekalongan Regency (Fig. 1). This irrigation primary-canal crosses a river that flows to the left. Before the landslide occurred, the rainfall intensity reached 153 mm/hour, the river water overflowed and hit the right side of the irrigation canal. [17]. Under these conditions, rainwater infiltration into the soil will most likely increase soil water content and pore pressure around the primary-canal embankment. All of this causes the soil bearing capacity to decrease resulting in landslides. To ascertain the causes of the landslide, it is important to carry out slope analysis at the landslide location by considering the intensity of rainfall.

**Fig. 1.** The primary canal in the North-Raman irrigation-area collapsed

Slope failure generally occurs on planes characterized by sloping surfaces that are influenced by gravitational

* Corresponding author: mas_mera@eng.unand.ac.id

forces. For example, canal embankment with extreme slopes so that the gravitational force exceeds the plane shear-angle, resulting in landslides [4]. There are two types of landslides, namely translational landslides and rotational landslides [9]. Translational landslides involve the movement of soil and rock along nearly flat or gently undulating slip planes. In contrast, rotational landslides occur along concave slip planes. This last type is common in Indonesia, especially in areas of contact between weathered soil zones and bedrock. Translational landslides usually have shallower slip planes and relatively thin weathering zones [9].

Two methods are commonly employed in slope stability analysis. They are the limit equilibrium method (LEM) and the finite element method (FEM). Both methods predict landslide heights, with findings generally indicating comparable results between FEM and LEM [14]. Geo-Slope or Plaxis 8.6 are software packages used for geotechnical numerical simulation based on LEM [17]. Both models are often useful to determine how sensitive the stability of a structure is to variation in a certain parameter. Previous researchers who used the Geo-Slope model to analyse slope stability stated that the results were satisfactory [14-16].

The present research objective is to carry out slope-stability analysis to determine the causes of collapse in primary-canal embankment in the North-Raman Irrigation-Area, Central Lampung, Pekalongan Regency using the Geo-Slope and Plaxis 8.6 numerical models. This study will provide essential information which is required to improve the design of hydraulic infrastructure affected by landslide-prone embankments.

## 2 Methods

### 2.1 Research site

This research was conducted at the location of the primary canal that collapsed (BRU-1B at STA 5,375) in the North-Raman Irrigation-Area, Pekalongan Regency, Ganti Mulyo Village, Central Lampung. Dimension of the irrigation canal is presented in Fig. 2. The location coordinates are 5°2'24.96" S - 105°21 '41.4" E. Fig. 3 shows the research site.

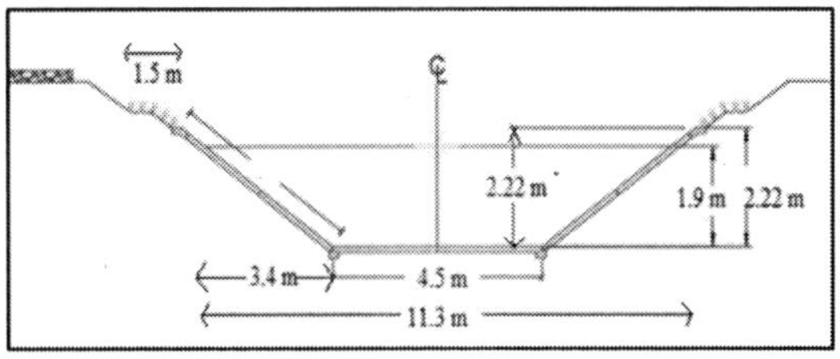


**Fig. 2**. Dimension of irrigation canal

### 2.2 Data collection

The data collected for the purposes of this study are rainfall data, geological data and soil properties. Daily rainfall data at the landslide location was obtained from BMKG Central Lampung. The rainfall station is the closest station to the landslide location. Geological data and soil properties were obtained from field surveys and the Geological Map of Central Lampung (Fig. 4). Fig. 5 depicts soil sampling (drill holes) carried out at BRU-1B at STA 5,375 to a depth of 10 meters. The required soil properties are soil unit weight $\gamma$ in kN/m$^3$, soil specific gravity $Gs$, effective cohesion $c'$ in kN/m$^2$, effective shear angle $\phi'$, void ratio, soil permeability, compression index $Cc$, consolidation coefficient $Cv$, average water content in percentage, and groundwater level.

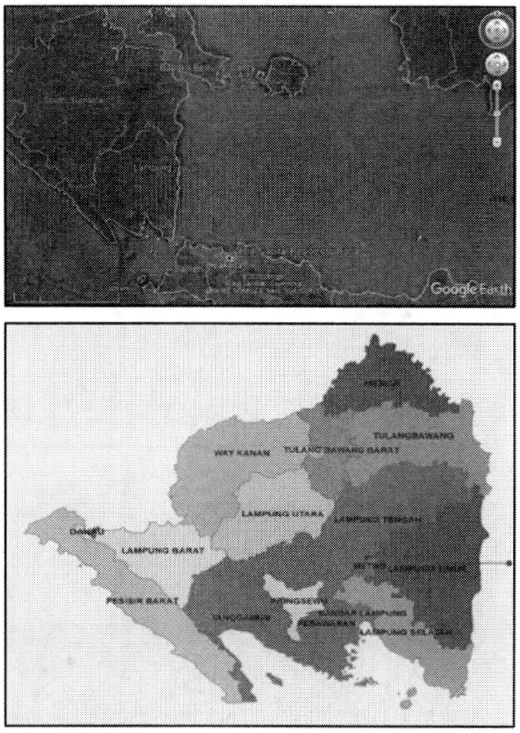


**Fig. 3**. Research site

### 2.3 The limit equilibrium method

The LEM is the theoretical basis of the Geo-slope model. This method is used to predict slope stability based on the comparison between supporting forces and driving forces acting on the slope. The results of this comparison are referred to as safety factors (*SF*). The *SF* values that represent slope-stability conditions are presented in Table 1. The simplest equilibrium method is the wedge method. The wedge method is one of the most basic approaches to assessing slope stability. This method involves dividing the slope into several vertical slices and analysing the forces acting on each slice. Overall stability is then determined by adding up the forces from all the slices and calculating a safety factor. The derivation of these forces is explained in Fig. 6 and Equation 1.

 https://doi.org/10.1051/e3sconf/202560413001

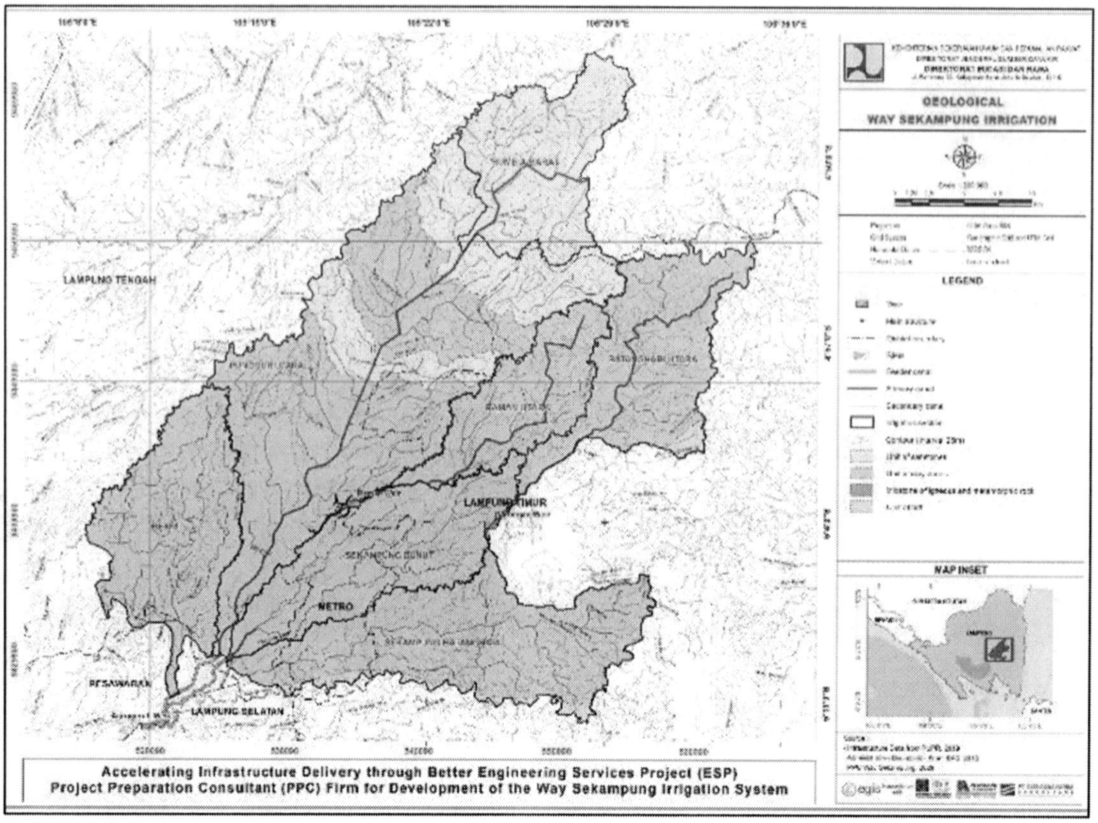


**Fig. 4**. Geological map in the Central Lampung

**Fig. 5**. Borehole sampling at BRU-1B di STA 5,375 up to 10 meters depth

**Table 1.** Relationship between *SF* values and landslide intensity

| No | *SF* Values | Landslide Intensity |
|---|---|---|
| 1 | < 1.07 | Landslides are common/frequent |
| 2 | 1.07 < *SF* < 1.5 | Landslides occur occasionally |
| 3 | > 1.5 | Landslides are rare |

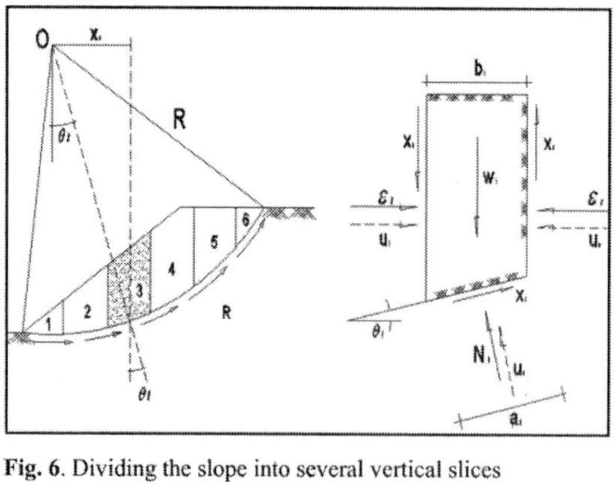


**Fig. 6**. Dividing the slope into several vertical slices

$$SF = \frac{\sum_{i=1}^{i=n}(c' a_i + N_i \phi)R}{\sum_{i=1}^{i=n}(W_i + sin\theta_i)R} \quad (1)$$

Where *SF* is safety factor, *R* is radius of the landslide circle (m), *c'* is effective cohesion (kN/m$^2$), *ϕ'* is effective shear angle ($^0$), *u* is pore water pressure, *Wi* is weight of the slide, $Ni = Wi \cos \theta i$ is normal force on each slide, $a_i$

is circle curve length (m) and $\theta i$ is slope angle of each slide.

## 3 Results and discussion

### 3.1 Landslide type

The primary canal in BRU-1B at STA 5,375 has a base width of 11.3 meters, height of 8 meters, and the width of the top of the embankment is 1.5 meters. The 25-meter length of this canal which consists of a right embankment and a 4.5-meter-wide base, collapsed into the river and rice fields. As a result, the North-Raman Irrigation-Area stopped operating. In an addition, 4000 hectares of rice fields cannot be irrigated by this irrigation canal. Visually, the landslide that occurred was identified as a translational landslide where the sliding surface formed a straight line. Fig. 7 shows the condition of the channel before and after the landslide.

**Fig. 7**. The primary canal: before landslide (top) and after landslide (bottom).

### 3.2 Void ratio

The void ratio in soil greatly influences its mechanical properties. The greater the pore number indicates the greater the soil's ability to hold water, thus affecting its stability and carrying capacity. In addition, the void ratio impacts soil permeability, namely the soil's ability to pass water or other fluids. The void ratio can reflect changes in the internal structure of the soil. A high void ratio allows water infiltration into the soil or piping. When piping occurs, fine particles are continuously lost in the flow, resulting in the volume of soil cavities increasing and the microstructure becoming looser so that the possibility of landslides is greater.

Fig. 8 presents the void ratio at the landslide location during the consolidation test. Based on the graph, the void ratio at the landslide location is 1.5. This value is quite high, indicating that the volume of voids is greater than the volume of soil grains. Therefore, it can be concluded that the high number of pores at the landslide location allows water infiltration into the soil, thereby reducing the stability and bearing capacity of the soil. The high void ratio at the landslide location causes the soil structure to become more porous and increases soil erodibility significantly.

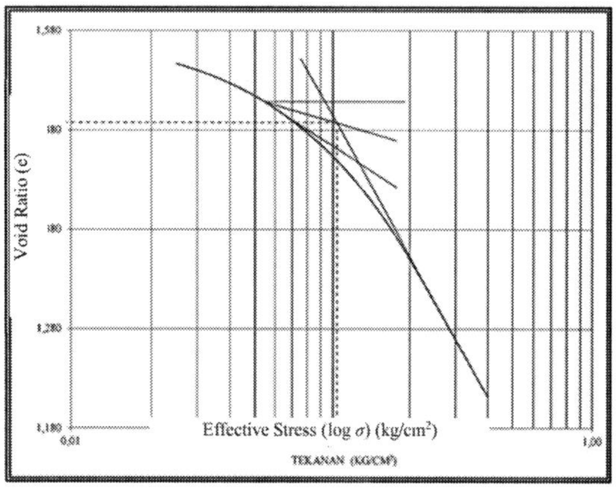


**Fig. 8**. Void ratio at the landslide location during the consolidation test

**Table 2.** Soil properties

| Soil parameters | | Depth of soil 2.5 m | 7 m |
|---|---|---|---|
| SPT | | 14 | 3 |
| Type of soils | | Silty clay (cm/s) | Silty clay |
| Soil permeability | | $2.6\times10^{-6}$ (cm/s) | $8.4\times10^{-5}$ (cm/s) |
| $\gamma_t$ | | 17.5 kN/m³ | 17.5 kN/m³ |
| Specific gravity (*Gs*) | | 2.699 gr/cm³ | 2.335 gr/cm³ |
| Groundwater level | | 7-8 m | |
| LL | | 86 % | 70.47 % |
| PL | | 50 % | 49.62 % |
| IP | | 35 % | 20.86 % |
| Particle size analyses | Gravels | 0 % | 11.25 % |
| | Sand | 11.6 % | 11.84 % |
| | Silt | 15.7 % | 56.34 % |
| | Clay | 73.24 % | 20.57 % |
| Consolida tion test | *Cc* | - | 0.411 |
| | *Cv* | - | $8.6\times10^{-4}$ cm³/s |
| Triaxial UU | $\phi'$ (°) | 0 | 5.954 |
| | *c'* | 0 | 0.144 kg/cm² |

### 3.3 Slope-stability analysis using geo-slope model

To analyse the impact of rainfall infiltration on slope stability using Geo-slope model, the following parameters are required: rainfall intensity in mm/h, slope angle in degrees, soil depth *Z*, ground water level in meters, soil type, soil permeability *k*, cohesion to effective stress *c'*, angle of internal friction to effective stress $\phi'$, and total unit weight of soil $\gamma_t$. By using these variables, the soil is assumed to be in the most critical and saturated condition due to high rainfall intensity.

The data required for the simulation is presented in Table 2. The data in this table are the results of laboratory tests. Testing was carried out using the undrained consolidated (UC) triaxial test method.

Fig. 9 shows the results of simulating the conditions of the irrigation canal-slope using Geo-slope model. The model results show that the safety factor is 0.97. Based on Table 1 regarding the relationship between *SF* values and slope conditions, the *SF* value below 1.07 indicates a high risk of landslides. Therefore, it can be concluded that the condition of the slope at the landslide location is classified as critical, where there is a high probability that frequent landslides will occur. By taking into account the *SF* and surface slip obtained from the Geo-slope model, these results are considered close to existing conditions.

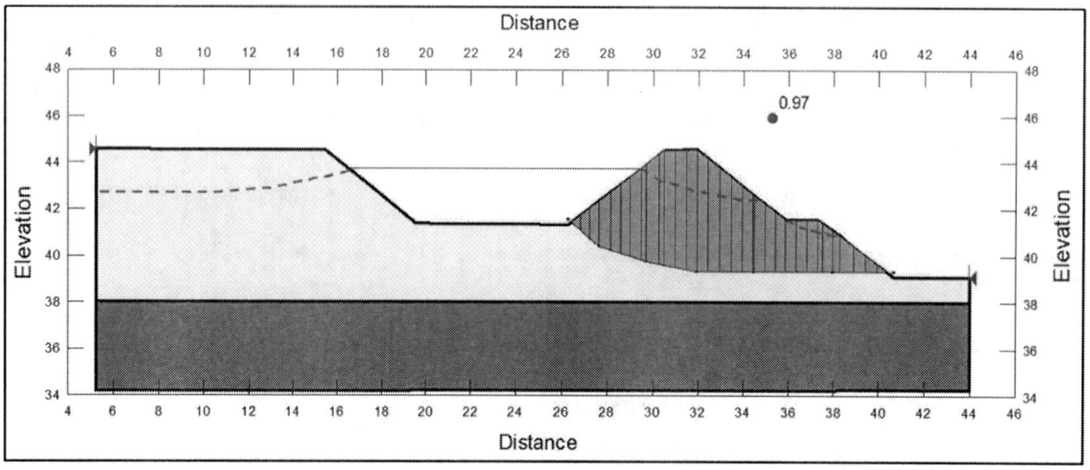


**Fig. 9**. Simulation results of irrigation canal-embankments using the Geo-slope model

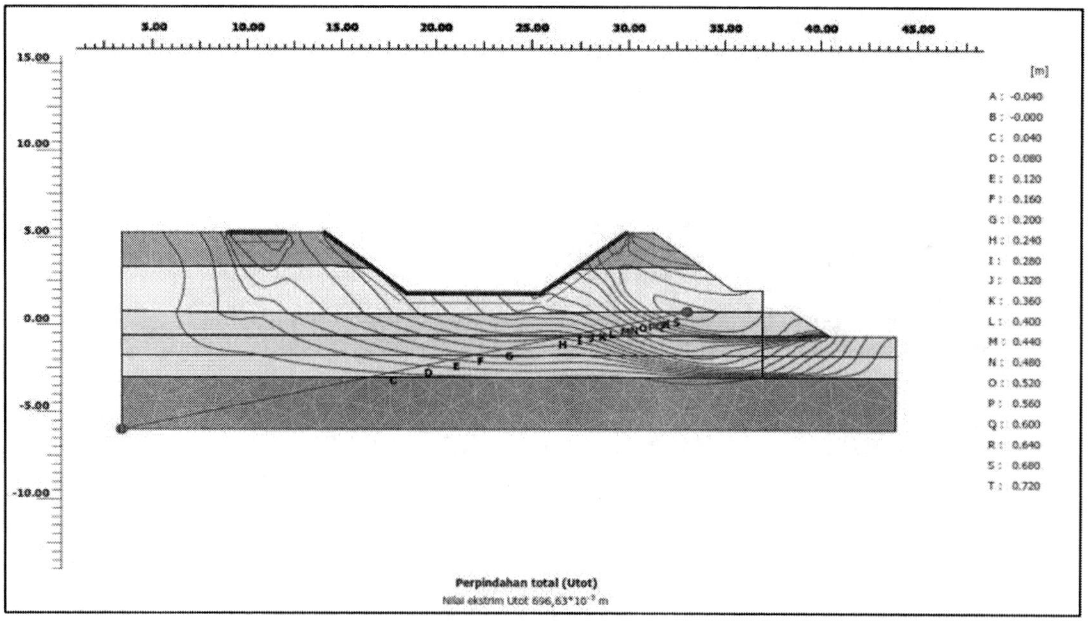


**Fig. 10**. Simulation results of irrigation canal-embankments using the plaxis 8.6 model

## 3.4 Slope-stability analysis using plaxis model

Based on simulation results using the Plaxis 8.6 model, the potential horizontal nodal displacement *Ux* is 0.69 meters (Fig. 10). This is in line with landslide conditions in the field, where the movement of landslide material reaches a distance of less than 1 meter. Positive values indicate movement to the right, while negative values indicate movement to the left. Soil elements that experience horizontal displacement are influenced by factors such as excess pore water pressure and the weight of the soil itself.

## 4 Conclusions

The results of the analysis of the slope stability of the primary-canal embankment in the Raman-North Irrigation-Area using two numerical models, namely the Geo-slope and Plaxis 8.6 models, show that the BRU-1B location is very prone to landslides. This can be indicated by a low safety factor (*SF* of 0.97 at the landslide location which is far below the allowable *SF*). This finding is also reinforced by laboratory results where void ratio (e) at the landslide location is classified as high, it is around 1.5. A high void ratio can cause water infiltration into the soil, thereby reducing soil stability and bearing capacity.

Acknowledgments are given to the Universitas Andalas and Universitas Teknokrat Indonesia (Hibah Hiputs: 006/UTI/LPPMI/E.1.1/VIII/2024) for providing facilities and financial assistance for this research.

## References

1. Chen, C., Y. Wang, J. Zhang, H. Zhang, H. Li, Q. Chen, A preliminary study of landslide dam failures induced by the combined influence of piping and overtopping. Journal of Hydrology **625**, 129984 (2023).
2. Chang, D. S., L.M. Zhang, Extended internal stability criteria for soils under seepage. Soils and Foundations **53**(4), 569-583 (2013).
3. Brunetti, M.T., S. Peruccacci, M. Rossi, S. Luciani S., D. Valigi, F. Guzzetti, Rainfall thresholds for the possible occurrence of landGeo-slopes in Italy, Natural Hazards and Earth Sistem Science (2010).
4. Baum, R. L., W.Z. Savage, J. W. Godt, Trigrs-A Fortran program for transient rainfall infiltration and grid-based regional slope-stability analysis, version 2.0. U.S. Geological Survey (2008).
5. Hardiyatmo, H.C., Penanganan tanah longsor dan erosi. Gajah Mada University Press, Yogyakarta (2012).
6. Muntohar A.S., Proposal ambang hujan untuk peringatan dini tanah longsor, Seminar/ workshop Application Research for Disaster and Humanitarian, University Club UGM Plaxis Versi 8 Manual Acuan (2009).
7. Rahardjo, H., T.T. Lee, E.C. Leong, R.B. Rezaur, Response of a residual soil slope to rainfall. Can. Geotech. J. **42**, 340–351 (2005).
8. Collison, A.J.C., M.G. Anderson, Using a combined slope hydrology/slope stability model to identify suitable conditions for landGeo-slope prevention by vegetation in the humid tropics. Earth Surface Processes and Landforms **21**, 737-747 (1996).
9. Fitri, L. Yao, D. Pratiwi, A. Phelia, F. Dewantoro, D. Safitri and K.N.A. Maulud, Effectiveness of a groundsill structure in reducing scouring problem at Cimadur River, Banten Province. IOP Conference Series: Earth and Environmental Science. 12026 (2021).
10. Fitri, H. Chen, L. Yao, R. Zheng, F. Rossi, Y. Yin, Evaluation of the groundsill's stability at downstream of "Citorek" Bridge in Cimadur River, Banten Province. IOP Conference Series: Earth and Environmental Science. 12029 (2021).
11. Fitri, M.S.N. Hadie, A. Agustina, D. Pratiwi, S. Susarman, G. Pramita and H.R.A. Salah, Analyses of flood peak discharge in Cimadur river basin, Banten Province, Indonesia. E3S Web of Conferences. 8006 (2021).
12. Pratiwi, D., A. Fitri, A. Phelia, N.A.A Adma, Kastamto, Analysis of urban flood using synthetic unit hydrograph (SUH) and flood mitigation strategies along way Halim River: a case study on Seroja street, Tanjung Senang District. E3S Web of Conferences. 7015 (2021).
13. Fredlund, D.G., H. Rahardjo, Soil mechanic for unsaturated soils", John Willey and Sons, 1993 Geo-Slope International Ltd. Stability Modelling with Slope/w 2007 version-An Engineering Methodology First edition (2012).
14. Rohmat H., Analisis stabilitas lereng pada longsor Desa Caok, Purwerejo, Jawa Tengah. Jurnal Sumber Daya Air **14**(1), 63-74 (2018).
15. Ciptaninga, K., Y. Yunus, S. M. Saleh, Analisis stabilitas lereng dengan konstruksi dinding penahan tanah tipe counterfort. Jurnal Arsip Rekayasa Sipil dan Perencanaan **1**(2),58-68 (2018).
16. Geo-Slope International Ltd., Stability modelling with slope/w 2007 version-An engineering methodology first edition. Geo Slope International Ltd: Canada (2012).
17. Badan Meteorologi, Klimatologi dan Geofisika (BMKG), Lampung Tengah (2021)

 https://doi.org/10.1051/e3sconf/202560413002

# Effectiveness of earthquake drains in mitigating liquefaction-induced settlement

*A'isyah* Salimah[1*], *Yelvi*[1], *Meidi* Andienti[1], *Widjojo* Adi Prakoso[2], and *Wiwik* Rahayu[2]

[1]Departement of Civil Engineering, Politeknik Negeri Jakarta, Depok, Indonesia
[2]Department of Civil Engineering, Faculty of Engineering, Universitas Indonesia, Depok, Indonesia

**Abstract.** This study evaluates the effectiveness of earthquake drains in mitigating liquefaction and examines their performance at specific frequencies in loose and medium-dense sediments. Two lab-scale shake table single-axis test series were conducted to assess this: one without mitigation and another using earthquake drains. Both models were instrumented and subjected to consistent shaking sequences at 1 Hz and 1.2 Hz frequencies. The results revealed reduced excess pore-water pressure generation around the drains due to the rapid dissipation of pore pressures during shaking. Additionally, the excess pore water pressure increased more slowly with the drains, as they allowed partial dissipation of pressure through drainage. Earthquake drains significantly reduced liquefaction potential, particularly in the middle area, where conditions shifted from liquefied to non-liquefied. Furthermore, ground treated with drains exhibited less differential settlement than untreated ground. The use of drains resulted in settlements being retrieved by 2% to 15% more than untreated conditions. The drains effectively mitigated the post-shaking liquefaction-induced settlement mechanisms. While this study demonstrates earthquake drains as a reliable and efficient measure for reducing liquefaction-induced settlement, further research is necessary to optimize their design and application.

## 1 Introduction

Liquefaction during earthquakes significantly threatens infrastructure stability, leading to severe ground deformations and structural failures. Indonesia is a country located in the Pacific Ring of Fire. This region is prone to earthquakes, volcanic eruptions, and tsunamis due to the meeting of three world tectonic plates: the Indo-Australian, Eurasian, and Pacific.

Sandy Poor (SP) soils, known as loose sand, store more water than well-graded soils. Unconsolidated and water-saturated sandy soils have a large porosity, allowing water to infiltrate easily, making them susceptible to liquefaction. Liquefaction occurs when the soil loses its shear strength due to an earthquake, and the bearing capacity of the soil drops suddenly. The pore water stress almost entirely replaces the total stress because the soil decreases in volume, pushing the ground surface upwards or flowing like mud, which causes the collapse of buildings.

Excess Pore Water Pressure (EPWP) is the additional pressure exerted by water within soil pores, above normal hydrostatic pressure, typically caused by external forces like seismic activity or rapid loading. This condition reduces the effective stress between soil particles, weakening the soil and potentially leading to liquefaction, where soil behaves like a liquid. Managing excess pore water pressure is crucial in geotechnical engineering to prevent ground deformations and structural failures, especially in earthquake-prone areas. Techniques such as earthquake drains can accelerate the dissipation of this pressure, enhancing soil stability and infrastructure resilience [1, 2].

In overcoming one of the failures in the Sandy Poor (SP) soil structure or loose sand, an experimental analysis of liquefaction potential using an earthquake drain with shaking table testing was conducted. An earthquake drain is a vertical channel that provides a drainage path that accumulates pore water pressure to dissipate ideally before the surrounding soil reaches a state of liquefaction [3, 4].

Other vertical drains, such as Prefabricated Vertical Drain (PVD), accelerate the soil. The consolidation process speeds up water dissipation from the soil [5]. The general characteristics of PVDs are that they are flat and are used in fine-grained soils with low water flow rates while preloading loads are applied [6]. When it is necessary to address the discharge of excess pore water pressure during earthquakes, PVDs alone may not be sufficient. Therefore, the method is changed from PVD to earthquake drains, which have a similar drainage function but offer a larger capacity for removing water from the ground. While PVDs can be effective for dissipation, they require a higher capacity to handle the increased flow rate during such events [7].

Earthquake drains can be applied in sandy soils placed under shallow foundations and are expected to effectively reduce pore water pressure and mitigate impacts caused

* Corresponding author: aisyah.salimah@sipil.pnj.ac.id

by liquefaction. Experimental analysis of liquefaction potential using earthquake drain is proactive research in improving the earthquake resistance of buildings and providing protection for residents and infrastructure in earthquake-prone areas [8].

In this research, relative density variations of 40% and 60% are employed, alongside vibration frequencies of 1 Hz and 1.2 Hz, to investigate the influence of these parameters on liquefaction potential. As highlighted by Salimah (2023), an increase in relative density percentage corresponds to a diminished likelihood of liquefaction and land subsidence, as denser soils exhibit greater resistance to shear stress and pore pressure build-up. Conversely, lower relative density percentages exacerbate the risk of liquefaction and subsidence, as looser soils are more susceptible to pore pressure accumulation during seismic activity. Liquefaction, which occurs when saturated soils lose their strength and behave as a viscous fluid due to elevated pore water pressure, poses significant risks to structural foundations and overall ground stability during earthquakes.

During liquefaction, water within the soil pores experiences an increase in excess pore water pressure, reducing the soil's effective stress to nearly zero. This causes the soil to lose its strength, behaving like a liquid and allows soil particles to settle back into a denser condition once the earthquake ends [9].

Several drainage techniques, such as earthquake drains, vertical drains, and sand drains, have been developed to facilitate the rapid dissipation of excess pore water pressure generated during seismic events to mitigate these hazards [10]. These systems work by enhancing soil drainage capacity, thus accelerating the stabilization of soil layers and minimizing the risk of liquefaction. By allowing water to escape more efficiently from the soil matrix, these drains play a critical role in reducing soil deformation, preserving structural integrity, and preventing catastrophic ground failure in earthquake-prone regions.

## 2 Research significance

This study demonstrates the effectiveness of earthquake drains in mitigating liquefaction in loose and medium-dense sediments during seismic events. This experiment hypothesizes that seismic drainage significantly reduces excess pore water pressure by facilitating rapid dissipation. The accelerated dissipation of pore water pressure prevents it from exceeding the effective stress of the soil, thereby reducing the liquefaction potential. The reduced liquefaction potential is also expected to decrease settlement events. These findings highlight the practical benefits of earthquake drains, providing a foundation for further research to optimize their design and application, hopefully enhancing infrastructure resilience in seismic regions.

## 3 Methods

The research was conducted at the Civil Engineering Laboratory, Politeknik Negeri Jakarta. The test box is 40 cm long and 40 cm wide, with a depth of 60 cm, where the depth of sand sediment is 40 cm. The test box is equipped with a one-way automatic drive to model the occurrence of an earthquake. Sample preparation using a rainer box system to drop sand into the sandbox, as seen in Fig. 1. This tool has a sand filter with a diameter of 0.6 mm. The sand rainer box is equipped with a pulley system to control the height of the sand drop, which helps regulate the relative density of the soil. Controlling the rainer box is essential to ensure the uniformity of the soil sediment [11]. Figure 2 shows a shaking table system using seismic control. Seismic input as 1 Hz and 1,2 Hz equal to 0,1 g and 0,18 g for this scale. Seismic input as 1 Hz frequency equal to 0,1 g for this scale. Seismic response was observed for 30 seconds for each test. Presentation of data 30 seconds before shaking, during 30 seconds of shaking, and 30 seconds after shaking, ensuring the observation covers the entire period before, during, and after the shaking event.

**Fig. 1.** Rainer box system

**Fig. 2.** Shaking table system

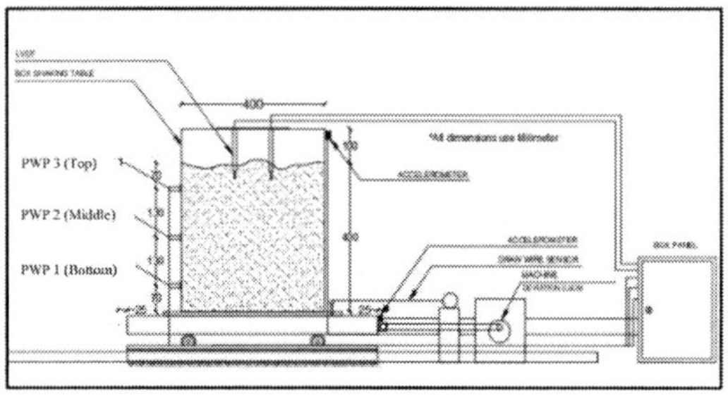


**Fig. 3.** Plan view of shaking table system

Figure 3 shows a plan view of the shaking table with the position of the sensor used in the measurement. Pressure transmitters are located at three levels to analyze the excess pore water pressure at each level and assess its significance. The data obtained is the value of the change in pore water pressure (Δuo) read in the layer at depths of 32 cm (bottom transmitter) for first elevation, 20 cm (middle transmitter) for second elevation, and 8 cm (top transmitter) for third elevation. An inverter is used in the machine motor to control acceleration, while an accelerometer sensor is placed on the table base to measure the actual acceleration. Linear Variable Differential Transformer (LVDT) used to measure settlement. There are 4 LVDT placed on the top of shallow foundation. Electromechanical transducer that can convert the rectilinear motion of an object to which it is coupled mechanically into a corresponding electrical signal.

This research used clean sand as a base material. Due to its purity and consistent grain size, silica sand is specifically used to ensure good results, making it ideal for applications requiring precision and durability, with particle sizes ranging from 0.15 mm to 0.5 mm. Properties of soil in Table 1 show uniform grain gradation Cu < 6, which means poorly graded soil. Study variations on shallow foundations include without and with mitigation using earthquake drains. Soil properties details are shown in Table 2.

**Table 1.** Soil properties

| Properties | Value | Unit |
|---|---|---|
| $\gamma d_{max}$ | 14,47 | $kg/m^3$ |
| $\gamma d_{min}$ | 13,33 | $kg/m^3$ |
| Gs | 2,65 | |
| $e_{max}$ | 0.950 | |
| $e_{min}$ | 0.796 | |
| $C_u$ | 2,056 | |
| $C_c$ | 0.794 | |
| c | 0.017 | |
| φ | 29.28 | ° |

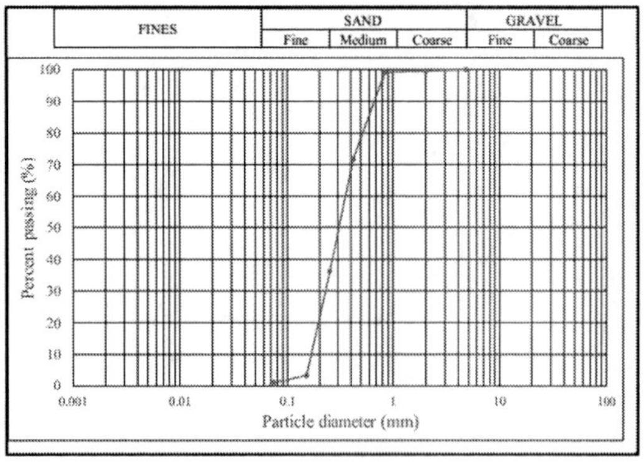


**Fig. 4.** Grain size distribution

Figure 4 shows the grain size gradation of the soil samples used. The distribution appears steep around the medium sand size, indicating that the sand grains are uniformly graded. This uniformity is essential to meet the requirements for preparing liquefaction soil samples.

**Table 2.** Experimental variations

| Mitigation | Symbol | Density relative (%) | | Frequency (Hz) |
|---|---|---|---|---|
| Shallow foundation | No ED | 40 | 60 | 1 |
| Shallow foundation with earthquake drain | ED | 40 | 60 | 1 |

Table 2 presents the experimental variations used in this study. The 'NO ED' symbol represents tests without mitigation, while the 'ED' symbol indicates tests with mitigation, where earthquake drains were used to reinforce the shallow foundation. The data analysed includes pressure transmitter readings from the top, middle, and bottom elevations.

# 4 Results

## 4.1 Excess pore water pressure (EPWP)

Excess Pore Water Pressure (EPWP) must be carefully managed in geotechnical engineering to prevent soil weakening and liquefaction, particularly in earthquake-prone areas. Techniques like earthquake drains help dissipate this pressure, improving soil stability and protecting infrastructure. In this study, EPWP measurements show the differences in pore pressure at various elevations before, during and post shaking. Figs. 5- 8 illustrates the variation in excess pore water pressure (EPWP) over time, comparing tests without earthquake drains (No ED) and with earthquake drains (ED) across different elevations (1, 2, and 3). The x-axis represents time, while the y-axis represents EPWP levels.

Solid line at Fig. 5 shows curves without earthquake drain used to mitigation. No ED 1 (blue solid line) shows the highest peak of EPWP, indicating significant pore pressure buildup and slow dissipation. The orange solid line reaches a moderate peak but is considerably lower than No ED 1, suggesting less EPWP buildup. No ED 3 (green solid line) shows the lowest EPWP, indicating minimal pressure buildup at this elevation. Dash line shows curves with earthquake drain used to mitigation. ED 1 (blue dashed line) shows a noticeable reduction in peak EPWP compared to No ED 1, demonstrating that earthquake drains effectively mitigate pore pressure. ED 2 (orange dashed line) also shows a reduction in EPWP compared to its No ED counterpart, though the reduction is less dramatic than in ED 1. ED 3 (green dashed line) shows the smallest peak EPWP and remains relatively flat, confirming effective mitigation at this elevation. In general, for relative densities 40% shows trend that EPWP rises rapidly 5 until 10 second after shaking begins and soon dissipate after end of shaking.

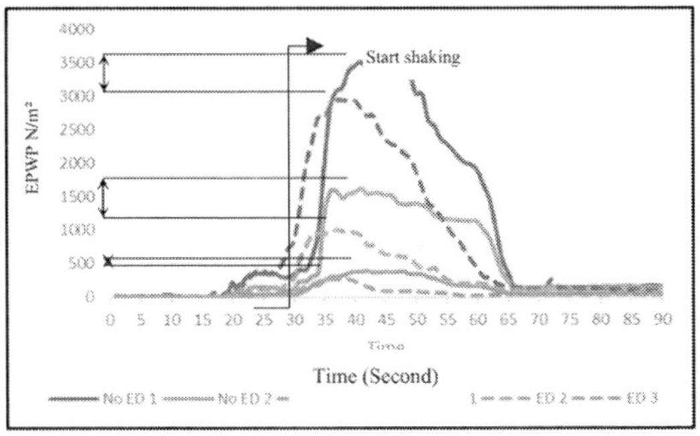


**Fig. 5.** EPWP for RD 40%, 1 Hz frequency

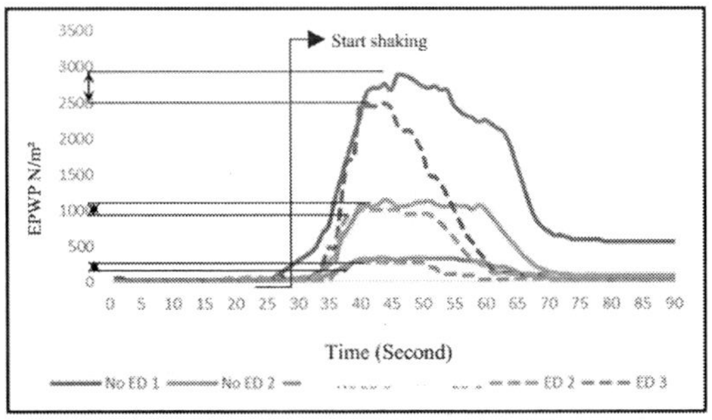


**Fig. 6.** EPWP for RD 60%, 1 Hz frequency

Similar to Fig. 5, Figure 6 illustrates the time series values of excess pore water pressure (EPWP) for sedimented soil with 60% relative density (RD), representing medium-density sand. The trend of the curve is comparable to that observed at 40% RD. However, upon closer examination, Figure 6 shows that EPWP increases 5 seconds longer than in Fig. 5 immediately after shaking begins, dissipating rapidly once shaking stops. This indicates that denser soil conditions result in a slightly longer duration for excess pore water pressure to dissipate.

Figure 7 shows the excess pore water pressure (EPWP) for 40% Relative Density (RD) at a frequency of 1.2 Hz. The solid line represents EPWP without mitigation, while the dashed line represents EPWP with earthquake drains as a mitigation measure. The curves demonstrate that the use of earthquake drains reduces EPWP effectively. Similarly, Figure 8 presents the EPWP results for 60% RD at the same frequency of 1.2 Hz, showing comparable outcomes. The use of earthquake drains accelerates water dissipation, thereby reducing pore water pressure during seismic events.

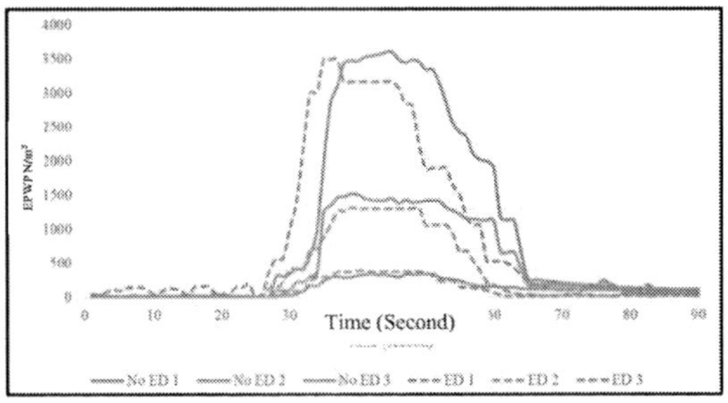


**Fig. 7.** EPWP for RD 40%, 1.2 Hz frequency

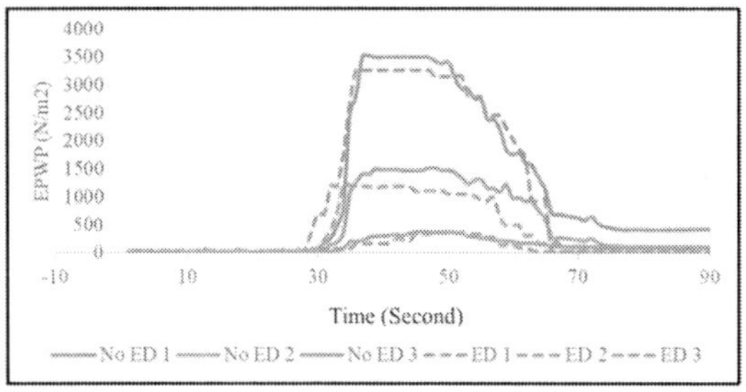


**Fig. 8.** EPWP for RD 60%, 1.2 Hz frequency

The key findings from Figs. 5-8 demonstrate earthquake drains significant role in reducing Excess Pore Water Pressure (EPWP) in soils subjected to dynamic loading. One of the most notable observations is the reduction of EPWP in both loose (40% relative density) and denser (60% relative density) soils. Earthquake drains allow excess water to dissipate from soil pores, thus preventing the buildup of excessive pressure that can weaken the soil structure during seismic activity.

The data further show the impact of relative density on EPWP. In looser soils with a relative density of 40%, the EPWP is higher due to the soil's tendency to compress more during loading. This compression causes a more significant accumulation of pore water pressure. In contrast, soils with a higher relative density of 60% exhibit lower initial EPWP. The denser arrangement of soil particles provides more excellent resistance to compression, resulting in less volume change and a more minor increase in pore water pressure. This demonstrates that soil density is critical in how pore water pressure develops under seismic conditions.

Moreover, earthquake drains effectively minimize the gap between EPWP and the effective stress within the soil. Since effective stress is the force that holds soil particles together, a smaller gap means that more stress is available to maintain the soil's strength. By controlling the rise in EPWP, earthquake drains help preserve soil stability during seismic events, reducing the risk of liquefaction and ground deformation. As a result, using earthquake drains enhances the seismic resistance of buildings and infrastructure, ensuring greater resilience and safety during earthquakes.

This study shows earthquake drains can effectively reduce the excess pore water pressure that occurs, thereby minimizing the gap between excess pore water pressure and effective soil stress. Which means that the earthquake drains play effectively in stabilizing the soil structure during an earthquake, thereby improving the safety and seismic resistance of the building.

## 4.2 Potential liquefaction

The ratio of excess pore water pressure to initial effective stress, denoted as $r_u$, plays a critical role in assessing liquefaction potential. The analysis of potential liquefaction ratio ($r_u$) values in relation to variations in Relative Density (RD) and earthquake frequency reveals that ru values increase significantly with higher RD and earthquake frequency. At lower relative densities, the ru values are greater, indicating that the soil is more susceptible to liquefaction. Additionally, higher earthquake frequencies can further elevate $r_u$ values, as faster seismic waves are more likely to trigger liquefaction. This analysis provides valuable insight into soil behavior under seismic conditions and its potential for liquefaction. The image below shows two graphs labeled Fig. 9 and 10, both depicting the potential liquefaction ($r_u$) over time for soil samples with varying Relative Densitiy (RD) and subjected to seismic frequencies. The x-axis represents time in seconds, while the y-axis represents the $r_u$ value. The $r_u > 1$ shows high potential liquefaction.

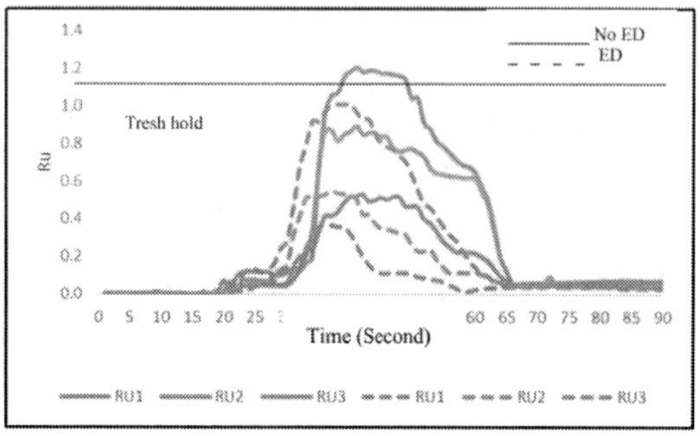


**Fig. 9.** Potential liquefaction ($r_u$) RD 40%, 1 Hz frequency

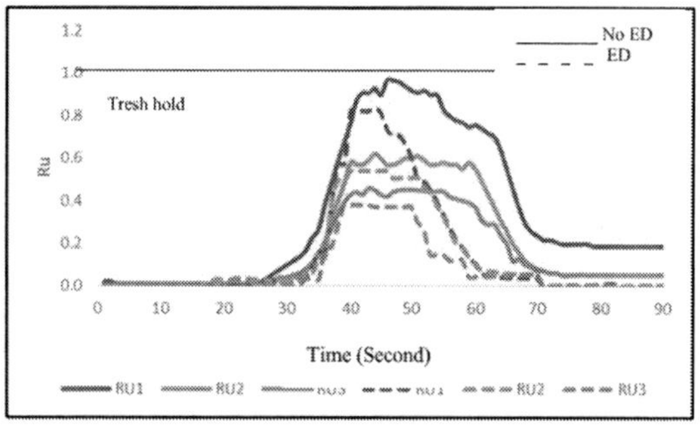


**Fig. 10.** Potential liquefaction ($r_u$) RD 60%, 1 Hz frequency

Figure 9 highlights a significant difference in the change of $r_u$ values between the shaking table tests with and without the use of earthquake drains. The solid line represents the ru values without mitigation, while the dashed line represents the $r_u$ values with earthquake drains. The dashed line is consistently lower than the solid line, demonstrating that earthquake drains effectively mitigate the increase in excess pore water pressure (EPWP).

Figure 10 follows a similar trend to Fig. 9, but the denser soil (60% RD) results in slightly lower ru values, indicating reduced liquefaction potential compared to the 40% RD case. Overall, the use of earthquake drains significantly reduces the $r_u$ values, particularly at the peak, effectively mitigating liquefaction potential.

A comparison between the two figures shows that Fig. 9 (40% RD) presents higher $r_u$ values, indicating that the soil is more prone to liquefaction compared to Fig. 10 (60% RD), where denser soil decreases the risk of liquefaction. In both figures, the effect of earthquake drains consistently lowers $r_u$ values, demonstrating their effectiveness in reducing liquefaction risk.

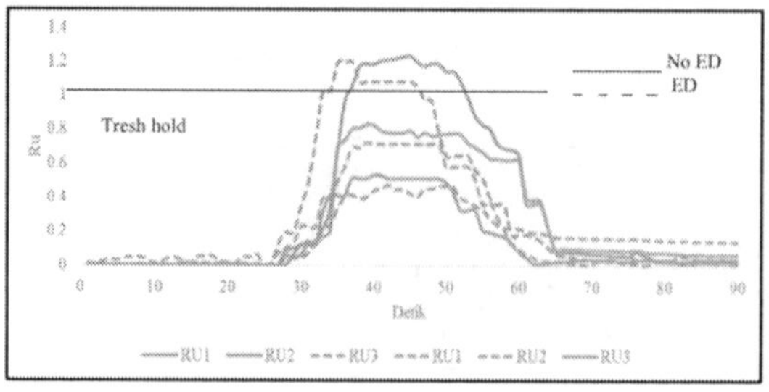


**Fig. 11.** Potential liquefaction ($r_u$) RD 40%, 1.2 Hz frequency

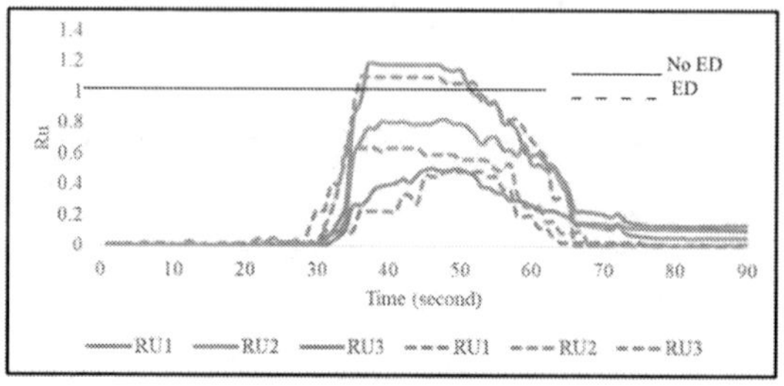


**Fig. 12.** Potential liquefaction ($r_u$) RD 60%, 1.2 Hz frequency

Figs. 11 and 12 show the $r_u$ values where the relative density of 40% loose shows a $r_u$ value higher than the relative density of 60%. The curves in both figures clearly show that higher relative densities and the use of earthquake drains significantly reduce liquefaction potential, with the denser soil in Fig. 12 showing a more pronounced reduction in $r_u$ values.

This shows that soil density conditions significantly affect the liquefaction potential. In addition, looking at Figs. 11 and 12 one by one, it shows that liquefaction occurs in the top of transmitter positions. In contrast, after mitigation using an earthquake drain, mitigation only occurs in the top layer, with a value that decreases from before, and the time of liquefaction is narrower. This shows that the earthquake drain can effectively quickly distribute pore water pressure so that liquefaction does not occur in specific layers.

### 4.3 Liquefaction-induced settlement

**Fig. 13.** Initial condition before shaking

The measurement of the foundation settlement that occurs using 4 LVDT sensors placed on the foundation will measure the settlement and be analyzed for settlement differences.

**Fig. 14.** Condition after shaking

Figure 13 shows the initial condition before shaking, where the sand is fully saturated. In contrast, Figure 14 illustrates the final condition after shaking. It is evident that water has risen to the surface, while the ground surface has subsided and become denser, causing the foundation to settle as it follows the ground's movement. This phenomenon is known as the change in relative density, which occurs immediately after the dissipation of water from soil pores due to cyclic loading.

Generally, settlement measure shows like the Figure 15, the denser the soil condition, the more the settlement will be reduced. Figure 15 shows the effect of using earthquake drains on settlement. The use can reduce the settlement by 2 % to 15% from the initial settlement without mitigation.

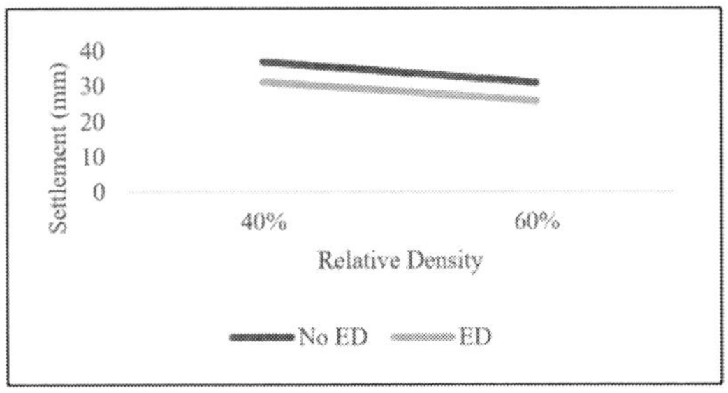


**Fig. 15.** EPWP for RD 60%, 1 Hz frequency

The subsidence value decreases as the relative density of the soil increases. Additionally, earthquake frequency plays a significant role in the extent of land subsidence—the stronger the earthquake, the greater the potential for subsidence. Besides these factors, the use of earthquake drains has been found to effectively mitigate the subsidence, reducing its severity even during seismic events.

## 5 Conclusion

The excess pore water pressure increase is highly dependent on soil depth, relative density, and seismic events. In samples with uniform relative density, the most significant excess pore water pressure occurs at the lowest position because pore water pressure trapped in deeper layers requires more time and pressure to escape. Lower relative density indicates looser soil, so if a seismic event occurs, the pore water pressure trying to rise will be more significant. Liquefaction potential is directly proportional to the increase in excess pore water pressure. However, in this assessment, a threshold determines whether the observed layer experiences liquefaction.

Mitigation using earthquake drains shows smaller excess pore water pressure values than without mitigation, and the liquefaction potential is also lower. Over time, earthquake drains can effectively dissipate water pressure from the lower layers directly to the upper layers, resulting in differences in the timing of excess pore water pressure increase. Earthquake drains can counteract the excess pressure that occurs. The reduction in excess pore water pressure positively impacts the soil-bearing capacity, helping to maintain the stability of the soil and the structures above it. Likewise, settlements can be reduced with mitigation using earthquake drains. The presence of earthquake drains (ED) consistently reduces EPWP across all elevations compared to the tests without mitigation (No ED), indicating the drains' success in dissipating excess pore water pressure during seismic events. This study demonstrates earthquake drains as a reliable and efficient measure to reduce liquefaction-induced settlement. However, further research is needed to optimize their design and application.

## References

1. Do, T.M., Laue, J., Mattsson, H. et al., Excess pore water pressure generation in fine granular materials under undrained cyclic triaxial loading. Geo-Engineering, **14**(8) (2023).
2. V. Maghvan, S., Imam, R., and Mc Cartney, J. S., Relative density effects on the bearing capacity of unsaturated sand. *Soils Found.* **59**(5), 1280-1291 (2019).
3. Salehi, Y., Shafaghat, A., and Khabbaz, H., A review on performance of stone columns as a ground improvement technique: Lessons learnt from past experiences and prospect for future development. Australian Geomechanics Journal. **57**(1), 71–91 (2022).
4. Brennan, A.J. and Madabhushi, G.S.P, Liquefaction and drainage in stratified soil. J. Geotech. Geoenviron. Eng. **131**(7), 876–885 (2005).
5. Kirkwood, P., and Dashti, S., Considerations for the mitigation of earthquake-induced soil liquefaction in urban environments. J. Geotech. Geoenviron. Eng. **144**(10), 1–15 (2018).
6. García-Torres S., Madabhushi, G.S.P., Performance of vertical drains in liquefaction mitigation under structures. Bulletin of Earthquake Engineering. **17**, 5849–5866 (2019).
7. J. Ramirez, S. Dashti, A. Liel, and B. Paramasivam, An experimental and numerical study of prefabricated vertical drains as a liquefaction countermeasure for mat-founded structures. Geo-Congress 2019: Earthquake Engineering and Soil Dynamics (GSP 308) (2019).

8. Brennan A.J., and Madabhushi, G.S.P., Effectiveness of vertical drains in mitigation of liquefaction. Soil Dynamics and Earthquake Engineering. **22**, 1059–1065 (2002).
9. Terzaghi, K., Peck, R. B., and Mesri, G., Soil Mechanics in Engineering Practice, (John Wiley and Sons, 1996).
10. Suravi Pal, Kousik Deb, Vertical composite drains for mitigating liquefaction hazard. Soil and Foundation. **59**(1), 196-207 (2019).
11. Salimah, A., Prakoso, W.A., et al., Uniformity of reconstitution of liquefiable sand specimen using a rainer box system. International Journal of GEOMATE. **26**(114), 9–16 (2024).

# Influence of soil structure interaction on the 4th story building

*Alex* Kurniawandy[1*], *Cindy* Dayu Tiara[1], *Riza* Aryanti[2], and *Muhammad* Aminsyah[2]

[1]Civil Engineering Department, Faculty of Engineering, Riau University, Pekanbaru, Riau, Indonesia
[2]Civil Engineering Department, Faculty of Engineering, Universitas Andalas, Padang, West Sumatra, Indonesia

**Abstract.** Design structure generally is carried out using fixed-based restraints, which assume soil is infinitely rigid. The soil has a certain rigidity and can deform a structure commonly called soil structure interaction (SSI). However, practitioners rarely consider interaction between soil and structural systems because of their high complexity and long computing time. The SSI reviews need to be carried out to obtain a more accurate analysis according to original conditions in the soil. The study aims to identify structural behavior with consideration of SSI in a case study of the 4th-story building in Pekanbaru. Structural analysis is done linearly using response spectrum analysis and nonlinear pushover analysis. Modeling was made into two types: model 1 using fixed base and model 2 using spring located on the points along the pile foundation as a representation of SSI. The analysis results show the natural period increases by 20% in the X direction and 18% in Y. The base shear decreases by 18% in the X direction and 16% in Y. The maximum displacement increases by 29% in the X direction and 24% in Y. Nonlinear pushover analysis results in an increase in the performance point value of structure with SSI modeling.

## 1 Introduction

Modeling structures using pinch supports assumes that the subgrade is rigid so that when the structural system is given an earthquake load, only the upper structure deforms while the soil does not move. This assumption is not in accordance with field conditions where, during an earthquake, the soil will experience movement. The process by which soil response and structural motion influence each other is called soil structure interaction [1-3].

Generally, structural analysis performed by practitioners rarely considers the interaction between soil and structure. However, soil-structure interaction has a prominent influence on heavy construction and construction built on soft soil. Therefore, it is necessary to review aspects of soil-structure interaction to obtain data closer to the field's original conditions.

Investigating the magnitude of the effect of soil-structure interaction (SSI) can be done by analyzing the behavior of the structure through linear analysis that occurs when the building is in elastic condition. The effect of soil-structure interaction (SSI) is closely related to dynamic loads. Therefore, linear analysis is performed with dynamic analysis using earthquake response spectrum data, which is considered more efficient in the computational process but can still provide a global picture and sufficient information about the structure's response to dynamic loads.

In addition, it is also necessary to consider the performance of structures involving SSIs to see if there is a change in structural performance when SSI are considered in the analysis. Structural performance analysis can be performed using advanced nonlinear pushover analysis, which is commonly used. This nonlinear pushover analysis is carried out to analyze the inelastic behavior of the structure against various earthquake intensities applied to the structure so that the performance of the structure in critical conditions can be known [4].

Therefore, the author will conduct a structural analysis by considering soil-structure interaction (SSI) to identify how much influence SSI has on structural response and structural performance. This research will discuss the influence of SSI on the existing 4th-story building in Pekanbaru City, Riau Province. This building uses reinforced concrete as its main material. The structural analysis was conducted using a linear dynamic analysis approach with earthquake response spectrum data and nonlinear pushover analysis. This research output compares base shear force, vibration period, displacement, and structural performance in structural analysis with and without SSI consideration.

## 2 Methodology

This research uses an existing structure as a case study with a structural system in the form of a moment-bearing frame system. This building is located in Pekanbaru City and is categorized as having medium soil (SD). Structural element data, loading data, and soil data are based on the detailed engineering design document of the building used as the object.

* Corresponding author: alexkurniawandy@eng.unri.ac.id

Soil structure interaction (SSI) modeling is performed using the indirect deformable piles interactive spings method by modeling the upper structure and lower structure in pile foundations into a single substructure. The soil is then modeled as a spring by inputting dynamic stiffness values that relate dynamic loads to displacements at several points of contact between the soil and foundation.

The soil spring stiffness in pile foundations consists of three types: the vertical direction spring at the base of the pile (end bearing), the horizontal direction lateral spring along the pile, and the vertical skin friction resistance spring (skin friction). An illustration the three types of springs can be seen in Fig. 1 below.

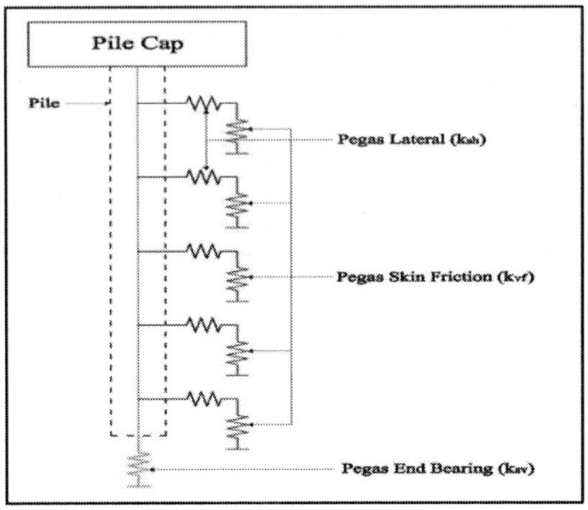


**Fig 1.** Illustration of soil-structure interaction in pile foundations

The soil spring stiffness in the horizontal and vertical directions can be calculated using the subgrade reaction modulus parameter ($K_s$) obtained through the Equation. (1) as follows [4]:

$$K_s = \frac{0.65}{D} \times \sqrt[12]{\frac{E_s D^4}{E_p I_p}} \times \frac{E_s}{1-\mu_s^2} \tag{1}$$

In which:
$K_s$ = modulus of subgrade reaction (kN/m$^3$)
$E_s$ = modulus of soil elasticity (kN/m$^2$)
$E_p$ = modulus of elasticity of the foundation (kN/m$^2$)
$I_p$ = foundation cross-section inertia (kg/cm$^2$)
$\mu_s$ = soil Poisson's ratio
$D$ = pole diameter or pole side dimension (m)

The vertical direction spring stiffness at the base of the pile ($K_{sv}$) can be calculated using the Equation (2) below:

$$K_{sv} = K_s \times A_p \tag{2}$$

In which:
$K_{sv}$ = end bearing spring stiffness (kN/m$^2$)
$A_p$ = cross-sectional area of the pile (m$^2$)

The horizontal direction spring stiffness along the pile ($K_{sh}$) can be calculated using the Equation (3) below:

$$K_{sh} = K_s \times D \times L \tag{3}$$

In which:
$K_{sh}$ = horizontal direction spring stiffness (kN/m$^2$)
$L$ = length of pile foundation (m)

The stiffness of the vertical skin friction resistance spring ($K_{vf}$) can be calculated using the Equation (4) below:

$$K_{vf} = 1.8 \times E_s \times \eta \times \lambda^{\left(0.5-\frac{\lambda}{\eta}\right)} \times \alpha \tag{4}$$

In which:
$K_{vf}$ = skin friction spring stiffness (kN/m$^2$)
$\eta$ = pile ratio
$\lambda$ = pile-soil stiffness ratio
$\alpha$ = vertical spring stiffness effect along the pile length

The values of pile ratio ($\eta$) and pile-soil stiffness ratio ($\lambda$) can be calculated using the Equations (5) and (6) below:

$$\eta = \frac{L}{D} \tag{5}$$

$$\lambda = \frac{E_p}{E_s} \tag{6}$$

The $\alpha$ value is an expression used to distribute the effect of soil vertical stiffness along the pile. The value of $\alpha = 1.00$ is used for the spring positioned at the base of the pile. While the value of $\alpha$ used for the spring along the pile can be calculated using the Equation (7) below:

$$\alpha = \frac{Z \times \Delta_z}{L^2} \tag{7}$$

In which:
$Z$ = depth of spring placement point (m)
$\Delta_z$ = distance between springs (m)

The spring stiffness values were calculated based on soil investigation by using Cone Penetration Test (CPT) at the construction site. Equations (1) to (7) are carried out to determine the stiffness and result of the calculation presented in Table 1 below.

**Table 1.** Calculation results of ground spring stiffness

| Soil Depth (m) | $K_{sh}$ (kN/m) | $K_{vf}$ (kN/m/m) | $K_{sv}$ (kN/m) |
|---|---|---|---|
| 1 | 12,124.81 | 0 | - |
| 2 | 26,308.18 | 0 | - |
| 3 | 62,249.65 | 0 | - |
| 4 | 154,325.83 | 0 | - |
| 5 | 327,005.03 | 0 | - |
| 6 | 291,731.83 | 0 | - |
| 7 | 380,461.25 | 0 | - |
| 8 | 398,413.57 | 0 | - |
| 9 | 160,706.13 | 0 | - |
| 10 | 75,843.19 | 0 | - |
| 11 | 452,636.27 | 0 | - |
| 12 | 340,524.42 | 0 | - |
| 13 | 398,413.57 | 0 | - |
| 14 | 264,384.39 | 0 | - |
| 15 | 371,370.33 | 0 | - |
| 16 | 692,899.53 | 0 | 12,991.87 |

## 3 Numerical model

This study uses two types of structural models, namely model 1 in the form of a top structure with a pinch support and model 2 in the form of a top structure and bottom structure modeled as a whole with a spring support placed at the nodal along the pile foundation as a representation of SSI. Model 1 (fixed base) and model 2 (SSI) used in this study are presented in Figs. 2 and 3.

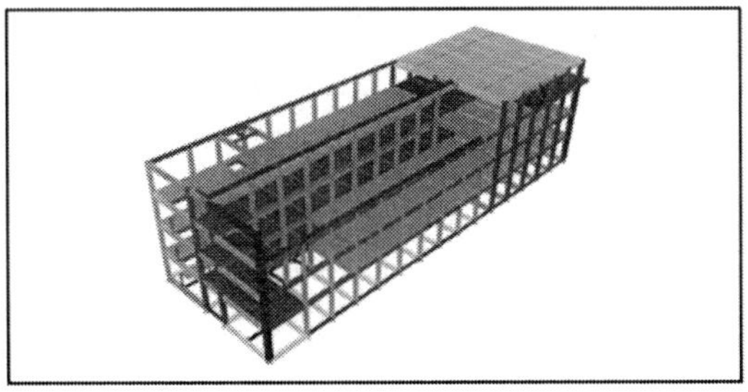

**Fig 2.** Model 1 fixed base

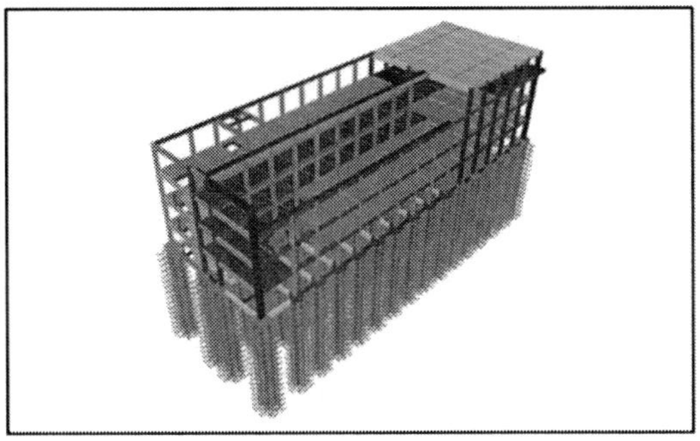

**Fig 3.** Model 2 soil structure interaction (SSI)

The structural analysis is carried out with a linear dynamic analysis approach with earthquake response spectrum data and further analysis in the form of nonlinear static analysis (pushover analysis) [5-7]. The plan response spectrum used is based on the Indonesian National Standard SNI 1726: 2019 based on the location of the building [8].

The building is located in Pekanbaru, and the design earthquake parameters include a short-period acceleration ($S_s$) of 0.453 g, and a 1-s period acceleration ($S_1$) of 0.281 g. The response spectrum graph is presented in Fig. 4 below.

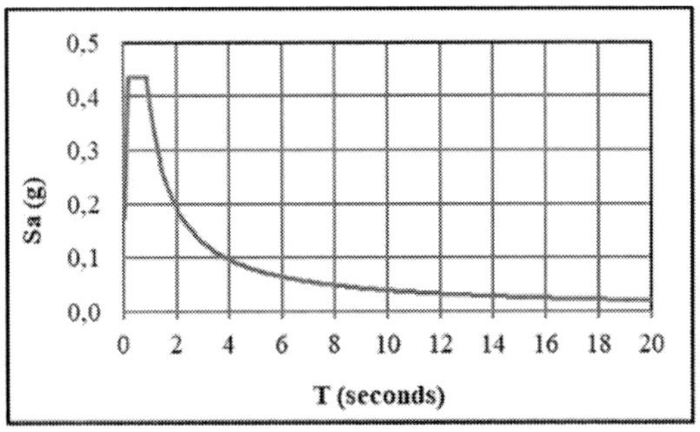


**Fig 4.** Design response spectral at the location of the existing building

Further analysis in the form of nonlinear pushover analysis was conducted based on ASCE 41 [9]. The nonlinear pushover analysis output in the form of a capacity curve is a graph of the relationship between the shear force applied to the structure and increased periodically with the displacement that occurs at the review point due to the application of the previous load.

## 4 Results and discussion

The results of the comparison of the two structural models show that the consideration of SSI in the existing structure produces an increase in the natural period of the structure, as can be seen in Table 2.

**Table 2.** Natural period of the structure

| Structure Model | Period (seconds) | |
|---|---|---|
| | X | Y |
| Model 1 (Fixed Base) | 0.98 | 0.90 |
| Model 2 (Soil-Structure Interaction) | 1.18 | 1.06 |

The results of the comparison of the natural periods of the two models are presented in Fig. 5 below.

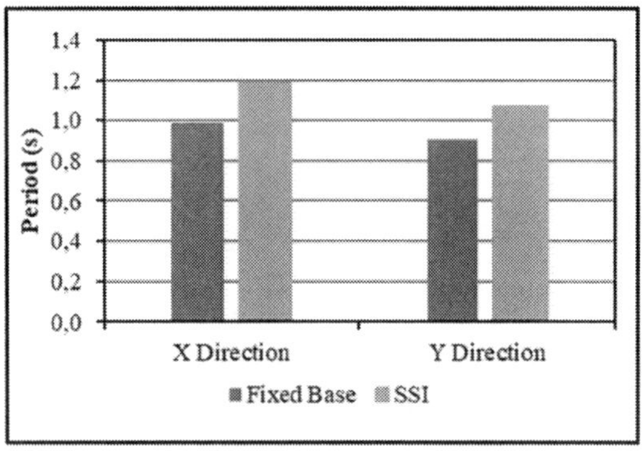


**Fig 5.** Comparison of natural period of structures

### 4.1 Shear force

The comparison of the two structural models shows that the consideration of SSI in the existing structure resulted in an increase in the base shear force of the structure, which can be seen in Table 3.

**Table 3.** Shear force of structure

| Structure Model | Shear Force (kN) | |
|---|---|---|
| | X | Y |
| Model 1 (Fixed Base) | 2,990.98 | 3,258.97 |
| Model 2 (Soil-Structure Interaction) | 2,466.50 | 2,740.16 |

The results of the comparison of the natural periods of the two models are presented in Fig. 5 above and Fig. 6 below.

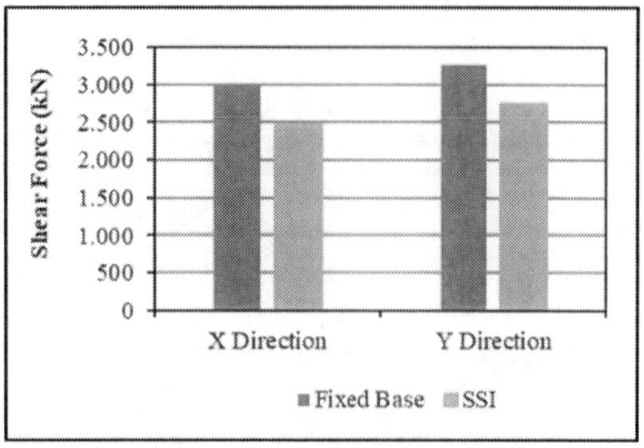


**Fig 6.** Comparison of structure base shear force

The results of the comparison of the two structural models show that the consideration of SSI in the existing structure results in changes in the distribution of shear forces received by each floor of the structure, which can be seen in Tables 5 and 6.

**Table 4.** Shear force distribution for fixed-based structure

| Story | Elevation (m) | Position | Shear Force (kN) | |
|---|---|---|---|---|
| | | | X dir. | Y dir. |
| 4 | 15.95 | Top | 751.35 | 810.58 |
| | | Bottom | 751.35 | 810.58 |
| 3 | 11.95 | Top | 1,790.35 | 1,954.51 |
| | | Bottom | 1,802.05 | 1,967.72 |
| 2 | 7.95 | Top | 2,579.53 | 2,825.12 |
| | | Bottom | 2,587.31 | 2,833.21 |
| 1 | 3.95 | Top | 2,987.90 | 3,256.41 |
| | | Bottom | 2,990.98 | 3,258.97 |
| Base | 0.0 | Top | 0 | 0 |
| | | Bottom | 0 | 0 |

As shown in Tables 4 and 5, the shear force distribution in the SSI model is higher than that in the fixed-base model.

**Table 5.** Shear force distribution for SSI structure

| Story | Elevation (m) | Position | Shear Force (kN) | |
|---|---|---|---|---|
| | | | X dir. | Y dir. |
| 4 | 15.95 | Top | 573.89 | 624.07 |
| | | Bottom | 573.89 | 624.07 |
| 3 | 11.95 | Top | 1,386.86 | 1,535.18 |
| | | Bottom | 1,396.26 | 1,546.06 |
| 2 | 7.95 | Top | 2,032.27 | 2,277.63 |
| | | Bottom | 2,039.14 | 2,285.15 |
| 1 | 3.95 | Top | 2,462.58 | 2,736.71 |
| | | Bottom | 2,466.51 | 2,740.16 |
| Base | 0.0 | Top | 2,488.27 | 2,763.71 |
| | | Bottom | 0.00 | 0.00 |

The results of the comparison of shear force distribution for each floor of the structure are presented in Figs. 7 and 8 below.

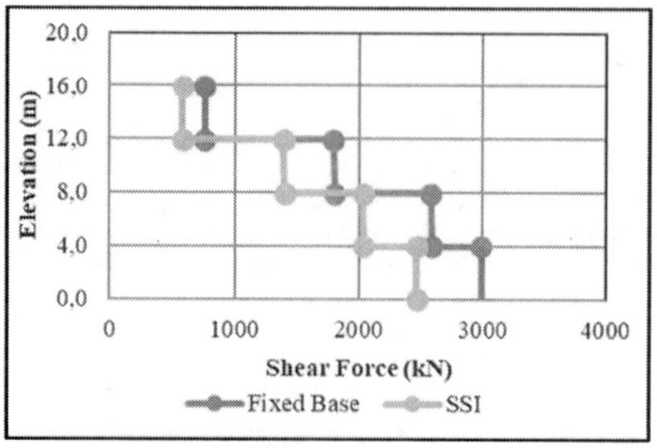


**Fig 7.** Shear force distribution in X-direction

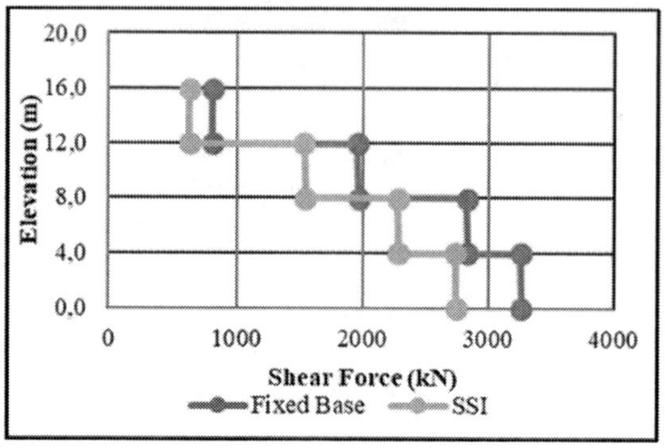


**Fig 8.** Shear force distribution in Y-direction

## 4.2 Maximum displacement

The results of the comparison of the two structural models show that the consideration of SSI in the existing structure results in an increase in the maximum displacement of each floor of the structure, which can be seen in Tables 6 and 7 below.

**Table 6.** Maximum displacement of model 1

| Story | Elevation (m) | Displacement (mm) | |
|---|---|---|---|
| | | X dir. | Y dir. |
| 4 | 15.95 | 25.96 | 33.32 |
| 3 | 11.95 | 20.52 | 27.89 |
| 2 | 7.95 | 14.13 | 17.24 |
| 1 | 3.95 | 6.05 | 7.27 |
| Base | 0.0 | 0.00 | 0.00 |

**Table 7.** Maximum displacement of model 2

| Story | Elevation (m) | Displacement (mm) | |
|---|---|---|---|
| | | X dir. | Y dir. |
| 4 | 15.95 | 29.85 | 37.37 |
| 3 | 11.95 | 25.21 | 31.96 |
| 2 | 7.95 | 18.89 | 22.21 |
| 1 | 3.95 | 10.57 | 12.11 |
| Base | 0.0. | 1.99 | 2.17 |

The results of the displacement comparison for each floor of the structure are presented in Figs. 9 and 10 below.

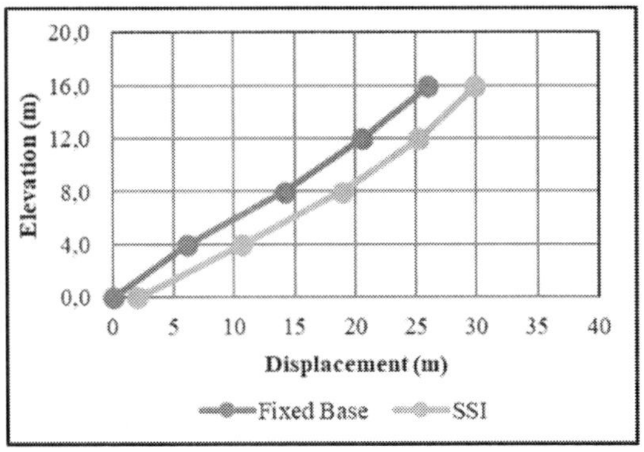


**Fig 9.** Maximum displacement in x-direction

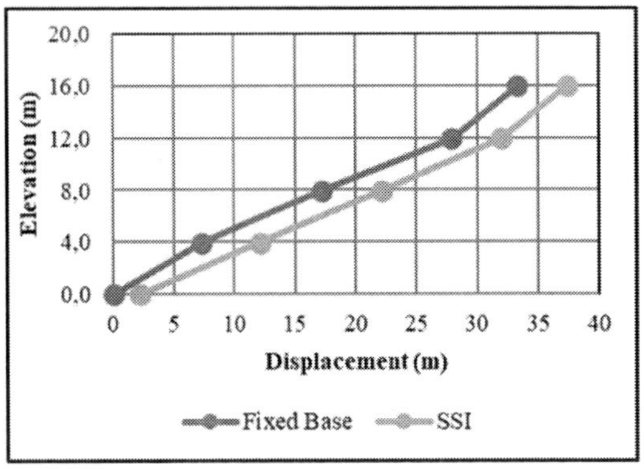


**Fig 10.** Maximum displacement in y-direction

It can be seen from the above figure that for the SSI model there is a shift in the restraint base of the structure. Then, the displacement of the SSI model is larger than that of the fixed-base model. The trend or shape of the displacement of each floor is almost identical for both models.

### 4.3 Story drift

The results of the comparison of the two structural models shows that the consideration of SSI in the existing structure results in changes in story drift, which can be seen from Tables 8-11.

**Table 8.** Story drift fixed base structure in x-direction

| Story | Floor Height h | Displ $\delta_X$ | Elastic Drift $\Delta_{elastic}$ | Inelastic Drift $\Delta_{inelastic}$ | Dritf Ratio X |
|---|---|---|---|---|---|
| | (mm) | (mm) | (mm) | (mm) | (%) |
| 4 | 4,000 | 25.96 | 5.45 | 19.99 | 0.50 |
| 3 | 4,000 | 20.52 | 6.38 | 23.42 | 0.59 |
| 2 | 4,000 | 14.12 | 8.07 | 29.61 | 0.74 |
| 1 | 4,250 | 6.05 | 6.05 | 22.19 | 0.52 |
| Base | 0 | - | - | - | - |

**Table 9.** Story drift fixed-base structure in y-direction

| Story | Floor Height h | Displ $\delta_Y$ | Elastic Drift $\Delta_{elastic}$ | Inelastic Drift $\Delta_{inelastic}$ | Dritf Ratio Y |
|---|---|---|---|---|---|
| | (mm) | (mm) | (mm) | (mm) | (%) |
| 4 | 4,000 | 33.30 | 5.42 | 19.87 | 0.50 |
| 3 | 4,000 | 27.88 | 10.65 | 39.05 | 0.98 |
| 2 | 4,000 | 17.23 | 9.97 | 36.54 | 0.91 |
| 1 | 4,250 | 7.27 | 7.27 | 26.65 | 0.63 |
| Base | 0 | - | - | - | - |

**Table 10.** Story drift SSI structure in x-direction

| Story | Floor Height h | Displ. $\delta_X$ | Total Drift $\Delta_{total}$ | Rotation $\theta$ | Displ. Rotation $\delta_\theta$ | Horizontal Drift $\Delta^*$ | Elastic Drift $\Delta_{elastic}$ | Inelastic Drift $\Delta_{inelastic}$ | Dritf Ratio X |
|---|---|---|---|---|---|---|---|---|---|
| | (mm) | (mm) | (mm) | (rad) | (mm) | (mm) | (mm) | (mm) | (%) |
| 4 | 4,000 | 29.85 | 4.646 | 0.000917 | 3.668 | 0.978 | 0.978 | 3.586 | 0.09 |
| 3 | 4,000 | 25.21 | 6.311 | 0.000917 | 3.668 | 2.643 | 2.643 | 9.691 | 0.24 |
| 2 | 4,000 | 18.89 | 8.317 | 0.000917 | 3.668 | 4.649 | 4.649 | 17.046 | 0.43 |
| 1 | 4,250 | 10.58 | 8.584 | 0.000917 | 3.897 | 4.687 | 4.687 | 17.185 | 0.40 |
| Base | 0.0 | 1.99 | 1.993 | 0.000000 | 0.000 | 1.993 | 1.993 | 7.308 | 0.05 |

**Table 11.** Story drift SSI structure (SSI) y-direction

| Story | Floor Height h | Displ. $\delta_Y$ | Total Drift $\Delta_{total}$ | Rotation $\theta$ | Displ. Rotation $\delta_\theta$ | Horizontal Drift $\Delta^*$ | Elastic Drift $\Delta_{elastic}$ | Inelastic Drift $\Delta_{inelastic}$ | Dritf Ratio Y |
|---|---|---|---|---|---|---|---|---|---|
| | (mm) | (mm) | (mm) | (rad) | (mm) | (mm) | (mm) | (mm) | (%) |
| 4 | 4,000 | 37.37 | 5.410 | 0.000812 | 3.248 | 2.162 | 2.162 | 7.927 | 0.20 |
| 3 | 4,000 | 31.96 | 9.751 | 0.000812 | 3.248 | 6.503 | 6.503 | 23.844 | 0.60 |
| 2 | 4,000 | 22.21 | 10.109 | 0.000812 | 3.248 | 6.861 | 6.861 | 25.157 | 0.63 |
| 1 | 4,250 | 12.11 | 9.933 | 0.000812 | 3.451 | 6.482 | 6.482 | 23.767 | 0.56 |
| Base | 0.0 | 2.16 | 2.168 | 0.000812 | 0.000 | 2.168 | 2.168 | 7.949 | 0.05 |

 https://doi.org/10.1051/e3sconf/202560413003

The story drift comparison results are also presented in Figs. 11 and 12 below.

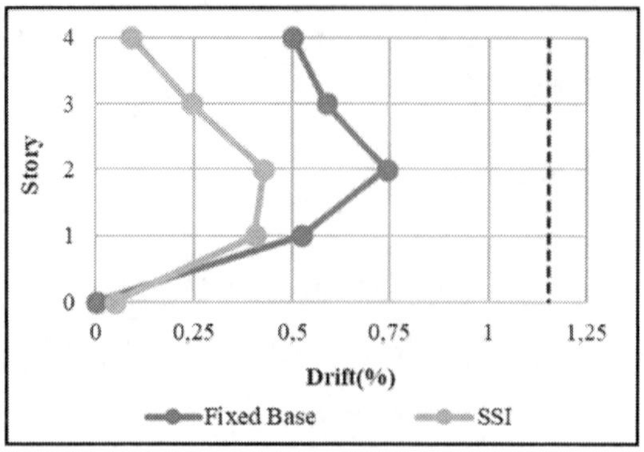


**Fig 11.** Story drift in x-direction

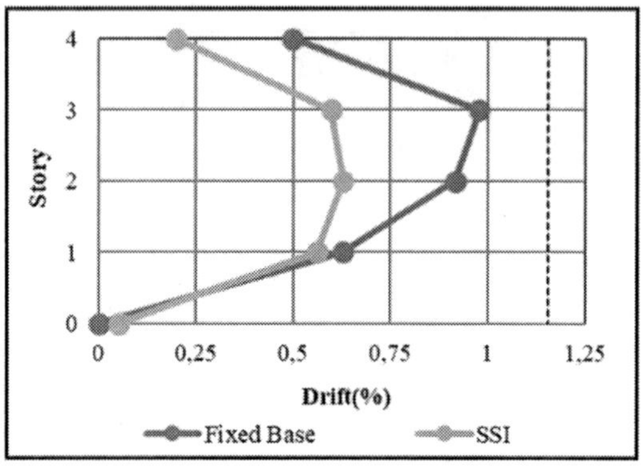


**Fig 12.** Story drift in y-direction

## 4.4 Capacity curve and performance points

The results of the comparison of the two structural models show that the consideration of SSI in the existing structure resulted in an increase in the performance point value. The comparison of the capacity curves of the two models in the X and Y directions is presented in Figs. 13 and 14.

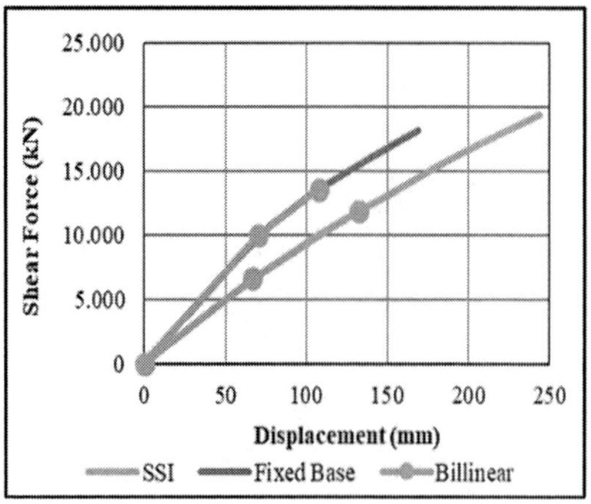


**Fig 13.** Capacity curves in x-direction

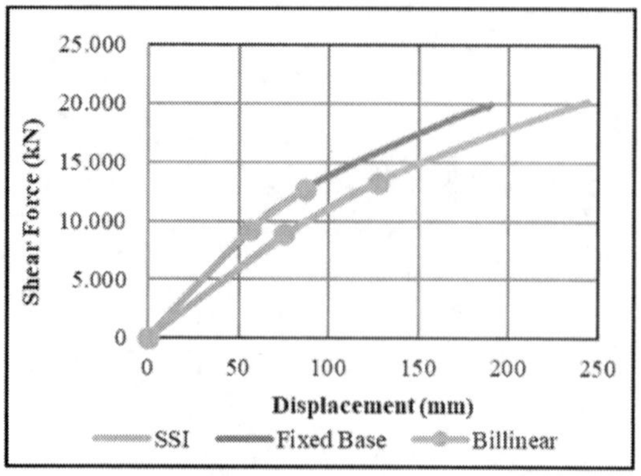


**Fig 14.** Capacity curves in y-direction

The performance points generated from the capacity curves for model 1 (fixed base) and model 2 (SSI) in the X and Y directions are shown in Table 12 as follows.

**Table 12.** Performance point

| Model | Directions | Displacement (mm) | Shear Force (kN) |
|---|---|---|---|
| Model 1 (Fixed Base) | X | 107.9 | 13,613.8 |
| | Y | 86.4 | 12,667.2 |
| Model 2 (SSI) | X | 132.7 | 11,906.1 |
| | Y | 127.2 | 13,297.5 |

From the table above, it can be seen that the displacement that occurs in the SSI model is greater than the fixed-base model at the performance point.

## 4.5 Plastic joint mechanism

The results of the comparison of the two structural models shows that the presence of SSI considerations in the existing structure results in changes in the plastic joint mechanism. The plastic hinge mechanisms produced through nonlinear pushover analysis on each structural model are presented in Figs. 15 to 18 below.

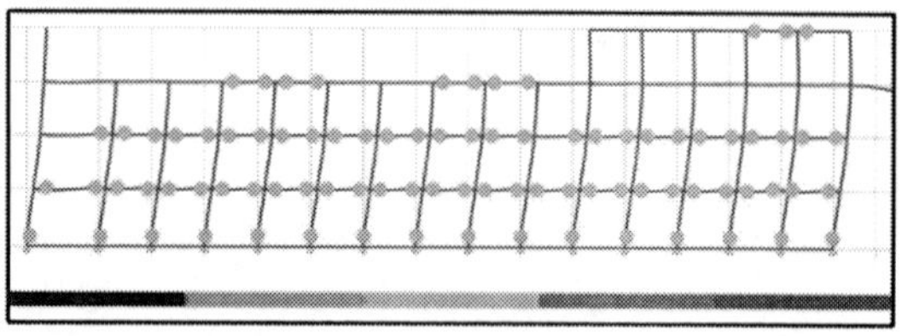

**Fig 15.** Plastic hinge mechanism of fixed base structure in x-direction

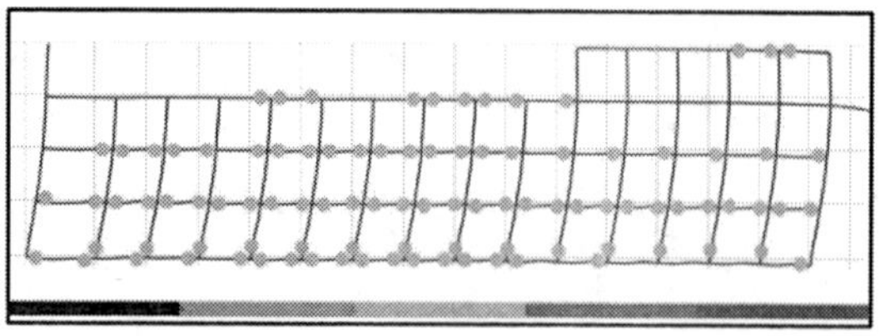

**Fig 16.** Plastic hinge mechanism of SSI structure in x-direction

From Figs. 15-18, there is no significant difference in the plastic joints that occur in either the fixed-base model or the SSI model. In the SSI model, there is plastification in the tie-beams at the base of the column because the base joint can be displaced.

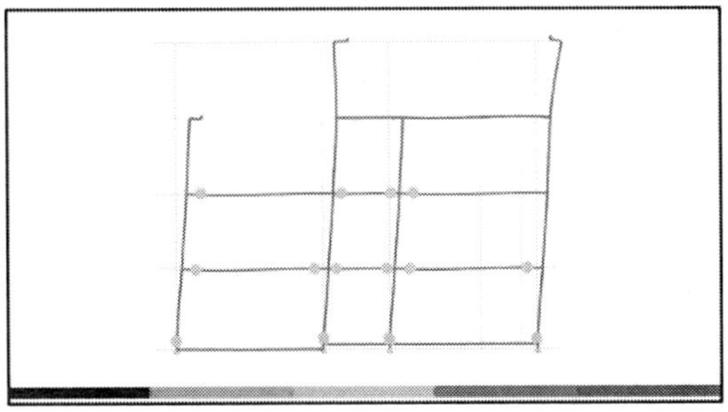

**Fig 17.** Plastic hinge mechanism fixed base structure in y-direction

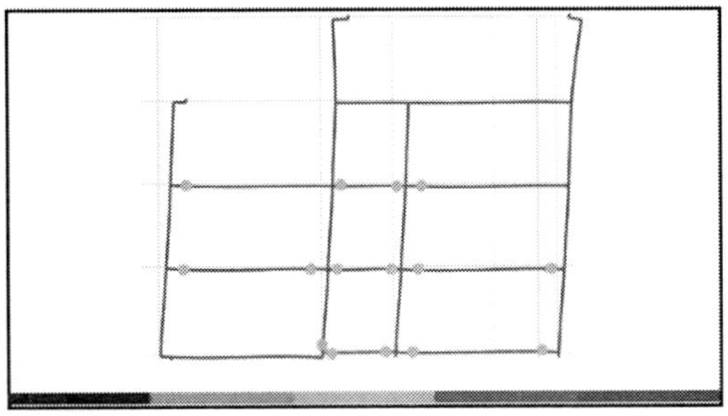

**Fig 18.** Plastic hinge mechanism SSI structure in y-direction

## 5 Conclusions

The effect of soil-structure interaction on a 4-story building located in Pekanbaru has been analyzed in this study. The conclusions obtained from this research are:

1. There is an increase in the natural period of the structure when SSI is considered. The increase in the SSI model is 20% in the X direction and 18% in the Y direction.
2. There is a decrease in the base shear force when SSI is considered by 18% in the X direction and 16% in the Y direction.
3. There is an increase in the maximum displacement of each floor when SSI is considered. With an average percentage of 29% in the X direction and 24% in the Y direction.
4. Nonlinear pushover analysis results in an increase in the performance point value of the structure with SSI modeling. The tie-beam at the base of the structure has plastification because the base joint can be displaced.

## References

1 F. M. Wani, J. Vemuri, C. Rajaram, D. V. Babu R, Effect of soil structure interaction on the dynamic response of reinforced concrete structures. Nat. Hazards Res., **2** (2022). https://doi.org/10.1016/j.nhres.2022.11.002.

2 S. Mokhles Al Samadi, Resonance effect on structures subjected to earthquake loads due to soil structure interaction (University of Jordan, Amman, 2014). https://doi.org/10.13140/RG.2.1.2380.1760.

3 A. Yanik, Y. Ulus, Soil–structure interaction consideration for base isolated structures under earthquake excitation. Buildings. **13** (2023). https://doi.org/10.3390/buildings13040915.

4 A. Kurniawandy, S. Nakazawa, A proposal of seismic index for existing buildings in Indonesia using pushover analysis. J. Eng. Technol. Sci., **52** (2020). https://doi.org/10.5614/j.eng.technol.sci.2020.52.3.2.

5 S. B. Magade, R. K. Ingle, Influence of clear edge distance and spacing of piles on failure of pile cap. Iran. J. Sci. Technol. - Trans. Civ. Eng., **44** (2020). https://doi.org/10.1007/s40996-019-00285-9.

6 A. Kurniawandy, S. Nakazawa, Seismic performance evaluation of existing building using seismic index method. MATEC Web Conf., **276** (2019). https://doi.org/10.1051/matecconf/201927601015.

7 A. Kurniawandy, S. Nakazawa, A. Hendry, Ridwan, R. Firdaus, Structural building screening and evaluation International Conference in Civil Engineering 2017. AIP Conf. Proc. 020031 (2017)

8 Indonesian code, *SNI 1726:2019* earthquake resistant design code for building and non-building of Indonesia (in Indonesian) (Standardization Agency of Indonesia, Jakarta, 2019)

9 American Society of Civil Engineers., ASCE/SEI 41-13 seismic evaluation and retrofit of existing building. (Reston, Virginia, 2013)

# Lessons learnt from community preparedness for Mount Merapi and Mount Marapi eruption disasters

*Bambang* Istijono[1*], *Fauzi*[1], *Muhammad* Al Giffari[1], *Dian* Hadiyansyah[2], *Andriani*[3], *Niken* Silmi Surjandari[4], *Anggih* Prastiyo[1], and *Marlon* de Luna Era[5]

[1]Graduate School, Universitas Andalas, Padang, Indonesia
[2]Energy and Mineral Resources Office, West Sumatra Government, Padang, Indonesia
[3]Civil Engineering Department, Faculty of Engineering, Universitas Andalas, Padang, Indonesia
[4]Civil Engineering Department, Faculty of Engineering, Universitas Sebelas Maret, Solo, Indonesia
[5]Behavioral Sciences Department, De La Salle University, Manila, Philippines

**Abstract.** Mount Merapi, located on the border of Central Java and the Special Region of Yogyakarta, and Mount Marapi in West Sumatra are active volcanoes. Eruption activities will have positive and negative impacts on the surrounding environment. The eruptions of Marapi on December 3, 2023, and Merapi on March 4, 2024, caused severe damage to road infrastructure, buildings, bridges, and settlements and resulted in casualties. This article describes the experiences of communities after the eruption of the two volcanoes. This research uses a combination of literature review and field observation methods to analyze how communities in the Mount Merapi and Mount Marapi areas prepare for disasters. The results of this study show that to increase community resilience to volcanic disasters and flash floods triggered by Mount Merapi and Mount Marapi, collaboration between the government, disaster agencies, and local residents is essential. Measures that can be taken are proactive evacuation protocols, infrastructure development such as sabo dams, and continuous education to increase awareness and preparedness. Recommendations include more intensive cooperation, educational initiatives, and strategic relocation from high-risk zones to effectively protect vulnerable communities.

## 1 Introduction

Indonesia is one of the countries that annually experience various natural disasters including volcanic eruptions. According to the Disaster Management Law No. 24 of 2007, "A disaster is an event or series of events that threaten and disrupt people's lives and livelihoods caused by both natural and non-natural factors as well as human factors resulting in human casualties, environmental damage, property losses, and psychological impacts". Volcanoes form when liquid magma, gas or other liquids emerge from the earth's crust. Indonesia is home to several active volcanoes, including Mount Agung in Bali, Mount Marapi in West Sumatra, Mount Merapi in Central Java, and Mount Bromo in East Java and others [1, 2].

Volcanic eruptions are one of Indonesia's most dangerous natural disasters, which will significantly impact society and even cost. According to Cassidy et al (2019) [3], a volcanic eruption is the process of releasing hot magma from the inside of the earth to the surface, which is indicated by eruptions that produce loose materials of various sizes or molten rock that forms lava or incandescent flow. Due to the dangers these eruptions various efforts are made to protect the community from these natural disasters. Mount Merapi and Mount Marapi are two of the most active volcanoes and frequently experience eruptions. The eruptions of Marapi and Merapi volcanoes not only produce volcanic ash and hot clouds, but also generate hazardous cold lava flows, severely disrupting the activities of the surrounding communities. Mount Merapi, standing at 2,968 meters above sea level is one of Indonesia most active volcanoes, located on the border of Central Java and the Special Region of Yogyakarta. This mountain is classified as a stratovolcano, distinguished by its conical shape and steep slopes, which are typical characteristic of this type of volcano. From 1906 to 2000, the Merapi experienced numerous eruptions [4]. Its activity has garnered significant public attention due to its scientific and cultural uniqueness. According to Nugraha et al (2019) [5], the current type of eruption of the Merapi is linked to the growth and collapse of the volcanic dome. This phenomenon has intensified since the 1800s. However, it's possible that dome growth-related eruptions occurred even earlier. Historical records indicate that more than 1,369 people died in the 1930 eruption of Mount Merapi. This eruption also destroyed many buildings and agricultural lands, and livestock also. In early 2021, Mount Merapi erupted again, causing cold lava flows and hot clouds, which forced many residents to evacuate in fear [1, 4].

* Corresponding author: bistijono@eng.unand.ac.id

Mount Marapi has erupted more than 500 times since 1770 and is one of the active volcanoes on the island of Sumatra [6]. According to the Geological Agency, due to limited research, there is insufficient data on this volcano in early December 2023, Mount Marapi erupted with a rumbling sound, and ash was ejected up to 3,000 meters above the crater. The Volcano Observation Post (PGA) in Bukittinggi monitors all volcanic activities, with seismograms recorded at eight seismic stations around the mountain. Monitoring Mount Marapi's seismic activity is critical for disaster warning and mitigation, with an alert status has been in place since August 3, 2011. The potential dangers of Mount Marapi include volcanic ash rain, lava flows and debris flows [7-9].

The eruption intervals of volcanoes range from 2 to 7 years, depending on the location of magma release. Volcanic eruptions such as those of Mount Merapi and Mount Marapi can be divided into several types. Central eruptions occur through the main crater, while the side eruptions occur on the slopes of the mountain. Crack eruptions occur along cracks or faults extending several kilometers, and eccentric eruptions occur in the same way as side eruptions. The process of volcanic eruptions varies but some produce molten lava, ash and solid material. A distinguishing feature of Mount Merapi's eruption is that they do not form a caldera (crater), instead they emit incandescent lava and hot clouds.

Volcanic eruptions are complex events that have a major impact on society and the environment. Adverse effects include casualties, infrastructure damage, and environmental pollution caused by hot clouds, lava flows, pyroclastic falls, toxic gases, and volcanic ash [10, 11]. Additionally, indirect dangers such as cold lava flows after rainfall have been observed following the eruptions of Mount Marapi on December 3, 2023, and Mount Merapi on March 4, 2024. Despite these hazards, eruptions can also yield benefits, such as thermal energy sources, more fertile soil, and mineral springs. Nevertheless, eruptions also produce sources of thermal energy, more fertile soil, and mineral springs [12].

Community preparedness is crucial to mitigate the risks and negative impacts of volcanic eruptions like those of Mount Merapi and Mount Marapi. Effective community preparedness can significantly reduce the impact of natural disasters. Local knowledge of eruption signs, evacuation routes and evacuation sites is essential. Enhancing early warning systems, evacuation procedures, and community response capabilities improves resilience and reduces the overall impact of volcanic disasters.

## 2 Research methodology

This research was carried out using literature study field observation method to analyze community preparedness for disaster around Mount Merapi and Mount Marapi. The qualitative research emphasizes on the condition of natural objects as critical instruments. Data collection was conducted using a triangulated approach with inductive data analysis. The research findings emphasize understanding and meaning over generalization [13, 14]. The method used to collect data and sources related to the research subject include literature study and field observation.

The literature study involves reviewing journal articles, newspapers and supporting books to understand the dynamics, obstacles and current learning opportunities. This approach helps identify ways to enhance the readiness of the community around the volcano.

The observation method involves making direct observations and recording the state or behavior of the target object to understand and interpret the relevant phenomena [15]. The results of the research are expected to enhance preparedness for volcanic eruptions and serves as a reference for policymakers and practitioners in developing effective interventions. This will help the community mitigate the negative impacts of volcanic disasters, which can be highly damaging and potentially deadly.

## 3 Results and discussion

Mount Marapi in West Sumatra suddenly erupted on Sunday 3 December 2023 at around 2:53 pm. At the time of eruption, the mountain was under Level II status, indicating an alert. The eruption produced volcanic ash that soared upwards in grey with a thick intensity. When the eruption occurred, dozens of climbers were trapped and as many as 23 people were reported dead [15]. The eruption of Mount Marapi gained widespread attention, because it occurred without any signs or symptoms beforehand. In fact, no volcanic earthquakes detected beforehand, and testimonies of the surviving climbers also stated that there were no signs of eruption at all.

Volcanic eruptions certainly have a negative impact on society as evidenced by the situation in Agam district. Based on data from Regional Disaster Management Agency (BPBD) of West Sumatra, the eruption resulted in 24 deaths, 22 injuries, 2 people being treated, and 709 people displaced. In addition to human casualties, regional infrastructure and livestock were also affected. The high number of affected individuals is partly due to a lack of understanding of how to protect themselves, the areas affected by the eruption of Mount Marapi have a high vulnerability to land movement disasters.

**Fig. 1.** The collapse of an elementary school and residential area in Nagari Koto Tuo due to flash floods following the eruption of Mount Marapi in West Sumatra.

**Fig. 2.** Infrastructure damaged by flash floods following the eruption of Mount Marapi in West Sumatera

The type of eruption of Mount Marapi in December 2023 differs significantly to the eruption in March 2024 which was a strombolian eruption. The eruption of Mount Marapi on Saturday 11 May 2024 triggered flash floods accompanied by mud or locally knowns as "galodo" leaving the affected victims traumatized. As felt by hundreds of victims at the Bukit Batabuah refugee post, at SDN 08 Kubang Duo, Koto Panjang, Canduang, Agam district, when the floods swept away their homes and other public facilities as shown in Figs. 1-3. The death toll from the flash floods and landslide in four areas in West Sumatra rose to 37 people [16].

Since the end of 2010, Mount Merapi in Central Java and the Special Region of Yogyakarta has been on an alert status or Level III, as designated by the Geological Agency through PVMBG. On 4 January 2021, lava domes, volcanoes, and hot clouds began to appear. At that time, the two lava domes of Mount Merapi had volumes of 1,598,700 $m^3$ and 2,267,400 $m^3$ respectively. This massive lava dome avalanche had the potential to cause hot clouds as far as 7 km to the southwest and 5 km to the south-southeast. The resulting lava dome collapse in the southwest did trigger hot clouds, and the eruption activity was considered very high. Lava flows were observed heading southwest, with the sound of their movement audible from the Kaliurang and Babadan Posts. The monitoring results showed that Mount Merapi's volcanic activity was still at an alert level. Potential hazards include lava flows, hot clouds in the south-southwest area, and ejections of volcanic material during explosive eruptions.

**Table 1.** Population data and the impact of the eruption of the Merapi volcano and the Marapi volcano

| Impact of eruptions | Mount Merapi | Mount Marapi |
|---|---|---|
| Eruption date | October 26, 2010 | December 3, 2023 |
| Casualties | 353 people | 70 people |
| Number of refugees | 350,000 people | 3,595 people |
| Infrastructure damage | High | Medium |
| Economic impact | Hundreds of million dollars | Local disadvantages |

Based on the Table 1 above, it is evident that the eruption of Mount Merapi in 2010 had a more significant impact than the eruption of Mount Marapi in December 2023, both in terms of the number of casualties, the number of refugees, infrastructure damage, and economic impact as shown in Fig. 3.

**Fig. 3.** Settlements destroyed by the eruption of Mount Merapi in Central Java

The Merapi volcano tends to exhibit higher volcanic activity and is more destructive than the Marapi volcano. In accordance with the Regulation of the Head of National Board for Disaster Management (BNPB) number 2 of 2012 concerning General Guidelines for Disaster Risk Assessment, efforts to reduce disaster risk involve understanding the relationship between threats, vulnerabilities, and capacities. These efforts include minimizing regional threats; reducing the vulnerability of threatened areas; and increasing the capacity of threatened areas. Therefore, to enhance the ability of the community to face and manage volcanic eruption disasters, various strategies and activities are implemented, examples are shown in Table 2 and 3.

**Table 2.** Availability of disaster management infrastructure

| Type | Mount Merapi (unit) | Mount Marapi (unit) |
|---|---|---|
| Sabo-dam | 272 | 2 |
| Observation post | 5 | 1 |
| Disaster Management Command post | 3 | - |
| Early warning system | 5 | 11 |
| Rivers that have the potential to channel cold lava flows | 15 | 25 |
| Evacuation routes | 3 | - |
| Temporary evacuation site | 6 | - |
| Livestock evacuation site | 3 | - |

The Klaten and Sleman Regency Governments implemented strategies to manage the threat of a Mount Merapi eruption during the COVID-19 pandemic. Despite the dual challenges, local governments and communities successfully controlled multi-hazard disasters. As a result, during the simultaneous occurrence of these two disasters, the community remained safe in evacuation camps, and no COVID-19 cases were reported among the evacuees.

One of the most effective approaches is the cooperation between the government, disaster management agencies and the community in minimizing the impact of volcanic eruptions. Enhancing preparedness through learning and education improves community knowledge and skills in disaster response, reduces casualties and infrastructure damage, and strengthens early warning and evacuation systems. There significant number of casualties resulting from volcanic eruptions indicates a low level of community preparedness. Therefore, educating communities on preparing for and managing eruption disasters is very crucial; for those living around volcanic areas.

### 3.1 Development of community knowledge in dealing with mountain eruption disasters

Indonesia is located in an area that is prone to natural disasters, including volcanic eruptions. Land use in these affected areas includes settlements, agriculture (gardens, fields, rice fields), and public facilities. The upper region is primarily covered by forests, some of which have been converted into fields/gardens. The central slopes are used for rice fields and gardens. According to the forecast map for land movements in West Sumatra, the disaster-prone area falls within a medium-high potential zone for landslides and materials flows [17]. The destruction of the slopes around Mount Marapi was not incidental. A number of factors contributed to this disaster. The steep slopes around the mountain provide an ideal path for unconsolidated volcanic material. The heavy rain exacerbated the situation by triggering erosion and mobilizing loose materials. Additionally, upstream land use changes, such as land clearing for agriculture or settlements, also contribute to increased surface erosion and accelerate the formation of material flows. Therefore, addressing natural disasters requires significant knowledge and education to enhance community preparedness for volcanic eruptions [1, 4]. When well-informed, people gain a better understanding of the causes, characteristics, and impacts of volcanic eruption [18, 19].

To effectively manage a volcanic eruption disaster, it is crucial to prioritize community education and knowledge dissemination. A comprehensive approach is needed to ensure that the community is well informed and prepared for potential eruptions.

1) The government and relevant agencies must ensure that the public is informed of the characteristics and potential dangers of eruptions from Mount Merapi and Mount Marapi, especially for those living nearby. To enhance public awareness, socialization and training activities must be carried out regularly. Various communication channels, both formal and informal, must be used to disseminate important information such as eruption history, early warning signs, danger zones, and evacuation routes.
2) The community should be encouraged to develop the household-level emergency plans, including supplies such as food, water, medicine, and communication equipment. In addition, the community should identify safe gathering places and cooperate with neighboring residents on evacuation plans similar to those in the Mount Merapi area. Regular training in self-rescue techniques, such as first aid and self-evacuation, is also essential.
3) The government should establish a disaster management system that incorporates local knowledge. The government can develop a more effective early warning system by leveraging public awareness of natural signs as self-warning indicators. Combining scientific and local knowledge will enhance the effectiveness of disaster warning and response systems. Strengthening local capacity is crucial. Training and forming rapid response teams, volunteers, and local community organizations can better prepare communities for disasters. A sense of shared responsibility will increase when communities actively participate in planning and implementing emergency response measures.

**Table 3.** Non-physical countermeasures

| Activities | Mount Merapi | Mount Marapi |
|---|---|---|
| Strengthening and implementing regulations, such as vulnerable areas and evacuation routes | The community is ready | Socialization to the community through students and 70 Walinagari |
| Identification of disaster management system | Facilities and infrastructure are ready | In EWS installation |
| Building community preparedness | Ready with the involvement of neighboring villages | Still in the socialization stage |
| Capacity building of disaster management agencies | Disaster resilient villages have been formed | Disaster resilient villages at the socialization level |
| Multi-stakeholder collaboration to build a better | Still continuing to be improved | Still continuing to be improved |

A holistic and participatory approach is expected to provide the community with adequate knowledge and preparedness in dealing with mountain eruption disasters. This will certainly increase the ability of the community to reduce disaster risk and accelerate recovery processes.

### 3.2 Cooperation between the government, disaster management agencies, and the community in preparing an evacuation system

The government plays significant role in coordinating disaster management efforts, while management institutions and the community play an active role in implementation.

The government is responsible for ensuring comprehensive planning, adequate budget allocation, and the availability of infrastructure to support evacuations.

The Regional Disaster Management Agency (BPBD) supports the Regent by coordinating, formulating policies, and implementing disaster management strategies. It ensures that the community has a clear emergency response plan, trained personnel and adequate equipment. This coordination helps ensure that all disaster management measures are carried out effectively and in an integrated manner [20]. Strong cooperation with local governments and communities in the Mount Merapi and Mount Marapi areas is strongly needed to ensure the rapid and effective mobilization of aid and resources during disasters. The community plays a crucial role in the evacuation process and the development of the evacuation system. From the very beginning, communities should be involved in the evacuation plan through socialization, training, and simulation. To improve community awareness, knowledge, and preparedness for volcanic eruptions, local communities can also participate in forming disaster preparedness teams that assist with evacuation and pre-disaster emergency management.

Evacuation maps are essential for identifying hazardous areas, safe locations, and evacuation routes, demonstrated by the communities in 70 Nagari on Mount Marapi. Preparing individuals, staff and organizations for emergency situations that require evacuation is known as evacuation simulations which teach people how to respond during a disaster. Various types of simulations, such as desktop simulation, field simulation, and full simulation, can be used. Developing evacuation systems and disaster management organizations comes with challenges and obstacles, highlighting the need for cooperation to address the issues.

1) Ensuring effective collaboration among all parties is a significant issue. While the government must act as the main coordinator, disaster management agencies such as BNBP and BPBD must contribute to developing evacuation strategies and infrastructure. The community must also be involved in evacuation planning and implementation processes.
2) Facing increasingly diverse and complex disaster phenomena presents a significant challenge. Therefore, a flexible evacuation system capable of managing volcanic eruption disasters is essential. Evacuation strategies and infrastructure must be continuedly maintained and improved.
3) Understanding spatial planning policies and local knowledge is crucial. Each community possesses unique local knowledge on managing volcanic eruptions, whether on Mount Merapi or Mount Marapi. However, they are often not involved in disaster management planning. Therefore, disaster management agencies must collaborate with local communities in developing efficient spatial planning policies and address potential losses from volcanic eruptions.
4) Overcoming the complexity of inter-institutional cooperation is essential. BNPB and BPBD must coordinate with other institutions, including ministries or national institutions, educational institutions and the community. The government must ensure that all institutions work effectively in creating a comprehensive evacuation system.

The explanation above is a strategic step to enhance disaster management capabilities. With clear regulations, active participation from disaster management institutions and community contributions, this approach will reduce the number of victims and accelerate recovery process, thereby mitigating the negative impacts of volcanic eruptions as shown in Fig. 4 and Fig. 5.

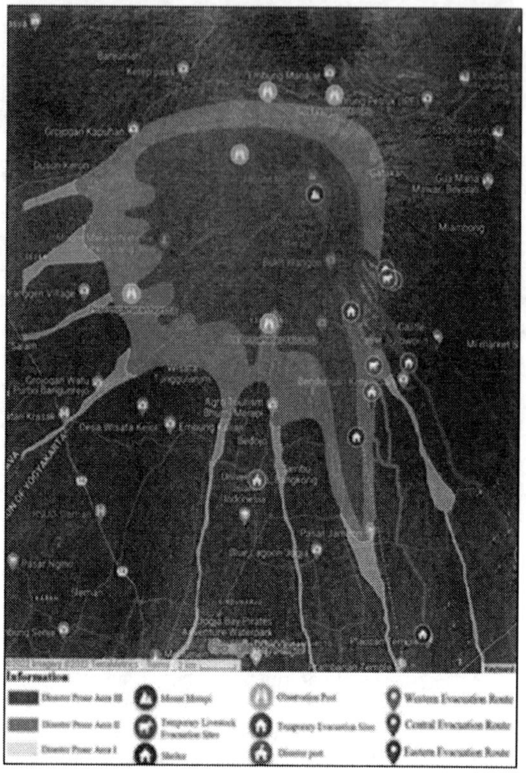

**Fig. 4.** Location map of ground motion and cold lava flow from the eruption of Mount Merapi in Central Java (source: BPBD Central Java, 2024)

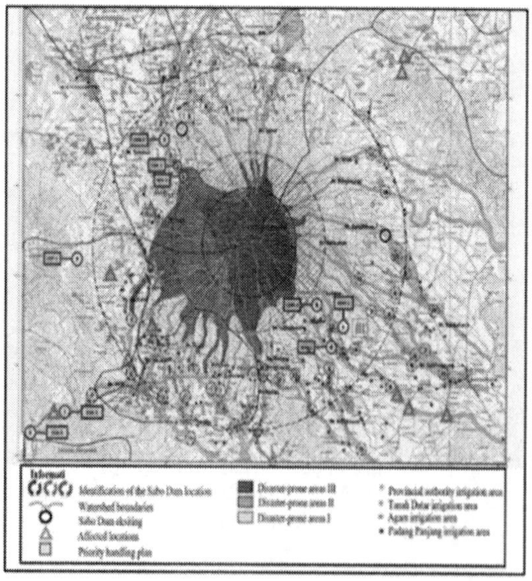

**Fig. 5.** Location map of ground motion and cold lava flow from the eruption of Mount Marapi in West Sumatra (source: BPBD West Sumatra, 2024)

### 3.3 Development of disaster mitigation

Disaster mitigation is a proactive measure taken before an emergency or disaster occurs to eliminate or reduce the impact and risk of harm [3, 18]. To mitigate the adverse impact of volcanic eruptions, efforts are underway to enhance disaster mitigation strategies for Mount Merapi and Mount Marapi. Both volcanoes remain active and experience significant explosive eruptions that pose severe risks to surrounding communities. Therefore, understanding and implementing effective mitigation strategies is essential.

The creation and provision of maps of areas prone to volcanic eruption disasters, including volcanic hazard risk zones and volcanic geological maps are essential to understand potential hazards. Location maps showing ground motion and cold lava flow from the eruption of Mount Merapi and Mount Marapi can be seen in Figs. 4 and 5. Governments and volcanological institutions must monitor volcanic activity using specialized equipment such as seismographs and tiltmeters [21, 22]. The government has established clear and organized procedures to deal with emergency situations such as the eruption of Mount Merapi. It is in the process of preparing for potential eruptions of Mount Marapi. These procedures include evacuation plans, provision of emergency facilities, and coordination with relevant parties.

An early warning system (EWS) has been installed in the Mount Merapi and Mount Marapi areas to monitor and anticipate potential disasters and volcanic eruptions. In the Mount Merapi area, the system includes 42 seismic stations, 10 GPS stations, 13 tiltmeter stations, 2 geochemical stations, 5 rain or cold lava monitoring stations and 34 camera stations.

Meanwhile, in the Mount Marapi area, there are 11 EWS designed to provide early warning of volcanic eruption hazards and 20 units of EWS sensors by BNBP to detect and alert for potential floods caused by volcanic eruptions. However, EWS in the Mount Merapi and Mount Marapi areas requires further evaluation to improve its effectiveness in disaster management. The existing EWS faces several challenges, including:
Enhance community socialization. Although 11 EWS units have been installed on Mount Marapi, additional socialization efforts are needed to ensure that the community fully understands the use of land, buildings, and other infrastructure around the volcano. This will help the community become more prepared and responsive to early warning signals. Expand the coverage of the early warning system. Although BNBP has installed 20 units of EWS sensors in rivers around Mount Marapi, there are still areas that are not covered. For example, the Siagarungguang trunk, which is upstream from Mount Marapi and downstream from Pasar Batusangkar, is a river with potential risk of flash floods and cold lava flow that require monitoring. Improves the speed and accuracy of the early warning system. Although sirens can provide rapid alerts, they are less effective in cases of hot clouds moving at speed of 100 km/h. Enhancing the early warning system to address these high-speed phenomena will improve its effectiveness.

Sabo dams are a robust alternative solution for managing lava flows and sediment from volcanic eruptions. Sabo dams are designed to capture sediment and lava, particularly after heavy rains. Observations in the Mount Marapi area, show only 2 sabo dams, which limits their effectiveness given the 25 rivers that flow downstream from Marapi and have potential flooding risks. In contrast, the Mount Merapi area benefits from 272 sabo dam units, which significantly help control and redirect volcanic material. To improve the disaster mitigations in the Mount Marapi area, increasing the number of sabo dams at several points is required. This will help prevent damage to infrastructure and the surrounding environment and mitigate the risk of lava floods caused by volcanic eruptions.

Strengthening human resources and support systems is also crucial for effective disaster mitigation. Personnels involved in disaster management should receive proper training and development. In addition, improved facilities and infrastructure are needed, such as evacuation routes and emergency services. The community must be aware of evacuation locations and understand what constitutes a safe area. It is vital that people are prepared with essential supplies, such as medicine and food, and that they follow guidance and instructions provided by the relevant authorities.

## 4 Conclusion and recommendations

### 4.1 Conclusion

The eruptions of Mount Merapi and Mount Marapi have had a great negative impact on the community, such as casualties, injuries, economic impacts, damage to infrastructure and livestock. The eruption of Mount Merapi in 2010 had a greater impact than the eruption of Mount Marapi in December 2023, highlighting the higher and more destructive volcanic activity of Mount Merapi. To improve the community's ability to deal with volcanic eruption disasters, cooperation among the government, disaster management agencies, and the community is essential.

The flash flood disaster or locally known as "galodo' that affected the districts of Agam, Tanah Datar, Padang Pariaman and the city of Padang Panjang was caused by lava flows following the eruption of Mount Marapi. To reduce disaster risk, several immediate actions are recommended:

1) Evacuation: Residents living along riverbanks upstream from Mount Marapi should evacuate to safer locations if it rains upstream. If there are signs of hazardous material flows, the community should quickly move away from the area and report the incident to authorities.
2) Infrastructure Development: Developing River flow control infrastructure, such as sabo dams, is crucial for mitigating the impact of flash floods. Additionally, repairing culverts, normalizing river channels, and improving water flow conditions are necessary to prevent flooding.

3) Coordination and Education: Effective coordination among local governments, relevant agencies, and the community is vital. Continuous socialization and education about soil movement and disaster mitigation efforts should be conducted to raise public awareness and preparedness for potential disaster threats.

### 4.2 Recommendations

There are several recommendations that can be made for the government, response agencies, and the community around the Mount Merapi and Mount Marapi areas:

1) Increasing cooperation between the government, disaster management agencies, and the community in dealing with volcanic eruption disasters.
2) Develop effective preparedness learning programs to improve community knowledge and skills.
3) Relocating of settlements in disaster-prone areas, intensively monitoring material flow, developing effective flood control infrastructure, and providing socialization and education on disaster mitigation are the main recommendations to protect and ensure the safety of communities around Mount Marapi.

We would like to thank Universitas Andalas for the Research Grant year 2024 number 332/UN.16.19/PT.01.03/PUJK/2024 date July 17, 2024.

## References

1. A. R. Prabowo, A. R. Dwicahyani, W. A. Jauhari, A. Aisyati, and P. W. Laksono, Development and application of humanistic logistics models for optimizing location-allocation problem solutions to volcanic eruption disaster (Case study: Volcanic eruption of Mount Merapi, Indonesia). Cogent Eng, **4**, 1 (2017). https://doi.org/10.1080/23311916.2017.1360541.
2. E. Purwaningsih, S. A. Liusti, E. Purnamasari, R. Ramadhan, and A. F. R. Nasution, The Mount Marapi eruption disaster evacuation path model using a local wisdom approach. International Journal of GEOMATE, **26**, 116, 64–71 (2024). https://doi.org/10.21660/2024.116.4353.
3. M. Cassidy et al., Explosive eruptions with little warning: Experimental petrology and volcano monitoring observations from the 2014 eruption of Kelud, Indonesia. Geochemistry, Geophysics, Geosystems, **20**, 8, 4218–4247 (2019). https://doi.org/10.1029/2018GC008161.
4. Rahayu, D. Priyo Ariyanto, S. Hartati, and J. Syamsiyah, Dampak erupsi Gunung Merapi terhadap lahan dan upaya-upaya pemulihannya. Caraka Tani-Jurnal Ilmu Ilmu Pertanian, **4**, 1, 61–72 (2014)
5. A. L. Nugraha, Hani'Ah, H. S. Firdaus, and S. Haeriah, Analysis of risk assessment of Mount Merapi eruption in settlement area of Sleman Regency', in IOP Conference Series: Earth and Environmental Science, Institute of Physics Publishing, (2019). https://doi.org/10.1088/1755-1315/313/1/012003.
6. D. Fiantis, Gusnidar, B. Malone, R. Pallasser, E. Van Ranst, and B. Minasny, Geochemical fingerprinting of volcanic soils used for wetland rice in West Sumatra, Indonesia. Geoderma Regional, **10**, 48–63 (2017). https://doi.org/10.1016/j.geodrs.2017.04.004.
7. D. Eka Putra, Supriatna, and Sitanala, Risiko erupsi Gunung Marapi di Kabupaten Tanah Datar. Jurnal Geosains Terapan, **2**, 3, 7–13 (2016)
8. E. A. D. C. Utami and D. Hermon, Analisis spasial kerentanan erupsi Gunung Marapi terhadap wilayah permukiman. Jurnal Pendidikan Tambusai, **8**, 1, 9310–9319 (2024)
9. F. Bari, B. Istijono, R. Yuhendra, A. Hakam, M. Noer, and T. Ophiyandri, Potential debris flow after earthquake in Mount Talamau Pasaman district and West Pasaman district. in IOP Conference Series: Earth and Environmental Science, Institute of Physics, (2023). https://doi.org/10.1088/1755-1315/1173/1/012069.
10. C. Michellier et al., Evaluating population vulnerability to volcanic risk in a data scarcity context: The case of Goma City, Virunga volcanic province (DR Congo). International Journal of Disaster Risk Reduction, **45** (2020). https://doi.org/10.1016/j.ijdrr.2019.101460.
11. J. Oldfield and A. Stevenson, After the fire: An ecological, phenomenological exploration of resilience-building following the Fuego volcanic eruption in Guatemala. Am J Community Psychol., (2024). https://doi.org/10.1002/ajcp.12748.
12. Pusat Vulkanologi dan Mitigasi Bencana Geologi, West Sumatra Mount Marapi eruption report. (Pusat Vulkanologi dan Mitigasi Bencana Geologi, Jakarta, 2024)
13. J. W. Creswell, Design research: qualitative, quantitative, and mixed methods approaches, fifth edition. Sage Publishing, Los Angeles, 2018)
14. Yin and Robert K., Qualitative research from start to finish, second edition (The Guilford Press, New York, 2016)
15. S. J. Taylor, R. Bogdan, and M. L. DeVault, Introduction to qualitative research methods, fourth edition. (Wiley, New Jersey, 2016)
16. BPBD Provinsi Sumatera Barat, Laporan bencana erupsi Gunung Marapi dan banjir galado di Sumatra Barat. Padang, (2024)
17. A. Andriani, B. Istijono, F. Akmal, and B. Martanto Adji, Identification of landslide hazard in residential area Kubang Tangah District, Sawahlunto. Journal of Applied Engineering and Technological Science, **5**, 2, 1154–1164 (2024)
18. I. Amri, S. R. Giyarsih, and D. Ruslanjari, Tsunami risk awareness, hazard warning knowledge, and intended evacuation behavior among beach users in Bantul, Indonesia. International Journal of Disaster Risk Reduction, **109**, 104594 (2024). https://doi.org/10.1016/j.ijdrr.2024.104594.

19. E. Noviana, O. Kurniaman, N. S. Sb, S. D. Nirmala, and R. S. Dewi, How to prepare disaster mitigation knowledge for prospective teachers in Elementary School?. AL-ISHLAH: Jurnal Pendidikan, **13**, 2, 1123–1134 (2021) https://doi.org/10.35445/alishlah.v13i2.496.
20. S. Menghoung, B. Istijono, F. A. Ismail, M. Noer, and R. F. Zakir, Build back better and stronger the 2022 earthquake-damaged houses in Talamau–Pasaman Barat District. in E3S Web of Conferences, **464** (2023). https://doi.org/10.1051/e3sconf/202346407006.
21. D. K. Chester, M. Degg, A. M. Duncan, and J. E. Guest, The increasing exposure of cities to the effects of volcanic eruptions: A global survey. Environmental Hazards, **2**, 3, 89–103 (2000). https://doi.org/10.3763/ehaz.2000.0214.
22. M. P. Poland and K. R. Anderson, Partly cloudy with a chance of lava flows: forecasting volcanic eruptions in the twenty-first century. J Geophys Res Solid Earth, **125**, (2020). https://doi/org/10.1029/2018JB016974

# Analysis of student preparedness behaviour in facing lava rainfall disasters

*Siti* Hadiyati Nur Hafida[1,2*], *Ayu* Dila Putri Yuliani[1], *Vina* Dahniar[2], *Aghnia* Mulia Iqlimah[1], *Rachma* Regita Sukamto[1], *Muhammad* Musiyam[1], and *Nasurullah*[3]

[1]Department of Geography Education, Faculty of Teacher Training and Education, Universitas Muhammadiyah Surakarta, Surakarta, Indonesia
[2]Disaster Mitigation Study Center, Universitas Muhammadiyah Surakarta, Surakarta, Indonesia
[3]Disaster Management Authority, Islamabad, Pakistan

**Abstract.** Students are vulnerable to natural disasters. This is further exacerbated by the large number of schools located in disaster-prone areas, so student preparedness behaviour needs to be continuously improved. This study aims to determine the level of preparedness, awareness, action, and effect students have in dealing with the volcanic eruption of Mount Merapi. This study used quantitative research; the instrument used was a questionnaire made by taking samples at SMK Muhammadiyah 1 Salam. The sampling method was stratified random sampling from students in grades 10, 11, and 12. The assessment of students' preparedness behaviour in dealing with rain lava was based on four variables: preparedness, awareness, action, and affect. Then, each variable was analyzed using descriptive statistics and multiple correlation analysis. The study results show that the relationship between preparedness, awareness, action, and affect shows a strong level of closeness with a value of 0.680. The study identifies critical gaps in knowledge and behavioural responses that must be addressed to enhance overall preparedness behaviour. The results underscore the importance of targeted educational interventions and community engagement in fostering a culture of resilience among students in disaster-prone areas.

## 1 Introduction

Indonesia is one of the countries with a high level of natural disasters. A natural disaster is a natural event that has a major impact on the human population [1]. Natural disasters that occur in Indonesia, such as landslides, floods, tsunamis, earthquakes, volcanic eruptions, and tornadoes. One of the natural disasters is eruption, and the most active volcano namely Mount Merapi. Mount Merapi with a height of 2968 meters above sea level is located between Sleman Regency, Magelang Regency, Boyolali Regency, and Klaten Regency. Mount Merapi is a stratovolcano with a high potential to erupt in Indonesia [2-4]. From 1672 to 2010, more than 80 eruptions were recorded, with a rest period of 1-18 years or an average of 4 years [5, 6].

Regencies in Central Java Province are areas prone to volcanic eruptions. The form of threat from volcanic eruptions is in the form of loss of life and damage to settlements, property, and objects [3]. The danger of the eruption of Mount Merapi will release volcanic material such as hot clouds, dust, and material (rocks, gravel, and sand). Hot clouds are in the form of toxic gas material with a very hot temperature of around 200-700 degrees Celsius, this hot cloud is very dangerous because it can burn anything in its path [4, 5, 7]. Volcanic dust released by the eruption of Mount Merapi will be dangerous to public health related to respiratory problems and irritation. Lava floods are a secondary hazard because when a volcano erupts, in addition to releasing clouds, it also releases material that falls down the slope [8, 9]. Volcanic material in the form of rocks, gravel, and sand will be pushed by runoff with its slope and will enter the river so that with high intensity it can cause a lava flood disaster [10].

The disaster-prone areas of Mount Merapi eruption consist of three areas, namely Disaster-Prone Area (DPA) I, has a mild eruption impact, and DPA II has a moderate risk of danger due to the eruption of Mount Merapi. Conversely, DPA III is the area with the highest risk of danger [11]. Each of these areas has different characteristics, so it is necessary to have an understanding to adjust to its characteristics. The area in DPA 1 that has the potential to be hit by cold lava is divided into two areas, namely the area prone to lava flows or floods and the area prone to ash fall. In the cold lava area, especially the area located along the river, valley, or downstream of the river that originates at the peak. SMK Muhammadiyah 1 Salam is located in DPA I, precisely in Jumoyo Village, which has a *Kali Putih* that originates at the peak of Mount Merapi, so that when Mount Merapi erupts, the *Kali Putih* area is prone to cold lava floods.

The volcanic eruption disaster in Salam District caused a lot of material losses, injuries, and fatalities due

* Corresponding author: shnh421@ums.ac.id

to the lack of knowledge, understanding, and skills in dealing with volcanic eruption disasters [12]. One of the activity sectors that felt the impact of the eruption of Mount Merapi was the education sector. There are 15 Senior High Schools (SMA) located on the slopes of Mount Merapi and are included in the DPA of Mount Merapi I to III. One of them is SMK Muhammadiyah 1 Salam, which is about 4.7 km from *Kali Putih*. Schools also receive special attention for disaster management activities, where students are one of the vulnerable groups in natural disaster [13-15]. For this reason, students in DPA such as SMK Muhammadiyah 1 Salam must be aware of the importance of preparedness behaviour for all members of their school to minimize the risk of loss in cold lava disasters.

Students are one of the vulnerable groups in natural disasters, this happens because of the lack of knowledge, understanding, and skills of students towards natural disasters. Through regular guidance from schools, it is expected that students always care about the environment are aware of disasters and increase protective actions that are useful for themselves and those around them to be applied directly and well in their lives. Therefore, the function of schools as a place to educate students both in terms of knowledge and behaviour is an effective means to provide and improve knowledge, understanding, and disaster skills, especially eruption disasters for students of SMK Muhammadiyah 1 Salam.

Students' preparedness in facing lava rainfall disasters is highly dependent on their understanding of the risks, actions taken, and emotional reactions to potential threats. Preparedness, which includes students' knowledge and abilities in facing disasters, is an important element in reducing disaster risk. However, this preparedness cannot be realized without adequate awareness of the dangers faced. This awareness must be developed through adequate education and experience.

In addition, students' attitudes when facing disasters are greatly influenced by the actions taken when a disaster occurs. Quick and appropriate actions can save lives and minimize the negative impacts of a disaster. On the other hand, the affect factor (emotion) or students' feelings towards the threat of disaster also influences how they respond to dangerous situations. Fear, anxiety, or panic can hinder their ability to act effectively when a disaster strikes.

This study aims to analyse students' preparedness behaviour in facing lava rainfall disasters by considering four main variables, namely preparedness, awareness, action, and affect. By understanding how these variables interact, it is hoped that the results of this study can provide appropriate recommendations to improve students' preparedness in facing disasters, especially in areas prone to lava rainfall, such as the slopes of Mount Merapi. In addition, this study seeks to assess the extent of the role of disaster education in increasing students' awareness and preparedness to face potential natural disaster threats.

## 2 Methodology

The research method used in this study is quantitative. According to Creswell [16], quantitative research methods are research based on the philosophy of post-positivism, which views that a symptom can be grouped, observed, measured, is causal, relatively fixed, and value-free. Quantitative research is research that requires a lot of use of numbers, starting from data collection, interpretation of the data, and the appearance of the results. At the conclusion stage of the research, it would be better if accompanied by pictures, tables, graphs, or other displays. This research was conducted at SMK Muhammadiyah 1 Salam, Salam District, Magelang Regency. The selection of the research location was because SMK Muhammadiyah 1 Salam is located in DPA I which is located close to the *Kali Putih* so if Mount Merapi erupts, the area will be affected.

The population used in this study consisted of students at SMK Muhammadiyah 1 Salam and the sample came from students in grades 10, 11, and 12 of SMK Muhammadiyah 1 Salam as many as 222 students. This technique is a sampling method that divides the population into smaller ones so that each member of the group has an equal chance of being selected as a sample [17]. Data collection techniques and instruments use questionnaires, and documentation. The questionnaire used will be related to the variable preparedness, awareness, action, and affect, where each variable will be analysed using descriptive statistics and multiple correlation tests. The multiple correlation is a statistical calculation that measures the degree of relationship between two or more variables, where the linear correlation must first be calculated [18, 19].

## 3 Results and discussion

### 3.1 Analysis of preparedness, awareness, action dan affect

Table 1 describes the results of preparedness caused by the rain lava disaster for students at SMK Muhammadiyah 1 Salam.

**Table 1.** Student preparedness in dealing with volcano eruption

| Indicators | Value |
|---|---|
| School Standard Operating Procedure | 85.14 |
| Emergency Number | 25.23 |
| Access to Information | 40.99 |
| Unexpected Rain Lava | 72.97 |
| Socialization of Disaster Mitigation | 39.36 |
| Risk Reduction | 87.39 |
| Early Preparation | 56.31 |
| Student Awareness | 38.29 |

Based on Table 1, the highest value is in the risk reduction indicator, where students understand that schools are prone to cold lava disasters must carry out annual disaster risk reduction activities. Disaster risk reduction is a planned step to strengthen students in efforts to reduce disaster risks and build a safe and resilient culture against disasters [20-22]. Furthermore, the lowest value is emergency number indicator, where only a small portion of students know about emergency numbers. Emergency numbers are one of the important indicators in preparing for rain lava disasters, emergency numbers can be contacted when something sudden happens such as a rain lava disaster that cannot be predicted when it will occur, as stated by Putra, et al., [23] having an important telephone number to ask for help in an emergency is an important aspect because through this, the community can coordinate across sectors when a disaster occurs. Emergency numbers that students need to save include telephone numbers for police stations, hospitals, the Regional Disaster Management Agency, SAR (Search and Rescue), and family telephone numbers/other emergency telephone numbers [24]. Saving emergency numbers can minimize the number of fatalities because they can be helped quickly.

Risk reduction helps protect vital infrastructure such as schools. When risk mitigation measures are implemented, such as building reinforcement and risk-based spatial planning, it helps reduce the potential for material damage that can incur significant costs in post-disaster recovery. Eruptions can cause massive economic losses, and risk reduction serves to reduce the economic impact by minimizing damage to infrastructure. Disaster risk reduction focuses not only on the physical aspects but also on building social and psychological resilience in communities. Risk reduction also includes environmental protection measures that can slow or prevent damage from eruptions. For example, rehabilitating and conserving watersheds and vegetation around volcanoes helps mitigate the effects of landslides and lahar floods. By better managing environmental risks, the impacts of eruptions can be minimized, both for local communities and ecosystems. In other words, risk reduction is key to protecting lives, reducing material losses, strengthening communities, and maintaining environmental sustainability in the face of volcanic eruption threats.

**Table 2.** Student awareness in dealing with volcano eruption

| Indicators | Value |
|---|---|
| Active in simulation | 50.90 |
| Participation in disaster simulation | 81.98 |
| Initial action | 68.02 |
| Signs of disaster | 45.95 |
| Local cultural/customary knowledge | 42.79 |

Table 2 describes the awareness caused by the rain lava disaster on students of SMK Muhammadiyah 1 Salam. Based on the table, the highest value lies in the participation in disaster simulation indicators. This can be interpreted that students realize when the rain lava disaster simulation activity is carried out, all school residents must participate, according to research by Yustisia et al., [25] that disaster simulations can affect students' knowledge and actions towards disaster management. When a rain-lava disaster occurs, students and all school residents already understand what they have to do, and this can minimize casualties. While in the lowest value lies in the local cultural/customary knowledge. If customary knowledge is linked to disaster studies, communication about disasters uses a distinctive language based on the past experiences of local people regarding the disaster itself, this language can be in the form of stories and oral stories, games, or dances [26]. However, less than 50% of students no longer use customary knowledge to be aware of the arrival of the rain lava disaster.

Students realize the importance of participating in evacuation simulations or disaster drills held by schools or communities. They understand that these drills are important to hone their readiness to face real emergencies. Through these simulation activities, students will eventually understand the evacuation routes at school and home that have been set for emergencies. They are aware of safe locations or shelters that have been designated in the event of a disaster. In addition, students understand the procedures for action that need to be taken when a disaster occurs, for example, how to act quickly but calmly when hearing a warning siren, or they know how to provide simple first aid to themselves or friends in an emergency. These activities show how student awareness can help them be more prepared for potential disasters, as well as how this plays an important role in creating a school and community environment that is more responsive and safer to disasters.

**Table 3.** Student actions in dealing with volcano eruption

| Indicators | Value |
|---|---|
| Evacuation preparation and equipment | 59.01 |
| Use of evacuation routes | 84.68 |
| Take shelter in a safe place | 88.74 |
| Quick, precise and active action | 80.63 |

Table 3 describes the actions of SMK Muhammadiyah 1 Salam students in response to the cold lava disaster. Based on the table, it is found that the highest value is in the indicator of taking shelter in a safe place, meaning that most students already know that when a rain lava disaster occurs, they take shelter in a safe place. Emergency shelter areas are safe locations for people after the evacuation process and have the function of ensuring the exchange of appropriate information for disaster victims [27]. Emergency shelters must respect the privacy of each individual and still allow for supervision to maintain a sense of security for all members of the community [28]. The lowest value is in the evacuation preparation and equipment indicator. Although the lowest, more than 50% of students have prepared equipment and important items that they will take when evacuating to a safe place. In line with research conducted by Kurniawati [29], that a simple anticipation in the event of a disaster is to store valuables. One way for people in disaster-prone areas to be alert and prepared is to urge them to store valuables in one bag so that when a disaster occurs, they can easily take them to an evacuation site. Similarly, students can do the same thing, namely preparing the valuables they bring to school in one place, for example when teaching and learning activities are finished, all items are put in a bag so that nothing is left behind during the evacuation.

One of the basic principles of disaster mitigation is having a clear and directed evacuation plan. For students, evacuating to a designated safe place reduces the risk of confusion and panic during a disaster. These safe places are usually designed and equipped with infrastructure that is more resistant to volcanic hazards and provides better protection than other places that are not reinforced. After a lahar rain occurs, secondary hazards such as mudflows or trapped volcanic material can occur. Students who do not immediately take shelter in a safe place may be trapped in more dangerous conditions after the initial phase of the disaster has passed. Therefore, it is important to ensure that students are immediately moved to a safe place that is free from these potential secondary hazards.

In emergencies such as lahar rain, students, especially young ones, can experience extreme stress or trauma. Taking shelter in a safe place, which is well-designed and organized, also helps create a sense of psychological safety. In this place, they can receive assistance from teachers or trained officers, which helps to reduce panic and increase students' emotional resilience. Thus, taking shelter in a safe place during a lava rain is not only important to protect students from the direct threat of lava but also helps them avoid secondary hazards, maintain physical health, and support psychological stability.

**Table 4.** The effect of the eruption felt by students

| Indicators | Value |
|---|---|
| Calming a traumatized friend | 72.07 |
| Monitoring aftershocks | 79.28 |
| Renovating school buildings | 82.88 |
| Impact of rain-induced lava disasters | 77.03 |

Table 4 describes the effect caused by the rain-lava disaster on students at SMK Muhammadiyah 1 Salam. Based on the table, the highest value is found in the indicator of schools renovating, which means that students agree that schools need to renovate buildings damaged by rain lava disasters. Based on research by Anisah and Sumarni [30], schools as formal education units should be safe places for children, where students spend more time at school, so schools need to renovate damaged buildings so as not to endanger students during the teaching and learning process. The lowest value is in the indicator of calming a traumatized friend. Even though it is the lowest value, most students can calm their friends who are afraid when the rain lava disaster occurs. According to Arifin [31], traumatic events due to disasters can cause Post-Traumatic Stress Disorder (PTSD) which interferes with sufferers in carrying out their daily lives, including disrupting their learning activities. It is very important to calm friends when a disaster occurs to minimize stress disorders that disrupt learning activities.

Buildings that are not structurally rehabilitated or reinforced are at risk of collapse, threatening the safety of students and staff. Renovating school buildings to earthquake-resistant construction standards is one of the primary steps to prevent damage or collapse from seismic activity associated with an eruption. School buildings that are not resistant to eruptions or natural disasters, such as ashfall, lava flows, and other volcanic materials, are particularly vulnerable to severe damage. Renovations that involve retrofitting roofs, walls, and ventilation to be resistant to volcanic ash help maintain school conditions and reduce the risk of permanent damage that could lead to prolonged school closures.

Schools that are rehabilitated to be more resistant to eruptions can continue to function even after a disaster has occurred. Disaster-resilient infrastructure allows schools to remain operational or recover quickly after an eruption, minimizing disruption to student learning. Renovations also often include the development of safe evacuation areas and emergency shelters, which are essential to maintaining educational continuity in disaster-prone areas. Renovating schools designed for disasters also serves as a reminder and education for the school community and surrounding community about the importance of disaster preparedness. Safe and resilient schools serve as models for students, teachers, and communities to be more aware of the importance of disaster mitigation, not only in school buildings but also in their daily lives. Renovating school buildings is a crucial step in ensuring physical safety, educational continuity, and community resilience in eruption-prone areas. Improving the strength of school infrastructure not only protects the building itself but also safeguards future generations and reduces the economic and social impacts of disasters.

### 3.2 The relationship between preparedness, awareness, action and affect

This study uses multiple correlation test, this test is used to determine the strength or weakness of the relationship between Awareness (X1), Action (X2), and effect (X3) on the Preparedness variable (Y). The correlation test uses the SPSS program with the results below.

**Table 5.** The relationship between awareness, action, and effect toward preparedness behaviour

| Variable Dependent | Variable Independent | B | Sig. |
|---|---|---|---|
| Preparedness | Constant (∝) | 1.502 | .000 |
| | Awareness | .569 | .000 |
| | Action | .328 | .007 |
| | Affect | .223 | .033 |

Based on the results of the SPSS output using multiple correlation tests, it is known that ∝ has a positive value of 1.502 which indicates a unidirectional influence between the awareness, action, effect, and preparedness behaviour. This shows that if all independent variables including awareness (X1), action (X2), effect (X3) have a value of 0 percent or do not change, then the preparedness behaviour value is 1.502

The results of the values in Table 5 show that the sig. value is 0.000 (<0.05) which can be concluded that the awareness (X1) have a significant relationship to the preparedness (Y) variable, but for action (X2), and effect (X3) (>0.05) variables didn't have a significant relationship to the preparedness (Y) variable. Preparedness and action have a close relationship where action and preparedness are very important to reduce the impact of natural disasters [32–34]. This is also similar to research conducted by Ayub et al., [35] which states that disaster awareness is directly related to preparedness behaviour or preparedness in facing disasters in students. Research conducted by Tursilowati [36] also stated that the results of the statistical tests carried out showed that the awareness characteristic variable had a significant value, which means that the influence of awareness on disaster preparedness will make the better level of awareness, and better the disaster preparedness.

## 4 Conclusion

Based on the study results above, the highest indicator of student preparedness is the school standard operating procedure, which has a value of 85.14%. This means that the school already has facilities supporting the safety of students vulnerable to rain-lava disasters, while the smallest value is in the emergency number. Awareness has the highest value in participation in disaster simulation, at 81.98%. This means that students are active in disaster simulation activities, while the signs of a disaster have the lowest value. This differs from participation in disaster simulation, which should explain the signs of a disaster and how to reduce the risk. Action has the highest value in the indicator of taking shelter in a safe place, which is 88.74%. Students already understand that they must take shelter in a safe place when a disaster occurs. At the same time, the smallest value is evacuation preparation and equipment. The effect has the highest value in school building renovation, with a value of 82.88%. This means that the school is fast in renovating buildings damaged by disasters. While the smallest value is calming friends during trauma, this may occur because the student's mentality is still lacking in dealing with critical situations. The relationship between preparedness, awareness, action, and affect shows a value of 0.680, so it has a strong level of closeness.

Analysing student preparedness behaviour in facing lava rainfall disasters highlights strengths and areas for improvement in current disaster preparedness efforts. While many students demonstrate basic knowledge of lava risks and appropriate responses, there is a clear need for more comprehensive and consistent education on the topic. The study emphasizes the role of schools as pivotal institutions in disaster risk reduction, mainly through integrating disaster education into the curriculum and active engagement with local disaster management agencies. Strengthening these efforts will enhance student preparedness and contribute to the broader community's resilience to lava disasters.

## References

1. S. Hardiyanto, and D. Pulungan. Komunikasi efektif sebagai upaya penanggulangan bencana alam di Kota Padang Sidimpuan. J. Interaksi, **3**, 30–39 (2019) http://doi.org/10.30596/interaksi.v3i1.2694
2. E. Febriati. and N. H. Muzaki. Pemanfaatan radar cuaca untuk mengidentifikasi sebaran abu vulkanik Gunung Merapi 3 Maret 2020. J. Widya Climago, **4**, (2022).
3. E. Hariyono. and S. Liliasari. The Characteristics of volcanic eruption in indonesia. in volcanoes - geological and geophysical setting, theoretical aspects and numerical modeling, applications to industry and their impact on the human health (ed. Aiello, G.) (InTech, 2018). http://doi.org/10.5772/intechopen.71449
4. S. Grémion. *et al.* Tracking the evolution of the summit lava dome of Merapi volcano between 2018 and 2019 using DEMs derived from TanDEM-X and Pléiades data. J. of Volcanology and Geothermal Research, **433**, 107732 (2023). http://doi.org/10.1016/j.jvolgeores.2022.107732
5. R. Gertisser. *et al.* Geological History, Chronology and magmatic evolution of merapi. in merapi volcano (eds. Gertisser, R., Troll, V. R., Walter, T. R., Nandaka, I. G. M. A. and Ratdomopurbo, A.) 137–193 (Springer International Publishing, Cham, 2023). http://doi.org/10.1007/978-3-031-15040-1_6
6. S. H. N. Hafida, N. K. M. Isa, M. H. Ibrahim, and J. Jumadi. The role of psychological adaptations in the forming of student's behavioural adaptations in dealing with volcano eruptions in disaster-prone areas of Mount Merapi. Disaster Advances, **15**, 18–26 (2022). http://doi.org/10.25303/1507da018026
7. G. Budiyanto. Land use planning for disaster-prone areas in Southern Region of Mount Merapi. Agrivita. J. Agr. Sci, **43**, (2021). http://doi.org/10.17503/agrivita.v1i1.2774
8. E. T. W. Mei, M. I. Sadali, and R. F. Putri. Persaudaraan sekolah (sister school) untuk pengurangan risiko bencana erupsi gunungapi. (Deepublish, Yogyakarta, 2022).
9. J. Ikhsan, D. Legono, A. Pamudji Rahardjo, P. Harsanto, and M. Fujita. Dynamics of lahar-affected river tributaries of the Progo river after the 2010 Mt. Merapi eruption. IOP Conf. Ser.: Earth Environ. Sci. **437**, 012009 (2020). http://doi.org/10.1088/1755-1315/437/1/012009
10. A. Purba, S. H. Sumantri, A. Kurnadi, and D. R Putra. Analisis kapasitas masyarakat terdampak erupsi Gunung Semeru. PENDIPA Journal of Science Education, **6**, 599–608 (2022). http://doi.org/10.33369/pendipa.6.2.599-608
11. S. H. N. Hafida, N. K. M. Isa, and N. Khotimah. Is psychological adaptability needed for high school students in disaster-prone areas of Mount Merapi Eruption? in IOP Conf. Series: Earth and Environmental Science, **986**, 012040 (2022). http://doi.org/10.1088/1755-1315/986/1/012040
12. L. SetiawanL. Studi fenomenologi: Kehidupan masyarakat paska erupsi Gunung Kelud tahun 2014 di Desa Puncu Kecamatan Puncu Kabupaten Kediri. Jurnal Penelitian Keperawatan, **4**, 88–100 (2018). http://doi.org/10.32660/jurnal.v4i2.319
13. D. Hermon, *et al.* A policy model of adaptation mitigation and social risks the volcano eruption disaster of Sinabung in Karo Regency - Indonesia. GEOMATE, **17**, (2019). http://doi.org/10.21660/2019.60.50944

14. E. D. Andita. *et al.* The influence of emotional support on the supportiveness of students at SMPN 6 Surakarta in Facing Flood Disasters. in IOP Conf. Series: Earth and Environmental Science, **1357**, 012039 (2024). http://doi.org/10.1088/1755-1315/1357/1/012039
15. A. Wijayanti, and S. H. N Hafida. Analysis of the psychological adaptability of high school students in disaster-prone areas of Mount Merapi. IOP Conf. Ser.: Earth Environ. Sci. **1190**, 012033 (2023). http://doi.org/10.1088/1755-1315/1190/1/012033
16. J. W. Creswell, and J. D. Creswell. Research design: qualitative, quantitative, and mixed methods approaches. (SAGE Publications, California, 2018).
17. E. Haute. van. Sampling Techniques. Sample types and sample size. in research methods in the social sciences: An a-z of key concepts (Oxford University Press, United Kingdom, 2021).
18. M. F Ikhwan. *et al.* Pearson correlation and multiple correlation analyses of the animal fat s-parameter. TEM Journal, **13**, 155–160 (2024). http://doi.org/10.18421/TEM131-15
19. I. Maryati, T. S. Sumartini, and D. Sofyan. Experiences of pearson formula in analysis regression. in IOP Conf. Series: Materials Science and Engineering, **1098,** 032088, (2021). http://doi.org/10.1088/1757-899X/1098/3/032088
20. R. Kamil, R. Yoseptry, and D. Rostini. Manajemen pendidikan mitigasi bencana untuk menumbuhkan kesiapsiagaan peserta didik sekolah menengah pertama sebagai usaha pengurangan risiko bencana di kawasan Sesar Lembang. NUSANTARA: Jurnal Ilmu Pengetahuan Sosial, **8**, 2144–2156 (2021). http://doi.org/10.31604/jips.v8i7.2021.2144-2156
21. N. Aghaei, H. Seyedin, and H. Sanaeinasab. Strategies for disaster risk reduction education: A systematic review. Journal of Education and Health Promotion, **7**, 1–8 (2018). http://doi.org/10.4103/jehp.jehp_31_18
22. A. Amri, D. K. Bird, K. Ronan, K. Haynes, and B. Towers. Disaster risk reduction education in Indonesia: challenges and recommendations for scaling up. Nat. Hazards Earth Syst. Sci. **17**, 595–612 (2017). https://doi.org/10.5194/nhess-17-595-2017
23. D. R. K. Putra, S. Maarif, and L. Kurniawan. Community preparedness in facing mount semeru eruption to support national security. International Journal of Humanities Education and Social Sciences (IJHESS), **2**, 1668–1675 (2023). http://doi.org/10.55227/ijhess.v2i5.448
24. A. I. Benardi, *et al.* Disaster preparedness in proximity of Merapi Volcano, Indonesia: Is there any relationship in knowledge and attitude of senior high school students?. International Journal of Safety and Security Engineering, **13**, 245–254 (2023). http://doi.org/10.18280/ijsse.130207
25. N. Yustisia, T. Aprilayutini, and T. A. Utama. Pengaruh simulasi menghadapi bencana gempa bumi terhadap kesiapsiagaan siswa SDN 86 Kota Bengkulu. Journal of Nursing and Public Health, **7**, 32–28 (2019). http://doi.org/10.37676/jnph.v7i2.888
26. M. N Fatanti, D. Rahmiati, and I.R Yustisia. Merawat tradisi lokal sebagai strategi pengurangan risiko bencana di Dusun Brau, Jawa Timur (Preserving local traditions as a strategy for disaster risk reduction in Brau Village, East Java). Jurnal IPTEK-KOM (Jurnal Ilmu Pengetahuan dan Teknologi Komunikasi), **21**, (2019). http://doi.org/10.17933/iptekkom.21.1.2019.75-91
27. Z. Y. İlerisoy, B.I Gökgöz, and A. Soyluk. Selection of emergency shelter areas using multi-criteria decision-making techniques: An assessment of the case of Erciş-Van, Turkey. Periodica Polytechnica Architecture, **53**, (2022). http://doi.org/10.3311/PPar.19410
28. T. Aryanti, and A. Muhlis. Disaster, gender, and space: Spatial vulnerability in post-disaster shelters. in IOP Conf. Series: Earth and Environmental Science, **447,** (2020). http://doi.org/10.1088/1755-1315/447/1/012012
29. D. Kurniawati. Komunikasi mitigasi bencana sebagai kewaspadaan masyarakat menghadapi bencana. Jurnal Simbolika Research and Learning in Communication Study, **6**, 51–58 (2020). http://doi.org/10.31289/simbollika.v6i1.3494
30. N. Anisah. Model sekolah aman bencana dalam upaya mewujudkan pendidikan karakter di MIN 1 Bantul. Literasi, **10**, 9–20 (2019). http://doi.org/10.21927/literasi.2019.10(1).9-20
31. S. Arifin. Peran komunikasi antarpribadi dalam proses pemulihan trauma: Pada remaja korban bencana. Perspektif, **1**, 534–548 (2022). http://doi.org/10.53947/perspekt.v1i5.185
32. J. Jagnoor, *et al.* Exploring the impact, response and preparedness to water-related natural disasters in the Barisal division of Bangladesh: a mixed methods study. BMJ Open, **9**, e026459 (2019). http://doi.org/10.1136/bmjopen-2018-026459
33. V. Cvetković, G. Roder, A. Öcal, P. Tarolli, and S. Dragićević. The role of gender in preparedness and response behaviours towards flood risk in Serbia. IJERPH, **15**, 2761 (2018). http://doi.org/10.3390/ijerph15122761
34. J. S. Becker, D. Paton, D. M. Johnston, K. R Ronan, and J. McClure. The role of prior experience in informing and motivating earthquake preparedness. International Journal of Disaster Risk Reduction, **22**, 179–193 (2017). http://doi.org/10.1016/j.ijdrr.2017.03.006
35. S. Ayub, Kosim, I. W. Gunada, and E. P Handayani. Kesadaran kebencanaan pada perilaku kesiapsiagaan siswa. ORBITA: Jurnal Pendidikan dan Ilmu Fisika, **6**, (2020). http://doi.org/10.31764/orbita.v6i2.3026
36. S. Y. Tursilowati. Determinan kesiapsiagaan bencana tsunami siswa di sekolah siaga bencana Kretek Bantul Yogyakarta. Jurnal Kesehatan Indra Husada, **8**, 231–240 (2020). http://doi.org/10.36973/jkih.v8i2.265

# Preliminary study site characterization using HVSR microtremor in West Bandung Basin

*Gita* Fitri Yulianto[1*], *Qashdina* Saimima Wakano[1], *Eko* Widi Santoso[2], *Mohamad* Ridwan[2], and *Budy* Santoso[1]

[1]Department of Geophysics, University of Padjadjaran, Sumedang, Indonesia
[2]Research Center for Geological Disaster, National Research and Innovation Agency, Central Jakarta, Indonesia

**Abstract.** Site characteristic become crucial factor in seismic hazard assessment correspond to the site response of earthquake motions. This study investigates site characterization in the western part of the Bandung Basin using the Horizontal-to-Vertical Spectral Ratio (HVSR) method. HVSR analysis was applied to the Rayleigh wave microtremor data that was conducted at observation sites i.e: Baros, Cibeber, Batujajar, and Giriasih. The analysis results revealed some soil parameters i.e: dominant period (Td) values ranging from 0.23 to 0.43 seconds, S-wave velocity structure down to a depth of 30 ($Vs_{30}$) in a range of 173.25 m/s in Cibeber to the highest was 281.76 m/s in Baros. The result of $Vs_{30}$ analysis were used to classify local site which is Baros site can be classified into the stiff soil (SD) category, Cibeber and Batujajar belong to the soft soil (SC) category, and Giriasih is classified as very soft soil (SE). The findings of this study provide valuable information for seismic hazard assessment, seismic site classification, and earthquake-resistant structural design in the western part of the Bandung Basin.

## 1 Introduction

An earthquake is a sudden release of energy that generates seismic waves which spreads in all directions, causing the ground to shake and crack, often causing damage to the affected area. These events are primarily triggered by various geological activities, including volcanic eruptions and tectonic plate movements [1]. As widely recognized, the Bandung region is particularly susceptible to earthquakes due to its underlying geological structure, characterized by active tectonic zones [2]. The seismic sources in Indonesia that have been well identified can be defined into shallow crustal that are located on land, and megathrust that are located in the ocean [3].

This study located on eastern part of the West Bandung basin, where many active faults around it and the closest one is Lembang Fault. The Lembang Fault is an active geological structure extending 29 kilometers in West-east direction located in the Northern Bandung City. It exhibits a slip rate of 2 mm/year [3-5]. This movement indicates that the Lembang Fault is capable of generating earthquakes with magnitudes ranging from 6.5 to 7.0 Mw, with a recurrence interval of approximately 500 years [5]. Beside the Lembang fault, there are many active faults around the study area, including: Cimandiri, Nyalindung and Rajamandala segment of active faults [3] which considered can have an impact if an earthquake occurs.

Given the high earthquake intensity and vulnerability of the West Bandung region, mitigation measures are crucial to minimize the potential impact of seismic events. To effectively implement these measures, a comprehensive understanding of site characteristics, site effect distribution and quantifying site response to seismic motion are essential. This study aims to address this need by employing the microtremor Horizontal-to-Vertical Spectral Ratio (HVSR) method to characterize sites in West Bandung. The HVSR analysis will provide valuable insights into the dominant frequency, seismic vulnerability index, $Vs_{30}$ and site class for the region. Analysis microtremor data by using HVSR method is very useful and has been widely applied for seismic microzonation [6].

## 2 Material and methods

### 2.1 Study area

The study area encompasses the western portion of the Bandung Basin, spanning from Cililin to Cihampelas, Batujajar, and Cimahi. This western region of the Bandung Basin is demarcated by Tertiary volcanic rocks and belongs to the Rajamandala Formation. The Rajamandala Formation is the sole non-volcanic sedimentary rock unit in the area, composed by limestone, mudstone, shale, and quartz sandstone [7, 8]. The study employed 14 single-station data measurement points with interval distance about 1 – 3 km to obtain general feature for the next study in more detail for this area, as illustrated in Fig. 1. The single station microtremor survey method using HVSR analysis developed by Nakamura (1989) is very possible to be carried out in densely populated or limited areas. The instrumentation used in the field work

* Corresponding author: gitafitry71@gmail.com

is a kind of seismograph using triaxial sensor developed by OYO Corp.

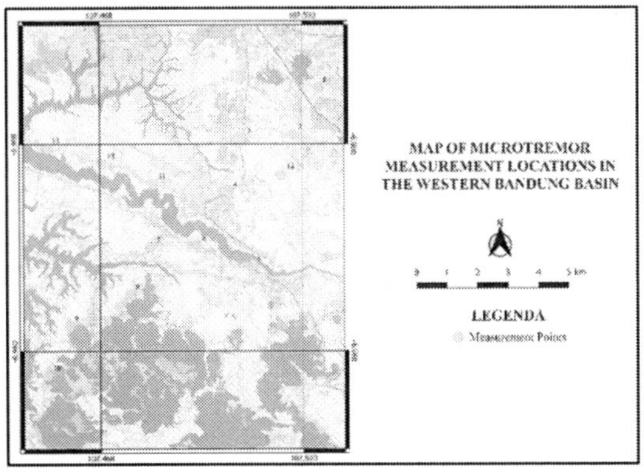


**Fig. 1.** Map of microtremor measurement locations in the Western Bandung Basin

### 2.2 HVSR method

This study utilized the single station microtremor method that was developed by Nakamura (1989). This method is relatively simple in operational because no require seismic source and requiring relatively short measurement times at each location.

The processing of microtremor data was conducted by using the HVSR method [9-11] involves employing specialized software like Geopsy to compute Fast Fourier Transform (FFT) and to generate HVSR curves by using mathematical equation 1 [12]:

$$HVSR = \frac{\sqrt{[(S_{North-South})^2 + (S_{East-West})^2]}}{S_{Vertical}} \quad (1)$$

The HVSR curves are describe a correlation beween dominant frequency ($f_0$) or Period ($T_0$) and Spectrum amplitude ($A_0$). Further analysis are calculate of some parameter such as seismic vulnerability index (Ivs), which serves as a quantitative measure of a site's susceptibility to earthquake damage.

Dominant frequency, denoted as $f_0$, represents the natural frequency of the subsurface layers that was influenced by the thickness and hardness of the soil layer. Consequently, the dominant frequency reflects subsurface characteristics. Dominant frequency is also related to shear-wave velocity and thickness in the subsurface [8]. Lower dominant frequencies can suggest increased seismic vulnerability and indicate thicker sediment. The dominant frequency can be estimated using the following empirical relationship [13]:

$$f_0 = \frac{V_s}{4h} \quad (2)$$

Where $f_0$ is the dominant frequency (Hz), $V_s$ is shear wave velocity (m/s), and h is the thickness of the sediment layer (m).

The dominant period is derived from the dominant frequency value (Fo), as shown in equation:

$$T_0 = \frac{1}{f_0} \quad (3)$$

where $T_0$ is the dominant period (s) and $f_0$ is the dominant frequency (Hz).

Seismic Vulnerability Index can indicate a potential of deformation and damage to the earthquake shaking. The correlation of some parameters i.e: seismic vulnerability index, dominant frequency and amplification factor can be shown by equation as follows [1]:

$$Kg = \frac{A_0^2}{f_0} \quad (4)$$

Where Kg is the seismic vulnerability index, Ao is the amplification factor, and fo is the natural frequency (Hz).

$Vs_{30}$ reflect soil hardness that one of the crucial parameters in application to engineering purpose. This parameter can be calculated by averaged over the top 30 meters of the subsurface. The relationship of $Vs_{30}$ and the amplification factor is crucial for understanding site-specific seismic response and predicting ground motion during earthquakes [14]. $Vs_{30}$ values can be determined using the following equation [15]:

$$Vs_{30} = \frac{30}{\sum_{i=1}^{N} \frac{h_i}{Vs_i}} \quad (5)$$

Where *Vs30* is average of *Vs* for the top of 30m, $h_i$ is thickness of layer i-th and $Vs_i$ is velocity of i-th.

## 3 Results and discussion

### 3.1 HVSR analysis results

Recorded microtremor data comprises seismic wave signals in three components: Vertical, North-South, and East-West. Data processing divided into some steps which is the first step is to identify and extract stationary signals from the noisy background. This process involves filtering and analyzing the data series to obtain the coherent microtremor signals that provide valuable information about the subsurface structure.

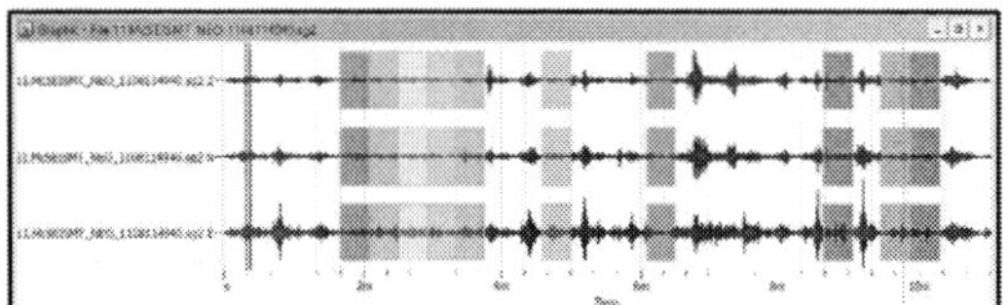

**Fig. 2.** Windowing of microtremor time series

The microtremor data in time series then divided into some segment consist of 20,48 s time window to select stationary data and noise present in the signal. The example of windowing process is shown in the site location 1 (Fig. 2). This window size selection is based on the characteristics of the noise and its impact on the data. Smoothing process was conducted by using Konno and Ohmachi method to removes high-frequency noise while preserving the underlying signal characteristics [16].

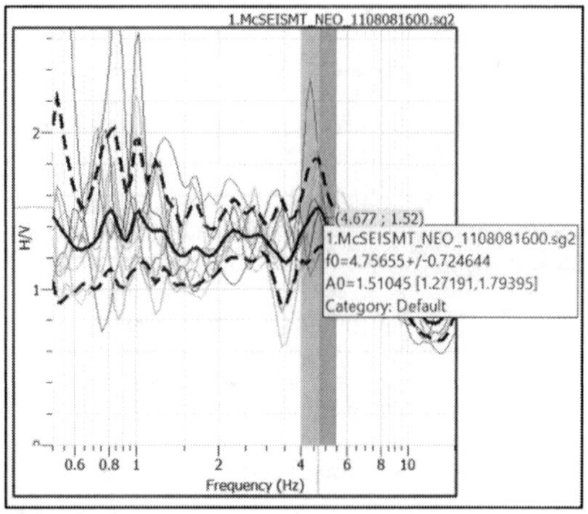


**Fig. 3.** Average horizontal-to-vertical spectral ratio curve (solid black line with standard deviation shown by dashed lines) statistically calculated from each time window's HVSR (colored lines).

Horizontal-to-vertical spectral ratio (H/V) curve (HVSR) describe distribution of amplitude of Fourier spectra ($A_0$) with respect to dominant frequency ($f_0$). As an example, shown in Fig. 3 for observation site 1, the extracted values for dominant frequency and amplitude were 4.677 Hz and 1.510, respectively.

The dominant frequency is determined from the peak amplitude of the H/V curve (represented by the solid black line without breaks). This dominant frequency reflects subsurface condition by referring to Kanai classification (1983) [17].

A comprehensive study was conducted to investigate the seismic response variations within the Western Bandung Basin, Indonesia. Utilizing the HVSR (Horizontal-to-Vertical Spectral Ratio) method, 14 sites were analyzed across the region, revealing a remarkable range of predominant frequencies ($f_0$) and amplification factors ($A_0$).

The highest dominant frequency value of 13.05 Hz was observed at the Batujajar research site. This finding indicates the presence of relatively hard and well-consolidated geological formations beneath the surface. This aligns with the geological history of the region, where sediments in Batujajar are relatively older compared to other areas in the basin. The extended time period has allowed for greater compaction and cementation, leading to increased stiffness and shear-wave velocity (Vs). Additionally, the relatively shallow depth to bedrock in Batujajar contributes to the higher $f_0$ value, as the stiffer bedrock material influences the overall seismic response.

The lowest dominant frequency value of 0.55 Hz was recorded in Cipatik, located at the foot of a mountain with steep and hilly topography. This low value suggests the presence of relatively soft and unconsolidated geological formations beneath the surface. This observation is consistent with the geological history of the area, where loose and easily compacted volcanic deposits dominate. The high water content in Cipatik's volcanic sediments further lowers Vs and amplifies the effect of seismic wave amplification.

The dominant period (T) is inversely proportional to the dominant frequency ($f_0$). In other words, as the frequency decreases, the period increases. While the relationship between dominant frequency/period and seismic vulnerability are shown by equation 5. An areas with higher dominant period values are considered to be at higher risk of earthquake damage, while areas with lower dominant period values are considered to be at lower risk [13].

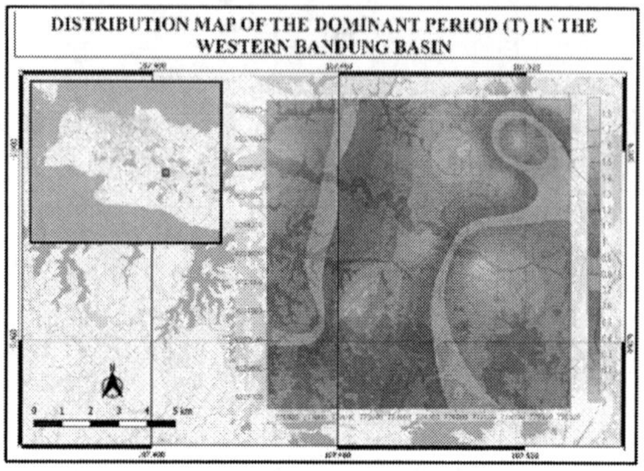


**Fig. 4.** Spatial distribution map of the dominant period (T) in the western Bandung Basin

A comprehensive analysis of dominant period values obtained from microtremor measurements at 14 locations within the western Bandung Basin unveils crucial insights into the region's soil characteristics and seismic response. The dominant period values are in a range from 0.07 to 1.8 seconds, indicating a predominant of low-period soils across the study area (Fig. 4).

Characterized by hard soil conditions, this category encompasses areas with dominant periods ranging from 0.07 to 0.09 seconds. These regions are likely underlain by hard bedrock or have a thin soil layer overlying bedrock. Due to the inherent rigidity of hard soils, seismic activity in these zones is typically low as the ground effectively dampens vibrations. Representing moderately hard soil conditions, this category encompasses areas with dominant periods ranging from 0.21 to 0.24 seconds. These regions exhibit thicker soil layers compared to Type I soils, potentially comprising deeper alluvial deposits or clay soils with a mix of denser materials. Such soils exhibit moderate responsiveness to seismic waves, with the potential for moderate amplification. Indicating soft soil conditions, this category encompasses areas with dominant periods ranging from 0.28 to 0.40 seconds. These regions feature thick sedimentary deposits, such as alluvial sediments or deep clay soils, which can lead to more significant seismic wave amplification, increasing the risk of earthquake-induced damage. These areas represent the accumulation of younger, less consolidated material. Characterized by very soft soil conditions, this category encompasses areas with dominant periods ranging from 0.45 to 0.62 seconds. These regions typically comprise extremely thick, loose sediments, such as marshlands or deep river deposits. Such areas possess a very high potential for seismic wave amplification, posing a substantial risk to infrastructure and buildings.

**Table 1.** Soil classification table based on microtremor dominant period values (modified from [17])

| Soil Classification | Periode (T) | Characteristic |
|---|---|---|
| Type I | 0.05 - 0.15 | Hard (tertiary or older rocks), consist of hard sandy rock and gravel. |
| Type II | 0.15 - 0.25 | Medium, consist of gravel, sandy hard clay, and loam. |
| Type III | 0.25 - 0.40 | Soft, consist of sand, sandy clay, and clay. |
| Type IV | > 0.40 | Very soft soil. |

Areas with dominant period values ranging from 1.2 to 1.8 seconds, represented by shades of red to light yellow on the map, are primarily located in the western and southeastern parts of the region. According to Kanai's soil classification [17], the identified range of 1.2 to 1.8 seconds falls under category IV that has a high potential for seismic amplification, which has the potential to have a greater impact on building damage. This category encompasses very soft soil conditions, typically composed of thick sedimentary deposits such as clays, sands, and alluvial materials. These sediments, due to their slow acoustic properties, exhibit longer dominant periods.

The analysis of Seismic Vulnerability Index (SVI) values derived from microtremor measurements revealed a distinct spatial variation across the 14 measurement points. In general, the SVI values ranged from 0.25 to 14.04, indicating significant heterogeneity in seismic vulnerability within the Western Bandung Basin as shown in Fig. 5.

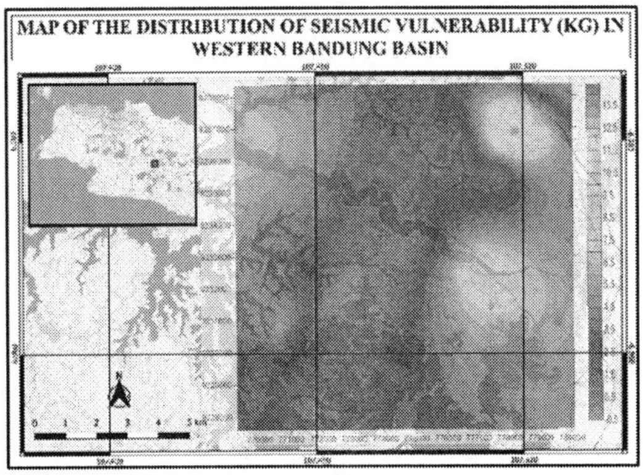


**Fig. 5.** Map of the spatial distribution of seismic vulnerability (kg) in the western Bandung Basin

The research points in the South Cimahi District show the highest seismic vulnerability index value, specifically 14.04. Referring to the geological map of the research area, these points feature hilly topography with steep slopes and are situated near an active fault, namely the Lembang Fault. The surface of this area consists of alluvial deposits, which increase the risk of strong ground shaking during earthquakes.

In contrast, the research points in the Cililin District exhibit the lowest seismic vulnerability index value, specifically 0.25. The soil in this area tends to be more compact and consolidated, such as bedrock or thin layers of soil overlying hard rock, naturally dampening seismic vibrations and reducing wave amplification. Additionally, this region is located far from seismic activity sources or active faults, thereby reducing earthquake risk. The area has relatively flat topography with minimal slopes, further the potential for seismic wave amplification.

### 3.2 1D S-wave velocity profiling

Site characterization include investigation of subsurface structure, thickness and hardness of soil/rock layers. Soil S-wave velocity is one of the important parameters for classifying soil types for engineering purposes [18]. Other parameters that can be used to determine the site class are based on the Standard Penetration Test (SPT) and soil shear strength [15].

Estimation of S-wave velocity was conducted by using inversion method to the HVSR curve by some soil parameters model including thickness, velocity and density. In this study, the ellipticity curve method of HVSR inversion with fitting is employed. The final model was considered by the smallest misfit between observed and calculated one. Some parameters that considered in the inversion process include compression-wave velocity (m/s), Poisson's ratio, shear-wave velocity (m/s), and density (kg/m³) that constructed by referring to local geology/geotechnical condition.

Since this study was intended for shallow layer survey to estimate $Vs_{30}$ and site class, then profile of S-wave velocity structure is constructed by modelling of subsurface down to a depth of 50 meters at each measurement location. The example of Vs profile were presented in Fig. 6 for four site resulting from inversion technique. It can be seen that the thickness and S-wave velocity are vary for each site that represent subsurface condition.

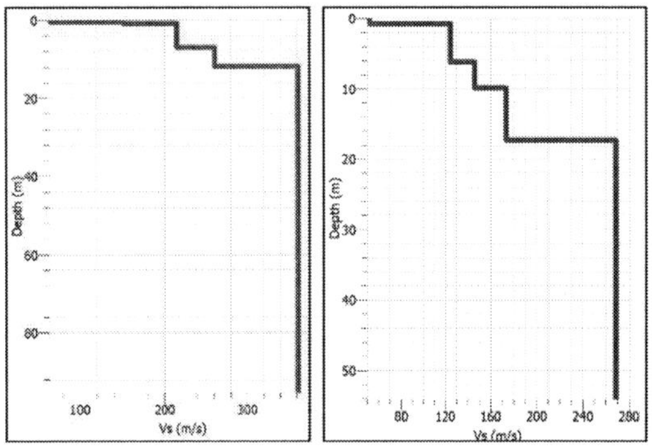


**Fig. 6.** The example of ground profiles generated from single microtremor data at site 1 and site 2

https://doi.org/10.1051/e3sconf/202560414001

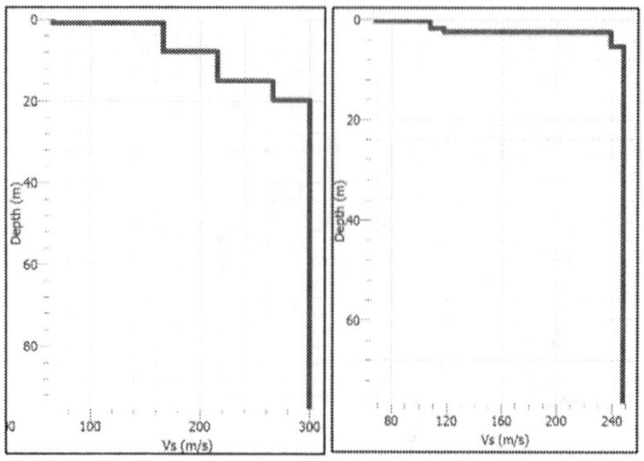


**Fig. 6.** Continue for site 3 and site 4

Site classification of this study area then was determined based on $Vs_{30}$ by referring to Indonesian National Standard (SNI-1726-2017) [15].

The Vs values for each layer are then calculated down to a depth of 30 meters to derive the $Vs_{30}$ value using the following equation and the example are shown in the location of Cimahi Tengah.

$$V_{S30} = \frac{30}{\sum_{i=1}^{N} \frac{h_i}{V_{S_i}}} \quad (6)$$

$$= \frac{30}{\frac{0.43}{66.73} + \frac{0.44}{150.88} + \frac{6.06}{215.88} + \frac{4.98}{260.80} + \frac{18.09}{362.18}} = 281.76\ m/s \quad (7)$$

Referring to result of $Vs_{30}$ calculation above, point 1 in Cimahi Tengah District has a $Vs_{30}$ about 281.76 m/s, indicating that this site can be classified into the SD (medium soil) category at a depth of 30 meters from the surface. Based on the calculation of $Vs_{30}$ values, it shows that the four points sampled in the Cimahi - Batujajar District area are predominantly of the SE (soft soil) to SD (medium soil) types.

The calculated $Vs_{30}$ values for the four location points range from 173.25 m/s to 281.76 m/s. Based on the standard $Vs_{30}$ site classification, in general the study area can be described in Fig. 7 by using kriging method which is $Vs_{30}$ are in a range of 173.25 m/s to 281.76 m/s.

The lowest $Vs_{30}$ value of 173.25 m/s was recorded at the observation point in Baros, Cimahi Tengah District. Based on the standard $Vs_{30}$ site classification table (Table 1.) [15], this value corresponds to the SE (Soft Soil) category. This classification indicates that the subsurface materials at this location are relatively loose and deformable, suggesting a higher susceptibility to ground motion amplification and liquefaction hazards during earthquakes. In contrast, higher $Vs_{30}$ values ranging from 213.30 m/s to 281.76 m/s were observed at the research points in Cibeber, Cimahi Selatan District, and Batujajar, Bandung Barat District. These values fall within the SD (Stiff Soil) category according to the $Vs_{30}$ site classification table. This classification suggests that the subsurface materials at these locations are moderately rigid and can transmit seismic waves more efficiently, implying a lower likelihood of severe ground motion amplification.

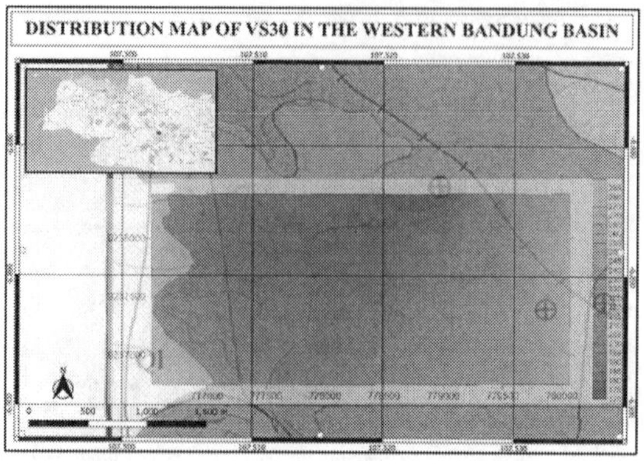


**Fig. 7.** Spatial distribution map of $Vs_{30}$ in the Western Bandung Basin

In contrast to the Cimahi site, higher $Vs_{30}$ values ranging from 213.30 m/s to 281.76 m/s were observed at the research points in Cibeber, Cimahi Selatan District, and Batujajar, Bandung Barat District. These values fall within the SD (Medium Soil). This classification suggests that the subsurface condition at these locations are moderately rigid. The smallest of $Vs_{30}$ was obtained at the Cibeber site in Cimahi Selatan area which is $Vs_{30}$ is about 173.25 m/s that can be classified into site class SE (soft soil). This location is typically associated with unconsolidated alluvial deposits.

The correlation of observed spatial variation in $Vs_{30}$ values and seismic site classifications to the topography of the region, in general, indicate the lowest $Vs_{30}$ (Baros site) is located at a lower elevation, while the location with higher $Vs_{30}$ values (Cibeber and Batujajar) are situated at higher elevations. This suggests a correlation between elevation and subsurface characteristics has a good agreement.

The high $Vs_{30}$ value of 281.76 m/s recorded at the Baros observation point indicates hard and dense soil conditions. This classification is typically associated with bedrock or compacted soils, such as sandstone or consolidated sedimentary rocks. The high shear wave velocity suggests that the subsurface materials at this location are rigid and efficient in transmitting seismic waves, implying a lower likelihood of severe ground motion amplification. The $Vs_{30}$ value of 213.30 m/s obtained at the Cibeber research point suggests softer soil conditions compared to Baros. This could be attributed to the presence of clay deposits or looser soils, which generally exhibit lower shear wave velocities than bedrock. The softer soil conditions at Cibeber may result in a higher susceptibility to ground motion amplification compared to Baros. The $Vs_{30}$ value of 221.31 m/s recorded at the Batujajar research point indicates intermediate soil conditions, falling between the hard and dense soils of Baros and the softer soils of Cibeber. This suggests that the subsurface materials at Batujajar may consist of a mixture of sedimentary rocks, volcanic rocks, or clayey materials, influencing the overall shear wave velocity.

Since $Vs_{30}$ reflect geological properties included type of soil/rock and geomorphology, therefore integrating local geological considerations are required for seismic hazard assessment. Study in more detail of site characteristic and geology local, and their effect to the earthquake motion are crucial to improve understanding related to seismic wave propagation and building damages.

## 4 Conclusion

The analysis of HVSR, dominant period, seismic vulnerability, HVSR inversion, and $Vs_{30}$ in the western part of the Bandung Basin shows a variation in $Vs_{30}$ values and seismic site classifications in the region. $Vs_{30}$ values range from 173.25 m/s to 281.76 m/s, with seismic site classifications ranging from SE (Soft Soil) to SD (Stiff Soil). This variation is influenced by local geological factors such as rock type, soil composition, and other geological conditions. Based on the result of $Vs_{30}$ calculation at entire observation site shows a good correlation to the topography.

This information is important for earthquake hazard assessment and building structure evaluation. S-wave velocity structure reflect site local condition that was related to the amplification phenomenon of the seismic motion. Deep understanding of site characterization as resulting from HVSR microtremor analysis is crucial for developing of seismic hazard analysis for this study area and implementing mitigation measures to minimize seismic risk in the western part of the Bandung Basin.

## References

1. Kurniawan, R., Eva, M. N., Marjiyono, and Sismanto. Pemetaan daerah rawan resiko gempa bumi menggunakan metode HVSR di kotamadya Denpasar dan sekitarnya, Bali. KURVATEK, **2**, 21-30, (2017).
2. Ambarwati, I. W., Feranie, S., and Tohari, A. Analisis Potensi Likuifaksi Di Wilayah Cekungan Bandung dengan Menggunakan Metode Uji Penetrasi Konus. RISET Geologi dan Pertambangan, **30**, 21-37. (2020).
3. Irsyam M, Cummins PR, Asrurifak M, Faizal L, Natawidjaja DH, Widiyantoro S, Meilano I, Triyoso W, Rudiyanto A, Hidayati S, Ridwan M, Hanifa NR, Syahbana AJ Development of the 2017 national seismic hazard maps of Indonesia. Earthq. Spectra **36**, 112–136. (2020)
4. Meilano, I., Abidin, H. Z., Andreas, H., Gumilar, I., Sarsito, D., Hanifa, R., Rino, Harjono, H, Kato, T, Kimata, F, and Fukuda, Y, Slip rate estimation of the Lembang fault West Java from geodetic observation: Journal of Disaster Research, **7**, 12–18. (2012)
5. Daryono, M. R. and Natawidjaja, D. H. Lembang Active fault study using high resolution digital elevationmodel (DEM)image: Poster session presented at Geotechnology Science Week, Research Centre for Geotechnology, Indonesia Institute of Science, Bandung, Indonesia. (2015)
6. Gosar, A., Study on the applicability of microtremor HVSR method to support seismic microzonation in the town of Idrija (W Slovenia), Nat. Hazards Earth Syst. Sci. Discuss. (2017)
7. Bronto, S., and Hartono, U. Potensi sumber daya geologi di daerah Cekungan Bandung dan sekitarnya. Jurnal Geologi Indonesia, **1**, 9-18, (2006).
8. Silitonga, P.H. Geological Map of the Bandung Quadrangle, Jawa, Scale 1:100.000, Geological Research and Development Centre. (2003)
9. Nakamura, Y. A Method for Dynamic Characteristics Estimation of Subsurface using Microtremor on the Ground Surface, Quarterly Report of Railway Technical Research Institute (RTRI), **30**(1). (1989),
10. Nakamura, Y, Clear Identification of Fundamental Idea of Nakamura's Technique and Its Applications, 12 WCEE 2656. (2000)
11. Nakamura, Y, On The H/V Spectrum, The 14 th World Conference on Earthquake Engineering October 12-17, (2008)
12. Rahman, A. S., Nugroho, H., Permana, D., Sativa, O., Riyanto, A., Sirait, R. K et al. Analisis nilai periode dominan tanah dan amplifikasi akibat gempabumi Tarutung M5.8 01 Oktober 2022 terhadap kerusakan infrastruktur berdasarkan metode HVSR. **4**, 28-33, (2023).
13. Putra, A. K., Yatini, Y., and Paripurno, E. T, Application of HVSR method on microtremor data for analysis of earthquake potential in Candipuro District, Lumajang, Indonesia. Journal of Geoscience, Engineering, Environment, and Technology, **8**, (2023).
14. Bustari, A. A., and Wibowo, N. B. Pemetaan sebaran nilai Vs30, faktor amplifikasi tanah, dan peak ground acceleration wilayah Bantul Timur. Cakrawala Jurnal Ilmiah Bidang Sains, **1**, 73-80. (2023).
15. (BSN) Badan Standardisasi Nasional Indonesia SNI-03-1726-2019: Tata cara perencanaan ketahanan gempa untuk struktur bangunan gedung dan non gedung. (2019)
16. Ambarsari, D. Analisis Mikrotremor Dengan Metode HVSR Untuk Mikrozonasi Kabupaten Gunungkidul Yogyakarta. Tugas akhir–sf, 141501. (2017).
17. Haerudin, N., Alami, F., and Yogi, I. B. S. The effect site analysis based on microtremor data using the Horizontal to Vertical Spectral Ratio (HVSR) method in the Bandar Lampung City. In Journal of Physics: Conference Series, **1572**, 1, 012075. (2020).
18. Kyaw, Z. L., Pramumijoyo, S., Husein, S., Fathani, T.F., Kiyono, J. Ground Motion in Yogyakarta City, Yogyakarta Special Province, Indonesia on Densely Microtremor Observations and Shear Wave Velocity, J. SE Asian Appl. Geol., Jan–Jun 2013, **5**(1), 10–20, (2013)

# Seismic performance evaluation of the Central Laboratory Universitas Andalas building with expansion joints

*Fauzan* [1*], *Geby* Aryo Agista[1], and *Zev* Al Jauhari[2]

[1]Department of Civil Engineering, Faculty of Engineering, Universitas Andalas, Indonesia
[2]Graduate School of Engineering, Toyohashi University of Technology, Japan

**Abstract.** The laboratory supports students' education and provides space for practice and research. One of the laboratories is the Central Laboratory Building of Universitas Andalas, located in Padang City, which has building expansion joints. Since Padang City is an earthquake-prone area, the planning of laboratory buildings must consider earthquake safety. An expansion joint is needed to ensure earthquake safety as a boundary for building structures. It is designed so that if a major earthquake occurs, the buildings are planned not to be damaged due to pounding so that potential losses can be minimized. This study evaluates the seismic performance of buildings with expansion joints based on the current Indonesian seismic codes. The study results show that the maximum displacement of the buildings (zones A, B, C, and D) does not exceed the distance between expansion joints. It prevents contact between them (pounding) during any seismic event, allowing the building to be declared safe during an earthquake. Additionally, the seismic performance of buildings complies with current Indonesian earthquake regulations, specifically regarding dynamic characteristics, inter-story drift, horizontal irregularity, and vertical irregularity.

## 1 Introduction

Padang City is prone to earthquakes, which have caused significant casualties in West Sumatra, Indonesia. This is due to its geographical location on the western coast of Sumatra Island, where there is a subduction zone and the meeting point of two tectonic plates, the Indo-Australian Plate and the Eurasian Plate [1]. Earthquakes result in various impacts, including material losses such as building damage and human lives lost. Efforts to reduce the risk of damage due to earthquakes need to consider the design and location of buildings, taking into account the characteristics of the earthquake and the local geology [2]. One way to ensure structural strength during an earthquake is to provide expansion joints in a tall and elongated building [3-5]. The design of the building's expansion joint system is very well used for buildings with elongated spans to avoid the domino effect from the earthquake load received. The vibrations caused by earthquake loads affect the behavior of building structures, especially in asymmetric buildings with T-shaped plans. Asymmetric buildings, where the center of gravity is not in the middle, can result in significant torsion when the building receives horizontal loads such as earthquake loads. Therefore, an evaluation of the building with expansion joints distance is necessary.

In recent years, many studies have been conducted on checking structural elements and evaluating building expansion joint distances, such as analyzing building structure expansion joint distances using a two-column dilatation system [6], analyzing expansion joint distances of L-layout buildings, and calculating reinforcement of beam and column elements around dilatation [7], Seismic performance and assessment of RC framed structure with geometric irregularities [8], building performance evaluation of A.N.S hotel building plan with and without dilatation in earthquake prone areas [9], Seismic performance evaluation of reinforced concrete building in Palu City [10]. One such building with expansion joints is the Central Laboratory Building of Universitas Andalas. This newly constructed building, inaugurated in 2023, is at the Universitas Andalas Campus, Limau Manis, Padang City, West Sumatra. It functions as the main laboratory facility at Universitas Andalas.

This study focused on evaluating the seismic performance of the central laboratory Universitas Andalas building plan with expansion joint according to the current Indonesian seismic codes, specifically regarding dynamic characteristics, inter-story drift, horizontal irregularity, and vertical irregularity.

## 2 Methodology

The Central Laboratory Building of Universitas Andalas is a support facility for students conducting practical activities and research. As shown in Fig. 1, the building is divided into zones A, B, C, and D.

In this study, seismic performance evaluation of building expansion joint is carried out by as-built drawing based on the current Indonesian seismic codes (SNI 1726:2019) [11] regarding the procedures for earthquake

* Corresponding author: fauzan@eng.unand.ac.id

resistance planning for building and non-building structures (SNI 2847:2019) [12], and minimum load for planning of buildings and other structures (SNI 1727:2020) [13], regarding structural concrete requirements for buildings using ETABS software.

**Fig. 1.** The Central Laboratory Building of Universitas Andalas

## 2.1 Building data

This study analyzes three models: Model 1 is a 3D structural model of laboratory buildings Zones A and B, Model 2 is a 3D structural model of laboratory building Zone C, and Model 3 is a 3D structural model of laboratory building Zone D. Figs. 2 and 3 show a plan view of the central laboratory building and detail of expansion joints.

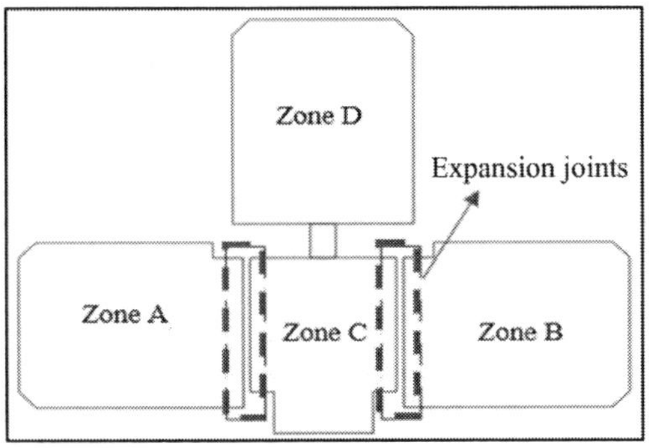


**Fig. 2.** Plan view of the Central Laboratory Building of Universitas Andalas

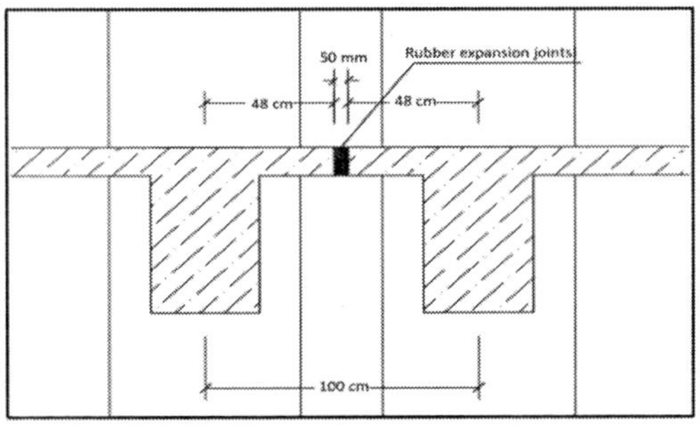


**Fig. 3.** Detail of building expansion joints

Table 1 shows the building data parameters used in the seismic performance evaluation of the Central Laboratory Building at Universitas Andalas. Based on the Detail Engineering Design (DED), the expansion joint between zones is 50 mm, as seen in Fig. 3.

**Table 1.** Building data of the laboratory building

| Parameter | Data |
|---|---|
| Building Name | Central Laboratory of Universitas Andalas |
| Location | Limau Manis, Padang City, West Sumatra |
| Building Function | Laboratory |
| Number of Floors | 3 Floors |
| Building Height | 12 m |
| Expansion Joint Distance | 50 mm |

This study obtained the concrete quality, reinforcing steel quality, dimensions of structural elements, and room functions from as-built drawings. The compressive strength of the existing concrete is 24.9 MPa, as specified in the plan drawings, and the specified weight density for normal concrete is 2,400 kg/m$^3$. The reinforcing steel quality of the longitudinal bar is categorized as minimum yield strength (fy = 400 MPa), minimum tensile strength (fu = 560 MPa and the reinforcing steel quality of the ties bar is fy = 240 MPa, fu = 400 MPa. Steel bars' weight density and modulus of elasticity are 7,850 kg/m$^3$ and 2.10$^5$ MPa, respectively. The structure includes primary columns and beams with reinforcement bars, detailed in Tables 2 and 3.

**Table 2.** Details of columns

| No. | Type | Section | | Flex. reinf. bar | Shear reinf. bar | |
|---|---|---|---|---|---|---|
| | | D | W | | S | M |
| 1 | K1 | 700 | 700 | 20D22 | Ø13-100 | Ø13-150 |
| 2 | K1-1 | 700 | 700 | 16D22 | Ø13-100 | Ø13-150 |
| 3 | K2 | Circle-500 | | 12S150 | Ø12-150 | Ø12-150 |
| 4 | K3 | 600 | 600 | 16D22 | Ø13-100 | Ø13-150 |
| 5 | K4 | 500 | 500 | 12D22 | Ø13-100 | Ø13-150 |

**Table 3.** Details of beams

| No. | Type | Section | | Flex bar | | Shear reinf. bar | |
|---|---|---|---|---|---|---|---|
| | | D | W | T | C | T | C |
| 1 | S1 | 700 | 400 | 6D19 | 6D19 | Ø10-100 | Ø10-150 |
| 3 | S2 | 500 | 300 | 5D16 | 5D16 | Ø10-100 | Ø10-150 |
| 4 | B1 | 700 | 400 | 10D19 | 5D19 | Ø10-100 | Ø10-150 |
| 6 | B2 | 700 | 400 | 10D19 | 5D19 | Ø10-100 | Ø10-150 |
| 7 | B3 | 500 | 300 | 5D19 | 3D19 | Ø10-100 | Ø10-150 |
| 8 | B4 | 400 | 300 | 5D16 | 3D16 | Ø10-100 | Ø10-150 |
| 9 | B5 | 300 | 200 | 3D13 | 5D16 | Ø10-100 | Ø10-150 |
| 10 | BS-1 | 500 | 300 | 5D19 | 3D19 | Ø10-100 | Ø10-150 |
| 11 | B6 | 350 | 250 | 5D16 | 3D16 | Ø10-100 | Ø10-150 |
| 12 | BD | 350 | 250 | 5D16 | 3D16 | Ø10-100 | Ø10-150 |

## 2.2 Structural modeling

The 3D structural modeling of the Central Laboratory Building at Universitas Andalas using ETABS software. The structural elements, including floor slabs, were modeled as thin shells with rigid diaphragms, and columns and beams were modeled as frame elements with fixed support boundary conditions. The fixed support boundary condition restricts all translation degrees of freedom for the assigned entities to zero. It models portions of the geometry connected to a rigid body. This study uses these rigid bodies for ground foundations, column-foundation joints, and beam-column joints. The 3D structural model of laboratory building Zones A, B, C, and D is illustrated in Figs. 4-6.

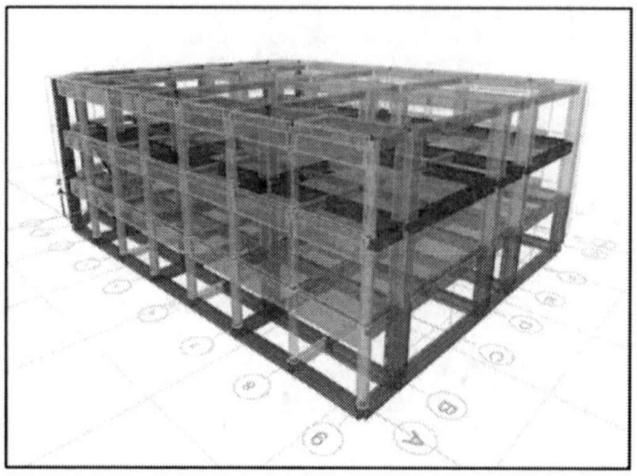

**Fig. 4.** 3D structural model of laboratory building zone A and B (model 1)

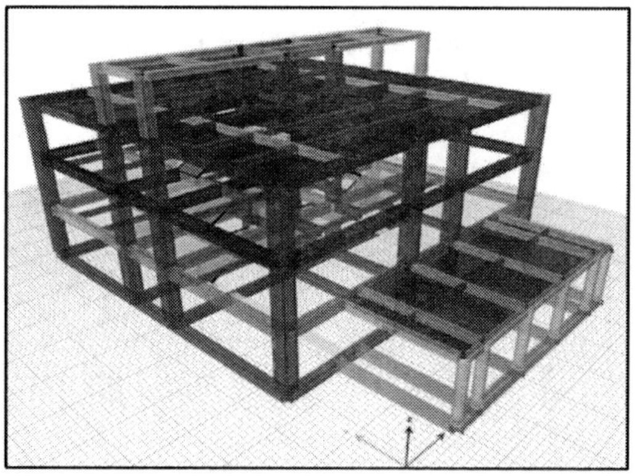

**Fig. 5.** 3D structural model of laboratory building zone C (model 2)

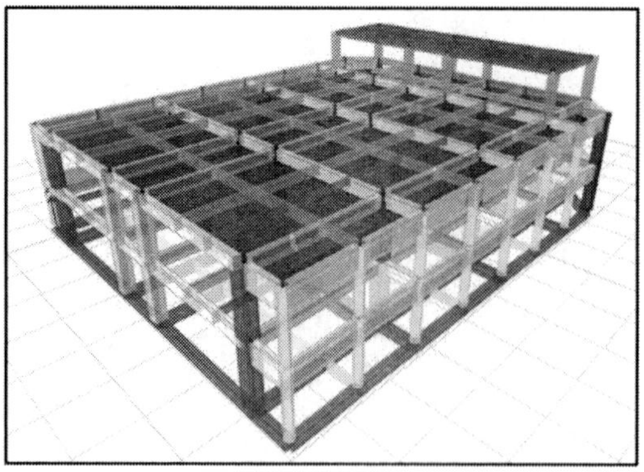

**Fig. 6.** 3D structural model of laboratory building zone D (model 3)

## 2.3 Loading analysis

### *2.3.1 Dead load*

Based on SNI 1727:2020 [10], the dead load is the weight of all building construction materials installed in this model, including the weight of the beam itself and comprising walls, floors, roofs, ceilings, stairs, fixed partition walls, finishes, building cladding, as well as other architectural and structural components, and installed service equipment as additional dead load (Tables 4-6).

**Table 4.** Additional dead load for first and second floors

| Type of load | Weight | Unit |
|---|---|---|
| Ceramic tile | 24 | kg/m$^2$ |
| Spacing | 20 | kg/m$^2$ |
| MEP | 50 | kg/m$^2$ |
| Plafond | 18 | kg/m$^2$ |
| Total | 112 | kg/m$^2$ |

**Table 5.** Additional dead load for roof floors

| Type of load | Weight | Unit |
|---|---|---|
| Waterproofing | 5 | kg/m$^2$ |
| Spacing | 20 | kg/m$^2$ |
| MEP | 50 | kg/m$^2$ |
| Plafond | 18 | kg/m$^2$ |
| Total | 93 | kg/m$^2$ |

**Table 6.** Additional dead load for stairs

| Type of load | Weight | Unit |
|---|---|---|
| Ceramic tile | 24 | kg/m$^2$ |
| Spacing | 20 | kg/m$^2$ |
| Total | 44 | kg/m$^2$ |

### *2.3.2 Live load*

Based on SNI 1727:2020, live load is caused by users and occupants of the building, structures not including construction loads, and environmental loads such as earthquake loads. The minimum distributed live loads are listed in Table 7.

**Table 7.** Live load

| Function | Weight | Unit |
|---|---|---|
| Floor system | | |
| Office room | 250 | kg/m$^2$ |
| Computer room | 480 | kg/m$^2$ |
| Corridor | 480 | kg/m$^2$ |
| Laboratory | 400 | kg/m$^2$ |
| Toilet | 250 | kg/m$^2$ |
| Roof deck | 250 | kg/m$^2$ |

### *2.3.3 Seismic load*

A seismic load is an equivalent static load that acts on a building or part of a building that mimics the influence of ground movement due to the earthquake. In this study, the earthquake loads using parameters based on the Indonesian seismic standard (SNI 1726:2019) with shelter building as risk category IV, priority factor (Ie) 1.5, the value of the response modification coefficient (R) is 8, the over-strength factor system (Ω) is 3, and the deflection magnification factor (Cd) is 5.5, the approach reinforcement parameter (Ct) is 0.0466, and the approach period parameter (x) is 0.9. The acceleration spectrum parameters are used in the response spectrum design of West Pasaman District, Indonesia, as shown in Fig. 7.

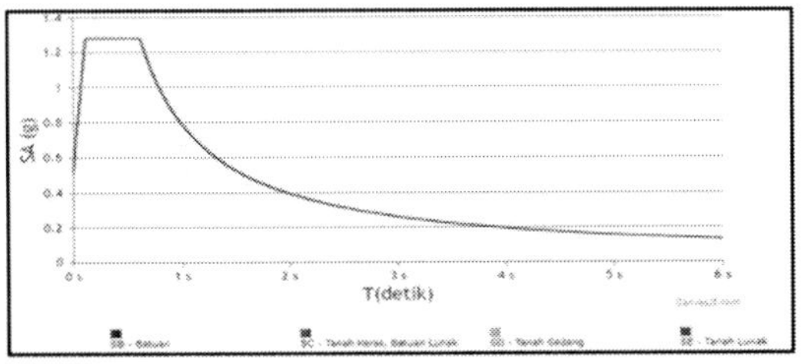


**Fig. 7.** Response spectrum of building's location

## 2.4 Load combinations

The load combinations are based on SNI 1726:2019 regarding the procedures for earthquake resistance planning for building and non-building structures, which are presented in Table 8. In ETABS, an envelope function in load combinations is used to determine the maximum and minimum values of response quantities (such as forces, moments, or displacements) from a set of load combinations. Instead of manually checking each load combination, an envelope automatically captures the extreme values (both positive and negative) across all the combinations included in the envelope.

**Table 8.** Additional dead load for stairs

| No | Load combination | Note |
|---|---|---|
| 1. | U = 1.4 D | Dead load |
| 2. | U = 1.2 D + 1.6 L | Dead and live load |
| 3. | U = 1.4 D + L ± 1.3 Ex ± 0.39 Ey | Dead, live, and earthquake loads |
| 4. | U = 1.4 D + L ±0.39 Ex ±1.3 Ey | |
| 5. | U = 0.7 D ±1.3 Ex ± 0.39 Ey | |
| 6. | U = 0.7 D ± 0.39 Ex ± 1.3 Ey | |
| 7. | Envelope | |

# 3 Results and discussion

## 3.1 Evaluation of the existing building's expansion joints

The expansion joint distance evaluation was conducted on the Central Laboratory Building of Universitas Andalas in Zones A, B, C, and D by calculating the maximum building displacement, as shown in Figs. 8-10.

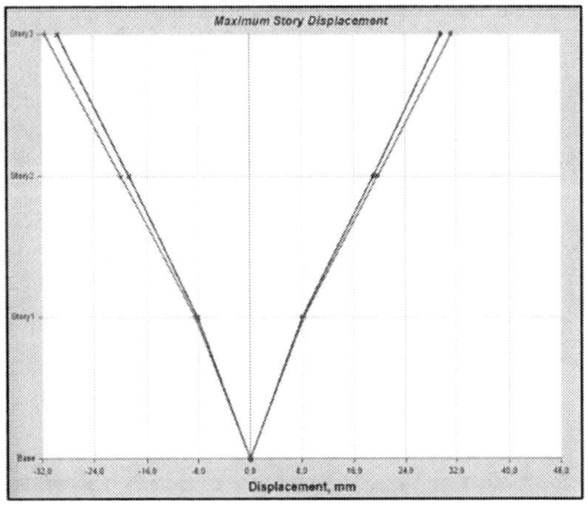


**Fig. 8.** Result of maximum displacement of model 1

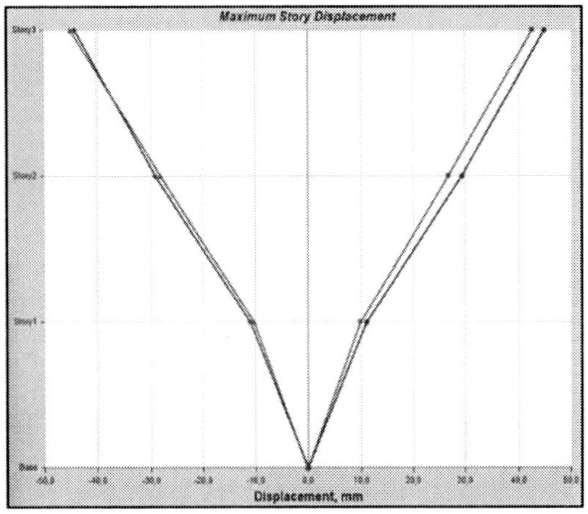


**Fig. 9.** Result of maximum displacement of model 2

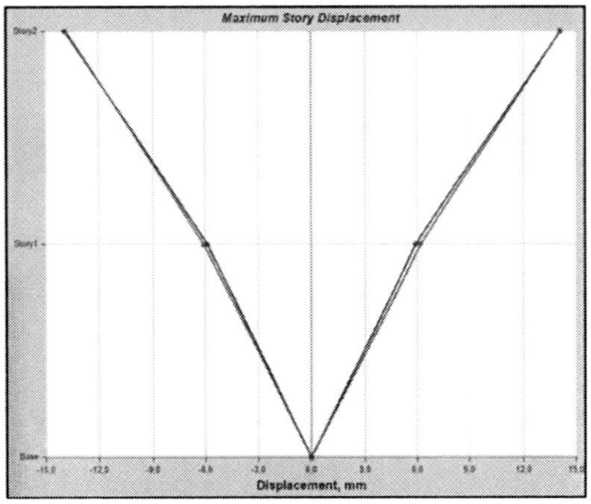


**Fig. 10.** Result of maximum displacement of model 3

**Table 9.** Maximum displacement of laboratory building

| Zone of laboratory | Model | X-direction (mm) | Y-direction (mm) |
|---|---|---|---|
| A and B | 1 | 29.253 | 31.221 |
| C | 2 | 44.880 | 42.662 |
| D | 3 | 14.110 | 14.098 |

Evaluation of the expansion joint distance was carried out on the Central Laboratory Building of Andalas University in Zone A and B with Zone C and Zone D with Zone C by calculating the maximum displacement of the building using the finite element method (FEM) of the ETABS computer program. The maximum value of the building displacement was obtained from the load combination envelope.

Table 9 shows the result of the maximum displacement in zones A, B, C, and D does not exceed the distance between expansion joints (50 mm), allowing the building to be declared safe in the event of an earthquake maximum (SNI 1726:2019). Therefore, the expansion joints distance in the building can be declared safe, and there will be no pounding during an earthquake.

## 3.2 Seismic performance of the laboratory building with expansion joints

### *3.2.1 Dynamic characteristic*

Period T is a fundamental property in the structural design process, especially in earthquake-resistant structures. The vibration period of the structure will determine the magnitude of the earthquake load that will be applied in the structural design. A structure's natural vibration period depends on its mass and stiffness to vibrate freely without external forces. Based on the Indonesian National Standard SNI 1726:2019 article 7.8.2, the fundamental period of the structure (T) in the appropriate direction must be obtained using the structural properties and deformation characteristics of the elements in the structural analysis. The period of Model 1 is 0.433 seconds, Model 2 is 0.58 seconds, and Model 3 is 0.369 seconds. Both models have met the building planning requirements for the shape mode, namely shape modes 1, 2, and 3 experience translation in the x and y directions, and shape mode 3 experiences rotation.

### *3.2.2 Inter-story drift*

The displacement that occurs in the structural elements around the expansion joints is reviewed from the structural analysis.

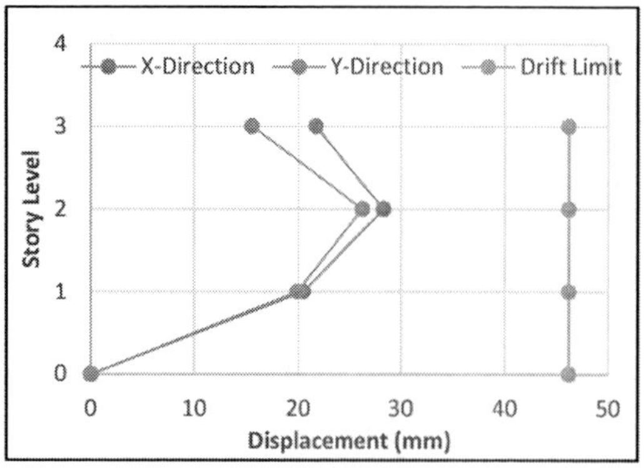


**Fig. 11.** Inter-story displacement of model 1

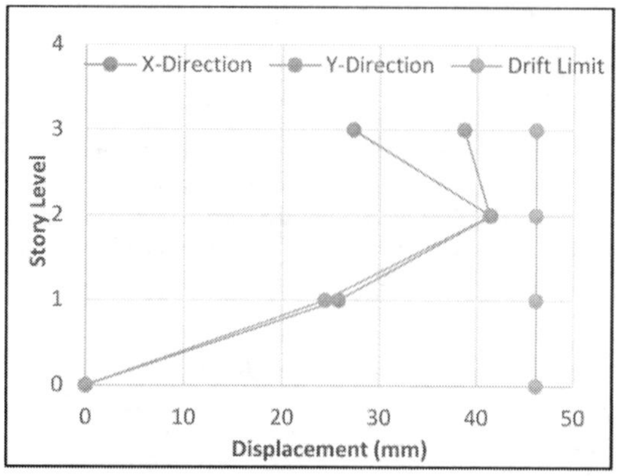


**Fig. 12.** Inter-story displacement of model 2

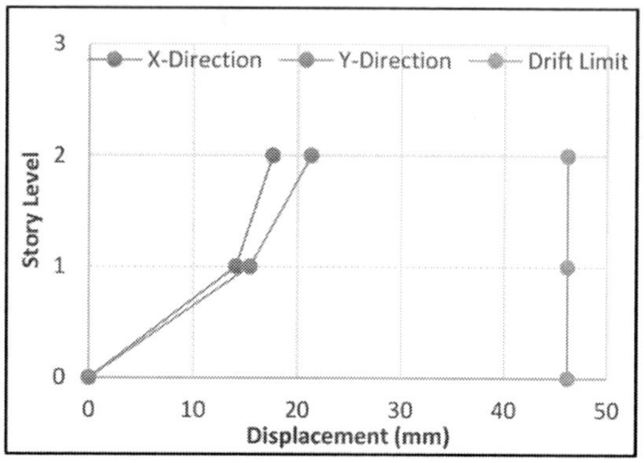


**Fig. 13.** Inter-story displacement of model 3

Inter-story drift, or story drift, is the difference in horizontal deflection between two adjacent building floors. It is calculated by subtracting the lower level's story displacement from the upper level's. Inter-story drift of permits is regulated in SNI 1726:2019. Inter-story drift for Models 1-3 can be seen in Figs. 11-13.

Figs. 11-13 shows that the value of X and Y displacement in each zone of the laboratory building does not exceed the permitted limits in Indonesian seismic codes (SNI 1726:2019), so the building has complied with the current seismic building codes.

### *3.2.3 Horizontal irregularity*

In this study, horizontal irregularity is conducted to torsional irregularity in SNI 1726:2019, which requires that the limit of torsion irregularity in the direction of the horizontal axis of the structure is 1.2 and in the vertical direction of the structure is 1.4. From the results of the structural analysis of the three models, it was found that the torsional irregularity of the building still met the requirements. The results of torsional horizontal irregularity are shown in Figs. 14-16.

Overall, the results of torsional irregularity in each zone of the laboratory building have met the current Indonesian seismic building codes. Still, on the 3rd floor of zone C, there is horizontal torsion irregularity. Torsional horizontal irregularity on the 3rd floor of zone C occurs when there is an uneven distribution of mass, stiffness, or strength across its floor plan, leading to a twisting or rotational response during lateral forces such as earthquakes, as shown in Fig. 17.

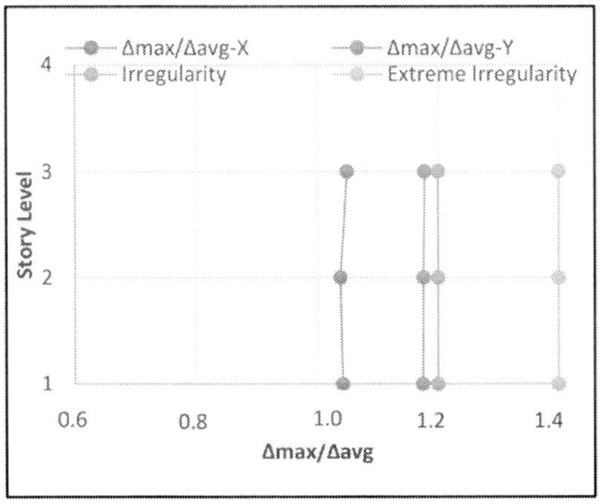


**Fig. 14.** Torsional irregularity of model 1 (Zone A and B)

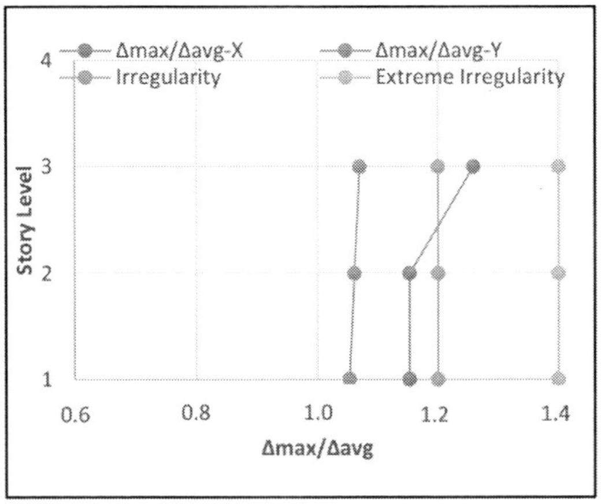


**Fig. 15.** Torsional irregularity of model 2 (Zone C)

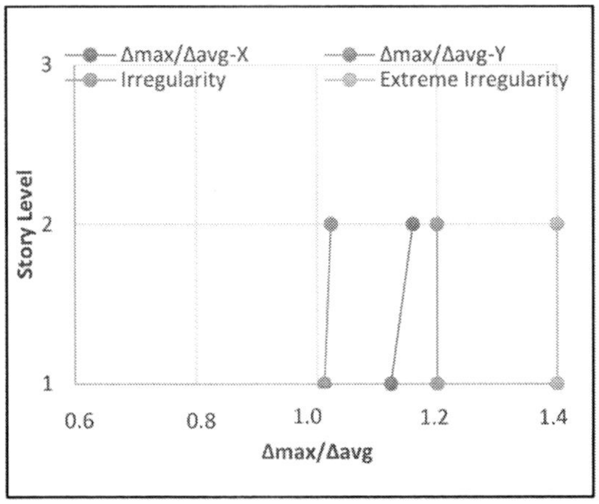


**Fig. 16.** Torsional irregularity of model 3 (Zone D)

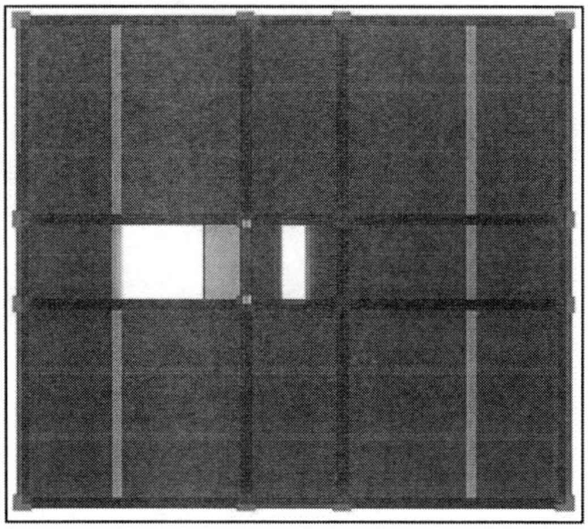

**Fig. 17.** Plan view of story level 3 of zone C (model 2)

### *3.2.4 Vertical irregularity*

Vertical stiffness irregularity in SNI 1726:2019 requires that soft story stiffness irregularity is defined as existing if there is a story whose lateral stiffness is less than 70% of the lateral stiffness of the story above it or less than 80% of the average stiffness of the three stories above it, as shown in Tables 10-12. From the tables, the results of stiffness irregularity in each zone of the laboratory building have met the current Indonesian seismic codes.

**Table 10.** The story stiffness of model 1

| Floor | X-direction | | Y-direction | |
|---|---|---|---|---|
| | **Story stiffness (kN/m)** | **Check** | **Story stiffness (kN/m)** | **Check** |
| 3 | 143337.81 | | 141970.076 | |
| 2 | 364256.359 | OK | 351665.881 | OK |
| 1 | 570341.574 | OK | 601582.539 | OK |

**Table 11.** The story stiffness of model 2

| Floor | X-direction | | Y-direction | |
|---|---|---|---|---|
| | **Story stiffness (kN/m)** | **Check** | **Story stiffness (kN/m)** | **Check** |
| 3 | 135682.15 | | 111180.873 | |
| 2 | 175168.994 | OK | 144031.375 | OK |
| 1 | 452396.146 | OK | 553805.072 | OK |

**Table 12.** The story stiffness of model 3

| Floor | X-direction | | Y-direction | |
|---|---|---|---|---|
| | **Story stiffness (kN/m)** | **Check** | **Story stiffness (kN/m)** | **Check** |
| 2 | 352534.236 | | 363073.94 | |
| 1 | 636144.765 | OK | 605493.7 | OK |

## 4 Conclusion

The Central Laboratory Building of Universitas Andalas, located in an earthquake-prone region, was thoroughly evaluated for its seismic performance using the current Indonesian seismic codes. The evaluation results indicated that the maximum displacement in zones A, B, C, and D did not exceed the building's expansion joint distance of 50 mm, ensuring the structure is safe against pounding during an earthquake. Additionally, the building's seismic performance, including dynamic characteristics, inter-story drift, horizontal irregularity, and vertical irregularity, all complied with the relevant seismic resistance codes.

The fundamental vibration periods for the different zones met the structural planning requirements, ensuring that the building remains dynamically stable during seismic events. Inter-story drift, which measures the difference in displacement between adjacent floors, was within permissible limits, preventing excessive deformation under seismic loads. However, minor torsional irregularities were observed on the third floor of Zone C, likely caused by uneven distribution of mass or stiffness across the building's layout. Despite this, vertical stiffness irregularity checks revealed no violations, indicating uniform structural stiffness across the building's levels.

Overall, the seismic evaluation confirmed that the building complies with Indonesian seismic regulations, ensuring it is well-prepared to withstand earthquakes with minimal safety and structural integrity risk.

The authors gratefully acknowledge the financial support provided by Universitas Andalas.

## References

1. S. M. Alif, K. E. Ching, T. Sagiya, W. N. Wahyuni, Determination of Euler pole parameters for Sundaland plate based on updated GNSS observations in Sumatra, Indonesia. Geosc. Lett. **11**, 1, 16 (2024).
2. A. Damiani, V. Poggi, C. Scaini, M. Kohrangi, P. Bazzurro, Impact of the uncertainty in the parameters of the earthquake occurrence model on loss estimates of urban building portfolios. Seism. Research Lett. **95**, 1, 135-149 (2024).
3. N. Khan and K. Kapoor, Performance Evaluation of A Structure With And Without Expansion Joints. *Introduction to Sustainable Solution Techniques in Civ. and Env. Eng. Sc.*, **255**. (2024)
4. I. Sydnaoui, R. B. N Mohamed, and M. A Kadir, The effects of seasonal thermal loads at expansion joints locations in Arabic area buildings. *Int. J of Adv. Sci. and Tech.*, **29** (2), 590-600 (2019)
5. M. Arun, T. Srinivas, and P. V. V. S. S. R. Krishna, Role of expansion joint in the study of seismic analysis of a multi-storied building. In E3S Web of Conferences, EDP Sciences. 309, 01115 (2021)
6. G. Caredda, N. Makoond, M. Buitrago, J. Sagaseta, M. Chryssanthopoulos, J. M. Adam, Enhancing building robustness through a fuse-based segmentation framework. Dev. in the Built Env. **19**, 100515 (2024).
7. J. Patil, S. Dhagdhage, P. Minde, Enhancing earthquake resilience in L-and T-shaped buildings using post-tensioned rocking shear walls. Innovative Infrastructure Solutions, **9**, 9, 352 (2024).
8. V. Mehta, M. H Chey, Seismic performance and assessment of RC framed structure with geometric irregularities. Asian Journal of Civil Engineering, **24**, 2, 479-496 (2023).
9. F. Nugroho, Building Performance Evaluation of ANS Hotel Building Plan with and Without Dilatation in Earthquake Prone Areas. ICTIS, **10**, (2016).
10. F. Nugroho, J. Tanjung, M. Maidiawati, R. Kurniawan, Seismic Performance Evaluation of Reinforced Concrete Building in Palu City, Bentang: Jurnal Teoritis dan Terapan Bidang Rekayasa Sipil, **12**, 2, 181-190 (2024).
11. Indonesian National Standardization Agency, SNI 1726:2019 Seismic Resistance Design Codes for Buildings and Other Structures, BSN, Jakarta, 2019, pp. 1-238.
12. Indonesian National Standardization Agency, SNI 2847:2019 Structural Concrete Requirements for Buildings, BSN, Jakarta, (2019).
13. Indonesian National Standardization Agency, SNI 1727:2020 Minimum Load for Planning of Buildings and Other Structures, BSN, Jakarta, (2020).

# Assessment of seismic performance of low-rise reinforced concrete frame structures with masonry infill using fragility curves

*Azia* Septriyeni Afrizal[1], *Jafril* Tanjung[1*], and *Jati* Sunaryati[1]

[1]Civil Engineering Department, Faculty of Engineering, Universitas Andalas, Padang, 25163, Indonesia

**Abstract.** The seismic performance of reinforced concrete frame structures with masonry infill is a critical area of study due to the complex interactions between the infill and the structural frame during seismic events. This study evaluates the impact of masonry infills on the seismic performance of RC frames by utilizing fragility curves, which are developed through pushover analysis. The presence of masonry infill can significantly enhance the stiffness and strength of reinforced concrete frames; however, it can also lead to unexpected failure modes and reduced ductility. The fragility curves generated provide a probabilistic assessment of damage at different seismic intensity levels, offering valuable insights into the vulnerabilities and resilience of these structures. The findings from this study highlight the importance of considering masonry infill in the design and assessment of RC frames for seismic resistance, and they contribute to the development of more effective seismic design strategies and retrofitting measures for buildings in earthquake-prone regions.

## 1 Introduction

Indonesia is located near the Pacific Ring of Fire and the Trans Asiatic Earthquake Belt, increasing the likelihood of an earthquake. West Sumatra Province is uniquely positioned in the very center of the Pacific Ring of Fire, a geologically active area renowned for its frequent seismic activity. This province lies at the critical intersection of two major continental plates: the Eurasian Plate and the Indo-Australian Plate, which together contribute to its geological complexity. Additionally, the Semangko Fault runs through the region, further intensifying the seismic potential of the area. Nearby, the Mentawai Fault also intersects with these tectonic plates, establishing a dynamic environment characterized by significant tectonic movements. As a result, all three of these features, West Sumatra Province, the Semangko Fault, and the Mentawai Fault, constitute active seismic regions that are continually influenced by the intricate interactions of the earth's tectonic forces. [1, 2].

Several major earthquakes have occurred in the earthquake-prone regions of West Sumatra Province, among which the earthquake that struck Padang City on September 30, 2009, stands out as particularly significant. This powerful seismic event registered a magnitude of 7.6 on the Richter scale and resulted in extensive damage throughout the city, leading not only to the collapse of numerous reinforced concrete buildings but also to widespread devastation that affected countless residents and disrupted the local infrastructure. The impact of this earthquake served as a stark reminder of the region's vulnerability to such natural disasters, highlighting the critical importance of implementing effective earthquake preparedness and response strategies to safeguard the community in the future [3].

In Indonesia, the most significant damage caused by earthquakes primarily affects simple structures, particularly because the majority of buildings are low-rise, comprising one or two-story houses. These houses are often constructed without the oversight of building and structural professionals, which leads to inadequate earthquake resistance and results in what are commonly referred to as non-engineered buildings. A prevalent practice in many countries, especially in low-rise reinforced concrete (RC) frames, is the use of masonry infill walls as partition walls. While infill is frequently overlooked in design considerations, extensive research has demonstrated that it can enhance both the lateral stiffness and strength of structural frames. Some studies further suggest that structures exhibiting greater lateral stiffness tend to experience reduced drift demands, thereby making them less susceptible to severe damage during seismic events. This insight underscores the importance of incorporating effective design practices that account for the contribution of infill in earthquake-prone areas, ultimately leading to improved resilience and safety for residents [4-6].

Brick masonry is frequently employed as an infill material in reinforced concrete (RC) structures, particularly in earthquake-prone regions such as West Sumatra, Indonesia. According to findings from a post-earthquake study conducted in the aftermath of a

* Corresponding author: jafriltanjung@eng.unand.ac.id

significant seismic event, as well as experimental outcomes from RC frames subjected to static lateral loading, the performance of these constructions is notably influenced by the presence of infilled brick masonry. While brick masonry infill can enhance the stiffness and lateral strength of the structure, providing critical support during seismic activity, it can simultaneously reduce the overall ductility of the frame. This reduction in ductility can pose significant risks to the residents, as it may lead to a greater likelihood of brittle failure during an earthquake, ultimately compromising the safety and integrity of the building. Therefore, it is crucial to carefully consider the implications of using brick masonry as infill in the design of RC structures, particularly in areas vulnerable to seismic events, to ensure the well-being of occupants [7].

Fired clay bricks are commonly utilized as construction materials for wall assemblies and as infill in reinforced concrete frame structures, particularly in earthquake-prone regions like West Sumatra, Indonesia. Post-earthquake field investigations have revealed that one of the significant causes of fatalities during seismic events is the collapse of brick walls, which can lead to catastrophic consequences for occupants and surrounding structures. It is widely believed that improper bricklaying techniques contribute significantly to these collapses, as inadequate alignment, insufficient bonding, and lack of attention to structural integrity can undermine the overall strength and stability of the walls. Consequently, addressing these construction practices is essential to enhancing the safety and resilience of brick masonry structures in the face of future earthquakes, thereby reducing the risk of injury and loss of life in vulnerable communities [8].

## 2 Methodology

This section aims to provide a comprehensive overview of the model structures utilized in the research study, the materials selected for construction, and the methodologies implemented to effectively achieve the study's objectives. By detailing these critical components, the intention is to offer readers a clear and thorough understanding of the foundational elements upon which the research is built, thereby enhancing the clarity and context of the findings presented. Through this detailed exploration, we hope to illuminate the rationale behind each choice made during the research process, illustrating how these decisions contribute to the overall integrity and validity of the study.

### 2.1 Model and material used

The buildings referred to as "ruko" in the city of Padang serve as the primary focus of evaluation in this study, representing typical structures found in the region. These ruko buildings are designed as two-story reinforced concrete structures, characterized by a floor-to-floor height of 4 meters. Additionally, the columns supporting these buildings are sized at 200x200 mm, while the beams are designed with a dimension of 200x250 mm. This specific architectural configuration not only reflects the common construction practices in Padang but also provides a basis for analyzing their structural performance and resilience under various conditions, particularly in the context of seismic activity. These structures are constructed using concrete of grade K-250, which provides a solid foundation, along with steel that possesses a strength of 400 MPa, ensuring robustness and stability. The building modeling, conducted using the SeismoStruct 2022 software, illustrates the performance of the structures both with and without masonry infill, and these comparative visualizations can be seen in Figs. 1-3.

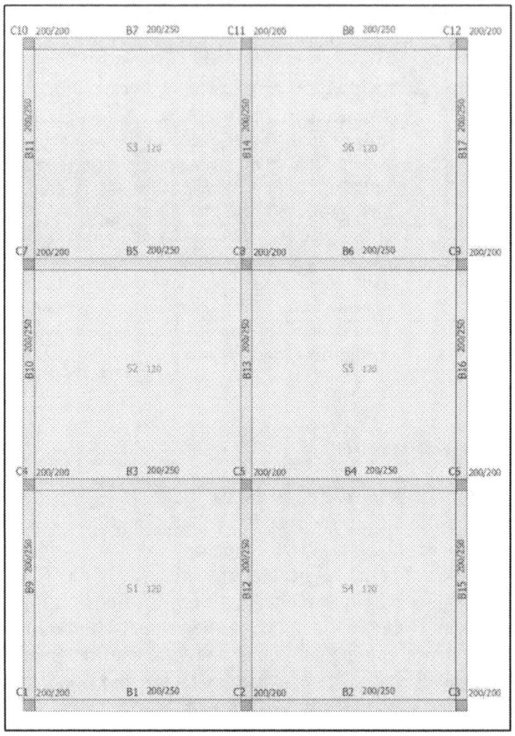


**Fig. 1.** The layout of the building.

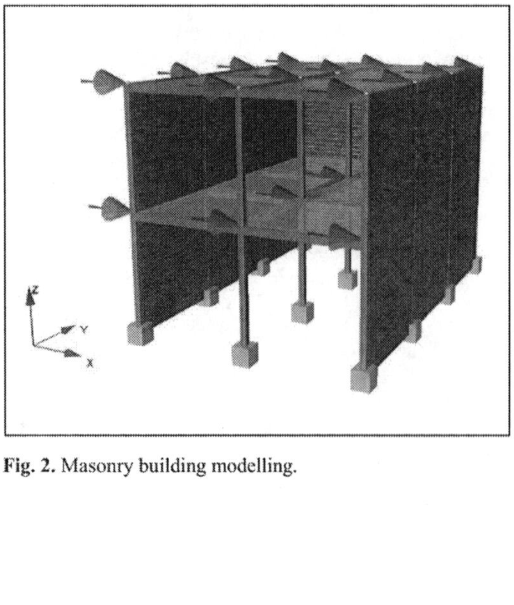


**Fig. 2.** Masonry building modelling.

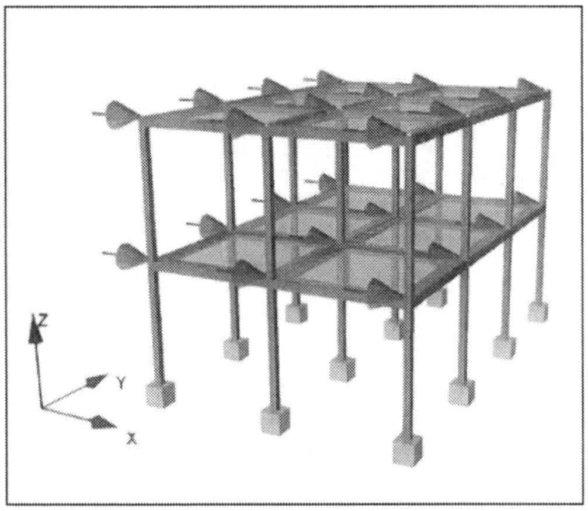


**Fig. 3.** Without masonry modelling.

## 2.2 Method

Pushover analysis is an advanced analytical technique in which the structural frame is subjected to a constant height-width partition while simultaneously applying a monotonically increasing lateral force until the desired level of displacement is achieved. This method involves a series of successive elastic assessments that are superimposed to approximate the global force-displacement curve of the structure as a whole. Initially, the gravity loads are applied to the model, which is then represented as either a two-dimensional or three-dimensional configuration. Furthermore, this model incorporates bilinear or trilinear load-deformation diagrams for all components that withstand lateral forces, allowing for a more accurate depiction of the structural behavior under seismic loading conditions. Through this comprehensive approach, engineers can effectively evaluate the performance and resilience of the structure, leading to informed design decisions that enhance its ability to withstand earthquakes [9].

HAZUS (Hazard United States) defines the fragility curve as a crucial analytical tool that illustrates the relationship between the likelihood of experiencing specific levels of damage known as Damage States (DS) in a building and the corresponding intensity of seismic activity, referred to as Intensity Measures (IM). This curve serves as a crucial framework for understanding the intricate relationship between varying levels of earthquake intensity and their potential impact on the structural integrity of buildings. By providing a visual representation of how different seismic forces can affect a structure's performance, the curve enables engineers and planners to more accurately assess risks associated with seismic activity. Consequently, this understanding allows them to implement effective mitigation strategies tailored to enhance the resilience and safety of structures in earthquake-prone areas. In doing so, it fosters informed decision-making that prioritizes both the protection of human life and the preservation of infrastructure in the face of natural disasters [10].

Fragility curves are very useful tools to determine the damage risk of bridge structures due to seismic load. The probability of exceedance in various damage states of ground motion is displayed by these curves. While there are many different kinds of distribution functions in mathematics, only three are commonly utilized in fragility analysis: uniform, lognormal, and normal distributions. The cumulative distribution function and the normal distribution function are frequently used by engineers. After that, the functions are created to have a lognormal distribution since it is dependable for analyzing the excess of risk and seismic hazard. Stated differently, equation (1) represents the standard equation utilized to generate the fragility curves or conditional exceedance.

$$Fragility = P\ [\ LS|\ IM = y\ ]\ , \tag{1}$$

Where LS is a limit states or damage states DS, IM is the intensity of ground motion, and y is a reached state due to ground motion intensity IM.

To effectively predict the probability of hazards that may arise both during and after seismic activity, fragility curves have been developed as essential tools in the realm of seismic risk assessment. These curves are widely utilized in engineering practices to mitigate potential damage and minimize energy loss during seismic events, thereby playing a crucial role in informing and supporting building codes and regulations tailored to specific geographic areas. The process of creating fragility curves typically begins with sophisticated methodologies, such as pushover analysis or nonlinear static analysis, which are vital for gaining a comprehensive understanding of how structures respond to varying levels of seismic loading. By employing these analytical techniques, engineers can better assess the vulnerability of different structural systems, ultimately leading to more resilient designs that enhance safety and performance in the face of earthquakes [11].

# 3 Results and discussion

The results of the fragility curves for the structure, both with and without masonry infill, are comprehensively illustrated in Figs. 4 and 5 for the x-direction, while the corresponding results for the y-direction are presented in Figs. 6 and 7. These figures provide a detailed visual representation of how the presence or absence of masonry infill significantly influences the structural performance and vulnerability during seismic events, enabling a clearer understanding of the varying risk profiles associated with each configuration. By meticulously comparing these curves, we can gain valuable insights into the effectiveness of masonry infill in enhancing the resilience of the structure under lateral loading conditions. This analysis not only highlights the protective benefits offered by masonry infill but also serves to inform future design considerations and risk mitigation strategies, ultimately contributing to the development of safer and more resilient buildings in earthquake-prone areas. Such findings are crucial for engineers and architects as they strive to create structures that can withstand the forces generated by seismic activity, thereby safeguarding both lives and property.

The fragility curve in the x-direction presented in the analysis clearly illustrates that the spectral displacement values for structures incorporating masonry infill are notably smaller when compared to those of structures that do not utilize such infill. This significant difference in displacement values indicates that the presence of masonry infill substantially enhances the overall stiffness and stability of the structures, effectively reducing their vulnerability to lateral forces during seismic events. Consequently, this finding underscores the critical role that masonry infill plays in improving structural performance and resilience, highlighting its importance as a fundamental design consideration in regions that are prone to earthquakes. Furthermore, by emphasizing the benefits of masonry infill, this analysis provides valuable insights for architects and engineers, encouraging them to prioritize this element in their designs to ensure that buildings can withstand the challenges posed by seismic activity, ultimately contributing to the safety and security of communities in high-risk areas.

The curve clearly illustrates that the spectral displacement values for structures incorporating masonry infill are significantly smaller when compared to those of structures that do not include such infill, indicating a marked improvement in performance under seismic loading conditions. Furthermore, it is evident from the detailed analysis that the spectral displacement values observed in the y-direction consistently exhibit lower readings than those measured in the x-direction, suggesting a directional dependency in the structural response to seismic forces. This discrepancy can be attributed to the inherent differences in the dimensions and geometric configurations of the buildings, which directly influence how each structure reacts to lateral loads during seismic events. Such insights not only underscore the critical importance of considering the presence of masonry infill but also highlight the necessity of evaluating the structural dimensions when assessing the overall seismic performance of buildings. Ultimately, these findings inform design practices aimed at enhancing the resilience of structures in earthquake-prone areas, ensuring that they can better withstand the forces of nature and protect the safety of their occupants.

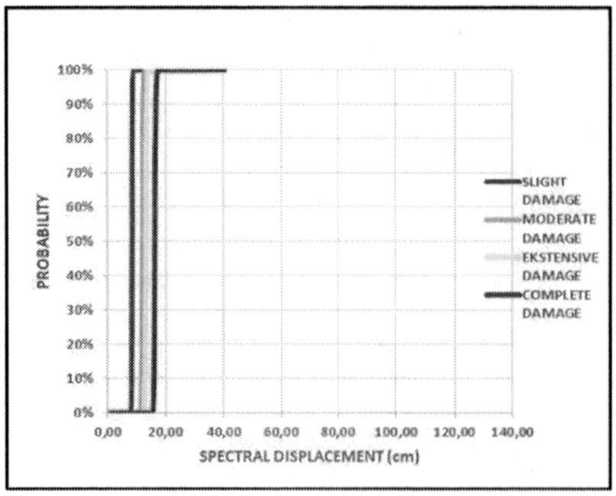


**Fig. 4.** Fragility curve x-direction with masonry infill.

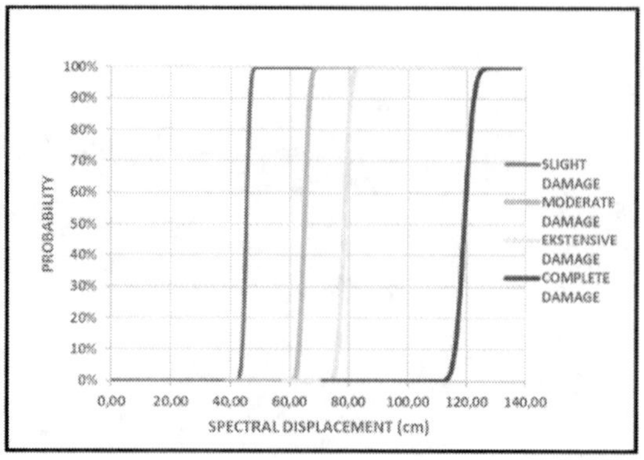


**Fig. 5.** Fragility curve x-direction without masonry infill.

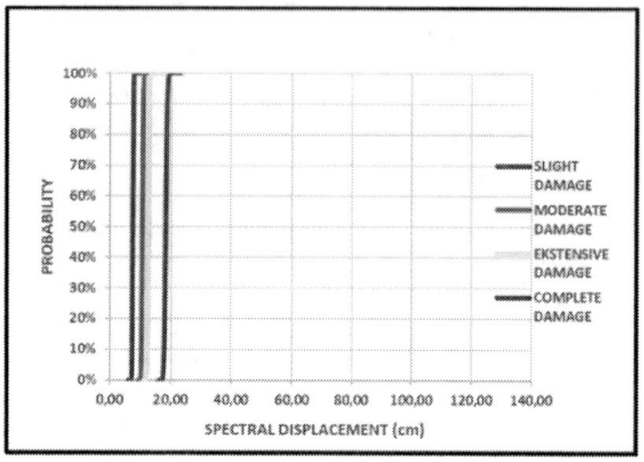


**Fig. 6.** Fragility curve y-direction with masonry infill.

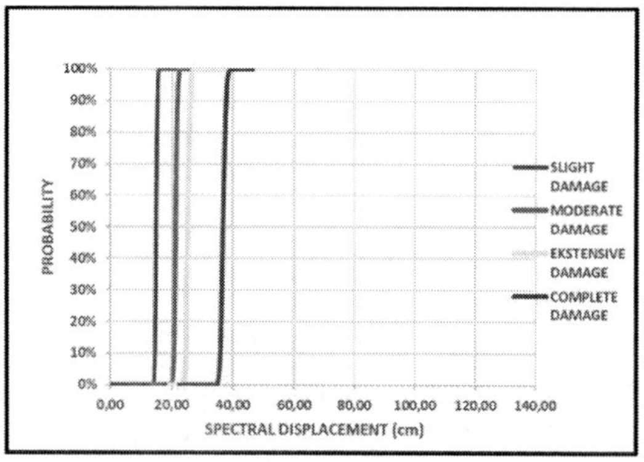


**Fig. 7.** Fragility curve y-direction without masonry infill.

Based on the results obtained from the comprehensive analysis of the building fragility curves, which compared structures with masonry infill to those without, the probabilities of various damage states can be effectively summarized as follows: these findings provide critical insights into the substantial influence that the presence of masonry infill has on the overall vulnerability of buildings during seismic events. By meticulously quantifying the likelihood of different damage scenarios, this analysis not only highlights the protective benefits conferred by masonry infill in enhancing structural performance but also underscores the urgent need for careful consideration of such essential structural elements throughout the design and construction processes. This approach is vital for improving the safety and resilience of buildings located in earthquake-prone areas, ensuring that they are better equipped to withstand the forces generated by seismic activity, thereby safeguarding the lives of occupants and the integrity of infrastructure:

### 3.1 In the x-direction:

In the assessment of damage conditions, the analysis reveals significant disparities in spectral displacement values between structures equipped with and without infill walls. For the slight damage condition, structures lacking infill walls sustain damage at a spectral displacement value of 45.67 cm, whereas those incorporating infill walls exhibit slight damage at a markedly lower displacement of only 8.84 cm. Moving on to moderate damage conditions, the findings indicate that structures without infill walls reach a critical spectral displacement of 63.92 cm, while those with infill walls incur moderate damage at just 12.20 cm. Furthermore, when considering extensive damage, the values are even more pronounced: structures without infill walls experience damage at 77.22 cm, in stark contrast to the 13.26 cm observed for structures with infill walls. Lastly, in the case of complete damage, structures devoid of infill walls withstand a substantial spectral displacement of 118.91 cm, while those with infill walls sustain complete damage at a significantly reduced level of 16.44 cm. These findings emphatically highlight the critical role that infill walls play in enhancing structural resilience across various damage scenarios, underscoring their importance in earthquake-resistant design strategies.

### 3.2 In the y-direction:

In the analysis of damage conditions, the findings reveal significant differences in spectral displacement values between structures with and without infill walls. For the slight damage condition, structures without infill walls experience damage at a spectral displacement value of 14.82 cm, while those with infill walls demonstrate this damage at a much lower value of 7.51 cm. Similarly, in the case of moderate damage, structures without infill walls reach a critical spectral displacement of 21.17 cm, whereas structures with infill walls sustain moderate damage at only 10.59 cm. Furthermore, for extensive damage conditions, the disparity is even more pronounced, with the threshold for structures lacking infill walls set at 24.84 cm, compared to just 12.52 cm for those that include infill walls. Finally, in scenarios of complete damage, structures without infill walls reach a critical spectral displacement of 36.65 cm, while those with infill walls incur complete damage at a significantly reduced value of 18.33 cm. These results clearly illustrate the protective benefits of masonry infill in enhancing the resilience of buildings against seismic forces.

## 4 Conclusion

Incorporating masonry infill into reinforced concrete (RC) structures is a prevalent and widely accepted practice in West Sumatra, where it is frequently employed not only to enhance the overall lateral stiffness and strength of buildings but also to significantly improve their ability to withstand the potentially devastating forces generated during seismic events. This integration of masonry infill serves to reinforce the structural framework, allowing buildings to better absorb and dissipate energy from lateral movements, ultimately contributing to their stability and resilience in the face of earthquakes. Moreover, this practice reflects a broader trend in the region, emphasizing the importance of adopting effective construction techniques that prioritize safety and structural integrity, especially in areas that are particularly vulnerable to seismic activity. Through the strategic use of masonry infill, builders and architects aim to create safer living and working environments, ensuring that communities are better protected against the impacts of natural disasters.

However, despite the critical support that this infill provides during seismic events, it can also inadvertently reduce the ductility of the structures, which poses significant risks to the safety and well-being of residents. Post-earthquake studies conducted in these regions have identified improper bricklaying techniques as a key contributing factor to the fatalities resulting from the collapse of brick walls during seismic activity. These alarming findings highlight the urgent need for the implementation of improved construction practices and a greater emphasis on adherence to building standards. It is essential to recognize that ensuring both strength and ductility in structural design is paramount for protecting lives and minimizing damage in the event of future earthquakes. This calls for a comprehensive approach to construction that not only addresses the materials used but also prioritizes skilled labor and rigorous quality control throughout the building process.

The results of the fragility curve analysis reveal compelling evidence that structures equipped with masonry infill exhibit significantly smaller spectral displacement values when compared to those that do not incorporate such infill. This finding strongly suggests that the inclusion of masonry infill within the structural design not only enhances the overall performance of the buildings but also significantly boosts their resilience during seismic events, enabling them to withstand lateral forces more effectively and reliably. Consequently, this indicates that structures that incorporate masonry infill are far more adept at mitigating the impacts of seismic activity than those lacking this crucial component, thereby reducing the potential for damage and enhancing safety. Such insights underscore the paramount importance of considering masonry infill during the design and construction phases of buildings, as it plays a vital role in improving their resistance to the forces generated by earthquakes.

Ultimately, this thoughtful integration of masonry infill into the design and construction of buildings contributes significantly to the safety and well-being of occupants residing in earthquake prone regions, thereby underscoring the urgent need for architects and engineers to prioritize this essential structural element in their design practices. By doing so, they can enhance the overall resilience of structures, ensuring that communities are better safeguarded against the devastating impacts of seismic hazards. This proactive approach not only mitigates the risks associated with potential earthquakes but also fosters a greater sense of security among residents, who can have increased confidence in the integrity of their homes and workplaces. Furthermore, as urban development continues to expand in areas susceptible to seismic activity, the emphasis on incorporating masonry infill becomes even more critical, as it plays a pivotal role in preserving both human life and infrastructure during natural disasters, ultimately contributing to the sustainable growth and stability of these communities in the long term.

## References

1. Masrilayanti, R. Kurniawan, A. L. Budi and S. H. Sourkan, IOP Conf., **1041**, 012003, (2021). https://doi.org/10.1088/1757-899X/1041/1/012003
2. L. Leilany, Jurnal Rivet, **02**, 01, (2022).
3. T. Al Fajri, R. Imani, and Z. Siregar, Jurnal Teknologi, **10**, 2, (2020)
4. R.S. Leksono, D. Iranata, and H. Kristijanto, Jurnal Teknik ITS, **1**, 1, (2012)
5. R. Suwal and R. Uprety, Journal of Engineering Issues and Solutions, **2**, 1, (2023). https://doi.org/10.3126/joeis.v2i1.49373
6. J. Monical and S. Pujol, Structures, **63**, 106345 (2024). https://doi.org/10.1016/j.istruc.2024.106345
7. J. Tanjung, F. A. Ismail, Maidiawati, and R. Putra, International Journal of GEOMATE, **18**, 66, (2020) https://doi.org/10.21660/2020.66.9488
8. J. Tanjung, F. A. Ismail, Maidiawati, and Y. Hayati, IOP Conf. **933**, 012028 (2020), https://doi.org/10.1088/1757-899X/933/1/012028
9. M. Saadi, D. Yahiaoui, N. Lahbari, and B. Tayeb, C.E.J. **7**, 07, (2021) http://dx.doi.org/10.28991/cej-2021-03091714
10. D. R. Silitonga, and I. Imran, J. Teknik Sipil, **26**, 1, (2019), https://doi.org/10.5614/jts.2019.26.1.7
11. M. Masrilayanti, A. P. Nasution, R. Kurniawan, J. Tanjung, and S. Sarmayenti, sciendo, **17**, 1, (2021), https://doi.org/10.2478/cee-2021-0022

# Identification of post-disaster housing reconstruction using a lean reconstruction approach (case study: Pasaman earthquake in 2022)

*Dyla* Midya Octavia[1,2] and *Taufika* Ophiyandri[1*]

[1]Department of Civil Engineering, Faculty of Engineering, Universitas Andalas, Padang, Indonesia
[2]Department of Civil Engineering, Adzkia University, Padang, Indonesia

**Abstract.** Housing reconstruction for affected communities is a priority need following a disaster. Mechanisms and strategic steps for implementation are needed to ensure that the handling is quicker and more targeted. This aligns with the lean principle of "Let the customer pull—do not make anything until it is needed, then make it quickly," and pursue perfection through continuous improvement. This research aims to identify the stages and mechanisms of housing reconstruction implementation in West Pasaman following the earthquake disaster in 2022 through a lean reconstruction approach. This research is part of a study on developing a lean reconstruction model for post-disaster housing reconstruction in Indonesia. The method used in this research is qualitative with a case study approach. The study results indicate that implementing post-disaster housing reconstruction in Pasaman has not met the lean principle, with several stages being repeated so that it can be classified as a wasteful process. The main problem that causes the repetition of the process is the problem of data collection. In addition, the distribution pattern of post-disaster housing reconstruction assistance has several weaknesses at various stages, so the final value of recipient satisfaction cannot be fully achieved.

## 1 Introduction

### 1.1 Background

Housing or housing reconstruction is the main factor in post-disaster recovery. It is a top priority in recovery efforts because decent housing is the foundation for community survival. Community participation, communication, resettlement, and cultural conformity have become recurring challenges and continue to cause housing reconstruction projects to fail [1]. Some beneficiaries modify or outright reject the housing provided or, in some cases, dismantle the house and sell its components.

This reconstruction process requires a holistic approach involving community participation and stakeholders and using environmentally friendly technology to ensure that buildings are resilient to future disasters.

On the other hand, housing reconstruction issues are not solely the result of limited technical capacity or construction engineering but also of factors such as inadequate planning and coordination, chaos among the parties involved, a scarcity of resources and supplies during the project, the simultaneous launch of a significant amount of aid, a lack of knowledge and experience in the provision and construction of post-disaster settlements, socio-cultural considerations, and a failure to consider the expectations and needs of the beneficiaries [2, 3].

There is more to post-disaster housing reconstruction than simply reconstructing. With all the challenges and problems mentioned above, A method is required to address, mitigate, or resolve the obstacles and issues associated with post-disaster home rebuilding. Lean construction is one methodology applicable to post-disaster house rehabilitation. The term "lean" originated in the manufacturing sector, referring to the lean production strategy implemented by Toyota, which provides high-quality, varied, and cost-effective output to satisfy consumer demands [4].

The aspiration to offer cheap homes, minimize construction waste, and enhance efficiency has resulted in adopting lean building principles. Lean construction projects are typically more manageable, safer, done more swiftly and economically, and exhibit superior quality. Implementing lean construction in-house reconstruction can enhance performance, expedite the process, mitigate environmental effects, and decrease material waste.

Prior studies have not specifically addressed the incorporation or implementation of lean principles in post-disaster home rebuilding. The study initially explored the feasibility of implementing lean methodologies in post-disaster reconstruction, demonstrating that approaches such as Pull Scheduling, Quick Mobilization, Lean Six Sigma, and Just in Time are

* Corresponding author: ophiyandri@eng.unand.ac.id

applicable in this context [5]. This indicates that lean thinking has significant potential for exploration within natural catastrophe management. Implementing lean principles in post-disaster rebuilding can facilitate a transition from a reactive to a proactive strategy in catastrophe management.

### 1.2 Research problems

The main issue in applying lean construction to post-disaster housing reconstruction is determining what kind of lean reconstruction model will be implemented. To develop a lean reconstruction model from the previously applied lean construction model, one can modify the lean construction techniques, combine several lean methods, or even create new strategies for lean reconstruction that utilize lean principles. To address this issue, it is necessary to identify the implementation of post-disaster reconstruction in Indonesia, one of which is Pasaman Regency, which experienced an earthquake in 2022 and whose reconstruction is still ongoing. Identification is carried out to understand what the stages of implementation are and what problems occur at each of those stages.

### 1.3 Research objectives

This research is part of developing a lean reconstruction model for post-disaster housing reconstruction in Indonesia. This research seeks to delineate the phases of post-disaster house reconstruction and the issues that may be addressed or mitigated by applying lean construction principles. In this paper, the research locus discussed is housing reconstruction implementation after the Pasaman Earthquake disaster in 2022. This research was conducted through a literature study on the report on implementing Rehabilitation-Reconstruction in West Pasaman Regency and other literature in journals and books. Furthermore, Interviews were conducted with stakeholders participating in post-disaster home rehabilitation at the regional level to ascertain the current implementation circumstances.

## 2 Literature review

### 2.1 Post-disaster reconstruction

The reconstruction phase is the most essential in disaster management. The principal challenges in the post-disaster rebuilding phase include inadequate planning, impulsive decision-making, disorganized recovery efforts, substandard built environments, material wastage, environmental degradation, protracted restoration timelines, and elevated prices [5]. Post-disaster rebuilding requires interdisciplinary participation, substantial resources, and many skill sets [8].

Reconstruction should rely on established procedures and rules that provide explicit institutional accountability for recovery efforts and promote public engagement. Recovery activities that focus on "building back better » can shape the fate of local communities and reduce the potential for damage due to future disasters [6].

Reconstruction is a long-term activity [7] that begins as soon as the emergency phase begins. A series of processes must be carried out from when the disaster occurs until the reconstruction ends [8] as shown in Fig. 1.

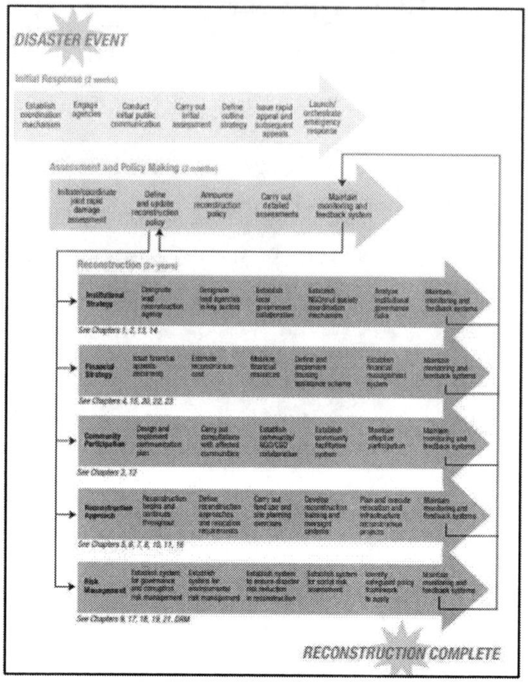


**Fig. 1.** Post-disaster response and reconstruction process

In the book "Handbook for Post-Disaster Housing and Community Reconstruction" [8], there are ten principles of post-disaster reconstruction:

a. Inclusive Policy: Empowering communities to rebuild their lives with a focus on vulnerable groups.
b. Immediate Reconstruction: Begin immediate reconstruction with new standards to improve building safety.
c. Community Participation: Communities should be involved as partners, not victims, to improve psychosocial recovery and community cohesion.
d. Realistic Financial Policy: Plan policies with sufficient and realistic funding integrated with other reconstruction activities.
e. Institutional and Coordination: Ensure effective policies and coordination before disasters to improve reconstruction outcomes.
f. Future Planning: Using reconstruction opportunities to plan for the future and preserve historic buildings.
g. Avoid Relocation: Relocation should be avoided except for safety, with participatory locations selected.
h. The Role of the Private Sector and NGOs: The contribution of NGOs and the private sector is crucial in reconstruction.

i. Assessment and Monitoring: Engage the community in evaluating and monitoring to improve reconstruction outcomes.
j. Sustainability: Reconstruction must be environmentally and economically sustainable without overriding environmental laws.

These principles aim to ensure an effective and sustainable reconstruction process.

## 2.2 Post-disaster housing reconstruction

Reconstruction of housing is an essential component of post-disaster recovery activities since it enables communities that have been affected to reestablish themselves. Post-disaster housing rehabilitation represents a significant obstacle in recovery efforts, particularly in developing nations [9]. Housing is essential for individuals as it pertains to their well-being. Housing serves as an asset that enhances family life in domains including health, education, economy, security, and social welfare. Facilitate social engagement, cultural pride, identification, political dynamics, and economic resources [2].

Several approaches can be used as models for post-disaster housing reconstruction. The selection of the best reconstruction approach to use is context-specific. It must consider the cost of reconstruction, improving housing and community security, livelihood recovery, political environment, cultural contexts, and the community's welfare, empowerment, and capacity goals. Conducting consultations with the community and evaluating capacity and requirements is essential before deciding which reconstruction approach to use. There are 5 approaches to post-disaster housing reconstruction that are generally carried out, including [8]:

a. Cash Approach (CA): offers unconditional financial aid without technical help.
b. Owner-Driven Reconstruction (ODR): offers conditional financial aid and regulatory and technical help to guarantee the improved reconstruction of the dwelling. Individuals who lost their residence received financial assistance or support for independent repairs.
c. Community-Driven Approach (CDA): Reconstruction is spearheaded by an institution that incorporates community involvement in decision-making and building processes.
d. Contractor-Driven Reconstruction In-Situ (CRIS): A designated contractor oversees reconstructing a damaged home at its original site before the disaster.
e. Contractor-Driven Reconstruction in Relocated Site (CRRS): Reconstruction is overseen by a contractor designated to construct a new residence at a different location.

The housing provision process, community engagement in housing design and oversight, the acquisition of building materials and skilled labor, and the government's involvement and policies contribute to enhancing and facilitating the post-disaster housing rehabilitation and reconstruction process in Indonesia [10].

The precision of beneficiary data in the housing allocation process is crucial to prevent redundant housing reconstructions. The government's capability is essential and must be strengthened from the initial phases of the home rehabilitation operation. Community involvement in housing rebuilding must be meticulously evaluated, analysed, and administered, contingent upon the contextual circumstances. Moreover, training and capacity building for communities before engaging in house construction would substantially enhance their involvement in the project [10]. Ill-informed communities can incite conflicts with contractors and builders, leading to building delays.

Furthermore, there is a need to raise awareness of earthquake safety problems among home builders, homeowners, and government authorities. Inspection and quality control must be performed throughout construction to ensure that the structure is created by the design and seismic safety criteria. [11].

Early community involvement, particularly in decision-making during the planning phase, is crucial for influencing future engagement. These results guide disaster recovery methods by promoting project concepts that integrate community involvement throughout the project's lifetime [12].

The experiences in Japan, Gujarat, and Bam illustrate that a crucial driver of success is the presence of a suitable institutional structure. A national specialized entity with representatives and units at the state and local government levels should enable centralized coordination and facilitate recovery and reconstruction initiatives. This case study demonstrates that the institutional framework and overarching principles for the design and execution of home rebuilding initiatives are mandated by legislative measures [1].

The housing reconstruction has been identified as the most arduous and complex initiative throughout the recovery period from the catastrophe. A community-based strategy is a procurement method applicable to home rebuilding. In this approach, the community not only serves as a beneficiary but also possesses the authority to oversee the construction phase [13]

During the pre-construction phase, twelve essential criteria have been recognized as Critical Success criteria (CSF) for community-based home rebuilding following a disaster. These factors include [14]: (a) Clarity and responsibility, (b) Suitable reconstruction policies/strategies, (c) Comprehension of community-oriented approaches, (d) Building public trust, (e) Capacity for facilitation, (f) Effective coordination and communication, (g) Sufficient financial resources, (h) Capability for implementation, (i) A notable degree of community engagement/control, (j) Support from governmental entities, (k) Participation of all community members, (l) Effective identification of beneficiaries

## 2.3 Lean reconstruction

Deliberately utilizing resources to cut down on waste and increase user value is what we mean when we talk about lean thinking. The success of the Toyota Company provides a clear illustration of the effectiveness of lean manufacturing. Because of this, there is a substantial demand for professionals knowledgeable in lean thinking and its implementation in fields other than manufacturing, such as logistics, distribution, services, retail, healthcare, construction, maintenance, and government. [15].

Lean construction emerged as a solution to the issues faced by the construction sector, characterized by inflated costs, prolonged schedules, and a culture that sometimes contravenes regulations [16, 17]. Lean construction derives from lean manufacturing, encompassing a multifaceted array of concepts such as continuous improvement, a flat organizational hierarchy, collaboration, waste elimination, resource efficiency, and cooperative supply chain management [18]. Lean Construction emerged as a solution to the issues faced in the construction sector, characterized by inflated expenditures, prolonged timelines, and a culture that often contravenes regulations [16, 17].

Lean construction is no longer novel; several studies have examined lean construction. The Last Planner System is a methodology within lean construction that has demonstrated efficacy in enhancing construction management practices across several dimensions. [19]. The use of BIM in workflow visualization to support the implementation of lean construction shows increased transparency of the process where the physical and managerial state of the site becomes more apparent, which has the potential to increase worker safety motivation and perception, thereby improving workflow and reducing waste as visualization helps reduce uncertainty and misunderstanding [20].

In addition to construction projects, lean construction is also implemented in housing development projects. Implementing lean construction can ensure effectiveness in housing development projects [21]. The results show increased efficiency, encouraging increased speed, reduced waste, and decreased work accidents. Lean construction will create better designs, improve communication and collaboration, and increase user satisfaction.

Over time, several studies began to discuss the application of lean construction to post-disaster reconstruction. Previous research has also identified lean construction strategies that can be implemented during the rebuilding of post-disaster housing, including, Pull Scheduling, Value Stream Mapping, Last Planner System, 5S (Sort, Straight, Shine, Standardize, and Sustain), and Lean Six Sigma. These five techniques need further study regarding the conditions for implementing housing reconstruction [22].

## 3 Results and discussion

### 3.1 Data collection

The data needed in this study is the existing condition of the post-disaster housing reconstruction implementation stages and the problems faced. Secondary data was obtained from institutional documents, including the Rehabilitation and Post-Earthquake Reconstruction Plan in West Pasaman Regency and Implementation Technical Instructions. The data is also supported by primary data from interviews conducted with parties implementing post-disaster housing reconstruction in the West Pasaman Regency Rules and other decisions. The following table (Table 1) shows a list of respondents interviewed in this study.

**Table 1.** List of interview respondents

| No. | Position | Institution |
|---|---|---|
| 1 | Chief Executive | West Pasaman Regional Disaster Management Agency |
| 2 | Head of Rehabilitation and Reconstruction Division | |
| 3 | Facilitator | |
| 4 | Beneficiary Communities | West Pasaman |

### 3.2 Results and analysis

#### *3.2.1 Disaster profile and impact*

The earthquake in West Pasaman Regency occurred on February 25, 2022. It had a severe impact on the Region. The 6.1-magnitude earthquake, accompanied by aftershocks, resulted in other aftershocks in the form of flash floods, liquefaction, and landslides, especially at the main points of the earthquake.

#### *3.2.2 Post-disaster housing reconstruction implementation policy*

The Post-Earthquake Rehabilitation and Reconstruction Plan (R3P) in West Pasaman Regency is contained in the Decree of the Regent of West Pasaman Number 188.45/704/BUP-PASBAR/2022. In the R3P, the recovery period set out in this rehabilitation and reconstruction plan is from 2022 to 2025. The priority for recovery is the residential sector because it directly impacts people's lives. For relocation, land needs will be provided between the district and provincial governments. In contrast, the government will give or finance stimulant funds to construct houses and residential infrastructure.

However, until now, the implementation of rehab in West Pasaman has not been based on this R3P because West Pasaman is still in an emergency transition period, while according to the provisions of the R3P, it is carried out during the recovery period (after the emergency harmonization period). Housing reconstruction is implemented during the emergency transition period for heavily damaged houses, and the BPBD of West Pasaman Regency handles it. Meanwhile, the relevant service, the West Pasaman Settlement Office, will carry out moderate and light damage. Housing reconstruction implementation during the emergency transition period uses BNPB's Ready to Use Fund (DSP) to accelerate reconstruction. Therefore, initially only 14 days, the emergency transition period continues to be extended until housing reconstruction is implemented. One of the provisions for using DSP is that it can only be used during the emergency transition period.

### *3.2.3 Implementation of housing reconstruction after the Pasaman earthquake*

Based on the analysis of institutional documents, such as technical instructions for implementation and interviews, Fig. 2 describes the existing conditions of the stages of post-disaster housing reconstruction implementation in the West Pasaman Regency.

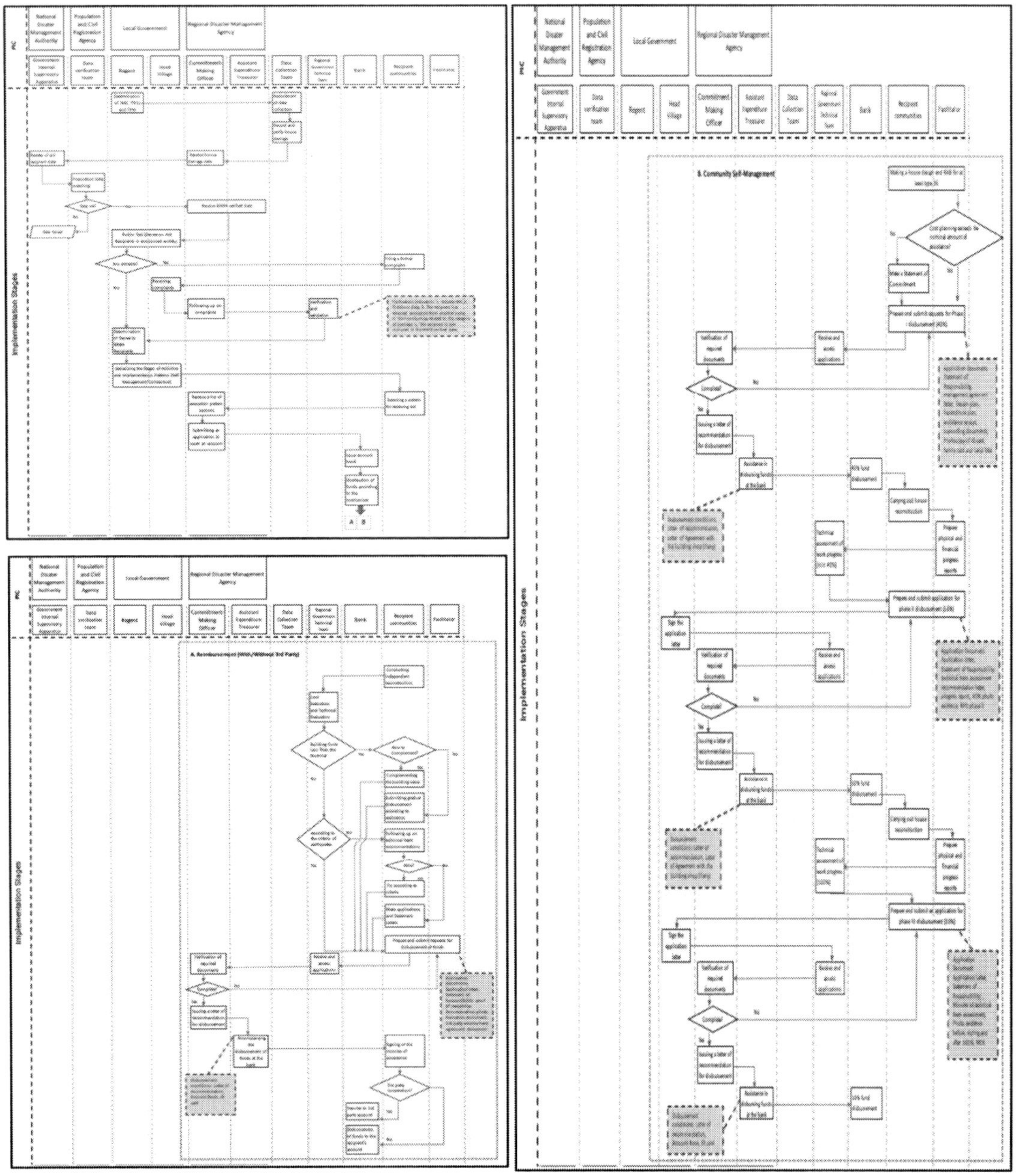

**Fig. 2.** Stages of housing reconstruction after the 2022 Pasaman earthquake

The implementation of housing reconstruction in West Pasaman Regency can be pretty slow if you compare the damage to previous disaster events, such as Lombok and Palu in 2018, which had a much more significant impact. In the implementation of post-disaster housing reconstruction in West Pasaman, several problems were found, and these problems were dominated by the problem of collecting data on aid recipients both at the beginning of the implementation and during the reconstruction. At the start of the implementation of the activity, which began with the collection of data on damaged houses, several problems were found, including:

a. There is a data collection team using rules that are not by National Disaster Management Authority (NDMA) provisions, where 50% of the damage is categorized as badly damaged; even though NDMA is listed for the category of severe damage, the level of damage is above 70%, so re-data collection or re-verification is carried out related to this.
b. As a result, the problem in point A caused protests in the community where the damage to their houses decreased from severely damaged to moderately damaged, and those that were initially moderately damaged became lightly damaged.
c. The data collection process also experienced interference from the community, which forced the categorization of houses at the highest level to be severely damaged).

Furthermore, in the implementation of activities, based on the stages of implementation in Fig. 2, it is known that the phases of reconstruction implementation do not support the acceleration of reconstruction. In addition, the community has complaints and dissatisfaction about the implementation procedures and reconstruction results.

The following is a description of the results of the identification of the implementation of post-disaster reconstruction in West Pasaman Regency:

a. The implementation procedure is unclear. The Government Internal Supervisory Apparatus of NDMA sends the data collection results of damaged houses to NDMA for review and matches them with population data. However, the type of data reviewed and the review indicators are unclear.
b. From the technical instructions, it is known that after the data on the list of beneficiaries is reviewed by NDMA, a public test of the data is carried out. If there are, the local government technical team will carry out complaints from the community, verification, and revalidation. After the public test, a Regent Decree was issued regarding the list of recipients of By Name by Address assistance. However, the procedure for determining the Decree based on verification and validation results is unclear. In addition, there are still many complaints from the public after the decree was issued.
c. The Regent's Decree should be the last used to assist. However, while monitoring activities, NDMA found indications of data errors, so it reviewed the data on the list of recipients again. The review results show that the number of aid recipients has decreased significantly. This, of course, has provoked protests from the community.
d. The technical instructions provide three (three) patterns or mechanisms for distributing assistance: Reimbursement, Self-Management, and Contractual. However, the contractual procedure is not described in detail, resulting in unclear procedures that will trigger problems in implementation.
e. The implementation of reconstruction with a reimbursement pattern is chosen by the community that has done repairs and reconstruction before the aid funds are distributed. In this pattern, the community must meet many requirements for the disbursement of funds. In addition, there is a technical evaluation where houses built independently by the community must meet the criteria for earthquake-safe and livable dwellings. However, the community did not know the requirements for these two things in advance because they were only conveyed during socialization. Meanwhile, quite a lot of people have repaired their houses before socialization was carried out.
f. The implementation pattern with self-management is a pattern that is generally chosen by people who do not have the financial ability to repair their houses before the disbursement of aid funds. This pattern is quite difficult for the community because of the requirements that must be met. One of them is that the community must prepare house design documents and cost budget plans as one of the requirements for the disbursement of funds. In making these documents, the community is indeed assisted by facilitators, but the limited number of facilitators will undoubtedly slow the completion of activities. The number and ability of facilitators largely determine the time it takes for this requirement.

## 4 Conclusion

The implementation of post-disaster housing reconstruction in Indonesia faces various problems that hinder its effectiveness and efficiency. Significant issues include poor data collection of aid recipients, resulting in repeated verification and disappointment in the community. In addition, complicated administrative requirements make it difficult for the community in an emergency, causing delays in the disbursement of funds. Socialization related to earthquake-safe and habitable housing standards was not carried out before the community reconstructed their homes independently. All of these problems are interrelated and contribute to the low level of community satisfaction and the failure to achieve the goal of post-disaster reconstruction, namely "building back better." Therefore, developing a lean reconstruction model for post-disaster housing reconstruction is essential to minimize existing problems and increase community satisfaction as beneficiaries.

The Research was funded by the Directorate of Research, Technology, and Community Service, Directorate General of Higher Education, Research, and Technology, Ministry of Education, Culture, Research, and Technology, on the Doctoral Dissertation Research Scheme, by the Research Contract Number: 041/E5/PG.02.00.PL/2024, fiscal Year 2024.

## References


1. A. A. Bilau, E. Witt, and I. Lill, A framework for managing post-disaster housing reconstruction. Procedia Economics and Finance, **21** (2015). https://doi.org/10.1016/s2212-5671(15)00182-3
2. S. Barakat, Housing reconstruction after conflict and disaster. Humanitarian Policy Group, **43** (2003)
3. C. H. Davidson, C. Johnson, G. Lizarralde, N. Dikmen, and A. Sliwinski, Truths and myths about community participation in post-disaster housing projects. Habitat International, **31** (2007). https://doi.org/10.1016/j.habitatint.2006.08.003
4. Lean Construction Institute. What is Lean? https://www.leanconstruction.org
5. Mojtahedi, and B. L. Oo, Possibility of applying lean in post-disaster reconstruction: An evaluation study. In Proceeding of the 20th Annual Conference of the International Group for Lean Construction (2012)
6. B. Rouhanizadeh, S. Kermanshachi, and T. J. Nipa, Identification, categorization, and weighting of barriers to timely post-disaster recovery process. ASCE, **4** (2019). http://toc.proceedings.com/49478webtoc.pdf
7. T. L. Moe, and P. Pathranarakul, An integrated approach to natural disaster management: Public project management and its critical success factors. Disaster Prevention and Management: An International Journal, **15** (2006). https://doi.org/10.1108/09653560610669882
8. A. K. Jha, J. D. Barenstein, P. M. Phelps, D. Pittet, and S. Sena, Safer homes, stronger communities: A handbook for reconstructing after natural disasters global facility for disaster reduction and recovery global facility for disaster. (World Bank Group, Washington D.C, 2010)
9. I. Ahmed, An overview of post-disaster permanent housing reconstruction in developing countries. International Journal of Disaster Resilience in the Built Environment, **2** (2011). https://doi.org/10.1108/17595901111149141
10. Azweri, C. Mutiawati, N. Al-Huda, and H. Mufiaty, Disaster recovery indicators of housing reconstruction: the story of post tsunami Aceh, Indonesia. International Journal of Disaster Management, **1** (2017)
11. K. S. Pribadi, D. Kusumastuti, S. A. H. Sagala, and R. Wimbardana, Post-disaster housing reconstruction in indonesia: Review and lessons from Aceh, Yogyakarta, West Java and West Sumatera earthquakes (Springer, Tokyo, 2014). https://doi.org/10.1007/978-4-431-54255-1_11
12. C. Venable, A. Opdyke, A. Javernick-Will, and A. Liel, Community participation in post-disaster shelter programs: Examining the evolution of participation in planning, design, and construction. (Construction Research Congress, New Orleans, 2018). https://doi.org/10.1061/9780784481295.02
13. T. Ophiyandri, R. Amaratunga, and C. Pathirage, Community based post-disaster housing reconstruction: Indonesian perspective (University of Salford, Salford, 2010). http://usir.salford.ac.uk/id/eprint/9761/
14. T. Ophiyandri, R. Amaratunga, and C. Pathirage and K. Keraminiyage, Critical success factors for community-based post-disaster housing reconstruction projects in the pre-construction stage in Indonesia. International Journal of Disaster Resilience in the Built Environment, **4** (2013). https://doi.org/10.1108/IJDRBE-03-2013-0005
15. C. Protzman, F. Whiton, J. Kerpchar, C. R. Lewandowski, S. Stenberg, P. Grounds, and J. Bond, The lean practitioner's field book (Productivity Press, Chennai, 2018). https://doi.org/10.4324/9781315373843-3
16. O. Salem, M. Asce, J. Solomon, A. Genaidy, and I. Minkarah, Lean construction: From theory to implementation. Journal of Management in Engineering, **22** (2006). https://doi.org/10.1061/ASCE0742-597X200622:4168
17. Z. K. Rybkowski, Last planner and its role as conceptual kanban. 18th Annual Conference of the International Group for Lean Construction, **1** (2010).
18. S. D. Green, The dark side of lean construction: Exploitation and ideology. In Proceedings of the 7th Annual Conference of the International Group for Lean Construction. Berkeley, California, USA, 26-28 Jul 1999. (1999).
19. R. F. Aziz, and S. M. Hafez, Applying lean thinking in construction and performance improvement. Alexandria Engineering Journal, **52** (2013). https://doi.org/10.1016/j.aej.2013.04.008
20. R. Sacks, M. Treckmann, and O. Rozenfeld, Visualization of workflow to support lean construction. Journal of Construction Engineering and Management, **135** (2009). https://doi.org/10.1061/ASCECO.1943-7862.0000102
21. A. Shastri, H. S. Chauhan, S. B. Chidapareddi, V. A. Chopade, and R. S. Kahane, Implementation of lean construction method, to ensure the effectiveness of a residential construction project. International Journal of Research in Engineering, Science and Management, **5** (2022)
22. D. M. Octavia, T. Ophiyandri,A. Suraji, and B. Hidayat, Identification of lean implementation in the post-disaster housing reconstruction. IOP Conference Series: Earth and Environmental Science, **1173** (2023). https://doi.org/10.1088/1755-1315/1173/1/012071

# Seismic vulnerability assessment of a hospital building with and without retrofitting using RC shear walls

*Fauzan*[1]*, *Geby* Aryo Agista[1], *Diva* Syandriaji[2], *Zev* Al Jauhari[3], and *Deara* Bima Suwarso[1]

[1]Department of Civil Engineering, Faculty of Engineering, Universitas Andalas, Indonesia
[2]Civil Engineering Department, Gifu University, Gifu, Japan
[3]Graduate School of Engineering, Toyohashi University of Technology, Japan

**Abstract.** Revisions to building regulations in Indonesia necessitate assessing building structures, especially hospital buildings. A structural assessment using the current building codes and vulnerability analysis is necessary to ensure the building's strength against the working loads. In this study, an assessment of the Third-Class Inpatient Building at the Sijunjung Regional General Hospital was carried out using the finite element method (ETABS V21). The structural assessment of the existing hospital building revealed that the inter-story drift values exceed the drift limit, and column and beam structural elements cannot withstand the working loads. The recommendation for retrofitting the building structure is to add RC shear walls to the existing building. The retrofitted building shows an increase in its structural capacity, and the inter-story drift and the strength of the structural elements have met the requirements based on the current Indonesian building codes. Furthermore, structural fragility curves of the hospital building were developed with and without retrofitting against earthquake loads. The vulnerability analysis results indicated that retrofitting the hospital building with shear walls reduces the probability of building damage due to earthquake loads by 52.69% at the level of extensive damage at a PGA of 0.60 g.

## 1 Introduction

Health development aims to enhance awareness, willingness, and the ability to live healthily for everyone, thereby improving public health and serving as an investment in developing socially and economically productive human resources. Hospitals are essential infrastructure needed to meet human health needs [1]. Sijunjung General Hospital, hereafter referred to as RSUD Sijunjung, is one of the hospitals in Sijunjung Regency that frequently undertakes development and improvements to enhance the quality and services of the hospital, including the construction of a Third-Class Inpatient Building at RSUD Sijunjung.

The building must be strong, rigid, and stable to withstand the loads imposed on its structure. Based on current building codes, routine inspections of building reliability, especially following updates to building codes, are necessary, as seen in the Assessment and Analysis of Office Building Retrofits in Riau Province, Indonesia [2]. Moreover, standards for building assessments continually evolve and improve over time. For example, earthquake resistance design procedures for buildings are regulated in SNI 1726:2012 [3], which was updated to SNI 1726:2019 [4]. Similarly, SNI 1727:2013, which specifies minimum loads for building and structural design [5], was revised to SNI 1727:2020 [6], and the requirements for structural concrete in buildings specified by the National Standardization Agency were updated from SNI 2847:2013 [7] to SNI 2847:2019 [8]. Visual observations revealed structural issues in columns, beams, and walls, leading to concrete quality testing to determine the building's suitability. Concrete quality was assessed using a hammer test, a non-destructive method outlined in SNI 4430:1997 [9]. This testing method involves applying an impact load to the concrete surface, with results obtained relatively quickly [10]. However, most tests indicated concrete quality above the planned 25 MPa.

Considering that hospital buildings are critical infrastructure that must remain safe during earthquakes and that numerous instances of structural element damage have been observed, coupled with the existing building's location in a seismically active area, it is essential to conduct a structural assessment. This assessment involves a structural response analysis to the loads applied, including seismic loads, using the ETABS V21 structural calculation software, following the current Indonesian National Standards (SNI) regulations [11-13].

## 2 Evaluation of the existing building

### 2.1 Existing building data

The existing building uses a reinforced concrete structure, a composite material with steel retrofitting embedded within the concrete. This configuration is designed to withstand tensile forces in the structure, resulting in

*Corresponding author: fauzan@eng.unand.ac.id

reinforced concrete having significant strength against both compressive and tensile forces [14].

Data related to structural analysis were obtained from existing as-built drawings and building design documents [15], as presented in Tables 1-3 and Figs. 1 and 2.

**Table 1.** Existing building data

| Name | Data |
|---|---|
| Building name | Third-Class Inpatient Building at Sijunjung Regional General Hospital |
| Function | Hospital |
| Structure type | Dual system |
| Amount floor | 4 floors |
| Building height | 12 m |
| Floor slab thickness | 13 cm |
| Building area | ± 5,035.82 m$^2$ |
| Column sections | K1(55x55), K2(40x50), and K3(40x40) |
| Beam sections | B1(35x60), B2(30x60), B3(30x50), B4(25x40), B1cor(30x50), B2cor(30x40), and B3cor(20x40) |
| Concrete quality | 25 MPa |
| Steel reinforcing bar quality | BJTS U-28<br>BJTP U-42 |

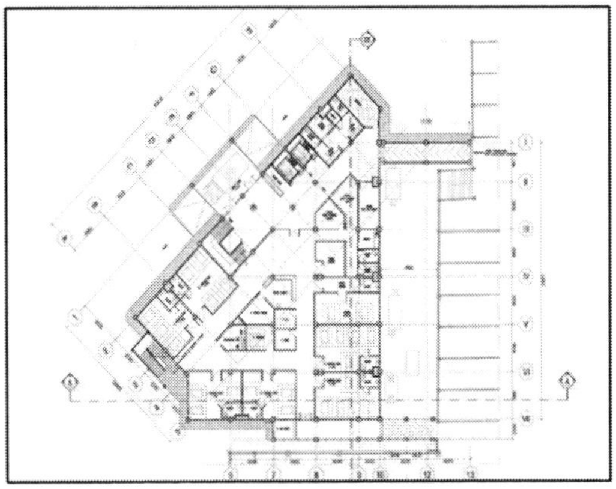

**Fig. 1.** The plan view of the third-class inpatient building at the Sijunjung regional general hospital

**Fig. 2.** The third-class inpatient building at Sijunjung regional general hospital

**Table 2.** Column reinforcement details

| No | Type | Section | | Flex. | Shear Reinf, | |
|---|---|---|---|---|---|---|
| | | Depth | Width | Reinf. | Support | Midspan |
| 1. | K1 | 550 | 550 | 12D19 | D13-150 | D13-150 |
| 2. | K2 | 400 | 500 | 10D19 | D10-150 | D10-150 |
| 3. | K3 | 400 | 500 | 8D19 | D10-150 | D10-150 |

**Table 3.** Beam reinforcement details

| No | Type | Section | | Support Area | | Midspan Area | |
|---|---|---|---|---|---|---|---|
| | | Depth | Width | Tensile | Compr | Tensile | Compr. |
| 1. | B1 | 350 | 600 | D19 | D16 | D19 | D16 |
| 2. | B2 | 300 | 600 | D19 | D13 | D19 | D13 |
| 3. | B3 | 300 | 500 | D19 | D13 | D19 | D13 |
| 4. | B4 | 250 | 400 | D16 | D10 | D16 | D10 |
| 5. | B1Cor | 300 | 500 | D16 | D16 | D16 | D16 |
| 6. | B2Cor | 300 | 400 | D16 | D16 | D16 | D16 |
| 7. | B3Cor | 200 | 400 | D13 | D13 | D13 | D13 |

## 2.2 Modeling structure

Structural modeling was performed using various applications, including the ETABS software [16]. The structural model is shown in Fig. 3.

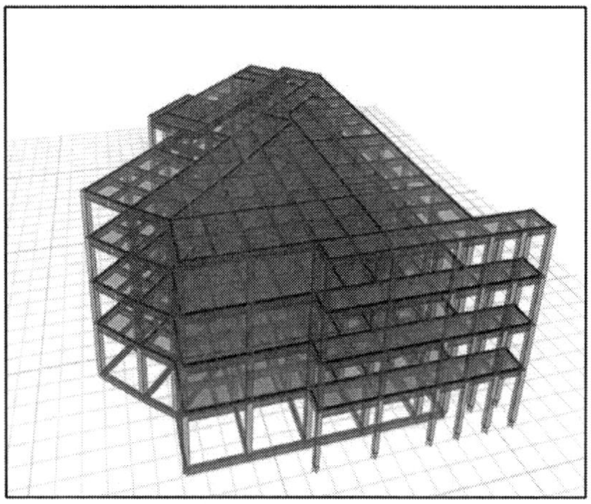

**Fig. 3.** 3D structural modeling of existing hospital building

## 2.3 Load analysis

### *2.3.1 Vertical load*

The dead loads include the weight of the structural elements calculated using ETABS V21.2 and additional superimposed dead loads (SIDL) on the floor slabs, roof slabs, and beams. The superimposed dead loads on the floor slabs include finishing materials, ceramics, ceiling loads, and Mechanical, Electrical, and Plumbing (MEP) components, while the beams include the weight of walls on the structure.

According to SNI 1727:2020, live loads must represent the maximum expected load due to occupancy and use of the building. The live loads for the third-class inpatient ward at Sijunjung Regional General Hospital are listed in Table 4 [6].

**Table 4.** Live load based on room function

| Function Room | Load (kN/m²) |
|---|---|
| Corridor | 4.79 |
| Corridor above the first floor | 3.83 |
| Operating room/ laboratory | 2.87 |
| Patient room | 1.92 |

### *2.3.2 Horizontal load*

The seismic loading is calculated based on SNI 1726:2019, which provides the guidelines for earthquake-resistant design of buildings and non-building structures [4]. According to these regulations and the building's utilization type, the structure is categorized under Risk Category IV with an importance factor (Ie) of 1.5. The structural type is a dual system of special reinforced concrete shear walls with a seismic load reduction factor (R) of 7, an overstrength factor (Ω) of 2.5, and a deflection amplification factor (Cd) of 5.5. Based on this structural type, a retrofitting parameter (Ct) of 0.0488 and a period parameter (x) of 0.75 are used. The acceleration parameters are derived from the earthquake data for Sijunjung Regency, with SDS and SD1 values of 0.658 g and 0.724 g, respectively. As shown in Fig. 4, these parameters are presented in a response spectrum graph.

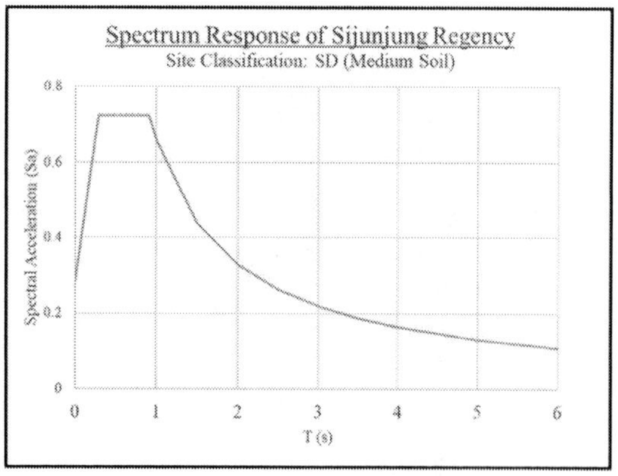


**Fig. 4.** Response spectrum design of Sijunjung regency, Indonesia

### *2.3.3 Load Combination*

SNI 1726:2019 regulates the combinations of dead, live, and earthquake loads. Table 5 presents the load combinations of the structure model.

**Table 5.** Load combinations

| No | Load combination | Note |
|---|---|---|
| 1. | U = 1.4 D | Dead load |
| 2. | U = 1.2 D + 1.6 L | Dead and live loads |
| 3. | U = 1.4 D + L ± 1.3 Ex ± 0.39 Ey | Dead, live, and earthquake loads |
| 4. | U = 1.4 D + L ±0.39 Ex ±1.3 Ey | |
| 5. | U = 0.7 D ±1.3 Ex ± 0.39 Ey | |
| 6. | U = 0.7 D ± 0.39 Ex ± 1.3 Ey | |

## 2.4 Inter-story drift

From the ETABS analysis results, the story forces are obtained, representing the inter-story drift values in millimeters. These values can then be calculated according to Table 6.

**Table 6.** Inter-story drift of the existing building

| Story | Inelastic Drift (mm) | | Drift Limit | Δ < Drift Limit |
|---|---|---|---|---|
| | ΔX | ΔY | | |
| 5 | 99.851 | 13.915 | 25.000 | NOT OK |
| 4 | 23.903 | 20.013 | 40.000 | OK |
| 3 | 50.538 | 22.132 | 40.000 | NOT OK |
| 2 | 63.906 | 20.324 | 40.000 | NOT OK |
| 1 | 68.281 | 15.077 | 50.000 | NOT OK |

The analysis results of inter-story drift in Table 6 indicate that the existing building does not comply with the Indonesian seismic codes as per SNI 1726:2019, chapter 7.8.6.

## 2.5 Structural capacity

### *2.5.1 Column capacity*

The column is a vertical compression member in a structural frame that supports the loads from beams [17]. The cross-sectional capacity of a column to resist the combination of axial force and bending moment can be represented by an interaction curve between these two forces, known as the column P-M interaction diagram.

Analysis of the interaction diagram indicates that columns K1, K2, and K3 in the Third-Class Inpatient Building at Sijunjung Regional General Hospital are not capable of withstanding the working load in both the X and Y directions, as the P-M values (points) have crossed the P-M interaction diagram line, as shown in Figs. 5-7.

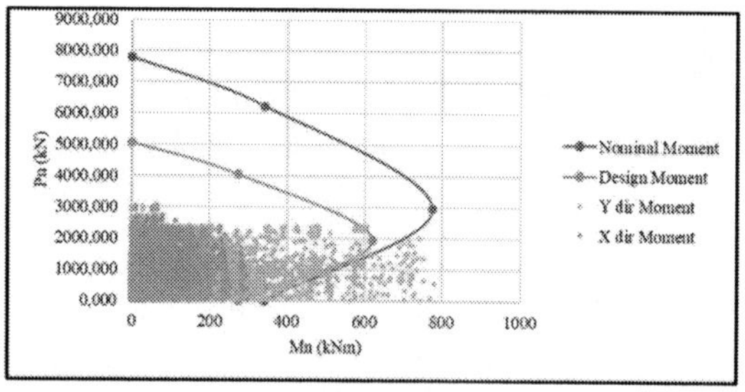


**Fig. 5.** P-M interaction diagram for column K1

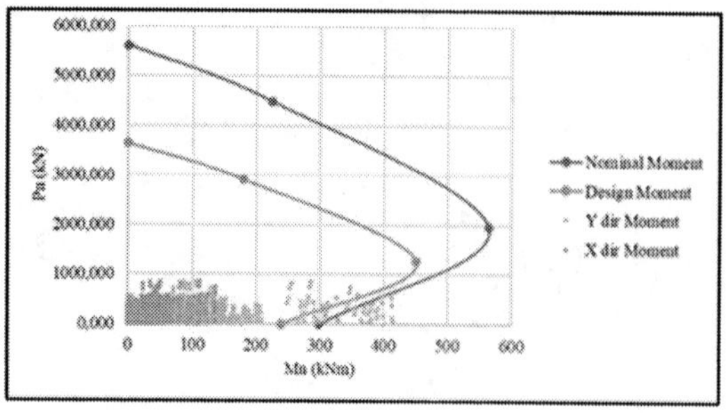


**Fig. 6** P-M interaction diagram for column K2

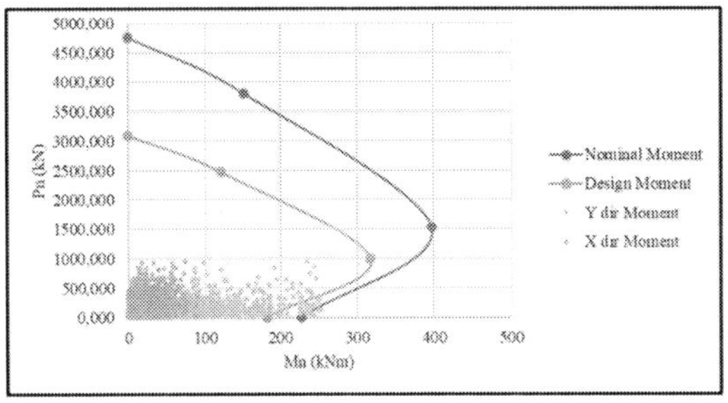


**Fig 7.** P-M interaction diagram for column K3

The results of the column's shear strength are shown in Table 7. The table compares the reduction factor nominal shear strength (ϕVn) and ultimate shear force (Vu).

**Table 7.** Shear capacity of columns

| Fl. | Code | Width | Depth | ΦVn | Vu | ΦVn ≥ Vu |
|---|---|---|---|---|---|---|
| 1 | K1 | 550 | 550 | 473.490 | 252.824 | OK |
| | K2 | 400 | 500 | 276.800 | 123.521 | OK |
| | K3 | 400 | 400 | 218.800 | 87.404 | OK |
| 2 | K1 | 550 | 550 | 473.490 | 302.292 | OK |
| | K2 | 400 | 500 | 276.800 | 112.086 | OK |
| | K3 | 400 | 400 | 218.800 | 122.086 | OK |
| 3 | K1 | 550 | 550 | 473.490 | 226.292 | OK |
| | K2 | 400 | 500 | 276.800 | 85.414 | OK |
| | K3 | 400 | 400 | 218.800 | 91.814 | OK |
| 4 | K1 | 550 | 550 | 473.490 | 141.254 | OK |
| | K2 | 400 | 500 | 276.800 | 48.847 | OK |
| | K3 | 400 | 400 | 218.800 | 57.283 | OK |
| 5 | K1 | 550 | 550 | 473.490 | 189.066 | OK |

The table above shows that the columns of the Third-Class Inpatient Building at Sijunjung Regional Hospital can resist the shear forces acting on the structure.

### *2.5.2 Beam Capacity*

The load-bearing capacity of beam analysis was carried out by reviewing all types of beams on each floor. Table 8 shows that the beam capacity is insufficient to withstand the forces acting on the structure, particularly for beams B1, B2, and the corridor beam.

## 3 Retrofitting the building with shear walls

### 3.1 Design of shear walls

Based on the structural evaluation, the appropriate retrofitting is using shear walls. Shear walls are structural walls made of reinforced concrete designed to resist shear and lateral forces caused by earthquakes. They must be designed to provide adequate stiffness to reduce inter-story drift caused by lateral loads such as earthquakes [18].

**Table 8.** Flexural and shear capacity of beams

| Fl. | Code | ΦMn | Mu | ΦMn ≥ Mu | ΦVn | Vu | ΦVn ≥ Vu. |
|---|---|---|---|---|---|---|---|
| 1 | B1 | 350.54 | 529.23 | NOT OK | 465.87 | 468.71 | NOT OK |
| | B2 | 292.74 | 425.73 | NOT OK | 310.50 | 381.50 | NOT OK |
| | B3 | 286.81 | 115.71 | OK | 372.60 | 102.70 | OK |
| | B4 | 109.91 | 105.01 | OK | 285.58 | 86.04 | OK |
| | B1cor | 171.15 | 228.32 | NOT OK | 258.75 | 185.37 | OK |
| | B2cor | 81.60 | 193.53 | NOT OK | 207.00 | 173.13 | OK |
| | B3cor | 37.15 | 68.80 | NOT OK | 134.48 | 77.63 | OK |
| 2 | B1 | 350.54 | 380.39 | NOT OK | 465.87 | 351.90 | OK |
| | B2 | 292.74 | 298.50 | NOT OK | 310.50 | 260.24 | OK |
| | B3 | 286.81 | 187.10 | OK | 372.60 | 129.95 | OK |
| | B4 | 109.91 | 98.06 | OK | 285.58 | 81.88 | OK |
| | B1cor | 171.15 | 216.47 | NOT OK | 258.75 | 176.92 | OK |
| | B2cor | 81.60 | 168.28 | NOT OK | 207.00 | 152.67 | OK |
| | B3cor | 37.15 | 59.43 | NOT OK | 134.48 | 68.11 | OK |
| 3 | B1 | 350.54 | 339.47 | OK | 465.87 | 350.50 | OK |
| | B2 | 292.74 | 300.71 | NOT OK | 310.50 | 253.16 | OK |
| | B3 | 286.81 | 175.32 | OK | 372.60 | 127.28 | OK |
| | B4 | 109.91 | 84.55 | OK | 285.58 | 70.99 | OK |
| | B1cor | 171.15 | 148.95 | OK | 258.75 | 124.13 | OK |
| | B2cor | 81.60 | 115.71 | NOT OK | 207.00 | 108.01 | OK |
| | B3cor | 37.15 | 42.67 | NOT OK | 134.48 | 51.59 | OK |
| 4 | B1 | 350.54 | 118.38 | OK | 465.87 | 102.93 | OK |
| | B2 | 292.74 | 162.66 | OK | 310.50 | 121.09 | OK |
| | B3 | 286.81 | 86.81 | OK | 372.60 | 74.54 | OK |
| | B4 | 109.91 | 49.74 | OK | 285.58 | 45.50 | OK |
| | B1cor | 171.15 | 75.72 | OK | 258.75 | 56.89 | OK |
| | B2cor | 81.60 | 61.98 | OK | 207.00 | 58.40 | OK |
| | B3cor | 37.15 | 39.44 | NOT OK | 134.48 | 48.82 | OK |
| 5 | B2 | 292.74 | 114.76 | OK | 310.50 | 74.73 | OK |
| | B3 | 286.81 | 198.42 | OK | 372.60 | 199.49 | OK |
| | B5 | 72.87 | 52.90 | OK | 134.48 | 27.87 | OK |

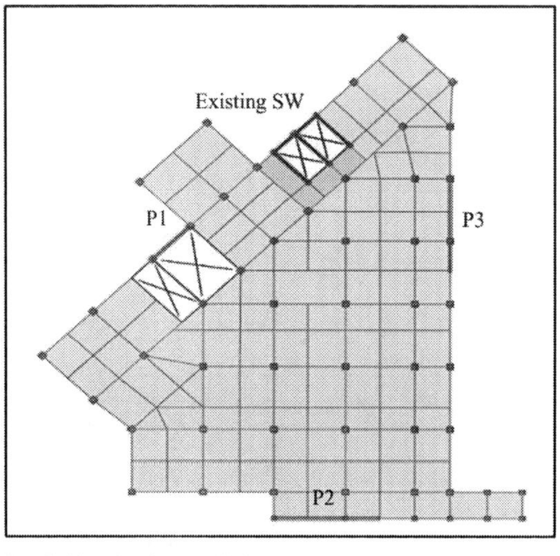


**Fig. 8.** Plan for shear wall placement

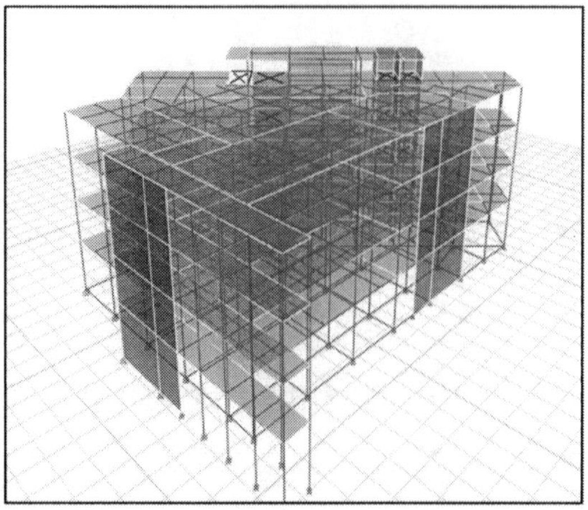

**Fig. 9.** 3D location of the shear wall

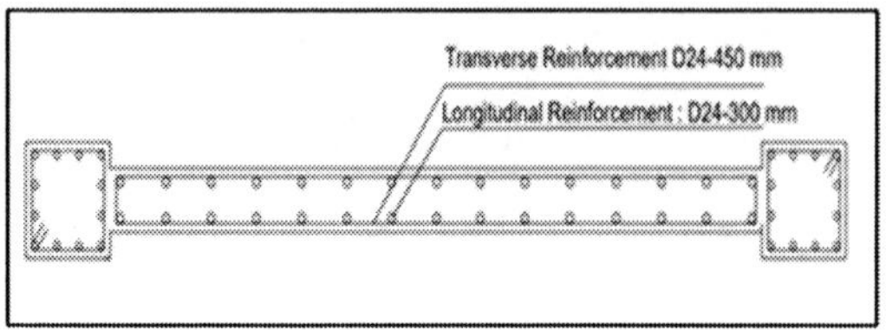


**Fig. 10.** Details of shear wall P1

The following are the recommended locations and details for the shear walls in the Third-Class inpatient ward at Sijunjung Regional General Hospital, as illustrated in Figs. 8-10.

The shear walls use transverse retrofitting of D24-300 and longitudinal retrofitting of D24-450 for the shear wall sections P1, P2, and P3. The shear walls, with a thickness of 300 mm, replace the existing building walls. Shear wall P1 has a length of 4.485 m, while shear walls P2 and P3 each measure 9 m [19].

## 3.2 Inter-story drift of the retrofitted building

Following the retrofitting with shear walls, the structural stiffness increases, leading to a reduction in inter-story drift values. The inter-story drift values post-retrofitting are presented in Table 9.

**Table 9**. Inter-story drift after retrofitting

| Story | Inelastic Drift (mm) | | Drift Limit | Δ after retrofitting < Drift Limit |
|---|---|---|---|---|
| | Δ max without | Δ max with | | |
| 5 | 99.851 | 20.141 | 25.000 | OK |
| 4 | 23.903 | 13.974 | 40.000 | OK |
| 3 | 50.538 | 13.790 | 40.000 | OK |
| 2 | 63.906 | 12.214 | 40.000 | OK |
| 1 | 68.281 | 8.246 | 50.000 | OK |

## 3.3 Structural capacity after retrofitting

### 3.3.1 Column capacity after retrofitting

The results of the structural capacity assessment after retrofitting with shear walls are presented in Figs. 11-13.

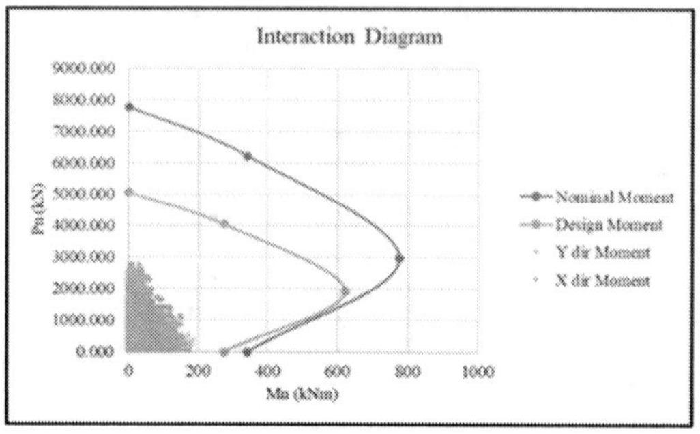


**Fig. 11.** P-M interaction diagram for column K1 after retrofitting

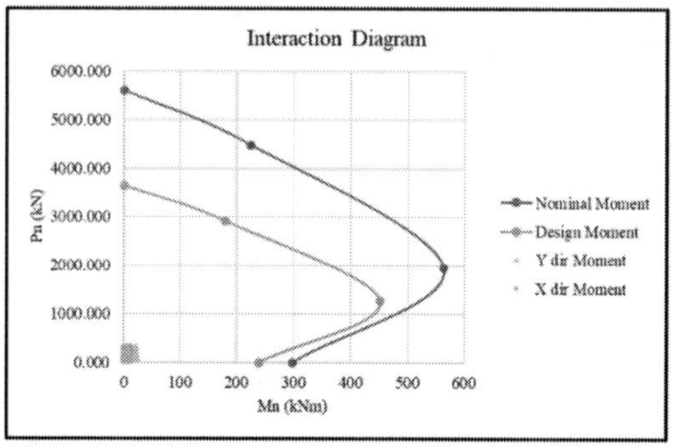


**Fig. 12.** P-M interaction diagram for column K2 after retrofitting

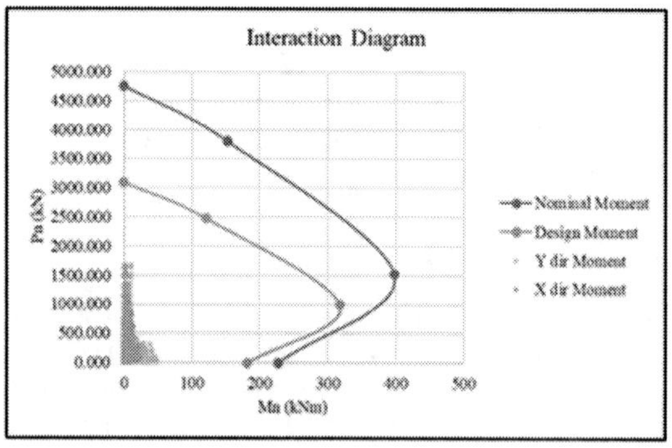


**Fig. 13.** P-M interaction diagram for column K3 after retrofitting

Analysis of the capacity checks in Figs. 11-13 indicate that columns K1, K2, and K3, which initially could not withstand the applied moments, exhibit enhanced strength and can support the structure following the retrofitting with shear walls.

### *3.3.2 Beam capacity of the retrofitted building*

**Table 10.** Beam capacity of beams the retrofitted building

| Fl | Code | ΦMn | Mu | ΦMn ≥ Mu | ΦVn | Vu | ΦVn ≥ Vu. |
|---|---|---|---|---|---|---|---|
| 1 | B1 | 350.54 | 222.07 | OK | 465.87 | 210.59 | OK |
| | B2 | 292.74 | 168.07 | OK | 310.50 | 157.00 | OK |
| | B3 | 286.81 | 123.46 | OK | 372.60 | 94.47 | OK |
| | B4 | 109.91 | 58.01 | OK | 285.58 | 59.52 | OK |
| | B1cor | 171.15 | 83.59 | OK | 258.75 | 74.86 | OK |
| | B2cor | 81.60 | 33.71 | OK | 207.00 | 38.28 | OK |
| | B3cor | 37.15 | 21.10 | OK | 134.48 | 27.49 | OK |
| 2 | B1 | 350.54 | 234.68 | OK | 465.87 | 253.18 | OK |
| | B2 | 292.74 | 108.79 | OK | 310.50 | 163.67 | OK |
| | B3 | 286.81 | 93.24 | OK | 372.60 | 106.70 | OK |
| | B4 | 109.91 | 60.27 | OK | 285.58 | 59.67 | OK |
| | B1cor | 171.15 | 99.45 | OK | 258.75 | 105.59 | OK |
| | B2cor | 81.60 | 42.44 | OK | 207.00 | 45.72 | OK |
| | B3cor | 37.15 | 21.81 | OK | 134.48 | 30.89 | OK |
| 3 | B1 | 350.54 | 245.34 | OK | 465.87 | 259.12 | OK |
| | B2 | 292.74 | 196.34 | OK | 310.50 | 175.75 | OK |
| | B3 | 286.81 | 136.11 | OK | 372.60 | 116.23 | OK |
| | B4 | 109.91 | 61.21 | OK | 285.58 | 60.56 | OK |
| | B1cor | 171.15 | 116.69 | OK | 258.75 | 121.06 | OK |
| | B2cor | 81.60 | 45.24 | OK | 207.00 | 48.17 | OK |
| | B3cor | 37.15 | 24.14 | OK | 134.48 | 32.04 | OK |
| 4 | B1 | 350.54 | 211.62 | OK | 465.87 | 258.14 | OK |
| | B2 | 292.74 | 130.67 | OK | 310.50 | 115.59 | OK |
| | B3 | 286.81 | 114.22 | OK | 372.60 | 95.56 | OK |
| | B4 | 109.91 | 61.21 | OK | 285.58 | 45.36 | OK |
| | B1cor | 171.15 | 92.56 | OK | 258.75 | 86.15 | OK |
| | B2cor | 81.60 | 34.56 | OK | 207.00 | 35.92 | OK |
| | B3cor | 37.15 | 21.62 | OK | 134.48 | 23.51 | OK |
| 5 | B2 | 292.74 | 108.86 | OK | 310.50 | 70.47 | OK |
| | B3 | 286.81 | 43.02 | OK | 372.60 | 32.52 | OK |
| | B5 | 72.87 | 35.14 | OK | 134.48 | 27.75 | OK |

Table 10 shows that the beams in the third-class inpatient building at Sijunjung Regional Hospital can resist the flexural and shear forces acting on the structure.

## 4 Development of fragility curve

Fragility curves enable engineers and decision-makers to comprehend the likelihood of different damage scenarios, which is crucial for risk assessment, emergency planning, and mitigation strategies. By identifying the most vulnerable aspects of a building, fragility curve analysis informs the implementation of retrofitting measures and enhances overall seismic resilience. This method provides a probabilistic approach to seismic risk assessment, offering a more nuanced perspective than deterministic methods, and is instrumental in improving the safety and preparedness of communities in earthquake-prone areas. The seismic vulnerability assessment can be illustrated in Fig. 14.

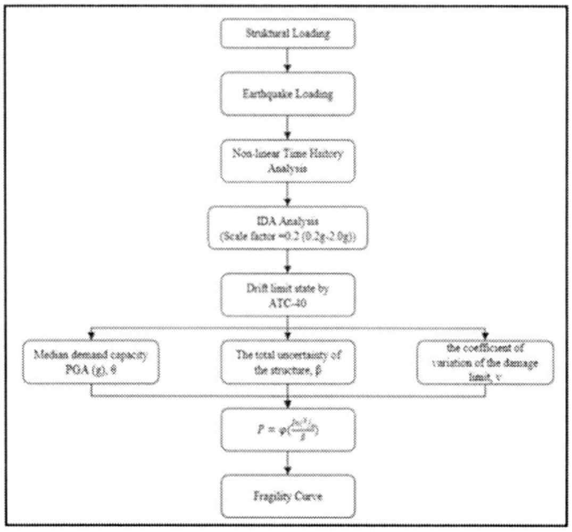


**Fig. 14.** Seismic vulnerability assessment flowchart

### 4.1 Non-linear time history analysis

In this study, non-linear time history analysis was performed using four types of earthquake data (Sijunjung Regency, Chi-Chi, Kobe, and Superstition Hills) with a scaling range of 0.2 to 2.0 and an interval of 0.2. Non-linear time history analysis is utilized as a parameter (EDP) to generate fragility curves with maximum drift values.

### 4.2 Incremental Dynamic Analysis (IDA)

Incremental Dynamic Analysis (IDA) is used to obtain displacements resulting from seismic loads with a scaling factor of 0.2. The structural response from the incremental dynamic analysis, in terms of displacements, can be observed for the X and Y earthquake directions, as shown in Table 11.

**Table 11.** Maximum displacement earthquake X and Y directions

| Sf | Eq. | X dir (mm) | Y dir (mm) | Eq. | X dir (mm) | Y dir (mm) | Eq. | X dir (mm) | Y dir (mm) |
|---|---|---|---|---|---|---|---|---|---|
| 0.2 | Chichi | 5.12 | 6.72 | Kobe | 4.27 | 6.66 | SH | 5.06 | 6.61 |
| 0.4 | Chichi | 10.24 | 13.43 | Kobe | 8.55 | 13.32 | SH | 10.13 | 13.22 |
| 0.6 | Chichi | 15.36 | 20.15 | Kobe | 12.82 | 19.98 | SH | 15.19 | 19.83 |
| 0.8 | Chichi | 20.49 | 26.86 | Kobe | 17.10 | 26.64 | SH | 20.25 | 26.44 |
| 1.0 | Chichi | 25.60 | 33.58 | Kobe | 21.37 | 33.30 | SH | 25.32 | 33.05 |
| 1.2 | Chichi | 30.72 | 40.29 | Kobe | 25.64 | 39.96 | SH | 30.38 | 39.66 |
| 1.4 | Chichi | 35.85 | 47.01 | Kobe | 29.91 | 46.62 | SH | 35.44 | 46.27 |
| 1.6 | Chichi | 40.97 | 53.72 | Kobe | 34.19 | 53.28 | SH | 40.51 | 52.88 |
| 1.8 | Chichi | 46.09 | 60.44 | Kobe | 38.46 | 59.94 | SH | 45.57 | 59.49 |
| 2.0 | Chichi | 51.21 | 67.15 | Kobe | 42.73 | 66.61 | SH | 50.63 | 66.10 |

Based on the structural analysis results using the IDA method, IDA curves are obtained with various variations according to the number of earthquake records used as seismic loads. In this study, the maximum displacement values listed in Table 11 can be used to determine the drift ratio (%) and intensity measure (PGA (g)) as variables for the IDA curves. Damage limits are analyzed based on ATC-40, as shown in Table 12. The IDA curves for the hospital building are illustrated in Fig. 15, and those after retrofitting are depicted in Fig. 16.

**Table 12.** The limit state by ATC-40

| Damages States | Drift Ratio (%) |
|---|---|
| Slight | 0 - 0.01 |
| Moderate | 0.01 – 0.02 |
| Extensive | 0.02 – 0.115 |
| Complete | 0.115 – 0.33 |

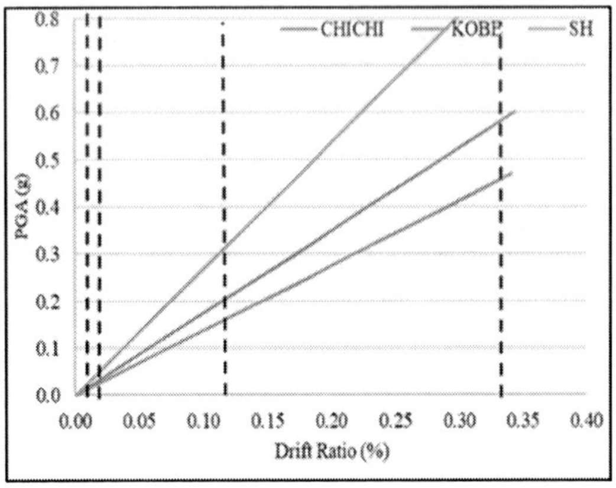


**Fig. 15.** Structural damage limit of the existing hospital building

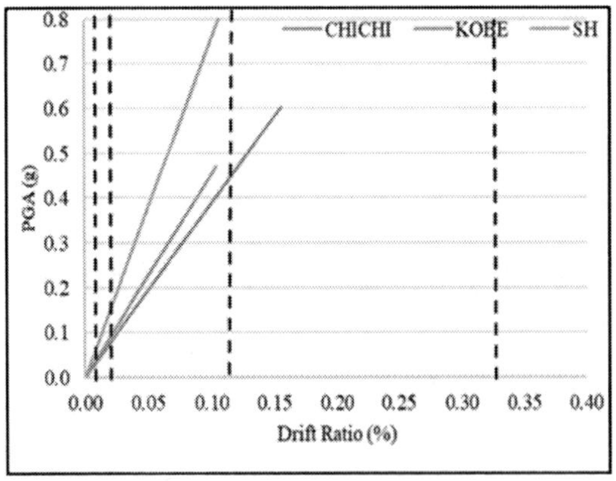


**Fig. 16.** Structural damage limit of the retrofitted hospital building

### 4.3 Seismic fragility curve

With the development of fragility curves for each building, a disaster risk map for a region is hoped to be created, providing an overview of the building fragility levels in that area.

Following the structural analysis results using the Incremental Dynamic Analysis (IDA) method, fragility curve parameters for various damage levels are obtained after retrofitting with shear walls, as shown in Tables 13 and 14. Subsequently, structural damage probabilities are determined to create seismic fragility curves, as illustrated in Figs. 17 and 18. The fragility curves show the probability of damage in percentage relative to PGA values.

**Table 13.** Parameter of fragility curve for an existing building

| Type of Earthquake | Slight PGA (g) | Moderate PGA (g) | Extensive PGA (g) | Complete PGA (g) |
|---|---|---|---|---|
| CHICHI | 0.030 | 0.058 | 0.351 | 1,006 |
| KOBE | 0.020 | 0.038 | 0.233 | 0.668 |
| SH | 0.029 | 0.055 | 0.334 | 0.959 |
| Total | 0.08 | 0.15 | 0.92 | 2.63 |
| N | 3 | 3 | 3 | 3 |
| σ | 0.006 | 0.011 | 0.064 | 0.183 |
| μ | 0.027 | 0.051 | 0.306 | 0.878 |
| v | 0.208 | 0.208 | 0.208 | 0.208 |
| Θ | 0.026 | 0.049 | 0.299 | 0.859 |
| b | 0.206 | 0.206 | 0.206 | 0.206 |

**Table 14.** Parameter of fragility curve for a retrofitted building

| Type of Earthquake | Slight PGA (g) | Moderate PGA (g) | Extensive PGA (g) | Complete PGA (g) |
|---|---|---|---|---|
| CHICHI | 0.053 | 0.100 | 0.607 | 1,741 |
| KOBE | 0.045 | 0.086 | 0.518 | 1,487 |
| SH | 0.075 | 0.143 | 0.866 | 2,486 |
| Total | 0.17 | 0.33 | 1.99 | 5.71 |
| N | 3 | 3 | 3 | 3 |
| σ | 0.016 | 0.030 | 0.181 | 0.520 |
| μ | 0.058 | 0.110 | 0.664 | 1,904 |
| v | 0.273 | 0.273 | 0.273 | 0.273 |
| Θ | 0.056 | 0.106 | 0.640 | 1,837 |
| b | 0.268 | 0.268 | 0.268 | 0.268 |

The seismic fragility curves illustrate the relationship between Peak Ground Acceleration (PGA) and the probability of different damage levels to the building. Derived from the building's exposure scale estimation, the fragility curves are depicted as continuous line curves, with line colors representing the damage levels due to earthquakes. The red line indicates the PGA values at the building location.

**Fig. 17.** Seismic fragility curves of the existing hospital building

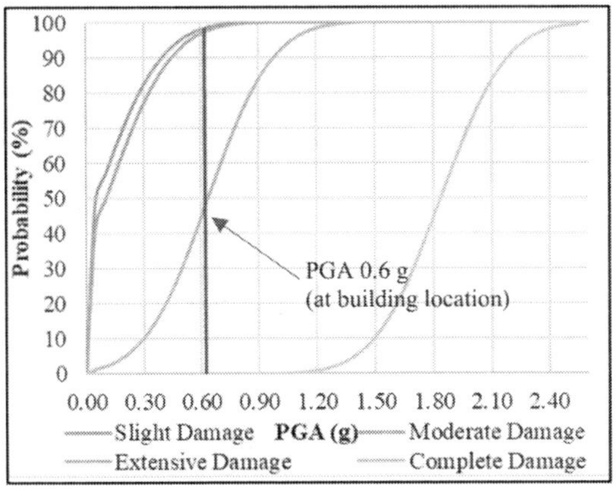


**Fig. 18.** Seismic fragility curves of the retrofitted hospital building

Fig. 17 shows that the building has a 100% probability of slight damage, a 100% probability of moderate damage, a 93% probability of extensive damage, and a 10% probability of collapse at a PGA of 0.60 g. After retrofitting, the probabilities of damage are reduced to 98% for slight damage, 97% for moderate damage, 44% for extensive damage, and 0% for the probability of collapse, as illustrated in Fig. 18.

## 5 Conclusion

Seismic vulnerability assessment of a hospital building with and without retrofitting using Reinforced Concrete (RC) shear walls indicates that the inter-story drift values of the structure exceed the drift limit. Furthermore, it was found that the column and beam elements are unable to support the loads applied to the structure based on seismic regulations (SNI 1726:2019), reinforced concrete structure regulations (SNI 2847:2019), and minimum load regulations (SNI 1727:2020). Therefore, the structure requires retrofitting with shear walls, using longitudinal retrofitting D24-300 mm and transverse retrofitting D24-450 mm, to withstand the forces acting on the structure.

The retrofitted building shows an increase in its structural capacity. The inter-story drift and the strength of the structural elements have met the requirements based on the current building codes, so this building is safe for hospital operations. The vulnerability analysis results indicated that retrofitting the hospital building with shear walls reduces the probability of building damage due to earthquake loads by 52.69% at the level of extensive damage at a PGA of 0.60 g (based on the Indonesian Seismic Map).

This research is funded by LPPM, Universitas Andalas under Grant RKAT 2024 (Contract No.: 149/UN16.19/PT.01.03/ PSS/2024).

## References

1. Ministry of Health of the Republic of Indonesia, Action Plan for the Health Systems and Strategies Center. Statute, (2022).
2. F. Fauzan, A. Maulana, and Z. Al Jauhari, Assessment and Retrofitting Analysis of an Office Building in Riau Province Indonesia based on The Current Building Codes, in AIP Conference Proceedings, American Institute of Physics, (2024).
3. Indonesian National Standardization Agency, SNI 1726:2012 Seismic Resistance Design Codes for Building and Other Structures, BSN, Jakarta, 1-138 (2012)
4. Indonesian National Standardization Agency, SNI 1726:2019 Seismic Resistance Design Codes for Building and Other Structures, BSN, Jakarta, pp.1-238 (2019)
5. National Standardization Agency of Indonesia, SNI 1727:2013 Minimum Loads for Building Design and Other Structures. Statute, (2013).
6. Indonesian National Standardization Agency, SNI 1727:2020 Minimum Load for Planning of Buildings and Other Structures, BSN, Jakarta, 1-302 (2020)
7. Indonesian National Standardization Agency, SNI 2847:2013 Structural Concrete Requirements for Buildings, BSN, Jakarta (2013)
8. Indonesian National Standardization Agency, SNI 2847:2019 Structural Concrete Requirements for Buildings, BSN, Jakarta, 1-695 (2019)
9. National Standardization Agency of Indonesia, SNI 4430:1997 Testing Method for Concrete Structural Elements Using Concrete Hammer N and NR. Statute, (1997).
10. J. Sumajouw, R. Pandaleke, and S. E. Wallah, Comparison of Compressive Strength Using Hammer Test on Reinforced Concrete Portal Specimens and Using Compression Testing Machine on Cube Specimens. J. Sipil Statik. **6**, pp. 941–948, (2018).
11. F. Fauzan and A. Hakam, Structural Evaluation of Damaged BRI Branch Office Building at Jalan Khatib Sulaiman Padang Due to Earthquake, International Journal of Civil Engineering and Technology. **8**, pp. 1170–1180, (2017).
12. F. Fauzan, F. A. Ismail, and Z. A. Jauhari, Seismic Retrofitting Analysis Using Concrete Jacketing and Shear Wall on Dental Hospital Building of Andalas University, in IOP Conference Series: Materials Science and Engineering. **602**, 012108, (2019).
13. F. Fauzan, N. M. Hanifah, E. W. Peratundhika, M. P. Monika, and Z. A. Jauhari, Structural Evaluation of 3-Story Dormitory Reinforced Concrete Building Considering Soil Liquefaction Potential, in E3S Web of Conferences. **156**, 05015, (2020).
14. R. Rifaldo, Feasibility Evaluation of Floor Slab and Beam Structure at the Kaliban School Project with 5 Floors, Bachelor Thesis, Batam Internasional University, Engineering Faculty, (2019).
15. Fauzan, A. Hakam, F. A. Ismail, Z. A. Jauhari, G. A. Agista, Seismic Damage Assessment and Retrofitting of a Hospital Building in West Pasaman, Indonesia. GEOMATE Journal, **27**, 120, 122–129 (2024).

16. M. A. Kurniawan, Evaluation of Structural Strength and Strengthening Recommendations for SDN 08 Campago Ipuh Bukittingi, Bachelor Thesis, Andalas University, Engineering Faculty, (2022).
17. A. I. S. Pratama, Structural Analysis Based on The Application of SNI 1726:2019 and SNI 2847:2019. J.Artesis. **2**, pp 1-12, (2022)
18. N. A. Usmat I, I. Imran, and M. A. Sultan, Analysis of Shear Wall Location on Building Structural Behavior Due to Earthquake Loads. J. Penelitian. **8**, 297, (2019).
19. D. W. Gawatre, T. S. Joshi, M. S. Kukade, P. S. Kurumkar, V. R. Kale, and M. A. R. Bodake, Analysis and Design of Shear Wall with Various Openings Criteria, International Journal of Trend in Scientific Research and Development. **2**, pp. 532-536, (2018).

# Landslide disaster mitigation in Sawahlunto City

*Tirza* Haqia Purnama[1], *Roni* Ekha Putera[1*], and *Hendri* Koeswara[1]

[1]Faculty of Social and Political Sciences, Universitas Andalas, Padang 25175, Indonesia

**Abstract.** The research aims to analyze the implementation of landslide disaster mitigation in Sawahlunto. Motivated by the high risk of landslides in Sawahlunto. Efforts are needed to reduce the risk of landslides by implementing the theory used of W.N. Carter's theory of non-structural and structural disaster mitigation. The method used is descriptive qualitative, with data collection techniques using interviews, documentation, and observation. This is concluded that the implementation of landslide disaster mitigation in Sawahlunto has been running, but the implementation has not been optimal. Several things, non-structural, cause this: In the legal framework, a lack of understanding of policies that contain appropriate mitigation as well as the role of each agency or institution; the disaster risk reduction forum (FPRB) has been formed at the level, but at the village level only has a disaster preparedness group (KSB); the lack of implementation of disaster mitigation measures by the community. Structural disaster mitigation activities are still inadequate, as seen from the unavailability of vertical evacuation buildings. Efforts that can be made to support the implementation of landslide disaster mitigation in Sawahlunto include reviewing mitigation policy documents, increasing the capacity of FPRB, and collaborating with various parties to support the implementation of disaster mitigation.

## 1 Introduction

Indonesia is ranked as the second most disaster-prone country in the world due to various geological, geographical, and climatic factors [1]. It is located in the Pacific Ring of Fire, an area of high seismic and volcanic activity due to the meeting of several large plates, including the Indo-Australian, Eurasian, and Pacific plates [2, 3]. In addition to natural factors, human activities such as deforestation and unplanned development also reduce the soil's water absorption capacity and cause erosion, increasing the risk of natural disasters. Global climate change also affects weather patterns in Indonesia, with extreme weather events such as tropical storms and extreme rainfall becoming more frequent and intense [4]. Therefore, containment and prevention efforts are essential to mitigate its impacts.

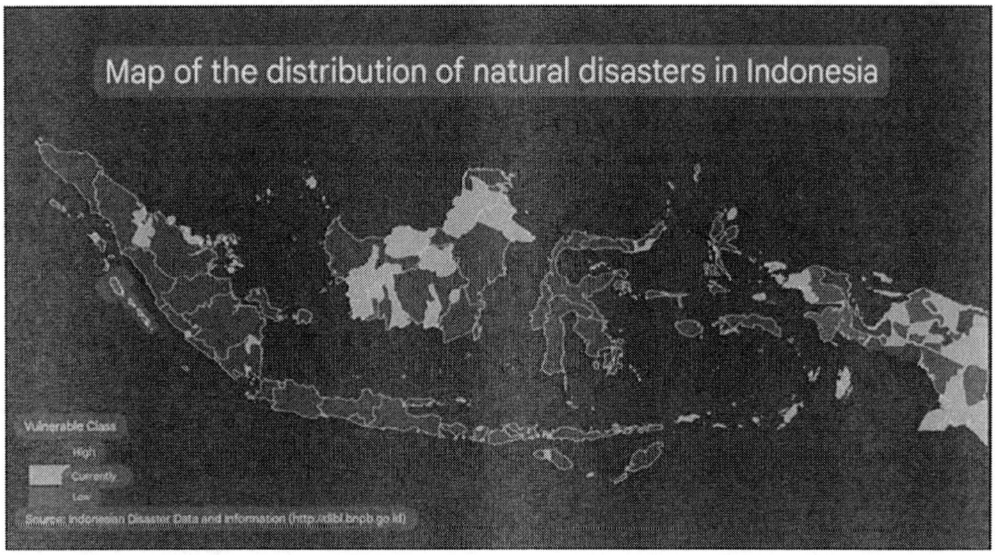


**Fig 1.** Map of the distribution of natural disasters in Indonesia

* Corresponding author : roniekhaputera@soc.unand.ac.id

Figure 1 shows Indonesia's high vulnerability to disasters. In addition, the islands of Sumatra, Java, Kalimantan, and Sulawesi, shown in red, are at the highest risk. Disasters are natural or non-natural disasters that cannot be predicted and cause casualties, environmental damage, and property losses [5, 6]. In addition, it is difficult to predict when exactly the disaster will occur.

The following data describes the impact and damage that occurred in Indonesia due to disasters in the period from January 2021 to May 2024, namely:

**Table 1.** Impact and damage from natural disasters in 2021 – 2024

| Year | Impact of natural disasters | | Natural disaster damage | | Total cases |
|---|---|---|---|---|---|
| | Died | Evacuate | Damaged house | Damaged facilities | |
| 2021 | 728 | 7,630,692 | 158,658 | 4,445 | 5,402 |
| 2022 | 858 | 6,144,534 | 95,403 | 1,983 | 3,544 |
| 2023 | 275 | 8,491,288 | 47,214 | 1,291 | 5,400 |
| 2024 | 239 | 3,526,054 | 37,287 | 706 | 783 |

From Table 1, the impact of various disasters in Indonesia from 2021 to 2024 is very concerning, causing losses and damage to different aspects such as houses and multiple facilities such as education, worship, and health. In addition, many casualties and people are displaced due to disasters. Based on this data, damage due to disasters is increasing yearly, and the total number of disasters continues to grow until May alone. In 2024, it has reached 783 disaster cases.

West Sumatra is a province located on the west coast of the island of Sumatra. It is one of Indonesia's regions that is very prone to natural disasters [7]. This vulnerability is due to West Sumatra's location on the Pacific Ring of Fire, which has high tectonic activity [8]. The topography of West Sumatra, which consists of hills and mountains, makes this area vulnerable to landslides, especially during the rainy season [7,9]. Heavy rainfall can cause soil erosion and increase the risk of landslides, destroying infrastructure, settlements, and agricultural land. The following table shows the disaster potential for each regency/city in West Sumatra.

**Tabel 2.** Distribution of disaster potential in West Sumatra

| Location | Disaster | | | | | |
|---|---|---|---|---|---|---|
| | Earthquake | Tsunami | Flood | Fire mountain | Landslides | Fire |
| Padang | Potential | Potential | Potential | | Potential | Potential |
| Pariaman | Potential | Potential | Potential | | Potential | Potential |
| Bukittinggi | Potential | | | Potential | Potential | Potential |
| Padang Panjang | Potential | | | Potential | Potential | Potential |
| Payakumbuh | | | Potential | Potential | | Potential |
| Sawahlunto | | | | | Potential | Potential |
| Solok | Potential | | Potential | | | Potential |
| Padang Pariaman | Potential | Potential | Potential | Potential | Potential | Potential |
| Agam | Potential | Potential | Potential | Potential | Potential | Potential |
| West Pasaman | Potential | Potential | Potential | Potential | Potential | Potential |
| Pasaman | Potential | | Potential | Potential | Potential | Potential |
| Limapuluh Kota | Potential | | Potential | Potential | Potential | Potential |
| Tanah Datar | Potential | | Potential | Potential | Potential | Potential |
| Solok | Potential | | Potential | Potential | Potential | Potential |
| South Solok | Potential | | Potential | Potential | Potential | Potential |
| South Coastal | Potential | Potential | Potential | | Potential | Potential |
| Sijunjung | | | Potential | | Potential | Potential |
| Dharmasraya | | | Potential | | Potential | Potential |
| Mentawai | Potential | Potential | | | Potential | Potential |

It can be seen in Table 2 that each region in West Sumatra province has more than one potential disaster. One is Sawahlunto, which has the potential for landslides and fires. Sawahlunto is an area that is vulnerable to natural and non-natural disasters [10] This is evidenced by the disasters that always occur in West Sumatra, as shown in Table 3 below.

 https://doi.org/10.1051/e3sconf/202560416001

**Table 3.** Landslide disaster in West Sumatra regency/city in 2017

| District/City | Landslide |
|---|---|
| Lima Puluh Kota | 8 |
| Agam | 15 |
| Dhamasraya | 1 |
| Mentawai | 1 |
| Padang Pariaman | 3 |
| Pasaman | 8 |
| West Pasaman | 9 |
| Pesisir Selatan | 7 |
| Sijunjung | 12 |
| Solok | 8 |
| Solok Selatan | 2 |
| Tanah Datar | 9 |
| Bukittinggi | 3 |
| Padang | 12 |
| Padang Panjang | 2 |
| Pariaman | 0 |
| Payakumbuh | 2 |
| Sawahlunto | 27 |
| Solok | 0 |
| Total | 139 |

Table 3 shows that Sawahlunto is the most frequent area for landslides; as of 2017, there have been 27. Sawahlunto has hilly contours and land slope conditions from 45°, 50° to 75°; this can potentially cause landslides in Sawahlunto if hit by continuous rain. Landslide-prone areas in Sawahlunto include all villages in the Silungkang sub-district, all towns in the Fresh Valley sub-district, all villages in the Barangin sub-district except Kolok Mudik and Kolok Nan Tuo, and several areas in the Talawi sub-district except Salak Village, Talawi Mudik, Talawi Hilir and Rantih. This is also evidenced by the number of landslides in Sawahlunto from 2018 to 2019, as shown in Table 4 below.

**Table 4.** Number of landslide in Sawahlunto

| Year | Landslides |
|---|---|
| 2018 | 123 |
| 2019 | 136 |

From Table 4, it can be seen that the most disastrous events in Sawahlunto were landslides; as many as 259 landslide events occurred from 2018 to 2019 due to the city's hilly topography with undulating contours. Sawahlunto is also prone to recent floods and landslides, which caused considerable losses, as in the news in Fig. 2.

Figure 2 shows that Sawahlunto recently experienced floods and landslides, which resulted in losses totaling 6 billion rupiah. Dedi Ardona, acting head of the Regional Disaster Management Agency (BPBD) of Sawahlunto, said the total loss was calculated from the agriculture, fisheries, houses/settlements, schools, roads, and irrigation sectors.

**Fig 2.** News of flood and landslide losses in Sawahlunto

Disaster management, which includes prevention/mitigation, preparedness, emergency response, rehabilitation, and reconstruction after a disaster occurs, is needed to avoid and reduce the impact of disaster risk [11]. To minimize the impact of disasters, prevention must be carried out at the pre-disaster stage through mitigation efforts. According to Law No. 24/2007, disaster mitigation is a series of efforts to reduce disaster risk through physical development and increasing awareness and ability to deal with disaster threats [12]

Several relevant previous studies can be used as references in this research. First, Haeril's research entitled "The Strategy of the Regional Disaster Management Agency (BPBD) in Natural Disaster Risk Reduction in Bima Regency." [13]. This research discusses the various strategies implemented by BPBD Bima District in reducing the risk of natural disasters, including preventive measures, preparedness, and emergency response. Haeril examines the effectiveness of BPBD programs and how they involve the community in mitigation efforts.

Second, a research entitled "Landslide Disaster Risk Reduction Strategy in Bogor Regency, West Java Province." [14]. This research explores the strategies implemented by BPBD Bogor Regency in reducing landslide risk, including the development of landslide retaining infrastructure, community education, and early warning systems. Nindy highlighted the importance of collaboration between the government, local communities, and non-governmental organizations in disaster management. The shortcoming of this research is the lack of long-term evaluation of the effectiveness of the strategies implemented, so the impact of the measures cannot be thoroughly measured.

Furthermore, research by Chintya entitled "Flood Disaster Management Strategy by the Regional Disaster Management Agency (BPBD) of Solok, West Sumatra." [15]. This study examines BPBD Solok's strategy in dealing with flood disasters, including river flow management, embankment construction, and the provision of evacuation facilities. We also discussed how the BPBD coordinated with related agencies and the community in a flood emergency. This research provides a comprehensive overview of the multi-faceted approach to flood management in urban areas.

Therefore, new research is needed to assess landslide mitigation in Sawahlunto. Understanding the strategies implemented and evaluating their effectiveness is expected to provide recommendations to improve preparedness and response to future landslides. Through this study, it is hoped to make a meaningful contribution to developing disaster mitigation policies and practices at the local level and support government and community efforts to create a safer and more resilient environment. This research provides a comprehensive overview of Sawahlunto in landslide mitigation and offers insights that can be applied to other areas with similar vulnerabilities.

## 2 Research method

This research uses a qualitative approach with a description method and is coupled with a literature study. This method aims to solve problems by describing the issues that occur factually, systematically, and accurately related to the properties, facts, and relationships between the phenomena to be investigated. The theory used to analyze this research problem is the theory of disaster mitigation by W.N. Carter, which will explore non-structural and structural disaster mitigation in Sawahlunto's landslide mitigation. In the nonstructural mitigation variable, there are six indicators, namely, legal framework, institutional establishment, raising public awareness, conducting training and education, providing incentives, and an early warning system.

## 3 Results and discussion

This research aims to analyze landslide mitigation efforts in Sawahlunto using W.N. Carter's Disaster Mitigation Theory, which includes structural and non-structural mitigation. Data was collected through field observation and review of related documents. The following are the main findings of this research:

### 3.1 Structural mitigation

#### *3.1.1 Construction and repair of waterways*

This effort aims to improve the capacity and efficiency of the drainage system in Sawahlunto. A good drainage system is essential to drain surface water quickly, especially during the rainy season, to prevent inundation that could potentially trigger landslides. Standing water not only damages infrastructure but also increases the risk of soil erosion, which can weaken slopes and trigger landslides.

To achieve this goal, the first step was to conduct a thorough survey and analysis to determine critical points vulnerable to waterlogging. This includes mapping areas that frequently experience drainage problems and identifying the leading causes of waterlogging, such as blocked or damaged drains. A comprehensive analysis will help design appropriate solutions, be it through repairing existing drains or constructing new, more effective drains.

Implementing the survey and analyzing the results involved various concrete actions. Improving existing waterways was made by widening or deepening the channels to increase capacity. In addition, new waterways were constructed to direct water flow more efficiently and reduce the risk of inundation. Regular cleaning of drains is also an essential part of this implementation to ensure the drains remain optimally functional and not clogged with sedimentation or trash. Through this approach, it is hoped that the drainage problem in Sawahlunto can be adequately addressed to minimize the risk of landslides due to waterlogging.

#### *3.1.2 Making gutters in the roof of a house*

This effort aims to rainwater from rooftops directly to drains or pools to pool water infiltration into the soil around the foundation's foundation. Water infiltration can lead to increased soil moisture and pore water pressure, weakening the foundation structure and increasing the risk of landslides, especially in vulnerable areas.

The implementation of this goal involves several concrete steps. First, communities are encouraged to install gutters on the roofs of their houses. These gutters collect rainwater and channel it directly into drains or rivers, reducing the risk of infiltration. Socialization and education on the importance of gutter installation were conducted through public campaigns, brochures, and social media to ensure that all residents understood its benefits.

In addition, local governments or related institutions can provide technical and material assistance to needy residents. Technical assistance can include consultations on the design and installation of effective gutters, while material assistance includes the provision of affordable gutters or subsidies for underprivileged families. This approach ensures that all levels of society can participate in the mitigation efforts so that overall landslide risk can be significantly reduced in Sawahlunto.

#### *3.1.3 Channeling household waste water to riol/sewer*

This effort aims to reduce wastewater infiltration into the soil, which can lead to increased pore water pressure and potential landslides. Household wastewater infiltrating into the soil can harm the environment and contribute to soil instability around residential areas. This increase in pore water pressure due to infiltration can weaken the soil structure and increase the risk of landslides, especially in areas with slopes.

The implementation of this objective requires a comprehensive approach. First, community outreach is conducted on the importance of proper household wastewater management. This includes information on the dangers of wastewater infiltration and ways to safely manage household waste, such as using proper septic systems or channeling waste directly into designated sewers. This educational campaign can be conducted through seminars, workshops, and brochure distribution in residential neighborhoods.

In addition to counseling, enforcement of regulations on wastewater management is also critical. Local governments should issue and implement rules governing the discharge of household wastewater, including regular inspections to ensure residents' compliance with the regulations. Sanctions or fines can be imposed for those who violate them to ensure compliance and awareness of the importance of proper waste management. In combination with education and law enforcement, this approach is expected to reduce wastewater infiltration into the ground, thereby reducing the risk of landslides and creating a safer environment in Sawahlunto.

#### *3.1.4 Routine cleaning of sedimentation and trash in waterways*

These efforts aim to prevent blockage of waterways that can cause flooding and accelerate the flow of water that can erode slopes. Blockage of waterways often results from sedimentation and accumulation of garbage, which not only blocks the flow of water but also increases the risk of flooding. This flooding, in turn, can accelerate soil erosion and result in landslides, especially in areas with steep slopes.

To achieve this goal, implementation included organizing gotong royong activities and scheduling regular cleaning of waterways, especially during the rainy season. Gotong royong involves active participation from the community in clearing waterways of garbage and sediments that block the flow of water. This activity clears the waterways and raises residents' awareness of the importance of keeping the environment clean.

Scheduling regular cleaning is also an essential part of this implementation. Local governments and relevant agencies should establish a regular cleaning schedule and ensure that it is carried out consistently, especially during the rainy season when water volumes increase. In addition, education on the negative impacts of blocked drains and the importance of community participation in cleaning activities should be continuously socialized. With a combination of gotong royong and regular cleaning, it is expected that waterways in Sawahlunto can function adequately, reduce the risk of flooding, and prevent soil erosion that can trigger landslides.

The results of BPBD's evaluation of flood and landslide disasters that have occurred in recent years show that the above measures need to be optimized and implemented thoughtfully. Proper preventive measures can reduce disaster risks and impacts. Acting Mayor of Sawahlunto Fauzan Hasan reminded the importance of going the extra mile in anticipating disaster risks rather than just coping when disasters occur. The commitment of all regional apparatus and the community is significant for the success of this mitigation effort. The circular letter for follow-up and accelerated implementation of BPBD recommendations is an essential first step in improving disaster preparedness and mitigation in Sawahlunto.

### 3.2 Non-structural mitigation

Place the figure as close as possible after the point where it is first referenced in the text. If there are many figures and tables, it might be necessary to place some before their text citation.

#### *3.2.1 Optimizing the role of disaster preparedness groups in village*

This effort aims to improve community preparedness by forming and training disaster preparedness groups that can respond quickly in the event of a disaster. Establishing disaster preparedness groups in each village and kelurahan aims to ensure that local units are ready to provide first aid and coordinate immediate evacuation in the event of a disaster. The training covers basic knowledge of different types of disasters, first aid skills, and safe and efficient evacuation procedures. Disaster simulations are conducted regularly to ensure that all members of the disaster preparedness group can apply their knowledge and skills in emergencies. In addition, disaster preparedness groups are equipped with essential equipment such as first aid kits, communication tools, and simple rescue equipment to support their operations.

The implementation of this program does not stop at training and simulation. The local government also conducts regular evaluations to assess the effectiveness and readiness of the disaster preparedness group. This evaluation includes an assessment of the team's response during simulations, the condition and completeness of equipment, and coordination between team members and related agencies. Based on the evaluation results, follow-up is carried out to correct deficiencies and increase the team's capacity. This may include additional training, equipment repair or replacement, and improvement of standard operating procedures (SOPs) for disaster management. With this comprehensive approach, disaster preparedness groups in Sawahlunto are expected to function optimally in minimizing disaster impacts and saving more lives.

#### *3.2.2 Legal framework*

The Sawahlunto Government is aware of the potential landslide hazard in its area. Therefore, it has taken proactive measures by establishing land use and development regulations in landslide-prone areas. These regulations aim to minimize the risk of landslides by controlling human activities that can aggravate the condition, such as development on steep slopes or deforestation.

#### *3.2.3 Institutional establishment*

Establishing a Regional Disaster Management Agency (BPBD) in Sawahlunto is essential in disaster mitigation efforts. BPBD is responsible for coordinating and implementing mitigation efforts, including risk monitoring, emergency management, and post-disaster rehabilitation. With the existence of BPBD, disaster management is expected to be done more structurally and effectively. BPBD also has a role in disaster-related data collection, risk analysis, and the preparation of comprehensive disaster management plans.

BPBDs also serve as communication and coordination centers between various government agencies, communities, and non-government organizations involved in disaster mitigation. This function is essential to ensure that information regarding disaster threats can be disseminated immediately and preventive actions can be taken quickly. In addition, BPBDs play a role in increasing community awareness and preparedness for disasters through various education and training programs. Collaboration between BPBD and multiple parties can improve Sawahlunto's resilience to disasters.

### *3.2.4 Raising public awareness*

Public awareness is critical in landslide mitigation. Sawahlunto has conducted public education and outreach campaigns on landslide hazards and preventive measures. The campaign includes various activities that aim to deliver information effectively and evenly to all levels of society. One method used was social media, which has a wide and fast reach. In addition, brochure distribution and seminars were also conducted to provide more in-depth and direct information to the community.

Social media, brochures, and seminars were educational tools to reach the broader community. The campaign aims to increase people's understanding of landslide risks and teach them how to protect themselves and their families. Through continuous education, the community is expected to be more aware of the early signs of landslides and know what actions to take during an emergency. Thus, the level of community preparedness can increase to minimize potential losses and casualties due to landslides. This effort shows a commitment to building a more resilient and disaster-prepared community.

## 4 Conclusion

In the context of Sawahlunto, the synergy between structural and non-structural measures has shown positive results in landslide mitigation. The construction of dams, retaining walls, improved drainage systems, and the application of modern technology for soil monitoring have helped significantly reduce landslide risk. On the other hand, non-structural efforts such as the implementation of legal frameworks, establishment of strong institutions, education and training, provision of incentives, and development of early warning systems have increased community awareness and preparedness in facing disasters.

However, challenges remain, particularly in funding, infrastructure maintenance, and increasing community awareness and participation on an ongoing basis. Limited funding may hamper efforts to maintain and repair the infrastructure that has been built. At the same time, community awareness and participation still need to be improved, requiring more effective education and communication strategies.

This study concludes that a holistic and adaptive mitigation approach is needed to reduce the impact of landslides effectively. Integration between structural and non-structural solutions should be continuously strengthened, considering local conditions and potential risks. Continuous regional capacity-building efforts, such as regular training and community education, and increased cooperation between the government, community, and private sector are crucial to achieving sustainable disaster mitigation.

Firstly, the increased funding by local government needs to seek additional sources of financing. Whether from the central government, international donors, or the private sector, to ensure the sustainability of disaster mitigation infrastructure maintenance and repair.

Infrastructure maintenance needs a regular program to ensure that dams, retaining walls, and drainage systems remain in good condition and effectively prevent landslides. Education and training programs need to be improved by involving more communities and using more interactive and engaging methods. Information technology and social media can be effective tools in disseminating information and raising awareness. They increase community participation, and the government needs to encourage active community participation in disaster mitigation efforts through more attractive incentives and rewards for citizens who contribute significantly.

Multisectoral cooperation: increased collaboration between the government, communities, and the private sector needs to be increased in disaster mitigation efforts. This partnership can cover various aspects, such as funding, training, and technology provision. Early warning systems need to be continuously developed and refined, ensuring that the information provided is accurate, timely, and easily accessible to all levels of society, and research and innovation further in disaster mitigation techniques should continue to be encouraged to find more effective and efficient solutions to the growing landslide risk.

The research funder is Universitas Andalas with grand number Nomor: 249/UN16.19/PT.01.03/PSS/2024.

## References

1. E. Suharini, D. Liesnoor S., and E. Kurniawan, Univers. J. Geosci. **2**, 195 (2014)
2. A. T. Fathani, R. E. Putera, A. Asrinaldi, T. R. Valentina, D. Hariyati, D. Holidin, and A. Agustiyara, E3S Web Conf. **464**, 05002 (2023)
3. P. Febri Wialdi, R. E. Putera, Kusdarini, R. Ariany, T. R. Valentina, and R. Dermawan, E3S Web Conf. **331**, 04001 (2021)
4. I. T. N. H. Ullum, A. Fitria, and W. Widodo, JSN J. Sains Nat. **2**, 40 (2024)
5. I. Setyawan, S. A. Sajidah, A. Sabila, A. P. Wijayanti, and K. S. Dewi, Proc. Int. Conf. Psychol. Stud. **4**, 423 (2023)

6. R. E. Putera, H. Nurasa, and Y. S. Sugandi, Bisnis Birokrasi J. **23**, (2018)
7. T. U. Vemileza Putri, I. K. Dewi, and L. Marlinda, Sci. Phys. Educ. J. **5**, 47 (2022)
8. M. Fuady, R. Munadi, and M. A. K. Fuady, IOP Conf. Ser. Mater. Sci. Eng. **1087**, 012011 (2021)
9. T. Ahmad, Roni Ekha Putera, and Hendri Koeswara, J. Soc. Policy Issues 45 (2024)
10. S. N. Fadhilah, R. E. Putera, D. Aromatica, and E. Hasyimi, Sawala J. Adm. Negara **10**, 196 (2022)
11. R. Fajria, R. E. Putera, and R. Ariany, E3S Web Conf. **464**, 10001 (2023)
12. N. Laframboise and B. Loko, IMF Work. Pap. **12**, 1 (2012)
13. H. Haeril, T. Irfadat, and M. Mas'ud, J. Stud. Ilmu Pemerintah. **3**, 1 (2022)
14. N. Pratiwi, IPDN (n.d.)
15. Chyntia Aulia and Geovani Meiwanda, J. Res. Dev. Public Policy **1**, 35 (2022)

# Geotechnical analysis of potential landslides on slopes in the Rimbo Panjang area of Padang City and mitigation

*Abileo* Fardan Fadlan[1], *Rina* Yuliet[1*], *Suhana* Koting[2], *Ainul* Mardhiah[1], *Bunga* Sucita Ikhsan[1], *Hafizh*[1], and *Raihan* Fadhlurahman[1]

[1]Department of Civil Engineering, Faculty of Engineering, Universitas Andalas, Padang 25163, Indonesia
[2]Department of Civil Engineering, Faculty of Engineering, Universiti Malaya, 50603 Pantai Valley, Kuala Lumpur, Malaysia

**Abstract.** The Lambung Bukik area of Padang City is a relocation area for the flash flood disaster in Batu Busuk Padang on 24 July 2012. This research analyzes the potential for slope failure in the Rimbo Panjang area, where many residents' houses are built on steep slopes. Geotechnical analysis was carried out to determine the soil characteristics in the Rimbo Panjang area, which was taken at four locations. The results of the classification show that the soil in the Rimbo Panjang area is inorganic silt - high plasticity (MH). The plasticity index value is 2.5 – 10.4%, indicating that the soil has low swelling potential. The activity value in inorganic silt soil is meager, namely in the range of 0.1 – 0.99, which is the mineral kaolinite and illite activity value. The activity values agree with the XRD test results, where the soil contains kaolinite and illite minerals with low swelling properties. The low swelling characteristic causes the slopes to have a low landslide potential. The safety factor value of the slopes is in the range of 1.79 – 2.71 > 1.5, so the slope can be said to be stable.

## 1 Introduction

Landslides are geological hazards usually associated with extreme events such as heavy rain [1]. Landslides are also influenced by the angle of the slope, where the steeper the slope, the greater the possibility of a landslide occurring [2]. Landslides can potentially damage the environment around the slopes [3-5]. Factors that cause natural slope instability include (1) changes in the slope profile due to additional loads or reduced strength; (2) an increase in pore water pressure, which causes swelling in cohesive soil; (3) a decrease in soil shear strength due to weathering, leaching, mineralogy changes, and cracks; vibrations caused by earthquakes, blasting or pile driving [6].

The Fukuoka-ken Seiho-oki earthquake that occurred in 2005 caused severe damage to residential areas on Genkai-jima Island. Most residential houses were built on excavation and embankment slopes that experienced landslide due to the earthquakes [7]. Apart from earthquakes, landslides caused by rainfall have caused deaths, destroyed infrastructure, and hampered economic development in many countries worldwide. Shallow landslides caused by rainfall are characterized by the movement of shallow soil layer materials along the slope. This movement can have a significant impact (depending on the potential energy), which can cause severe damage to both the built and natural environments [8-10].

High and prolonged rainfall around March 2021 caused landslides in front of the Science Techno Park building (SPT) Universitas Andalas. Landslides failed in the soil retaining wall structure, which was close to the slope where the landslide occurred. Figure 1 shows that the failure of the existing retaining wall is not caused by the dimensions of the retaining wall being unsafe against overturning, shearing, and bearing capacity but rather due to the occurrence of landslides on steep slopes that have not been provided with a retaining wall and are close to the existing retaining wall [11].

**Fig. 1.** Damage to gravity walls

Figure 2 shows the failure of gravity walls in student dormitories due to high rainfall.

* Corresponding author: rina@eng.unand.ac.id

**Fig. 2.** Damage to gravity walls in student dormitories

The flash flood on 24 July 2012 resulted in several houses on the banks of the Batu Busuk River being damaged and uninhabitable. Due to the high risk of a flood disaster that might happen again, the Padang City government relocated the affected residents to the Rimbo Panjang area, which is close to the location of the flash flood disaster. The Rimbo Panjang area is in a location that is safe from flooding. Until 2023, the quality of infrastructure in the Rimbo Panjang relocation area has yet to improve significantly [12]. Based on this, it is necessary to conduct a geotechnical analysis to determine the potential for landslides on slopes in the Rimbo Panjang area as a relocation area.

From the results of field visits to the Rimbo Panjang relocation area, it was found that many residents' houses were located on steep natural slopes with slopes greater than 45°, as shown in Fig. 3. These natural slopes are not equipped with soil-retaining structures, which is dangerous for houses built on these slopes. Steep slopes and high and prolonged rainfall are factors that cause landslides. Based on this, a problem-solving approach needs to be carried out using geotechnical analysis of potential slope failures in the Rimbo Panjang area and handling landslides.

**Fig. 3.** Residential conditions in Rimbo Panjang

## 2 Research methods

### 2.1 Study area

The Rimbo Panjang area is located in Lambung Bukit, Padang City, directly bordering Solok Regency. This area is dominated by protected forest areas and community forests, covering almost 82% of the total area, the rest of which are rice fields, gardens, and fields. In general, Rimbo Panjang has a wet tropical climate, is strongly influenced by the west wind, and has a very short dry month with an average rainfall of 347.5 mm/month. Rimbo Panjang is located in Pauh District at 10 m - 1600m above sea level. Most of the topography of this area has an average land slope of 40% (steep to very steep) [13].

### 2.2 Data sources

This research uses data field investigations and data from previous service reports.

### 2.3 Soil sampling

In this research, a location survey was carried out to describe the sampling area. Soil sampling was carried out at four locations in the Rimbo Panjang area (Figure 4) using a hoe for disturbed soil and a sampling tube for undisturbed soil. Soil sampling was carried out for four days. The coordinates of points 1, 2, 3, and 4 on the map, respectively, are -0.910068, 100.450842; -0.910476, 100.450720; -0.910379, 100.448805; and -0.910038, 100.450124 as shown in Fig. 4.

**Fig. 4.** Soil sampling area in Rimbo Panjang

### 2.4 Laboratory testing

Laboratory tests were carried out to describe the soil material at the research location. The following tests are carried out: (1) test the soil index properties or soil classification; (2) test mechanical properties; (3) SEM and XRD.

### 2.5 Slope stability analysis

In this research, stability analysis was carried out on natural slopes in the Rimbo Panjang area. Natural slopes are formed due to natural activities such as erosion, tectonic movements, and so on. The material that forms the slope tends to slip due to its own weight and external forces resisted by the shear strength of the soil of the material. Disturbances to stability occur when the shear resistance of the soil cannot compensate for the forces that cause sliding on the landslide surface. Natural slopes that have been stable for years can experience landslides. For slope stability analysis, the safety factors required by SNI 8460:2017 consider the costs and consequences of slope failure and the level of uncertainty in the analysis conditions.

## 3 Results and discussion

### 3.1 Geotechnical analysis

Figure 5 shows the particle size distribution curve of soil samples taken at Rimbo Panjang research areas. The sieve analysis results show that the slope material in the Rimbo Panjang area consists of 60% - 67% silt, 22% - 24% clay, 9% - 15% sand, and 0 - 2% gravel. This value shows that the silt soil type dominates the slope material.

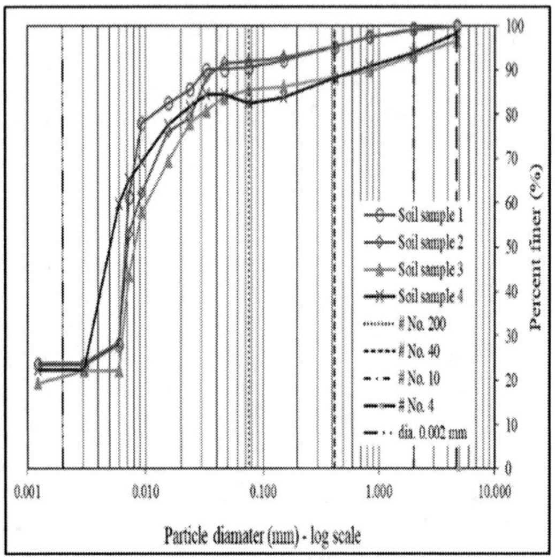


**Fig. 5.** Particle-size distribution for soil samples

The results of tests on soil index properties and soil classification at four locations in the Rimbo Panjang area can be seen in Table 1. From Table 1, it can be seen that the water content value of the original soil is in the range of 35.6% - 58.5%, the volume weight is in the range of 1.4 - 1.6 gr/cm3, and the average specific gravity is 2.6. The plasticity index (PI) value of the soil in the research area is meager, namely in the range of 2.5 – 10.4%, where, according to Chen (1988), soil with a PI < 12% has a low degree of swelling and according to Snethen et al. (1977) if PI < 25% has a low swelling potential classification as well. Based on the USCS, the soil in the Rimbo Panjang area can be classified as high plasticity inorganic silt soil with the group symbol MH. The activity value in inorganic silt soil is meager, namely in the range of 0.1 – 0.5, where, according to Skempton (1953), soil that has an activity value between 0.33 – 0.46 usually contains the mineral kaolinite and soil that has an activity value of 0.99 contains the mineral illite. From the XRD test results, the soil in the Rimbo Panjang area also contains the minerals kaolinite and illite.

**Table 1.** Index properties and classification of soils

| Description | Soil Sample | | | |
|---|---|---|---|---|
| | 1 | 2 | 3 | 4 |
| Moisture content, w (%) | 58.5 | 53.2 | 56.4 | 35.6 |
| Unit weight, g (gr/cm$^3$) | 1.6 | 1.5 | 1.5 | 1.4 |
| Specific gravity | 2.6 | 2.6 | 2.6 | 2.6 |
| Mechanical analysis of soil: | | | | |
| % passing sieve No. 200 | 90.6 | 91.9 | 85.6 | 82.5 |
| Gravel (%) | 0.0 | 0.1 | 3,37 | 1.8 |
| Sand (%) | 9.4 | 8.0 | 11.0 | 15.7 |
| Silt (%) | 67.1 | 68.0 | 63.4 | 60.0 |
| Clay (%) | 23.4 | 23.8 | 22.2 | 22.4 |
| Atterberg Limits: | | | | |
| Liquid Limit, LL (%) | 55.6 | 54.0 | 51.7 | 55.7 |
| Plastic Limit, PL (%) | 45.5 | 44.2 | 49.2 | 45.4 |
| Plasticity Index, PI (%) | 10.1 | 9.8 | 2.5 | 10.4 |
| Classification of soil (USCS) | Inorganic silt high plasticity (MH) | | | |
| Activity (A) | 0.43 | 0.41 | 0.11 | 0.46 |

Figure 6 shows the relationship between normal and shear stress from soil samples taken at four locations in the research area.

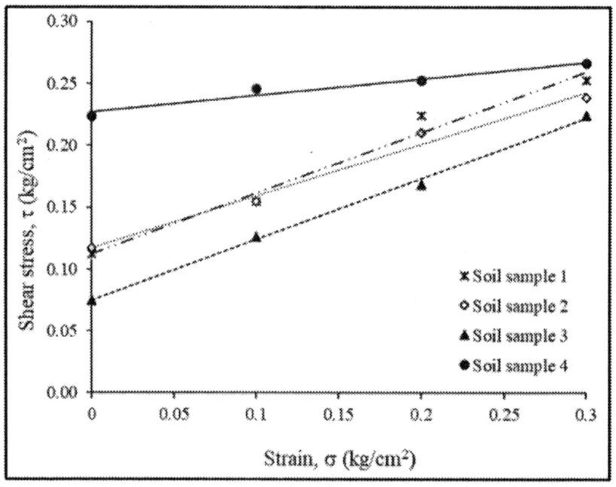


**Fig. 6.** Direct shear plot of soil samples

The results of shear strength tests at four locations showed internal friction angle values in the range of 6.01°-26.16° and cohesion values in the range of 0.08-0.23 kg/cm$^2$ (Table 2). The shear strength of soils is the most critical aspect of geotechnical engineering.

**Table 2.** Shear strength parameters

| Parameters | Soil sample | | | |
|---|---|---|---|---|
| | 1 | 2 | 3 | 4 |
| Cohesion, c (kg/cm$^2$) | 0.11 | 0.12 | 0.08 | 0.23 |
| Angle of friction, $\phi$ (°) | 26.11 | 22.84 | 26.16 | 6.01 |

Figure 7 shows the results of the Scanning Electron Microscope (SEM) test, where you can see the arrangement of groups of colloid-sized particles with variations in grain size with grain diameters smaller than 0.002 mm (2 μm) and grain variations greater than 0.002 mm.

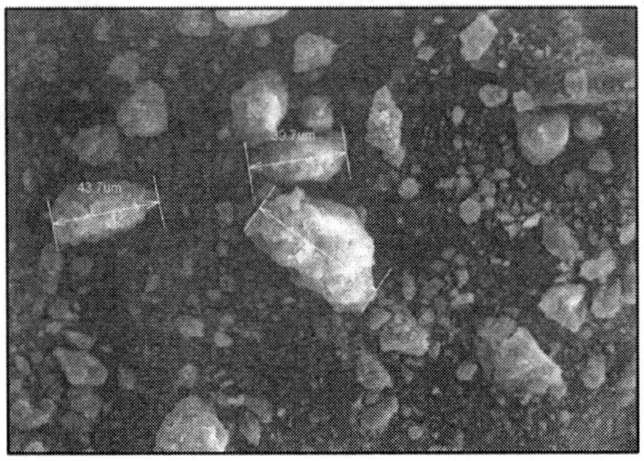

**Fig. 7.** Scanning electron micrograph of a specimen

Figure 8 shows the results of the XRD test on soil in the Rimbo Panjang area of Padang city. The graph shows that the soil in the Rimbo Panjang area quantitatively contains the minerals gibbsite, kaolinite, illite, cristobalite, and quartz.

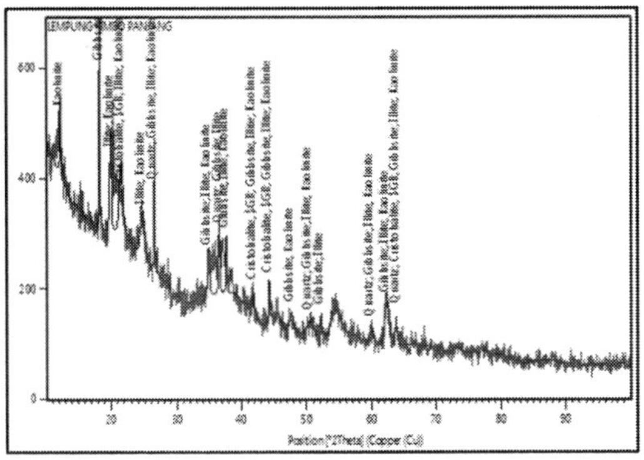


**Fig. 8.** X-ray diffraction of a specimen

Table 3 shows the minerals that dominate quantitatively in the soil in the Rimbo Panjang area.

**Table 3.** Rimbo Panjang x-ray diffraction (XRD) test results

| Mineral name | Compound name | Chemical formula |
|---|---|---|
| Gibbsite | Aluminum Hydroxide | $Al(OH)_3$ |
| Kaolinite | Aluminum Silicate Hydroxide | $Al_2(Si_2O_5)(OH)_4$ |
| Illite | Potassium Aluminum Silicate Hydroxide | $K(Al_4Si_2O_9(OH)_3)$ |
| Cristobalite | Silicon Oxide | $SiO_2$ |
| Quartz | Silicon Oxide | $SiO_2$ |

The XRD test results show that the minerals found in the soil in the Rimbo Panjang area, such as illite and kaolinite, are minerals whose ability to absorb water is small, so their swelling properties are also small. When these minerals absorb water, the soil volume will increase by less. Because little water is absorbed, changes in soil volume are also small. Small volume changes do not affect the stability of the slope and do not damage the strength of the building structure that occupies the soil.

### 3.2 Slope stability analysis

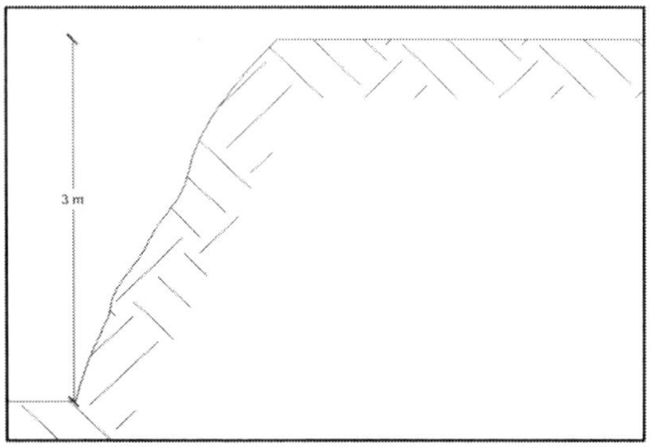


**Fig. 9.** Existing slope conditions

Soil slope stability analysis is generally based on the limit equilibrium approach, plastic limit theory, and numerical methods such as the finite element method. This research uses a limit-balance approach. Figure 9 shows the condition of the existing slope.

Figure 10 shows a circular landslide surface with the soil mass above divided into several vertical slices. By using the bishop method, slope stability analysis can be calculated.

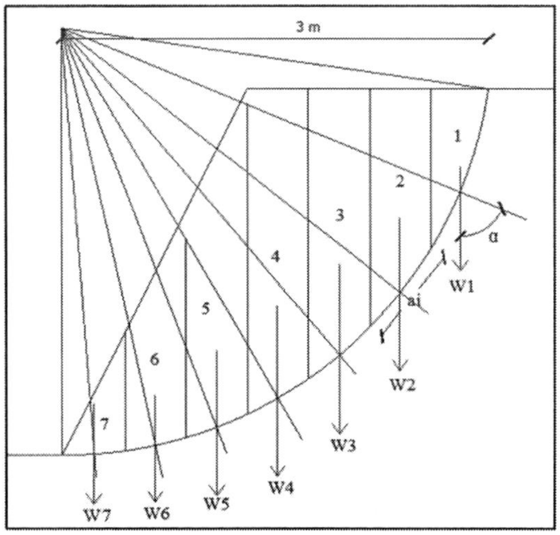


**Fig. 10.** Circular slip surface with vertical slices.

Based on the calculation results for the three failure areas that were tried, the minimum safety factor values were obtained as in Table 4. From Table 4, it can be seen that the minimum safety factor values at the 4 locations are in the range of 1.59 – 2.71; this shows that the slope is stable where FS > 1.5

**Table 4.** Safety factors of the slopes in the Rimbo Panjang area.

| Location | Factors of Safety |
|---|---|
| 1 | 2.56 |
| 2 | 2.32 |
| 3 | 1.59 |
| 4 | 2.71 |

### 3.3 Landslide mitigation

Forms of management and control to mitigate landslide disasters include identifying the causes of landslides and determining the level of landslide vulnerability or susceptibility of an area, which are assessed based on significant factors. The causes and characteristics of slope failure must be understood first before initiating landslide mitigation measures. Many methods have been used to stabilize slopes. Factors that need to be considered in slope stabilization [13]: (1) What is the purpose of stabilizing the slope (2) How much time is available (3) How accessible is the site, and what type of construction equipment can be mobilized there? Access considerations may limit the stabilization methods that can be used; and (4) How much will the repair cost? It makes no sense to stabilize a slope when the costs exceed the benefits. Several methods commonly used for slope stabilization are (1) Drainage; (2) Excavations and buttress fill; (3) Retaining structures (prestressed anchors and anchored, gravity walls, MSE walls, and soil nailed walls, gabions); (4) Reinforcing piles and drilled shafts; (5) Injection methods; (6) Vegetation; (7) Thermal treatment; (8) Bridging; and (9) Removal and replacement of the sliding mass. For slopes in the Rimbo Panjang area, it is recommended that Gabions as landslide protection.

## 4 Conclusion

The results of geotechnical analysis in the Rimbo Panjang area show that the soil in the area is classified as high plasticity inorganic silt soil with the symbol MH group. Soil types are dominated by silt soil in the range of 60% – 67% and clay soil in the range of 22.2% – 23.8%. The soil has a low swelling based on the plasticity index (PI) value, which is 2.5 – 10.4%. The activity value in inorganic silt soil is meager, namely in the range 0.1 – 0.5, which is the activity value for the mineral kaolinite, and the activity value is 0.99 for soil containing the mineral illite. This activity value is in close agreement with the XRD test results, where the soil in the Rimbo Panjang area contains kaolinite and illite minerals, which have a low level of expansion. The low level of development affects the instability of natural slopes, so the slopes in the Rimbo Panjang area have a low potential for landslides. The stability analysis results show that the safety factor value is 1.59 – 2.71 > 1.5, so the slopes in the Rimbo Panjang area are stable. Landslide mitigation must be carried out considering the many factors that cause landslides, such as those caused by excessive infiltration of rainwater into the subsurface layer of the soil, thereby increasing pore water pressure and causing a reduction in the effective shear strength of the soil. Even though the slopes are stable and do not have the potential for landslides, landslide mitigation in the Panjang forest area must be carried out using retaining walls.

This article is part of the research project under PSS scheme funded by LPPM Universitas Andalas in 2024, with contract number 242/UN16.19/PT.01.03/PSS/2024, dated 17 July 2024.

## References

1. O. B. Akintan, J. A. Olusola, O. P. Imole and M. O. Adeyemi, Geotechnical and GIS-based environmental factors and vulnerability studies of the Okemesi landslide, Nigeria, Regional Sustainability, **4**, 249-260 (2023).
2. R. Yuvaraj and B. Dolui, Geographical assessment of landslide susceptibility using statistical approach, Quaternary Sci. Advances, **11** (2023).
3. S. Lee, Current and Future Status of GIS-based Landslide Susceptibility Mapping: A Literature Review, Korean Journal of Remote Sensing, **35**(1), 179-193 (2019).
4. B. Youssef, I. Bouskri, B. Brahim, S. Kader, I. Brahim, B. Abdelkrim and V. Spalevic, The contribution of the frequency ratio model and the prediction rate for the analysis of landslide risk in the Tizi N'tichka area on the national road (RN9) linking Marrakech and Ouarzazate, Catena, **232**, (2023).
5. S. Patil and S. S. Panhalkar, Remote sensing and GIS-based landslide susceptibility mapping using LNRF method in part of Western Ghats of India, Quaternary Science Advances, **11** (2023).
6. W. Wu, S. Guo and Z. Shao, Landslide risk evaluation and its causative factors in typical mountain, Ecological Indicators, **154** (2023).
7. R. Yuliet, A. Hakam, H. G. Putra and R. L. Alfian, Evaluasi kegagalan struktur dinding penahan tanah gravitasi pada lereng tanah kohesif, Jurnal Ilmiah Poli Rekayasa, **18**(2), 53-58 (2023).
8. D. T. S. U. Andalas, Pembuatan Peta Photo Daerah Relokasi Pasca Bencana Rimbo Panjang, Kelurahan Lambung Bukit, Kota Padang, Padang, (2023).
9. Y. Julianto, A. Hakam and R. Yuliet, Analisa stabilitas lereng dan perencanaan penstabilan lereng (Studi kasus Polres Arosuka Kabupaten Solok), Jurnal Teknik Sipil ITP, **7**(1), 34-42 (2020).
10. Badan Standarisasi Nasional, SNI 8460:2017 Persyaratan Perancangan Geoteknik, Jakarta: BSN, (2017).
11. https://bappeda.padang.go.id/uploads/audios/bapped a_padang_62fd097772ce8.pdf, RPJPD Kota Padang Tahun 2005-2025., (Bappeda Padang)

12. T. A. Kobayashi, N. Y. Asufuku, H. E. Nagase, G. Chen, K. Kasama, A. Hirooka, H. Wada, Y. Noyama and H. Uchida, Damage to Residential Retaining Walls at the Genkai-Jima Island Induced by the 2005 Fukuoka-Ken Seiho-Oki Earthquake, Soils and Foundations, **46**(6), 793-804 (2006).
13. H. G. Putra, R. Yuliet and H. Oktavian, Analisis kelongsoran lereng batu apit jalan alternatif Padang-Bukitinggi daerah Sicincin Malalak STA 5+350, Jurnal Rekayasa Sipil, **9**(1),10-23 (2013)

# Cultural disaster: a threat to the preservation of Minangkabau culture

*Oktavianus* Oktavianus[1*], *Nopriyasman* Nopriyasman[2], and *Indirawati* Zahid[3]

[1]Department of Linguistics, Faculty of Humanities, Universitas Andalas, Padang, Indonesia
[2]Department of History, Faculty of Humanities, Universitas Andalas, Padang, Indonesia
[3]Department of Language, Academy of Malay Studies, Universiti Malaya, Malaysia

**Abstract.** Culture is something essential and a marker of the identity of an ethnic group. Therefore, culture must always be part of human life. However, due to various factors, the culture of a tribe is often neglected and which lead to cultural disasters. In this regard, this study examined cultural disaster in Minangkabau. The study used a qualitative and quantitative approach. The respondents were the younger generation of Minangkabau who were studying at university. Discussions with cultural figures, religious leaders, educators, and community leaders were also conducted to know their perspectives on cultural disasters. The study results showed that there has indeed been a shift in Minangkabau cultural values which could lead to a cultural disaster. The lack of instilling cultural values and the ignorance of the collective awareness regarding the importance of knowledge and implementation of cultural values lead to a cultural disaster. The result of the study indicated a potential threat of a cultural disaster in Minangkabau, as indicated by the lack of basic knowledge of the young generation about cultural values in Minangkabau expressions, kinship terms, traditional arts, and culinary. The learning process of Minangkabau cultures can be one of the ways to anticipate the Minangkabau cultural disaster.

## 1 Introduction

Minangkabau is one of the tribes in Indonesia that adheres to a maternal lineage in which the determination of *suku* (clan) is from the mother's line. In this system, Minangkabau culture regards a child as descended from his or her mother's *suku* (clan) [1]. The Minang matriliny has a broad implication for Minangkabau culture, such as the marriage system, the position of Minangkabau men and women in the kinship system and the family, the status and management of inheritance, the use of addressing terms, the way to behave and treat people in the kindship system and even in public spaces.

Minangkabau, as a tribe and region, is rich in culture. Minangkabau is rich in traditional arts such as *Saluang, Rabab, Randai, Pancak Silat,* and various types of *Permainan Anak Nagari* (Anak Nagari Performances) customs, traditional expressions [2], and proverbs [3-5]. Minangkabau is also rich in various Minangkabau culinary delights and specialties. *Rumah Makan Minangkabau*, which people mostly call *Rumah Makan Padang*, spread in West Sumatra, throughout Indonesia, and even abroad [6]. The owners of *Rumah Makan Padang* even appear with all their creativity in cooking and the menu names [7].

The cultural richness of the Minangkabau tribe should be part of everyday life in all actions, attitudes, and behaviors, individually and in groups. Minangkabau people should show their ethnicity as Minangkabau people. Minangkabau people should understand and internalize Minangkabau values within themselves. *The urang Minang* (Minangkabau people) should implement the concept of *Adat Basandi Syarak-Syarak Basandi kitabullah* (Adat is based on religion, religion is based on *kitabullah* (Quran) in their life. Minangkabau people also need to embed the principles of *Alam Terkembang Jadi Guru* (Humans learn from Nature) in all aspects of their lives.

Minangkabau culture is accommodative to other cultures. Cultural contact with other cultures makes the Minangkabau culture dynamic. Many factors cause Minangkabau culture to come into contact with other cultures in the archipelago. Inter-ethnic marriages, the Minangkabau people's pattern of migration for various reasons, and the situation demands in this era of globalization, which make Minangkabau culture have to adapt to the times, are among the factors that make Minangkabau culture have to be dynamic and flexible [8]. The following expressions can be a reflection of the dynamism of Minangkabau culture.

(1) *Tagang bajelo-jelo, kandua badantiang-dantiang,*
(Tight strings are coiled, loose strings are very tight)

(2) *Di mano bumi dipijak, di situ langik dijujuang*
(When in Rome, do what the Romans do)

The expression in 1 state that in doing something, there is room for negotiation and not to state something

* Corresponding author: oktavianus@hum.unand.ac.id

very rigidly. The expression in 2 contains the meaning stating that adaptation to the environment where someone stays, and lives are necessary, especially in building harmonization between the plurality of tribes in Indonesia who have mixed with the Minangkabau tribe both in Minangkabau and overseas.

The openness of Minangkabau culture to the demands of the situation and globalization does not necessarily shift the core values of Minangkabau. The Minangkabau people must maintain the core values of their culture. The strength and the specificity of Minangkabau cultural values are reflected in the following expressions.

(3) *Indak lakang di paneh*
(Not damaged by heat)
*Indak lapuak di ujan.*
(Not decayed by rain)
*Dibubuik indak mati.*
(Pulled out, not dead)
*Dianjak indak layua.*
(Transplanted, not wilted)

Minangkabau people must continuously revitalize and preserve Minangkabau's tangible and intangible cultural heritage because the cultural heritage is the marker of Minangkabau's ethnicity. Minangkabau politeness, wisdom, courtesy, hospitality, discipline, responsibility, and phatic communion are among Minangkabau's core values that must be learned, adapted, and implemented in life by the Minangkabau people.

Violating the fundamental values of *Minangkabau* culture in all its aspects can lead to cultural erosion and eruption. Hamzah quotes the opinion of Dr. Pudentia, Chair of the Oral Tradition Association, stating that cultural disasters are far more devastating than natural disasters [9]. Natural disasters can even come from cultural disasters. Controlling natural disasters has difficulties. However, controlling cultural disasters has its level of complexity. Cultural disasters have long-term effects. Natural disasters are physically real, but cultural disasters are abstract and sometimes intangible.

The ulama and community leaders state their anxiety about cultural disasters in Minangkabau. The noble values behind *Adat Basandi Syarak-Syarak Basandi Kitabullah* (Adat is based on religion, and religion is based on *kitabullah* (Quran)) have become increasingly distant from the Minangkabau younger generation. If allowed, they will face a bleak future in the Minangkabau because they allow these values to fade [10].

Observing the current dynamics of Minangkabau culture, this research is an effort to examine the potential for cultural disasters in Minangkabau society. The study focused on two problems: What are the threats to a cultural disaster in Minangkabau? What strategic steps can be taken to minimize the threats of cultural disasters in Minangkabau?

This study aims to explore and determine the potential for cultural disasters in Minangkabau culture and society. This study also attempts to formulate strategic steps for the Minangkabau people to overcome and minimize cultural disasters in their regions. This study will benefit the future preservation of Minangkabau culture and maintain the authenticity and preservation of the Minangkabau core values, which seek the characteristics of a matrilineal society. Therefore, Minangkabau, as one of the ethnic groups, can keep their ethnicity for a long time.

## 2 Literature review

Many studies have been carried out on Minangkabau culture. In general, we can put the studies about Minangkabau culture into two groups, namely studies that merely reveal the essence of Minangkabau culture and studies about Minangkabau culture and Islam, which is stated in the expression *Adat Basandi Syarak-Syarak Basandi Kitabullah* (Adat is based on religion and religion is based on *kitabullah* (Quran)). These principles have become the proper reference for strengthening and revitalizing the Minangkabau culture for all Minangkabau people.

Studies on the Minangkabau culture was carried out by cultural experts, community leaders, and scientists from various universities in West Sumatra, other regions in Indonesia, and even in the world. Foreign researchers have even carried out studies on Minangkabau [11]. These studies have taken place from the past to the present. The results of this study uncover the uniqueness of Minangkabau culture. They can shape the character of the Minangkabau people as one of the ethnicities that also color the world's culture.

The experts have studied Minangkabau culture through various forms of Minangkabau expressions. This study has explored the cultural values contained in Minangkabau proverbs and expressions. This study has revealed that strengthening the character of the Minangkabau people will occur if the cultural values contained in various forms of expression are implemented optimally in all aspects of life [12, 13]. Otherwise, a cultural disaster will occur. Violating core values such as empathy, sympathy, togetherness, politeness, responsibility, honesty, caring, toughness, and fairness can trigger a cultural disaster.

The experts have also described how Minangkabau people behave, show their work ethic, and build their perspectives [14-17]. In the 1970s, Idrus Hakimy Datuak Rajo Pangulu even had a program broadcasted by Radio Republic of Indonesia in Padang every Tuesday evening after evening prayers with the topic *Mangaji Adaik* (Learning *Adaik*). Kato studied the patterns and attitudes of Migrant Minangkabau people [18, 19].

The significance of Minangkabau culture for every Minangkabau person, even for human culture in general, has been discussed extensively by Hamka in his books published repeatedly in 2014 and 2015 [20-22]. The content of these books is significant for developing a positive community culture. The experts have also studied Hamka's thoughts about Islam and cultural values. These studies stated that Islamic and cultural values can be combined to strengthen human cultures, and They must be implemented in every aspect of human life [23-25].

These studies will provide knowledge to the Minangkabau people, especially the younger generation, so that we can expect cultural strengthening and not a cultural disaster. However, bad events that occur in

Minangkabau land, such as gambling, sexual harassment, theft, corruption, and lack of understanding of the Minangkabau cultural order, are indicators of an impending cultural disaster.

## 3 Method

The research was conducted in Minangkabau West Sumatra Province. We carried out this study using a combination of qualitative and quantitative approaches. This research explored the perceptions, knowledge, and understanding of the young Minangkabau generation regarding essential matters of Minangkabau culture. The researchers sent 25 questions to respondents. The 25 questions included questions regarding the respondent's knowledge of Minangkabau culture in general, such as the kinship system, customs, traditional arts, traditional clothing, knowledge of Minangkabau expressions and proverbs, knowledge and skill of culinary and types of Minangkabau specialty, and the respondent's involvement in various Minangkabau cultural activities.

This study used two groups of respondents. The first group is 50 respondents the age of 20s who were not in the process of studying *Masyarakat dan Kebudayaan Minangkabau* (Minangkabau Culture and Society), with the assumption that they have studied Minangkabau culture from school and their families. The second group consisted of 72 Minangkabau youth taking *Masyarakat dan Kebudayaan Minangkabau* courses (Minangkabau Culture and Society) at Fakultas Ilmu Budaya, Universitas Andalas. The reason for selecting young Minangkabau at this level of education is because they are the spearheads of the guardians and preservers of Minangkabau culture. They are also young people who have been in contact with globalization, so through their views, we can see the existence of Minangkabau culture today. Do they still care about Minangkabau culture, or are they more likely to be swept away by globalization?

We also discussed with cultural figures, experts, community leaders, and several families. We conducted the discussions to learn their opinions and perceptions about Minangkabau culture today. We also discussed their roles and participation in anticipating cultural disasters in Minangkabau society and culture.

## 4 Results and discussion

Based on the analysis of data collected through respondents and discussions with experts, as well as the observations on cultural behaviour and language loyalty in Minangkabau people, there are concerns and threats for cultural disaster in Minangkabau society. Many factors influence it, such as a lack of basic knowledge of various aspects of Minangkabau cultures, cultural pride, and globalization that expose cultures from other regions in other countries. In the following section, we observe the indicators of threats from the responses of the young Minangkabau generation regarding various aspects of Minangkabau culture.

### 4.1 Basic knowledge of Minangkabau culture

#### *4.1.1 Minangkabau language and expression*

Minangkabau language is the first language and mother tongue of most Minangkabau people. At this time, the linguistic situation in Minangkabau society is already heterogeneous. The era of globalization, with all its effects and the presence of social media in all its types, influences the existence of the Minangkabau language today. National and foreign languages are competitors to the Minangkabau language in various communication domains. As a result, the cultural values contained in the Minangkabau language are not widely recognized by the Minangkabau people. The level of the use of the Minangkabau language is presented in Fig. 1.

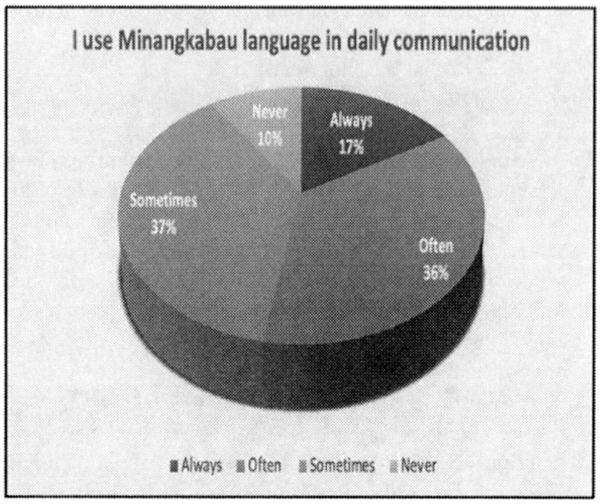


**Fig.1** The use of Minangkabau language

Figure 1 shows that only 17% of respondents said they always use the Minangkabau language when speaking. As many as 36% answered that they often use the Minangkabau language, 37% answered that they sometimes used the Minangkabau language, and 10% never used the Minangkabau language in their daily speech. This figure indicates the potency of the threat to the preservation of Minangkabau culture among young Minangkabau people.

Language has two critical roles in culture: language as part of culture and language as a tool to develop the culture. Clydekaymaben Kluckhohn 1953 formulated seven elements of culture [26, 27]. One element of culture is language. Referring to this idea, we can enrich and strengthen our knowledge of the culture through language mastery. The development and maintenance of culture is closely related to the development and maintenance of language. Language extinction can lead to cultural extinction.

The Minangkabau people have thousands of expressions and proverbs that contain cultural values as a guide to action and behavior. Researchers and experts have documented these expressions and proverbs in books and dictionaries. However, there is a tendency for the younger generation to no longer understand those traditional expressions, as in Fig. 2.

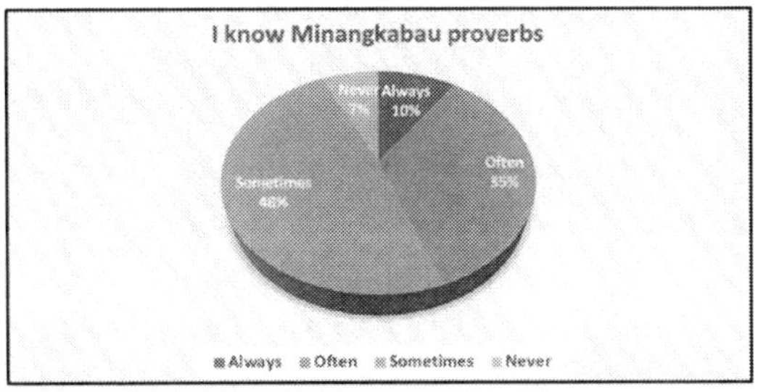


**Fig. 2**. Familiarity with Minangkabau proverbs

Figure 2 shows that only 10% of respondents always recognized Minangkabau proverbs. As many as 35% of the respondents answered that they were often familiar with Minangkabau proverbs and expressions. As many as 48% of the respondents responded that they were sometimes familiar with Minangkabau proverbs and expressions, and 7% were unfamiliar with Minangkabau proverbs.

The range of percentages in Figure 2 must give us an ideal picture of the younger generation's perception of Minangkabau proverbs and expressions. In discussions with respondents, they revealed that they heard proverbs spoken by older people and at traditional events, but they rarely used them. The practical use of the Minangkabau language for daily communication has fulfilled their communication needs so they no longer need to use proverbs and expressions that require much thought to create those proverbs and expressions. Except for forms of proverbs whose source of inspiration is close to their lives, they will use metaphors to express emotions in interactions. Regarding understanding Minangkabau proverbs, we can see the illustration in Fig. 3.

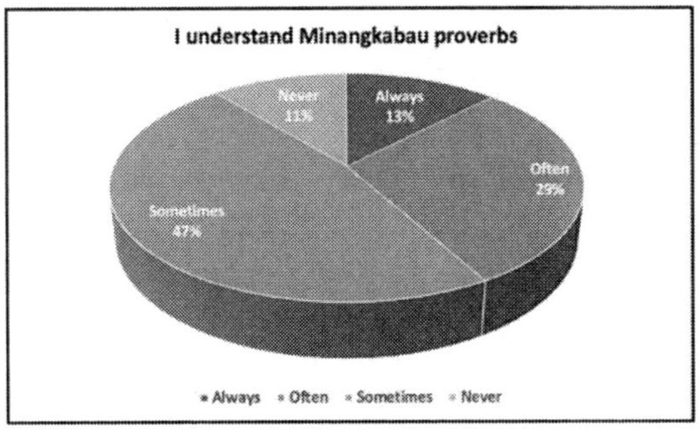


**Fig. 3**. Undertanding Minangkabau proverbs

Figure 3 shows that only 13% of respondents answered that they always understand Minangkabau proverbs. As many as 29% of the respondents said they often understand Minangkabau proverbs. As many as 47% of the respondents said they sometimes understand Minangkabau proverbs. As many as 11% answered that they never understood Minangkabau proverbs. This percentage range also did not portray the future of Minangkabau proverbs among the younger generation. These findings appeared to align with the results of the research conducted by Oktavianus 2003-2005. Of the 240 respondents surveyed based on age group 20-25, with elementary school to universities in all city districts in West Sumatra, the study's results showed different things. The understanding of the Minangkabau proverbs in rural areas was relatively higher than in urban areas. The level of understanding of Minangkabau proverbs in rural areas was in the range of 60-65%, while in urban areas, it was in the range of 25-30%. These findings were almost comparable to current conditions.

Observations in the field based on interactions with the younger generation of Minangkabau both in classes on courses related to Minangkabau and in various public spaces such as markets, schools, universities, hospitals, places of worship, airports, and even in the family realm showed that there was a tendency not to use a complex language like proverbs and expressions. The younger generation's language provides less support for the survival of the Minangkabau language at a certain level. Language has its ecology, such as physical, social, and cultural ecology [28]. If the language speakers learn from language ecology, they will gain a specific language containing specific proverbs. Minangkabau expressions and proverbs or metaphors are complex languages constructed through sources of inspiration from the environment. Closeness to Minangkabau language ecology is one of the ways to learn the Minangkabau language.

### 4.1.2 Kinship terms

The Minangkabau tribe is one of the tribes that adhere to a kinship system based on maternal lineage. We name such a kinship system as matrilineal. The matrilineal kinship system has main elements such as the mother's lineage, marriage must take place outside one's group, and the mother's role in educating, securing, and family welfare [17]. Minangkabau tribe must carry out marriage across or outside their clan. The mother's role as guardian of wealth means protecting an inheritance based on the mother's lineage, and the Minangkabau man cannot bring *Harato Pusako* to his wife's house when he is married.

The matrilineal kinship system has far-reaching implications in Minangkabau culture. Many kinships terms function as markers and labels for each component of relations, both within a clan/tribe and across the clans or tribes. The placement and use of the term must be correct. Mistakes in the placement and use of these terms can have harmful and even fatal consequences in Minangkabau culture. The survey results for the first group of respondents are related to two fundamental questions, as shown in Fig. 4.

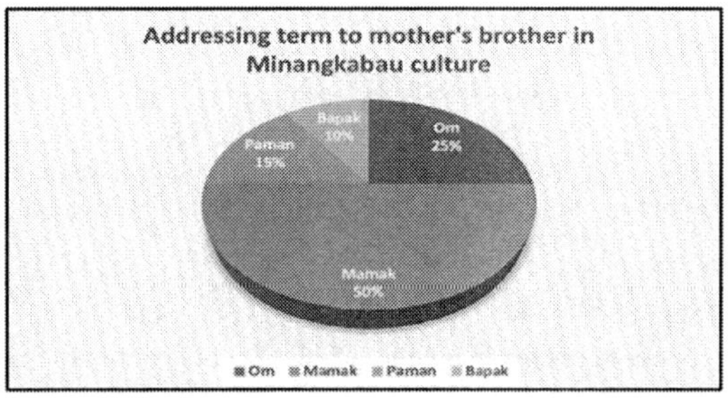


**Fig. 4**. Understanding the kinship terms

As many as 50% of respondents answered that addressing terms for their mother's brother as *Mamak*, 25% as *Paman*, and 15% as *Om*—10% as *Bapak*. As many as 50% of respondents answered correctly. The 15% answer for the use of *Om* is interesting to note. *Mamak* is different from *Paman* and even *Om*. In Minangkabau culture, *Mamak* has his role and responsibilities, which are not the same as those of *Om* and *Paman*. The fatal mistake is when *Mamak* is called *Bapak* during daily interaction. The calling of *Mamak* as *Bapak* has cultural implications in Minangkabau. The Minangkabau cultural leaders agree to categorize it as a cultural disaster.

Then, the understanding of the terms for addressing the father's brother is shown in Fig. 5.

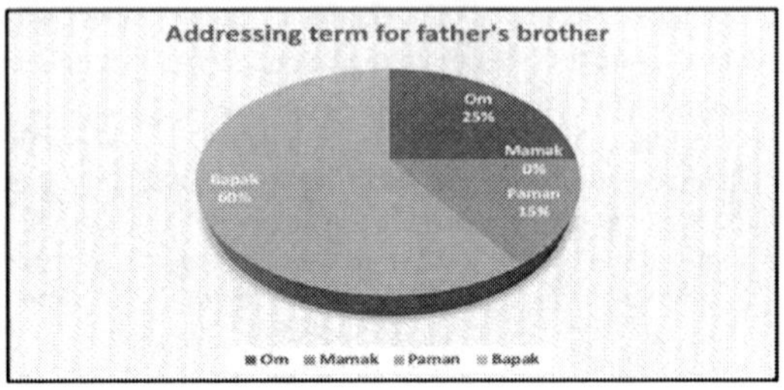


**Fig. 5**. Understanding the kinship terms

**Table 1**. Basic knowledge of kinship term system

| No | Questions about kinship terms | Responses (%) | |
|---|---|---|---|
| | | True | False |
| 1 | Kinship term for the father's sisters | 98.6% | 1.4% |
| 2 | Kinship term for the husband of older or younger sisters | 94.4% | 5.6% |
| 3 | Kinship term for the relationship between a sister's husband and the sister's brother | 87.5% | 12.5% |
| 4 | Kinship term for a sensitive relationship in Minangkabau | 72.2% | 27.8 |
| 5 | Kinship term for the relationship between the older sister's husband and the younger sister's husband. | 88.9% | 11.1 |
| 6 | The name of the children's *Suku* if a man from Suku Caniago marries a woman from Suku Melayu | 90.3% | 9.7% |
| 7 | The name of the children's Suku if a Minangkabau woman from Suku Caniago marries a man from Suku Batak. | 44.4% | 55.6% |

As many as 60% of respondents answered that addressing terms for the father's brother was *Mamak*, 25% answered that it was *Om*, and 15% stated that it was *Paman*. As many as 60% of respondents answered correctly. The 25% answer as *Paman* and 15% *Om* are interesting. They get this addressing term from their mother, relatives, playmates, and social media. Someone needs to remind and correct them that these address terms are inappropriate. They realize they call their mother's brother and father's brother but do not care about appropriately using the terms.

Evaluation at the end of the semester to the 72 respondents who took *Manusia dan Kebudayaan Minangkabau* course showed that 93.1% of the respondents could answer that the addressing term for their mother's brother was *Mamak*, while addressing term for their father's brother was *Bapak*. After the learning process related to Minangkabau culture, we can observe the basic knowledge of respondents to Minangkabau culture in Table 1.

The answers to questions (1) and (2) were 98.6% and 94.4%. Respondents have a good understanding when given illustrations. This kind of learning can reconfirm what they already knew briefly before. The answer to question (3) is 87,5%. Most respondents felt unfamiliar with the terms *Urang Sumando* (sister's husband), *Mamak Rumah* (Wife's brother), and *Urang Kampuang* (Wife's brother). For question number 4, the correct answer from respondents was only 72.2%. Respondents needed to recognize kinship terms such as *Urang Sumando* (sister's husband), *Urang Kampung*, or *Mamak Rumah* (sister's brother). In this case, the sensitive kinship relationship lies in the relationship between the *Urang Sumando* (the sister's husband) and the *Mamak Rumah* (the wife's brother). In some areas, *Mamak Rumah* is called *Urang Kampuang* because when a man is married, his wife's village becomes his village, too. Another implication is that in the communication and interaction between *Urang Kampung* (the wife's brother) and *Urang Sumando* (the sister's wife), the ideal language they can use is *kiasan* (implied meaning) [3, 13].

In terms of the use of language, Minangkabau has the concepts of *Kato nan Ampek*, namely *Kato Mandaki, Kato Manurun, Kato Malereang*, and *Kato Mandata* [2]. Minangkabau people should use *Kato Mandaki* to talk to older people. *Kato Manurun* is used to talk to those whose position is lower than the speaker's, such as parents to their children. *Kato Mandata* is used to talk to friends. Minangkabau people use *Kato Malereang* or *Kiasan* (Indirect language) to talk to those who have sensitive kinship relationships, such as *Mamak Rumah* (wife's brother) to *Urang Sumando* (sister's husband).

The answers to questions (5) and (6) were 88.9% and 90.3%. Respondents generally understand the terms *Pambayan, Dunsanak Pambayan* kinship relations in Minangkabau. These addressing terms are used to name the relationship between the older sister's husband and the younger sister's husband in case one nuclear family has two or three married women.

In some parts of Minangkabau, people are named *Pambayan* as *Kawan Sarumah*, though they do not live in the same house. However, some respondents still need to learn these addressing terms. In Minangkabau culture, people can refer *Pambayan* or *Dunsanak Pambayan* sometimes to a sensitive kinship relationship. Economic inequality and social status can also lead to conflict in specific contexts. Therefore, *Mamak Rumah* and *Mintuo* must be wise in treating *Urang Sumando* or *Minantu* if in their *kaum* there are *Minantu* who is *Badunsanak Pambayan*.

The answer to question 7 was 44.4% correct and 55,6% wrong. Inter-ethnic marriages cause the Minangkabau young generation to have more knowledge of their identity. The mistake of using kinship terms for mother's brother and father's brother even comes from this group. It became worse when they lived in *perantauan* [18]. The lack of attention of the nuclear family to Minangkabau culture causes the younger generation's Minangkabau knowledge to erode increasingly. It, of course, could lead to a cultural disaster.

### *4.1.3 Anak nagari performances*

Minangkabau has traditional arts such as *Saluang, Rabab, Randai, Pancak Silek, Tari Minangkabau, Takok-Taki.* Minangkabau singing is spread and famous everywhere. Regarding religion, Minangkabau has *Salawek Dulang* and *Dikia /Rabanna*. These all include cultural elements that the Minangkabau people must preserve as cultural assets. This study also asked several questions about the Minangkabau generation's initial perception of traditional art and games for *Anak Nagari*. We can see the results of the survey in Fig. 6.

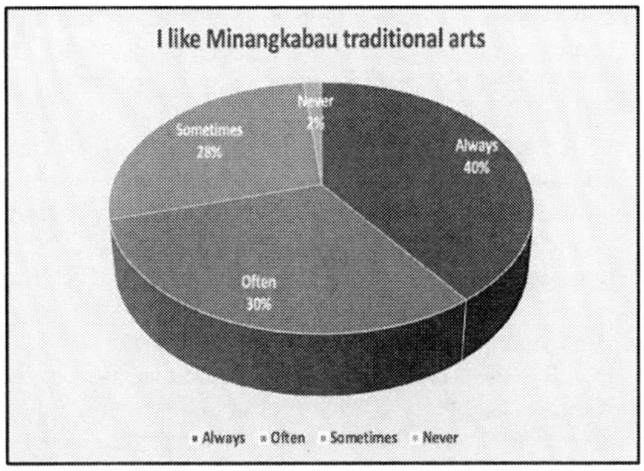


**Fig. 6.** Responses to Minangkabau traditional arts

Figure 5 indicates that 40% of the younger generation of Minangkabau people have always liked traditional Minangkabau arts. As many as 30% said often, and 28% said sometimes. As many as 2% of the respondents stated that they do not like (never) Minangkabau traditional arts. The percentage of the answers as *often* and *sometimes* indicated that respondents must pay more attention to traditional Minangkabau arts. The observation in the fields confirmed that there was a tendency for the younger generation not to listen to *Saluang*, *Rabab*, and *Randai*. The contributing factors include the topic and the story's content, which they consider less relevant to the current time. It is challenging for the Minangkabau people to revitalize Minangkabau traditional arts through younger generations.

### *4.1.4 Culinary and cultural activities*

Minangkabau is rich in cultural activities and various culinary delights. The preservation of Minangkabau culture will also be seen from the extent to which Minangkabau cuisine is typical and the extent to which the young Minangkabau generation wants to involve themselves in various cultural activities. We can observe the Minangkabau generation's perception of culinary and cultural activities in Table 2.

**Table 2.** Questions on culinary and cultural activity

| No | Culinary and cultural activity | Responses | | | |
|---|---|---|---|---|---|
| | | Always | Often | Some times | Never |
| 1 | I like Minangkabau's Culinary | 75% | 20% | 4.5% | 0% |
| 2 | I can cook specific Minangkabau food | 8% | 29.2% | 43.1% | 19.4% |
| 3 | I participate in Minangkabau cultural activities | 4.3% | 20.8 | 61.1% | 13.9% |

Table 2 shows us that the level of respondents' liking for Minangkabau cuisine is relatively high. As many as 95% of respondents answered that they *always* and *often* like Minangkabau culinary delights. There were no respondents who stated that they did not like Minangkabau culinary delights. It is a positive indicator for the development and sustainability of the Minangkabau culinary industry. However, the number of respondents who were constantly good at cooking Minangkabau food was 37.2%. It indicates that the majority of respondents need to possess cooking skills. It is a challenge to maintain Minangkabau culture from the culinary sector. Among the contributing factors is the ease of getting various types of food in public places, so there is no need to cook anymore. An unfortunate consequence of this phenomenon is that people must pass on Minangkabau cooking skills to most young people.

The involvement of the younger generation in cultural activities, who answered *always* and *often* was also low, namely 25.1%. This figure is also relatively low. This low figure is due to, among other things, the younger generation's lack of skills in various cultural events, their dislike and disinterest, and the emergence of alternative games and entertainment presented via the internet. Trends of this kind also pose a threat of a gradual cultural disaster.

### 4.2 Cultural disaster in Minangkabau

Nowadays, we widely hear people use the term *Cultural Disaster*. The word *disaster* collocates with *Nature* to create the term *Natural Disaster*. A sudden calamitous disaster can bring significant damage, loss, and destruction (to the environment) (https://www.merriam-webster.com/dictionary/disaster). However, we can use the word *disaster* to refer to cultural disaster, which means culture's damage, loss, and destruction.

Some findings of this study can be indicators of a cultural disaster in the future. A lack of understanding of the cultural values contained in thousands of Minangkabau expressions and proverbs can weaken Minangkabau culture, ultimately leading to Minangkabau culture becoming extinct. Minangkabau customs and culture, as outlined in the *Adat Basandi Syarak-Syarak Basandi Kitabullah* (Adat is based on religion, religion is based on *kitabullah* (Quran)), completed by the principle of *Alam Takambang jadi guru* (Humans learn from Nature) have regulated human relations in two dimensions, namely, human relations with the Creator and human relations with Allah's creations. The Islamic religion is the religion of most of the Minangkabau tribe and has even incorporated two essential elements in life: *Ihsan* toward Allah and *Ihsan* all His creations.

The lack of understanding of the kinship system in Minangkabau can also lead to cultural disasters. The kinship system is not just a term but has broad cultural implications. To maintain harmony in Minangkabau culture, we must treat everyone who adheres to a particular kinship system in a certain way. For example, *Urang Sumando* must treat the *Mamak Rumah*, *Bapak*, *Adiak*, *Orang Tuo*, *Pambayan*, and *Mintuo* differently in social and cultural interactions. If he successfully, does it, people call him *Urang Sumando Niniak Mamak* (Good Urang Sumando). However, if he fails to behave appropriately with those he interacts with, people call him *Urang Sumando Langau Ijau* (Bad Urang Sumando). An ideal language in this situation is the contextual language [29, 30]. If someone fails to treat another person appropriately, people ridicule, scold, or label him with the expressions in Table 3.

**Table 3**. Negative expressions

| No | Expressions |
|---|---|
| (1) | *Ndak tau di adaik*<br>(Stupid) |
| (2) | *Indak bapatang sawah di inyo*<br>(Stupid) |
| (3) | *Indak tau di atah takunyah*<br>(Stupid) |
| (4) | *Ongok*<br>(Stupid) |
| (5) | *Indak bapangana*<br>(Stupid) |
| (6) | *Barangin*<br>(Stupid) |
| (7) | *Agak barangin*<br>(Rather stupid) |
| (8) | *Ndak amuah baraja*<br>(Stone headed) |
| (9) | *Ndak bautak*<br>(Stupid) |

The lack of knowledge of the younger generation from various cultural activities is also a threat to Minangkabau culture and can lead to cultural disaster. Cultural activities require skills and knowledge. The younger generation of Minangkabau must learn it. These cultural activities can have multiple roles, such as entertainment, economic empowerment, and preserving Minangkabau culture. Formatting cultures can be transformed into tourism destinations, as seen in Bali and other tourist destinations in Indonesia.

### 4.3 Cultural preservation

The government and related parties have made much effort to preserve and maintain. The government has enacted Law of the Republic of Indonesia Number 5 of 2017 concerning the Advancement of Culture. Article 4 of Law Number 5 of 2017 states that the aims of advancing culture cover developing the noble values of national culture, enriching cultural diversity, strengthening national identity, strengthening national unity and integrity, enlightening the nation's life, improving the nation's image, realizing civil society, improving people's welfare, preserving the nation's cultural heritage, and influencing the direction of development of world civilization. Article 5 of Law Number 5 of 2017 states that the objects of cultural advancement include oral traditions, manuscripts, customs, rites, traditional knowledge, traditional technology, art, language, folk games, and traditional sports [31]. The government has even established cultural institutions throughout Indonesia as a follow-up to this law. We expect this effort to be part of preservation steps to avoid cultural disasters.

However, top-down efforts will only be successful if bottom-up efforts in collective awareness from all related components also support the government's actions. In the context of advancing Minangkabau culture, all components must build collective awareness to advance Minangkabau culture in all aspects. Minangkabau children exist and stay in all sectors of life, such as at home and in places where they study, such as schools, universities, madrasas, places of worship, and various public places. Cultural concern and caring for each other must be maintained and revitalized continuously to avoid cultural disasters. Minangkabau is rich in oral stories containing cultural values to teach the lesson [32]. Besides, the Minangkabau people are also rich in local wisdom to preserve their culture and avoid cultural disasters that can lead to natural disasters [33]. Humans must understand the philosophy of life in order to live happily, understand and solve the secrets of life by referring to religion and considering socio-cultural aspects [34].

## 5 Conclusion

Based on the study conducted on the current condition of Minangkabau culture, this research can derive seven conclusions. *First*, the results of data analysis show that the threat of a cultural disaster in Minangkabau could occur. *Second*, we can see this indicator from the tendency of the young generation's low understanding of Minangkabau culture as the object of this study. *Third*, the government has carried out efforts to preserve Minangkabau culture by enacting cultural advancement laws and other top-down efforts by the government. *Fourth*, these efforts will succeed if the relevant parties have collective and cultural awareness. Strengthening the Minangkabau language with all the values contained in expressions and proverbs is a step that language teachers, linguists, and other related parties must do continuously.

*Fifth*, society must continuously involve the Minangkabau young generation in various cultural events. *Sixth*, Minangkabau cultural learning patterns and strategies must constantly be updated to adapt to the current situation. We can utilize current technological advances to advance Minangkabau culture, so we do not just blame globalization for triggering cultural disasters. *Seventh*, the digitalization of various languages related to the Minangkabau culture is a positive step, so globalization is not always the scapegoat for the decline of a culture. Globalization spreads technology quickly in order to help human beings survive.

This research was conducted with the help and support of respondents, experts, and Fakultas Ilmu Budaya Universitas Andalas. Therefore, it is our pleasure to thank and appreciate the related parties. This research was conducted with a team of three researchers. Oktavianus, as the first author, designed the research and collected and analyzed the data. Nopriyasman, as the team members, collected and analyzed the data. The three of us work together to formulate the draft and final version of the article.

## References

1. R. Yatim. ADAT: The Legacy of Minangkabau. (Kuala Lumpur: Yayasan Warisan Negara, 2015).
2. A. B. Dt. Madjo Indo. Kato pusako:papatah, patitih, mamang, pantun ajaran, dan filsafat Minangkabau. (Jakarta. PT. Rora Karya, 1999).
3. 3.A.A Navis. Alam Terkembang Jadi Guru: Adat dan Kebudayaan Minangkabau. (Jakarta: Grafiti Press, 1984).
4. A. Nafis. Peribahasa Minanngkabau. (Jakarta: Penerbit Intermasa, 1996).
5. I. Fanany and R. Fanany. Wisdom of the Malay Proverbs. (Kuala Lumpur: Dewan Bahasa dan Pustaka, 2003).
6. O. Oktavianus. Landscape of Cultural Identity at Rumah Makan Minang. JATI: Journal of Southeast Asian Studies, **24**(2), 155-175, (2019). https://doi.org/10.22452/jati.vol24no2.9
7. A. Hadi. Menu Restoran Padang dengan Nama Partai Yang Lagi Trending. Harian Kami.Com (2024). https://www.hariankami.com/gaya-hidup-kami/23611955165/menu-restoran-padang-dengan-nama-partai-yang-lagi-trending. Retrieved, 16/05:2024. 21:45.
8. M. Naim. Budaya Minangkabau di Era Globalisasi. (Artikel Media Online Palanta Minang, 2008). https://palantaminang.wordpress.com/2008/04/01/budaya-minangkabau-di-era-globalisasi/
9. M. Hamzah. Bencana Alam Kecil Dibandingkan Bencana Budaya. (Berita Satu, 2012). https://www.beritasatu.com/news/73825/bencana-alam-kecil-dibandingkan-bencana-budaya
10. M. Abidin. Tiga Sepilin: Surau Solusi untuk Bang-sa. (Yogyakarta: GrePublishing, 2016).
11. F.K Errington. Manners And Meaning in West Sumatra: The Social Context of Consciousness. (New Haven and London: Yale University Press, 1984).
12. O. Oktavianus. Bertutur Berkias dalam Bahasa Minangkabau. Denpasar. Disertasi Doktor di Universitas Udayana, (2005).
13. O. Oktavianus. Kiasan dalam Bahasa Minang-kabau. (Padang: Minangkabau Press, 2022).
14. I. Hakimy, I. Rangkaian Mustika Adat Basandi Syarak di Minangkabau. (Bandung: Bandung: Remaja Rosdakarya, 1991).
15. I. Hakimy. 1000 Pepatah-Petitih-Mamangan-Bi-dal-Pantun-Gurindam. Bandung: Remaja Karya, 1996).
16. I. Hakimy. Pokok-Pokok Pengetahuan Adat Alam Minangkabau. (Bandung: Remaja Rosdakarya, 1994).
17. A. M.S. Adat Minangkabau: Pola dan Tujuan Hidup Orang Minang. (Jakarta: Citra Harta Prima, 2016).
18. T. Kato. Adat Minangkabau dan Merantau dalam Perspektif Sejarah. (Jakarta: Penerbit Balai Pustaka, 2005).
19. M. Naim. Merantau: Pola Migrasi Suku Minang-kabau. (Yogyakarta: RajaGrafindo Persada, 2012).
20. Hamka. Peribadi Hebat. (Jakarta: Gema Insani, 2014).
21. Hamka. Falsafah Hidup: Memecahkan Rahasia Kehidupan Berdasarkan Tuntunan Al Qur'an dan As Sunnah. (Jakarta: Penerbit Republika, 2015).
22. Hamka. Lembaga Hidup: Ikhtiar Sepenuh Hati Memenuhi Ragam Kewajiban untuk Hidup Sesuai Ketetapan Ilahi. (Jakarta: Penerbit Republika, 2015).
23. F. Ferdinal, O. Oktavianus and I. Zahid. Exploring the Beauty of Islamic Values Through Metaphorical Expressions in Literary Work. Afkar: Jurnal Akidah; Pemikiran Islam, **25**(2), 421–458, (2023). https://doi.org/10.22452/afkar.vol25no2.13
24. O. Oktavianus. Lanskap Linguistik Metafora Ba-hasa Minangkabau dalam Dakwah Agama Islam. (Prosiding Seminar Nasional Bahasa, Sastra dan Budaya. Fakultas Ilmu Budaya Unud, 2023). https://ejournal1.unud.ac.id/index.php/snbsb/article/view/755/595.
25. O. Oktavianus, F. Ferdinal and I Zahid. Metafora Leksikal dan Nilai Budaya dalam Novel Hamka Di Bawah Lindungan Ka'bah. Pertanika: Social Sciences and Humanities, 32(S1): 111 – 137, (2024). https://doi.org/10.47836/pjssh.32.S1.05.
26. M.J. Herskovits and C. Kluckhohn. A Biographical Memoir. (Washington D.C: National Academy of Science, 1964).
27. Koentjaraningrat. Kebudayaan, Mentalitas dan Pembangunan. (Jakarta: Gramedia, 1993).
28. A. Fill and P. Muhlhausler, (Eds). The Eco-linguistics Readers: Language, Ecology, and Environment. (London: Continuum, 2001).
29. Thomas, J.A. Meaning in Interaction: Introduction to Pragmatics (1st Edition). Routledge: Tylor and Francis), 1995).
30. G. Yule. Pragmatics. Oxford: Oxford University Press, 2000).

31. Undang-Undang Republik Indonesia No. 5 Tahun 2017 tentang Pemajuan Kebudayaan. https://kebudayaan.kemdikbud.go.id/bpcbsumbar/wp-content/uploads/sites/28/2019/04/UU-No-5-Tahun-2017-tentang-Pemajuan-Kebudayaan.pdf
32. N. Nopriyasman, G. Asnan, A. Fauzi, I.P. Hastuti, A. H. Ritonga, V. Kurniawan, and R. Mairiska, Reading Indigenous signs: The wisdom of Nagari communities toward natural disaster in Pasaman Barat. International Journal of Disaster Risk Reduction (IJDRR), **107**(1), 1-6, (2024). https://doi.org/10.1016/j.ijdrr.2024.104497.
33. K. Anwar, O. Oktavianus, V.K. Khanh., and V.Q.A. Khanh. Disaster mitigation through oral stories: a case study around Singkarak Lake. In E3S Web of Conferences 464, 16002. EDP Sciences, (2023). https://doi.org/10.1051/e3sconf/202346416002
34. 34. Hamka. Dari Hati ke Hati: Tentang Agama, Sosial-Budaya dan Politik. (Jakarta: Pustaka Panjimas, 2005)

# Design of MSE Wall with geotextile reinforcement for temporary mitigation of landslide in Padang Panjang – Sicincin road section (STA 64+100)

*Muhammad* Saddam[1], *Azwin*[1], *Vivi* Anggraini[2], and *Rina* Yuliet[1*]

[1]Department of Civil Engineering, Faculty of Engineering, Universitas Andalas, Padang 25163, Indonesia
[2]Department of Civil Engineering, School of Engineering, Subang Jaya, Malaysia

**Abstract.** The Padang Panjang - Sicincin roads section was cut off due to a flash flood on Saturday, 11 May 2024. This caused a landslide in the area resulting in the national road access being cut off, and traffic flow was forced to be diverted to several alternative routes. The study aims to obtain the total length of geotextile at any depth (L), the vertical spacing of the layers ($S_v$), and the lap length ($l_l$) in MSE wall heights at STA 64 + 100. The computation method for external and internal stability (pull out and break), uses the Load Resistance Factor Design (LRFD). From the calculation results for the effective height ($H_e$) of the MSE wall of 8 m, the reinforcement length is 6 m, the embedment depth is 0.6 m, the vertical spacing ($S_v$) is 0.8 m from the internal stability, and the fold length ($l_l$) is 1 m. The design results obtained MSE wall dimensions that are safe for external and internal stability. These dimensions can be used for temporary landslide handling but not long-term. For long-term stability, adding facing elements like secant piles, cantilever walls, and other forms is recommended to avoid erosion due to river water flow.

## 1 Introduction

The main road section on the Padang Panjang – Sicincin city was cut off due to flash floods that caused landslides on the area's road slopes, so traffic flow was forced to be diverted to several alternative routes. From the results of the field visit, information was obtained that STA 64-100 was one of the road sections that experienced landslides, so the STA 64+100 road section was selected for this study, as shown in Fig. 1. The design of the MSE wall associated with the geotextile as the reinforcement was carried out to temporarily mitigate landslides on the Padang Panjang - Sicincin boundary at (STA 64+100). The design results of the optimal MSE wall dimensions can be used as a reference for designing landslide mitigation in other areas under the same conditions.

The MSE walls consist of a face cover and steel or geosynthetic reinforcements tied to the face cover and installed in layers in a granular soil embankment that quickly drains water. The combination of reinforcement and granular soil embankment produces an internally stable composite structure [1]

The MSE walls are an alternative to conventional retaining walls, such as gravity retaining walls and cantilever walls, which are widely used for road construction. MSE walls are prone to landslides and retain soil on slopes close to vertical to minimize the width of the road embankment area, especially on high embankments

**Fig. 1.** Landslide location in Padang Panjang-Sicincin Road section (STA64+100) [2]

This study can provide information on the design of MSE walls by the Indonesian National Standard (SNI) 8460-2017. Other researchers have used SNI 8460-2017 to design MSE walls [3]. Research on the design of MSE walls with Geotextile reinforcement has been widely conducted, but many still need to refer to SNI 8460-2017; this can be seen from the references used by researchers [4-7]. After the publication of SNI 8460-2017, there are still researchers who have not used SNI 8460-2017 for the

* Corresponding author: rina@eng.unand.ac.id

design of MSE walls [8], and there are researchers who use other standards from outside Indonesia, such as the FHWA standard [9]. Research on slope landslides and handling of slope landslides with Retaining Walls without geotextiles has been conducted by the research team since 2012 and has been published in several national journals [10-13].

The primary purpose of using reinforcement is to increase the stability of steep slopes for slope improvement and to increase the stability of the surface. Fig. 2 shows three failure modes in reinforced slopes, namely internal, external, and combined failure.

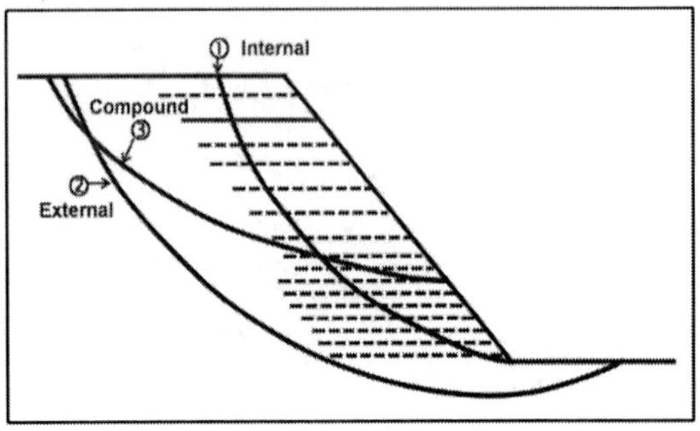


**Fig. 2.** Failure modes for reinforced soil slopes (FHWA) [14]

## 2 Research methods

### 2.1 Field survey

The field survey was carried out to describe the area experiencing the landslide and to measure the geometry of the existing slope.

### 2.2 Data sources

The secondary data obtained is information about the dominant material used for the MSE wall, namely, 100 kN woven geotextile, sand, and gravel mixture as material backfill taken from the Sicincin quarry area and compacted with a relative compaction of 96.86% ($\gamma_r$ = 23.6 kN/m$^3$).

### 2.3 Analysis of MSE Walls

#### *2.3.1 Establish project requirements*

In this section, the geometry of the MSE wall is determined, including the effective height of the MSE wall, the length of the MSE wall, service life, type of reinforcement, earthquakes, and the presence of groundwater. According to SNI 8460 2017, the required reinforcement length is L $\geq$ 0.7 $H_e$, where $H_e$ is the effective height of the MSE wall. Regardless of the effective height, L must be $\geq$ 2.5 m (Fig. 3). The length of the reinforcement must be the same for the entire height of the wall except for certain things, such as adding length and reducing the length of the reinforcement. The minimum reinforcement length also increases due to the work of external loads and the softness of the foundation soil. The vertical spacing for MSE walls with geotextile reinforcement is 0.2 m – 1.25 m. This spacing can change with depth.

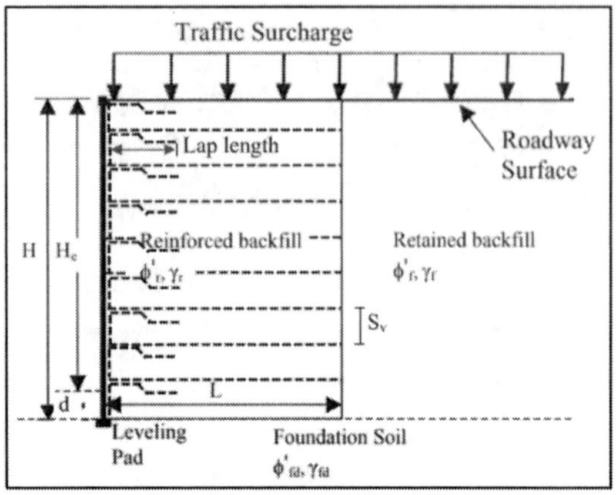


**Fig. 3.** The parameters for analysis of an MSE wall [14]

#### *2.3.2 Evaluate project parameters*

The soil parameters of shear strength and unit weight of embankment on reinforcement, embankment parameters behind reinforcement, and foundation soil parameters are required for MSE wall design requirements. The bearing resistance factor of the foundation for consideration of service limit and consideration of strength limit is assumed to be 359 kN/m$^2$ and 502 kN/m$^2$, respectively. Traffic and off-road load (q) are determined based on the road class I [2].

#### *2.3.3 Estimate the depth of embedment and length of reinforcement*

The facing element in this study is a flexible facing element made of geosynthetic material. The geosynthetic facing element is installed by bending the geosynthetic 180° and returning it to the back to form a sloped surface. The minimum embedment of the facing element (d) for the slope in front of the horizontal wall is $H_e/20 > 0.35$ m [2, 14].

#### *2.3.4 Estimate unfactored loads*

The computation of the stability of the MSE wall is carried out by considering the mass of the MSE wall as a rigid body. The active earth pressure coefficient can be calculated as follows:

$$K_a = \tan^2\left(45 - \frac{f'_f}{2}\right) \tag{1}$$

The vertical forces:

$$V_1 = g_r HL \tag{2}$$

$$V_s = q_t\, L \tag{3}$$

The horizontal forces:

$$F_1 = \frac{1}{2} K_{af}\, g_f\, H^2 \tag{4}$$

$$F_2 = K_{af}\, q\, H \tag{5}$$

The components of the forces acting on the MSE wall can be seen in Fig. 4.

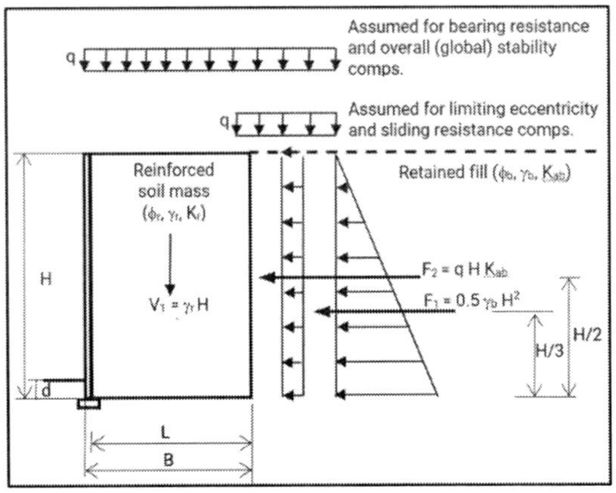


**Fig. 4.** Components of forces and moments for external stability [2, 14]

### *2.3.5 Load factors*

Table 1 shows the load factors, and Table 2 shows the resistance factors used in the LRFD method.

**Table 1.** The load factors [14]

| Load combination | Load factors | | |
|---|---|---|---|
| | EV | EH | LL |
| Strength I (max.) | 1.35 | 1.5 | 1.75 |
| Strength I (min.) | 1 | 0.9 | 1.75 |
| Service I | 1 | 1 | 1 |

**Table 2.** The resistance factors [14]

| Item | Resistance factors |
|---|---|
| Sliding ($\phi_s$) | 1 |
| Bearing ($\phi_b$) | 0.65 |
| Tensile ($\phi_t$) | 0.65 |
| Pullout ($\phi_p$) | 0.9 |

### *2.3.6 Evaluate the external stability*

In the LRFD method, the soil load is symbolized by EV and EH, while the live load is symbolized as LL. The forces acting can be used to calculate the external stability of the MSE wall.

### *2.3.7 Evaluate internal stability*

Internal failure of MSE walls can occur in two different failure modes, but both cause significant movements in the MSE wall structure, leading to the collapse of the wall structure. The two internal failure modes are:

1. Failure of the reinforcement material, namely excessive elongation or breaking of the
2. reinforcement, due to high tensile forces on the reinforcement. (FS break > 1.5)
3. Failure due to removing reinforcement from the embankment due to the high tensile force on the reinforcement (FS pullout > 1.5).

## 3 Results and discussion

The following project requirements were set for designing the MSE wall:

- Height of wall, $H_e$ = 8 m
- Length of wall = 20 m
- Design life = 75 years
- Type reinforcement = geotextile woven
- No seismic considerations
- No groundwater influence

Soil parameters used in MSE wall analysis:

- Reinforced backfill, $\phi'_r = 43°$, $\gamma_r = 23.6$ kN/m$^3$
- Retained backfill, $\phi'_f = 43°$, $\gamma_f = 23.6$ kN/m$^3$
- Parameters of soil, $\phi'_{fd} = 43°$, $\gamma_{fd} = 23.6$ kN/m$^3$
- Live load surcharge = 10 kN/m$^2$ (Table 1)
- Traffic surcharge = 15 kN/m$^2$ (Table 1)

Minimum embedment depth (d) = $H_e/20$; for this design, assume d = 0.6 m; thus, H = $H_e$ + d = 8.6 m. The minimum initial length is assumed to be 0.7 H or 6 m.

Based on equation (1), the lateral earth pressure coefficient is:

$$K_{ar} = \tan^2\left(45 - \frac{34^o}{2}\right) = 0.189 \tag{6}$$

$$K_{af} = \tan^2\left(45 - \frac{30^o}{2}\right) = 0.189 \tag{7}$$

Calculate $V_1$ using equation (2) and $V_s$ using equation (3). The results of the calculation of vertical forces and moments can be seen in Table 3.

**Table 3.** Unfactored vertical forces and moments

| | Force (kN/m) | Moment arm (m) | Moment (kN.m) | LRFD load type |
|---|---|---|---|---|
| $V_1$ | 1,218 | 3 | 3,653 | EV |
| $V_s$ | 150 | 3 | 450 | LL |

Calculate $F_1$ using equation (4) and $F_2$ using equation (5). The results of the calculation of horizontal forces and moments can be seen in Table 4

**Table 4.** Unfactored horizontal forces and moments

| | Force (kN/m) | Moment arm (m) | Moment (kN.m) | LRFD load type |
|---|---|---|---|---|
| $F_1$ | 165 | 2.87 | 473 | EH |
| $F_2$ | 41 | 4.30 | 175 | LL |

Table 5, Table 6, and Table 7 show the computation of external stability (sliding resistance, overturning/limiting eccentricity, and bearing capacity resistance).

**Table 5.** Computations of sliding resistance

| Item | Strength I (max) | Strength I (min) |
|---|---|---|
| Lateral load ($H_m$), kN/m | 319 | 220 |
| Vertical load ($V_1$), kN/m | 1,644 | 1,218 |
| Nominal sliding resistance ($V_{Nm}$), kN/m | 1,533 | 1,136 |
| Sliding resistance ($V_{Fm}$), kN/m | 1,533 | 1,136 |
| Is $V_{Fm} > H_m$? | YES | YES |
| Capacity Demand Ratio (CDR) | 4.8 | 5.2 |
| Minimum sliding resistance ($V_{Fmin}$), kN/m | 1,136 | |
| Maximum lateral load (Hmmax), kN/m | 319 | |
| Is $V_F$min > Hmmax | YES | |
| Critical Capacity Demand Ratio ($CDR_{critis}$) | 3.56 | |

**Table 6.** Computations of limiting eccentricity

| Item | Strength I (max) | Strength I (min) |
|---|---|---|
| Total vertical load ($V_A$), kN/m | 1,644 | 1,218 |
| Resisting moments ($M_{RA}$), kN-m/m | 4,932 | 3,653 |
| Overturning moments (M), kN-m/m | 1,015 | 732 |
| Net moment ($M_A$), kN-m/m | 3,917 | 2,922 |
| Location of the resultant force (a), m | 2.38 | 2.40 |
| Eccentricity ($e_L$), m | 0.63 | 0.61 |
| Limiting eccentricity (e), m | 1.5 | 1.5 |
| Is $e_L \leq e$? | Yes | Yes |
| Calculated $e_L$/L | 0.10 | 0.10 |
| Maximum overturning moments ($M_{OA\text{-}C}$), kN-m/m | 1,015 | |
| Minimum resisting moments ($M_{RA\text{-}C}$), kN-m/m | 3,653 | |
| Critical net moments ($M_{A\text{-}C}$), kN-m/m | 2,638 | |
| Minimum vertical force ($V_{AC}$), kN/m | 1,218 | |
| Location of the resultant force (a), m | 2.17 | |
| Eccentricity from center of wall base: ($e_L$), m | 0.844 | |
| Limiting eccentricity, (e), m | 1.51 | |
| Is $e_L \leq e$? | Yes | |
| Effective width of base (B'), m | 4.33 | |
| Calculated $e_L$/L | 0.14 | |

**Table 7.** Computations of bearing resistance.

| Item | Strength I (max) | Strength I (min) | Service I |
|---|---|---|---|
| Vertical load ($\Sigma V$), kN/m | 1,906 | 1,480 | 1,368 |
| Resisting moments ($M_{RA}$), kN-m/m | 5,719 | 4,441 | 4,103 |
| Overturning Moments ($M_{OA}$), kN-m/m | 1,015 | 732 | 648 |
| Net moment ($M_A$), kN-m/m | 4,704 | 3,709 | 3,455 |
| Eccentricity ($e_L$), m | 0.54 | 0.50 | 0.48 |
| Limiting eccentricity (e), m | 1.51 | 1.51 | 1.00 |
| Is $e_L \leq e$? | YES | YES | YES |
| Effective width of base (B'), m | 4.93 | 5.01 | 5.05 |
| Bearing stress ($\sigma_V$), kN/m$^2$ | 386 | 295 | 271 |
| Bearing resistance ($q_{nf\text{-}str}$ for strength or $q_{nf\text{-}ser}$ for service), (kN/m$^2$) | 502 | 359 | 359 |
| Is $\sigma_V < q_{nf\text{-}str}$ or $q_{nf\text{-}ser}$? | YES | YES | YES |
| Capacity Demand Ratio (CDR) | 1.30 | 1.22 | 1.33 |
| Minimum resisting moments ($M_{RA\text{-}C}$), kN-m/m | 4,441 | | |
| Maximum overturning moments ($M_{OA\text{-}C}$), kN-m/m | 1,015 | | |
| Net moment ($M_{A\text{-}C}$), kN-m/m | 3,425 | | |
| Minimum vertical force ($\Sigma V_C$), kN/m | 1,480 | | |
| Eccentricity ($e_L$), m | 0.7 | | |
| Limiting eccentricity (e), m | 1.51 | | |
| Is eL ≤ e? | YES | | |
| Critical Capacity Demand Ratio ($CDR_{critis}$) | 1.57 | | |

The results of external stability calculations in Table 5, Table 6, and Table 7 show that the MSE wall designed with a total height of H = 8.6 m and a reinforcement length of 6 m based on critical values has a capacity demand ratio of 3.56 for shear, 1.57 for bearing capacity respectively. This value is greater than the minimum standard, which is > 1. This shows that the MSE Wall has qualified external stability. For internal stability, the safety factor against pulling out ($FS_P$) = 1.5, and the safety factor against breaking or breaking ($FS_B$) = 1.5. Based on these values, the calculation results obtained a vertical spacing (Sv) of 0.8 m; this value is by SNI requirements, which is in the range of 0.2 m - 1.25 m. If divided by the height of the MSE wall, the number of layers of geotextile is 8.6 / 0.8 ≈ 10 layers of geotextile. For the lap length from the calculation results, the value obtained is $l_l$ = 0.1 m < 1 m, so a fold length of 1 m is used as the minimum requirement.

## 4 Conclusion

From the design results for an effective height of 8 m, the embedment depth (d) is 6 m, so the total wall height is (H) 8.6 m. With these dimensions, the MSE wall is safe for external stability. The vertical spacing ($S_v$) is 0.8 m, and the lap length ($l_l$) is 1 m, which is safe for internal stability. These dimensions can be used for temporary landslide handling but not long-term stability. For long-term stability, adding facing elements like secant piles, cantilever walls, and other forms are recommended to avoid erosion due to river water flow.

This article is part of the research project under the PSS scheme funded by LPPM Universitas Andalas in 2024, with contract number 244/UN16.19/PT.01.03/PSS/2024, dated 17 July 2024.

## References

1. Badan Standarisasi Nasional, SNI 8460:2017 Persyaratan perancangan geoteknik, (BSN, Jakarta, 2017)
2. Dinas PUPR and HKI infrastruktur, Transisi darurat kepemulihan bencana alam ruas batas kota Padang Panjang - Sicincin (N.040). (2024).
3. A. A. Chandra and S. Rindi, Analisis stabilitas lereng pada ruas jalan Abe-Arso Sta 3+700 dengan perkuatan geotekstil. CRANE: Civil Engineering Research Jounal, **2**, 2, 50-60 (2021)
4. H. Setiawan, Design of retaining walls with geotextile reinforcement (case study of Donggala ring road). Infrastruktur, **2** (2012)
5. F. N. Azizah, N. S. Surjandari and S. As'ad, Penggunaan geotekstil pada Lereng Sungai Gajah Putih Surakarta. Matriks Teknik Sipil, **2**, 1, 140-146 (2014)
6. R. R. Legrans, Pengaruh jenis tanah terhadap kestabilan dinding mse dengan perkuatan geotekstil di daerah Reklamasi Malalayang. Tekno, **14**, 66, 34-43 (2016)
7. I. Prasetyo, B. Setiawan and R. H. Dananjaya, Alisis stabilitas lereng bertingkat dengan perkuatan geotekstil menggunakan metode elemen hingga. Matriks Teknik Sipil, **5**, 922-926 (2017)
8. M. R. Ramadhani, Pengaruh kekuatan, jumlah lapisan, dan jarak geotekstil untuk perkuatan tanah timbunan di atas tanah lunak. Application to the Civil Engineering in the Field of Geotechnical Engineering Program, Master Thesis, Lampung, 2021
9. C. Aziza and A. Suhendra, Analisis deformasi lateral mse wall dengan perkuatan geogrid terhadap variasi jenis material timbunan. Jurnal Mitra Teknik Sipil, **5**, 1, 153-168 (2022)
10. R. Yuliet, A. Hakam, H. G. Putra and R. L. Alfian, Evaluasi kegagalan struktur dinding penahan tanah gravitasi pada lereng tanah kohesif. Jurnal Ilmiah Poli Rekayasa, **18**, 2, 53-58 (2023)
11. R. Yuliet, A. Hakam, I. Firtsilova and F. Anafwil, Pembuatan program aplikasi konstruksi dinding penahan tanah. Jurnal Rekayasa Sipil, **8**, 2, 75-84 (2012)
12. H. G. Putra, R. Yuliet and H. Oktavian, Analisis kelongsoran lereng batu apit jalan alternatif Padang - Bukitinggi Daerah Sicincin Malalak STA 5+350. Jurnal Rekayasa Sipil, **9**, 1, 10-23 (2013)
13. Y. Julianto, A. Hakam and R. Yuliet, Analisa stabilitas lereng dan Perencanaan penstabilan lereng (Studi kasus polres Arosuka Kabupaten Solok). Jurnal Teknik Sipil ITP, **7**, 1, 34-42 (2020)
14. R. R. Berg, B. R. Christopher and N. C. Samtani, Design and construction of mechanically stabilized earth walls and reinforced soil slopes – volume ii (Federal Highway Administration U.S. Department Transportation, Washington, D.C., 2009)

# The incidence and characteristics of acute respiratory infection (ARI) after the earthquake in West Pasaman and Pasaman Districts, West Sumatra Province in 2022

*Isnaini*[1,2*], *Finny* Fitry Yani[1,2], *Yusri* Dianne Jurnalis[1,2], and *Rinang* Mariko[1,2]

[1]Department of Child Health, Faculty of Medicine, Universitas Andalas, Padang, Indonesia
[2]Department of Pediatric, Dr. M. Djamil General Hospital, Padang, Indonesia

**Abstract:** In February 2022, an earthquake struck Pasaman City in West Sumatra and caused outbreaks of infectious diseases such as acute respiratory infections (ARI). Method: This study used an ecological design by collecting data on the population in West Pasaman and Pasaman Regency in 2022. 13 subjects in West Pasaman and 16 subjects in Pasaman were obtained. ARI cases were mostly found in the female (51.7%) and in the >5 years old group (55.2%). The nutritional status was mostly found in the undernourished category (79.3%). There was no significant relationship between the location of natural disasters and fever or cough in ARI patients (>0.05). Most subjects had a cold (75.9%). Most ARI patients from West Pasaman do not have families with smoking habits (84.6%), while most ARI patients from Pasaman have families with smoking habits (87.5%). There is a significant relationship between the location of natural disasters and families with smoking habits in ARI patients (<0.05). The prevalence of post-earthquake ARI in West Pasaman District and Pasaman Regency was higher among females, particularly in over 5 years old. The affected individuals had low nutritional status, with most of them had cold.

## 1 Introduction

Indonesia is very vulnerable to natural disasters due to its geographic, geological, climatic, and population characteristics. Between 2004 and 2018, Indonesia had eleven earthquakes with a magnitude above 7 on the Richter Scale (SR) [1, 2]. West Sumatra a province in Indonesia is located in close proximity to the subduction zone, which is responsible for the formation of the Mentawai and Sumatra faults. This geographical positioning makes it one of the Indonesian provinces that is highly susceptible to earthquake activity [3]. In February 2022, an earthquake tragedy struck Pasaman City in West Sumatra, resulting in seven fatalities, several injuries, and damage to 1,765 structures [4]. Figs. 1 and 2 shows the conditions of the earthquake victims.

Natural disasters not only result in a high death toll and destruction of infrastructure, but they also present a variety of health risks, such as the spread of infectious illnesses, which are particularly dangerous for children, the elderly, and refugees [5–7]. After natural disasters, outbreaks of infectious diseases such as acute respiratory infections (ARI) are often found in refugee camps [8]. Following the 2009 earthquake in Sumatra and the 2010 earthquake in Haiti, the most cases that were documented were acute respiratory infections (16.3%). In the 10 days following the 2013 Lushan earthquake in Sichuan Province, China, ARI was also a prevalent infectious condition among children [9].

Acute respiratory infections (ARIs) are illnesses that invest symptoms such as fever, coughing, difficulty swallowing, and difficulty breathing. They are a leading cause of illness and mortality in children under the age of five worldwide. Acute respiratory infections (ARI) encompass both upper and lower respiratory tract infections, with the common cold and influenza being the most widespread types. The leading cause of mortality among the 5.4 million children under the age of five who died in 2017 was Acute Respiratory Infection (ARI). Furthermore, the mortality rate for ARI in impoverished countries is two to six times higher compared to industrialized nations [10–12]. Age, gender, nutritional status, exclusive breastfeeding, socioeconomic level, population density, indoor pollution, passive smoking, and other variables can all have an impact on this [11].

Overcrowding in refugee camps, poor air ventilation, dirty drinking water, inadequate personal hygiene, and the temporary shutdown of emergency response and local health services are some of the variables that may make people more susceptible to contracting ARI after a catastrophe [6, 7]. Apart from environmental variables, public health awareness and knowledge are also thought to be important determinants of the prevalence of infectious illnesses during crisis occurrences [5]. In light of this context, the author will investigate the incidence and characteristics of ARI after the earthquake disaster in West Pasaman and Pasaman Districts, West Sumatra Province in 2022.

* Corresponding author: isnainialpis@gmail.com

**Fig. 1.** Tents for earthquake victims

**Fig. 2.** Medical examination of children affected by the earthquake

## 2 Method

This study used an ecological design by collecting data on the population. Case data describes the cumulative number of ARI and describes specific individual characteristic that affected by the earthquake in West Pasaman Regency and Pasaman Regency in 2022. The data were collected by The Disaster Team of Department of Child Health, Faculty of Medicine, Universitas Andalas and The Indonesian Pediatric Society Disaster Team West Sumatra Branch, data collected from February 2022 to April 2022.

## 3 Results

During the research period, 13 subjects in West Pasaman and 16 subjects in Pasaman were obtained. ARI cases were mostly found in the female gender (51.7%) and the age of children in the >5 years group (55.2%). There was no significant relationship between gender and disaster location in ARI cases. The relationship between age and disaster location in ARI cases could not be analyzed. In 2022, according to the Central Bureau of Statistics (BPS) of West Pasaman Regency, the population aged 0-19 years was 163,317 people while Pasaman Regency had a population aged 0-19 years of 109,070 people. Thus, the prevalence of ARI in children in West Pasaman District is 0.008% and in Pasaman District is 0.014%. The characteristics of the study subjects can be seen in Table 1.

**Table 1.** Research subject characteristics

| Variable | Location | | Total | p value |
|---|---|---|---|---|
| | West Pasaman n=13 | Pasaman n=16 | | |
| Gender<br>Male<br>Female | <br>5 (38.5%)<br>8 (61.5%) | <br>9 (56.3%)<br>7 (43.8%) | <br>14 (48.3%)<br>15 (51.7%) | 0,462 |
| Age (year)<br>1-2<br>2-3<br>3-4<br>4-5<br>>5 | <br>2 (15.4%)<br>2 (15.4%)<br>1 (7.7%)<br>1 (7.7%)<br>7 (53.8%) | <br>3 (18.8%)<br>2 (12.5%)<br>1 (6.3%)<br>0 (0%)<br>9 (56.3%) | <br>5 (17.2%)<br>4 (13.8%)<br>2 (6.9%)<br>1 (3.4%)<br>16 (55.2%) | - |

The nutritional status of the subjects was mostly found in the undernourished category (79.3%). Both earthquake disaster locations had more subjects with poor nutritional status than good nutrition (92.3% vs 7.7%, and 68.6% vs 31.3%). The results of bivariate analysis showed no significant relationship between the location of the earthquake and the nutritional status of patients with ARI (p value>0.05). The nutritional status of the study subjects can be seen in Table 2.

**Table 2.** Research subject nutritional status

| Variable | Location | | Total | p value |
|---|---|---|---|---|
| | West Pasaman n=13 | Pasaman n=16 | | |
| Good<br>Deficient<br>Poor | 1(7.7%)<br>12(9.3%)<br>0(0%) | 5(31.3%)<br>11(68.6%)<br>0(0%) | 6(20.7%)<br>23(79.3% )<br>0 (0%) | 0,183 |

Clinical manifestation of the subject including fever are investigated in the study. This study shown that most subjects did not have fever (65.5%). There was no significant relationship between the location of natural disasters and fever in ARI patients (>0.05). Other manifestation such as cough are seen in 13 subjects (44.5%). There was no significant relationship between the location of natural disasters and cough in patients with ARI (>0.05). Most subjects had a cold (75.9%). There was no significant relationship between the location of the natural disaster and fever in patients with ARI (>0.05). The clinical manifestation can be seen in Table 3.

**Table 3.** Research subject clinical manifestation

| Variable | Location | | Total | p value |
|---|---|---|---|---|
| | West Pasaman n=13 | Pasaman n=16 | | |
| Fever<br>Yes<br>None | 6(46.2%)<br>7(53.8%) | 4(25%)<br>12(75%) | 10(34.5%)<br>19(65.5%) | 1.000 |
| Cough<br>Yes<br>None | 7(53.8%)<br>6(46.2%) | 6(37.5%)<br>10(62.5%) | 13(44.5%)<br>16(55.2%) | 0.614 |
| Cold<br>Yes<br>None | 11(84.6%)<br>2(15.4%) | 11(68.8%)<br>5(31.3%) | 22(75.9%)<br>7(24.1%) | 0.292 |

Family smoking is a risk factor assessed in this study. Most ARI patients from West Pasaman do not have families with smoking habits (84.6%), while most ARI patients from Pasaman have families with smoking habits (87.5%). There is a significant relationship between the location of natural disasters and families with smoking habits in ARI patients (<0.05). The description of families with smoking habits in the subject can be seen in Table 4.

**Table 4.** Research description of families with smoking habits

| Variable | Location | Total | p value | |
|---|---|---|---|---|
| | West Pasaman n=13 | Pasaman n=16 | | |
| Yes | 2(15.4%) | 14(87.5%) | 16(55.2%) | 0.00 |
| None | 11(84.6%) | 2(12.5%) | 13(44.8%) | |

## 4 Discussion

Earthquakes inflict significant harm on both the environment and civilization, particularly on the health system. Following a tragedy, there are three subsequent phases that take place. Phase 1 is known as the impact phase and lasts for 4 days. During this stage, individuals who have been affected by the catastrophe are moved to a different location, saved from danger, and promptly provided with medical care for any injuries sustained. Common issues experienced during this stage include hypothermia, hyperthermia, and dehydration. period 2, often known as the post-impact period, occurs from day 4 to week 4. During this stage, the occurrence of infectious illnesses starts. Pathogens can be conveyed by water, air, or soil. Airborne infections encompass a range of illnesses caused by viruses (such as influenza, RSV, adenovirus), bacteria (including Streptococcus pneumoniae, pertussis, TB, legionella, Mycoplasma pneumoniae), and other pathogens including measles, varicella, and Neisseria meningitidis. period 3 is the recuperation period, which occurs after week 4. During this stage, the prevalent infectious illnesses are those characterized by a prolonged incubation time [13].

This study finds that female patients with ARI was 51.7%, which was higher than the proportion of male patients (48.3%), but the difference between these percentages was not significant. There was no discernible correlation between gender and the occurrence of natural catastrophes in individuals with ARI. The findings of this investigation diverge from the study conducted by Shah [14], which shown a higher incidence of acute respiratory infections (ARI) in males compared to women, both for viral-induced ARI (52.6% vs. 41.4%) and bacterial-induced ARI (55.1% vs. 44.9%). However, the data from this study also revealed a very small disparity in prevalence. Ahmad and Al Subaie [15, 16] both observed similar findings, with a male-to-female incidence ratio of 60% to 40%.

The study revealed that the age group with the largest number of patients with acute respiratory infections (ARIs) was above 5 years old in both catastrophe areas, accounting for 53.8% and 56.3% respectively. The bivariate analysis does not allow for the assessment of the link between age and the occurrence of natural disasters in patients with ARI. This finding aligns with the study conducted by Fitry Yani [17] which revealed that the majority of cases of Acute Respiratory Infection (ARI) occurred in individuals older than 5 years (50%) after the seismic catastrophe in West Sumatra. This finding is consistent with the study on the earthquake catastrophe in Nepal in 2015, which indicated that 65% of children above the age of 6 had upper respiratory tract infections [18].

Post-earthquake catastrophes can lead to a high prevalence of upper respiratory tract infections among youngsters [14, 17, 19]. Children are more susceptible to respiratory tract infections due to their underdeveloped immune systems in comparison to adults. Furthermore, shelters provided for individuals affected by disasters typically have high levels of humidity, congestion, and inadequate ventilation, so promoting the transfer of viruses and bacteria and facilitating the spread of diseases. Hassen demonstrated that shelters with inadequate ventilation resulted in a 4.3-fold increase in the likelihood of children developing acute respiratory infections (ARI). The conditions outside the shelter are particularly unfavorable as the surrounding air is contaminated as a result of the accident. During the nighttime, children may be subjected to frigid air as a result of insufficient shelter conditions. These circumstances can elevate a child's susceptibility to upper respiratory tract infection [14, 18, 20]. There is no correlation between the gender and age of a kid and the location of a natural catastrophe in terms of the risk of acquiring upper respiratory tract infection. Essentially, all children have an equal likelihood of having this condition following a disaster.

According to the data collected from this study, the majority of the individuals had a nutritional status that fell into the undernourished group (79.3%). The earthquake catastrophe sites had a higher prevalence of individuals with inadequate nutrition compared to those with adequate nutrition (92.3% vs 7.7%, and 68.6% vs 31.3%). Nevertheless, the bivariate analysis conducted in this study revealed that there is no statistically significant correlation between the earthquake's location and the nutritional state of patients with ARI, as shown by a p-value greater than 0.05. In a research conducted by Tazinya [21] it was shown that undernourished children had a considerably greater proportion of acute respiratory infections (ARI) compared to well-nourished children, with an odds ratio of 3.01 (95% confidence interval: 1.66-5.43). The difference in proportions was statistically significant ($P < 0.05$). Nevertheless, in the multivariate analysis, nutritional status did not show a significant impact when compared to other risk variables ($p = 0.06$). The variation in statistical methodology employed across different research may account for the observed discrepancy, as several investigations have consistently demonstrated a substantial impact of malnutrition on acute respiratory infections (ARI). Ronaldi's research on children impacted by the 2018 North Lombok earthquake [13] revealed that there is no correlation between the nutritional health of children and the incidence of sickness following the earthquake (with a significance level of $q>0.005$).

Conversely, Maria's research [27] demonstrated a clear and statistically significant correlation between inadequate nutritional status and the occurrence of acute respiratory infections (ARI). Malnutrition leads to airway anomalies that disrupt the normal functioning of the airway, namely its ability to guard against disease-causing substances. In children with inadequate nutritional status, impaired physiological processes can lead to the entry of disease-causing substances into the lungs through the airway.

The study found no statistically significant correlation between the occurrence of natural disasters and the presence of fever in individuals with acute respiratory infections ($p>0.05$). According to Chen's research [23], there was no statistically significant correlation ($P=0.559$) between the presence of viral, bacterial, or bacterial co-infection with fever symptoms in babies and children with acute respiratory infections (ARI). newborns diagnosed with viruses had a significantly greater incidence of fever ($P < 0.001$) compared to newborns who were not infected with viruses. Chen [22] found that 51% of patients were diagnosed with at least one respiratory infection. The most frequently detected viruses were Human rhinovirus (HRV) (23%) and Respiratory syncytial virus (RSV) (22.7%). In contrast, Khan conducted a research in Pakistan with 314 children, where 89% of the observed clinical signs of Acute Respiratory Infection (ARI) were fever, followed by cough and rhinorrhea.

This study discovered that coughing, which is one of the symptoms of Acute Respiratory Infection (ARI), was not influenced by the location of the natural catastrophe. A literature review documenting the clinical manifestations of patients affected by the 2004 Indian Ocean tsunami tragedy revealed that around 325 out of 1021 patients (32%) presented with minor cough symptoms [7]. In their respective studies, Khan and Desouza [23, 24] discovered that cough symptoms manifested in 79% of the patients. During the earthquake, the evacuation circumstances were characterized by extreme overcrowding, with individuals cramped together in a single tent [9]. The presence of overcrowded tents and inadequate sanitation in emergency situations might heighten the likelihood of fast transmission of airborne illnesses, as patients may cough or sneeze [7, 13]. Following the earthquake, a multitude of pulmonary contaminants were prevalent in the atmosphere.

This study concluded that there was no substantial correlation between colds and the geographical location of the accident. In their study, Khan [23] discovered that 61.8% of patients experienced symptoms of a cold. In their study, Desouza [24] discovered that 80.72% of patients experienced symptoms of a cold. There is compelling data indicating the occurrence of a flu pandemic following an earthquake, which peaks one week after the seismic event and exhibits a twentyfold increase compared to pre-earthquake levels. The influenza virus primarily targets specific demographics, particularly youngsters [25].

A strong correlation was observed between the geographical occurrence of the natural calamity and the smoking behaviors of families in patients with Acute Respiratory Infection (ARI) (p-value $<0.05$). The prevalence of smoking parents is a contributing factor that is more frequently observed in one of the catastrophe regions. According to a study conducted by Maria, children whose parents smoke are more likely to have acute respiratory infections (ARI) compared to toddlers whose parents do not smoke. Parents who are exposed to acute respiratory infections (ARI) in newborns as a result of being exposed to cigarette smoke for a duration of 30 minutes might experience harm to endothelial cells and blood clotting cells, leading to the constriction and rigidity of blood vessels [24].

The study revealed that environmental variables, including being exposed to wood smoke, cigarette smoke, and living with someone who coughs, were found to have a substantial impact on the occurrence of acute respiratory infections (ARI) ($P < 0.05$). According to this study, passive smoking was identified as a significant risk factor for acute respiratory infections (ARI), increasing the chances by 4.67 (1.91-11.40) compared to children who were not exposed to secondhand smoke [21].

Tobacco use is a prevalent contributing factor to Acute Respiratory Infections (ARI) [26]. Exposure to cigarette smoke is associated with cough complaints, upper respiratory tract infections, and hospitalizations related to respiratory issues in children. Passive smoking has emerged as a significant health concern due to the prevalence of individuals spending a majority of their time in enclosed spaces. Secondhand smoke, equally detrimental as direct smoking, leads to significant health issues in youngsters. It exacerbates and simplifies the occurrence of respiratory disorders in both the upper and lower respiratory tracts, beginning in early childhood. Exposure to secondhand smoke has a substantial impact on the hospitalization rates and healthcare expenses of children, primarily owing to respiratory ailments. Furthermore, inadequate airflow and excessively congested environments during catastrophes can contribute to the spread of upper respiratory tract diseases [25].

Medical masks, including those for children, are distributed to individuals at the disaster scene. Chaabna [27] asserts that the utilization of medical masks is efficacious in preventing respiratory infections. It is strongly advised to wear medical masks in community settings when there is widespread transmission of the virus, and it is not feasible to maintain physical distance. Furthermore, at the location of the catastrophe, children get health screenings. Children exhibiting signs of Acute Respiratory Infection (ARI) get pharmacological treatment. Chiu (2022) states that it is crucial to supply children with medical equipment during catastrophes that is similar to the supplies provided to adults. Pediatricians are recommended to provide parents with information and knowledge on catastrophe emergency planning [28].

## 5 Conclusion

The occurrence of Acute Respiratory Infection (ARI) following the earthquake in West Pasaman District was 0.008%, whereas in Pasaman District it was 0.14%. The prevalence of post-earthquake Acute Respiratory Infection (ARI) in West Pasaman District and Pasaman Regency was higher among females, particularly in the age range of over 5 years. The affected individuals had low nutritional status, with most of them not experiencing fever but exhibiting signs of cough and cold. Additionally, the majority of the participants did not have a family member who smoked.

Raising early awareness of the occurrence of Acute Respiratory Infections (ARI) is crucial in order to avert epidemics. It is imperative to provide education on hygiene and healthy living practices, as well as improve infrastructure, in order to prevent Acute Respiratory Infections (ARI) following the earthquake, particularly in the West Pasaman and Pasaman Regency.

## References

1. M. Fuady, R. Munadi, and M. A. K. Fuady, Disaster mitigation in Indonesia: between plans and reality. IOP Conf. Ser. Mater. Sci. Eng., **1**. 012011, (2021). https://doi.org/10.1088/1757-899X/1087/1/012011
2. T. Simanjuntak and M. Ririmasse, Archaeology of disaster in Indonesia: where are we now?. Ber. Sedimentol., **47**, 3, (2021). https://doi.org/10.51835/bsed.2021.47.3.351
3. R. P. Yanti, Suharsono, I. R. Palupi, and W. Hidayat, Preventive toward earthquake's disaster in West Sumatera Based on geophysic analysis. J. Dialog Penanggulangan Bencana, **8**,1, (2017)
4. P. Maulida, R. Putra, and A. Kurniawan, Studi pergeseran koseismik gempa Pasaman M6.1 2022 menggunakan data pengamatan gps harian. Geoid, **18**, 1 (2022).
5. D. N. Pascapurnama, A. Murakami, H. Chagan-Yasutan, T. Hattori, H. Sasaki, and S. Egawa, Integrated health education in disaster risk reduction: Lesson learned from disease outbreak following natural disasters in Indonesia. Int. J. Disaster Risk Reduct., **29**, pp. 94–102, (2018). https://doi.org/10.1016/j.ijdrr.2017.07.01
6. M. Mavrouli, S. Mavroulis, E. Lekkas, and A. Tsakris, The impact of earthquakes on public health: A narrative review of infectious diseases in the post-disaster period aiming to disaster risk reduction. Microorganisms, . **11**,2, (2023), https://doi.org/10.3390/microorganisms11020419
7. M. Mavrouli, S. Mavroulis, E. Lekkas, and A. Tsakris, Respiratory infections following earthquake-induced tsunamis: transmission risk factors and lessons learned for disaster risk management. Int. J. Environ. Res. Public Health, **18**,9, (2021). https://doi.org/10.3390/ijerph-18094952
8. T. Kawano, Y. Tsugawa, K. NIshiyama, H. Morita, O. Yamamura, and K. Hasegawa, Shelter crowding and increased incidence of acute respiratory infection in evacuees following the Great Eastern Japan Earthquake and tsunami. Epidemiol. Infect.,**144**, 4, (2016), https://doi.org/10.1017/S0950268815001715
9. M. Mavrouli, S. Mavroulis, E. Lekkas, and A. Tsakris, Earthquake-Triggered Respiratory Infectious Diseases. Encyclopedia, 2023. https://encyclopedia.pub/entry/41156 (accessed Dec. 15, (2023)
10. H. Dagne, Z. Andualem, B. Dagnew, and A. A. Taddese, Acute respiratory infection and its associated factors among children under-five years attending pediatrics ward at University of Gondar Comprehensive Specialized Hospital, Northwest Ethiopia: institution-based cross-sectional study. BMC Pediatr., **20**,1, (2020). https://doi.org/10.1186/s12887-020-1997-2
11. F. Ujunwa and C. Ezeonu, Risk factors for acute respiratory tract infections in under-five children in Enugu Southeast Nigeria. Ann. Med. Health Sci. Res.,**4**,1, (2014). https://doi.org/10.4103/2141-9248.126610
12. F. Mir et al., Risk Factors for acute respiratory infections in children between 0 and 23 months of age in a Peri-Urban District in Pakistan: A matched case–control study. Front. Pediatr., **9**,1 (2022). https://doi.org/10.3389/fped.2021.704545
13. M. Ronaldi, I. B. G. Suwibawa, K. T. Kadafi, N. E. Dewi, J. A. Effendi, and A. B. Pulungan, The impacts of 2018 earthquakes to the pediatric population in NorthLombok Field Hospital, West Nusa Tenggara, Indonesia. Curr Pediatr Res, (2020)
14. N. Shah, M. A. Abro, M. A. Abro, A. Khan, F. Anwar, and H. Akhtar, Disease pattern in earthquake affected areas of Pakistan: data from Kaghan valley. J. Ayub Med. Coll. Abbottabad, **22**, 3, (2010)
15. S. S. Al Subaie and M. M. Al Saadi, Features associated with severe disease in hospitalized children with 2009 influenza A (H1N1) infection at a university hospital in Riyadh, Saudi Arabia. Ann. Saudi Med., **32**,1, (2012). https://doi.org/10.5144/0256-4947.2012.53
16. A. S. Ahmad, Clinical Presentation and Outcome in Hospitalized Patients of 2009 Pandemic Influenza A (H1N1) viral infection in Oman. Oman Med. J., **26**,5, pp. 329–336 (2011). https://doi.org/10.5001/omj.2011.82
17. F. F. Yani, E. Elfitrimelly, and D. Basir, Acute

respiratory infection in children after earthquake on September 30th, 2009 and its related factors in Malalak and Maninjau villages at Agam District West Sumatera. Paediatr. Respir. Rev., **11**,111, (2010). https://doi.org/10.1016/S1526-0542(10)70172-X

18. J. Wang et al., 2015 Nepal Earthquake: Analysis of child rescue and treatment by a field hospital. Disaster Med. Public Health Prep., **10**, 5, (2016). https://doi.org/10.1017/dmp.2016.22.
19. S. Goyet et al., Post-earthquake health-service support, Nepal. Bull. World Health Organ., **96**, 4, (2018). https://doi.org/10.2471/BLT.17.205666
20. S. Hassen et al., Determinants of acute respiratory infection (ARI) among under-five children in rural areas of Legambo District, South Wollo Zone, Ethiopia: A matched case–control study. Int. J. Infect. Dis., . 96, (2020). https://doi.org/10.1016/j.ijid.2020.-05.012.
21. A. A. Tazinya, G. E. Halle-Ekane, L. T. Mbuagbaw, M. Abanda, J. Atashili, and M. T. Obama, Risk factors for acute respiratory infections in children under five years attending the Bamenda Regional Hospital in Cameroon. BMC Pulm. Med., . **18**, 1, (2018). https://doi.org/10.1186/s12890-018-0579-7
22. J. Chen et al., Epidemiology and clinical characteristics of acute respiratory tract infections among hospitalized infants and young children in Chengdu, West China, 2009–2014. BMC Pediatr., . **18**, 1, (2018). https://doi.org/10.1186/s12887-018-1203-y.
23. E. A. Khan, M. H. Raja, S. Chaudhry, T. Zahra, S. Naeem, and M. Anwar, Outcome of upper respiratory tract infections in healthy children: Antibiotic stewardship in treatment of acute upper respiratory tract infections. Pakistan J. Med. Sci., . **36**, 4, (2020). https://doi.org/10.12669/pjms.36.4.1420.
24. J. Desouza, M. Patil, I. Thomas, and J. Desouza, Current clinical profiles of acute respiratory tract infections in children between 2 months to 5 years. Int. J. Contemp. Pediatr., . **11**, 2, (2024). https://doi.org/10.18203/2349-3291.ijcp20240089.
25. M. Y. Yavuz, M. A. Türk, B. Oral2, and A. C. Beyan, Earthquake and respiratory pollutants. Eurasian J. Pulmonol (2023). https://doi.org/10.14744/ejp.2022.6003
26. M. Thomas and P. A. Boma, Upper Respiratory Tract Infection. StatPearls [Internet]. Treasure Island (FL): StatPearls Publishing, (2023)
27. L. Maria, M. Simanjuntak, T. Silangit, and J. M. Siahaan, Determinants of Acute Respiratory Infection in Children Under Five in Simalingkar, Medan, North Sumatera. J. Epidemiol. Public Heal., **5**, 1, (2020). https://doi.org/10.26911/jepublichealth-.2020.05.01.03
28. S. Aslan, O. Bostan Gayret, M. Erol, S. Mandel Isikli, O. Buke, and A. Ozel, Determination of the Relation Between Passive Cigarette Smoking in Children and Respiratory Tract Infections by Evaluation of Urine Cotinine/Creatinine Levels. Med. Bull. Haseki, **60**, 3, (2022). https://doi.org/10.4274/haseki.galenos.2022.8045

# A conceptual framework for the process protocol of integration quality, safety, health and sustainable construction management systems for infrastructure projects in Indonesia

*Utami* Dewi Arman[1], *Akhmad* Suraji[2*], and *Benny* Hidayat[2]

[1]Postgraduate Program, Universitas Andalas, Padang, West Sumatra, Indonesia
[2]Department of Civil Engineering, Faculty of Engineering, Universitas Andalas, Padang, West Sumatra, Indonesia

**Abstract.** Construction accidents on infrastructure projects in Indonesia from 2017-2021 have increased by 5.65% from previous years. Construction accidents that result in property damage, death or injury to both workers and the public, pollution and environmental damage cause sustainable development goals not to be achieved. Construction accidents are not only focused on the impacts, but it is necessary to explore how the incident occurred. This is the basis for the need to integrate quality, safety, health and sustainability in construction. The current construction safety management system is not yet integrated with other management systems. This paper aims to develop a conceptual framework in the process of integrating construction quality, safety, health and sustainability management systems. Development of a conceptual framework based on literature review, ISO standards and existing construction safety planning guidelines. The development of this conceptual framework is part of a research proposal that is being formulated to develop a process protocol for integrating process of construction quality, safety, health and sustainability management systems in infrastructure projects. This process protocol can simplify the complexity and overlapping of the processes of each management system so that it is easier to understand and to implement by all project stakeholders.

## 1 Introduction

### 1.1 Backgrounds

According BPJSTK (2021), construction accidents in Indonesia as of 2021 have reached 234,270 cases and this number has increased by 5.65% from the previous year of 221,740 cases. Several cases of infrastructure project accidents such as girder collapses, crane overturns, formwork collapses and recent material landslides cannot be separated from technical, human resource and organizational failures in managing projects.

Accidents need to be analysed from how the incident occurred, not only from the results of the incident. In the opinion of [1, 2], structural failure occurs due to the development of deformation and structural fracture because the structure is not designed properly or is not built according to design, apart from that it is also indicated by other failures such as the use of inappropriate materials. Structural failures often occur due to human errors made during the design process. Errors in the design process could be viewed from personnel factors are caused by the level of ability, experience, accountability, or personality type of the design consultant. Meanwhile, if we look at organizational factors, it is caused by lack of education or training, lack of competent resources, professionalism, and use of technology. If we view from project-related factors, human error can be caused by inaccuracies in determining the scope of work and there is no integration between design and other consideration factors [3]. Lack of workmanship in a construction can result in defects resulting in changes in appearance and other examples because lack of attention to foundation design can cause ground movement, which is characterized by cracking of brick walls, changes in construction and possibly also by sagging roofs and changes in the shape of the building. Work that results in a defect can be described as work that does not meet the description or requirements of the contract, in particular the drawings or specifications, with all that implies, quality, workmanship, performance, or design [4].

Failure can also occur during the expected service life of a structure, for example the collapse of bridges and other structures due to lack of maintenance after years of construction [5]. Whittington et al (1992) stated that accidents are initiated by a series of failures from accident events which combine individual and organizational factors as causes of construction accidents. This theory also assumes that accidents in construction projects result from failures initiated by workers, management in the field, project management and organizational policies. Reason (1991) also stated that failures in the system can trigger unsafe conditions and behaviour in operational areas, for example operators commit negligence and

* Corresponding author: akhmad.suraji@eng.unand.ac.id

violations in dangerous areas which allows incidents and accidents to occur [6].

Construction accidents result in increased direct and indirect costs. Direct costs include insurance costs, workers' medical costs, and costs of loss or damage to property, while indirect costs include increased transportation costs and fines for late completion [7-10]. Work accidents have an impact not only on workers but also on their future and their families. Efforts to prevent construction accidents can be mitigated by planning an effective construction safety management system [11]. Construction accidents in the form of property damage, death or injury to workers and the public, pollution and environmental damage cause sustainable construction goals not to be achieved. The concept of sustainable construction is still a global issue to be discussed in the construction sector.

In addition to having to meet reliability, safety and health standards, construction projects must also meet sustainability standard. Some previous researchers [12-15] incorporated construction sustainable indicators into construction management practices to achieve project success. Construction sustainability assessments need to be reviewed throughout the project life cycle starting from the pre-construction, construction, operation, maintenance and demolition phases [16]. Factors related to occupational safety and health [17-19], factors related to construction quality [20, 21], socio-technical considerations [22] and lean construction principles [23] can be taken into account as factors that influence project success to create sustainable construction.

By reviewing how defects and failures occurred and even causing construction accidents which result in loss of reputation, damage to assets and property, as well as environmental damage, this becomes the basis for considering the need to integrate quality, safety, health, environmental and construction sustainability management systems. It resulted a few research questions which are:

1. What is a more appropriate integration approach in integrating elements in the quality, safety, health and sustainability management system into an integrated management system?
2. How to integrate processes that are fragmented in the quality, occupational safety and health, environmental and sustainability so that it become integrated processes?
3. What are the duties and responsibilities of each organization in implementing an integrated management system at the conceptual, planning, design, selection, implementation, supervision, operational, maintenance and demolition phase?
4. What are the integrated risks that impact buildings, workers, environmental and social factors?
5. How can a more comprehensive risk assessment methodology be developed?
6. How are the integrated documents in the implementation of the integrated management system at the conceptual, planning, design, selection, construction, supervision, operational and maintenance and demolition phases?

So therefore, this study aims to develop a conceptual framework in the implementation of integrating construction quality, safety, health and sustainability management systems.

## 1.2 Integrated management system

Integrated management system is a system that integrates various management functions needed to support business operations into one comprehensive management system involving organizational structure, resources and procedures including planning, implementation, monitoring, and controlling aimed to be able to improve the company's business performance. The integrated management system must also adopt the Plan-Do-Check-Action concept as an effort to continuously improve as a management system. Seghezzi explained that there are three different ways of system integration, including addition, merger and integration. In "addition", the partial systems for quality and environmental management are separated and described in separate documents but their content is made comparable. In "merging", work instructions are fully integrated, but not procedures and manuals, an overall management system is created but partial systems are still visible. In "integration", companies can select or develop a basic management system as their general system and include all partial systems in it" [24].

Wilkinson and Dale also describe a four-level model for integrating management systems. The first level of management systems and functions is still separate. The second level is a combination of systems based on the relationships identified between one management system and other management systems. Documentation is combined and integration into each function is still required. The third level involves integrating selected parts of each management system with other certified management systems, but without using the identified relationships. The fourth level is to integrate certified and non-certified systems with the overall management system with policies and objectives aligned and supporting the organization's overall strategy, policies and business objectives. Kirkby stated that there are three possible management system models, including separate, parallel and integrated management system models. The first level is "separate", where management systems cover their own distinct areas for each set of requirements. The second level is "parallel" where the management system utilizes standard areas and all common elements such as management reviews and internal audits are directed to the same system. The final level is "integrated" which combines all standards into one common management system. Karapetrovic (2002) concluded that there were three types of organizations that were found to only integrate documentation, equalize core processes, objectives and resources, and finally have all parts of the management system integrated in one management system. The process is summarized where two levels can be defined, the first is "partial integration" which ranges from simple collaboration to the alignment and compatibility of objectives, processes and resources of separate management systems and the second is "full integration" in where management systems lose their unique identity, resulting in incomplete integration into a multifunctional integrated management system. The same opinion from Beckmerhagen et al (2003) suggested system integration at three levels. The first level, "harmonization", namely harmonizing management systems, an organization partially integrates documentation. The second level, "cooperation" denotes joint system improvement using integrated audits and resources. The third level is "integration" meaning that one management system and other management systems are combined into a comprehensive integrated management system [25, 26].

It is possible to integrate quality and safety management systems, which has been done in previous studies. Theoretical studies were carried out by looking at the similarities in the ISO 9001 and OSHA 18000 management standard clauses [27-29] while empirical studies were carried out by investigating the relationship between factors that influence quality and safety. Quality and safety are interconnected with each other and the presence of defects or rework carries a risk of work accidents [30-33]. Some elements that can be integrated are implementation objectives, policies, management commitment, organizational structure and responsibilities, human resources, training, processes, work instructions, investigations, communication, company safety climate and culture, audits and documentation [34]. The benefits of integrating quality with safety are that the implementation of the management system becomes more optimal, there is increased competitiveness, and resource utilization becomes more efficient. To integrate quality and environmental management systems, which has been done in previous studies.

The similarity or suitability of the clauses in the ISO 9000 and ISO 14000 standards is the basis for integrating these two management systems. Based on the similarity or suitability of the clauses, the elements that can be integrated include policies, goals and objectives, processes, work instructions, audits, management reviews, corrective actions, communication and documentation. The reduction in waste from production results is due to the "zero defects" concept with the implemented quality management system which can further improve the company's reputation [35]. The benefits of integrating quality and environmental management systems can improve company performance, increase productivity, reduce costs and waste, increase customer satisfaction, improve company reputation, reduce duplication of procedures, save on resource use [36].

Quality, safety, health and environmental (QHSE) management systems make it possible to be integrated due to the suitability of elements in the ISO 9001, ISO 14001 and OHSAS 18001 standards [37-41] and ISO 45001 [42]. Some previous researchers also proposed efforts to QHSE management systems to realize sustainable construction based on ISO 9001, ISO 14001, OHSAS 18001 and ISO 45001 [43, 44]. Elements that have similarities between systems include policies, goals and objectives, organizational structure responsibilities, documentation, planning, records, resources and infrastructure, control and action, workforce, communication, continuous improvement, corrective action, process control, maintenance and calibration, integration management guidelines, training, operational procedures, nonconformities, as well as internal and external audit policies. The benefits of QHSE integration are increasing the effectiveness of the management system, reducing costs, goals can be aligned, eliminating conflict of responsibilities between departments and simplifying management standards increasing work productivity, reducing resources, internal audits can be unified, increasing training, reducing duplication of documentation, reducing bureaucracy, system better communication, the company's competitiveness increases and can increase compliance with regulations and the impact of integration which reduces accident incidents, environmental related issues and improves quality.

Integration also has a significant impact on the sustainability of construction project management, especially reducing economic, social and environmental risks, organizing good, transparent and accountable organizations, optimizing resources, implementing rewards and punishments and developing new technological innovations. Meanwhile, the difficulty in integrating these three management systems is that each management system has different targets and the safety and health management system is "forced" to conform to the other two management systems. The success of integration is influenced by the company's internal will, ability to make improvements and external factors that influence the process. Leadership commitment and awareness are the most important factors within a company to achieve increased sustainability. Policy and planning are the most important factors in the success of implementing the integration process.

Two integration models were proposed by [44], namely clause-based and process-based integration models. However, the process-based integration model in his paper is not explained for the reason that it is considered not much different from the clause-based integration model and process integration also depends on the management system implemented by the organization. This means that the integration process is not described and explained by him and is a matter for future research in developing an integration process that can be standardized.

Some previous researchers adopted the risk management framework into the implementation of management system integration. Zeng developed a quality, safety, occupational health and environmental risk assessment methodology for contractors in implementing an integrated management system based on the Plan-Do-Check-Action concept. This integrated risk assessment was carried out using the Failure Mode Effect and Analysis (FMEA) method. Williams and Bertsch integrated risk management systems with quality and concluded that factors that influence quality simply can be assessed as identified risks such as risks that exist at the operational stage. Quality does not play a role in managing unidentified risks. The risk management system can also be integrated with other management systems. Sánchez and Lopez identified, classified and assessed construction sustainability risks. Sustainability aspects include environmental aspects (land, water, air, biodiversity, resources, energy), social aspects (culture, accessibility, participation, security, public facilities and social integration) and economic aspects (costs, technical requirements, bureaucracy, economics). social, and heritage/historical heritage) [45]. However, not many discuss the adoption of the risk management process into an integrated management system starting from conceptual, planning, implementation, operations, maintenance to the demolition stage. It is important to examine how risk management is implemented throughout the project life cycle.

### 1.3 A review of construction safety management system guidelines in Indonesia

The current construction safety management system guidelines still separate a few management systems such as quality, safety and health, environmental and sustainability management systems where the processes, tasks and responsibilities of the organization and documents are still fragmented so that these elements become complex and overlapping. By referring to the management system integration theory according to Seghezzi, it can be assessed that Indonesia's construction safety management system is currently not yet an integrated management system among its partial management systems. The construction safety management system only "adding" or "merging" of its partial management systems. It is necessary to integrate and to simplify processes, organizational responsibilities and documents in implementing an integrated construction safety management system.

This also requires a comprehensive safety risk assessment methodology starting by including design safety risks at the pre-construction stage, work safety risks at the implementation stage, safety risks at the operational and maintenance stages and safety risks at the demolition stage. By considering sustainability standard which are one of the construction safety standards, it is necessary to identify construction sustainability risks as well by examining failure risks in economic, social and environmental sustainability as hazards or residual safety risks that need to be identified and evaluated so that the construction safety risk assessment is more comprehensive and integrated.

Besides that, the implementation of the construction safety management system in Ministerial of Public Works and Housing Regulation No. 10 of 2021 is explained only at the conceptual and planning, design, selection, implementation and supervision phases. It is necessary to transform into an integrated phases from the pre-construction, construction, operation, maintenance to demolition stages that can be implemented and complied with by all project stakeholders, including clients, contractors, sub-contractor, feasibility study and planning consultants, design consultants, supervisory consultants, suppliers and the community.

## 2 Research methodology

This study is qualitative research with deductive reasoning. It was begun by identifying problem statements by analysing regulation and reviewing literatures of previous studies related to integrated management system.

To develop a conceptual framework was conducted by analysing Ministerial of Public Works and Housing Regulation No. 10 of 2021 concerning Construction Safety Management System Guideline. The next step is to analysis the conformity of the clauses among ISO 9001:2015, ISO 45001:2018, ISO 14001:2015, ISO 31000:2018, ISO 37101:2016 and ISO 55001:2014. ISO 9001:2015 focuses on increasing customer satisfaction, ISO 45001:2018 focuses on eliminating hazards and minimizing the risk of work accidents, ISO 14001:2018 focuses on achieving environmental values for stakeholders and society, ISO 31000:2018 provides a risk management framework for any industrial sector. ISO 37101:2016 focuses on achieving sustainable development targets, meanwhile ISO 55001:2014 focuses on achieving asset management targets.

To identify the duties and responsibilities of organizations in the study using the 2015 Construction Design and Management (CDM) regulations which are regulation-based operational guidelines that help project stakeholders including clients, design consultants, principal design consultants, sub-contractor and principal contractor in carrying out the construction process starting from the pre-construction until the project is completed and equipped with rules that determine the duties and obligations of stakeholders in implementing occupational safety and health management.

Meanwhile, to identify safety risks at the pre-construction, construction, operational, maintenance and demolition phases in this study, the Construction Hazard Assessment Implication Review (CHAIR) method is used. CHAIR is a method that facilitates discussion with the aim of inviting the project stakeholders involved to jointly identify integrated risks that have the potential to occur during pre-construction, construction, maintenance, renovation and demolition in developing a safer design.

## 3 Results and discussion

Developing a conceptual framework is an important first step in conducting research. By seeing at the current guideline of implementing construction safety management systems in Indonesia, where construction safety management system is not yet integrated. The construction safety management system is part of the construction work implementation management system to ensure construction safety. Construction safety standards include construction security, safety, health and sustainability standards. This means that the construction safety management system combines quality, safety, health, environmental and construction sustainability management systems. However, implementation is still fragmented.

Furthermore, the construction safety management system is transformed into an integrated management system. Ministerial of Public Works and Housing Regulation No. 10 of 2021 concerns construction safety system guideline, describes the implementation of a fragmented construction safety management system with other management systems combined and of course some elements of this system are duplicated and overlap with each other. The duties and responsibilities of each organization as well as implementation documents are also not clearly defined. Therefore, it requires a process protocol that is lean, clearly defined throughout the project cycle and easy to implement and understand by all project stakeholders.

Protocols can be considered as references or guidelines that must be adhered to by parties involved in organizing these activities. By using process modeling, some complex processes can be easily broken down into processes that are easy to identify and easy to understand. A process protocol is a modeling tool that is capable of representing stakeholder interests in a process, providing flexibility and clarity that allows activities to be formulated into a framework called standardization. This framework encourages users to understand processes more easily, problems can be identified, improvements can be made in communication and coordination, control and management of resources in order to adopt a shared vision.

The development of construction quality, safety, health and sustainability management system integration process protocols is based on an analysis of regulations, guidelines and ISO standards. the conformity of the clauses in ISO 9001:2015, ISO 45001:2018, ISO 14001:2015, ISO 31000:2018, ISO 37101:2016 and ISO 55001:2014. Some of the compatibility clauses among ISO are as follows goals and objectives, leadership, organizational responsibility and authority, risk management planning, resources, communication, documentation, operational procedures, control, audit and continuous improvement. Besides that, the state of the art management system elements that allow for integration are as follows goals and objectives, leadership, organizational structure, planning, procedures, use of resources, audits, documentation and management reviews.

The duties and responsibilities of organizations in this study using the 2015 Construction Design and Management (CDM) regulation which are regulation-based operational guidelines that help project stakeholders including clients, design consultants, principal design consultants, sub-contractor and principal contractor in carrying out the construction process starting from the pre-construction until the project is completed and equipped with rules that determine the duties and obligations of stakeholders in implementing occupational safety and health management. Meanwhile, safety risks at the pre-construction, construction, operational, maintenance and demolition stages in this study, the Construction Hazard Assessment Implication Review (CHAIR) method is used. It could be facilitated by discussion with the aim of inviting the project stakeholders involved to jointly identify integrated risks that have the potential to occur during construction, maintenance, renovation and demolition in developing a design.

Safety plan on feasibility study and planning, safety plan on design, construction safety plan, supervision safety plan, quality program for supervising, construction work quality plan, environmental control and monitoring plan as well as work traffic management plan also needs to be analyzed to formulate an integration process starting from pre-construction to demotion phase. The following is the conceptual framework of the research shown in Fig. 1.

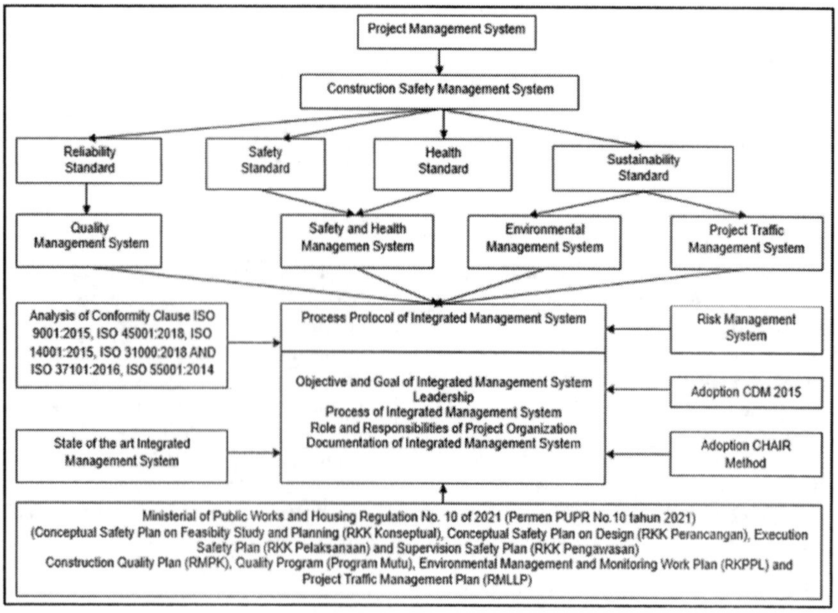


**Fig. 1.** Research conceptual framework

## 4 Conclusion

The current construction safety management system cannot be said to be an integrated management system in terms of goals and objectives, leadership, organizational duties and responsibilities, processes and documentation which are still fragmented between partial management systems, giving rise to overlapping and duplication. The construction safety management system only "adding or combining" partial management systems such as quality, occupational safety and health, environmental and construction sustainability management systems. Implementation of the construction safety management system is carried out throughout the project life cycle, starting from the pre-construction, construction, operational and maintenance stages to demolition. Meanwhile, the implementation of the construction safety management in the guidelines is only at the conceptual and planning, design, selection, implementation and supervision stages. The need to review the implementation of the construction safety management system at the operational, maintenance and demolition stages. Opportunities and impacts of risks in construction safety planning in safety management guidelines require a more comprehensive methodology and need to include safety risks in design components, risks at the maintenance and demolition/demolition stages. In addition, considering that sustainability standards are part of construction safety standards, it is necessary to include economic, social and environmental sustainability indicators as residual risks that need to be identified. Existing studies related to integrated management systems have not been able to provide solutions to research problems, so they require a novelty to integrate quality, safety, health and construction sustainability management systems with different framework. A conceptual framework as rationale thinking for developing a lean process protocol to integrate quality, safety, health, and construction sustainability management systems in the future research

## References

1. Eldukair. Analysis of Recent U.S. Structural and Construction Failures. Journal of Performance of Constructed Facilities. **5**, 1, 57-73. (1991).
2. K. Wardhana, and F.C.Hadipriono. Study of Recent Building Failures in the United States. Journal of Performance of Construction Facilities.2003.**17**:151-158 (2003). https://doi.org/10.1061/(ASCE)08873828(2003)17:3(151)
3. P.E.D, Love, R. Lopez, Y.M Goh and C.M Tam. What Goes up, Shouldn't Come down: Learning from Construction and Engineering Failures. The Twelfth East Asia-Pacific Conference on Structural Engineering and Construction. Procedia Engineering. **14,** 844–850 (2011). http://dx.doi.org/10.1016/j.proeng.2011.07.107
4. A. Atkinson. The management of construction failures and defects. In: Stephenson, P (Ed.), 13th Annual ARCOM Conference, 15-17 September 199, King's College, Cambridge. Association of Researchers in Construction Management. **1**, 342-52 (1997)
5. K.L. Carper. Structural Failures During Construction. Journal of Performance of Constructed Facilities. **1,** 3, 132-144 (1987).
6. A. Suraji, A.R Duff and S. J. Peckitt. Development of Causal Model of Construction Accident Causation. Journal of Construction Engineering and Management **127**(4) 337-344 (2001). https://doi.org/10.1227/01.NEU.0000210260.55124.A4
7. S.S Hosseinian, and Z.J Torghabeh. Major Theories of Construction Accident Causation Models: A Literature Review. International Journal of Advances in Engineering and Technology (IJAET). **4**, 2, 53-66, Sept 2012 (2012)
8. O.S Williams. R.A Hamid. M.S.Misnan. Accident Causal Factors on the Building Construction Sites: A Review. International Journal of Built Environment and Sustainability IJBES **5**(1) 78-92 (2018) http://dx.doi.org/10.11113/ijbes.v5.n1.248
9. D. Sukamani and J. Wang. SEM Model for Investigating Factor of an Accident Affecting Safety Performance in Construction Sites in Nepal. Engineering Letters. **28**. 3. (2020)
10. M. Alkaissy et al. Safety Management in Construction: 20 Years of Risk Modeling. Safety Science. **129,** 104805. (2020) https://doi.org/10.1016/j.ssci.2020.104805
11. W. A. Asanka. M. Ranasinghe. Study on The Impacts on Accidents on Construction Projects. 6th International Conference on Structural Engineering and Construction Management (CSECM) 2015. Kandy. Sri Lanka. 11th – 13th December 2015. 58-67 (2015).
12. A. Presley and L. Meade. Benchmarking for Sustainability: An Application to The Sustainable Construction Industry. Benchmarking: An International Journal. **17**(3) 435–451. (2010). http://dx.doi.org/10.1108/14635771011049380
13. S Banihashemi. M.R Hosseini. H. Golizadeh. S. Sankaran. Critical Success Factors (CSFs) for Integration of Sustainability into Construction Project Management Practices in Developing Countries. International Journal of Project Management. **35,** 1103–1119 (2017). http://dx.doi.org/10.1016/j.ijproman.2017.01.014
14. A Goel. LS Ganesh. A Kaur. Sustainability Integration in The Management of Construction Projects: A Morphological Analysis of Over Two Decades' research Literature. Journal of Cleaner Production. (2019). http://dx.doi.org/10.1016/j.jclepro.2019.117676
15. M. Stanitsas. K. Kirytopoulos and V. Leopoulos. Integrating Sustainability Indicators into Project

Management: The Case of Construction Industry. J. of Cleaner Production. **279,** 123774. (2021). https://doi.org/10.1016/j.jclepro.2020.123774

16. J.P Reyes. J.T San-José. J. Cuadrado, and R. Sancibrian, Health and Safety Criteria for Determining the Sustainable Value of Construction Projects. Safety Science. **62,** 221–232 (2014). http://dx.doi.org/10.1016/j.ssci.2013.08.023
17. J. Hinze. R. Godfrey, and J. Sullivan. Integration of construction worker safety and health in assessment of sustainable construction. Journal of Construction Engineering and Management. ASCE. June 2013. (2013). http://dx.doi.org/10.1061/(ASCE)CO.19437862.0000651
18. M. Mapar et al. Sustainability Indicators for Municipalities of Megacities: Integrating Health.Safety and Environmental Performance. Ecological Indicators. **83,** 271–291 (2017). http://dx.doi.org/10.1016/j.ecolind.2017.08.012
19. R. S. Mohandes and X. Zhang. Developing a holistic occupational health and safety risk assessment model: an application to a case of sustainable construction project. Journal of Cleaner Production. **291,** 125934 (2021). https://doi.org/10.1016/j.jclepro.2021.125934
20. M. I. Eraqi. Integrated Quality Management and Sustainability for Enhancing the Competitiveness of Tourism in Egypt. International Journal Services and Operations Management. **5**(1) 14–28 (2009).
21. C. Kuei, and M.H. Lu. Integrating quality management principles into sustainability management. Total Quality Management and Business Excellence. **24**(1-2) 62-78 (2013). http://dx.doi.org/10.1080/14783363.2012.669536
22. A. Chaudhuri. J. Jayaram. A Socio-technical View of Performance Impact of Integrated Quality and Sustainability Strategies. International Journal of Production Research. (2019). https://doi.org/10.1080/00207543.2018.1492162
23. F. Emuze, and J. Smallwood. The Integration of Health and Safety (H&S). Lean and Sustainability in Construction: A literature Review. Proceedings IGLC-21. July 2013. Fortaleza. Brazil
24. A. Griffith. P. Stephenson, and K. Bhutto. Integrated Management System for Construction Quality. Safety and Environment: A Framework for IMS. International Journal of Construction Management. **5**(2) 51-60 (2005). https://doi.org/10.1080/15623599.2005.10773074
25. G. Wilkinson and B.G. Dale. Integrated management systems: an examination of the concept and theory. TQM Magazine. **11**(2) 95-104 (1999)
26. M Bernardo. M Casadesus. S Karapetrovic, and I Heras. How integrated are environmental. quality and other standardized management systems? an empirical study. Journal of Cleaner Production **17,** 742–750 (2009) http://dx.doi.org/10.1016/j.jclepro.2008.11.003
27. S.G Herrero et al. From the traditional concept of safety management to safety integrated with quality. J. of Safety Research. **33**(1-20) (2002). https://doi.org/10.1016/S0022-4375(02)00008-7
28. L.S Pheng and C.Y Pong. Integrating ISO 9001 and OHSAS 18001 for Construction. J. of Cons. Eng. and Man. **129**(3) (2003). http://dx.doi.org/10.1061/(ASCE)07339364(2003)129:3(338)
29. T.W Loushine et al. Quality and safety management in construction. The Total Quality Management Journal. **17**(9), 1171–1212 (2006). http://dx.doi.org/10.1080=14783360600750469
30. J. Wanberg et al. Relationship between construction safety and quality performance. J. Cons. Eng. Man. **139** (2013). http://dx.doi.org/10.1061/(ASCE)CO.1943-7862.0000732
31. P.E.D Love et al. The symbiotic nature of safety and quality in construction: incidents and rework non-conformance. Safety Science. **79**(55–62) (2015). http://dx.doi.org/10.1016/j.ssci.2015.05.009
32. P.E.D. Love et al. Quality and safety in construction: creating a no-harm environment. J. Con. Eng. Man. 05016006 (2016). http://dx.doi.org/10.1061/(ASCE)CO.1943-7862.0001133
33. P.E.D. Love et al. Unearthing the nature and interplay of quality and safety in constructio projects: an empirical study. Safety Science. **103** 270-279 (2018). https://doi.org/10.1016/j.ssci.2017.11.026
34. I.M Shohet. HH Wei. MJ Skibniewski. B Tak, and M Revivi. Integrated communication. control. and command of construction safety and quality. Journal Construction Engineering Management. **145**(9) 04019051 (2019). http://dx.doi.org/10.1061/(ASCE)CO.19437862.0001679
35. I.A. Aboulnaga. Integrating Quality and Environmental Management as Competitive Business Strategy for 21st Century. Environmental Management and Health. **9**(2) 65–71 (1998)
36. S.X. Zeng. P. Tian, and J.J. Shi. Implementing Integration of ISO 9001 and ISO 14001 for Construction. Managerial Auditing Journal. **20**(4) 394-407 (2005). http://dx.doi.org/10.1108/02686900510592070
37. Y.J. Shen and D.H.T. Walker. Integrating OHS. EMS and QM with Constructability Principles When Construction Planning a Design and Construct Project Case Study. The TQM Magazine. **13**(4) 247–259 (2001). http://dx.doi.org/10.1108/09544780110392241
38. K. Bhutto. A Griffith, and P. Stephenson. Integration of quality. health and safety and environment management systems in contractor organisations. In: Khosrowshahi. F (Ed.). 20th Annual ARCOM

Conference. 1-3 September 2004. Heriot Watt University. Association of Researchers in Construction Management. **2**. 1211 (2004)

39. A.R.A. Hamid. B. Singh. W.Z.W. Yusof, and A.K.T. Yang. Integration of Safety. Health. Environment and Quality (SHEQ) Management System in Construction: A Review. Jurnal Kejuruteraan Awam. **16**(1) 24-37 (2004).
40. G. Santos. F. Mendes, and J. Barbosa. Certification and integration of management systems: the experience of Portuguese small and medium enterprises. Journal of Cleaner Production. **19** 1965-1974 (2011). http://dx.doi.org/10.1016/j.jclepro.2011.06.017
41. K.M.P. Carvalho. F. Picchi. G. Camarini, and M.Q.O. Edna. Benefits in the implementation of safety, health, environmental and quality integrated system. IACSIT International Journal of Engineering and Technology. **7**(4) (2015). http://dx.doi.org/10.7763/IJET.2015.V7.814
42. A.H. Laksana. R. Arifuddin. S. Burhanuddin, and M.A. Abdurrahman. Integration conceptual framework of quality management system-occupational safety and health and environment (QHSE) at PT. Wijaya Karya. IOP Conf. Series: Earth and Env. Sci. **419**(012147) (2020). http://dx.doi.org/10.1088/17551315/419/1/012147
43. A. Griffith. Management systems for sustainable construction integrating environmental. quality and safety management systems. I. J. Env. Tech. and Man. **2**(1/2/3) (2002). http://dx.doi.org/10.1504/IJETM.2002.000781
44. M. Rofi'udin and Y. Latief. Integration process models of quality, safety, health and environmental management systems to achieve sustainability construction. Proceedings of the International Conference on Industrial Engineering and Operations Management (Bandung. Indonesia. March 6-8. 2018)
45. G. Fernández-Sánchez and F. Rodríguez-López. A methodology to identify sustainability indicators in construction project management-application to infrastructure projects in Spain. Ecological Indicators. **10**. 1193–1201 (2010). http://dx.doi.org/10.1016/j.ecolind.2010.04.009

https://doi.org/10.1051/e3sconf/202560416007

# Management of inclusive data for Merapi volcano eruption risk areas using participatory action research (PAR) with the perspective of women, the elderly, and persons with disabilities

*Meilinarti*[1]*, *Eko* Swasto[2], *Ida* Ngurah[1], *Dewi* Candraningrum[3], and *Kania* Bening Rahmayna[4]

[1]Yayasan Plan International Indonesia, South Jakarta, Indonesia
[2]Klaten Disability Committee, Klaten, Indonesia
[3]Faculty of Teacher Training and Education, Universitas Muhammadiyah Surakarta, Indonesia
[4]Faculty of Social and Communication Sciences, Universitas Kristen Satya Wacana Salatiga, Salatiga, Indonesia

**Abstract.** This research, conducted by Plan Indonesia, YAKKUM Emergency Unit (YEU), and the Klaten Association of Disabled People (PPDK), with support from ELRHA, UKAid, the Norwegian Ministry of Foreign Affairs, and Plan International, aimed to foster inclusive research by integrating Views from the Frontline (VFL) guidelines with the Feminist Participatory Action Research (FPAR) method. The study involved women, people with disabilities, including youth and children with disabilities, as well as the elderly. Research participants included people with disabilities, the elderly, health cadres, and elderly cadres, and was conducted from May to December 2023 in the disaster-prone areas of Kemalang and Manisrenggo Districts, Central Java. Data analysis was done through triangulation, utilizing multi-party meetings to verify findings based on various times, places, and participants. The research identified types of threats related to the Merapi volcano eruption and needs based on the Humanitarian Inclusion Standard, focusing on gender, age, and ability. Key recommendations include integrating disaster risk management approaches with community-led action plans, enhancing inclusion, promoting local leadership, and developing a national advocacy strategy to increase community involvement in disaster risk reduction efforts, particularly for women, the elderly, and people with disabilities.

## 1 Introduction

This paper was developed by Plan Indonesia, YAKKUM Emergency Unit (YEU), Klaten Disability Association (PPDK) with support from Enhancing Learning and Research for Humanitarian Assistance (ELRHA), UKAid, the Norwegian Ministry of Foreign Affairs and Plan International in order to promote inclusive research involving women, people with disabilities, including young people and children with disabilities, and the elderly (Lansia) through the integration of the Views from the Frontline (VFL) guidelines with the Feminist Participatory Action Research (FPAR) method involving the research process by persons with disabilities, the elderly, health cadres, and elderly cadres.

The development of this guideline was carried out during the period from May to December 2023 in Kemalang District and Manisrenggo District which are areas of Disaster Risk Area 3 and Disaster-Prone Area 2. During this development, consultations were carried out involving representatives of the village government, people with disabilities, the elderly, health cadres, elderly cadres, village volunteers, academics and related agencies in the Klaten Regency government such as the Regional Disaster Management Agency (BPBD), the AP3KB Social Service (Women's Empowerment, Child Protection, Population Control, and Family Planning Service), the Population and Civil Registration Service (Dukcapil), the Ministry of Communication and Information (Kominfo), and the Central Statistics Agency (BPS).

Structural gender inequalities entail that disasters affect women, girls, persons with disabilities, and the elderly in diverse ways than they influence boys, men, and or so-called 'normal people' (givingcompass.org). The vulnerability of women multiplies when they are in a lower socioeconomic group, predominantly in the Global South, in this case risk-area Klaten close to mount Merapi slopes. This vulnerability influences preparedness, evacuation, response, number of deaths and recovery.

The causes for this vulnerability can frequently be tracked to the roles women hold in society and existing gender and cultural standards where they exist. This can comprise the tasks and chores women and girls carry out, the clothing they wear, the way they anticipated to behave, etc. This vulnerability is also part of the intersectional nature of the gendered effect of disasters [1]. For example, women and girls in the disaster-prone area tend to have higher rates of poverty, greater encounters accessing health care, shrinking access to disaster mitigation and preparedness. They often live in women-led households and are more likely to be casualty of the natural disaster.

* Corresponding author: meilinarti.meilinarti@plan-international.org

Before a natural disaster, women and girls usually have the chief responsibility for caring for a home and the persons in it including children, older family members and people with disabilities. Their caregiving responsibilities may prevent their ability to evacuate. After a disaster, women will likely be accountable for caring for the sick and injured while still maintaining their daily house chores. And if the major breadwinner is killed during the disaster, women often need to search for outside employment and, especially in an international context, girls are pulled from school to take care of the household. In all countries, ferocity against women and girls is a factor post-disaster. Sexual assaults, physical abuses, and human trafficking multiply after a catastrophe. Stress is worsened in crowded shelters or momentary housing which may lead to violent behaviour, including sexual attack.

In cultural communities that require modest clothing, women and girls may find it harder to run away from danger (i.e. running from lava eruption, tsunami, or a collapsed building) because of the barriers their clothing may create. Additionally, modest dress and/or cultural norms may mean females engage in different cultural and recreational activities. This could mean girls may not be taught how to swim or to climb trees. This creates barriers that make it difficult to take care of themselves when trying to volcano eruption or survive flooding. During recovery, women and girls may have to work harder to carry out the functions of daily living for their families. This can include lining up for relief supplies, having to travel farther to access water or cooking in challenging conditions. These activities often happen during the day, limiting the access women and girls have to education or outside employment [2].

The aims of this paper is to develop a guideline to promote inclusive data involving women, girls, people with disabilities, including young people and children with disabilities, and the elderly (Lansia) through the integration of the Views from the Frontline (VFL) guidelines with the Feminist Participatory Action Research (FPAR) method involving the research process by people with disabilities, the elderly, health cadres, and elderly cadres.

## 2 Methods

Methods being applied in this research are FPAR, GEDSI, DDEI, and VFL. First, FPAR (Feminist Participatory Action Research, herein after abbreviated as FPAR), is a method that is consistent with the values and objectives of PLAN. PLAN uses methods that hold human rights justice objectives to support and advance women's rights movements and enhance women's human rights. FPAR generates novel forms of collaborative associations that are essential to empowering women and consolidating their voices and promoting agency.

FPAR is a political choice (as is all research) that begins with the belief that knowledge, data, and expertise is gendered, constructed to create privileged authority, and that women have expertise that should inform policy decision-making. Our primary goal in developing the FPAR research paper is to change systems and structures to improve the lives of marginalized women. By 'change' we mean 'structural change' or the change in structures and other systems of marginalization that largely eliminate women's agency and role in disaster response. Participants are not the objects of research but rather the subjects of inquiry who set the agenda, participate in data collection and analysis, and control the use of results, including deciding what actions to take or what direction to take in the future (apwld.org).

Second, GEDSI (Gender, Disability, and Social Inclusion) is also a component that is at the heart of the intersectionality applied in this paper. This GEDSI analysis will help researchers to observe five domains: overview of laws, policies, regulations, and institutional practices; cultural norms and beliefs; gender roles, responsibilities, and time use; access to and control over assets and resources; and patterns of power and decision-making related to the lives of women, older people, people with disabilities, young people, sexual minorities, and those marginalized in disaster events.

Third, Data-Driven Approach to Equity and Inclusion (DDEI). DDEI is the approach and method used in this guide to measure, analyse, and improve diversity, equity, and inclusion in data collection. In a data-driven approach, decisions are made based on data, not intuition. Following a data-driven approach in this paper offers a measurable method. This is because data-driven strategies use real facts and information, and not just instincts in disaster situations. This strategy is a diversity strategy, which is an action-based plan designed to reduce disparities and create a culture of inclusion in research.

Fourth, VFL (Views from the Frontline), is an innovative participatory monitoring method designed to strengthen public accountability in policy implementation in KRB (Disaster-Prone Areas) by reporting on the experiences of the most vulnerable communities for the implementation of disaster risk reduction at the local level [3]. VFL collects various perspectives from communities (women, disabilities, elderly), local governments, and civil society organizations that are most affected by disasters. VFL is a strategy for stakeholders in strengthening community resilience.

## 3 Results and discussion

Collaboration between communities, disaster-affected people, collectives, universities, policymakers, and others is commonplace in research today [4] and produces important information that can be used to address the social issues affecting those involved. Benefit of using this approach includes ensuring authenticity, impartiality, and shared ownership of the work, as exemplified in this research.

However, there were challenges in conducting and writing this paper. One challenge was providing a platform for community co-researcher—also referred to as co-researchers [5] —to accurately share the data they contributed to the collaborative paper. In other words, they can struggle to critically inform the stories they narrate about the issues they experience are most important. This paper addresses this by giving pen and paper directly to people (especially women, disabled, and older people) to write about their experiences. The perspectives of marginalized groups (including many of the fellow researchers in this paper) are often devalued, ignored, and marginalized in society and even intentionally by the state. Their voices are routinely silenced and ignored when they try to raise their concerns. [6] discuss how the marginalization of discriminated groups is a control mechanism used to misrepresent public opinion. Another form of silencing is the practice of reproducing hegemonic discourses and positivist views. This is a problem that occurs within academia, as [6] allude to. Burns, Hyde, Killett, [7] have revealed how this silencing also happens within research bodies.

Researchers then choose to conduct Participatory Research, or other derivative forms such as Participatory Action Research (PAR) or participatory organizational research, as one way to lessen the problem of the erasure of fellow researchers' voices [7]. This approach is reiterated when this paper uses PAR and Feminist Participatory Action Research (FPAR) principles in a paper-writing project to address the issue of unheard and unseen voices of fellow researchers (such as those affected by the Central Java volcanic eruption of Mount Merapi).

This paper project was a Community Based Participatory Action Research (CBPAR) project, which included focus groups focused on "what was the disaster response experience?" from women living in the volcanic eruption risk area. The focus groups obtained information about the challenges faced by women due to the eruption. The PLAN team was aware of this issue and ensured that it was emphasized when presenting the paper.

Based on [8] perspectives on some of the potential benefits and challenges of community-government / university / organization co-creation in community-based research. [8] writes about the empowerment aspect of community-based participatory research. He notes how this is a collaborative approach that involves each partner throughout the research process. Also noted is how the research "increases community capacity, broader stakeholder participation in decision-making, and promotes social justice" (p. 1). More specifically, [8] asserts that empowerment through research transpires when all researchers take control of the development of the paper, and also when they gain social influence through confidence in their abilities.

This paper discusses the value of inclusive data relationships by elevating the voices of co-researchers (community members/affected residents) and provides suggestions on how to care about their participation and boost the knowledge they share. In addition, it discusses the various aspects and stages of participatory research. It then provides a detailed explanation of the experiences of women, disabilities, and older people, especially in relation to Sexual and Reproductive Health and Rights (SRHR) (www.unicef.org), monitored by information on the methods and concepts that allow this inclusive data to achieve amplification and authenticity of the voices of co-researchers that are published and broadcast publicly (collaborations.miami.edu). The paper then discovers the idea of power relations with and how this form of power enables the research team to create a balanced relationship that values each contribution to the process of writing this paper. The paper exposes the voices of affected residents, written by them, by revealing some of the encounters and limitations of their disaster response achievements so far.

This paper is a catalyst for advocacy for inclusive, equitable, and participatory data. It proves how authoritative and optimistic the impact of our voices can be on the weeks we experience. This also confirms that the approach we use in this research supports preserve the stories shared by our co-researchers, the residents affected by the Merapi volcanic eruption.

States that the meaning of power-with is finding ways to share power that permit it to advance and cultivate within a group [9, 10] narrate power-with as a form of power that assists build strength among a group and incorporates mutual interests. Power-with also amplifies individual knowledge and skills. Power-with builds relationships between different groups, which aids them facilitate social change. Advocacy groups, such as PLAN Indonesia, YEU, PPDK (Perkumpulan Penyandang Disabilitas Klaten), and other teams, tend to use power-with to construct allies and multiply support for their goals [10, 11] explains that those who use this form of power prioritize commitment and co-creation, value experiential knowledge, and consider that everyone learns from each other. This research was made possible thanks to social interconnection coordinated by, among others, PLAN Indonesia.

One of the principles of FPAR is the validity of all knowledge because everyone teaches and studies from each other and has knowledge to distribute and grow. Power-with indicates participation, another important principle of FPAR. The "with" in power-with emphasizes collaboration and inclusion. Participation allows input from all members of the research team to contribute in a meaningful way. No research can be successful without equal contribution. In doing so, this research paper creates a secure climate with co-researchers to accelerate dialogue. This paper pursues to ensure that co-researchers' concerns and inputs are conveyed in a genuine way. This is primarily to fulfill the primary objectives of this paper, which is to convey the voices of the co-researchers who wrote in this paper without negotiation. Authenticity, novelty, and feminist and inclusive partisanship have been the driving powers behind the writing of this paper.

Research Stage being applied in this research is as follows. First, Research Code of Ethics. The Research Team must receive an explanation and training on PSEAH (Protection from Sexual Abuse and Harassment) at the beginning of the activity stage. The Research Team will sign a statement of agreement on PSEAH during the activity. Components that are part of the code of ethics include: a. Respecting and adapting to local cultural culture, dressing appropriately and politely, smiling, greeting, and greeting in a friendly manner. b. Being tolerant of participants and informants who will carry out their worship. c. Not committing sexual harassment, physical or verbal violence. d. Disciplined and punctual. e. Maintaining privacy/personal matters and the rights of others. f. Understanding duties, functions, and work procedures. g. Being open to criticism and suggestions. h. Communicating with Plan Indonesia, YEU, and PPDK staff, between members of the Research Team, and participants and informants of the researcher/limited discussion using Indonesian or other appropriate language of instruction. i. Separating personal relationships with participants and informants in terms of job responsibilities. j. Must not influence the opinions and answers of participants and informants. k. Must not falsify data. l. Maintain the confidentiality of participant and informant data and information provided by participants and informants only for the benefit of the RESEARCH TEAM in collaboration with Plan Indonesia, YEU, and PPDK. m. Ensure that the RESEARCH TEAM process is carried out transparently, accountably, efficiently, and effectively. n. Must not do things that are SARA (Ethnicity, Religion, Race, and Inter-Group, namely views or actions based on sentimental thoughts regarding self-identity concerning descent, religion, nationality, or ethnicity and group). o. Not be gender biased or have prejudices against certain genders, ages, and disabilities that result in injustice or gender-based violence. p. Do not intimidate and do not promise anything to participants and informants. q. Be neutral and provide equal treatment to participants and informants, considering their special needs. r. Do not expect any reward or to receive any compensation from participants and informants in any form. s. Implement health protocols to prevent the transmission of COVID-19 or the risk of certain diseases in accordance with announcements or regulations from the government wherever they are

Second, obligations of the Research Team: a. willing to participate in research team briefings and other activities according to the activity schedule in the direction of the research coordinator team. b. Receive capacity building related to PSEAH, Disaster Risk Reduction (DRR), Humanitarian Inclusion Standards and Feminist Participatory Action Research Approach (FPAR) from the research coordinator team. c. Conduct research activities based on the specified targets, namely women and men with disabilities and the elderly, their companions, female heads of families, and/or women who are pregnant, breastfeeding or have children and toddlers and heads of families who have wives who are pregnant, breastfeeding or have children and toddlers in the agreed area. d. Committed to coordinating with the team coordinator, government, and organizers through community companions from supporting institutions. e. Willing to comply with the code of ethics in data collection, signing a statement of having received information and willingness (informed consent), as well as a statement letter to be involved as a research team and committed to following the process in full. f. Able to explain the background, objectives, and description of the survey process to participants and informants.

Third result is Community Socialization and Consultation. Based on the FPAR/PTPF method, all PTPF activities must begin with community socialization and consultation before the research is designed and started. This research requires a series of socialization and consultation with various stakeholders. At this stage, PTPF will only consult with women, to ensure that there is a discussion on power relations and gender issues that may not be easy to discuss in front of men, so that they can be explored in depth. The objectives of pre-research community socialization and consultation should include: a. Gaining support, collaboration, and ownership of the community/stakeholders. b. Forming a research team with the research community. The selection of the research team is based on a consultation process with the community and village government with the following criteria: 1. Have knowledge of the profile of the research location in this case the village in the research location. 2. Women, the elderly and the disabled who are willing to be involved in the research are strongly encouraged to be involved. 3. Actively involved in community groups, especially women, such as PKK members, elderly or health Posyandu cadres, disability groups, village disaster preparedness volunteers, people with disabilities or their companions, the elderly, from the research location. 4. Having been involved in data collection or surveys will provide added value to the research team. 5. Often involved in activities with the elderly and/or disabled. 6. Can communicate well in the language used at the research location in this case Javanese and Indonesian. 7. Have disaster experience in this context, the eruption of Mount Merapi. c. Have disaster experience in this context that has occurred and experienced themselves while living in the village where the research is located. d. Identify the problems that are the focus of the research project. e. Design research methods, questions, and objectives. f. Develop an advocacy plan for long-term change (sustainable during the research process).

Fourth is Research Process. At this stage, the research team collects data based on the set of questions and activities carried out including: 1. Ensure that the data collection process is in accordance with the research target through coordination and approval with the local Village Government. Management of activities at the research location including official notification to the village, hamlet, RT/RW which is the research location must be carried out before the activity takes place. Including activities to ensure the place and time of the discussion, accommodation needs, and transportation of elderly or disabled participants and their companions, consumption needs and stationery supplies for the discussion, especially considering the needs of women's groups. 2. Preparing additional support needs or assistive devices needed by the research team or informants (people being recorded) in the research area, especially those at risk or not easily accessible for the disabled or elderly, to ensure that all data represents local conditions and nothing is left behind as a manifestation of the VFL spirit. 3. Conducting consultations with target informants based on data collection management through the choice of agreed data collection methods through Focus Group Discussions (FGD) and Interviews. Target Informants are people who know the daily situation, are willing, and able to communicate with the research team. This method is carried out through the following divisions: 4. Checking the documentation of information from the interview/interview process and FGD and ensuring that all processes and information are managed properly and pay attention to aspects of information confidentiality. Each research team activity consists of a Research Team or interviewer, discussion note taker, and officers for administration and logistics.

Fifth, analysis of research results shows the following. The analysis stage is carried out by triangulating data through multi-party meetings for checking provided by research participants and informants based on time, space, and different people. The analysis components include: 1. Threats in the context of developing this guideline are in accordance with the identification of information on the types of threats known and agreed upon with village stakeholders together with the research team, participation related to disaster risk reduction based on the profiles of researchers and informants from the local location [12]. 2. Identification of needs based on Humanitarian Inclusion Standards associated with data on gender, age, and ability (disability, children and the elderly, especially women). The components of the Humanitarian Inclusion Standard are as follows: Seven Sectors of the Humanitarian Inclusion Standard: 1. Protection Inclusion Standard, 2. Drinking Water and Environmental Sanitation Inclusion Standard, 3. Food Security and Livelihood Inclusion Standard, 4. Nutrition Inclusion Standard, 5. Temporary Shelter, Settlement, and Household Equipment Inclusion Standard, 6. Health Inclusion Standard, 7. Education Inclusion Standard, 3. Accommodate recommended actions to meet needs including the nine key HIS (Humanitarian Inclusion Standards) and other needs for children, youth, elderly and disabled people especially women in Disaster Management.

The results of the analysis show that strengthening community inclusion is a vital factor in data inclusion. In the interviews managed in this research, in communities most at risk of disasters, this data was openly accessible and disaggregated by community, age, gender, disability, and village location. This data is at present being used to update development results and strategy. The FPAR, GEDSI, DDEI and VFL procedures have improved the inclusion and collaboration of communities, local government and civil society groups in decision-making in the KRB. For example, communities have worked with local civil society groups to create community commitment alliances to encourage for and take collective action to decrease disaster risk.

The results also show local community-led action planning (women, older people, and persons with disabilities) is crucial in data inclusion. The development of local action plans in communities means that the basic needs of communities at risks of disaster are now being addressed. To date, local action plans have been changed and realized. These plans address dangers to lives and revenues such as poverty, water insecurity, battle and obstacles to risk reduction such as exclusion of communities by local government. Each plan identifies central activities to be undertaken, by whom and the resources needed. For example, food security in disaster situations. This research is able to inspire the accomplishment of local community-led planning in disaster-prone areas.

Championing community leadership (women, elderly, disabled) had been promoted during the field research. Around the world, Plan Indonesia and its partners are developing a program by nurturing community leadership so that communities at risk of disaster have the resources and strength to make real changes (www.gndr.org). For example, communities have procured the scheme to tackle water shortages in disaster-prone areas. With the support and collaboration of local governments, this research paper will be an intervention in producing local leaders in disaster-prone situations as part of disaster risk mitigation and reduction.

Advocating for national-level participation is also a must agenda to be foreseen to make data inclusion successful. A national advocacy plan can be developed through the research results of this paper, with multi-stakeholder consultations at the country level. These plans can be used to guide national campaigns and initiatives aimed at building systematic community involvement in development plans, decisions, and actions, especially disaster risk reduction. In the Merapi volcanic eruption area, for example, several community organizations have voiced the voices of communities affected by the eruption and have sought to strengthen disaster risk reduction policies through advocacy and dialogue with various stakeholders.

## 4 Conclusion

This research paper combines multiple perspectives, insights, and methods to strengthen policymaking. This integrated research approach to disaster risk management will bridge the gap between the urgent needs of affected people, science, policymaking, and implementation. Therefore, this paper outlines a series of recommendations for strengthening the capacity of all stakeholders in Disaster Risk Reduction [13].

Understanding disaster risk is a challenge not only for affected people, but for all relevant DRR stakeholders. It requires a comprehensive understanding of the complex interconnections and interdependencies between different hazards, as well as the complex dimensions of vulnerability and exposure – encompassing the root causes and drivers of disaster risk [14]. This paper can help understand disaster risk in a more inclusive way than previous disaster research guides that were only quantitative and exclusive, and often left out the voices of the most vulnerable people in disaster-affected areas [15]. This paper is a step towards investing in disaster risk reduction for community resilience by enhancing disaster preparedness for effective response and in post-disaster recovery, rehabilitation, and reconstruction.

## References

1. Bofu-Tawamba, Ndana; Ruby Bright, Stephanie Clohesy, Christine Grumm et al. The Uprising of Women in Philanthropy, Routledge, (2024).
2. Shuttleworth, Russel; Linda R. Mona. The Routledge Handbook of Disability and Sexuality, Routledge, (2020)

3. López-Carresi, Alejandro; Maureen Fordham, Ben Wisner, Ilan Kelman, JC Gaillard. Disaster Management - International lessons in risk reduction, response and recovery, Routledge, (2013).
4. Begun, A. L., Berger, L. K., Otto-Salaj, L. L., and Rose, S. J. Developing effective social work university-community research collaborations. Social Work, **55**(1), 54–62. (2010). https://doi.org/10.1093/sw/55.1.54
5. Walmsley, J., and Mannan, H. Parents as co-researchers: A participatory action research initiative involving parents of people with intellectual disabilities in Ireland. British Journal of Learning Disabilities, **37**(4), 271–276. (2009). https://doi.org/10.1111/j.1468-3156.2009.00582.x
6. Gibson, A. N., and Hughes-Hassell, S. We will not be silent: Amplifying marginalized voices in LIS education and research. The Library Quarterly, **87**(4), 317–329. (2017). https://doi.org/10.1086/693488
7. Burns, D., Hyde, P., Killett, A., Poland, F., and Gray, R. Participatory organizational research: Examining voice in the co-production of knowledge. British Journal of Management, **25**(1), 133–144. (2014). https://doi.org/10.1111/j.1467-8551.2012.00841.x
8. Tremblay, C. Community-based participatory research (CBPR) as a tool for empowerment and public policy. Victoria, BC: Office of Community-Based Research, University of Victoria. (2009).
9. Tett, L. Partnerships, community groups and social inclusion. Studies in Continuing Education, 27(1), 1–15. (2005). https://doi.org/10.1080/01580370500056364
10. Vene Klasen, L., Miller, V., and Ruthraff, J. V. A new weave of power, people, and politics: The action guide for advocacy and citizen and participation. Development in Practice, **14**(1), 294–296. (2004).
11. Terez, T. Power over vs. power with. (2010). Retrieved from https://www.slideshare.net/SwaamTech/power-over-vs-power-with
12. Agastya, Ida Bagus Oka; Eko Teguh Paripurno, Bambang Prastistho, Helmy Murwanto, Johan Danu Prasetya, Arif Rianto Budi Nugraho. "Model media learning for disaster risk reduction in the Batur UNESCO Global Geopark", International Journal of Geoheritage and Parks, (2024)
13. Giustiniani, Flavia Zorzi; Emanuele Sommario, Federico Casolari, Giulio Bartolini. Routledge Handbook of Human Rights and Disasters, Routledge, (2018).
14. Alcántara-Ayala, Irasema; Christopher Gomez, Ksenia Chmutina, Dewald van Niekerk et al. Disaster Risk, Routledge, (2022).
15. Battersby, Stephen. Clay's Handbook of Environmental Health, Routledge, (2022)

# AUTHOR INDEX